Triangles and Angles

Right Triangle
Triangle has one 90°
(right) angle.

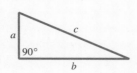

Pythagorean Formula
(*for right triangles*)
$c^2 = a^2 + b^2$

Right Angle
Measure is 90°.

Isosceles Triangle
Two sides are equal.
$AB = BC$

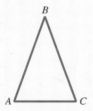

Straight Angle
Measure is 180°.

Equilateral Triangle
All sides are equal.
$AB = BC = CA$

Complementary Angles
The sum of the measures of
two complementary
angles is 90°.

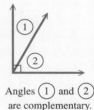

Angles ① and ②
are complementary.

Sum of the Angles of Any Triangle
$A + B + C = 180°$

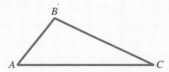

Supplementary Angles
The sum of the
measures of two
supplementary
angles is 180°.

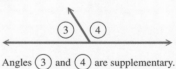

Angles ③ and ④ are supplementary.

Similar Triangles
Corresponding angles are
equal; corresponding sides
are proportional.
$A = D, B = E, C = F$
$$\frac{AB}{DE} = \frac{AC}{DF} = \frac{BC}{EF}$$

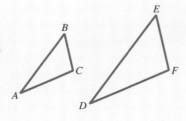

Vertical Angles
Vertical angles have
equal measures.

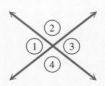

Angle ① = Angle ③

Angle ② = Angle ④

Algebra for College Students

FIFTH EDITION

Algebra for College Students

FIFTH EDITION

Margaret L. Lial
American River College

John Hornsby
University of New Orleans

Terry McGinnis

PEARSON

Addison
Wesley

Boston   San Francisco   New York
London   Toronto   Sydney   Tokyo   Singapore   Madrid
Mexico City   Munich   Paris   Cape Town   Hong Kong   Montreal

Publisher: Greg Tobin

Editor in Chief: Maureen O'Connor

Project Editor: Suzanne Alley

Assistant Editor: Jolene Lehr

Managing Editor: Ron Hampton

Production Supervisor: Kathleen A. Manley

Production Services: Elm Street Publishing Services, Inc.

Compositor: Beacon Publishing Services

Associate Media Producer: Sara Anderson

Software Development: David Malone and John O'Brien

Marketing Manager: Dona Kenly

Marketing Coordinator: Lindsay Skay

Prepress Supervisor: Caroline Fell

Manufacturing Buyer: Evelyn Beaton

Text Designer: Susan Carsten Raymond

Cover Designer: Dennis Schaefer

Cover Photograph: Daryl Benson/Masterfile

For permission to use copyrighted material, grateful acknowledgment is made to the copyright holders on page I-11, which is hereby made part of this copyright page.

Library of Congress Cataloging-in-Publication Data
Lial, Margaret L.
    Algebra for College Students.—5th ed. / Margaret L. Lial, John Hornsby, Terry McGinnis.
       p.   cm.
    Includes index.
    ISBN 0-321-16830-5 (alk. paper)
    1. Algebra.   I. Hornsby, John.   II. McGinnis, Terry.   III. Title.

  QA154.3.L53   2004
  512.9—dc21

                                       2003051792

2 3 4 5 6 7 8 9 10—QWT—07 06 05 04

Contents

v

List of Applications

Preface

The fifth edition of *Algebra for College Students* continues our ongoing commitment to provide the best possible text and supplements package to help instructors teach and students succeed. To that end, we have tried to address the diverse needs of today's students through a more open design, updated figures and graphs, helpful features, careful explanations of topics, and a comprehensive package of supplements and study aids. We have also taken special care to respond to the suggestions of users and reviewers and have added many new examples and exercises based on their feedback. Students who have never studied algebra—as well as those who require further review of basic algebraic concepts before taking additional courses in mathematics, business, science, nursing, or other fields—will benefit from the text's student-oriented approach.

After many years of benefiting from her behind-the-scenes assistance, we are pleased to welcome Terry McGinnis as coauthor of this series, which includes this text as well as the following books:

- *Beginning Algebra,* Ninth Edition, by Lial, Hornsby, and McGinnis
- *Intermediate Algebra,* Ninth Edition, by Lial, Hornsby, and McGinnis
- *Beginning and Intermediate Algebra,* Third Edition, by Lial, Hornsby, and McGinnis

WHAT'S NEW IN THIS EDITION?

We believe students and instructors will welcome the following new features.

New Real-Life Applications   We are always on the lookout for interesting data to use in real-life applications. As a result, we have included many new or updated examples and exercises throughout the text that focus on real-life applications of mathematics. These applied problems provide a modern flavor that will appeal to and motivate students. (See pp. 156, 169, and 258.) A comprehensive List of Applications appears at the beginning of the text. (See pp. xi–xiv.)

New Figures and Photos   Today's students are more visually oriented than ever. Thus, we have made a concerted effort to add mathematical figures, diagrams, tables, and graphs whenever possible. (See pp. 102, 216, and 668.) Many of the graphs use a style similar to that seen by students in today's print and electronic media. Photos have been incorporated to enhance applications in examples and exercises. (See pp. 45, 465, and 542.)

Increased Emphasis on Problem Solving   Introduced in Chapter 2, our six-step problem-solving method has been refined and integrated throughout the text. The six steps, *Read, Assign a Variable, Write an Equation, Solve, State the Answer,* and *Check,* are emphasized in boldface type and repeated in examples and exercises to reinforce the problem-solving process for students. (See pp. 69, 247, and 433.) Special boxes that include additional problem-solving information and tips are interspersed throughout the text. (See pp. 60, 68, and 82.)

XV

Chapter Openers   New chapter openers feature real-world applications of mathematics that are relevant to students and tied to specific material within the chapters. Examples of topics include retirement planning, college tuition, and the U.S. budget. (See pp. 1, 49, and 299—Chapters 1, 2, and 5.)

Now Try Exercises   To actively engage students in the learning process, each example now concludes with a reference to one or more parallel exercises from the corresponding exercise set. In this way, students are able to immediately apply and reinforce the concepts and skills presented in the examples. (See pp. 18, 153, and 575.)

Summary Exercises   Based on user feedback, we have more than doubled the number of in-chapter summary exercises. These special exercise sets provide students with the all-important *mixed* review problems they need to master topics. Summaries of solution methods or additional examples may be included. (See pp. 90, 121, and 500.)

Glossary   A comprehensive glossary of key terms from throughout the text is included at the back of the book. (See pp. G-1 to G-9.)

WHAT FAMILIAR FEATURES HAVE BEEN RETAINED?

We have retained the popular features of previous editions of the text, some of which follow.

Learning Objectives   Each section begins with clearly stated, numbered objectives, and the included material is directly keyed to these objectives so that students know exactly what is covered in each section. (See pp. 32, 270, and 491.)

Cautions and Notes   One of the most popular features of previous editions, **CAUTION** and **NOTE** boxes warn students about common errors and emphasize important ideas throughout the exposition. (See pp. 357, 398, and 469.) There are more of these in the fifth edition than in the fourth, and the new text design makes them easier to spot.

Connections   Connections boxes have been streamlined. They continue to provide connections to the real world or to other mathematical concepts, historical background, and thought-provoking questions for writing or class discussion. (See pp. 75, 250, and 812.)

Ample and Varied Exercise Sets   The text contains a wealth of exercises to provide students with opportunities to practice, apply, connect, and extend the algebraic skills they are learning. Numerous illustrations, tables, graphs, and photos have been added to the exercise sets to help students visualize the problems they are solving. Problem types include writing ✐, estimation, graphing calculator ▦, and challenging exercises that go beyond the examples, as well as applications and multiple-choice, matching, true/false, and fill-in-the-blank problems. (See pp. 21, 157, and 310.)

Relating Concepts Exercises   These sets of exercises help students tie together topics and develop problem-solving skills as they compare and contrast ideas, identify and describe patterns, and extend concepts to new situations. (See pp. 80, 209, and 543.) These exercises make great collaborative activities for pairs or small groups of students.

Technology Insights Exercises   We assume that all students of this text have access to scientific calculators. *While graphing calculators are not required for this text,* some students may go on to courses that use them. For this reason, we have included Technology Insights exercises in selected exercise sets. These exercises provide an opportunity for students to interpret typical results seen on graphing calculator screens. Actual calculator screens from the Texas Instruments TI-83 Plus graphing calculator are featured. (See pp. 147, 385, and 516.)

Group Activities   Appearing at the end of each chapter, these real-data activities allow students to apply the mathematical content of the chapter in a collaborative setting. (See pp. 122, 519, and 581.)

Ample Opportunity for Review   Each chapter concludes with an extensive Chapter Summary that features Key Terms, New Symbols, Test Your Word Power, and a Quick Review of each section's content with additional examples. A comprehensive set of Chapter Review Exercises, keyed to individual sections, is included, as are Mixed Review Exercises and a Chapter Test. Beginning with Chapter 2, each chapter concludes with a set of Cumulative Review Exercises that cover material going back to Chapter 1. (See pp. 124–136 and 520–531.)

WHAT CONTENT CHANGES HAVE BEEN MADE?

We have worked hard to fine-tune and polish presentations of topics throughout the text based on user and reviewer feedback. Some of the content changes include the following:

- Former Section 1.2 on operations of real numbers has been split into two sections, with exponents, square roots, and order of operations now covered in Section 1.3.

- In Chapter 3, the material on equations of lines in Section 3.3 has been reordered to cover slope-intercept form first. The presentation on functions in Section 3.5 has been rewritten.

- The material on systems of linear equations has been pulled together in Chapter 4. It includes former Sections 11.1–11.3, 12.1, 12.3, and 12.4.

- Chapter 5 covers exponents and polynomials, including polynomial division. Factoring is covered in a separate Chapter 6.

- The section on rational exponents and radicals has been split into two sections. Section 8.1 includes radical expressions and graphs, and Section 8.2 covers rational exponents.

- Chapter 10 focuses on additional graphs of relations and functions (former Sections 8.1–8.4 and 8.7).

- Inverse, exponential, and logarithmic functions are now covered in Chapter 11, followed by polynomial and rational functions in Chapter 12.

- Chapter 13 pulls all the material on conic sections, nonlinear inequalities, systems of inequalities, and linear programming together in one chapter. It includes former Sections 8.5, 8.6, 11.4, and 11.5.

- Three new appendices cover An Introduction to Calculators, Properties of Matrices (former Section 12.2), and Matrix Inverses (former Section 12.5).

WHAT SUPPLEMENTS ARE AVAILABLE?

Our extensive supplements package includes an Annotated Instructor's Edition, testing materials, solutions manuals, tutorial software, videotapes, and a state-of-the-art Web site. For more information about any of the following supplements, please contact your Addison-Wesley sales consultant.

For the Student

Student's Solutions Manual (ISBN 0-321-17329-5)   The *Student's Solutions Manual* by Jeff Cole, Anoka-Ramsey Community College, provides detailed solutions to the odd-numbered section and summary exercises and to all Relating Concepts, Chapter Review, Chapter Test, and Cumulative Review exercises.

Addison-Wesley Math Tutor Center   The Addison-Wesley Math Tutor Center is staffed by qualified college mathematics instructors who tutor students on examples and exercises from the textbook. Tutoring is provided via toll-free telephone, toll-free fax, e-mail, and the Internet. Interactive Web-based technology allows students and tutors to view and listen to live instruction—in real-time over the Internet. The Math Tutor Center is accessed through a registration number that can be packaged with a new textbook or purchased separately. (*Note:* MyMathLab students obtain access to the Math Tutor Center through their MyMathLab access code.)

InterAct Math® Tutorial Software (ISBN 0-321-16831-3)   Available on CD-ROM, this interactive tutorial software provides algorithmically generated practice exercises that are correlated at the objective level to the odd-numbered exercises in the text. Every exercise in the program is accompanied by an example and a guided solution designed to involve students in the solution process. Selected problems also include a video clip to help students visualize concepts. The software tracks student activity and scores and can generate printed summaries of students' progress.

MathXL   MathXL is an on-line testing, homework, and tutorial system that uses algorithmically generated exercises correlated to the textbook. Students can take chapter tests and receive personalized study plans that diagnose weaknesses and link students to areas they need to study and retest. Students can also work unlimited practice problems and receive tutorial instruction for areas in which they need improvement. MathXL can be packaged with new copies of *Algebra for College Students,* Fifth Edition. Please contact your Addison-Wesley sales representative for details.

Videotape Series (ISBN 0-321-17326-0)   This series of videotapes, created specifically for *Algebra for College Students,* Fifth Edition, features an engaging team of lecturers who provide comprehensive lessons on every objective in the text. The videos include a stop-the-tape feature that encourages students to pause the video, work through the example presented on their own, and then resume play to watch the video instructor go over the solution.

Digital Video Tutor (ISBN 0-321-17354-6)   This supplement provides the entire set of videotapes for the text in digital format on CD-ROM, making it easy and convenient for students to watch video segments from a computer, either at home or on campus. Available for purchase with the text at minimal cost, the Digital Video Tutor is ideal for distance learning and supplemental instruction.

MyMathLab   MyMathLab is a complete on-line course for Addison-Wesley mathematics textbooks that provides interactive, multimedia instruction correlated to the textbook content. MyMathLab is easily customizable to suit the needs of students and instructors and provides a comprehensive and efficient on-line course-management system that allows for diagnosis, assessment, and tracking of students' progress.

MyMathLab features:

- Chapter and section folders in the on-line course mirror the textbook Table of Contents and contain a wide range of multimedia instruction, including video lectures, tutorial software, and electronic supplements.

- Actual pages of the textbook are loaded into MyMathLab, and as students work through a section of on-line text, they can link to multimedia resources—such as video and audio clips, tutorial exercises, and interactive animations—that are correlated directly to examples and exercises in the text.

- Hyperlinks take the user directly to on-line testing, diagnosis, tutorials, and tracking in MathXL—Addison-Wesley's tutorial and testing system for mathematics and statistics.

- Print supplements are available on-line, side-by-side with the textbook.

For more information, visit our Web site at www.mymathlab.com or contact your Addison-Wesley sales representative for a live demonstration.

For the Instructor

CLASSROOM EXAMPLE

Find each power of i.
(a) i^{28} (b) i^{19} (c) i^{-9}
(d) i^{-22}
Answer: (a) 1 (b) $-i$
(c) $-i$ (d) -1

TEACHING TIP Show examples of real numbers and imaginary numbers written in standard form. For example, $3 = 3 + 0i$ and $4i = 0 + 4i$.

Annotated Instructor's Edition (ISBN 0-321-16835-6) For immediate access, the Annotated Instructor's Edition provides answers to all text exercises in the margin or next to the corresponding exercise, as well as Classroom Examples (formerly Chalkboard Examples) and Teaching Tips, printed in blue for easy visibility. Based on user feedback, we have increased the number of Classroom Examples and Teaching Tips.

Exercises designed for writing 🖊 and graphing calculator 📊 use are indicated in both the Student Edition and the Annotated Instructor's Edition.

Instructor's Solutions Manual (ISBN 0-321-17327-9) The *Instructor's Solutions Manual*, by Jeff Cole, Anoka-Ramsey Community College, provides complete solutions to all text exercises.

Answer Book (ISBN 0-321-17325-2) The *Answer Book* provides answers to all the exercises in the text.

Printed Test Bank (ISBN 0-321-17328-7) Written by Jon Becker, Indiana University Northwest, the *Printed Test Bank* contains two diagnostic pretests, four free-response and two multiple-choice test forms per chapter, and two final exams. Additional practice exercises for almost every objective of every section of the text are also included. A conversion guide from the fourth to the fifth edition is also included.

Adjunct Support Manual (ISBN 0-321-19743-7) This manual includes resources designed to help both new and adjunct faculty with course preparation and classroom management, as well as offers helpful teaching tips.

Adjunct Support Center The Adjunct Support Center offers consultation on suggested syllabi, helpful tips on using the textbook support package, assistance with textbook content, and advice on classroom strategies from qualified mathematics instructors with over fifty years of combined teaching experience. The Adjunct Support Center is available Sunday through Thursday evenings from 5 P.M. to midnight. Phone: 1-800-435-4084; E-mail: adjunctsupport@awl.com; Fax: 1-877-262-9774.

TestGen with QuizMaster (ISBN 0-321-16833-X)   TestGen enables instructors to build, edit, print, and administer tests using a computerized bank of questions developed to cover all the objectives of the text. Instructors can modify test bank questions or add new questions by using the built-in question editor, which allows users to create graphs, import graphics, insert math notation, and insert variable numbers or text. Tests can be printed or administered on-line via the Web or other network. TestGen comes packaged with Quiz-Master, which allows students to take tests on a local area network. The software is available on a dual-platform Windows/Macintosh CD-ROM.

MathXL   MathXL is an on-line testing, homework, and tutorial system that uses algorithmically generated exercises correlated to the textbook. Instructors can assign tests and homework provided by Addison-Wesley or create and customize their own tests and homework assignments. Instructors can also track their students' results and tutorial work in an on-line gradebook. Students can take chapter tests and receive personalized study plans that diagnose weaknesses and link students to areas they need to study and retest. Students can also work unlimited practice problems and receive tutorial instruction for areas in which they need improvement. MathXL can be packaged with new copies of *Algebra for College Students,* Fifth Edition. Please contact your Addison-Wesley sales representative for details.

MyMathLab   MyMathLab is a complete on-line course for Addison-Wesley mathematics textbooks that provides interactive, multimedia instruction correlated to the textbook content. MyMathLab is easily customizable to suit the needs of students and instructors and provides a comprehensive and efficient on-line course-management system that allows for diagnosis, assessment, and tracking of students' progress.

MyMathLab features:

- Chapter and section folders in the on-line course mirror the textbook Table of Contents and contain a wide range of multimedia instruction, including video lectures, tutorial software, and electronic supplements.

- Actual pages of the textbook are loaded into MyMathLab, and as students work through a section of on-line text, they can link to multimedia resources—such as video and audio clips, tutorial exercises, and interactive animations—that are correlated directly to examples and exercises in the text.

- Hyperlinks take the user directly to on-line testing, diagnosis, tutorials, and tracking in MathXL—Addison-Wesley's tutorial and testing system for mathematics and statistics.

- Instructors can create, copy, edit, assign, and track all tests and homework for their course as well as track students' results and practice work.

- With push-button ease, instructors can remove, hide, or annotate Addison-Wesley preloaded content, add their own course documents, or change the order in which material is presented.

- Using the communication tools found in MyMathLab, instructors can hold on-line office hours, host a discussion board, create communication groups within their class, send e-mail, and maintain a course calendar.
- Print supplements are available on-line, side-by-side with the textbook.

For more information, visit our Web site at www.mymathlab.com or contact your Addison-Wesley sales representative for a live demonstration.

ACKNOWLEDGMENTS

The comments, criticisms, and suggestions of users, nonusers, instructors, and students have positively shaped this textbook over the years, and we are most grateful for the many responses we have received. The feedback gathered for this revision of the text was particularly helpful, and we especially wish to thank the following individuals who provided invaluable suggestions:

Mary Kay Abbey, *Montgomery College*
Marwan Abu-Sawwa, *Florida Community College, Jacksonville*
Jose Alvarado, *University of Texas, Pan American*
Sonya Armstrong, *West Virginia State College*
Rajappa Asthagiri, *Miami University, Middletown*
Mary Lou Baker, *Columbia State Community College*
Dixilee Blackinton, *Weber State University*
Bob Bohac, *North Idaho College*
Lisa Cuneo, *Pennsylvania State University, DuBois*
Charles Curtis, *Missouri Southern State College*
Jo Dobbin, *Fayetteville Technical Community College*
Sharon Edgmon, *Bakersfield College*
Lucy Edwards, *Las Positas College*
Joe Howe, *St. Charles Community College*
Matthew Hudock, *St. Philip's College*
Dale Hughes, *Johnson County Community College*
Judy Kasabian, *El Camino College*
Nancy Ketchum, *Moberly Area Community College*
Tony Masci, *Notre Dame College*
Timothy McLendon, *East Central College*
Kausha Miller, *Lexington Community College*
Kathy Nickell, *College of DuPage*
Marilyn Platt, *Gaston College*
Joan Prymas, *Herkimer County Community College*
Nelissa Rutishauser, *Mohawk Valley Community College*
Bettie Truitt, *Black Hawk College*
Tony Vavra, *West Virginia Northern Community College*
Angela Walters, *Capitol College*
Jackie Wing, *Angelina College*
Mary Wolyniak, *Broome Community College*

Over the years, we have come to rely on an extensive team of experienced professionals. Our sincere thanks go to these dedicated individuals at Addison-Wesley, who worked long and hard to make this revision a success: Maureen O'Connor, Suzanne Alley, Dona Kenly, Kathy Manley, Dennis Schaefer, Susan Raymond, Jolene Lehr, Lindsay Skay, and Sara Anderson.

Abby Tanenbaum did an outstanding job checking the answers to exercises and also provided invaluable assistance during the production process. Steven Pusztai provided his customary excellent production work. Thanks are due Jeff Cole, who supplied accurate, helpful solutions manuals, and Jon Becker, who provided the comprehensive *Printed Test Bank.* We are most grateful to Paul Van Erden for yet another accurate, useful index; Becky Troutman for preparing the comprehensive List of Applications; and Shannon d'Hemecourt and Perian Herring for accuracy checking page proofs.

As an author team, we are committed to the goal stated earlier in this Preface—to provide the best possible text and supplements package to help instructors teach and students succeed. We are most grateful to all those over the years who have aspired to this goal with us. As we continue to work toward it, we would welcome any comments or suggestions you might have. Please feel free to send your comments via e-mail to math@aw.com.

Margaret L. Lial
John Hornsby
Terry McGinnis

Feature Walkthrough

Linear Equations, Inequalities, and Applications

2

Tuition for full-time students at both public and private 4-year universities in the United States has risen steadily from 1990 through 2000. Over this 10-year period, tuition increased 85% at public colleges and 86.6% at private institutions. (*Source:* National Center for Education Statistics, U.S. Department of Education.) In Sections 2.2 and 2.3, we discuss the use of percent—one of the most common everyday applications of mathematics—and find the year 2000 costs for both public and private universities in Exercises 39 and 40 of Section 2.3.

Chapter Opener

Each chapter opens with an application and section outline. The application in the opener is tied to specific material within the chapter.

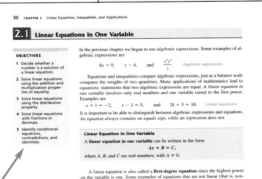

Learning Objectives

Each section opens with a highlighted list of clearly stated, numbered learning objectives. These learning objectives are reinforced throughout the section by restating the learning objective where appropriate so that students always know exactly what is being covered.

Notes

Important ideas are emphasized in *Note* boxes that appear throughout the text.

Cautions

Students are warned of common errors through the use of *Caution* boxes that are found throughout the text.

locate the point (3, 1). The line through these two points is the required graph. See Figure 25. $\left(\text{Verify that the point obtained using } -\frac{2}{3} \text{ as the slope is also on this line.}\right)$

FIGURE 25

Now Try Exercise 25.

NOTE The slope-intercept form of a linear equation is the most useful for several reasons. Every linear equation (of a nonvertical line) has a *unique* (one and only one) slope-intercept form. In Section 3.5 we study *linear functions*, which are defined using slope-intercept form. Also, this is the form we use when graphing a line with a graphing calculator. (See Section 3.1, Example 6.)

FIGURE 5

CAUTION Because absolute value represents distance, and distance is always positive (or 0), *the absolute value of a number is always positive (or 0).*

Classroom Examples and Teaching Tips

The Annotated Instructor's Edition provides answers to all text exercises and Group Activities in color in the margin or next to the corresponding exercise. In addition, *Classroom Examples* and *Teaching Tips* are included to assist instructors in creating examples to use in class that are different from what students have in their textbooks. *Teaching Tips* offer guidance on presenting the material at hand.

New! Now Try Exercises

Now Try exercises are found after each example to encourage active learning. This feature asks students to work exercises in the exercise sets that parallel the example just studied.

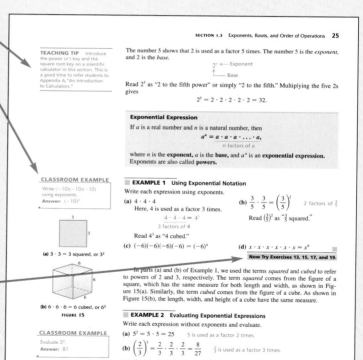

Connections

Connections boxes provide connections to the real world or to other mathematical concepts, historical background, and thought-provoking questions for writing or class discussion.

Writing Exercises

Writing exercises abound in the Lial series through the *Connections* boxes and also in the exercise sets. Some writing exercises require only short written answers, and others require lengthier journal-type responses where students are asked to fully explain terminology, procedures and methods, document their understanding using examples, or make connections between topics.

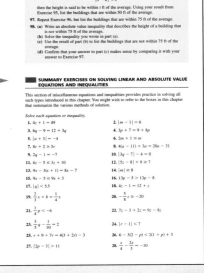

Problem Solving

The Lial *six-step problem-solving* method is clearly explained in Chapter 2 and is then continually reinforced in examples and exercises throughout the text to aid students in solving application problems.

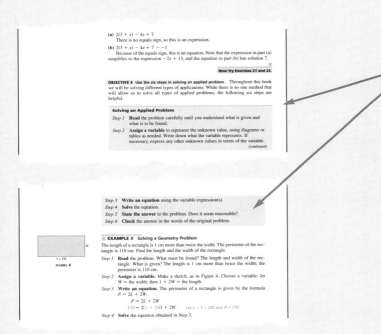

Summary Exercises

Summary Exercises appear in many chapters to provide students with *mixed* practice problems needed to master topics.

Ample and Varied Exercise Sets

Algebra students require a large number of varied practice exercises to master the material they have just learned. This text contains thousands of exercises including summary and review exercises, numerous conceptual and writing exercises, and challenging exercises that go beyond the examples. Multiple-choice, matching, true/false, and completion exercises help to provide variety. Exercises suitable for graphing calculator use are marked with an icon.

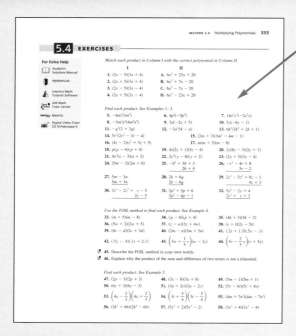

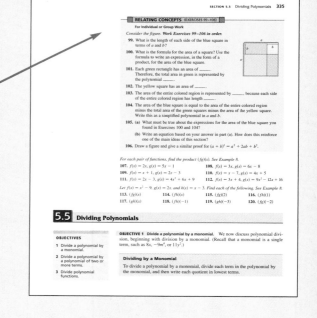

Relating Concepts

Found in selected exercise sets, these exercises tie together topics and highlight the relationships among various concepts and skills. For example, they may show how algebra and geometry are related or how a graph of a linear equation in two variables is related to the solution of the corresponding linear equation in one variable. These sets of exercises make great collaborative activities for small groups of students.

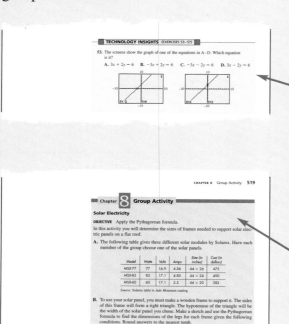

Technology Insights

Technology Insights exercises are found in selected exercise sets throughout the text. These exercises illustrate the power of graphing calculators and provide an opportunity for students to interpret typical results seen on graphing calculator screens. (A graphing calculator is *not* required to complete these exercises.)

Group Activities

Appearing at the end of each chapter, these activities allow students to work collaboratively to solve a problem related to the chapter material.

Ample Opportunity for Review

One of the most popular features of the Lial textbooks is the extensive and well thought-out end-of-chapter material. At the end of each chapter, students will find:

Key Terms and *New Symbols* that are keyed back to the appropriate section for easy reference and study.

Test Your Word Power helps students understand and master mathematical vocabulary; key terms from the chapter are presented with four possible definitions in multiple-choice format.

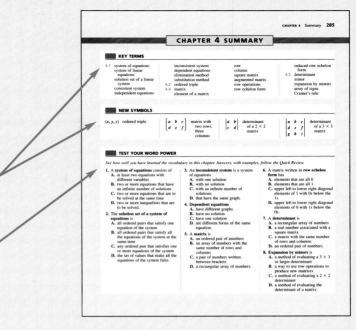

Quick Review sections give students not only the main concepts from the chapter (referenced back to the appropriate section) but also an adjacent example of each concept.

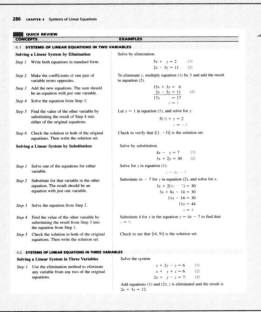

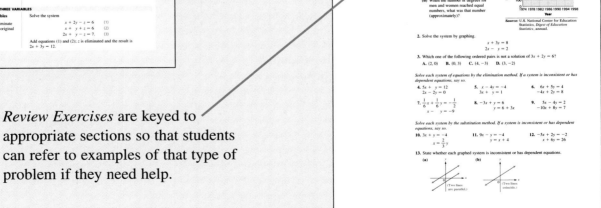

Review Exercises are keyed to appropriate sections so that students can refer to examples of that type of problem if they need help.

Mixed Review Exercises require students to solve problems without the help of section references.

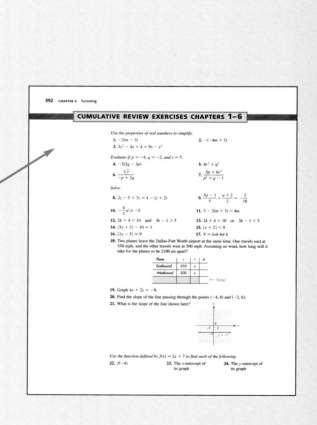

Chapter Tests help students practice for the real thing.

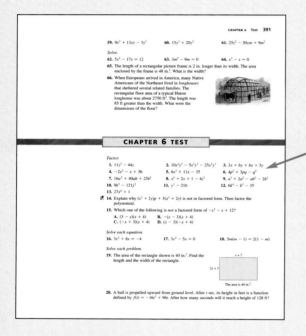

Cumulative Review Exercises gather various types of exercises from preceding chapters to help students remember and retain what they are learning throughout the course.

Algebra for College Students

FIFTH EDITION

Review of the Real Number System

Social Security is the largest source of income for elderly Americans. It is projected that in about 30 years, however, there will be twice as many older Americans as there are today, and the excess revenues now accumulating in Social Security's trust funds will be exhausted. (*Source:* Social Security Administration.)

To supplement their retirement incomes, more and more Americans have begun investing in mutual funds, pension plans, and other means of savings. In Exercise 93 of Section 1.3, we relate the concepts of this chapter to the percent of U.S. households investing in mutual funds.

1

1.1 Basic Concepts

In this chapter we review some of the basic symbols and rules of algebra.

OBJECTIVE 1 Write sets using set notation. A **set** is a collection of objects called the **elements** or **members** of the set. In algebra, the elements of a set are usually numbers. Set braces, { }, are used to enclose the elements. For example, 2 is an element of the set {1, 2, 3}. Since we can count the number of elements in the set {1, 2, 3}, it is a *finite set*.

In our study of algebra, we refer to certain sets of numbers by name. The set

$$N = \{1, 2, 3, 4, 5, 6, \ldots\}$$

is called the **natural numbers** or the **counting numbers.** The three dots show that the list continues in the same pattern indefinitely. We cannot list all of the elements of the set of natural numbers, so it is an *infinite set*.

When 0 is included with the set of natural numbers, we have the set of **whole numbers,** written

$$W = \{0, 1, 2, 3, 4, 5, 6, \ldots\}.$$

A set containing no elements, such as the set of whole numbers less than 0, is called the **empty set,** or **null set,** usually written $\emptyset$ or { }.

CAUTION Do not write $\{\emptyset\}$ for the empty set; $\{\emptyset\}$ is a set with one element, $\emptyset$. Use the notation $\emptyset$ or { } for the empty set.

To write the fact that 2 is an element of the set {1, 2, 3}, we use the symbol $\in$ (read "is an element of").

$$2 \in \{1, 2, 3\}$$

The number 2 is also an element of the set of natural numbers N, so we may write

$$2 \in N.$$

To show that 0 is *not* an element of set N, we draw a slash through the symbol $\in$.

$$0 \notin N$$

Two sets are equal if they contain exactly the same elements. For example, {1, 2} = {2, 1}, because the sets contain the same elements. (Order doesn't matter.) On the other hand, {1, 2} $\neq$ {0, 1, 2} ($\neq$ means "is not equal to") since one set contains the element 0 while the other does not.

In algebra, letters called **variables** are often used to represent numbers or to define sets of numbers. For example,

$$\{x \mid x \text{ is a natural number between 3 and 15}\}$$

(read "the set of all elements x such that x is a natural number between 3 and 15") defines the set

$$\{4, 5, 6, 7, \ldots, 14\}.$$

The notation $\{x \,|\, x$ is a natural number between 3 and 15$\}$ is an example of **set-builder notation.**

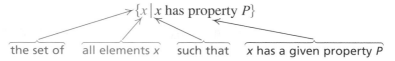

$$\{x \,|\, x \text{ has property } P\}$$

the set of   all elements x such that x has a given property P

▦ EXAMPLE 1   Listing the Elements in Sets

List the elements in each set.

(a) $\{x \,|\, x$ is a natural number less than 4$\}$
   The natural numbers less than 4 are 1, 2, and 3. This set is $\{1, 2, 3\}$.

(b) $\{y \,|\, y$ is one of the first five even natural numbers$\} = \{2, 4, 6, 8, 10\}$

(c) $\{z \,|\, z$ is a natural number greater than or equal to 7$\}$
   The set of natural numbers greater than or equal to 7 is an infinite set, written with three dots as $\{7, 8, 9, 10, \ldots\}$.   ▦

Now Try Exercise 1.

▦ EXAMPLE 2   Using Set-Builder Notation to Describe Sets

Use set-builder notation to describe each set.

(a) $\{1, 3, 5, 7, 9\}$
   There are often several ways to describe a set with set-builder notation. One way to describe this set is

$$\{y \,|\, y \text{ is one of the first five odd natural numbers}\}.$$

(b) $\{5, 10, 15, \ldots\}$
   This set can be described as $\{x \,|\, x$ is a multiple of 5 greater than 0$\}$.   ▦

Now Try Exercises 13 and 15.

OBJECTIVE 2 Use number lines.   A good way to get a picture of a set of numbers is to use a **number line.** To construct a number line, choose any point on a horizontal line and label it 0. Next, choose a point to the right of 0 and label it 1. The distance from 0 to 1 establishes a scale that can be used to locate more points, with positive numbers to the right of 0 and negative numbers to the left of 0. The number 0 is neither positive nor negative. A number line is shown in Figure 1.

FIGURE 1

The set of numbers identified on the number line in Figure 1, including positive and negative numbers and 0, is part of the set of **integers,** written

$$I = \{\ldots, -3, -2, -1, 0, 1, 2, 3, \ldots\}.$$

Each number on a number line is called the **coordinate** of the point that it labels, while the point is the **graph** of the number. Figure 2 shows a number line with several selected points graphed on it.

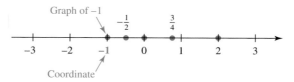

FIGURE 2

The fractions $-\frac{1}{2}$ and $\frac{3}{4}$, graphed on the number line in Figure 2, are examples of rational numbers. A **rational number** can be expressed as the quotient of two integers, with denominator not 0. Rational numbers can also be written in decimal form, either as terminating decimals such as $\frac{3}{5} = .6$, $\frac{1}{8} = .125$, or $\frac{11}{4} = 2.75$, or as repeating decimals such as $\frac{1}{3} = .33333\ldots$ or $\frac{3}{11} = .272727\ldots$. A repeating decimal is often written with a bar over the repeating digit(s). Using this notation, $.2727\ldots$ is written $.\overline{27}$.

Decimal numbers that neither terminate nor repeat are *not* rational, and thus are called **irrational numbers.** Many square roots are irrational numbers; for example, $\sqrt{2} = 1.4142136\ldots$ and $-\sqrt{7} = -2.6457513\ldots$ repeat indefinitely without pattern. $\left(\text{Some square roots }are\text{ rational: }\sqrt{16} = 4, \sqrt{100} = 10,\text{ and so on.}\right)$ Another irrational number is π, the ratio of the distance around or circumference of a circle to its diameter.

Some of the rational and irrational numbers just discussed are graphed on the number line in Figure 3. The rational numbers together with the irrational numbers make up the set of **real numbers.** Every point on a number line corresponds to a real number, and every real number corresponds to a point on the number line.

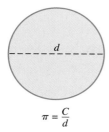

$\pi = \dfrac{C}{d}$

FIGURE 3

OBJECTIVE 3 Know the common sets of numbers. The following sets of numbers will be used throughout the rest of this text.

Sets of Numbers	
Natural numbers or counting numbers	$\{1, 2, 3, 4, 5, 6, \ldots\}$
Whole numbers	$\{0, 1, 2, 3, 4, 5, 6, \ldots\}$
Integers	$\{\ldots, -3, -2, -1, 0, 1, 2, 3, \ldots\}$

Rational numbers	$\left\{ \dfrac{p}{q} \,\middle|\, p \text{ and } q \text{ are integers, } q \neq 0 \right\}$
	Examples: $\frac{4}{1}$ or 4, 1.3, $-\frac{9}{2}$ or $-4\frac{1}{2}$, $\frac{16}{8}$ or 2, $\sqrt{9}$ or 3, $.\overline{6}$
Irrational numbers	$\{x \mid x \text{ is a real number that is not rational}\}$
	Examples: $\sqrt{3}, -\sqrt{2}, \pi$
Real numbers	$\{x \mid x \text{ is represented by a point on a number line}\}$*

The relationships among these various sets of numbers are shown in Figure 4; in particular, the figure shows that the set of real numbers includes both the rational and irrational numbers. Every real number is either rational or irrational. Also, notice that the integers are elements of the set of rational numbers and that whole numbers and natural numbers are elements of the set of integers.

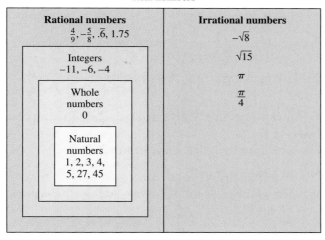

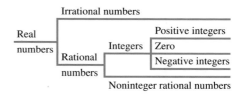

FIGURE 4 The Real Numbers

EXAMPLE 3 Identifying Examples of Number Sets

Which numbers in

$$\left\{ -8, -\sqrt{6}, -\frac{9}{64}, 0, .5, \frac{2}{3}, 1.\overline{12}, \sqrt{3}, 2 \right\}$$

are elements of each set?

(a) Integers
    -8, 0, and 2 are integers.

(b) Rational numbers
    -8, $-\frac{9}{64}$, 0, .5, $\frac{2}{3}$, $1.\overline{12}$, and 2 are rational numbers.

(c) Irrational numbers
    $-\sqrt{6}$ and $\sqrt{3}$ are irrational numbers.

(d) Real numbers
    All the numbers in the given set are real numbers.

Now Try Exercise 25.

---

*An example of a number that is not a coordinate of a point on a number line is $\sqrt{-1}$. This number, called an *imaginary number,* is discussed in Chapter 8.

▨ **EXAMPLE 4   Determining Relationships between Sets of Numbers**

Decide whether each statement is *true* or *false.*

(a) All irrational numbers are real numbers.

This is true. As shown in Figure 4, the set of real numbers includes all irrational numbers.

(b) Every rational number is an integer.

This statement is false. Although some rational numbers are integers, other rational numbers, such as $\frac{2}{3}$ and $-\frac{1}{4}$, are not.   ▨

Now Try Exercise 27.

OBJECTIVE 4   Find additive inverses.   Look again at the number line in Figure 1. For each positive number, there is a negative number on the opposite side of 0 that lies the same distance from 0. These pairs of numbers are called *additive inverses, negatives,* or *opposites* of each other. For example, 5 is the additive inverse of -5, and -5 is the additive inverse of 5.

Additive Inverse

For any real number a, the number $-a$ is the **additive inverse** of a.

Change the sign of a number to get its additive inverse. The sum of a number and its additive inverse is always 0.

The symbol "$-$" can be used to indicate any of the following:

1. a negative number, such as -9 or -15;

2. the additive inverse of a number, as in "-4 is the additive inverse of 4";

3. subtraction, as in $12 - 3$.

In the expression $-(-5)$, the symbol "$-$" is being used in two ways: the first $-$ indicates the additive inverse of -5, and the second indicates a negative number, -5. Since the additive inverse of -5 is 5, then $-(-5) = 5$. This example suggests the following property.

Number	Additive Inverse
6	-6
-4	4
$\frac{2}{3}$	$-\frac{2}{3}$
-8.7	8.7
0	0

For any real number a, $-(-a) = a.$

Numbers written with positive or negative signs, such as $+4$, $+8$, -9, and -5, are called **signed numbers.** A positive number can be called a signed number even though the positive sign is usually left off. The table in the margin shows the additive inverses of several signed numbers. The number 0 is its own additive inverse.

OBJECTIVE 5   Use absolute value.   Geometrically, the **absolute value** of a number a, written $|a|$, is the distance on the number line from 0 to a. For example, the absolute value of 5 is the same as the absolute value of -5 because each number lies five units from 0. See Figure 5. That is,

$$|5| = 5 \qquad \text{and} \qquad |-5| = 5.$$

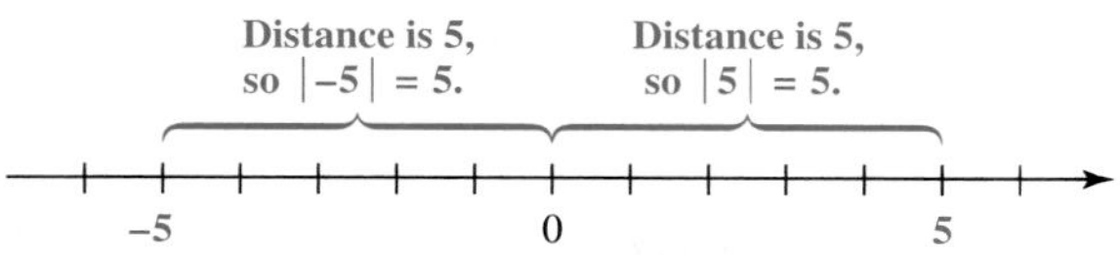

Distance is 5,
so $|{-}5| = 5$.

Distance is 5,
so $|5| = 5.$

FIGURE 5

CAUTION   Because absolute value represents distance, and distance is always positive (or 0), *the absolute value of a number is always positive (or 0).*

The formal definition of absolute value follows.

Absolute Value

$$|a| = \begin{cases} a & \text{if } a \text{ is positive or 0} \\ -a & \text{if } a \text{ is negative} \end{cases}$$

The second part of this definition, $|a| = -a$ if a is negative, requires careful thought. If a is a *negative* number, then $-a$, the additive inverse or opposite of a, is a positive number, so $|a|$ is positive. For example, if $a = -3$, then

$$|a| = |-3| = -(-3) = 3. \qquad |a| = -a \text{ if } a \text{ is negative.}$$

EXAMPLE 5   Evaluating Absolute Value Expressions

Find the value of each expression.

(a) $|13| = 13$ **(b)** $|-2| = -(-2) = 2$ **(c)** $|0| = 0$

(d) $-|8|$

Evaluate the absolute value first. Then find the additive inverse.

$$-|8| = -(8) = -8$$

(e) $-|-8|$

Work as in part (d): $|-8| = 8$, so

$$-|-8| = -(8) = -8.$$

(f) $|-2| + |5|$

Evaluate each absolute value first, then add.

$$|-2| + |5| = 2 + 5 = 7$$

Now Try Exercises 43, 47, 49, and 53.

Absolute value is useful in applications comparing size without regard to sign.

EXAMPLE 6   Comparing Rates of Change in Industries

The projected annual rates of employment change (in percent) in some of the fastest growing and most rapidly declining industries from 1994 through 2005 are shown in the table on the next page.

Industry (1994–2005)	Percent Rate of Change
Health services	5.7
Computer and data processing services	4.9
Child day care services	4.3
Footware, except rubber and plastic	−6.7
Household audio and video equipment	−4.2
Luggage, handbags, and leather products	−3.3

Source: U.S. Bureau of Labor Statistics.

What industry in the list is expected to see the greatest change? the least change?

We want the greatest *change,* without regard to whether the change is an increase or a decrease. Look for the number in the list with the largest absolute value. That number is found in footware, since $|-6.7| = 6.7$. Similarly, the least change is in the luggage, handbags, and leather products industry: $|-3.3| = 3.3$.

Now Try Exercise 59.

OBJECTIVE 6   Use inequality symbols.   The statement $4 + 2 = 6$ is an **equation;** it states that two quantities are equal. The statement $4 \neq 6$ (read "4 is not equal to 6") is an **inequality,** a statement that two quantities are *not* equal. When two numbers are not equal, one must be less than the other. The symbol $<$ means "is less than." For example,

$$8 < 9, \quad -6 < 15, \quad -6 < -1, \quad \text{and} \quad 0 < \frac{4}{3}.$$

The symbol $>$ means "is greater than." For example,

$$12 > 5, \quad 9 > -2, \quad -4 > -6, \quad \text{and} \quad \frac{6}{5} > 0.$$

Notice that in each case, the symbol "points" toward the smaller number.

The number line in Figure 6 shows the graphs of the numbers 4 and 9. We know that $4 < 9$. On the graph, 4 is to the left of 9. The smaller of two numbers is always to the left of the other on a number line.

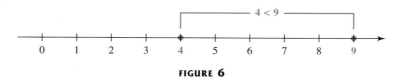

FIGURE 6

Inequalities on a Number Line

On a number line,

$$a < b \text{ if } a \text{ is to the left of } b; \qquad a > b \text{ if } a \text{ is to the right of } b.$$

We can use a number line to determine order. As shown on the number line in Figure 7, -6 is located to the left of 1. For this reason, $-6 < 1$. Also, $1 > -6$. From the same number line, $-5 < -2$, or $-2 > -5$.

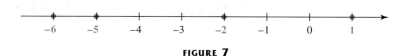

FIGURE 7

CAUTION   Be careful when ordering negative numbers. Since -5 is to the left of -2 on the number line in Figure 7, $-5 < -2$, or $-2 > -5$. In each case, the symbol points to -5, the smaller number.

The following table summarizes results about positive and negative numbers in both words and symbols.

Words	Symbols
Every negative number is less than 0.	If a is negative, then $a < 0$.
Every positive number is greater than 0.	If a is positive, then $a > 0$.
0 is neither positive nor negative.	

In addition to the symbols $\neq$, $<$, and $>$, the symbols $\leq$ and $\geq$ are often used.

INEQUALITY SYMBOLS

Symbol	Meaning	Example
$\neq$	is not equal to	$3 \neq 7$
$<$	is less than	$-4 < -1$
$>$	is greater than	$3 > -2$
$\leq$	is less than or equal to	$6 \leq 6$
$\geq$	is greater than or equal to	$-8 \geq -10$

The following table shows several inequalities and why each is true.

Inequality	Why It Is True
$6 \leq 8$	$6 < 8$
$-2 \leq -2$	$-2 = -2$
$-9 \geq -12$	$-9 > -12$
$-3 \geq -3$	$-3 = -3$
$6 \cdot 4 \leq 5(5)$	$24 < 25$

Notice the reason why $-2 \leq -2$ is true. With the symbol $\leq$, if *either* the $<$ part *or* the $=$ part is true, then the inequality is true. This is also the case with the $\geq$ symbol.

In the last row of the table, recall that the dot in $6 \cdot 4$ indicates the product 6×4, or 24, and $5(5)$ means 5×5, or 25. Thus, the inequality $6 \cdot 4 \leq 5(5)$ becomes $24 \leq 25$, which is true.

OBJECTIVE 7   Graph sets of real numbers.   Inequality symbols and variables are used to write sets of real numbers. For example, the set $\{x \mid x > -2\}$ consists of all the real numbers greater than -2. On a number line, we show the elements of this set (the set of all real numbers to the right of -2) by drawing an arrow from -2 to the right. We use a parenthesis at -2 to indicate that -2 is *not* an element of the given set. The result, shown in Figure 8, is the graph of the set $\{x \mid x > -2\}$.

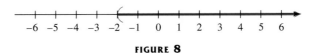

FIGURE 8

The set of numbers greater than -2 is an example of an **interval** on the number line. To write intervals, we use **interval notation.** Using this notation, we write the interval of all numbers greater than -2 as $(-2, \infty)$. The infinity symbol ∞ does not indicate a number; it shows that the interval includes all real numbers greater than -2. The left parenthesis indicates that -2 is not included. A parenthesis is *always* used next to the infinity symbol in interval notation. The set of all real numbers is written in interval notation as $(-\infty, \infty)$.

▨ EXAMPLE 7   Graphing an Inequality Written in Interval Notation

Write $\{x \mid x < 4\}$ in interval notation and graph the interval.

The interval is written $(-\infty, 4)$. The graph is shown in Figure 9. Since the elements of the set are all real numbers *less than* 4, the graph extends to the left.

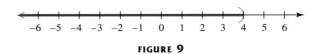

FIGURE 9

Now Try Exercise 101.

The set $\{x \mid x \le -6\}$ includes all real numbers less than or equal to -6. To show that -6 is part of the set, a square bracket is used at -6, as shown in Figure 10. In interval notation, this set is written $(-\infty, -6]$.

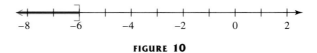

FIGURE 10

▨ EXAMPLE 8   Graphing an Inequality Written in Interval Notation

Write $\{x \mid x \ge -4\}$ in interval notation and graph the interval.

This set is written in interval notation as $[-4, \infty)$. The graph is shown in Figure 11. We use a square bracket at -4 since -4 is part of the set.

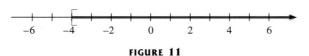

FIGURE 11

Now Try Exercise 103.

NOTE   In a previous course you may have graphed $\{x \mid x > -2\}$ using an open circle instead of a parenthesis at -2. Also, you may have graphed $\{x \mid x \geq -4\}$ using a solid dot instead of a bracket at -4.

It is common to graph sets of numbers that are *between* two given numbers. For example, the set $\{x \mid -2 < x < 4\}$ includes all real numbers between -2 and 4, but not the numbers -2 and 4 themselves. This set is written in interval notation as $(-2, 4)$. The graph has a heavy line between -2 and 4 with parentheses at -2 and 4. See Figure 12. The inequality $-2 < x < 4$ is read "-2 is less than x and x is less than 4," or "x is between -2 and 4."

FIGURE 12

▨ EXAMPLE 9   Graphing a Three-Part Inequality

Write $\{x \mid 3 < x \leq 10\}$ in interval notation and graph the interval.

Use a parenthesis at 3 and a square bracket at 10 to get $(3, 10]$ in interval notation. The graph is shown in Figure 13. Read the inequality $3 < x \leq 10$ as "3 is less than x and x is less than or equal to 10," or "x is between 3 and 10, excluding 3 and including 10."

FIGURE 13

Now Try Exercise 109.

1.1 EXERCISES

Write each set by listing its elements. See Example 1.

1. $\{x \mid x$ is a natural number less than 6$\}$

2. $\{m \mid m$ is a natural number less than 9$\}$

3. $\{z \mid z$ is an integer greater than 4$\}$

4. $\{y \mid y$ is an integer greater than 8$\}$

5. $\{z \mid z$ is an integer less than or equal to 4$\}$

6. $\{p \mid p$ is an integer less than 3$\}$

7. $\{a \mid a$ is an even integer greater than 8$\}$

8. $\{k \mid k$ is an odd integer less than 1$\}$

9. $\{x \mid x$ is an irrational number that is also rational$\}$

10. $\{r \mid r$ is a number that is both positive and negative$\}$

11. $\{p \mid p$ is a number whose absolute value is 4$\}$

12. $\{w \mid w$ is a number whose absolute value is 7$\}$

Write each set using set-builder notation. See Example 2. (More than one description is possible.)

13. $\{2, 4, 6, 8\}$

14. $\{11, 12, 13, 14\}$

15. $\{4, 8, 12, 16, \ldots\}$

16. $\{\ldots, -6, -3, 0, 3, 6, \ldots\}$

17. A student claimed that $\{x \mid x$ is a natural number greater than 3$\}$ and $\{y \mid y$ is a natural number greater than 3$\}$ actually name the same set, even though different variables are used. Was this student correct?

18. A student claimed that $\{\emptyset\}$ and $\emptyset$ name the same set. Was this student correct?

Graph the elements of each set on a number line.

19. $\{-3, -1, 0, 4, 6\}$

20. $\{-4, -2, 0, 3, 5\}$

21. $\left\{-\dfrac{2}{3}, 0, \dfrac{4}{5}, \dfrac{12}{5}, \dfrac{9}{2}, 4.8\right\}$

22. $\left\{-\dfrac{6}{5}, -\dfrac{1}{4}, 0, \dfrac{5}{6}, \dfrac{13}{4}, 5.2, \dfrac{11}{2}\right\}$

✍ **23.** Explain the difference between the graph of a number and the coordinate of a point.

✍ **24.** Explain why the real numbers $.36$ and $.\overline{36}$ have different points as graphs on a number line.

*Which elements of each set are **(a)** natural numbers, **(b)** whole numbers, **(c)** integers, **(d)** rational numbers, **(e)** irrational numbers, **(f)** real numbers? See Example 3.*

25. $\left\{-8, -\sqrt{5}, -.6, 0, \dfrac{3}{4}, \sqrt{3}, \pi, 5, \dfrac{13}{2}, 17, \dfrac{40}{2}\right\}$

26. $\left\{-9, -\sqrt{6}, -.7, 0, \dfrac{6}{7}, \sqrt{7}, 4.\overline{6}, 8, \dfrac{21}{2}, 13, \dfrac{75}{5}\right\}$

Decide whether each statement is true *or* false. *If false, tell why. See Example 4.*

27. Every integer is a whole number.

28. Every natural number is an integer.

29. Every irrational number is an integer.

30. Every integer is a rational number.

31. Every natural number is a whole number.

32. Some rational numbers are irrational.

33. Some rational numbers are whole numbers.

34. Some real numbers are integers.

35. The absolute value of any number is the same as the absolute value of its additive inverse.

36. The absolute value of any nonzero number is positive.

*Give **(a)** the additive inverse and **(b)** the absolute value of each number. See the discussion of additive inverses and Example 5.*

37. 6   **38.** 8   **39.** -12 **40.** -15 **41.** $\dfrac{6}{5}$ **42.** $.13$

Find the value of each expression. See Example 5.

43. $|-8|$ **44.** $|-11|$ **45.** $\left|\dfrac{3}{2}\right|$ **46.** $\left|\dfrac{7}{4}\right|$

47. $-|5|$ **48.** $-|17|$ **49.** $-|-2|$ **50.** $-|-8|$

51. $-|4.5|$ **52.** $-|12.6|$ **53.** $|-2| + |3|$ **54.** $|-16| + |12|$

55. $|-9| - |-3|$ **56.** $|-10| - |-5|$

57. $|-1| + |-2| - |-3|$ **58.** $|-6| + |-4| - |-10|$

Solve each problem. See Example 6.

59. The table shows the percent change in population from 1990 through 1999 for some of the largest cities in the United States.

City	Percent Change
New York	1.4
Los Angeles	4.2
Chicago	.6
Philadelphia	−10.6
Houston	8.7
Detroit	−6.1

Source: U.S. Bureau of the Census.

(a) Which city had the greatest change in population? What was this change? Was it an increase or a decline?

(b) Which city had the smallest change in population? What was this change? Was it an increase or a decline?

60. The table gives the net trade balance, in millions of dollars, for selected U.S. trade partners for April 2002.

Country	Trade Balance (in millions of dollars)
Germany	−2815
China	−7552
Netherlands	823
France	−951
Turkey	96
Australia	373

Source: U.S. Bureau of the Census.

A negative balance means that imports exceeded exports, while a positive balance means that exports exceeded imports.

(a) Which country had the greatest discrepancy between exports and imports? Explain.

(b) Which country had the smallest discrepancy between exports and imports? Explain.

Sea level refers to the surface of the ocean. The depth of a body of water such as an ocean or sea can be expressed as a negative number, representing average depth in feet below sea level. On the other hand, the altitude of a mountain can be expressed as a positive number, indicating its height in feet above sea level. The table gives selected depths and heights.

Body of Water	Average Depth in Feet (as a negative number)	Mountain	Altitude in Feet (as a positive number)
Pacific Ocean	−12,925	McKinley	20,320
South China Sea	−4,802	Point Success	14,158
Gulf of California	−2,375	Matlalcueyetl	14,636
Caribbean Sea	−8,448	Rainier	14,410
Indian Ocean	−12,598	Steele	16,644

Source: World Almanac and Book of Facts, 2002.

61. List the bodies of water in order, starting with the deepest and ending with the shallowest.

62. List the mountains in order, starting with the shortest and ending with the tallest.

63. *True* or *false:* The absolute value of the depth of the Pacific Ocean is greater than the absolute value of the depth of the Indian Ocean.

64. *True* or *false:* The absolute value of the depth of the Gulf of California is greater than the absolute value of the depth of the Caribbean Sea.

Use the number line to answer true *or* false *to each statement.*

$$-6\ -5\ -4\ -3\ -2\ -1\quad 0\quad 1\quad 2\quad 3\quad 4\quad 5\quad 6$$

65. $-6 < -2$ **66.** $-4 < -3$ **67.** $-4 > -3$ **68.** $-2 > -1$

69. $3 > -2$ **70.** $5 > -3$ **71.** $-3 \geq -3$ **72.** $-4 \leq -4$

Rewrite each statement with $>$ so that it uses $<$ instead; rewrite each statement with $<$ so that it uses $>$.

73. $6 > 2$ **74.** $4 > 1$ **75.** $-9 < 4$ **76.** $-5 < 1$

77. $-5 > -10$ **78.** $-8 > -12$ **79.** $0 < x$ **80.** $-2 < x$

Use an inequality symbol to write each statement.

81. 7 is greater than y.

82. -4 is less than 12.

83. 5 is greater than or equal to 5.

84. -3 is less than or equal to -3.

85. $3t - 4$ is less than or equal to 10.

86. $5x + 4$ is greater than or equal to 19.

87. $5x + 3$ is not equal to 0.

88. $6x + 7$ is not equal to -3.

89. t is between -3 and 5.

90. r is between -4 and 12.

91. $3x$ is between -3 and 4, including -3 and excluding 4.

92. $5y$ is between -2 and 6, excluding -2 and including 6.

First simplify each side of the inequality. Then tell whether the resulting statement is true *or* false.

93. $-6 < 7 + 3$ **94.** $-7 < 4 + 2$ **95.** $2 \cdot 5 \geq 4 + 6$

96. $8 + 7 \leq 3 \cdot 5$ **97.** $-|-3| \geq -3$ **98.** $-|-5| \leq -5$

99. $-8 > -|-6|$ **100.** $-9 > -|-4|$

Write each set using interval notation and graph the interval. See Examples 7–9.

101. $\{x \mid x > -1\}$ **102.** $\{x \mid x < 5\}$ **103.** $\{x \mid x \leq 6\}$

104. $\{x \mid x \geq -3\}$ **105.** $\{x \mid 0 < x < 3.5\}$ **106.** $\{x \mid -4 < x < 6.1\}$

107. $\{x \mid 2 \leq x \leq 7\}$ **108.** $\{x \mid -3 \leq x \leq -2\}$ **109.** $\{x \mid -4 < x \leq 3\}$

110. $\{x \mid 3 \leq x < 6\}$ **111.** $\{x \mid 0 < x \leq 3\}$ **112.** $\{x \mid -1 \leq x < 6\}$

The graph on the next page shows egg production in millions of eggs in selected states for 1998 and 1999. Use this graph to work Exercises 113–116.

113. In 1999, which states had production greater than 500 million eggs?

114. In which states was 1999 egg production less than 1998 egg production?

115. If x represents 1999 egg production for Texas (TX) and y represents 1999 egg production for Ohio (OH), which is true: $x < y$ or $x > y$?

116. If x represents 1999 egg production for Indiana (IN) and y represents 1999 egg production for Pennsylvania (PA), write an equation or inequality that compares the production in these two states.

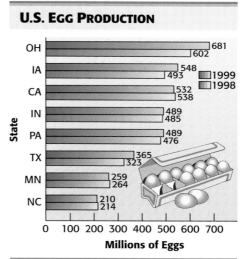

U.S. Egg Production

Source: Iowa Agricultural Statistics.

✍ **117.** List the sets of numbers introduced in this section. Give a short explanation, including three examples, for each set.

✍ **118.** List at least five symbols introduced in this section, and give a true statement involving each one.

1.2   Operations on Real Numbers

OBJECTIVES

1 Add real numbers.

2 Subtract real numbers.

3 Find the distance between two points on a number line.

4 Multiply real numbers.

5 Divide real numbers.

In this section we review the rules for adding, subtracting, multiplying, and dividing real numbers.

OBJECTIVE 1 Add real numbers.   Recall that the answer to an addition problem is called the **sum.** The rules for adding real numbers follow.

Adding Real Numbers

Like signs   To add two numbers with the *same* sign, add their absolute values. The sign of the answer (either $+$ or $-$) is the same as the sign of the two numbers.

Unlike signs   To add two numbers with *different* signs, subtract the smaller absolute value from the larger. The sign of the answer is the same as the sign of the number with the larger absolute value.

▓ **EXAMPLE 1**   Adding Two Negative Numbers

Find each sum.

(a) $-12 + (-8)$

First find the absolute values.

$$|-12| = 12 \qquad \text{and} \qquad |-8| = 8$$

Because -12 and -8 have the *same* sign, add their absolute values. Both numbers are negative, so the answer is negative.

$$-12 + (-8) = -(12 + 8) = -(20) = -20$$

(b) $-6 + (-3) = -(|-6| + |-3|) = -(6 + 3) = -9$

(c) $-1.2 + (-.4) = -(1.2 + .4) = -1.6$

(d) $-\dfrac{5}{6} + \left(-\dfrac{1}{3}\right) = -\left(\dfrac{5}{6} + \dfrac{1}{3}\right) = -\left(\dfrac{5}{6} + \dfrac{2}{6}\right) = -\dfrac{7}{6}$

Now Try Exercise 11.

▨ EXAMPLE 2   Adding Numbers with Different Signs

Find each sum.

(a) $-17 + 11$

First find the absolute values.

$$|-17| = 17 \qquad \text{and} \qquad |11| = 11$$

Because -17 and 11 have *different* signs, subtract their absolute values.

$$17 - 11 = 6$$

The number -17 has a larger absolute value than 11, so the answer is negative.

$$-17 + 11 = -6$$

Negative because $|-17| > |11|$

(b) $4 + (-1)$

Subtract the absolute values, 4 and 1. Because 4 has the larger absolute value, the sum must be positive.

$$4 + (-1) = 4 - 1 = 3$$

Positive because $|4| > |-1|$

(c) $-9 + 17 = 17 - 9 = 8$

(d) $-16 + 12$

The absolute values are 16 and 12. Subtract the absolute values. The negative number has the larger absolute value, so the answer is negative.

$$-16 + 12 = -(16 - 12) = -4$$

(e) $-\dfrac{4}{5} + \dfrac{2}{3}$

Write each fraction with a common denominator.

$$\frac{4}{5} = \frac{4 \cdot 3}{5 \cdot 3} = \frac{12}{15} \qquad \text{and} \qquad \frac{2}{3} = \frac{2 \cdot 5}{3 \cdot 5} = \frac{10}{15}$$

$$-\frac{4}{5} + \frac{2}{3} = -\frac{12}{15} + \frac{10}{15}$$

$$= -\left(\frac{12}{15} - \frac{10}{15}\right) \qquad -\frac{12}{15} \text{ has the larger absolute value.}$$

$$= -\frac{2}{15} \qquad \text{Subtract.}$$

(f) $-2.3 + 5.6 = 3.3$

Now Try Exercises 13, 15, and 17.

OBJECTIVE 2 Subtract real numbers. Recall that the answer to a subtraction problem is called the **difference.** Thus, the difference between 6 and 4 is 2. To see how subtraction should be defined, compare the following two statements.

$$6 - 4 = 2$$
$$6 + (-4) = 2$$

Similarly, $9 - 3 = 6$ and $9 + (-3) = 6$ so that $9 - 3 = 9 + (-3)$. To subtract 3 from 9, we add the additive inverse of 3 to 9. These examples suggest the following rule for subtraction.

Subtraction

For all real numbers a and b,

$$a - b = a + (-b).$$

That is, change the sign of the second number and add.

EXAMPLE 3 Subtracting Real Numbers

Find each difference.

Change to addition.
Change sign of second number.

(a) $6 - 8 = 6 + (-8) = -2$

Changed
Sign changed

(b) $-12 - 4 = -12 + (-4) = -16$

(c) $-10 - (-7) = -10 + [-(-7)]$ This step is often omitted.
$$= -10 + 7$$
$$= -3$$

(d) $-2.4 - (-8.1) = -2.4 + 8.1 = 5.7$

(e) $\dfrac{8}{3} - \left(-\dfrac{5}{3}\right) = \dfrac{8}{3} + \dfrac{5}{3} = \dfrac{13}{3}$

Now Try Exercises 19, 23, 25, and 27.

When working a problem that involves both addition and subtraction, add and subtract in order from left to right. Work inside brackets or parentheses first.

EXAMPLE 4   Adding and Subtracting Real Numbers

Perform the indicated operations.

(a) $15 - (-3) - 5 - 12 = (15 + 3) - 5 - 12$ Work from left to right.
$$= 18 - 5 - 12$$
$$= 13 - 12$$
$$= 1$$

(b) $-9 - [-8 - (-4)] + 6 = -9 - [-8 + 4] + 6$ Work inside brackets.
$$= -9 - [-4] + 6$$
$$= -9 + 4 + 6$$
$$= -5 + 6$$
$$= 1$$

Now Try Exercises 39 and 41.

OBJECTIVE 3 Find the distance between two points on a number line.   The number line in Figure 14 shows several points. To find the distance between the points 4 and 7, we subtract: $7 - 4 = 3$. Since distance is always positive (or 0), we must be careful to subtract in such a way that the answer is positive (or 0). Or, to avoid this problem altogether, we can find the absolute value of the difference. Then the distance between 4 and 7 is either

$$|7 - 4| = |3| = 3 \qquad \text{or} \qquad |4 - 7| = |-3| = 3.$$

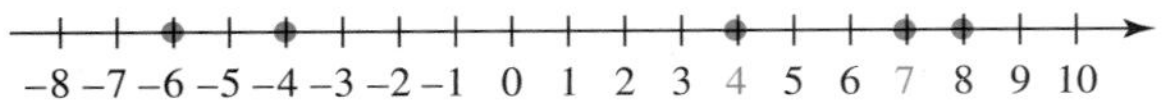

FIGURE 14

Distance

The **distance** between two points on a number line is the absolute value of the difference between the numbers.

EXAMPLE 5   Finding Distance between Points on the Number Line

Find the distance between each pair of points from Figure 14.

(a) 8 and -4

Find the absolute value of the difference of the numbers, taken in either order.

$$|8 - (-4)| = 12 \qquad \text{or} \qquad |-4 - 8| = 12$$

(b) -4 and -6

$$|-4 - (-6)| = 2 \qquad \text{or} \qquad |-6 - (-4)| = 2$$

Now Try Exercise 51.

OBJECTIVE 4 Multiply real numbers.   The answer to a multiplication problem is called the **product.** For example, 24 is the product of 8 and 3. The rules for finding signs of products of real numbers are given next.

> **Multiplying Real Numbers**
>
> *Like signs*   The product of two numbers with the *same* sign is positive.
>
> *Unlike signs*   The product of two numbers with *different* signs is negative.

◾ EXAMPLE 6   Multiplying Real Numbers

Find each product.

(a) $-3(-9) = 27$ Same sign; product is positive.

(b) $-.5(-.4) = .2$ **(c)** $-\dfrac{3}{4}\left(-\dfrac{5}{3}\right) = \dfrac{5}{4}$

(d) $6(-9) = -54$ Different signs; product is negative.

(e) $-.05(.3) = -.015$ **(f)** $\dfrac{2}{3}(-3) = -2$ **(g)** $-\dfrac{5}{8}\left(\dfrac{12}{13}\right) = -\dfrac{15}{26}$ ◾

> **Now Try Exercises 61, 65, 67, and 73.**

OBJECTIVE 5 Divide real numbers.   Earlier, we defined subtraction in terms of addition. Now we define division in terms of multiplication. The result of dividing one number by another is called the **quotient.** The quotient of two real numbers $a \div b$ ($b \neq 0$) is the real number q such that $q \cdot b = a$. That is,

$$a \div b = q \quad \text{only if} \quad q \cdot b = a.$$

For example, $36 \div 9 = 4$ since $4 \cdot 9 = 36$. Similarly, $35 \div (-5) = -7$ since $-7(-5) = 35$. The quotient $a \div b$ can also be denoted $\frac{a}{b}$. Thus, $35 \div (-5)$ can be written $\frac{35}{-5}$. As above, $\frac{35}{-5} = -7$ since -7 answers the question, "What number multiplied by -5 gives the product 35?" Now consider $\frac{5}{0}$. There is *no* number whose product with 0 gives 5. On the other hand, $\frac{0}{0}$ would be satisfied by *every* real number, because any number multiplied by 0 gives 0. When dividing, we always want a *unique* quotient, and therefore ***division by 0 is undefined.*** Thus,

$$\frac{15}{0} \text{ is undefined} \qquad \text{and} \qquad -\frac{1}{0} \text{ is undefined.}$$

CAUTION   Division by 0 is undefined. However, dividing 0 by a nonzero number gives the quotient 0. For example,

$$\frac{6}{0} \text{ is undefined,} \quad \text{but} \quad \frac{0}{6} = 0 \quad \text{(since } 0 \cdot 6 = 0\text{).}$$

Be careful when 0 is involved in a division problem.

Recall that $\frac{a}{b} = a \cdot \frac{1}{b}$. Thus, dividing by b is the same as multiplying by $\frac{1}{b}$. If $b \neq 0$, then $\frac{1}{b}$ is the **reciprocal** (or *multiplicative inverse*) of b. When multiplied, reciprocals have a product of 1. The table on the next page gives several numbers and their reciprocals. There is no reciprocal for 0 because there is no number that can be multiplied by 0 to give a product of 1.

Number	Reciprocal	
$-\frac{2}{5}$	$-\frac{5}{2}$	$-\frac{2}{5}\left(-\frac{5}{2}\right) = 1$
-6	$-\frac{1}{6}$	$-6\left(-\frac{1}{6}\right) = 1$
$\frac{7}{11}$	$\frac{11}{7}$	$\frac{7}{11}\left(\frac{11}{7}\right) = 1$
$.05$	20	$.05(20) = 1$
0	None	

CAUTION A number and its additive inverse have *opposite* signs; however, a number and its reciprocal always have the *same* sign.

The preceding discussion suggests the following definition of division.

Division

For all real numbers a and b (where $b \neq 0$),

$$a \div b = \frac{a}{b} = a \cdot \frac{1}{b}.$$

That is, multiply the first number by the reciprocal of the second number.

Since division is defined as multiplication by the reciprocal, the rules for signs of quotients are the same as those for signs of products.

Dividing Real Numbers

Like signs The quotient of two nonzero real numbers with the *same* sign is positive.

Unlike signs The quotient of two nonzero real numbers with *different* signs is negative.

EXAMPLE 7 Dividing Real Numbers

Find each quotient.

(a) $\dfrac{-12}{4} = -12 \cdot \dfrac{1}{4} = -3$ $\qquad \frac{a}{b} = a \cdot \frac{1}{b}$

(b) $\dfrac{6}{-3} = 6\left(-\dfrac{1}{3}\right) = -2$ $\qquad$ The reciprocal of -3 is $-\frac{1}{3}$.

(c) $\dfrac{-\dfrac{2}{3}}{-\dfrac{5}{9}} = -\dfrac{2}{3} \cdot \left(-\dfrac{9}{5}\right) = \dfrac{6}{5}$ $\qquad$ The reciprocal of $-\frac{5}{9}$ is $-\frac{9}{5}$.

Now Try Exercises 75, 77, and 87.

The rules for multiplication and division suggest the following results.

Equivalent Forms of a Fraction

The fractions $\dfrac{-x}{y}$, $\dfrac{x}{-y}$, and $-\dfrac{x}{y}$ are equivalent. (Assume $y \neq 0$.)

Example: $\dfrac{-4}{7} = \dfrac{4}{-7} = -\dfrac{4}{7}$.

The fractions $\dfrac{x}{y}$ and $\dfrac{-x}{-y}$ are equivalent.

Example: $\dfrac{4}{7} = \dfrac{-4}{-7}$.

The forms $\frac{x}{-y}$ and $\frac{-x}{-y}$ are not used very often.

Every fraction has three signs: the sign of the numerator, the sign of the denominator, and the sign of the fraction itself. Changing any two of these three signs does not change the value of the fraction. Changing only one sign, or changing all three, *does* change the value.

1.2 EXERCISES

For Extra Help

Student's
Solutions Manual

MyMathLab

InterAct Math
Tutorial Software

AW Math
Tutor Center

MathXL

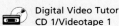
Digital Video Tutor
CD 1/Videotape 1

Complete each statement and give an example.

1. The sum of a positive number and a negative number is 0 if _____.

2. The sum of two positive numbers is a _____ number.

3. The sum of two negative numbers is a _____ number.

4. The sum of a positive number and a negative number is negative if _____.

5. The sum of a positive number and a negative number is positive if _____.

6. The difference between two positive numbers is negative if _____.

7. The difference between two negative numbers is negative if _____.

8. The product of two numbers with like signs is _____.

9. The product of two numbers with unlike signs is _____.

10. The quotient formed by any nonzero number divided by 0 is _____, and the quotient formed by 0 divided by any nonzero number is _____.

Add or subtract as indicated. See Examples 1–3.

11. $-6 + (-13)$ **12.** $-8 + (-15)$ **13.** $13 + (-4)$

14. $19 + (-13)$ **15.** $-\dfrac{7}{3} + \dfrac{3}{4}$ **16.** $-\dfrac{5}{6} + \dfrac{3}{8}$

17. $-2.3 + .45$ **18.** $-.238 + 4.55$ **19.** $-6 - 5$

20. $-8 - 13$ **21.** $8 - (-13)$ **22.** $13 - (-22)$

23. $-16 - (-3)$ **24.** $-21 - (-8)$ **25.** $-12.31 - (-2.13)$

26. $-15.88 - (-9.22)$ **27.** $\dfrac{9}{10} - \left(-\dfrac{4}{3}\right)$ **28.** $\dfrac{3}{14} - \left(-\dfrac{1}{4}\right)$

29. $|-8 - 6|$ **30.** $|-7 - 9|$ **31.** $-|-4 + 9|$

32. $-|-5 + 7|$ **33.** $-2 - |-4|$ **34.** $9 - |-13|$

Perform the indicated operations. See Example 4.

35. $-7 + 5 - 9$ **36.** $-12 + 13 - 19$ **37.** $6 - (-2) + 8$

38. $7 - (-3) + 12$ **39.** $-9 - 4 - (-3) + 6$ **40.** $-10 - 5 - (-12) + 8$

41. $-8 - (-12) - (2 - 6)$ **42.** $-3 + (-14) + (-5 + 3)$ **43.** $-.382 + 4 - .6$

44. $3 - 2.94 - (-.63)$ **45.** $\left(-\dfrac{5}{4} - \dfrac{2}{3}\right) + \dfrac{1}{6}$ **46.** $\left(-\dfrac{5}{8} + \dfrac{1}{4}\right) - \left(-\dfrac{1}{4}\right)$

47. $-\dfrac{3}{4} - \left(\dfrac{1}{2} - \dfrac{3}{8}\right)$ **48.** $\dfrac{7}{5} - \left(\dfrac{9}{10} - \dfrac{3}{2}\right)$

49. $|-11| - |-5| - |7| + |-2|$ **50.** $|-6| + |-3| - |4| - |-8|$

The number line has several points labeled. Find the distance between each pair of points. See Example 5.

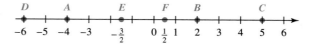

51. *A* and *B* **52.** *A* and *C* **53.** *D* and *F* **54.** *E* and *C*

✍ **55.** Give an example of a difference between two negative numbers that is equal to 5. State the rule for determining the sign of the answer after subtraction has been changed to addition.

✍ **56.** Give an example of a sum of a positive number and a negative number that is equal to 4. State the rule for determining the sign of the answer when adding two numbers with different signs.

✍ **57.** A statement that is often heard is "Two negatives give a positive." When is this true? When is it not true? Give a more precise statement that conveys this message.

✍ **58.** Explain why the reciprocal of a nonzero number must have the same sign as the number.

Multiply. See Example 6.

59. $5(-7)$ **60.** $6(-6)$ **61.** $-8(-5)$ **62.** $-10(-4)$

63. $-10\left(-\dfrac{1}{5}\right)$ **64.** $-\dfrac{1}{2}(-12)$ **65.** $\dfrac{3}{4}(-16)$ **66.** $\dfrac{4}{5}(-35)$

67. $-\dfrac{5}{2}\left(-\dfrac{12}{25}\right)$ **68.** $-\dfrac{9}{7}\left(-\dfrac{35}{36}\right)$ **69.** $-\dfrac{3}{8}\left(-\dfrac{24}{9}\right)$ **70.** $-\dfrac{2}{11}\left(-\dfrac{99}{4}\right)$

71. $-2.4(-2.45)$ **72.** $-3.45(-2.14)$ **73.** $3.4(-3.14)$ **74.** $5.66(-2.1)$

Divide where possible. See Example 7.

75. $\dfrac{-14}{2}$ **76.** $\dfrac{-26}{13}$ **77.** $\dfrac{-24}{-4}$ **78.** $\dfrac{-36}{-9}$ **79.** $\dfrac{100}{-25}$

80. $\dfrac{300}{-60}$ **81.** $\dfrac{0}{-8}$ **82.** $\dfrac{0}{-10}$ **83.** $\dfrac{5}{0}$ **84.** $\dfrac{12}{0}$

85. $-\dfrac{10}{17} \div \left(-\dfrac{12}{5}\right)$ **86.** $-\dfrac{22}{23} \div \left(-\dfrac{33}{4}\right)$ **87.** $\dfrac{\frac{12}{13}}{-\frac{4}{3}}$ **88.** $\dfrac{\frac{5}{6}}{-\frac{1}{30}}$

89. $-\dfrac{27.72}{13.2}$ **90.** $\dfrac{-126.7}{36.2}$ **91.** $\dfrac{-100}{-.01}$ **92.** $\dfrac{-50}{-.05}$

Solve each problem.

93. The highest temperature ever recorded in Juneau, Alaska, was 90°F. The lowest temperature ever recorded there was -22°F. What is the difference between these two temperatures? (*Source: World Almanac and Book of Facts,* 2002.)

94. On August 10, 1936, a temperature of 120°F was recorded in Arkansas. On February 13, 1905, Arkansas recorded a temperature of -29°F. What is the difference between these two temperatures? (*Source: World Almanac and Book of Facts,* 2002.)

95. The Standard and Poor's 500, an index measuring the performance of 500 leading stocks, had an annual return of 37.58% in 1995. For 2000, its annual return was -9.10%. Find the difference between these two percents. (*Source:* Legg Mason Wood Walker, Inc.)

96. When George W. Bush took office in January 2001, the U.S. federal budget was at a record surplus of $236 billion. It was at a record deficit of $-$255$ billion when his father, George H. W. Bush, left office in January 1993. Find the difference between these two amounts. (*Source: Economic Report of the President,* 2001.)

The table shows Social Security finances (in billions of dollars). Use this table to work Exercises 97 and 98.

Year	Tax Revenue	Cost of Benefits
2000	538	409
2010*	916	710
2020*	1479	1405
2030*	2041	2542

*Projected
Source: Social Security Board of Trustees.

97. Find the difference between Social Security tax revenue and cost of benefits for each year shown in the table.

 **98.** Interpret your answer for 2030.

Use the graph of California exports to work Exercises 99–102.

99. What is the difference between the January and February changes?

100. What is the difference between the changes in April and May?

101. Which of the following is the best estimate of the difference between the October and November changes?

    A. −6000   **B.** −6500

    C. −7000   **D.** −7500

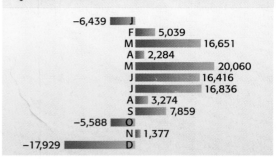

CALIFORNIA EXPORTS
Change in number of 20-ft containers exported each month in 1997 versus 1996.

J −6,439
F 5,039
M 16,651
A 2,284
M 20,060
J 16,416
J 16,836
A 3,274
S 7,859
O −5,588
N 1,377
D −17,929

Source: USA Today, June 23, 1998.

102. Which of the following is the best estimate of the difference between the November and December changes?

    A. 17,000   **B.** 18,000   **C.** 19,000   **D.** 20,000

RELATING CONCEPTS (EXERCISES 103–106)

For Individual or Group Work

In Section 1.1 we discussed the meanings of a < b, a = b, and a > b. Choose two numbers a and b such that a < b. **Work Exercises 103–106 in order.**

103. Find the difference $a - b$.

104. How does the answer in Exercise 103 compare to 0? (Is it greater than, less than, or equal to 0?)

105. Repeat Exercise 103 with different values for a and b.

106. How does the answer in Exercise 105 compare to 0? Based on your observations in these exercises, complete the following statement: If $a < b$, then $a - b$ _____ 0.

1.3   Exponents, Roots, and Order of Operations

OBJECTIVES

1 Use exponents.

2 Find square roots.

3 Use the order of operations.

4 Evaluate algebraic expressions for given values of variables.

Two or more numbers whose product is a third number are **factors** of that third number. For example, 2 and 6 are factors of 12 since $2 \cdot 6 = 12$. Other integer factors of 12 are 1, 3, 4, 12, −1, −2, −3, −4, −6, and −12.

OBJECTIVE 1 Use exponents. In algebra, we use *exponents* as a way of writing products of repeated factors. For example, the product $2 \cdot 2 \cdot 2 \cdot 2 \cdot 2$ is written

$$\underbrace{2 \cdot 2 \cdot 2 \cdot 2 \cdot 2}_{5 \text{ factors of } 2} = 2^5.$$

The number 5 shows that 2 is used as a factor 5 times. The number 5 is the *exponent,* and 2 is the *base.*

$$2^5 \longleftarrow \text{Exponent}$$
$$\text{Base}$$

Read 2^5 as "2 to the fifth power" or simply "2 to the fifth." Multiplying the five 2s gives

$$2^5 = 2 \cdot 2 \cdot 2 \cdot 2 \cdot 2 = 32.$$

Exponential Expression

If a is a real number and n is a natural number, then

$$a^n = \underbrace{a \cdot a \cdot a \cdot \ldots \cdot a,}_{n \text{ factors of } a}$$

where n is the **exponent,** a is the **base,** and a^n is an **exponential expression.** Exponents are also called **powers.**

◼ **EXAMPLE 1** Using Exponential Notation

Write each expression using exponents.

(a) $4 \cdot 4 \cdot 4$
Here, 4 is used as a factor 3 times.

$$\underbrace{4 \cdot 4 \cdot 4}_{3 \text{ factors of } 4} = 4^3$$

Read 4^3 as "4 cubed."

(b) $\dfrac{3}{5} \cdot \dfrac{3}{5} = \left(\dfrac{3}{5}\right)^2$ 2 factors of $\frac{3}{5}$

Read $\left(\frac{3}{5}\right)^2$ as "$\frac{3}{5}$ squared."

(c) $(-6)(-6)(-6)(-6) = (-6)^4$

(d) $x \cdot x \cdot x \cdot x \cdot x \cdot x = x^6$ ◼

Now Try Exercises 13, 15, 17, and 19.

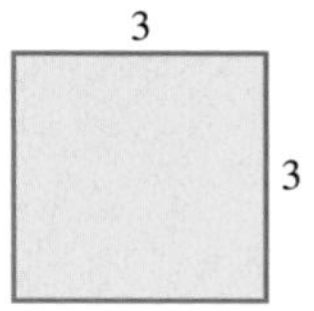

(a) $3 \cdot 3 = 3$ squared, or 3^2

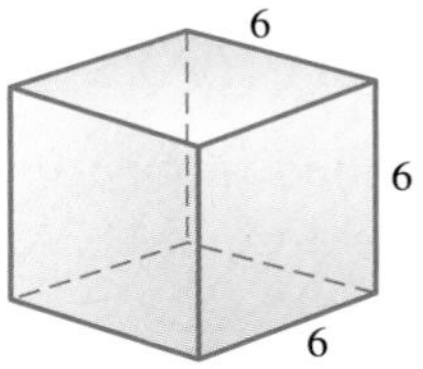

(b) $6 \cdot 6 \cdot 6 = 6$ cubed, or 6^3

FIGURE 15

In parts (a) and (b) of Example 1, we used the terms *squared* and *cubed* to refer to powers of 2 and 3, respectively. The term *squared* comes from the figure of a square, which has the same measure for both length and width, as shown in Figure 15(a). Similarly, the term *cubed* comes from the figure of a cube. As shown in Figure 15(b), the length, width, and height of a cube have the same measure.

◼ **EXAMPLE 2** Evaluating Exponential Expressions

Write each expression without exponents and evaluate.

(a) $5^2 = 5 \cdot 5 = 25$ 5 is used as a factor 2 times.

(b) $\left(\dfrac{2}{3}\right)^3 = \dfrac{2}{3} \cdot \dfrac{2}{3} \cdot \dfrac{2}{3} = \dfrac{8}{27}$ $\frac{2}{3}$ is used as a factor 3 times.

(c) $2^6 = 2 \cdot 2 \cdot 2 \cdot 2 \cdot 2 \cdot 2 = 64$ ◼

Now Try Exercises 21 and 27.

Be careful when evaluating an exponential expression with a negative sign.

▓ **EXAMPLE 3** Evaluating Exponential Expressions with Negative Signs

Evaluate each exponential expression.

(a) $(-3)^5 = (-3)(-3)(-3)(-3)(-3) = -243$

(b) $(-2)^6$

The exponent 6 applies to the number -2.

$$(-2)^6 = (-2)(-2)(-2)(-2)(-2)(-2) = 64 \qquad \text{The base is } -2.$$

(c) -2^6

Since there are no parentheses, the exponent 6 applies *only* to the number 2, not to -2.

$$-2^6 = -(2 \cdot 2 \cdot 2 \cdot 2 \cdot 2 \cdot 2) = -64 \qquad \text{The base is 2.} \qquad ▓$$

Now Try Exercises 29, 31, and 33.

Example 3 suggests the following generalizations.

> The product of an *odd* number of negative factors is negative.
> The product of an *even* number of negative factors is positive.

CAUTION As shown in Examples 3(b) and (c), it is important to distinguish between $-a^n$ and $(-a)^n$.

$$-a^n = -1\underbrace{(a \cdot a \cdot a \cdot \ldots \cdot a)}_{n \text{ factors of } a} \qquad \text{The base is } a.$$

$$(-a)^n = \underbrace{(-a)(-a) \cdot \ldots \cdot (-a)}_{n \text{ factors of } -a} \qquad \text{The base is } -a.$$

OBJECTIVE 2 **Find square roots.** As we saw in Example 2(a), $5^2 = 5 \cdot 5 = 25$, so 5 squared is 25. The opposite (inverse) of squaring a number is called taking its **square root.** For example, a square root of 25 is 5. Another square root of 25 is -5 since $(-5)^2 = 25$; thus, 25 has two square roots, 5 and -5.

We write the positive or *principal* square root of a number with the symbol $\sqrt{}$, called a **radical sign.** For example, the positive or principal square root of 25 is written $\sqrt{25} = 5$. The negative square root of 25 is written $-\sqrt{25} = -5$. Since the square of any nonzero real number is positive, *the square root of a negative number, such as* $\sqrt{-25}$, *is not a real number.*

▓ **EXAMPLE 4** Finding Square Roots

Find each square root that is a real number.

(a) $\sqrt{36} = 6$ since 6 is positive and $6^2 = 36$.

(b) $\sqrt{0} = 0$ since $0^2 = 0$.   **(c)** $\sqrt{\dfrac{9}{16}} = \dfrac{3}{4}$ since $\left(\dfrac{3}{4}\right)^2 = \dfrac{9}{16}$.

(d) $\sqrt{.16} = .4$ since $(.4)^2 = .16$. **(e)** $\sqrt{100} = 10$ since $10^2 = 100$.

(f) $-\sqrt{100} = -10$ since the negative sign is outside the radical sign.

(g) $\sqrt{-100}$ is not a real number because the negative sign is inside the radical sign. No *real number* squared equals -100.

Notice the difference among the expressions in parts (e), (f), and (g). Part (e) is the positive or principal square root of 100, part (f) is the negative square root of 100, and part (g) is the square root of -100, which is not a real number.

Now Try Exercises 37, 41, 43, and 47.

CAUTION The symbol $\sqrt{\ }$ is used only for the *positive* square root, except that $\sqrt{0} = 0$. The symbol $-\sqrt{\ }$ is used for the negative square root.

OBJECTIVE 3 Use the order of operations. To simplify an expression such as $5 + 2 \cdot 3$, what should we do first—add 5 and 2, or multiply 2 and 3? When an expression involves more than one operation symbol, we use the following **order of operations.**

Order of Operations

1. Work separately above and below any **fraction bar.**
2. If **grouping symbols** such as **parentheses (   )**, **square brackets [   ]**, or **absolute value bars | |** are present, start with the innermost set and work outward.
3. Evaluate all **powers, roots,** and **absolute values.**
4. Do any **multiplications** or **divisions** in order, working from left to right.
5. Do any **additions** or **subtractions** in order, working from left to right.

EXAMPLE 5   Using the Order of Operations

Simplify.

(a) $5 + 2 \cdot 3$

First multiply and then add.

$$5 + 2 \cdot 3 = 5 + 6 \qquad \text{Multiply.}$$
$$= 11 \qquad \text{Add.}$$

(b) $24 \div 3 \cdot 2 + 6$

Multiplications and divisions are done *in the order in which they appear from left to right,* so divide first.

$$24 \div 3 \cdot 2 + 6 = 8 \cdot 2 + 6 \qquad \text{Divide.}$$
$$= 16 + 6 \qquad \text{Multiply.}$$
$$= 22 \qquad \text{Add.}$$

Now Try Exercises 53 and 57.

EXAMPLE 6   Using the Order of Operations

Simplify.

(a) $10 \div 5 + 2|3 - 4|$

Evaluate the absolute value first.

$$10 \div 5 + 2|3 - 4| = 10 \div 5 + 2|-1|$$ Subtract inside the absolute value bars.

$$= 10 \div 5 + 2 \cdot 1$$ Take the absolute value.

$$= 2 + 2$$ Divide and multiply.

$$= 4$$ Add.

(b) $4 \cdot 3^2 + 7 - (2 + 8)$

Work inside the parentheses first.

$$4 \cdot 3^2 + 7 - (2 + 8) = 4 \cdot 3^2 + 7 - 10$$ Add inside parentheses.

$$= 4 \cdot 9 + 7 - 10$$ Evaluate powers.

$$= 36 + 7 - 10$$ Multiply.

$$= 43 - 10$$ Add.

$$= 33$$ Subtract.

(c) $\dfrac{1}{2} \cdot 4 + (6 \div 3 - 7)$

Work inside the parentheses, dividing before subtracting.

$$\frac{1}{2} \cdot 4 + (6 \div 3 - 7) = \frac{1}{2} \cdot 4 + (2 - 7)$$ Divide inside parentheses.

$$= \frac{1}{2} \cdot 4 + (-5)$$ Subtract inside parentheses.

$$= 2 + (-5)$$ Multiply.

$$= -3$$ Add.

Now Try Exercises 65 and 71.

EXAMPLE 7   Using the Order of Operations

Simplify $\dfrac{5 + (-2^3)(2)}{6 \cdot \sqrt{9} - 9 \cdot 2}$.

Work separately above and below the fraction bar.

$$\frac{5 + (-2^3)(2)}{6 \cdot \sqrt{9} - 9 \cdot 2} = \frac{5 + (-8)(2)}{6 \cdot 3 - 9 \cdot 2}$$ Evaluate powers and roots.

$$= \frac{5 - 16}{18 - 18}$$ Multiply.

$$= \frac{-11}{0}$$ Subtract.

Since division by 0 is undefined, the given expression is undefined.

Now Try Exercise 75.

OBJECTIVE 4 Evaluate algebraic expressions for given values of variables. Any collection of numbers, variables, operation symbols, and grouping symbols, such as

$$6ab, \qquad 5m - 9n, \qquad \text{and} \qquad -2(x^2 + 4y),$$

is called an **algebraic expression.** Algebraic expressions have different numerical values for different values of the variables. We can evaluate such expressions by *substituting* given values for the variables.

Algebraic expressions are used in problem solving. For example, if movie tickets cost \$7 each, the amount in dollars you pay for x tickets can be represented by the algebraic expression $7x$. We can substitute different numbers of tickets to get the costs to purchase those tickets.

▎ EXAMPLE 8 Evaluating Algebraic Expressions

Evaluate each expression if $m = -4$, $n = 5$, $p = -6$, and $q = 25$.

(a) $5m - 9n$

Replace m with -4 and n with 5.

$$5m - 9n = 5(-4) - 9(5) = -20 - 45 = -65$$

(b) $\dfrac{m + 2n}{4p} = \dfrac{-4 + 2(5)}{4(-6)} = \dfrac{-4 + 10}{-24} = \dfrac{6}{-24} = -\dfrac{1}{4}$

(c) $-3m^3 - n^2(\sqrt{q}) = -3(-4)^3 - (5)^2(\sqrt{25})$ Substitute; $m = -4$, $n = 5$, and $q = 25$.

$$= -3(-64) - 25(5)$$ Evaluate powers and roots.

$$= 192 - 125$$ Multiply.

$$= 67$$ Subtract.

Now Try Exercises 79 and 85.

CAUTION To avoid errors when evaluating expressions, use parentheses around any negative numbers that are substituted for variables, as shown in Example 8.

1.3 EXERCISES

Decide whether each statement is true *or* false. *If false, correct the statement so it is true.*

1. $-4^6 = (-4)^6$

2. $-4^7 = (-4)^7$

3. $\sqrt{16}$ is a positive number.

4. $3 + 5 \cdot 6 = 3 + (5 \cdot 6)$

5. $(-2)^7$ is a negative number.

6. $(-2)^8$ is a positive number.

7. The product of 8 positive factors and 8 negative factors is positive.

8. The product of 3 positive factors and 3 negative factors is positive.

9. In the exponential expression -3^5, -3 is the base.

10. $\sqrt{a}$ is positive for all positive numbers a.

11. Evaluate each exponential expression.

   (a) 8^2 **(b)** -8^2

   (c) $(-8)^2$ **(d)** $-(-8)^2$

12. Evaluate each exponential expression.

   (a) 4^3 **(b)** -4^3

   (c) $(-4)^3$ **(d)** $-(-4)^3$

Write each expression using exponents. See Example 1.

13. $10 \cdot 10 \cdot 10 \cdot 10$ **14.** $8 \cdot 8 \cdot 8$ **15.** $\frac{3}{4} \cdot \frac{3}{4} \cdot \frac{3}{4} \cdot \frac{3}{4} \cdot \frac{3}{4}$

16. $\frac{1}{2} \cdot \frac{1}{2}$ **17.** $(-9)(-9)(-9)$ **18.** $(-4)(-4)(-4)(-4)$

19. $z \cdot z \cdot z \cdot z \cdot z \cdot z \cdot z$ **20.** $a \cdot a \cdot a \cdot a \cdot a$

Evaluate each expression. See Examples 2 and 3.

21. 4^2 **22.** 2^4 **23.** $.28^3$ **24.** $.91^3$

25. $\left(\frac{1}{5}\right)^3$ **26.** $\left(\frac{1}{6}\right)^4$ **27.** $\left(\frac{4}{5}\right)^4$ **28.** $\left(\frac{7}{10}\right)^3$

29. $(-5)^3$ **30.** $(-2)^5$ **31.** $(-2)^8$ **32.** $(-3)^6$

33. -3^6 **34.** -4^6 **35.** -8^4 **36.** -10^3

Find each square root. If it is not a real number, say so. See Example 4.

37. $\sqrt{81}$ **38.** $\sqrt{64}$ **39.** $\sqrt{169}$ **40.** $\sqrt{225}$

41. $-\sqrt{400}$ **42.** $-\sqrt{900}$ **43.** $\sqrt{\frac{100}{121}}$ **44.** $\sqrt{\frac{225}{169}}$

45. $-\sqrt{.49}$ **46.** $-\sqrt{.64}$ **47.** $\sqrt{-36}$ **48.** $\sqrt{-121}$

49. Match each square root with the appropriate value or description.

  (a) $\sqrt{144}$ **(b)** $\sqrt{-144}$ **(c)** $-\sqrt{144}$

  A. -12 **B.** 12 **C.** Not a real number

50. Explain why $\sqrt{-900}$ is not a real number.

51. If a is a positive number, is $-\sqrt{-a}$ positive, negative, or not a real number?

52. If a is a positive number, is $-\sqrt{a}$ positive, negative, or not a real number?

Simplify each expression. Use the order of operations. See Examples 5–7.

53. $12 + 3 \cdot 4$ **54.** $15 + 5 \cdot 2$ **55.** $6 \cdot 3 - 12 \div 4$

56. $9 \cdot 4 - 8 \div 2$ **57.** $10 + 30 \div 2 \cdot 3$ **58.** $12 + 24 \div 3 \cdot 2$

59. $-3(5)^2 - (-2)(-8)$ **60.** $-9(2)^2 - (-3)(-2)$ **61.** $5 - 7 \cdot 3 - (-2)^3$

62. $-4 - 3 \cdot 5 + 6^2$ **63.** $-7\left(\sqrt{36}\right) - (-2)(-3)$ **64.** $-8\left(\sqrt{64}\right) - (-3)(-7)$

65. $6|4 - 5| - 24 \div 3$ **66.** $-4|2 - 4| + 8 \cdot 2$ **67.** $|-6 - 5|(-8) + 3^2$

68. $(-6 - 3)|-2 - 3| \div 9$ **69.** $6 + \frac{2}{3}(-9) - \frac{5}{8} \cdot 16$ **70.** $7 - \frac{3}{4}(-8) + 12 \cdot \frac{5}{6}$

71. $-14\left(-\frac{2}{7}\right) \div (2 \cdot 6 - 10)$ **72.** $-12\left(-\frac{3}{4}\right) - (6 \cdot 5 \div 3)$

73. $\dfrac{\left(-5 + \sqrt{4}\right)(-2^2)}{-5 - 1}$ **74.** $\dfrac{\left(-9 + \sqrt{16}\right)(-3^2)}{-4 - 1}$

75. $\dfrac{2(-5) + (-3)(-2)}{-8 + 3^2 - 1}$ **76.** $\dfrac{3(-4) + (-5)(-8)}{2^3 - 2 - 6}$

77. $\dfrac{5 - 3\left(\dfrac{-5 - 9}{-7}\right) - 6}{-9 - 11 + 3 \cdot 7}$ **78.** $\dfrac{-4\left(\dfrac{12 - (-8)}{3 \cdot 2 + 4}\right) - 5(-1 - 7)}{-9 - (-7) - [-5 - (-8)]}$

Evaluate each expression if a = −3, b = 64, and c = 6. See Example 8.

79. $3a + \sqrt{b}$ **80.** $-2a - \sqrt{b}$ **81.** $\sqrt{b} + c - a$ **82.** $\sqrt{b} - c + a$

83. $4a^3 + 2c$ **84.** $-3a^4 - 3c$ **85.** $\dfrac{2c + a^3}{4b + 6a}$ **86.** $\dfrac{3c + a^2}{2b - 6c}$

Solve each problem.

Residents of Linn County, Iowa in the Cedar Rapids Community School District can use the expression

$$(v \times .5485 - 4850) \div 1000 \times 31.44$$

to determine their property taxes, where v is home value. (*Source: The Gazette,* August 19, 2000.) Use the expression to calculate the amount of property taxes to the nearest dollar that the owner of a home with each of the following values would pay. Follow the order of operations.

87. $100,000 **88.** $150,000 **89.** $200,000

The Blood Alcohol Concentration (BAC) of a person who has been drinking is given by the expression

number of oz × % alcohol × .075 ÷ body weight in lb − hr of drinking × .015.

(*Source:* Lawlor, J., *Auto Math Handbook: Mathematical Calculations, Theory, and Formulas for Automotive Enthusiasts,* HP Books, 1991.)

90. Suppose a policeman stops a 190-lb man who, in 2 hr, has ingested four 12-oz beers (48 oz), each having a 3.2% alcohol content.

(a) Substitute the values in the formula, and write the expression for the man's BAC.
(b) Calculate the man's BAC to the nearest thousandth. Follow the order of operations.

91. Find the BAC to the nearest thousandth for a 135-lb woman who, in 3 hr, has drunk three 12-oz beers (36 oz), each having a 4.0% alcohol content.

92. (a) Calculate the BACs in Exercises 90 and 91 if each person weighs 25 lb more and the rest of the variables stay the same. How does increased weight affect a person's BAC?
(b) Predict how decreased weight would affect the BAC of each person in Exercises 90 and 91. Calculate the BACs if each person weighs 25 lb less and the rest of the variables stay the same.

93. An approximation of the percent of U.S. households investing in mutual funds during the years 1980 through 2000 can be obtained by substituting a given year for x in the expression

$$2.2023x - 4356.6$$

and then evaluating. (*Source:* Investment Company Institute.) Approximate the percent of U.S. households investing in mutual funds in each year. Round answers to the nearest tenth of a percent.

(a) 1980 (b) 1990 (c) 2000
(d) How has the percent of households investing in mutual funds changed from 1980 to 2000?

94. An approximation of federal spending on education in billions of dollars from 1997 through 2001 can be obtained using the expression

$$3.31714x - 6597.86,$$

where x represents the year. (*Source:* U.S. Department of Education.)

(a) Use this expression to complete the table. Round answers to the nearest tenth.

Year	Education Spending (in billions of dollars)
1997	26.5
1998	29.8
1999	_____
2000	_____
2001	_____

(b) Describe the trend in the amount of federal spending on education during these years.

1.4 Properties of Real Numbers

OBJECTIVES

1 Use the distributive property.

2 Use the inverse properties.

3 Use the identity properties.

4 Use the commutative and associative properties.

5 Use the multiplication property of 0.

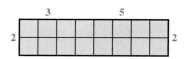

Area of left part is $2 \cdot 3 = 6$.
Area of right part is $2 \cdot 5 = 10$.
Area of total rectangle is $2(3 + 5) = 16$.

FIGURE 16

The study of any object is simplified when we know the properties of the object. For example, a property of water is that it freezes when cooled to 0°C. Knowing this helps us to predict the behavior of water.

The study of numbers is no different. The basic properties of real numbers studied in this section reflect results that occur consistently in work with numbers, so they have been generalized to apply to expressions with variables as well.

OBJECTIVE 1 Use the distributive property. Notice that

$$2(3 + 5) = 2 \cdot 8 = 16$$

and

$$2 \cdot 3 + 2 \cdot 5 = 6 + 10 = 16,$$

so

$$2(3 + 5) = 2 \cdot 3 + 2 \cdot 5.$$

This idea is illustrated by the divided rectangle in Figure 16. Similarly,

$$-4[5 + (-3)] = -4(2) = -8$$

and

$$-4(5) + (-4)(-3) = -20 + 12 = -8,$$

so

$$-4[5 + (-3)] = -4(5) + (-4)(-3).$$

These arithmetic examples are generalized to *all* real numbers as the **distributive property of multiplication with respect to addition,** or simply the **distributive property.**

Distributive Property

For any real numbers a, b, and c,

$$a(b + c) = ab + ac \qquad \text{and} \qquad (b + c)a = ba + ca.$$

The distributive property can also be written

$$ab + ac = a(b + c) \qquad \text{and} \qquad ba + ca = (b + c)a.$$

It can be extended to more than two numbers as well.

$$a(b + c + d) = ab + ac + ad$$

This property is important because it provides a way to rewrite a *product* $a(b + c)$ as a sum $ab + ac$, or a *sum* as a product.

NOTE When we rewrite $a(b + c)$ as $ab + ac$, we sometimes refer to the process as "removing parentheses."

EXAMPLE 1   Using the Distributive Property

Use the distributive property to rewrite each expression.

(a) $3(x + y)$
Use the first form of the property to rewrite this product as a sum.

$$3(x + y) = 3x + 3y$$

(b) $-2(5 + k) = -2(5) + (-2)(k)$
$$= -10 - 2k$$

(c) $4x + 8x$
Use the second form of the property to rewrite this sum as a product.

$$4x + 8x = (4 + 8)x = 12x$$

(d) $3r - 7r = 3r + (-7r)$ Definition of subtraction
$$= [3 + (-7)]r \qquad \text{Distributive property}$$
$$= -4r$$

(e) $5p + 7q$
Because there is no common number or variable here, we cannot use the distributive property to rewrite the expression.

(f) $6(x + 2y - 3z) = 6x + 6(2y) + 6(-3z)$
$$= 6x + 12y - 18z$$

Now Try Exercises 11, 13, 15, and 19.

As illustrated in Example 1(d), the distributive property can also be used for subtraction, so

$$a(b - c) = ab - ac.$$

OBJECTIVE 2 Use the inverse properties. In Section 1.1 we saw that the additive inverse of a number a is $-a$ and that the sum of a number and its additive inverse is 0. For example, 3 and -3 are additive inverses, as are -8 and 8. The number 0 is its own additive inverse. In Section 1.2, we saw that two numbers with a product of 1 are reciprocals. As mentioned there, another name for reciprocal is *multiplicative inverse*. This is similar to the idea of an additive inverse. Thus, 4 and $\frac{1}{4}$ are multiplicative inverses, as are $-\frac{2}{3}$ and $-\frac{3}{2}$. (Recall that reciprocals have the same sign.) We can extend these properties of arithmetic, the **inverse properties** of addition and multiplication, to the real numbers of algebra.

Inverse Properties

For any real number a, there is a single real number $-a$, such that
$$a + (-a) = 0 \qquad \text{and} \qquad -a + a = 0.$$

The inverse "undoes" addition with the result 0.

For any *nonzero* real number a, there is a single real number $\frac{1}{a}$ such that
$$a \cdot \frac{1}{a} = 1 \qquad \text{and} \qquad \frac{1}{a} \cdot a = 1.$$

The inverse "undoes" multiplication with the result 1.

OBJECTIVE 3 Use the identity properties. The numbers 0 and 1 each have a special property. Zero is the only number that can be added to any number to get that number. That is, adding 0 leaves the identity of a number unchanged. For this reason, 0 is called the **identity element for addition** or the **additive identity.** In a similar way, multiplying by 1 leaves the identity of any number unchanged, so 1 is the **identity element for multiplication** or the **multiplicative identity.** The following **identity properties** summarize this discussion and extend these properties from arithmetic to algebra.

Identity Properties

For any real number a,
$$a + 0 = 0 + a = a.$$

Start with a number a; add 0. The answer is "identical" to a.

Also,
$$a \cdot 1 = 1 \cdot a = a.$$

Start with a number a; multiply by 1. The answer is "identical" to a.

EXAMPLE 2 Using the Identity Property $1 \cdot a = a$

Simplify each expression.

(a) $12m + m = 12m + 1m$ Identity property

$\qquad\qquad\;\; = (12 + 1)m$ Distributive property

$\qquad\qquad\;\; = 13m$ Add inside parentheses.

(b) $y + y = 1y + 1y$ Identity property

$\quad\quad\quad = (1 + 1)y$ Distributive property

$\quad\quad\quad = 2y$ Add inside parentheses.

(c) $-(m - 5n) = -1(m - 5n)$ Identity property

$\quad\quad\quad\quad\quad = -1(m) + (-1)(-5n)$ Distributive property

$\quad\quad\quad\quad\quad = -m + 5n$ Multiply.

Now Try Exercises 21 and 23.

Expressions such as $12m$ and $5n$ from Example 2 are examples of *terms.* A **term** is a number or the product of a number and one or more variables. Terms with exactly the same variables raised to exactly the same powers are called **like terms.** Some examples of like terms are

$$5p \text{ and } -21p \quad\quad -6x^2 \text{ and } 9x^2. \quad\quad \text{Like terms}$$

Some examples of unlike terms are

$$3m \text{ and } 16x \quad\quad 7y^3 \text{ and } -3y^2. \quad\quad \text{Unlike terms}$$

The numerical factor in a term is called the **numerical coefficient,** or just the **coefficient.** For example, in the term $9x^2$, the coefficient is 9.

OBJECTIVE 4 Use the commutative and associative properties. Simplifying expressions as in parts (a) and (b) of Example 2 is called **combining like terms.** Only like terms may be combined. To combine like terms in an expression such as

$$-2m + 5m + 3 - 6m + 8,$$

we need two more properties. From arithmetic, we know that

$$3 + 9 = 12 \quad\quad \text{and} \quad\quad 9 + 3 = 12.$$

Also,

$$3 \cdot 9 = 27 \quad\quad \text{and} \quad\quad 9 \cdot 3 = 27.$$

Furthermore, notice that

$$(5 + 7) + (-2) = 12 + (-2) = 10$$

and

$$5 + [7 + (-2)] = 5 + 5 = 10.$$

Also,

$$(5 \cdot 7)(-2) = 35(-2) = -70$$

and

$$(5)[7 \cdot (-2)] = 5(-14) = -70.$$

These arithmetic examples can be extended to algebra.

Commutative and Associative Properties

For any real numbers a, b, and c,

$$a + b = b + a$$
and
$$ab = ba.$$

Commutative properties

Reverse the order of two terms or factors.

Also,

$$a + (b + c) = (a + b) + c$$
and
$$a(bc) = (ab)c.$$

Associative properties

Shift parentheses among three terms or factors; order stays the same.

The commutative properties are used to change the *order* of the terms or factors in an expression. Think of commuting from home to work and then from work to home. The associative properties are used to *regroup* the terms or factors of an expression. Remember, to *associate* is to be part of a group.

EXAMPLE 3 **Using the Commutative and Associative Properties**

Simplify $-2m + 5m + 3 - 6m + 8$.

$$-2m + 5m + 3 - 6m + 8$$
$$= (-2m + 5m) + 3 - 6m + 8 \qquad \text{Order of operations}$$
$$= (-2 + 5)m + 3 - 6m + 8 \qquad \text{Distributive property}$$
$$= 3m + 3 - 6m + 8$$

By the order of operations, the next step would be to add $3m$ and 3, but they are unlike terms. To get $3m$ and $-6m$ together, use the associative and commutative properties. Begin by inserting parentheses and brackets according to the order of operations.

$$[(3m + 3) - 6m] + 8$$
$$= [3m + (3 - 6m)] + 8 \qquad \text{Associative property}$$
$$= [3m + (-6m + 3)] + 8 \qquad \text{Commutative property}$$
$$= [(3m + [-6m]) + 3] + 8 \qquad \text{Associative property}$$
$$= (-3m + 3) + 8 \qquad \text{Combine like terms.}$$
$$= -3m + (3 + 8) \qquad \text{Associative property}$$
$$= -3m + 11 \qquad \text{Add.}$$

In practice, many of these steps are not written down, but you should realize that the commutative and associative properties are used whenever the terms in an expression are rearranged to combine like terms.

Now Try Exercise 27.

EXAMPLE 4 **Using the Properties of Real Numbers**

Simplify each expression.

(a) $5y - 8y - 6y + 11y$
$$= (5 - 8 - 6 + 11)y \qquad \text{Distributive property}$$
$$= 2y \qquad \text{Combine like terms.}$$

(b) $3x + 4 - 5(x + 1) - 8$

$$= 3x + 4 - 5x - 5 - 8 \qquad \text{Distributive property}$$
$$= 3x - 5x + 4 - 5 - 8 \qquad \text{Commutative property}$$
$$= -2x - 9 \qquad \text{Combine like terms.}$$

(c) $8 - (3m + 2) = 8 - 1(3m + 2) \qquad \text{Identity property}$

$$= 8 - 3m - 2 \qquad \text{Distributive property}$$
$$= 6 - 3m \qquad \text{Combine like terms.}$$

(d) $3x(5)(y) = [3x(5)]y \qquad \text{Order of operations}$

$$= [3(x \cdot 5)]y \qquad \text{Associative property}$$
$$= [3(5x)]y \qquad \text{Commutative property}$$
$$= [(3 \cdot 5)x]y \qquad \text{Associative property}$$
$$= (15x)y \qquad \text{Multiply.}$$
$$= 15(xy) \qquad \text{Associative property}$$
$$= 15xy$$

As previously mentioned, many of these steps are not usually written out.

Now Try Exercises 29 and 31.

OBJECTIVE 5 Use the multiplication property of 0.   The additive identity property gives a special property of 0, namely that $a + 0 = a$ for any real number a. The **multiplication property of 0** gives a special property of 0 that involves multiplication: The product of any real number and 0 is 0.

Multiplication Property of 0

For any real number a,

$$a \cdot 0 = 0 \qquad \text{and} \qquad 0 \cdot a = 0.$$

1.4 EXERCISES

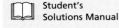

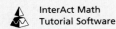
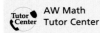
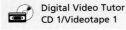
Choose the correct response in Exercises 1–4.

1. The identity element for addition is

  A. $-a$ **B.** 0   **C.** 1   **D.** $\dfrac{1}{a}$.

2. The identity element for multiplication is

  A. $-a$ **B.** 0   **C.** 1   **D.** $\dfrac{1}{a}$.

3. The additive inverse of a is

  A. $-a$ **B.** 0   **C.** 1   **D.** $\dfrac{1}{a}$.

4. The multiplicative inverse of a, where $a \neq 0$, is

  A. $-a$ **B.** 0   **C.** 1   **D.** $\dfrac{1}{a}$.

Complete each statement.

5. The multiplication property of 0 says that the _____ of 0 and any real number is _____.

6. The commutative property is used to change the _____ of two terms or factors.

7. The associative property is used to change the _____ of three terms or factors.

8. Like terms are terms with the _____ variables raised to the _____ powers.

9. When simplifying an expression, only _____ terms can be combined.

10. The coefficient in the term $-8yz^2$ is _____.

Simplify each expression. See Examples 1 and 2.

11. $2(m + p)$ **12.** $3(a + b)$ **13.** $-12(x - y)$ **14.** $-10(p - q)$

15. $5k + 3k$ **16.** $6a + 5a$ **17.** $7r - 9r$ **18.** $4n - 6n$

19. $-8z + 4w$ **20.** $-12k + 3r$ **21.** $a + 7a$ **22.** $s + 9s$

23. $-(2d - f)$ **24.** $-(3m - n)$

Simplify each expression. See Examples 1–4.

25. $-12y + 4y + 3 + 2y$ **26.** $-5r - 9r + 8r - 5$

27. $-6p + 5 - 4p + 6 + 11p$ **28.** $-8x - 12 + 3x - 5x + 9$

29. $3(k + 2) - 5k + 6 + 3$ **30.** $5(r - 3) + 6r - 2r + 4$

31. $-2(m + 1) - (m - 4)$ **32.** $6(a - 5) - (a + 6)$

33. $.25(8 + 4p) - .5(6 + 2p)$ **34.** $.4(10 - 5x) - .8(5 + 10x)$

35. $-(2p + 5) + 3(2p + 4) - 2p$ **36.** $-(7m - 12) - 2(4m + 7) - 8m$

37. $2 + 3(2z - 5) - 3(4z + 6) - 8$ **38.** $-4 + 4(4k - 3) - 6(2k + 8) + 7$

Complete each statement so that the indicated property is illustrated. Simplify each answer, if possible.

39. $5x + 8x = $ _____
(distributive property)

40. $9y - 6y = $ _____
(distributive property)

41. $5(9r) = $ _____
(associative property)

42. $-4 + (12 + 8) = $ _____
(associative property)

43. $5x + 9y = $ _____
(commutative property)

44. $-5 \cdot 7 = $ _____
(commutative property)

45. $1 \cdot 7 = $ _____
(identity property)

46. $-12x + 0 = $ _____
(identity property)

47. $-\dfrac{1}{4}ty + \dfrac{1}{4}ty = $ _____
(inverse property)

48. $-\dfrac{9}{8}\left(-\dfrac{8}{9}\right) = $ _____
(inverse property)

49. $8(-4 + x) = $ _____
(distributive property)

50. $3(x - y + z) = $ _____
(distributive property)

51. $0(.875x + 9y - 88z) = $ _____
(multiplication property of 0)

52. $0(35t^2 - 8t + 12) = $ _____
(multiplication property of 0)

53. Give an "everyday" example of a commutative operation and of one that is not commutative.

54. Give an "everyday" example of inverse operations.

The distributive property can be used to mentally perform calculations. For example, calculate $38 \cdot 17 + 38 \cdot 3$ as follows.

$$38 \cdot 17 + 38 \cdot 3 = 38(17 + 3) \qquad \text{Distributive property}$$
$$= 38(20)$$
$$= 760$$

Use the distributive property to calculate each value mentally.

55. $96 \cdot 19 + 4 \cdot 19$

56. $27 \cdot 60 + 27 \cdot 40$

57. $58 \cdot \dfrac{3}{2} - 8 \cdot \dfrac{3}{2}$

58. $8.75(15) - 8.75(5)$

59. $4.31(69) + 4.31(31)$

60. $\dfrac{8}{5}(17) + \dfrac{8}{5}(13)$

RELATING CONCEPTS (EXERCISES 61–66)

For Individual or Group Work

When simplifying the expression $3x + 4 + 2x + 7$ to $5x + 11$, several important steps are usually done mentally. **Work Exercises 61–66 in order,** *providing the property that justifies each statement in the given simplification. (These steps could be done in other orders.)*

61. $3x + 4 + 2x + 7 = (3x + 4) + (2x + 7)$

62. $ = 3x + (4 + 2x) + 7$

63. $ = 3x + (2x + 4) + 7$

64. $ = (3x + 2x) + (4 + 7)$

65. $ = (3 + 2)x + (4 + 7)$

66. $ = 5x + 11$

---

67. Write a paragraph explaining the properties introduced in this section. Give examples.

68. Explain how the distributive property is used to combine like terms. Give an example.

69. By the distributive property, $a(b + c) = ab + ac$. This property is more completely named the distributive property of multiplication with respect to addition. Is there a distributive property of addition with respect to multiplication? That is, does

$$a + (b \cdot c) = (a + b)(a + c)$$

for all real numbers, a, b, and c? To find out, try some sample values of a, b, and c.

70. Are there *any* different numbers that satisfy the statement $a - b = b - a$? Give an example if your answer is yes.

| Chapter | 1 | Group Activity |

How Americans Spend Their Money

OBJECTIVE   Construct and read bar graphs and circle graphs.

Graphs and tables are a great way of presenting information. They allow you to make comparisons and approximations as well as draw conclusions more quickly than reading a page of text with the same information.

    Listed in the table are common personal consumption expenditures of Americans during the years 1993 through 1999.

Personal Consumption Expenditures (in billions of dollars)

Category	1993	1994	1995	1996	1997	1998	1999
Food and Tobacco	$733.4	$761.7	$783.8	$805.2	$832.3	$900.2	$963.8
Clothing, Accessories, Jewelry	298.1	312.7	323.4	338.0	353.3	368.3	397.2
Personal Care	65.1	68.4	71.9	75.0	79.4	80.5	86.0
Housing	672.8	712.7	750.3	787.4	829.8	858.2	906.2
Household Operation	504.1	535.0	562.8	592.8	620.7	643.8	682.5
Medical Care	785.5	826.1	871.6	912.4	957.3	1040.9	1102.6
Personal Business	357.4	370.4	389.1	416.2	459.1	533.7	586.2
Transportation	504.0	542.2	572.3	611.6	636.4	648.6	705.5
Recreation	340.2	370.2	402.5	432.3	462.9	489.8	534.9
Education and Research	98.5	104.7	112.2	119.7	129.4	139.4	148.9
Religious and Welfare Activities	121.3	131.2	139.8	151.1	157.6	162.6	170.2
TOTAL	$4480.4	$4735.3	$4979.7	$5241.7	$5518.2	$5866.0	$6284.0

Source: U.S. Bureau of Economic Analysis.

A. Have each person in the group construct a bar graph for one category of personal consumption expenditures over the 7-yr period shown in the table.

B. Have each person in the group construct a circle graph for one year shown in the table. Include all eleven categories of personal consumption expenditures. Use approximate values, as needed.

C. As a group, examine the graphs you have constructed.

    1. Discuss any conclusions you can draw from them.

    2. What do the graphs tell you about how personal consumption of Americans changed over the 7 yr?

    3. Write a paragraph that summarizes the group's conclusions.

CHAPTER **1** SUMMARY

KEY TERMS

1.1 set
elements (members)
empty set (null set)
variable
set-builder notation
number line
coordinate
graph
additive inverse
(negative, opposite)

signed numbers
absolute value
equation
inequality
interval
interval notation
1.2 sum
difference
product
quotient

reciprocal
(multiplicative
inverse)
1.3 factors
exponent (power)
base
exponential expression
square root
algebraic expression

1.4 identity element for
addition
identity element for
multiplication
term
like terms
coefficient (numerical
coefficient)
combining like terms

NEW SYMBOLS

$\{a, b\}$ set containing
the elements a
and b

$\emptyset$ or { } empty set

$\in$ is an element of
(a set)

$\notin$ is not an
element of

$\neq$ is not equal to

$\{x \mid x \text{ has property } P\}$
set-builder notation

$|x|$ absolute value of x

$<$ is less than

$\leq$ is less than or equal
to

$>$ is greater than

$\geq$ is greater than or
equal to

∞ infinity

$-\infty$ negative
infinity

$(-\infty, \infty)$ set of all real
numbers

(a, ∞) the interval
$\{x \mid x > a\}$

$(-\infty, a)$ the interval
$\{x \mid x < a\}$

$(a, b]$ the interval
$\{x \mid a < x \leq b\}$

a^m m factors of a

$\sqrt{}$ radical sign

$\sqrt{a}$ positive (or
principal) square
root of a

TEST YOUR WORD POWER

See how well you have learned the vocabulary in this chapter. Answers, with examples, follow the Quick Review.

1. The **empty set** is a set
  A. with 0 as its only element
  B. with an infinite number of
  elements
  C. with no elements
  D. of ideas.

2. A **variable** is
  A. a symbol used to represent an
  unknown number
  B. a value that makes an equation
  true
  C. a solution of an equation
  D. the answer in a division problem.

3. The **absolute value** of a number is
  A. the graph of the number
  B. the reciprocal of the number
  C. the opposite of the number
  D. the distance between 0 and the
  number on a number line.

4. The **reciprocal** of a nonzero number
  a is
  A. a
  B. $\frac{1}{a}$
  C. $-a$
  D. 1.

5. A **factor** is
  A. the answer in an addition problem
  B. the answer in a multiplication
  problem
  C. one of two or more numbers that
  are added to get another number
  D. any number that divides evenly
  into a given number.

6. An **exponential expression** is
  A. a number that is a repeated factor
  in a product
  B. a number or a variable written
  with an exponent

  C. a number that shows how many
  times a factor is repeated in a
  product
  D. an expression that involves
  addition.

7. A **term** is
  A. a numerical factor
  B. a number or a product of numbers
  and variables raised to powers
  C. one of several variables with the
  same exponents
  D. a sum of numbers and variables
  raised to powers.

8. A **numerical coefficient** is
  A. the numerical factor in a term
  B. the number of terms in an
  expression
  C. a variable raised to a power
  D. the variable factor in a term.

QUICK REVIEW

CONCEPTS	EXAMPLES

1.1 BASIC CONCEPTS

Sets of Numbers

Natural Numbers
$\{1, 2, 3, 4, \ldots\}$

10, 25, 143

Whole Numbers
$\{0, 1, 2, 3, 4, \ldots\}$

0, 8, 47

Integers
$\{\ldots, -2, -1, 0, 1, 2, \ldots\}$

$-22, -7, 0, 4, 9$

Rational Numbers
$\left\{\dfrac{p}{q} \,\middle|\, p \text{ and } q \text{ are integers, } q \neq 0\right\}$
(all terminating or repeating decimals)

$-\dfrac{2}{3}, -.14, 0, 6, \dfrac{5}{8}, .33333\ldots$

Irrational Numbers
$\{x \mid x \text{ is a real number that is not rational}\}$
(all nonterminating, nonrepeating decimals)

$\pi, .125469\ldots, \sqrt{3}, -\sqrt{22}$

Real Numbers
$\{x \mid x \text{ is represented by a point on a number line}\}$
(all rational and irrational numbers)

$-3, .7, \pi, -\dfrac{2}{3}$

Absolute Value $|a| = \begin{cases} a & \text{if } a \text{ is positive or 0} \\ -a & \text{if } a \text{ is negative} \end{cases}$

$|12| = 12$
$|-12| = 12$

1.2 OPERATIONS ON REAL NUMBERS

Addition

Like Signs: Add the absolute values. The answer has the same sign as the two numbers.

$-2 + (-7) = -(2 + 7) = -9$

Unlike Signs: Subtract the smaller absolute value from the larger. The answer has the sign of the number with the larger absolute value.

$-5 + 8 = 8 - 5 = 3$
$-12 + 4 = -(12 - 4) = -8$

Subtraction
Change the sign of the second number and add.

$-5 - (-3) = -5 + 3 = -2$

Multiplication and Division

Like Signs: The answer is positive when multiplying or dividing two numbers with the same sign.

$-3(-8) = 24 \qquad \dfrac{-15}{-5} = 3$

Unlike Signs: The answer is negative when multiplying or dividing two numbers with different signs.

$-7(5) = -35 \qquad \dfrac{-24}{12} = -2$

CONCEPTS	EXAMPLES

1.3 EXPONENTS, ROOTS, AND ORDER OF OPERATIONS

The product of an even number of negative factors is positive. The product of an odd number of negative factors is negative.

$(-5)^2$ is positive: $(-5)^2 = (-5)(-5) = 25$

$(-5)^3$ is negative: $(-5)^3 = (-5)(-5)(-5) = -125$

Order of Operations

1. Work separately above and below any fraction bar.
2. If parentheses, brackets, or absolute value bars are present, start with the innermost set and work outward.
3. Evaluate all exponents, roots, and absolute values.
4. Multiply or divide in order from left to right.
5. Add or subtract in order from left to right.

$$\frac{12 + 3}{5 \cdot 2} = \frac{15}{10} = \frac{3}{2}$$

$$(-6)[2^2 - (3 + 4)] + 3 = (-6)[2^2 - 7] + 3$$
$$= (-6)[4 - 7] + 3$$
$$= (-6)[-3] + 3$$
$$= 18 + 3$$
$$= 21$$

1.4 PROPERTIES OF REAL NUMBERS

For real numbers a, b, and c:

Distributive Property

$a(b + c) = ab + ac$

$12(4 + 2) = 12 \cdot 4 + 12 \cdot 2$

Inverse Properties

$a + (-a) = 0$ and $-a + a = 0$

$a \cdot \dfrac{1}{a} = 1$ and $\dfrac{1}{a} \cdot a = 1$

$5 + (-5) = 0 \qquad -12 + 12 = 0$

$5 \cdot \dfrac{1}{5} = 1 \qquad -\dfrac{1}{3}(-3) = 1$

Identity Properties

$a + 0 = 0 + a = a$ and $a \cdot 1 = 1 \cdot a = a$

$-32 + 0 = -32 \qquad 17.5 \cdot 1 = 17.5$

Commutative Properties

$a + b = b + a$ and $ab = ba$

$9 + (-3) = -3 + 9 \qquad 6(-4) = (-4)6$

Associative Properties

$a + (b + c) = (a + b) + c$ and $a(bc) = (ab)c$

$7 + (5 + 3) = (7 + 5) + 3 \qquad -4(6 \cdot 3) = (-4 \cdot 6)3$

Multiplication Property of 0

$a \cdot 0 = 0$ and $0 \cdot a = 0$

$4 \cdot 0 = 0 \qquad 0(-3) = 0$

Answers to Test Your Word Power

1. C; *Example:* The set of whole numbers less than 0 is the empty set, written $\emptyset$.   **2.** A; *Examples: a, b, c*   **3.** D; *Examples:* $|2| = 2$ and $|-2| = 2$ **4.** B; *Examples:* 3 is the reciprocal of $\frac{1}{3}$; $-\frac{5}{2}$ is the reciprocal of $-\frac{2}{5}$.   **5.** D; *Examples:* 2 and 5 are factors of 10 since both divide evenly (without remainder) into 10; other integer factors of 10 are $-10, -5, -2, -1, 1,$ and 10.   **6.** B; *Examples:* 3^4 and x^{10} **7.** B; *Examples:* $6, \frac{x}{2}, -4ab^2$ **8.** A; *Examples:* The term $8z$ has numerical coefficient 8, and $-10x^3y$ has numerical coefficient -10.

CHAPTER 1 REVIEW EXERCISES

If you need help with any of these Review Exercises, look in the section indicated in brackets.

[1.1]   *Graph the elements of each set on a number line.*

1. $\left\{-4, -1, 2, \dfrac{9}{4}, 4\right\}$

2. $\left\{-5, -\dfrac{11}{4}, -.5, 0, 3, \dfrac{13}{3}\right\}$

Find the value of each expression.

3. $|-16|$

4. $-|-4|$

5. $|-8| - |-3|$

Let set $S = \left\{-9, -\dfrac{4}{3}, -\sqrt{4}, -.25, 0, .\overline{35}, \dfrac{5}{3}, \sqrt{7}, \sqrt{-9}, \dfrac{12}{3}\right\}$. Simplify the elements of S as necessary, and then list the elements that belong to the specified set.

6. Whole numbers

7. Integers

8. Rational numbers

9. Real numbers

Write each set by listing its elements.

10. $\{x \mid x$ is a natural number between 3 and 9$\}$

11. $\{y \mid y$ is a whole number less than 4$\}$

Write true *or* false *for each inequality.*

12. $4 \cdot 2 \le |12 - 4|$

13. $2 + |-2| > 4$

14. $4(3 + 7) > -|40|$

The graph shows the percent change in car sales from January 2000 to January 2001 for various automakers. Use this graph to work Exercises 15–18.

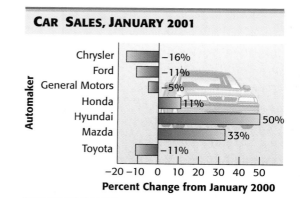

CAR SALES, JANUARY 2001

Source: Automakers.

15. Which automaker had the greatest change in sales? What was that change?

16. Which automaker had the smallest change in sales? What was that change?

17. *True* or *false:* The absolute value of the percent change for Honda was greater than the absolute value of the percent change for Toyota.

18. *True* or *false:* The percent change for Hyundai was more than four times greater than the percent change for Honda.

Write each set in interval notation and graph the interval.

19. $\{x \mid x < -5\}$

20. $\{x \mid -2 < x \le 3\}$

[1.2]   *Add or subtract as indicated.*

21. $-\dfrac{5}{8} - \left(-\dfrac{7}{3}\right)$

22. $-\dfrac{4}{5} - \left(-\dfrac{3}{10}\right)$

23. $-5 + (-11) + 20 - 7$

24. $-9.42 + 1.83 - 7.6 - 1.9$

25. $-15 + (-13) + (-11)$

26. $-1 - 3 - (-10) + (-7)$

27. $\dfrac{3}{4} - \left(\dfrac{1}{2} - \dfrac{9}{10}\right)$

28. $-|-12| - |-9| + (-4) - |10|$

29. Telescope Peak, altitude 11,049 ft, is next to Death Valley, 282 ft below sea level. Find the difference between these altitudes. (*Source: World Almanac and Book of Facts*, 2002.)

Multiply or divide as indicated.

30. $2(-5)(-3)(-3)$

31. $-\dfrac{3}{7}\left(-\dfrac{14}{9}\right)$

32. $\dfrac{75}{-5}$

33. $\dfrac{-2.3754}{-.74}$

34. Which one of the following is undefined: $\dfrac{5}{7-7}$ or $\dfrac{7-7}{5}$?

[1.3] *Evaluate each expression.*

35. 10^4

36. $\left(\dfrac{3}{7}\right)^3$

37. $(-5)^3$

38. -5^3

Find each square root. If it is not a real number, say so.

39. $\sqrt{400}$

40. $\sqrt{\dfrac{64}{121}}$

41. $-\sqrt{.81}$

42. $\sqrt{-64}$

Use the order of operations to simplify each expression.

43. $-14\left(\dfrac{3}{7}\right) + 6 \div 3$

44. $-\dfrac{2}{3}[5(-2) + 8 - 4^3]$

45. $\dfrac{-5(3^2) + 9\left(\sqrt{4}\right) - 5}{6 - 5(-2)}$

Evaluate each expression if $k = -4$, $m = 2$, and $n = 16$.

46. $4k - 7m$

47. $-3\sqrt{n} + m + 5k$

48. $\dfrac{4m^3 - 3n}{7k^2 - 10}$

49. The following expression for *body mass index* (BMI) can help determine ideal body weight.

$704 \times$ (weight in pounds) $\div$ (height in inches)^2

A BMI of 19 to 25 corresponds to a healthy weight. (*Source: Washington Post.*)

(a) Derek Jeter is 6 ft 3 in. tall and weighs 195 lb. (*Source:* www.mlb.com) Find his BMI (to the nearest whole number).

(b) Calculate your BMI.

[1.4]   *Use the properties of real numbers to simplify each expression.*

50. $2q + 19q$ 　　　　　　　**51.** $13z - 17z$ 　　　　　　　**52.** $-m + 6m$

53. $5p - p$ 　　　　　　　**54.** $-2(k + 3)$ 　　　　　　　**55.** $6(r + 3)$

56. $9(2m + 3n)$ 　　　　　　　　　**57.** $-(-p + 6q) - (2p - 3q)$

58. $-3y + 6 - 5 + 4y$ 　　　　　　　　**59.** $2a + 3 - a - 1 - a - 2$

60. $-3(4m - 2) + 2(3m - 1) - 4(3m + 1)$

Complete each statement so that the indicated property is illustrated. Simplify each answer, if possible.

61. $2x + 3x =$ _____ 　　　　**62.** $-4 \cdot 1 =$ _____
　　　　　(distributive property) 　　　　　　　　　　(identity property)

63. $2(4x) =$ _____ 　　　　**64.** $-3 + 13 =$ _____
　　　　　(associative property) 　　　　　　　　　　(commutative property)

65. $-3 + 3 =$ _____ 　　　　**66.** $5(x + z) =$ _____
　　　　　(inverse property) 　　　　　　　　　　(distributive property)

67. $0 + 7 =$ _____ 　　　　**68.** $8 \cdot \dfrac{1}{8} =$ _____
　　　　　(identity property) 　　　　　　　　　　(inverse property)

MIXED REVIEW EXERCISES*

The table gives revenue and expenditures (both in millions of dollars) for Yahoo!, Inc., for three different years.

Year	Revenue	Expenditures
1996	19.07	21.39
1997	67.41	90.29
1998	203.20	177.61

Source: www.quote.com

Determine the absolute value of the difference between revenue and expenditures for each year, and tell whether the company made a profit (i.e., was "in the black") or experienced a loss (i.e., was "in the red"). (These descriptions go back to the days when bookkeepers used black ink to represent gains and red ink to represent losses. This convention is still used. For example, Yahoo! and on-line brokers display stock gains in black and losses in red.)

69. 1996 　　　　　　　　**70.** 1997 　　　　　　　　**71.** 1998

Perform the indicated operations.

72. $\left(-\dfrac{4}{5}\right)^4$ 　　　　　　　　　　　　**73.** $-\dfrac{5}{8}(-40)$

74. $-25\left(-\dfrac{4}{5}\right) + 3^3 - 32 \div \sqrt{4}$ 　　　　　　**75.** $-8 + |-14| + |-3|$

---

*The order of exercises in this final group does not correspond to the order in which topics occur in the chapter. This random ordering should help you prepare for the chapter test in yet another way.

76. $\dfrac{6 \cdot \sqrt{4} - 3 \cdot \sqrt{16}}{-2 \cdot 5 + 7(-3) - 10}$

77. $-\sqrt{25}$

78. $-\dfrac{10}{21} \div \left(-\dfrac{5}{14}\right)$

79. $.8 - 4.9 - 3.2 + 1.14$

80. -3^2

81. $\dfrac{-38}{-19}$

82. $-2(k - 1) + 3k - k$

83. $-\sqrt{-100}$

84. $-(3k - 4h)$

85. $-4.6(2.48)$

86. $-\dfrac{2}{3}(-15) + (2^4 - 8 \div 4)$

87. $-2x + 5 - 4x - 1$

88. $-\dfrac{2}{3} - \left(\dfrac{1}{6} - \dfrac{5}{9}\right)$

89. Evaluate $-m(3k^2 + 5m)$ if $k = -4$ and $m = 2$.

90. To evaluate $(3 + 2)^2$, should you work within the parentheses first, or should you square 3 and square 2 and then add?

CHAPTER **1** TEST

1. Graph $\left\{-3, .75, \dfrac{5}{3}, 5, 6.3\right\}$ on a number line.

Let $A = \left\{-\sqrt{6}, -1, -.5, 0, 3, \sqrt{25}, 7.5, \frac{24}{2}, \sqrt{-4}\right\}$. First simplify each element as needed, and then list the elements from A that belong to each set.

2. Whole numbers

3. Integers

4. Rational numbers

5. Real numbers

Write each set in interval notation and graph the interval.

6. $\{x \mid x < -3\}$

7. $\{y \mid -4 < y \le 2\}$

Perform the indicated operations.

8. $-6 + 14 + (-11) - (-3)$

9. $10 - 4 \cdot 3 + 6(-4)$

10. $7 - 4^2 + 2(6) + (-4)^2$

11. $\dfrac{10 - 24 + (-6)}{\sqrt{16}(-5)}$

12. $\dfrac{-2[3 - (-1 - 2) + 2]}{\sqrt{9}(-3) - (-2)}$

13. $\dfrac{8 \cdot 4 - 3^2 \cdot 5 - 2(-1)}{-3 \cdot 2^3 + 1}$

The table shows the heights in feet of some selected mountains and the depths in feet (as negative numbers) of some selected ocean trenches.

Mountain	Height	Trench	Depth
Foraker	17,400	Philippine	−32,995
Wilson	14,246	Cayman	−24,721
Pikes Peak	14,110	Java	−23,376

Source: World Almanac and Book of Facts, 2002.

14. What is the difference between the height of Mt. Foraker and the depth of the Philippine Trench?

15. What is the difference between the height of Pikes Peak and the depth of the Java Trench?

16. How much deeper is the Cayman Trench than the Java Trench?

Find each square root. If the number is not real, say so.

17. $\sqrt{196}$　　　　　　　**18.** $-\sqrt{225}$　　　　　　　**19.** $\sqrt{-16}$

20. For the expression $\sqrt{a}$, under what conditions will its value be **(a)** positive, **(b)** not real, **(c)** 0?

21. Evaluate $\dfrac{8k + 2m^2}{r - 2}$ if $k = -3$, $m = -3$, and $r = 25$.

22. Use the properties of real numbers to simplify $-3(2k - 4) + 4(3k - 5) - 2 + 4k$.

23. How does the subtraction sign affect the terms $-4r$ and 6 when simplifying $(3r + 8) - (-4r + 6)$? What is the simplified form?

Match each statement in Column I with the appropriate property in Column II. Answers may be used more than once.

I	**II**
24. $6 + (-6) = 0$	**A.** Distributive property
25. $-2 + (3 + 6) = (-2 + 3) + 6$	**B.** Inverse property
26. $5x + 15x = (5 + 15)x$	**C.** Identity property
27. $13 \cdot 0 = 0$	**D.** Associative property
28. $-9 + 0 = -9$	**E.** Commutative property
29. $4 \cdot 1 = 4$	**F.** Multiplication property of 0
30. $(a + b) + c = (b + a) + c$	

Linear Equations, Inequalities, and Applications

2

Tuition for full-time students at both public and private 4-year universities in the United States has risen steadily from 1990 through 2000. Over this 10-year period, tuition increased 85% at public colleges and 86.6% at private institutions. (*Source:* National Center for Education Statistics, U.S. Department of Education.) In Sections 2.2 and 2.3, we discuss the use of percent—one of the most common everyday applications of mathematics—and find the year 2000 costs for both public and private universities in Exercises 39 and 40 of Section 2.3.

2.1 Linear Equations in One Variable

In the previous chapter we began to use *algebraic expressions.* Some examples of algebraic expressions are

$$8x + 9, \qquad y - 4, \qquad \text{and} \qquad \frac{x^3y^8}{z}. \qquad \text{Algebraic expressions}$$

Equations and inequalities compare algebraic expressions, just as a balance scale compares the weights of two quantities. Many applications of mathematics lead to *equations,* statements that two algebraic expressions are equal. A *linear equation in one variable* involves only real numbers and one variable raised to the first power. Examples are

$$x + 1 = -2, \qquad x - 3 = 5, \qquad \text{and} \qquad 2k + 5 = 10. \qquad \text{Linear equations}$$

It is important to be able to distinguish between algebraic expressions and equations. *An equation always contains an equals sign, while an expression does not.*

Linear Equation in One Variable

A **linear equation in one variable** can be written in the form

$$Ax + B = C,$$

where A, B, and C are real numbers, with $A \neq 0$.

A linear equation is also called a **first-degree equation** since the highest power on the variable is one. Some examples of equations that are not linear (that is, *nonlinear*) are

$$x^2 + 3y = 5, \qquad \frac{8}{x} = -22, \qquad \text{and} \qquad \sqrt{x} = 6. \qquad \text{Nonlinear equations}$$

OBJECTIVE 1 Decide whether a number is a solution of a linear equation. If the variable in an equation can be replaced by a real number that makes the statement true, then that number is a **solution** of the equation. For example, 8 is a solution of the equation $x - 3 = 5$, since replacing x with 8 gives a true statement. An equation is *solved* by finding its **solution set,** the set of all solutions. The solution set of the equation $x - 3 = 5$ is $\{8\}$.

Equivalent equations are equations that have the same solution set. To solve an equation, we usually start with the given equation and replace it with a series of simpler equivalent equations. For example,

$$5x + 2 = 17, \qquad 5x = 15, \qquad \text{and} \qquad x = 3$$

are all equivalent since each has the solution set $\{3\}$.

OBJECTIVE 2 Solve linear equations using the addition and multiplication properties of equality. Two important properties that are used in producing equivalent equations are the **addition** and **multiplication properties of equality.**

Addition and Multiplication Properties of Equality

Addition Property of Equality

For all real numbers A, B, and C, the equations
$$A = B \qquad \text{and} \qquad A + C = B + C$$
are equivalent.

That is, the same number may be added to each side of an equation without changing the solution set.

Multiplication Property of Equality

For all real numbers A and B, and for $C \neq 0$, the equations
$$A = B \qquad \text{and} \qquad AC = BC$$
are equivalent.

That is, each side of an equation may be multiplied by the same nonzero number without changing the solution set.

Because subtraction and division are defined in terms of addition and multiplication, respectively, these properties can be extended: The same number may be subtracted from each side of an equation, and each side of an equation may be divided by the same nonzero number, without changing the solution set.

■ **EXAMPLE 1   Using the Addition and Multiplication Properties to Solve a Linear Equation**

Solve $4x - 2x - 5 = 4 + 6x + 3$.

The goal is to use the addition and multiplication properties to get x alone on one side of the equation. First, combine like terms on each side of the equation to obtain
$$2x - 5 = 7 + 6x.$$

Next, use the addition property to get the terms with x on the same side of the equation and the remaining terms (the numbers) on the other side. One way to do this is to first add 5 to each side.

$$
\begin{array}{ll}
2x - 5 + 5 = 7 + 6x + 5 & \text{Add 5.} \\
2x = 12 + 6x & \text{Combine like terms.} \\
2x - 6x = 12 + 6x - 6x & \text{Subtract } 6x. \\
-4x = 12 & \text{Combine like terms.} \\
\dfrac{-4x}{-4} = \dfrac{12}{-4} & \text{Divide by } -4. \\
x = -3 &
\end{array}
$$

To be sure that -3 is the solution, check by substituting for x in the *original* equation.

$$
\begin{array}{lll}
\textit{Check:} \qquad 4x - 2x - 5 = 4 + 6x + 3 & & \text{Original equation} \\
4(-3) - 2(-3) - 5 = 4 + 6(-3) + 3 & \text{?} & \text{Let } x = -3. \\
-12 + 6 - 5 = 4 - 18 + 3 & \text{?} & \text{Multiply.} \\
-11 = -11 & & \text{True}
\end{array}
$$

The true statement indicates that $\{-3\}$ is the solution set. ■

Now Try Exercise 13.

NOTE Notice that in Example 1 the equality symbols are aligned in a column. Do not use more than one equality symbol in a horizontal line of work when solving an equation.

We use the following steps to solve a linear equation in one variable. (Some equations may not require all these steps.)

Solving a Linear Equation in One Variable

Step 1 **Clear fractions.** Eliminate any fractions by multiplying each side by the least common denominator.

Step 2 **Simplify each side separately.** Use the distributive property to clear parentheses and combine like terms as needed.

Step 3 **Isolate the variable terms on one side.** Use the addition property to get all terms with variables on one side of the equation and all numbers on the other.

Step 4 **Isolate the variable.** Use the multiplication property to get an equation with just the variable (with coefficient 1) on one side.

Step 5 **Check.** Substitute the proposed solution into the original equation.

OBJECTIVE 3 Solve linear equations using the distributive property. In Example 1 we did not use Step 1 or the distributive property in Step 2 as given in the box. Many equations, however, will require one or both of these steps.

EXAMPLE 2 Using the Distributive Property to Solve a Linear Equation

Solve $2(k - 5) + 3k = k + 6$.

Step 1 Since there are no fractions in this equation, Step 1 does not apply.

Step 2 Use the distributive property to simplify and combine terms on the left side of the equation.

$$2(k - 5) + 3k = k + 6$$
$$2k - 10 + 3k = k + 6 \qquad \text{Distributive property}$$
$$5k - 10 = k + 6 \qquad \text{Combine like terms.}$$

Step 3 Next, use the addition property of equality.

$$5k - 10 + 10 = k + 6 + 10 \qquad \text{Add 10.}$$
$$5k = k + 16 \qquad \text{Combine like terms.}$$
$$5k - k = k + 16 - k \qquad \text{Subtract } k.$$
$$4k = 16 \qquad \text{Combine like terms.}$$

Step 4 Use the multiplication property of equality to get just k on the left.

$$\frac{4k}{4} = \frac{16}{4} \qquad \text{Divide by 4.}$$
$$k = 4$$

Step 5 Check that the solution set is {4} by substituting 4 for k in the original equation.

Now Try Exercise 15.

NOTE Because of space limitations, we will not always show the check when solving an equation. To be sure that your solution is correct, you should *always* check your work.

OBJECTIVE 4 Solve linear equations with fractions or decimals. When fractions or decimals appear as coefficients in equations, our work can be made easier if we multiply each side of the equation by the least common denominator (LCD) of all the fractions. This is an application of the multiplication property of equality, and it produces an equivalent equation with integer coefficients.

EXAMPLE 3   Solving a Linear Equation with Fractions

Solve $\dfrac{x + 7}{6} + \dfrac{2x - 8}{2} = -4$.

Start by eliminating the fractions. Multiply both sides by the LCD, 6.

Step 1 $\qquad 6\left(\dfrac{x + 7}{6} + \dfrac{2x - 8}{2}\right) = 6(-4)$

Step 2 $\quad 6\left(\dfrac{x + 7}{6}\right) + 6\left(\dfrac{2x - 8}{2}\right) = 6(-4)$ $\qquad$ Distributive property

$\qquad\qquad\qquad x + 7 + 3(2x - 8) = -24$ $\qquad$ Multiply.

$\qquad\qquad\qquad x + 7 + 6x - 24 = -24$ $\qquad$ Distributive property

$\qquad\qquad\qquad\qquad\quad 7x - 17 = -24$ $\qquad$ Combine like terms.

Step 3 $\qquad\qquad 7x - 17 + 17 = -24 + 17$ $\qquad$ Add 17.

$\qquad\qquad\qquad\qquad\qquad 7x = -7$ $\qquad$ Combine like terms.

Step 4 $\qquad\qquad\qquad\quad \dfrac{7x}{7} = \dfrac{-7}{7}$ $\qquad$ Divide by 7.

$\qquad\qquad\qquad\qquad\qquad x = -1$

Step 5 Check by substituting -1 for x in the original equation.

$$\dfrac{x + 7}{6} + \dfrac{2x - 8}{2} = -4$$

$$\dfrac{-1 + 7}{6} + \dfrac{2(-1) - 8}{2} = -4 \quad ? \qquad \text{Let } x = -1.$$

$$\dfrac{6}{6} + \dfrac{-10}{2} = -4 \quad ?$$

$$1 - 5 = -4 \quad ?$$

$$-4 = -4 \qquad \text{True}$$

The solution checks, so the solution set is $\{-1\}$.

Now Try Exercise 41.

In later sections we solve problems involving interest rates and concentrations of solutions. These problems involve percents that are converted to decimals. The equations that are used to solve such problems involve decimal coefficients. We can clear these decimals by multiplying by a power of 10 that allows us to obtain integer coefficients.

■ **EXAMPLE 4   Solving a Linear Equation with Decimals**

Solve $.06x + .09(15 - x) = .07(15)$.

Since each decimal number is given in hundredths, multiply both sides of the equation by 100. (This is done by moving the decimal points two places to the right.) To multiply the second term, $.09(15 - x)$, by 100, remember the associative property: To multiply three terms, first multiply any two of them. Here we will multiply $100(.09)$ first to get 9, so the product $100(.09)(15 - x)$ becomes $9(15 - x)$.

$$.06x + .09(15 - x) = .07(15)$$

$$.06x + .09(15 - x) = .07(15) \qquad \text{Multiply each term by 100.}$$

$$6x + 9(15 - x) = 7(15)$$

$$6x + 9(15) - 9x = 105 \qquad \text{Distributive property; multiply.}$$

$$-3x + 135 = 105 \qquad \text{Combine like terms.}$$

$$-3x + 135 - 135 = 105 - 135 \qquad \text{Subtract 135.}$$

$$-3x = -30$$

$$\frac{-3x}{-3} = \frac{-30}{-3} \qquad \text{Divide by } -3.$$

$$x = 10$$

Check to verify that the solution set is $\{10\}$.

Now Try Exercise 45.

OBJECTIVE 5   Identify conditional equations, contradictions, and identities.   All the preceding equations had solution sets containing one element; for example, $2(k - 5) + 3k = k + 6$ has solution set $\{4\}$. Some equations that appear to be linear have no solutions, while others have an infinite number of solutions. The table gives the names of these types of equations.

Type of Equation	Number of Solutions	Indication When Solving
Conditional	One	Final line is $x =$ a number. (See Example 5(a).)
Contradiction	None; solution set $\emptyset$	Final line is false, such as $0 = 1$. (See Example 5(c).)
Identity	Infinite; solution set {all real numbers}	Final line is true, such as $0 = 0$. (See Example 5(b).)

■ **EXAMPLE 5   Recognizing Conditional Equations, Identities, and Contradictions**

Solve each equation. Decide whether it is a *conditional equation*, an *identity*, or a *contradiction*.

(a) $5x - 9 = 4(x - 3)$

$$5x - 9 = 4x - 12 \qquad \text{Distributive property}$$
$$5x - 9 - 4x = 4x - 12 - 4x \qquad \text{Subtract } 4x.$$
$$x - 9 = -12 \qquad \text{Combine like terms.}$$
$$x - 9 + 9 = -12 + 9 \qquad \text{Add 9.}$$
$$x = -3$$

The solution set, $\{-3\}$, has only one element, so $5x - 9 = 4(x - 3)$ is a conditional equation.

(b) $5x - 15 = 5(x - 3)$

Use the distributive property to clear parentheses on the right side.

$$5x - 15 = 5x - 15$$
$$0 = 0 \qquad \text{Subtract } 5x \text{ and add 15.}$$

The final line, $0 = 0$, indicates that the solution set is {all real numbers}, and the equation $5x - 15 = 5(x - 3)$ is an identity. (*Note:* The first step yielded $5x - 15 = 5x - 15$, which is true for all values of x. We could have identified the equation as an identity at that point.)

(c) $\qquad 5x - 15 = 5(x - 4)$

$$5x - 15 = 5x - 20 \qquad \text{Distributive property}$$
$$5x - 15 - 5x = 5x - 20 - 5x \qquad \text{Subtract } 5x.$$
$$-15 = -20 \qquad \text{False}$$

Since the result, $-15 = -20$, is *false,* the equation has no solution. The solution set is $\emptyset$, so the equation $5x - 15 = 5(x - 4)$ is a contradiction.

Now Try Exercises 53, 57, and 59.

2.1 EXERCISES

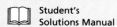
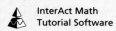

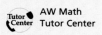

1. Which equations are linear equations in x?

A. $3x + x - 1 = 0$ **B.** $8 = x^2$ **C.** $6x + 2 = 9$ **D.** $\dfrac{1}{2}x - \dfrac{1}{x} = 0$

2. Which of the equations in Exercise 1 are nonlinear equations in x? Explain why.

3. Decide whether 6 is a solution of $3(x + 4) = 5x$ by substituting 6 for x. If it is not a solution, explain why.

4. Use substitution to decide whether -2 is a solution of $5(x + 4) - 3(x + 6) = 9(x + 1)$. If it is not a solution, explain why.

5. The equation $4[x + (2 - 3x)] = 2(4 - 4x)$ is an identity. Let x represent the number of letters in your last name. Is this number a solution of this equation? Check your answer.

6. The expression $.06(10 - x)(100)$ is equivalent to which of the following?

A. $.06 - .06x$ **B.** $60 - 6x$ **C.** $6 - 6x$ **D.** $6 - .06x$

7. Identify each as an *expression* or an *equation.*

(a) $3x = 6$ **(b)** $3x + 6$
(c) $5x + 6(x - 3) = 12x + 6$ **(d)** $5x + 6(x - 3) - (12x + 6)$

8. Explain why $6x + 9 = 6x + 8$ cannot have a solution. (No work is necessary.)

Solve and check each equation. See Examples 1 and 2.

9. $7x + 8 = 1$

10. $5x - 4 = 21$

11. $7x - 5x + 15 = x + 8$

12. $2x + 4 - x = 4x - 5$

13. $12w + 15w - 9 + 5 = -3w + 5 - 9$

14. $-4t + 5t - 8 + 4 = 6t - 4$

15. $2(x + 3) = -4(x + 1)$

16. $4(t - 9) = 8(t + 3)$

17. $3(2w + 1) - 2(w - 2) = 5$

18. $4(x - 2) + 2(x + 3) = 6$

19. $2x + 3(x - 4) = 2(x - 3)$

20. $6x - 3(5x + 2) = 4(1 - x)$

21. $6p - 4(3 - 2p) = 5(p - 4) - 10$

22. $-2k - 3(4 - 2k) = 2(k - 3) + 2$

23. $-[2z - (5z + 2)] = 2 + (2z + 7)$

24. $-[6x - (4x + 8)] = 9 + (6x + 3)$

25. $-3m + 6 - 5(m - 1) = 4m - (2m - 4) - 9m + 5$

26. $4(k + 2) - 8k - 5 = -3k + 9 - 2(k + 6)$

27. $-[3x - (2x + 5)] = -4 - [3(2x - 4) - 3x]$

28. $2[-(x - 1) + 4] = 5 + [-(6x - 7) + 9x]$

29. $-(9 - 3a) - (4 + 2a) - 3 = -(2 - 5a) + (-a) + 1$

30. $-(-2 + 4x) - (3 - 4x) + 5 = -(-3 + 6x) + x + 1$

31. To solve the linear equation

$$\frac{8x}{3} - \frac{2x}{4} = -13,$$

we are allowed to multiply each side by the least common denominator of all the fractions in the equation. What is this least common denominator?

32. To solve the linear equation

$$.05x + .12(x + 5000) = 940,$$

we multiply each side by a power of 10 so that all coefficients are integers. What is the smallest power of 10 that will accomplish this goal?

33. Suppose that in solving the equation

$$\frac{1}{3}x + \frac{1}{2}x = \frac{1}{6}x,$$

you begin by multiplying each side by 12, rather than the *least* common denominator, 6. Would you get the correct solution anyway? Explain.

34. What is the final line in the check for the solution of the equation in Example 4?

Solve each equation. See Examples 3 and 4.

35. $-\dfrac{5}{9}k = 2$

36. $\dfrac{3}{11}z = -5$

37. $\dfrac{6}{5}x = -1$

38. $-\dfrac{7}{8}r = 6$

39. $\dfrac{m}{2} + \dfrac{m}{3} = 5$

40. $\dfrac{x}{5} - \dfrac{x}{4} = 1$

41. $\dfrac{x - 10}{5} + \dfrac{2}{5} = -\dfrac{x}{3}$

42. $\dfrac{2r - 3}{7} + \dfrac{3}{7} = -\dfrac{r}{3}$

43. $\dfrac{4t + 1}{3} = \dfrac{t + 5}{6} + \dfrac{t - 3}{6}$

44. $\dfrac{2x + 5}{5} = \dfrac{3x + 1}{2} + \dfrac{-x + 7}{2}$

45. $.05x + .12(x + 5000) = 940$

46. $.09k + .13(k + 300) = 61$

47. $.02(50) + .08r = .04(50 + r)$

48. $.20(14{,}000) + .14t = .18(14{,}000 + t)$

49. $.05x + .10(200 - x) = .45x$

50. $.08x + .12(260 - x) = .48x$

51. A student tried to solve the equation $8x = 7x$ by dividing each side by x, obtaining $8 = 7$. He gave the solution set as $\emptyset$. Why is this incorrect?

52. Explain the distinction between a conditional equation, an identity, and a contradiction.

Decide whether each equation is conditional, *an* identity, *or a* contradiction. *Give the solution set. See Example 5.*

53. $-2p + 5p - 9 = 3(p - 4) - 5$

54. $-6k + 2k - 11 = -2(2k - 3) + 4$

55. $6x + 2(x - 2) = 9x + 4$

56. $-4(x + 2) = -3(x + 5) - x$

57. $-11m + 4(m - 3) + 6m = 4m - 12$

58. $3p - 5(p + 4) + 9 = -11 + 15p$

59. $7[2 - (3 + 4r)] - 2r = -9 + 2(1 - 15r)$

60. $4[6 - (1 + 2m)] + 10m = 2(10 - 3m) + 8m$

Decide whether each pair of equations is equivalent. If not equivalent, explain why.

61. $5x = 10$ and $\dfrac{5x}{x + 2} = \dfrac{10}{x + 2}$

62. $x + 1 = 9$ and $\dfrac{x + 1}{8} = \dfrac{9}{8}$

63. $x = -3$ and $\dfrac{x}{x + 3} = \dfrac{-3}{x + 3}$

64. $m = 1$ and $\dfrac{m + 1}{m - 1} = \dfrac{2}{m - 1}$

65. $k = 4$ and $k^2 = 16$

66. $p^2 = 36$ and $p = 6$

2.2 Formulas

OBJECTIVES

1 Solve a formula for a specified variable.

2 Solve applied problems using formulas.

3 Solve percent problems.

A **mathematical model** is an equation or inequality that describes a real situation. Models for many applied problems already exist; they are called *formulas*. A **formula** is a mathematical equation in which variables are used to describe a relationship. Some formulas that we will be using are

$$d = rt, \qquad I = prt, \qquad \text{and} \qquad P = 2L + 2W.$$

A list of some common formulas used in algebra is given inside the covers of the book.

OBJECTIVE 1 Solve a formula for a specified variable. In some applications, the appropriate formula may be solved for a different variable than the one to be found. For example, the formula $I = prt$ says that interest on a loan or investment equals principal (amount borrowed or invested) times rate (annually in percent) times time at interest (in years). To determine how long it will take for an investment at a stated interest rate to earn a predetermined amount of interest, it would help to first solve the formula for t. This process is called **solving for a specified variable.**

The steps used in the following examples are very similar to those used in solving linear equations. When you are solving for a specified variable, the key is to treat that variable as if it were the only one; treat all other variables like numbers (constants). The following additional suggestions may be helpful.

Solving for a Specified Variable

Step 1 Get all terms containing the specified variable on one side of the equation and all terms without that variable on the other side.

Step 2 If necessary, use the distributive property to combine the terms with the specified variable.* The result should be the product of a sum or difference and the variable.

Step 3 Divide both sides by the factor that is the coefficient of the specified variable.

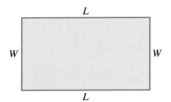

Perimeter, *P*, distance around a rectangle, is given by
$P = 2L + 2W.$

FIGURE 1

■ **EXAMPLE 1** Solving for a Specified Variable

Solve the formula $P = 2L + 2W$ for *W*.

This formula gives the relationship between the perimeter (distance around) a rectangle, *P*, the length of the rectangle, *L*, and the width of the rectangle, *W*. See Figure 1.

Solve the formula for *W* by getting *W* alone on one side of the equals sign. To begin, subtract 2*L* from both sides.

$$P = 2L + 2W$$

Step 1
$$P - 2L = 2L + 2W - 2L \qquad \text{Subtract } 2L.$$
$$P - 2L = 2W$$

Step 2 is not needed here.

Step 3
$$\frac{P - 2L}{2} = \frac{2W}{2} \qquad \text{Divide both sides by 2.}$$
$$\frac{P - 2L}{2} = W \quad \text{or} \quad W = \frac{P - 2L}{2}$$

Now Try Exercise 9.

CAUTION In Step 3 of Example 1, you cannot simplify the fraction by dividing 2 into the term 2*L*. The subtraction in the numerator must be done before the division.

$$\frac{P - 2L}{2} \neq P - L$$

■ **EXAMPLE 2** Solving a Formula with Parentheses

The formula for the perimeter of a rectangle is sometimes written in the equivalent form $P = 2(L + W)$. Solve this form for *W*.

---

*Using the distributive property to write $ab + ac$ as $a(b + c)$ is called *factoring*. See Chapter 6.

One way to begin is to use the distributive property on the right side of the equation to get $P = 2L + 2W$, which we would then solve as in Example 1. Another way to begin is to divide by the coefficient 2.

$$P = 2(L + W)$$

$$\frac{P}{2} = L + W \qquad \text{Divide by 2.}$$

$$\frac{P}{2} - L = W \quad \text{or} \quad W = \frac{P}{2} - L \qquad \text{Subtract } L.$$

We can show that this result is equivalent to our result in Example 1 by multiplying L by $\frac{2}{2}$.

$$\frac{P}{2} - \frac{2}{2}(L) = W \qquad \frac{2}{2} = 1, \text{ so } L = \frac{2}{2}(L).$$

$$\frac{P}{2} - \frac{2L}{2} = W$$

$$\frac{P - 2L}{2} = W \qquad \text{Subtract fractions.}$$

The final line agrees with the result in Example 1.

Now Try Exercise 15.

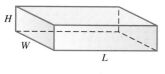

FIGURE 2

A rectangular solid has the shape of a box, but is solid. See Figure 2. The labels H, W, and L represent the height, width, and length of the figure, respectively. The surface area of any solid three-dimensional figure is the total area of its surface. For a rectangular solid, the surface area A is

$$A = 2HW + 2LW + 2LH.$$

EXAMPLE 3 **Using the Distributive Property to Solve for a Specified Variable**

Given the surface area, height, and width of a rectangular solid, write a formula for the length.

To solve for the length L, treat L as the only variable and treat all other variables as constants.

$$A = 2HW + 2LW + 2LH$$

$$A - 2HW = 2LW + 2LH \qquad \text{Subtract } 2HW.$$

$$A - 2HW = L(2W + 2H) \qquad \text{Use the distributive property on the right side.}$$

$$\frac{A - 2HW}{2W + 2H} = L \qquad \text{Divide by } 2W + 2H.$$

or $$L = \frac{A - 2HW}{2W + 2H}$$

Now Try Exercise 21.

CAUTION The most common error in working a problem like Example 3 is not using the distributive property correctly. We must write the expression so that the specified variable is a *factor;* then we can divide by its coefficient in the final step.

OBJECTIVE 2 Solve applied problems using formulas. The next example uses the distance formula, $d = rt$, which relates d, the distance traveled, r, the rate or speed, and t, the travel time.

▉ EXAMPLE 4 Finding Average Speed

Janet Branson found that usually it took her $\frac{3}{4}$ hr each day to drive a distance of 15 mi to work. What was her speed?

Find the formula for speed (rate) r by solving $d = rt$ for r.

$$d = rt$$

$$\frac{d}{t} = \frac{rt}{t} \qquad \text{Divide by } t.$$

$$\frac{d}{t} = r \quad \text{or} \quad r = \frac{d}{t}$$

Notice that only Step 3 was needed to solve for r in this example. Now find Janet's speed by substituting the given values of d and t into this formula.

$$r = \frac{d}{t}$$

$$r = \frac{15}{\dfrac{3}{4}} \qquad \text{Let } d = 15, \, t = \frac{3}{4}.$$

$$r = 15 \cdot \frac{4}{3} \qquad \text{Multiply by the reciprocal of } \frac{3}{4}.$$

$$r = 20$$

Her speed averaged 20 mph. (That is, at times she may have traveled a little faster or a little slower than 20 mph, but overall her speed was 20 mph.) ▉

Now Try Exercise 25.

▉ PROBLEM SOLVING ▉

As seen in Example 4, it may be convenient to first solve for a specific unknown variable before substituting the given values. This is particularly useful when we wish to substitute several different values for the same variable. For example, an economics class might need to solve the equation $I = prt$ for r to find rates that produce specific amounts of interest for various principals and times.

OBJECTIVE 3 Solve percent problems. An important everyday use of mathematics involves the concept of percent. Percent is written with the symbol %. The word

percent means "per one hundred." One percent means "one per one hundred" or "one one-hundredth."

$$1\% = .01 \qquad \text{or} \qquad 1\% = \frac{1}{100}$$

Solving a Percent Problem

Let a represent a partial amount of b, the base, or whole amount. Then the following formula can be used to solve a percent problem.

$$\frac{\textbf{amount}}{\textbf{base}} = \frac{a}{b} = \textbf{percent (represented as a decimal)}$$

For example, if a class consists of 50 students and 32 are males, then the percent of males in the class is

$$\frac{\text{amount}}{\text{base}} = \frac{a}{b}$$
$$= \frac{32}{50} \qquad \text{Let } a = 32,\ b = 50.$$
$$= .64 \quad \text{or} \quad 64\%.$$

▨ EXAMPLE 5   Solving Percent Problems

(a) A 50-L mixture of acid and water contains 10 L of acid. What is the percent of acid in the mixture?

The given amount of the mixture is 50 L, and the part that is acid (percentage) is 10 L. Let x represent the percent of acid. Then, the percent of acid in the mixture is

$$x = \frac{10}{50}$$
$$x = .20 \quad \text{or} \quad 20\%.$$

(b) If a savings account balance of \$3550 earns 8% interest in one year, how much interest is earned?

Let x represent the amount of interest earned (that is, the part of the whole amount invested). Since $8\% = .08$, the equation is

$$\frac{x}{3550} = .08 \qquad \tfrac{a}{b} = \text{percent}$$
$$x = .08(3550) \qquad \text{Multiply by 3550.}$$
$$x = 284.$$

The interest earned is \$284.

Now Try Exercises 37 and 39.

Graphs sometimes represent the percents of a whole amount that satisfy certain conditions.

▨ **EXAMPLE 6** Interpreting Percents from a Graph

The country of origin of immigrants to the United States is shifting from mainly European in the 19th century to Hispanic (from Mexico and Central America) and Asian. In 2000, the foreign-born population of the United States from all countries was 28,379,000 to the nearest thousand. Use the graph in Figure 3 to determine how many came from Asia.

FOREIGN-BORN POPULATION OF THE UNITED STATES, 2000

Other 44.6%
Mexico 27.6%
Cuba 3.4%
Central America 5.1%
Asia 17.1%
Great Britain 2.2%

Source: U.S. Bureau of the Census.

FIGURE 3

According to the graph, 17.1% of the foreign-born population came from Asia. Let x represent the required number of immigrants from Asia.

$$\frac{x}{28,379,000} = .171$$
$$x = .171(28,379,000)$$
$$x = 4,852,809$$

Therefore, 4,853,000 to the nearest thousand immigrants came from Asia in 2000.

▨

Now Try Exercise 47.

2.2 **EXERCISES**

RELATING CONCEPTS (EXERCISES 1–6)

For Individual or Group Work

Consider the following equations:

First Equation	**Second Equation**
$x = \dfrac{5x + 8}{3}$	$t = \dfrac{bt + k}{c} \quad (c \neq 0).$

Solving the second equation for t requires the same logic as solving the first equation for x. When solving for t, we treat all other variables as though they were constants. **Work Exercises 1–6 in order,** *to see the "parallel logic" of solving for x and solving for t.*

1. (a) Clear the first equation of fractions by multiplying each side by 3.
   (b) Clear the second equation of fractions by multiplying each side by c.

2. (a) Get the terms involving x on the left side of the first equation by subtracting $5x$ from each side.
   (b) Get the terms involving t on the left side of the second equation by subtracting bt from each side.

3. (a) Combine like terms on the left side of the first equation. What property allows us to write $3x - 5x$ as $(3 - 5)x = -2x$?
   (b) Write the expression on the left side of the second equation so that t is a factor. What property allows us to do this?

4. (a) Divide each side of the first equation by the coefficient of x.
   (b) Divide each side of the second equation by the coefficient of t.

5. Look at your answer for the second equation. What restriction must be placed on the variables? Why is this necessary?

6. Write a short paragraph summarizing what you have learned in this group of exercises.

Solve each formula for the specified variable. See Examples 1 and 2.

7. $I = prt$ for r (simple interest)

8. $d = rt$ for t (distance)

9. $P = 2L + 2W$ for L
   (perimeter of a rectangle)

10. $A = bh$ for b
    (area of a parallelogram)

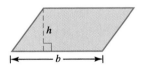

11. $V = LWH$ for W
    (volume of a rectangular solid)

12. $P = a + b + c$ for b
    (perimeter of a triangle)

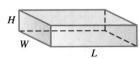

13. $C = 2\pi r$ for r
    (circumference of a circle)

14. $A = \dfrac{1}{2}bh$ for h (area of a triangle)

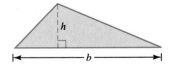

15. $A = \dfrac{1}{2}h(B + b)$ for B
    (area of a trapezoid)

16. $S = 2\pi rh + 2\pi r^2$ for h
    (surface area of a right circular cylinder)

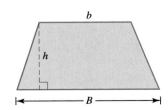

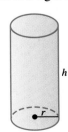

17. $F = \dfrac{9}{5}C + 32$ for C

(Celsius to Fahrenheit)

18. $C = \dfrac{5}{9}(F - 32)$ for F

(Fahrenheit to Celsius)

19. When a formula is solved for a particular variable, several different equivalent forms may be possible. If we solve $A = \frac{1}{2}bh$ for h, one possible correct answer is

$$h = \dfrac{2A}{b}.$$

Which one of the following is *not* equivalent to this?

A. $h = 2\left(\dfrac{A}{b}\right)$ **B.** $h = 2A\left(\dfrac{1}{b}\right)$ **C.** $h = \dfrac{A}{\dfrac{1}{2}b}$ **D.** $h = \dfrac{\dfrac{1}{2}A}{b}$

20. Suppose the formula

$$A = 2HW + 2LW + 2LH$$

is solved for L as follows.

$$A = 2HW + 2LW + 2LH$$
$$A - 2LW - 2HW = 2LH$$
$$\dfrac{A - 2LW - 2HW}{2H} = L$$

While there are no algebraic errors here, what is wrong with the final equation, if we are interested in solving for L?

Solve each equation for the specified variable. Use the distributive property to factor as necessary. See Example 3.

21. $2k + ar = r - 3y$ for r

22. $4s + 7p = tp - 7$ for p

23. $w = \dfrac{3y - x}{y}$ for y

24. $c = \dfrac{-2t + 4}{t}$ for t

Solve each problem. See Example 4.

25. In 1998 Jeff Gordon won the World 600 (mile) race with a speed of 136.424 mph. Find his time to the nearest thousandth. (*Source: Sports Illustrated 1999 Sports Almanac.*)

26. In 1975, rain shortened the Indianapolis 500 race to 435 mi. It was won by Bobby Unser, who averaged 149.213 mph. What was his time to the nearest thousandth? (*Source: Sports Illustrated 1998 Sports Almanac.*)

27. Faye Korn traveled from Kansas City to Louisville, a distance of 520 mi, in 10 hr. Find her rate in miles per hour.

28. The distance from Melbourne to London is 10,500 mi. If a jet averages 500 mph between the two cities, what is its travel time in hours?

29. As of 2001, the highest temperature ever recorded in Chicago was 40°C. Find the corresponding Fahrenheit temperature. (*Source: World Almanac and Book of Facts,* 2002.)

30. The lowest temperature recorded in Salt Lake City in 1997 was 8°F. Find the corresponding Celsius temperature. (*Source: World Almanac and Book of Facts,* 1999.)

31. The base of the Great Pyramid of Cheops is a square whose perimeter is 920 m. What is the length of each side of this square? (*Source: Atlas of Ancient Archaeology.*)

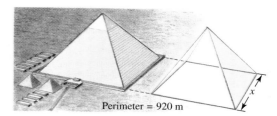

Perimeter = 920 m

32. The Peachtree Plaza Hotel in Atlanta is in the shape of a cylinder with radius 46 m and height 220 m. Find its volume to the nearest tenth. (*Hint:* Use the π key on your calculator.)

33. The circumference of a circle is 480π in. What is its radius? What is its diameter?

34. The radius of a circle is 2.5 in. What is the diameter of the circle? What is its circumference?

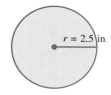

$r = 2.5$ in.

35. A cord of wood contains 128 ft³ (cubic feet) of wood. If a stack of wood is 4 ft wide and 4 ft high, how long must it be if it contains exactly 1 cord?

36. Give one set of possible dimensions for a stack of wood that contains 1.5 cords. (See Exercise 35.)

Solve each problem. See Example 5.

37. A mixture of alcohol and water contains a total of 36 oz of liquid. There are 9 oz of pure alcohol in the mixture. What percent of the mixture is water? What percent is alcohol?

38. A mixture of acid and water is 35% acid. If the mixture contains a total of 40 L, how many liters of pure acid are in the mixture? How many liters of pure water are in the mixture?

39. A real estate agent earned $6300 commission on a property sale of $210,000. What is her rate of commission?

40. A certificate of deposit for 1 yr pays $221 simple interest on a principal of $3400. What is the interest rate being paid on this deposit?

When a consumer loan is paid off ahead of schedule, the finance charge is smaller than if the loan were paid off over its scheduled life. By one method, called the rule of 78, the amount of unearned interest (finance charge that need not be paid) is given by

$$u = f \cdot \frac{k(k + 1)}{n(n + 1)},$$

where u is the amount of unearned interest (money saved) when a loan scheduled to run n payments is paid off k payments ahead of schedule. The total scheduled finance charge is f. Use this formula to solve Exercises 41–44.

41. Rhonda Alessi bought a new Ford and agreed to pay it off in 36 monthly payments. The total finance charge was $700. Find the unearned interest if she paid the loan off 4 payments ahead of schedule.

42. Charles Vosburg bought a car and agreed to pay it off in 36 monthly payments. The total finance charge on the loan was $600. With 12 payments remaining, Charles decided to pay the loan in full. Find the amount of unearned interest.

43. The finance charge on a loan taken out by Vic Denicola is $380.50. If there were 24 equal monthly installments needed to repay the loan, and the loan is paid in full with 8 months remaining, find the amount of unearned interest.

44. Adrian Ortega is scheduled to repay a loan in 24 equal monthly installments. The total finance charge on the loan is $450. With 9 payments remaining, he decides to repay the loan in full. Find the amount of unearned interest.

Exercises 45 and 46 deal with winning percentage in the standings of sports teams. Winning percentage (Pct.) is commonly expressed as a decimal rounded to the nearest thousandth. To find the winning percentage of a team, divide the number of wins (W) by the total number of games played (W + L).

45. At the start of play on September 15, 2000, the standings of the Central Division of the American League were as shown. Find the winning percentage of each team.

   (a) Chicago   **(b)** Cleveland
   (c) Detroit

	W	L	Pct.
Chicago	87	58	
Cleveland	77	65	
Detroit	71	74	
Kansas City	68	78	.466
Minnesota	63	82	.434

46. Repeat Exercise 45 for the following standings for the Eastern Division of the National League.

   (a) Atlanta   **(b)** New York
   (c) Florida

	W	L	Pct.
Atlanta	86	60	
New York	84	62	
Florida	69	76	
Montreal	61	84	.421
Philadelphia	60	85	.414

Television networks have been losing viewers to cable programming since 1982, as the two graphs show. Use these graphs to answer Exercises 47–50. See Example 6.

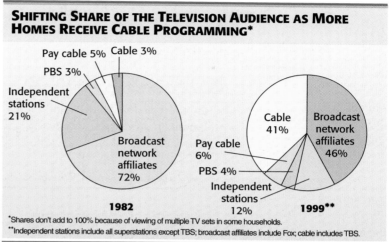

SHIFTING SHARE OF THE TELEVISION AUDIENCE AS MORE HOMES RECEIVE CABLE PROGRAMMING*

Pay cable 5% Cable 3%
PBS 3%
Independent stations 21%
Broadcast network affiliates 72%
1982

Cable 41%
Broadcast network affiliates 46%
Pay cable 6%
PBS 4%
Independent stations 12%
1999**

*Shares don't add to 100% because of viewing of multiple TV sets in some households.
**Independent stations include all superstations except TBS; broadcast affiliates include Fox; cable includes TBS.

Source: Nielsen Media Research, National Cable Television Association Report, Spring, 2000.

47. In a typical group of 50,000 television viewers, how many would have watched cable in 1982?

48. In 1982, how many of a typical group of 110,000 viewers watched independent stations?

49. How many of a typical group of 35,000 viewers watched cable in 1999?

50. In a typical group of 65,000 viewers, how many watched independent stations in 1999?

An average middle-income family will spend $160,140 to raise a child born in 1999 from birth to age 17. The graph shows the percent spent for various categories. Use the graph to answer Exercises 51 and 52. See Example 6.

51. To the nearest dollar, how much will be spent to provide housing for the child?

52. (a) To the nearest dollar, how much will be spent for health care?
   (b) Use your answer from part (a) to decide how much will be spent for transportation.

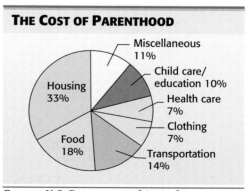

THE COST OF PARENTHOOD

Miscellaneous 11%
Child care/ education 10%
Health care 7%
Clothing 7%
Transportation 14%
Food 18%
Housing 33%

Source: U.S. Department of Agriculture.

2.3 Applications of Linear Equations

OBJECTIVE 1 Translate from words to mathematical expressions. Producing a mathematical model of a real situation often involves translating verbal statements into mathematical statements.

PROBLEM SOLVING

Usually there are key words and phrases in a verbal problem that translate into mathematical expressions involving addition, subtraction, multiplication, and division. Translations of some commonly used expressions follow.

Translating from Words to Mathematical Expressions

Verbal Expression	Mathematical Expression (where x and y are numbers)
Addition	
The **sum** of a number and 7	$x + 7$
6 **more than** a number	$x + 6$
3 **plus** a number	$3 + x$
24 **added to** a number	$x + 24$
A number **increased by** 5	$x + 5$
The **sum** of two numbers	$x + y$
Subtraction	
2 **less than** a number	$x - 2$
12 **minus** a number	$12 - x$
A number **decreased by** 12	$x - 12$
A number **subtracted from** 10	$10 - x$
The **difference between** two numbers	$x - y$
Multiplication	
16 **times** a number	$16x$
A number **multiplied by** 6	$6x$
$\frac{2}{3}$ **of** a number (used with fractions and percent)	$\frac{2}{3}x$
Twice (2 times) a number	$2x$
The **product** of two numbers	xy
Division	
The **quotient** of 8 and a number	$\frac{8}{x} \ (x \neq 0)$
A number **divided by** 13	$\frac{x}{13}$
The **ratio** of two numbers or the **quotient** of two numbers	$\frac{x}{y} \ (y \neq 0)$

CAUTION Because subtraction and division are not commutative operations, it is important to correctly translate expressions involving them. For example, "2 less than a number" is translated as $x - 2$, *not* $2 - x$. "A number subtracted from 10" is expressed as $10 - x$, *not* $x - 10$.

For division, it is understood that the number by which we are dividing is the denominator, and the number that is divided is the numerator. For example, "a number divided by 13" and "13 divided into x" both translate as $\frac{x}{13}$. Similarly, "the quotient of x and y" is translated as $\frac{x}{y}$.

OBJECTIVE 2 Write equations from given information. The symbol for equality, $=$, is often indicated by the word *is*. In fact, because equal mathematical expressions represent names for the same number, any words that indicate the idea of "sameness" translate to $=$.

EXAMPLE 1 Translating Words into Equations

Translate each verbal sentence into an equation.

Verbal Sentence	Equation
Twice a number, decreased by 3, is 42.	$2x - 3 = 42$
The product of a number and 12, decreased by 7, is 105.	$12x - 7 = 105$
The quotient of a number and the number plus 4 is 28.	$\dfrac{x}{x+4} = 28$
The quotient of a number and 4, plus the number, is 10.	$\dfrac{x}{4} + x = 10$

Now Try Exercises 7 and 17.

OBJECTIVE 3 Distinguish between expressions and equations. To distinguish between algebraic expressions and equations, remember that an expression translates as a phrase. An equation includes the $=$ symbol and translates as a sentence.

EXAMPLE 2 Distinguishing between Expressions and Equations

Decide whether each is an *expression* or an *equation*.

(a) $2(3 + x) - 4x + 7$

There is no equals sign, so this is an expression.

(b) $2(3 + x) - 4x + 7 = -1$

Because of the equals sign, this is an equation. Note that the expression in part (a) simplifies to the expression $-2x + 13$, and the equation in part (b) has solution 7.

Now Try Exercises 21 and 23.

OBJECTIVE 4 Use the six steps in solving an applied problem. Throughout this book we will be solving different types of applications. While there is no one method that will allow us to solve all types of applied problems, the following six steps are helpful.

Solving an Applied Problem

Step 1 **Read** the problem carefully until you understand what is given and what is to be found.

Step 2 **Assign a variable** to represent the unknown value, using diagrams or tables as needed. Write down what the variable represents. If necessary, express any other unknown values in terms of the variable.

(continued)

Step 3 **Write an equation** using the variable expression(s).

Step 4 **Solve** the equation.

Step 5 **State the answer** to the problem. Does it seem reasonable?

Step 6 **Check** the answer in the words of the original problem.

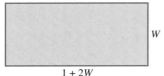

$1 + 2W$

FIGURE 4

■ EXAMPLE 3 Solving a Geometry Problem

The length of a rectangle is 1 cm more than twice the width. The perimeter of the rectangle is 110 cm. Find the length and the width of the rectangle.

Step 1 **Read** the problem. What must be found? The length and width of the rectangle. What is given? The length is 1 cm more than twice the width; the perimeter is 110 cm.

Step 2 **Assign a variable.** Make a sketch, as in Figure 4. Choose a variable: let W = the width; then $1 + 2W$ = the length.

Step 3 **Write an equation.** The perimeter of a rectangle is given by the formula $P = 2L + 2W$.

$$P = 2L + 2W$$
$$110 = 2(1 + 2W) + 2W \qquad \text{Let } L = 1 + 2W \text{ and } P = 110.$$

Step 4 **Solve** the equation obtained in Step 3.

$$110 = 2(1 + 2W) + 2W$$
$$110 = 2 + 4W + 2W \qquad \text{Distributive property}$$
$$110 = 2 + 6W \qquad \text{Combine like terms.}$$
$$110 - 2 = 2 + 6W - 2 \qquad \text{Subtract 2.}$$
$$108 = 6W$$
$$\frac{108}{6} = \frac{6W}{6} \qquad \text{Divide by 6.}$$
$$18 = W$$

Step 5 **State the answer.** The width of the rectangle is 18 cm and the length is $1 + 2(18) = 37$ cm.

Step 6 **Check** the answer by substituting these dimensions into the words of the original problem.

Now Try Exercise 29.

■ EXAMPLE 4 Finding Unknown Numerical Quantities

Two outstanding major league pitchers in recent years are Roger Clemens and Greg Maddux. Between 1984 and 1999, they pitched a total of 916 games. Clemens pitched 44 more games than Maddux. How many games did each player pitch? (*Source: Who's Who in Baseball,* 2000.)

Step 1 **Read** the problem. We are asked to find the number of games each player pitched.

Step 2   **Assign a variable** to represent the number of games of one of the men.

Let m = the number of games for Maddux.

We must also find the number of games for Clemens. Since he pitched 44 more games than Maddux,

$m + 44$ = Clemens' number of games.

Step 3   **Write an equation.** The sum of the numbers of games is 916, so

Maddux's games + Clemens' games = Total

$$m + (m + 44) = 916.$$

Step 4   **Solve** the equation.

$$m + (m + 44) = 916$$
$$2m + 44 = 916 \qquad \text{Combine like terms.}$$
$$2m = 872 \qquad \text{Subtract 44.}$$
$$m = 436 \qquad \text{Divide by 2.}$$

Step 5   **State the answer.** Since m represents the number of Maddux's games, Maddux pitched 436 games. Also, $m + 44 = 436 + 44 = 480$ is the number of games pitched by Clemens.

Step 6   **Check.** 480 is 44 more than 436, and the sum of 436 and 480 is 916. The conditions of the problem are satisfied, and our solution checks.

Now Try Exercise 33.

CAUTION   A common error in solving applied problems is forgetting to answer all the questions asked in the problem. In Example 4, we were asked for the number of games for *each* player, so there was extra work in Step 5 in order to find Clemens' number.

OBJECTIVE 5   Solve further percent problems.   Recall from the previous section that percent means "per one hundred," so 5% means .05, 14% means .14, and so on.

EXAMPLE 5   Solving a Percent Problem

In 2002 there were 301 long-distance area codes in the United States. This was an increase of 250% over the number when the area code plan originated in 1947. How many area codes were there in 1947? (*Source:* SBC Telephone Directory.)

Step 1   **Read** the problem. We are given that the number of area codes increased by 250% from 1947 to 2002, and there were 301 area codes in 2002. We must find the original number of area codes.

Step 2   **Assign a variable.** Let x represent the number of area codes in 1947. Since $250\% = 250(.01) = 2.5$, $2.5x$ represents the number of codes added since then.

Step 3   **Write an equation** from the given information.

the number in 1947 + the increase = 301

$$x + 2.5x = 301$$

Step 4   **Solve** the equation.

$$1x + 2.5x = 301 \qquad \text{Identity property}$$
$$3.5x = 301 \qquad \text{Combine like terms.}$$
$$x = 86 \qquad \text{Divide by 3.5.}$$

Step 5   **State the answer.** There were 86 area codes in 1947.

Step 6   **Check** that the increase, $301 - 86 = 215$, is 250% of 86.

Now Try Exercise 39.

CAUTION   Watch for two common errors that occur in solving problems like the one in Example 5.

1. Do not try to find 250% of 301 and subtract that amount from 301. The 250% should be applied to *the amount in 1947, not the amount in 2002.*

2. Do not write the equation as

$$x + 2.5 = 301.$$

The percent must be multiplied by some amount; in this case, the amount is the number of area codes in 1947, giving $2.5x$.

OBJECTIVE 6   Solve investment problems.   We use linear equations to solve certain types of investment problems. The investment problems in this chapter deal with *simple interest.* In most real-world applications, *compound interest* (covered in a later chapter) is used.

EXAMPLE 6   Solving an Investment Problem

After winning the state lottery, Mark LeBeau has $40,000 to invest. He will put part of the money in an account paying 4% interest and the remainder into stocks paying 6% interest. His accountant tells him that the total annual income from these investments should be $2040. How much should he invest at each rate?

Step 1   **Read** the problem again.

Step 2   **Assign a variable.**

Let $x =$ the amount to invest at 4%;

then $40,000 - x =$ the amount to invest at 6%.

The formula for interest is $I = prt$. Here the time, t, is 1 yr. Use a table to organize the given information.

Principal	Rate (as a decimal)	Interest	
x	.04	.04x	
40,000 − x	.06	.06(40,000 − x)	
40,000		2040	← Total

Step 3 **Write an equation.** The last column of the table gives the equation.

interest at 4% + interest at 6% = total interest

$$.04x \quad + \quad .06(40{,}000 - x) \quad = \quad 2040$$

Step 4 **Solve** the equation. We do so without clearing decimals.

$.04x + .06(40{,}000) - .06x = 2040$	Distributive property
$-.02x + 2400 = 2040$	Combine like terms; multiply.
$-.02x = -360$	Subtract 2400.
$x = 18{,}000$	Divide by $-.02$.

Step 5 **State the answer.** Mark should invest $18,000 of the money at 4% and $40,000 − $18,000 = $22,000 at 6%.

Step 6 **Check** by finding the annual interest at each rate; they should total $2040.

Now Try Exercise 43.

NOTE In Example 6, we chose to let the variable represent the amount invested at 4%. Students often ask, "Can I let the variable represent the other unknown?" The answer is yes. The equation will be different, but in the end the answers will be the same.

OBJECTIVE 7 Solve mixture problems.   Mixture problems involving rates of concentration can be solved with linear equations.

EXAMPLE 7 Solving a Mixture Problem

A chemist must mix 8 L of a 40% acid solution with some 70% solution to get a 50% solution. How much of the 70% solution should be used?

Step 1 **Read** the problem. The problem asks for the amount of 70% solution to be used.

Step 2 **Assign a variable.** Let $x =$ the number of liters of 70% solution to be used. The information in the problem is illustrated in Figure 5.

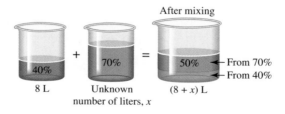

FIGURE 5

Use the given information to complete a table as shown on the next page.

Number of Liters	Percent (as a decimal)	Liters of Pure Acid
8	.40	.40(8) = 3.2
x	.70	.70x
8 + x	.50	.50(8 + x)

Sum must equal

The numbers in the last column were found by multiplying the strengths and the numbers of liters. The number of liters of pure acid in the 40% solution plus the number of liters in the 70% solution must equal the number of liters in the 50% solution.

Step 3 **Write an equation.**

$$3.2 + .70x = .50(8 + x)$$

Step 4 **Solve.** $3.2 + .70x = 4 + .50x$ Distributive property

$\qquad\qquad\qquad .20x = .8$ Subtract 3.2 and .50x.

$\qquad\qquad\qquad\quad x = 4$ Divide by .20.

Step 5 **State the answer.** The chemist should use 4 L of the 70% solution.

Step 6 **Check.** 8 L of 40% solution plus 4 L of 70% solution is 8(.40) + 4(.70) = 6 L of acid. Similarly, 8 + 4 or 12 L of 50% solution has 12(.50) = 6 L of acid in the mixture. The total amount of pure acid is 6 L both before and after mixing, so the answer checks.

Now Try Exercise 51.

In some mixture problems, you will need to use the fact that the percent of acid in pure water is 0%. Similarly, pure acid is 100% acid.

EXAMPLE 8   Solving a Mixture Problem When One Ingredient Is Pure

The octane rating of gasoline is a measure of its antiknock qualities. For a standard fuel, the octane rating is the percent of isooctane. How many liters of pure isooctane should be mixed with 200 L of 94% isooctane, referred to as 94 octane, to get a mixture that is 98% isooctane?

Step 1 **Read** the problem. The problem asks for the amount of pure isooctane.

Step 2 **Assign a variable.** Let x = the number of liters of pure (100%) isooctane. Fill in a table with the given information. Recall that 100% = 100(.01) = 1.

Number of Liters	Percent (as a decimal)	Liters of Pure Isooctane
x	1	x
200	.94	.94(200)
x + 200	.98	.98(x + 200)

Step 3 **Write an equation.** The equation comes from the last column of the table, as in Example 7.

$$x + .94(200) = .98(x + 200)$$

Step 4 **Solve.**

$x + .94(200) = .98x + .98(200)$	Distributive property
$x + 188 = .98x + 196$	Multiply.
$.02x = 8$	Subtract $.98x$ and 188.
$x = 400$	Divide by .02.

Step 5 **State the answer.** 400 L of isooctane are needed.

Step 6 **Check** by showing that $400 + .94(200) = .98(400 + 200)$.

Now Try Exercise 53.

| **CONNECTIONS** |

Probably the most famous study of problem-solving techniques was developed by George Polya (1888–1985). Among his many publications was the modern classic *How to Solve It.* In this book, Polya proposed a four-step process for problem solving.

Polya's Four-Step Process for Problem Solving

1. **Understand the problem.** You must first decide what you are to find.

2. **Devise a plan.** Here are some strategies that may prove useful.

   Problem-Solving Strategies
   If a formula applies, use it.
   Write an equation and solve it.
   Draw a sketch.
   Make a table or a chart.
   Look for a pattern.
   Use trial and error.
   Work backward.

   We used the first of these strategies in the previous section. In this section we used the next three strategies.

3. **Carry out the plan.** This is where the algebraic techniques you are learning in this book can be helpful.

4. **Look back and check.** Is your answer reasonable? Does it answer the question that was asked?

For Discussion or Writing

Compare Polya's four steps with the six steps for problem solving given earlier. Which of our steps correspond with each of Polya's steps?

2.3 EXERCISES

In each of the following, **(a)** *translate as an expression and* **(b)** *translate as an equation or inequality. Use x to represent the number.*

1. (a) 12 more than a number _____
  (b) 12 is more than a number. _____

2. (a) 3 less than a number _____
  (b) 3 is less than a number. _____

3. (a) 4 smaller than a number _____
  (b) 4 is smaller than a number. _____

4. (a) 6 greater than a number _____
  (b) 6 is greater than a number. _____

5. Which one of the following is *not* a valid translation of "20% of a number"?

  A. $.20x$ **B.** $.2x$

  C. $\dfrac{x}{5}$ **D.** $20x$

6. Explain why $13 - x$ is *not* a correct translation of "13 less than a number."

Translate each verbal phrase into a mathematical expression. Use x to represent the unknown number. See Example 1.

7. Twice a number, decreased by 13

8. The product of 6 and a number, decreased by 12

9. 12 increased by three times a number

10. 12 more than one-half of a number

11. The product of 8 and 12 less than a number

12. The product of 9 more than a number and 6 less than the number

13. The quotient of three times a number and 7

14. The quotient of 6 and five times a nonzero number

Use the variable x for the unknown, and write an equation representing the verbal sentence. Then solve the problem. See Example 1.

15. The sum of a number and 6 is -31. Find the number.

16. The sum of a number and -4 is 12. Find the number.

17. If the product of a number and -4 is subtracted from the number, the result is 9 more than the number. Find the number.

18. If the quotient of a number and 6 is added to twice the number, the result is 8 less than the number. Find the number.

19. When $\frac{2}{3}$ of a number is subtracted from 12, the result is 10. Find the number.

20. When 75% of a number is added to 6, the result is 3 more than the number. Find the number.

Decide whether each is an expression *or an* equation. *See Example 2.*

21. $5(x + 3) - 8(2x - 6)$

22. $-7(z + 4) + 13(z - 6)$

23. $5(x + 3) - 8(2x - 6) = 12$

24. $-7(z + 4) + 13(z - 6) = 18$

25. $\dfrac{r}{2} - \dfrac{r + 9}{6} - 8$

26. $\dfrac{r}{2} - \dfrac{r + 9}{6} = 8$

27. In your own words, list the six steps suggested for solving an applied problem.

28. In a recent year the two most popular places where book buyers shopped were large chain bookstores and small chain/independent bookstores. In a sample of book buyers, 70 more shopped at large chain bookstores than at small chain/independent bookstores. A total of 442 book buyers shopped at these two types of stores. Complete the problem-solving steps to find how many buyers shopped at each type of bookstore. (*Source:* Book Industry Study Group.)

Step 1   We are asked to find _____.

Step 2   Let x = the number of book buyers at large chain bookstores.
Then $x - 70 =$ _____.

Step 3   _____ + _____ = 442

Step 4 $x =$ _____

Step 5   There were _____ large chain bookstore shoppers and _____ small chain/independent shoppers.

Step 6   The number of _____ was _____ more than the number of _____ and the total number of these shoppers was _____.

Use the six-step problem-solving method to solve each problem. See Examples 3 and 4.

29. The John Hancock Center in Chicago has a rectangular base. The length of the base measures 65 ft less than twice the width. The perimeter of this base is 860 ft. What are the dimensions of the base?

30. The John Hancock Center (Exercise 29) tapers as it rises. The top floor is rectangular and has perimeter 520 ft. The width of the top floor measures 20 ft more than one-half its length. What are the dimensions of the top floor?

The perimeter of the top floor is 520 ft.

$\frac{1}{2}L + 20$

L

$2W - 65$ W

The perimeter of the base is 860 ft.

31. The Bermuda Triangle supposedly causes trouble for aircraft pilots. It has a perimeter of 3075 mi. The shortest side measures 75 mi less than the middle side, and the longest side measures 375 mi more than the middle side. Find the lengths of the three sides.

32. The Vietnam Veterans Memorial in Washington, D.C., is in the shape of two sides of an isosceles triangle. If the two walls of equal length were joined by a straight line of 438 ft, the perimeter of the resulting triangle would be 931.5 ft. Find the lengths of the two walls. (*Source:* Pamphlet obtained at Vietnam Veterans Memorial.)

438 ft

33. The two companies with top revenues in the Fortune 500 list for 2002 were Wal-Mart and Exxon Mobil. Their revenues together totaled $412 million. Exxon Mobil revenues were $28 million less than Wal-Mart revenues. What were the revenues for each corporation? (*Source:* www.fortune.com/lists/F500/)

34. In a recent year, video rental revenue was $.27 billion more than twice video sales revenue. Together, these revenues amounted to $9.81 billion. What was the revenue from each of these sources? (*Source:* Paul Kagan Associates, Inc.)

35. In the 1996 presidential election, Bill Clinton and Bob Dole together received 538 electoral votes. Clinton received 220 more votes than Dole. How many votes did each candidate receive? (*Source:* Congressional Quarterly, Inc.)

36. Ted Williams and Rogers Hornsby were two great hitters. Together they got 5584 hits in their careers. Hornsby got 276 more hits than Williams. How many base hits did each get? (*Source:* Neft, D. S. and R. M. Cohen, *The Sports Encyclopedia: Baseball,* St. Martins Griffin; New York, 1997.)

Solve each percent problem. See Example 5.

37. Composite scores on the ACT exam rose from 20.6 in 1990 to 21.0 in 1998. What percent increase was this? (*Source:* The American College Testing Program.)

38. In 1998, the number of participants in the ACT exam was 995,000. Earlier, in 1990, a total of 817,000 took the exam. What percent increase was this? (*Source:* The American College Testing Program.)

39. In 1990, the average tuition for public 4-year universities in the United States was $2035 for full-time students. By 2000, it had risen approximately 85%. To the nearest dollar, what was the approximate cost in 2000? (*Source:* National Center for Education Statistics, U.S. Department of Education.)

40. In 1990, the average tuition for private 4-year universities in the United States was $10,348 for full-time students. By 2000, it had risen approximately 86.6%. To the nearest dollar, what was the approximate cost in 2000? (*Source:* National Center for Education Statistics, U.S. Department of Education.)

41. At the end of a day, Jeff Hornsby found that the total cash register receipts at the motel where he works amounted to $2725. This included the 9% sales tax charged. Find the amount of the tax.

42. Fino Roverato sold his house for $159,000. He got this amount knowing that he would have to pay a 6% commission to his agent. What amount did he have after the agent was paid?

Solve each investment problem. See Example 6.

43. Carter Fenton earned $12,000 last year by giving tennis lessons. He invested part at 3% simple interest and the rest at 4%. In one year, he earned a total of $440 in interest. How much did he invest at each rate?

Principal	Rate (as a decimal)	Interest
x	.03	
	.04	

44. Melissa Wright won $60,000 on a slot machine in Las Vegas. She invested part at 2% simple interest and the rest at 3%. In one year, she earned a total of $1600 in interest. How much was invested at each rate?

Principal	Rate (as a decimal)	Interest
x	.02	

45. Michael Pellissier invested some money at 4.5% simple interest and $1000 less than twice this amount at 3%. His total annual income from the interest was $1020. How much was invested at each rate?

46. Holly Rioux invested some money at 3.5% simple interest, and $5000 more than 3 times this amount at 4%. In one year, she earned $1440 in interest. How much did she invest at each rate?

47. Jerry and Lucy Keefe have $29,000 invested in stocks paying 5%. How much additional money should they invest in certificates of deposit paying 2% so that the total return on the two investments is 3%?

48. Ron Hampton placed $15,000 in an account paying 6%. How much additional money should he deposit at 4% so that the total return on the two investments is 5.5%?

Solve each problem involving rates of concentration and mixtures. See Examples 7 and 8.

49. Ten liters of a 4% acid solution must be mixed with a 10% solution to get a 6% solution. How many liters of the 10% solution are needed?

Liters of Solution	Percent (as a decimal)	Liters of Pure Acid
10	.04	
x	.10	
	.06	

50. How many liters of a 14% alcohol solution must be mixed with 20 L of a 50% solution to get a 30% solution?

Liters of Solution	Percent (as a decimal)	Liters of Pure Alcohol
x	.14	
	.50	

51. In a chemistry class, 12 L of a 12% alcohol solution must be mixed with a 20% solution to get a 14% solution. How many liters of the 20% solution are needed?

52. How many liters of a 10% alcohol solution must be mixed with 40 L of a 50% solution to get a 40% solution?

53. How much pure dye must be added to 4 gal of a 25% dye solution to increase the solution to 40%? (*Hint:* Pure dye is 100% dye.)

54. How much water must be added to 6 gal of a 4% insecticide solution to reduce the concentration to 3%? (*Hint:* Water is 0% insecticide.)

55. Randall Albritton wants to mix 50 lb of nuts worth $2 per lb with some nuts worth $6 per lb to make a mixture worth $5 per lb. How many pounds of $6 nuts must he use?

56. Lee Ann Spahr wants to mix tea worth 2¢ per oz with 100 oz of tea worth 5¢ per oz to make a mixture worth 3¢ per oz. How much 2¢ tea should be used?

57. Why is it impossible to mix candy worth $4 per lb and candy worth $5 per lb to obtain a final mixture worth $6 per lb?

58. Write an equation based on the following problem, solve the equation, and explain why the problem has no solution.

> How much 30% acid should be mixed with 15 L of 50% acid to obtain a mixture that is 60% acid?

RELATING CONCEPTS (EXERCISES 59–63)

For Individual or Group Work

Consider each problem.

Problem A
Jack has $800 invested in two accounts. One pays 5% interest per year and the other pays 10% interest per year. The amount of yearly interest is the same as he would get if the entire $800 was invested at 8.75%. How much does he have invested at each rate?

Problem B
Jill has 800 L of acid solution. She obtained it by mixing some 5% acid with some 10% acid. Her final mixture of 800 L is 8.75% acid. How much of each of the 5% and 10% solutions did she use to get her final mixture?

In Problem A, let x represent the amount invested at 5% interest, and in Problem B, let y represent the amount of 5% acid used. **Work Exercises 59–63 in order.**

59. (a) Write an expression in x that represents the amount of money Jack invested at 10% in Problem A.
  (b) Write an expression in y that represents the amount of 10% acid solution Jill used in Problem B.

60. (a) Write expressions that represent the amount of interest Jack earns per year at 5% and at 10%.
  (b) Write expressions that represent the amount of pure acid in Jill's 5% and 10% acid solutions.

61. (a) The sum of the two expressions in part (a) of Exercise 60 must equal the total amount of interest earned in one year. Write an equation representing this fact.
  (b) The sum of the two expressions in part (b) of Exercise 60 must equal the amount of pure acid in the final mixture. Write an equation representing this fact.

62. (a) Solve Problem A.
  (b) Solve Problem B.

63. Explain the similarities between the processes used in solving Problems A and B.

2.4   Further Applications of Linear Equations

OBJECTIVES

1 Solve problems about different denominations of money.

2 Solve problems about uniform motion.

3 Solve problems about angles.

There are three common applications of linear equations that we did not discuss in Section 2.3: money problems, uniform motion problems, and problems involving the angles of a triangle.

OBJECTIVE 1 Solve problems about different denominations of money.   These problems are very similar to the simple interest problems in Section 2.3.

PROBLEM SOLVING

In problems involving money, use the basic fact that

$$\begin{bmatrix} \text{number of monetary} \\ \text{units of the same kind} \end{bmatrix} \times [\text{denomination}] = \begin{bmatrix} \text{total monetary} \\ \text{value} \end{bmatrix}.$$

For example, 30 dimes have a monetary value of $30(.10) = 3.00$ dollars.
Fifteen 5-dollar bills have a value of $15(5) = 75$ dollars.

EXAMPLE 1   Solving a Money Denomination Problem

For a bill totaling $5.65, a cashier received 25 coins consisting of nickels and quarters. How many of each type of coin did the cashier receive?

Step 1   **Read** the problem. The problem asks that we find the number of nickels and the number of quarters the cashier received.

Step 2   **Assign a variable.**

Let x represent the number of nickels;
then $25 - x$ represents the number of quarters.

We can organize the information in a table as we did with investment problems.

	Number of Coins	Denomination	Value
Nickels	x	.05	.05x
Quarters	$25 - x$	.25	.25$(25 - x)$
			5.65 ← Total

Step 3   **Write an equation.** From the last column of the table,
$$.05x + .25(25 - x) = 5.65.$$

Step 4 **Solve.**

$$.05x + .25(25 - x) = 5.65$$
$$5x + 25(25 - x) = 565 \qquad \text{Multiply by 100.}$$
$$5x + 625 - 25x = 565 \qquad \text{Distributive property}$$
$$-20x = -60 \qquad \text{Subtract 625; combine terms.}$$
$$x = 3 \qquad \text{Divide by } -20.$$

Step 5 **State the answer.** The cashier has 3 nickels and $25 - 3 = 22$ quarters.

Step 6 **Check.** The cashier has $3 + 22 = 25$ coins, and the value of the coins is $\$.05(3) + \$.25(22) = \$5.65$, as required.

Now Try Exercise 11.

CAUTION Be sure that your answer is reasonable when working problems like Example 1. Because you are dealing with a number of coins, the correct answer can neither be negative nor a fraction.

OBJECTIVE 2 Solve problems about uniform motion.

PROBLEM SOLVING

Uniform motion problems use the distance formula, $d = rt$. In this formula, **when rate (or speed) is given in miles per hour, time must be given in hours.** To solve such problems, **draw a sketch** to illustrate what is happening in the problem, and make a table to summarize the given information.

EXAMPLE 2 Solving a Motion Problem (Motion in Opposite Directions)

Two cars leave the same place at the same time, one going east and the other west. The eastbound car averages 40 mph, while the westbound car averages 50 mph. In how many hours will they be 300 mi apart?

Step 1 **Read** the problem. We are looking for the time it takes for the two cars to be 300 mi apart.

Step 2 **Assign a variable.** A sketch shows what is happening in the problem: The cars are going in *opposite* directions. See Figure 6.

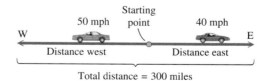

Starting point
50 mph   40 mph
W   E
Distance west   Distance east

Total distance = 300 miles

FIGURE 6

Let *x* represent the time traveled by each car. Summarize the information of the problem in a table. When the expressions for rate and time are entered, *fill in each distance by multiplying rate by time* using the formula $d = rt$.

	Rate	Time	Distance
Eastbound Car	40	x	40x
Westbound Car	50	x	50x
			300 ← Total

Step 3 **Write an equation.** From the sketch in Figure 6, the sum of the two distances is 300.

$$40x + 50x = 300$$

Step 4 **Solve.** $90x = 300$ Combine like terms.

$$x = \frac{300}{90}$$ Divide by 90.

$$x = \frac{10}{3}$$ Lowest terms

Step 5 **State the answer.** The cars travel $\frac{10}{3} = 3\frac{1}{3}$ hr, or 3 hr 20 min.

Step 6 **Check.** The eastbound car traveled $40\left(\frac{10}{3}\right) = \frac{400}{3}$ mi, and the westbound car traveled $50\left(\frac{10}{3}\right) = \frac{500}{3}$ mi, for a total of $\frac{400}{3} + \frac{500}{3} = \frac{900}{3} = 300$ mi, as required.

Now Try Exercise 21.

CAUTION It is a common error to write 300 as the distance for each car in Example 2. Three hundred miles is the *total* distance traveled.

As in Example 2, in general, the equation for a problem involving motion in opposite directions is of the form

partial distance + partial distance = total distance.

EXAMPLE 3 Solving a Motion Problem (Motion in the Same Direction)

Jeff Bezzone can bike to work in $\frac{3}{4}$ hr. When he takes the bus, the trip takes $\frac{1}{4}$ hr. If the bus travels 20 mph faster than Jeff rides his bike, how far is it to his workplace?

Step 1 **Read** the problem. We must find the distance between Jeff's home and his workplace.

Step 2 **Assign a variable.** Although the problem asks for a distance, it is easier here to let x be Jeff's speed when he rides his bike to work. Then the speed of the bus is $x + 20$. For the trip by bike,

$$d = rt = x \cdot \frac{3}{4} = \frac{3}{4}x,$$

and by bus,

$$d = rt = (x + 20) \cdot \frac{1}{4} = \frac{1}{4}(x + 20).$$

Summarize this information in a table.

	Rate	Time	Distance	
Bike	x	$\dfrac{3}{4}$	$\dfrac{3}{4}x$	⎫
Bus	$x + 20$	$\dfrac{1}{4}$	$\dfrac{1}{4}(x + 20)$	⎬ Same

Step 3   **Write an equation.** The key to setting up the correct equation is to understand that the distance in each case is the same. See Figure 7.

Home   Workplace

FIGURE 7

Since the distance is the same in each case,

$$\frac{3}{4}x = \frac{1}{4}(x + 20).$$

Step 4   **Solve** the equation. First multiply each side by 4.

$$4\left(\frac{3}{4}x\right) = 4\left(\frac{1}{4}\right)(x + 20) \qquad \text{Multiply by 4.}$$
$$3x = x + 20 \qquad \text{Multiply; identity property}$$
$$2x = 20 \qquad \text{Subtract } x.$$
$$x = 10 \qquad \text{Divide by 2.}$$

Step 5   **State the answer.** The required distance is

$$d = \frac{3}{4}x = \frac{3}{4}(10) = \frac{30}{4} = 7.5 \text{ mi.}$$

Step 6   **Check** by finding the distance using

$$d = \frac{1}{4}(x + 20) = \frac{1}{4}(10 + 20) = \frac{30}{4} = 7.5 \text{ mi,}$$

the same result.

Now Try Exercise 25.

As in Example 3, the equation for a problem involving motion in the same direction is often of the form

$$\text{one distance} = \text{other distance.}$$

> **NOTE**   In Example 3 it was easier to let the variable represent a quantity other than the one that we were asked to find. This is the case in some problems. It takes practice to learn when this approach is the best, and practice means working lots of problems!

OBJECTIVE 3   Solve problems about angles.   An important result of Euclidean geometry (the geometry of the Greek mathematician Euclid) is that the sum of the angle measures of any triangle is 180°. This property is used in the next example.

■ EXAMPLE 4   Finding Angle Measures

Find the value of x, and determine the measure of each angle in Figure 8.

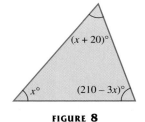

FIGURE 8

Step 1   **Read** the problem. We are asked to find the measure of each angle.

Step 2   **Assign a variable.** Let x represent the measure of one angle.

Step 3   **Write an equation.** The sum of the three measures shown in the figure must be 180°.

$$x + (x + 20) + (210 - 3x) = 180$$

Step 4   **Solve.**

$-x + 230 = 180$	Combine like terms.
$-x = -50$	Subtract 230.
$x = 50$	Divide by -1.

Step 5   **State the answer.** One angle measures 50°, another measures $x + 20 = 50 + 20 = 70°$, and the third measures $210 - 3x = 210 - 3(50) = 60°$.

Step 6   **Check.** Since $50° + 70° + 60° = 180°$, the answers are correct.   ■

Now Try Exercise 31.

We discuss another way to solve the problems in this section in Chapter 4.

2.4   EXERCISES

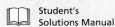

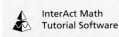
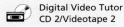
Solve each problem.

1. What amount of money is found in a coin hoard containing 38 nickels and 26 dimes?

2. The distance between Cape Town, South Africa, and Miami is 7700 mi. If a jet averages 480 mph between the two cities, what is its travel time in hours?

3. Tri Phong traveled from Chicago to Des Moines, a distance of 300 mi, in 5 hr. What was his rate in miles per hour?

4. A square has perimeter 40 in. What would be the perimeter of an equilateral triangle whose sides each measure the same length as the side of the square?

✐ *Write a short explanation in Exercises 5–8.*

5. Read over Example 3 in this section. The solution of the equation is 10. Why is *10 mph* not the answer to the problem?

6. Suppose that you know that two angles of a triangle have equal measures, and the third angle measures 36°. Explain in a few words the strategy you would use to find the measures of the equal angles without actually writing an equation.

7. In a problem about the number of coins of different denominations, would an answer that is a fraction be reasonable? What about a negative number?

8. In a motion problem the rate is given as x mph and the time is given as 30 min. What variable expression represents the distance in miles?

Solve each problem. See Example 1.

9. Otis Taylor has a box of coins that he uses when playing poker with his friends. The box currently contains 44 coins, consisting of pennies, dimes, and quarters. The number of pennies is equal to the number of dimes, and the total value is $4.37. How many of each denomination of coin does he have in the box?

Number of Coins	Denomination	Value
x	.01	.01x
x		
	.25	
		4.37 ← Total

10. Nana Nantambu found some coins while looking under her sofa pillows. There were equal numbers of nickels and quarters, and twice as many half-dollars as quarters. If she found $2.60 in all, how many of each denomination of coin did she find?

Number of Coins	Denomination	Value
x	.05	.05x
x		
2x	.50	
		2.60 ← Total

11. Kim Falgout's daughter, Madeline, has a piggy bank with 47 coins. Some are quarters, and the rest are half-dollars. If the total value of the coins is $17.00, how many of each denomination does she have?

12. John Joslyn has a jar in his office that contains 39 coins. Some are pennies, and the rest are dimes. If the total value of the coins is $2.64, how many of each denomination does he have?

13. Dave Bowers collects U.S. gold coins. He has a collection of 41 coins. Some are $10 coins, and the rest are $20 coins. If the face value of the coins is $540, how many of each denomination does he have?

14. In the nineteenth century, the United States minted two-cent and three-cent pieces. Frances Steib has three times as many three-cent pieces as two-cent pieces, and the face value of these coins is $2.42. How many of each denomination does she have?

15. A total of 550 people attended a Kenny Loggins concert. Floor tickets cost $40 each, while balcony tickets cost $28 each. If a total of $20,800 was collected, how many of each type of ticket were sold?

16. The Delgado Community College production of *The Music Man* was a big success. For opening night, 410 tickets were sold. Students paid $3 each, while nonstudents paid $7 each. If a total of $1650 was collected, how many students and how many nonstudents attended?

In Exercises 17–20, find the rate based on the information provided. Use a calculator and round your answers to the nearest hundredth. All events were at the 2000 Summer Olympics in Sidney, Australia. (Source: http://espn.go.com/oly/summer00)

	Event	Participant	Distance	Time
17.	100-m hurdles, Women	Olga Shishigina, Kazakhstan	100 m	12.65 sec
18.	400-m hurdles, Women	Irina Privalova, Russia	400 m	53.02 sec
19.	400-m hurdles, Men	Angelo Taylor, USA	400 m	47.50 sec
20.	400-m dash, Men	Michael Johnson, USA	400 m	43.84 sec

Solve each problem. See Examples 2 and 3.

21. Two steamers leave a port on a river at the same time, traveling in opposite directions. Each is traveling 22 mph. How long will it take for them to be 110 mi apart?

	Rate	Time	Distance
First Steamer		t	
Second Steamer	22		
			110

22. A train leaves Kansas City, Kansas, and travels north at 85 km per hr. Another train leaves at the same time and travels south at 95 km per hr. How long will it take before they are 315 km apart?

	Rate	Time	Distance
First Train	85	t	
Second Train			
			315

23. Agents Mulder and Scully are driving to Georgia to investigate "Big Blue," a giant aquatic reptile reported to inhabit one of the local lakes. Mulder leaves Washington at 8:30 A.M. and averages 65 mph. His partner, Scully, leaves at 9:00 A.M., following the same path and averaging 68 mph. At what time will Scully catch up with Mulder?

	Rate	Time	Distance
Mulder			
Scully			

24. Lois and Clark are covering separate stories and have to travel in opposite directions. Lois leaves the *Daily Planet* at 8:00 A.M. and travels at 35 mph. Clark leaves at 8:15 A.M. and travels at 40 mph. At what time will they be 140 mi apart?

	Rate	Time	Distance
Lois			
Clark			

25. Latrella can get to school in 15 min if she rides her bike. It takes her 45 min if she walks. Her speed when walking is 10 mph slower than her speed when riding. What is her speed when she rides?

	Rate	Time	Distance
Riding			
Walking			

26. When Dewayne drives his car to work, the trip takes 30 min. When he rides the bus, it takes 45 min. The average speed of the bus is 12 mph less than his speed when driving. Find the distance he travels to work.

	Rate	Time	Distance
Car			
Bus			

27. Johnny leaves Memphis to visit his cousin, Anne Hoffman, who lives in the town of Hornsby, TN, 80 mi away. He travels at an average speed of 50 mph. One-half hour later, Anne leaves to visit Johnny, traveling at an average speed of 60 mph. How long after Anne leaves will it be before they meet?

28. On an automobile trip, Aimee Cardella maintained a steady speed for the first two hours. Rush-hour traffic slowed her speed by 25 mph for the last part of the trip. The entire trip, a distance of 125 mi, took $2\frac{1}{2}$ hr. What was her speed during the first part of the trip?

Find the measure of each angle in the triangles shown. (Be sure to substitute your value of x into each angle expression.) See Example 4.

29.
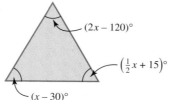

30.
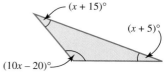

31.
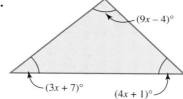

32.
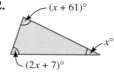

RELATING CONCEPTS (EXERCISES 33–36)

For Individual or Group Work

Consider the following two figures. **Work Exercises 33–36 in order.**

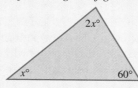

FIGURE A

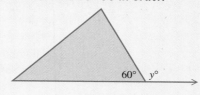

FIGURE B

33. Solve for the measures of the unknown angles in Figure A.

34. Solve for the measure of the unknown angle marked $y°$ in Figure B.

35. Add the measures of the two angles you found in Exercise 33. How does the sum compare to the measure of the angle you found in Exercise 34?

36. From Exercises 33–35, make a conjecture (an educated guess) about the relationship among the angles marked ①, ②, and ③ in the figure shown here.

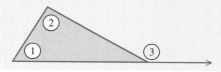

*In Exercises 37 and 38, the angles marked with variable expressions are called **vertical angles**. It is shown in geometry that vertical angles have equal measures. Find the measure of each angle.*

37.

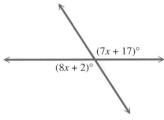

$(7x + 17)°$

$(8x + 2)°$

38.

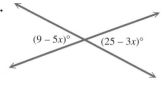

$(9 - 5x)°$ $(25 - 3x)°$

39. Two angles whose sum is 90° are called **complementary angles.** Find the measures of the complementary angles shown in the figure.

$(5x - 1)°$

$(2x)°$

40. Two angles whose sum is 180° are called **supplementary angles.** Find the measures of the supplementary angles shown in the figure.

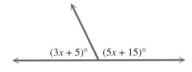

$(3x + 5)°$ $(5x + 15)°$

Another type of application often studied in introductory and intermediate algebra courses involves consecutive integers. Consecutive integers are integers that follow each other in counting order, such as 8, 9, and 10. Suppose we wish to solve the following problem:

Find three consecutive integers such that the sum of the first and third, increased by 3, is 50 more than the second.

Let x represent the first of the unknown integers. Then $x + 1$ will be the second, and $x + 2$ will be the third. The equation we need can be found by going back to the words of the original problem.

Sum of the first and third	increased by 3	is	50 more than the second.
↓	↓	↓	↓
$x + (x + 2)$	$+ 3$	$=$	$(x + 1) + 50$

The solution of this equation is 46, meaning that the first integer is $x = 46$, the second is $x + 1 = 47$, and the third is $x + 2 = 48$. The three integers are 46, 47, and 48. Check by substituting these numbers back into the words of the original problem.

Solve each problem involving consecutive integers.

41. Find three consecutive integers such that the sum of the first and twice the second is 17 more than twice the third.

42. Find four consecutive integers such that the sum of the first three is 54 more than the fourth.

43. If I add my current age to the age I will be next year on this date, the sum is 103 years. How old will I be 10 years from today?

44. Two pages facing each other in this book have 189 as the sum of their page numbers. What are the two page numbers?

▬ SUMMARY EXERCISES ON SOLVING APPLIED PROBLEMS

The applications that follow are of the various types introduced in this chapter. Use the strategies you have developed to solve each problem.

1. The length of a rectangle is 3 in. more than its width. If the length were decreased by 2 in. and the width were increased by 1 in., the perimeter of the resulting rectangle would be 24 in. Find the dimensions of the original rectangle.

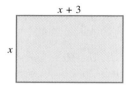

2. A farmer wishes to enclose a rectangular region with 210 m of fencing in such a way that the length is twice the width and the region is divided into two equal parts, as shown in the figure. What length and width should be used?

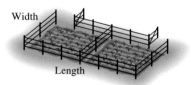

3. An electronics store offered a videodisc player for $255. This was the sale price, after the regular price had been discounted 40%. What was the regular price?

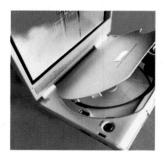

4. After a discount of 30%, the sale price of *The Parents' Guide to Kids' Sports* was $6.27. What was the regular price of the book? (Give your answer to the nearest 5¢.)

5. An amount of money is invested at 4% annual simple interest, and twice that amount is invested at 5%. The total annual interest is $112. How much is invested at each rate?

6. An amount of money is invested at 3% annual simple interest, and $2000 more than that amount is invested at 4%. The total annual interest is $920. How much is invested at each rate?

7. In the 1940 presidential election, Franklin Roosevelt and Wendell Willkie together received 531 electoral votes. Roosevelt received 367 more votes than Willkie in the landslide. How many votes did each man receive? (*Source:* Congressional Quarterly, Inc.)

8. Two of the highest paid business executives in a recent year were Mike Eisner, chairman of Disney, and Ed Horrigan, vice chairman of RJR Nabisco. Together their salaries totaled $61.8 million. Eisner earned $18.4 million more than Horrigan. What was the salary for each executive?

9. Atlanta and Cincinnati are 440 mi apart. John leaves Cincinnati, driving toward Atlanta at an average speed of 60 mph. Pat leaves Atlanta at the same time, driving toward Cincinnati in her antique auto, averaging 28 mph. How long will it take them to meet?

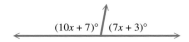

10. Moses Tanui from Kenya won the 1996 men's Boston marathon with a rate of 12.19 mph. The women's race was won by Uta Pippig from Germany, who ran at 10.69 mph. Pippig's time was .3 hr longer than Tanui's. Find their winning times. (*Source: Universal Almanac.*)

11. A pharmacist has 20 L of a 10% drug solution. How many liters of 5% solution must be added to get a mixture that is 8%?

12. A certain metal is 20% tin. How many kilograms of this metal must be mixed with 80 kg of a metal that is 70% tin to get a metal that is 50% tin?

13. A cashier has a total of 126 bills in fives and tens. The total value of the money is $840. How many of each type bill does he have?

14. The hit movie *Titanic* earned more in Europe than in the United States. As of September 1998, the average cost for a movie ticket in the United States was $4.59. In Great Britain, it cost $10.59 and in Germany, $8.42. If 74 million tickets were sold in Great Britain and Germany, how many would have to be sold in each country to earn $686 million? Round answers to the nearest million. (*Source: Parade, September 13, 1998.*)

15. Find the measure of each angle.

16. Find the measure of each marked angle.

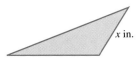

17. The sum of the smallest and largest of three consecutive integers is 32 more than the middle integer. What are the three integers?

18. If the smaller of two consecutive odd integers is doubled, the result is 7 more than the larger of the two integers. Find the two integers.

19. The perimeter of a triangle is 34 in. The middle side is twice as long as the shortest side. The longest side is 2 in. less than three times the shortest side. Find the lengths of the three sides.

20. The perimeter of a rectangle is 43 in. more than the length. The width is 10 in. Find the length of the rectangle.

2.5 Linear Inequalities in One Variable

OBJECTIVES

1 Solve linear inequalities using the addition property.

2 Solve linear inequalities using the multiplication property.

3 Solve linear inequalities with three parts.

4 Solve applied problems using linear inequalities.

In Section 1.1, we used interval notation to write solution sets of inequalities, using a parenthesis to indicate that an endpoint is not included and a square bracket to indicate that an endpoint is included. We summarize the various types of intervals here.

Interval Notation

Type of Interval	Set	Interval Notation	Graph
Open interval	$\{x \mid a < x\}$	(a, ∞)	
	$\{x \mid a < x < b\}$	(a, b)	
	$\{x \mid x < b\}$	$(-\infty, b)$	
	$\{x \mid x \text{ is a real number}\}$	$(-\infty, \infty)$	
Half-open interval	$\{x \mid a \le x\}$	$[a, \infty)$	
	$\{x \mid a < x \le b\}$	$(a, b]$	
	$\{x \mid a \le x < b\}$	$[a, b)$	
	$\{x \mid x \le b\}$	$(-\infty, b]$	
Closed interval	$\{x \mid a \le x \le b\}$	$[a, b]$	

An inequality says that two expressions are *not* equal. Solving inequalities is similar to solving equations.

Linear Inequality in One Variable

A **linear inequality in one variable** can be written in the form
$$Ax + B < C,$$
where A, B, and C are real numbers, with $A \ne 0$.

(Throughout this section we give definitions and rules only for $<$, but they are also valid for $>$, $\le$, and $\ge$.) Examples of linear inequalities include

$$x + 5 < 2, \qquad x - 3 \ge 5, \qquad \text{and} \qquad 2k + 5 \le 10. \qquad \text{Linear inequalities}$$

OBJECTIVE 1 Solve linear inequalities using the addition property. We solve an inequality by finding all numbers that make the inequality true. Usually, an inequality has an infinite number of solutions. These solutions, like solutions of equations, are found by producing a series of simpler equivalent inequalities. **Equivalent inequalities** are inequalities with the same solution set. We use the addition and multiplication properties of inequality to produce equivalent inequalities.

> **Addition Property of Inequality**
>
> For all real numbers A, B, and C, the inequalities
> $$A < B \qquad \text{and} \qquad A + C < B + C$$
> are equivalent.
>   That is, adding the same number to each side of an inequality does not change the solution set.

As with equations, the addition property can be used to *subtract* the same number from each side of an inequality.

■ **EXAMPLE 1   Using the Addition Property of Inequality**

Solve $x - 7 < -12$.

Add 7 to each side.

$$x - 7 + 7 < -12 + 7 \qquad \text{Add 7.}$$
$$x < -5$$

Check: Substitute -5 for x in the equation $x - 7 = -12$. The result should be a true statement.

$$x - 7 = -12$$
$$-5 - 7 = -12 \qquad ? \qquad \text{Let } x = -5.$$
$$-12 = -12 \qquad \text{True}$$

This shows that -5 is the boundary point. Now test a number on each side of -5 to verify that numbers *less than* -5 make the inequality true. Choose -4 and -6.

$$x - 7 < -12$$

$-4 - 7 < -12$? Let $x = -4$.	$-6 - 7 < -12$? Let $x = -6$.
$-11 < -12$ False	$-13 < -12$ True
-4 is not in the solution set.	-6 is in the solution set.

The check confirms that $(-\infty, -5)$, graphed in Figure 9, is the correct solution set.

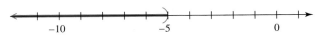

FIGURE 9

Now Try Exercise 9.

EXAMPLE 2   Using the Addition Property of Inequality

Solve $14 + 2m \le 3m$ and graph the solution set.

First, subtract $2m$ from each side.

$$14 + 2m \le 3m$$
$$14 + 2m - 2m \le 3m - 2m \qquad \text{Subtract } 2m.$$
$$14 \le m \qquad \text{Combine like terms.}$$

The inequality $14 \le m$ (14 is less than or equal to m) can also be written $m \ge 14$ (m is greater than or equal to 14). Notice that in each case, the inequality symbol points to the smaller number, 14.

Check:
$$14 + 2m = 3m$$
$$14 + 2(14) = 3(14) \qquad ? \qquad \text{Let } m = 14.$$
$$42 = 42 \qquad \text{True}$$

So 14 satisfies the equality part of $\le$. Choose 10 and 15 as test points.

$$14 + 2m < 3m$$

$14 + 2(10) < 3(10)$? Let $m = 10$.	$14 + 2(15) < 3(15)$? Let $m = 15$.
$34 < 30$ False	$44 < 45$ True
10 is not in the solution set.	15 is in the solution set.

The check confirms that $[14, \infty)$ is the correct solution set. See Figure 10.

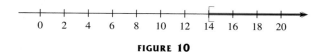

FIGURE 10

Now Try Exercise 21.

CAUTION   Errors often occur in graphing inequalities when the variable term is on the right side. (This is probably due to the fact that we read from left to right.) To guard against such errors, it is a good idea to rewrite these inequalities so that the variable is on the left, as discussed in Example 2.

OBJECTIVE 2 Solve linear inequalities using the multiplication property.   Solving an inequality such as $3x \le 15$ requires dividing each side by 3, using the *multiplication property of inequality*, which is a little more involved than the multiplication property of *equality*. To see how the multiplication property of inequality works, start with the true statement

$$-2 < 5.$$

Multiply each side by, say, 8.

$$-2(8) < 5(8) \qquad \text{Multiply by 8.}$$
$$-16 < 40 \qquad \text{True}$$

This gives a true statement. Start again with $-2 < 5$, and multiply each side by -8.

$$-2(-8) < 5(-8) \qquad \text{Multiply by } -8.$$
$$16 < -40 \qquad \text{False}$$

The result, $16 < -40$, is false. To make it true, we must change the direction of the inequality symbol to get

$$16 > -40. \quad \text{True}$$

As these examples suggest, multiplying each side of an inequality by a *negative* number reverses the direction of the inequality symbol. The same is true for dividing by a negative number since division is defined in terms of multiplication.

Multiplication Property of Inequality

For all real numbers A, B, and C, with $C \neq 0$,

(a) the inequalities

$$A < B \qquad \text{and} \qquad AC < BC$$

are equivalent **if $C > 0$;**

(b) the inequalities

$$A < B \qquad \text{and} \qquad AC > BC$$

are equivalent **if $C < 0$.**

That is, each side of an inequality may be multiplied (or divided) by a *positive* number without changing the direction of the inequality symbol. *Multiplying (or dividing) by a **negative** number requires that we reverse the inequality symbol.*

CAUTION   Remember to reverse the direction of the inequality symbol when multiplying or dividing by a *negative* number.

■ EXAMPLE 3   Using the Multiplication Property of Inequality

Solve each inequality and graph the solution set.

(a) $5m \leq -30$

Use the multiplication property to divide each side by 5. Since $5 > 0$, do *not* reverse the inequality symbol.

$$5m \leq -30$$

$$\frac{5m}{5} \leq \frac{-30}{5} \qquad \text{Divide by 5.}$$

$$m \leq -6$$

Check that the solution set is the interval $(-\infty, -6]$, graphed in Figure 11.

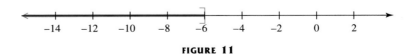

FIGURE 11

(b) $-4k \leq 32$

Divide each side by -4. Since $-4 < 0$, reverse the inequality symbol.

$$-4k \leq 32$$

$$\frac{-4k}{-4} \geq \frac{32}{-4} \qquad \text{Divide by } -4 \text{ and reverse the symbol.}$$

$$k \geq -8$$

Check the solution set. Figure 12 shows the graph of the solution set, $[-8, \infty)$.

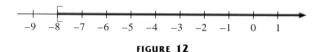

FIGURE 12

Now Try Exercises 13 and 17.

The steps used in solving a linear inequality are given here.

Solving a Linear Inequality

Step 1 **Simplify each side separately.** Use the distributive property to clear parentheses and combine like terms as needed.

Step 2 **Isolate the variable terms on one side.** Use the addition property of inequality to get all terms with variables on one side of the inequality and all numbers on the other side.

Step 3 **Isolate the variable.** Use the multiplication property of inequality to change the inequality to the form $x < k$ or $x > k$.

Remember: Reverse the direction of the inequality symbol *only* when *multiplying or dividing each side of an inequality by a **negative number.***

EXAMPLE 4 Solving a Linear Inequality Using the Distributive Property

Solve $-3(x + 4) + 2 \geq 7 - x$ and graph the solution set.

Step 1 $\qquad -3x - 12 + 2 \geq 7 - x \qquad$ Distributive property

$\qquad\qquad -3x - 10 \geq 7 - x$

Step 2 $\qquad -3x - 10 + x \geq 7 - x + x \qquad$ Add x.

$\qquad\qquad -2x - 10 \geq 7$

$\qquad -2x - 10 + 10 \geq 7 + 10 \qquad$ Add 10.

$\qquad\qquad -2x \geq 17$

Step 3 $\qquad\qquad \dfrac{-2x}{-2} \leq \dfrac{17}{-2} \qquad$ Divide by -2; change $\geq$ to $\leq$.

$\qquad\qquad x \leq -\dfrac{17}{2}$

Figure 13 shows the graph of the solution set, $\left(-\infty, -\frac{17}{2}\right]$.

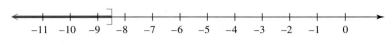

FIGURE 13

Now Try Exercise 23.

NOTE   In Step 2 of Example 4, if we add $3x$ to both sides of the inequality, we have

$$-3x - 10 + 3x \geq 7 - x + 3x \qquad \text{Add } 3x.$$
$$-10 \geq 2x + 7$$
$$-10 - 7 \geq 2x + 7 - 7 \qquad \text{Subtract } 7.$$
$$-17 \geq 2x$$
$$-\frac{17}{2} \geq x. \qquad \text{Divide by } 2.$$

The final line is read "$-\frac{17}{2}$ is greater than or equal to x," which means the same thing as "x is less than or equal to $-\frac{17}{2}$." Thus, the solution set is the same.

▦ EXAMPLE 5   Solving a Linear Inequality with Fractions

Solve $-\frac{2}{3}(r - 3) - \frac{1}{2} < \frac{1}{2}(5 - r)$ and graph the solution set.

To clear fractions, multiply each side by the least common denominator, 6.

$$-\frac{2}{3}(r - 3) - \frac{1}{2} < \frac{1}{2}(5 - r)$$

$$6\left[-\frac{2}{3}(r - 3) - \frac{1}{2}\right] < 6\left[\frac{1}{2}(5 - r)\right] \qquad \text{Multiply by 6.}$$

$$6\left[-\frac{2}{3}(r - 3)\right] - 6\left(\frac{1}{2}\right) < 6\left[\frac{1}{2}(5 - r)\right] \qquad \text{Distributive property}$$

$$-4(r - 3) - 3 < 3(5 - r)$$

Step 1 $\qquad -4r + 12 - 3 < 15 - 3r \qquad \text{Distributive property}$

$$-4r + 9 < 15 - 3r$$

Step 2 $\qquad -4r + 9 + 3r < 15 - 3r + 3r \qquad \text{Add } 3r.$

$$-r + 9 < 15$$

$$-r + 9 - 9 < 15 - 9 \qquad \text{Subtract 9.}$$

$$-r < 6$$

Step 3   To solve for r, multiply each side of the inequality by -1. Since -1 is negative, change the direction of the inequality symbol.

$$-1(-r) > -1(6) \qquad \text{Multiply by } -1, \text{ change } < \text{ to } >.$$

$$r > -6$$

Check that the solution set is $(-6, \infty)$. See the graph in Figure 14 on the next page.

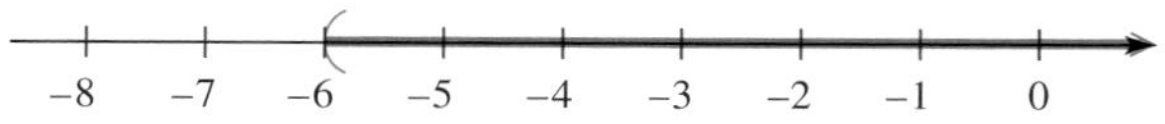

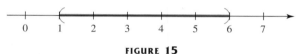

FIGURE 14

Now Try Exercise 29.

OBJECTIVE 3 **Solve linear inequalities with three parts.**   For some applications, it is necessary to work with an inequality such as

$$3 < x + 2 < 8,$$

where $x + 2$ is *between* 3 and 8. To solve this inequality, we subtract 2 from each of the three parts of the inequality, giving

$$3 - 2 < x + 2 - 2 < 8 - 2$$
$$1 < x < 6.$$

Thus, x must be between 1 and 6, so $x + 2$ will be between 3 and 8. The solution set, $(1, 6)$, is graphed in Figure 15.

FIGURE 15

CAUTION   When inequalities have three parts, the order of the parts is important. It would be *wrong* to write an inequality as $8 < x + 2 < 3$, since this would imply that $8 < 3$, a false statement. In general, three-part inequalities are written so that the symbols point in the same direction and both point toward the smaller number.

EXAMPLE 6   Solving a Three-Part Inequality

Solve $-2 \le -3k - 1 \le 5$ and graph the solution set.

Begin by adding 1 to each of the three parts to isolate the variable term in the middle.

$$-2 + 1 \le -3k - 1 + 1 \le 5 + 1 \qquad \text{Add 1 to each part.}$$

$$-1 \le -3k \le 6$$

$$\frac{-1}{-3} \ge \frac{-3k}{-3} \ge \frac{6}{-3} \qquad \text{Divide each part by } -3; \text{ reverse the inequality symbols.}$$

$$\frac{1}{3} \ge k \ge -2$$

$$-2 \le k \le \frac{1}{3} \qquad \text{Rewrite in the order on the number line.}$$

Check that the solution set is $\left[-2, \frac{1}{3}\right]$, as shown in Figure 16.

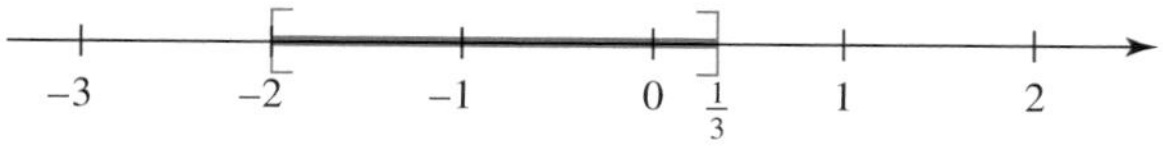

FIGURE 16

Now Try Exercise 51.

Examples of the types of solution sets to be expected from solving linear equations or linear inequalities are shown below.

Solution Sets of Linear Equations and Inequalities

Equation or Inequality	Typical Solution Set	Graph of Solution Set
Linear equation $5x + 4 = 14$	$\{2\}$	(number line with point at 2)
Linear inequality $5x + 4 < 14$	$(-\infty, 2)$	(number line shaded left of 2)
Linear inequality $5x + 4 > 14$	$(2, \infty)$	(number line shaded right of 2)
Three-part inequality $-1 \leq 5x + 4 \leq 14$	$[-1, 2]$	(number line shaded from -1 to 2)

OBJECTIVE 4 Solve applied problems using linear inequalities. In addition to the familiar "is less than" and "is greater than," the expressions "is no more than" and "is at least" also indicate inequalities. Expressions for inequalities sometimes appear in applied problems. The table shows how to interpret these expressions.

Word Expression	Interpretation
a is at least b	$a \geq b$
a is no less than b	$a \geq b$
a is at most b	$a \leq b$
a is no more than b	$a \leq b$

In Examples 7 and 8, we show how to solve applied problems with inequalities. We use the six problem-solving steps, changing Step 3 from "Write an equation" to "Write an inequality."

▨ EXAMPLE 7 Using a Linear Inequality to Solve a Rental Problem

A rental company charges $15 to rent a chain saw, plus $2 per hr. Al Ghandi can spend no more than $35 to clear some logs from his yard. What is the *maximum* amount of time he can use the rented saw?

Step 1 **Read** the problem again.

Step 2 **Assign a variable.** Let $h = $ the number of hours he can rent the saw.

Step 3 **Write an inequality.** He must pay $15, plus $2h$, to rent the saw for h hours, and this amount must be *no more than* $35.

$$
\underbrace{15 + 2h}_{\substack{\text{Cost of} \\ \text{renting}}} \quad \underbrace{\leq}_{\substack{\text{is no} \\ \text{more than}}} \quad \underbrace{35}_{\text{35 dollars.}}
$$

Step 4 **Solve.**

$$2h \leq 20 \qquad \text{Subtract 15.}$$

$$h \leq 10 \qquad \text{Divide by 2.}$$

Step 5 **State the answer.** He can use the saw for a maximum of 10 hr. (Of course, he may use it for less time, as indicated by the inequality $h \leq 10$.)

Step 6 **Check.** If Al uses the saw for 10 hr, he will spend $15 + 2(10) = 35$ dollars, the maximum amount.

Now Try Exercise 65.

EXAMPLE 8 Finding an Average Test Score

Martha has scores of 88, 86, and 90 on her first three algebra tests. An average score of at least 90 will earn an A in the class. What possible scores on her fourth test will earn her an A average?

Let x represent the score on the fourth test. Her average score must be at least 90. To find the average of four numbers, add them and then divide by 4.

$$\underbrace{\frac{88 + 86 + 90 + x}{4}}_{\text{Average}} \quad \underbrace{\geq}_{\substack{\text{is at}\\\text{least}}} \quad \underbrace{90}_{90.}$$

$$\frac{264 + x}{4} \geq 90 \qquad \text{Add the scores.}$$

$$264 + x \geq 360 \qquad \text{Multiply by 4.}$$

$$x \geq 96 \qquad \text{Subtract 264.}$$

She must score 96 or more on her fourth test.

Check: $\qquad \dfrac{88 + 86 + 90 + 96}{4} = \dfrac{360}{4} = 90$

A score of 96 or more will give an average of at least 90, as required.

Now Try Exercise 63.

2.5 EXERCISES

Match each inequality in Column I with the correct graph or interval notation in Column II.

I **II**

1. $x \leq 3$

A. (number line with open circle at 3, shaded right; marks at 0 and 3)

2. $x > 3$

B. (number line with open circle at 3, shaded left; marks at 0 and 3)

3. $x < 3$ **C.** $(3, \infty)$

4. $x \geq 3$ **D.** $(-\infty, 3]$

5. $-3 \leq x \leq 3$ **E.** $(-3, 3)$

6. $-3 < x < 3$ **F.** $[-3, 3]$

7. Explain how to determine whether to use parentheses or brackets when graphing the solution set of an inequality.

✐ **8.** Describe the steps used to solve a linear inequality. Explain when it is necessary to reverse the inequality symbol.

Solve each inequality. Give the solution set in both interval and graph forms. See Examples 1–5.

9. $x + 4 \geq 20$ **10.** $t + 40 \geq 50$ **11.** $3k - 1 > 20$ **12.** $5z - 6 < 64$

13. $4x < 16$ **14.** $2m > 10$ **15.** $-\dfrac{3}{4}r \geq 30$ **16.** $-1.5y \leq -\dfrac{9}{2}$

17. $-1.3m \geq -5.2$ **18.** $-2.5y \leq -1.25$ **19.** $\dfrac{2k - 5}{-4} > 5$ **20.** $\dfrac{3z - 2}{-5} < 6$

21. $6x - 4 \geq -2x$ **22.** $-2m + 8 \leq 2m$

23. $-(4 + r) + 2 - 3r < -14$ **24.** $-(9 + k) - 5 + 4k \geq 4$

25. $-3(z - 6) > 2z - 2$ **26.** $-2(x + 4) \leq 6x + 16$

27. $\dfrac{2}{3}(3k - 1) \geq \dfrac{3}{2}(2k - 3)$ **28.** $\dfrac{7}{5}(10m - 1) < \dfrac{2}{3}(6m + 5)$

29. $-\dfrac{1}{4}(p + 6) + \dfrac{3}{2}(2p - 5) < 10$ **30.** $\dfrac{3}{5}(k - 2) - \dfrac{1}{4}(2k - 7) \leq 3$

31. $3(2x - 4) - 4x < 2x + 3$ **32.** $7(4 - x) + 5x < 2(16 - x)$

33. $8\left(\dfrac{1}{2}x + 3\right) < 8\left(\dfrac{1}{2}x - 1\right)$ **34.** $10x + 2(x - 4) < 12x - 10$

RELATING CONCEPTS (EXERCISES 35–39)

For Individual or Group Work

Work Exercises 35–39 in order.

35. Solve the linear equation $5(x + 3) - 2(x - 4) = 2(x + 7)$ and graph the solution set on a number line.

36. Solve the linear inequality $5(x + 3) - 2(x - 4) > 2(x + 7)$ and graph the solution set on a number line.

37. Solve the linear inequality $5(x + 3) - 2(x - 4) < 2(x + 7)$ and graph the solution set on a number line.

38. Graph all the solution sets of the equation and inequalities in Exercises 35–37 on the same number line. What set do you obtain?

39. Based on the results of Exercises 35–37, complete the following using a conjecture (educated guess): The solution set of $-3(x + 2) = 3x + 12$ is $\{-3\}$, and the solution set of $-3(x + 2) < 3x + 12$ is $(-3, \infty)$. Therefore the solution set of $-3(x + 2) > 3x + 12$ is _____.

40. Which is the graph of $-2 < x$?

A.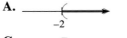

B.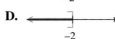

C.

D.

Solve each inequality. Give the solution set in both interval and graph forms. See Example 6.

41. $-4 < x - 5 < 6$

42. $-1 < x + 1 < 8$

43. $-9 \leq k + 5 \leq 15$

44. $-4 \leq m + 3 \leq 10$

45. $-6 \leq 2z + 4 \leq 16$

46. $-15 < 3p + 6 < -12$

47. $-19 \leq 3x - 5 \leq 1$

48. $-16 < 3t + 2 < -10$

49. $-1 \leq \dfrac{2x - 5}{6} \leq 5$

50. $-3 \leq \dfrac{3m + 1}{4} \leq 3$

51. $4 \leq 5 - 9x < 8$

52. $4 \leq 3 - 2x < 8$

Find the unknown numbers in each description.

53. Six times a number is between -12 and 12.

54. Half a number is between -3 and 2.

55. When 1 is added to twice a number, the result is greater than or equal to 7.

56. If 8 is subtracted from a number, then the result is at least 5.

57. One third of a number is added to 6, giving a result of at least 3.

58. Three times a number, minus 5, is no more than 7.

The July 14th weather forecast by time of day for the 2000 U.S. Olympic Track and Field Trials, held July 14–23, 2000, in Sacramento, California, is shown in the figure. Use this graph to work Exercises 59–62.

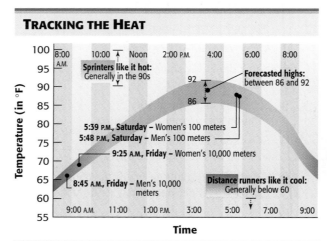

Source: Accuweather, Bee research.

59. Sprinters prefer Fahrenheit temperatures in the 90s. Using the upper boundary of the forecast, in what time period is the temperature expected to be at least 90°F?

60. Distance runners prefer cool temperatures. During what time period are temperatures predicted to be no more than 70°F? Use the lower forecast boundary.

61. What range of temperatures is predicted for the Women's 100-m event?

62. What range of temperatures is forecast for the Men's 10,000-m event?

Solve each problem. See Examples 7 and 8.

63. Margaret Westmoreland earned scores of 90 and 82 on her first two tests in English Literature. What score must she make on her third test to keep an average of 84 or greater?

64. Jacques d'Hemecourt scored 92 and 96 on his first two tests in Methods in Teaching Mathematics. What score must he make on his third test to keep an average of 90 or greater?

65. A couple wishes to rent a car for one day while on vacation. Avis automobile rental wants $35.00 per day and 14¢ per mile, while Enterprise wants $34.00 per day and 16¢ per mile. After how many miles would the price to rent from Enterprise exceed the price to rent from Avis?

66. Jane and Terry Brandsma went to Long Island for a week. They needed to rent a car, so they checked out two rental firms. Avis wanted $28 per day, with no mileage fee. Downtown Toyota wanted $108 per week and 14¢ per mile. How many miles would they have to drive before the Avis price is less than the Toyota price?

A product will produce a profit only when the revenue R from selling the product exceeds the cost C of producing it. In Exercises 67 and 68, find the smallest whole number of units x that must be sold for the business to show a profit for the item described.

67. Peripheral Visions, Inc. finds that the cost to produce x studio-quality videotapes is $C = 20x + 100$, while the revenue produced from them is $R = 24x$ (C and R in dollars).

68. Speedy Delivery finds that the cost to make x deliveries is $C = 3x + 2300$, while the revenue produced from them is $R = 5.50x$ (C and R in dollars).

69. Suppose that $4 < x < 1$. What can you say about x?

70. What is wrong with writing a statement of inequality as $10 > 5 < 8$? What three-part inequality would avoid this problem?

71. A BMI (body mass index) between 19 and 25 is considered healthy. Use the formula

$$\text{BMI} = \frac{704 \times (\text{weight in pounds})}{(\text{height in inches})^2}$$

to find the weight range w, to the nearest pound, that gives a healthy BMI for each height. (*Source: Washington Post.*)

(a) 72 in.   **(b)** Your height in inches

72. To achieve the maximum benefit from exercising, the heart rate in beats per minute should be in the target heart rate zone (THR). For a person aged A, the formula is

$$.7(220 - A) \le \text{THR} \le .85(220 - A).$$

Find the THR to the nearest whole number for each age. (*Source:* Hockey, Robert V., *Physical Fitness: The Pathway to Healthful Living,* Times Mirror/Mosby College Publishing, 1989.)

(a) 35   **(b)** Your age

2.6 Set Operations and Compound Inequalities

OBJECTIVES

1 Find the intersection of two sets.

2 Solve compound inequalities with the word *and*.

3 Find the union of two sets.

4 Solve compound inequalities with the word *or*.

The table shows symptoms of an underactive thyroid and an overactive thyroid.

Underactive Thyroid	Overactive Thyroid
Sleepiness, s	Insomnia, i
Dry hands, d	Moist hands, m
Intolerance of cold, c	Intolerance of heat, h
Goiter, g	Goiter, g

Source: *The Merck Manual of Diagnosis and Therapy*, 16th Edition, Merck Research Laboratories, 1992.

Let N be the set of symptoms for an underactive thyroid, and let O be the set of symptoms for an overactive thyroid. Suppose we are interested in the set of symptoms that are found in *both* sets *N and O*. In this section we discuss the use of the words *and* and *or* as they relate to sets and inequalities.

OBJECTIVE 1 Find the intersection of two sets.   The intersection of two sets is defined using the word *and*.

Intersection of Sets

For any two sets A and B, the **intersection** of A and B, symbolized $A \cap B$, is defined as follows:

$$A \cap B = \{x \mid x \text{ is an element of } A \text{ and } x \text{ is an element of } B\}.$$

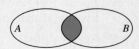

EXAMPLE 1   Finding the Intersection of Two Sets

Let $A = \{1, 2, 3, 4\}$ and $B = \{2, 4, 6\}$. Find $A \cap B$.

The set $A \cap B$ contains those elements that belong to both A *and* B: the numbers 2 and 4. Therefore,

$$A \cap B = \{1, 2, 3, 4\} \cap \{2, 4, 6\}$$
$$= \{2, 4\}.$$

Now Try Exercise 7.

A **compound inequality** consists of two inequalities linked by a connective word such as *and* or *or*. Examples of compound inequalities are

$$x + 1 \le 9 \quad \text{and} \quad x - 2 \ge 3$$

and

$$2x > 4 \quad \text{or} \quad 3x - 6 < 5.$$

OBJECTIVE 2 Solve compound inequalities with the word *and*.   Use the following steps.

Solving a Compound Inequality with *and*

Step 1 Solve each inequality in the compound inequality individually.

Step 2 Since the inequalities are joined with *and,* the solution set of the compound inequality will include all numbers that satisfy both inequalities in Step 1 (the intersection of the solution sets).

■ **EXAMPLE 2** Solving a Compound Inequality with *and*

Solve the compound inequality

$$x + 1 \le 9 \quad \text{and} \quad x - 2 \ge 3.$$

Step 1 Solve each inequality in the compound inequality individually.

$$x + 1 \le 9 \qquad \text{and} \qquad x - 2 \ge 3$$
$$x + 1 - 1 \le 9 - 1 \quad \text{and} \quad x - 2 + 2 \ge 3 + 2$$
$$x \le 8 \qquad \text{and} \qquad x \ge 5$$

Step 2 Because the inequalities are joined with the word *and,* the solution set will include all numbers that satisfy both inequalities in Step 1 at the same time. Thus, the compound inequality is true whenever $x \le 8$ and $x \ge 5$ are both true. The top graph in Figure 17 shows $x \le 8$, and the bottom graph shows $x \ge 5$.

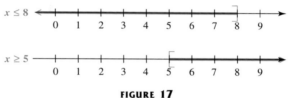

FIGURE 17

Find the intersection of the two graphs in Figure 17 to get the solution set of the compound inequality. The intersection of the two graphs in Figure 18 shows that the solution set in interval notation is $[5, 8]$.

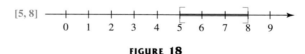

FIGURE 18

Now Try Exercise 27.

■ **EXAMPLE 3** Solving a Compound Inequality with *and*

Solve the compound inequality

$$-3x - 2 > 5 \quad \text{and} \quad 5x - 1 \le -21.$$

Step 1 Solve each inequality separately.

$$-3x - 2 > 5 \qquad \text{and} \quad 5x - 1 \le -21$$
$$-3x > 7 \qquad \text{and} \qquad 5x \le -20$$
$$x < -\frac{7}{3} \qquad \text{and} \qquad x \le -4$$

The graphs of $x < -\frac{7}{3}$ and $x \leq -4$ are shown in Figure 19.

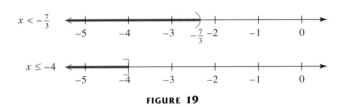

FIGURE 19

Step 2 Now find all values of x that satisfy both conditions; that is, the real numbers that are less than $-\frac{7}{3}$ and also less than or equal to -4. As shown by the graph in Figure 20, the solution set is $(-\infty, -4]$.

FIGURE 20

Now Try Exercise 31.

EXAMPLE 4   Solving a Compound Inequality with *and*

Solve $x + 2 < 5$ and $x - 10 > 2$.

First solve each inequality separately.

$$x + 2 < 5 \quad \text{and} \quad x - 10 > 2$$
$$x < 3 \quad \text{and} \quad x > 12$$

The graphs of $x < 3$ and $x > 12$ are shown in Figure 21.

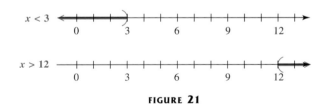

FIGURE 21

There is no number that is both less than 3 *and* greater than 12, so the given compound inequality has no solution. The solution set is ∅. See Figure 22.

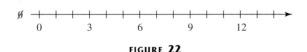

FIGURE 22

Now Try Exercise 25.

OBJECTIVE 3 Find the union of two sets.   The union of two sets is defined using the word *or*.

Union of Sets

For any two sets A and B, the **union** of A and B, symbolized $A \cup B$, is defined as follows:

$$A \cup B = \{x \mid x \text{ is an element of } A \textbf{ or } x \text{ is an element of } B\}.$$

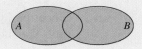

EXAMPLE 5   Finding the Union of Two Sets

Let $A = \{1, 2, 3, 4\}$ and $B = \{2, 4, 6\}$. Find $A \cup B$.

Begin by listing all the elements of set A: 1, 2, 3, 4. Then list any additional elements from set B. In this case the elements 2 and 4 are already listed, so the only additional element is 6. Therefore,

$$
\begin{aligned}
A \cup B &= \{1, 2, 3, 4\} \cup \{2, 4, 6\} \\
&= \{1, 2, 3, 4, 6\}.
\end{aligned}
$$

The union consists of all elements in either A *or* B (or both).

Now Try Exercise 13.

In Example 5, notice that although the elements 2 and 4 appeared in both sets A and B, they are written only once in $A \cup B$.

OBJECTIVE 4 Solve compound inequalities with the word *or*.   Use the following steps.

Solving a Compound Inequality with *or*

Step 1   Solve each inequality in the compound inequality individually.

Step 2   Since the inequalities are joined with *or*, the solution set of the compound inequality includes all numbers that satisfy either one of the two inequalities in Step 1 (the union of the solution sets).

EXAMPLE 6   Solving a Compound Inequality with *or*

Solve $6x - 4 < 2x$ or $-3x \le -9$.

Step 1   Solve each inequality separately.

$$
\begin{aligned}
6x - 4 &< 2x \quad &\text{or} \quad -3x &\le -9 \\
4x &< 4 \\
x &< 1 \quad &\text{or} \quad x &\ge 3
\end{aligned}
$$

The graphs of these two inequalities are shown in Figure 23 on the next page.

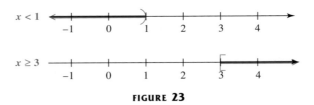

FIGURE 23

Step 2 Since the inequalities are joined with *or*, find the union of the two solution sets. The union is shown in Figure 24 and is written

$$(-\infty, 1) \cup [3, \infty).$$

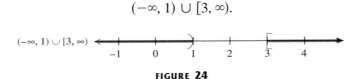

FIGURE 24

Now Try Exercise 43.

CAUTION When inequalities are used to write the solution set in Example 6, it *should* be written as

$$x < 1 \quad \text{or} \quad x \geq 3,$$

which keeps the numbers 1 and 3 in their order on the number line. Writing $3 \leq x < 1$ would imply that $3 \leq 1$, which is **FALSE.** There is no other way to write the solution set of such a union.

EXAMPLE 7 Solving a Compound Inequality with *or*

Solve $-4x + 1 \geq 9$ or $5x + 3 \leq -12$.

First we solve each inequality separately.

$$-4x + 1 \geq 9 \quad \text{or} \quad 5x + 3 \leq -12$$
$$-4x \geq 8 \quad \text{or} \quad 5x \leq -15$$
$$x \leq -2 \quad \text{or} \quad x \leq -3$$

The graphs of these two inequalities are shown in Figure 25.

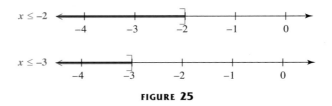

FIGURE 25

By taking the union, we obtain the interval $(-\infty, -2]$. It is graphed in Figure 26.

FIGURE 26

Now Try Exercise 37.

■ EXAMPLE 8   Applying Intersection and Union

The five highest domestic grossing films (adjusted for inflation) are listed in the table.

Five All-Time Highest Grossing Films

Film	Admissions	Gross Income
Gone with the Wind	200,605,313	$972,900,000
Star Wars	178,119,595	$863,900,000
The Sound of Music	142,415,376	$690,700,000
E.T.	135,987,938	$659,500,000
The Ten Commandments	131,000,000	$635,400,000

Source: New York Times Almanac, 2001.

List the elements of the following sets.

(a) The set of top five films with admissions greater than 180,000,000 *and* gross greater than $800,000,000

The only film that satisfies both conditions is *Gone with the Wind,* so the set is

$$\{Gone\ with\ the\ Wind\}.$$

(b) The set of top five films with admissions less than 170,000,000 *or* gross greater than $700,000,000

Here, a film that satisfies at least one of the conditions is in the set. This set includes all five films:

$$\{Gone\ with\ the\ Wind,\ Star\ Wars,\ The\ Sound\ of\ Music,\ E.T.,\ The\ Ten\ Commandments\}.$$

Now Try Exercise 69.

2.6   EXERCISES

For Extra Help

 Student's
Solutions Manual

 MyMathLab

 InterAct Math
Tutorial Software

Tutor Center AW Math
Tutor Center

MathXL MathXL

Digital Video Tutor
CD 2/Videotape 3

Decide whether each statement is true *or* false. *If it is false, explain why.*

1. The union of the solution sets of $x + 1 = 5$, $x + 1 < 5$, and $x + 1 > 5$ is $(-\infty, \infty)$.

2. The intersection of the sets $\{x \mid x \geq 7\}$ and $\{x \mid x \leq 7\}$ is $\emptyset$.

3. The union of the sets $(-\infty, 8)$ and $(8, \infty)$ is $\{8\}$.

4. The intersection of the sets $(-\infty, 8]$ and $[8, \infty)$ is $\{8\}$.

5. The intersection of the set of rational numbers and the set of irrational numbers is $\{0\}$.

6. The union of the set of rational numbers and the set of irrational numbers is the set of real numbers.

Let $A = \{1, 2, 3, 4, 5, 6\}$, $B = \{1, 3, 5\}$, $C = \{1, 6\}$, and $D = \{4\}$. Specify each set. See Examples 1 and 5.

7. $B \cap A$　　　　**8.** $A \cap B$　　　　**9.** $A \cap D$　　　　**10.** $B \cap C$

11. $B \cap \emptyset$　　　　**12.** $A \cap \emptyset$　　　　**13.** $A \cup B$　　　　**14.** $B \cup D$

✒ **15.** Give an example of intersection applied to a real-life situation.

✒ **16.** A compound inequality uses one of the words *and* or *or*. Explain how you will determine whether to use *intersection* or *union* when graphing the solution set.

Two sets are specified by graphs. Graph the intersection of the two sets.

17.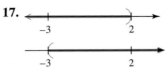

18.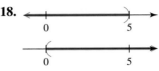

19.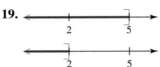

20.

For each compound inequality, give the solution set in both interval and graph forms. See Examples 2–4.

21. $x < 2$ and $x > -3$ **22.** $x < 5$ and $x > 0$ **23.** $x \le 2$ and $x \le 5$

24. $x \ge 3$ and $x \ge 6$ **25.** $x \le 3$ and $x \ge 6$ **26.** $x \le -1$ and $x \ge 3$

27. $x - 3 \le 6$ and $x + 2 \ge 7$ **28.** $x + 5 \le 11$ and $x - 3 \ge -1$

29. $-3x > 3$ and $x + 3 > 0$ **30.** $-3x < 3$ and $x + 2 < 6$

31. $3x - 4 \le 8$ and $-4x + 1 \ge -15$ **32.** $7x + 6 \le 48$ and $-4x \ge -24$

Two sets are specified by graphs. Graph the union of the two sets.

33.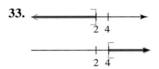

34.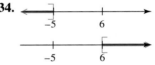

35.

36.

For each compound inequality, give the solution set in both interval and graph forms. See Examples 6 and 7.

37. $x \le 1$ or $x \le 8$ **38.** $x \ge 1$ or $x \ge 8$

39. $x \ge -2$ or $x \ge 5$ **40.** $x \le -2$ or $x \le 6$

41. $x \ge -2$ or $x \le 4$ **42.** $x \ge 5$ or $x \le 7$

43. $x + 2 > 7$ or $1 - x > 6$ **44.** $x + 1 > 3$ or $x + 4 < 2$

45. $x + 1 > 3$ or $-4x + 1 > 5$ **46.** $3x < x + 12$ or $x + 1 > 10$

Express each set in the simplest interval form. (Hint: Graph each set and look for the intersection or union.)

47. $(-\infty, -1] \cap [-4, \infty)$ **48.** $[-1, \infty) \cap (-\infty, 9]$

49. $(-\infty, -6] \cap [-9, \infty)$ **50.** $(5, 11] \cap [6, \infty)$

51. $(-\infty, 3) \cup (-\infty, -2)$ **52.** $[-9, 1] \cup (-\infty, -3)$

53. $[3, 6] \cup (4, 9)$ **54.** $[-1, 2] \cup (0, 5)$

For each compound inequality, decide whether intersection *or* union *should be used. Then give the solution set in both interval and graph forms. See Examples 2, 3, 4, 6, and 7.*

55. $x < -1$ and $x > -5$

56. $x > -1$ and $x < 7$

57. $x < 4$ or $x < -2$

58. $x < 5$ or $x < -3$

59. $-3x \leq -6$ or $-3x \geq 0$

60. $2x - 6 \leq -18$ and $2x \geq -18$

61. $x + 1 \geq 5$ and $x - 2 \leq 10$

62. $-8x \leq -24$ or $-5x \geq 15$

RELATING CONCEPTS (EXERCISES 63–68)

For Individual or Group Work

The figures represent the backyards of neighbors Luigi, Maria, Than, and Joe. Find the area and the perimeter of each yard. Suppose that each resident has 150 ft of fencing and enough sod to cover 1400 ft² of lawn. Give the name or names of the residents whose yards satisfy each description. **Work Exercises 63–68 in order.**

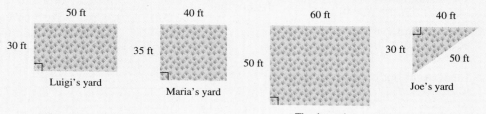

Luigi's yard — 50 ft by 30 ft

Maria's yard — 40 ft by 35 ft

Than's yard — 60 ft by 50 ft

Joe's yard — 40 ft, 30 ft, 50 ft

63. The yard can be fenced *and* the yard can be sodded.

64. The yard can be fenced *and* the yard cannot be sodded.

65. The yard cannot be fenced *and* the yard can be sodded.

66. The yard cannot be fenced *and* the yard cannot be sodded.

67. The yard can be fenced *or* the yard can be sodded.

68. The yard cannot be fenced *or* the yard can be sodded.

Use the graphs to answer Exercises 69 and 70. See Example 8.

69. In which years did the number of players with 30–39 home runs exceed 20 *and* the number of players with 40 or more home runs exceed 45?

70. In which years were the number of players with 20–29 home runs less than 20 *or* the number of players with 30–39 home runs at least 20?

GOING, GOING, GONE
Home runs have been flying out of major-league ballparks at an increasing rate. A breakdown:

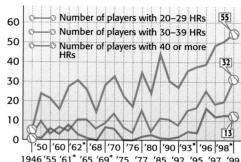

*Indicates expansion year

Source: Bee research.

2.7 Absolute Value Equations and Inequalities

OBJECTIVES

1 Use the distance definition of absolute value.

2 Solve equations of the form $|ax + b| = k$, for $k > 0$.

3 Solve inequalities of the form $|ax + b| < k$ and of the form $|ax + b| > k$, for $k > 0$.

4 Solve absolute value equations that involve rewriting.

5 Solve equations of the form $|ax + b| = |cx + d|$.

6 Solve special cases of absolute value equations and inequalities.

In a production line, quality is controlled by randomly choosing items from the line and checking to see how selected measurements vary from the optimum measure. These differences are sometimes positive and sometimes negative, so they are expressed with absolute value. For example, a machine that fills quart milk cartons might be set to release 1 qt (32 oz) plus or minus 2 oz per carton. Then the number of ounces in each carton should satisfy the *absolute value inequality* $|x - 32| \le 2$, where x is the number of ounces.

OBJECTIVE 1 Use the distance definition of absolute value. In Chapter 1 we saw that the absolute value of a number x, written $|x|$, represents the distance from x to 0 on the number line. For example, the solutions of $|x| = 4$ are 4 and -4, as shown in Figure 27.

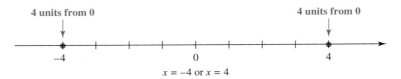

FIGURE 27

Because absolute value represents distance from 0, it is reasonable to interpret the solutions of $|x| > 4$ to be all numbers that are *more* than 4 units from 0. The set $(-\infty, -4) \cup (4, \infty)$ fits this description. Figure 28 shows the graph of the solution set of $|x| > 4$. Because the graph consists of two separate intervals, the solution set is described using *or* as

$$x < -4 \quad \text{or} \quad x > 4.$$

FIGURE 28

The solution set of $|x| < 4$ consists of all numbers that are *less* than 4 units from 0 on the number line. Another way of thinking of this is to think of all numbers *between* -4 and 4. This set of numbers is given by $(-4, 4)$, as shown in Figure 29. Here, the graph shows that $-4 < x < 4$, which means $x > -4$ *and* $x < 4$.

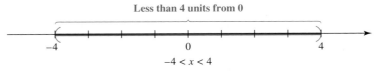

FIGURE 29

The equation and inequalities just described are examples of **absolute value equations and inequalities.** They involve the absolute value of a variable expression and generally take the form

$$|ax + b| = k, \qquad |ax + b| > k, \qquad \text{or} \qquad |ax + b| < k,$$

where k is a positive number. From Figures 27–29, we see that

$$|x| = 4 \text{ has the same solution set as } x = -4 \text{ or } x = 4,$$
$$|x| > 4 \text{ has the same solution set as } x < -4 \text{ or } x > 4,$$
$$|x| < 4 \text{ has the same solution set as } x > -4 \text{ and } x < 4.$$

Thus, we solve an absolute value equation or inequality by first rewriting it as an equivalent statement without absolute value bars. Notice that, except for special cases, the solution set of an absolute value *equation* includes exactly *two points*. However, the solution set of an absolute value *inequality* includes one or more *intervals*. We summarize these facts in the next box.

Solving Absolute Value Equations and Inequalities

Let k be a positive real number, and p and q be real numbers.

1. To solve $|ax + b| = k,$ solve the compound equation

$$ax + b = k \quad \text{or} \quad ax + b = -k.$$

The solution set is usually of the form $\{p, q\}$, which includes two numbers.

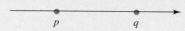

2. To solve $|ax + b| > k,$ solve the compound inequality

$$ax + b > k \quad \text{or} \quad ax + b < -k.$$

The solution set is of the form $(-\infty, p) \cup (q, \infty)$, which consists of two separate intervals.

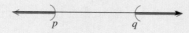

3. To solve $|ax + b| < k,$ solve the compound inequality

$$-k < ax + b < k.$$

The solution set is of the form (p, q), a single interval.

NOTE Some people prefer to write the compound statements in parts 1 and 2 of the preceding summary as

$$ax + b = k \quad \text{or} \quad -(ax + b) = k$$

and

$$ax + b > k \quad \text{or} \quad -(ax + b) > k.$$

These forms are equivalent to those we give in the summary and produce the same results.

OBJECTIVE 2 **Solve equations of the form** $|ax + b| = k$, **for** $k > 0$. The next example shows how we use a compound equation to solve a typical absolute value equation. Remember that because absolute value refers to distance from the origin, each absolute value equation will have two parts.

■ EXAMPLE 1   Solving an Absolute Value Equation

Solve $|2x + 1| = 7$.

For $|2x + 1|$ to equal 7, $2x + 1$ must be 7 units from 0 on the number line. This can happen only when $2x + 1 = 7$ or $2x + 1 = -7$. This is the first case in the preceding summary. Solve this compound equation as follows.

$$2x + 1 = 7 \quad \text{or} \quad 2x + 1 = -7$$
$$2x = 6 \quad \text{or} \quad 2x = -8$$
$$x = 3 \quad \text{or} \quad x = -4$$

Check by substitution in the original absolute value equation to verify that the solution set is $\{-4, 3\}$. The graph is shown in Figure 30.

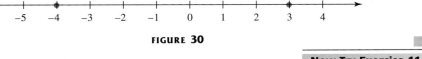

FIGURE 30

Now Try Exercise 11.

OBJECTIVE 3 **Solve inequalities of the form** $|ax + b| < k$ **and of the form** $|ax + b| > k$, **for** $k > 0$.

■ EXAMPLE 2   Solving an Absolute Value Inequality with >

Solve $|2x + 1| > 7$.

By part 2 of the summary, this absolute value inequality is rewritten as

$$2x + 1 > 7 \quad \text{or} \quad 2x + 1 < -7,$$

because $2x + 1$ must represent a number that is *more* than 7 units from 0 on either side of the number line. Now, solve the compound inequality.

$$2x + 1 > 7 \quad \text{or} \quad 2x + 1 < -7$$
$$2x > 6 \quad \text{or} \quad 2x < -8$$
$$x > 3 \quad \text{or} \quad x < -4$$

Check these solutions. The solution set is $(-\infty, -4) \cup (3, \infty)$. See Figure 31. Notice that the graph consists of two intervals.

FIGURE 31

Now Try Exercise 25.

EXAMPLE 3 Solving an Absolute Value Inequality with <

Solve $|2x + 1| < 7$.

The expression $2x + 1$ must represent a number that is less than 7 units from 0 on either side of the number line. Another way of thinking of this is to realize that $2x + 1$ must be between -7 and 7. As part 3 of the summary shows, this is written as the three-part inequality

$$-7 < 2x + 1 < 7.$$

We solved such inequalities in Section 2.5 by working with all three parts at the same time.

$$-7 < 2x + 1 < 7$$
$$-8 < 2x < 6 \qquad \text{Subtract 1 from each part.}$$
$$-4 < x < 3 \qquad \text{Divide each part by 2.}$$

Check that the solution set is $(-4, 3)$, so the graph consists of the single interval shown in Figure 32.

FIGURE 32

Now Try Exercise 39.

Look back at Figures 30, 31, and 32, with the graphs of $|2x + 1| = 7$, $|2x + 1| > 7$, and $|2x + 1| < 7$. If we find the union of the three sets, we get the set of all real numbers. This is because for any value of x, $|2x + 1|$ will satisfy one and only one of the following: it is equal to 7, greater than 7, or less than 7.

CAUTION When solving absolute value equations and inequalities of the types in Examples 1, 2, and 3, remember the following.

1. The methods described apply when the constant is alone on one side of the equation or inequality and is *positive*.

2. Absolute value equations and absolute value inequalities in the form $|ax + b| > k$ translate into "or" compound statements.

3. Absolute value inequalities in the form $|ax + b| < k$ translate into "and" compound statements, which may be written as three-part inequalities.

4. An "or" statement *cannot* be written in three parts. It would be incorrect to write

$$-7 > 2x + 1 > 7$$

in Example 2, because this would imply that $-7 > 7$, which is *false*.

OBJECTIVE 4 Solve absolute value equations that involve rewriting. Sometimes an absolute value equation or inequality requires some rewriting before it can be set up as a compound statement, as shown in the next example.

EXAMPLE 4 Solving an Absolute Value Equation That Requires Rewriting

Solve $|x + 3| + 5 = 12$.

First get the absolute value alone on one side of the equals sign by subtracting 5 from each side.

$$|x + 3| + 5 - 5 = 12 - 5 \qquad \text{Subtract 5.}$$
$$|x + 3| = 7$$

Now use the method shown in Example 1.

$$x + 3 = 7 \quad \text{or} \quad x + 3 = -7$$
$$x = 4 \quad \text{or} \qquad x = -10$$

Check that the solution set is $\{4, -10\}$ by substituting into the original equation. ∎

Now Try Exercise 63.

We use a similar method to solve an absolute value *inequality* that requires rewriting.

OBJECTIVE 5 **Solve equations of the form $|ax + b| = |cx + d|$.** By definition, for two expressions to have the same absolute value, they must either be equal or be negatives of each other.

Solving $|ax + b| = |cx + d|$

To solve an absolute value equation of the form

$$|ax + b| = |cx + d|,$$

solve the compound equation

$$ax + b = cx + d \quad \text{or} \quad ax + b = -(cx + d).$$

EXAMPLE 5 Solving an Equation with Two Absolute Values

Solve $|z + 6| = |2z - 3|$.

This equation is satisfied either if $z + 6$ and $2z - 3$ are equal to each other or if $z + 6$ and $2z - 3$ are negatives of each other.

$$z + 6 = 2z - 3 \quad \text{or} \quad z + 6 = -(2z - 3)$$

Solve each equation.

$$z + 6 = 2z - 3 \quad \text{or} \quad z + 6 = -2z + 3$$
$$6 + 3 = 2z - z \qquad\qquad 3z = -3$$
$$9 = z \qquad\qquad\quad \text{or} \qquad z = -1$$

The solution set is $\{9, -1\}$. ∎

Now Try Exercise 71.

OBJECTIVE 6 Solve special cases of absolute value equations and inequalities.   When a typical absolute value equation or inequality involves a *negative constant or 0* alone on one side, use the properties of absolute value to solve. Keep in mind the following.

1. The absolute value of an expression can never be negative: $|a| \geq 0$ for all real numbers a.

2. The absolute value of an expression equals 0 only when the expression is equal to 0.

The next two examples illustrate these special cases.

▇ EXAMPLE 6   Solving Special Cases of Absolute Value Equations

Solve each equation.

(a) $|5r - 3| = -4$

Since the absolute value of an expression can never be negative, there are no solutions for this equation. The solution set is ∅.

(b) $|7x - 3| = 0$

The expression $7x - 3$ will equal 0 *only* if

$$7x - 3 = 0.$$

The solution of this equation is $\frac{3}{7}$. The solution set is $\left\{\frac{3}{7}\right\}$. It consists of only one element that checks by substitution in the original equation.

Now Try Exercises 79 and 81.

▇ EXAMPLE 7   Solving Special Cases of Absolute Value Inequalities

Solve each inequality.

(a) $|x| \geq -4$

The absolute value of a number is never negative. For this reason, $|x| \geq -4$ is true for *all* real numbers. The solution set is $(-\infty, \infty)$.

(b) $|k + 6| - 3 < -5$

Add 3 to both sides to get the absolute value expression alone on one side.

$$|k + 6| < -2$$

There is no number whose absolute value is less than -2, so this inequality has no solution. The solution set is ∅.

(c) $|m - 7| + 4 \leq 4$

Adding -4 to both sides gives

$$|m - 7| \leq 0.$$

The value of $|m - 7|$ will never be less than 0. However, $|m - 7|$ will equal 0 when $m = 7$. Therefore, the solution set is $\{7\}$.

Now Try Exercises 85 and 91.

CONNECTIONS

Absolute value is used to find the relative error of a measurement in science, engineering, manufacturing, and other fields. If x_t represents the expected value of a measurement and x represents the actual measurement, then the *relative error in x* equals the absolute value of the difference between x_t and x divided by x_t. That is,

$$\text{relative error in } x = \left| \frac{x_t - x}{x_t} \right|.$$

In many situations in the work world, the relative error must be less than some predetermined amount. For example, suppose a machine filling *quart* milk cartons is set for a relative error no greater than .05. Here $x_t = 32$ oz, the relative error $= .05$ oz, and we must find x, given

$$\left| \frac{32 - x}{32} \right| = \left| 1 - \frac{x}{32} \right| \le .05.$$

For Discussion or Writing

With this tolerance level, how many ounces may a carton contain?

 EXERCISES

Match each absolute value equation or inequality in Column I with the graph of its solution set in Column II.

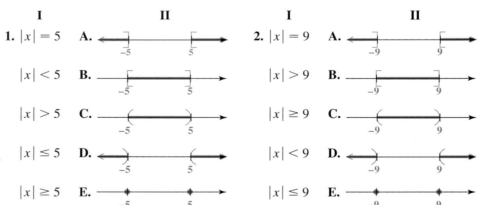

1. $|x| = 5$

$|x| < 5$

$|x| > 5$

$|x| \le 5$

$|x| \ge 5$

2. $|x| = 9$

$|x| > 9$

$|x| \ge 9$

$|x| < 9$

$|x| \le 9$

3. Explain when to use *and* and when to use *or* if you are solving an absolute value equation or inequality of the form $|ax + b| = k$, $|ax + b| < k$, or $|ax + b| > k$, where k is a positive number.

4. How many solutions will $|ax + b| = k$ have if

(a) $k = 0$; (b) $k > 0$; (c) $k < 0$?

Solve each equation. See Example 1.

5. $|x| = 12$ **6.** $|k| = 14$ **7.** $|4x| = 20$ **8.** $|5x| = 30$

9. $|y - 3| = 9$ **10.** $|p - 5| = 13$ **11.** $|2x - 1| = 11$ **12.** $|2y + 3| = 19$

13. $|4r - 5| = 17$ **14.** $|5t - 1| = 21$ **15.** $|2y + 5| = 14$ **16.** $|2x - 9| = 18$

17. $\left| \dfrac{1}{2}x + 3 \right| = 2$ **18.** $\left| \dfrac{2}{3}q - 1 \right| = 5$

19. $\left| 1 + \dfrac{3}{4}k \right| = 7$ **20.** $\left| 2 - \dfrac{5}{2}m \right| = 14$

Solve each inequality and graph the solution set. See Example 2.

21. $|x| > 3$ **22.** $|y| > 5$ **23.** $|k| \geq 4$ **24.** $|r| \geq 6$

25. $|r + 5| \geq 20$ **26.** $|3x - 1| \geq 8$ **27.** $|t + 2| > 10$ **28.** $|4x + 1| \geq 21$

29. $|3 - x| > 5$ **30.** $|5 - x| > 3$ **31.** $|-5x + 3| \geq 12$ **32.** $|-2x - 4| \geq 5$

33. The graph of the solution set of $|2x + 1| = 9$ is given here.

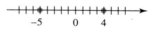

-5   0   4

Without actually doing the algebraic work, graph the solution set of each inequality, referring to the graph above.

(a) $|2x + 1| < 9$ (b) $|2x + 1| > 9$

34. The graph of the solution set of $|3x - 4| < 5$ is given here.

$-\dfrac{1}{3}$ 3

Without actually doing the algebraic work, graph the solution set of the following, referring to the graph above.

(a) $|3x - 4| = 5$ (b) $|3x - 4| > 5$

Solve each inequality and graph the solution set. See Example 3. (Hint: Compare your answers to those in Exercises 21–32.)

35. $|x| \leq 3$ **36.** $|y| \leq 5$ **37.** $|k| < 4$ **38.** $|r| < 6$

39. $|r + 5| \leq 20$ **40.** $|3x - 1| < 8$ **41.** $|t + 2| \leq 10$ **42.** $|4x + 1| < 21$

43. $|3 - x| \leq 5$ **44.** $|5 - x| \leq 3$ **45.** $|-5x + 3| \leq 12$ **46.** $|-2x - 4| \leq 5$

Decide which method you should use to solve each absolute value equation or inequality. Find the solution set and in Exercises 47–58, graph the solution set. See Examples 1–3.

47. $|-4 + k| > 9$ **48.** $|-3 + t| > 8$ **49.** $|r + 5| > 20$ **50.** $|2x - 1| < 7$

51. $|7 + 2z| = 5$ **52.** $|9 - 3p| = 3$ **53.** $|3r - 1| \leq 11$ **54.** $|2s - 6| \leq 6$

55. $|-6x - 6| \leq 1$ **56.** $|-2x - 6| \leq 5$ **57.** $|2x - 1| \geq 7$ **58.** $|-4 + k| \leq 9$

59. $|x| - 1 = 4$ **60.** $|x + 3| = 10$ **61.** $|x + 2| = 3$ **62.** $|x - 4| = 1$

Solve each equation or inequality. Give the solution set in set notation for equations and in interval notation for inequalities. See Example 4.

63. $|x + 4| + 1 = 2$ **64.** $|x + 5| - 2 = 12$ **65.** $|2x + 1| + 3 > 8$

66. $|6x - 1| - 2 > 6$ **67.** $|x + 5| - 6 \leq -1$ **68.** $|r - 2| - 3 \leq 4$

69. $|2 - x| > 3$ **70.** $|4 - x| < 1$

Solve each equation. See Example 5.

71. $|3x + 1| = |2x + 4|$ **72.** $|7x + 12| = |x - 8|$

73. $\left| m - \dfrac{1}{2} \right| = \left| \dfrac{1}{2}m - 2 \right|$ **74.** $\left| \dfrac{2}{3}r - 2 \right| = \left| \dfrac{1}{3}r + 3 \right|$

75. $|6x| = |9x + 1|$ **76.** $|13x| = |2x + 1|$

77. $|2p - 6| = |2p + 11|$ **78.** $|3x - 1| = |3x + 9|$

Solve each equation or inequality. See Examples 6 and 7.

79. $|12t - 3| = -8$ **80.** $|13w + 1| = -3$ **81.** $|4x + 1| = 0$

82. $|6r - 2| = 0$ **83.** $|2q - 1| = -6$ **84.** $|8n + 4| = -4$

85. $|x + 5| > -9$ **86.** $|x + 9| > -3$ **87.** $|7x + 3| \leq 0$

88. $|4x - 1| \leq 0$ **89.** $|5x - 2| = 0$ **90.** $|4 + 7x| = 0$

91. $|10z + 7| + 3 < 1$ **92.** $|4x + 1| - 2 < -5$

93. The 1998 recommended daily intake (RDI) of calcium for females aged 19–50 is 1000 mg. (*Source: World Almanac and Book of Facts, 2000.*) Actual vitamin needs vary from person to person. Write an absolute value inequality, with x representing the RDI, to express the RDI plus or minus 100 mg/day and solve it.

94. The average clotting time of blood is 7.45 sec with a variation of plus or minus 3.6 sec. Write this statement as an absolute value inequality with x representing the time and solve it.

RELATING CONCEPTS (EXERCISES 95–98)

For Individual or Group Work

The ten tallest buildings in Kansas City, Missouri, are listed along with their heights.

Building	Height (in feet)
One Kansas City Place	632
AT&T Town Pavilion	590
Hyatt Regency	504
Kansas City Power and Light	476
City Hall	443
Fidelity Bank and Trust Building	433
1201 Walnut	427
Federal Office Building	413
Commerce Tower	407
City Center Square	404

Source: World Almanac and Book of Facts, 2001.

*Use this information to **work Exercises 95–98 in order.***

95. To find the average of a group of numbers, we add the numbers and then divide by the number of items added. Use a calculator to find the average of the heights.

96. Let k represent the average height of these buildings. If a height x satisfies the inequality

$$|x - k| < t,$$

then the height is said to be within t ft of the average. Using your result from Exercise 95, list the buildings that are within 50 ft of the average.

97. Repeat Exercise 96, but list the buildings that are within 75 ft of the average.

98. **(a)** Write an absolute value inequality that describes the height of a building that is *not* within 75 ft of the average.
(b) Solve the inequality you wrote in part (a).
(c) Use the result of part (b) to list the buildings that are not within 75 ft of the average.
(d) Confirm that your answer to part (c) makes sense by comparing it with your answer to Exercise 97.

SUMMARY EXERCISES ON SOLVING LINEAR AND ABSOLUTE VALUE EQUATIONS AND INEQUALITIES

This section of miscellaneous equations and inequalities provides practice in solving all such types introduced in this chapter. You might wish to refer to the boxes in this chapter that summarize the various methods of solution.

Solve each equation or inequality.

1. $4z + 1 = 49$

2. $|m - 1| = 6$

3. $6q - 9 = 12 + 3q$

4. $3p + 7 = 9 + 8p$

5. $|a + 3| = -4$

6. $2m + 1 \leq m$

7. $8r + 2 \geq 5r$

8. $4(a - 11) + 3a = 20a - 31$

9. $2q - 1 = -7$

10. $|3q - 7| - 4 = 0$

11. $6z - 5 \leq 3z + 10$

12. $|5z - 8| + 9 \geq 7$

13. $9x - 3(x + 1) = 8x - 7$

14. $|m| \geq 8$

15. $9x - 5 \geq 9x + 3$

16. $13p - 5 > 13p - 8$

17. $|q| < 5.5$

18. $4z - 1 = 12 + z$

19. $\frac{2}{3}x + 8 = \frac{1}{4}x$

20. $-\frac{5}{8}z \geq -20$

21. $\frac{1}{4}p < -6$

22. $7z - 3 + 2z = 9z - 8z$

23. $\frac{3}{5}q - \frac{1}{10} = 2$

24. $|r - 1| < 7$

25. $r + 9 + 7r = 4(3 + 2r) - 3$

26. $6 - 3(2 - p) < 2(1 + p) + 3$

27. $|2p - 3| > 11$

28. $\frac{x}{4} - \frac{2x}{3} = -10$

29. $|5a + 1| \le 0$

30. $5z - (3 + z) \ge 2(3z + 1)$

31. $-2 \le 3x - 1 \le 8$

32. $-1 \le 6 - x \le 5$

33. $|7z - 1| = |5z + 3|$

34. $|p + 2| = |p + 4|$

35. $|1 - 3x| \ge 4$

36. $\dfrac{1}{2} \le \dfrac{2}{3} r \le \dfrac{5}{4}$

37. $-(m + 4) + 2 = 3m + 8$

38. $\dfrac{p}{6} - \dfrac{3p}{5} = p - 86$

39. $-6 \le \dfrac{3}{2} - x \le 6$

40. $|5 - x| < 4$

41. $|x - 1| \ge -6$

42. $|2r - 5| = |r + 4|$

43. $8q - (1 - q) = 3(1 + 3q) - 4$

44. $8t - (t + 3) = -(2t + 1) - 12$

45. $|r - 5| = |r + 9|$

46. $|r + 2| < -3$

47. $2x + 1 > 5$ or $3x + 4 < 1$

48. $1 - 2x \ge 5$ and $7 + 3x \ge -2$

| Chapter | **2** | **Group Activity** |

Comparing Long-Distance Costs

OBJECTIVE Write an inequality to solve an applied problem.

Cellular phones are popular tools for both local and long-distance phone calls. Frequently, rate plans include long-distance telephoning as an option. The plans vary among different companies and often offer a limited number of "anytime" minutes.

   Consider the following pricing schemes for regular and cellular phones.

- The long-distance plan for an *in-home* phone costs $6.95 per month plus $.05 per min for long-distance calls both within your state or between states, with no limit to the number of minutes of call time.

- One option for a *cellular* phone is a flat monthly fee of $59.99 that includes 450 min of "anytime" local or long-distance calls.

Note: Basic phone rates are *not* included in the in-home plan, but since you intend to have an in-home phone anyway, you can disregard those costs. Also, calls in excess of the limits for the cellular plan are expensive: $.35 per minute over the maximum. You do *not* expect to exceed the number of minutes included in the basic cellular rate plan, so do not worry about those extra charges.

We might ask "Which plan is more economical?"

A. To answer this question, let x represent the number of minutes of long-distance calls in a month.

   1. Write an expression that represents the monthly costs for the in-home rate plan.

   2. Write the expression that represents the monthly cost for the cellular rate plan.

   3. How many minutes of long-distance calls would you have to make in one month with the in-home phone to exceed the cost of the cellular phone plan? Write a linear inequality that states that the in-home rate plan costs more than the cellular rate plan.

   4. Solve the linear inequality and answer the question posed in Problem 3. What does your answer mean in terms of comparing phone costs?

B. Analyze your answers.

   1. Compare phone costs for the two plans.

   2. Suppose you use the cellular phone plan for 450 min (the maximum number of minutes without incurring excess charges). How much more money would you pay compared to the in-home plan?

C. There were actually two different rate plans for *in-home* long distance. Another plan costs $3.95 per month plus $.07 per minute. How many minutes of long distance would you have to call in one month before the $.05 plan costs less than the $.07 plan?

CHAPTER 2 SUMMARY

KEY TERMS

2.1 linear (first-degree)
equation in one
variable
solution
solution set
equivalent equations

conditional equation
contradiction
identity
2.2 mathematical model
formula
percent

2.5 inequality
linear inequality in
one variable
equivalent inequalities
2.6 intersection
compound inequality

union
2.7 absolute value
equation
absolute value
inequality

NEW SYMBOLS

1° one degree   ∩ set intersection   ∪ set union

TEST YOUR WORD POWER

See how well you have learned the vocabulary in this chapter. Answers, with examples, follow the Quick Review.

1. An **algebraic expression** is
   A. an expression that uses any of the four basic operations or the operation of raising to powers or taking roots on any collection of variables and numbers
   B. an expression that contains fractions
   C. an equation that uses any of the four basic operations or the operation of raising to powers or taking roots on any collection of variables and numbers
   D. an equation in algebra.

2. An **equation** is
   A. an algebraic expression
   B. an expression that contains fractions
   C. an expression that uses any of the four basic operations or the operation of raising to powers or taking roots on any collection of variables and numbers
   D. a statement that two algebraic expressions are equal.

3. A **solution set** is the set of numbers that
   A. make an expression undefined
   B. make an equation false

   C. make an equation true
   D. make an expression equal to 0.

4. The **intersection** of two sets A and B is the set of elements that belong
   A. to both A and B
   B. to either A or B, or both
   C. to either A or B, but not both
   D. to just A.

5. The **union** of two sets A and B is the set of elements that belong
   A. to both A and B
   B. to either A or B, or both
   C. to either A or B, but not both
   D. to just B.

QUICK REVIEW

CONCEPTS	EXAMPLES

2.1 LINEAR EQUATIONS IN ONE VARIABLE

Addition and Multiplication Properties of Equality

The same number may be added to (or subtracted from) each side of an equation to obtain an equivalent equation. Similarly, the same nonzero number may be multiplied by or divided into each side of an equation to obtain an equivalent equation.

If $2x + 5 = 10$,

then
$$2x + 5 + (-5) = 10 + (-5),$$
$$2x + 5 - 5 = 10 - 5,$$
$$\frac{1}{2}(2x + 5) = \frac{1}{2}(10),$$
$$\frac{2x + 5}{2} = \frac{10}{2}.$$

CONCEPTS	EXAMPLES
Solving a Linear Equation in One Variable	Solve $4(8 - 3t) = 32 - 8(t + 2)$.

Step 1 Clear fractions.

Step 2 Simplify each side separately.

Step 3 Isolate the variable terms on one side.

$$32 - 12t = 32 - 8t - 16$$
$$32 - 12t = 16 - 8t$$

$$32 - 12t + 12t = 16 - 8t + 12t$$
$$32 = 16 + 4t$$
$$32 - 16 = 16 + 4t - 16$$
$$16 = 4t$$

Step 4 Isolate the variable.

$$\frac{16}{4} = \frac{4t}{4}$$
$$4 = t$$

Step 5 Check.

The solution set is $\{4\}$. This can be checked by substituting 4 for t in the original equation.

2.2 FORMULAS

Solving a Formula for a Specified Variable

Solve $A = \frac{1}{2}bh$ for h.

Step 1 Get all terms with the specified variable on one side and all terms without that variable on the other side.

Step 2 If necessary, use the distributive property to combine terms with the specified variable.

Step 3 Divide both sides by the factor that is the coefficient of the specified variable.

$$A = \frac{1}{2}bh$$

$$2A = 2\left(\frac{1}{2}bh\right)$$

$$2A = bh$$

$$\frac{2A}{b} = h \qquad \text{Divide by } b.$$

2.3 APPLICATIONS OF LINEAR EQUATIONS

Solving an Applied Problem

How many liters of 30% alcohol solution and 80% alcohol solution must be mixed to obtain 100 L of 50% alcohol solution?

Step 1 Read the problem.

Let $\quad x =$ number of liters of 30% solution needed; then $100 - x =$ number of liters of 80% solution needed.

Step 2 Assign a variable.

Summarize the information of the problem in a table.

Liters of Solution	Percent (as a decimal)	Liters of Pure Alcohol
x	.30	.30x
$100 - x$	.80	.80$(100 - x)$
100	.50	.50(100)

(continued)

CONCEPTS	EXAMPLES

Step 3   Write an equation.

The equation is

$$.30x + .80(100 - x) = .50(100).$$

Step 4   Solve the equation.

The solution of the equation is 60.

Step 5   State the answer.

60 L of 30% solution and $100 - 60 = 40$ L of 80% solution are needed.

Step 6   Check.

$$.30(60) + .80(100 - 60) = 50 \text{ is true.}$$

2.4 FURTHER APPLICATIONS OF LINEAR EQUATIONS

To solve a uniform motion problem, draw a sketch and make a table. Use the formula $d = rt$.

Two cars start from towns 400 mi apart and travel toward each other. They meet after 4 hr. Find the speed of each car if one travels 20 mph faster than the other.

Let x = speed of the slower car in miles per hour;
then $x + 20$ = speed of the faster car.

Use the information in the problem and $d = rt$ to complete a table.

	Rate	Time	Distance	
Slower Car	x	4	4x	
Faster Car	x + 20	4	4(x + 20)	
			400	← Total

A sketch shows that the sum of the distances, $4x$ and $4(x + 20)$, must be 400.

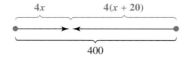

The equation is

$$4x + 4(x + 20) = 400.$$

Solving this equation gives $x = 40$. The slower car travels 40 mph, and the faster car travels $40 + 20 = 60$ mph.

Problems involving denominations of money and mixture problems are solved using methods similar to the one used for the mixture problem shown in the example for Section 2.3.

2.5 LINEAR INEQUALITIES IN ONE VARIABLE

Solving a Linear Inequality in One Variable

Step 1   Simplify each side of the inequality by clearing parentheses and combining like terms.

Solve $3(x + 2) - 5x \leq 12$.

$$3x + 6 - 5x \leq 12$$
$$-2x + 6 \leq 12$$

CONCEPTS	EXAMPLES

Step 2 Use the addition property of inequality to get all terms with variables on one side and all terms without variables on the other side.

$$-2x \leq 6$$

Step 3 Use the multiplication property of inequality to write the inequality in the form $x < k$ or $x > k$.

$$\frac{-2x}{-2} \geq \frac{6}{-2}$$
$$x \geq -3$$

If an inequality is multiplied or divided by a *negative* number, the inequality symbol *must be reversed.*

The solution set $[-3, \infty)$ is graphed here.

2.6 SET OPERATIONS AND COMPOUND INEQUALITIES

Solving a Compound Inequality

Solve $x + 1 > 2$ and $2x < 6$.

Step 1 Solve each inequality in the compound inequality individually.

$$x + 1 > 2 \quad \text{and} \quad 2x < 6$$
$$x > 1 \quad \text{and} \quad x < 3$$

Step 2 If the inequalities are joined with *and,* then the solution set is the intersection of the two individual solution sets.

The solution set is $(1, 3)$.

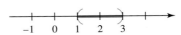

If the inequalities are joined with *or,* then the solution set is the union of the two individual solution sets.

Solve $x \geq 4$ or $x \leq 0$.
The solution set is $(-\infty, 0] \cup [4, \infty)$.

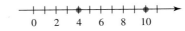

2.7 ABSOLUTE VALUE EQUATIONS AND INEQUALITIES

Solving Absolute Value Equations and Inequalities
Let k be a positive number.
To solve $|ax + b| = k$, solve the compound equation

$$ax + b = k \quad \text{or} \quad ax + b = -k.$$

Solve $|x - 7| = 3$.
$$x - 7 = 3 \quad \text{or} \quad x - 7 = -3$$
$$x = 10 \quad \text{or} \quad x = 4$$

The solution set is $\{4, 10\}$.

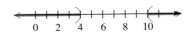

To solve $|ax + b| > k$, solve the compound inequality

$$ax + b > k \quad \text{or} \quad ax + b < -k.$$

Solve $|x - 7| > 3$.
$$x - 7 > 3 \quad \text{or} \quad x - 7 < -3$$
$$x > 10 \quad \text{or} \quad x < 4$$

The solution set is $(-\infty, 4) \cup (10, \infty)$.

(continued)

CONCEPTS	EXAMPLES
To solve $\lvert ax + b \rvert < k$, solve the compound inequality $$-k < ax + b < k.$$	Solve $\lvert x - 7 \rvert < 3$. $$-3 < x - 7 < 3$$ $$4 < x < 10 \qquad \text{Add 7.}$$ The solution set is $(4, 10)$. (number line from 0 to 10, open interval between 4 and 10)
To solve an absolute value equation of the form $$\lvert ax + b \rvert = \lvert cx + d \rvert,$$ solve the compound equation $$ax + b = cx + d \quad \text{or} \quad ax + b = -(cx + d).$$	Solve $\lvert x + 2 \rvert = \lvert 2x - 6 \rvert$. $$x + 2 = 2x - 6 \quad \text{or} \quad x + 2 = -(2x - 6)$$ $$x = 8 \qquad \text{or} \qquad x = \frac{4}{3}$$ The solution set is $\left\{ \frac{4}{3}, 8 \right\}$.

Answers to Test Your Word Power

1. A; *Examples:* $\frac{3y - 1}{2}$, $6 + \sqrt{2x}$, $4a^3b - c$ **2.** D; *Examples:* $2a + 3 = 7$, $3y = -8$, $x^2 = 4$ **3.** C; *Example:* $\{8\}$ is the solution set of $2x + 5 = 21$.   **4.** A; *Example:* If $A = \{2, 4, 6, 8\}$ and $B = \{1, 2, 3\}$, $A \cap B = \{2\}$.   **5.** B; *Example:* Using the sets A and B from Answer 4, $A \cup B = \{1, 2, 3, 4, 6, 8\}$.

CHAPTER 2 REVIEW EXERCISES

[2.1] *Solve each equation.*

1. $-(8 + 3z) + 5 = 2z + 6$

2. $-\dfrac{3}{4}x = -12$

3. $\dfrac{2q + 1}{3} - \dfrac{q - 1}{4} = 0$

4. $5(2x - 3) = 6(x - 1) + 4x$

Decide whether the given equation is conditional, *an* identity, *or a* contradiction. *Give the solution set.*

5. $7r - 3(2r - 5) + 5 + 3r = 4r + 20$

6. $8p - 4p - (p - 7) + 9p + 6 = 12p - 7$

7. $-2r + 6(r - 1) + 3r - (4 - r) = -(r + 5) - 5$

[2.2] *Solve each formula for the specified variable.*

8. $V = LWH$ for H

9. $A = \dfrac{1}{2}(B + b)h$ for b

Solve each equation for x.

10. $M = -\dfrac{1}{4}(x + 3y)$

11. $P = \dfrac{3}{4}x - 12$

12. Give the steps you would use to solve $-2x + 5 = 7$.

[2.2, 2.3]   *Solve each problem.*

13. A rectangular solid has a volume of 180 ft³. Its length is 6 ft and its width is 5 ft. Find its height.

14. The total number of AIDS cases reported in 1997 was 58,443. In 1998, this figure had decreased to 48,269. What approximate percent decrease did this represent? (*Source:* U.S. Centers for Disease Control.)

15. Find the simple interest rate that Francesco Castellucio is earning, if a principal of $30,000 earns $7800 interest in 4 yr.

16. If the Fahrenheit temperature is 77°, what is the corresponding Celsius temperature?

17. The circle graph shows the projected racial composition of the U.S. workforce in the year 2006. The projected total number of people in the workforce for that year is 148,847,000. How many of these will be in the Hispanic category?

18. Refer to Exercise 17. How many people in the workplace in 2006 will be in the category labeled "Asian and other"?

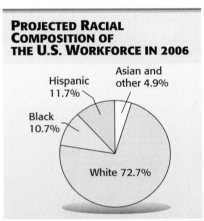

PROJECTED RACIAL COMPOSITION OF THE U.S. WORKFORCE IN 2006

Hispanic 11.7%
Asian and other 4.9%
Black 10.7%
White 72.7%

Source: U.S. Bureau of Labor Statistics.

Write each phrase as a mathematical expression, using x as the variable.

19. One-third of a number, subtracted from 9

20. The product of 4 and a number, divided by 9 more than the number

Solve each problem.

21. The length of a rectangle is 3 m less than twice the width. The perimeter of the rectangle is 42 m. Find the length and width of the rectangle.

22. In a triangle with two sides of equal length, the third side measures 15 in. less than the sum of the two equal sides. The perimeter of the triangle is 53 in. Find the lengths of the three sides.

23. A candy clerk has three times as many kilograms of chocolate creams as peanut clusters. The clerk has 48 kg of the two candies altogether. How many kilograms of peanut clusters does the clerk have?

24. How many liters of a 20% solution of a chemical should be mixed with 15 L of a 50% solution to get a 30% mixture?

25. How much water should be added to 30 L of a 40% acid solution to reduce it to a 30% solution?

Liters of Solution	Percent (as a decimal)	Liters of Pure Acid
	.40	
x	.00	
	.30	

26. Anna Mae Wood invested some money at 6% and $4000 less than this amount at 4%. Find the amount invested at each rate if her total annual interest income is $840.

Principal	Rate (as a decimal)	Interest
x	.06	
	.04	

[2.4]

27. A grocery store clerk has $3.50 in dimes and quarters in her cash drawer. The number of dimes is 1 less than twice the number of quarters. How many of each denomination are there?

28. When Jim emptied his pockets one evening, he found he had 19 nickels and dimes with a total value of $1.55. How many of each denomination did he have?

29. Which choice is the best *estimate* for the average speed of a trip of 405 mi that lasted 8.2 hr?

A. 50 mph   **B.** 30 mph   **C.** 60 mph   **D.** 40 mph

30. (a) A driver averaged 53 mph and took 10 hr to travel from Memphis to Chicago. What is the distance between Memphis and Chicago?
(b) A small plane traveled from Warsaw to Rome, averaging 164 mph. The trip took 2 hr. What is the distance from Warsaw to Rome?

31. A passenger train and a freight train leave a town at the same time and go in opposite directions. They travel at 60 mph and 75 mph, respectively. How long will it take for them to be 297 mi apart?

	Rate	Time	Distance
Passenger Train	60	x	
Freight Train	75	x	

32. Two cars leave towns 230 km apart at the same time, traveling directly toward one another. One car travels 15 km per hr slower than the other. They pass one another 2 hr later. What are their speeds?

	Rate	Time	Distance
Faster Car	x	2	
Slower Car	x − 15	2	

33. An automobile averaged 45 mph for the first part of a trip and 50 mph for the second part. If the entire trip took 4 hr and covered 195 mi, for how long was the rate 45 mph?

34. An 85-mi trip to the beach took the Valenzuela family 2 hr. During the second hour, a rainstorm caused them to average 7 mph less than they traveled during the first hour. Find their average rate for the first hour.

35. Find the measure of each angle in the triangle.

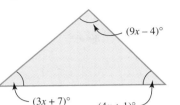

$(9x - 4)°$

$(3x + 7)°$ $(4x + 1)°$

36. Find the measure of each marked angle.

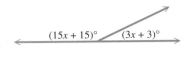

$(15x + 15)°$ $(3x + 3)°$

[2.5] *Solve each inequality. Express the solution set in interval form.*

37. $-\dfrac{2}{3}k < 6$

38. $-5x - 4 \geq 11$

39. $\dfrac{6a + 3}{-4} < -3$

40. $5 - (6 - 4k) \geq 2k - 7$

41. $8 \leq 3z - 1 < 14$

42. $\dfrac{5}{3}(m - 2) + \dfrac{2}{5}(m + 1) > 1$

43. The perimeter of a rectangular playground must be no greater than 120 m. The width of the playground must be 22 m. Find the possible lengths of the playground.

44. The hit movie *Titanic* earned more in Europe than in the United States. The average movie ticket in London, for example, costs the equivalent of $10.59. (*Source: Parade,* September 13, 1998.) A student group from the United States is touring London and wishes to see the movie there. If $1000 is available to purchase tickets and the group receives a $50 discount from the tour company, how many tickets can be purchased?

45. To pass Algebra, a student must have an average of at least 70 on five tests. On the first four tests, a student has scores of 75, 79, 64, and 71. What possible scores on the fifth test would guarantee a passing grade in the class?

46. While solving the inequality

$$10x + 2(x - 4) < 12x - 13,$$

a student did all the work correctly and obtained the statement $-8 < -13$. The student did not know what to do at this point, because the variable "disappeared." How would you explain to the student the interpretation of this result?

[2.6] *Let $A = \{a, b, c, d\}$, $B = \{a, c, e, f\}$, and $C = \{a, e, f, g\}$. Find each set.*

47. $A \cap B$ **48.** $A \cap C$ **49.** $B \cup C$ **50.** $A \cup C$

Solve each compound inequality. Give the solution set in both interval and graph forms.

51. $x > 6$ and $x < 9$

52. $x + 4 > 12$ and $x - 2 < 12$

53. $x > 5$ or $x \leq -3$

54. $x \geq -2$ or $x < 2$

55. $x - 4 > 6$ and $x + 3 \leq 10$

56. $-5x + 1 \geq 11$ or $3x + 5 \geq 26$

Express each union or intersection in simplest interval form.

57. $(-3, \infty) \cap (-\infty, 4)$

58. $(-\infty, 6) \cap (-\infty, 2)$

59. $(4, \infty) \cup (9, \infty)$

60. $(1, 2) \cup (1, \infty)$

[2.7] *Solve each absolute value equation.*

61. $|x| = 7$ **62.** $|x + 2| = 9$ **63.** $|3k - 7| = 8$

64. $|z - 4| = -12$ **65.** $|2k - 7| + 4 = 11$ **66.** $|4a + 2| - 7 = -3$

67. $|3p + 1| = |p + 2|$ **68.** $|2m - 1| = |2m + 3|$

Solve each absolute value inequality. Give the solution set in interval form.

69. $|p| < 14$

70. $|-t + 6| \leq 7$

71. $|2p + 5| \leq 1$

72. $|x + 1| \geq -3$

▨ MIXED REVIEW EXERCISES*

Solve.

73. $5 - (6 - 4k) > 2k - 5$

74. $ak + bt = 6t - sk$ for k

75. $x < 3$ and $x \geq -2$

76. $\dfrac{4x + 2}{4} + \dfrac{3x - 1}{8} = \dfrac{x + 6}{16}$

77. $|3k + 6| \geq 0$

78. $-5r \geq -10$

79. A newspaper recycling collection bin is in the shape of a box, 1.5 ft wide and 5 ft long. If the volume of the bin is 75 ft³, find the height.

80. The sum of the smallest and largest of three consecutive integers is 47 more than the middle integer. What are the integers?

81. $|3x + 2| + 4 = 9$

82. $.05x + .03(1200 - x) = 42$

83. $|m + 3| \leq 13$

84. $\dfrac{3}{4}(a - 2) - \dfrac{1}{3}(5 - 2a) < -2$

85. $-4 < 3 - 2k < 9$

86. $-.3x + 2.1(x - 4) \leq -6.6$

87. The complement of an angle measures 10° less than one-fifth of its supplement. Find the measure of the angle.

---

*The order of exercises in this final group does not correspond to the order in which topics occur in the chapter. This random ordering should help you prepare for the chapter test.

88. A loan has a finance charge of $450. The loan was scheduled to run for 24 mo. Find the unearned interest if the loan is paid off with 5 payments left. (Refer to Section 2.2, Exercises 41–44.)

89. To qualify for a company pension plan, an employee must average at least $1000 per month in earnings. During the first four months of the year, an employee made $900, $1200, $1040, and $760. What possible amounts earned during the fifth month will qualify the employee?

90. $|5r - 1| > 14$

91. $x \geq -2$ or $x < 4$

92. How many liters of a 20% solution of a chemical should be mixed with 10 L of a 50% solution to get a 40% mixture?

93. $|m - 1| = |2m + 3|$

94. $\dfrac{3x}{5} - \dfrac{x}{2} = 3$

95. $|m + 3| \leq 1$

96. $|3k - 7| = 4$

97. In the 2000 presidential election, Al Gore received five fewer electoral votes than George W. Bush. A total of 537 electoral votes were cast. How many electoral votes did each candidate receive? (*Source:* www.cnn.com)

98. $5(2x - 7) = 2(5x + 3)$

In Exercises 99 and 100, sketch the graph of each solution set.

99. $x > 6$ and $x < 8$

100. $-5x + 1 \geq 11$ or $3x + 5 \geq 26$

101. The 2000 median weekly earnings of full-time workers by occupation were as shown in the table.

Weekly Earnings of Full-time Workers (in dollars)

Occupation	Men	Women
Managerial/Professional	994	709
Technical/Sales/Administrative Support	655	452
Service	357	316
Operators/Fabricators/Laborers	487	351

Source: U.S. Bureau of Labor Statistics.

List the elements of each set.

(a) The set of occupations with median earnings for men less than $900 and for women greater than $500

(b) The set of occupations with median earnings for men greater than $600 or for women less than $400

CHAPTER **2** TEST

Solve each equation.

1. $3(2x - 2) - 4(x + 6) = 3x + 8 + x$

2. $.08x + .06(x + 9) = 1.24$

3. $\dfrac{x + 6}{10} + \dfrac{x - 4}{15} = \dfrac{x + 2}{6}$

4. Decide whether the equation

$$3x - (2 - x) + 4x + 2 = 8x + 3$$

is *conditional*, an *identity*, or a *contradication*. Give its solution set.

5. Solve $-16t^2 + vt - S = 0$ for v.

6. Solve $ar + 2 = 3r - 6t$ for r.

Solve each problem.

7. The 1997 Daytona 500 (mile) race was won by Jeff Gordon, who averaged 148.295 mph. What was Gordon's time?

8. A certificate of deposit pays $2281.25 in simple interest for 1 yr on a principal of $36,500. What is the rate of interest?

9. Of the 38,159 offices, stations, and branches of the U.S. Postal Service in 1998, 27,952 were actually classified as post offices. What percent to the nearest tenth were classified as post offices? (*Source:* U.S. Postal Service.)

10. Tyler McGinnis invested some money at 3% simple interest and some at 5% simple interest. The total amount of his investments was $28,000, and the interest he earned during the first year was $1240. How much did he invest at each rate?

11. Two cars leave from the same point at the same time, traveling in opposite directions. One travels 15 mph slower than the other. After 6 hr, they are 630 mi apart. Find the rate of each car.

12. Find the measure of each angle.

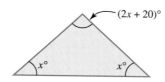

Solve each inequality. Give the solution set in both interval and graph forms.

13. $4 - 6(x + 3) \le -2 - 3(x + 6) + 3x$

14. $-\dfrac{4}{7}x > -16$

15. $-6 \le \dfrac{4}{3}x - 2 \le 2$

Solve each problem.

16. A student must have an average of at least 80 on the four tests in a course to get a B. The student had scores of 83, 76, and 79 on the first three tests. What minimum score on the fourth test would guarantee a B in the course?

17. A product will break even or produce a profit only if the revenue R (in dollars) from selling the product is at least equal to the cost C (in dollars) of producing it. Suppose that the cost to produce x units of carpet is $C = 50x + 5000$, while the revenue is $R = 60x$. For what values of x is R at least equal to C?

18. Let $A = \{1, 2, 5, 7\}$ and $B = \{1, 5, 9, 12\}$. Find

    (a) $A \cap B$ **(b)** $A \cup B$.

19. Solve each compound inequality.

    (a) $3k \geq 6$ and $k - 4 < 5$ **(b)** $-4x \leq -24$ or $4x - 2 < 10$

Solve each absolute value equation or inequality.

20. $|4x - 3| = 7$ **21.** $|5 - 6x| > 12$ **22.** $|7 - x| \leq -1$

23. $|3 - 5x| = |2x + 8|$ **24.** $|-3x + 4| - 4 < -1$

25. If $k < 0$, what is the solution set of

    (a) $|5x + 3| < k$ **(b)** $|5x + 3| > k$ **(c)** $|5x + 3| = k$?

CUMULATIVE REVIEW EXERCISES CHAPTERS 1–2

From now on, each chapter will conclude with a set of cumulative review exercises designed to cover the major topics from the beginning of the course. This feature allows the student to constantly review topics that have been introduced up to that point.

Let $A = \left\{-8, -\frac{2}{3}, -\sqrt{6}, 0, \frac{4}{5}, 9, \sqrt{36}\right\}$. Simplify the elements of A as necessary and then list the elements that belong to the set.

1. Natural numbers   **2.** Whole numbers   **3.** Integers

4. Rational numbers   **5.** Irrational numbers   **6.** Real numbers

Add or subtract, as indicated.

7. $-\dfrac{4}{3} - \left(-\dfrac{2}{7}\right)$ **8.** $|-4| - |2| + |-6|$

9. $(-2)^4 + (-2)^3$ **10.** $\sqrt{25} - 5(-1)^0$

Evaluate each expression.

11. $(-3)^5$ **12.** $\left(\dfrac{6}{7}\right)^3$ **13.** $\left(-\dfrac{2}{3}\right)^3$ **14.** -4^6

15. Which one of the following is not a real number: $-\sqrt{36}$ or $\sqrt{-36}$?

16. Which one of the following is undefined: $\dfrac{4 - 4}{4 + 4}$ or $\dfrac{4 + 4}{4 - 4}$?

Evaluate if $a = 2$, $b = -3$, and $c = 4$.

17. $-3a + 2b - c$ **18.** $-8(a^2 + b^3)$ **19.** $\dfrac{3a^3 - b}{4 + 3c}$

Use the properties of real numbers to simplify each expression.

20. $-7r + 5 - 13r + 12$ **21.** $-(3k + 8) - 2(4k - 7) + 3(8k + 12)$

Identify the property of real numbers illustrated by each equation.

22. $(a + b) + 4 = 4 + (a + b)$

23. $4x + 12x = (4 + 12)x$

Solve each equation.

24. $-4x + 7(2x + 3) = 7x + 36$

25. $-\dfrac{3}{5}x + \dfrac{2}{3}x = 2$

26. $.06x + .03(100 + x) = 4.35$

27. $P = a + b + c$ for b

28. $4(2x - 6) + 3(x - 2) = 11x + 1$

29. $\dfrac{2}{3}x + \dfrac{5}{8}x = \dfrac{31}{24}x$

Solve each inequality. Give the solution set in both interval and graph forms.

30. $3 - 2(x + 7) \le -x + 3$

31. $-4 < 5 - 3x \le 0$

32. $2x + 1 > 5$ or $2 - x > 2$

33. $|-7k + 3| \ge 4$

Solve each problem.

34. Karl Manley invested some money at 7% interest and the same amount at 10%. His total interest for the year was $150 less than one-tenth of the total amount he invested. How much did he invest at each rate?

35. A dietician must use three foods, A, B, and C, in a diet. He must include twice as many grams of food A as food C, and 5 g of food B. The three foods must total at most 24 g. What is the largest amount of food C that the dietician can use?

36. Laurie Reilly got scores of 88 and 78 on her first two tests. What score must she make on her third test to keep an average of 80 or greater?

37. How much pure alcohol should be added to 7 L of 10% alcohol to increase the concentration to 30% alcohol?

38. A coin collection contains 29 coins. It consists of pennies, nickels, and quarters. The number of quarters is 4 less than the number of nickels, and the face value of the collection is $2.69. How many of each denomination are there in the collection?

39. Clark's rule, a formula used in reducing drug dosage according to weight from the recommended adult dosage to a child dosage, is

$$\frac{\text{weight of child in pounds}}{150} \times \text{adult dose} = \text{child's dose}.$$

Find a child's dosage if the child weighs 55 lb and the recommended adult dosage is 120 mg.

40. Since 1975, the number of daily newspapers has steadily declined. According to the table,

(a) by how much did the number of daily newspapers decrease between 1990 and 1998?

(b) by what *percent* did the number of daily newspapers decrease from 1990 to 1998?

Year	Number of Daily Newspapers
1975	1756
1980	1745
1985	1676
1990	1611
1995	1533
1996	1520
1997	1509
1998	1489

Source: Statistical Abstract of the United States, 1999.

Graphs, Linear Equations, and Functions

3

Graphs are widely used in the media because they present a lot of information in an easy-to-understand form. As the saying goes, "A picture is worth a thousand words." It is important to be able to read graphs correctly and understand how to use the data they provide. In Section 3.2, Example 8, we use a graph to find the average rate of change each year from 1997 to 2001 in the number of U.S. households owning more than one personal computer.

3.1 The Rectangular Coordinate System

The line graph in Figure 1 shows profits for cable television station CNBC from 1994 through 1999.

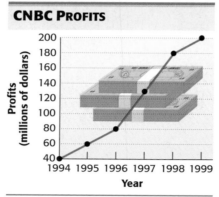

CNBC PROFITS

Source: Fortune, May 24, 1999, p. 142.

FIGURE 1

The line graph represents information based on a method for locating a point in a plane developed by René Descartes, a 17th-century French mathematician. It is said that Descartes, who was lying in bed ill, was watching a fly crawl about on the ceiling near a corner of the room. It occurred to him that the location of the fly on the ceiling could be described by determining its distances from the two adjacent walls. In this chapter we use this insight to plot points and graph linear equations in two variables whose graphs are straight lines.

OBJECTIVE 1 Plot ordered pairs. Each of the pairs of numbers $(3, 1)$, $(-5, 6)$, and $(4, -1)$ is an example of an **ordered pair;** that is, a pair of numbers written within parentheses in which the order of the numbers is important. We graph an ordered pair using two perpendicular number lines that intersect at their 0 points, as shown in Figure 2. The common 0 point is called the **origin.** The position of any point in this plane is determined by referring to the horizontal number line, the **x-axis,** and the vertical number line, the **y-axis.** The first number in the ordered pair indicates the position relative to the *x*-axis, and the second number indicates the position relative to the *y*-axis. The *x*-axis and the *y*-axis make up a **rectangular** (or **Cartesian,** for Descartes) **coordinate system.**

To locate, or **plot,** the point on the graph that corresponds to the ordered pair $(3, 1)$, we move three units from 0 to the right along the *x*-axis, and then one unit up parallel to the *y*-axis. The point corresponding to the ordered pair $(3, 1)$ is labeled *A* in Figure 3. Additional points are labeled *B−E.* The phrase "the point corresponding to the ordered pair $(3, 1)$" is often abbreviated as "the point $(3, 1)$." The numbers in an ordered pair are called the **coordinates** of the corresponding point.

We can relate this method of locating ordered pairs to the line graph in Figure 1. We move along the horizontal axis to a year, then up parallel to the vertical axis to find profit for that year. Thus, we can write the ordered pair $(1998, 180)$ to indicate that in 1998, profit was $180 million.

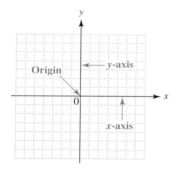

FIGURE 2

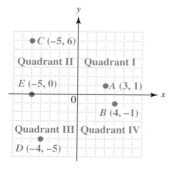

FIGURE 3

The four regions of the graph, shown in Figure 3, are called **quadrants I, II, III, and IV,** reading counterclockwise from the upper right quadrant. The points on the *x*-axis and *y*-axis do not belong to any quadrant. For example, point *E* in Figure 3 belongs to no quadrant.

Now Try Exercises 3, 9, 13, 15, and 21.

OBJECTIVE 2 Find ordered pairs that satisfy a given equation. Each solution to an equation with two variables, such as $2x + 3y = 6$, includes two numbers, one for each variable. To keep track of which number goes with which variable, we write the solutions as ordered pairs. (If *x* and *y* are used as the variables, the *x*-value is given first.) For example, we can show that $(6, -2)$ is a solution of $2x + 3y = 6$ by substitution.

$$2x + 3y = 6$$
$$2(6) + 3(-2) = 6 \quad ? \qquad \text{Let } x = 6, y = -2.$$
$$12 - 6 = 6 \quad ?$$
$$6 = 6 \qquad \text{True}$$

Because the ordered pair $(6, -2)$ makes the equation true, it is a solution. On the other hand, $(5, 1)$ is *not* a solution of the equation $2x + 3y = 6$ because

$$2(5) + 3(1) = 10 + 3 = 13 \neq 6.$$

To find ordered pairs that satisfy an equation, select any number for one of the variables, substitute it into the equation for that variable, and then solve for the other variable. Two other ordered pairs satisfying $2x + 3y = 6$ are $(0, 2)$ and $(3, 0)$. Since any real number could be selected for one variable and would lead to a real number for the other variable, linear equations in two variables have an infinite number of ordered-pair solutions.

EXAMPLE 1 Completing Ordered Pairs

Complete the table of ordered pairs for $2x + 3y = 6$.

x	*y*	
-3		← Represents the ordered pair $(-3, \_\_)$
	-4	← Represents the ordered pair $(\_\_, -4)$

First let $x = -3$ and substitute into the equation to find *y*. Then let $y = -4$ and substitute to find *x*.

$$2x + 3y = 6 \qquad\qquad 2x + 3y = 6$$
$$2(-3) + 3y = 6 \quad \text{Let } x = -3. \qquad 2x + 3(-4) = 6 \quad \text{Let } y = -4.$$
$$-6 + 3y = 6 \qquad\qquad 2x - 12 = 6$$
$$3y = 12 \qquad\qquad 2x = 18$$
$$y = 4 \qquad\qquad x = 9$$

The ordered pair is $(-3, 4)$. The ordered pair is $(9, -4)$.

These pairs lead to the following completed table.

x	*y*
-3	4
9	-4

Now Try Exercise 23(a).

OBJECTIVE 3 Graph lines. Since an equation in two variables like $2x + 3y = 6$ is satisfied by an infinite number of ordered pairs, how might we express the solution set of such an equation? The **graph of an equation** is the set of points corresponding to *all* ordered pairs that satisfy the equation. It gives a "picture" of the equation.

To graph an equation, we plot a number of ordered pairs that satisfy the equation until we have enough points to suggest the shape of the graph. For example, to graph $2x + 3y = 6$, we plot all ordered pairs found in Objective 2 and Example 1. These points, shown in a table of values and plotted in Figure 4(a), appear to lie on a straight line. If all the ordered pairs that satisfy the equation $2x + 3y = 6$ were graphed, they would form the straight line shown in Figure 4(b).

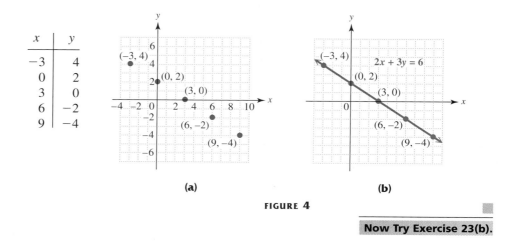

x	y
-3	4
0	2
3	0
6	-2
9	-4

(a)　　　　　　**(b)**

FIGURE 4

Now Try Exercise 23(b).

The equation $2x + 3y = 6$ is called a **first-degree equation** because it has no term with a variable to a power greater than 1.

The graph of any first-degree equation in two variables is a straight line.

Since first-degree equations with two variables have straight-line graphs, they are called *linear equations in two variables.* (We discussed linear equations in one variable in Chapter 2.)

Linear Equation in Two Variables

A **linear equation in two variables** can be written in the form

$$Ax + By = C,$$

where A, B, and C are real numbers (A and B not both 0). This form is called **standard form.**

OBJECTIVE 4 Find x- and y-intercepts. A straight line is determined if any two different points on the line are known, so finding two different points is enough to graph the line. Two useful points for graphing are the x- and y-intercepts. The **x-intercept** is the point (if any) where the line intersects the x-axis; likewise, the **y-intercept** is the point (if any) where the line intersects the *y*-axis.* In Figure 4(b), the *y*-value of the point where the line intersects the *x*-axis is 0. Similarly, the *x*-value of the point where the line intersects the *y*-axis is 0. This suggests a method for finding the *x*- and *y*-intercepts.

Finding Intercepts

In the equation of a line, let $y = 0$ to find the *x*-intercept; let $x = 0$ to find the *y*-intercept.

■ EXAMPLE 2   Finding Intercepts

Find the *x*- and *y*-intercepts of $4x - y = -3$ and graph the equation.

We find the *x*-intercept by letting $y = 0$.

$$4x - 0 = -3 \qquad \text{Let } y = 0.$$
$$4x = -3$$
$$x = -\frac{3}{4} \qquad \text{x-intercept is } \left(-\tfrac{3}{4}, 0\right).$$

For the *y*-intercept, we let $x = 0$.

$$4(0) - y = -3 \qquad \text{Let } x = 0.$$
$$-y = -3$$
$$y = 3 \qquad \text{y-intercept is } (0, 3).$$

The intercepts are the two points $\left(-\tfrac{3}{4}, 0\right)$ and $(0, 3)$. We show these ordered pairs in the table next to Figure 5 and use them to draw the graph.

x	y
$-\frac{3}{4}$	0
0	3

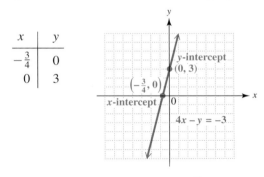

FIGURE 5

Now Try Exercise 33.

NOTE While two points, such as the two intercepts in Figure 5, are sufficient to graph a straight line, it is a good idea to use a third point to guard against errors. Verify by substitution that $(-2, -5)$ also lies on the graph of $4x - y = -3$.

---

*Some texts define an intercept as a number, not a point. For example, "*y*-intercept (0, 4)" would be given as "*y*-intercept 4."

OBJECTIVE 5 Recognize equations of vertical and horizontal lines. A graph can fail to have an *x*-intercept or a *y*-intercept, which is why the phrase "if any" was added when discussing intercepts.

EXAMPLE 3 Graphing a Horizontal Line

Graph $y = 2$.

Writing $y = 2$ as $0x + 1y = 2$ shows that any value of *x*, including $x = 0$, gives $y = 2$, making the *y*-intercept $(0, 2)$. Since *y* is always 2, there is no value of *x* corresponding to $y = 0$, so the graph has no *x*-intercept. The graph, shown with a table of ordered pairs in Figure 6, is a horizontal line.

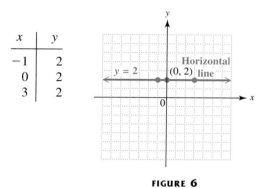

x	y
−1	2
0	2
3	2

FIGURE 6

Now Try Exercise 39.

EXAMPLE 4 Graphing a Vertical Line

Graph $x + 1 = 0$.

The form $1x + 0y = -1$ shows that every value of *y* leads to $x = -1$, making the *x*-intercept $(-1, 0)$. No value of *y* makes $x = 0$, so the graph has no *y*-intercept. The only way a straight line can have no *y*-intercept is to be vertical, as shown in Figure 7.

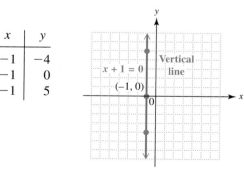

x	y
−1	−4
−1	0
−1	5

FIGURE 7

Now Try Exercise 43.

CAUTION To avoid confusing equations of horizontal and vertical lines remember that

1. An equation with only the variable x will always intersect the *x-axis* and thus will be *vertical*.
2. An equation with only the variable y will always intersect the *y-axis* and thus will be *horizontal*.

Some lines have both the x- and y-intercepts at the origin.

■ **EXAMPLE 5** **Graphing a Line That Passes through the Origin**

Graph $x + 2y = 0$.

    Find the intercepts.

$x + 2y = 0$			$x + 2y = 0$		
$x + 2(0) = 0$	Let $y = 0$.		$0 + 2y = 0$	Let $x = 0$.	
$x + 0 = 0$			$2y = 0$		
$x = 0$	*x*-intercept is $(0, 0)$.		$y = 0$	*y*-intercept is $(0, 0)$.	

Both intercepts are the same point, $(0, 0)$, which means that the graph passes through the origin. To find another point so that we can graph the line, we choose any nonzero number for x or y. If we choose $x = 4$ and solve for y, then

$$x + 2y = 0$$
$$4 + 2y = 0 \qquad \text{Let } x = 4.$$
$$2y = -4$$
$$y = -2.$$

This gives the ordered pair $(4, -2)$. These two points lead to the graph shown in Figure 8. As a check, verify that $(-2, 1)$ also lies on the line.

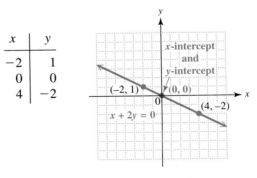

x	y
-2	1
0	0
4	-2

FIGURE 8

To find the additional point, we could have chosen any number (except 0) for y instead of x.

Now Try Exercise 45.

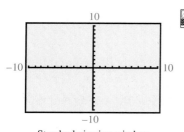

Standard viewing window

FIGURE 9

OBJECTIVE 6 Use a graphing calculator to graph an equation. When graphing by hand, we first set up a rectangular coordinate system, then plot points and draw the graph. Similarly, when graphing with a graphing calculator, we first tell the calculator how to set up a rectangular coordinate system. This involves choosing the minimum and maximum x- and y-values that will determine the viewing screen. In the screen shown in Figure 9, we chose minimum x- and y-values of -10 and maximum x- and y-values of 10. The *scale* on each axis determines the distance between the tick marks; in the screen shown, the scale is 1 for both axes. We refer to this as the *standard viewing window*.

To graph an equation, we usually need to solve the equation for y in order to enter it into the calculator. Once the equation is graphed, we can use the calculator to find the intercepts or any other point on the graph easily.

EXAMPLE 6 Graphing a Linear Equation with a Graphing Calculator and Finding the Intercepts

Use a graphing calculator to graph $4x - y = 3$.

Because we want to be able to see the intercepts on the screen, we use them to determine an appropriate window. Here, the x-intercept is $(.75, 0)$ and the y-intercept is $(0, -3)$. Although many choices are possible, we choose the standard viewing window. We must solve the equation for y to enter it into the calculator.

$$4x - y = 3$$
$$-y = -4x + 3 \qquad \text{Subtract } 4x.$$
$$y = 4x - 3 \qquad \text{Multiply by } -1.$$

The graph is shown in Figures 10 and 11, which also give the intercepts at the bottoms of the screens.

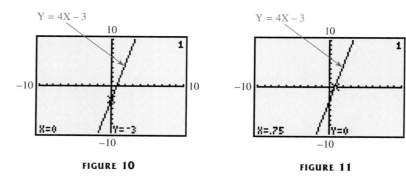

FIGURE 10   **FIGURE 11**

Some calculators have the capability of locating the x-intercept (called "Root" or "Zero"). Consult your owner's manual.

Now Try Exercise 59.

3.1 **EXERCISES**

In Exercises 1 and 2, answer each question by locating ordered pairs on the graphs.

1. The graph shows the percent of women in math or computer science professions.

   (a) If (x, y) represents a point on the graph, what does x represent? What does y represent?

   (b) In what decade (10-yr period) did the percent of women in math or computer science professions decrease?

   (c) Write an ordered pair (x, y) that gives the approximate percent of women in math or computer science professions in 1990.

   (d) What does the ordered pair (2000, 30) mean in the context of this graph?

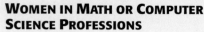

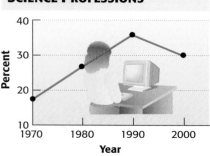

Source: U.S. Bureau of the Census and Bureau of Labor Statistics.

2. The graph indicates federal government tax revenues in billions of dollars.

   (a) If (x, y) represents a point on the graph, what does x represent? What does y represent?

   (b) Estimate revenue in 1996.

   (c) Write an ordered pair (x, y) that gives approximate federal tax revenues in 1995.

   (d) What does the ordered pair (1998, 1720) mean in the context of this graph?

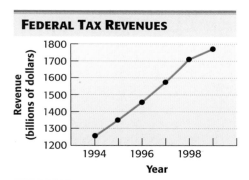

Source: U.S. Office of Management and Budget.

Fill in each blank with the correct response.

3. The point with coordinates $(0, 0)$ is called the _____ of a rectangular coordinate system.

4. For any value of x, the point $(x, 0)$ lies on the _____-axis.

5. To find the x-intercept of a line, we let _____ equal 0 and solve for _____; to find the y-intercept, we let _____ equal 0 and solve for _____.

6. The equation _____ = 4 has a horizontal line as its graph.
        (x or y)

7. To graph a straight line, we must find a minimum of _____ points.

8. The point (_____, 4) is on the graph of $2x - 3y = 0$.

Name the quadrant, if any, in which each point is located.

9. (a) $(1, 6)$ **(b)** $(-4, -2)$ **10. (a)** $(-2, -10)$ **(b)** $(4, 8)$
    (c) $(-3, 6)$ **(d)** $(7, -5)$ **(c)** $(-9, 12)$ **(d)** $(3, -9)$
    (e) $(-3, 0)$ **(e)** $(0, -8)$

11. Use the given information to determine the possible quadrants in which the point (x, y) must lie.

    (a) $xy > 0$ **(b)** $xy < 0$ **(c)** $\dfrac{x}{y} < 0$ **(d)** $\dfrac{x}{y} > 0$

12. What must be true about the coordinates of any point that lies along an axis?

Plot each point on a rectangular coordinate system.

13. $(2, 3)$ **14.** $(-1, 2)$ **15.** $(-3, -2)$ **16.** $(1, -4)$ **17.** $(0, 5)$

18. $(-2, -4)$ **19.** $(-2, 4)$ **20.** $(3, 0)$ **21.** $(-2, 0)$ **22.** $(3, -3)$

*In Exercises 23–28, **(a)** complete the given table for each equation, and then **(b)** graph the equation. See Example 1 and Figure 4.*

23. $x - y = 3$

x	y
0	
	0
5	
2	

24. $x - y = 5$

x	y
0	
	0
1	
3	

25. $x + 2y = 5$

x	y
0	
	0
2	
	2

26. $x + 3y = -5$

x	y
0	
	0
1	
	-1

27. $4x - 5y = 20$

x	y
0	
	0
2	
	-3

28. $6x - 5y = 30$

x	y
0	
	0
3	
	-2

29. Explain why the graph of $x + y = k$ cannot pass through quadrant III if $k > 0$.

30. Explain how to determine the intercepts and graph of the linear equation $4x - 3y = 12$.

31. A student attempted to graph $4x + 5y = 0$ by finding intercepts. She first let $x = 0$ and found y; then she let $y = 0$ and found x. In both cases, the resulting point was $(0, 0)$. She knew that she needed at least two points to graph the line, but was unsure what to do next because finding intercepts gave her only one point. Explain to her what to do next.

32. What is the equation of the x-axis? What is the equation of the y-axis?

Find the x- and y-intercepts. Then graph each equation. See Examples 2–5.

33. $2x + 3y = 12$ **34.** $5x + 2y = 10$ **35.** $x - 3y = 6$

36. $x - 2y = -4$ **37.** $\dfrac{2}{3}x - 3y = 7$ **38.** $\dfrac{5}{7}x + \dfrac{6}{7}y = -2$

39. $y = 5$ **40.** $y = -3$ **41.** $x = 2$

42. $x = -3$ **43.** $x + 4 = 0$ **44.** $y + 2 = 0$

45. $x + 5y = 0$ **46.** $x - 3y = 0$ **47.** $2x = 3y$

48. $4y = 3x$ **49.** $-\dfrac{2}{3}y = x$ **50.** $3y = -\dfrac{4}{3}x$

A linear equation can be used as a model to describe real data in some cases. Exercises 51 and 52 are based on this idea.

51. Track qualifying records at North Carolina Motor Speedway from 1965–1998 are approximated by the linear equation $y = 1.22x + 118$, where y is the speed (in miles per hour) in year x. In the equation, $x = 0$ corresponds to 1965, $x = 10$ corresponds to 1975, and so on. Use the equation to approximate the speed of the 1995 winner, Hut Stricklin. (*Source:* NASCAR.)

52. According to Families USA Foundation, the national average family health care cost in dollars between 1980 and 2000 can be approximated by the linear equation $y = 382.75x + 1742$, where $x = 0$ corresponds to 1980 and $x = 20$ corresponds to 2000. Based on this equation, find the national average health care cost in 2000.

TECHNOLOGY INSIGHTS   (EXERCISES 53–57)

53. The screens show the graph of one of the equations in A–D. Which equation is it?

    A. $3x + 2y = 6$ **B.** $-3x + 2y = 6$ **C.** $-3x - 2y = 6$ **D.** $3x - 2y = 6$

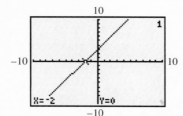

54. The table of ordered pairs was generated by a graphing calculator with a TABLE feature.

    (a) What is the x-intercept?
    (b) What is the y-intercept?
    (c) Which equation corresponds to this table of values?

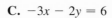

        A. $Y_1 = 2X - 3$ **B.** $Y_1 = -2X - 3$
        C. $Y_1 = 2X + 3$ **D.** $Y_1 = -2X + 3$

55. Refer to the equation in Exercise 52. A portion of its graph is shown on the accompanying screen, along with the coordinates of a point on the line displayed at the bottom. How is this point interpreted in the context of the model?

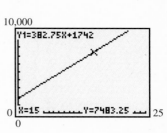

(continued)

56. The screens each show the graph of $x + y = 15$ (which was entered as $y = -x + 15$). However, different viewing windows are used. Which window would be more useful for this graph? Why?

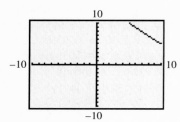 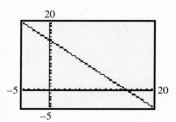

57. The screen shows the graph of $x + 2y = 0$ from Example 5. In what form should you enter the equation into the calculator?

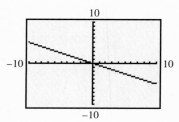

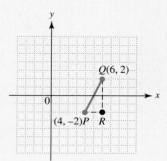 *Graph each equation using a graphing calculator. Use a standard viewing window. See Example 6.*

58. $4x - y = 10$ **59.** $5x + 2y = -10$ **60.** $3x + 4y = -6$

61. $3.6x - y = -5.8$ **62.** $y - 4.2 = 1.5x$

RELATING CONCEPTS (EXERCISES 63–68)

For Individual or Group Work

If the endpoints of a line segment are known, then the coordinates of the midpoint of the segment can be found. The figure shows the coordinates of the points P and Q. Let $\overline{PQ}$ represent the line segment with endpoints at P and Q. To derive a formula for the midpoint of $\overline{PQ}$, work Exercises 63–68 in order.

63. In the figure, R is the point with the same x-coordinate as Q and the same y-coordinate as P. Write the ordered pair that corresponds to R.

64. From the graph, determine the coordinates of the midpoint of $\overline{PR}$.

65. From the graph, determine the coordinates of the midpoint of $\overline{QR}$.

66. The x-coordinate of the midpoint M of $\overline{PQ}$ is the x-coordinate of the midpoint of $\overline{PR}$ and the y-coordinate is the y-coordinate of the midpoint of $\overline{QR}$. Write the ordered pair that corresponds to M.

67. The average of two numbers is found by dividing their sum by 2. Find the average of the x-coordinates of points P and Q. Find the average of the y-coordinates of points P and Q.

68. Comparing your answers to Exercises 66 and 67, what connection is there between the coordinates of P and Q and the coordinates of M?

The result of the preceding Relating Concepts exercises leads to the **midpoint formula.**

Midpoint Formula

If the endpoints of a line segment PQ are (x_1, y_1) and (x_2, y_2), then its midpoint M is

$$\left(\frac{x_1 + x_2}{2}, \frac{y_1 + y_2}{2} \right).$$

For example, the midpoint of the segment with endpoints $(4, -3)$ and $(6, -1)$ is

$$\left(\frac{4 + 6}{2}, \frac{-3 + (-1)}{2} \right) = \left(\frac{10}{2}, \frac{-4}{2} \right) = (5, -2).$$

Use the midpoint formula to find the midpoint of each segment with the given endpoints.

69. $(-8, 4)$ and $(-2, -6)$ **70.** $(5, 2)$ and $(-1, 8)$ **71.** $(3, -6)$ and $(6, 3)$

72. $(-10, 4)$ and $(7, 1)$ **73.** $(-9, 3)$ and $(9, 8)$ **74.** $(4, -3)$ and $(-1, 3)$

75. $(2.5, 3.1)$ and $(1.7, -1.3)$ **76.** $(6.2, 5.8)$ and $(1.4, -.6)$

3.2   The Slope of a Line

OBJECTIVES

1 Find the slope of a line given two points on the line.

2 Find the slope of a line given an equation of the line.

3 Graph a line given its slope and a point on the line.

4 Use slopes to determine whether two lines are parallel, perpendicular, or neither.

5 Solve problems involving average rate of change.

Slope (steepness) is used in many practical ways. The slope of a highway (sometimes called the *grade*) is often given as a percent. For example, a 10% $\left(\text{or } \frac{10}{100} = \frac{1}{10}\right)$ slope means the highway rises 1 unit for every 10 horizontal units. Stairs and roofs have slopes too, as shown in Figure 12.

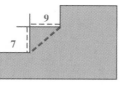   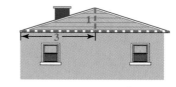

Slope is $\frac{7}{9}$.   Slope (or pitch) is $\frac{1}{3}$.

FIGURE 12

In each example mentioned, slope is the ratio of vertical change, or **rise,** to horizontal change, or **run.** A simple way to remember this is to think "slope is rise over run."

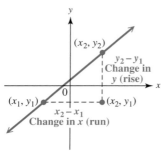

FIGURE 13

OBJECTIVE 1 Find the slope of a line given two points on the line.   To get a formal definition of the slope of a line, we designate two different points on the line. To differentiate between the points, we write them as (x_1, y_1) and (x_2, y_2). See Figure 13. (The small numbers 1 and 2 in these ordered pairs are called *subscripts*. Read (x_1, y_1) as "*x*-sub-one, *y*-sub-one.")

As we move along the line in Figure 13 from (x_1, y_1) to (x_2, y_2), the *y*-value changes (vertically) from y_1 to y_2, an amount equal to $y_2 - y_1$. As *y* changes from y_1 to y_2, the value of *x* changes (horizontally) from x_1 to x_2 by the amount $x_2 - x_1$. The ratio of the change in *y* to the change in *x* (the rise over the run) is called the *slope* of the line, with the letter *m* traditionally used for slope.

Slope Formula

The **slope** of the line through the distinct points (x_1, y_1) and (x_2, y_2) is

$$m = \frac{\text{rise}}{\text{run}} = \frac{\text{change in } y}{\text{change in } x} = \frac{y_2 - y_1}{x_2 - x_1} \quad (x_1 \neq x_2).$$

▨ **EXAMPLE 1   Finding the Slope of a Line**

Find the slope of the line through the points $(2, -1)$ and $(-5, 3)$.

If $(2, -1) = (x_1, y_1)$ and $(-5, 3) = (x_2, y_2)$, then

$$m = \frac{y_2 - y_1}{x_2 - x_1} = \frac{3 - (-1)}{-5 - 2} = \frac{4}{-7} = -\frac{4}{7}.$$

See Figure 14. If the pairs are reversed so that $(2, -1) = (x_2, y_2)$ and $(-5, 3) = (x_1, y_1)$, the slope is the same.

$$m = \frac{-1 - 3}{2 - (-5)} = \frac{-4}{7} = -\frac{4}{7}$$

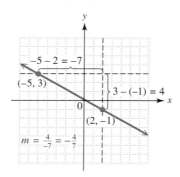

FIGURE 14

Now Try Exercise 19.

Example 1 suggests that the slope is the same no matter which point we consider first. Also, using similar triangles from geometry, we can show that the slope is the same no matter which two different points on the line we choose.

CAUTION   In calculating slope, be careful to subtract the y-values and the x-values in the *same order*.

Correct   **Incorrect**

$$\frac{y_2 - y_1}{x_2 - x_1} \quad \text{or} \quad \frac{y_1 - y_2}{x_1 - x_2} \qquad\qquad \frac{y_2 - y_1}{x_1 - x_2} \;\;\text{or}\;\; \frac{y_1 - y_2}{x_2 - x_1}$$

Also, remember that the change in y is the *numerator* and the change in x is the *denominator*.

OBJECTIVE 2   Find the slope of a line given an equation of the line.   When an equation of a line is given, one way to find the slope is to use the definition of slope by first finding two different points on the line.

EXAMPLE 2   Finding the Slope of a Line

Find the slope of the line $4x - y = -8$.

The intercepts can be used as the two different points needed to find the slope. Let $y = 0$ to find that the x-intercept is $(-2, 0)$. Then let $x = 0$ to find that the y-intercept is $(0, 8)$. Use these two points in the slope formula. The slope is

$$m = \frac{\text{rise}}{\text{run}} = \frac{8 - 0}{0 - (-2)} = \frac{8}{2} = 4.$$

Now Try Exercise 29.

EXAMPLE 3   Finding Slopes of Horizontal and Vertical Lines

Find the slope of each line.

(a) $y = 2$

Figure 6 in Section 3.1 shows that the graph of $y = 2$ is a horizontal line. To find the slope, select two different points on the line, such as $(3, 2)$ and $(-1, 2)$, and use the slope formula.

$$m = \frac{\text{rise}}{\text{run}} = \frac{2 - 2}{3 - (-1)} = \frac{0}{4} = 0$$

In this case, the *rise* is 0, so the slope is 0.

(b) $x = -1$

As shown in Figure 7 (Section 3.1), the graph of $x = -1$ or $x + 1 = 0$ is a vertical line. Two points that satisfy the equation $x = -1$ are $(-1, 5)$ and $(-1, -4)$. If we use these two points to try to find the slope, we obtain

$$m = \frac{\text{rise}}{\text{run}} = \frac{-4 - 5}{-1 - (-1)} = \frac{-9}{0}.$$

Since division by 0 is undefined, the slope is undefined. This is why the definition of slope includes the restriction $x_1 \neq x_2$.

Now Try Exercises 35 and 37.

Generalizing from Example 3, we can make the following statements about slopes of horizontal and vertical lines.

Slopes of Horizontal and Vertical Lines

The slope of a horizontal line is 0; the slope of a vertical line is undefined.

The slope of a line can also be found directly from its equation. Look again at the equation $4x - y = -8$ from Example 2. Solve this equation for y.

$$4x - y = -8 \qquad \text{Equation from Example 2}$$
$$-y = -4x - 8 \qquad \text{Subtract } 4x.$$
$$y = 4x + 8 \qquad \text{Multiply by } -1.$$

Notice that the slope, 4, found using the slope formula in Example 2 is the same number as the coefficient of x in the equation $y = 4x + 8$. We will see in the next section that this always happens, *as long as the equation is solved for y.*

■ EXAMPLE 4 Finding the Slope from an Equation

Find the slope of the graph of $3x - 5y = 8$.

Solve the equation for y.

$$3x - 5y = 8$$
$$-5y = -3x + 8 \qquad \text{Subtract } 3x.$$
$$y = \frac{3}{5}x - \frac{8}{5} \qquad \text{Divide by } -5.$$

The slope is given by the coefficient of x, so the slope is $\frac{3}{5}$.

■

Now Try Exercise 31.

OBJECTIVE 3 Graph a line given its slope and a point on the line. Example 5 shows how to graph a straight line by using the slope and one point on the line.

■ EXAMPLE 5 Using the Slope and a Point to Graph Lines

Graph each line.

(a) With slope $\frac{2}{3}$ passing through the point $(-1, 4)$

First locate the point $P(-1, 4)$ on a graph as shown in Figure 15. Then use the slope to find a second point. From the slope formula,

$$m = \frac{\text{change in } y}{\text{change in } x} = \frac{2}{3},$$

so move *up* 2 units and then 3 units to the *right* to locate another point on the graph (labeled R). The line through $(-1, 4)$ and R is the required graph.

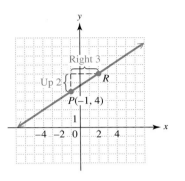

FIGURE 15

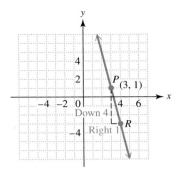

FIGURE 16

(b) Through $(3, 1)$ with slope -4

Start by locating the point $P(3, 1)$ on a graph. Find a second point R on the line by writing the slope -4 as $\frac{-4}{1}$ and using the slope formula.

$$m = \frac{\text{change in } y}{\text{change in } x} = \frac{-4}{1}$$

Move *down* 4 units from $(3, 1)$, and then move 1 unit to the *right*. Draw a line through this second point R and $(3, 1)$, as shown in Figure 16.

The slope also could be written as

$$m = \frac{\text{change in } y}{\text{change in } x} = \frac{4}{-1}.$$

In this case the second point R is located *up* 4 units and 1 unit to the *left*. Verify that this approach also produces the line in Figure 16.

Now Try Exercises 41 and 43.

In Example 5(a), the slope of the line is the *positive* number $\frac{2}{3}$. The graph of the line in Figure 15 goes up (rises) from left to right. The line in Example 5(b) has *negative* slope, -4. As Figure 16 shows, its graph goes down (falls) from left to right. These facts suggest the following generalization.

A positive slope indicates that the line goes *up* from left to right;
a negative slope indicates that the line goes *down* from left to right.

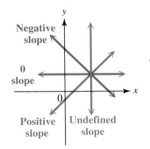

FIGURE 17

Figure 17 shows lines of positive, 0, negative, and undefined slopes.

OBJECTIVE 4 Use slopes to determine whether two lines are parallel, perpendicular, or neither. The slopes of a pair of parallel or perpendicular lines are related in a special way. Recall that the slope of a line measures the steepness of the line. Since parallel lines have equal steepness, their slopes must be equal; also, lines with the same slope are parallel.

Slopes of Parallel Lines

Two nonvertical lines with the same slope are parallel;
two nonvertical parallel lines have the same slope.

EXAMPLE 6   **Determining Whether Two Lines Are Parallel**

Are the lines L_1, through $(-2, 1)$ and $(4, 5)$, and L_2, through $(3, 0)$ and $(0, -2)$, parallel?

The slope of L_1 is

$$m_1 = \frac{5 - 1}{4 - (-2)} = \frac{4}{6} = \frac{2}{3}.$$

The slope of L_2 is

$$m_2 = \frac{-2 - 0}{0 - 3} = \frac{-2}{-3} = \frac{2}{3}.$$

Because the slopes are equal, the two lines are parallel.

Now Try Exercise 49.

To see how the slopes of perpendicular lines are related, consider a nonvertical line with slope $\frac{a}{b}$. If this line is rotated $90°$, the vertical change and the horizontal change are reversed and the slope is $-\frac{b}{a}$, since the horizontal change is now negative. See Figure 18. Thus, the slopes of perpendicular lines have product -1 and are negative reciprocals of each other. For example, if the slopes of two lines are $\frac{3}{4}$ and $-\frac{4}{3}$, then the lines are perpendicular because $\frac{3}{4}\left(-\frac{4}{3}\right) = -1$.

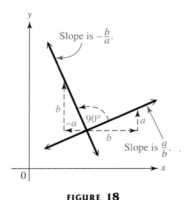

FIGURE 18

Slopes of Perpendicular Lines

If neither is vertical, perpendicular lines have slopes that are negative reciprocals; that is, their product is -1. Also, lines with slopes that are negative reciprocals are perpendicular.

EXAMPLE 7   **Determining Whether Two Lines Are Perpendicular**

Are the lines with equations $2y = 3x - 6$ and $2x + 3y = -6$ perpendicular?

Find the slope of each line by first solving each equation for y.

$$2y = 3x - 6 \qquad\qquad 2x + 3y = -6$$

$$y = \frac{3}{2}x - 3 \qquad\qquad 3y = -2x - 6$$

↑
Slope

$$y = -\frac{2}{3}x - 2$$

↑
Slope

Since the product of the slopes of the two lines is $\frac{3}{2}\left(-\frac{2}{3}\right) = -1$, the lines are perpendicular.

Now Try Exercise 51.

NOTE In Example 7, alternatively, we could have found the slope of each line by using intercepts and the slope formula.

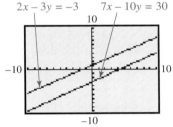

$2x - 3y = -3$ $7x - 10y = 30$

The graphs are *not* parallel, though they may appear to be.

FIGURE 19

We must be careful when interpreting calculator graphs of parallel and perpendicular lines. For example, the graphs of the equations in Figure 19 appear to be parallel. However, checking their slopes algebraically, we find that

$$2x - 3y = -3 \qquad\qquad 7x - 10y = 30$$

$$-3y = -2x - 3 \qquad\qquad -10y = -7x + 30$$

$$y = \frac{2}{3}x + 1 \qquad\qquad y = \frac{7}{10}x - 3.$$

Since the slopes $\frac{2}{3}$ and $\frac{7}{10}$ are not equal, the lines are *not* parallel.

Figure 20(a) shows graphs of the perpendicular lines from Example 7. As graphed in the standard viewing window, the lines do not appear to be perpendicular. However, if we use a *square viewing window* as in Figure 20(b), we get a more realistic view. (Many graphing calculators can set a square window automatically. See your owner's manual.) These two cases indicate that we cannot rely completely on what we see on a calculator screen—we must understand the mathematical concepts as well.

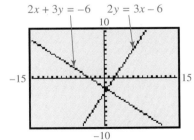

$2x + 3y = -6$ $2y = 3x - 6$

In the standard window, the lines *do not* appear to be perpendicular.

(a)

$2x + 3y = -6$ $2y = 3x - 6$

In the square window, the lines *do* appear to be perpendicular.

(b)

FIGURE 20

OBJECTIVE 5 Solve problems involving average rate of change. We know that the slope of a line is the ratio of the vertical change in *y* to the horizontal change in *x*. Thus, slope gives the *average rate of change* in *y* per unit of change in *x*, where the value of *y* depends on the value of *x*. The next examples illustrate this idea. We assume a linear relationship between *x* and *y*.

EXAMPLE 8 Interpreting Slope as Average Rate of Change

The graph in Figure 21 approximates the percent of U.S. households owning multiple personal computers in the years 1997 through 2001. Find the average rate of change in percent per year.

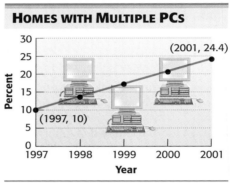

HOMES WITH MULTIPLE PCs

(2001, 24.4)

(1997, 10)

Year

Source: The Yankee Group.

FIGURE 21

To determine the average rate of change, we need two pairs of data. From the graph, if *x* = 1997, then *y* = 10 and if *x* = 2001, then *y* = 24.4, so we have the ordered pairs (1997, 10) and (2001, 24.4). By the slope formula,

$$\text{average rate of change} = \frac{\text{change in } y}{\text{change in } x} = \frac{24.4 - 10}{2001 - 1997} = \frac{14.4}{4} = 3.6.$$

This means that the number of U.S. households owning multiple computers *increased* by 3.6% each year from 1997 to 2001.

Now Try Exercise 69.

EXAMPLE 9 Interpreting Slope as Average Rate of Change

In 1997, sales of VCRs numbered 16.7 million. In 2002, estimated sales of VCRs were 13.3 million. Find the average rate of change, in millions, per year. (*Source: The Gazette,* June 22, 2002.)

To use the slope formula, we need two ordered pairs. Here, if *x* = 1997, then *y* = 16.7 and if *x* = 2002, then *y* = 13.3, which gives the ordered pairs (1997, 16.7) and (2002, 13.3). (Note that *y* is in millions.)

$$\text{average rate of change} = \frac{13.3 - 16.7}{2002 - 1997} = \frac{-3.4}{5} = -.68$$

Sales of VCRs

(1997, 16.7)

(2002, 13.3)

Year

FIGURE 22

The graph in Figure 22 confirms that the line through the ordered pairs falls from left to right and therefore has negative slope. Thus, sales of VCRs *decreased* by .68 million each year from 1997 to 2002.

Now Try Exercise 71.

3.2 EXERCISES

1. A ski slope drops 30 ft for every horizontal 100 ft. Which of the following express its slope? (There are several correct choices.)

    A. $-.3$ **B.** $-\dfrac{3}{10}$ **C.** $-3\dfrac{1}{3}$

    D. $-\dfrac{30}{100}$ **E.** $-\dfrac{10}{3}$

2. A hill has slope $-.05$. How many feet in the vertical direction correspond to a run of 50 ft?

3. Match each situation in (a)–(d) with the most appropriate graph in A–D.

    (a) Sales rose sharply during the first quarter, leveled off during the second quarter, and then rose slowly for the rest of the year.

    (b) Sales fell sharply during the first quarter, and then rose slowly during the second and third quarters before leveling off for the rest of the year.

    (c) Sales rose sharply during the first quarter, and then fell to the original level during the second quarter before rising steadily for the rest of the year.

    (d) Sales fell during the first two quarters of the year, leveled off during the third quarter, and rose during the fourth quarter.

    A.
    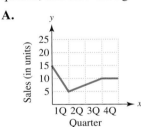

    B.
    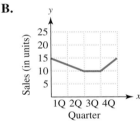

    C.
    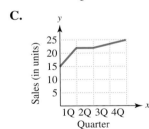

    D.
    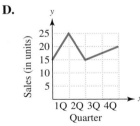

Determine the slope of each line segment in the given figure.

    4. *AB*   **5.** *BC*   **6.** *CD*

    7. *DE*   **8.** *EF*

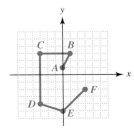

Calculate the value of each slope m using the slope formula. See Example 1.

9. $m = \dfrac{6 - 2}{5 - 3}$

10. $m = \dfrac{5 - 7}{-4 - 2}$

11. $m = \dfrac{4 - (-1)}{-3 - (-5)}$

12. $m = \dfrac{-6 - 0}{0 - (-3)}$

13. $m = \dfrac{-5 - (-5)}{3 - 2}$

14. $m = \dfrac{7 - (-2)}{-3 - (-3)}$

 15. Which of the following forms of the slope formula are correct? Explain.

A. $\dfrac{y_1 - y_2}{x_2 - x_1}$ **B.** $\dfrac{y_1 - y_2}{x_1 - x_2}$ **C.** $\dfrac{x_2 - x_1}{y_2 - y_1}$ **D.** $\dfrac{y_2 - y_1}{x_2 - x_1}$

Find the slope of the line through each pair of points. See Example 1.

16. $(-2, -3)$ and $(-1, 5)$

17. $(-4, 3)$ and $(-3, 4)$

18. $(-4, 1)$ and $(2, 6)$

19. $(-3, -3)$ and $(5, 6)$

20. $(2, 4)$ and $(-4, 4)$

21. $(-6, 3)$ and $(2, 3)$

Find the slope of each line.

22.

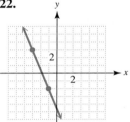

23.

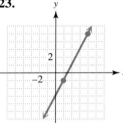

24.

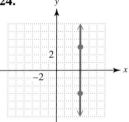

Based on the figure shown here, determine which line satisfies the given description.

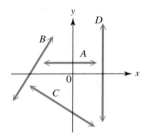

25. The line has positive slope.

26. The line has negative slope.

27. The line has slope 0.

28. The line has undefined slope.

Find the slope of the line and sketch the graph. See Examples 1–4.

29. $x + 2y = 4$

30. $x + 3y = -6$

31. $5x - 2y = 10$

32. $4x - y = 4$

33. $y = 4x$

34. $y = -3x$

35. $x - 3 = 0$

36. $y + 5 = 0$

37. $y = -4$

Graph the line described. See Example 5.

38. Through $(-4, 2)$; $m = \dfrac{1}{2}$

39. Through $(-2, -3)$; $m = \dfrac{5}{4}$

40. Through $(0, -2)$; $m = -\dfrac{2}{3}$

41. Through $(0, -4)$; $m = -\dfrac{3}{2}$

42. Through $(-1, -2)$; $m = 3$

43. Through $(-2, -4)$; $m = 4$

44. $m = 0$; through $(2, -5)$

45. Undefined slope; through $(-3, 1)$

46. Undefined slope; through $(-4, 1)$ **47.** $m = 0$; through $(5, 3)$

48. If a line has slope $-\frac{4}{9}$, then any line parallel to it has slope _____, and any line perpendicular to it has slope _____.

Decide whether each pair of lines is parallel, perpendicular, *or* neither. *See Examples 6 and 7.*

49. The line through $(4, 6)$ and $(-8, 7)$ and the line through $(-5, 5)$ and $(7, 4)$

50. The line through $(15, 9)$ and $(12, -7)$ and the line through $(8, -4)$ and $(5, -20)$

51. $2x + 5y = -7$ and $5x - 2y = 1$ **52.** $x + 4y = 7$ and $4x - y = 3$

53. $2x + y = 6$ and $x - y = 4$ **54.** $4x - 3y = 6$ and $3x - 4y = 2$

55. $3x = y$ and $2y - 6x = 5$ **56.** $x = 6$ and $6 - x = 8$

57. $2x + 5y = -8$ and $6 + 2x = 5y$ **58.** $4x + y = 0$ and $5x - 8 = 2y$

59. $4x - 3y = 8$ and $4y + 3x = 12$ **60.** $2x = y + 3$ and $2y + x = 3$

Solve each problem.

61. The upper deck at Comiskey Park in Chicago has produced, among other complaints, displeasure with its steepness. It is 160 ft from home plate to the front of the upper deck and 250 ft from home plate to the back. The top of the upper deck is 63 ft above the bottom. What is its slope? (Consider the slope as a positive number here.)

62. When designing the FleetCenter arena in Boston, architects designed the ramps leading up to the entrances so that circus elephants would be able to walk up the ramps. The maximum grade (or slope) that an elephant will walk on is 13%. Suppose that such a ramp was constructed with a horizontal run of 150 ft. What would be the maximum vertical rise the architects could use?

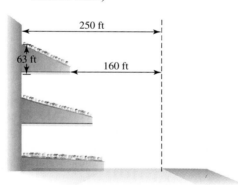

Find and interpret the average rate of change illustrated in each graph.

63.

64.

65.

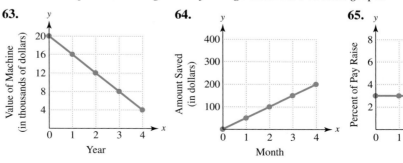

66. If the graph of a linear equation rises from left to right, then the average rate of change
is _____. If the graph of a linear equation falls from left to right, then
  (positive/negative)
the average rate of change is _____.
                          (positive/negative)

Solve each problem. See Examples 8 and 9.

67. The table gives book publishers'
approximate net dollar sales (in
millions) from 1995 through 2000.

Book Publishers' Sales

Year	Sales (in millions)
1995	19,000
1996	20,000
1997	21,000
1998	22,000
1999	23,000
2000	24,000

Source: Book Industry Study
Group.

(a) Find the average rate of change for
1995–1996, 1995–1999, and
1998–2000.

(b) What do you notice about your
answers in part (a)? What does this
tell you?

68. The table gives the number of cellular
telephone subscribers (in thousands)
from 1994 through 1999.

Cellular Telephone Subscribers

Year	Subscribers (in thousands)
1994	24,134
1995	33,786
1996	44,043
1997	55,312
1998	69,209
1999	86,047

Source: Cellular Telecommunications
Industry Association, Washington, D.C.,
State of the Cellular Industry (Annual).

(a) Find the average rate of change in
subscribers for 1994–1995,
1995–1996, and so on.

(b) Is the average rate of change in
successive years approximately the
same? If the ordered pairs in the
table were plotted, could an
approximately straight line be
drawn through them?

69. Merck pharmaceutical company research and development expenditures
(in millions of dollars) in recent
years are closely approximated by
the graph.

(a) Use the given ordered pairs to
determine the average rate of
change in these expenditures per
year.

(b) Explain how a positive rate of
change is interpreted in this
situation.

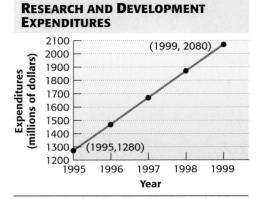

**RESEARCH AND DEVELOPMENT
EXPENDITURES**

Source: Merck & Co., Inc. 1999 Annual Report.

70. The graph provides a good
approximation of the number of
food stamp recipients (in millions)
from 1994 through 1998.

   (a) Use the given ordered pairs to
find the average rate of change
in food stamp recipients per
year during this period.

   (b) Interpret what a negative slope
means in this situation.

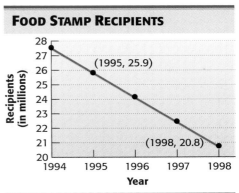

FOOD STAMP RECIPIENTS

Source: U.S. Bureau of the Census.

71. When introduced in 1997, a DVD player sold for about $500. In 2002, the average
price was $155. Find and interpret the average rate of change in price per year.
(*Source: The Gazette,* June 22, 2002.)

72. In 1997 when DVD players entered the market, .349 million (that is, 349,000) were
sold. In 2002, sales of DVD players reached 15.5 million (estimated). Find and interpret
the average rate of change in sales, in millions, per year. Round your answer to the
nearest hundredth. (*Source: The Gazette,* June 22, 2002.)

TECHNOLOGY INSIGHTS (EXERCISES 73 AND 74)

73. The graphing calculator screen
shows two lines. One is the graph
of $y_1 = -2x + 3$ and the other is
the graph of $y_2 = 3x - 4$. Which
is which?

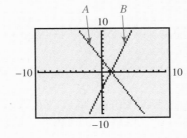

74. The graphing calculator screen
shows two lines. One is the graph
of $y_1 = 2x - 5$ and the other is
the graph of $y_2 = 4x - 5$. Which
is which?

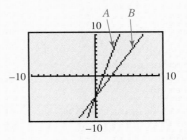

Solve each problem, using your knowledge of the slopes of parallel and perpendicular lines.

75. Show that $(-13, -9)$, $(-11, -1)$, $(2, -2)$, and $(4, 6)$ are the vertices of a parallelogram.
(*Hint:* A parallelogram is a four-sided figure with opposite sides parallel.)

76. Is the figure with vertices at $(-11, -5)$, $(-2, -19)$, $(12, -10)$, and $(3, 4)$ a parallelogram?
Is it a rectangle? (*Hint:* A rectangle is a parallelogram with a right angle.)

RELATING CONCEPTS (EXERCISES 77–82)

For Individual or Group Work

Three points that lie on the same straight line are said to be **collinear.** *Consider the points A(3, 1), B(6, 2), and C(9, 3).* **Work Exercises 77–82 in order.**

77. Find the slope of segment AB.

78. Find the slope of segment BC.

79. Find the slope of segment AC.

80. If slope of segment AB = slope of segment BC = slope of segment AC, then A, B, and C are collinear. Use the results of Exercises 77–79 to show that this statement is satisfied.

81. Use the slope formula to determine whether the points $(1, -2)$, $(3, -1)$, and $(5, 0)$ are collinear.

82. Repeat Exercise 81 for the points $(0, 6)$, $(4, -5)$, and $(-2, 12)$.

3.3 Linear Equations in Two Variables

OBJECTIVES

1 Write an equation of a line given its slope and y-intercept.

2 Graph a line using its slope and y-intercept.

3 Write an equation of a line given its slope and a point on the line.

4 Write an equation of a line given two points on the line.

5 Write an equation of a line parallel or perpendicular to a given line.

6 Write an equation of a line that models real data.

7 Use a graphing calculator to solve linear equations in one variable.

OBJECTIVE 1 Write an equation of a line given its slope and y-intercept. In the previous section we found the slope of a line from the equation of the line by solving the equation for y. For example, we found that the slope of the line with equation $y = 4x + 8$ is 4, the coefficient of x. What does the number 8 represent?

To find out, suppose a line has slope m and y-intercept $(0, b)$. We can find an equation of this line by choosing another point (x, y) on the line, as shown in Figure 23. Using the slope formula,

$$m = \frac{y - b}{x - 0}$$

$$m = \frac{y - b}{x}$$

$$mx = y - b \qquad \text{Multiply by } x.$$

$$mx + b = y \qquad \text{Add } b.$$

$$y = mx + b. \qquad \text{Rewrite.}$$

FIGURE 23

This last equation is called the *slope-intercept form* of the equation of a line, because we can identify the slope and y-intercept at a glance. Thus, in the line with equation $y = 4x + 8$, the number 8 indicates that the y-intercept is $(0, 8)$.

Slope-Intercept Form

The **slope-intercept form** of the equation of a line with slope m and y-intercept $(0, b)$ is

$$y = mx + b.$$

Slope y-intercept is $(0, b)$.

EXAMPLE 1   Using the Slope-Intercept Form to Find an Equation of a Line

Find an equation of the line with slope $-\frac{4}{5}$ and y-intercept $(0, -2)$.

Here $m = -\frac{4}{5}$ and $b = -2$. Substitute these values into the slope-intercept form.

$$y = mx + b \qquad \text{Slope-intercept form}$$

$$y = -\frac{4}{5}x - 2 \qquad m = -\frac{4}{5};\ b = -2$$

Now Try Exercise 19.

OBJECTIVE 2   Graph a line using its slope and y-intercept.   If the equation of a line is written in slope-intercept form, we can use the slope and y-intercept to obtain its graph.

EXAMPLE 2   Graphing Lines Using Slope and y-Intercept

Graph each line using the slope and y-intercept.

(a) $y = 3x - 6$

Here $m = 3$ and $b = -6$. Plot the y-intercept $(0, -6)$. The slope 3 can be interpreted as

$$m = \frac{\text{rise}}{\text{run}} = \frac{\text{change in } y}{\text{change in } x} = \frac{3}{1}.$$

From $(0, -6)$, move *up* 3 units and to the *right* 1 unit, and plot a second point at $(1, -3)$. Join the two points with a straight line to obtain the graph in Figure 24.

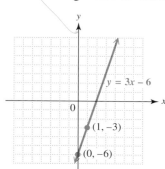

FIGURE 24

(b) $3y + 2x = 9$

Write the equation in slope-intercept form by solving for y.

$$3y + 2x = 9$$

$$3y = -2x + 9 \qquad \text{Subtract } 2x.$$

$$y = -\frac{2}{3}x + 3 \qquad \text{Slope-intercept form}$$

Slope ⟶   ⟵ y-intercept is $(0, 3)$.

To graph this equation, plot the y-intercept $(0, 3)$. The slope can be interpreted as either $\frac{-2}{3}$ or $\frac{2}{-3}$. Using $\frac{-2}{3}$, move from $(0, 3)$ *down* 2 units and to the *right* 3 units to locate the point $(3, 1)$. The line through these two points is the required graph. See Figure 25. $\left(\text{Verify that the point obtained using } \frac{2}{-3} \text{ as the slope is also on this line.}\right)$

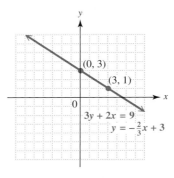

FIGURE 25

Now Try Exercise 25.

NOTE The slope-intercept form of a linear equation is the most useful for several reasons. Every linear equation (of a nonvertical line) has a *unique* (one and only one) slope-intercept form. In Section 3.5 we study *linear functions*, which are defined using slope-intercept form. Also, this is the form we use when graphing a line with a graphing calculator. (See Section 3.1, Example 6.)

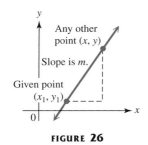

FIGURE 26

OBJECTIVE 3 Write an equation of a line given its slope and a point on the line. Let m represent the slope of a line and (x_1, y_1) represent a given point on the line. Let (x, y) represent any other point on the line. See Figure 26. Then by the slope formula,

$$m = \frac{y - y_1}{x - x_1}$$

$$m(x - x_1) = y - y_1 \qquad \text{Multiply each side by } x - x_1.$$

$$y - y_1 = m(x - x_1). \qquad \text{Rewrite.}$$

This last equation is the *point-slope form* of the equation of a line.

Point-Slope Form

The **point-slope form** of the equation of a line with slope m passing through the point (x_1, y_1) is

$$\overset{\text{Slope}}{\underset{\text{Given point}}{y - y_1 = m(x - x_1).}}$$

To use this form to write the equation of a line, we need to know the coordinates of a point (x_1, y_1) and the slope m of the line.

EXAMPLE 3 Using the Point-Slope Form

Find an equation of the line with slope $\frac{1}{3}$ passing through the point $(-2, 5)$.

Use the point-slope form of the equation of a line, with $(x_1, y_1) = (-2, 5)$ and $m = \frac{1}{3}$.

$$y - y_1 = m(x - x_1) \qquad \text{Point-slope form}$$

$$y - 5 = \frac{1}{3}[x - (-2)] \qquad y_1 = 5,\, m = \tfrac{1}{3},\, x_1 = -2$$

$$y - 5 = \frac{1}{3}(x + 2)$$

$$3y - 15 = x + 2 \qquad \text{Multiply by 3.}$$

$$-x + 3y = 17 \qquad \text{Subtract } x;\text{ add 15.}$$

In Section 3.1, we defined *standard form* for a linear equation as

$$Ax + By = C,$$

where A, B, and C are real numbers. Most often, however, A, B, and C are integers. In this case, let us agree that integers A, B, and C have no common factor (except 1) and $A \geq 0$. For example, the final equation in Example 3, $-x + 3y = 17$, is written in standard form as $x - 3y = -17$.

NOTE The definition of "standard form" is not standard from one text to another. Any linear equation can be written in many different (all equally correct) forms. For example, the equation $2x + 3y = 8$ can be written as

$$2x = 8 - 3y, \qquad 3y = 8 - 2x, \qquad x + \frac{3}{2}y = 4, \qquad 4x + 6y = 16,$$

and so on. In addition to writing it in the form $Ax + By = C$ with $A \geq 0$, let us agree that the form $2x + 3y = 8$ is preferred over any multiples of each side, such as $4x + 6y = 16$. (To write $4x + 6y = 16$ in standard form, divide each side by 2.)

Now Try Exercise 31.

OBJECTIVE 4 Write an equation of a line given two points on the line. To find an equation of a line when two points on the line are known, first use the slope formula to find the slope of the line. Then use the slope with either of the given points and the point-slope form of the equation of a line.

EXAMPLE 4 Finding an Equation of a Line Given Two Points

Find an equation of the line passing through the points $(-4, 3)$ and $(5, -7)$. Write the equation in standard form.

First find the slope by using the slope formula.

$$m = \frac{-7 - 3}{5 - (-4)} = -\frac{10}{9}$$

Use either $(-4, 3)$ or $(5, -7)$ as (x_1, y_1) in the point-slope form of the equation of a line. If you choose $(-4, 3)$, then $-4 = x_1$ and $3 = y_1$.

$$y - y_1 = m(x - x_1)$$ Point-slope form

$$y - 3 = -\frac{10}{9}[x - (-4)]$$ $y_1 = 3, m = -\frac{10}{9}, x_1 = -4$

$$y - 3 = -\frac{10}{9}(x + 4)$$

$$9y - 27 = -10x - 40$$ Multiply by 9; distributive property.

$$10x + 9y = -13$$ Standard form

Verify that if $(5, -7)$ were used, the same equation would result.

Now Try Exercise 49.

A horizontal line has slope 0. Using point-slope form, the equation of a horizontal line through the point (a, b) is

$$y - y_1 = m(x - x_1)$$

$$y - b = 0(x - a)$$ $y_1 = b, m = 0, x_1 = a$

$$y - b = 0$$

$$y = b.$$

Notice that point-slope form does not apply to a vertical line, since the slope of a vertical line is undefined. A vertical line through the point (a, b) has equation $x = a$.

In summary, horizontal and vertical lines have the following special equations.

Equations of Horizontal and Vertical Lines

The horizontal line through the point (a, b) has equation $y = b.$
The vertical line through the point (a, b) has equation $x = a.$

Now Try Exercises 41 and 43.

OBJECTIVE 5 **Write an equation of a line parallel or perpendicular to a given line.** As mentioned in the previous section, parallel lines have the same slope and perpendicular lines have slopes that are negative reciprocals of each other.

EXAMPLE 5 **Finding Equations of Parallel or Perpendicular Lines**

Find an equation of the line passing through the point $(-4, 5)$ and **(a)** parallel to the line $2x + 3y = 6$; **(b)** perpendicular to the line $2x + 3y = 6$. Write each equation in slope-intercept form.

(a) We find the slope of the line $2x + 3y = 6$ by solving for y.

$$2x + 3y = 6$$

$$3y = -2x + 6$$ Subtract $2x$.

$$y = -\frac{2}{3}x + 2$$ Divide by 3.

$$\underset{\text{Slope}}{\uparrow}$$

The slope is given by the coefficient of x, so $m = -\frac{2}{3}$.
See the figure.

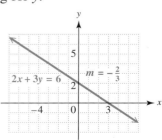

The required equation of the line through $(-4, 5)$ and parallel to $2x + 3y = 6$ must also have slope $-\frac{2}{3}$. To find this equation, we use the point-slope form, with $(x_1, y_1) = (-4, 5)$ and $m = -\frac{2}{3}$.

$$y - 5 = -\frac{2}{3}[x - (-4)] \qquad y_1 = 5,\ m = -\tfrac{2}{3},\ x_1 = -4$$

$$y - 5 = -\frac{2}{3}(x + 4)$$

$$y - 5 = -\frac{2}{3}x - \frac{8}{3} \qquad \text{Distributive property}$$

$$y = -\frac{2}{3}x - \frac{8}{3} + \frac{15}{3} \qquad \text{Add } 5 = \tfrac{15}{3}.$$

$$y = -\frac{2}{3}x + \frac{7}{3} \qquad \text{Combine like terms.}$$

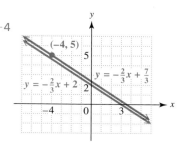

We did not clear fractions after the substitution step here because we want the equation in slope-intercept form—that is, solved for y. Both lines are shown in the figure.

(b) To be perpendicular to the line $2x + 3y = 6$, a line must have a slope that is the negative reciprocal of $-\frac{2}{3}$, which is $\frac{3}{2}$. We use $(-4, 5)$ and slope $\frac{3}{2}$ in the point-slope form to get the equation of the perpendicular line shown in the figure.

$$y - 5 = \frac{3}{2}[x - (-4)] \qquad y_1 = 5,\ m = \tfrac{3}{2},\ x_1 = -4$$

$$y - 5 = \frac{3}{2}(x + 4)$$

$$y - 5 = \frac{3}{2}x + 6 \qquad \text{Distributive property}$$

$$y = \frac{3}{2}x + 11 \qquad \text{Add 5.}$$

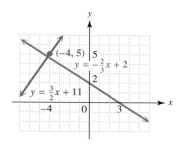

Now Try Exercises 61 and 65.

A summary of the various forms of linear equations follows.

Forms of Linear Equations

Equation	Description	When to Use
$y = mx + b$	**Slope-Intercept Form** Slope is m. y-intercept is $(0, b)$.	The slope and y-intercept can be easily identified and used to quickly graph the equation.
$y - y_1 = m(x - x_1)$	**Point-Slope Form** Slope is m. Line passes through (x_1, y_1).	This form is ideal for finding the equation of a line if the slope and a point on the line or two points on the line are known.

(continued)

Equation	Description	When to Use
Ax + By = C	**Standard Form** (A, B, and C integers, A ≥ 0) Slope is $-\frac{A}{B}$ $(B \neq 0)$. x-intercept is $\left(\frac{C}{A}, 0\right)$ $(A \neq 0)$. y-intercept is $\left(0, \frac{C}{B}\right)$ $(B \neq 0)$.	The x- and y-intercepts can be found quickly and used to graph the equation. Slope must be calculated.
y = b	**Horizontal Line** Slope is 0. y-intercept is (0, b).	If the graph intersects only the y-axis, then y is the only variable in the equation.
x = a	**Vertical Line** Slope is undefined. x-intercept is (a, 0).	If the graph intersects only the x-axis, then x is the only variable in the equation.

OBJECTIVE 6 Write an equation of a line that models real data. We can use the information presented in this section to write equations of lines that mathematically describe, or *model,* real data if the given set of data changes at a fairly constant rate. In this case, the data fit a linear pattern, and the rate of change is the slope of the line.

■ **EXAMPLE 6** Determining a Linear Equation to Describe Real Data

Suppose it is time to fill your car with gasoline. At your local station, 89-octane gas is selling for $1.60 per gal.

(a) Write an equation that describes the cost y to buy x gal of gas.

Experience has taught you that the total price you pay is determined by the number of gallons you buy multiplied by the price per gallon (in this case, $1.60). As you pump the gas, two sets of numbers spin by: the number of gallons pumped and the price for that number of gallons.

The table uses ordered pairs to illustrate this situation.

Number of Gallons Pumped	Price of This Number of Gallons
0	0($1.60) = $0.00
1	1($1.60) = $1.60
2	2($1.60) = $3.20
3	3($1.60) = $4.80
4	4($1.60) = $6.40

If we let x denote the number of gallons pumped, then the total price y in dollars can be found by the linear equation

Total price ⎤   ⎡ Number of gallons

$$y = 1.60x.$$

Theoretically, there are infinitely many ordered pairs (x, y) that satisfy this equation, but here we are limited to nonnegative values for x, since we cannot have a negative number of gallons. There is also a practical maximum value for x in this situation, which varies from one car to another. What determines this maximum value?

(b) You can also get a car wash at the gas station if you pay an additional $3.00. Write an equation that defines the price for gas and a car wash.

Since an additional $3.00 will be charged, you pay $1.60x + 3.00$ dollars for x gallons of gas and a car wash, or

$$y = 1.6x + 3. \qquad \text{Delete unnecessary 0s.}$$

(c) Interpret the ordered pairs (5, 11) and (10, 19) in relation to the equation from part (b).

The ordered pair (5, 11) indicates that the price of 5 gal of gas and a car wash is $11.00. Similarly, (10, 19) indicates that the price of 10 gal of gas and a car wash is $19.00.

Now Try Exercises 69 and 73.

NOTE In Example 6(a), the ordered pair (0, 0) satisfied the equation, so the linear equation has the form $y = mx$, where $b = 0$. If a realistic situation involves an initial charge plus a charge per unit as in Example 6(b), the equation has the form $y = mx + b$, where $b \neq 0$.

EXAMPLE 7 Finding an Equation of a Line That Models Data

Average annual tuition and fees for in-state students at public 4-year colleges are shown in the table for selected years and graphed as ordered pairs of points in the *scatter diagram* in Figure 27, where $x = 0$ represents 1990, $x = 4$ represents 1994, and so on, and y represents the cost in dollars.

Year	Cost (in dollars)
1990	2035
1994	2820
1996	3151
1998	3486
2000	3774

Source: U.S. National Center for Education Statistics.

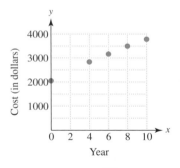

FIGURE 27

(a) Find an equation that models the data.

Since the points in Figure 27 lie approximately on a straight line, we can write a linear equation that models the relationship between year x and cost y. We choose two data points, (0, 2035) and (10, 3774), to find the slope of the line.

$$m = \frac{3774 - 2035}{10 - 0} = \frac{1739}{10} = 173.9$$

The slope 173.9 indicates that the cost of tuition and fees for in-state students at public 4-year colleges increased by about $174 per year from 1990 to 2000. We use this slope, the y-intercept (0, 2035), and the slope-intercept form to write an equation of the line. Thus,

$$y = 173.9x + 2035.$$

(b) Use the equation from part (a) to approximate the cost of tuition and fees at public 4-year colleges in 2002.

The value $x = 12$ corresponds to the year 2002, so we substitute 12 for x in the equation.

$$y = 173.9x + 2035$$
$$y = 173.9(12) + 2035$$
$$y = 4121.8$$

According to the model, average tuition and fees for in-state students at public 4-year colleges in 2002 were about $4122.

Now Try Exercise 79.

NOTE In Example 7, if we had chosen different data points, we would have gotten a slightly different equation. However, all such equations should be similar.

EXAMPLE 8 Finding an Equation of a Line That Models Data

Retail spending (in billions of dollars) on prescription drugs in the United States is shown in the graph in Figure 28.

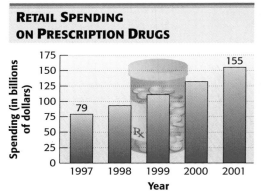

RETAIL SPENDING ON PRESCRIPTION DRUGS

Source: American Institute for Research analysis of Scott-Levin data.

FIGURE 28

(a) Write an equation that models the data.

The data shown in the bar graph increase linearly; that is, we could draw a straight line through the tops of any two bars that would be close to the top of each bar. We can use the data and the point-slope form of the equation of a line to get an equation that models the relationship between year x and spending on prescription drugs y. If we let $x = 7$ represent 1997, $x = 8$ represent 1998, and so on, the given data for 1997 and 2001 can be written as the ordered pairs $(7, 79)$ and $(11, 155)$. The slope of the line through these two points is

$$m = \frac{155 - 79}{11 - 7} = \frac{76}{4} = 19.$$

Thus, retail spending on prescription drugs increased by about $19 billion per year. Using this slope, one of the points, say $(7, 79)$, and the point-slope form, we obtain

$$y - y_1 = m(x - x_1) \qquad \text{Point-slope form}$$
$$y - 79 = 19(x - 7) \qquad (x_1, y_1) = (7, 79); \ m = 19$$
$$y - 79 = 19x - 133 \qquad \text{Distributive property}$$
$$y = 19x - 54. \qquad \text{Slope-intercept form}$$

Thus, retail spending y (in billions of dollars) on prescription drugs in the United States in year x can be approximated by the equation $y = 19x - 54$.

(b) Use the equation from part (a) to predict retail spending on prescription drugs in the United States in 2004. (Assume a constant rate of change.)

Since $x = 7$ represents 1997 and 2004 is 7 yr after 1997, $x = 14$ represents 2004. We substitute 14 for x in the equation.

$$y = 19x - 54 = 19(14) - 54 = 212$$

According to the model, $212 billion will be spent on prescription drugs in 2004.

Now Try Exercise 81.

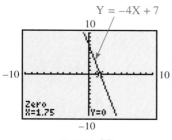

$Y = -4X + 7$

FIGURE 29

OBJECTIVE 7 Use a graphing calculator to solve linear equations in one variable. Figure 29 shows the graph of $Y = -4X + 7$. From the values at the bottom of the screen, we see that when $X = 1.75$, $Y = 0$. This means that $X = 1.75$ satisfies the equation $-4X + 7 = 0$, a linear equation in one variable. Therefore, the solution set of $-4X + 7 = 0$ is $\{1.75\}$. We can verify this algebraically by substitution. (The word "Zero" indicates that the x-intercept has been located.)

EXAMPLE 9 Solving an Equation with a Graphing Calculator

Use a graphing calculator to solve $-2x - 4(2 - x) = 3x + 4$.

We must write the equation as an equivalent equation with 0 on one side.

$$-2x - 4(2 - x) - 3x - 4 = 0 \qquad \text{Subtract } 3x \text{ and } 4.$$

Then we graph $Y = -2X - 4(2 - X) - 3X - 4$ to find the x-intercept. The standard viewing window cannot be used because the x-intercept does not lie in the interval $[-10, 10]$. As seen in Figure 30, the x-intercept of the graph is $(-12, 0)$, and thus the solution (or zero) of the equation is -12. The solution set is $\{-12\}$.

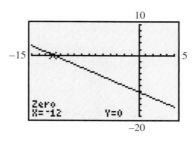

FIGURE 30

Now Try Exercise 83.

3.3 EXERCISES

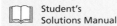
1. The following equations all represent the same line. Which one is in standard form as defined in the text?

A. $3x - 2y = 5$ **B.** $2y = 3x - 5$ **C.** $\dfrac{3}{5}x - \dfrac{2}{5}y = 1$ **D.** $3x = 2y + 5$

2. Which equation is in point-slope form?

A. $y = 6x + 2$ **B.** $4x + y = 9$ **C.** $y - 3 = 2(x - 1)$ **D.** $2y = 3x - 7$

3. Which equation in Exercise 2 is in slope-intercept form?

4. Write the equation $y + 2 = -3(x - 4)$ in slope-intercept form.

5. Write the equation from Exercise 4 in standard form.

6. Write the equation $10x - 7y = 70$ in slope-intercept form.

Match each equation with the graph that it most closely resembles. (Hint: Determine the signs of m and b to help you make your decision.)

7. $y = 2x + 3$ **A.**   **B.**   **C.**

8. $y = -2x + 3$

9. $y = -2x - 3$

10. $y = 2x - 3$ **D.**   **E.**   **F.**

11. $y = 2x$

12. $y = -2x$

13. $y = 3$ **G.**   **H.**

14. $y = -3$

Find the equation in slope-intercept form of the line satisfying the given conditions. See Example 1.

15. $m = 5;\ b = 15$ **16.** $m = -2;\ b = 12$

17. $m = -\dfrac{2}{3};\ b = \dfrac{4}{5}$ **18.** $m = -\dfrac{5}{8};\ b = -\dfrac{1}{3}$

19. Slope $\dfrac{2}{5}$; y-intercept $(0, 5)$ **20.** Slope $-\dfrac{3}{4}$; y-intercept $(0, 7)$

Write an equation in slope-intercept form of the line shown in each graph. (Hint: Use the indicated points to find the slope.)

21.

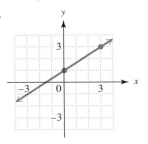

22.

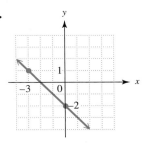

*For each equation, **(a)** write it in slope-intercept form, **(b)** give the slope of the line, **(c)** give the y-intercept, and **(d)** graph the line. See Example 2.*

23. $-x + y = 4$　　**24.** $-x + y = 6$　　**25.** $6x + 5y = 30$

26. $3x + 4y = 12$　　**27.** $4x - 5y = 20$　　**28.** $7x - 3y = 3$

29. $x + 2y = -4$　　**30.** $x + 3y = -9$

Find an equation of the line that satisfies the given conditions. Write the equation in standard form. See Example 3.

31. Through $(-2, 4)$; slope $-\dfrac{3}{4}$ 　　**32.** Through $(-1, 6)$; slope $-\dfrac{5}{6}$

33. Through $(5, 8)$; slope -2 　　**34.** Through $(12, 10)$; slope 1

35. Through $(-5, 4)$; slope $\dfrac{1}{2}$ 　　**36.** Through $(7, -2)$; slope $\dfrac{1}{4}$

37. x-intercept $(3, 0)$; slope 4 　　**38.** x-intercept $(-2, 0)$; slope -5

✎ **39.** In your own words, list all the forms of linear equations in two variables and describe when each form should be used.

✎ **40.** Explain why the point-slope form of an equation cannot be used to find the equation of a vertical line.

Write an equation of the line that satisfies the given conditions.

41. Through $(9, 5)$; slope 0 　　**42.** Through $(-4, -2)$; slope 0

43. Through $(9, 10)$; undefined slope 　　**44.** Through $(-2, 8)$; undefined slope

45. Through $(.5, .2)$; vertical 　　**46.** Through $\left(\dfrac{5}{8}, \dfrac{2}{9}\right)$; vertical

47. Through $(-7, 8)$; horizontal 　　**48.** Through $(2, 7)$; horizontal

Find an equation of the line passing through the given points. Write the equation in standard form. See Example 4.

49. $(3, 4)$ and $(5, 8)$ 　　**50.** $(5, -2)$ and $(-3, 14)$

51. $(6, 1)$ and $(-2, 5)$ 　　**52.** $(-2, 5)$ and $(-8, 1)$

53. $\left(-\dfrac{2}{5}, \dfrac{2}{5}\right)$ and $\left(\dfrac{4}{3}, \dfrac{2}{3}\right)$ 　　**54.** $\left(\dfrac{3}{4}, \dfrac{8}{3}\right)$ and $\left(\dfrac{2}{5}, \dfrac{2}{3}\right)$

55. $(2, 5)$ and $(1, 5)$ 　　**56.** $(-2, 2)$ and $(4, 2)$

57. $(7, 6)$ and $(7, -8)$

58. $(13, 5)$ and $(13, -1)$

59. $(1, -3)$ and $(-1, -3)$

60. $(-4, -6)$ and $(5, -6)$

Find an equation of the line satisfying the given conditions. Write the equation in slope-intercept form. See Example 5.

61. Through $(7, 2)$; parallel to $3x - y = 8$

62. Through $(4, 1)$; parallel to $2x + 5y = 10$

63. Through $(-2, -2)$; parallel to $-x + 2y = 10$

64. Through $(-1, 3)$; parallel to $-x + 3y = 12$

65. Through $(8, 5)$; perpendicular to $2x - y = 7$

66. Through $(2, -7)$; perpendicular to $5x + 2y = 18$

67. Through $(-2, 7)$; perpendicular to $x = 9$

68. Through $(8, 4)$; perpendicular to $x = -3$

Write an equation in the form $y = mx$ for each situation. Then give the three ordered pairs associated with the equation for x-values 0, 5, and 10. See Example 6(a).

69. x represents the number of hours traveling at 45 mph, and y represents the distance traveled (in miles).

70. x represents the number of compact discs sold at \$16 each, and y represents the total cost of the discs (in dollars).

71. x represents the number of gallons of gas sold at \$1.50 per gal, and y represents the total cost of the gasoline (in dollars).

72. x represents the number of days a videocassette is rented at \$3.50 per day, and y represents the total charge for the rental (in dollars).

✎ *For each situation, **(a)** write an equation in the form $y = mx + b$; **(b)** find and interpret the ordered pair associated with the equation for $x = 5$; and **(c)** answer the question. See Examples 6(b) and 6(c).*

73. A membership to the Midwest Athletic Club costs \$99 plus \$39 per month. (*Source:* Midwest Athletic Club.) Let x represent the number of months selected. How much does the first year's membership cost?

74. For a family membership, the athletic club in Exercise 73 charges a membership fee of \$159 plus \$60 for each additional family member after the first. Let x represent the number of additional family members. What is the membership fee for a four-person family?

75. A cell phone plan includes 900 anytime minutes for \$50 per month, plus a one-time activation fee of \$25. A Nokia 5165 cell phone is included at no additional charge. (*Source:* U.S. Cellular.) Let x represent the number of months of service. If you sign a 2-yr contract, how much will this cell phone plan cost? (Assume that you never use more than the allotted number of minutes.)

76. Another cell phone plan includes 450 anytime minutes for $35 per month, plus $19.95 for a Nokia 5165 cell phone and $25 for a one-time activation fee. (*Source:* U.S. Cellular.) Let *x* represent the number of months of service. If you sign a 1-yr contract, how much will this cell phone package cost? (Assume that you never use more than the allotted number of minutes.)

77. A rental car costs $50 plus $.20 per mile. Let *x* represent the number of miles driven, and *y* represent the total charge to the renter. How many miles was the car driven if the renter paid $84.60?

78. There is a $30 fee to rent a chain saw, plus $6 per day. Let *x* represent the number of days the saw is rented and *y* represent the charge to the user in dollars. If the total charge is $138, for how many days is the saw rented?

Solve each problem. In part (a), give equations in slope-intercept form. See Examples 7 and 8. (Source for Exercises 79 and 80: Jupiter Media Metrix.)

79. The percent of households that access the Internet by high-speed broadband is shown in the graph, where the year 2000 corresponds to *x* = 0.

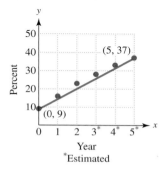

  (a) Use the ordered pairs from the graph to write an equation that models the data. What does the slope tell us in the context of this problem?
  (b) Use the equation from part (a) to predict the percent of U.S. households that will access the Internet by broadband in 2006. Round your answer to the nearest percent.

80. The percent of U.S. households that access the Internet by dial-up is shown in the graph, where the year 2000 corresponds to *x* = 0.

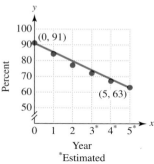

  (a) Use the ordered pairs from the graph to write an equation that models the data. What does the slope tell us in the context of this problem?
  (b) Use the equation from part (a) to predict the percent of U.S. households that will access the Internet by dial-up in 2006. Round your answer to the nearest percent.

81. The number of post offices in the United States is shown in the bar graph.

  (a) Use the information given for the years 1995 and 2000, letting *x* = 5 represent 1995, *x* = 10 represent 2000, and *y* represent the number of post offices, to write an equation that models the data.
  (b) Use the equation to approximate the number of post offices in 1998. How does this result compare to the actual value, 27,952?

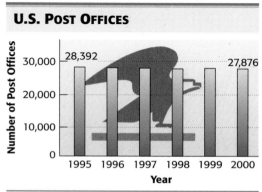

U.S. POST OFFICES

Source: U.S. Postal Service, *Annual Report of the Postmaster General.*

82. Median household income of African-Americans is shown in the bar graph.

(a) Use the information given for the years 1995 and 1999, letting $x = 5$ represent 1995, $x = 9$ represent 1999, and y represent the median income, to write an equation that models median household income.

(b) Use the equation to approximate the median income for 1997. How does your result compare to the actual value, \$25,050?

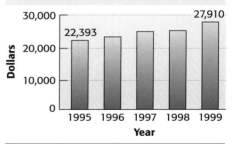

MEDIAN HOUSEHOLD INCOME FOR AFRICAN–AMERICANS

Source: U.S. Bureau of the Census.

TECHNOLOGY INSIGHTS (EXERCISES 83–88)

In Exercises 83–88, do the following.

(a) Simplify and rewrite the equation so that the right side is 0. Then replace 0 with y.

(b) The graph of the equation for y is shown with each exercise. Use the graph to determine the solution of the given equation. See Example 9.

(c) Solve the equation using the methods of Chapter 2.

83. $2x + 7 - x = 4x - 2$

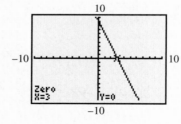

84. $7x - 2x + 4 - 5 = 3x + 1$

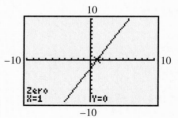

85. $3(2x + 1) - 2(x - 2) = 5$

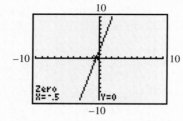

86. $4x - 3(4 - 2x) = 2(x - 3) + 6x + 2$

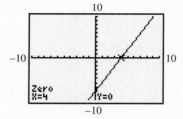

87. The graph of y_1 is shown in the standard viewing window. Which is the only choice that could possibly be the solution of the equation $y_1 = 0$?

  A. -15 **B.** 0 **C.** 5 **D.** 15

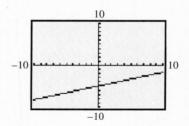

88. (a) Solve $-2(x - 5) = -x - 2$ using the methods of Chapter 2.

✐ **(b)** Explain why the standard viewing window of a graphing calculator cannot graphically support the solution found in part (a). What minimum and maximum x-values would make it possible for the solution to be seen?

RELATING CONCEPTS (EXERCISES 89–94)

For Individual or Group Work

In Section 2.2 we learned how formulas can be applied to problem solving. **Work Exercises 89–94 in order,** *to see how the formula that relates Celsius and Fahrenheit temperatures is derived.*

89. There is a linear relationship between Celsius and Fahrenheit temperatures. When $C = 0°$, $F = $ _____°, and when $C = 100°$, $F = $ _____°.

90. Think of ordered pairs of temperatures (C, F), where C and F represent corresponding Celsius and Fahrenheit temperatures. The equation that relates the two scales has a straight-line graph that contains the two points determined in Exercise 89. What are these two points?

91. Find the slope of the line described in Exercise 90.

92. Now think of the point-slope form of the equation in terms of C and F, where C replaces x and F replaces y. Use the slope you found in Exercise 91 and one of the two points determined earlier, and find the equation that gives F in terms of C.

93. To obtain another form of the formula, use the equation you found in Exercise 92 and solve for C in terms of F.

✐ **94.** The equation found in Exercise 92 is graphed on the graphing calculator screen shown here. Interpret the display at the bottom, in the context of this group of exercises.

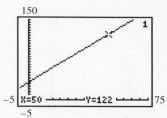

3.4 Linear Inequalities in Two Variables

OBJECTIVES

1. Graph linear inequalities in two variables.

2. Graph the intersection of two linear inequalities.

3. Graph the union of two linear inequalities.

4. Use a graphing calculator to solve linear inequalities in one variable.

OBJECTIVE 1   Graph linear inequalities in two variables.   In Chapter 2 we graphed linear inequalities in one variable on the number line. In this section we graph linear inequalities in two variables on a rectangular coordinate system.

Linear Inequality in Two Variables

An inequality that can be written as

$$Ax + By < C \qquad \text{or} \qquad Ax + By > C,$$

where A, B, and C are real numbers and A and B are not both 0, is a **linear inequality in two variables.**

The symbols $\leq$ and $\geq$ may replace $<$ and $>$ in the definition.

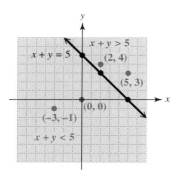

FIGURE 31

Consider the graph in Figure 31. The graph of the line $x + y = 5$ divides the points in the rectangular coordinate system into three sets: those points that lie on the line itself and satisfy the equation $x + y = 5$ [like $(0, 5)$, $(2, 3)$, and $(5, 0)$], those that lie in the half-plane above the line and satisfy the inequality $x + y > 5$ [like $(5, 3)$ and $(2, 4)$], and those that lie in the half-plane below the line and satisfy the inequality $x + y < 5$ [like $(0, 0)$ and $(-3, -1)$]. The graph of the line $x + y = 5$ is called the **boundary line** for the inequalities $x + y > 5$ and $x + y < 5$. Graphs of linear inequalities in two variables are *regions* in the real number plane that may or may not include boundary lines.

To graph a linear inequality in two variables, follow these steps.

Graphing a Linear Inequality

Step 1 **Draw the graph of the straight line that is the boundary.** Make the line solid if the inequality involves $\leq$ or $\geq$; make the line dashed if the inequality involves $<$ or $>$.

Step 2 **Choose a test point.** Choose any point not on the line and substitute the coordinates of this point in the inequality.

Step 3 **Shade the appropriate region.** Shade the region that includes the test point if it satisfies the original inequality; otherwise, shade the region on the other side of the boundary line.

EXAMPLE 1 Graphing a Linear Inequality

Graph $3x + 2y \geq 6$.

Step 1 First graph the line $3x + 2y = 6$. The graph of this line, the boundary of the graph of the inequality, is shown in Figure 32.

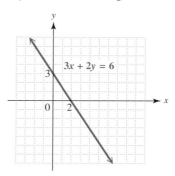

FIGURE 32

Step 2 The graph of the inequality $3x + 2y \geq 6$ includes the points of the line $3x + 2y = 6$ and either the points *above* the line $3x + 2y = 6$ or the points *below* that line. To decide which, select any point not on the boundary line $3x + 2y = 6$ as a test point. The origin, $(0, 0)$, is often a good choice because the substitution is easy. Substitute the values from the test point $(0, 0)$ for x and y in the inequality $3x + 2y > 6$.

$$3(0) + 2(0) > 6 \qquad ?$$
$$0 > 6 \qquad \text{False}$$

Step 3   Because the result is false, $(0, 0)$ does *not* satisfy the inequality, and so the solution set includes all points on the other side of the line. This region is shaded in Figure 33.

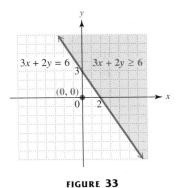

FIGURE 33

Now Try Exercise 7.

If the inequality is written in the form $y > mx + b$ or $y < mx + b$, then the inequality symbol indicates which half-plane to shade.

If $y > mx + b$, then shade above the boundary line;

if $y < mx + b$, then shade below the boundary line.

This method works *only* if the inequality is solved for y.

EXAMPLE 2   Graphing a Linear Inequality

Graph $x - 3y < 4$.

First graph the boundary line, shown in Figure 34. The points of the boundary line do not belong to the inequality $x - 3y < 4$ (because the inequality symbol is $<$, not $\leq$). For this reason, the line is dashed. Now solve the inequality for y.

$$x - 3y < 4$$
$$-3y < -x + 4 \qquad \text{Subtract } x.$$
$$y > \frac{1}{3}x - \frac{4}{3} \qquad \text{Multiply by } -\frac{1}{3}; \text{ change } < \text{ to } >.$$

Because of the *is greater than* symbol, shade *above* the line. As a check, choose a test point not on the line, say $(1, 2)$, and substitute for x and y in the original inequality.

$$1 - 3(2) < 4 \qquad ?$$
$$-5 < 4 \qquad \text{True}$$

This result agrees with the decision to shade above the line. The solution set, graphed in Figure 34, includes only those points in the shaded half-plane (not those on the line).

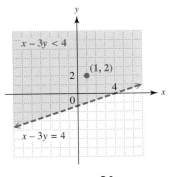

FIGURE 34

Now Try Exercise 9.

OBJECTIVE 2 Graph the intersection of two linear inequalities.   In Section 2.6, we used the words *and* and *or* to solve compound inequalities. In that section, the inequalities had one variable. We can extend those ideas to include inequalities in two variables. A pair of inequalities joined with the word *and* is interpreted as the intersection of the solution sets of the inequalities. The graph of the intersection of two or more inequalities is the region of the plane where all points satisfy all of the inequalities at the same time.

EXAMPLE 3 Graphing the Intersection of Two Inequalities

Graph $2x + 4y \geq 5$ and $x \geq 1$.

To begin, we graph each of the two inequalities $2x + 4y \geq 5$ and $x \geq 1$ separately. The graph of $2x + 4y \geq 5$ is shown in Figure 35(a), and the graph of $x \geq 1$ is shown in Figure 35(b).

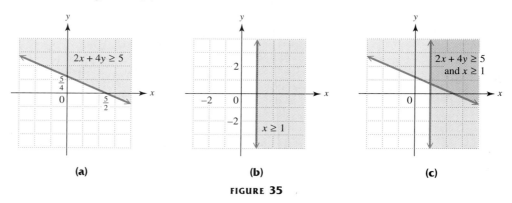

(a)   (b)   (c)

FIGURE 35

In practice, the graphs in Figures 35(a) and (b) are graphed on the same axes. Then we use heavy shading to identify the intersection of the graphs, as shown in Figure 35(c).

To check, we use a test point from each of the four regions formed by the intersection of the boundary lines. Verify that only ordered pairs in the heavily shaded region satisfy both inequalities.

Now Try Exercise 19.

OBJECTIVE 3 Graph the union of two linear inequalities.   When two inequalities are joined by the word *or*, we must find the union of the graphs of the inequalities. The graph of the union of two inequalities includes all of the points that satisfy either inequality.

EXAMPLE 4   Graphing the Union of Two Inequalities

Graph $2x + 4y \geq 5$ or $x \geq 1$.

The graphs of the two inequalities are shown in Figures 35(a) and (b) in Example 3. The graph of the union is shown in Figure 36.

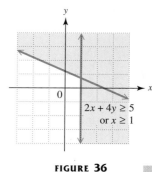

FIGURE 36

Now Try Exercise 29.

| **CONNECTIONS** |

Suppose a factory can have *no more than* 200 workers on a shift, but must have *at least* 100 and must manufacture *at least* 3000 units at minimum cost. The managers need to know how many workers should be on a shift in order to produce the required units at minimal cost. *Linear programming* is a method for finding the optimal (best possible) solution that meets all the conditions for such problems. The first step in solving linear programming problems with two variables is to express the conditions (constraints) as inequalities, graph the system of inequalities, and identify the region that satisfies all the inequalities at once.

For Discussion or Writing

Let *x* represent the number of workers and *y* represent the number of units manufactured.

1. Write three inequalities expressing the conditions given in the problem.

2. Graph the inequalities from Item 1 and shade the intersection.

3. The cost per worker is $50 per day and the cost to manufacture 1 unit is $100. Write an expression representing the total daily cost, *C*.

4. Find values of *x* and *y* for several points in or on the boundary of the shaded region. Include any "corner points."

5. Of the values of *x* and *y* that you chose in Item 4, which gives the least cost when substituted in the cost equation from Item 3? What does your answer mean in terms of the given problem? Is your answer reasonable? Explain.

OBJECTIVE 4 Use a graphing calculator to solve linear inequalities in one variable. Recall from Section 3.3 that the *x*-intercept of the graph of the line $y = mx + b$ indicates the solution of the equation $mx + b = 0$. We can extend this observation to find solutions of the associated inequalities $mx + b > 0$ and $mx + b < 0$. The solution set of $mx + b > 0$ is the set of all *x*-values for which the graph of $y = mx + b$ is *above* the *x*-axis. (We consider points above because the symbol is $>$.) On the other hand, the solution set of $mx + b < 0$ is the set of all *x*-values for which the graph of $y = mx + b$ is *below* the *x*-axis. (We consider points below because the symbol is $<$.)

For example, in Figure 37 the *x*-intercept of $y = 3x - 9$ is $(3, 0)$. Therefore,

the solution set of $3x - 9 = 0$ is $\{3\}$.

Because the graph of *y* lies above the *x*-axis for *x*-values greater than 3,

the solution set of $3x - 9 > 0$ is $(3, \infty)$.

Because the graph lies below the *x*-axis for *x*-values less than 3,

the solution set of $3x - 9 < 0$ is $(-\infty, 3)$.

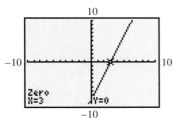

FIGURE 37

To solve the equation $-2(3x + 1) = -2x + 18$ and the associated inequalities $-2(3x + 1) > -2x + 18$ and $-2(3x + 1) < -2x + 18$, we must rewrite the equation so that the right side equals 0:

$$-2(3x + 1) + 2x - 18 = 0.$$

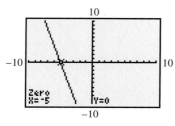

FIGURE 38

Graphing

$$y = -2(3x + 1) + 2x - 18$$

yields the x-intercept $(-5, 0)$, as shown in Figure 38. Because the graph of y lies *above* the x-axis for x-values less than -5,

the solution set of $-2(3x + 1) > -2x + 18$ is $(-\infty, -5)$.

Because the graph of y lies *below* the x-axis for x-values greater than -5,

the solution set of $-2(3x + 1) < -2x + 18$ is $(-5, \infty)$.

3.4 EXERCISES

For Extra Help

 Student's
Solutions Manual

 MyMathLab

 InterAct Math
Tutorial Software

 AW Math
Tutor Center

 MathXL

Digital Video Tutor
CD 3/Videotape 4

In Exercises 1–4, fill in the first blank with either solid *or* dashed. *Fill in the second blank with* either above *or* below.

1. The boundary of the graph of $y \leq -x + 2$ will be a _____ line, and the shading will be _____ the line.

2. The boundary of the graph of $y < -x + 2$ will be a _____ line, and the shading will be _____ the line.

3. The boundary of the graph of $y > -x + 2$ will be a _____ line, and the shading will be _____ the line.

4. The boundary of the graph of $y \geq -x + 2$ will be a _____ line, and the shading will be _____ the line.

5. How is the boundary line $Ax + By = C$ used in graphing either $Ax + By < C$ or $Ax + By > C$?

6. Describe the two methods discussed in the text for deciding which region is the solution set of a linear inequality in two variables.

Graph each linear inequality in two variables. See Examples 1 and 2.

7. $x + y \leq 2$ **8.** $x + y \leq -3$ **9.** $4x - y < 4$

10. $3x - y < 3$ **11.** $x + 3y \geq -2$ **12.** $x + 4y \geq -3$

13. $x + y > 0$ **14.** $x + 2y > 0$ **15.** $x - 3y \leq 0$

16. $x - 5y \leq 0$ **17.** $y < x$ **18.** $y \leq 4x$

Graph each compound inequality. See Example 3.

19. $x + y \leq 1$ and $x \geq 1$ **20.** $x - y \geq 2$ and $x \geq 3$

21. $2x - y \geq 2$ and $y < 4$ **22.** $3x - y \geq 3$ and $y < 3$

23. $x + y > -5$ and $y < -2$ **24.** $6x - 4y < 10$ and $y > 2$

Use the method described in Section 2.7 to write each inequality as a compound inequality, and graph its solution set in the rectangular coordinate plane.

25. $|x| < 3$ **26.** $|y| < 5$ **27.** $|x + 1| < 2$ **28.** $|y - 3| < 2$

Graph each compound inequality. See Example 4.

29. $x - y \geq 1$ or $y \geq 2$

30. $x + y \leq 2$ or $y \geq 3$

31. $x - 2 > y$ or $x < 1$

32. $x + 3 < y$ or $x > 3$

33. $3x + 2y < 6$ or $x - 2y > 2$

34. $x - y \geq 1$ or $x + y \leq 4$

TECHNOLOGY INSIGHTS (EXERCISES 35–42)

Match each inequality with its calculator graph. (Hint: Use the slope, y-intercept, and inequality symbol in making your choice.)

35. $y \leq 3x - 6$

36. $y \geq 3x - 6$

37. $y \leq -3x - 6$

38. $y \geq -3x - 6$

A.
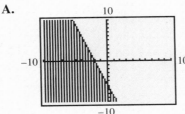

B.
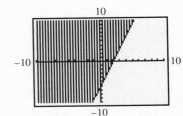

C.
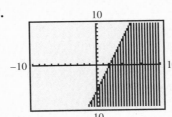

D.
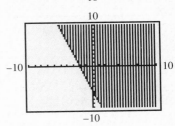

The graph of a linear equation $y = mx + b$ is shown on a graphing calculator screen, along with the x-value of the x-intercept of the line. Use the screen to solve (a) $y = 0$, (b) $y < 0$, and (c) $y > 0$. See Objective 4.

39.
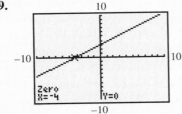

40.
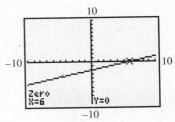

41.
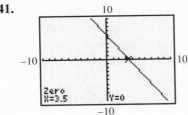

42.
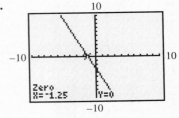

Solve the equation in part (a) and the associated inequalities in parts (b) and (c) using the methods of Chapter 2. Then graph the left side as y in the standard viewing window of a graphing calculator and explain how the graph supports your answers in parts (a)–(c).

43. (a) $5x + 3 = 0$
   (b) $5x + 3 > 0$
   (c) $5x + 3 < 0$

44. (a) $6x + 3 = 0$
   (b) $6x + 3 > 0$
   (c) $6x + 3 < 0$

45. (a) $-8x - (2x + 12) = 0$
   (b) $-8x - (2x + 12) \geq 0$
   (c) $-8x - (2x + 12) \leq 0$

46. (a) $-4x - (2x + 18) = 0$
   (b) $-4x - (2x + 18) \geq 0$
   (c) $-4x - (2x + 18) \leq 0$

3.5   Introduction to Functions

OBJECTIVES

1 Define and identify relations and functions.

2 Find domain and range.

3 Identify functions defined by graphs and equations.

4 Use function notation.

5 Identify linear functions.

We often describe one quantity in terms of another. Consider the following.

- The amount of your paycheck if you are paid hourly depends on the number of hours you worked.

- The cost at the gas station depends on the number of gallons of gas you pumped into your car.

- The distance traveled by a car moving at a constant speed depends on the time traveled.

We can use ordered pairs to represent these corresponding quantities. For example, we indicate the relationship between the amount of your paycheck and hours worked by writing ordered pairs in which the first number represents hours worked and the second number represents paycheck amount in dollars. Then the ordered pair (5, 40) indicates that when you work 5 hr, your paycheck is $40. Similarly, the ordered pairs (10, 80) and (20, 160) show that working 10 hr results in an $80 paycheck and working 20 hr results in a $160 paycheck. In this example, what would the ordered pair (40, 320) indicate?

Since the amount of your paycheck *depends* on the number of hours worked, your paycheck amount is called the *dependent variable,* and the number of hours worked is called the *independent variable.* Generalizing, if the value of the variable *y* depends on the value of the variable *x*, then *y* is the **dependent variable** and *x* is the **independent variable.**

Independent variable ⌐ ⌐Dependent variable
          ↓ ↓
        (x, y)

OBJECTIVE 1 Define and identify relations and functions. Since we can write related quantities using ordered pairs, a set of ordered pairs such as

$$\{(5, 40), (10, 80), (20, 160), (40, 320)\}$$

is called a *relation.*

Relation

A **relation** is a set of ordered pairs.

A special kind of relation, called a *function,* is very important in mathematics and its applications.

Function

A **function** is a relation in which, for each value of the first component of the ordered pairs, there is *exactly one value* of the second component.

■ **EXAMPLE 1 Determining Whether Relations Are Functions**

Tell whether each relation defines a function.

$$F = \{(1, 2), (-2, 4), (3, -1)\}$$
$$G = \{(-2, -1), (-1, 0), (0, 1), (1, 2), (2, 2)\}$$
$$H = \{(-4, 1), (-2, 1), (-2, 0)\}$$

Relations F and G are functions, because for each different x-value there is exactly one y-value. Notice that in G, the last two ordered pairs have the same y-value (1 is paired with 2, and 2 is paired with 2). This does not violate the definition of function, since the first components (x-values) are different and each is paired with only one second component (y-value).

In relation H, however, the last two ordered pairs have the *same x-value* paired with *two different y-values* (-2 is paired with both 1 and 0), so H is a relation but not a function. ***In a function, no two ordered pairs can have the same first component and different second components.***

Different y-values

$$H = \{(-4, 1), (-2, 1), (-2, 0)\} \qquad \text{Not a function}$$

Same x-value

Now Try Exercises 5 and 7.

In a function, there is *exactly one* value of the dependent variable, the second component, for each value of the independent variable, the first component. This is what makes functions so important in applications.

NOTE The relation from the beginning of this section representing hours worked and corresponding paycheck amount is a function since each *x*-value is paired with exactly one *y*-value. You would not be happy, for example, if you and a coworker each worked 20 hr at the same hourly rate and your paycheck was $160 while his was $200.

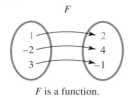

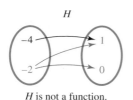

F is a function.

H is not a function.

FIGURE 39

Relations and functions can also be expressed as a correspondence or *mapping* from one set to another, as shown in Figure 39 for function *F* and relation *H* from Example 1. The arrow from 1 to 2 indicates that the ordered pair (1, 2) belongs to *F*— each first component is paired with exactly one second component. In the mapping for set *H*, which is not a function, the first component −2 is paired with two different second components, 1 and 0.

Since relations and functions are sets of ordered pairs, we can represent them using tables and graphs. A table and graph for function *F* is shown in Figure 40.

Finally, we can describe a relation or function using a rule that tells how to determine the dependent variable for a specific value of the independent variable. The rule may be given in words: the dependent variable is twice the independent variable. Usually the rule is an equation:

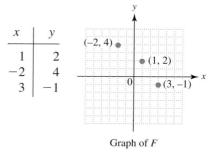

x	*y*
1	2
−2	4
3	−1

Graph of *F*

FIGURE 40

$$y = 2x.$$

Dependent variable / Independent variable

This is the most efficient way to define a relation or function.

NOTE Another way to think of a function relationship is to think of the independent variable as an input and the dependent variable as an output. This is illustrated by the input-output (function) machine for the function defined by $y = 2x$.

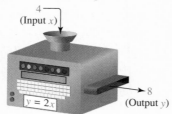

Function machine

OBJECTIVE 2 Find domain and range.

Domain and Range

In a relation, the set of all values of the independent variable (*x*) is the **domain;** the set of all values of the dependent variable (*y*) is the **range.**

EXAMPLE 2 Finding Domains and Ranges of Relations

Give the domain and range of each relation. Tell whether the relation defines a function.

(a) $\{(3, -1), (4, 2), (4, 5), (6, 8)\}$

The domain, the set of x-values, is $\{3, 4, 6\}$; the range, the set of y-values, is $\{-1, 2, 5, 8\}$. This relation is not a function because the same x-value 4 is paired with two different y-values, 2 and 5.

(b)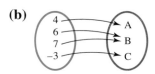

The domain of this relation is

$$\{4, 6, 7, -3\};$$

the range is

$$\{A, B, C\}.$$

This mapping defines a function—each x-value corresponds to exactly one y-value.

(c)

x	y
-5	2
0	2
5	2

This is a table of ordered pairs, so the domain is the set of x-values $\{-5, 0, 5\}$ and the range is the set of y-values $\{2\}$. The table defines a function because each different x-value corresponds to exactly one y-value (even though it is the same y-value).

Now Try Exercises 11, 13, and 15.

As mentioned previously, the graph of a relation is the graph of its ordered pairs. The graph gives a picture of the relation, which can be used to determine its domain and range.

EXAMPLE 3 Finding Domains and Ranges from Graphs

Give the domain and range of each relation.

(a)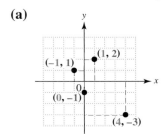

The domain is the set of x-values,

$$\{-1, 0, 1, 4\}.$$

The range is the set of y-values,

$$\{-3, -1, 1, 2\}.$$

(b)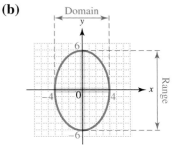

The x-values of the points on the graph include all numbers between -4 and 4, inclusive. The y-values include all numbers between -6 and 6, inclusive. Using interval notation,

the domain is $[-4, 4]$;
the range is $[-6, 6]$.

(c)

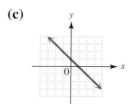

(d)

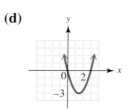

The arrowheads indicate that the line extends indefinitely left and right, as well as up and down. Therefore, both the domain and the range include all real numbers, written $(-\infty, \infty)$.

The arrowheads indicate that the graph extends indefinitely left and right, as well as upward. The domain is $(-\infty, \infty)$. Because there is a least y-value, -3, the range includes all numbers greater than or equal to -3, written $[-3, \infty)$.

Now Try Exercises 17 and 19.

Since relations are often defined by equations, such as $y = 2x + 3$ and $y^2 = x$, we must sometimes determine the domain of a relation from its equation. In this book, we assume the following agreement on the domain of a relation.

Agreement on Domain

Unless specified otherwise, the domain of a relation is assumed to be all real numbers that produce real numbers when substituted for the independent variable.

To illustrate this agreement, since any real number can be used as a replacement for x in $y = 2x + 3$, the domain of this function is the set of all real numbers. As another example, the function defined by $y = \frac{1}{x}$ has all real numbers except 0 as domain, since y is undefined if $x = 0$. In general, the domain of a function defined by an algebraic expression is all real numbers, except those numbers that lead to division by 0 or an even root of a negative number.

OBJECTIVE 3 Identify functions defined by graphs and equations. Most of the relations we have seen in the examples are functions—that is, each x-value corresponds to exactly one y-value. Since each value of x leads to only one value of y in a function, any vertical line drawn through the graph of a function must intersect the graph in at most one point. This is the *vertical line test* for a function.

Vertical Line Test

If every vertical line intersects the graph of a relation in no more than one point, then the relation represents a function.

For example, the graph shown in Figure 41(a) is not the graph of a function since a vertical line intersects the graph in more than one point. The graph in Figure 41(b) does represent a function.

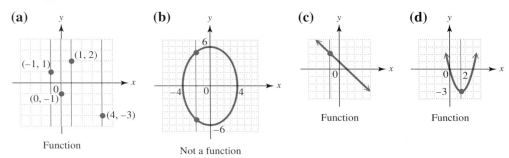

Not a function–the same
x-value corresponds to
four different y-values.

(a)

Function–each x-value
corresponds to only one
y-value.

(b)

FIGURE 41

◾ EXAMPLE 4   Using the Vertical Line Test

Use the vertical line test to determine whether each relation graphed in Example 3 is
a function.

(a)

Function

(b)

Not a function

(c)

Function

(d)

Function

The graphs in (a), (c), and (d) represent functions. The graph of the relation in
(b) fails the vertical line test, since the same x-value corresponds to two different
y-values; therefore, it is not the graph of a function.

Now Try Exercise 21.

NOTE   Graphs that do not represent functions are still relations. Remember
that all equations and graphs represent relations and that all relations have a
domain and range.

The vertical line test is a simple method for identifying a function defined by a
graph. It is more difficult to decide whether a relation defined by an equation is a
function. The next example gives some hints that may help.

◾ EXAMPLE 5   Identifying Functions from Their Equations

Decide whether each relation defines a function and give the domain.

(a) $y = x + 4$

In the defining equation (or rule), $y = x + 4$, y is always found by adding 4 to x.
Thus, each value of x corresponds to just one value of y and the relation defines a
function; x can be any real number, so the domain is $\{x \mid x$ is a real number$\}$ or $(-\infty, \infty)$.

(b) $y = \sqrt{2x - 1}$

For any choice of x in the domain, there is exactly one corresponding value for y (the radical is a nonnegative number), so this equation defines a function. Refer to the agreement on domain stated previously. Since the equation involves a square root, the quantity under the radical sign cannot be negative. Thus,

$$2x - 1 \geq 0$$
$$2x \geq 1$$
$$x \geq \frac{1}{2},$$

and the domain of the function is $\left[\frac{1}{2}, \infty\right)$.

(c) $y^2 = x$

The ordered pairs $(16, 4)$ and $(16, -4)$ both satisfy this equation. Since one value of x, 16, corresponds to two values of y, 4 and -4, this equation does not define a function. Because x is equal to the square of y, the values of x must always be nonnegative. The domain of the relation is $[0, \infty)$.

(d) $y \leq x - 1$

By definition, y is a function of x if every value of x leads to exactly one value of y. In this example, a particular value of x, say 1, corresponds to many values of y. The ordered pairs $(1, 0)$, $(1, -1)$, $(1, -2)$, $(1, -3)$, and so on, all satisfy the inequality. For this reason, an inequality never defines a function. Any number can be used for x, so the domain is the set of real numbers, $(-\infty, \infty)$.

(e) $y = \dfrac{5}{x - 1}$

Given any value of x in the domain, we find y by subtracting 1, then dividing the result into 5. This process produces exactly one value of y for each value in the domain, so this equation defines a function. The domain includes all real numbers except those that make the denominator 0. We find these numbers by setting the denominator equal to 0 and solving for x.

$$x - 1 = 0$$
$$x = 1$$

Thus, the domain includes all real numbers except 1. In interval notation this is written as $(-\infty, 1) \cup (1, \infty)$.

Now Try Exercises 27, 29, and 35.

In summary, three variations of the definition of function are given here.

Variations of the Definition of Function

1. A **function** is a relation in which, for each value of the first component of the ordered pairs, there is exactly one value of the second component.

2. A **function** is a set of ordered pairs in which no first component is repeated.

3. A **function** is a rule or correspondence that assigns exactly one range value to each domain value.

OBJECTIVE 4 Use function notation.   When a function f is defined with a rule or an equation using x and y for the independent and dependent variables, we say "y is a function of x" to emphasize that y *depends on x*. We use the notation

$$y = f(x),$$

called **function notation,** to express this and read $f(x)$ as "f of x." (In this special notation the parentheses do not indicate multiplication.) The letter f stands for *function.* For example, if $y = 9x - 5$, we can name this function f and write

$$f(x) = 9x - 5.$$

Note that $f(x)$ *is just another name for the dependent variable y.* For example, if $y = f(x) = 9x - 5$ and $x = 2$, then we find y, or $f(2)$, by replacing x with 2.

$$y = f(2)$$
$$= 9 \cdot 2 - 5$$
$$= 18 - 5$$
$$= 13.$$

The statement "if $x = 2$, then $y = 13$" represents the ordered pair $(2, 13)$ and is abbreviated with function notation as

$$f(2) = 13.$$

Read $f(2)$ as "f of 2" or "f at 2." Also,

$$f(0) = 9 \cdot 0 - 5 = -5 \qquad \text{and} \qquad f(-3) = 9(-3) - 5 = -32.$$

These ideas and the symbols used to represent them can be illustrated as follows.

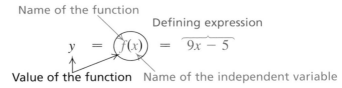

Name of the function

Defining expression

Value of the function   Name of the independent variable

CAUTION   The symbol $f(x)$ *does not* indicate "f times x," but represents the y-value for the indicated x-value. As just shown, $f(2)$ is the y-value that corresponds to the x-value 2.

▌ EXAMPLE 6   Using Function Notation

Let $f(x) = -x^2 + 5x - 3$. Find the following.

(a) $f(2)$

Replace x with 2.

$$f(x) = -x^2 + 5x - 3$$
$$f(2) = -2^2 + 5 \cdot 2 - 3$$
$$= -4 + 10 - 3$$
$$= 3$$

Thus, $f(2) = 3$; the ordered pair $(2, 3)$ belongs to f.

(b) $f(q)$

Replace x with q.

$$f(x) = -x^2 + 5x - 3$$
$$f(q) = -q^2 + 5q - 3$$

The replacement of one variable with another is important in later courses.

Now Try Exercises 41 and 45.

Sometimes letters other than f, such as g, h, or capital letters F, G, and H are used to name functions.

▨ EXAMPLE 7   Using Function Notation

Let $g(x) = 2x + 3$. Find and simplify $g(a + 1)$.

$$g(x) = 2x + 3$$
$$g(a + 1) = 2(a + 1) + 3 \qquad \text{Replace } x \text{ with } a + 1.$$
$$= 2a + 2 + 3$$
$$= 2a + 5$$

Now Try Exercise 49.

Functions can be evaluated in a variety of ways, as shown in Example 8.

▨ EXAMPLE 8   Using Function Notation

For each function, find $f(3)$.

(a) $f(x) = 3x - 7$

$$f(3) = 3(3) - 7 \qquad \text{Replace } x \text{ with } 3.$$
$$f(3) = 2$$

(b) $f = \{(-3, 5), (0, 3), (3, 1), (6, -1)\}$

We want $f(3)$, the y-value of the ordered pair where $x = 3$. As indicated by the ordered pair $(3, 1)$, when $x = 3$, $y = 1$, so $f(3) = 1$.

(c)

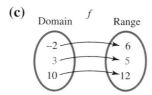

The domain element 3 is paired with 5 in the range, so $f(3) = 5$.

(d)

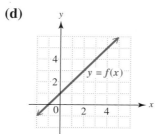

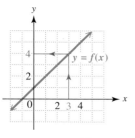

FIGURE 42

To evaluate $f(3)$, find 3 on the x-axis. See Figure 42. Then move up until the graph of f is reached. Moving horizontally to the y-axis gives 4 for the corresponding y-value. Thus, $f(3) = 4$.

Now Try Exercises 53, 55, and 57.

If a function f is defined by an equation with x and y, not with function notation, use the following steps to find $f(x)$.

Finding an Expression for $f(x)$

Step 1 Solve the equation for y.

Step 2 Replace y with $f(x)$.

EXAMPLE 9 Writing Equations Using Function Notation

Rewrite each equation using function notation. Then find $f(-2)$ and $f(a)$.

(a) $y = x^2 + 1$

This equation is already solved for y. Since $y = f(x)$,
$$f(x) = x^2 + 1.$$

To find $f(-2)$, let $x = -2$.
$$f(-2) = (-2)^2 + 1$$
$$= 4 + 1$$
$$= 5$$

Find $f(a)$ by letting $x = a$: $f(a) = a^2 + 1$.

(b) $x - 4y = 5$

First solve $x - 4y = 5$ for y. Then replace y with $f(x)$.
$$x - 4y = 5$$
$$x - 5 = 4y$$
$$y = \frac{x - 5}{4} \quad \text{so} \quad f(x) = \frac{1}{4}x - \frac{5}{4}$$

Now find $f(-2)$ and $f(a)$.
$$f(-2) = \frac{1}{4}(-2) - \frac{5}{4} = -\frac{7}{4} \qquad \text{Let } x = -2.$$
$$f(a) = \frac{1}{4}a - \frac{5}{4} \qquad \text{Let } x = a.$$

Now Try Exercise 59.

OBJECTIVE 5 Identify linear functions. Our first two-dimensional graphing was of straight lines. Linear equations (except for vertical lines with equations $x = a$) define *linear functions*.

Linear Function

A function that can be defined by

$$f(x) = mx + b$$

for real numbers m and b is a **linear function.**

Recall from Section 3.3 that m is the slope of the line and $(0, b)$ is the y-intercept. In Example 9(b), we wrote the equation $x - 4y = 5$ as the linear function defined by

$$f(x) = \frac{1}{4}x - \frac{5}{4}.$$

Slope⌐   ⌐y-intercept is $\left(0, -\frac{5}{4}\right)$.

$f(x) = \frac{1}{4}x - \frac{5}{4}$

$-\frac{5}{4}$ $m = \frac{1}{4}$

To graph this function, plot the y-intercept and use the definition of slope as $\frac{\text{rise}}{\text{run}}$ to find a second point on the line. Draw the straight line through the points to obtain the graph shown in the margin.

A linear function defined by $f(x) = b$ (whose graph is a horizontal line) is sometimes called a **constant function.** The domain of any linear function is $(-\infty, \infty)$. The range of a nonconstant linear function is $(-\infty, \infty)$, while the range of the constant function defined by $f(x) = b$ is $\{b\}$.

Now Try Exercise 67.

3.5 EXERCISES

✎ **1.** In your own words, define a function and give an example.

✎ **2.** In your own words, define the domain of a function and give an example.

3. In an ordered pair of a relation, is the first element the independent or the dependent variable?

4. Give an example of a relation that is not a function, having domain $\{-3, 2, 6\}$ and range $\{4, 6\}$. (There are many possible correct answers.)

Tell whether each relation defines a function. See Example 1.

5. $\{(5, 1), (3, 2), (4, 9), (7, 6)\}$ **6.** $\{(8, 0), (5, 4), (9, 3), (3, 8)\}$

7. $\{(2, 4), (0, 2), (2, 5)\}$ **8.** $\{(9, -2), (-3, 5), (9, 2)\}$

9. $\{(-3, 1), (4, 1), (-2, 7)\}$ **10.** $\{(-12, 5), (-10, 3), (8, 3)\}$

Decide whether each relation defines a function and give the domain and range. See Examples 1–4.

11. $\{(1, 1), (1, -1), (0, 0), (2, 4), (2, -4)\}$ **12.** $\{(2, 5), (3, 7), (4, 9), (5, 11)\}$

13.

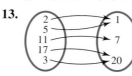

14.

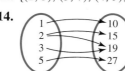

15.

x	y
1	5
1	2
1	-1
1	-4

16.

x	y
4	-3
2	-3
0	-3
-2	-3

17.

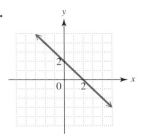

18.

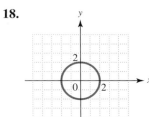

19.

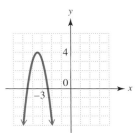

20.

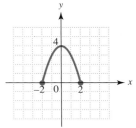

21.

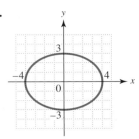

22.

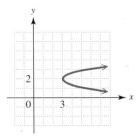

Decide whether each relation defines y as a function of x. Give the domain. See Example 5.

23. $y = x^2$ **24.** $y = x^3$ **25.** $x = y^6$ **26.** $x = y^4$

27. $y = 2x - 6$ **28.** $y = -6x + 8$ **29.** $x + y < 4$ **30.** $x - y < 3$

31. $y = \sqrt{x}$ **32.** $y = -\sqrt{x}$ **33.** $xy = 1$ **34.** $xy = -3$

35. $y = \sqrt{4x + 2}$ **36.** $y = \sqrt{9 - 2x}$ **37.** $y = \dfrac{2}{x - 9}$ **38.** $y = \dfrac{-7}{x - 16}$

39. Choose the correct response: The notation $f(3)$ means

   A. the variable f times 3 or $3f$.
   B. the value of the dependent variable when the independent variable is 3.
   C. the value of the independent variable when the dependent variable is 3.
   D. f equals 3.

40. Give an example of a function from everyday life. (*Hint:* Fill in the blanks: _____ depends on _____, so _____ is a function of _____.)

Let $f(x) = -3x + 4$ *and* $g(x) = -x^2 + 4x + 1$. *Find the following. See Examples 6 and 7.*

41. $f(0)$ **42.** $f(-3)$ **43.** $g(-2)$ **44.** $g(10)$

45. $f(p)$ **46.** $g(k)$ **47.** $f(-x)$ **48.** $g(-x)$

49. $f(x + 2)$ **50.** $f(a + 4)$ **51.** $f(2m - 3)$ **52.** $f(3t - 2)$

For each function, find (a) $f(2)$ and (b) $f(-1)$. See Example 8.

53. $f = \{(-1, 3), (4, 7), (0, 6), (2, 2)\}$ **54.** $f = \{(2, 5), (3, 9), (-1, 11), (5, 3)\}$

55.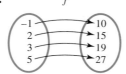

56.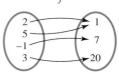

57.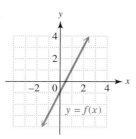

58.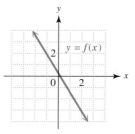

An equation that defines y as a function of x is given. (a) Solve for y in terms of x and replace y with the function notation $f(x)$. (b) Find $f(3)$. See Example 9.

59. $x + 3y = 12$ **60.** $x - 4y = 8$ **61.** $y + 2x^2 = 3$

62. $y - 3x^2 = 2$ **63.** $4x - 3y = 8$ **64.** $-2x + 5y = 9$

65. Fill in each blank with the correct response.

The equation $2x + y = 4$ has a straight _____ as its graph. One point that lies on the graph is (3, _____). If we solve the equation for y and use function notation, we obtain $f(x) =$ _____. For this function, $f(3) =$ _____, meaning that the point (_____, _____) lies on the graph of the function.

66. Which of the following defines a linear function?

     A. $y = \dfrac{x - 5}{4}$ **B.** $y = \dfrac{1}{x}$ **C.** $y = x^2$ **D.** $y = \sqrt{x}$

Graph each linear function. Give the domain and range. See Objective 5.

67. $f(x) = -2x + 5$ **68.** $g(x) = 4x - 1$ **69.** $h(x) = \dfrac{1}{2}x + 2$

70. $F(x) = -\dfrac{1}{4}x + 1$ **71.** $G(x) = 2x$ **72.** $H(x) = -3x$

73. $g(x) = -4$ **74.** $f(x) = 5$

75. Suppose that a package weighing x lb costs $f(x)$ dollars to mail to a given location, where

$$f(x) = 2.75x.$$

     (a) What is the value of $f(3)$?

✎ **(b)** In your own words, describe what 3 and the value $f(3)$ mean in part (a), using the terminology *independent variable* and *dependent variable*.

   (c) How much would it cost to mail a 5-lb package? Interpret this question and its answer using function notation.

76. Suppose that a Yellow Cab driver charges $1.50 per mile.

   (a) Fill in the table with the correct response for the price $f(x)$ he charges for a trip of x mi.

   (b) The linear function that gives a rule for the amount charged is $f(x) = $ _____.

   (c) Graph this function for the domain {0, 1, 2, 3}.

x	$f(x)$
0	
1	
2	
3	

Forensic scientists use the lengths of certain bones to calculate the height of a person. Two bones often used are the tibia (t), the bone from the ankle to the knee, and the femur (r), the bone from the knee to the hip socket. A person's height (h) is determined from the lengths of these bones using functions defined by the following formulas. All measurements are in centimeters.

For men: $h(r) = 69.09 + 2.24r$ or $h(t) = 81.69 + 2.39t$

For women: $h(r) = 61.41 + 2.32r$ or $h(t) = 72.57 + 2.53t$

77. Find the height of a man with a femur measuring 56 cm.

78. Find the height of a man with a tibia measuring 40 cm.

79. Find the height of a woman with a femur measuring 50 cm.

80. Find the height of a woman with a tibia measuring 36 cm.

Femur

Tibia

Federal regulations set standards for the size of the quarters of marine mammals. A pool to house sea otters must have a volume of "the square of the sea otter's average adult length (in meters) multiplied by 3.14 and by .91 meter." If x represents the sea otter's average adult length and f(x) represents the volume (in cubic meters) of the corresponding pool size, this formula can be written as

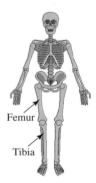

$$f(x) = .91(3.14)x^2.$$

Find the volume of the pool for each adult sea otter length (in meters). Round answers to the nearest hundredth.

81. .8   **82.** 1.0   **83.** 1.2   **84.** 1.5

85. The graph shows the daily megawatts of electricity used on a record-breaking summer day in Sacramento, California.

ELECTRICITY USE

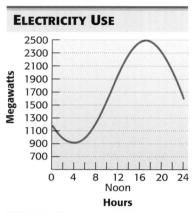

Source: Sacramento Municipal Utility District.

(a) Is this the graph of a function?
(b) What is the domain?
(c) Estimate the number of megawatts used at 8 A.M.
(d) At what time was the most electricity used? the least electricity?
(e) Call this function f. What is $f(12)$? What does it mean?

86. Refer to the graph to answer the questions.

GALLONS OF WATER IN A POOL AT TIME *t*

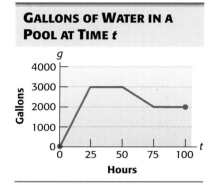

(a) What numbers are possible values of the independent variable? the dependent variable?
(b) For how long is the water level increasing? decreasing?
(c) How many gallons of water are in the pool after 90 hr?
(d) Call this function f. What is $f(0)$? What does it mean?
(e) What is $f(25)$? What does it mean?

TECHNOLOGY INSIGHTS (EXERCISES 87 AND 88)

87. The calculator screen shows the graph of a linear function $y = f(x)$, along with the display of coordinates of a point on the graph. Use function notation to write what the display indicates.

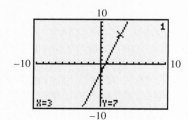

88. The table was generated by a graphing calculator for a linear function $Y_1 = f(X)$. Use the table to work parts (a)–(e).

(a) What is $f(2)$?
(b) If $f(X) = -3.7$, what is the value of X?
(c) What is the slope of the line?
(d) What is the y-intercept of the line?
(e) Find the expression for $f(X)$.

X	Y₁
0	3.5
1	2.3
2	1.1
3	-.1
4	-1.3
5	-2.5
6	-3.7

X=0

3.6 | Variation

$C = 2\pi r$

Certain types of functions are very common, especially in business and the physical sciences. These are functions where y depends on a multiple of x, or y depends on a number divided by x. In such situations, y is said to *vary directly as* x (in the first case) or *vary inversely as* x (in the second case). For example, by the distance formula, the distance traveled varies directly as the rate (or speed) and the time. Formulas for area and volume are other familiar examples of *direct variation*.

On the other hand, the force required to keep a car from skidding on a curve varies inversely as the radius of the curve. Another example of *inverse variation* is how travel time is inversely proportional to rate or speed.

OBJECTIVE 1 Write an equation expressing direct variation. The circumference of a circle is given by the formula $C = 2\pi r$, where r is the radius of the circle. See the figure. Circumference is always a constant multiple of the radius. (C is always found by multiplying r by the constant 2π.) Thus,

> As the *radius increases*, the *circumference increases*.

The reverse is also true.

> As the *radius decreases*, the *circumference decreases*.

Because of this, the circumference is said to *vary directly* as the radius.

Direct Variation

y varies directly as x if there exists a real number k such that

$$y = kx.$$

Also, y is said to be **proportional to** x. The number k is called the **constant of variation.** In direct variation, for $k > 0$, as the value of x increases, the value of y also increases. Similarly, as x decreases, y decreases.

OBJECTIVE 2 Find the constant of variation, and solve direct variation problems. The direct variation equation $y = kx$ defines a linear function, where the constant of variation k is the slope of the line. For example, we wrote the equation

$$y = 1.60x$$

to describe the cost y to buy x gal of gas in Example 6 of Section 3.3. The cost varies directly as, or is proportional to, the number of gallons of gas purchased. That is, as the number of gallons of gas increases, cost increases; also, as the number of gallons of gas decreases, cost decreases. The constant of variation k is 1.60, the cost of 1 gal of gas.

▧ **EXAMPLE 1** Finding the Constant of Variation and the Variation Equation

Steven Pusztai is paid an hourly wage. One week he worked 43 hr and was paid $795.50. How much does he earn per hour?

Let h represent the number of hours he works and P represent his corresponding pay. Then, P varies directly as h, so

$$P = kh.$$

Here, k represents Steven's hourly wage. Since $P = 795.50$ when $h = 43$,

$$795.50 = 43k$$
$$k = 18.50. \quad \text{Use a calculator.}$$

His hourly wage is $18.50, and P and h are related by

$$P = 18.50h.$$

Now Try Exercise 31.

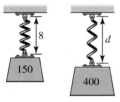

FIGURE 43

▧ **EXAMPLE 2** Solving a Direct Variation Problem

Hooke's law for an elastic spring states that the distance a spring stretches is proportional to the force applied. If a force of 150 newtons* stretches a certain spring 8 cm, how much will a force of 400 newtons stretch the spring? See Figure 43.

If d is the distance the spring stretches and f is the force applied, then $d = kf$ for some constant k. Since a force of 150 newtons stretches the spring 8 cm, use these values to find k.

$$d = kf \qquad \text{Variation equation}$$
$$8 = k \cdot 150 \qquad \text{Let } d = 8 \text{ and } f = 150.$$
$$k = \frac{8}{150} \qquad \text{Find } k.$$
$$k = \frac{4}{75}$$

Substitute $\frac{4}{75}$ for k in the variation equation $d = kf$ to get

$$d = \frac{4}{75}f.$$

For a force of 400 newtons,

$$d = \frac{4}{75}(400) \qquad \text{Let } f = 400.$$
$$= \frac{64}{3}.$$

The spring will stretch $\frac{64}{3}$ cm if a force of 400 newtons is applied.

Now Try Exercise 35.

---

*A newton is a unit of measure of force used in physics.

In summary, use the following steps to solve a variation problem.

Solving a Variation Problem

Step 1   Write the variation equation.

Step 2   Substitute the initial values and solve for k.

Step 3   Rewrite the variation equation with the value of k from Step 2.

Step 4   Substitute the remaining values, solve for the unknown, and find the required answer.

The direct variation equation $y = kx$ is a linear equation. However, other kinds of variation involve other types of equations. For example, one variable can be proportional to a power of another variable.

Direct Variation as a Power

y **varies directly as the nth power of x** if there exists a real number k such that

$$y = kx^n.$$

$A = \pi r^2$

An example of direct variation as a power is the formula for the area of a circle, $A = \pi r^2$. Here, π is the constant of variation, and the area varies directly as the *square* of the radius.

■ EXAMPLE 3   Solving a Direct Variation Problem

The distance a body falls from rest varies directly as the square of the time it falls (disregarding air resistance). If a skydiver falls 64 ft in 2 sec, how far will she fall in 8 sec?

Step 1   If d represents the distance the skydiver falls and t the time it takes to fall, then d is a function of t, and, for some constant k,

$$d = kt^2.$$

Step 2   To find the value of k, use the fact that the skydiver falls 64 ft in 2 sec.

$$d = kt^2 \qquad \text{Variation equation}$$
$$64 = k(2)^2 \qquad \text{Let } d = 64 \text{ and } t = 2.$$
$$k = 16 \qquad \text{Find } k.$$

Step 3   Using 16 for k, the variation equation becomes

$$d = 16t^2.$$

Step 4   Now let $t = 8$ to find the number of feet the skydiver will fall in 8 sec.

$$d = 16(8)^2 \qquad \text{Let } t = 8.$$
$$= 1024$$

The skydiver will fall 1024 ft in 8 sec.

Now Try Exercise 37.

As pressure on trash increases, volume of trash decreases.

FIGURE 44

OBJECTIVE 3 Solve inverse variation problems.   In direct variation, where $k > 0$, as x increases, y increases. Similarly, as x decreases, y decreases. Another type of variation is *inverse variation*. With inverse variation, where $k > 0$, as one variable increases, the other variable decreases. For example, in a closed space, volume decreases as pressure increases, as illustrated by a trash compactor. See Figure 44. As the compactor presses down, the pressure on the trash increases; in turn, the trash occupies a smaller space.

Inverse Variation

y varies inversely as x if there exists a real number k such that

$$y = \frac{k}{x}.$$

Also, **y varies inversely as the nth power of x** if there exists a real number k such that

$$y = \frac{k}{x^n}.$$

The inverse variation equation also defines a function. Since x is in the denominator, these functions are *rational functions*. (See Chapter 7.) Another example of inverse variation comes from the distance formula. In its usual form, the formula is

$$d = rt.$$

Dividing each side by r gives

$$t = \frac{d}{r}.$$

Here, t (time) varies inversely as r (rate or speed), with d (distance) serving as the constant of variation. For example, if the distance between Chicago and Des Moines is 300 mi, then

$$t = \frac{300}{r}$$

and the values of r and t might be any of the following.

$\left.\begin{array}{l} r = 50, t = 6 \\ r = 60, t = 5 \\ r = 75, t = 4 \end{array}\right\}$ As *r* increases, *t* decreases. $\left.\begin{array}{l} r = 30, t = 10 \\ r = 25, t = 12 \\ r = 20, t = 15 \end{array}\right\}$ As *r* decreases, *t* increases.

If we *increase* the rate (speed) we drive, time *decreases*. If we *decrease* the rate (speed) we drive, what happens to time?

EXAMPLE 4   Solving an Inverse Variation Problem

In the manufacturing of a certain medical syringe, the cost of producing the syringe varies inversely as the number produced. If 10,000 syringes are produced, the cost is $2 per syringe. Find the cost per syringe to produce 25,000 syringes.

Let x = the number of syringes produced,

and c = the cost per syringe.

Here, as production increases, cost decreases and as production decreases, cost increases. Since c varies inversely as x, there is a constant k such that

$$c = \frac{k}{x}.$$

Find k by replacing c with 2 and x with 10,000.

$$2 = \frac{k}{10{,}000}$$

$$20{,}000 = k \qquad\qquad \text{Multiply by 10,000.}$$

Since $c = \frac{k}{x}$,

$$c = \frac{20{,}000}{25{,}000} = .80. \qquad \text{Let } k = 20{,}000 \text{ and } x = 25{,}000.$$

The cost per syringe to make 25,000 syringes is $.80.

Now Try Exercise 39.

EXAMPLE 5 Solving an Inverse Variation Problem

The weight of an object above Earth varies inversely as the square of its distance from the center of Earth. A space shuttle in an elliptical orbit has a maximum distance from the center of Earth (*apogee*) of 6700 mi. Its minimum distance from the center of Earth (*perigee*) is 4090 mi. See Figure 45. If an astronaut in the shuttle weighs 57 lb at its apogee, what does the astronaut weigh at its perigee?

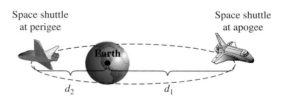

Space shuttle at perigee   Space shuttle at apogee

Earth

d_2 d_1

FIGURE 45

If w is the weight and d is the distance from the center of Earth, then

$$w = \frac{k}{d^2}$$

for some constant k. At the apogee the astronaut weighs 57 lb, and the distance from the center of Earth is 6700 mi. Use these values to find k.

$$57 = \frac{k}{(6700)^2} \qquad \text{Let } w = 57 \text{ and } d = 6700.$$

$$k = 57(6700)^2$$

Then the weight at the perigee with $d = 4090$ mi is

$$w = \frac{57(6700)^2}{(4090)^2} \approx 153 \text{ lb.} \qquad \text{Use a calculator.}$$

Now Try Exercise 43.

OBJECTIVE 4 Solve joint variation problems. It is common for one variable to depend on several others. If one variable varies directly as the *product* of several other variables (perhaps raised to powers), the first variable is said to *vary jointly* as the others.

Joint Variation

y **varies jointly as** *x* **and** *z* if there exists a real number *k* such that

$$y = kxz.$$

CAUTION Note that *and* in the expression "*y* varies directly as *x and z*" translates as the product $y = kxz$. The word *and* does not indicate addition here.

EXAMPLE 6 Solving a Joint Variation Problem

The interest on a loan or an investment is given by the formula $I = prt$. Here, for a given principal *p*, the interest earned *I* varies jointly as the interest rate *r* and the time *t* the principal is left at interest. If an investment earns \$100 interest at 5% for 2 yr, how much interest will the same principal earn at 4.5% for 3 yr?

We use the formula $I = prt$, where *p* is the constant of variation because it is the same for both investments. For the first investment, we have $I = 100$, $r = .05$, and $t = 2$, so

$$I = prt$$
$$100 = p(.05)(2) \qquad \text{Let } I = 100, r = .05, \text{ and } t = 2.$$
$$100 = .1p$$
$$\frac{100}{.1} = p$$
$$p = 1000.$$

Now we find *I* when $p = 1000$, $r = .045$, and $t = 3$.

$$I = 1000(.045)(3) \qquad \text{Let } p = 1000, r = .045, \text{ and } t = 3.$$
$$I = 135$$

The interest will be \$135.

Now Try Exercise 45.

OBJECTIVE 5 Solve combined variation problems. There are many combinations of direct and inverse variation. Example 7 shows a typical **combined variation** problem.

EXAMPLE 7 Solving a Combined Variation Problem

Body mass index, or BMI, is used by physicians to assess a person's level of fatness. A BMI from 19 through 25 is considered desirable. BMI varies directly as an individual's weight in pounds and inversely as the square of the individual's height in inches. A person who weighs 118 lb and is 64 in. tall has a BMI of 20. (The BMI is rounded to the nearest whole number.) Find the BMI of a person who weighs 165 lb with a height of 70 in. (*Source: Washington Post.*)

Let B represent the BMI, w the weight, and h the height. Then

$$B = \frac{kw}{h^2}.$$ ←——— BMI varies directly as the weight.
←——— BMI varies inversely as the square of the height.

To find k, let $B = 20$, $w = 118$, and $h = 64$.

$$20 = \frac{k(118)}{64^2}$$

$$k = \frac{20(64^2)}{118} \qquad \text{Multiply by } 64^2; \text{ divide by 118.}$$

$$k \approx 694 \qquad \text{Use a calculator.}$$

Now find B when $k = 694$, $w = 165$, and $h = 70$.

$$B = \frac{694(165)}{70^2} \approx 23 \qquad \text{Nearest whole number}$$

The person's BMI is 23.

Now Try Exercise 47.

3.6 EXERCISES

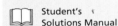

Use personal experience or intuition to determine whether the situation suggests direct *or* inverse *variation.*

1. The number of different lottery tickets you buy and your probability of winning that lottery

2. The rate and the distance traveled by a pickup truck in 3 hr

3. The amount of pressure put on the accelerator of a car and the speed of the car

4. The number of days from now until December 25 and the magnitude of the frenzy of Christmas shopping

5. Your age and the probability that you believe in Santa Claus

6. The surface area of a balloon and its diameter

7. The number of days until the end of the baseball season and the number of home runs that Barry Bonds has

8. The amount of gasoline you pump and the amount you pay

Determine whether each equation represents direct, inverse, joint, *or* combined *variation.*

9. $y = \dfrac{3}{x}$ **10.** $y = \dfrac{8}{x}$ **11.** $y = 10x^2$ **12.** $y = 2x^3$

13. $y = 3xz^4$ **14.** $y = 6x^3z^2$ **15.** $y = \dfrac{4x}{wz}$ **16.** $y = \dfrac{6x}{st}$

17. For $k > 0$, if y varies directly as x, when x increases, y _____, and when x decreases, y _____.

18. For $k > 0$, if y varies inversely as x, when x increases, y _____, and when x decreases, y _____.

Solve each problem.

19. If x varies directly as y, and $x = 9$ when $y = 3$, find x when $y = 12$.

20. If x varies directly as y, and $x = 10$ when $y = 7$, find y when $x = 50$.

21. If a varies directly as the square of b, and $a = 4$ when $b = 3$, find a when $b = 2$.

22. If h varies directly as the square of m, and $h = 15$ when $m = 5$, find h when $m = 7$.

23. If z varies inversely as w, and $z = 10$ when $w = .5$, find z when $w = 8$.

24. If t varies inversely as s, and $t = 3$ when $s = 5$, find s when $t = 5$.

25. If m varies inversely as p^2, and $m = 20$ when $p = 2$, find m when $p = 5$.

26. If a varies inversely as b^2, and $a = 48$ when $b = 4$, find a when $b = 7$.

27. p varies jointly as q and r^2, and $p = 200$ when $q = 2$ and $r = 3$. Find p when $q = 5$ and $r = 2$.

28. f varies jointly as g^2 and h, and $f = 50$ when $g = 4$ and $h = 2$. Find f when $g = 3$ and $h = 6$.

29. Explain the difference between inverse variation and direct variation.

30. What is meant by the constant of variation in a direct variation problem? If you were to graph the linear equation $y = kx$ for some nonnegative constant k, what role would the value of k play in the graph?

Solve each problem involving variation. See Examples 1–7.

31. Todd bought 8 gal of gasoline and paid $13.59. To the nearest tenth of a cent, what is the price of gasoline per gallon?

32. Melissa gives horseback rides at Shadow Mountain Ranch. A 2.5-hr ride costs $50.00. What is the price per hour?

33. The volume of a can of tomatoes is proportional to the height of the can. If the volume of the can is 300 cm³ when its height is 10.62 cm, find the volume of a can with height 15.92 cm.

34. The force required to compress a spring is proportional to the change in length of the spring. If a force of 20 newtons is required to compress a certain spring 2 cm, how much force is required to compress the spring from 20 cm to 8 cm?

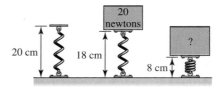

35. The weight of an object on Earth is directly proportional to the weight of that same object on the moon. A 200-lb astronaut would weigh 32 lb on the moon. How much would a 50-lb dog weigh on the moon?

36. The pressure exerted by a certain liquid at a given point varies directly as the depth of the point beneath the surface of the liquid. The pressure at 30 m is 80 newtons per m². What pressure is exerted at 50 m?

37. For a body falling freely from rest (disregarding air resistance), the distance the body falls varies directly as the square of the time. If an object is dropped from the top of a tower 576 ft high and hits the ground in 6 sec, how far did it fall in the first 4 sec?

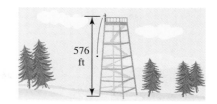

38. The amount of water emptied by a pipe varies directly as the square of the diameter of the pipe. For a certain constant water flow, a pipe emptying into a canal will allow 200 gal of water to escape in an hour. The diameter of the pipe is 6 in. How much water would a 12-in. pipe empty into the canal in an hour, assuming the same water flow?

39. Over a specified distance, speed varies inversely with time. If a Dodge Viper on a test track goes a certain distance in one-half minute at 160 mph, what speed is needed to go the same distance in three-fourths minute?

40. For a constant area, the length of a rectangle varies inversely as the width. The length of a rectangle is 27 ft when the width is 10 ft. Find the width of a rectangle with the same area if the length is 18 ft.

41. The frequency of a vibrating string varies inversely as its length. That is, a longer string vibrates fewer times in a second than a shorter string. Suppose a piano string 2 ft long vibrates 250 cycles per sec. What frequency would a string 5 ft long have?

42. The current in a simple electrical circuit varies inversely as the resistance. If the current is 20 amps when the resistance is 5 ohms, find the current when the resistance is 7.5 ohms.

43. The amount of light (measured in foot-candles) produced by a light source varies inversely as the square of the distance from the source. If the illumination produced 1 m from a light source is 768 foot-candles, find the illumination produced 6 m from the same source.

44. The force with which Earth attracts an object above Earth's surface varies inversely with the square of the distance of the object from the center of Earth. If an object 4000 mi from the center of Earth is attracted with a force of 160 lb, find the force of attraction if the object were 6000 mi from the center of Earth.

45. For a given interest rate, simple interest varies jointly as principal and time. If $2000 left in an account for 4 yr earned interest of $280, how much interest would be earned in 6 yr?

46. The collision impact of an automobile varies jointly as its mass and the square of its speed. Suppose a 2000-lb car traveling at 55 mph has a collision impact of 6.1. What is the collision impact of the same car at 65 mph?

47. The force needed to keep a car from skidding on a curve varies inversely as the radius of the curve and jointly as the weight of the car and the square of the speed. If 242 lb of force keep a 2000-lb car from skidding on a curve of radius 500 ft at 30 mph, what force would keep the same car from skidding on a curve of radius 750 ft at 50 mph?

48. The maximum load that a cylindrical column with a circular cross section can hold varies directly as the fourth power of the diameter of the cross section and inversely as the square of the height. A 9-m column 1 m in diameter will support 8 metric tons. How many metric tons can be supported by a column 12 m high and $\frac{2}{3}$ m in diameter?

9 m

1 m

Load = 8 metric tons

49. The number of long-distance phone calls between two cities in a certain time period varies jointly as the populations of the cities, p_1 and p_2, and inversely as the distance between them. If 80,000 calls are made between two cities 400 mi apart, with populations of 70,000 and 100,000, how many calls are made between cities with populations of 50,000 and 75,000 that are 250 mi apart?

50. A body mass index from 27 through 29 carries a slight risk of weight-related health problems, while one of 30 or more indicates a great increase in risk. Use your own height and weight and the information in Example 7 to determine your BMI and whether you are at risk.

51. Natural gas provides 35.8% of U.S. energy. (*Source:* U.S. Energy Department.) The volume of gas varies inversely as the pressure and directly as the temperature. (Temperature must be measured in *Kelvin* (K), a unit of measurement used in physics.) If a certain gas occupies a volume of 1.3 L at 300 K and a pressure of 18 newtons per cm², find the volume at 340 K and a pressure of 24 newtons per cm².

52. The maximum load of a horizontal beam that is supported at both ends varies directly as the width and the square of the height and inversely as the length between the supports. A beam 6 m long, .1 m wide, and .06 m high supports a load of 360 kg. What is the maximum load supported by a beam 16 m long, .2 m wide, and .08 m high?

Exercises 53 and 54 describe weight-estimation formulas that fishermen have used over the years. Girth *is the distance around the body of the fish.* (*Source: Sacramento Bee,* November 9, 2000.)

53. The weight of a bass varies jointly as its girth and the square of its length. A prize-winning bass weighed in at 22.7 lb and measured 36 in. long with 21 in. girth. How much would a bass 28 in. long with 18 in. girth weigh?

54. The weight of a trout varies jointly as its length and the square of its girth. One angler caught a trout that weighed 10.5 lb and measured 26 in. long with 18 in. girth. Find the weight of a trout that is 22 in. long with 15 in. girth.

RELATING CONCEPTS (EXERCISES 55–62)

For Individual or Group Work

A routine activity such as pumping gasoline can be related to many of the concepts studied in this chapter. Suppose that premium unleaded costs $1.25 per gallon. **Work Exercises 55–62 in order.**

55. Zero gallons of gasoline cost $0.00, while 1 gallon costs $1.25. Represent these two pieces of information as ordered pairs of the form (gallons, price).

56. Use the information from Exercise 55 to find the slope of the line on which the two points lie.

57. Write the slope-intercept form of the equation of the line on which the two points lie.

58. Using function notation, if $f(x) = ax + b$ represents the line from Exercise 57, what are the values of a and b?

59. How does the value of a from Exercise 58 relate to gasoline in this situation? With relationship to the line, what do we call this number?

60. Why does the equation from Exercise 57 satisfy the conditions for direct variation? In the context of variation, what do we call the value of a?

61. The graph of the equation from Exercise 57 is shown in the calculator screen. How is the display at the bottom of the screen interpreted in the context of these exercises?

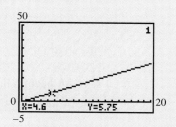

62. The table was generated by a graphing calculator, with Y_1 entered as the equation from Exercise 57. Interpret the entry for X = 12 in the context of these exercises.

X	Y₁
7	8.75
8	10
9	11.25
10	12.5
11	13.75
12	15
13	16.25

X=12

Chapter **3** **Group Activity**

Choosing an Energy Source (or How to Get Hot Water)

OBJECTIVE Write and graph linear functions that model given data.

There are many different ways to heat water. In this activity you will look at three different energy sources that may be used to provide heat for a 40-gal home water tank.

Have each student in your group choose one of the three types of water heaters listed in the table.

Type of Hot Water Heater	Size (in gallons)	Price	Operating Cost per Month (manufacturer's estimate)	Hot Water Temperature
Kenmore Economizer 6 —Electric	40	$139.99	$35.00	120°–130°
Kenmore Economizer 6 —Natural Gas	40	$139.99	$13.25	120°–130°
Sunbather Water Heater —Solar	40	$950.00	$0.00	*

Source: Jade Mountain 1999.

A. Using data for the water heater you selected, find a linear equation that represents total cost y of heating water with respect to time. Let x represent number of months. Write the equation in slope-intercept form.

B. Graph your equation using domain [0, 60] and range [0, 1000].

C. As a group, compare the graphs of your equations.

    1. What are the y-intercepts?

    2. How do the slopes compare? Which is the steepest? Which has 0 slope?

D. Discuss other factors to consider when choosing each type of water heater. Which of these water heaters would you choose to heat your home?

---

*No temperature listed but the ad says "Best for warm climates, preheating water, summer only in cold places, or when hot water needed only in afternoons and evenings."

CHAPTER 3 SUMMARY

KEY TERMS

3.1 ordered pair
origin
x-axis
y-axis
rectangular
(Cartesian)
coordinate system
plot
coordinate
quadrant

graph of an equation
first-degree equation
linear equation in two
variables
standard form
x-intercept
y-intercept
3.2 rise
run
slope

3.3 slope-intercept form
point-slope form
3.4 linear inequality in
two variables
boundary line
3.5 dependent variable
independent variable
relation
function
domain

range
function notation
linear function
constant function
3.6 vary directly
proportional
constant of variation
vary inversely
vary jointly
combined variation

NEW SYMBOLS

(a, b) ordered pair

x_1 a specific value of the
variable *x* (read
"*x*-sub-one")

m slope

f(x) function of *x* (read
"*f* of *x*")

TEST YOUR WORD POWER

See how well you have learned the vocabulary in this chapter. Answers, with examples, follow the Quick Review.

1. An **ordered pair** is a pair of numbers written
   A. in numerical order between brackets
   B. between parentheses or brackets
   C. between parentheses in which order is important
   D. between parentheses in which order does not matter.

2. A **linear equation in two variables** is an equation that can be written in the form
   A. $Ax + By < C$
   B. $ax = b$
   C. $y = x^2$
   D. $Ax + By = C$.

3. An **intercept** is
   A. the point where the *x*-axis and *y*-axis intersect
   B. a pair of numbers written between parentheses in which order matters

   C. one of the four regions determined by a rectangular coordinate system
   D. the point where a graph intersects the *x*-axis or the *y*-axis.

4. The **slope** of a line is
   A. the measure of the run over the rise of the line
   B. the distance between two points on the line
   C. the ratio of the change in *y* to the change in *x* along the line
   D. the horizontal change compared to the vertical change between two points on the line.

5. In a relationship between two variables *x* and *y*, the **independent variable** is
   A. *x*, if *x* depends on *y*
   B. *x*, if *y* depends on *x*
   C. either *x* or *y*
   D. the larger of *x* and *y*.

6. In a relationship between two variables *x* and *y*, the **dependent variable** is
   A. *y*, if *y* depends on *x*
   B. *y*, if *x* depends on *y*
   C. either *x* or *y*
   D. the smaller of *x* and *y*.

7. A **relation** is
   A. a set of ordered pairs
   B. the ratio of the change in *y* to the change in *x* along a line
   C. the set of all possible values of the independent variable
   D. all the second components of a set of ordered pairs.

8. A **function** is
   A. the numbers in an ordered pair
   B. a set of ordered pairs in which each *x*-value corresponds to exactly one *y*-value
   C. a pair of numbers written between parentheses in which order matters
   D. the set of all ordered pairs that satisfy an equation.

(continued)

9. The **domain** of a function is
   A. the set of all possible values of the dependent variable y
   B. a set of ordered pairs
   C. the difference between the x-values
   D. the set of all possible values of the independent variable x.

10. The **range** of a function is
   A. the set of all possible values of the dependent variable y
   B. a set of ordered pairs
   C. the difference between the y-values
   D. the set of all possible values of the independent variable x.

◼ QUICK REVIEW

CONCEPTS	EXAMPLES

3.1 THE RECTANGULAR COORDINATE SYSTEM

Finding Intercepts

To find the x-intercept, let $y = 0$.

To find the y-intercept, let $x = 0$.

The graph of $2x + 3y = 12$ has

x-intercept $(6, 0)$

and y-intercept $(0, 4)$.

3.2 THE SLOPE OF A LINE

If $x_2 \neq x_1$, then

$$m = \frac{\text{rise}}{\text{run}} = \frac{\text{change in } y}{\text{change in } x} = \frac{y_2 - y_1}{x_2 - x_1}.$$

A vertical line has undefined slope.

A horizontal line has 0 slope.

Parallel lines have equal slopes.

For $2x + 3y = 12$,

$$m = \frac{4 - 0}{0 - 6} = -\frac{2}{3}.$$

$x = 3$ has undefined slope.

$y = -5$ has $m = 0$.

$y = 2x + 3$	$4x - 2y = 6$
$m = 2$	$-2y = -4x + 6$
	$y = 2x - 3$
	$m = 2$

These lines are parallel.

The slopes of perpendicular lines are negative reciprocals with a product of -1.

$y = 3x - 1$	$x + 3y = 4$
$m = 3$	$3y = -x + 4$
	$y = -\dfrac{1}{3}x + \dfrac{4}{3}$
	$m = -\dfrac{1}{3}$

These lines are perpendicular.

3.3 LINEAR EQUATIONS IN TWO VARIABLES

Slope-Intercept Form

$y = mx + b$

$y = 2x + 3$ $m = 2$, y-intercept is $(0, 3)$.

Point-Slope Form

$y - y_1 = m(x - x_1)$

$y - 3 = 4(x - 5)$ $(5, 3)$ is on the line, $m = 4$.

CONCEPTS	EXAMPLES

Standard Form

$Ax + By = C$ (A, B, C integers, $A \geq 0$)

$$2x - 5y = 8$$

Horizontal Line

$y = b$

$$y = 4$$

Vertical Line

$x = a$

$$x = -1$$

3.4 LINEAR INEQUALITIES IN TWO VARIABLES

Graphing a Linear Inequality

Step 1   Draw the graph of the line that is the boundary. Make the line solid if the inequality involves $\leq$ or $\geq$; make the line dashed if the inequality involves $<$ or $>$.

Step 2   Choose any point not on the line as a test point. Substitute the coordinates in the inequality.

Step 3   Shade the region that includes the test point if the test point satisfies the original inequality; otherwise, shade the region on the other side of the boundary line.

Graph $2x - 3y \leq 6$.

Draw the graph of $2x - 3y = 6$. Use a solid line because of $\leq$.

Choose $(1, 2)$.

$$2(1) - 3(2) = 2 - 6 \leq 6 \qquad \text{True}$$

Shade the side of the line that includes $(1, 2)$.

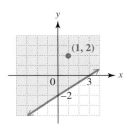

3.5 INTRODUCTION TO FUNCTIONS

A **function** is a set of ordered pairs such that for each first component there is one and only one second component. The set of first components is called the **domain,** and the set of second components is called the **range.**

$y = f(x) = x^2$ defines a function f, with domain $(-\infty, \infty)$ and range $[0, \infty)$.

To evaluate a function using function notation (that is, $f(x)$ notation) for a given value of x, substitute the value wherever x appears.

If $f(x) = x^2 - 7x + 12$, then

$$f(1) = 1^2 - 7(1) + 12 = 6.$$

To write an equation that defines a function in function notation,

Step 1   Solve the equation for y.

Step 2   Replace y with $f(x)$.

Write $2x + 3y = 12$ using function notation.

$$3y = -2x + 12 \qquad \text{Subtract 2x.}$$

$$y = -\frac{2}{3}x + 4 \qquad \text{Divide by 3.}$$

$$f(x) = -\frac{2}{3}x + 4$$

CONCEPTS	EXAMPLES

3.6 VARIATION

If there is some constant k such that:

$y = kx^n$, then y varies directly as x^n.

$y = \dfrac{k}{x^n}$, then y varies inversely as x^n.

$y = kxz$, then y varies jointly as x and z.

The area of a circle varies directly as the square of the radius.

$$A = kr^2 \qquad \text{Here, } k = \pi.$$

Pressure varies inversely as volume.

$$p = \dfrac{k}{V}$$

For a given principal, interest varies jointly as interest rate and time.

$$I = krt \qquad k \text{ is the given principal.}$$

Answers to Test Your Word Power

1. C; *Examples:* (0, 3), (3, 8), (4, 0)   **2.** D; *Examples:* $3x + 2y = 6$, $x = y - 7$, $4x = y$ **3.** D; *Example:* In Figure 4(b) of Section 3.1, the x-intercept is (3, 0) and the y-intercept is (0, 2).   **4.** C; *Example:* The line through (3, 6) and (5, 4) has slope $\frac{4-6}{5-3} = \frac{-2}{2} = -1$.
5. B; *Example:* See Answer 6, which follows.   **6.** A; *Example:* When borrowing money, the amount you borrow (independent variable) determines the size of your payments (dependent variable).   **7.** A; *Example:* The set $\{(2, 0), (4, 3), (6, 6), (8, 9)\}$ defines a relation.
8. B; The relation given in Answer 7 is a function since the x-value of each ordered pair corresponds to exactly one y-value.
9. D; *Example:* In the function in Answer 7, the domain is the set of x-values, $\{2, 4, 6, 8\}$.   **10.** A; *Example:* In the function in Answer 7, the range is the set of y-values, $\{0, 3, 6, 9\}$.

CHAPTER 3 REVIEW EXERCISES

[3.1] *Complete the table of ordered pairs for each equation. Then graph the equation.*

1. $3x + 2y = 10$

x	y
0	
	0
2	
	-2

2. $x - y = 8$

x	y
2	
	-3
3	
	-2

Find the x- and y-intercepts and then graph each equation.

3. $4x - 3y = 12$

4. $5x + 7y = 28$

5. $2x + 5y = 20$

6. $x - 4y = 8$

✎ **7.** Explain how the signs of the x- and y-coordinates of a point determine the quadrant in which the point lies.

[3.2] *Find the slope of each line.*

8. Through $(-1, 2)$ and $(4, -5)$

9. Through $(0, 3)$ and $(-2, 4)$

10. $y = 2x + 3$

11. $3x - 4y = 5$

12. $x = 5$

13. Parallel to $3y = 2x + 5$

14. Perpendicular to $3x - y = 4$

15. Through $(-1, 5)$ and $(-1, -4)$

16. $y + 6 = 0$

17.

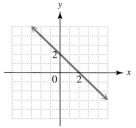

Tell whether each line has positive, negative, 0, *or* undefined *slope.*

18.

19.

20.

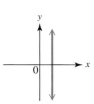

21.

22. If a walkway rises 2 ft for every 10 ft on the horizontal, which of the following express its slope (or grade)? (There are several correct choices.)

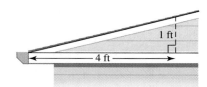

  A. .2   **B.** $\dfrac{2}{10}$ **C.** $\dfrac{1}{5}$

  D. 20%   **E.** 5   **F.** $\dfrac{20}{100}$

  G. 500%   **H.** $\dfrac{10}{2}$

23. If the pitch of a roof is $\frac{1}{4}$, how many feet in the horizontal direction correspond to a rise of 3 ft?

24. Family income in the United States has steadily increased for many years (primarily due to inflation). In 1970 the median family income was about $10,000 a year. In 1999 it was about $49,000 a year. Find the average rate of change of median family income to the nearest dollar over that period. (*Source:* U.S. Bureau of the Census.)

[3.3] *Find an equation for each line, if possible. Write the equation in slope-intercept form.*

25. Slope $-\dfrac{1}{3}$; y-intercept $(0, -1)$

26. Slope 0; y-intercept $(0, -2)$

27. Slope $-\dfrac{4}{3}$; through $(2, 7)$

28. Slope 3; through $(-1, 4)$

29. Vertical; through $(2, 5)$

30. Through $(2, -5)$ and $(1, 4)$

31. Through $(-3, -1)$ and $(2, 6)$

32. The line pictured in Exercise 17

33. Parallel to $4x - y = 3$ and through $(7, -1)$

34. Perpendicular to $2x - 5y = 7$ and through $(4, 3)$

35. The Midwest Athletic Club (Section 3.3, Exercises 73 and 74) offers two special membership plans. (*Source:* Midwest Athletic Club.) For each plan, write a linear equation in slope-intercept form and give the cost y in dollars of a 1-yr membership. Let x represent the number of months.

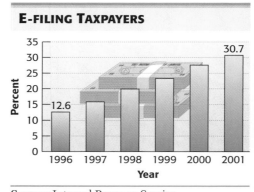

(a) Executive VIP/Gold membership: $159 fee plus $57 per month

(b) Executive Regular/Silver membership: $159 fee plus $47 per month

36. The percent of tax returns filed electronically for the years 1996–2001 is shown in the graph.

(a) Use the information given for the years 1996 and 2001, letting $x = 6$ represent 1996, $x = 11$ represent 2001, and y represent the percent of returns filed electronically to find a linear equation that models the data. Write the equation in slope-intercept form. Interpret the slope of this equation.

(b) Use your equation from part (a) to predict the percent of tax returns that will be filed electronically in 2005. (Assume a constant rate of change.)

E-FILING TAXPAYERS

Source: Internal Revenue Service.

[3.4] *Graph the solution set of each inequality or compound inequality.*

37. $3x - 2y \le 12$

38. $5x - y > 6$

39. $x \ge 2$

40. $2x + y \le 1$ and $x \ge 2y$

41. $x \ge 2$ or $y \ge 2$

[3.5]   *In Exercises 42–45, give the domain and range of each relation. Identify any functions.*

42. $\{(-4, 2), (-4, -2), (1, 5), (1, -5)\}$

43.

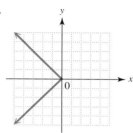

44.

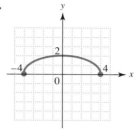

45.

Determine whether each equation or inequality defines y as a function of x. Give the domain in each case. Identify any linear functions.

46. $y = 3x - 3$

47. $y < x + 2$

48. $y = |x|$

49. $y = \sqrt{4x + 7}$

50. $x = y^2$

51. $y = \dfrac{7}{x - 6}$

52. Explain the test that allows us to determine whether a graph is that of a function.

Given $f(x) = -2x^2 + 3x - 6$, find each function value or expression.

53. $f(0)$

54. $f(2.1)$

55. $f\left(-\dfrac{1}{2}\right)$

56. $f(k)$

57. The table shows profits for cable television station CNBC from 1994 through 1999.

(a) Does the table define a function?
(b) What are the domain and range?
(c) Call this function f. Give two ordered pairs that belong to f.
(d) Find $f(1994)$. What does it mean?
(e) If $f(x) = 180$, what does x equal?

Year	Profit (in millions of dollars)
1994	40
1995	60
1996	80
1997	130
1998	180
1999	200

Source: *Fortune*, May 24, 1999, p. 142.

58. The equation $2x^2 - y = 0$ defines y as a function of x. Rewrite it using $f(x)$ notation, and find $f(3)$.

59. Suppose that $2x - 5y = 7$ defines a function. If $y = f(x)$, which one of the following defines the same function?

A. $f(x) = \dfrac{7 - 2x}{5}$

B. $f(x) = \dfrac{-7 - 2x}{5}$

C. $f(x) = \dfrac{-7 + 2x}{5}$

D. $f(x) = \dfrac{7 + 2x}{5}$

60. Can the graph of a linear function have undefined slope? Explain.

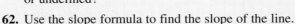

RELATING CONCEPTS (EXERCISES 61–72)

For Individual or Group Work

Refer to the straight-line graph and **work Exercises 61–72** *in order.*

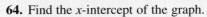

61. By just looking at the graph, how can you tell whether the slope is positive, negative, 0, or undefined?

62. Use the slope formula to find the slope of the line.

63. What is the slope of any line parallel to the line shown? perpendicular to the line shown?

64. Find the *x*-intercept of the graph.

65. Find the *y*-intercept of the graph.

66. Use function notation to write the equation of the line. Use f to designate the function.

67. Find $f(8)$.

68. If $f(x) = -8$, what is the value of x?

69. Graph the solution set of $f(x) \geq 0$.

70. What is the solution set of $f(x) = 0$?

71. What is the solution set of $f(x) < 0$? (Use the graph and the result of Exercise 70.)

72. What is the solution set of $f(x) > 0$? (Use the graph and the result of Exercise 70.)

[3.6]

73. In which one of the following does *y* vary inversely as *x*?

    A. $y = 2x$ **B.** $y = \dfrac{x}{3}$ **C.** $y = \dfrac{3}{x}$ **D.** $y = x^2$

Solve each problem.

74. For the subject in a photograph to appear in the same perspective in the photograph as in real life, the viewing distance must be properly related to the amount of enlargement. For a particular camera, the viewing distance varies directly as the amount of enlargement. A picture taken with this camera that is enlarged 5 times should be viewed from a distance of 250 mm. Suppose a print 8.6 times the size of the negative is made. From what distance should it be viewed?

75. The frequency (number of vibrations per second) of a vibrating guitar string varies inversely as its length. That is, a longer string vibrates fewer times in a second than a shorter string. Suppose a guitar string .65 m long vibrates 4.3 times per sec. What frequency would a string .5 m long have?

76. The volume of a rectangular box of a given height is proportional to its width and length. A box with width 2 ft and length 4 ft has volume 12 ft^3. Find the volume of a box with the same height that is 3 ft wide and 5 ft long.

CHAPTER **3** TEST

1. Complete the table of ordered pairs for the equation $2x - 3y = 12$.

x	y
1	
3	
	-4

Find the x- and y-intercepts, and graph each equation.

2. $3x - 2y = 20$ **3.** $y = 5$ **4.** $x = 2$

5. Find the slope of the line through the points $(6, 4)$ and $(-4, -1)$.

6. Describe how the graph of a line with undefined slope is situated in a rectangular coordinate system.

Determine whether each pair of lines is parallel, perpendicular, *or* neither.

7. $5x - y = 8$ and $5y = -x + 3$ **8.** $2y = 3x + 12$ and $3y = 2x - 5$

 9. In 1980, there were 119,000 farms in Iowa. As of 2001, there were 93,500. Find and interpret the average rate of change in the number of farms per year. Round your answer to the nearest whole number. (*Source:* Iowa Agricultural Statistics Service.)

Find an equation of each line, and write it in slope-intercept form.

10. Through $(4, -1)$; $m = -5$ **11.** Through $(-3, 14)$; horizontal

12. Through $(-7, 2)$ and parallel to $3x + 5y = 6$

13. Through $(-7, 2)$ and perpendicular to $y = 2x$

14. Through $(-2, 3)$ and $(6, -1)$

15. Which one of the following has positive slope and negative y-coordinate for its y-intercept?

A.   **B.**   **C.**   **D.**

16. The bar graph shows median household income for Hispanics.

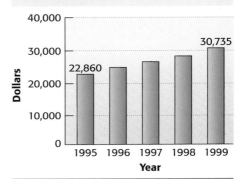

MEDIAN HOUSEHOLD INCOME FOR HISPANICS

Source: U.S. Bureau of the Census.

(a) Use the information for the years 1995 and 1999 to find an equation that models the data. Let $x = 5$ represent 1995, $x = 9$ represent 1999, and y represent the median income. Write the equation in slope-intercept form.

(b) Use the equation from part (a) to approximate median household income for 1997 to the nearest dollar. How does your result compare to the actual value, $26,628?

Graph each inequality or compound inequality.

17. $3x - 2y > 6$

18. $y < 2x - 1$ and $x - y < 3$

19. Which one of the following is the graph of a function?

A. 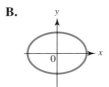   **B.**   **C.** 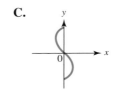   **D.**

20. Which of the following does not define a function?

A. $\{(0, 1), (-2, 3), (4, 8)\}$ **B.** $y = 2x - 6$ **C.** $y = \sqrt{x + 2}$ **D.**

x	y
0	1
3	2
0	2
6	3

21. Give the domain and range of the relation shown in **(a)** choice A of Problem 19 and **(b)** choice A of Problem 20.

22. If $f(x) = -x^2 + 2x - 1$, find $f(1)$ and $f(a)$.

23. Graph the linear function defined by $f(x) = \frac{2}{3}x - 1$. What is its domain and range?

Solve each problem.

24. The current in a simple electrical circuit is inversely proportional to the resistance. If the current is 80 amps when the resistance is 30 ohms, find the current when the resistance is 12 ohms.

25. The force of the wind blowing on a vertical surface varies jointly as the area of the surface and the square of the velocity. If a wind blowing at 40 mph exerts a force of 50 lb on a surface of 500 ft², how much force will a wind of 80 mph place on a surface of 2 ft²?

CUMULATIVE REVIEW EXERCISES CHAPTERS 1–3

Decide whether each statement is always true, sometimes true, *or* never true. *If the statement is* sometimes true, *give examples where it is true and where it is false.*

1. The absolute value of a negative number equals the additive inverse of the number.

2. The quotient of two integers with nonzero denominator is a rational number.

3. The sum of two negative numbers is positive.

4. The sum of a positive number and a negative number is 0.

Perform each operation.

5. $-|-2| - 4 + |-3| + 7$

6. $(-.8)^2$

7. $\sqrt{-64}$

8. $-\dfrac{2}{3}\left(-\dfrac{12}{5}\right)$

Simplify.

9. $-(-4m + 3)$

10. $3x^2 - 4x + 4 + 9x - x^2$

11. $\dfrac{3\sqrt{16} - (-1)7}{4 + (-6)}$

12. Write $-3 < x \le 5$ in interval notation.

13. Is $\sqrt{\dfrac{-2 + 4}{-5}}$ a real number?

Evaluate each expression if $p = -4$, $q = -2$, and $r = 5$.

14. $-3(2q - 3p)$

15. $|p|^3 - |q^3|$

16. $\dfrac{\sqrt{r}}{-p + 2q}$

Solve.

17. $2z - 5 + 3z = 4 - (z + 2)$

18. $\dfrac{3a - 1}{5} + \dfrac{a + 2}{2} = -\dfrac{3}{10}$

19. $V = \dfrac{1}{3}\pi r^2 h$ for h

Solve each problem.

20. The lowest temperature ever recorded in Allagash, Maine, was $-55°C$, on January 14, 1999. What was the corresponding Fahrenheit temperature? (*Source:* National Climatic Data Center.)

21. If each side of a square were increased by 4 in., the perimeter would be 8 in. less than twice the perimeter of the original square. Find the length of a side of the original square.

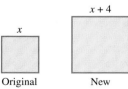

Original square   New square

22. Two planes leave the Dallas–Fort Worth airport at the same time. One travels east at 550 mph, and the other travels west at 500 mph. Assuming no wind, how long will it take for the planes to be 2100 mi apart?

West ←   Airport   → East

Solve. Write each solution set in interval notation and graph it.

23. $-4 < 3 - 2k < 9$

24. $-.3x + 2.1(x - 4) \le -6.6$

25. $x > 6$ and $x < 8$

26. $-5x + 1 \ge 11$ or $3x + 5 > 26$

Solve.

27. $|2k - 7| + 4 = 11$

28. $|3m + 6| \ge 0$

29. How are the solution sets of a linear equation and the two associated inequalities related?

30. Complete the table of ordered pairs at the right for the equation $3x - 4y = 12$.

x	y
0	
	0
2	

31. Find the x- and y-intercepts of the line with equation $3x + 5y = 12$ and graph the line.

32. Consider the points $A(-2, 1)$ and $B(3, -5)$.

    (a) Find the slope of the line AB.

    (b) Find the slope of a line perpendicular to line AB.

33. Graph the inequality $-2x + y < -6$.

Write an equation for each line. Express the equation in slope-intercept form.

34. Slope $-\dfrac{3}{4}$; y-intercept $(0, -1)$

35. Horizontal; through $(2, -2)$

36. Through $(4, -3)$ and $(1, 1)$

37. Give the domain and range of the relation. Does it define a function? Explain.

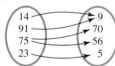

38. For the function defined by
$$f(x) = -4x + 10,$$

    (a) find the domain and range.

    (b) what is $f(-3)$?

39. Use the information in the graph to find and interpret the average rate of change in millions of U.S. cell phone subscribers per year from 1992 to 2000.

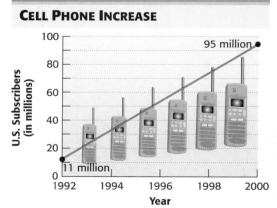

Source: Cellular Telecommunications Industry Association, Intel Corp.

40. The cost of a pizza varies directly as the square of its radius. If a pizza with a 7-in. radius costs $6.00, how much should a pizza with a 9-in. radius cost?

Systems of Linear Equations

4

Americans continue to flock to movie theaters in record numbers. Box office grosses reached $8.14 billion in 2001, the first time they exceeded $8 billion, as more than 1.4 billion movie tickets were sold. Surprisingly, the top three movies of the year were family films—*Harry Potter and the Sorcerer's Stone, Shrek,* and *Monsters, Inc.* not only attracted scores of adults with kids, but also unaccompanied adults, perhaps wishing to get away from it all for a few hours. (*Source:* ACNielsen EDI.) In Exercise 19 of the Chapter 4 Review Exercises, we use systems of linear equations to find out just how much money these top films earned.

4.1 Systems of Linear Equations in Two Variables

The worldwide personal computer market share for different manufacturers has varied, with first one, then another obtaining a larger share. As shown in Figure 1, Hewlett-Packard's share rose from 1995 through 1998, while Packard Bell, NEC saw its share decline. The graphs intersect at the point when the two companies had the same market share.

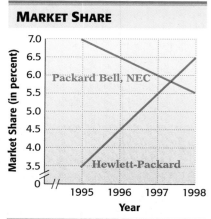

MARKET SHARE

Source: Intelliquest; IDC.

FIGURE 1

We could use a linear equation to model the graph of Hewlett-Packard's market share and another linear equation to model the graph of Packard Bell, NEC's market share. Such a set of equations is called a **system of equations,** in this case a **system of linear equations.** The point where the graphs in Figure 1 intersect is a solution of each of the individual equations. It is also the solution of the system of linear equations.

OBJECTIVE 1 Decide whether an ordered pair is a solution of a linear system. The **solution set of a linear system** of equations contains all ordered pairs that satisfy all the equations of the system at the same time.

EXAMPLE 1 Deciding Whether an Ordered Pair Is a Solution

Decide whether the given ordered pair is a solution of the system.

(a) $x + y = 6$
$4x - y = 14$; $(4, 2)$

Replace x with 4 and y with 2 in each equation of the system.

$$
\begin{array}{c|c}
x + y = 6 & 4x - y = 14 \\
4 + 2 = 6 \quad ? & 4(4) - 2 = 14 \quad ? \\
6 = 6 \quad \text{True} & 14 = 14 \quad \text{True}
\end{array}
$$

Since $(4, 2)$ makes both equations true, $(4, 2)$ is a solution of the system.

(b) $3x + 2y = 11$; $(-1, 7)$
$\ x + 5y = 36$

$$3x + 2y = 11$$
$$3(-1) + 2(7) = 11 \quad ?$$
$$-3 + 14 = 11$$
$$11 = 11 \qquad \text{True}$$

$$x + 5y = 36$$
$$-1 + 5(7) = 36 \quad ?$$
$$-1 + 35 = 36$$
$$34 = 36 \qquad \text{False}$$

The ordered pair $(-1, 7)$ is not a solution of the system, since it does not make *both* equations true.

Now Try Exercises 11 and 13.

OBJECTIVE 2 Solve linear systems by graphing. One way to find the solution set of a linear system of equations is to graph each equation and find the point where the graphs intersect.

EXAMPLE 2 Solving a System by Graphing

Solve the system of equations by graphing.

$$x + y = 5 \qquad (1)$$
$$2x - y = 4 \qquad (2)$$

When we graph these linear equations as shown in Figure 2, the graph suggests that the point of intersection is the ordered pair $(3, 2)$.

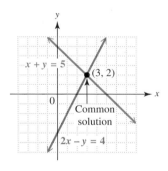

FIGURE 2

To be sure that $(3, 2)$ is a solution of *both* equations, we check by substituting 3 for x and 2 for y in each equation.

$$x + y = 5 \qquad (1)$$
$$3 + 2 = 5 \quad ?$$
$$5 = 5 \qquad \text{True}$$

$$2x - y = 4 \qquad (2)$$
$$2(3) - 2 = 4 \quad ?$$
$$6 - 2 = 4 \quad ?$$
$$4 = 4 \qquad \text{True}$$

Since $(3, 2)$ makes both equations true, $\{(3, 2)\}$ is the solution set of the system.

Now Try Exercise 15.

There are three possibilities for the solution set of a linear system in two variables.

Graphs of Linear Systems in Two Variables

1. The two graphs intersect in a single point. The coordinates of this point give the only solution of the system. In this case the system is **consistent,** and the equations are **independent.** This is the most common case. See Figure 3(a).

2. The graphs are parallel lines. In this case the system is **inconsistent;** that is, there is no solution common to both equations of the system, and the solution set is ∅. See Figure 3(b).

3. The graphs are the same line. In this case the equations are **dependent,** since any solution of one equation of the system is also a solution of the other. The solution set is an infinite set of ordered pairs representing the points on the line. See Figure 3(c).

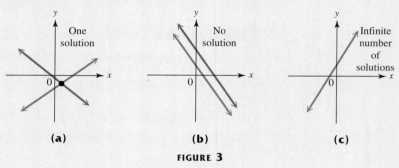

FIGURE 3

OBJECTIVE 3 **Solve linear systems (with two equations and two variables) by elimination.** While it is possible to find the solution of a system of equations by graphing, it can be difficult to read exact coordinates, especially if they are not integers, from a graph. Because of this, we usually use algebraic methods to solve systems. One such method, called the **elimination method,** involves combining the two equations of the system so that one variable is *eliminated*. This is done using the following fact.

$$\text{If} \quad a = b \text{ and } c = d, \quad \text{then} \quad a + c = b + d.$$

The general method of solving a system by the elimination method is summarized as follows.

Solving a Linear System by Elimination

Step 1   **Write both equations in standard form $Ax + By = C$.**

Step 2   **Make the coefficients of one pair of variable terms opposites.** Multiply one or both equations by appropriate numbers so that the sum of the coefficients of either the x- or y-terms is 0.

Step 3   **Add** the new equations to eliminate a variable. The sum should be an equation with just one variable.

Step 4   **Solve** the equation from Step 3 for the remaining variable.

> *Step 5* **Find the other value.** Substitute the result of Step 4 into either of the original equations and solve for the other variable.
>
> *Step 6* **Check** the solution in both of the original equations. Then write the solution set.

EXAMPLE 3 Solving a System by Elimination

Solve the system

$$5x - 2y = 4 \qquad (1)$$
$$2x + 3y = 13. \qquad (2)$$

Step 1 Both equations are in standard form.

Step 2 Suppose that you wish to eliminate the variable x. One way to do this is to multiply equation (1) by 2 and equation (2) by -5.

$$10x - 4y = 8 \qquad \text{2 times each side of equation (1)}$$
$$-10x - 15y = -65 \qquad \text{-5 times each side of equation (2)}$$

Step 3 Now add.

$$\begin{array}{r} 10x - 4y = 8 \\ -10x - 15y = -65 \\ \hline -19y = -57 \end{array}$$

Step 4 Solve for y. $\qquad y = 3 \qquad$ Divide by -19.

Step 5 To find x, substitute 3 for y in either equation (1) or (2). Substituting in equation (2) gives

$$2x + 3y = 13 \qquad (2)$$
$$2x + 3(3) = 13 \qquad \text{Let } y = 3.$$
$$2x + 9 = 13$$
$$2x = 4 \qquad \text{Subtract 9.}$$
$$x = 2. \qquad \text{Divide by 2.}$$

Step 6 The solution is (2, 3). To check, substitute 2 for x and 3 for y in both equations (1) and (2).

$$\begin{array}{ll} 5x - 2y = 4 \quad (1) & 2x + 3y = 13 \quad (2) \\ 5(2) - 2(3) = 4 \quad ? & 2(2) + 3(3) = 13 \quad ? \\ 10 - 6 = 4 \quad ? & 4 + 9 = 13 \quad ? \\ 4 = 4 \quad \text{True} & 13 = 13 \quad \text{True} \end{array}$$

The solution set is $\{(2, 3)\}$.

Now Try Exercise 19.

EXAMPLE 4 Solving a System with Fractional Coefficients

Solve the system

$$5x - 2y = 4 \qquad (1)$$
$$\frac{1}{2}x + \frac{3}{4}y = \frac{13}{4}. \qquad (2)$$

If an equation in a system has fractional coefficients, as in equation (2), first multiply by the least common denominator to clear the fractions.

$$4\left(\frac{1}{2}x + \frac{3}{4}y\right) = 4 \cdot \frac{13}{4} \qquad \text{Multiply equation (2) by the LCD, 4.}$$

$$4 \cdot \frac{1}{2}x + 4 \cdot \frac{3}{4}y = 4 \cdot \frac{13}{4} \qquad \text{Distributive property}$$

$$2x + 3y = 13 \qquad \text{Equivalent to equation (2)}$$

The system of equations becomes

$$5x - 2y = 4 \qquad (1)$$
$$2x + 3y = 13, \qquad \text{Equation (2) with fractions cleared}$$

which is identical to the system we solved in Example 3. The solution set is $\{(2, 3)\}$. To confirm this, check the solution in both equations (1) and (2).

Now Try Exercise 31.

NOTE If an equation in a system contains decimal coefficients, it is best to first clear the decimals by multiplying by an appropriate power of ten, depending on the number of decimal places. Then solve the system. For example, we multiply *each side* of the equation

$$.5x + .75y = 3.25$$

by 100 to get the equivalent equation

$$50x + 75y = 325.$$

OBJECTIVE 4 Solve special systems. As we saw in Figures 3(b) and (c), some systems of linear equations have no solution or an infinite number of solutions. Examples 5 and 6 show how to recognize these systems when solving algebraically.

EXAMPLE 5 Solving a System of Dependent Equations

Solve the system

$$2x - y = 3 \qquad (1)$$
$$6x - 3y = 9. \qquad (2)$$

We multiply equation (1) by -3, and then add the result to equation (2).

$$-6x + 3y = -9 \qquad \text{-3 times each side of equation (1)}$$
$$\underline{6x - 3y = 9} \qquad (2)$$
$$0 = 0 \qquad \text{True}$$

Adding these equations gives the true statement $0 = 0$. In the original system, we could get equation (2) from equation (1) by multiplying equation (1) by 3. Because of this, equations (1) and (2) are equivalent and have the same graph, as shown in Figure 4. The equations are dependent. The solution set is the set of all points on the line with equation $2x - y = 3$, written

$$\{(x, y) \mid 2x - y = 3\}$$

and read "the set of all ordered pairs (x, y), such that $2x - y = 3$."

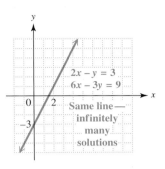

FIGURE 4

Now Try Exercise 23.

NOTE When a system has dependent equations and an infinite number of solutions, as in Example 5, either equation of the system could be used to write the solution set. We prefer to use an equation (in standard form) with coefficients that are integers having no common factor (except 1). Other texts may express such solutions differently.

EXAMPLE 6 Solving an Inconsistent System

Solve the system

$$x + 3y = 4 \qquad (1)$$
$$-2x - 6y = 3. \qquad (2)$$

Multiply equation (1) by 2, and then add the result to equation (2).

$$\begin{array}{rl}
2x + 6y = 8 & \text{Equation (1) multiplied by 2} \\
\underline{-2x - 6y = 3} & \text{(2)} \\
0 = 11 & \text{False}
\end{array}$$

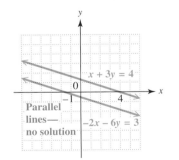

FIGURE 5

The result of the addition step is a false statement, which indicates that the system is inconsistent. As shown in Figure 5, the graphs of the equations of the system are parallel lines. There are no ordered pairs that satisfy both equations, so there is no solution for the system; the solution set is $\emptyset$.

Now Try Exercise 25.

The results of Examples 5 and 6 are generalized as follows.

Special Cases of Linear Systems

If both variables are eliminated when a system of linear equations is solved,

1. there are infinitely many solutions if the resulting statement is *true;*
2. there is no solution if the resulting statement is *false.*

Slopes and y-intercepts can be used to decide if the graphs of a system of equations are parallel lines or if they coincide. In Example 5, writing each equation in slope-intercept form shows that both lines have slope 2 and y-intercept $(0, -3)$, so the graphs are the same line and the system has an infinite solution set.

In Example 6, both equations have slope $-\frac{1}{3}$ but y-intercepts $\left(0, \frac{4}{3}\right)$ and $\left(0, -\frac{1}{2}\right)$, showing that the graphs are two distinct parallel lines. Thus, the system has no solution.

OBJECTIVE 5 **Solve linear systems (with two equations and two variables) by substitution.** Linear systems can also be solved algebraically by the **substitution method.** This method is most useful for solving linear systems in which one variable has coefficient 1 or -1.

The substitution method is summarized as follows.

Solving a Linear System by Substitution

Step 1 **Solve one of the equations for either variable.** If one of the variable terms has coefficient 1 or -1, choose it since the substitution method is usually easier this way.

Step 2 **Substitute** for that variable in the other equation. The result should be an equation with just one variable.

Step 3 **Solve** the equation from Step 2.

Step 4 **Find the other value.** Substitute the result from Step 3 into the equation from Step 1 to find the value of the other variable.

Step 5 **Check** the solution in both of the original equations. Then write the solution set.

EXAMPLE 7 Solving a System by Substitution

Solve the system

$$3x + 2y = 13 \qquad (1)$$
$$4x - y = -1. \qquad (2)$$

Step 1 To use the substitution method, first solve one of the equations for either x or y. Since the coefficient of y in equation (2) is -1, it is easiest to solve for y in equation (2).

$$4x - y = -1 \qquad (2)$$
$$-y = -1 - 4x \qquad \text{Subtract } 4x.$$
$$y = 1 + 4x \qquad \text{Multiply by } -1.$$

Step 2 Substitute $1 + 4x$ for y in equation (1).

$$3x + 2y = 13 \qquad (1)$$
$$3x + 2(1 + 4x) = 13 \qquad \text{Let } y = 1 + 4x.$$

Step 3 Solve for x.

$$3x + 2 + 8x = 13 \qquad \text{Distributive property}$$
$$11x = 11 \qquad \text{Combine terms; subtract 2.}$$
$$x = 1 \qquad \text{Divide by 11.}$$

Step 4 Now solve for *y*. Since $y = 1 + 4x$,

$$y = 1 + 4(1) = 5. \qquad \text{Let } x = 1.$$

Step 5 Check the solution (1, 5) in both equations (1) and (2).

$3x + 2y = 13$	(1)		$4x - y = -1$	(2)
$3(1) + 2(5) = 13$	?		$4(1) - 5 = -1$	?
$3 + 10 = 13$	?		$4 - 5 = -1$	?
$13 = 13$	True		$-1 = -1$	True

The solution set is {(1, 5)}.

Now Try Exercise 43.

EXAMPLE 8 Solving a System by Substitution

Solve the system

$$\frac{2}{3}x - \frac{1}{2}y = \frac{7}{6} \qquad (1)$$

$$3x - 2y = 6. \qquad (2)$$

This system will be easier to solve if you clear the fractions in equation (1). Multiply by the LCD, 6.

$$6 \cdot \frac{2}{3}x - 6 \cdot \frac{1}{2}y = 6 \cdot \frac{7}{6}$$

$$4x - 3y = 7 \qquad (3)$$

Now the system consists of equations (2) and (3). To use the substitution method, one equation must be solved for one of the two variables. Solve equation (2) for *x*.

$$3x = 2y + 6$$

$$x = \frac{2y + 6}{3}$$

Substitute $\frac{2y + 6}{3}$ for *x* in equation (3).

$$4x - 3y = 7 \qquad (3)$$

$$4\left(\frac{2y + 6}{3}\right) - 3y = 7 \qquad \text{Let } x = \frac{2y + 6}{3}.$$

$$3\left[4\left(\frac{2y + 6}{3}\right)\right] - 3(3y) = 3(7) \qquad \text{Multiply by 3 to clear the fraction.}$$

$$4(2y + 6) - 9y = 21$$

$$8y + 24 - 9y = 21 \qquad \text{Distributive property}$$

$$24 - y = 21 \qquad \text{Combine terms.}$$

$$-y = -3 \qquad \text{Subtract 24.}$$

$$y = 3 \qquad \text{Multiply by } -1.$$

Since $x = \frac{2y + 6}{3}$ and $y = 3$,

$$x = \frac{2(3) + 6}{3} = \frac{6 + 6}{3} = 4.$$

A check verifies that the solution set is {(4, 3)}.

Now Try Exercise 49.

NOTE While the substitution method is not usually the best choice for solving a system like the one in Example 8, it is sometimes necessary to use it to solve a system of *nonlinear* equations.

 OBJECTIVE 6 **Recognize how a graphing calculator is used to solve a linear system.** In Example 2 we showed how to solve the system

$$x + y = 5$$
$$2x - y = 4$$

by graphing the two lines and finding their point of intersection. We can also do this with a graphing calculator.

EXAMPLE 9 **Finding the Solution Set of a System from a Graphing Calculator Screen**

The solution set of the system

$$x + y = 5$$
$$2x - y = 4$$

can be found from the calculator screen in Figure 6. The two lines were graphed by solving the first equation to get $y = 5 - x$ and the second to get $y = 2x - 4$. The coordinates of their point of intersection are displayed at the bottom of the screen, indicating that the solution set is $\{(3, 2)\}$. (Compare this graph to the one found in Figure 2.)

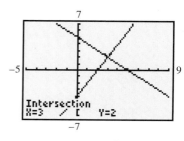

FIGURE 6

Now Try Exercise 59.

4.1 EXERCISES

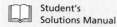

Fill in the blanks with the correct responses.

1. If $(3, -6)$ is a solution of a linear system in two variables, then substituting _____ for x and _____ for y leads to true statements in *both* equations.

2. A solution of a system of independent linear equations in two variables is a(n) _____.

3. If the solution process leads to a false statement such as $0 = 5$ when solving a system, the solution set is _____.

4. If the solution process leads to a true statement such as $0 = 0$ when solving a system, the system has _____ equations.

5. If the two lines forming a system have the same slope and different y-intercepts, the system has _____ solution(s).
(how many?)

6. If the two lines forming a system have different slopes, the system has _____ solution(s).
(how many?)

7. Which ordered pair could possibly be a solution of the graphed system of equations? Why?

    A. $(3, 3)$
    B. $(-3, 3)$
    C. $(-3, -3)$
    D. $(3, -3)$

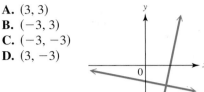

8. Which ordered pair could possibly be a solution of the graphed system of equations? Why?

    A. $(3, 0)$
    B. $(-3, 0)$
    C. $(0, 3)$
    D. $(0, -3)$

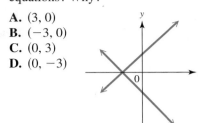

9. Match each system with the correct graph.

    (a) $x + y = 6$ **(b)** $x + y = -6$ **(c)** $x + y = 0$ **(d)** $x + y = 0$
           $x - y = 0$ $x - y = 0$ $x - y = -6$ $x - y = 6$

A.

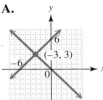

B.

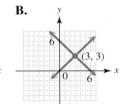

C.

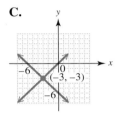

D.

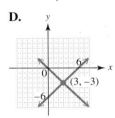

Decide whether the given ordered pair is a solution of the given system. See Example 1.

10. $x + y = 6$
    $x - y = 4$; $(5, 1)$

11. $x - y = 17$
    $x + y = -1$; $(8, -9)$

12. $2x - y = 8$
    $3x + 2y = 20$; $(5, 2)$

13. $3x - 5y = -12$
    $x - y = 1$; $(-1, 2)$

Solve each system by graphing. See Example 2.

14. $x + y = 4$
    $2x - y = 2$

15. $x + y = -5$
    $-2x + y = 1$

16. $x - 4y = -4$
    $3x + y = 1$

Solve each system by elimination. If the system is inconsistent or has dependent equations, say so. See Examples 3–6.

17. $2x - 5y = 11$
    $3x + y = 8$

18. $-2x + 3y = 1$
    $-4x + y = -3$

19. $3x + 4y = -6$
    $5x + 3y = 1$

20. $4x + 3y = 1$
    $3x + 2y = 2$

21. $3x + 3y = 0$
    $4x + 2y = 3$

22. $8x + 4y = 0$
    $4x - 2y = 2$

23. $7x + 2y = 6$
    $-14x - 4y = -12$

24. $x - 4y = 2$
    $4x - 16y = 8$

25. $5x - 5y = 3$
    $x - y = 12$

26. $2x - 3y = 7$
    $-4x + 6y = 14$

27. $x - \dfrac{1}{2}y = 2$
    $-x + \dfrac{2}{5}y = -\dfrac{8}{5}$

28. $\dfrac{3}{2}x + y = 3$
    $\dfrac{2}{3}x + \dfrac{1}{3}y = 1$

29. $x + y = 0$
$2x - 2y = 0$

30. $3x + 3y = 0$
$-2x - y = 0$

31. $\dfrac{1}{2}x + \dfrac{1}{3}y = -\dfrac{1}{3}$
$\dfrac{1}{2}x + 2y = -7$

32. $\dfrac{1}{5}x + y = \dfrac{6}{5}$
$\dfrac{1}{10}x + \dfrac{1}{3}y = \dfrac{5}{6}$

Write each equation in slope-intercept form and then tell how many solutions the system has. Do not actually solve.

33. $3x + 7y = 4$
$6x + 14y = 3$

34. $-x + 2y = 8$
$4x - 8y = 1$

35. $2x = -3y + 1$
$6x = -9y + 3$

36. $5x = -2y + 1$
$10x = -4y + 2$

37. Suppose that two linear equations are graphed on the same set of coordinate axes. Sketch what the graph might look like if the system has the given description.

(a) The system has a single solution. **(b)** The system has no solution.
(c) The system has infinitely many solutions.

38. Assuming you want to minimize the amount of work required, tell whether you would use the substitution or elimination method to solve each system. Explain your answers. Then solve the system.

(a) $6x - y = 5$
$y = 11x$

(b) $3x + y = -7$
$x - y = -5$

(c) $3x - 2y = 0$
$9x + 8y = 7$

Solve each system by substitution. If the system is inconsistent or has dependent equations, say so. See Examples 5–8.

39. $4x + y = 6$
$y = 2x$

40. $2x - y = 6$
$y = 5x$

41. $3x - 4y = -22$
$-3x + y = 0$

42. $-3x + y = -5$
$x + 2y = 0$

43. $-x - 4y = -14$
$2x - y = 1$

44. $-3x - 5y = -17$
$4x = y - 8$

45. $5x - 4y = 9$
$3 - 2y = -x$

46. $6x - y = -9$
$4 + 7x = -y$

47. $x = 3y + 5$
$x = \dfrac{3}{2}y$

48. $x = 6y - 2$
$x = \dfrac{3}{4}y$

49. $\dfrac{1}{2}x + \dfrac{1}{3}y = 3$
$-3x + y = 0$

50. $\dfrac{1}{4}x - \dfrac{1}{5}y = 9$
$5x - y = 0$

51. $y = 2x$
$4x - 2y = 0$

52. $x = 3y$
$3x - 9y = 0$

53. $x = 5y$
$5x - 25y = 5$

54. $y = -4x$
$8x + 2y = 4$

TECHNOLOGY INSIGHTS (EXERCISES 55–58)

55. The table shown was generated by a graphing calculator. The functions defined by Y_1 and Y_2 are linear. Based on the table, what are the coordinates of the point of intersection of the graphs?

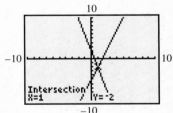

56. The functions defined by Y_1 and Y_2 in the table are linear.

   (a) Use the methods of Chapter 3 to find the equation for Y_1.

   (b) Use the methods of Chapter 3 to find the equation for Y_2.

   (c) Solve the system of equations formed by Y_1 and Y_2.

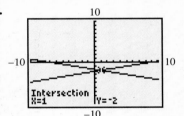

57. The solution set of the system

$$y_1 = 3x - 5$$
$$y_2 = -4x + 2$$

is $\{(1, -2)\}$. Using slopes and y-intercepts, determine which one of the two calculator graphs is the appropriate one for this system.

A.

B.

58. Which one of the ordered pairs listed could be the only possible solution of the system whose graphs are shown in the standard viewing window of a graphing calculator?

   A. $(15, -15)$ **B.** $(15, 15)$

   C. $(-15, 15)$ **D.** $(-15, -15)$

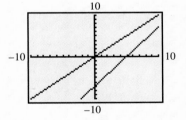

*For each system **(a)** solve by elimination or substitution and **(b)** use a graphing calculator to support your result. In part (b), be sure to solve each equation for y first. See Example 9.*

59. $x + y = 10$
    $2x - y = 5$

60. $6x + y = 5$
    $-x + y = -9$

61. $3x - 2y = 4$
    $3x + y = -2$

62. $2x - 3y = 3$
    $2x + 2y = 8$

Answer the questions in Exercises 63–66 by observing the graphs provided.

63. Annette Hebert compared the monthly payments she would incur for two types of mortgages: fixed-rate and variable-rate. Her observations led to the following graphs.

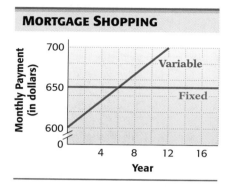

MORTGAGE SHOPPING

(a) For which years would the monthly payment be more for the fixed-rate mortgage than for the variable-rate mortgage?

(b) In what year would the payments be the same, and what would those payments be?

64. The figure shows graphs that represent supply and demand for a certain brand of low-fat frozen yogurt at various prices per half-gallon (in dollars).

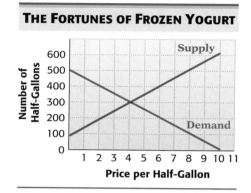

THE FORTUNES OF FROZEN YOGURT

(a) At what price does supply equal demand?

(b) For how many half-gallons does supply equal demand?

(c) What are the supply and demand at a price of $2 per half-gallon?

65. The graph shows network share (the percentage of TV sets in use) for the early evening news programs for the three major broadcast networks from 1986 through 2000.

(a) Between what years did the ABC early evening news dominate?

(b) During what year did ABC's dominance end? Which network equaled ABC's share that year? What was that share?

(c) During what years did ABC and CBS have equal network share? What was the share for each of these years?

(d) Which networks most recently had equal share? Write their share as an ordered pair of the form

(year, share).

✎ (e) Describe the general trend in viewership for the three major networks during these years.

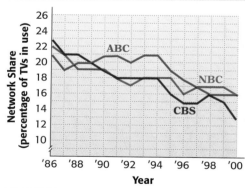

WHO'S WATCHING THE EVENING NEWS?

Source: Nielsen Media Research.

66. The graph at the top of the next page shows how the production of vinyl LPs, audiocassettes, and compact discs (CDs) changed over the years from 1986 through 1998.

(a) In what year did cassette production and CD production reach equal levels? What was that level?

(b) Express the point of intersection of the graphs of LP production and CD production as an ordered pair of the form

(year, production level).

(c) Between what years did cassette production first stabilize and remain fairly constant?

🖉 **(d)** Describe the trend in CD production from 1986 through 1998. If a straight line were used to approximate its graph, would the line have positive, negative, or 0 slope?

🖉 **(e)** If a straight line were used to approximate the graph of cassette production from 1990 through 1998, would the line have positive, negative, or 0 slope? Explain.

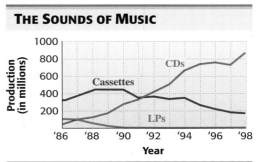

THE SOUNDS OF MUSIC

Source: Recording Industry Association of America.

Use the graph given at the beginning of this section (repeated here) to work Exercises 67–70.

67. For which years was Hewlett-Packard's share less than Packard Bell, NEC's share?

68. Estimate the year in which market share for Hewlett-Packard and Packard Bell, NEC was the same. About what was this share?

69. If $x = 0$ represents 1995 and $x = 3$ represents 1998, the market shares y (in percent) of these companies are closely modeled by the linear equations in the following system.

$$y = -.5x + 7 \qquad \text{Packard Bell, NEC}$$
$$y = x + 3.5 \qquad \text{Hewlett-Packard}$$

Solve this system. Round values to the nearest tenth as necessary.

70. Using your solution from Exercise 69, in what month and year did these companies have the same market share? What was that share?

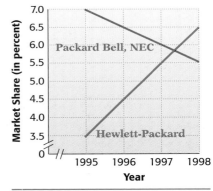

MARKET SHARE

Source: Intelliquest; IDC.

A system such as

$$\frac{3}{x} + \frac{4}{y} = \frac{5}{2}$$
$$\frac{5}{x} - \frac{3}{y} = \frac{7}{4}$$

can be solved by elimination. One way to do this is to let $p = \frac{1}{x}$ and $q = \frac{1}{y}$. Substitute, solve for p and q, and then find x and y. $\left(Hint: \frac{3}{x} = 3 \cdot \frac{1}{x} = 3p.\right)$ *Use this method to solve each system.*

71. $\dfrac{3}{x} + \dfrac{4}{y} = \dfrac{5}{2}$

$\dfrac{5}{x} - \dfrac{3}{y} = \dfrac{7}{4}$

72. $\dfrac{4}{x} - \dfrac{9}{y} = -1$

$-\dfrac{7}{x} + \dfrac{6}{y} = -\dfrac{3}{2}$

73. $\dfrac{2}{x} - \dfrac{5}{y} = \dfrac{3}{2}$

$\dfrac{4}{x} + \dfrac{1}{y} = \dfrac{4}{5}$

74. $\dfrac{2}{x} + \dfrac{3}{y} = \dfrac{11}{2}$

$-\dfrac{1}{x} + \dfrac{2}{y} = -1$

Solve by any method. Assume that a and b represent nonzero constants.

75. $ax + by = 2$
$-ax + 2by = 1$

76. $2ax - y = 3$
$y = 5ax$

77. $3ax + 2y = 1$
$-ax + y = 2$

78. $ax + by = c$
$ax - 2by = c$

RELATING CONCEPTS (EXERCISES 79–82)

For Individual or Group Work

Work Exercises 79–82 in order, to see the connections between systems of linear equations and the graphs of linear functions.

79. Use elimination or substitution to solve the system

$$3x + y = 6 \quad \text{(1)}$$
$$-2x + 3y = 7. \quad \text{(2)}$$

80. For equation (1) in the system of Exercise 79, solve for y and rename it $f(x)$. What special kind of function is f?

81. For equation (2) in the system of Exercise 79, solve for y and rename it $g(x)$. What special kind of function is g?

82. Use the result of Exercise 79 to fill in the blanks with the appropriate responses:

Because the graphs of f and g are straight lines that are neither parallel nor coincide, they intersect in exactly _____ point. The coordinates of the point are (_____, _____). Using function notation, this is given by $f($_____$) = $ _____ and $g($_____$) = $ _____.

🖊 **83.** Explain how to solve a system using the elimination method. Make up a system of your own to illustrate the steps.

🖊 **84.** Explain how to solve a system using the substitution method. Make up a system of your own to illustrate the steps.

4.2   Systems of Linear Equations in Three Variables

OBJECTIVES

1 Understand the geometry of systems of three equations in three variables.

A solution of an equation in three variables, such as

$$2x + 3y - z = 4,$$

is called an **ordered triple** and is written (x, y, z). For example, the ordered triple $(0, 1, -1)$ is a solution of the equation, because

$$2(0) + 3(1) - (-1) = 0 + 3 + 1 = 4.$$

Verify that another solution of this equation is $(10, -3, 7)$.

In the rest of this chapter, the term *linear equation* is extended to equations of the form

$$Ax + By + Cz + \cdots + Dw = K,$$

where not all the coefficients A, B, C, ..., D equal 0. For example,

$$2x + 3y - 5z = 7 \quad \text{and} \quad x - 2y - z + 3u - 2w = 8$$

are linear equations, the first with three variables and the second with five variables.

OBJECTIVE 1   Understand the geometry of systems of three equations in three variables.   Consider the solution of a system of linear equations in three variables, such as

$$\begin{aligned} 4x + 8y + z &= 2 \\ x + 7y - 3z &= -14 \\ 2x - 3y + 2z &= 3. \end{aligned}$$

Theoretically, a system of this type can be solved by graphing. However, the graph of a linear equation with three variables is a *plane*, not a line. Since the graph of each equation of the system is a plane, which requires three-dimensional graphing, this method is not practical. However, it does illustrate the number of solutions possible for such systems, as shown in Figure 7.

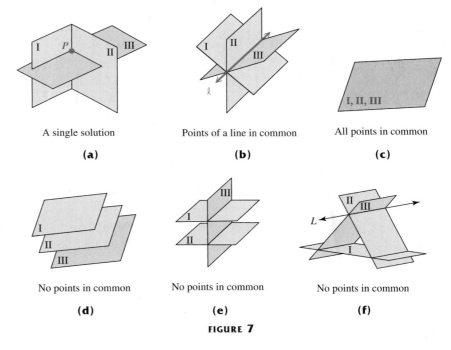

A single solution	Points of a line in common	All points in common
(a)	**(b)**	**(c)**

No points in common	No points in common	No points in common
(d)	**(e)**	**(f)**

FIGURE 7

Figure 7 illustrates the following cases.

Graphs of Linear Systems in Three Variables

1. The three planes may meet at a single, common point that is the solution of the system. See Figure 7(a). (continued)

2 Solve linear systems (with three equations and three variables) by elimination.

3 Solve linear systems (with three equations and three variables) where some of the equations have missing terms.

4 Solve special systems (with three equations and three variables).

2. The three planes may have the points of a line in common so that the infinite set of points that satisfy the equation of the line is the solution of the system. See Figure 7(b).

3. The three planes may coincide so that the solution of the system is the set of all points on a plane. See Figure 7(c).

4. The planes may have no points common to all three so that there is no solution of the system. See Figures 7(d), (e), and (f).

OBJECTIVE 2 **Solve linear systems (with three equations and three variables) by elimination.** Since graphing to find the solution set of a system of three equations in three variables is impractical, these systems are solved with an extension of the elimination method, summarized as follows.

Solving a Linear System in Three Variables

Step 1 **Eliminate a variable.** Use the elimination method to eliminate any variable from any two of the original equations. The result is an equation in two variables.

Step 2 **Eliminate the same variable again.** Eliminate the *same* variable from any *other* two equations. The result is an equation in the same two variables as in Step 1.

Step 3 **Eliminate a different variable and solve.** Use the elimination method to eliminate a second variable from the two equations in two variables that result from Steps 1 and 2. The result is an equation in one variable that gives the value of that variable.

Step 4 **Find a second value.** Substitute the value of the variable found in Step 3 into either of the equations in two variables to find the value of the second variable.

Step 5 **Find a third value.** Use the values of the two variables from Steps 3 and 4 to find the value of the third variable by substituting into an appropriate equation.

Step 6 **Check** the solution in all of the original equations. Then write the solution set.

EXAMPLE 1 **Solving a System in Three Variables**

Solve the system

$$4x + 8y + z = 2 \qquad (1)$$
$$x + 7y - 3z = -14 \qquad (2)$$
$$2x - 3y + 2z = 3. \qquad (3)$$

Step 1 As before, the elimination method involves eliminating a variable from the sum of two equations. The choice of which variable to eliminate is arbitrary.

Suppose we decide to begin by eliminating z. We multiply equation (1) by 3 and then add the result to equation (2).

$$
\begin{array}{ll}
12x + 24y + 3z = 6 & \text{Multiply each side of (1) by 3.} \\
\underline{x + 7y - 3z = -14} & \text{(2)} \\
13x + 31y = -8 & \text{Add. \quad (4)}
\end{array}
$$

Step 2   Equation (4) has only two variables. To get another equation without z, we multiply equation (1) by -2 and add the result to equation (3). It is essential at this point to *eliminate the same variable, z.*

$$
\begin{array}{ll}
-8x - 16y - 2z = -4 & \text{Multiply each side of (1) by } -2. \\
\underline{2x - 3y + 2z = 3} & \text{(3)} \\
-6x - 19y = -1 & \text{Add. \quad (5)}
\end{array}
$$

Step 3   Now we solve the system of equations (4) and (5) for x and y. This step is possible only if the *same* variable is eliminated in Steps 1 and 2.

$$
\begin{array}{ll}
78x + 186y = -48 & \text{Multiply each side of (4) by 6.} \\
\underline{-78x - 247y = -13} & \text{Multiply each side of (5) by 13.} \\
-61y = -61 & \text{Add.} \\
y = 1 &
\end{array}
$$

Step 4   Now we substitute 1 for y in either equation (4) or (5). Choosing (5) gives

$$
\begin{array}{ll}
-6x - 19y = -1 & \text{(5)} \\
-6x - 19(1) = -1 & \text{Let } y = 1. \\
-6x - 19 = -1 & \\
-6x = 18 & \\
x = -3. &
\end{array}
$$

Step 5   We substitute -3 for x and 1 for y in any one of the three original equations to find z. Choosing (1) gives

$$
\begin{array}{ll}
4x + 8y + z = 2 & \text{(1)} \\
4(-3) + 8(1) + z = 2 & \text{Let } x = -3 \text{ and } y = 1. \\
-4 + z = 2 & \\
z = 6. &
\end{array}
$$

Step 6   It appears that the ordered triple $(-3, 1, 6)$ is the only solution of the system. We must check that the solution satisfies all three equations of the system. For equation (1),

$$
\begin{array}{ll}
4x + 8y + z = 2 & \text{(1)} \\
4(-3) + 8(1) + 6 = 2 & \text{?} \\
-12 + 8 + 6 = 2 & \text{?} \\
2 = 2. & \text{True}
\end{array}
$$

Because $(-3, 1, 6)$ also satisfies equations (2) and (3), the solution set is $\{(-3, 1, 6)\}$.

Now Try Exercise 3.

OBJECTIVE 3 Solve linear systems (with three equations and three variables) where some of the equations have missing terms. When this happens, one elimination step can be omitted.

■ EXAMPLE 2 Solving a System of Equations with Missing Terms

Solve the system

$$6x - 12y = -5 \quad (1)$$
$$8y + z = 0 \quad (2)$$
$$9x - z = 12. \quad (3)$$

Since equation (3) is missing the variable y, eliminate y using equations (1) and (2).

$$
\begin{array}{ll}
12x - 24y \quad\quad = -10 & \text{Multiply each side of (1) by 2.} \\
\underline{\quad\quad 24y + 3z = \quad 0} & \text{Multiply each side of (2) by 3.} \\
12x \quad\quad + 3z = -10 & \text{Add.} \quad (4)
\end{array}
$$

Use this result, together with equation (3), to eliminate z. Multiply equation (3) by 3.

$$
\begin{array}{ll}
27x - 3z = \quad 36 & \text{Multiply each side of (3) by 3.} \\
\underline{12x + 3z = -10} & (4) \\
39x \quad\quad = \quad 26 & \text{Add.}
\end{array}
$$

$$x = \frac{26}{39} = \frac{2}{3}$$

Substituting into equation (3) gives

$$9x - z = 12 \quad (3)$$
$$9\left(\frac{2}{3}\right) - z = 12 \qquad \text{Let } x = \tfrac{2}{3}.$$
$$6 - z = 12$$
$$z = -6.$$

Substituting -6 for z in equation (2) gives

$$8y + z = 0 \quad (2)$$
$$8y - 6 = 0 \qquad \text{Let } z = -6.$$
$$8y = 6$$
$$y = \frac{3}{4}.$$

Check in each of the original equations of the system to verify that the solution set of the system is $\left\{\left(\tfrac{2}{3}, \tfrac{3}{4}, -6\right)\right\}$.

Now Try Exercise 21.

OBJECTIVE 4 Solve special systems (with three equations and three variables). Linear systems with three variables may be inconsistent or may include dependent equations. The next examples illustrate these cases.

EXAMPLE 3   Solving an Inconsistent System with Three Variables

Solve the system

$$2x - 4y + 6z = 5 \qquad (1)$$
$$-x + 3y - 2z = -1 \qquad (2)$$
$$x - 2y + 3z = 1. \qquad (3)$$

Eliminate x by adding equations (2) and (3) to get the equation

$$y + z = 0.$$

Now, *eliminate x again,* using equations (1) and (3).

$$\begin{array}{ll} -2x + 4y - 6z = -2 & \text{Multiply each side of (3) by } -2. \\ \underline{2x - 4y + 6z = 5} & (1) \\ 0 = 3 & \text{False} \end{array}$$

The resulting false statement indicates that equations (1) and (3) have no common solution. Thus, the system is inconsistent and the solution set is $\emptyset$. The graph of this system would show these two planes parallel to one another.

Now Try Exercise 29.

NOTE   If you get a false statement when adding as in Example 3, you do not need to go any further with the solution. Since two of the three planes are parallel, it is not possible for the three planes to have any common points.

EXAMPLE 4   Solving a System of Dependent Equations with Three Variables

Solve the system

$$2x - 3y + 4z = 8 \qquad (1)$$
$$-x + \frac{3}{2}y - 2z = -4 \qquad (2)$$
$$6x - 9y + 12z = 24. \qquad (3)$$

Multiplying each side of equation (1) by 3 gives equation (3). Multiplying each side of equation (2) by -6 also gives equation (3). Because of this, the equations are dependent. All three equations have the same graph, as illustrated in Figure 7(c). The solution set is written

$$\{(x, y, z) \mid 2x - 3y + 4z = 8\}.$$

Although any one of the three equations could be used to write the solution set, we use the equation in standard form with coefficients that are integers with no common factor (except 1), as we did in Section 4.1.

Now Try Exercise 33.

We can extend the method discussed in this section to solve larger systems. For example, to solve a system of four equations in four variables, eliminate a variable from three pairs of equations to get a system of three equations in three unknowns. Then proceed as shown above.

4.2 EXERCISES

1. Explain what the following statement means: The solution set of the system

$$2x + y + z = 3$$
$$3x - y + z = -2$$
$$4x - y + 2z = 0$$

is $\{(-1, 2, 3)\}$.

2. The two equations

$$x + y + z = 6$$
$$2x - y + z = 3$$

have a common solution of $(1, 2, 3)$. Which equation would complete a system of three linear equations in three variables having solution set $\{(1, 2, 3)\}$?

A. $3x + 2y - z = 1$ **B.** $3x + 2y - z = 4$
C. $3x + 2y - z = 5$ **D.** $3x + 2y - z = 6$

Solve each system of equations. See Example 1.

3. $2x - 5y + 3z = -1$
   $x + 4y - 2z = 9$
   $x - 2y - 4z = -5$

4. $x + 3y - 6z = 7$
   $2x - y + z = 1$
   $x + 2y + 2z = -1$

5. $3x + 2y + z = 8$
   $2x - 3y + 2z = -16$
   $x + 4y - z = 20$

6. $-3x + y - z = -10$
   $-4x + 2y + 3z = -1$
   $2x + 3y - 2z = -5$

7. $2x + 5y + 2z = 0$
   $4x - 7y - 3z = 1$
   $3x - 8y - 2z = -6$

8. $5x - 2y + 3z = -9$
   $4x + 3y + 5z = 4$
   $2x + 4y - 2z = 14$

9. $x + 2y + z = 4$
   $2x + y - z = -1$
   $x - y - z = -2$

10. $x - 2y + 5z = -7$
    $-2x - 3y + 4z = -14$
    $-3x + 5y - z = -7$

11. $\dfrac{1}{3}x + \dfrac{1}{6}y - \dfrac{2}{3}z = -1$

    $-\dfrac{3}{4}x - \dfrac{1}{3}y - \dfrac{1}{4}z = 3$

    $\dfrac{1}{2}x + \dfrac{3}{2}y + \dfrac{3}{4}z = 21$

12. $\dfrac{2}{3}x - \dfrac{1}{4}y + \dfrac{5}{8}z = 0$

    $\dfrac{1}{5}x + \dfrac{2}{3}y - \dfrac{1}{4}z = -7$

    $-\dfrac{3}{5}x + \dfrac{4}{3}y - \dfrac{7}{8}z = -5$

13. $-x + 2y + 6z = 2$
    $3x + 2y + 6z = 6$
    $x + 4y - 3z = 1$

14. $2x + y + 2z = 1$
    $x + 2y + z = 2$
    $x - y - z = 0$

15. $x + y - z = -2$
    $2x - y + z = -5$
    $-x + 2y - 3z = -4$

16. $x + 2y + 3z = 1$
    $-x - y + 3z = 2$
    $-6x + y + z = -2$

Solve each system of equations. See Example 2.

17. $2x - 3y + 2z = -1$
    $x + 2y + z = 17$
    $2y - z = 7$

18. $2x - y + 3z = 6$
    $x + 2y - z = 8$
    $2y + z = 1$

19. $4x + 2y - 3z = 6$
$x - 4y + z = -4$
$-x + 2z = 2$

20. $2x + 3y - 4z = 4$
$x - 6y + z = -16$
$-x + 3z = 8$

21. $2x + y = 6$
$3y - 2z = -4$
$3x - 5z = -7$

22. $4x - 8y = -7$
$4y + z = 7$
$-8x + z = -4$

23. $-5x + 2y + z = 5$
$-3x - 2y - z = 3$
$-x + 6y = 1$

24. $x + y - z = 0$
$2y - z = 1$
$2x + 3y - 4z = -4$

25. $4x - z = -6$
$\dfrac{3}{5}y + \dfrac{1}{2}z = 0$
$\dfrac{1}{3}x + \dfrac{2}{3}z = -5$

26. $5x - 2z = 8$
$4y + 3z = -9$
$\dfrac{1}{2}x + \dfrac{2}{3}y = -1$

27. Using your immediate surroundings, give an example of three planes that

(a) intersect in a single point;
(b) do not intersect;
(c) intersect in infinitely many points.

28. Suppose that a system has infinitely many ordered triple solutions of the form (x, y, z) such that

$$x + y + 2z = 1.$$

Give three specific ordered triples that are solutions of the system.

Solve each system of equations. If the system is inconsistent or has dependent equations, say so. See Examples 1, 3, and 4.

29. $2x + 2y - 6z = 5$
$-3x + y - z = -2$
$-x - y + 3z = 4$

30. $-2x + 5y + z = -3$
$5x + 14y - z = -11$
$7x + 9y - 2z = -5$

31. $-5x + 5y - 20z = -40$
$x - y + 4z = 8$
$3x - 3y + 12z = 24$

32. $x + 4y - z = 3$
$-2x - 8y + 2z = -6$
$3x + 12y - 3z = 9$

33. $2x + y - z = 6$
$4x + 2y - 2z = 12$
$-x - \dfrac{1}{2}y + \dfrac{1}{2}z = -3$

34. $2x - 8y + 2z = -10$
$-x + 4y - z = 5$
$\dfrac{1}{8}x - \dfrac{1}{2}y + \dfrac{1}{8}z = -\dfrac{5}{8}$

35. $x + y - 2z = 0$
$3x - y + z = 0$
$4x + 2y - z = 0$

36. $2x + 3y - z = 0$
$x - 4y + 2z = 0$
$3x - 5y - z = 0$

Extend the method of this section to solve each system. Express the solution in the form (x, y, z, w).

37. $x + y + z - w = 5$
$2x + y - z + w = 3$
$x - 2y + 3z + w = 18$
$-x - y + z + 2w = 8$

38. $3x + y - z + 2w = 9$
$x + y + 2z - w = 10$
$x - y - z + 3w = -2$
$-x + y - z + w = -6$

39. $3x + y - z + w = -3$
$2x + 4y + z - w = -7$
$-2x + 3y - 5z + w = 3$
$5x + 4y - 5z + 2w = -7$

40. $x - 3y + 7z + w = 11$
$2x + 4y + 6z - 3w = -3$
$3x + 2y + z + 2w = 19$
$4x + y - 3z + w = 22$

RELATING CONCEPTS (EXERCISES 41–50)

For Individual or Group Work

Suppose that on a distant planet a function of the form

$$f(x) = ax^2 + bx + c \quad (a \neq 0)$$

describes the height in feet of a projectile x sec after it has been projected upward. **Work Exercises 41–50 in order,** *to see how this can be related to a system of three equations in three variables a, b, and c.*

41. After 1 sec, the height of a certain projectile is 128 ft. Thus, $f(1) = 128$. Use this information to find one equation in the variables a, b, and c. (*Hint:* Substitute 1 for x and 128 for $f(x)$.)

42. After 1.5 sec, the height is 140 ft. Find a second equation in a, b, and c.

43. After 3 sec, the height is 80 ft. Find a third equation in a, b, and c.

44. Write a system of three equations in a, b, and c, based on your answers in Exercises 41–43. Solve the system.

45. What is the function f for this particular projectile?

46. In the function f written in Exercise 45, the _____ of the projectile is a function of the _____ elapsed since it was projected.

47. What was the initial height of the projectile? (*Hint:* Find $f(0)$.)

48. The projectile reaches its maximum height in 1.625 sec. Find its maximum height.

49. In Chapter 10 we discuss graphs of functions of the form $f(x) = ax^2 + bx + c$ ($a \neq 0$). Use a system of equations to find the values of a, b, and c for the function of this form that satisfies $f(1) = 2$, $f(-1) = 0$, and $f(-2) = 8$. Then write the expression for $f(x)$.

50. The accompanying table was generated by a graphing calculator for a function $Y_1 = aX^2 + bX + c$. Use any three points shown to find the values of a, b, and c. Then write the expression for Y_1.

X	Y1
1	8
2	15
3	24
4	35
5	48
6	63
7	80

X=1

---

51. Discuss why it is necessary to eliminate the same variable in the first two steps of the elimination method with three equations and three variables.

52. In Step 3 of the elimination method for solving systems in three variables, does it matter which variable is eliminated? Explain.

4.3   Applications of Systems of Linear Equations

Many applied problems involve more than one unknown quantity. Although some problems with two unknowns can be solved using just one variable, it is often easier to use two variables. To solve a problem with two unknowns, we must write two equations that relate the unknown quantities. The system formed by the pair of equations can then be solved using the methods of this chapter.

The following steps, based on the six-step problem-solving method first introduced in Chapter 2, give a strategy for solving applied problems using more than one variable.

Solving an Applied Problem by Writing a System of Equations

Step 1   **Read** the problem carefully until you understand what is given and what is to be found.

Step 2   **Assign variables** to represent the unknown values, using diagrams or tables as needed. *Write down* what each variable represents.

Step 3   **Write a system of equations** that relates the unknowns.

Step 4   **Solve** the system of equations.

Step 5   **State the answer** to the problem. Does it seem reasonable?

Step 6   **Check** the answer in the words of the original problem.

OBJECTIVE 1   Solve geometry problems using two variables.   Problems about the perimeter of a geometric figure often involve two unknowns and can be solved using systems of equations.

EXAMPLE 1   Finding the Dimensions of a Soccer Field

Unlike football, where the dimensions of a playing field cannot vary, a rectangular soccer field may have a width between 50 and 100 yd and a length between 50 and 100 yd. Suppose that one particular field has a perimeter of 320 yd. Its length measures 40 yd more than its width. What are the dimensions of this field? (*Source: Microsoft Encarta Encyclopedia 2000.*)

Step 1   **Read** the problem again. We are asked to find the dimensions of the field.

Step 2   **Assign variables.** Let L = the length and W = the width. Figure 8 shows a soccer field with the length labeled L and the width labeled W.

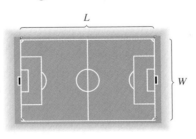

FIGURE 8

Step 3 **Write a system of equations.** Because the perimeter is 320 yd, we find one equation by using the perimeter formula:

$$2L + 2W = 320.$$

Because the length is 40 yd more than the width, we have

$$L = W + 40.$$

The system is

$$2L + 2W = 320 \qquad (1)$$
$$L = W + 40. \qquad (2)$$

Step 4 **Solve** the system of equations. Since equation (2) is solved for L, we can use the substitution method. We substitute $W + 40$ for L in equation (1), and solve for W.

$2L + 2W = 320$	(1)
$2(W + 40) + 2W = 320$	Let $L = W + 40$.
$2W + 80 + 2W = 320$	Distributive property
$4W + 80 = 320$	Combine terms.
$4W = 240$	Subtract 80.
$W = 60$	Divide by 4.

Let $W = 60$ in the equation $L = W + 40$ to find L.

$$L = 60 + 40 = 100$$

Step 5 **State the answer.** The length is 100 yd, and the width is 60 yd. The answer is reasonable, since both dimensions are within the ranges given in the problem.

Step 6 **Check.** The perimeter of this soccer field is

$$2(100) + 2(60) = 320 \text{ yd},$$

and the length, 100 yd, is 40 yd more than the width, since

$$100 - 40 = 60.$$

The answer is correct.

Now Try Exercise 3.

OBJECTIVE 2 Solve money problems using two variables. Professional sport ticket prices increase annually. Average per-ticket prices in three of the four major sports (football, basketball, and hockey) now exceed $30.00.

EXAMPLE 2 Solving a Problem about Ticket Prices

During recent National Hockey League and National Basketball Association seasons, two hockey tickets and one basketball ticket purchased at their average prices would have cost $110.40. One hockey ticket and two basketball tickets would have cost $106.32. What were the average ticket prices for the two sports? (*Source:* Team Marketing Report, Chicago.)

Step 1 **Read** the problem again. There are two unknowns.

Step 2 **Assign variables.** Let h represent the average price for a hockey ticket and b represent the average price for a basketball ticket.

Step 3 **Write a system of equations.** Because two hockey tickets and one basketball ticket cost a total of $110.40, one equation for the system is

$$2h + b = 110.40.$$

By similar reasoning, the second equation is

$$h + 2b = 106.32.$$

Therefore, the system is

$$2h + b = 110.40 \qquad (1)$$
$$h + 2b = 106.32. \qquad (2)$$

Step 4 **Solve** the system of equations. To eliminate h, multiply equation (2) by -2 and add.

$$\begin{array}{ll} 2h + b = 110.40 & (1) \\ \underline{-2h - 4b = -212.64} & \text{Multiply each side of (2) by } -2. \\ -3b = -102.24 & \text{Add.} \\ b = 34.08 & \text{Divide by } -3. \end{array}$$

To find the value of h, let $b = 34.08$ in equation (2).

$$\begin{array}{ll} h + 2b = 106.32 & (2) \\ h + 2(34.08) = 106.32 & \text{Let } b = 34.08. \\ h + 68.16 = 106.32 & \text{Multiply.} \\ h = 38.16 & \text{Subtract } 68.16. \end{array}$$

Step 5 **State the answer.** The average price for one basketball ticket was $34.08. For one hockey ticket, the average price was $38.16.

Step 6 **Check** that these values satisfy the conditions stated in the problem.

Now Try Exercise 11.

OBJECTIVE 3 **Solve mixture problems using two variables.** We solved mixture problems earlier using one variable. For many mixture problems it seems more natural to use more than one variable and a system of equations.

EXAMPLE 3 **Solving a Mixture Problem**

How many ounces each of 5% hydrochloric acid and 20% hydrochloric acid must be combined to get 10 oz of solution that is 12.5% hydrochloric acid?

Step 1 **Read** the problem. Two solutions of different strengths are being mixed together to get a specific amount of a solution with an "in-between" strength.

Step 2 **Assign variables.** Let x represent the number of ounces of 5% solution and y represent the number of ounces of 20% solution. Use a table to summarize the information from the problem.

Ounces of Solution	Percent (as a decimal)	Ounces of Pure Acid
x	5% = .05	.05x
y	20% = .20	.20y
10	12.5% = .125	(.125)10

Figure 9 illustrates what is happening in the problem.

Ounces of solution: x + y = 10

Ounces of pure acid: $.05x$ + $.20y$ = $.125(10)$

— **FIGURE 9**

Step 3 **Write a system of equations.** When the x oz of 5% solution and the y oz of 20% solution are combined, the total number of ounces is 10, so

$$x + y = 10. \qquad (1)$$

The ounces of acid in the 5% solution ($.05x$) plus the ounces of acid in the 20% solution ($.20y$) should equal the total ounces of acid in the mixture, which is $(.125)10$, or 1.25. That is,

$$.05x + .20y = 1.25. \qquad (2)$$

Notice that these equations can be quickly determined by reading down in the table or using the labels in Figure 9.

Step 4 **Solve** the system of equations (1) and (2). Eliminate x by first multiplying equation (2) by 100 to clear it of decimals and then multiplying equation (1) by -5.

$$
\begin{array}{rl}
5x + 20y = & 125 \qquad \text{Multiply each side of (2) by 100.} \\
\underline{-5x - 5y = -50} & \qquad \text{Multiply each side of (1) by } -5. \\
15y = & 75 \qquad \text{Add.} \\
y = & 5
\end{array}
$$

Because $y = 5$ and $x + y = 10$, x is also 5.

Step 5 **State the answer.** The desired mixture will require 5 oz of the 5% solution and 5 oz of the 20% solution.

Step 6 **Check** that these values satisfy both equations of the system.

Now Try Exercise 17.

| CONNECTIONS |

Problems that can be solved by writing a system of equations have been of interest historically. The following problem appeared in a Hindu work that dates back to about 850 A.D.

> The mixed price of 9 citrons (a lemonlike fruit shown in the photo) and 7 fragrant wood apples is 107; again, the mixed price of 7 citrons and 9 fragrant wood apples is 101. O you arithmetician, tell me quickly the price of a citron and the price of a wood apple here, having distinctly separated those prices well.

For Discussion or Writing

What do you think is meant by "the mixed price" in the problem quoted above? Write a system of equations for this problem. (You will be asked to solve it in Exercise 35.)

OBJECTIVE 4 Solve distance-rate-time problems using two variables. Motion problems require the distance formula, $d = rt$, where d is distance, r is rate (or speed), and t is time. These applications often lead to systems of equations, as in the next example.

■ EXAMPLE 4 Solving a Motion Problem

A car travels 250 km in the same time that a truck travels 225 km. If the speed of the car is 8 km per hr faster than the speed of the truck, find both speeds.

Step 1 **Read** the problem again. Given the distances traveled, we need to find the speed of each vehicle.

Step 2 **Assign variables.**

$$\text{Let} \quad x = \text{the speed of the car,}$$
$$\text{and} \quad y = \text{the speed of the truck.}$$

As in Example 3, a table helps organize the information. Fill in the given information for each vehicle (in this case, distance) and use the assigned variables for the unknown speeds (rates).

	d	r	t
Car	250	x	
Truck	225	y	

The table shows nothing about time. To get an expression for time, solve the distance formula, $d = rt$, for t.

$$\frac{d}{r} = t$$

The two times can be written as $\frac{250}{x}$ and $\frac{225}{y}$.

Step 3 **Write a system of equations.** The problem states that the car travels 8 km per hr faster than the truck. Since the two speeds are x and y,

$$x = y + 8. \qquad (1)$$

Both vehicles travel for the same time, so from the table,

$$\frac{250}{x} = \frac{225}{y}.$$

This is not a linear equation. However, multiplying each side by xy gives

$$250y = 225x,$$

which is linear. The system is

$$x = y + 8$$
$$250y = 225x. \qquad (2)$$

Step 4 **Solve** the system of equations by substitution. Replace x with $y + 8$ in equation (2).

$250y = 225x$	(2)
$250y = 225(y + 8)$	Let $x = y + 8$.
$250y = 225y + 1800$	Distributive property
$25y = 1800$	Subtract $225y$.
$y = 72$	Divide by 25.

Because $x = y + 8$, the value of x is $72 + 8 = 80$.

Step 5 **State the answer.** The car's speed is 80 km per hr, and the truck's speed is 72 km per hr.

Step 6 **Check.** This is especially important since one of the equations had variable denominators.

$$\text{Car:} \quad t = \frac{d}{r} = \frac{250}{80} = 3.125$$

$$\text{Truck:} \quad t = \frac{d}{r} = \frac{225}{72} = 3.125$$

Times are equal.

Since $80 - 72 = 8$, the conditions of the problem are satisfied.

Now Try Exercise 27.

OBJECTIVE 5 **Solve problems with three variables using a system of three equations.** To solve such problems, we extend the method used for two unknowns. Since three variables are used, three equations are necessary to find a solution.

■ **EXAMPLE 5** **Solving a Problem Involving Prices**

At Panera Bread, a loaf of honey wheat bread costs $2.40, a loaf of pumpernickel bread costs $3.35, and a loaf of French bread costs $2.10. On a recent day, three times as many loaves of honey wheat were sold as pumpernickel. The number of loaves of French bread sold was 5 less than the number of loaves of honey wheat sold. Total receipts for these breads were $56.90. How many loaves of each type of bread were sold? (*Source:* Panera Bread menu.)

Step 1 **Read** the problem again. There are three unknowns in this problem.

Step 2 **Assign variables** to represent the three unknowns.

Let $\quad x = $ the number of loaves of honey wheat,

$\quad\quad y = $ the number of loaves of pumpernickel,

and $\quad z = $ the number of loaves of French bread.

Step 3 **Write a system of three equations** using the information in the problem. Since three times as many loaves of honey wheat were sold as pumpernickel,

$$x = 3y, \quad \text{or} \quad x - 3y = 0. \quad (1)$$

Also,

Number of loaves of French bread	equals	5 less than the number of loaves of honey wheat.
↓	↓	↓
z	$=$	$x - 5,$

so $\quad\quad\quad\quad\quad\quad x - z = 5. \quad (2)$

Multiplying the cost of a loaf of each kind of bread by the number of loaves of that kind sold and adding gives the total receipts.

$$2.40x + 3.35y + 2.10z = 56.90$$

Multiply each side of this equation by 100 to clear it of decimals.

$$240x + 335y + 210z = 5690 \quad (3)$$

Step 4 **Solve** the system of three equations using the method shown in Section 4.2. Solving the system

$$x - 3y = 0 \qquad (1)$$
$$x - z = 5 \qquad (2)$$
$$240x + 335y + 210z = 5690 \qquad (3)$$

leads to

$$x = 12, \quad y = 4, \quad \text{and} \quad z = 7.$$

Step 5 **State the answer.** The solution is $(12, 4, 7)$, so 12 loaves of honey wheat, 4 loaves of pumpernickel, and 7 loaves of French bread were sold.

Step 6 **Check.** Since $12 = 3 \cdot 4$, the number of loaves of honey wheat is three times the number of loaves of pumpernickel. Also, $12 - 7 = 5$, so the number of loaves of French bread is 5 less than the number of loaves of honey wheat. Multiply the appropriate cost per loaf by the number of loaves sold and add the results to check that total receipts were $56.90.

Now Try Exercise 45.

EXAMPLE 6 **Solving a Business Production Problem**

A company produces three color television sets, models X, Y, and Z. Each model X set requires 2 hr of electronics work, 2 hr of assembly time, and 1 hr of finishing time. Each model Y requires 1, 3, and 1 hr of electronics, assembly, and finishing time, respectively. Each model Z requires 3, 2, and 2 hr of the same work, respectively. There are 100 hr available for electronics, 100 hr available for assembly, and 65 hr available for finishing per week. How many of each model should be produced each week if all available time must be used?

Step 1 **Read** the problem again. There are three unknowns.

Step 2 **Assign variables.**

Let $x =$ the number of model X produced per week,

$y =$ the number of model Y produced per week,

and $z =$ the number of model Z produced per week.

We organize the information in a table.

	Each Model X	Each Model Y	Each Model Z	Totals
Hours of Electronics Work	2	1	3	100
Hours of Assembly Time	2	3	2	100
Hours of Finishing Time	1	1	2	65

Step 3 **Write a system of three equations.** The x model X sets require $2x$ hr of electronics, the y model Y sets require $1y$ (or y) hr of electronics, and the

z model Z sets require $3z$ hr of electronics. Since 100 hr are available for electronics,

$$2x + y + 3z = 100. \qquad (1)$$

Similarly, from the fact that 100 hr are available for assembly,

$$2x + 3y + 2z = 100, \qquad (2)$$

and the fact that 65 hr are available for finishing leads to the equation

$$x + y + 2z = 65. \qquad (3)$$

Again, notice the advantage of setting up a table. By reading across, we can easily determine the coefficients and constants in the equations of the system.

Step 4 **Solve** the system

$$\begin{aligned}
2x + y + 3z &= 100 \\
2x + 3y + 2z &= 100 \\
x + y + 2z &= 65
\end{aligned}$$

to find $x = 15$, $y = 10$, and $z = 20$.

Step 5 **State the answer.** The company should produce 15 model X, 10 model Y, and 20 model Z sets per week.

Step 6 **Check** that these values satisfy the conditions of the problem.

Now Try Exercise 47.

4.3 EXERCISES

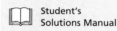
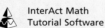

Solve each problem. See Example 1.

1. During the 2000 Major League Baseball season, the St. Louis Cardinals played 162 games. They won 28 more games than they lost. What was their win–loss record that year?

2. Refer to Exercise 1. During the same 162-game season, the Chicago Cubs lost 32 more games than they won. What was the team's win–loss record?

2000 MLB FINAL STANDINGS NATIONAL LEAGUE CENTRAL

Team	W	L
St. Louis	___	___
Cincinnati	85	77
Milwaukee	73	89
Houston	72	90
Pittsburgh	69	93
Chicago	___	___

Source: www.mlb.com

3. Venus and Serena measured a tennis court and found that it was 42 ft longer than it was wide and had a perimeter of 228 ft. What were the length and the width of the tennis court?

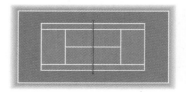

4. Shaq and Kobe found that the width of their basketball court was 44 ft less than the length. If the perimeter was 288 ft, what were the length and the width of their court?

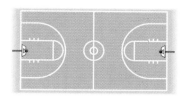

5. The two biggest U.S. companies in terms of revenue in 2000 were ExxonMobil and General Motors. ExxonMobil's revenue was $29 billion more than that of General Motors. Total revenue for the two companies was $399 billion. What was the revenue for each company? (*Source:* Bridge News, MarketGuide.com)

6. The top two U.S. trading partners during the first four months of 2000 were Canada and Mexico. Exports and imports with Mexico were $57 billion less than those with Canada. Total exports and imports involving these two countries were $211 billion. How much were U.S. exports and imports with each country? (*Source:* U.S. Bureau of the Census.)

In Exercises 7 and 8, find the measures of the angles marked x and y. Remember that (1) the sum of the measures of the angles of a triangle is 180°, (2) supplementary angles have a sum of 180°, and (3) vertical angles have equal measures.

7.

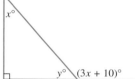

8.

The Fan Cost Index (FCI) represents the cost of four average-price tickets, four small soft drinks, two small beers, four hot dogs, parking for one car, two game programs, and two souvenir caps to a sporting event. For example, in a recent year, the FCI for Major League Baseball was $105.63. This was by far the least for the four major professional sports. (Source: Team Marketing Report, Chicago.)

Use the concept of FCI in Exercises 9 and 10. See Example 2.

9. The FCI prices for the National Hockey League and the National Basketball Association totaled $423.12. The hockey FCI was $16.36 more than that of basketball. What were the FCIs for these sports?

10. The FCI prices for Major League Baseball and the National Football League totaled $311.03. The football FCI was $105.87 more than that of baseball. What were the FCIs for these sports?

Solve each problem. See Example 2.

11. Andrew McGinnis works at Wendy's Old Fashioned Hamburgers. During one particular lunch hour, he sold 15 single hamburgers and 10 double hamburgers, totaling $63.25. Another lunch hour, he sold 30 singles and 5 doubles, totaling $78.65. How much did each type of burger cost? (*Source:* Wendy's Old Fashioned Hamburgers menu.)

12. Tokyo and New York are among the most expensive cities worldwide for business travelers. Using average costs per day for each city (which includes room, meals, laundry, and two taxi fares), 2 days in Tokyo and 3 days in New York cost $2015. Four days in Tokyo and 2 days in New York cost $2490. What is the average cost per day for each city? (*Source:* ECA International.)

The formulas p = br (percentage = base × rate) and I = prt (simple interest = principal × rate × time) are used in the applications in Exercises 17–24. In general, we are using

$$\text{portion} = \text{whole} \times \text{percent}.$$

To prepare to use these formulas, answer the questions in Exercises 13 and 14.

13. If a container of liquid contains 60 oz of solution, what is the number of ounces of pure acid if the given solution contains the following acid concentrations?

(a) 10%   **(b)** 25%   **(c)** 40%   **(d)** 50%

14. If $5000 is invested in an account paying simple annual interest, how much interest will be earned during the first year at the following rates?

(a) 2%   **(b)** 3%   **(c)** 4%   **(d)** 3.5%

15. If a pound of turkey costs $.99, how much will *x* pounds cost?

16. If a ticket to the movie *Eight Legged Freaks* costs $8 and *y* tickets are sold, how much is collected from the sale?

Solve each problem. See Example 3.

17. How many gallons each of 25% alcohol and 35% alcohol should be mixed to get 20 gal of 32% alcohol?

Gallons of Solution	Percent (as a decimal)	Gallons of Pure Alcohol
x	25% = .25	
y	35% = .35	
20	32% =	

18. How many liters each of 15% acid and 33% acid should be mixed to get 120 L of 21% acid?

Liters of Solution	Percent (as a decimal)	Liters of Pure Acid
x	15% = .15	
y	33% =	
120	21% =	

19. Pure acid is to be added to a 10% acid solution to obtain 54 L of a 20% acid solution. What amounts of each should be used?

20. A truck radiator holds 36 L of fluid. How much pure antifreeze must be added to a mixture that is 4% antifreeze to fill the radiator with a mixture that is 20% antifreeze?

21. A party mix is made by adding nuts that sell for $2.50 per kg to a cereal mixture that sells for $1 per kg. How much of each should be added to get 30 kg of a mix that will sell for $1.70 per kg?

	Number of Kilograms	Price per Kilogram	Value
Nuts	x	2.50	
Cereal	y	1.00	
Mixture		1.70	

22. A popular fruit drink is made by mixing fruit juices. Such a drink with 50% juice is to be mixed with another drink that is 30% juice to get 200 L of a drink that is 45% juice. How much of each should be used?

	Liters of Drink	Percent (as a decimal)	Liters of Pure Juice
50% Juice	x	.50	
30% Juice	y	.30	
Mixture		.45	

23. A total of $3000 is invested, part at 2% simple interest and part at 4%. If the total annual return from the two investments is $100, how much is invested at each rate?

Principal	Rate (as a decimal)	Interest
x	.02	.02x
y	.04	.04y
3000		100

24. An investor will invest a total of $15,000 in two accounts, one paying 4% annual simple interest, and the other 3%. If he wants to earn $550 annual interest, how much should he invest at each rate?

Principal	Rate (as a decimal)	Interest
x	.04	
y	.03	
15,000		

The formula d = rt (distance = rate × time) is used in the applications in Exercises 27–30. To prepare to use this formula, answer the questions in Exercises 25 and 26.

25. If the speed of a killer whale is 25 mph and the whale swims for y hr, how many miles does the whale travel?

26. If the speed of a boat in still water is 10 mph, and the speed of the current of a river is x mph, what is the speed of the boat

Upstream (against the current)

Downstream (with the current)

    (a) going upstream (that is, against the current, which slows the boat down);

    (b) going downstream (that is, with the current, which speeds the boat up)?

Solve each problem. See Example 4.

27. A train travels 150 km in the same time that a plane covers 400 km. If the speed of the plane is 20 km per hr less than 3 times the speed of the train, find both speeds.

	r	t	d
Train	x		150
Plane	y		400

28. A freight train and an express train leave towns 390 km apart, traveling toward one another. The freight train travels 30 km per hr slower than the express train. They pass one another 3 hr later. What are their speeds?

	r	t	d
Freight Train	x	3	
Express Train	y	3	

29. In his motorboat, Bill Ruhberg travels upstream at top speed to his favorite fishing spot, a distance of 36 mi, in 2 hr. Returning, he finds that the trip downstream, still at top speed, takes only 1.5 hr. Find the speed of Bill's boat and the speed of the current.

	r	t	d
Upstream	$x - y$	2	
Downstream	$x + y$		

30. Traveling for 3 hr into a steady headwind, a plane flies 1650 mi. The pilot determines that flying *with* the same wind for 2 hr, he could make a trip of 1300 mi. Find the speed of the plane and the speed of the wind.

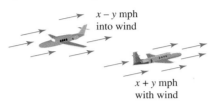

$x - y$ mph
into wind

$x + y$ mph
with wind

Use the problem-solving techniques of this section to solve each problem with two variables. See Examples 1–4.

31. At age 61, rock icon Tina Turner generated the most revenue on the concert circuit in 2000. Turner and second-place 'N Sync together took in $157 million from ticket sales. If 'N Sync took in $3.8 million less than Turner, how much did each generate? (*Source:* Pollstar.)

32. Carol Britz plans to mix pecan clusters that sell for $3.60 per lb with chocolate truffles that sell for $7.20 per lb to get a mixture that she can sell in Valentine boxes for $4.95 per lb. How much of the $3.60 clusters and the $7.20 truffles should she use to create 80 lb of the mix?

	Number of Pounds	Price per Pound	Value
Pecan Clusters	x		
Chocolate Truffles	y		
Valentine Mixture	80		

33. Tickets to a production of *King Lear* at the University of Miami cost $5 for general admission or $4 with a student ID. If 184 people paid to see a performance and $812 was collected, how many of each type of ticket were sold?

34. At a business meeting at Panera Bread, the bill for two cappuccinos and three house lattes was $10.95. At another table, the bill for one cappuccino and two house lattes was $6.65. How much did each type of beverage cost? (*Source:* Panera Bread menu.)

35. The mixed price of 9 citrons and 7 fragrant wood apples is 107; again, the mixed price of 7 citrons and 9 fragrant wood apples is 101. O you arithmetician, tell me quickly the price of a citron and the price of a wood apple here, having distinctly separated those prices well. (*Source:* Hindu work, A.D. 850.)

36. Braving blizzard conditions on the planet Hoth, Luke Skywalker sets out at top speed in his snow speeder for a rebel base 4800 mi away. He travels into a steady headwind and makes the trip in 3 hr. Returning, he finds that the trip back, still at top speed but now with a tailwind, takes only 2 hr. Find the top speed of Luke's snow speeder and the speed of the wind.

	r	t	d
Into Headwind			
With Tailwind			

Solve each problem involving three variables. See Examples 5 and 6. (In Exercises 37–40, remember that the sum of the measures of the angles of a triangle is 180°.)

37. In the figure, $z = x + 10$ and $x + y = 100$. Determine a third equation involving x, y, and z, and then find the measures of the three angles.

38. In the figure, x is 10 less than y and 20 less than z. Write a system of equations and find the measures of the three angles.

39. In a certain triangle, the measure of the second angle is $10°$ more than three times the first. The third angle measure is equal to the sum of the measures of the other two. Find the measures of the three angles.

40. The measure of the largest angle of a triangle is $12°$ less than the sum of the measures of the other two. The smallest angle measures $58°$ less than the largest. Find the measures of the angles.

41. The perimeter of a triangle is 70 cm. The longest side is 4 cm less than the sum of the other two sides. Twice the shortest side is 9 cm less than the longest side. Find the length of each side of the triangle.

42. The perimeter of a triangle is 56 in. The longest side measures 4 in. less than the sum of the other two sides. Three times the shortest side is 4 in. more than the longest side. Find the lengths of the three sides.

43. In a random sample of 100 Americans of voting age, 10 more Americans identify themselves as Independents than Republicans. Six fewer Americans identify themselves as Republicans than Democrats. Assuming that all of those sampled are Republican, Democrat, or Independent, how many of those in the sample identify themselves with each political affiliation? (*Source:* The Gallup Organization.)

44. In the 2000 Summer Olympics in Sydney, Australia, the United States earned 14 more gold medals than silver. The number of bronze medals earned was 17 less than twice the number of silver medals. The United States earned a total of 97 medals. How many of each kind of medal did the United States earn? (*Source: The Gazette,* October 2, 2000.)

45. Tickets for one show on the Harlem Globetrotters' 75th Anniversary Tour cost $10, $18, or, for VIP seats, $30. So far, five times as many $18 tickets have been sold as VIP tickets. The number of $10 tickets equals the number of $18 tickets plus twice the number of VIP tickets. Sales of these tickets total $9500. How many of each kind of ticket have been sold? (*Source:* www.ticketmaster.com)

46. Three kinds of tickets are available for a *Prosthetic Forehead* concert: "up close," "in the middle," and "far out." "Up close" tickets cost $10 more than "in the middle" tickets, while "in the middle" tickets cost $10 more than "far out" tickets. Twice the cost of an "up close" ticket is $20 more than 3 times the cost of a "far out" ticket. Find the price of each kind of ticket.

47. A hardware supplier manufactures three kinds of clamps, types A, B, and C. Production restrictions require it to make 10 units more type C clamps than the total of the other types and twice as many type B clamps as type A. The shop must produce a total of 490 units of clamps per day. How many units of each type can be made per day?

48. A Mardi Gras trinket manufacturer supplies three wholesalers, A, B, and C. The output from a day's production is 320 cases of trinkets. She must send wholesaler A three times as many cases as she sends B, and she must send wholesaler C 160 cases less than she provides A and B together. How many cases should she send to each wholesaler to distribute the entire day's production to them?

49. A plant food is to be made from three chemicals. The mix must include 60% of the first and second chemicals. The second and third chemicals must be in a ratio of 4 to 3 by weight. How much of each chemical is needed to make 750 kg of the plant food?

50. How many ounces of 5% hydrochloric acid, 20% hydrochloric acid, and water must be combined to get 10 oz of solution that is 8.5% hydrochloric acid, if the amount of water used must equal the total amount of the other two solutions?

51. During a recent National Hockey League regular season, the Dallas Stars played 82 games. Together, their wins and losses totaled 74. They tied 18 fewer games than they lost. How many wins, losses, and ties did they have that year?

Team	GP	W	L	T	GF	GA	Pts
Dallas	82	__	__	__	252	198	104
Detroit	82	38	26	18	253	197	94
Phoenix	82	38	37	7	240	243	83
St. Louis	82	36	35	11	236	239	83
Chicago	82	34	35	13	223	210	81
Toronto	82	30	44	8	230	273	68

Source: Sports Illustrated Sports Almanac.

52. During a recent National Hockey League season, the Boston Bruins played 82 games. Their losses and ties totaled 56, and they had 21 fewer wins than losses. How many wins, losses, and ties did they have that year?

Team	GP	W	L	T	GF	GA	Pts
Buffalo	82	40	30	12	237	208	92
Pittsburgh	82	38	36	8	285	280	84
Ottawa	82	31	36	15	226	234	77
Montreal	82	31	36	15	249	276	77
Hartford	82	32	39	11	226	256	75
Boston	82	__	__	__	234	300	61

Source: Sports Illustrated Sports Almanac.

4.4   Solving Systems of Linear Equations by Matrix Methods

[A]

$$\begin{bmatrix} -1 & 0 \\ 1 & -2 \end{bmatrix}$$

[B]

$$\begin{bmatrix} 8 & -1 & -3 \\ 2 & 1 & 6 \\ 0 & 5 & -3 \\ 5 & 9 & 7 \end{bmatrix}$$

FIGURE 10

OBJECTIVE 1   Define a matrix.   An ordered array of numbers such as

$$\text{Rows} \begin{bmatrix} 2 & 3 & 5 \\ 7 & 1 & 2 \end{bmatrix}$$

Columns

is called a **matrix.** The numbers are called **elements** of the matrix. Matrices (the plural of *matrix*) are named according to the number of **rows** and **columns** they contain. The rows are read horizontally, and the columns are read vertically. For example, the first row in the preceding matrix is 2 3 5 and the first column is $\frac{2}{7}$. This matrix is a 2×3 (read "two by three") matrix because it has 2 rows and 3 columns. The number of rows is given first, and then the number of columns. Two other examples follow.

$$\begin{bmatrix} -1 & 0 \\ 1 & -2 \end{bmatrix} \quad \begin{array}{l} 2 \times 2 \\ \text{matrix} \end{array} \qquad \begin{bmatrix} 8 & -1 & -3 \\ 2 & 1 & 6 \\ 0 & 5 & -3 \\ 5 & 9 & 7 \end{bmatrix} \quad \begin{array}{l} 4 \times 3 \\ \text{matrix} \end{array}$$

A **square matrix** is one that has the same number of rows as columns. The 2×2 matrix is a square matrix.

Figure 10 shows how a graphing calculator displays the preceding two matrices. Work with matrices is made much easier by using technology when available. Consult your owner's manual for details.

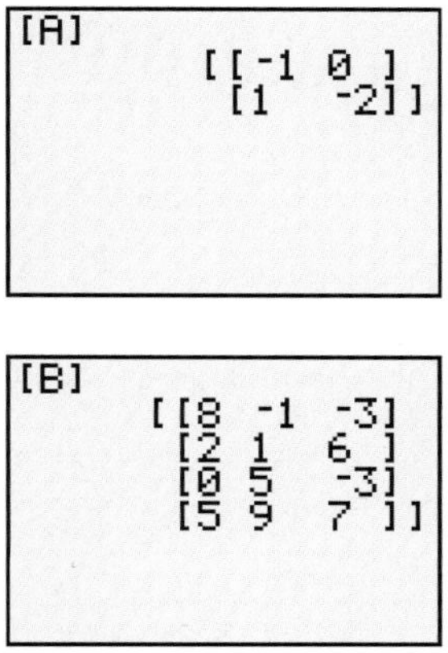

In this section, we discuss a matrix method of solving linear systems that is really just a very structured way of using the elimination method. The advantage of this new method is that it can be done by a graphing calculator or a computer, allowing large systems of equations to be solved easily.

OBJECTIVE 2   Write the augmented matrix for a system.   To begin, we write an *augmented matrix* for the system. An **augmented matrix** has a vertical bar that separates the columns of the matrix into two groups. For example, to solve the system

$$\begin{aligned} x - 3y &= 1 \\ 2x + y &= -5, \end{aligned}$$

start with the augmented matrix

$$\left[\begin{array}{cc|c} 1 & -3 & 1 \\ 2 & 1 & -5 \end{array} \right].$$

Place the coefficients of the variables to the left of the bar, and the constants to the right. The bar separates the coefficients from the constants. The matrix is just a shorthand way of writing the system of equations, so the rows of the augmented matrix can be treated the same as the equations of a system of equations.

We know that exchanging the position of two equations in a system does not change the system. Also, multiplying any equation in a system by a nonzero number does not change the system. Comparable changes to the augmented matrix of a system of equations produce new matrices that correspond to systems with the same solutions as the original system.

The following **row operations** produce new matrices that lead to systems having the same solutions as the original system.

> ### Matrix Row Operations
>
> **1.** Any two rows of the matrix may be interchanged.
> **2.** The elements in any row may be multiplied by any nonzero real number.
> **3.** Any row may be changed by adding to the elements of the row the product of a real number and the corresponding elements of another row.

Examples of these row operations follow.

Row operation 1:

$$\begin{bmatrix} 2 & 3 & 9 \\ 4 & 8 & -3 \\ 1 & 0 & 7 \end{bmatrix} \quad \text{becomes} \quad \begin{bmatrix} 1 & 0 & 7 \\ 4 & 8 & -3 \\ 2 & 3 & 9 \end{bmatrix}.$$

Interchange row 1 and row 3.

Row operation 2:

$$\begin{bmatrix} 2 & 3 & 9 \\ 4 & 8 & -3 \\ 1 & 0 & 7 \end{bmatrix} \quad \text{becomes} \quad \begin{bmatrix} 6 & 9 & 27 \\ 4 & 8 & -3 \\ 1 & 0 & 7 \end{bmatrix}.$$

Multiply the numbers in row 1 by 3.

Row operation 3:

$$\begin{bmatrix} 2 & 3 & 9 \\ 4 & 8 & -3 \\ 1 & 0 & 7 \end{bmatrix} \quad \text{becomes} \quad \begin{bmatrix} 0 & 3 & -5 \\ 4 & 8 & -3 \\ 1 & 0 & 7 \end{bmatrix}.$$

Multiply the numbers in row 3 by −2; add them to the corresponding numbers in row 1.

The third row operation corresponds to the way we eliminated a variable from a pair of equations in the previous sections.

OBJECTIVE 3 **Use row operations to solve a system with two equations.** Row operations can be used to rewrite a matrix until it is the matrix of a system where the solution is easy to find. The goal is a matrix in the form

$$\left[\begin{array}{cc|c} 1 & a & b \\ 0 & 1 & c \end{array}\right] \quad \text{or} \quad \left[\begin{array}{ccc|c} 1 & a & b & c \\ 0 & 1 & d & e \\ 0 & 0 & 1 & f \end{array}\right]$$

for systems with two or three equations, respectively. Notice that there are 1s down the diagonal from upper left to lower right and 0s below the 1s. A matrix written this way is said to be in **row echelon form.** When these matrices are rewritten as systems of equations, the value of one variable is known, and the rest can be found by substitution. The following examples illustrate this method.

EXAMPLE 1 Using Row Operations to Solve a System with Two Variables

Use row operations to solve the system

$$x - 3y = 1$$
$$2x + y = -5.$$

We start with the augmented matrix of the system.

$$\begin{bmatrix} 1 & -3 & \big| & 1 \\ 2 & 1 & \big| & -5 \end{bmatrix}$$

Now we use the various row operations to change this matrix into one that leads to a system that is easier to solve.

It is best to work by columns. We start with the first column and make sure that there is a 1 in the first row, first column position. There is already a 1 in this position. Next, we get 0 in every position below the first. To get a 0 in row two, column one, we use the third row operation and add to the numbers in row two the result of multiplying each number in row one by -2. (We abbreviate this as $-2R_1 + R_2$.) Row one remains unchanged.

$$\begin{bmatrix} 1 & -3 & \big| & 1 \\ 2 + 1(-2) & 1 + -3(-2) & \big| & -5 + 1(-2) \end{bmatrix}$$

Original number -2 times number
from row two   from row one

$$\begin{bmatrix} 1 & -3 & \big| & 1 \\ 0 & 7 & \big| & -7 \end{bmatrix} \quad -2R_1 + R_2$$

The matrix now has a 1 in the first position of column one, with 0 in every position below the first.

Now we go to column two. A 1 is needed in row two, column two. We get this 1 by using the second row operation, multiplying each number of row two by $\frac{1}{7}$.

$$\begin{bmatrix} 1 & -3 & \big| & 1 \\ 0 & 1 & \big| & -1 \end{bmatrix} \quad \frac{1}{7}R_2$$

This augmented matrix leads to the system of equations

$$\begin{array}{ccc} 1x - 3y = 1 & & x - 3y = 1 \\ 0x + 1y = -1 & \text{or} & y = -1. \end{array}$$

From the second equation, $y = -1$. We substitute -1 for y in the first equation to get

$$x - 3y = 1$$
$$x - 3(-1) = 1$$
$$x + 3 = 1$$
$$x = -2.$$

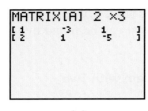

(a)

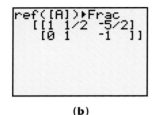

(b)

FIGURE 11

The solution set of the system is $\{(-2, -1)\}$. Check this solution by substitution in both equations of the system.

Now Try Exercise 3.

If the augmented matrix of the system in Example 1 is entered as matrix [A] in a graphing calculator (Figure 11(a)) and the row echelon form of the matrix is found (Figure 11(b)), then the system becomes

$$x + \frac{1}{2}y = -\frac{5}{2}$$
$$y = -1.$$

While this system looks different from the one we obtained in Example 1, it is equivalent, since its solution set is also $\{(-2, -1)\}$.

OBJECTIVE 4 Use row operations to solve a system with three equations. A linear system with three equations is solved in a similar way. We use row operations to get 1s down the diagonal from left to right and all 0s below each 1.

EXAMPLE 2   Using Row Operations to Solve a System with Three Variables

Use row operations to solve the system

$$x - y + 5z = -6$$
$$3x + 3y - z = 10$$
$$x + 3y + 2z = 5.$$

Start by writing the augmented matrix of the system.

$$\begin{bmatrix} 1 & -1 & 5 & | & -6 \\ 3 & 3 & -1 & | & 10 \\ 1 & 3 & 2 & | & 5 \end{bmatrix}$$

This matrix already has 1 in row one, column one. Next get 0s in the rest of column one. First, add to row two the results of multiplying each number of row one by -3. This gives the matrix

$$\begin{bmatrix} 1 & -1 & 5 & | & -6 \\ 0 & 6 & -16 & | & 28 \\ 1 & 3 & 2 & | & 5 \end{bmatrix}. \qquad -3R_1 + R_2$$

Now add to the numbers in row three the results of multiplying each number of row one by -1.

$$\begin{bmatrix} 1 & -1 & 5 & | & -6 \\ 0 & 6 & -16 & | & 28 \\ 0 & 4 & -3 & | & 11 \end{bmatrix} \qquad -1R_1 + R_3$$

Introduce 1 in row two, column two by multiplying each number in row two by $\frac{1}{6}$.

$$\begin{bmatrix} 1 & -1 & 5 & \bigm| & -6 \\ 0 & 1 & -\frac{8}{3} & \bigm| & \frac{14}{3} \\ 0 & 4 & -3 & \bigm| & 11 \end{bmatrix} \qquad \frac{1}{6}R_2$$

To obtain 0 in row three, column two, add to row three the results of multiplying each number in row two by -4.

$$\begin{bmatrix} 1 & -1 & 5 & \bigm| & -6 \\ 0 & 1 & -\frac{8}{3} & \bigm| & \frac{14}{3} \\ 0 & 0 & \frac{23}{3} & \bigm| & -\frac{23}{3} \end{bmatrix} \qquad -4R_2 + R_3$$

Finally, obtain 1 in row three, column three by multiplying each number in row three by $\frac{3}{23}$.

$$\begin{bmatrix} 1 & -1 & 5 & \bigm| & -6 \\ 0 & 1 & -\frac{8}{3} & \bigm| & \frac{14}{3} \\ 0 & 0 & 1 & \bigm| & -1 \end{bmatrix} \qquad \frac{3}{23}R_3$$

This final matrix gives the system of equations

$$x - y + 5z = -6$$
$$y - \frac{8}{3}z = \frac{14}{3}$$
$$z = -1.$$

Substitute -1 for z in the second equation, $y - \frac{8}{3}z = \frac{14}{3}$, to find that $y = 2$. Finally, substitute 2 for y and -1 for z in the first equation, $x - y + 5z = -6$, to determine that $x = 1$. The solution set of the original system is $\{(1, 2, -1)\}$. Check by substitution.

Now Try Exercise 15.

OBJECTIVE 5 Use row operations to solve special systems. In the final example we show how to recognize inconsistent systems or systems with dependent equations when solving these systems with row operations.

EXAMPLE 3 Recognizing Inconsistent Systems or Dependent Equations

Use row operations to solve each system.

(a) $\quad 2x - 3y = 8$
$\quad\;\; -6x + 9y = 4$

$$\begin{bmatrix} 2 & -3 & \bigm| & 8 \\ -6 & 9 & \bigm| & 4 \end{bmatrix} \qquad \text{Write the augmented matrix.}$$

$$\begin{bmatrix} 1 & -\frac{3}{2} & \bigm| & 4 \\ -6 & 9 & \bigm| & 4 \end{bmatrix} \qquad \frac{1}{2}R_1$$

$$\begin{bmatrix} 1 & -\frac{3}{2} & \bigm| & 4 \\ 0 & 0 & \bigm| & 28 \end{bmatrix} \qquad 6R_1 + R_2$$

The corresponding system of equations is

$$x - \frac{3}{2}y = 4$$

$$0 = 28, \qquad \text{False}$$

which has no solution and is inconsistent. The solution set is $\emptyset$.

(b) $-10x + 12y = 30$

$\qquad 5x - 6y = -15$

$$\begin{bmatrix} -10 & 12 & | & 30 \\ 5 & -6 & | & -15 \end{bmatrix} \qquad \text{Write the augmented matrix.}$$

$$\begin{bmatrix} 1 & -\frac{6}{5} & | & -3 \\ 5 & -6 & | & -15 \end{bmatrix} \qquad -\frac{1}{10}R_1$$

$$\begin{bmatrix} 1 & -\frac{6}{5} & | & -3 \\ 0 & 0 & | & 0 \end{bmatrix} \qquad -5R_1 + R_2$$

The corresponding system is

$$x - \frac{6}{5}y = -3$$

$$0 = 0, \qquad \text{True}$$

which has dependent equations. Using the second equation of the original system, we write the solution set as

$$\{(x, y) \,|\, 5x - 6y = -15\}.$$

Now Try Exercises 11 and 13.

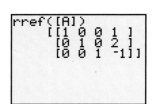

FIGURE 12

| **CONNECTIONS** |

An extension of the matrix method described in this section involves transforming an augmented matrix into **reduced row echelon form.** This form has 1s down the main diagonal and 0s above and below this diagonal. For example, the matrix for the system in Example 2 could be transformed into the matrix

$$\begin{bmatrix} 1 & 0 & 0 & | & 1 \\ 0 & 1 & 0 & | & 2 \\ 0 & 0 & 1 & | & -1 \end{bmatrix} \qquad \text{which gives the equivalent system} \qquad \begin{aligned} x &= 1 \\ y &= 2 \\ z &= -1. \end{aligned}$$

The calculator screens in Figure 12 indicate how easily this transformation can be obtained using technology.

For Discussion or Writing

1. Write the reduced row echelon form for the matrix of the system in Example 1.
2. If transforming to reduced row echelon form leads to all 0s in the final row, what kind of system is represented?

4.4 EXERCISES

1. Consider the matrix $\begin{bmatrix} -2 & 3 & 1 \\ 0 & 5 & -3 \\ 1 & 4 & 8 \end{bmatrix}$ and answer the following.

  (a) What are the elements of the second row?
  (b) What are the elements of the third column?
  (c) Is this a square matrix? Explain why or why not.
  (d) Give the matrix obtained by interchanging the first and third rows.
  (e) Give the matrix obtained by multiplying the first row by $-\frac{1}{2}$.
  (f) Give the matrix obtained by multiplying the third row by 3 and adding to the first row.

2. Give the dimensions of each matrix.

(a) $\begin{bmatrix} 3 & -7 \\ 4 & 5 \\ -1 & 0 \end{bmatrix}$

(b) $\begin{bmatrix} 4 & 9 & 0 \\ -1 & 2 & -4 \end{bmatrix}$

(c)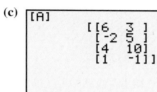

(d)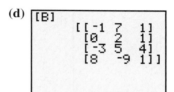

Complete the steps in the matrix solution of each system by filling in the boxes. Give the final system and the solution set. See Example 1.

3. $4x + 8y = 44$
$2x - y = -3$

$\begin{bmatrix} 4 & 8 & | & 44 \\ 2 & -1 & | & -3 \end{bmatrix}$

$\begin{bmatrix} 1 & \blacksquare & | & \blacksquare \\ 2 & -1 & | & -3 \end{bmatrix} \quad \frac{1}{4}R_1$

$\begin{bmatrix} 1 & 2 & | & 11 \\ 0 & \blacksquare & | & \blacksquare \end{bmatrix} \quad -2R_1 + R_2$

$\begin{bmatrix} 1 & 2 & | & 11 \\ 0 & 1 & | & \blacksquare \end{bmatrix} \quad -\frac{1}{5}R_2$

4. $2x - 5y = -1$
$3x + y = 7$

$\begin{bmatrix} 2 & -5 & | & -1 \\ 3 & 1 & | & 7 \end{bmatrix}$

$\begin{bmatrix} 1 & -\dfrac{5}{2} & | & \blacksquare \\ 3 & 1 & | & 7 \end{bmatrix} \quad \frac{1}{2}R_1$

$\begin{bmatrix} 1 & -\dfrac{5}{2} & | & -\dfrac{1}{2} \\ 0 & \blacksquare & | & \blacksquare \end{bmatrix} \quad -3R_1 + R_2$

$\begin{bmatrix} 1 & -\dfrac{5}{2} & | & -\dfrac{1}{2} \\ 0 & 1 & | & \blacksquare \end{bmatrix} \quad \frac{2}{17}R_2$

Use row operations to solve each system. See Examples 1 and 3.

5. $x + y = 5$
$x - y = 3$

6. $x + 2y = 7$
$x - y = -2$

7. $2x + 4y = 6$
$3x - y = 2$

8. $4x + 5y = -7$
$x - y = 5$

9. $3x + 4y = 13$
$2x - 3y = -14$

10. $5x + 2y = 8$
$3x - y = 7$

11. $-4x + 12y = 36$
$x - 3y = 9$

12. $2x - 4y = 8$
$-3x + 6y = 5$

13. $2x + y = 4$
$4x + 2y = 8$

14. $-3x - 4y = 1$
$6x + 8y = -2$

Complete the steps in the matrix solution of each system by filling in the boxes. Give the final system and the solution set. See Example 2.

15. $x + y - z = -3$
$2x + y + z = 4$
$5x - y + 2z = 23$

$$\begin{bmatrix} 1 & 1 & -1 & | & -3 \\ 2 & 1 & 1 & | & 4 \\ 5 & -1 & 2 & | & 23 \end{bmatrix}$$

$$\begin{bmatrix} 1 & 1 & -1 & | & -3 \\ 0 & \blacksquare & \blacksquare & | & \blacksquare \\ 0 & \blacksquare & \blacksquare & | & \blacksquare \end{bmatrix} \quad \begin{array}{l} -2R_1 + R_2 \\ -5R_1 + R_3 \end{array}$$

$$\begin{bmatrix} 1 & 1 & -1 & | & -3 \\ 0 & 1 & \blacksquare & | & \blacksquare \\ 0 & -6 & 7 & | & 38 \end{bmatrix} \quad -1R_2$$

$$\begin{bmatrix} 1 & 1 & -1 & | & -3 \\ 0 & 1 & -3 & | & -10 \\ 0 & 0 & \blacksquare & | & \blacksquare \end{bmatrix} \quad 6R_2 + R_3$$

$$\begin{bmatrix} 1 & 1 & -1 & | & -3 \\ 0 & 1 & -3 & | & -10 \\ 0 & 0 & 1 & | & \blacksquare \end{bmatrix} \quad -\frac{1}{11}R_3$$

16. $2x + y + 2z = 11$
$2x - y - z = -3$
$3x + 2y + z = 9$

$$\begin{bmatrix} 2 & 1 & 2 & | & 11 \\ 2 & -1 & -1 & | & -3 \\ 3 & 2 & 1 & | & 9 \end{bmatrix}$$

$$\begin{bmatrix} 1 & \blacksquare & \blacksquare & | & \blacksquare \\ 2 & -1 & -1 & | & -3 \\ 3 & 2 & 1 & | & 9 \end{bmatrix} \quad \frac{1}{2}R_1$$

$$\begin{bmatrix} 1 & \frac{1}{2} & 1 & | & \frac{11}{2} \\ 0 & \blacksquare & \blacksquare & | & \blacksquare \\ 0 & \blacksquare & \blacksquare & | & \blacksquare \end{bmatrix} \quad \begin{array}{l} -2R_1 + R_2 \\ -3R_1 + R_3 \end{array}$$

$$\begin{bmatrix} 1 & \frac{1}{2} & 1 & | & \frac{11}{2} \\ 0 & 1 & \blacksquare & | & \blacksquare \\ 0 & \frac{1}{2} & -2 & | & -\frac{15}{2} \end{bmatrix} \quad -\frac{1}{2}R_2$$

$$\begin{bmatrix} 1 & \frac{1}{2} & 1 & | & \frac{11}{2} \\ 0 & 1 & \frac{3}{2} & | & 7 \\ 0 & 0 & \blacksquare & | & \blacksquare \end{bmatrix} \quad -\frac{1}{2}R_2 + R_3$$

$$\begin{bmatrix} 1 & \frac{1}{2} & 1 & | & \frac{11}{2} \\ 0 & 1 & \frac{3}{2} & | & 7 \\ 0 & 0 & 1 & | & \blacksquare \end{bmatrix} \quad -\frac{4}{11}R_3$$

Use row operations to solve each system. See Examples 2 and 3.

17. $x + y - 3z = 1$
$2x - y + z = 9$
$3x + y - 4z = 8$

18. $2x + 4y - 3z = -18$
$3x + y - z = -5$
$x - 2y + 4z = 14$

19. $x + y - z = 6$
$2x - y + z = -9$
$x - 2y + 3z = 1$

20. $x + 3y - 6z = 7$
$2x - y + 2z = 0$
$x + y + 2z = -1$

21. $x - y = 1$
$y - z = 6$
$x + z = -1$

22. $x + y = 1$
$2x - z = 0$
$y + 2z = -2$

23. $x - 2y + z = 4$
$3x - 6y + 3z = 12$
$-2x + 4y - 2z = -8$

24. $4x + 8y + 4z = 9$
$x + 3y + 4z = 10$
$5x + 10y + 5z = 12$

25. $x + 2y + 3z = -2$
$2x + 4y + 6z = -5$
$x - y + 2z = 6$

26. $x + 3y + z = 1$
$2x + 6y + 2z = 2$
$3x + 9y + 3z = 3$

27. Write a short explanation of each term. Include examples.

    (a) Matrix   **(b)** Row of a matrix
    (c) Column of a matrix   **(d)** Square matrix
    (e) Augmented matrix   **(f)** Row operations on a matrix

28. Compare the use of the third row operation on a matrix and the elimination method of solving a system of linear equations. Give examples.

The augmented matrix for the system in Exercise 3 is shown in the graphing calculator screen on the left as matrix [A]. The screen in the middle shows the row echelon form for [A]. Compare it to the matrix shown in the answer section for Exercise 3. The screen on the right shows the reduced row echelon form, and from this it can be determined by inspection that the solution set of the system is $\{(1, 5)\}$.

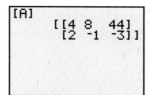 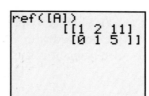 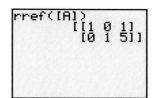

Use a graphing calculator and either one of the two matrix methods illustrated to solve each system.

29. $4x + y = 5$
$2x + y = 3$

30. $5x + 3y = 7$
$7x - 3y = -19$

31. $5x + y - 3z = -6$
$2x + 3y + z = 5$
$-3x - 2y + 4z = 3$

32. $x + y + z = 3$
$3x - 3y - 4z = -1$
$x + y + 3z = 11$

33. $x + z = -3$
$y + z = 3$
$x + y = 8$

34. $x - y = -1$
$-y + z = -2$
$x + z = -2$

4.5 Determinants and Cramer's Rule

Recall from Section 4.4 that an ordered array of numbers within square brackets is called a *matrix* (plural *matrices*). Matrices are named according to the number of rows and columns they contain. A *square matrix* has the same number of rows and columns.

$$\begin{bmatrix} -1 & 0 \\ 1 & -2 \end{bmatrix} \quad \begin{array}{l} 2 \times 2 \\ \text{square matrix} \end{array}$$

Associated with every *square matrix* is a real number called the **determinant** of the matrix. A determinant is symbolized by the entries of the matrix placed between two vertical bars, such as

$$\begin{vmatrix} 2 & 3 \\ 7 & 1 \end{vmatrix} \quad \begin{array}{l} 2 \times 2 \\ \text{determinant} \end{array} \qquad \begin{vmatrix} 7 & 4 & 3 \\ 0 & 1 & 5 \\ 6 & 0 & 1 \end{vmatrix}. \quad \begin{array}{l} 3 \times 3 \\ \text{determinant} \end{array}$$

Like matrices, determinants are named according to the number of rows and columns they contain.

CAUTION Matrices are enclosed with square brackets, while determinants are denoted with vertical bars. Also, a matrix is an *array of numbers,* but its determinant is a single number.

OBJECTIVE 1 **Evaluate 2 × 2 determinants.** As mentioned above, the value of a determinant is a *real number.* The value of the 2 × 2 determinant

$$\begin{vmatrix} a & b \\ c & d \end{vmatrix}$$

is defined as follows.

Value of a 2 × 2 Determinant

$$\begin{vmatrix} a & b \\ c & d \end{vmatrix} = ad - bc$$

▨ **EXAMPLE 1** **Evaluating a 2 × 2 Determinant**

Evaluate the determinant.

$$\begin{vmatrix} -1 & -3 \\ 4 & -2 \end{vmatrix}$$

Here $a = -1$, $b = -3$, $c = 4$, and $d = -2$, so

$$\begin{vmatrix} -1 & -3 \\ 4 & -2 \end{vmatrix} = -1(-2) - (-3)4 = 2 + 12 = 14.$$

Now Try Exercise 3.

A 3×3 determinant can be evaluated in a similar way.

Value of a 3 × 3 Determinant

$$\begin{vmatrix} a_1 & b_1 & c_1 \\ a_2 & b_2 & c_2 \\ a_3 & b_3 & c_3 \end{vmatrix} = \begin{aligned} &(a_1b_2c_3 + b_1c_2a_3 + c_1a_2b_3) \\ &- (a_3b_2c_1 + b_3c_2a_1 + c_3a_2b_1) \end{aligned}$$

This rule for evaluating a 3×3 determinant is hard to remember. A method for calculating a 3×3 determinant that is easier to use is based on the rule. Rearranging terms and using the distributive property gives

$$\begin{vmatrix} a_1 & b_1 & c_1 \\ a_2 & b_2 & c_2 \\ a_3 & b_3 & c_3 \end{vmatrix} = a_1(b_2c_3 - b_3c_2) - a_2(b_1c_3 - b_3c_1) + a_3(b_1c_2 - b_2c_1). \tag{1}$$

Each of the quantities in parentheses represents a 2×2 determinant that is the part of the 3×3 determinant remaining when the row and column of the multiplier are eliminated, as shown below.

$$a_1(b_2c_3 - b_3c_2) \qquad \begin{vmatrix} a_1 & b_1 & c_1 \\ a_2 & b_2 & c_2 \\ a_3 & b_3 & c_3 \end{vmatrix}$$

$$a_2(b_1c_3 - b_3c_1) \qquad \begin{vmatrix} a_1 & b_1 & c_1 \\ a_2 & b_2 & c_2 \\ a_3 & b_3 & c_3 \end{vmatrix}$$

$$a_3(b_1c_2 - b_2c_1) \qquad \begin{vmatrix} a_1 & b_1 & c_1 \\ a_2 & b_2 & c_2 \\ a_3 & b_3 & c_3 \end{vmatrix}$$

These 2×2 determinants are called **minors** of the elements in the 3×3 determinant. In the determinant above, the minors of a_1, a_2, and a_3 are, respectively,

$$\begin{vmatrix} b_2 & c_2 \\ b_3 & c_3 \end{vmatrix}, \qquad \begin{vmatrix} b_1 & c_1 \\ b_3 & c_3 \end{vmatrix}, \qquad \text{and} \qquad \begin{vmatrix} b_1 & c_1 \\ b_2 & c_2 \end{vmatrix}.$$

OBJECTIVE 2 Use expansion by minors to evaluate 3 × 3 determinants. A 3×3 determinant can be evaluated by multiplying each element in the first column by its minor and combining the products as indicated in equation (1). This is called **expansion of the determinant by minors** about the first column.

■ EXAMPLE 2 Evaluating a 3 × 3 Determinant

Evaluate the determinant using expansion by minors about the first column.

$$\begin{vmatrix} 1 & 3 & -2 \\ -1 & -2 & -3 \\ 1 & 1 & 2 \end{vmatrix}$$

In this determinant, $a_1 = 1$, $a_2 = -1$, and $a_3 = 1$. Multiply each of these numbers by its minor, and combine the three terms using the definition. Notice that the second term in the definition is *subtracted*.

$$\begin{vmatrix} 1 & 3 & -2 \\ -1 & -2 & -3 \\ 1 & 1 & 2 \end{vmatrix} = 1\begin{vmatrix} -2 & -3 \\ 1 & 2 \end{vmatrix} - (-1)\begin{vmatrix} 3 & -2 \\ 1 & 2 \end{vmatrix} + 1\begin{vmatrix} 3 & -2 \\ -2 & -3 \end{vmatrix}$$

$$= 1[-2(2) - (-3)1] + 1[3(2) - (-2)1]$$
$$\quad + 1[3(-3) - (-2)(-2)]$$
$$= 1(-1) + 1(8) + 1(-13)$$
$$= -1 + 8 - 13$$
$$= -6$$

Now Try Exercise 9.

To obtain equation (1), we could have rearranged terms in the definition of the determinant and used the distributive property to factor out the three elements of the second or third column or of any of the three rows. Therefore, expanding by minors about any row or any column results in the same value for a 3 × 3 determinant. To determine the correct signs for the terms of other expansions, the following **array of signs** is helpful.

Array of Signs for a 3 × 3 Determinant

$$\begin{array}{ccc} + & - & + \\ - & + & - \\ + & - & + \end{array}$$

The signs alternate for each row and column beginning with a + in the first row, first column position. For example, if the expansion is to be about the second column, the first term would have a minus sign associated with it, the second term a plus sign, and the third term a minus sign.

■ **EXAMPLE 3** Evaluating a 3 × 3 Determinant

Evaluate the determinant of Example 2 using expansion by minors about the second column.

$$\begin{vmatrix} 1 & 3 & -2 \\ -1 & -2 & -3 \\ 1 & 1 & 2 \end{vmatrix} = -3\begin{vmatrix} -1 & -3 \\ 1 & 2 \end{vmatrix} + (-2)\begin{vmatrix} 1 & -2 \\ 1 & 2 \end{vmatrix} - 1\begin{vmatrix} 1 & -2 \\ -1 & -3 \end{vmatrix}$$

$$= -3(1) - 2(4) - 1(-5)$$
$$= -3 - 8 + 5$$
$$= -6$$

As expected, the result is the same as in Example 2.

Now Try Exercise 15.

OBJECTIVE 3 Use a graphing calculator to evaluate determinants. The graphing calculator function det(A) assigns to each square matrix [A] one and only one real number, the determinant of A.

■ **EXAMPLE 4** Evaluating Determinants Using a Graphing Calculator

Evaluate the determinants in Examples 1 and 2 using a graphing calculator.

Figure 13 shows how a graphing calculator displays the correct value for the determinant in Example 1. Similarly, Figure 14 supports the result of Example 2.

```
[A]
      [[-1 -3]
       [4  -2]]
det([A])
               14
```

```
[B]
      [[1  3 -2]
       [-1 -2 -3]
       [1  1  2]]
det([B])
               -6
```

FIGURE 13   **FIGURE 14**

Now Try Exercise 23.

| **CONNECTIONS** |

Determinants of larger dimensions (such as 4 × 4) can be evaluated by extending the concepts presented thus far. However, because of the tedious calculations and chance for error, they are usually evaluated by computer or graphing calculator. For example, the determinant

$$\begin{vmatrix} -1 & -2 & 3 & 2 \\ 0 & 1 & 4 & -2 \\ 3 & -1 & 4 & 0 \\ 2 & 1 & 0 & 3 \end{vmatrix}$$

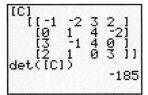

```
[C]
  [[-1 -2  3  2]
   [0   1  4 -2]
   [3  -1  4  0]
   [2   1  0  3]]
det([C])
            -185
```

FIGURE 15

is equal to -185, as shown in the graphing calculator screen in Figure 15.

For Discussion or Writing

1. Use the array of signs

$$\begin{array}{cccc} + & - & + & - \\ - & + & - & + \\ + & - & + & - \\ - & + & - & + \end{array}$$

to evaluate the preceding determinant by hand, expanding about the fourth row.

2. Explain how finding a determinant illustrates the function concept.

OBJECTIVE 4 **Understand the derivation of Cramer's rule.** Determinants can be used to solve a system of the form

$$a_1 x + b_1 y = c_1 \qquad (1)$$
$$a_2 x + b_2 y = c_2. \qquad (2)$$

The result will be a formula that can be used to solve any system of two equations with two variables. To get this general solution, we eliminate y and solve for x by first multiplying each side of equation (1) by b_2 and each side of equation (2) by $-b_1$. Then we add these results and solve for x.

$a_1 b_2 x + b_1 b_2 y = c_1 b_2$	Multiply equation (1) by b_2.
$\underline{-a_2 b_1 x - b_1 b_2 y = -c_2 b_1}$	Multiply equation (2) by $-b_1$.
$(a_1 b_2 - a_2 b_1)x = c_1 b_2 - c_2 b_1$	

$$x = \frac{c_1 b_2 - c_2 b_1}{a_1 b_2 - a_2 b_1} \qquad (\text{if } a_1 b_2 - a_2 b_1 \neq 0)$$

To solve for y, we multiply each side of equation (1) by $-a_2$ and each side of equation (2) by a_1 and add.

$-a_1 a_2 x - a_2 b_1 y = -a_2 c_1$	Multiply equation (1) by $-a_2$.
$\underline{a_1 a_2 x + a_1 b_2 y = a_1 c_2}$	Multiply equation (2) by a_1.
$(a_1 b_2 - a_2 b_1)y = a_1 c_2 - a_2 c_1$	

$$y = \frac{a_1 c_2 - a_2 c_1}{a_1 b_2 - a_2 b_1} \qquad (\text{if } a_1 b_2 - a_2 b_1 \neq 0)$$

Both numerators and the common denominator of these values for x and y can be written as determinants because

$$a_1 c_2 - a_2 c_1 = \begin{vmatrix} a_1 & c_1 \\ a_2 & c_2 \end{vmatrix},$$

$$c_1 b_2 - c_2 b_1 = \begin{vmatrix} c_1 & b_1 \\ c_2 & b_2 \end{vmatrix},$$

and $\qquad\qquad\qquad a_1 b_2 - a_2 b_1 = \begin{vmatrix} a_1 & b_1 \\ a_2 & b_2 \end{vmatrix}.$

Using these results, the solutions for x and y become

$$x = \frac{\begin{vmatrix} c_1 & b_1 \\ c_2 & b_2 \end{vmatrix}}{\begin{vmatrix} a_1 & b_1 \\ a_2 & b_2 \end{vmatrix}} \quad \text{and} \quad y = \frac{\begin{vmatrix} a_1 & c_1 \\ a_2 & c_2 \end{vmatrix}}{\begin{vmatrix} a_1 & b_1 \\ a_2 & b_2 \end{vmatrix}}, \quad \begin{vmatrix} a_1 & b_1 \\ a_2 & b_2 \end{vmatrix} \neq 0.$$

For convenience, we denote the three determinants in the solution as

$$\begin{vmatrix} a_1 & b_1 \\ a_2 & b_2 \end{vmatrix} = D, \qquad \begin{vmatrix} c_1 & b_1 \\ c_2 & b_2 \end{vmatrix} = D_x, \qquad \text{and} \qquad \begin{vmatrix} a_1 & c_1 \\ a_2 & c_2 \end{vmatrix} = D_y.$$

Notice that the elements of D are the four coefficients of the variables in the given system; the elements of D_x are obtained by replacing the coefficients of x by the respective constants; the elements of D_y are obtained by replacing the coefficients of y by the respective constants.

These results are summarized as **Cramer's rule.**

Cramer's Rule for 2 × 2 Systems

Given the system

$$a_1x + b_1y = c_1$$
$$a_2x + b_2y = c_2 \quad \text{with } a_1b_2 - a_2b_1 = D \neq 0,$$

then

$$x = \frac{\begin{vmatrix} c_1 & b_1 \\ c_2 & b_2 \end{vmatrix}}{\begin{vmatrix} a_1 & b_1 \\ a_2 & b_2 \end{vmatrix}} = \frac{D_x}{D} \quad \text{and} \quad y = \frac{\begin{vmatrix} a_1 & c_1 \\ a_2 & c_2 \end{vmatrix}}{\begin{vmatrix} a_1 & b_1 \\ a_2 & b_2 \end{vmatrix}} = \frac{D_y}{D}.$$

OBJECTIVE 5 Apply Cramer's rule to solve linear systems. To use Cramer's rule to solve a system of equations, find the three determinants, D, D_x, and D_y, and then write the necessary quotients for x and y.

CAUTION As indicated in the box, Cramer's rule does not apply if $D = a_1b_2 - a_2b_1 = 0$. When $D = 0$, the system is inconsistent or has dependent equations. For this reason, it is a good idea to evaluate D first.

EXAMPLE 5 Using Cramer's Rule to Solve a 2 × 2 System

Use Cramer's rule to solve the system

$$5x + 7y = -1$$
$$6x + 8y = 1.$$

By Cramer's rule, $x = \dfrac{D_x}{D}$ and $y = \dfrac{D_y}{D}$. As previously mentioned, it is a good idea to find D first since if $D = 0$, Cramer's rule does not apply. If $D \neq 0$, then find D_x and D_y.

$$D = \begin{vmatrix} 5 & 7 \\ 6 & 8 \end{vmatrix} = 5(8) - 7(6) = -2$$

$$D_x = \begin{vmatrix} -1 & 7 \\ 1 & 8 \end{vmatrix} = -1(8) - 7(1) = -15$$

$$D_y = \begin{vmatrix} 5 & -1 \\ 6 & 1 \end{vmatrix} = 5(1) - (-1)6 = 11$$

From Cramer's rule,

$$x = \frac{D_x}{D} = \frac{-15}{-2} = \frac{15}{2} \qquad \text{and} \qquad y = \frac{D_y}{D} = \frac{11}{-2} = -\frac{11}{2}.$$

The solution set is $\left\{\left(\frac{15}{2}, -\frac{11}{2}\right)\right\}$, as can be verified by checking in the given system.

Now Try Exercise 27.

Because graphing calculators can evaluate determinants, Cramer's rule can be applied using them. Figure 16 shows how the work of Example 5 is accomplished, with D the determinant of matrix [A], D_x the determinant of matrix [B], and D_y the determinant of matrix [C].

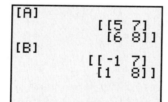

FIGURE 16

In a similar manner, Cramer's rule can be applied to systems of three equations with three variables.

Cramer's Rule for 3 × 3 Systems

Given the system

$$a_1 x + b_1 y + c_1 z = d_1$$
$$a_2 x + b_2 y + c_2 z = d_2$$
$$a_3 x + b_3 y + c_3 z = d_3$$

with

$$D_x = \begin{vmatrix} d_1 & b_1 & c_1 \\ d_2 & b_2 & c_2 \\ d_3 & b_3 & c_3 \end{vmatrix}, \qquad D_y = \begin{vmatrix} a_1 & d_1 & c_1 \\ a_2 & d_2 & c_2 \\ a_3 & d_3 & c_3 \end{vmatrix},$$

$$D_z = \begin{vmatrix} a_1 & b_1 & d_1 \\ a_2 & b_2 & d_2 \\ a_3 & b_3 & d_3 \end{vmatrix}, \qquad D = \begin{vmatrix} a_1 & b_1 & c_1 \\ a_2 & b_2 & c_2 \\ a_3 & b_3 & c_3 \end{vmatrix} \neq 0,$$

then

$$x = \frac{D_x}{D}, \qquad y = \frac{D_y}{D}, \qquad \text{and} \qquad z = \frac{D_z}{D}.$$

EXAMPLE 6   Using Cramer's Rule to Solve a 3 × 3 System

Use Cramer's rule to solve the system

$$\begin{aligned} x + y - z + 2 &= 0 \\ 2x - y + z + 5 &= 0 \\ x - 2y + 3z - 4 &= 0. \end{aligned}$$

To use Cramer's rule, first rewrite the system in the form

$$\begin{aligned} x + y - z &= -2 \\ 2x - y + z &= -5 \\ x - 2y + 3z &= 4. \end{aligned}$$

Expand by minors about row 1 to find D.

$$\begin{aligned} D &= \begin{vmatrix} 1 & 1 & -1 \\ 2 & -1 & 1 \\ 1 & -2 & 3 \end{vmatrix} \\ &= 1 \begin{vmatrix} -1 & 1 \\ -2 & 3 \end{vmatrix} - 1 \begin{vmatrix} 2 & 1 \\ 1 & 3 \end{vmatrix} + (-1) \begin{vmatrix} 2 & -1 \\ 1 & -2 \end{vmatrix} \\ &= 1(-1) - 1(5) - 1(-3) \\ &= -3 \end{aligned}$$

Expanding D_x by minors about row 1 gives

$$\begin{aligned} D_x &= \begin{vmatrix} -2 & 1 & -1 \\ -5 & -1 & 1 \\ 4 & -2 & 3 \end{vmatrix} \\ &= -2 \begin{vmatrix} -1 & 1 \\ -2 & 3 \end{vmatrix} - 1 \begin{vmatrix} -5 & 1 \\ 4 & 3 \end{vmatrix} + (-1) \begin{vmatrix} -5 & -1 \\ 4 & -2 \end{vmatrix} \\ &= -2(-1) - 1(-19) - 1(14) \\ &= 7. \end{aligned}$$

In the same way, $D_y = -22$ and $D_z = -21$, so that

$$x = \frac{D_x}{D} = \frac{7}{-3} = -\frac{7}{3}, \qquad y = \frac{D_y}{D} = \frac{-22}{-3} = \frac{22}{3}, \qquad z = \frac{D_z}{D} = \frac{-21}{-3} = 7.$$

Check that the solution set is $\left\{\left(-\frac{7}{3}, \frac{22}{3}, 7\right)\right\}$.

Now Try Exercise 33.

As mentioned earlier, Cramer's rule does not apply when $D = 0$. The next example illustrates this case.

EXAMPLE 7 Determining When Cramer's Rule Does Not Apply

Use Cramer's rule to solve the system

$$\begin{aligned} 2x - 3y + 4z &= 8 \\ 6x - 9y + 12z &= 24 \\ x + 2y - 3z &= 5. \end{aligned}$$

First, find D.

$$D = \begin{vmatrix} 2 & -3 & 4 \\ 6 & -9 & 12 \\ 1 & 2 & -3 \end{vmatrix}$$

$$= 2\begin{vmatrix} -9 & 12 \\ 2 & -3 \end{vmatrix} - 6\begin{vmatrix} -3 & 4 \\ 2 & -3 \end{vmatrix} + 1\begin{vmatrix} -3 & 4 \\ -9 & 12 \end{vmatrix}$$

$$= 2(3) - 6(1) + 1(0)$$

$$= 0$$

Since $D = 0$ here, Cramer's rule does not apply and we must use another method to solve the system. Multiplying each side of the first equation by 3 shows that the first two equations have the same solution set, so this system has dependent equations and an infinite solution set.

Now Try Exercise 35.

4.5 EXERCISES

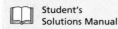

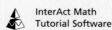

1. Which one of the following is the expression for the determinant $\begin{vmatrix} -2 & -3 \\ 4 & -6 \end{vmatrix}$?

A. $-2(-6) + (-3)4$ **B.** $-2(-6) - 3(4)$

C. $-3(4) - (-2)(-6)$ **D.** $-2(-6) - (-3)4$

2. Evaluate $\begin{vmatrix} 0 & 0 \\ 3 & -4 \end{vmatrix}$ and $\begin{vmatrix} 0 & 1 & 2 \\ 0 & -3 & 4 \\ 0 & 2 & 6 \end{vmatrix}$ and make a conjecture (educated guess) about the

value of a determinant that has all 0s in a row or a column.

Evaluate each determinant. See Example 1.

3. $\begin{vmatrix} -2 & 5 \\ -1 & 4 \end{vmatrix}$

4. $\begin{vmatrix} 3 & -6 \\ 2 & -2 \end{vmatrix}$

5. $\begin{vmatrix} 1 & -2 \\ 7 & 0 \end{vmatrix}$

6. $\begin{vmatrix} -5 & -1 \\ 1 & 0 \end{vmatrix}$

7. $\begin{vmatrix} 0 & 4 \\ 0 & 4 \end{vmatrix}$

8. $\begin{vmatrix} 8 & -3 \\ 0 & 0 \end{vmatrix}$

Evaluate each determinant by expansion by minors about the first column. See Example 2.

9. $\begin{vmatrix} -1 & 2 & 4 \\ -3 & -2 & -3 \\ 2 & -1 & 5 \end{vmatrix}$

10. $\begin{vmatrix} 2 & -3 & -5 \\ 1 & 2 & 2 \\ 5 & 3 & -1 \end{vmatrix}$

11. $\begin{vmatrix} 1 & 0 & -2 \\ 0 & 2 & 3 \\ 1 & 0 & 5 \end{vmatrix}$

12. $\begin{vmatrix} 2 & -1 & 0 \\ 0 & -1 & 1 \\ 1 & 2 & 0 \end{vmatrix}$

13. Explain in your own words how to evaluate a 2 × 2 determinant. Illustrate with an example.

14. Explain in your own words how to evaluate a 3 × 3 determinant. Illustrate with an example.

Evaluate each determinant by expansion by minors about any row or column. (Hint: The work is easier if you choose a row or a column with 0s.) See Example 3.

15. $\begin{vmatrix} 4 & 4 & 2 \\ 1 & -1 & -2 \\ 1 & 0 & 2 \end{vmatrix}$

16. $\begin{vmatrix} 3 & -1 & 2 \\ 1 & 5 & -2 \\ 0 & 2 & 0 \end{vmatrix}$

17. $\begin{vmatrix} 3 & 5 & -2 \\ 1 & -4 & 1 \\ 3 & 1 & -2 \end{vmatrix}$

18. $\begin{vmatrix} 0 & 0 & 3 \\ 4 & 0 & -2 \\ 2 & -1 & 3 \end{vmatrix}$

19. $\begin{vmatrix} 3 & 0 & -2 \\ 1 & -4 & 1 \\ 3 & 1 & -2 \end{vmatrix}$

20. $\begin{vmatrix} 1 & 1 & 2 \\ 5 & 5 & 7 \\ 3 & 3 & 1 \end{vmatrix}$

21. Explain why a determinant with a row or column of 0s has a value of 0.

Use a graphing calculator with matrix capabilities to find each determinant. See Example 4.

22. $\begin{vmatrix} .68 & .94 \\ .31 & -.56 \end{vmatrix}$

23. $\begin{vmatrix} 1.5 & 2.6 & 9.3 \\ 5.2 & -1.4 & 8.6 \\ 0 & .7 & 1.2 \end{vmatrix}$

24. $\begin{vmatrix} \sqrt{5} & \sqrt{2} & -\sqrt{3} \\ \sqrt{7} & -\sqrt{6} & \sqrt{10} \\ -\sqrt{5} & -\sqrt{2} & \sqrt{17} \end{vmatrix}$ (To as many places as the calculator shows)

25. Consider the system
$$4x + 3y - 2z = 1$$
$$7x - 4y + 3z = 2$$
$$-2x + y - 8z = 0.$$

Match each determinant in parts (a)–(d) with its correct representation from choices A–D.

(a) D

(b) D_x

(c) D_y

(d) D_z

A. $\begin{vmatrix} 1 & 3 & -2 \\ 2 & -4 & 3 \\ 0 & 1 & -8 \end{vmatrix}$ **B.** $\begin{vmatrix} 4 & 3 & 1 \\ 7 & -4 & 2 \\ -2 & 1 & 0 \end{vmatrix}$

C. $\begin{vmatrix} 4 & 1 & -2 \\ 7 & 2 & 3 \\ -2 & 0 & -8 \end{vmatrix}$ **D.** $\begin{vmatrix} 4 & 3 & -2 \\ 7 & -4 & 3 \\ -2 & 1 & -8 \end{vmatrix}$

26. For the system
$$x + 3y - 6z = 7$$
$$2x - y + z = 1$$
$$x + 2y + 2z = -1,$$

$D = -43$, $D_x = -43$, $D_y = 0$, and $D_z = 43$. What is the solution set of the system?

Use Cramer's rule to solve each linear system in two variables. See Example 5.

27. $3x + 5y = -5$
  $-2x + 3y = 16$

28. $5x + 2y = -3$
  $4x - 3y = -30$

29. $8x + 3y = 1$
  $6x - 5y = 2$

30. $3x - y = 9$
  $2x + 5y = 8$

31. $2x + 3y = 4$
  $5x + 6y = 7$

32. $4x + 5y = 6$
  $7x + 8y = 9$

Use Cramer's rule where applicable to solve each linear system in three variables. See Examples 6 and 7.

33. $2x + 3y + 2z = 15$
  $x - y + 2z = 5$
  $x + 2y - 6z = -26$

34. $x - y + 6z = 19$
  $3x + 3y - z = 1$
  $x + 9y + 2z = -19$

35. $2x - 3y + 4z = 8$
  $6x - 9y + 12z = 24$
  $-4x + 6y - 8z = -16$

36. $7x + y - z = 4$
  $2x - 3y + z = 2$
  $-6x + 9y - 3z = -6$

37. $3x + 5z = 0$
  $2x + 3y = 1$
  $-y + 2z = -11$

38. $-x + 2y = 4$
  $3x + y = -5$
  $2x + z = -1$

39. $x - 3y = 13$
  $2y + z = 5$
  $-x + z = -7$

40. $-5x - y = -10$
  $3x + 2y + z = -3$
  $-y - 2z = -13$

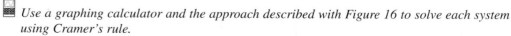 *Use a graphing calculator and the approach described with Figure 16 to solve each system using Cramer's rule.*

41. $x + 2y + z = 10$
  $2x - y - 3z = -20$
  $-x + 4y + z = 18$

42. $2x + y + 3z = 1$
  $x - 2y + z = -3$
  $-3x + y - 2z = -4$

43. $-8w + 4x - 2y + z = -28$
$-w + x - y + z = -10$
$w + x + y + z = -4$
$27w + 9x + 3y + z = 2$

44. $5w + 2x - 3y + z = 4.7$
$-2w + x + 2y - z = -3.2$
$w + 3x - y + 2z = 2.1$
$2w + x - 5y + 3z = 3.4$

Solve each system for x and y using Cramer's rule. Assume a and b are nonzero constants.

45. $bx + y = a^2$
$ax + y = b^2$

46. $ax + by = \dfrac{b}{a}$
$x + y = \dfrac{1}{b}$

47. $b^2x + a^2y = b^2$
$ax + by = a$

48. $x + \dfrac{1}{b}y = b$
$\dfrac{1}{a}x + y = a$

There is another method for evaluating a 3 × 3 determinant. Refer to Example 2, and copy the first two columns to the right of the original determinant to obtain

$$\begin{vmatrix} 1 & 3 & -2 \\ -1 & -2 & -3 \\ 1 & 1 & 2 \end{vmatrix} \begin{matrix} 1 & 3 \\ -1 & -2 \\ 1 & 1 \end{matrix}.$$

Multiply along the diagonals as shown, placing the product at the end of the arrow.

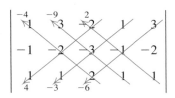

Add the top numbers: $-4 - 9 + 2 = -11.$

Add the bottom numbers: $4 - 3 - 6 = -5.$

Find the *difference* between these sums to obtain the final answer:

$$-11 - (-5) = -6.$$

Use this method to find each determinant in the indicated exercise.

49. Exercise 15

50. Exercise 16

51. Exercise 17

52. Exercise 18

53. Exercise 19

54. Exercise 20

Solve each equation by finding an expression for the determinant on the left, and then solving using the methods of Chapter 2.

55. $\begin{vmatrix} 4 & x \\ 2 & 3 \end{vmatrix} = 8$

56. $\begin{vmatrix} 5 & 3 \\ x & x \end{vmatrix} = 20$

57. $\begin{vmatrix} x & 4 \\ x & -3 \end{vmatrix} = 0$

58. Look at the coefficients and constants in the systems in Exercises 31 and 32. Notice that in both cases, the six numbers are consecutive integers. Make up a system having this same pattern for its coefficients and constants, and solve it using Cramer's rule. Compare the solutions in Exercises 31, 32, and here. What do you notice?

59. Use Cramer's rule to prove that the following system has solution set $\{(-1, 2)\}$.

$$ax + (a + 1)y = a + 2$$
$$(a + 3)x + (a + 4)y = a + 5, \quad \text{where } D \neq 0.$$

60. Under what conditions can a system *not* be solved using Cramer's rule?

RELATING CONCEPTS (EXERCISES 61–68)

For Individual or Group Work

In this section we have seen how determinants can be used to solve systems of equations. There are other applications of determinants. Here, we show how a determinant can be used to find the area of a triangle if we know the coordinates of its vertices.

Suppose that $A(x_1, y_1)$, $B(x_2, y_2)$, and $C(x_3, y_3)$ are the coordinates of the vertices of triangle ABC in the coordinate plane. Then it can be shown that the area of the triangle is given by the absolute value of

$$\frac{1}{2} \begin{vmatrix} x_1 & y_1 & 1 \\ x_2 & y_2 & 1 \\ x_3 & y_3 & 1 \end{vmatrix}.$$

Work Exercises 61–64 in order.

61. Sketch triangle *ABC* in the coordinate plane, given that the coordinates of *A* are $(0, 0)$, of *B* are $(-3, -4)$, and of *C* are $(2, -2)$.

62. Write the determinant expression described above that gives the area of triangle *ABC* described in Exercise 57.

63. Evaluate the absolute value of the determinant expression in Exercise 58 to find the area.

64. Use the determinant expression described above to find the area of the triangle with vertices at $(3, 8)$, $(-1, 4)$, and $(0, 1)$.

Here is yet another application of determinants. Recall the formula for slope and the point-slope form of the equation of a line from Chapter 3. Use these formulas to ***work Exercises 65–68 in order*** *and see how a determinant can be used in writing the equation of a line.*

65. Write the expression for the slope of a line passing through the points (x_1, y_1) and (x_2, y_2).

66. Using the expression from Exercise 65 as *m*, and the point (x_1, y_1), write the point-slope form of the equation of the line.

67. Using the equation obtained in Exercise 66, multiply both sides by $x_2 - x_1$, and write the equation so that 0 is on the right side.

68. Consider the *determinant equation*

$$\begin{vmatrix} x & y & 1 \\ x_1 & y_1 & 1 \\ x_2 & y_2 & 1 \end{vmatrix} = 0.$$

Expand by minors on the left and show that this determinant equation yields the same result that you obtained in Exercise 67.

Several theorems are useful when calculating determinants. These theorems are true for square matrices of any size.

Determinant Theorems

1. If every element in a row (or column) of matrix A is 0, then $|A| = 0$.

2. If the rows of matrix A are the corresponding columns of matrix B, then $|B| = |A|$.

3. If any two rows (or columns) of matrix A are interchanged to form matrix B, then $|B| = -|A|$.

4. Suppose matrix B is formed by multiplying every element of a row (or column) of matrix A by the real number k. Then $|B| = k \cdot |A|$.

5. If two rows (or columns) of matrix A are identical, then $|A| = 0$.

6. Changing a row (or column) of a matrix by adding to it a constant times another row (or column) does not change the determinant of the matrix.

Use the determinant theorems to find the value of each determinant.

69. $\begin{vmatrix} 1 & 0 & 0 \\ 1 & 0 & 1 \\ 3 & 0 & 0 \end{vmatrix}$

70. $\begin{vmatrix} -1 & 2 & 4 \\ 4 & -8 & -16 \\ 3 & 0 & 5 \end{vmatrix}$

71. $\begin{vmatrix} 6 & 8 & -12 \\ -1 & 0 & 2 \\ 4 & 0 & -8 \end{vmatrix}$

72. $\begin{vmatrix} 4 & 8 & 0 \\ -1 & -2 & 1 \\ 2 & 4 & 3 \end{vmatrix}$

73. $\begin{vmatrix} -4 & 1 & 4 \\ 2 & 0 & 1 \\ 0 & 2 & 4 \end{vmatrix}$

74. $\begin{vmatrix} 6 & 3 & 2 \\ 1 & 0 & 2 \\ 5 & 7 & 3 \end{vmatrix}$

Use the appropriate determinant theorem(s) to tell why each statement is true. Do not evaluate the determinants.

75. $\begin{vmatrix} 4 & -2 \\ 3 & 8 \end{vmatrix} = \begin{vmatrix} 4 & 3 \\ -2 & 8 \end{vmatrix}$

76. $\begin{vmatrix} 2 & 1 & 6 \\ 3 & 0 & 2 \\ 4 & 1 & 8 \end{vmatrix} = \begin{vmatrix} 2 & 3 & 4 \\ 1 & 0 & 1 \\ 6 & 2 & 8 \end{vmatrix}$

77. $\begin{vmatrix} -1 & 8 & 9 \\ 0 & 2 & 1 \\ 3 & 2 & 0 \end{vmatrix} = -\begin{vmatrix} 8 & -1 & 9 \\ 2 & 0 & 1 \\ 2 & 3 & 0 \end{vmatrix}$

78. $\begin{vmatrix} 2 & 6 \\ 3 & 5 \end{vmatrix} = -\begin{vmatrix} 3 & 5 \\ 2 & 6 \end{vmatrix}$

79. $-\dfrac{1}{2}\begin{vmatrix} 5 & -8 & 2 \\ 3 & -6 & 9 \\ 2 & 4 & 4 \end{vmatrix} = \begin{vmatrix} 5 & 4 & 2 \\ 3 & 3 & 9 \\ 2 & -2 & 4 \end{vmatrix}$

80. $3\begin{vmatrix} 6 & 0 & 2 \\ 4 & 1 & 3 \\ 2 & 8 & 6 \end{vmatrix} = \begin{vmatrix} 6 & 0 & 2 \\ 4 & 3 & 3 \\ 2 & 24 & 6 \end{vmatrix}$

81. $\begin{vmatrix} 3 & -4 \\ 2 & 5 \end{vmatrix} = \begin{vmatrix} 3 & -4 \\ 5 & 1 \end{vmatrix}$

82. $\begin{vmatrix} -1 & 6 \\ 3 & -5 \end{vmatrix} = \begin{vmatrix} -1 & 5 \\ 3 & -2 \end{vmatrix}$

Chapter **4** **Group Activity**

Top Concert Tours

OBJECTIVE   Use systems of equations to analyze data.

This activity will use systems of equations to analyze the top-grossing concert tours of 1997 and 2000.

- In 1997 the top-grossing North American concert tours were The Rolling Stones and U2. Together they grossed $169.2 million. The Rolling Stones grossed $9.4 million more than U2. The two groups performed a total of 79 shows, and U2 performed 13 more shows than The Rolling Stones. (*Source:* Pollstar.)

- In 2000 the top-grossing North American concert tours were Tina Turner and 'N Sync. Together they grossed $156.6 million. Tina Turner grossed $3.8 million more than 'N Sync. The two groups performed 181 shows, and 'N Sync performed 9 less shows than Tina Turner. (*Source:* Pollstar.)

Using the given information, complete the table. One student should complete the information for 1997 and the other student for 2000.

A. Write a system of equations to find the total gross for each group.

B. Write a system of equations to find the number of shows per year for each group.

C. Use the information from Exercises A and B to find the gross per show.

Year	Group	Total Gross	Shows per Year	Gross per Show
1997	Rolling Stones			
1997	U2			
2000	Tina Turner			
2000	'N Sync			

D. Once you have completed the table, compare the results. Answer these questions.

   1. What differences do you notice between 1997 and 2000?

   2. Which group had the highest total gross?

   3. Which group had the highest gross per show?

CHAPTER 4 SUMMARY

KEY TERMS

4.1 system of equations
system of linear
equations
solution set of a linear
system
consistent system
independent equations

inconsistent system
dependent equations
elimination method
substitution method
4.2 ordered triple
4.4 matrix
element of a matrix

row
column
square matrix
augmented matrix
row operations
row echelon form

reduced row echelon
form
4.5 determinant
minor
expansion by minors
array of signs
Cramer's rule

NEW SYMBOLS

(x, y, z) ordered triple

$$\begin{bmatrix} a & b & c \\ d & e & f \end{bmatrix}$$ matrix with
two rows,
three
columns

$$\begin{vmatrix} a & b \\ c & d \end{vmatrix}$$ determinant
of a 2×2
matrix

$$\begin{vmatrix} a & b & c \\ d & e & f \\ g & h & i \end{vmatrix}$$ determinant
of a 3×3
matrix

TEST YOUR WORD POWER

See how well you have learned the vocabulary in this chapter. Answers, with examples, follow the Quick Review.

1. A **system of equations** consists of
  A. at least two equations with
  different variables
  B. two or more equations that have
  an infinite number of solutions
  C. two or more equations that are to
  be solved at the same time
  D. two or more inequalities that are
  to be solved.

2. The **solution set of a system of
equations** is
  A. all ordered pairs that satisfy one
  equation of the system
  B. all ordered pairs that satisfy all
  the equations of the system at the
  same time
  C. any ordered pair that satisfies one
  or more equations of the system
  D. the set of values that make all the
  equations of the system false.

3. An **inconsistent system** is a system
of equations
  A. with one solution
  B. with no solution
  C. with an infinite number of
  solutions
  D. that have the same graph.

4. **Dependent equations**
  A. have different graphs
  B. have no solution
  C. have one solution
  D. are different forms of the same
  equation.

5. A **matrix** is
  A. an ordered pair of numbers
  B. an array of numbers with the
  same number of rows and
  columns
  C. a pair of numbers written
  between brackets
  D. a rectangular array of numbers.

6. A matrix written in **row echelon
form** has
  A. elements that are all 0
  B. elements that are all 1
  C. upper left to lower right diagonal
  elements of 1 with 0s below the
  1s
  D. upper left to lower right diagonal
  elements of 0 with 1s below the
  0s.

7. A **determinant** is
  A. a rectangular array of numbers
  B. a real number associated with a
  square matrix
  C. a matrix with the same number
  of rows and columns
  D. an ordered pair of numbers.

8. **Expansion by minors** is
  A. a method of evaluating a 3×3
  or larger determinant
  B. a way to use row operations to
  produce new matrices
  C. a method of evaluating a 2×2
  determinant
  D. a method of evaluating the
  determinant of a matrix.

QUICK REVIEW	
CONCEPTS	**EXAMPLES**

4.1 SYSTEMS OF LINEAR EQUATIONS IN TWO VARIABLES

Solving a Linear System by Elimination

Step 1 Write both equations in standard form.

Step 2 Make the coefficients of one pair of variable terms opposites.

Step 3 Add the new equations. The sum should be an equation with just one variable.

Step 4 Solve the equation from Step 3.

Step 5 Find the value of the other variable by substituting the result of Step 4 into either of the original equations.

Step 6 Check the solution in both of the original equations. Then write the solution set.

Solve by elimination.

$$5x + y = 2 \quad (1)$$
$$2x - 3y = 11 \quad (2)$$

To eliminate y, multiply equation (1) by 3 and add the result to equation (2).

$$15x + 3y = 6$$
$$\underline{2x - 3y = 11} \quad (2)$$
$$17x = 17$$
$$x = 1$$

Let $x = 1$ in equation (1), and solve for y.

$$5(1) + y = 2$$
$$y = -3$$

Check to verify that $\{(1, -3)\}$ is the solution set.

Solving a Linear System by Substitution

Step 1 Solve one of the equations for either variable.

Step 2 Substitute for that variable in the other equation. The result should be an equation with just one variable.

Step 3 Solve the equation from Step 2.

Step 4 Find the value of the other variable by substituting the result from Step 3 into the equation from Step 1.

Step 5 Check the solution in both of the original equations. Then write the solution set.

Solve by substitution.

$$4x - y = 7 \quad (1)$$
$$3x + 2y = 30 \quad (2)$$

Solve for y in equation (1).

$$y = 4x - 7$$

Substitute $4x - 7$ for y in equation (2), and solve for x.

$$3x + 2(4x - 7) = 30$$
$$3x + 8x - 14 = 30$$
$$11x - 14 = 30$$
$$11x = 44$$
$$x = 4$$

Substitute 4 for x in the equation $y = 4x - 7$ to find that $y = 9$.

Check to see that $\{(4, 9)\}$ is the solution set.

4.2 SYSTEMS OF LINEAR EQUATIONS IN THREE VARIABLES

Solving a Linear System in Three Variables

Step 1 Use the elimination method to eliminate any variable from any two of the original equations.

Solve the system

$$x + 2y - z = 6 \quad (1)$$
$$x + y + z = 6 \quad (2)$$
$$2x + y - z = 7. \quad (3)$$

Add equations (1) and (2); z is eliminated and the result is $2x + 3y = 12$.

CONCEPTS	EXAMPLES
Step 2 Eliminate the *same* variable from any *other* two equations.	Eliminate z again by adding equations (2) and (3) to get $3x + 2y = 13$. Now solve the system

$$2x + 3y = 12 \quad (4)$$
$$3x + 2y = 13. \quad (5)$$

Step 3   Eliminate a second variable from the two equations in two variables that result from Steps 1 and 2. The result is an equation in one variable that gives the value of that variable.

To eliminate x, multiply equation (4) by -3 and equation (5) by 2.

$$\begin{array}{r} -6x - 9y = -36 \\ \underline{6x + 4y = \quad 26} \\ -5y = -10 \\ y = 2 \end{array}$$

Step 4   Substitute the value of the variable found in Step 3 into either of the equations in two variables to find the value of the second variable.

Let $y = 2$ in equation (4).

$$2x + 3(2) = 12$$
$$2x + 6 = 12$$
$$2x = 6$$
$$x = 3$$

Step 5   Use the values of the two variables from Steps 3 and 4 to find the value of the third variable by substituting into an appropriate equation.

Let $y = 2$ and $x = 3$ in any of the original equations to find $z = 1$.

Step 6   Check the solution in all of the original equations. Then write the solution set.

Check. The solution set is $\{(3, 2, 1)\}$.

4.3   APPLICATIONS OF SYSTEMS OF LINEAR EQUATIONS

Use the six-step problem-solving method.

Step 1   Read the problem carefully.

The perimeter of a rectangle is 18 ft. The length is 3 ft more than twice the width. What are the dimensions of the rectangle?

Step 2   Assign variables.

Let x represent the length and y represent the width.

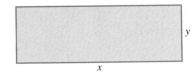

Step 3   Write a system of equations that relates the unknowns.

From the perimeter formula, one equation is $2x + 2y = 18$. From the problem, another equation is $x = 3 + 2y$.

Step 4   Solve the system.

Solve the system

$$2x + 2y = 18$$
$$x = 3 + 2y$$

to find that $x = 7$ and $y = 2$.   (continued)

CONCEPTS	EXAMPLES

Steps 5 and 6 State the answer and check.

The length is 7 ft, and the width is 2 ft. Since the perimeter is

$$2(7) + 2(2) = 18, \quad \text{and} \quad 3 + 2(2) = 7,$$

the answer checks.

4.4 SOLVING SYSTEMS OF LINEAR EQUATIONS BY MATRIX METHODS

Matrix Row Operations

1. Any two rows of the matrix may be interchanged.

$$\begin{bmatrix} 1 & 5 & 7 \\ 3 & 9 & -2 \\ 0 & 6 & 4 \end{bmatrix} \text{ becomes } \begin{bmatrix} 3 & 9 & -2 \\ 1 & 5 & 7 \\ 0 & 6 & 4 \end{bmatrix}$$ Interchange R_1 and R_2.

2. The elements in any row may be multiplied by any nonzero real number.

$$\begin{bmatrix} 1 & 5 & 7 \\ 3 & 9 & -2 \\ 0 & 6 & 4 \end{bmatrix} \text{ becomes } \begin{bmatrix} 1 & 5 & 7 \\ 1 & 3 & -\frac{2}{3} \\ 0 & 6 & 4 \end{bmatrix}$$ $\frac{1}{3}R_2$

3. Any row may be changed by adding to the elements of the row the product of a real number and the elements of another row.

$$\begin{bmatrix} 1 & 5 & 7 \\ 3 & 9 & -2 \\ 0 & 6 & 4 \end{bmatrix} \text{ becomes } \begin{bmatrix} 1 & 5 & 7 \\ 0 & -6 & -23 \\ 0 & 6 & 4 \end{bmatrix}$$ $-3R_1 + R_2$

A system can be solved by matrix methods. Write the augmented matrix and use row operations to obtain a matrix in row echelon form.

Solve using row operations: $\begin{array}{l} x + 3y = 7 \\ 2x + y = 4. \end{array}$

$$\begin{bmatrix} 1 & 3 & | & 7 \\ 2 & 1 & | & 4 \end{bmatrix}$$ Augmented matrix

$$\begin{bmatrix} 1 & 3 & | & 7 \\ 0 & -5 & | & -10 \end{bmatrix}$$ $-2R_1 + R_2$

$$\begin{bmatrix} 1 & 3 & | & 7 \\ 0 & 1 & | & 2 \end{bmatrix}$$ $-\frac{1}{5}R_2$

$$\begin{array}{l} x + 3y = 7 \\ y = 2 \end{array}$$

When $y = 2$, $x + 3(2) = 7$, so $x = 1$. The solution set is $\{(1, 2)\}$.

4.5 DETERMINANTS AND CRAMER'S RULE

Value of a 2 × 2 Determinant

$$\begin{vmatrix} a & b \\ c & d \end{vmatrix} = ad - bc$$

Evaluate.

$$\begin{vmatrix} 3 & 4 \\ -2 & 6 \end{vmatrix} = 3(6) - 4(-2) = 26$$

Determinants larger than 2 × 2 are evaluated by expansion by minors about a column or row.

Array of Signs for a 3 × 3 Determinant

$$\begin{array}{ccc} + & - & + \\ - & + & - \\ + & - & + \end{array}$$

Evaluate by expanding about the second column.

$$\begin{vmatrix} 2 & -3 & -2 \\ -1 & -4 & -3 \\ -1 & 0 & 2 \end{vmatrix} = -(-3)\begin{vmatrix} -1 & -3 \\ -1 & 2 \end{vmatrix} + (-4)\begin{vmatrix} 2 & -2 \\ -1 & 2 \end{vmatrix}$$

$$- 0\begin{vmatrix} 2 & -2 \\ -1 & -3 \end{vmatrix}$$

$$= 3(-5) - 4(2) - 0(-8)$$

$$= -15 - 8 + 0 = -23$$

CONCEPTS	EXAMPLES

Cramer's Rule for 2 × 2 Systems

Given the system

$$a_1x + b_1y = c_1$$
$$a_2x + b_2y = c_2$$

where $a_1b_2 - a_2b_1 = D \neq 0$,

$$\text{then} \quad x = \frac{\begin{vmatrix} c_1 & b_1 \\ c_2 & b_2 \end{vmatrix}}{\begin{vmatrix} a_1 & b_1 \\ a_2 & b_2 \end{vmatrix}} = \frac{D_x}{D}$$

$$\text{and} \quad y = \frac{\begin{vmatrix} a_1 & c_1 \\ a_2 & c_2 \end{vmatrix}}{\begin{vmatrix} a_1 & b_1 \\ a_2 & b_2 \end{vmatrix}} = \frac{D_y}{D}.$$

Cramer's Rule for 3 × 3 Systems

Given the system

$$a_1x + b_1y + c_1z = d_1$$
$$a_2x + b_2y + c_2z = d_2$$
$$a_3x + b_3y + c_3z = d_3$$

with

$$D = \begin{vmatrix} a_1 & b_1 & c_1 \\ a_2 & b_2 & c_2 \\ a_3 & b_3 & c_3 \end{vmatrix} \neq 0, \quad D_x = \begin{vmatrix} d_1 & b_1 & c_1 \\ d_2 & b_2 & c_2 \\ d_3 & b_3 & c_3 \end{vmatrix},$$

$$D_y = \begin{vmatrix} a_1 & d_1 & c_1 \\ a_2 & d_2 & c_2 \\ a_3 & d_3 & c_3 \end{vmatrix}, \quad D_z = \begin{vmatrix} a_1 & b_1 & d_1 \\ a_2 & b_2 & d_2 \\ a_3 & b_3 & d_3 \end{vmatrix},$$

$$\text{then} \quad x = \frac{D_x}{D}, \quad y = \frac{D_y}{D}, \quad \text{and} \quad z = \frac{D_z}{D}.$$

Solve using Cramer's rule.

$$x - 2y = -1$$
$$2x + 5y = 16$$

$$x = \frac{\begin{vmatrix} -1 & -2 \\ 16 & 5 \end{vmatrix}}{\begin{vmatrix} 1 & -2 \\ 2 & 5 \end{vmatrix}} = \frac{-5 + 32}{5 + 4} = \frac{27}{9} = 3$$

$$y = \frac{\begin{vmatrix} 1 & -1 \\ 2 & 16 \end{vmatrix}}{\begin{vmatrix} 1 & -2 \\ 2 & 5 \end{vmatrix}} = \frac{16 + 2}{5 + 4} = \frac{18}{9} = 2$$

The solution set is $\{(3, 2)\}$.

Solve using Cramer's rule.

$$3x + 2y + z = -5$$
$$x - y + 3z = -5$$
$$2x + 3y + z = 0$$

Using the method of expansion by minors, it can be shown that $D_x = 45$, $D_y = -30$, $D_z = 0$, and $D = -15$. Therefore,

$$x = \frac{D_x}{D} = \frac{45}{-15} = -3,$$

$$y = \frac{D_y}{D} = \frac{-30}{-15} = 2,$$

$$z = \frac{D_z}{D} = \frac{0}{-15} = 0.$$

The solution set is $\{(-3, 2, 0)\}$.

Answers to Test Your Word Power

1. C; *Example:* $\begin{array}{l} 3x - y = 3 \\ 2x + y = 7 \end{array}$ **2.** B; *Example:* The ordered pair $(2, 3)$ satisfies both equations of the system in Answer 1, so $\{(2, 3)\}$ is the solution set of the system.   **3.** B; *Example:* The equations of two parallel lines form an inconsistent system; their graphs never intersect, so the system has no solution.   **4.** D; *Example:* The equations $4x - y = 8$ and $8x - 2y = 16$ are dependent because their graphs are the same line.   **5.** D; *Examples:* $\begin{bmatrix} 3 & -1 & 0 \\ 4 & 2 & 1 \end{bmatrix}, \begin{bmatrix} 1 & 2 \\ 4 & 3 \end{bmatrix}$ **6.** C; *Example:* $\begin{bmatrix} 1 & -3 & 2 & 1 \\ 0 & 1 & 4 & -1 \\ 0 & 0 & 1 & 2 \end{bmatrix}$ **7.** B; *Examples:* $\begin{vmatrix} 1 & 2 \\ 4 & 3 \end{vmatrix}, \begin{vmatrix} 1 & 4 & 0 \\ 3 & -2 & -1 \\ 0 & 5 & 0 \end{vmatrix}$

8. A; *Example:* See Section 4.5, Example 3 on page 273.

CHAPTER **4** REVIEW EXERCISES

[4.1]

1. The graph shows the trends during the years 1974 through 1996 relating to bachelor's degrees awarded in the United States.

   (a) Between what years shown on the horizontal axis did the number of degrees for men and women reach equal numbers?

   (b) When the number of degrees for men and women reached equal numbers, what was that number (approximately)?

BACHELOR'S DEGREES IN THE U.S.

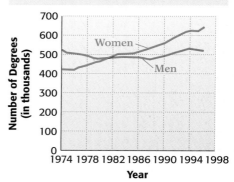

Source: U.S. National Center for Education Statistics, *Digest of Education Statistics*, annual.

2. Solve the system by graphing.

$$x + 3y = 8$$
$$2x - y = 2$$

3. Which one of the following ordered pairs is not a solution of $3x + 2y = 6$?

   A. $(2, 0)$ **B.** $(0, 3)$ **C.** $(4, -3)$ **D.** $(3, -2)$

Solve each system of equations by the elimination method. If a system is inconsistent or has dependent equations, say so.

4. $5x + y = 12$
   $2x - 2y = 0$

5. $x - 4y = -4$
   $3x + y = 1$

6. $6x + 5y = 4$
   $-4x + 2y = 8$

7. $\dfrac{1}{6}x + \dfrac{1}{6}y = -\dfrac{1}{2}$
   $x - y = -9$

8. $-3x + y = 6$
   $y = 6 + 3x$

9. $5x - 4y = 2$
   $-10x + 8y = 7$

Solve each system by the substitution method. If a system is inconsistent or has dependent equations, say so.

10. $3x + y = -4$
    $x = \dfrac{2}{3}y$

11. $9x - y = -4$
    $y = x + 4$

12. $-5x + 2y = -2$
    $x + 6y = 26$

13. State whether each graphed system is inconsistent or has dependent equations.

   (a)

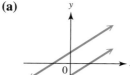

   (Two lines are parallel.)

   (b)

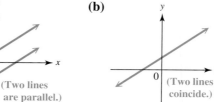

   (Two lines coincide.)

14. Explain why the system

$$y = 3x + 2$$
$$y = 3x - 4$$

has $\emptyset$ as its solution set without doing any algebraic work but answering based on your knowledge of the graphs of the two lines.

[4.2]   *Solve each system. If a system is inconsistent or has dependent equations, say so.*

15. $2x + 3y - z = -16$
$x + 2y + 2z = -3$
$-3x + y + z = -5$

16. $4x - y = 2$
$3y + z = 9$
$x + 2z = 7$

17. $3x - y - z = -8$
$4x + 2y + 3z = 15$
$-6x + 2y + 2z = 10$

[4.3]   *Solve each problem using a system of equations.*

18. A regulation National Hockey League ice rink has perimeter 570 ft. The length is 30 ft longer than twice the width. What are the dimensions of an NHL ice rink? (*Source: Microsoft Encarta Encyclopedia 2000.*)

19. The two top-grossing movies of 2001 were *Harry Potter and the Sorcerer's Stone* and *Shrek. Shrek* grossed $26 million less than *Harry Potter and the Sorcerer's Stone,* and together the two films took in $562 million. (*Source:* ACNielsen EDI.)

(a) How much did each of these movies earn?
(b) If *Shrek* earned $27 million more than *Monsters, Inc.,* how much did *Monsters, Inc.* earn? (*Hint:* Use your answer from part (a).)
(c) What is the total amount these top three films earned?

20. A plane flies 560 mi in 1.75 hr traveling with the wind. The return trip later against the same wind takes the plane 2 hr. Find the speed of the plane and the speed of the wind.

	r	t	d
With Wind	$x + y$	1.75	
Against Wind		2	

21. Sweet's Candy Store is offering a special mix for Valentine's Day. Ms. Sweet will mix some $2-per-lb nuts with some $1-per-lb chocolate candy to get 100 lb of mix, which she will sell at $1.30 per lb. How many pounds of each should she use?

	Number of Pounds	Price per Pound	Value
Nuts	x		
Chocolate	y		
Mixture	100		

22. The sum of the measures of the angles of a triangle is 180°. The largest angle measures 10° less than the sum of the other two. The measure of the middle-sized angle is the average of the other two. Find the measures of the three angles.

23. Maria Gonzales sells real estate. On three recent sales, she made 10% commission, 6% commission, and 5% commission. Her total commissions on these sales were $17,000, and she sold property worth $280,000. If the 5% sale amounted to the sum of the other two, what were the three sales prices?

24. How many liters each of 8%, 10%, and 20% hydrogen peroxide should be mixed together to get 8 L of 12.5% solution, if the amount of 8% solution used must be 2 L more than the amount of 20% solution used?

25. In the great baseball year of 1961, Yankee teammates Mickey Mantle, Roger Maris, and John Blanchard combined for 136 home runs. Mantle hit 7 fewer than Maris. Maris hit 40 more than Blanchard. What were the home run totals for each player? (*Source:* Neft, David S. and Richard M. Cohen, *The Sports Encyclopedia: Baseball 1997.*)

[4.4]   *Solve each system of equations using row operations.*

26. $2x + 5y = -4$
$4x - y = 14$

27. $6x + 3y = 9$
$-7x + 2y = 17$

28. $x + 2y - z = 1$
$3x + 4y + 2z = -2$
$-2x - y + z = -1$

29. $x + 3y = 7$
$3x + z = 2$
$y - 2z = 4$

[4.5]   *Evaluate each determinant.*

30. $\begin{vmatrix} 2 & -9 \\ 8 & 4 \end{vmatrix}$

31. $\begin{vmatrix} 7 & 0 \\ 5 & -3 \end{vmatrix}$

32. $\begin{vmatrix} 2 & 10 & 4 \\ 0 & 1 & 3 \\ 0 & 6 & -1 \end{vmatrix}$

33. $\begin{vmatrix} -1 & 7 & 2 \\ 3 & 0 & 5 \\ -1 & 2 & 6 \end{vmatrix}$

34. Under what conditions can a system *not* be solved using Cramer's rule?

35. Why can't the system $\begin{aligned} 3x + 2y + z &= 0 \\ -x + y - 3z &= 1 \end{aligned}$ be solved using Cramer's rule?

Use Cramer's rule to solve each system of equations.

36. $3x - 4y = -32$
$2x + y = -3$

37. $-4x + 3y = -12$
$2x + 6y = 15$

38. $4x + y + z = 11$
$x - y - z = 4$
$y + 2z = 0$

39. $-x + 3y - 4z = 4$
$2x + 4y + z = -14$
$3x - y + 2z = -8$

MIXED REVIEW EXERCISES

Solve by any method.

40. $\dfrac{2}{3}x + \dfrac{1}{6}y = \dfrac{19}{2}$

$\dfrac{1}{3}x - \dfrac{2}{9}y = 2$

41. $2x + 5y - z = 12$
$-x + y - 4z = -10$
$-8x - 20y + 4z = 31$

42. $x = 7y + 10$
$2x + 3y = 3$

43. $x + 4y = 17$
$-3x + 2y = -9$

44. $-7x + 3y = 12$
$5x + 2y = 8$

45. $2x - 5y = 8$
$3x + 4y = 10$

46. To make a 10% acid solution for chemistry class, Xavier wants to mix some 5% solution with 10 L of 20% solution. How many liters of 5% solution should he use?

Liters of Solution	Percent (as a decimal)	Liters of Pure Acid

47. In the 2000 Summer Olympics in Sydney, Australia, the top medal-winning countries were the United States, Russia, and China, with a combined total of 244 medals. The United States won 9 more medals than Russia, while China won 29 fewer medals than Russia. How many medals did each country win? (*Source: The Gazette,* October 2, 2000.)

RELATING CONCEPTS (EXERCISES 48–52)

For Individual or Group Work

Thus far in this text we have studied only linear *equations. In later chapters we will study the graphs of other kinds of equations. One such graph is a* circle, *which has an equation of the form*

$$x^2 + y^2 + ax + by + c = 0.$$

It is a fact from geometry that given three *noncollinear* points *(that is, points that do not all lie on the same straight line), there will be a circle that contains them. For example, the points* (4, 2),

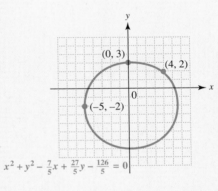

$(-5, -2),$ *and* (0, 3) *lie on the circle whose equation is shown in the figure.* **Work Exercises 48–52 in order,** *to find an equation of the circle passing through the points* (2, 1), (−1, 0), *and* (3, 3).

48. Let $x = 2$ and $y = 1$ in the equation $x^2 + y^2 + ax + by + c = 0$ to find an equation in a, b, and c.

49. Let $x = -1$ and $y = 0$ to find a second equation in a, b, and c.

50. Let $x = 3$ and $y = 3$ to find a third equation in a, b, and c.

51. Solve the system of equations formed by your answers in Exercises 48–50 to find the values of a, b, and c. What is the equation of the circle?

52. Explain why the relation whose graph is a circle is not a function.

CHAPTER 4 TEST

Hank Aaron and Babe Ruth are the all-time home run leaders in the major leagues. Use the graphs to answer Exercises 1 and 2.

1. Was there any year in Babe Ruth's career that he had more home runs than Hank Aaron in the same year of Aaron's career? Explain.

2. After 15 yr, which player had more home runs? Who had fewer?

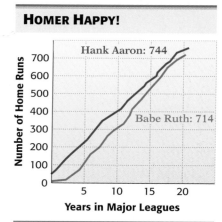

HOMER HAPPY!

Hank Aaron: 744

Babe Ruth: 714

Number of Home Runs

Years in Major Leagues

Source: ESPN 1998 Sports Almanac; CNN/*Sports Illustrated* and Major League Baseball web sites.

3. Use a graph to solve the system $\begin{aligned} x + y &= 7 \\ x - y &= 5. \end{aligned}$

Solve each system by elimination. If a system is inconsistent or has dependent equations, say so.

4. $3x + y = 12$
   $2x - y = 3$

5. $-5x + 2y = -4$
   $6x + 3y = -6$

6. $3x + 4y = 8$
   $8y = 7 - 6x$

7. $3x + 5y + 3z = 2$
   $6x + 5y + z = 0$
   $3x + 10y - 2z = 6$

8. $4x + y + z = 11$
   $x - y - z = 4$
   $y + 2z = 0$

Solve each system by substitution. If a system is inconsistent or has dependent equations, say so.

9. $2x - 3y = 24$
   $y = -\dfrac{2}{3}x$

10. $12x - 5y = 8$
   $3x = \dfrac{5}{4}y + 2$

Solve each problem using a system of equations.

11. Julia Roberts is one of the biggest box office stars in Hollywood. As of July 2001, her two top-grossing domestic films, *Pretty Woman* and *Runaway Bride,* together earned $330.7 million. If *Runaway Bride* grossed $26.1 million less than *Pretty Woman,* how much did each film gross? (*Source:* ACNielsen EDI.)

12. Two cars start from points 420 mi apart and travel toward each other. They meet after 3.5 hr. Find the average speed of each car if one travels 30 mph slower than the other.

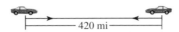

13. A chemist needs 12 L of a 40% alcohol solution. She must mix a 20% solution and a 50% solution. How many liters of each will be required to obtain what she needs?

14. A local electronics store will sell 7 AC adaptors and 2 rechargeable flashlights for $86, or 3 AC adaptors and 4 rechargeable flashlights for $84. What is the price of a single AC adaptor and a single rechargeable flashlight?

15. The owner of a tea shop wants to mix three kinds of tea to make 100 oz of a mixture that will sell for $.83 per oz. He uses Orange Pekoe, which sells for $.80 per oz, Irish Breakfast, for $.85 per oz, and Earl Grey, for $.95 per oz. If he wants to use twice as much Orange Pekoe as Irish Breakfast, how much of each kind of tea should he use?

Solve each system using row operations.

16. $3x + 2y = 4$
$5x + 5y = 9$

17. $x + 3y + 2z = 11$
$3x + 7y + 4z = 23$
$5x + 3y - 5z = -14$

Evaluate each determinant.

18. $\begin{vmatrix} 6 & 8 \\ 2 & -7 \end{vmatrix}$

19. $\begin{vmatrix} 2 & 0 & 8 \\ -1 & 7 & 9 \\ 12 & 5 & -3 \end{vmatrix}$

Solve each system using Cramer's rule.

20. $2x - 3y = -33$
$4x + 5y = 11$

21. $x + y - z = -4$
$2x - 3y - z = 5$
$x + 2y + 2z = 3$

22. Use any method described in this chapter to solve the system

$$4x - 2y = -8$$
$$3y - 5z = 14$$
$$2x + z = -10.$$

CUMULATIVE REVIEW EXERCISES CHAPTERS 1–4

Evaluate.

1. $(-3)^4$

2. -3^4

3. $-(-3)^4$

4. $\sqrt{.49}$

5. $-\sqrt{.49}$

6. $\sqrt{-.49}$

7. $\sqrt[3]{64}$

8. $\sqrt[3]{-64}$

Evaluate if $x = -4$, $y = 3$, and $z = 6$.

9. $|2x| + 3y - z^3$

10. $-5(x^3 - y^3)$

11. Which property of real numbers justifies the statement $5 + (3 \cdot 6) = 5 + (6 \cdot 3)$?

Solve each equation.

12. $7(2x + 3) - 4(2x + 1) = 2(x + 1)$

13. $|6x - 8| = 4$

14. $ax + by = cx + d$ for x

15. $.04x + .06(x - 1) = 1.04$

Solve each inequality.

16. $\dfrac{2}{3}x + \dfrac{5}{12}x \le 20$

17. $|3x + 2| \le 4$

18. $|12t + 7| \ge 0$

19. A recent survey measured public recognition of the most popular contemporary advertising slogans. Complete the results shown in the table if 2500 people were surveyed.

Slogan (product or company)	Percent Recognition (nearest tenth of a percent)	Actual Number Who Recognized Slogan (nearest whole number)
Please Don't Squeeze the . . . (Charmin)	80.4%	
The Breakfast of Champions (Wheaties)	72.5%	
The King of Beers (Budweiser)		1570
Like a Good Neighbor (State Farm)		1430

[Other slogans included "You're in Good Hands" (Allstate), "Snap, Crackle, Pop" (Rice Krispies), and "The Un-Cola" (7-Up).]

Source: Department of Integrated Marketing Communications, Northwestern University.

Solve each problem.

20. A triangle has an area of 42 m². The base is 14 m long. Find the height of the triangle.

21. A jar contains only pennies, nickels, and dimes. The number of dimes is 1 more than the number of nickels, and the number of pennies is 6 more than the number of nickels. How many of each denomination can be found in the jar, if the total value is $4.80?

22. Two angles of a triangle have the same measure. The measure of the third angle is 4° less than twice the measure of each of the equal angles. Find the measures of the three angles.

Measures are in degrees.

In Exercises 23–27, point A has coordinates $(-2, 6)$ and point B has coordinates $(4, -2)$.

23. What is the equation of the horizontal line through A?

24. What is the equation of the vertical line through B?

25. What is the slope of line AB?

26. What is the slope of a line perpendicular to line AB?

27. What is the standard form of the equation of line AB?

28. Graph the line having slope $\frac{2}{3}$ and passing through the point $(-1, -3)$.

29. Graph the inequality $-3x - 2y \le 6$.

30. Given that $f(x) = x^2 + 3x - 6$, find each of the following.

    (a) $f(-3)$ **(b)** $f(a)$

31. If y varies directly as x and $y = 5$ when $x = 12$, find y when $x = 42$.

Solve by any method.

32. $-2x + 3y = -15$
$\quad\ 4x -\ \ y = 15$

33. $x - 3y = 7$
$\quad\ 2x - 6y = 14$

34. $x + y + z = 10$
$\quad\ x - y - z = 0$
$\quad -x + y - z = -4$

In Exercises 35–38, solve each problem using a system of equations. Use the method of your choice: elimination, substitution, row operations, or Cramer's rule.

35. Mabel Johnston bought apples and oranges at DeVille's Grocery. She bought 6 lb of fruit. Oranges cost $.90 per lb, while apples cost $.70 per lb. If she spent a total of $5.20, how many pounds of each kind of fruit did she buy?

36. Two of the best-selling toys in a recent year were Tickle Me Elmo and Snacktime Kid. Based on their average retail prices, Elmo cost $8.63 less than Kid, and together they cost $63.89. What was the average retail price for each toy? (*Source:* NPD Group, Inc.)

37. Kenneth and Peggy are planning to move, and they need some cardboard boxes. They can buy 10 small and 20 large boxes for $65, or 6 small and 10 large boxes for $34. Find the cost of each size of box.

38. At the Chalmette Nut Shop, 6 lb of peanuts and 12 lb of cashews cost $60, while 3 lb of peanuts and 4 lb of cashews cost $22. Find the cost of each type of nut.

The graph shows a company's costs to produce computer parts and the revenue from the sale of those computer parts.

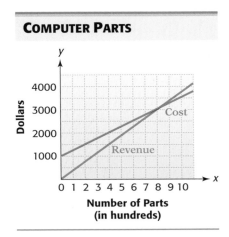

39. At what production level does the cost equal the revenue? What is the revenue at that point?

40. Profit is revenue less cost. Estimate the profit on the sale of 1100 parts.

41. Find the value of the determinant $\begin{vmatrix} 3 & 1 & 0 \\ -2 & 0 & 2 \\ 0 & -1 & 1 \end{vmatrix}$.

42. Use Cramer's rule to solve the system with $D = 28$, $D_x = 35$, $D_y = 56$, and $D_z = 42$.

Exponents, Polynomials, and Polynomial Functions

Just how much is a trillion? A trillion, which is written 1,000,000,000,000, is an incredibly huge number. A trillion is a million million or a thousand billion. A trillion seconds would last more than 31,000 yr, that is, 310 centuries. To be a trillionaire, a person would need a stack of $1000 bills over 67 mi high.

The U.S. budget first exceeded $1 trillion in 1987 and topped $2 trillion for the 2003 fiscal year. (*Source: The Gazette*, February 5, 2002.) In Exercise 151 of Section 5.1, we use exponents to write large numbers like the amount of the U.S. budget using scientific notation.

5.1 Integer Exponents and Scientific Notation

Recall that we use exponents to write products of repeated factors. For example,

$$2^5 \text{ is defined as } 2 \cdot 2 \cdot 2 \cdot 2 \cdot 2 = 32.$$

The number 5, the *exponent,* shows that the *base* 2 appears as a factor 5 times. The quantity 2^5 is called an *exponential* or a *power.* We read 2^5 as "2 to the fifth power" or "2 to the fifth."

OBJECTIVE 1 Use the product rule for exponents.   There are several useful rules that simplify work with exponents. For example, the product $2^5 \cdot 2^3$ can be simplified as follows.

$$\overbrace{2^5 \cdot 2^3 = (2 \cdot 2 \cdot 2 \cdot 2 \cdot 2)(2 \cdot 2 \cdot 2) = 2^8}^{5 + 3 = 8}$$

This result, that products of exponential expressions with the *same base* are found by adding exponents, is generalized as the **product rule for exponents.**

> **Product Rule for Exponents**
>
> If m and n are natural numbers and a is any real number, then
> $$a^m \cdot a^n = a^{m+n}.$$
>
> That is, when multiplying powers of like bases, keep the same base and add the exponents.

To see that the product rule is true, use the definition of an exponent as follows.

$$a^m = \underbrace{a \cdot a \cdot a \cdot \ldots \cdot a}_{a \text{ appears as a factor } m \text{ times.}} \qquad a^n = \underbrace{a \cdot a \cdot a \cdot \ldots \cdot a}_{a \text{ appears as a factor } n \text{ times.}}$$

From this,

$$a^m \cdot a^n = \underbrace{a \cdot a \cdot a \cdot \ldots \cdot a}_{m \text{ factors}} \underbrace{a \cdot a \cdot a \cdot \ldots \cdot a}_{n \text{ factors}}$$

$$= \underbrace{a \cdot a \cdot a \cdot \ldots \cdot a}_{(m+n) \text{ factors}}$$

$$a^m \cdot a^n = a^{m+n}.$$

EXAMPLE 1 Using the Product Rule for Exponents

Apply the product rule for exponents, if possible, in each case.

(a) $3^4 \cdot 3^7 = 3^{4+7} = 3^{11}$

(b) $5^3 \cdot 5 = 5^3 \cdot 5^1 = 5^{3+1} = 5^4$

(c) $y^3 \cdot y^8 \cdot y^2 = y^{3+8+2} = y^{13}$

(d) $(5y^2)(-3y^4) = 5(-3)y^2y^4$ Associative and commutative properties

$\qquad\qquad\quad = -15y^{2+4}$ Product rule

$\qquad\qquad\quad = -15y^6$

(e) $(7p^3q)(2p^5q^2) = 7(2)p^3p^5qq^2 = 14p^8q^3$

(f) $x^2 \cdot y^4$

   The product rule does not apply because the bases are not the same.   ▇

Now Try Exercises 7, 11, 13, and 15.

CAUTION   Be careful in problems like Example 1(a) not to multiply the bases. Notice that $3^4 \cdot 3^7 \neq 9^{11}$. Remember to keep the *same* base and add the exponents.

OBJECTIVE 2   Define 0 and negative exponents.   So far we have discussed only positive exponents. How should we define a 0 exponent? Suppose we multiply 4^2 by 4^0. Extending the product rule, we should have

$$4^2 \cdot 4^0 = 4^{2+0} = 4^2.$$

For the product rule to hold true, 4^0 must equal 1, and so we define a^0 this way for any nonzero real number a.

Zero Exponent

If a is any nonzero real number, then

$$a^0 = 1.$$

The expression **0^0 is undefined.***

▇ **EXAMPLE 2   Using 0 as an Exponent**

Evaluate each expression.

(a) $6^0 = 1$ **(b)** $(-6)^0 = 1$ Base is -6.

(c) $-6^0 = -(6^0) = -(1) = -1$ Base is 6.   **(d)** $-(-6)^0 = -1$

(e) $5^0 + 12^0 = 1 + 1 = 2$ **(f)** $(8k)^0 = 1, \quad k \neq 0$ ▇

Now Try Exercises 17 and 25.

   How should we define a negative exponent? Extending the product rule again,

$$8^2 \cdot 8^{-2} = 8^{2+(-2)} = 8^0 = 1.$$

This indicates that 8^{-2} is the reciprocal of 8^2. But $\frac{1}{8^2}$ is the reciprocal of 8^2, and a number can have only one reciprocal. Therefore, it is reasonable to conclude that $8^{-2} = \frac{1}{8^2}$. We can generalize and make the following definition.

Negative Exponent

For any natural number n and any nonzero real number a,

$$a^{-n} = \frac{1}{a^n}.$$

---

*In advanced treatments, 0^0 is called an *indeterminant form*.

With this definition, the expression a^n is meaningful for any integer exponent n and any nonzero real number a. The product rule is valid for integers m and n.

CAUTION *A negative exponent does not indicate a negative number;* negative exponents lead to reciprocals. For example,

$$3^{-2} = \frac{1}{3^2} = \frac{1}{9}. \qquad \text{Not negative}$$

On the other hand,

$$-3^{-2} = -\frac{1}{3^2} = -\frac{1}{9} \qquad \text{Negative}$$

because of the negative sign preceding the base.

EXAMPLE 3 Using Negative Exponents

In parts (a)–(f), write each expression with only positive exponents.

(a) $2^{-3} = \dfrac{1}{2^3}$

(b) $6^{-1} = \dfrac{1}{6^1} = \dfrac{1}{6}$

(c) $(5z)^{-3} = \dfrac{1}{(5z)^3}, \quad z \neq 0$
↑
Base is 5z.

(d) $5z^{-3} = 5\left(\dfrac{1}{z^3}\right) = \dfrac{5}{z^3}, \quad z \neq 0$
↑
Base is z.

(e) $-m^{-2} = -\dfrac{1}{m^2}, \quad m \neq 0$
(What is the base here?)

(f) $(-m)^{-2} = \dfrac{1}{(-m)^2}, \quad m \neq 0$
(What is the base here?)

In parts (g) and (h), evaluate each expression.

(g) $3^{-1} + 4^{-1} = \dfrac{1}{3} + \dfrac{1}{4} = \dfrac{4}{12} + \dfrac{3}{12} = \dfrac{7}{12}$ $\qquad \frac{1}{3} \cdot \frac{4}{4} = \frac{4}{12}; \frac{1}{4} \cdot \frac{3}{3} = \frac{3}{12}$

(h) $5^{-1} - 2^{-1} = \dfrac{1}{5} - \dfrac{1}{2} = \dfrac{2}{10} - \dfrac{5}{10} = -\dfrac{3}{10}$

Now Try Exercises 33, 37, 39, 41, and 45.

CAUTION In Example 3(g), note that $3^{-1} + 4^{-1} \neq (3 + 4)^{-1}$. The expression on the left is equal to $\frac{7}{12}$, as shown in the example, while the expression on the right is $7^{-1} = \frac{1}{7}$. Similar reasoning can be applied to part (h).

EXAMPLE 4 Using Negative Exponents

Evaluate each expression.

(a) $\dfrac{1}{2^{-3}} = \dfrac{1}{\frac{1}{2^3}} = 1 \div \dfrac{1}{2^3} = 1 \cdot \dfrac{2^3}{1} = 2^3 = 8$

Multiply by the reciprocal.

(b) $\dfrac{2^{-3}}{3^{-2}} = \dfrac{\frac{1}{2^3}}{\frac{1}{3^2}} = \dfrac{1}{2^3} \div \dfrac{1}{3^2} = \dfrac{1}{2^3} \cdot \dfrac{3^2}{1} = \dfrac{3^2}{2^3} = \dfrac{9}{8}$

Now Try Exercises 51 and 53.

Example 4 suggests the following generalizations.

Special Rules for Negative Exponents

If $a \neq 0$ and $b \neq 0$, then $\qquad \dfrac{1}{a^{-n}} = a^n \qquad$ and $\qquad \dfrac{a^{-n}}{b^{-m}} = \dfrac{b^m}{a^n}.$

OBJECTIVE 3 Use the quotient rule for exponents. A quotient, such as $\dfrac{a^8}{a^3}$, can be simplified in much the same way as a product. (In all quotients of this type, assume that the denominator is not 0.) Using the definition of an exponent,

$$\frac{a^8}{a^3} = \frac{a \cdot a \cdot a \cdot a \cdot a \cdot a \cdot a \cdot a}{a \cdot a \cdot a} = a \cdot a \cdot a \cdot a \cdot a = a^5.$$

Notice that $8 - 3 = 5$. In the same way,

$$\frac{a^3}{a^8} = \frac{a \cdot a \cdot a}{a \cdot a \cdot a \cdot a \cdot a \cdot a \cdot a \cdot a} = \frac{1}{a^5} = a^{-5}.$$

Here $3 - 8 = -5$. These examples suggest the following **quotient rule for exponents.**

Quotient Rule for Exponents

If a is any nonzero real number and m and n are integers, then

$$\frac{a^m}{a^n} = a^{m-n}.$$

That is, when dividing powers of like bases, keep the same base and subtract the exponent of the denominator from the exponent of the numerator.

EXAMPLE 5 Using the Quotient Rule for Exponents

Apply the quotient rule for exponents, if possible, and write each result using only positive exponents.

Numerator exponent
Denominator exponent

(a) $\dfrac{3^7}{3^2} = 3^{7-2} = 3^5$
Minus sign

(b) $\dfrac{p^6}{p^2} = p^{6-2} = p^4, \quad p \neq 0$

(c) $\dfrac{k^7}{k^{12}} = k^{7-12} = k^{-5} = \dfrac{1}{k^5}, \quad k \neq 0$

(d) $\dfrac{2^7}{2^{-3}} = 2^{7-(-3)} = 2^{7+3} = 2^{10}$

(e) $\dfrac{8^{-2}}{8^5} = 8^{-2-5} = 8^{-7} = \dfrac{1}{8^7}$

(f) $\dfrac{6}{6^{-1}} = \dfrac{6^1}{6^{-1}} = 6^{1-(-1)} = 6^2$

(g) $\dfrac{z^{-5}}{z^{-8}} = z^{-5-(-8)} = z^3, \quad z \neq 0$

(h) $\dfrac{a^3}{b^4}, \quad b \neq 0$

The quotient rule does not apply because the bases are different. ▮

Now Try Exercises 63, 67, 75, and 77.

CAUTION As seen in Example 5, be very careful when working with quotients that involve negative exponents in the denominator. Always be sure to write the numerator exponent, then a minus sign, and then the denominator exponent.

OBJECTIVE 4 Use the power rules for exponents. The expression $(3^4)^2$ can be simplified as

$$(3^4)^2 = 3^4 \cdot 3^4 = 3^{4+4} = 3^8,$$

where $4 \cdot 2 = 8$. This example suggests the first **power rule for exponents.** The other two power rules can be demonstrated with similar examples.

Power Rules for Exponents

If a and b are real numbers and m and n are integers, then

(a) $(a^m)^n = a^{mn},$ **(b)** $(ab)^m = a^m b^m,$ and **(c)** $\left(\dfrac{a}{b}\right)^m = \dfrac{a^m}{b^m} \quad (b \neq 0).$

That is,

(a) to raise a power to a power, multiply exponents;

(b) to raise a product to a power, raise each factor to that power; and

(c) to raise a quotient to a power, raise the numerator and the denominator to that power.

EXAMPLE 6 Using the Power Rules for Exponents

Use one or more power rules to simplify each expression.

(a) $(p^8)^3 = p^{8 \cdot 3} = p^{24}$

(b) $\left(\dfrac{2}{3}\right)^4 = \dfrac{2^4}{3^4} = \dfrac{16}{81}$

(c) $(3y)^4 = 3^4 y^4 = 81y^4$

(d) $(6p^7)^2 = 6^2 p^{7 \cdot 2} = 6^2 p^{14} = 36p^{14}$

(e) $\left(\dfrac{-2m^5}{z}\right)^3 = \dfrac{(-2)^3 m^{5 \cdot 3}}{z^3} = \dfrac{(-2)^3 m^{15}}{z^3} = \dfrac{-8m^{15}}{z^3}, \quad z \neq 0$

Now Try Exercises 79, 81, 83, and 87.

The reciprocal of a^n is $\dfrac{1}{a^n} = \left(\dfrac{1}{a}\right)^n$. Also, by definition, a^n and a^{-n} are reciprocals since

$$a^n \cdot a^{-n} = a^n \cdot \dfrac{1}{a^n} = 1.$$

Thus, since both are reciprocals of a^n,

$$a^{-n} = \left(\dfrac{1}{a}\right)^n.$$

Some examples of this result are

$$6^{-3} = \left(\dfrac{1}{6}\right)^3 \quad \text{and} \quad \left(\dfrac{1}{3}\right)^{-2} = 3^2.$$

This discussion can be generalized as follows.

More Special Rules for Negative Exponents

If $a \neq 0$ and $b \neq 0$ and n is an integer, then

$$a^{-n} = \left(\dfrac{1}{a}\right)^n \quad \text{and} \quad \left(\dfrac{a}{b}\right)^{-n} = \left(\dfrac{b}{a}\right)^n.$$

That is, any nonzero number raised to the negative nth power is equal to the reciprocal of that number raised to the nth power.

EXAMPLE 7 Using Negative Exponents with Fractions

Write each expression with only positive exponents and then evaluate.

(a) $\left(\dfrac{3}{7}\right)^{-2} = \left(\dfrac{7}{3}\right)^2 = \dfrac{49}{9}$ **(b)** $\left(\dfrac{4}{5}\right)^{-3} = \left(\dfrac{5}{4}\right)^3 = \dfrac{125}{64}$

Now Try Exercise 55.

The definitions and rules of this section are summarized here.

Definitions and Rules for Exponents

For all integers m and n and all real numbers a and b, the following rules apply.

Product Rule	$a^m \cdot a^n = a^{m+n}$
Quotient Rule	$\dfrac{a^m}{a^n} = a^{m-n} \quad (a \neq 0)$
Zero Exponent	$a^0 = 1 \quad (a \neq 0)$
Negative Exponent	$a^{-n} = \dfrac{1}{a^n} \quad (a \neq 0)$

(continued)

Power Rules	$(a^m)^n = a^{mn}$	$(ab)^m = a^m b^m$

$$\left(\frac{a}{b}\right)^m = \frac{a^m}{b^m} \quad (b \neq 0)$$

Special Rules	$\dfrac{1}{a^{-n}} = a^n \quad (a \neq 0)$	$\dfrac{a^{-n}}{b^{-m}} = \dfrac{b^m}{a^n} \quad (a, b \neq 0)$

$$a^{-n} = \left(\frac{1}{a}\right)^n \quad (a \neq 0) \qquad \left(\frac{a}{b}\right)^{-n} = \left(\frac{b}{a}\right)^n \quad (a, b \neq 0)$$

OBJECTIVE 5 Simplify exponential expressions. With the rules of exponents developed so far in this section, we can simplify expressions that involve one or more rules.

EXAMPLE 8 Using the Definitions and Rules for Exponents

Simplify each expression so that no negative exponents appear in the final result. Assume all variables represent nonzero real numbers.

(a) $3^2 \cdot 3^{-5} = 3^{2+(-5)} = 3^{-3} = \dfrac{1}{3^3}$ or $\dfrac{1}{27}$

(b) $x^{-3} x^{-4} x^2 = x^{-3+(-4)+2} = x^{-5} = \dfrac{1}{x^5}$

(c) $(4^{-2})^{-5} = 4^{(-2)(-5)} = 4^{10}$ **(d)** $(x^{-4})^6 = x^{(-4)6} = x^{-24} = \dfrac{1}{x^{24}}$

(e) $\dfrac{x^{-4}y^2}{x^2 y^{-5}} = \dfrac{x^{-4}}{x^2} \cdot \dfrac{y^2}{y^{-5}}$

$= x^{-4-2} \cdot y^{2-(-5)}$

$= x^{-6} y^7$

$= \dfrac{y^7}{x^6}$

(f) $(2^3 x^{-2})^{-2} = (2^3)^{-2} \cdot (x^{-2})^{-2}$

$= 2^{-6} x^4$

$= \dfrac{x^4}{2^6}$ or $\dfrac{x^4}{64}$

(g) $\left(\dfrac{3x^2}{y}\right)^2 \left(\dfrac{4x^3}{y^{-2}}\right)^{-1} = \dfrac{3^2 (x^2)^2}{y^2} \cdot \dfrac{y^{-2}}{4x^3}$ Combination of rules

$= \dfrac{9x^4}{y^2} \cdot \dfrac{y^{-2}}{4x^3}$ Power rule

$= \dfrac{9}{4} x^{4-3} y^{-2-2}$ Quotient rule

$= \dfrac{9x}{4y^4}$

Now Try Exercises 91, 93, 107, and 117.

NOTE There is often more than one way to simplify expressions like those in Example 8. For instance, we could simplify Example 8(e) as follows.

$$\frac{x^{-4}y^2}{x^2y^{-5}} = \frac{y^5y^2}{x^4x^2} \qquad \text{Use } \frac{a^{-n}}{b^{-m}} = \frac{b^m}{a^n}.$$

$$= \frac{y^7}{x^6} \qquad \text{Product rule}$$

OBJECTIVE 6 **Use the rules for exponents with scientific notation.** Scientists often use numbers that are very large or very small. For example, the number of one-celled organisms that will sustain a whale for a few hours is 400,000,000,000,000, and the shortest wavelength of visible light is approximately .0000004 m. It is simpler to write these numbers using *scientific notation.*

In scientific notation, a number is written with the decimal point after the first nonzero digit and multiplied by a power of 10, as indicated in the following definition.

Scientific Notation

A number is written in **scientific notation** when it is expressed in the form

$$a \times 10^n$$

where $1 \le |a| < 10$, and n is an integer.

For example, in scientific notation,

$$8000 = 8 \times 1000 = 8 \times 10^3.$$

When using scientific notation, it is customary to use a times sign $\times$ instead of a multiplication dot. The following numbers are not in scientific notation.

$$.230 \times 10^4 \qquad\qquad 46.5 \times 10^{-3}$$

.230 is less than 1.   46.5 is greater than 10.

To write a number in scientific notation, use the following steps. (If the number is negative, ignore the negative sign, go through these steps, and then attach a negative sign to the result.)

Converting to Scientific Notation

Step 1 **Position the decimal point.** Place a caret, ^, to the right of the first nonzero digit, where the decimal point will be placed.

Step 2 **Determine the numeral for the exponent.** Count the number of digits from the decimal point to the caret. This number gives the absolute value of the exponent on 10.

Step 3 **Determine the sign for the exponent.** Decide whether multiplying by 10^n should make the result of Step 1 larger or smaller. The exponent should be positive to make the result larger; it should be negative to make the result smaller.

It is helpful to remember that for $n \ge 1$, $10^{-n} < 1$ and $10^n \ge 10$.

▧ **EXAMPLE 9** Writing Numbers in Scientific Notation

Write each number in scientific notation.

(a) 820,000

Step 1 Place a caret to the right of the 8 (the first nonzero digit) to mark the new location of the decimal point.

$$8{\scriptstyle\wedge}20,000$$

Step 2 Count from the decimal point, which is understood to be after the last 0, to the caret.

$$8.20,000. \quad \leftarrow \text{Decimal point}$$

Count 5 places.

Step 3 Since the number 8.2 is to be made larger, the exponent on 10 is positive.

$$820,000 = 8.2 \times 10^5$$

(b) .0000072

Count from left to right.

$$.000007.2$$

6 places

Since the number 7.2 is to be made smaller, the exponent on 10 is negative.

$$.0000072 = 7.2 \times 10^{-6}$$

(c) $-.0000462 = -4.62 \times 10^{-5}$

Count 5 places.

▧

Now Try Exercises 127, 131, and 133.

To convert a number written in scientific notation to standard notation, just work in reverse.

Converting from Scientific Notation

Multiplying a number by a positive power of 10 makes the number larger, so move the decimal point to the right if n is positive in 10^n.

Multiplying by a negative power of 10 makes a number smaller, so move the decimal point to the left if n is negative.

If n is 0, leave the decimal point where it is.

▧ **EXAMPLE 10** Converting from Scientific Notation to Standard Notation

Write each number in standard notation.

(a) 6.93×10^7

$$6.9300000 \qquad \text{Attach 0s as necessary.}$$

7 places

We moved the decimal point 7 places to the right. (It was necessary to attach five 0s.)

$$6.93 \times 10^7 = 69,300,000$$

(b) 4.7×10^{-6}

$$\underset{\underset{\text{6 places}}{\wedge\wedge\wedge\wedge\wedge}}{000004.7} \qquad \text{Attach 0s as necessary.}$$

We moved the decimal point 6 places to the left.

$$4.7 \times 10^{-6} = .0000047$$

(c) $-1.083 \times 10^{0} = -1.083 \times 1 = -1.083$

Now Try Exercises 135, 137, and 139.

When problems require operations with numbers that are very large and/or very small, it is often advantageous to write the numbers in scientific notation first, and then perform the calculations using the rules for exponents.

EXAMPLE 11 Using Scientific Notation in Computation

Evaluate $\dfrac{1,920,000 \times .0015}{.000032 \times 45,000}$.

First, express all numbers in scientific notation.

$$\frac{1,920,000 \times .0015}{.000032 \times 45,000} = \frac{1.92 \times 10^{6} \times 1.5 \times 10^{-3}}{3.2 \times 10^{-5} \times 4.5 \times 10^{4}}$$

Next, use the commutative and associative properties and the rules for exponents to simplify the expression.

$$\frac{1,920,000 \times .0015}{.000032 \times 45,000} = \frac{1.92 \times 1.5 \times 10^{6} \times 10^{-3}}{3.2 \times 4.5 \times 10^{-5} \times 10^{4}}$$

$$= \frac{1.92 \times 1.5 \times 10^{3}}{3.2 \times 4.5 \times 10^{-1}}$$

$$= \frac{1.92 \times 1.5}{3.2 \times 4.5} \times 10^{4}$$

$$= .2 \times 10^{4}$$

$$= (2 \times 10^{-1}) \times 10^{4}$$

$$= 2 \times 10^{3}$$

$$= 2000$$

Now Try Exercise 149.

EXAMPLE 12 Using Scientific Notation to Solve Problems

In 1990, the national health care expenditure was $695.6 billion. By 2000, this figure had risen by a factor of 1.9; that is, it almost doubled in only 10 yr. (*Source:* U.S. Centers for Medicare & Medicaid Services.)

(a) Write the 1990 health care expenditure using scientific notation.

$$695.6 \text{ billion} = 695.6 \times 10^{9} = (6.956 \times 10^{2}) \times 10^{9}$$

$$= 6.956 \times 10^{11} \qquad \text{Product rule}$$

In 1990, the expenditure was 6.956×10^{11}.

(b) What was the expenditure in 2000?

Multiply the result in part (a) by 1.9.

$$(6.956 \times 10^{11}) \times 1.9 = (1.9 \times 6.956) \times 10^{11} \qquad \text{Commutative and associative properties}$$

$$= 13.216 \times 10^{11} \qquad \text{Round to three decimal places.}$$

The 2000 expenditure was about \$1,321,600,000,000, over \$1 trillion.

Now Try Exercise 153.

5.1 EXERCISES

For Extra Help

Student's
Solutions Manual

MyMathLab

InterAct Math
Tutorial Software

AW Math
Tutor Center

MathXL

Digital Video Tutor
CD 5/Videotape 6

Decide whether each expression has been simplified correctly. If not, correct it.

1. $(ab)^2 = ab^2$

2. $y^2 \cdot y^6 = y^{12}$

3. $\left(\dfrac{4}{a}\right)^3 = \dfrac{4^3}{a} \quad (a \neq 0)$

4. $xy^0 = 0 \quad (y \neq 0)$

5. State the product rule for exponents in your own words. Give an example.

6. Your friend evaluated $4^5 \cdot 4^2$ as 16^7. Explain to him why his answer is incorrect.

Apply the product rule for exponents, if possible, in each case. See Example 1.

7. $13^4 \cdot 13^8$

8. $9^6 \cdot 9^4$

9. $x^3 \cdot x^5 \cdot x^9$

10. $y^4 \cdot y^5 \cdot y^6$

11. $(-3w^5)(9w^3)$

12. $(-5x^2)(3x^4)$

13. $(2x^2y^5)(9xy^3)$

14. $(8s^4t)(3s^3t^5)$

15. $r^2 \cdot s^4$

16. $p^3 \cdot q^2$

In Exercises 17 and 18, match the expression in Column I with its equivalent expression in Column II. Choices may be used once, more than once, or not at all.* See Example 2.

	I	II		I	II
17.	**(a)** 9^0	**A.** 0	**18.**	**(a)** $2x^0$	**A.** 0
	(b) -9^0	**B.** 1		**(b)** $-2x^0$	**B.** 1
	(c) $(-9)^0$	**C.** -1		**(c)** $(2x)^0$	**C.** -1
	(d) $-(-9)^0$	**D.** 9		**(d)** $(-2x)^0$	**D.** 2
		E. -9			**E.** -2

Evaluate. Assume all variables represent nonzero real numbers.* See Example 2.

19. 25^0

20. 14^0

21. -7^0

22. -10^0

23. $(-15)^0$

24. $(-20)^0$

25. $3^0 + (-3)^0$

26. $5^0 + (-5)^0$

27. $-3^0 + 3^0$

28. $-5^0 + 5^0$

29. $-4^0 - m^0$

30. $-8^0 - k^0$

---

*The authors thank Mitchel Levy of Broward Community College for his suggestions for these exercises.

In Exercises 31 and 32, match the expression in Column I with its equivalent expression in Column II. Choices may be used once, more than once, or not at all.* See Example 3.

	I	**II**
31. (a) 4^{-2}	**A.** 16
(b) -4^{-2}	**B.** $\dfrac{1}{16}$
(c) $(-4)^{-2}$	**C.** -16
(d) $-(-4)^{-2}$	**D.** $-\dfrac{1}{16}$

	I	**II**
32. (a) 5^{-3}	**A.** 125
(b) -5^{-3}	**B.** -125
(c) $(-5)^{-3}$	**C.** $\dfrac{1}{125}$
(d) $-(-5)^{-3}$	**D.** $-\dfrac{1}{125}$

Write each expression with only positive exponents. Assume all variables represent nonzero real numbers. In Exercises 45–48, simplify each expression. See Example 3.

33. 5^{-4} **34.** 7^{-2} **35.** 8^{-1} **36.** 12^{-1}

37. $(4x)^{-2}$ **38.** $(5t)^{-3}$ **39.** $4x^{-2}$ **40.** $5t^{-3}$

41. $-a^{-3}$ **42.** $-b^{-4}$ **43.** $(-a)^{-4}$ **44.** $(-b)^{-6}$

45. $5^{-1} + 6^{-1}$ **46.** $2^{-1} + 8^{-1}$ **47.** $8^{-1} - 3^{-1}$ **48.** $6^{-1} - 4^{-1}$

49. Consider the expressions $-a^n$ and $(-a)^n$. In some cases they are equal and in some cases they are not. Using $n = 2, 3, 4, 5,$ and 6 and $a = 2$, draw a conclusion as to when they are equal and when they are opposites.

50. Your friend thinks that $(-3)^{-2}$ is a negative number. Why is she incorrect?

Evaluate each expression. See Examples 4 and 7.

51. $\dfrac{1}{4^{-2}}$ **52.** $\dfrac{1}{3^{-3}}$ **53.** $\dfrac{2^{-2}}{3^{-3}}$ **54.** $\dfrac{3^{-3}}{2^{-2}}$

55. $\left(\dfrac{2}{3}\right)^{-3}$ **56.** $\left(\dfrac{3}{2}\right)^{-3}$ **57.** $\left(\dfrac{4}{5}\right)^{-2}$ **58.** $\left(\dfrac{5}{4}\right)^{-2}$

In Exercises 59 and 60, match the expression in Column I with its equivalent expression in Column II. Choices may be used once, more than once, or not at all.*

	I	**II**
59. (a) $\left(\dfrac{1}{3}\right)^{-1}$	**A.** $\dfrac{1}{3}$
(b) $\left(-\dfrac{1}{3}\right)^{-1}$	**B.** 3
(c) $-\left(\dfrac{1}{3}\right)^{-1}$	**C.** $-\dfrac{1}{3}$
(d) $-\left(-\dfrac{1}{3}\right)^{-1}$	**D.** -3

	I	**II**
60. (a) $\left(\dfrac{2}{5}\right)^{-2}$	**A.** $\dfrac{25}{4}$
(b) $\left(-\dfrac{2}{5}\right)^{-2}$	**B.** $-\dfrac{25}{4}$
(c) $-\left(\dfrac{2}{5}\right)^{-2}$	**C.** $\dfrac{4}{25}$
(d) $-\left(-\dfrac{2}{5}\right)^{-2}$	**D.** $-\dfrac{4}{25}$

61. State the quotient rule for exponents in your own words. Give an example.

62. State the three power rules for exponents in your own words. Give examples.

---

*The authors thank Mitchel Levy of Broward Community College for his suggestions for these exercises.

Apply the quotient rule for exponents, if applicable, and write each result using only positive exponents. Assume all variables represent nonzero real numbers. See Example 5.

63. $\dfrac{4^8}{4^6}$ **64.** $\dfrac{5^9}{5^7}$ **65.** $\dfrac{x^{12}}{x^8}$ **66.** $\dfrac{y^{14}}{y^{10}}$

67. $\dfrac{r^7}{r^{10}}$ **68.** $\dfrac{y^8}{y^{12}}$ **69.** $\dfrac{6^4}{6^{-2}}$ **70.** $\dfrac{7^5}{7^{-3}}$

71. $\dfrac{6^{-3}}{6^7}$ **72.** $\dfrac{5^{-4}}{5^2}$ **73.** $\dfrac{7}{7^{-1}}$ **74.** $\dfrac{8}{8^{-1}}$

75. $\dfrac{r^{-3}}{r^{-6}}$ **76.** $\dfrac{s^{-4}}{s^{-8}}$ **77.** $\dfrac{x^3}{y^2}$ **78.** $\dfrac{y^5}{t^3}$

Use one or more power rules to simplify each expression. Assume all variables represent nonzero real numbers. See Example 6.

79. $(x^3)^6$ **80.** $(y^5)^4$ **81.** $\left(\dfrac{3}{5}\right)^3$ **82.** $\left(\dfrac{4}{3}\right)^2$

83. $(4t)^3$ **84.** $(5t)^4$ **85.** $(-6x^2)^3$ **86.** $(-2x^5)^5$

87. $\left(\dfrac{-4m^2}{t}\right)^3$ **88.** $\left(\dfrac{-5n^4}{r^2}\right)^3$

Simplify each expression so that no negative exponents appear in the final result. Assume all variables represent nonzero real numbers. See Examples 1–8.

89. $3^5 \cdot 3^{-6}$ **90.** $4^4 \cdot 4^{-6}$ **91.** $a^{-3}a^2a^{-4}$ **92.** $k^{-5}k^{-3}k^4$

93. $(k^2)^{-3}k^4$ **94.** $(x^3)^{-4}x^5$ **95.** $-4r^{-2}(r^4)^2$ **96.** $-2m^{-1}(m^3)^2$

97. $(5a^{-1})^4(a^2)^{-3}$ **98.** $(3p^{-4})^2(p^3)^{-1}$ **99.** $(z^{-4}x^3)^{-1}$ **100.** $(y^{-2}z^4)^{-3}$

101. $7k^2(-2k)(4k^{-5})^0$ **102.** $3a^2(-5a^{-6})(-2a)^0$ **103.** $\dfrac{(p^{-2})^0}{5p^{-4}}$

104. $\dfrac{(m^4)^0}{9m^{-3}}$ **105.** $\dfrac{(3pq)q^2}{6p^2q^4}$ **106.** $\dfrac{(-8xy)y^3}{4x^5y^4}$

107. $\dfrac{4a^5(a^{-1})^3}{(a^{-2})^{-2}}$ **108.** $\dfrac{12k^{-2}(k^{-3})^{-4}}{6k^5}$ **109.** $\dfrac{(-y^{-4})^2}{6(y^{-5})^{-1}}$

110. $\dfrac{2(-m^{-1})^{-4}}{9(m^{-3})^2}$ **111.** $\dfrac{(2k)^2m^{-5}}{(km)^{-3}}$ **112.** $\dfrac{(3rs)^{-2}}{3^2r^2s^{-4}}$

113. $\dfrac{(2k)^2k^3}{k^{-1}k^{-5}}(5k^{-2})^{-3}$ **114.** $\dfrac{(3r^2)^2r^{-5}}{r^{-2}r^3}(2r^{-6})^2$ **115.** $\left(\dfrac{3k^{-2}}{k^4}\right)^{-1} \cdot \dfrac{2}{k}$

116. $\left(\dfrac{7m^{-2}}{m^{-3}}\right)^{-2} \cdot \dfrac{m^3}{4}$ **117.** $\left(\dfrac{2p}{q^2}\right)^3\left(\dfrac{3p^4}{q^{-4}}\right)^{-1}$ **118.** $\left(\dfrac{5z^3}{2a^2}\right)^{-3}\left(\dfrac{8a^{-1}}{15z^{-2}}\right)^{-3}$

119. $\dfrac{2^2y^4(y^{-3})^{-1}}{2^5y^{-2}}$ **120.** $\dfrac{3^{-1}m^4(m^2)^{-1}}{3^2m^{-2}}$

121. $\dfrac{(2m^2p^3)^2(4m^2p)^{-2}}{(-3mp^4)^{-1}(2m^3p^4)^3}$ **122.** $\dfrac{(-5y^3z^4)^2(2yz^5)^{-2}}{10(y^4z)^3(3y^3z^2)^{-1}}$

123. $\dfrac{(-3y^3x^3)(-4y^4x^2)(x^2)^{-4}}{18x^3y^2(y^3)^3(x^3)^{-2}}$ **124.** $\dfrac{(2m^3x^2)^{-1}(3m^4x)^{-3}}{(m^2x^3)^3(m^2x)^{-5}}$

125. $\left(\dfrac{p^2 q^{-1}}{2p^{-2}}\right)^2 \cdot \left(\dfrac{p^3 \cdot 4q^{-2}}{3q^{-5}}\right)^{-1} \cdot \left(\dfrac{pq^{-5}}{q^{-2}}\right)^3$ **126.** $\left(\dfrac{a^6 b^{-2}}{2a^{-2}}\right)^{-1} \cdot \left(\dfrac{6a^{-2}}{5b^{-4}}\right)^2 \cdot \left(\dfrac{2b^{-1} a^2}{3b^{-2}}\right)^{-1}$

Write each number in scientific notation. See Example 9.

127. 530 **128.** 1600 **129.** .830 **130.** .0072

131. .00000692 **132.** .875 **133.** $-38{,}500$ **134.** $-976{,}000{,}000$

Write each number in standard notation. See Example 10.

135. 7.2×10^4 **136.** 8.91×10^2 **137.** 2.54×10^{-3}

138. 5.42×10^{-4} **139.** -6×10^4 **140.** -9×10^3

141. 1.2×10^{-5} **142.** 2.7×10^{-6}

Find each value. See Example 11.

143. $\dfrac{12 \times 10^4}{2 \times 10^6}$ **144.** $\dfrac{16 \times 10^5}{4 \times 10^8}$ **145.** $\dfrac{3 \times 10^{-2}}{12 \times 10^3}$

146. $\dfrac{5 \times 10^{-3}}{25 \times 10^2}$ **147.** $\dfrac{.05 \times 1600}{.0004}$ **148.** $\dfrac{.003 \times 40{,}000}{.00012}$

149. $\dfrac{20{,}000 \times .018}{300 \times .0004}$ **150.** $\dfrac{840{,}000 \times .03}{.00021 \times 600}$

Solve each problem. See Example 12.

151. The U.S. budget didn't pass **\$1,000,000,000** until 1917. Seventy years later (in 1987) it exceeded **\$1,000,000,000,000** for the first time. President George W. Bush's budget request for fiscal 2003 was **\$2,128,000,000,000.** If stacked in dollar bills, this amount would stretch **144,419** mi, almost two-thirds of the distance to the moon. Write the four boldfaced numbers in scientific notation. (*Source: The Gazette,* February 5, 2002.)

152. In 1970, Wal-Mart had **1500** employees. In 1997, Wal-Mart became the largest private employer in the United States, with **680,000** employees. In 1999, Wal-Mart became the largest private employer in the world, with **1,100,000** employees. By 2007, the company is expected to have **2,200,000** employees. Write these four numbers in scientific notation. (*Source:* Wal-Mart.)

153. On October 28, 1998, IBM announced a computer capable of 3.9×10^8 operations per second. This was 15,000 times faster than the normal desktop computer at that time. What was the number of operations that the normal desktop could do? (*Source:* IBM.)

154. In the early years of the Powerball Lottery, a player would choose five numbers from 1 through 49 and one number from 1 through 42. It can be shown that there are about 8.009×10^7 different ways to do this. Suppose that a group of 2000 persons decided to purchase tickets for all these numbers and each ticket cost \$1.00. How much should each person have expected to pay? (*Source:* www.powerball.com)

155. The speed of light is approximately 3×10^{10} cm per sec. How long will it take light to travel 9×10^{12} cm?

156. The average distance from Earth to the sun is 9.3×10^7 mi. How long would it take a rocket, traveling at 2.9×10^3 mph, to reach the sun?

157. A *light-year* is the distance that light travels in one year. Find the number of miles in a light-year if light travels 1.86×10^5 mi per sec.

158. Use the information given in the previous two exercises to find the number of minutes necessary for light from the sun to reach Earth.

159. **(a)** The planet Mercury has an average distance from the sun of 3.6×10^7 mi, while the average distance of Venus to the sun is 6.7×10^7 mi. How long would it take a spacecraft traveling at 1.55×10^3 mph to travel from Venus to Mercury? (Give your answer in hours, in standard notation.)

   (b) Use the information from part (a) to find the number of days it would take the spacecraft to travel from Venus to Mercury. Round your answer to the nearest whole number of days.

160. When the distance between the centers of the moon and Earth is 4.60×10^8 m, an object on the line joining the centers of the moon and Earth exerts the same gravitational force on each when it is 4.14×10^8 m from the center of Earth. How far is the object from the center of the moon at that point?

TECHNOLOGY INSIGHTS (EXERCISES 161–164)

The screen on the left shows how a graphing calculator displays 250,000 and .000000034 in scientific notation. When put in scientific mode, it will calculate and display results as shown in the screen on the right.

```
250000
            2.5E5
.000000034
           3.4E-8
```

```
(2.5E5)*(2E-3)
             5E2
(1.25E-4)/(5E-9)

           2.5E4
```

Predict the result the calculator will give for each screen. (Use the usual scientific notation to write your answers.)

161. $(1.5\text{E}12)*(5\text{E}^-3)$

162. $(3.2\text{E}^-5)*(3\text{E}12)$

163. $(8.4\text{E}14)/(2.1\text{E}^-3)$

164. $(2.5\text{E}10)/(2\text{E}^-3)$

5.2   Adding and Subtracting Polynomials

OBJECTIVE 1 Know the basic definitions for polynomials.   Just as whole numbers are the basis of arithmetic, *polynomials* are fundamental in algebra. To understand polynomials, we review several words from Chapter 1. A *term* is a number, a variable, or the product or quotient of a number and one or more variables raised to powers. Examples of terms include

$$4x, \quad \frac{1}{2}m^5 \left(\text{or } \frac{m^5}{2}\right), \quad -7z^9, \quad 6x^2z, \quad \frac{5}{3x^2}, \quad \text{and} \quad 9. \qquad \text{Terms}$$

The number in the product is called the *numerical coefficient,* or just the *coefficient.* In the term $8x^3$, the coefficient is 8. In the term $-4p^5$, it is -4. The coefficient of the term k is understood to be 1. The coefficient of $-r$ is -1. In the term $\frac{x}{3}$, the coefficient is $\frac{1}{3}$ since $\frac{x}{3} = \frac{1x}{3} = \frac{1}{3}x$. More generally, any factor in a term is the coefficient of the product of the remaining factors. For example, $3x^2$ is the coefficient of y in the term $3x^2y$, and $3y$ is the coefficient of x^2 in $3x^2y$.

Recall that any combination of variables or constants (numerical values) joined by the basic operations of addition, subtraction, multiplication, and division (except by 0), or raising to powers or taking roots is called an *algebraic expression.* The simplest kind of algebraic expression is a *polynomial.*

> **Polynomial**
>
> A **polynomial** is a term or a finite sum of terms in which all variables have whole number exponents and no variables appear in denominators.

Examples of polynomials include

$$3x - 5, \quad 4m^3 - 5m^2p + 8, \quad \text{and} \quad -5t^2s^3. \qquad \text{Polynomials}$$

Even though the expression $3x - 5$ involves subtraction, it is a sum of terms since it could be written as $3x + (-5)$.

Some examples of expressions that are not polynomials are

$$x^{-1} + 3x^{-2}, \quad \sqrt{9 - x}, \quad \text{and} \quad \frac{1}{x}. \qquad \text{Not polynomials}$$

The first of these is not a polynomial because it has negative integer exponents, the second because it involves a variable under a radical, and the third because it contains a variable in the denominator.

Most of the polynomials used in this book contain only one variable. A polynomial containing only the variable x is called a **polynomial in x.** A polynomial in one variable is written in **descending powers** of the variable if the exponents on the variable decrease from left to right. For example,

$$x^5 - 6x^2 + 12x - 5$$

is a polynomial in descending powers of x. The term -5 in this polynomial can be thought of as $-5x^0$, since $-5x^0 = -5(1) = -5$.

■ EXAMPLE 1 Writing Polynomials in Descending Powers

Write each polynomial in descending powers of the variable.

(a) $y - 6y^3 + 8y^5 - 9y^4 + 12$

Write the polynomial as

$$8y^5 - 9y^4 - 6y^3 + y + 12.$$

(b) $-2 + m + 6m^2 - 4m^3$ is written as $-4m^3 + 6m^2 + m - 2$. ■

Now Try Exercise 1.

Some polynomials with a specific number of terms are so common that they are given special names. A polynomial with exactly three terms is a **trinomial,** and a polynomial with exactly two terms is a **binomial.** A single-term polynomial is a **monomial.** The table gives examples.

Type of Polynomial	Examples
Monomial	$5x, \quad 7m^9, \quad -8, \quad x^2y^2$
Binomial	$3x^2 - 6, \quad 11y + 8, \quad 5a^2b + 3a$
Trinomial	$y^2 + 11y + 6, \quad 8p^3 - 7p + 2m, \quad -3 + 2k^5 + 9z^4$
None of these	$p^3 - 5p^2 + 2p - 5, \quad -9z^3 + 5c^3 + 2m^5 + 11r^2 - 7r$

OBJECTIVE 2 Find the degree of a polynomial. The **degree of a term** with one variable is the exponent on the variable. For example, the degree of $2x^3$ is 3, the degree of $-x^4$ is 4, and the degree of $17x$ (that is, $17x^1$) is 1. The degree of a term with more than one variable is defined to be the sum of the exponents on the variables. For example, the degree of $5x^3y^7$ is 10, because $3 + 7 = 10$.

The greatest degree of any term in a polynomial is called the **degree of the polynomial.** In most cases, we will be interested in finding the degree of a polynomial in one variable. For example, $4x^3 - 2x^2 - 3x + 7$ has degree 3, because the greatest degree of any term is 3 (the degree of $4x^3$).

The table shows several polynomials and their degrees.

Polynomial	Degree
$9x^2 - 5x + 8$	2
$17m^9 + 18m^{14} - 9m^3$	14
$5x$	1, because $5x = 5x^1$
-2	0, because $-2 = -2x^0$ (Any nonzero constant has degree 0.)
$5a^2b^5$	7, because $2 + 5 = 7$
$x^3y^9 + 12xy^4 + 7xy$	12, because the degrees of the terms are 12, 5, and 2; 12 is the greatest.

NOTE The number 0 has no degree, since 0 times a variable to any power is 0.

Now Try Exercises 21, 25, and 27.

OBJECTIVE 3 Add and subtract polynomials. We use the distributive property to simplify polynomials by combining terms. For example,

$$x^3 + 4x^2 + 5x^2 - 1 = x^3 + (4 + 5)x^2 - 1 \quad \text{Distributive property}$$
$$= x^3 + 9x^2 - 1.$$

On the other hand, the terms in the polynomial $4x + 5x^2$ cannot be combined. As these examples suggest, only terms containing exactly the same variables to the same powers may be combined. Recall that such terms are called *like terms*.

CAUTION Remember that only *like terms* can be combined.

EXAMPLE 2 Combining Like Terms

Combine like terms.

(a) $-5y^3 + 8y^3 - y^3 = (-5 + 8 - 1)y^3 \qquad$ Distributive property
$$= 2y^3$$

(b) $6x + 5y - 9x + 2y = 6x - 9x + 5y + 2y \qquad$ Associative and commutative properties

$$= -3x + 7y \qquad$$ Combine like terms.

Since $-3x$ and $7y$ are unlike terms, no further simplification is possible.

(c) $5x^2y - 6xy^2 + 9x^2y + 13xy^2 = 5x^2y + 9x^2y - 6xy^2 + 13xy^2$
$$= 14x^2y + 7xy^2$$

Now Try Exercises 31, 37, and 43.

We use the following rule to add two polynomials.

Adding Polynomials

To add two polynomials, combine like terms.

Polynomials can be added horizontally or vertically, as seen in the next example.

EXAMPLE 3   Adding Polynomials

Add: $(3a^5 - 9a^3 + 4a^2) + (-8a^5 + 8a^3 + 2)$.

Use the commutative and associative properties to rearrange the polynomials so that like terms are together. Then use the distributive property to combine like terms.

$$(3a^5 - 9a^3 + 4a^2) + (-8a^5 + 8a^3 + 2)$$
$$= 3a^5 - 8a^5 - 9a^3 + 8a^3 + 4a^2 + 2$$
$$= -5a^5 - a^3 + 4a^2 + 2 \qquad \text{Combine like terms.}$$

Add these same two polynomials vertically by placing like terms in columns.

$$\begin{array}{r} 3a^5 - 9a^3 + 4a^2 \\ -8a^5 + 8a^3 + 2 \\ \hline -5a^5 - a^3 + 4a^2 + 2 \end{array}$$

Now Try Exercises 51 and 65.

In Chapter 1, we defined subtraction of real numbers as

$$a - b = a + (-b).$$

That is, we add the first number and the negative (or opposite) of the second. We can give a similar definition for subtraction of polynomials by defining the **negative of a polynomial** as that polynomial with the sign of every coefficient changed.

Subtracting Polynomials

To subtract two polynomials, add the first polynomial and the negative of the *second* polynomial.

EXAMPLE 4   Subtracting Polynomials

Subtract: $(-6m^2 - 8m + 5) - (-5m^2 + 7m - 8)$.

Change every sign in the second polynomial and add.

$$(-6m^2 - 8m + 5) - (-5m^2 + 7m - 8)$$
$$= -6m^2 - 8m + 5 + 5m^2 - 7m + 8$$
$$= -6m^2 + 5m^2 - 8m - 7m + 5 + 8 \qquad \text{Rearrange terms.}$$
$$= -m^2 - 15m + 13 \qquad \text{Combine like terms.}$$

Check by adding the sum, $-m^2 - 15m + 13$, to the second polynomial. The result should be the first polynomial.

To subtract these two polynomials vertically, write the first polynomial above the second, lining up like terms in columns.

$$\begin{array}{r} -6m^2 - 8m + 5 \\ -5m^2 + 7m - 8 \end{array}$$

Change all the signs in the second polynomial and add.

$$\begin{array}{r} -6m^2 - 8m + 5 \\ + 5m^2 - 7m + 8 \qquad \text{Change all signs.} \\ \hline -m^2 - 15m + 13 \qquad \text{Add in columns.} \end{array}$$

Now Try Exercises 61 and 69.

5.2 EXERCISES

Write each polynomial in descending powers of the variable. See Example 1.

1. $2x^3 + x - 3x^2 + 4$ **2.** $3y^2 + y^4 - 2y^3 + y$ **3.** $4p^3 - 8p^5 + p^7$

4. $q^2 + 3q^4 - 2q + 1$ **5.** $-m^3 + 5m^2 + 3m^4 + 10$ **6.** $4 - x + 3x^2$

Give the numerical coefficient and the degree of each term.

7. $7z$ **8.** $3r$ **9.** $-15p^2$ **10.** $-27k^3$ **11.** x^4

12. y^6 **13.** $\dfrac{t}{6}$ **14.** $\dfrac{m}{4}$ **15.** $-mn^5$ **16.** $-a^5b$

Identify each polynomial as a monomial, binomial, trinomial, *or* none of these. *Also give the degree.*

17. 25 **18.** 5 **19.** $7m - 22$

20. $6x + 15$ **21.** $-7y^6 + 11y^8$ **22.** $12k^2 - 9k^5$

23. $-5m^3 + 6m - 9m^2$ **24.** $4z^2 - 11z + 2$

25. $-6p^4q - 3p^3q^2 + 2pq^3 - q^4$ **26.** $8s^3t - 4s^2t^2 + 2st^3 + 9$

27. Which one of the following is a trinomial in descending powers, having degree 6?

A. $5x^6 - 4x^5 + 12$ **B.** $6x^5 - x^6 + 4$
C. $2x + 4x^2 - x^6$ **D.** $4x^6 - 6x^4 + 9x^2 - 8$

28. Give an example of a polynomial of four terms in the variable x, having degree 5, written in descending powers, lacking a fourth-degree term.

Combine terms. See Example 2.

29. $5z^4 + 3z^4$ **30.** $8r^5 - 2r^5$ **31.** $-m^3 + 2m^3 + 6m^3$

32. $3p^4 + 5p^4 - 2p^4$ **33.** $x + x + x + x + x$ **34.** $z - z - z + z$

35. $m^4 - 3m^2 + m$ **36.** $5a^5 + 2a^4 - 9a^3$ **37.** $y^2 + 7y - 4y^2$

38. $2c^2 - 4 + 8 - c^2$ **39.** $2k + 3k^2 + 5k^2 - 7$ **40.** $4x^2 + 2x - 6x^2 - 6$

41. $n^4 - 2n^3 + n^2 - 3n^4 + n^3$ **42.** $2q^3 + 3q^2 - 4q - q^3 + 5q^2$

43. $3ab^2 + 7a^2b - 5ab^2 + 13a^2b$ **44.** $6m^2n - 8mn^2 + 3mn^2 - 7m^2n$

45. $4 - (2 + 3m) + 6m + 9$ **46.** $8a - (3a + 4) - (5a - 3)$

47. $6 + 3p - (2p + 1) - (2p + 9)$ **48.** $4x - 8 - (-1 + x) - (11x + 5)$

 49. Define *polynomial* in your own words. Give examples. Include the words *term*, *monomial*, *binomial*, and *trinomial* in your explanation.

 50. Write a paragraph explaining how to add and subtract polynomials. Give examples.

Add or subtract as indicated. See Examples 3 and 4.

51. $(5x^2 + 7x - 4) + (3x^2 - 6x + 2)$ **52.** $(4k^3 + k^2 + k) + (2k^3 - 4k^2 - 3k)$

53. $(6t^2 - 4t^4 - t) + (3t^4 - 4t^2 + 5)$ **54.** $(3p^2 + 2p - 5) + (7p^2 - 4p^3 + 3p)$

55. $(y^3 + 3y + 2) + (4y^3 - 3y^2 + 2y - 1)$ **56.** $(2x^5 - 2x^4 + x^3 - 1) + (x^4 - 3x^3 + 2)$

57. $(3r + 8) - (2r - 5)$ **58.** $(2d + 7) - (3d - 1)$

59. $(2a^2 + 3a - 1) - (4a^2 + 5a + 6)$ **60.** $(q^4 - 2q^2 + 10) - (3q^4 + 5q^2 - 5)$

61. $(z^5 + 3z^2 + 2z) - (4z^5 + 2z^2 - 5z)$ **62.** $(5t^3 - 3t^2 + 2t) - (4t^3 + 2t^2 + 3t)$

63. Add.
$$21p - 8$$
$$-9p + 4$$

64. Add.
$$15m - 9$$
$$4m + 12$$

65. Add.
$$-12p^2 + 4p - 1$$
$$3p^2 + 7p - 8$$

66. Add.
$$-6y^3 + 8y + 5$$
$$9y^3 + 4y - 6$$

67. Subtract.
$$12a + 15$$
$$7a - 3$$

68. Subtract.
$$-3b + 6$$
$$2b - 8$$

69. Subtract.
$$6m^2 - 11m + 5$$
$$-8m^2 + 2m - 1$$

70. Subtract.
$$-4z^2 + 2z - 1$$
$$3z^2 - 5z + 2$$

71. Add.
$$12z^2 - 11z + 8$$
$$5z^2 + 16z - 2$$
$$-4z^2 + 5z - 9$$

72. Add.
$$-6m^3 + 2m^2 + 5m$$
$$8m^3 + 4m^2 - 6m$$
$$-3m^3 + 2m^2 - 7m$$

73. Add.
$$6y^3 - 9y^2 + 8$$
$$4y^3 + 2y^2 + 5y$$

74. Add.
$$-7r^8 + 2r^6 - r^5$$
$$3r^6 + 5$$

75. Subtract.
$$-5a^4 + 8a^2 - 9$$
$$6a^3 - a^2 + 2$$

76. Subtract.
$$-2m^3 + 8m^2$$
$$m^4 - m^3 + 2m$$

Perform the indicated operations. See Examples 2–4.

77. Subtract $4y^2 - 2y + 3$ from $7y^2 - 6y + 5$.

78. Subtract $-(-4x + 2z^2 + 3m)$ from $[(2z^2 - 3x + m) + (z^2 - 2m)]$.

79. $(-4m^2 + 3n^2 - 5n) - [(3m^2 - 5n^2 + 2n) + (-3m^2) + 4n^2]$

80. $[-(4m^2 - 8m + 4m^3) - (3m^2 + 2m + 5m^3)] + m^2$

81. $[-(y^4 - y^2 + 1) - (y^4 + 2y^2 + 1)] + (3y^4 - 3y^2 - 2)$

82. $[2p - (3p - 6)] - [(5p - (8 - 9p)) + 4p]$

83. $-[3z^2 + 5z - (2z^2 - 6z)] + [(8z^2 - [5z - z^2]) + 2z^2]$

84. $5k - (5k - [2k - (4k - 8k)]) + 11k - (9k - 12k)$

5.3   Polynomial Functions

OBJECTIVES

1 Recognize and evaluate polynomial functions.

2 Use a polynomial function to model data.

3 Add and subtract polynomial functions.

4 Graph basic polynomial functions.

OBJECTIVE 1   Recognize and evaluate polynomial functions.   In Chapter 3 we studied linear (first-degree polynomial) functions, defined as $f(x) = mx + b$. Now we consider more general polynomial functions.

Polynomial Function

A **polynomial function of degree n** is defined by

$$f(x) = a_n x^n + a_{n-1}x^{n-1} + \cdots + a_1 x + a_0,$$

for real numbers $a_n, a_{n-1}, \ldots, a_1,$ and a_0, where $a_n \neq 0$ and n is a whole number.

Another way of describing a polynomial function is to say that it is a function defined by a polynomial in one variable, consisting of one or more terms. It is usually written in descending powers of the variable, and its degree is the degree of the polynomial that defines it.

Suppose that we consider the polynomial $3x^2 - 5x + 7$, so

$$f(x) = 3x^2 - 5x + 7.$$

If $x = -2$, then $f(x) = 3x^2 - 5x + 7$ takes on the value

$$\begin{aligned} f(-2) &= 3(-2)^2 - 5(-2) + 7 \qquad \text{Let } x = -2.\\ &= 3(4) + 10 + 7\\ &= 29. \end{aligned}$$

Thus, $f(-2) = 29$ and the ordered pair $(-2, 29)$ belongs to f.

■ EXAMPLE 1   Evaluating Polynomial Functions

Let $f(x) = 4x^3 - x^2 + 5$. Find each value.

(a) $f(3)$

$$\begin{aligned} f(x) &= 4x^3 - x^2 + 5\\ f(3) &= 4(3)^3 - 3^2 + 5 \qquad \text{Substitute 3 for } x.\\ &= 4(27) - 9 + 5 \qquad \text{Order of operations}\\ &= 108 - 9 + 5\\ &= 104 \end{aligned}$$

(b) $f(-4) = 4(-4)^3 - (-4)^2 + 5 \qquad \text{Let } x = -4.$
$$\begin{aligned} &= 4(-64) - 16 + 5\\ &= -267 \end{aligned}$$

■

Now Try Exercise 3.

While f is the most common letter used to represent functions, recall that other letters such as g and h are also used. The capital letter P is often used for polynomial functions. Note that the function defined as $P(x) = 4x^3 - x^2 + 5$ yields the same ordered pairs as the function f in Example 1.

OBJECTIVE 2   Use a polynomial function to model data.   Polynomial functions can be used to approximate data. They are usually valid for small intervals, and they allow us to predict (with caution) what might happen for values just outside the intervals. These intervals are often periods of years, as shown in Example 2.

■ EXAMPLE 2   Using a Polynomial Model to Approximate Data

The number of U.S. households estimated to see and pay at least one bill on-line each month during the years 2000 through 2006 can be modeled by the polynomial function defined by

$$P(x) = .808x^2 + 2.625x + .502,$$

where $x = 0$ corresponds to the year 2000, $x = 1$ corresponds to 2001, and so on, and $P(x)$ is in millions. Use this function to approximate the number of households expected to pay at least one bill on-line each month in 2004.

Since $x = 4$ corresponds to 2004, we must find $P(4)$.

$$P(x) = .808x^2 + 2.625x + .502$$
$$P(4) = .808(4)^2 + 2.625(4) + .502 \quad \text{Let } x = 4.$$
$$= 23.93 \quad \text{Evaluate.}$$

Thus, in 2004 about 23.93 million households are expected to pay at least one bill on-line each month.

Now Try Exercise 9.

OBJECTIVE 3 Add and subtract polynomial functions. The operations of addition, subtraction, multiplication, and division are also defined for functions. For example, businesses use the equation "profit equals revenue minus cost," written using function notation as

$$P(x) = R(x) - C(x),$$

where x is the number of items produced and sold. Thus, the profit function is found by subtracting the cost function from the revenue function.

We define the following **operations on functions.**

Adding and Subtracting Functions

If $f(x)$ and $g(x)$ define functions, then

$$(f + g)(x) = f(x) + g(x) \quad \text{Sum}$$

and

$$(f - g)(x) = f(x) - g(x). \quad \text{Difference}$$

In each case, the domain of the new function is the intersection of the domains of $f(x)$ and $g(x)$.

EXAMPLE 3 Adding and Subtracting Functions

For the polynomial functions defined by

$$f(x) = x^2 - 3x + 7 \quad \text{and} \quad g(x) = -3x^2 - 7x + 7,$$

find **(a)** the sum and **(b)** the difference.

(a) $(f + g)(x) = f(x) + g(x)$ Use the definition.
$$= (x^2 - 3x + 7) + (-3x^2 - 7x + 7) \quad \text{Substitute.}$$
$$= -2x^2 - 10x + 14 \quad \text{Add the polynomials.}$$

(b) $(f - g)(x) = f(x) - g(x)$ Use the definition.
$$= (x^2 - 3x + 7) - (-3x^2 - 7x + 7) \quad \text{Substitute.}$$
$$= (x^2 - 3x + 7) + (3x^2 + 7x - 7) \quad \text{Change subtraction to addition.}$$
$$= 4x^2 + 4x \quad \text{Add.}$$

Now Try Exercise 15.

EXAMPLE 4 Adding and Subtracting Functions

For the functions defined by

$$f(x) = 10x^2 - 2x \quad \text{and} \quad g(x) = 2x,$$

find each of the following.

(a) $(f + g)(2)$

$$
\begin{aligned}
(f + g)(2) &= f(2) + g(2) && \text{Use the definition.} \\
&= [10(2)^2 - 2(2)] + 2(2) && \text{Substitute.} \\
&= 40
\end{aligned}
$$

Alternatively, we could first find $(f + g)(x)$.

$$
\begin{aligned}
(f + g)(x) &= f(x) + g(x) && \text{Use the definition.} \\
&= (10x^2 - 2x) + 2x && \text{Substitute.} \\
&= 10x^2 && \text{Combine like terms.}
\end{aligned}
$$

Then,

$$(f + g)(2) = 10(2)^2 = 40. \qquad \text{The result is the same.}$$

(b) $(f - g)(x)$ and $(f - g)(1)$

$$
\begin{aligned}
(f - g)(x) &= f(x) - g(x) && \text{Use the definition.} \\
&= (10x^2 - 2x) - 2x && \text{Substitute.} \\
&= 10x^2 - 4x && \text{Combine like terms.}
\end{aligned}
$$

Then,

$$(f - g)(1) = 10(1)^2 - 4(1) = 6. \qquad \text{Substitute.}$$

Confirm that $f(1) - g(1)$ gives the same result.

Now Try Exercises 17 and 19.

OBJECTIVE 4 Graph basic polynomial functions. Functions were introduced in Section 3.5. Recall that each input (or x-value) of a function results in one output (or y-value). The simplest polynomial function is the **identity function,** defined by $f(x) = x$. The domain (set of x-values) of this function is all real numbers, $(-\infty, \infty)$, and it pairs each real number with itself. Therefore, the range (set of y-values) is also $(-\infty, \infty)$. Its graph is a straight line, as first seen in Chapter 3. (Notice that a *linear function* is a specific kind of polynomial function.) Figure 1 shows the graph of $f(x) = x$ and a table of selected ordered pairs.

x	$f(x) = x$
-2	-2
-1	-1
0	0
1	1
2	2

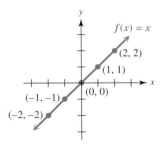

FIGURE 1

Another polynomial function, defined by $f(x) = x^2$, is the **squaring function.** For this function, every real number is paired with its square. The input can be any real number, so the domain is $(-\infty, \infty)$. Since the square of any real number is non-negative, the range is $[0, \infty)$. Its graph is a *parabola*. Figure 2 shows the graph and a table of selected ordered pairs.

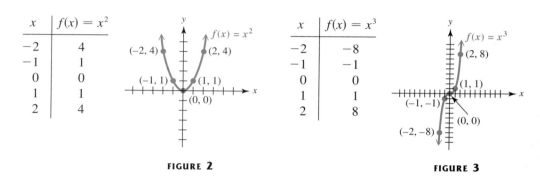

x	$f(x) = x^2$
-2	4
-1	1
0	0
1	1
2	4

FIGURE 2

x	$f(x) = x^3$
-2	-8
-1	-1
0	0
1	1
2	8

FIGURE 3

The **cubing function** is defined by $f(x) = x^3$. Every real number is paired with its cube. The domain and the range are both $(-\infty, \infty)$. Its graph is neither a line nor a parabola. See Figure 3 and the table of ordered pairs.

■ **EXAMPLE 5** **Graphing Variations of the Identity, Squaring, and Cubing Functions**

Graph each function by creating a table of ordered pairs. Give the domain and range of each function by observing the graphs.

(a) $f(x) = 2x$

To find each range value, multiply the domain value by 2. Plot the points and join them with a straight line. See Figure 4. Both the domain and the range are $(-\infty, \infty)$.

x	$f(x) = 2x$
-2	-4
-1	-2
0	0
1	2
2	4

FIGURE 4

(b) $f(x) = -x^2$

For each input x, square it and then take its opposite. Plotting and joining the points gives a parabola that opens down. It is a *reflection* of the graph of the squaring function. See the table and Figure 5 on the next page. The domain is $(-\infty, \infty)$, and the range is $(-\infty, 0]$.

x	$f(x) = -x^2$
-2	-4
-1	-1
0	0
1	-1
2	-4

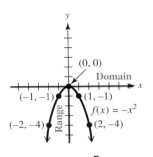

FIGURE 5

x	$f(x) = x^3 - 2$
-2	-10
-1	-3
0	-2
1	-1
2	6

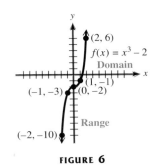

FIGURE 6

(c) $f(x) = x^3 - 2$

For this function, cube the input and then subtract 2 from the result. The graph is that of the cubing function *shifted* 2 units down. See the table and Figure 6. The domain and range are both $(-\infty, \infty)$.

Now Try Exercises 31, 33, and 35.

5.3 EXERCISES

For Extra Help

 Student's Solutions Manual

 MyMathLab

 InterAct Math Tutorial Software

 AW Math Tutor Center

*Math*XP MathXL

Digital Video Tutor CD 5/Videotape 6

For each polynomial function, find **(a)** $f(-1)$ *and* **(b)** $f(2)$. *See Example 1.*

1. $f(x) = 6x - 4$

2. $f(x) = -2x + 5$

3. $f(x) = x^2 - 3x + 4$

4. $f(x) = 3x^2 + x - 5$

5. $f(x) = 5x^4 - 3x^2 + 6$

6. $f(x) = -4x^4 + 2x^2 - 1$

7. $f(x) = -x^2 + 2x^3 - 8$

8. $f(x) = -x^2 - x^3 + 11x$

Solve each problem. See Example 2.

9. The number of airports in the United States during the period from 1970 through 1997 can be approximated by the polynomial function defined by

$$f(x) = -6.77x^2 + 445.34x + 11,279.82,$$

where $x = 0$ represents 1970, $x = 1$ represents 1971, and so on. Use this function to approximate the number of airports in each given year. (*Source:* U.S. Federal Aviation Administration.)

(a) 1970   **(b)** 1985   **(c)** 1997

10. The number of cases commenced by U.S. Courts of Appeals during the period from 1990 through 1998 can be approximated by the polynomial function defined by

$$f(x) = -145.32x^2 + 2610.84x + 41,341.13,$$

where $x = 0$ represents 1990, $x = 1$ represents 1991, and so on. Use this function to approximate the number of cases commenced in each given year. (*Source:* Administrative Office of the U.S. Courts, *Statistical Tables for the Federal Judiciary,* annual.)

(a) 1993   **(b)** 1995   **(c)** 1997

11. The number of bank debit cards issued during the period from 1990 through 2000 can be modeled by the polynomial function defined by

$$P(x) = -.31x^3 + 5.8x^2 - 15x + 9,$$

where $x = 0$ corresponds to the year 1990, $x = 1$ corresponds to 1991, and so on, and $P(x)$ is in millions. Use this function to approximate the number of bank debit cards issued in each given year. Round answers to the nearest million. (*Source: Statistical Abstract of the United States, 2000.*)

(a) 1990   (b) 1996   (c) 1999

12. The number of military personnel on active duty in the United States during the period from 1990 through 1998 is approximated by the polynomial function defined by

$$P(x) = .0045x^3 - .072x^2 + .19x + 3.7,$$

where $x = 0$ corresponds to 1990, $x = 1$ corresponds to 1991, and so on, and $P(x)$ is in millions. Based on this model, how many military personnel were on active duty in each given year? Round answers to the nearest tenth of a million. (*Source:* U.S. Department of Defense.)

(a) 1990   (b) 1994   (c) 1998

For each pair of functions, find **(a)** $(f + g)(x)$ *and* **(b)** $(f - g)(x)$. *See Example 3.*

13. $f(x) = 5x - 10$, $g(x) = 3x + 7$ **14.** $f(x) = -4x + 1$, $g(x) = 6x + 2$

15. $f(x) = 4x^2 + 8x - 3$, $g(x) = -5x^2 + 4x - 9$

16. $f(x) = 3x^2 - 9x + 10$, $g(x) = -4x^2 + 2x + 12$

Let $f(x) = x^2 - 9$, $g(x) = 2x$, *and* $h(x) = x - 3$. *Find each of the following. See Example 4.*

17. $(f + g)(x)$ **18.** $(f - g)(x)$ **19.** $(f + g)(3)$

20. $(f - g)(-3)$ **21.** $(f - h)(x)$ **22.** $(f + h)(x)$

23. $(f - h)(-3)$ **24.** $(f + h)(-2)$ **25.** $(g + h)(-10)$

26. $(g - h)(10)$ **27.** $(g - h)(-3)$ **28.** $(g + h)\left(\dfrac{1}{3}\right)$

29. Construct two polynomial functions defined by $f(x)$, a polynomial of degree 3, and $g(x)$, a polynomial of degree 4. Find $(f - g)(x)$ and $(g - f)(x)$. Use your answers to decide whether subtraction of polynomial functions is a commutative operation. Explain.

30. Find two polynomial functions defined by $f(x)$ and $g(x)$ such that

$$(f + g)(x) = 3x^3 - x + 3.$$

Graph each function. Give the domain and range. See Example 5.

31. $f(x) = -2x + 1$ **32.** $f(x) = 3x + 2$ **33.** $f(x) = -3x^2$

34. $f(x) = \dfrac{1}{2}x^2$ **35.** $f(x) = x^3 + 1$ **36.** $f(x) = -x^3 + 2$

5.4 Multiplying Polynomials

OBJECTIVE 1 Multiply terms.   Recall that the product of the two terms $3x^4$ and $5x^3$ is found by using the commutative and associative properties, along with the rules for exponents.

$$(3x^4)(5x^3) = 3 \cdot 5 \cdot x^4 \cdot x^3$$
$$= 15x^{4+3}$$
$$= 15x^7$$

EXAMPLE 1 Multiplying Monomials

Find each product.

(a) $-4a^3(3a^5) = -4(3)a^3 \cdot a^5 = -12a^8$

(b) $2m^2z^4(8m^3z^2) = 2(8)m^2 \cdot m^3 \cdot z^4 \cdot z^2 = 16m^5z^6$

Now Try Exercises 5 and 7.

OBJECTIVE 2 Multiply any two polynomials.   We use the distributive property to extend this process to find the product of any two polynomials.

EXAMPLE 2 Multiplying Polynomials

Find each product.

(a) $-2(8x^3 - 9x^2)$

$$-2(8x^3 - 9x^2) = -2(8x^3) - 2(-9x^2) \qquad \text{Distributive property}$$
$$= -16x^3 + 18x^2$$

(b) $5x^2(-4x^2 + 3x - 2) = 5x^2(-4x^2) + 5x^2(3x) + 5x^2(-2)$
$$= -20x^4 + 15x^3 - 10x^2$$

(c) $(3x - 4)(2x^2 + x)$

Use the distributive property to multiply each term of $2x^2 + x$ by $3x - 4$.

$$(3x - 4)(2x^2 + x) = (3x - 4)(2x^2) + (3x - 4)(x)$$

Here $3x - 4$ has been treated as a single expression so that the distributive property could be used. Now use the distributive property two more times.

$$= 3x(2x^2) + (-4)(2x^2) + (3x)(x) + (-4)(x)$$
$$= 6x^3 - 8x^2 + 3x^2 - 4x$$
$$= 6x^3 - 5x^2 - 4x$$

(d) $2x^2(x + 1)(x - 3) = 2x^2[(x + 1)(x) + (x + 1)(-3)]$
$$= 2x^2[x^2 + x - 3x - 3]$$
$$= 2x^2(x^2 - 2x - 3)$$
$$= 2x^4 - 4x^3 - 6x^2$$

Now Try Exercises 11, 13, and 19.

It is often easier to multiply polynomials by writing them vertically. To find the product $(3x - 4)(2x^2 + x)$ from Example 2(c) vertically, proceed as follows. (Notice how this process is similar to that of finding the product of two numbers, such as 24×78.)

1. Multiply x and $3x - 4$.

$$
\begin{array}{r}
3x \phantom{{}- 4} - 4 \\
2x^2 + x \\
\hline
\end{array}
$$
$$x(3x - 4) \rightarrow 3x^2 - 4x$$

2. Multiply $2x^2$ and $3x - 4$. Line up like terms of the products in columns.

$$
\begin{array}{r}
3x \phantom{{}- 4} - 4 \\
2x^2 + x \\
\hline
3x^2 - 4x \\
\end{array}
$$
$$2x^2(3x - 4) \rightarrow 6x^3 - 8x^2$$

3. Combine like terms.

$$ 6x^3 - 5x^2 - 4x$$

▌ EXAMPLE 3   Multiplying Polynomials Vertically

Find each product.

(a) $(5a - 2b)(3a + b)$

$$
\begin{array}{r}
5a \phantom{{}- 2b} - 2b \\
3a + b \\
\hline
5ab - 2b^2 \quad \leftarrow b(5a - 2b) \\
15a^2 - 6ab \phantom{{}- 2b^2} \quad \leftarrow 3a(5a - 2b) \\
\hline
15a^2 - ab - 2b^2 \quad \text{Combine like terms.}
\end{array}
$$

(b) $(3m^3 - 2m^2 + 4)(3m - 5)$

$$
\begin{array}{r}
3m^3 - 2m^2 + 4 \\
3m - 5 \\
\hline
-15m^3 + 10m^2 \phantom{{}+ 12m} - 20 \qquad -5(3m^3 - 2m^2 + 4) \\
9m^4 - 6m^3 \phantom{{}+ 10m^2} + 12m \phantom{{}- 20} \qquad 3m(3m^3 - 2m^2 + 4) \\
\hline
9m^4 - 21m^3 + 10m^2 + 12m - 20 \qquad \text{Combine like terms.}
\end{array}
$$

Now Try Exercise 25.

NOTE We can also use a rectangle to model polynomial multiplication. For example, to find the product

$$(5a - 2b)(3a + b),$$

from Example 3(a), label a rectangle with each term as shown here. Then put the product of each pair of monomials in the appropriate box.

	$3a$	b
$5a$		
$-2b$		

	$3a$	b
$5a$	$15a^2$	$5ab$
$-2b$	$-6ab$	$-2b^2$

The product of the original binomials is the sum of these four monomial products.

$$
\begin{aligned}
(5a - 2b)(3a + b) &= 15a^2 + 5ab - 6ab - 2b^2 \\
&= 15a^2 - ab - 2b^2
\end{aligned}
$$

OBJECTIVE 3 Multiply binomials. When working with polynomials, the product of two binomials occurs repeatedly. There is a shortcut method for finding these products. Recall that a binomial has just two terms, such as $3x - 4$ or $2x + 3$. We can find the product of these binomials using the distributive property as follows.

$$(3x - 4)(2x + 3) = 3x(2x + 3) - 4(2x + 3)$$
$$= 3x(2x) + 3x(3) - 4(2x) - 4(3)$$
$$= 6x^2 + 9x - 8x - 12$$

Before combining like terms to find the simplest form of the answer, we check the origin of each of the four terms in the sum. First, $6x^2$ is the product of the two *first* terms.

$$(3x - 4)(2x + 3) \qquad 3x(2x) = 6x^2 \qquad \text{First terms}$$

To get $9x$, the *outer* terms are multiplied.

$$(3x - 4)(2x + 3) \qquad 3x(3) = 9x \qquad \text{Outer terms}$$

The term $-8x$ comes from the *inner* terms.

$$(3x - 4)(2x + 3) \qquad -4(2x) = -8x \qquad \text{Inner terms}$$

Finally, -12 comes from the *last* terms.

$$(3x - 4)(2x + 3) \qquad -4(3) = -12 \qquad \text{Last terms}$$

The product is found by combining these four results.

$$(3x - 4)(2x + 3) = 6x^2 + 9x + (-8x) + (-12)$$
$$= 6x^2 + x - 12$$

To keep track of the order of multiplying these terms, we use the initials FOIL (First, Outer, Inner, Last). All the steps of the FOIL method can be done as follows. Try to do as many of these steps as possible mentally.

$$\begin{array}{cc} \text{First} \quad \text{Last} & 6x^2 \qquad -12 \\ (3x - 4)(2x + 3) & (3x - 4)(2x + 3) \\ \text{Inner} & -8x \\ \text{Outer} & 9x \\ & \overline{x} \quad \text{Add.} \end{array}$$

EXAMPLE 4   Using the FOIL Method

Use the FOIL method to find each product.

(a) $(4m - 5)(3m + 1)$

First terms $\qquad (4m - 5)(3m + 1) \qquad 4m(3m) = 12m^2$

Outer terms $\qquad (4m - 5)(3m + 1) \qquad 4m(1) = 4m$

Inner terms $\qquad (4m - 5)(3m + 1) \qquad -5(3m) = -15m$

Last terms $\qquad (4m - 5)(3m + 1) \qquad -5(1) = -5$

Simplify by combining the four terms.

$$\begin{matrix} & \text{F} & \text{O} & \text{I} & \text{L} \\ (4m - 5)(3m + 1) = & 12m^2 + 4m - 15m - 5 \end{matrix}$$
$$= 12m^2 - 11m - 5$$

The procedure can be written in compact form as follows.

Combine these four results to get $12m^2 - 11m - 5$.

$$\begin{matrix} \text{First} & \text{Outer} & \text{Inner} & \text{Last} \\ \downarrow & \downarrow & \downarrow & \downarrow \end{matrix}$$

(b) $(6a - 5b)(3a + 4b) = 18a^2 + 24ab - 15ab - 20b^2$
$$= 18a^2 + 9ab - 20b^2$$

(c) $(2k + 3z)(5k - 3z) = 10k^2 - 6kz + 15kz - 9z^2$ FOIL
$$= 10k^2 + 9kz - 9z^2$$

Now Try Exercises 35 and 39.

OBJECTIVE 4 **Find the product of the sum and difference of two terms.** Some types of binomial products occur frequently. The product of the sum and difference of the same two terms, x and y, is

$$(x + y)(x - y) = x^2 - xy + xy - y^2 \quad \text{FOIL}$$
$$= x^2 - y^2.$$

Product of the Sum and Difference of Two Terms

The **product of the sum and difference of the two terms x and y** is the difference of the squares of the terms.

$$(x + y)(x - y) = x^2 - y^2$$

EXAMPLE 5 Multiplying the Sum and Difference of Two Terms

Find each product.

(a) $(p + 7)(p - 7) = p^2 - 7^2$
$$= p^2 - 49$$

(b) $(2r + 5)(2r - 5) = (2r)^2 - 5^2$
$$= 2^2 r^2 - 25$$
$$= 4r^2 - 25$$

(c) $(6m + 5n)(6m - 5n) = (6m)^2 - (5n)^2$
$$= 36m^2 - 25n^2$$

(d) $2x^3(x + 3)(x - 3) = 2x^3(x^2 - 9)$
$$= 2x^5 - 18x^3$$

Now Try Exercises 47 and 51.

OBJECTIVE 5 **Find the square of a binomial.** Another special binomial product is the *square of a binomial.* To find the square of $x + y$, or $(x + y)^2$, multiply $x + y$ by itself.

$$(x + y)(x + y) = x^2 + xy + xy + y^2$$
$$= x^2 + 2xy + y^2$$

A similar result is true for the square of a difference.

Square of a Binomial

The **square of a binomial** is the sum of the square of the first term, twice the product of the two terms, and the square of the last term.

$$(x + y)^2 = x^2 + 2xy + y^2$$
$$(x - y)^2 = x^2 - 2xy + y^2$$

EXAMPLE 6 **Squaring Binomials**

Find each product.

(a) $(m + 7)^2 = m^2 + 2 \cdot m \cdot 7 + 7^2$
$$= m^2 + 14m + 49$$

(b) $(p - 5)^2 = p^2 - 2 \cdot p \cdot 5 + 5^2$
$$= p^2 - 10p + 25$$

(c) $(2p + 3v)^2 = (2p)^2 + 2(2p)(3v) + (3v)^2$
$$= 4p^2 + 12pv + 9v^2$$

(d) $(3r - 5s)^2 = (3r)^2 - 2(3r)(5s) + (5s)^2$
$$= 9r^2 - 30rs + 25s^2$$

Now Try Exercises 59 and 63.

CAUTION As the products in the formula for the square of a binomial show,
$$(x + y)^2 \neq x^2 + y^2.$$

More generally,
$$(x + y)^n \neq x^n + y^n \quad (n \neq 1).$$

We can use the patterns for the special products with more complicated products, as the following example shows.

EXAMPLE 7 Multiplying More Complicated Binomials

Find each product.

(a) $[(3p - 2) + 5q][(3p - 2) - 5q]$

$$= (3p - 2)^2 - (5q)^2 \qquad \text{Product of sum and difference of terms}$$

$$= 9p^2 - 12p + 4 - 25q^2 \qquad \text{Square both quantities.}$$

(b) $[(2z + r) + 1]^2 = (2z + r)^2 + 2(2z + r)(1) + 1^2 \qquad \text{Square of a binomial}$

$$= 4z^2 + 4zr + r^2 + 4z + 2r + 1 \qquad \begin{array}{l}\text{Square again; use the}\\\text{distributive property.}\end{array}$$

(c) $(x + y)^3 = (x + y)^2(x + y)$

$$= (x^2 + 2xy + y^2)(x + y) \qquad \text{Square } x + y.$$

$$= x^3 + 2x^2y + xy^2 + x^2y + 2xy^2 + y^3$$

$$= x^3 + 3x^2y + 3xy^2 + y^3$$

(d) $(2a + b)^4 = (2a + b)^2(2a + b)^2$

$$= (4a^2 + 4ab + b^2)(4a^2 + 4ab + b^2) \qquad \text{Square } 2a + b.$$

$$= 16a^4 + 16a^3b + 4a^2b^2 + 16a^3b + 16a^2b^2$$

$$\quad + 4ab^3 + 4a^2b^2 + 4ab^3 + b^4$$

$$= 16a^4 + 32a^3b + 24a^2b^2 + 8ab^3 + b^4$$

Now Try Exercises 69, 73, and 77.

OBJECTIVE 6 Multiply polynomial functions. In the previous section we introduced operations on functions and saw how functions can be added and subtracted. Functions can also be multiplied.

Multiplying Functions

If $f(x)$ and $g(x)$ define functions, then

$$(fg)(x) = f(x) \cdot g(x). \qquad \text{Product}$$

The domain of the product function is the intersection of the domains of $f(x)$ and $g(x)$.

EXAMPLE 8 Multiplying Polynomial Functions

For $f(x) = 3x + 4$ and $g(x) = 2x^2 + x$, find $(fg)(x)$ and $(fg)(-1)$.

$$(fg)(x) = f(x) \cdot g(x) \qquad \text{Use the definition.}$$

$$= (3x + 4)(2x^2 + x)$$

$$= 6x^3 + 3x^2 + 8x^2 + 4x \qquad \text{FOIL}$$

$$= 6x^3 + 11x^2 + 4x \qquad \text{Combine like terms.}$$

Then

$$(fg)(-1) = 6(-1)^3 + 11(-1)^2 + 4(-1) \qquad \text{Let } x = -1.$$

$$= -6 + 11 - 4$$

$$= 1.$$

(What does $f(-1) \cdot g(-1)$ equal?)

Now Try Exercises 113 and 115.

5.4   EXERCISES

Match each product in Column I with the correct polynomial in Column II.

I	**II**
1. $(2x - 5)(3x + 4)$	**A.** $6x^2 + 23x + 20$
2. $(2x + 5)(3x + 4)$	**B.** $6x^2 + 7x - 20$
3. $(2x - 5)(3x - 4)$	**C.** $6x^2 - 7x - 20$
4. $(2x + 5)(3x - 4)$	**D.** $6x^2 - 23x + 20$

Find each product. See Examples 1–3.

5. $-8m^3(3m^2)$ **6.** $4p^2(-5p^4)$ **7.** $14x^2y^3(-2x^5y)$

8. $-5m^3n^4(4m^2n^5)$ **9.** $3x(-2x + 5)$ **10.** $5y(-6y - 1)$

11. $-q^3(2 + 3q)$ **12.** $-3a^4(4 - a)$ **13.** $6k^2(3k^2 + 2k + 1)$

14. $5r^3(2r^2 - 3r - 4)$ **15.** $(2m + 3)(3m^2 - 4m - 1)$

16. $(4z - 2)(z^2 + 3z + 5)$ **17.** $m(m + 5)(m - 8)$

18. $p(p - 6)(p + 4)$ **19.** $4z(2z + 1)(3z - 4)$ **20.** $2y(8y - 3)(2y + 1)$

21. $4x^3(x - 3)(x + 2)$ **22.** $2y^5(y - 8)(y + 2)$ **23.** $(2y + 3)(3y - 4)$

24. $(5m - 3)(2m + 6)$

25. $\begin{array}{r} -b^2 + 3b + 3 \\ 2b + 4 \\ \hline \end{array}$ **26.** $\begin{array}{r} -r^2 - 4r + 8 \\ 3r - 2 \\ \hline \end{array}$

27. $\begin{array}{r} 5m - 3n \\ 5m + 3n \\ \hline \end{array}$ **28.** $\begin{array}{r} 2k + 6q \\ 2k - 6q \\ \hline \end{array}$ **29.** $\begin{array}{r} 2z^3 - 5z^2 + 8z - 1 \\ 4z + 3 \\ \hline \end{array}$

30. $\begin{array}{r} 3z^4 - 2z^3 + z - 5 \\ 2z - 5 \\ \hline \end{array}$ **31.** $\begin{array}{r} 2p^2 + 3p + 6 \\ 3p^2 - 4p - 1 \\ \hline \end{array}$ **32.** $\begin{array}{r} 5y^2 - 2y + 4 \\ 2y^2 + y + 3 \\ \hline \end{array}$

Use the FOIL method to find each product. See Example 4.

33. $(m + 5)(m - 8)$ **34.** $(p - 6)(p + 4)$ **35.** $(4k + 3)(3k - 2)$

36. $(5w + 2)(2w + 5)$ **37.** $(z - w)(3z + 4w)$ **38.** $(s + t)(2s - 5t)$

39. $(6c - d)(2c + 3d)$ **40.** $(2m - n)(3m + 5n)$ **41.** $(.2x + 1.3)(.5x - .1)$

42. $(.5y - .4)(.1y + 2.1)$ **43.** $\left(3w + \dfrac{1}{4}z\right)(w - 2z)$ **44.** $\left(5r - \dfrac{2}{3}y\right)(r + 5y)$

✎ **45.** Describe the FOIL method in your own words.

✎ **46.** Explain why the product of the sum and difference of two terms is not a trinomial.

Find each product. See Example 5.

47. $(2p - 3)(2p + 3)$ **48.** $(3x - 8)(3x + 8)$ **49.** $(5m - 1)(5m + 1)$

50. $(6y + 3)(6y - 3)$ **51.** $(3a + 2c)(3a - 2c)$ **52.** $(5r - 4s)(5r + 4s)$

53. $\left(4x - \dfrac{2}{3}\right)\left(4x + \dfrac{2}{3}\right)$ **54.** $\left(3t + \dfrac{5}{4}\right)\left(3t - \dfrac{5}{4}\right)$ **55.** $(4m + 7n^2)(4m - 7n^2)$

56. $(2k^2 + 6h)(2k^2 - 6h)$ **57.** $(5y^3 + 2)(5y^3 - 2)$ **58.** $(3x^3 + 4)(3x^3 - 4)$

Find each square. See Example 6.

59. $(y - 5)^2$

60. $(a - 3)^2$

61. $(2p + 7)^2$

62. $(3z + 8)^2$

63. $(4n + 3m)^2$

64. $(5r + 7s)^2$

65. $\left(k - \dfrac{5}{7}p\right)^2$

66. $\left(q - \dfrac{3}{4}r\right)^2$

67. How do the expressions $(x + y)^2$ and $x^2 + y^2$ differ?

68. Find the product $101 \cdot 99$ using the special product rule $(x + y)(x - y) = x^2 - y^2$.

Find each product. See Example 7.

69. $[(5x + 1) + 6y]^2$

70. $[(3m - 2) + p]^2$

71. $[(2a + b) - 3]^2$

72. $[(4k + h) - 4]^2$

73. $[(2a + b) - 3][(2a + b) + 3]$

74. $[(m + p) + 5][(m + p) - 5]$

75. $[(2h - k) + j][(2h - k) - j]$

76. $[(3m - y) + z][(3m - y) - z]$

77. $(y + 2)^3$

78. $(z - 3)^3$

79. $(5r - s)^3$

80. $(x + 3y)^3$

81. $(q - 2)^4$

82. $(r + 3)^4$

Find each product.

83. $(2a + b)(3a^2 + 2ab + b^2)$

84. $(m - 5p)(m^2 - 2mp + 3p^2)$

85. $(4z - x)(z^3 - 4z^2x + 2zx^2 - x^3)$

86. $(3r + 2s)(r^3 + 2r^2s - rs^2 + 2s^3)$

87. $(m^2 - 2mp + p^2)(m^2 + 2mp - p^2)$

88. $(3 + x + y)(-3 + x - y)$

89. $ab(a + b)(a + 2b)(a - 3b)$

90. $mp(m - p)(m - 2p)(2m + p)$

In Exercises 91–94, two expressions are given. Replace x with 3 and y with 4 to show that, in general, the two expressions do not equal each other.

91. $(x + y)^2$; $x^2 + y^2$

92. $(x + y)^3$; $x^3 + y^3$

93. $(x + y)^4$; $x^4 + y^4$

94. $(x + y)^5$; $x^5 + y^5$

Find the area of each figure. Express it as a polynomial in descending powers of the variable x. Refer to the formulas on the inside covers of this book if necessary.

95.

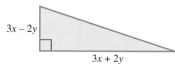

$3x - 2y$

$3x + 2y$

96.

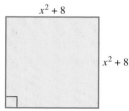

$x^2 + 8$

$x^2 + 8$

97.

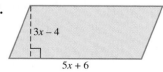

$3x - 4$

$5x + 6$

98.

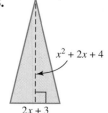

$x^2 + 2x + 4$

$2x + 3$

RELATING CONCEPTS (EXERCISES 99–106)

For Individual or Group Work

Consider the figure. **Work Exercises 99–106 in order.**

99. What is the length of each side of the blue square in terms of *a* and *b*?

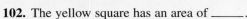

100. What is the formula for the area of a square? Use the formula to write an expression, in the form of a product, for the area of the blue square.

101. Each green rectangle has an area of _____. Therefore, the total area in green is represented by the polynomial _____.

102. The yellow square has an area of _____.

103. The area of the entire colored region is represented by _____, because each side of the entire colored region has length _____.

104. The area of the blue square is equal to the area of the entire colored region minus the total area of the green squares minus the area of the yellow square. Write this as a simplified polynomial in *a* and *b*.

105. **(a)** What must be true about the expressions for the area of the blue square you found in Exercises 100 and 104?

(b) Write an equation based on your answer in part (a). How does this reinforce one of the main ideas of this section?

106. Draw a figure and give a similar proof for $(a + b)^2 = a^2 + 2ab + b^2$.

For each pair of functions, find the product $(fg)(x)$. *See Example 8.*

107. $f(x) = 2x, g(x) = 5x - 1$

108. $f(x) = 3x, g(x) = 6x - 8$

109. $f(x) = x + 1, g(x) = 2x - 3$

110. $f(x) = x - 7, g(x) = 4x + 5$

111. $f(x) = 2x - 3, g(x) = 4x^2 + 6x + 9$

112. $f(x) = 3x + 4, g(x) = 9x^2 - 12x + 16$

Let $f(x) = x^2 - 9$, $g(x) = 2x$, *and* $h(x) = x - 3$. *Find each of the following. See Example 8.*

113. $(fg)(x)$

114. $(fh)(x)$

115. $(fg)(2)$

116. $(fh)(1)$

117. $(gh)(x)$

118. $(fh)(-1)$

119. $(gh)(-3)$

120. $(fg)(-2)$

5.5 Dividing Polynomials

OBJECTIVES

1 Divide a polynomial by a monomial.

2 Divide a polynomial by a polynomial of two or more terms.

3 Divide polynomial functions.

OBJECTIVE 1 Divide a polynomial by a monomial. We now discuss polynomial division, beginning with division by a monomial. (Recall that a monomial is a single term, such as $8x$, $-9m^4$, or $11y^2$.)

Dividing by a Monomial

To divide a polynomial by a monomial, divide each term in the polynomial by the monomial, and then write each quotient in lowest terms.

■ **EXAMPLE 1** Dividing a Polynomial by a Monomial

Divide.

(a) $\dfrac{15x^2 - 12x + 6}{3} = \dfrac{15x^2}{3} - \dfrac{12x}{3} + \dfrac{6}{3}$ Divide each term by 3.

$\qquad\qquad\qquad = 5x^2 - 4x + 2$ Write in lowest terms.

Check this answer by multiplying it by the divisor, 3. You should get $15x^2 - 12x + 6$ as the result.

$$3\underbrace{(5x^2 - 4x + 2)}_{} = \underbrace{15x^2 - 12x + 6}_{}$$

$\qquad\qquad$ Divisor Quotient$\qquad$ Original polynomial

(b) $\dfrac{5m^3 - 9m^2 + 10m}{5m^2} = \dfrac{5m^3}{5m^2} - \dfrac{9m^2}{5m^2} + \dfrac{10m}{5m^2}$ Divide each term by $5m^2$.

$\qquad\qquad\qquad = m - \dfrac{9}{5} + \dfrac{2}{m}$ Write in lowest terms.

This result is not a polynomial. (Why?) The quotient of two polynomials need not be a polynomial.

(c) $\dfrac{8xy^2 - 9x^2y + 6x^2y^2}{x^2y^2} = \dfrac{8xy^2}{x^2y^2} - \dfrac{9x^2y}{x^2y^2} + \dfrac{6x^2y^2}{x^2y^2}$

$\qquad\qquad\qquad = \dfrac{8}{x} - \dfrac{9}{y} + 6$

Now Try Exercises 5, 9, and 11.

OBJECTIVE 2 Divide a polynomial by a polynomial of two or more terms. The process for dividing one polynomial by another polynomial that is not a monomial is similar to that for dividing whole numbers.

■ **EXAMPLE 2** Dividing a Polynomial by a Polynomial

Divide $\dfrac{2m^2 + m - 10}{m - 2}$.

Write the problem as if dividing whole numbers, making sure that both polynomials are written in descending powers of the variables.

$$m - 2 \overline{)2m^2 + m - 10}$$

Divide the first term of $2m^2 + m - 10$ by the first term of $m - 2$. Since $\dfrac{2m^2}{m} = 2m$, place this result above the division line.

$$\begin{array}{r} 2m \\ m - 2 \overline{)2m^2 + m - 10} \end{array} \longleftarrow \text{Result of } \tfrac{2m^2}{m}$$

Multiply $m - 2$ and $2m$, and write the result below $2m^2 + m - 10$.

$$\begin{array}{r} 2m \\ m - 2 \overline{)2m^2 + m - 10} \\ \underline{2m^2 - 4m} \end{array} \longleftarrow 2m(m - 2) = 2m^2 - 4m$$

Now subtract $2m^2 - 4m$ from $2m^2 + m$. Do this by mentally changing the signs on $2m^2 - 4m$ and *adding*.

$$
\begin{array}{r}
2m \phantom{{}- 10} \\
m - 2{\overline{\smash{)}2m^2 + \phantom{{}}m - 10}} \\
\underline{2m^2 - 4m} \phantom{{}- 10} \\
5m \phantom{{}- 10}
\end{array}
$$
$\longleftarrow$ Subtract. The difference is $5m$.

Bring down -10 and continue by dividing $5m$ by m.

$$
\begin{array}{r}
2m + 5 \phantom{{}- 10} \\
m - 2{\overline{\smash{)}2m^2 + \phantom{{}}m - 10}} \\
\underline{2m^2 - 4m} \phantom{{}- 10} \\
5m - 10 \\
\underline{5m - 10} \\
0
\end{array}
$$
$\longleftarrow \frac{5m}{m} = 5$

$\longleftarrow$ Bring down -10.

$\longleftarrow 5(m - 2) = 5m - 10$

$\longleftarrow$ Subtract. The difference is 0.

Finally, $(2m^2 + m - 10) \div (m - 2) = 2m + 5$. Check by multiplying $m - 2$ and $2m + 5$. The result should be $2m^2 + m - 10$.

Now Try Exercise 17.

EXAMPLE 3   Dividing a Polynomial with a Missing Term

Divide $3x^3 - 2x + 5$ by $x - 3$.

Make sure that $3x^3 - 2x + 5$ is in descending powers of the variable. Add a term with 0 coefficient as a placeholder for the missing x^2-term.

$$
x - 3{\overline{\smash{)}3x^3 + 0x^2 - 2x + 5}}
$$

Missing term

Start with $\dfrac{3x^3}{x} = 3x^2$.

$$
\begin{array}{r}
3x^2 \phantom{{}- 2x + 5} \\
x - 3{\overline{\smash{)}3x^3 + 0x^2 - 2x + 5}} \\
\underline{3x^3 - 9x^2} \phantom{{}- 2x + 5}
\end{array}
$$
$\longleftarrow \frac{3x^3}{x} = 3x^2$

$\longleftarrow 3x^2(x - 3)$

Subtract by mentally changing the signs on $3x^3 - 9x^2$ and adding.

$$
\begin{array}{r}
3x^2 \phantom{{}- 2x + 5} \\
x - 3{\overline{\smash{)}3x^3 + 0x^2 - 2x + 5}} \\
\underline{3x^3 - 9x^2} \phantom{{}- 2x + 5} \\
9x^2 \phantom{{}- 2x + 5}
\end{array}
$$
$\longleftarrow$ Subtract.

Bring down the next term.

$$
\begin{array}{r}
3x^2 \phantom{{}- 2x + 5} \\
x - 3{\overline{\smash{)}3x^3 + 0x^2 - 2x + 5}} \\
\underline{3x^3 - 9x^2} \phantom{{}- 2x + 5} \\
9x^2 - 2x \phantom{{}+ 5}
\end{array}
$$
$\longleftarrow$ Bring down $-2x$.

In the next step, $\dfrac{9x^2}{x} = 9x$.

$$
\begin{array}{r}
3x^2 + \;\; 9x \qquad \longleftarrow \frac{9x^2}{x} = 9x \\
x - 3 \overline{)\,3x^3 + 0x^2 - \;\; 2x + 5\,} \\
\underline{3x^3 - 9x^2} \qquad\qquad \\
9x^2 - \;\; 2x \qquad \\
\underline{9x^2 - 27x} \qquad \longleftarrow 9x(x - 3) \\
25x + 5 \longleftarrow \text{Subtract; bring down 5.}
\end{array}
$$

Finally, $\frac{25x}{x} = 25$.

$$
\begin{array}{r}
3x^2 + \;\; 9x + 25 \longleftarrow \frac{25x}{x} = 25 \\
x - 3 \overline{)\,3x^3 + 0x^2 - \;\; 2x + \;\; 5\,} \\
\underline{3x^3 - 9x^2} \qquad\qquad\qquad \\
9x^2 - \;\; 2x \qquad\quad \\
\underline{9x^2 - 27x} \qquad\quad \\
25x + \;\; 5 \quad \\
\underline{25x - 75} \longleftarrow 25(x - 3) \\
80 \longleftarrow \text{Remainder}
\end{array}
$$

Write the remainder, 80, as the numerator of the fraction $\frac{80}{x - 3}$. In summary,

$$
\frac{3x^3 - 2x + 5}{x - 3} = 3x^2 + 9x + 25 + \frac{80}{x - 3}.
$$

Check by multiplying $x - 3$ and $3x^2 + 9x + 25$ and adding 80 to the result. You should get $3x^3 - 2x + 5$.

Now Try Exercise 33.

CAUTION Remember to include $\dfrac{\text{remainder}}{\text{divisor}}$ as part of the answer.

EXAMPLE 4 Dividing by a Polynomial with a Missing Term

Divide $6r^4 + 9r^3 + 2r^2 - 8r + 7$ by $3r^2 - 2$.

The polynomial $3r^2 - 2$ has a missing term. Write it as $3r^2 + 0r - 2$ and divide as usual.

$$
\begin{array}{r}
2r^2 + 3r + \;\; 2 \\
3r^2 + 0r - 2 \overline{)\,6r^4 + 9r^3 + 2r^2 - 8r + \;\; 7\,} \\
\underline{6r^4 + 0r^3 - 4r^2} \qquad\qquad\qquad \\
9r^3 + 6r^2 - 8r \qquad \\
\underline{9r^3 + 0r^2 - 6r} \qquad \\
6r^2 - 2r + \;\; 7 \\
\underline{6r^2 + 0r - \;\; 4} \\
-2r + 11
\end{array}
$$

Missing term ⟶

Since the degree of the remainder, $-2r + 11$, is less than the degree of the divisor, $3r^2 - 2$, the division process is now finished. The result is written

$$
2r^2 + 3r + 2 + \frac{-2r + 11}{3r^2 - 2}.
$$

Now Try Exercise 37.

CAUTION   Remember the following steps when dividing a polynomial by a polynomial of two or more terms.

1. Be sure the terms in both polynomials are in descending powers.

2. Write any missing terms with 0 placeholders.

◼ **EXAMPLE 5**   **Performing a Division with a Fractional Coefficient in the Quotient**

Divide $2p^3 + 5p^2 + p - 2$ by $2p + 2$.

$$\frac{3p^2}{2p} = \frac{3}{2}p$$

$$
\begin{array}{r}
p^2 + \dfrac{3}{2}p - 1 \\
2p + 2 \overline{)\,2p^3 + 5p^2 + \ p - 2} \\
\underline{2p^3 + 2p^2 \qquad\quad} \\
3p^2 + \ p \\
\underline{3p^2 + 3p} \\
-2p - 2 \\
\underline{-2p - 2} \\
0
\end{array}
$$

Since the remainder is 0, the quotient is $p^2 + \dfrac{3}{2}p - 1$.

◼

Now Try Exercise 39.

OBJECTIVE 3   Divide polynomial functions.   In the preceding sections, we used operations on functions to add, subtract, and multiply polynomial functions. We now define the quotient of two functions.

> **Dividing Functions**
>
> If $f(x)$ and $g(x)$ define functions, then
>
> $$\left(\frac{f}{g}\right)(x) = \frac{f(x)}{g(x)}. \qquad \text{Quotient}$$
>
> The domain of the quotient function is the intersection of the domains of $f(x)$ and $g(x)$, excluding any values of x where $g(x) = 0$.

◼ **EXAMPLE 6**   **Dividing Polynomial Functions**

For $f(x) = 2x^2 + x - 10$ and $g(x) = x - 2$, find $\left(\frac{f}{g}\right)(x)$ and $\left(\frac{f}{g}\right)(-3)$.

$$\left(\frac{f}{g}\right)(x) = \frac{f(x)}{g(x)} = \frac{2x^2 + x - 10}{x - 2}$$

This quotient was found in Example 2, with m replacing x. The result here is $2x + 5$, so

$$\left(\frac{f}{g}\right)(x) = 2x + 5, \quad x \neq 2.$$

Then

$$\left(\frac{f}{g}\right)(-3) = 2(-3) + 5 = -1. \qquad \text{Let } x = -3.$$

(Which is easier to find here—$\left(\frac{f}{g}\right)(-3)$ or $\frac{f(-3)}{g(-3)}$?)

Now Try Exercises 57 and 59.

5.5 EXERCISES

Complete each statement with the correct word or words.

1. We find the quotient of two monomials by using the _____ rule for _____.

2. To divide a polynomial by a monomial, divide _____ of the polynomial by the _____.

3. When dividing polynomials that are not monomials, first write them in _____.

4. If a polynomial in a division problem has a missing term, insert a term with _____ as a placeholder.

Divide. See Example 1.

5. $\dfrac{15x^3 - 10x^2 + 5}{5}$

6. $\dfrac{27m^4 - 18m^3 + 9m}{9}$

7. $\dfrac{9y^2 + 12y - 15}{3y}$

8. $\dfrac{80r^2 - 40r + 10}{10r}$

9. $\dfrac{15m^3 + 25m^2 + 30m}{5m^2}$

10. $\dfrac{64x^3 - 72x^2 + 12x}{8x^3}$

11. $\dfrac{14m^2n^2 - 21mn^3 + 28m^2n}{14m^2n}$

12. $\dfrac{24h^2k + 56hk^2 - 28hk}{16h^2k^2}$

13. $\dfrac{8wxy^2 + 3wx^2y + 12w^2xy}{4wx^2y}$

14. $\dfrac{12ab^2c + 10a^2bc + 18abc^2}{6a^2bc}$

Complete the division.

15.
$$\begin{array}{r} r^2 \\ 3r-1\overline{)3r^3 - 22r^2 + 25r - 6} \\ \underline{3r^3 - r^2 } \\ -21r^2 \end{array}$$

16.
$$\begin{array}{r} 3b^2 \\ 2b-5\overline{)6b^3 - 7b^2 - 4b - 40} \\ \underline{6b^3 - 15b^2 } \\ 8b^2 \end{array}$$

Divide. See Examples 2–5.

17. $\dfrac{y^2 + 3y - 18}{y + 6}$

18. $\dfrac{q^2 + 4q - 32}{q - 4}$

19. $\dfrac{3t^2 + 17t + 10}{3t + 2}$

20. $\dfrac{2k^2 - 3k - 20}{2k + 5}$

21. $\dfrac{p^2 + 2p + 20}{p + 6}$

22. $\dfrac{x^2 + 11x + 16}{x + 8}$

23. $\dfrac{3m^3 + 5m^2 - 5m + 1}{3m - 1}$

24. $\dfrac{8z^3 - 6z^2 - 5z + 3}{4z + 3}$

25. $(2z^3 - 5z^2 + 6z - 15) \div (2z - 5)$

26. $(3p^3 + p^2 + 18p + 6) \div (3p + 1)$

27. $(4x^3 + 9x^2 - 10x + 3) \div (4x + 1)$

28. $(10z^3 - 26z^2 + 17z - 13) \div (5z - 3)$

29. $\dfrac{6x^3 - 19x^2 + 14x - 15}{3x^2 - 2x + 4}$

30. $\dfrac{8m^3 - 18m^2 + 37m - 13}{2m^2 - 3m + 6}$

31. $(x^3 + 2x - 3) \div (x - 1)$

32. $(2x^3 - 11x^2 + 28) \div (x - 5)$

33. $(3x^3 - x + 4) \div (x - 2)$

34. $(3k^3 + 9k - 14) \div (k - 2)$

35. $\dfrac{4k^4 + 6k^3 + 3k - 1}{2k^2 + 1}$

36. $\dfrac{6y^4 + 4y^3 + 4y - 6}{3y^2 + 2y - 3}$

37. $(x^4 - 4x^3 + 5x^2 - 3x + 2) \div (x^2 + 3)$

38. $(3t^4 + 5t^3 - 8t^2 - 13t + 2) \div (t^2 - 5)$

39. $(2p^3 + 7p^2 + 9p + 2) \div (2p + 2)$

40. $(3a^2 - 11a + 17) \div (2a + 6)$

41. $\dfrac{p^3 - 1}{p - 1}$

42. $\dfrac{8a^3 + 1}{2a + 1}$

Divide.

43. $\left(2x^2 - \dfrac{7}{3}x - 1\right) \div (3x + 1)$

44. $\left(m^2 + \dfrac{7}{2}m + 3\right) \div (2m + 3)$

45. $\left(3a^2 - \dfrac{23}{4}a - 5\right) \div (4a + 3)$

46. $\left(3q^2 + \dfrac{19}{5}q - 3\right) \div (5q - 2)$

Solve each problem.

47. The volume of a box is $2p^3 + 15p^2 + 28p$. The height is p and the length is $p + 4$; find the width.

48. Suppose a car travels a distance of $(2m^3 + 15m^2 + 13m - 63)$ km in $(2m + 9)$ hr. Find its rate of speed.

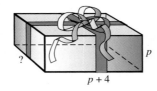

49. For $P(x) = x^3 - 4x^2 + 3x - 5$, find $P(-1)$. Then divide $P(x)$ by $D(x) = x + 1$. Compare the remainder to $P(-1)$. What do these results suggest?

50. Let $P(x) = 4x^3 - 8x^2 + 13x - 2$, and $D(x) = 2x - 1$. Use division to find polynomials $Q(x)$ and $R(x)$ so that $P(x) = Q(x) \cdot D(x) + R(x)$.

For each pair of functions, find the quotient $\left(\dfrac{f}{g}\right)(x)$ and give any x-values that are not in the domain of the quotient function. See Example 6.

51. $f(x) = 10x^2 - 2x$, $g(x) = 2x$

52. $f(x) = 18x^2 - 24x$, $g(x) = 3x$

53. $f(x) = 2x^2 - x - 3$, $g(x) = x + 1$

54. $f(x) = 4x^2 - 23x - 35$, $g(x) = x - 7$

55. $f(x) = 8x^3 - 27$, $g(x) = 2x - 3$

56. $f(x) = 27x^3 + 64$, $g(x) = 3x + 4$

Let $f(x) = x^2 - 9$, $g(x) = 2x$, and $h(x) = x - 3$. Find each of the following. See Example 6.

57. $\left(\dfrac{f}{g}\right)(x)$

58. $\left(\dfrac{f}{h}\right)(x)$

59. $\left(\dfrac{f}{g}\right)(2)$

60. $\left(\dfrac{f}{h}\right)(1)$

61. $\left(\dfrac{h}{g}\right)(x)$

62. $\left(\dfrac{f}{h}\right)(-3)$

63. $\left(\dfrac{h}{g}\right)(3)$

64. $\left(\dfrac{f}{g}\right)(-1)$

Chapter **5** **Group Activity**

Comparing Mathematical Models

OBJECTIVE Use polynomial models to approximate and compare data.

The number of individuals (in thousands) covered by private or governmental health insurance in the United States from 1994 through 1998 can be modeled by the polynomial function defined by

$$P(x) = 1206x + 222{,}448, \qquad \text{Linear model}$$

where x represents years since 1994. The number of individuals (in thousands) can also be modeled by

$$Q(x) = 12x^2 + 1159x + 222{,}471 \qquad \text{Quadratic model}$$

and $\qquad R(x) = 104x^3 - 613x^2 + 2054x + 222{,}347, \qquad$ Cubic model

again where x represents years since 1994.

A. Use each model to determine the number of individuals (in thousands) covered by private or governmental health insurance in the United States for each year from 1994 through 1998. Complete the table.

	Number of Individuals Insured (in thousands)		
Year	Using P(x)	Using Q(x)	Using R(x)
1994			
1995			
1996			
1997			
1998			

B. The actual data are given in the table. Compare the data you entered in the table in part A to the actual data. Which model provides the best approximation of (that is, *best fits*) the actual data? Explain your answer.

Year	Number of Individuals Insured (in thousands)
1994	222,387
1995	223,733
1996	225,077
1997	225,646
1998	227,462

Source: U.S. Bureau of the Census, Health Insurance Historical Table 1.

C. Use the models to make predictions.

  1. Predict the number of individuals (in thousands) who were covered by private or governmental health insurance in 1999 using P, Q, and R.

  2. The actual number of individuals (in thousands) insured in 1999 was 231,533. How do your predictions compare to the actual number? Which model gave the best approximation?

  3. Use each model to predict the number of individuals (in thousands) insured in 2002. What do you notice about your predictions?

  4. Discuss the validity of using these models to predict the number of individuals (in thousands) covered by private or governmental insurance in years beyond 1998.

CHAPTER **5** SUMMARY

KEY TERMS

5.2 term
  numerical coefficient
    (coefficient)
  algebraic expression
  polynomial
	polynomial in x
	descending powers
	trinomial
	binomial
	monomial
		degree of a term
		degree of a
		  polynomial
		negative of a
		  polynomial
			5.3 polynomial function
			  identity function
			  squaring function
			  cubing function

TEST YOUR WORD POWER

See how well you have learned the vocabulary in this chapter. Answers, with examples, follow the Quick Review.

1. A **polynomial** is an algebraic expression made up of
  A. a term or a finite product of terms with positive coefficients and exponents
  B. the sum of two or more terms with whole number coefficients and exponents
  C. the product of two or more terms with positive exponents
  D. a term or a finite sum of terms with real coefficients and whole number exponents.

2. A **monomial** is a polynomial with
  A. only one term
  B. exactly two terms
  C. exactly three terms
  D. more than three terms.

3. A **binomial** is a polynomial with
  A. only one term
  B. exactly two terms
  C. exactly three terms
  D. more than three terms.

4. A **trinomial** is a polynomial with
  A. only one term
  B. exactly two terms
  C. exactly three terms
  D. more than three terms.

5. **FOIL** is a method for
  A. adding two binomials
  B. adding two trinomials
  C. multiplying two binomials
  D. multiplying two trinomials.

QUICK REVIEW

CONCEPTS	EXAMPLES

5.1 INTEGER EXPONENTS AND SCIENTIFIC NOTATION

CONCEPTS	EXAMPLES
Definitions and Rules for Exponents For all integers m and n and all real numbers a and b:	Apply the rules of exponents.
Product Rule: $\quad a^m \cdot a^n = a^{m+n}$	$3^4 \cdot 3^2 = 3^6$
Quotient Rule: $\quad \dfrac{a^m}{a^n} = a^{m-n} \quad (a \neq 0)$	$\dfrac{2^5}{2^3} = 2^2$
Zero Exponent: $\quad a^0 = 1 \quad (a \neq 0)$	$27^0 = 1, \quad (-5)^0 = 1$
Negative Exponent: $\quad a^{-n} = \dfrac{1}{a^n} \quad (a \neq 0)$	$5^{-2} = \dfrac{1}{5^2}$
Power Rules: $\quad (a^m)^n = a^{mn}$ $\qquad\qquad\quad (ab)^m = a^m b^m$ $\qquad\qquad\quad \left(\dfrac{a}{b}\right)^n = \dfrac{a^n}{b^n} \quad (b \neq 0)$	$(6^3)^4 = 6^{12}$ $(5p)^4 = 5^4 p^4$ $\left(\dfrac{2}{3}\right)^5 = \dfrac{2^5}{3^5}$

(continued)

CONCEPTS	EXAMPLES

Special Rules for Negative Exponents:

$$\frac{1}{a^{-n}} = a^n \qquad (a \neq 0)$$

$$\frac{a^{-n}}{b^{-m}} = \frac{b^m}{a^n} \qquad (a, b \neq 0)$$

$$a^{-n} = \left(\frac{1}{a}\right)^n \quad (a \neq 0)$$

$$\left(\frac{a}{b}\right)^{-n} = \left(\frac{b}{a}\right)^n \quad (a, b \neq 0)$$

$$\frac{1}{3^{-2}} = 3^2$$

$$\frac{5^{-3}}{4^{-6}} = \frac{4^6}{5^3}$$

$$4^{-3} = \left(\frac{1}{4}\right)^3$$

$$\left(\frac{4}{7}\right)^{-2} = \left(\frac{7}{4}\right)^2$$

Scientific Notation

A number is in scientific notation when it is written as a product of a number between 1 and 10 (inclusive of 1) and an integer power of 10.

Write 23,500,000,000 in scientific notation.

$$23,500,000,000 = 2.35 \times 10^{10}$$

Write 4.3×10^{-6} in standard notation.

$$4.3 \times 10^{-6} = .0000043$$

5.2   ADDING AND SUBTRACTING POLYNOMIALS

Add or subtract polynomials by combining like terms.

$$(x^2 - 2x + 3) + (2x^2 - 8) = 3x^2 - 2x - 5$$

$$(5x^4 + 3x^2) - (7x^4 + x^2 - x) = -2x^4 + 2x^2 + x$$

5.3   POLYNOMIAL FUNCTIONS

The graph of $f(x) = x$ is a line, and the graph of $f(x) = x^2$ is a parabola. These graphs and the graph of $f(x) = x^3$ define the identity, squaring, and cubing functions, respectively.

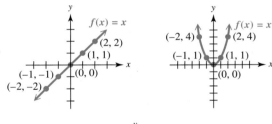

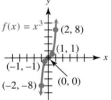

5.4   MULTIPLYING POLYNOMIALS

To multiply two polynomials, multiply each term of one by each term of the other.

$$(x^3 + 3x)(4x^2 - 5x + 2)$$

$$= 4x^5 + 12x^3 - 5x^4 - 15x^2 + 2x^3 + 6x$$

$$= 4x^5 - 5x^4 + 14x^3 - 15x^2 + 6x$$

CONCEPTS	EXAMPLES

To multiply two binomials, use the **FOIL method.** Multiply the **First** terms, the **Outer** terms, the **Inner** terms, and the **Last** terms. Then add these products.

$$(2x + 3)(x - 7) = 2x(x) + 2x(-7) + 3x + 3(-7)$$
$$= 2x^2 - 14x + 3x - 21$$
$$= 2x^2 - 11x - 21$$

Special Products
$$(x + y)(x - y) = x^2 - y^2$$
$$(x + y)^2 = x^2 + 2xy + y^2$$
$$(x - y)^2 = x^2 - 2xy + y^2$$

$$(3m + 8)(3m - 8) = 9m^2 - 64$$
$$(5a + 3b)^2 = 25a^2 + 30ab + 9b^2$$
$$(2k - 1)^2 = 4k^2 - 4k + 1$$

5.5   DIVIDING POLYNOMIALS

Dividing by a Monomial
To divide a polynomial by a monomial, divide each term in the polynomial by the monomial, and then write each fraction in lowest terms.

Dividing by a Polynomial
Use the long division process.

$$\frac{2x^3 - 4x^2 + 6x - 8}{2x} = \frac{2x^3}{2x} - \frac{4x^2}{2x} + \frac{6x}{2x} - \frac{8}{2x}$$
$$= x^2 - 2x + 3 - \frac{4}{x}$$

Divide $\dfrac{m^3 - m^2 + 2m + 5}{m + 1}$.

$$
\begin{array}{r}
m^2 - 2m + 4 \\
m + 1 \overline{\smash{)}\, m^3 - m^2 + 2m + 5} \\
\underline{m^3 + m^2} \\
-2m^2 + 2m \\
\underline{-2m^2 - 2m} \\
4m + 5 \\
\underline{4m + 4} \\
1 \leftarrow \text{Remainder}
\end{array}
$$

The answer is $m^2 - 2m + 4 + \dfrac{1}{m + 1}$.

Answers to Test Your Word Power
1. D; *Example:* $5x^3 + 2x^2 - 7$ **2.** A; *Examples:* $-4, 2x^3, 15a^2b$ **3.** B; *Example:* $3t^3 + 5t$ **4.** C; *Example:* $2a^2 - 3ab + b^2$
5. C; *Example:* $(m + 4)(m - 3) = m(m) - 3m + 4m + 4(-3) = m^2 + m - 12$
          F   O   I   L

CHAPTER 5 REVIEW EXERCISES

[5.1] *Simplify. Write answers with only positive exponents. Assume all variables represent nonzero real numbers.*

1. 4^3

2. $\left(\dfrac{1}{3}\right)^4$

3. $(-5)^3$

4. $\dfrac{2}{(-3)^{-2}}$

5. $\left(\dfrac{2}{3}\right)^{-4}$

6. $\left(\dfrac{5}{4}\right)^{-2}$

7. $5^{-1} + 6^{-1}$

8. $(5 + 6)^{-1}$

9. $-3^0 + 3^0$

10. $(3^{-4})^2$

11. $(x^{-4})^{-2}$

12. $(xy^{-3})^{-2}$

13. $(z^{-3})^3 z^{-6}$

14. $(5m^{-3})^2(m^4)^{-3}$

15. $\dfrac{(3r)^2 r^4}{r^{-2} r^{-3}}(9r^{-3})^{-2}$

16. $\left(\dfrac{5z^{-3}}{z^{-1}}\right)\dfrac{5}{z^2}$

17. $\left(\dfrac{6m^{-4}}{m^{-9}}\right)^{-1}\left(\dfrac{m^{-2}}{16}\right)$

18. $\left(\dfrac{3r^5}{5r^{-3}}\right)^{-2}\left(\dfrac{9r^{-1}}{2r^{-5}}\right)^3$

19. $(-3x^4 y^3)(4x^{-2} y^5)$

20. $\dfrac{6m^{-4} n^3}{-3mn^2}$

21. $\dfrac{(5p^{-2} q)(4p^5 q^{-3})}{2p^{-5} q^5}$

22. $\left(\dfrac{a^{-2} b^{-1}}{3a^2}\right)^{-2}\left(\dfrac{b^{-2} \cdot 3a^4}{2b^{-3}}\right)^{-2}\left(\dfrac{a^{-4} b^5}{a^3}\right)^{-2}$

23. Explain the difference between the expressions $(-6)^0$ and -6^0.

24. Give an example to show that, in general, $(2a)^{-3}$ is not equal to $\dfrac{2}{a^3}$ by choosing a specific value for a.

25. Is $\left(\dfrac{a}{b}\right)^{-1} = \dfrac{a^{-1}}{b^{-1}}$ for all a and $b \neq 0$? If not, explain.

26. Is $(ab)^{-1} = ab^{-1}$ for all a and $b \neq 0$? If not, explain.

27. Give an example to show that $(x^2 + y^2)^2 \neq x^4 + y^4$ by choosing specific values for x and y.

Write in scientific notation.

28. 13,450

29. .0000000765

30. .138

31. In 2000, the total population of the United States was **281,400,000.** Of this amount, **50,454** Americans were centenarians, that is, age **100** or older. Write the three boldfaced numbers using scientific notation. (*Source:* U.S. Bureau of the Census.)

Write each number in standard notation.

32. 1.21×10^6

33. 5.8×10^{-3}

Use scientific notation to compute. Give answers in both scientific notation and standard notation.

34. $\dfrac{16 \times 10^4}{8 \times 10^8}$

35. $\dfrac{6 \times 10^{-2}}{4 \times 10^{-5}}$

36. $\dfrac{.0000000164}{.0004}$

37. $\dfrac{.0009 \times 12,000,000}{400,000}$

38. In a recent year, the estimated population of Luxembourg was 3.92×10^5. The population density was 400 people per mi^2. Based on this information, what is the area of Luxembourg to the nearest square mile?

39. The population of Fresno, California, is approximately 3.45×10^5. The population density is 5449 per mi^2.

   (a) Write the population density in scientific notation.
   (b) To the nearest square mile, what is the area of Fresno?

[5.2] *Give the numerical coefficient of each term.*

40. $14p^5$

41. $-z$

42. $\dfrac{x}{10}$

43. $504p^3r^5$

For each polynomial, (a) write in descending powers, (b) identify as a monomial, binomial, trinomial, *or* none of these, *and (c) give the degree.*

44. $9k + 11k^3 - 3k^2$

45. $14m^6 + 9m^7$

46. $-5y^4 + 3y^3 + 7y^2 - 2y$

47. $-7q^5r^3$

48. Give an example of a polynomial in the variable x such that it has degree 5, is lacking a third-degree term, and is in descending powers of the variable.

Add or subtract as indicated.

49. Add.
$$3x^2 - 5x + 6$$
$$-4x^2 + 2x - 5$$

50. Subtract.
$$-5y^3 \qquad + 8y - 3$$
$$4y^2 + 2y + 9$$

51. $(4a^3 - 9a + 15) - (-2a^3 + 4a^2 + 7a)$

52. $(3y^2 + 2y - 1) + (5y^2 - 11y + 6)$

53. Find the perimeter of the triangle.

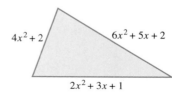

[5.3]

54. For the polynomial function defined by $f(x) = -2x^2 + 5x + 7$, find

    (a) $f(-2)$ **(b)** $f(3)$.

55. For $f(x) = 2x + 3$ and $g(x) = 5x^2 - 3x + 2$, find each of the following.

    (a) $(f + g)(x)$ **(b)** $(f - g)(x)$ **(c)** $(f + g)(-1)$ **(d)** $(f - g)(-1)$

56. The number of people, in millions, enrolled in Health Maintenance Organizations (HMOs) during the period from 1990 through 2000 can be modeled by the polynomial function defined by

$$f(x) = .241x^2 + 3.26x + 30.0,$$

where $x = 0$ corresponds to 1990, $x = 1$ corresponds to 1991, and so on. Use this model to approximate the number of people enrolled in each given year. (*Source:* Interstudy; U.S. National Center for Health Statistics.)

    (a) 1990   **(b)** 1995   **(c)** 2000

Graph each polynomial function defined as follows.

57. $f(x) = -2x + 5$ **58.** $f(x) = x^2 - 6$ **59.** $f(x) = -x^3 + 1$

[5.4] *Find each product.*

60. $-6k(2k^2 + 7)$ **61.** $(3m - 2)(5m + 1)$

62. $(7y - 8)(2y + 3)$ **63.** $(3w - 2t)(2w - 3t)$

64. $(2p^2 + 6p)(5p^2 - 4)$ **65.** $(3q^2 + 2q - 4)(q - 5)$

66. $(3z^3 - 2z^2 + 4z - 1)(3z - 2)$ **67.** $(6r^2 - 1)(6r^2 + 1)$

68. $\left(z + \dfrac{3}{5}\right)\left(z - \dfrac{3}{5}\right)$ **69.** $(4m + 3)^2$ **70.** $(2n - 10)^2$

[5.5] *Divide.*

71. $\dfrac{4y^3 - 12y^2 + 5y}{4y}$ **72.** $\dfrac{2p^3 + 9p^2 + 27}{2p - 3}$

73. $\dfrac{5p^4 + 15p^3 - 33p^2 - 9p + 18}{5p^2 - 3}$

▎ MIXED REVIEW EXERCISES

74. Match each expression (a)–(i) in Column I with its equivalent expression A–I in Column II. Choices may be used once, more than once, or not at all.

I

    (a) 4^{-2} **(f)** -4^0

    (b) -4^2 **(g)** $-4^0 + 4^0$

    (c) 4^0 **(h)** $-4^0 - 4^0$

    (d) $(-4)^0$ **(i)** $4^{-2} + 4^{-1}$

    (e) $(-4)^{-2}$

II

    A. $\dfrac{1}{16}$ **F.** $\dfrac{5}{16}$

    B. 0 **G.** -16

    C. 1 **H.** -2

    D. $-\dfrac{1}{16}$ **I.** none of these

    E. -1

Perform the indicated operations, then simplify. Write answers with only positive exponents. Assume all variables represent nonzero real numbers.

75. $(4x + 1)(2x - 3)$

76. $\dfrac{6^{-1}y^3(y^2)^{-2}}{6y^{-4}(y^{-1})}$

77. 5^{-3}

78. $(y^6)^{-5}(2y^{-3})^{-4}$

79. $7p^5(3p^4 + p^3 + 2p^2)$

80. $(2x - 9)^2$

81. $\dfrac{(-z^{-2})^3}{5(z^{-3})^{-1}}$

82. $-(-3)^2$

83. $\dfrac{8x^2 - 23x + 2}{x - 3}$

84. $\dfrac{(5z^2x^3)^2(2zx^2)^{-1}}{(-10zx^{-3})^{-2}(3z^{-1}x^{-4})^2}$

85. $[(3m - 5n) + p][(3m - 5n) - p]$

86. $\dfrac{20y^3x^3 + 15y^4x + 25yx^4}{10yx^2}$

87. $(2k - 1) - (3k^2 - 2k + 6)$

88. $(-5 + 11w) + (6 + 5w) + (-15 - 8w^2)$

CHAPTER **5** TEST

1. Match the expression in Column I with its equivalent expression from Column II. Choices may be used once, more than once, or not at all.

I	**II**
(a) 7^{-2}	**A.** 1
(b) 7^0	**B.** $\dfrac{1}{9}$
(c) -7^0	**C.** $\dfrac{1}{49}$
(d) $(-7)^0$	**D.** -1
(e) -7^2	**E.** -49
(f) $7^{-1} + 2^{-1}$	**F.** $\dfrac{9}{14}$
(g) $(7 + 2)^{-1}$	**G.** $\dfrac{2}{7}$
(h) $\dfrac{7^{-1}}{2^{-1}}$	**H.** 0
(i) $(-7)^{-2}$	**I.** none of these

Simplify. Write answers with only positive exponents. Assume all variables represent nonzero real numbers.

2. $(3x^{-2}y^3)^{-2}(4x^3y^{-4})$

3. $\dfrac{36r^{-4}(r^2)^{-3}}{6r^4}$

4. $\left(\dfrac{4p^2}{q^4}\right)^3\left(\dfrac{6p^8}{q^{-8}}\right)^{-2}$

5. $(-2x^4y^{-3})^0(-4x^{-3}y^{-8})^2$

6. Write 9.1×10^{-7} in standard form.

7. Use scientific notation to simplify $\dfrac{2{,}500{,}000 \times .00003}{.05 \times 5{,}000{,}000}$. Write the answer in both scientific notation and standard notation.

8. If $f(x) = -2x^2 + 5x - 6$ and $g(x) = 7x - 3$, find each of the following.

(a) $f(4)$ (b) $(f + g)(x)$ (c) $(f - g)(x)$ (d) $(f - g)(-2)$

9. Graph the function defined by $f(x) = -2x^2 + 3$.

10. The number of medical doctors, in thousands, in the United States during the period from 1990 through 1999 can be modeled by the polynomial function defined by

$$f(x) = .178x^2 + 19.1x + 614,$$

where $x = 0$ corresponds to 1990, $x = 1$ corresponds to 1991, and so on. Use this model to approximate the number of doctors to the nearest thousand in each given year. (*Source:* American Medical Association.)

(a) 1990   (b) 1996   (c) 1999

Perform the indicated operations.

11. $(4x^3 - 3x^2 + 2x - 5) - (3x^3 + 11x + 8) + (x^2 - x)$

12. $(5x - 3)(2x + 1)$

13. $(2m - 5)(3m^2 + 4m - 5)$

14. $(6x + y)(6x - y)$

15. $(3k + q)^2$

16. $[2y + (3z - x)][2y - (3z - x)]$

17. $\dfrac{16p^3 - 32p^2 + 24p}{4p^2}$

18. $(x^3 + 3x^2 - 4) \div (x - 1)$

19. If $f(x) = x^2 + 3x + 2$ and $g(x) = x + 1$, find (a) $(fg)(x)$ and (b) $(fg)(-2)$.

20. Use $f(x)$ and $g(x)$ from Problem 19 to find (a) $\left(\frac{f}{g}\right)(x)$ and (b) $\left(\frac{f}{g}\right)(-2)$.

CUMULATIVE REVIEW EXERCISES CHAPTERS 1–5

Match each number in Column I with the choice or choices of sets of numbers in Column II to which the number belongs.

I	II
1. 34	**A.** Natural numbers
2. 0	**B.** Whole numbers
3. 2.16	**C.** Integers
4. $-\sqrt{36}$	**D.** Rational numbers
5. $\sqrt{13}$	**E.** Irrational numbers
6. $-\dfrac{4}{5}$	**F.** Real numbers

Evaluate.

7. $9 \cdot 4 - 16 \div 4$

8. $-|8 - 13| - |-4| + |-9|$

Solve.

9. $-5(8 - 2z) + 4(7 - z) = 7(8 + z) - 3$ **10.** $3(x + 2) - 5(x + 2) = -2x - 4$

11. $A = p + prt$ for t

12. $2(m + 5) - 3m + 1 > 5$

13. $|3x - 1| = 2$

14. $|3z + 1| \geq 7$

15. A recent survey polled teens about the most important inventions of the twentieth century. Complete the results shown in the table if 1500 teens were surveyed.

Most Important Invention	Percent	Actual Number
Personal computer		480
Pacemaker	26%	
Wireless communication	18%	
Television		150

Source: Lemelson-MIT Program.

16. Find the measure of each angle of the triangle.

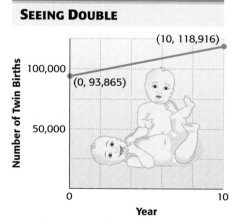

Find the slope of each line described.

17. Through $(-4, 5)$ and $(2, -3)$

18. Horizontal, through $(4, 5)$

Find an equation of each line. Write the equation in slope-intercept form.

19. Through $(4, -1)$, $m = -4$

20. Through $(0, 0)$ and $(1, 4)$

Graph each equation or inequality.

21. $-3x + 4y = 12$

22. $y \leq 2x - 6$

23. $3x + 2y < 0$

24. The graph shows the annual number of twin births in the United States for selected years.

(a) Use the information given on the graph to find and interpret the average rate of change in the number of twin births per year.

(b) If $x = 0$ represents 1990, use your answer from part (a) to write an equation of the line in slope-intercept form that models the annual number of twin births for the years 1990 through 2000.

(c) Use the equation from part (b) to approximate the number of twin births in 2002.

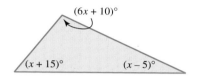

SEEING DOUBLE

Source: National Center for Health Statistics.

25. Give the domain and range of the relation $\{(-4, -2), (-1, 0), (2, 0), (5, 2)\}$. Does this relation define a function?

26. One source of renewable energy is wind, although as of 2000, it provided less than 5 trillion BTUs in the United States. (*Source:* U.S. Energy Information Administration, *Annual Energy Review.*) The force of the wind blowing on a vertical surface varies jointly as the area of the surface and the square of the velocity. If a wind of 40 mph exerts a force of 50 lb on a surface of $\frac{1}{2}$ ft^2, how much force will a wind of 80 mph place on a surface of 2 ft^2?

Solve each system.

27. $3x - 4y = 1$
    $2x + 3y = 12$

28. $3x - 2y = 4$
    $-6x + 4y = 7$

29. $x + 3y - 6z = 7$
    $2x - y + z = 1$
    $x + 2y + 2z = -1$

Solve each problem using a system of equations.

30. The Star-Spangled Banner that flew over Fort McHenry during the War of 1812 had a perimeter of 144 ft. Its length measured 12 ft more than its width. Find the dimensions of this flag, which is displayed in the Smithsonian Institution's Museum of American History in Washington, D.C. (*Source:* National Park Service brochure.)

31. A chemist needs 9 L of a 20% solution of alcohol. She has a 15% solution on hand, as well as a 30% solution. How many liters of the 15% solution and the 30% solution should she mix to get the 20% solution she needs?

Simplify. Write answers with only positive exponents. Assume all variables represent positive real numbers.

32. $\left(\dfrac{2m^3n}{p^2}\right)^3$

33. $\dfrac{x^{-6}y^3z^{-1}}{x^7y^{-4}z}$

34. $(2m^{-2}n^3)^{-3}$

Perform the indicated operations.

35. $2(3x^2 - 8x + 1) - 4(x^2 - 3x - 9)$

36. $(3x + 2y)(5x - y)$

37. $(8m + 5n)(8m - 5n)$

38. $(x + 2y)(x^2 - 2xy + 4y^2)$

39. $\dfrac{m^3 - 3m^2 + 5m - 3}{m - 1}$

40. In 2000, the population of the United States was 281.4 million. (*Source:* U.S. Bureau of the Census.)

    (a) Write the 2000 population using scientific notation.
    (b) Write $1 trillion, that is, $1,000,000,000,000, using scientific notation.
    (c) Using your answers from parts (a) and (b), calculate how much each person in the United States in the year 2000 would have had to contribute in order to make someone a trillionaire. Write this amount in standard notation to the nearest dollar.

Factoring

6

Factoring is used to solve quadratic equations, which have many useful applications. An important one is to express the distance a falling or propelled object travels in a specific time. Such equations are used in astronomy and the space program to describe the motion of objects in space. In Section 6.5 and the Group Activity, we use the concepts of this chapter to explore the heights of propelled objects.

6.1   Greatest Common Factors; Factoring by Grouping

Writing a polynomial as the product of two or more simpler polynomials is called **factoring** the polynomial. For example, the product of $3x$ and $5x - 2$ is $15x^2 - 6x$, and $15x^2 - 6x$ can be factored as the product $3x(5x - 2)$.

$$3x(5x - 2) = 15x^2 - 6x \qquad \text{Multiplying}$$
$$15x^2 - 6x = 3x(5x - 2) \qquad \text{Factoring}$$

Notice that both multiplying and factoring use the distributive property, but in opposite directions. Factoring "undoes" or reverses multiplying.

OBJECTIVE 1   Factor out the greatest common factor.   The first step in factoring a polynomial is to find the *greatest common factor* for the terms of the polynomial. The **greatest common factor (GCF)** is the largest term that is a factor of all terms in the polynomial. For example, the greatest common factor for $8x + 12$ is 4, since 4 is the largest term that is a factor of (divides into) both $8x$ and 12. Using the distributive property,

$$8x + 12 = 4(2x) + 4(3) = 4(2x + 3).$$

As a check, multiply 4 and $2x + 3$. The result should be $8x + 12$. Using the distributive property this way is called *factoring out the greatest common factor.*

■ **EXAMPLE 1   Factoring Out the Greatest Common Factor**

Factor out the greatest common factor.

(a) $9z - 18$

Since 9 is the GCF, factor 9 from each term.

$$9z - 18 = 9 \cdot z - 9 \cdot 2 = 9(z - 2)$$

(b) $56m + 35p = 7(8m + 5p)$

(c) $2y + 5$ There is no common factor other than 1.

(d) $12 + 24z = 12 \cdot 1 + 12 \cdot 2z$
$$= 12(1 + 2z) \qquad \text{12 is the GCF.}$$

■

Now Try Exercise 5.

CAUTION   In Example 1(d), remember to write the factor 1. Always check answers by multiplying.

■ **EXAMPLE 2   Factoring Out the Greatest Common Factor**

Factor out the greatest common factor.

(a) $9x^2 + 12x^3$

The numerical part of the greatest common factor is 3, the largest number that divides into both 9 and 12. For the variable parts, x^2 and x^3, use the least exponent that appears on x; here the least exponent is 2. The GCF is $3x^2$.

$$9x^2 + 12x^3 = 3x^2(3) + 3x^2(4x) \qquad \text{GCF} = 3x^2$$
$$= 3x^2(3 + 4x)$$

(b) $32p^4 - 24p^3 + 40p^5 = 8p^3(4p) + 8p^3(-3) + 8p^3(5p^2) \qquad \text{GCF} = 8p^3$
$$= 8p^3(4p - 3 + 5p^2)$$

(c) $3k^4 - 15k^7 + 24k^9 = 3k^4(1 - 5k^3 + 8k^5)$

(d) $24m^3n^2 - 18m^2n + 6m^4n^3$

The numerical part of the GCF is 6. Here 2 is the least exponent that appears on m, while 1 is the least exponent on n. The GCF is $6m^2n$.

$$24m^3n^2 - 18m^2n + 6m^4n^3 = 6m^2n(4mn) + 6m^2n(-3) + 6m^2n(m^2n^2)$$
$$= 6m^2n(4mn - 3 + m^2n^2)$$

(e) $25x^2y^3 + 30y^5 - 15x^4y^7 = 5y^3(5x^2 + 6y^2 - 3x^4y^4)$

Now Try Exercises 9 and 13.

A greatest common factor need not be a monomial. The next example shows a binomial greatest common factor.

EXAMPLE 3 Factoring Out a Binomial Factor

Factor out the greatest common factor.

(a) $(x - 5)(x + 6) + (x - 5)(2x + 5)$

The greatest common factor here is $x - 5$.

$$(x - 5)(x + 6) + (x - 5)(2x + 5) = (x - 5)[(x + 6) + (2x + 5)]$$
$$= (x - 5)(x + 6 + 2x + 5)$$
$$= (x - 5)(3x + 11)$$

(b) $z^2(m + n)^2 + x^2(m + n)^2 = (m + n)^2(z^2 + x^2)$

(c) $p(r + 2s)^2 - q(r + 2s)^3 = (r + 2s)^2[p - q(r + 2s)]$
$$= (r + 2s)^2(p - qr - 2qs)$$

(d) $(p - 5)(p + 2) - (p - 5)(3p + 4)$
$$= (p - 5)[(p + 2) - (3p + 4)]$$
$$= (p - 5)[p + 2 - 3p - 4] \qquad \text{Be careful with signs.}$$
$$= (p - 5)[-2p - 2]$$
$$= (p - 5)[-2(p + 1)] \text{ or } -2(p - 5)(p + 1) \qquad \text{Look for a common factor.}$$

Now Try Exercises 21 and 25.

When the coefficient of the term of greatest degree is negative, it is sometimes preferable to factor out the -1 that is understood along with the GCF.

EXAMPLE 4 Factoring Out a Negative Common Factor

Factor $-a^3 + 3a^2 - 5a$ in two ways.

First, a could be used as the common factor, giving

$$-a^3 + 3a^2 - 5a = a(-a^2) + a(3a) + a(-5)$$
$$= a(-a^2 + 3a - 5).$$

Because of the leading negative sign, $-a$ could be used as the common factor.

$$-a^3 + 3a^2 - 5a = -a(a^2) + (-a)(-3a) + (-a)(5)$$
$$= -a(a^2 - 3a + 5)$$

Either answer is correct.

Now Try Exercise 33.

> **N O T E**   Example 4 showed two ways of factoring a polynomial. Sometimes there may be a reason to prefer one of these forms over the other, but both are correct.

OBJECTIVE 2   **Factor by grouping.**   Sometimes the *individual terms* of a polynomial have a greatest common factor of 1, but it still may be possible to factor the polynomial by using a process called *factoring by grouping*. We usually factor by grouping when a polynomial has more than three terms. For example, to factor the polynomial

$$ax - ay + bx - by,$$

group the terms as follows.

Terms with common factor a Terms with common factor b

$$(ax - ay) + (bx - by)$$

Then factor $ax - ay$ as $a(x - y)$ and factor $bx - by$ as $b(x - y)$ to get

$$(ax - ay) + (bx - by) = a(x - y) + b(x - y).$$

On the right, the common factor is $x - y$. The final factored form is

$$ax - ay + bx - by = (x - y)(a + b).$$

EXAMPLE 5   **Factoring by Grouping**

Factor $3x - 3y - ax + ay$.

Grouping terms gives

$$(3x - 3y) + (-ax + ay) = 3(x - y) + a(-x + y).$$

There is no simple common factor here. However, if we factor out $-a$ instead of a in the second group of terms, we get

$$3(x - y) - a(x - y).$$

Now factor out the common binomial factor $(x - y)$ to obtain

$$(x - y)(3 - a).$$

Check by multiplying: $(x - y)(3 - a) = 3x - ax - 3y + ay$
$$= 3x - 3y - ax + ay$$

Now Try Exercise 43.

> **N O T E**   In Example 5, different grouping would lead to the product
> $$(a - 3)(y - x).$$
> Verify by multiplying that this is also correct.

Use the following steps to factor by grouping.

Factoring by Grouping

Step 1   **Group terms.** Collect the terms into groups so that each group has a common factor.

Step 2   **Factor within the groups.** Factor out the common factor in each group.

Step 3   **Factor the entire polynomial.** If each group now has a common factor, factor it out. If not, try a different grouping.

▋ EXAMPLE 6   Factoring by Grouping

Factor $6ax + 12bx + a + 2b$ by grouping.

$$6ax + 12bx + a + 2b = (6ax + 12bx) + (a + 2b) \qquad \text{Group terms.}$$

Now factor $6x$ from the first group, and use the identity property of multiplication to introduce the factor 1 in the second group.

$$(6ax + 12bx) + (a + 2b) = 6x(a + 2b) + 1(a + 2b)$$
$$= (a + 2b)(6x + 1) \qquad \text{Factor out } a + 2b.$$

Again, as in Example 1(d), remember to write the 1. Check by multiplying.

Now Try Exercise 41.

▋ EXAMPLE 7   Rearranging Terms before Factoring by Grouping

Factor $p^2q^2 - 10 - 2q^2 + 5p^2$.

Neither the first two terms nor the last two terms have a common factor except the identity element 1. Rearrange and group the terms as follows.

$$(p^2q^2 - 2q^2) + (5p^2 - 10) \qquad \text{Rearrange and group the terms.}$$
$$= q^2(p^2 - 2) + 5(p^2 - 2) \qquad \text{Factor out the common factors.}$$
$$= (p^2 - 2)(q^2 + 5) \qquad \text{Factor out } p^2 - 2.$$

Check by multiplying.

Now Try Exercise 53.

CAUTION   In Example 7, do not stop at the step

$$q^2(p^2 - 2) + 5(p^2 - 2).$$

This expression is *not in factored form* because it is a *sum* of two terms, $q^2(p^2 - 2)$ and $5(p^2 - 2)$, not a *product*.

6.1 EXERCISES

Factor out the greatest common factor. Simplify the factors, if possible. See Examples 1–4.

1. $12m + 60$ **2.** $15r - 27$ **3.** $8k^3 + 24k$

4. $9z^4 + 81z$ **5.** $xy - 5xy^2$ **6.** $5h^2j + hj$

7. $-4p^3q^4 - 2p^2q^5$ **8.** $-3z^5w^2 - 18z^3w^4$

9. $21x^5 + 35x^4 + 14x^3$ **10.** $6k^3 - 36k^4 + 48k^5$

11. $10t^5 - 8t^4 - 2t^3$ **12.** $6p^3 - 3p^2 + 9p^4$

13. $15a^2c^3 - 25ac^2 + 5ac$ **14.** $15y^3z^3 + 27y^2z^4 - 36yz^5$

15. $16z^2n^6 + 64zn^7 - 32z^3n^3$ **16.** $5r^3s^5 + 10r^2s^2 - 15r^4s^2$

17. $-27m^3p^5 + 36m^4p^3 - 72m^5p^4$ **18.** $-50r^4t^2 + 80r^3t^3 - 90r^2t^4$

19. $14a^3b^2 + 7a^2b - 21a^5b^3 + 42ab^4$ **20.** $12km^3 - 24k^3m^2 + 36k^2m^4 - 60k^4m^3$

21. $(m - 4)(m + 2) + (m - 4)(m + 3)$ **22.** $(z - 5)(z + 7) + (z - 5)(z + 9)$

23. $(2z - 1)(z + 6) - (2z - 1)(z - 5)$ **24.** $(3x + 2)(x - 4) - (3x + 2)(x + 8)$

25. $5(2 - x)^2 - (2 - x)^3 + 4(2 - x)$ **26.** $3(5 - x)^4 + 2(5 - x)^3 - (5 - x)^2$

27. $4(3 - x)^2 - (3 - x)^3 + 3(3 - x)$ **28.** $2(t - s) + 4(t - s)^2 - (t - s)^3$

29. $15(2z + 1)^3 + 10(2z + 1)^2 - 25(2z + 1)$

30. $6(a + 2b)^2 - 4(a + 2b)^3 + 12(a + 2b)^4$

31. $5(m + p)^3 - 10(m + p)^2 - 15(m + p)^4$

32. $-9a^2(p + q) - 3a^3(p + q)^2 + 6a(p + q)^3$

Factor each polynomial twice. First use a common factor with a positive coefficient, and then use a common factor with a negative coefficient. See Example 4.

33. $-r^3 + 3r^2 + 5r$ **34.** $-t^4 + 8t^3 - 12t$ **35.** $-12s^5 + 48s^4$

36. $-16y^4 + 64y^3$ **37.** $-2x^5 + 6x^3 + 4x^2$ **38.** $-5a^3 + 10a^4 - 15a^5$

Factor by grouping. See Examples 5–7.

39. $mx + 3qx + my + 3qy$ **40.** $2k + 2h + jk + jh$ **41.** $10m + 2n + 5mk + nk$

42. $3ma + 3mb + 2ab + 2b^2$ **43.** $4 - 2q - 6p + 3pq$ **44.** $20 + 5m + 12n + 3mn$

45. $p^2 - 4zq + pq - 4pz$ **46.** $r^2 - 9tw + 3rw - 3rt$ **47.** $2xy - 8y + 3x - 12$

48. $6y^2 + 9y + 4xy + 6x$ **49.** $m^3 + 4m^2 - 6m - 24$ **50.** $2a^3 + a^2 - 14a - 7$

51. $-3a^3 - 3ab^2 + 2a^2b + 2b^3$ **52.** $-16m^3 + 4m^2p^2 - 4mp + p^3$

53. $4 + xy - 2y - 2x$ **54.** $2ab^2 - 4 - 8b^2 + a$

55. $8 + 9y^4 - 6y^3 - 12y$ **56.** $x^3y^2 - 3 - 3y^2 + x^3$

57. $1 - a + ab - b$ **58.** $2ab^2 - 8b^2 + a - 4$

Factor out the variable that is raised to the smaller exponent. (For example, in Exercise 59, factor out m^{-5}.)

59. $3m^{-5} + m^{-3}$ **60.** $k^{-2} + 2k^{-4}$

61. $3p^{-3} + 2p^{-2}$ **62.** $-5q^{-3} + 8q^{-2}$

✏️ **63.** When directed to factor the polynomial $4x^2y^5 - 8xy^3$ completely, a student wrote $2xy^3(2xy^2 - 4)$. When the teacher did not give him full credit, he complained because when his answer is multiplied out, the result is the original polynomial. Was the teacher justified in her grading? Why or why not?

64. Refer to Exercise 58. One form of the answer is $(2b^2 + 1)(a - 4)$. Give two other acceptable factored forms of $2ab^2 - 8b^2 + a - 4$.

65. Which one of the following is an example of a polynomial in factored form?

A. $3x^2y^3 + 6x^2(2x + y)$ **B.** $5(x + y)^2 - 10(x + y)^3$

C. $(-2 + 3x)(5y^2 + 4y + 3)$ **D.** $(3x + 4)(5x - y) - (3x + 4)(2x - 1)$

6.2   Factoring Trinomials

OBJECTIVES

1 Factor trinomials when the coefficient of the squared term is 1.

2 Factor trinomials when the coefficient of the squared term is not 1.

3 Use an alternative method for factoring trinomials.

4 Factor by substitution.

OBJECTIVE 1 Factor trinomials when the coefficient of the squared term is 1. We begin by finding the product of $x + 3$ and $x - 5$.

$$(x + 3)(x - 5) = x^2 - 5x + 3x - 15$$
$$= x^2 - 2x - 15$$

We see by this result that the factored form of $x^2 - 2x - 15$ is $(x + 3)(x - 5)$.

$$\text{Factored form} \rightarrow (x + 3)(x - 5) \underset{\text{Factoring}}{\overset{\text{Multiplication}}{=}} x^2 - 2x - 15 \leftarrow \text{Product}$$

Since multiplying and factoring are operations that "undo" each other, factoring trinomials involves using FOIL backwards. As shown here, the x^2-term came from multiplying x and x, and -15 came from multiplying 3 and -5.

Product of x and x is x^2.

$$(x + 3)(x - 5) = x^2 - 2x - 15$$

Product of 3   and -5 is -15.

We found the $-2x$ in $x^2 - 2x - 15$ by multiplying the outer terms, then the inner terms, and adding.

Outer terms: $x(-5) = -5x$

$$(x + 3)(x - 5)$$

Add to get $-2x$.

Inner terms: $3 \cdot x = 3x$

Based on this example, use the following steps to factor a trinomial $x^2 + bx + c$, with 1 as the coefficient of the squared term.

Factoring $x^2 + bx + c$

Step 1 **Find pairs whose product is c.** Find all pairs of integers whose product is the third term of the trinomial, c.

Step 2 **Find pairs whose sum is b.** Choose the pair whose sum is the coefficient of the middle term, b.

If there are no such integers, the polynomial cannot be factored. A polynomial that cannot be factored with integer coefficients is **prime.**

■ **EXAMPLE 1** Factoring Trinomials in $x^2 + bx + c$ Form

Factor each polynomial.

(a) $y^2 + 2y - 35$

Step 1 Find pairs of numbers whose product is -35.

$$-35(1)$$
$$35(-1)$$
$$7(-5)$$
$$5(-7)$$

Step 2 Write sums of those numbers.

$$-35 + 1 = -34$$
$$35 + (-1) = 34$$
$$7 + (-5) = 2 \quad \leftarrow \text{Coefficient of the middle term}$$
$$5 + (-7) = -2$$

The required numbers are 7 and -5, so

$$y^2 + 2y - 35 = (y + 7)(y - 5).$$

Check by finding the product of $y + 7$ and $y - 5$.

(b) $r^2 + 8r + 12$

Look for two numbers with a product of 12 and a sum of 8. Of all pairs of numbers having a product of 12, only the pair 6 and 2 has a sum of 8. Therefore,

$$r^2 + 8r + 12 = (r + 6)(r + 2).$$

Because of the commutative property, it would be equally correct to write $(r + 2)(r + 6)$. Check by multiplying.

Now Try Exercises 5 and 7.

■ **EXAMPLE 2** Recognizing a Prime Polynomial

Factor $m^2 + 6m + 7$.

Look for two numbers whose product is 7 and whose sum is 6. Only two pairs of integers, 7 and 1 and -7 and -1, give a product of 7. Neither of these pairs has a sum of 6, so $m^2 + 6m + 7$ cannot be factored with integer coefficients and is prime.

Now Try Exercise 9.

We use a similar process to factor a trinomial that has more than one variable.

▨ **EXAMPLE 3** **Factoring a Trinomial in Two Variables**

Factor $p^2 + 6ap - 16a^2$.

Look for two expressions whose product is $-16a^2$ and whose sum is $6a$. The quantities $8a$ and $-2a$ have the necessary product and sum, so

$$p^2 + 6ap - 16a^2 = (p + 8a)(p - 2a).$$

▨

Now Try Exercise 11.

Sometimes a trinomial will have a common factor that should be factored out first.

▨ **EXAMPLE 4** **Factoring a Trinomial with a Common Factor**

Factor $16y^3 - 32y^2 - 48y$.

Start by factoring out the greatest common factor, $16y$.

$$16y^3 - 32y^2 - 48y = 16y(y^2 - 2y - 3)$$

To factor $y^2 - 2y - 3$, look for two integers whose product is -3 and whose sum is -2. The necessary integers are -3 and 1, so

$$16y^3 - 32y^2 - 48y = 16y(y - 3)(y + 1).$$

▨

Now Try Exercise 31.

CAUTION   When factoring, always look for a common factor first. Remember to write the common factor as part of the answer.

OBJECTIVE 2 **Factor trinomials when the coefficient of the squared term is not 1.**   We can use a generalization of the method shown in Objective 1 to factor a trinomial of the form $ax^2 + bx + c$, where $a \neq 1$. To factor $3x^2 + 7x + 2$, for example, we first identify the values of a, b, and c.

$$ax^2 + bx + c$$
$$\downarrow \quad \downarrow \quad \downarrow$$
$$3x^2 + 7x + 2$$
$$a = 3, \quad b = 7, \quad c = 2$$

The product ac is $3 \cdot 2 = 6$, so we must find integers having a product of 6 and a sum of 7 (since the middle term has coefficient 7). The necessary integers are 1 and 6, so we write $7x$ as $1x + 6x$, or $x + 6x$, giving

$$3x^2 + 7x + 2 = 3x^2 + \underline{x + 6x} + 2.$$
$$x + 6x = 7x$$

Now we factor by grouping.

$$3x^2 + x + 6x + 2 = (3x^2 + x) + (6x + 2)$$
$$= x(3x + 1) + 2(3x + 1)$$
$$3x^2 + 7x + 2 = (3x + 1)(x + 2)$$

EXAMPLE 5 Factoring a Trinomial in $ax^2 + bx + c$ Form

Factor $12r^2 - 5r - 2$.

Since $a = 12$, $b = -5$, and $c = -2$, the product ac is -24. The two integers whose product is -24 and whose sum is -5 are -8 and 3.

$$12r^2 - 5r - 2 = 12r^2 + 3r - 8r - 2 \qquad \text{Write } -5r \text{ as } 3r - 8r.$$
$$= 3r(4r + 1) - 2(4r + 1) \qquad \text{Factor by grouping.}$$
$$= (4r + 1)(3r - 2) \qquad \text{Factor out the common factor.} \blacksquare$$

Now Try Exercise 19.

OBJECTIVE 3 Use an alternative method for factoring trinomials. Alternatively, trying repeated combinations and using FOIL is helpful when the product ac is large. This method is shown using the two polynomials we just factored.

EXAMPLE 6 Factoring Trinomials in $ax^2 + bx + c$ Form

Factor each trinomial.

(a) $3x^2 + 7x + 2$

To factor this polynomial, we must find the correct numbers to put in the blanks.

$$3x^2 + 7x + 2 = (\underline{}x + \underline{})(\underline{}x + \underline{})$$

Addition signs are used since all the signs in the polynomial indicate addition. The first two expressions have a product of $3x^2$, so they must be $3x$ and $1x$ or x.

$$3x^2 + 7x + 2 = (3x + \underline{})(x + \underline{})$$

The product of the two last terms must be 2, so the numbers must be 2 and 1. There is a choice. The 2 could be used with the $3x$ or with the x. Only one of these choices can give the correct middle term, $7x$. We use FOIL to try each one.

$$\overset{3x}{\overbrace{(3x + 2)(x + 1)}} \qquad \overset{6x}{\overbrace{(3x + 1)(x + 2)}}$$
$$\underset{2x}{\underbrace{}} \qquad \underset{x}{\underbrace{}}$$

$$3x + 2x = 5x \qquad\qquad 6x + x = 7x$$

Wrong middle term   Correct middle term

Therefore, $3x^2 + 7x + 2 = (3x + 1)(x + 2)$.

(b) $12r^2 - 5r - 2$

To reduce the number of trials, we note that the trinomial has no common factor. This means that neither of its factors can have a common factor. We should keep this in mind as we choose factors. We try 4 and 3 for the two first terms.

$$12r^2 - 5r - 2 = (4r\underline{})(3r\underline{})$$

We do not know what signs to use yet. The factors of -2 are -2 and 1 or -1 and 2. We try both possibilities to see if we obtain the correct middle term, $-5r$.

$$(4r - 2)(3r + 1) \qquad \overset{8r}{\overbrace{(4r - 1)(3r + 2)}}$$
$$\qquad\qquad\qquad\qquad \underset{-3r}{\underbrace{}}$$

Wrong: $4r - 2$ has a $8r - 3r = 5r$
common factor of 2.   Wrong middle term

The middle term on the right is $5r$, instead of the $-5r$ that is needed. We get $-5r$ by interchanging the signs in the factors.

$$\overset{\displaystyle -8r}{(4r + 1)(3r - 2)}$$
$$\underset{\displaystyle 3r}{}$$

$$-8r + 3r = -5r$$
Correct middle term

Thus, $12r^2 - 5r - 2 = (4r + 1)(3r - 2)$.

Now Try Exercise 21.

NOTE As shown in Example 6(b), if the terms of a polynomial have no common factor (except 1), then none of the terms of its factors can have a common factor. Remembering this will eliminate some potential factors.

This alternative method of factoring a trinomial $ax^2 + bx + c$, $a \neq 1$, is summarized here.

Factoring $ax^2 + bx + c$

Step 1 **Find pairs whose product is a.** Write all pairs of integer factors of the coefficient of the squared term, a.

Step 2 **Find pairs whose product is c.** Write all pairs of integer factors of the last term, c.

Step 3 **Choose inner and outer terms.** Use FOIL and various combinations of the factors from Steps 1 and 2 until the necessary middle term is found.

If no such combinations exist, the trinomial is prime.

EXAMPLE 7 Factoring a Trinomial in Two Variables

Factor $18m^2 - 19mx - 12x^2$.

There is no common factor (except 1). Follow the steps to factor the trinomial. There are many possible factors of both 18 and -12. Try 6 and 3 for 18 and -3 and 4 for -12.

$$(6m - 3x)(3m + 4x) \qquad (6m + 4x)(3m - 3x)$$
Wrong: common factor   Wrong: common factors

Since 6 and 3 do not work as factors of 18, try 9 and 2 instead, with 3 and -4 as factors of -12.

$$(9m + 3x)(2m - 4x) \qquad \overset{\displaystyle 27mx}{(9m - 4x)(2m + 3x)}$$
Wrong: common factors $\underset{\displaystyle -8mx}{}$

$$27mx + (-8mx) = 19mx$$
Wrong middle term

The result on the right differs from the correct middle term only in sign, so interchange the signs in the factors. Check by multiplying.

$$18m^2 - 19mx - 12x^2 = (9m + 4x)(2m - 3x)$$

Now Try Exercise 23.

EXAMPLE 8 Factoring $ax^2 + bx + c, a < 0$

Factor $-3x^2 + 16x + 12$.

While it is possible to factor this polynomial directly, it is helpful to first factor out -1. Then proceed as in the earlier examples.

$$-3x^2 + 16x + 12 = -1(3x^2 - 16x - 12)$$
$$= -1(3x + 2)(x - 6)$$
$$= -(3x + 2)(x - 6)$$

This factored form can be written in other ways. Two of them are

$$(-3x - 2)(x - 6) \quad \text{and} \quad (3x + 2)(-x + 6).$$

Verify that these both give the original trinomial when multiplied.

Now Try Exercise 33.

EXAMPLE 9 Factoring a Trinomial with a Common Factor

Factor $16y^3 + 24y^2 - 16y$.

$$16y^3 + 24y^2 - 16y = 8y(2y^2 + 3y - 2) \qquad \text{GCF} = 8y$$
$$= 8y(2y - 1)(y + 2) \qquad \text{Remember the common factor.}$$

Now Try Exercise 39.

OBJECTIVE 4 **Factor by substitution.** Sometimes we can factor a more complicated polynomial by making a substitution of a variable for an expression.

EXAMPLE 10 Factoring a Polynomial Using Substitution

Factor $2(x + 3)^2 + 5(x + 3) - 12$.

Since the binomial $x + 3$ appears to powers 2 and 1, we let the substitution variable represent $x + 3$. We may choose any letter we wish except x. We choose y to equal $x + 3$.

$$2(x + 3)^2 + 5(x + 3) - 12 = 2y^2 + 5y - 12 \qquad \text{Let } y = x + 3.$$
$$= (2y - 3)(y + 4) \qquad \text{Factor.}$$

Now we replace y with $x + 3$ and simplify to get

$$2(x + 3)^2 + 5(x + 3) - 12 = [2(x + 3) - 3][(x + 3) + 4]$$
$$= (2x + 6 - 3)(x + 7)$$
$$= (2x + 3)(x + 7).$$

Now Try Exercise 49.

CAUTION Remember to make the final substitution of $x + 3$ for y in Example 10.

EXAMPLE 11 **Factoring a Trinomial in $ax^4 + bx^2 + c$ Form**

Factor $6y^4 + 7y^2 - 20$.

The variable y appears to powers in which the larger exponent is twice the smaller exponent. In a case such as this, let the substitution variable equal the smaller power. Here, let $m = y^2$. Then $m^2 = (y^2)^2 = y^4$, and the given trinomial becomes

$$6m^2 + 7m - 20,$$

which is factored as

$$6m^2 + 7m - 20 = (3m - 4)(2m + 5).$$

Since $m = y^2$,

$$6y^4 + 7y^2 - 20 = (3y^2 - 4)(2y^2 + 5).$$

Now Try Exercise 59.

NOTE Some students feel comfortable enough about factoring to factor polynomials like the one in Example 11 directly, without using the substitution method.

6.2 EXERCISES

For Extra Help

Student's
Solutions Manual

MyMathLab

InterAct Math
Tutorial Software

AW Math
Tutor Center

MathXL

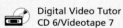
Digital Video Tutor
CD 6/Videotape 7

1. Which one of the following is *not* a valid way of starting the process of factoring $12x^2 + 29x + 10$?

  A. $(12x \quad)(x \quad)$ **B.** $(4x \quad)(3x \quad)$
  C. $(6x \quad)(2x \quad)$ **D.** $(8x \quad)(4x \quad)$

2. Which one of the following is the completely factored form of $2x^6 - 5x^5 - 3x^4$?

  A. $x^4(2x + 1)(x - 3)$ **B.** $x^4(2x - 1)(x + 3)$
  C. $(2x^5 + x^4)(x - 3)$ **D.** $x^3(2x^2 + x)(x - 3)$

3. Which one of the following is *not* a factored form of $-x^2 + 16x - 60$?

  A. $(x - 10)(-x + 6)$ **B.** $(-x - 10)(x + 6)$
  C. $(-x + 10)(x - 6)$ **D.** $-(x - 10)(x - 6)$

4. Which one of the following is the completely factored form of $4x^2 - 4x - 24$?

  A. $4(x - 2)(x + 3)$ **B.** $4(x + 2)(x + 3)$
  C. $4(x + 2)(x - 3)$ **D.** $4(x - 2)(x - 3)$

Factor each trinomial. See Examples 1–9.

  5. $y^2 + 7y - 30$ **6.** $z^2 + 2z - 24$ **7.** $p^2 + 15p + 56$ **8.** $k^2 - 11k + 30$
  9. $-m^2 + 11m - 60$ **10.** $-p^2 + 12p + 27$ **11.** $a^2 - 2ab - 35b^2$

12. $z^2 + 8zw + 15w^2$

13. $y^2 - 3yq - 15q^2$

14. $k^2 - 11hk + 28h^2$

15. $x^2y^2 + 11xy + 18$

16. $p^2q^2 - 5pq - 18$

17. $-6m^2 - 13m + 15$

18. $-15y^2 + 17y + 18$

19. $10x^2 + 3x - 18$

20. $8k^2 + 34k + 35$

21. $20k^2 + 47k + 24$

22. $27z^2 + 42z - 5$

23. $15a^2 - 22ab + 8b^2$

24. $15p^2 + 24pq + 8q^2$

25. $36m^2 - 60m + 25$

26. $25r^2 - 90r + 81$

27. $40x^2 + xy + 6y^2$

28. $14c^2 - 17cd - 6d^2$

29. $6x^2z^2 + 5xz - 4$

30. $8m^2n^2 - 10mn + 3$

31. $24x^2 + 42x + 15$

32. $36x^2 + 18x - 4$

33. $-15a^2 - 70a + 120$

34. $-12a^2 - 10a + 42$

35. $-11x^3 + 110x^2 - 264x$

36. $-9k^3 - 36k^2 + 189k$

37. $2x^3y^3 - 48x^2y^4 + 288xy^5$

38. $6m^3n^2 - 24m^2n^3 - 30mn^4$

39. $6a^3 + 12a^2 - 90a$

40. $3m^4 + 6m^3 - 72m^2$

41. $13y^3 + 39y^2 - 52y$

42. $4p^3 + 24p^2 - 64p$

43. $12p^3 - 12p^2 + 3p$

44. $45t^3 + 60t^2 + 20t$

45. When a student was given the polynomial $4x^2 + 2x - 20$ to factor completely on a test, the student lost some credit when her answer was $(4x + 10)(x - 2)$. She complained to her teacher that when we multiply $(4x + 10)(x - 2)$, we get the original polynomial. Write a short explanation of why she lost some credit for her answer, even though the product is indeed $4x^2 + 2x - 20$.

46. When factoring the polynomial $-4x^2 - 29x + 24$, Margo obtained $(-4x + 3)(x + 8)$, while Steve got $(4x - 3)(-x - 8)$. Who is correct? Explain your answer.

Factor each trinomial. See Example 10.

47. $12p^6 - 32p^3r + 5r^2$

48. $2y^6 + 7xy^3 + 6x^2$

49. $10(k + 1)^2 - 7(k + 1) + 1$

50. $4(m - 5)^2 - 4(m - 5) - 15$

51. $3(m + p)^2 - 7(m + p) - 20$

52. $4(x - y)^2 - 23(x - y) - 6$

53. $a^2(a + b)^2 - ab(a + b)^2 - 6b^2(a + b)^2$

54. $m^2(m - p) + mp(m - p) - 2p^2(m - p)$

55. $p^2(p + q) + 4pq(p + q) + 3q^2(p + q)$

56. $2k^2(5 - y) - 7k(5 - y) + 5(5 - y)$

57. $z^2(z - x) - zx(x - z) - 2x^2(z - x)$

58. $r^2(r - s) - 5rs(s - r) - 6s^2(r - s)$

Factor each trinomial. See Example 11.

59. $p^4 - 10p^2 + 16$

60. $k^4 + 10k^2 + 9$

61. $2x^4 - 9x^2 - 18$

62. $6z^4 + z^2 - 1$

63. $16x^4 + 16x^2 + 3$

64. $9r^4 + 9r^2 + 2$

6.3 Special Factoring

OBJECTIVES

1 Factor a difference of squares.

2 Factor a perfect square trinomial.

3 Factor a difference of cubes.

4 Factor a sum of cubes.

OBJECTIVE 1 Factor a difference of squares. The special products introduced in Section 5.4 are used in reverse when factoring. Recall that the product of the sum and difference of two terms leads to a **difference of squares,** a pattern that occurs often when factoring.

Difference of Squares

$$x^2 - y^2 = (x + y)(x - y)$$

EXAMPLE 1 Factoring Differences of Squares

Factor each polynomial.

(a) $4a^2 - 64$

There is a common factor of 4.

$$4a^2 - 64 = 4(a^2 - 16) \qquad \text{Factor out the common factor.}$$
$$= 4(a + 4)(a - 4) \qquad \text{Factor the difference of squares.}$$

$$\begin{array}{ccccccc} A^2 & - & B^2 & = & (A\ + & B) & (A\ - & B) \end{array}$$

(b) $16m^2 - 49p^2 = (4m)^2 - (7p)^2 = (4m + 7p)(4m - 7p)$

$$\begin{array}{ccccccc} A^2 & - & B^2 & = & (A\ + & B) & (A\ - & B) \end{array}$$

(c) $81k^2 - (a + 2)^2 = (9k)^2 - (a + 2)^2 = (9k + a + 2)(9k - (a + 2))$
$$= (9k + a + 2)(9k - a - 2)$$

We could have used the method of substitution here.

(d) $x^4 - 81 = (x^2 + 9)(x^2 - 9) \qquad \text{Factor the difference of squares.}$
$$= (x^2 + 9)(x + 3)(x - 3) \qquad \text{Factor } x^2 - 9.$$

Now Try Exercises 9, 13, and 15.

CAUTION Assuming no greatest common factor except 1, it is not possible to factor (with real numbers) a *sum* of squares such as $x^2 + 25$. In particular, $x^2 + y^2 \neq (x + y)^2$.

OBJECTIVE 2 Factor a perfect square trinomial. Two other special products from Section 5.4 lead to the following rules for factoring.

Perfect Square Trinomial

$$x^2 + 2xy + y^2 = (x + y)^2$$
$$x^2 - 2xy + y^2 = (x - y)^2$$

Because the trinomial $x^2 + 2xy + y^2$ is the square of $x + y$, it is called a **perfect square trinomial.** In this pattern, both the first and the last terms of the trinomial must be perfect squares. In the factored form, twice the product of the first and the last terms must give the middle term of the trinomial. It is important to understand these patterns in terms of words, since they occur with many different symbols (other than x and y).

$$4m^2 + 20m + 25 \qquad\qquad p^2 - 8p + 64$$

Perfect square trinomial   Not a perfect square trinomial;
                                                    middle term should be $\pm 16p$.

EXAMPLE 2   Factoring Perfect Square Trinomials

Factor each polynomial.

(a) $144p^2 - 120p + 25$

Here $144p^2 = (12p)^2$ and $25 = 5^2$. The sign on the middle term is $-$, so if $144p^2 - 120p + 25$ is a perfect square trinomial, the factored form will have to be

$$(12p - 5)^2.$$

Take twice the product of the two terms to see if this is correct.

$$2(12p)(-5) = -120p$$

This is the middle term of the given trinomial, so

$$144p^2 - 120p + 25 = (12p - 5)^2.$$

(b) $4m^2 + 20mn + 49n^2$

If this is a perfect square trinomial, it will equal $(2m + 7n)^2$. By the pattern in the box, if multiplied out, this squared binomial has a middle term of $2(2m)(7n) = 28mn$, which *does not equal* $20mn$. Verify that this trinomial cannot be factored by the methods of the previous section either. It is prime.

(c) $(r + 5)^2 + 6(r + 5) + 9 = [(r + 5) + 3]^2$
$$= (r + 8)^2,$$

since $2(r + 5)(3) = 6(r + 5)$, the middle term.

(d) $m^2 - 8m + 16 - p^2$

Since there are four terms, use factoring by grouping. The first three terms here form a perfect square trinomial. Group them together, and factor as follows.

$$(m^2 - 8m + 16) - p^2 = (m - 4)^2 - p^2$$

The result is the difference of squares. Factor again to get

$$(m - 4)^2 - p^2 = (m - 4 + p)(m - 4 - p).$$

Now Try Exercises 23, 25, and 33.

Perfect square trinomials, of course, can be factored using the general methods shown earlier for other trinomials. The patterns given here provide "shortcuts."

OBJECTIVE 3 Factor a difference of cubes.   A **difference of cubes,** $x^3 - y^3$, can be factored as follows.

Difference of Cubes

$$x^3 - y^3 = (x - y)(x^2 + xy + y^2)$$

We could check this pattern by finding the product of $x - y$ and $x^2 + xy + y^2$.

▌**EXAMPLE 3** **Factoring Differences of Cubes**

Factor each polynomial.

(a) $m^3 - 8 = m^3 - 2^3 = (m - 2)(m^2 + 2m + 2^2) = (m - 2)(m^2 + 2m + 4)$

Check:

$$\overset{\displaystyle m^3 \qquad -8}{(m - 2)(m^2 + 2m + 4)}$$

$$-2m$$

↑ Opposite of the product of the cube roots gives the middle term.

(b) $27x^3 - 8y^3 = (3x)^3 - (2y)^3$

$$= (3x - 2y)[(3x)^2 + (3x)(2y) + (2y)^2]$$

$$= (3x - 2y)(9x^2 + 6xy + 4y^2)$$

(c) $1000k^3 - 27n^3 = (10k)^3 - (3n)^3$

$$= (10k - 3n)[(10k)^2 + (10k)(3n) + (3n)^2]$$

$$= (10k - 3n)(100k^2 + 30kn + 9n^2)$$

Now Try Exercises 37 and 39.

OBJECTIVE 4 Factor a sum of cubes. While an expression of the form $x^2 + y^2$ (a sum of squares) cannot be factored with real numbers, a **sum of cubes** is factored as follows.

Sum of Cubes

$$x^3 + y^3 = (x + y)(x^2 - xy + y^2)$$

To verify this result, find the product of $x + y$ and $x^2 - xy + y^2$. Compare this pattern with the pattern for a difference of cubes.

NOTE The sign of the second term in the binomial factor of a sum or difference of cubes is *always the same* as the sign in the original polynomial. In the trinomial factor, the first and last terms are *always positive;* the sign of the middle term is *the opposite of* the sign of the second term in the binomial factor.

EXAMPLE 4 Factoring Sums of Cubes

Factor each polynomial.

(a) $r^3 + 27 = r^3 + 3^3 = (r + 3)(r^2 - 3r + 3^2)$
$$= (r + 3)(r^2 - 3r + 9)$$

(b) $27z^3 + 125 = (3z)^3 + 5^3 = (3z + 5)[(3z)^2 - (3z)(5) + 5^2]$
$$= (3z + 5)(9z^2 - 15z + 25)$$

(c) $(x + 2)^3 + t^3 = [(x + 2) + t][(x + 2)^2 - (x + 2)t + t^2]$
$$= (x + 2 + t)(x^2 + 4x + 4 - xt - 2t + t^2)$$

Now Try Exercises 41 and 43.

CAUTION A common error is to think that the xy-term has a coefficient of 2 when factoring the sum or difference of cubes. Since there is no coefficient of 2, expressions of the form $x^2 + xy + y^2$ and $x^2 - xy + y^2$ usually cannot be factored further.

The special types of factoring in this section are summarized here. *These should be memorized.*

Special Types of Factoring	
Difference of Squares	$x^2 - y^2 = (x + y)(x - y)$
Perfect Square Trinomial	$x^2 + 2xy + y^2 = (x + y)^2$
	$x^2 - 2xy + y^2 = (x - y)^2$
Difference of Cubes	$x^3 - y^3 = (x - y)(x^2 + xy + y^2)$
Sum of Cubes	$x^3 + y^3 = (x + y)(x^2 - xy + y^2)$

6.3 EXERCISES

For Extra Help

 Student's Solutions Manual

 MyMathLab

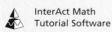 InterAct Math Tutorial Software

Tutor Center AW Math Tutor Center

*Math*XP MathXL

 Digital Video Tutor CD 6/Videotape 7

1. Which of the following binomials are differences of squares?

   A. $64 - m^2$ **B.** $2x^2 - 25$ **C.** $k^2 + 9$ **D.** $4z^4 - 49$

2. Which of the following binomials are sums or differences of cubes?

   A. $64 + y^3$ **B.** $125 - p^6$ **C.** $9x^3 + 125$ **D.** $(x + y)^3 - 1$

3. Which of the following trinomials are perfect squares?

   A. $x^2 - 8x - 16$ **B.** $4m^2 + 20m + 25$
   C. $9z^4 + 30z^2 + 25$ **D.** $25a^2 - 45a + 81$

4. Of the twelve polynomials listed in Exercises 1–3, which ones can be factored using the methods of this section?

5. The binomial $9x^2 + 81$ is an example of the sum of two squares that can be factored. Under what conditions can the sum of two squares be factored?

6. Insert the correct signs in the blanks.

   (a) $8 + t^3 = (2 \_\_ t)(4 \_\_ 2t \_\_ t^2)$ **(b)** $z^3 - 1 = (z \_\_ 1)(z^2 \_\_ z \_\_ 1)$

Factor each polynomial. See Examples 1–4.

7. $p^2 - 16$ | **8.** $k^2 - 9$ | **9.** $25x^2 - 4$

10. $36m^2 - 25$ | **11.** $18a^2 - 98b^2$ | **12.** $32c^2 - 98d^2$

13. $64m^4 - 4y^4$ | **14.** $243x^4 - 3t^4$ | **15.** $(y + z)^2 - 81$

16. $(h + k)^2 - 9$ | **17.** $16 - (x + 3y)^2$ | **18.** $64 - (r + 2t)^2$

19. $(p + q)^2 - (p - q)^2$ | **20.** $(a + b)^2 - (a - b)^2$ | **21.** $k^2 - 6k + 9$

22. $x^2 + 10x + 25$ | **23.** $4z^2 + 4zw + w^2$ | **24.** $9y^2 + 6yz + z^2$

25. $16m^2 - 8m + 1 - n^2$ | **26.** $25c^2 - 20c + 4 - d^2$ | **27.** $4r^2 - 12r + 9 - s^2$

28. $9a^2 - 24a + 16 - b^2$ | **29.** $x^2 - y^2 + 2y - 1$ | **30.** $-k^2 - h^2 + 2kh + 4$

31. $98m^2 + 84mn + 18n^2$ | | **32.** $80z^2 - 40zw + 5w^2$

33. $(p + q)^2 + 2(p + q) + 1$ | | **34.** $(x + y)^2 + 6(x + y) + 9$

35. $(a - b)^2 + 8(a - b) + 16$ | | **36.** $(m - n)^2 + 4(m - n) + 4$

37. $8x^3 - y^3$ | **38.** $z^3 - 125p^3$ | **39.** $64g^3 - 27h^3$

40. $27a^3 - 8b^3$ | **41.** $24n^3 + 81p^3$ | **42.** $250x^3 + 16y^3$

43. $(y + z)^3 + 64$ | **44.** $(p - q)^3 + 125$ | **45.** $(a + b)^3 - (a - b)^3$

46. $k^6 + (k + 3)^3$ | **47.** $m^6 - 125$ | **48.** $27r^6 + 1$

49. $1000x^9 - 27$ | **50.** $729p^9 - 64$ | **51.** $125y^6 + z^6$

RELATING CONCEPTS (EXERCISES 52–57)

For Individual or Group Work

The binomial $x^6 - y^6$ may be considered either as a difference of squares or a difference of cubes. **Work Exercises 52–57 in order.**

52. Factor $x^6 - y^6$ by first factoring as a difference of squares. Then factor further by considering one of the factors as a sum of cubes and the other factor as a difference of cubes.

53. Based on your answer in Exercise 52, fill in the blank with the correct factors so that $x^6 - y^6$ is factored completely:

$$x^6 - y^6 = (x - y)(x + y) \underline{\hspace{4cm}}.$$

54. Factor $x^6 - y^6$ by first factoring as a difference of cubes. Then factor further by considering one of the factors as a difference of squares.

55. Based on your answer in Exercise 54, fill in the blank with the correct factor so that $x^6 - y^6$ is factored:

$$x^6 - y^6 = (x - y)(x + y) \underline{\hspace{4cm}}.$$

56. Notice that the factor you wrote in the blank in Exercise 55 is a fourth-degree polynomial, while the two factors you wrote in the blank in Exercise 53 are both second-degree polynomials. What must be true about the product of the two factors you wrote in the blank in Exercise 53? Verify this.

57. If you have a choice of factoring as a difference of squares or a difference of cubes, how should you start to more easily obtain the complete factored form of the polynomial? Base the answer on your results in Exercises 52–56 and the methods of factoring explained in this section.

In some cases, the method of factoring by grouping can be combined with the methods of special factoring discussed in this section. For example, to factor $8x^3 + 4x^2 + 27y^3 - 9y^2$, we proceed as follows.

$$8x^3 + 4x^2 + 27y^3 - 9y^2 = (8x^3 + 27y^3) + (4x^2 - 9y^2)$$

Associative and commutative properties

$$= (2x + 3y)(4x^2 - 6xy + 9y^2) + (2x + 3y)(2x - 3y)$$

Factor within groups.

$$= (2x + 3y)[(4x^2 - 6xy + 9y^2) + (2x - 3y)]$$

Factor out the greatest common factor, $2x + 3y$.

$$= (2x + 3y)(4x^2 - 6xy + 9y^2 + 2x - 3y)$$

Combine terms.

In problems such as this, how we choose to group in the first step is essential to factoring correctly. If we reach a "dead end," then we should group differently and try again.

Use the method just described to factor each polynomial.

58. $27x^3 + 9x^2 + y^3 - y^2$

59. $125p^3 + 25p^2 + 8q^3 - 4q^2$

60. $1000k^3 + 20k - m^3 - 2m$

61. $27a^3 + 15a - 64b^3 - 20b$

62. $y^4 + y^3 + y + 1$

63. $8t^4 - 24t^3 + t - 3$

64. $10x^2 + 5x^3 - 10y^2 + 5y^3$

65. $64m^2 - 512m^3 - 81n^2 + 729n^3$

6.4   A General Approach to Factoring

OBJECTIVES

1 Factor out any common factor.

2 Factor binomials.

3 Factor trinomials.

4 Factor polynomials of more than three terms.

In this section, we summarize and apply the factoring methods presented in the preceding sections. A polynomial is completely factored when the polynomial is in the following form.

1. The polynomial is written as a product of prime polynomials with integer coefficients.

2. None of the polynomial factors can be factored further, except that a monomial factor need not be factored completely.

Factoring a Polynomial

Step 1   **Factor out any common factor.**

Step 2   **If the polynomial is a binomial,** check to see if it is the difference of squares, the difference of cubes, or the sum of cubes.

If the polynomial is a trinomial, check to see if it is a perfect square trinomial. If it is not, factor as in Section 6.2.

If the polynomial has more than three terms, try to factor by grouping.

Step 3   **Check** the factored form by multiplying.

OBJECTIVE 1   Factor out any common factor.   This step is always the same, regardless of the number of terms in the polynomial.

EXAMPLE 1 Factoring Out a Common Factor

Factor each polynomial.

(a) $9p + 45 = 9(p + 5)$ **(b)** $8m^2p^2 + 4mp = 4mp(2mp + 1)$

(c) $5x(a + b) - y(a + b) = (a + b)(5x - y)$

Now Try Exercises 13 and 23.

OBJECTIVE 2 Factor binomials. Use one of the following rules.

For a **binomial** (two terms), check for the following.

Difference of squares	$x^2 - y^2 = (x + y)(x - y)$
Difference of cubes	$x^3 - y^3 = (x - y)(x^2 + xy + y^2)$
Sum of cubes	$x^3 + y^3 = (x + y)(x^2 - xy + y^2)$

EXAMPLE 2 Factoring Binomials

Factor each polynomial, if possible.

(a) $64m^2 - 9n^2 = (8m)^2 - (3n)^2$ Difference of squares

$= (8m + 3n)(8m - 3n)$

(b) $8p^3 - 27 = (2p)^3 - 3^3$ Difference of cubes

$= (2p - 3)[(2p)^2 + (2p)(3) + 3^2]$

$= (2p - 3)(4p^2 + 6p + 9)$

(c) $100m^3 + 1 = (10m)^3 + 1^3$ Sum of cubes

$= (10m + 1)[(10m)^2 - (10m)(1) + 1^2]$

$= (10m + 1)(100m^2 - 10m + 1)$

(d) $25m^2 + 121$ is prime. It is the sum of squares.

Now Try Exercises 7, 11, and 29.

OBJECTIVE 3 Factor trinomials.

For a **trinomial** (three terms), decide if it is a perfect square trinomial of the form

$$x^2 + 2xy + y^2 = (x + y)^2, \quad \text{or} \quad x^2 - 2xy + y^2 = (x - y)^2,$$

or, if not, use the methods of Section 6.2.

EXAMPLE 3 Factoring Trinomials

Factor each trinomial.

(a) $p^2 + 10p + 25 = (p + 5)^2$ Perfect square trinomial

(b) $49z^2 - 42z + 9 = (7z - 3)^2$ Perfect square trinomial

(c) $y^2 - 5y - 6 = (y - 6)(y + 1)$

The numbers -6 and 1 have a product of -6 and a sum of -5.

(d) $r^2 + 18r + 72 = (r + 6)(r + 12)$

(e) $2k^2 - k - 6 = (2k + 3)(k - 2)$

Use either method from Section 6.2.

(f) $28z^2 + 6z - 10 = 2(14z^2 + 3z - 5)$ Factor out the common factor.

$\qquad\qquad\quad = 2(7z + 5)(2z - 1)$

Now Try Exercises 9, 19, and 41.

OBJECTIVE 4 Factor polynomials of more than three terms.   Try factoring by grouping.

EXAMPLE 4 Factoring Polynomials with More Than Three Terms

Factor each polynomial.

(a) $xy^2 - y^3 + x^3 - x^2y = (xy^2 - y^3) + (x^3 - x^2y)$

$\qquad\qquad\qquad\qquad = y^2(x - y) + x^2(x - y)$

$\qquad\qquad\qquad\qquad = (x - y)(y^2 + x^2)$

(b) $20k^3 + 4k^2 - 45k - 9 = (20k^3 + 4k^2) - (45k + 9)$ Be careful with signs.

$\qquad\qquad\qquad\qquad\quad = 4k^2(5k + 1) - 9(5k + 1)$

$\qquad\qquad\qquad\qquad\quad = (5k + 1)(4k^2 - 9)$ $5k + 1$ is a common factor.

$\qquad\qquad\qquad\qquad\quad = (5k + 1)(2k + 3)(2k - 3)$ Difference of squares

(c) $4a^2 + 4a + 1 - b^2 = (4a^2 + 4a + 1) - b^2$ Associative property

$\qquad\qquad\qquad\qquad = (2a + 1)^2 - b^2$ Perfect square trinomial

$\qquad\qquad\qquad\qquad = (2a + 1 + b)(2a + 1 - b)$ Difference of squares

(d) $8m^3 + 4m^2 - n^3 - n^2$

First, notice that the terms must be rearranged before grouping because

$$(8m^3 + 4m^2) - (n^3 + n^2) = 4m^2(2m + 1) - n^2(n + 1),$$

which cannot be factored further. Write the polynomial as follows.

$8m^3 + 4m^2 - n^3 - n^2 = (8m^3 - n^3) + (4m^2 - n^2)$ Group the cubes and squares.

$\qquad\qquad\qquad\qquad = (2m - n)(4m^2 + 2mn + n^2) + (2m - n)(2m + n)$

Factor each group.

$\qquad\qquad\qquad\qquad = (2m - n)(4m^2 + 2mn + n^2 + 2m + n)$

Factor out the common factor $2m - n$.

Now Try Exercises 21 and 45.

6.4 EXERCISES

Factor each polynomial. See Examples 1–4.

1. $100a^2 - 9b^2$

2. $10r^2 + 13r - 3$

3. $3p^4 - 3p^3 - 90p^2$

4. $k^4 - 16$

5. $3a^2pq + 3abpq - 90b^2pq$

6. $49z^2 - 16$

7. $225p^2 + 256$

8. $18m^3n + 3m^2n^2 - 6mn^3$

9. $6b^2 - 17b - 3$

10. $k^2 - 6k - 16$

11. $x^3 - 1000$

12. $6t^2 + 19tu - 77u^2$

13. $4(p + 2) + m(p + 2)$

14. $40p - 32r$

15. $9m^2 - 45m + 18m^3$

16. $4k^2 + 28kr + 49r^2$

17. $54m^3 - 2000$

18. $mn - 2n + 5m - 10$

19. $9m^2 - 30mn + 25n^2$

20. $2a^2 - 7a - 4$

21. $kq - 9q + kr - 9r$

22. $56k^3 - 875$

23. $16z^3x^2 - 32z^2x$

24. $9r^2 + 100$

25. $x^2 + 2x - 35$

26. $9 - a^2 + 2ab - b^2$

27. $x^4 - 625$ ·

28. $2m^2 - mn - 15n^2$

29. $p^3 + 64$

30. $48y^2z^3 - 28y^3z^4$

31. $64m^2 - 625$

32. $14z^2 - 3zk - 2k^2$

33. $12z^3 - 6z^2 + 18z$

34. $225k^2 - 36r^2$

35. $256b^2 - 400c^2$

36. $z^2 - zp - 20p^2$

37. $1000z^3 + 512$

38. $64m^2 - 25n^2$

39. $10r^2 + 23rs - 5s^2$

40. $12k^2 - 17kq - 5q^2$

41. $24p^3q + 52p^2q^2 + 20pq^3$

42. $32x^2 + 16x^3 - 24x^5$

43. $48k^4 - 243$

44. $14x^2 - 25xq - 25q^2$

45. $m^3 + m^2 - n^3 - n^2$

46. $64x^3 + y^3 - 16x^2 + y^2$

47. $x^2 - 4m^2 - 4mn - n^2$

48. $4r^2 - s^2 - 2st - t^2$

49. $18p^5 - 24p^3 + 12p^6$

50. $k^2 - 6k + 16$

51. $2x^2 - 2x - 40$

52. $27x^3 - 3y^3$

53. $(2m + n)^2 - (2m - n)^2$

54. $(3k + 5)^2 - 4(3k + 5) + 4$

55. $50p^2 - 162$

56. $y^2 + 3y - 10$

57. $12m^2rx + 4mnrx + 40n^2rx$

58. $18p^2 + 53pr - 35r^2$

59. $21a^2 - 5ab - 4b^2$

60. $x^2 - 2xy + y^2 - 4$

61. $x^2 - y^2 - 4$

62. $(5r + 2s)^2 - 6(5r + 2s) + 9$

63. $(p + 8q)^2 - 10(p + 8q) + 25$

64. $z^4 - 9z^2 + 20$

65. $21m^4 - 32m^2 - 5$

66. $(x - y)^3 - (27 - y)^3$

67. $(r + 2t)^3 + (r - 3t)^3$

68. $16x^3 + 32x^2 - 9x - 18$

69. $x^5 + 3x^4 - x - 3$

70. $x^{16} - 1$

71. $m^2 - 4m + 4 - n^2 + 6n - 9$

72. $x^2 + 4 + x^2y + 4y$

6.5 Solving Equations by Factoring

OBJECTIVES

1 Learn and use the zero-factor property.

2 Solve applied problems that require the zero-factor property.

In the previous four sections, we factored polynomial *expressions.* Now we can use factoring to solve polynomial *equations.* In Chapter 2, we developed methods for solving linear, or first-degree, equations. Solving higher-degree polynomial equations requires other methods.

OBJECTIVE 1 Learn and use the zero-factor property. Some polynomial equations can be solved by factoring. Solving equations by factoring depends on a special property of the number 0, called the **zero-factor property.**

> **Zero-Factor Property**
>
> If two numbers have a product of 0, then at least one of the numbers must be 0. That is, if $ab = 0$, then either $a = 0$ or $b = 0$.

To prove the zero-factor property, we first assume $a \neq 0$. (If a does equal 0, then the property is proved already.) If $a \neq 0$, then $\frac{1}{a}$ exists, and both sides of $ab = 0$ can be multiplied by $\frac{1}{a}$ to get

$$\frac{1}{a} \cdot ab = \frac{1}{a} \cdot 0$$
$$b = 0.$$

Thus, if $a \neq 0$, then $b = 0$, and the property is proved.

CAUTION If $ab = 0$, then $a = 0$ or $b = 0$. However, if $ab = 6$, for example, it is not necessarily true that $a = 6$ or $b = 6$; in fact, it is very likely that *neither $a = 6$ nor $b = 6$. The zero-factor property works only for a product equal to 0.*

EXAMPLE 1 Using the Zero-Factor Property to Solve an Equation

Solve $(x + 6)(2x - 3) = 0$.

Here the product of $x + 6$ and $2x - 3$ is 0. By the zero-factor property, this can be true only if

$$x + 6 = 0 \qquad \text{or} \qquad 2x - 3 = 0.$$

Solve these two equations.

$$x + 6 = 0 \qquad \text{or} \quad 2x - 3 = 0$$
$$x = -6 \quad \text{or} \qquad 2x = 3$$
$$x = \frac{3}{2}$$

Check these two solutions by substitution in the original equation.

If $x = -6$, then

$$(x + 6)(2x - 3) = 0$$
$$(-6 + 6)[2(-6) - 3] = 0 \quad ?$$
$$0(-15) = 0 \quad ?$$
$$0 = 0. \quad \text{True}$$

If $x = \frac{3}{2}$, then

$$(x + 6)(2x - 3) = 0$$
$$\left(\frac{3}{2} + 6\right)\left(2 \cdot \frac{3}{2} - 3\right) = 0 \quad ?$$
$$\frac{15}{2}(0) = 0 \quad ?$$
$$0 = 0. \quad \text{True}$$

Both solutions check; the solution set is $\left\{-6, \frac{3}{2}\right\}$.

Now Try Exercise 5.

Since the product $(x + 6)(2x - 3)$ equals $2x^2 + 9x - 18$, the equation of Example 1 has a squared term and is an example of a *quadratic equation*. A quadratic equation has degree 2.

Quadratic Equation

An equation that can be written in the form

$$ax^2 + bx + c = 0,$$

where a, b, and c are real numbers, with $a \neq 0$, is a **quadratic equation.** This form is called **standard form.**

Quadratic equations are discussed in more detail in Chapter 9.

The steps involved in solving a quadratic equation by factoring are summarized here.

Solving a Quadratic Equation by Factoring

Step 1 **Write in standard form.** Rewrite the equation if necessary so that one side is 0.

Step 2 **Factor** the polynomial.

Step 3 **Use the zero-factor property.** Set each variable factor equal to 0.

Step 4 **Find the solution(s).** Solve each equation formed in Step 3.

Step 5 **Check** each solution in the *original* equation.

EXAMPLE 2 Solving Quadratic Equations by Factoring

Solve each equation.

(a) $2x^2 + 3x = 2$

Step 1
$$2x^2 + 3x = 2$$
$$2x^2 + 3x - 2 = 0 \qquad \text{Standard form}$$

Step 2
$$(2x - 1)(x + 2) = 0 \qquad \text{Factor.}$$

Step 3 $2x - 1 = 0$ or $x + 2 = 0$ Zero-factor property

Step 4 $2x = 1$ or $x = -2$ Solve each equation.

$$x = \frac{1}{2}$$

Step 5   Check each solution in the original equation.

If $x = \frac{1}{2}$, then

$$2x^2 + 3x = 2$$

$$2\left(\frac{1}{2}\right)^2 + 3\left(\frac{1}{2}\right) = 2 \quad ?$$

$$2\left(\frac{1}{4}\right) + \frac{3}{2} = 2 \quad ?$$

$$\frac{1}{2} + \frac{3}{2} = 2 \quad ?$$

$$2 = 2. \quad \text{True}$$

If $x = -2$, then

$$2x^2 + 3x = 2$$

$$2(-2)^2 + 3(-2) = 2 \quad ?$$

$$2(4) - 6 = 2 \quad ?$$

$$8 - 6 = 2 \quad ?$$

$$2 = 2. \quad \text{True}$$

Because both solutions check, the solution set is $\left\{\frac{1}{2}, -2\right\}$.

(b) $4x^2 = 4x - 1$

$$4x^2 - 4x + 1 = 0 \qquad \text{Standard form}$$

$$(2x - 1)^2 = 0 \qquad \text{Factor.}$$

$$2x - 1 = 0 \qquad \text{Zero-factor property}$$

$$2x = 1$$

$$x = \frac{1}{2}$$

There is only one solution because the trinomial is a perfect square. The solution set is $\left\{\frac{1}{2}\right\}$.

Now Try Exercises 11 and 29.

■ **EXAMPLE 3**   **Solving a Quadratic Equation with a Missing Term**

Solve $5z^2 - 25z = 0$.

This quadratic equation has a missing term. Comparing it with the standard form $ax^2 + bx + c = 0$ shows that $c = 0$. The zero-factor property can still be used.

$$5z^2 - 25z = 0$$

$$5z(z - 5) = 0 \qquad \text{Factor.}$$

$$5z = 0 \quad \text{or} \quad z - 5 = 0 \qquad \text{Zero-factor property}$$

$$z = 0 \quad \text{or} \qquad z = 5$$

The solutions are 0 and 5, as can be verified by substituting in the original equation. The solution set is $\{0, 5\}$.

Now Try Exercise 19.

CAUTION   Remember to include the solution 0 when writing the solution set of the equation in Example 3.

EXAMPLE 4 Solving an Equation That Requires Rewriting

Solve $(2q + 1)(q + 1) = 2(1 - q) + 6$.

Write the equation in standard form $ax^2 + bx + c = 0$ by first multiplying on each side.

$$(2q + 1)(q + 1) = 2(1 - q) + 6$$

$$2q^2 + 3q + 1 = 2 - 2q + 6$$

$$2q^2 + 5q - 7 = 0 \qquad \text{Standard form}$$

$$(2q + 7)(q - 1) = 0 \qquad \text{Factor.}$$

$$2q + 7 = 0 \quad \text{or} \quad q - 1 = 0 \qquad \text{Zero-factor property}$$

$$2q = -7 \quad \text{or} \qquad q = 1$$

$$q = -\frac{7}{2}$$

Check that the solution set is $\left\{-\frac{7}{2}, 1\right\}$.

Now Try Exercise 35.

The zero-factor property can be extended to solve certain polynomial equations of degree 3 or higher, as shown in the next example.

EXAMPLE 5 Solving an Equation of Degree 3

Solve $-x^3 + x^2 = -6x$.

Start by adding $6x$ to each side to get 0 on the right side.

$$-x^3 + x^2 + 6x = 0$$

To make the factoring step easier, multiply each side by -1.

$$x^3 - x^2 - 6x = 0$$

$$x(x^2 - x - 6) = 0 \qquad \text{Factor out } x.$$

$$x(x - 3)(x + 2) = 0 \qquad \text{Factor the trinomial.}$$

Use the zero-factor property, extended to include the three variable factors.

$$x = 0 \quad \text{or} \quad x - 3 = 0 \quad \text{or} \quad x + 2 = 0$$

$$x = 3 \quad \text{or} \qquad x = -2$$

Check that the solution set is $\{0, 3, -2\}$.

Now Try Exercise 37.

OBJECTIVE 2 Solve applied problems that require the zero-factor property. The next example shows an application that leads to a quadratic equation. We continue to use the six-step problem-solving method introduced in Chapter 2.

EXAMPLE 6 Using a Quadratic Equation in an Application

Some surveyors are surveying a lot that is in the shape of a parallelogram. They find that the longer sides of the parallelogram are each 8 m longer than the distance

between them. The area of the lot is 48 m². Find the length of the longer sides and the distance between them.

Step 1 **Read** the problem again. There will be two answers.

Step 2 **Assign a variable.** Let x represent the distance between the longer sides. Then $x + 8$ is the length of each longer side. See Figure 1.

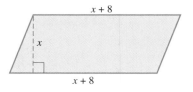

FIGURE 1

Step 3 **Write an equation.** The area of a parallelogram is given by $A = bh$, where b is the length of the longer side and h is the distance between the longer sides. Here $b = x + 8$ and $h = x$.

$$A = bh$$
$$48 = (x + 8)x \qquad \text{Let } A = 48, b = x + 8, h = x.$$

Step 4 **Solve.**

$$48 = x^2 + 8x \qquad\qquad \text{Distributive property}$$
$$0 = x^2 + 8x - 48 \qquad\qquad \text{Standard form}$$
$$0 = (x + 12)(x - 4) \qquad\qquad \text{Factor.}$$
$$x + 12 = 0 \quad \text{or} \quad x - 4 = 0 \qquad \text{Zero-factor property}$$
$$x = -12 \quad \text{or} \qquad x = 4$$

Step 5 **State the answer.** A distance cannot be negative, so reject -12 as a solution. The only possible solution is 4, so the distance between the longer sides is 4 m. The length of the longer sides is $4 + 8 = 12$ m.

Step 6 **Check.** The length of the longer sides is 8 m more than the distance between them, and the area is $4 \cdot 12 = 48$ m² as required, so the answer checks.

Now Try Exercise 57.

CAUTION When applications lead to quadratic equations, a solution of the equation may not satisfy the physical requirements of the problem, as in Example 6. Reject such solutions.

A function defined by a quadratic polynomial is called a *quadratic function*. In Chapter 10 we investigate quadratic functions in detail. The next example uses such a function.

EXAMPLE 7 Using a Quadratic Function in an Application

Quadratic functions are used to describe the height a falling object or a propelled object reaches in a specific time. For example, if a toy rocket is launched vertically

upward from ground level with an initial velocity of 128 ft per sec, then its height in feet after t sec is a function defined by

$$h(t) = -16t^2 + 128t,$$

if air resistance is neglected. After how many seconds will the rocket be 220 ft above the ground?

We must let $h(t) = 220$ and solve for t.

$$220 = -16t^2 + 128t \qquad \text{Let } h(t) = 220.$$
$$16t^2 - 128t + 220 = 0 \qquad \text{Standard form}$$
$$4t^2 - 32t + 55 = 0 \qquad \text{Divide by 4.}$$
$$(2t - 11)(2t - 5) = 0 \qquad \text{Factor.}$$
$$2t - 11 = 0 \quad \text{or} \quad 2t - 5 = 0 \qquad \text{Zero-factor property}$$
$$t = 5.5 \quad \text{or} \qquad t = 2.5$$

The rocket will reach a height of 220 ft twice: on its way up at 2.5 sec and again on its way down at 5.5 sec.

Now Try Exercise 65.

CONNECTIONS

In Section 5.3 we saw that the graph of $f(x) = x^2$ is a parabola. In general, the graph of $f(x) = ax^2 + bx + c$, $a \neq 0$, is a parabola, and the x-intercepts of its graph give the real number solutions of the equation $ax^2 + bx + c = 0$. In the screens, we show how a graphing calculator can locate these x-intercepts (called *zeros* of the function) for $Y_1 = f(X) = 2X^2 + 3X - 2$. Notice that this quadratic expression was found on the left side of the equation in Example 2(a) earlier in this section, where the equation was written in standard form.

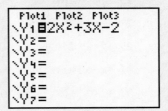

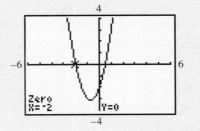
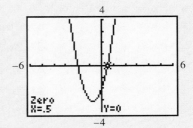

The x-intercepts (zeros) given with the graphs are the same as the solutions found in Example 2(a). This method of graphical solution can be used for any type of equation.

6.5 EXERCISES

 1. Explain in your own words how the zero-factor property is used in solving a quadratic equation.

 2. One of the following equations is *not* in proper form for using the zero-factor property. Which one is it? Explain why it is not in proper form.

A. $(x + 2)(x - 6) = 0$ **B.** $x(3x - 7) = 0$
C. $3t(t + 8)(t - 9) = 0$ **D.** $y(y - 3) + 6(y - 3) = 0$

Solve each equation using the zero-factor property. See Example 1.

3. $(x - 5)(x + 10) = 0$ **4.** $(x + 3)(x + 7) = 0$

5. $(2k - 5)(3k + 8) = 0$ **6.** $(3q - 4)(2q + 5) = 0$

Solve each equation. See Examples 2–4.

7. $m^2 - 3m - 10 = 0$ **8.** $x^2 + x - 12 = 0$ **9.** $z^2 + 9z + 18 = 0$

10. $x^2 - 18x + 80 = 0$ **11.** $2x^2 = 7x + 4$ **12.** $2x^2 = 3 - x$

13. $15k^2 - 7k = 4$ **14.** $3c^2 + 3 = -10c$

15. $2x^2 - 12 - 4x = x^2 - 3x$ **16.** $3p^2 + 9p + 30 = 2p^2 - 2p$

17. $(5z + 1)(z + 3) = -2(5z + 1)$ **18.** $(3x + 1)(x - 3) = 2 + 3(x + 5)$

19. $4p^2 + 16p = 0$ **20.** $2a^2 - 8a = 0$

21. $6m^2 - 36m = 0$ **22.** $-3m^2 + 27m = 0$

23. $-3m^2 + 27 = 0$ **24.** $-2a^2 + 8 = 0$

25. $4p^2 - 16 = 0$ **26.** $9x^2 - 81 = 0$

27. $-x^2 = 9 - 6x$ **28.** $-m^2 - 8m = 16$

29. $9k^2 + 24k + 16 = 0$ **30.** $4m^2 - 20m + 25 = 0$

31. $(x - 3)(x + 5) = -7$ **32.** $(x + 8)(x - 2) = -21$

33. $(2x + 1)(x - 3) = 6x + 3$ **34.** $(3x + 2)(x - 3) = 7x - 1$

35. $(x + 3)(x - 6) = (2x + 2)(x - 6)$ **36.** $(2x + 1)(x + 5) = (x + 11)(x + 3)$

Solve each equation. See Example 5.

37. $2x^3 - 9x^2 - 5x = 0$ **38.** $6x^3 - 13x^2 - 5x = 0$

39. $9t^3 = 16t$ **40.** $25x^3 = 64x$

41. $2r^3 + 5r^2 - 2r - 5 = 0$ **42.** $2p^3 + p^2 - 98p - 49 = 0$

43. $-x^3 + 6x^2 + 9x - 54 = 0$ **44.** $6t^3 + 5t^2 - 6t - 5 = 0$

45. $x^3 - 3x^2 - 4x + 12 = 0$

46. A student tried to solve the equation in Exercise 39 by first dividing each side by t, obtaining $9t^2 = 16$. She then solved the resulting equation by the zero-factor property to get the solution set $\left\{-\frac{4}{3}, \frac{4}{3}\right\}$. What was incorrect about her procedure?

47. Without actually solving each equation, determine which one of the following has 0 in its solution set.

A. $4x^2 - 25 = 0$ **B.** $x^2 + 2x - 3 = 0$ **C.** $6x^2 + 9x + 1 = 0$ **D.** $x^3 + 4x^2 = 3x$

Solve each equation.

48. $2(x - 1)^2 - 7(x - 1) - 15 = 0$

49. $4(2k + 3)^2 - (2k + 3) - 3 = 0$

50. $5(3a - 1)^2 + 3 = -16(3a - 1)$

51. $2(m + 3)^2 = 5(m + 3) - 2$

52. $(2k - 3)^2 = 16k^2$

53. $9p^2 = (5p + 2)^2$

Solve each problem. See Examples 6 and 7.

54. A garden has an area of 320 ft². Its length is 4 ft more than its width. What are the dimensions of the garden?

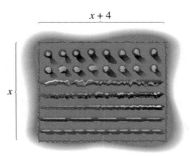

$x + 4$

x

55. A square mirror has sides measuring 2 ft less than the sides of a square painting. If the difference between their areas is 32 ft², find the lengths of the sides of the mirror and the painting.

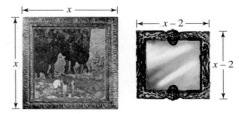

x

x

$x - 2$

$x - 2$

56. A sign has the shape of a triangle. The length of the base is 3 m less than the height. What are the measures of the base and the height, if the area is 44 m²?

h

Yard Sale Today

$h - 3$

57. The base of a parallelogram is 7 ft more than the height. If the area of the parallelogram is 60 ft², what are the measures of the base and the height?

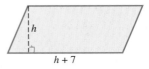

h

$h + 7$

58. A farmer has 300 ft of fencing and wants to enclose a rectangular area of 5000 ft². What dimensions should she use?

59. A rectangular landfill has an area of 30,000 ft². Its length is 200 ft more than its width. What are the dimensions of the landfill?

60. Find two consecutive integers such that the sum of their squares is 61.

61. Find two consecutive integers such that their product is 72.

62. A box with no top is to be constructed from a piece of cardboard whose length measures 6 in. more than its width. The box is to be formed by cutting squares that measure 2 in. on each side from the four corners, and then folding up the sides. If the volume of the box will be 110 in.^{3}, what are the dimensions of the piece of cardboard?

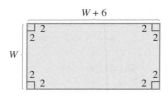

63. The surface area of the box with open top shown in the figure is 161 in.^{2}. Find the dimensions of the base. (*Hint:* The surface area of the box is a function defined by $S(x) = x^2 + 16x$.)

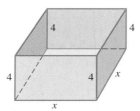

64. Refer to Example 7. After how many seconds will the rocket be 240 ft above the ground? 112 ft above the ground?

65. If an object is propelled upward with an initial velocity of 64 ft per sec from a height of 80 ft, then its height in feet t sec after it is propelled is a function defined by

$$f(t) = -16t^2 + 64t + 80.$$

How long after it is propelled will it hit the ground? (*Hint:* When it hits the ground, its height is 0 ft.)

66. If a baseball is dropped from a helicopter 625 ft above the ground, then its distance in feet from the ground t sec later is a function defined by

$$f(t) = -16t^2 + 625.$$

How long after it is dropped will it hit the ground?

67. If a rock is dropped from a building 576 ft high, then its distance in feet from the ground t sec later is a function defined by

$$f(t) = -16t^2 + 576.$$

How long after it is dropped will it hit the ground?

TECHNOLOGY INSIGHTS (EXERCISES 68–71)

As shown in the Connections box following Example 7, the solutions of the quadratic equation $ax^2 + bx + c = 0$ $(a \neq 0)$ are represented on the graph of the quadratic function $f(x) = ax^2 + bx + c$ by the x-intercepts.

Solve each equation using the zero-factor property, and confirm that your solutions correspond to the x-intercepts (zeros) shown on the accompanying graphing calculator screens.

68. $2x^2 - 7x - 4 = 0$

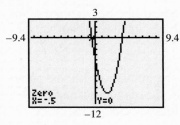 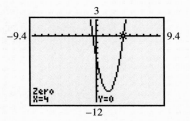

69. $2x^2 + 7x - 15 = 0$

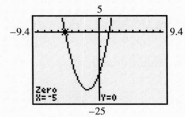 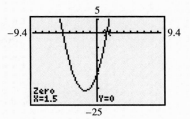

70. $-x^2 + 3x = -10$

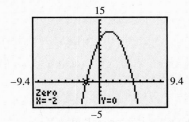 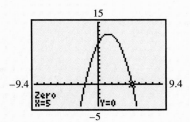

71. $-x^2 + x = -12$

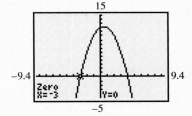 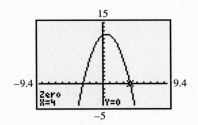

Chapter **Group Activity**

Finding the Height of a Propelled Object

OBJECTIVE Use a quadratic function to model the height of a propelled object.

An important application of quadratic functions deals with the height of a propelled object as a function of time elapsed after it is propelled. We saw instances of this in Example 7 of Section 6.5 and the corresponding exercises. Now we consider the general case. If air resistance is neglected, the height $f(x)$ (in feet) of an object propelled directly upward from an initial height s_0 ft with initial velocity (speed) v_0 ft per sec is

$$f(x) = -16x^2 + v_0 x + s_0,$$

where x is the number of seconds after the object is propelled. The coefficient of x^2, -16, is a constant based on the gravitational force of Earth. This constant varies on other surfaces, such as the moon or other planets.

A. Create a quadratic model for the following situation.

> A ball is thrown directly upward from an initial height of 100 ft with an initial velocity of 80 ft per sec.

1. Define a function that describes the height of the ball in terms of time x.

2. Use a graphing calculator or calculate several ordered pairs and plot them to graph your function. Use domain $[0, 10]$ and range $[-100, 300]$.*

B. Use your model to get information about the flight of the ball.

1. Use your graph to estimate the maximum height of the ball and when it reaches that height.

2. After how many seconds will the ball reach the ground (height 0 ft)? Estimate this answer from the graph and check it in the equation.

3. Use the graph to estimate the time interval when the height of the ball is greater than 150 ft. Check your estimate by substituting it into your function.

C. Answer the following questions.

1. Why must x be a nonnegative number?

2. What does it mean if $s_0 = 0$?

---

*It is easy to misinterpret the graph of this function. It does not define the *path* followed by the ball; it defines height as a function of time.

CHAPTER **6** SUMMARY

■ KEY TERMS

6.1 factoring
greatest common
factor (GCF)

6.2 prime polynomial
6.3 difference of squares
perfect square
trinomial

difference of cubes
sum of cubes
6.5 quadratic equation

standard form of a
quadratic equation

■ TEST YOUR WORD POWER

See how well you have learned the vocabulary in this chapter. Answers, with examples, follow the Quick Review.

1. Factoring is
   A. a method of multiplying
   polynomials
   B. the process of writing a
   polynomial as a product
   C. the answer in a multiplication
   problem
   D. a way to add the terms of a
   polynomial.

2. Factoring by grouping is used to
   factor
   A. out the GCF
   B. a sum or difference of squares
   C. by the substitution method
   D. a polynomial that has more than
   three terms.

3. A **quadratic equation** is a polyno-
   mial equation of
   A. degree 1
   B. degree 2

   C. degree 3
   D. degree 4.

4. The **zero-factor property** is used to
   A. factor a perfect square trinomial
   B. factor by grouping
   C. solve a polynomial equation of
   degree 2 or more
   D. solve a linear equation.

■ QUICK REVIEW

CONCEPTS	EXAMPLES

6.1 GREATEST COMMON FACTORS; FACTORING BY GROUPING

The Greatest Common Factor

The product of the largest common numerical factor and each common variable raised to the least exponent that appears on that variable in any term is the greatest common factor of the terms of the polynomial.

Factor $4x^2y - 50xy^2$.
The greatest common factor is $2xy$.
$$4x^2y - 50xy^2 = 2xy(2x - 25y)$$

Factoring by Grouping

Step 1 Group the terms so that each group has a common factor.

Step 2 Factor out the common factor in each group.

Step 3 If the groups now have a common factor, factor it out. If not, try a different grouping.

Factor by grouping.
$$5a - 5b - ax + bx = (5a - 5b) + (-ax + bx)$$
$$= 5(a - b) - x(a - b)$$
$$= (a - b)(5 - x)$$

(continued)

CONCEPTS	EXAMPLES

6.2 FACTORING TRINOMIALS

To factor a trinomial, choose factors of the first term and factors of the last term. Then, place them in a pair of parentheses of this form:

$$(\qquad)(\qquad).$$

Try various combinations of the factors until the correct middle term of the trinomial is found.

Factor $15x^2 + 14x - 8$.

The factors of 15 are 5 and 3, and 15 and 1.

The factors of -8 are -4 and 2, 4 and -2, -1 and 8, and 1 and -8.

Various combinations of these factors lead to the correct factorization

$$15x^2 + 14x - 8 = (5x - 2)(3x + 4).$$

Check by multiplying, using the FOIL method.

6.3 SPECIAL FACTORING

Difference of Squares

$$x^2 - y^2 = (x + y)(x - y)$$

Perfect Square Trinomials

$$x^2 + 2xy + y^2 = (x + y)^2$$
$$x^2 - 2xy + y^2 = (x - y)^2$$

Difference of Cubes

$$x^3 - y^3 = (x - y)(x^2 + xy + y^2)$$

Sum of Cubes

$$x^3 + y^3 = (x + y)(x^2 - xy + y^2)$$

Factor.

$$\begin{aligned}
4m^2 - 25n^2 &= (2m)^2 - (5n)^2 \\
&= (2m + 5n)(2m - 5n)
\end{aligned}$$

$$9y^2 + 6y + 1 = (3y + 1)^2$$
$$16p^2 - 56p + 49 = (4p - 7)^2$$

$$8 - 27a^3 = (2 - 3a)(4 + 6a + 9a^2)$$

$$64z^3 + 1 = (4z + 1)(16z^2 - 4z + 1)$$

6.4 A GENERAL APPROACH TO FACTORING

Step 1   Factor out any common factors.

Step 2   For a binomial, check for the difference of squares, the difference of cubes, or the sum of cubes.

For a trinomial, see if it is a perfect square. If not, factor as in Section 6.2.

For more than three terms, try factoring by grouping.

Step 3   Check the factored form by multiplying.

Factor.

$$\begin{aligned}
ak^3 + 2ak^2 - 9ak - 18a &= a(k^3 + 2k^2 - 9k - 18) \\
&= a[(k^3 + 2k^2) - (9k + 18)] \\
&= a[k^2(k + 2) - 9(k + 2)] \\
&= a[(k + 2)(k^2 - 9)] \\
&= a(k + 2)(k - 3)(k + 3)
\end{aligned}$$

CONCEPTS	EXAMPLES

6.5 SOLVING EQUATIONS BY FACTORING

Step 1 Rewrite the equation if necessary so that one side is 0.

Step 2 Factor the polynomial.

Step 3 Set each factor equal to 0.

Step 4 Solve each equation from Step 3.

Step 5 Check each solution.

Solve.

$$2x^2 + 5x = 3$$
$$2x^2 + 5x - 3 = 0 \qquad \text{Standard form}$$
$$(2x - 1)(x + 3) = 0$$

$$2x - 1 = 0 \quad \text{or} \quad x + 3 = 0$$
$$2x = 1 \qquad\qquad x = -3$$
$$x = \frac{1}{2}$$

A check verifies that the solution set is $\left\{-3, \frac{1}{2}\right\}$.

Answers to Test Your Word Power

1. B; *Example:* $x^2 - 5x - 14 = (x - 7)(x + 2)$ **2.** D; *Example:* $x^2 + 5x + xy + 5y = x(x + 5) + y(x + 5) = (x + 5)(x + y)$

3. B; *Examples:* $y^2 - 3y + 2 = 0, x^2 - 9 = 0, 2m^2 = 6m + 8$ **4.** C; *Example:* Use the zero-factor property to write $(x + 4)(x - 2) = 0$ as $x + 4 = 0$ or $x - 2 = 0$, then solve each linear equation to find the solution set $\{-4, 2\}$.

CHAPTER 6 REVIEW EXERCISES

[6.1]   *Factor out the greatest common factor.*

1. $12p^2 - 6p$

2. $21x^2 + 35x$

3. $12q^2b + 8qb^2 - 20q^3b^2$

4. $6r^3t - 30r^2t^2 + 18rt^3$

5. $(x + 3)(4x - 1) - (x + 3)(3x + 2)$

6. $(z + 1)(z - 4) + (z + 1)(2z + 3)$

Factor by grouping.

7. $4m + nq + mn + 4q$

8. $x^2 + 5y + 5x + xy$

9. $2m + 6 - am - 3a$

10. $x^2 + 3x - 3y - xy$

[6.2]   *Factor completely.*

11. $3p^2 - p - 4$

12. $6k^2 + 11k - 10$

13. $12r^2 - 5r - 3$

14. $10m^2 + 37m + 30$

15. $10k^2 - 11kh + 3h^2$

16. $9x^2 + 4xy - 2y^2$

17. $24x - 2x^2 - 2x^3$

18. $6b^3 - 9b^2 - 15b$

19. $y^4 + 2y^2 - 8$

20. $2k^4 - 5k^2 - 3$

21. $p^2(p + 2)^2 + p(p + 2)^2 - 6(p + 2)^2$

22. $3(r + 5)^2 - 11(r + 5) - 4$

23. When asked to factor $x^2y^2 - 6x^2 + 5y^2 - 30$, a student gave the following incorrect answer: $x^2(y^2 - 6) + 5(y^2 - 6)$. Why is this answer incorrect? What is the correct answer?

24. If the area of this rectangle is represented by $4p^2 + 3p - 1$, what is the width in terms of p?

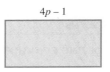

$4p - 1$

[6.3] *Factor completely.*

25. $16x^2 - 25$

26. $9t^2 - 49$

27. $36m^2 - 25n^2$

28. $x^2 + 14x + 49$

29. $9k^2 - 12k + 4$

30. $r^3 + 27$

31. $125x^3 - 1$

32. $m^6 - 1$

33. $x^8 - 1$

34. $x^2 + 6x + 9 - 25y^2$

35. $(a + b)^3 - (a - b)^3$

36. $x^5 - x^3 - 8x^2 + 8$

[6.5] *Solve each equation.*

37. $x^2 - 8x + 16 = 0$

38. $(5x + 2)(x + 1) = 0$

39. $p^2 - 5p + 6 = 0$

40. $q^2 + 2q = 8$

41. $6z^2 = 5z + 50$

42. $6r^2 + 7r = 3$

43. $8k^2 + 14k + 3 = 0$

44. $-4m^2 + 36 = 0$

45. $6x^2 + 9x = 0$

46. $(2x + 1)(x - 2) = -3$

47. $(r + 2)(r - 2) = (r - 2)(r + 3) - 2$

48. $2x^3 - x^2 - 28x = 0$

49. $-t^3 - 3t^2 + 4t + 12 = 0$

50. $(r + 2)(5r^2 - 9r - 18) = 0$

Solve each problem.

51. A triangular wall brace has the shape of a right triangle. One of the perpendicular sides is 1 ft longer than twice the other. The area enclosed by the triangle is 10.5 ft². Find the shorter of the perpendicular sides.

x

$2x + 1$

The area is 10.5 ft².

52. A rectangular parking lot has a length 20 ft more than its width. Its area is 2400 ft². What are the dimensions of the lot?

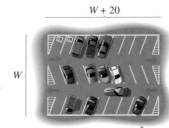

$W + 20$

W

The area is 2400 ft².

A rock is propelled directly upward from ground level. After t sec, its height is given by $f(t) = -16t^2 + 256t$ (if air resistance is neglected).

53. When will the rock return to the ground?

54. After how many seconds will it be 240 ft above the ground?

55. Why does the question in Exercise 54 have two answers?

MIXED REVIEW EXERCISES

Factor completely.

56. $30a + am - am^2$

57. $16 - 81k^2$

58. $8 - a^3$

59. $9x^2 + 13xy - 3y^2$ **60.** $15y^3 + 20y^2$ **61.** $25z^2 - 30zm + 9m^2$

Solve.

62. $5x^2 - 17x = 12$ **63.** $3m^2 - 9m = 0$ **64.** $x^3 - x = 0$

65. The length of a rectangular picture frame is 2 in. longer than its width. The area enclosed by the frame is 48 in.^{2}. What is the width?

66. When Europeans arrived in America, many Native Americans of the Northeast lived in *longhouses* that sheltered several related families. The rectangular floor area of a typical Huron longhouse was about 2750 ft^2. The length was 85 ft greater than the width. What were the dimensions of the floor?

CHAPTER 6 TEST

Factor.

1. $11z^2 - 44z$ **2.** $10x^2y^5 - 5x^2y^3 - 25x^5y^3$ **3.** $3x + by + bx + 3y$

4. $-2x^2 - x + 36$ **5.** $6x^2 + 11x - 35$ **6.** $4p^2 + 3pq - q^2$

7. $16a^2 + 40ab + 25b^2$ **8.** $x^2 + 2x + 1 - 4z^2$ **9.** $a^3 + 2a^2 - ab^2 - 2b^2$

10. $9k^2 - 121j^2$ **11.** $y^3 - 216$ **12.** $6k^4 - k^2 - 35$

13. $27x^6 + 1$

14. Explain why $(x^2 + 2y)p + 3(x^2 + 2y)$ is not in factored form. Then factor the polynomial.

15. Which one of the following is *not* a factored form of $-x^2 - x + 12$?

    A. $(3 - x)(x + 4)$ **B.** $-(x - 3)(x + 4)$

    C. $(-x + 3)(x + 4)$ **D.** $(x - 3)(-x + 4)$

Solve each equation.

16. $3x^2 + 8x = -4$ **17.** $3x^2 - 5x = 0$ **18.** $5m(m - 1) = 2(1 - m)$

Solve each problem.

19. The area of the rectangle shown is 40 in.^{2}. Find the length and the width of the rectangle.

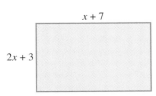

$x + 7$

$2x + 3$

The area is 40 in.^{2}.

20. A ball is propelled upward from ground level. After t sec, its height in feet is a function defined by $f(t) = -16t^2 + 96t$. After how many seconds will it reach a height of 128 ft?

CUMULATIVE REVIEW EXERCISES CHAPTERS **1–6**

Use the properties of real numbers to simplify.

1. $-2(m - 3)$

2. $-(-4m + 3)$

3. $3x^2 - 4x + 4 + 9x - x^2$

Evaluate if $p = -4$, $q = -2$, and $r = 5$.

4. $-3(2q - 3p)$

5. $8r^2 + q^2$

6. $\dfrac{\sqrt{r}}{-p + 2q}$

7. $\dfrac{5p + 6r^2}{p^2 + q - 1}$

Solve.

8. $2z - 5 + 3z = 4 - (z + 2)$

9. $\dfrac{3a - 1}{5} + \dfrac{a + 2}{2} = -\dfrac{3}{10}$

10. $-\dfrac{4}{3}d \geq -5$

11. $3 - 2(m + 3) < 4m$

12. $2k + 4 < 10$ and $3k - 1 > 5$

13. $2k + 4 > 10$ or $3k - 1 < 5$

14. $|5x + 3| - 10 = 3$

15. $|x + 2| < 9$

16. $|2x - 5| \geq 9$

17. $V = lwh$ for h

18. Two planes leave the Dallas-Fort Worth airport at the same time. One travels east at 550 mph, and the other travels west at 500 mph. Assuming no wind, how long will it take for the planes to be 2100 mi apart?

Plane	r	t	d
Eastbound	550	x	
Westbound	500	x	

← Total

19. Graph $4x + 2y = -8$.

20. Find the slope of the line passing through the points $(-4, 8)$ and $(-2, 6)$.

21. What is the slope of the line shown here?

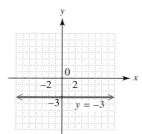

Use the function defined by $f(x) = 2x + 7$ to find each of the following.

22. $f(-4)$

23. The x-intercept of its graph

24. The y-intercept of its graph

Solve each system.

25. $3x - 2y = -7$
       $2x + 3y = 17$

26. $2x + 3y - 6z = 5$
       $8x - y + 3z = 7$
       $3x + 4y - 3z = 7$

Perform the indicated operations. In Exercises 27 and 28, assume that variables represent nonzero real numbers.

27. $(3x^2y^{-1})^{-2}(2x^{-3}y)^{-1}$

28. $\dfrac{5m^{-2}y^3}{3m^{-3}y^{-1}}$

29. $(3x^3 + 4x^2 - 7) - (2x^3 - 8x^2 + 3x)$

30. $(7x + 3y)^2$

31. $(2p + 3)(5p^2 - 4p - 8)$

Factor.

32. $16w^2 + 50wz - 21z^2$

33. $4x^2 - 4x + 1 - y^2$

34. $4y^2 - 36y + 81$

35. $100x^4 - 81$

36. $8p^3 + 27$

Solve.

37. $(p - 1)(2p + 3)(p + 4) = 0$

38. $9q^2 = 6q - 1$

39. A sign is to have the shape of a triangle with a height 3 ft greater than the length of the base. How long should the base be if the area is to be 14 ft^2?

40. A game board has the shape of a rectangle. The longer sides are each 2 in. longer than the distance between them. The area of the board is 288 in.^{2}. Find the length of the longer sides and the distance between them.

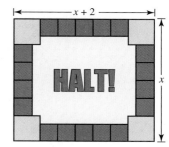

Rational Expressions and Functions

7

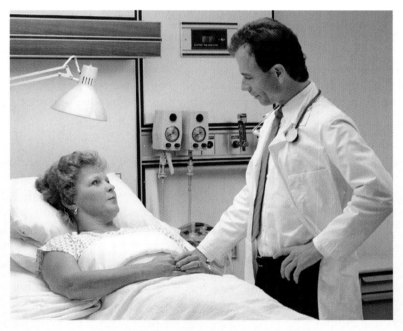

In 1999, 16 of every 100 Americans had no health insurance coverage. (If you happen to be one such person, you should be very concerned, considering the exorbitant cost of health care today.) The population at that time was about 274,000,000 Americans. (*Source:* U.S. Bureau of the Census.) In Example 4 of Section 7.5, we use proportions to determine the number of Americans without health insurance.

7.1 Rational Expressions and Functions; Multiplying and Dividing

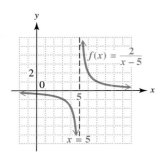

FIGURE 1

OBJECTIVE 1 Define rational expressions. In arithmetic, a rational number is the quotient of two integers, with the denominator not 0. In algebra, a **rational expression** or *algebraic fraction* is the quotient of two polynomials, again with the denominator not 0. For example,

$$\frac{x}{y}, \quad \frac{-a}{4}, \quad \frac{m+4}{m-2}, \quad \frac{8x^2-2x+5}{4x^2+5x}, \quad \text{and} \quad x^5 \left(\text{or } \frac{x^5}{1} \right)$$

are all rational expressions. In other words, rational expressions are the elements of the set

$$\left\{ \frac{P}{Q} \,\middle|\, P, Q \text{ polynomials, with } Q \neq 0 \right\}.$$

OBJECTIVE 2 Define rational functions and describe their domains. A function that is defined by a rational expression is called a **rational function** and has the form

$$f(x) = \frac{P(x)}{Q(x)},$$

where $Q(x) \neq 0$.

The domain of a rational function includes all real numbers except those that make $Q(x)$—that is, the denominator—equal to 0. For example, the domain of

$$f(x) = \frac{2}{x-5}$$

includes all real numbers except 5, because 5 would make the denominator equal to 0.

Figure 1 shows a graph of the function defined by $f(x) = \frac{2}{x-5}$. Notice that the graph does not exist when $x = 5$. It does not intersect the dashed vertical line whose equation is $x = 5$. This line is an *asymptote*. We discuss graphs of rational functions in more detail in Section 7.4.

EXAMPLE 1 Finding Numbers That Are Not in the Domains of Rational Functions

Find all numbers that are not in the domain of each rational function, and then state the domain.

(a) $f(x) = \dfrac{3}{7x-14}$

The only values that cannot be used are those that make the denominator 0. To find these values, set the denominator equal to 0 and solve the resulting equation.

$$7x - 14 = 0$$
$$7x = 14 \qquad \text{Add 14.}$$
$$x = 2 \qquad \text{Divide by 7.}$$

The number 2 cannot be used as a replacement for x; the domain of f includes all real numbers except 2.

(b) $g(x) = \dfrac{3 + x}{x^2 - 4x + 3}$

Set the denominator equal to 0, and solve the equation.

$$x^2 - 4x + 3 = 0$$
$$(x - 3)(x - 1) = 0 \qquad \text{Factor.}$$
$$x - 3 = 0 \quad \text{or} \quad x - 1 = 0 \qquad \text{Zero-factor property}$$
$$x = 3 \quad \text{or} \quad x = 1$$

The domain of g includes all real numbers except 3 and 1.

(c) $h(x) = \dfrac{8x + 2}{3}$

The denominator, 3, can never be 0, so the domain includes all real numbers.

(d) $f(x) = \dfrac{2}{x^2 + 4}$

Setting $x^2 + 4$ equal to 0 leads to $x^2 = -4$. There is no real number whose square is -4. Therefore, any real number can be used, and as in part (c), the domain includes all real numbers.

Now Try Exercises 11, 15, 17, and 19.

OBJECTIVE 3 Write rational expressions in lowest terms. In arithmetic, we write the fraction $\frac{15}{20}$ in lowest terms by dividing the numerator and denominator by 5 to get $\frac{3}{4}$. We write rational expressions in lowest terms in a similar way, using the **fundamental property of rational numbers.**

> **Fundamental Property of Rational Numbers**
>
> If $\frac{a}{b}$ is a rational number and if c is any nonzero real number, then
>
> $$\frac{a}{b} = \frac{ac}{bc}.$$
>
> That is, the numerator and denominator of a rational number may either be multiplied or divided by the same nonzero number without changing the value of the rational number.

Since $\frac{c}{c}$ is equivalent to 1, the fundamental property is based on the identity property of multiplication.

A rational expression is a quotient of two polynomials. Since the value of a polynomial is a real number for every value of the variable for which it is defined, any statement that applies to rational numbers will also apply to rational expressions. We use the following steps to write rational expressions in lowest terms.

> **Writing a Rational Expression in Lowest Terms**
>
> *Step 1* **Factor** both numerator and denominator to find their greatest common factor (GCF).
>
> *Step 2* **Apply the fundamental property.**

■ **EXAMPLE 2** Writing Rational Expressions in Lowest Terms

Write each rational expression in lowest terms.

(a) $\dfrac{8k}{16} = \dfrac{k \cdot 8}{2 \cdot 8} = \dfrac{k}{2} \cdot 1 = \dfrac{k}{2}$

Here, the GCF of the numerator and denominator is 8. We then applied the fundamental property.

(b) $\dfrac{8 + k}{16}$

The numerator cannot be factored, so this expression cannot be simplified further and is in lowest terms.

(c) $\dfrac{a^2 - a - 6}{a^2 + 5a + 6} = \dfrac{(a - 3)(a + 2)}{(a + 3)(a + 2)}$ Factor the numerator and the denominator.

$\qquad = \dfrac{a - 3}{a + 3} \cdot 1$ $\dfrac{a + 2}{a + 2} = 1$

$\qquad = \dfrac{a - 3}{a + 3}$ Lowest terms

(d) $\dfrac{y^2 - 4}{2y + 4} = \dfrac{(y + 2)(y - 2)}{2(y + 2)} = \dfrac{y - 2}{2}$

(e) $\dfrac{x^3 - 27}{x - 3} = \dfrac{(x - 3)(x^2 + 3x + 9)}{x - 3}$ Factor the difference of cubes.

$\qquad = x^2 + 3x + 9$ Lowest terms

(f) $\dfrac{pr + qr + ps + qs}{pr + qr - ps - qs} = \dfrac{(pr + qr) + (ps + qs)}{(pr + qr) - (ps + qs)}$ Group terms.

$\qquad = \dfrac{r(p + q) + s(p + q)}{r(p + q) - s(p + q)}$ Factor within groups.

$\qquad = \dfrac{(p + q)(r + s)}{(p + q)(r - s)}$ Factor by grouping.

$\qquad = \dfrac{r + s}{r - s}$ Lowest terms

■

Now Try Exercises 27, 31, 35, 43, and 47.

CAUTION Be careful! When using the fundamental property of rational numbers, only common *factors* may be divided. For example,

$$\frac{y - 2}{2} \neq y \quad \text{and} \quad \frac{y - 2}{2} \neq y - 1$$

because the 2 in $y - 2$ is not a *factor* of the numerator. To see this, replace y with a number and evaluate the fraction. For example, if $y = 5$, then

$$\frac{y - 2}{2} = \frac{5 - 2}{2} = \frac{3}{2}.$$

This does not equal y or $y - 1 = 5 - 1 = 4$. *Remember to factor before writing a fraction in lowest terms.*

In the rational expression from Example 2(c),

$$\frac{a^2 - a - 6}{a^2 + 5a + 6}, \quad \text{or} \quad \frac{(a - 3)(a + 2)}{(a + 3)(a + 2)},$$

a can take any value except -3 or -2 since these values make the denominator 0. In the simplified rational expression

$$\frac{a - 3}{a + 3},$$

a cannot equal -3. Because of this,

$$\frac{a^2 - a - 6}{a^2 + 5a + 6} = \frac{a - 3}{a + 3}$$

for all values of a except -3 or -2. From now on such statements of equality will be made with the understanding that they apply only for those real numbers that make neither denominator equal 0. We will no longer state such restrictions.

▪ EXAMPLE 3   Writing Rational Expressions in Lowest Terms

Write each rational expression in lowest terms.

(a) $\dfrac{m - 3}{3 - m}$

In this rational expression, the numerator and denominator are opposites. The given expression can be written in lowest terms by writing the denominator as $-1(m - 3)$, giving

$$\frac{m - 3}{3 - m} = \frac{m - 3}{-1(m - 3)} = \frac{1}{-1} = -1.$$

The numerator could have been rewritten instead to get the same result.

(b) $\dfrac{r^2 - 16}{4 - r} = \dfrac{(r + 4)(r - 4)}{4 - r}$

$$= \frac{(r + 4)(r - 4)}{-1(r - 4)} \qquad \text{Write } 4 - r \text{ as } -1(r - 4).$$

$$= \frac{r + 4}{-1} \qquad\qquad\quad \text{Fundamental property}$$

$$= -(r + 4) \quad \text{or} \quad -r - 4 \qquad \text{Lowest terms} \qquad\qquad ▪$$

Now Try Exercises 49 and 51.

As shown in Examples 3(a) and (b), the quotient

$$\frac{a}{-a} \quad (a \neq 0)$$

can be simplified as

$$\frac{a}{-a} = \frac{a}{-1(a)} = \frac{1}{-1} = -1.$$

The following statement summarizes this result.

In general, if the numerator and the denominator of a rational expression are opposites, then the expression equals -1.

Based on this result,

$$\frac{q - 7}{7 - q} = -1 \quad \text{and} \quad \frac{-5a + 2b}{5a - 2b} = -1.$$

However,

$$\frac{r - 2}{r + 2}$$

cannot be simplified further since the numerator and the denominator are *not* opposites.

OBJECTIVE 4 Multiply rational expressions. To multiply rational expressions, follow these steps. (In practice, we usually simplify before multiplying.)

Multiplying Rational Expressions

Step 1 **Factor** all numerators and denominators as completely as possible.

Step 2 **Apply the fundamental property.**

Step 3 **Multiply** remaining factors in the numerator and remaining factors in the denominator. Leave the denominator in factored form.

Step 4 **Check** to be sure the product is in lowest terms.

▨ EXAMPLE 4 Multiplying Rational Expressions

Multiply.

(a) $\dfrac{5p - 5}{p} \cdot \dfrac{3p^2}{10p - 10} = \dfrac{5(p - 1)}{p} \cdot \dfrac{3p \cdot p}{2 \cdot 5(p - 1)}$ Factor.

$\qquad\qquad\qquad = \dfrac{1}{1} \cdot \dfrac{3p}{2}$ Lowest terms

$\qquad\qquad\qquad = \dfrac{3p}{2}$ Multiply.

(b) $\dfrac{k^2 + 2k - 15}{k^2 - 4k + 3} \cdot \dfrac{k^2 - k}{k^2 + k - 20} = \dfrac{(k + 5)(k - 3)}{(k - 3)(k - 1)} \cdot \dfrac{k(k - 1)}{(k + 5)(k - 4)}$

$\qquad\qquad\qquad\qquad\qquad = \dfrac{k}{k - 4}$

(c) $(p - 4) \cdot \dfrac{3}{5p - 20} = \dfrac{p - 4}{1} \cdot \dfrac{3}{5p - 20}$ Write $p - 4$ as $\frac{p - 4}{1}$.

$\qquad\qquad\qquad = \dfrac{p - 4}{1} \cdot \dfrac{3}{5(p - 4)}$ Factor.

$\qquad\qquad\qquad = \dfrac{3}{5}$

(d) $\dfrac{x^2 + 2x}{x + 1} \cdot \dfrac{x^2 - 1}{x^3 + x^2} = \dfrac{x(x + 2)}{x + 1} \cdot \dfrac{(x + 1)(x - 1)}{x^2(x + 1)}$ Factor.

$$= \dfrac{(x + 2)(x - 1)}{x(x + 1)}$$ Multiply; lowest terms

(e) $\dfrac{x - 6}{x^2 - 12x + 36} \cdot \dfrac{x^2 - 3x - 18}{x^2 + 7x + 12} = \dfrac{x - 6}{(x - 6)^2} \cdot \dfrac{(x + 3)(x - 6)}{(x + 3)(x + 4)}$ Factor.

$$= \dfrac{1}{x + 4}$$ Lowest terms

Remember to include 1 in the numerator when all other factors are eliminated using the fundamental property.

Now Try Exercises 71, 73, and 77.

OBJECTIVE 5 **Find reciprocals for rational expressions.** The rational numbers $\frac{a}{b}$ and $\frac{c}{d}$ are reciprocals of each other if they have a product of 1. The **reciprocal** of a rational expression is defined in the same way: Two rational expressions are reciprocals of each other if they have a product of 1. Recall that 0 has no reciprocal. The table shows several rational expressions and their reciprocals. In the first two cases, check that the product of the rational expression and its reciprocal is 1.

The examples in the table suggest the following procedure.

Rational Expression	Reciprocal
$\dfrac{5}{k}$	$\dfrac{k}{5}$
$\dfrac{m^2 - 9m}{2}$	$\dfrac{2}{m^2 - 9m}$
$\dfrac{0}{4}$	undefined

Finding the Reciprocal

To find the reciprocal of a nonzero rational expression, invert the rational expression.

OBJECTIVE 6 **Divide rational expressions.** Dividing rational expressions is like dividing rational numbers.

Dividing Rational Expressions

To divide two rational expressions, *multiply* the first by the reciprocal of the second.

EXAMPLE 5 Dividing Rational Expressions

Divide.

(a) $\dfrac{2z}{9} \div \dfrac{5z^2}{18} = \dfrac{2z}{9} \cdot \dfrac{18}{5z^2}$ Multiply by the reciprocal of the divisor.

$$= \dfrac{2z}{9} \cdot \dfrac{2 \cdot 9}{5z^2}$$ Factor.

$$= \dfrac{4}{5z}$$ Multiply; lowest terms

(b) $\dfrac{m^2pq^3}{mp^4} \div \dfrac{m^5p^2q}{mpq^2}$

Use the definitions of division and multiplication and the properties of exponents.

$$\dfrac{m^2pq^3}{mp^4} \div \dfrac{m^5p^2q}{mpq^2} = \dfrac{m^2pq^3}{mp^4} \cdot \dfrac{mpq^2}{m^5p^2q} \qquad \text{Multiply by the reciprocal.}$$

$$= \dfrac{m^3p^2q^5}{m^6p^6q} \qquad \text{Properties of exponents}$$

$$= \dfrac{q^4}{m^3p^4} \qquad \text{Properties of exponents}$$

(c) $\dfrac{8k - 16}{3k} \div \dfrac{3k - 6}{4k^2} = \dfrac{8k - 16}{3k} \cdot \dfrac{4k^2}{3k - 6} \qquad \text{Multiply by the reciprocal.}$

$$= \dfrac{8(k - 2)}{3k} \cdot \dfrac{4k^2}{3(k - 2)} \qquad \text{Factor.}$$

$$= \dfrac{32k}{9} \qquad \text{Multiply; lowest terms}$$

(d) $\dfrac{5m^2 + 17m - 12}{3m^2 + 7m - 20} \div \dfrac{5m^2 + 2m - 3}{15m^2 - 34m + 15}$

$$= \dfrac{5m^2 + 17m - 12}{3m^2 + 7m - 20} \cdot \dfrac{15m^2 - 34m + 15}{5m^2 + 2m - 3} \qquad \text{Definition of division}$$

$$= \dfrac{(5m - 3)(m + 4)}{(m + 4)(3m - 5)} \cdot \dfrac{(3m - 5)(5m - 3)}{(5m - 3)(m + 1)} \qquad \text{Factor.}$$

$$= \dfrac{5m - 3}{m + 1} \qquad \text{Lowest terms}$$

Now Try Exercises 63, 79, and 85.

7.1 EXERCISES

Rational expressions often can be written in lowest terms in seemingly different ways. For example,

$$\dfrac{y - 3}{-5} \qquad and \qquad \dfrac{-y + 3}{5}$$

look different, but we get the second quotient by multiplying the first by -1 in both the numerator and denominator. To practice recognizing equivalent rational expressions, match the expressions in Exercises 1–6 with their equivalents in choices A–F.

1. $\dfrac{x - 3}{x + 4}$ **2.** $\dfrac{x + 3}{x - 4}$ **3.** $\dfrac{x - 3}{x - 4}$ **4.** $\dfrac{x + 3}{x + 4}$ **5.** $\dfrac{3 - x}{x + 4}$ **6.** $\dfrac{x + 3}{4 - x}$

A. $\dfrac{-x - 3}{4 - x}$ **B.** $\dfrac{-x - 3}{-x - 4}$ **C.** $\dfrac{3 - x}{-x - 4}$ **D.** $\dfrac{-x + 3}{-x + 4}$ **E.** $\dfrac{x - 3}{-x - 4}$ **F.** $\dfrac{-x - 3}{x - 4}$

✐ **7.** In Example 1(a), we showed that the domain of the rational function defined by $f(x) = \frac{3}{7x - 14}$ does not include 2. Explain in your own words why this is so. In general, how do we find the value or values excluded from the domain of a rational function?

✐ **8.** The domain of the rational function defined by $g(x) = \frac{x + 1}{x^2 + 3}$ includes all real numbers. Explain.

Find all numbers that are not in the domain of each function. See Example 1.

9. $f(x) = \dfrac{x}{x - 7}$

10. $f(x) = \dfrac{x}{x + 3}$

11. $f(x) = \dfrac{6x - 5}{7x + 1}$

12. $f(x) = \dfrac{8x - 3}{2x + 7}$

13. $f(x) = \dfrac{12x + 3}{x}$

14. $f(x) = \dfrac{9x + 8}{x}$

15. $f(x) = \dfrac{3x + 1}{2x^2 + x - 6}$

16. $f(x) = \dfrac{2x + 4}{3x^2 + 11x - 42}$

17. $f(x) = \dfrac{x + 2}{14}$

18. $f(x) = \dfrac{x - 9}{26}$

19. $f(x) = \dfrac{2x^2 - 3x + 4}{3x^2 + 8}$

20. $f(x) = \dfrac{9x^2 - 8x + 3}{4x^2 + 1}$

21. (a) Identify the two *terms* in the numerator and the two *terms* in the denominator of the rational expression $\dfrac{x^2 + 4x}{x + 4}$.

✐ **(b)** Describe the steps you would use to write this rational expression in lowest terms. (*Hint:* It simplifies to x.)

22. Only one of the following rational expressions can be simplified. Which one is it?

A. $\dfrac{x^2 + 2}{x^2}$ **B.** $\dfrac{x^2 + 2}{2}$ **C.** $\dfrac{x^2 + y^2}{y^2}$ **D.** $\dfrac{x^2 - 5x}{x}$

23. Only one of the following rational expressions is *not* equivalent to $\frac{x - 3}{4 - x}$. Which one is it?

A. $\dfrac{3 - x}{x - 4}$ **B.** $\dfrac{x + 3}{4 + x}$ **C.** $-\dfrac{3 - x}{4 - x}$ **D.** $-\dfrac{x - 3}{x - 4}$

24. Which two of the following rational expressions equal -1?

A. $\dfrac{2x + 3}{2x - 3}$ **B.** $\dfrac{2x - 3}{3 - 2x}$ **C.** $\dfrac{2x + 3}{3 + 2x}$ **D.** $\dfrac{2x + 3}{-2x - 3}$

Write each rational expression in lowest terms. See Example 2.

25. $\dfrac{x^2(x + 1)}{x(x + 1)}$

26. $\dfrac{y^3(y - 4)}{y^2(y - 4)}$

27. $\dfrac{(x + 4)(x - 3)}{(x + 5)(x + 4)}$

28. $\dfrac{(2x + 7)(x - 1)}{(2x + 3)(2x + 7)}$

29. $\dfrac{4x(x + 3)}{8x^2(x - 3)}$

30. $\dfrac{5y^2(y + 8)}{15y(y - 8)}$

31. $\dfrac{3x + 7}{3}$

32. $\dfrac{4x - 9}{4}$

33. $\dfrac{6m + 18}{7m + 21}$

34. $\dfrac{5r - 20}{3r - 12}$

35. $\dfrac{3z^2 + z}{18z + 6}$

36. $\dfrac{2x^2 - 5x}{16x - 40}$

37. $\dfrac{2t + 6}{t^2 - 9}$

38. $\dfrac{5s - 25}{s^2 - 25}$

39. $\dfrac{x^2 + 2x - 15}{x^2 + 6x + 5}$

40. $\dfrac{y^2 - 5y - 14}{y^2 + y - 2}$

41. $\dfrac{8x^2 - 10x - 3}{8x^2 - 6x - 9}$

42. $\dfrac{12x^2 - 4x - 5}{8x^2 - 6x - 5}$

43. $\dfrac{a^3 + b^3}{a + b}$

44. $\dfrac{r^3 - s^3}{r - s}$

45. $\dfrac{2c^2 + 2cd - 60d^2}{2c^2 - 12cd + 10d^2}$

46. $\dfrac{3s^2 - 9st - 54t^2}{3s^2 - 6st - 72t^2}$

47. $\dfrac{ac - ad + bc - bd}{ac - ad - bc + bd}$

48. $\dfrac{2xy + 2xw + y + w}{2xy + y - 2xw - w}$

Write each rational expression in lowest terms. See Example 3.

49. $\dfrac{7 - b}{b - 7}$

50. $\dfrac{r - 13}{13 - r}$

51. $\dfrac{x^2 - y^2}{y - x}$

52. $\dfrac{m^2 - n^2}{n - m}$

53. $\dfrac{(a - 3)(x + y)}{(3 - a)(x - y)}$

54. $\dfrac{(8 - p)(x + 2)}{(p - 8)(x - 2)}$

55. $\dfrac{5k - 10}{20 - 10k}$

56. $\dfrac{7x - 21}{63 - 21x}$

57. $\dfrac{a^2 - b^2}{a^2 + b^2}$

58. $\dfrac{p^2 + q^2}{p^2 - q^2}$

59. Explain in a few words how to multiply rational expressions. Give an example.

60. Explain in a few words how to divide rational expressions. Give an example.

Multiply or divide as indicated. See Examples 4 and 5.

61. $\dfrac{x^3}{3y} \cdot \dfrac{9y^2}{x^5}$

62. $\dfrac{a^4}{5b^2} \cdot \dfrac{25b^4}{a^3}$

63. $\dfrac{5a^4b^2}{16a^2b} \div \dfrac{25a^2b}{60a^3b^2}$

64. $\dfrac{s^3t^2}{10s^2t^4} \div \dfrac{8s^4t^2}{5t^6}$

65. $\dfrac{(-3mn)^2 \cdot (4m^2n)^3}{16m^2n^4(mn^2)^3} \div \dfrac{24(m^2n^2)^4}{(3m^2n^3)^2}$

66. $\dfrac{(-4a^2b^3)^2 \cdot (3a^2b^4)^2}{(2a^2b^3)^4 \cdot (3a^3b)^2} \div \dfrac{(ab)^4}{(a^2b^3)^2}$

67. $\dfrac{(x + 2)(x + 1)}{(x + 3)(x - 2)} \cdot \dfrac{(x + 3)(x + 4)}{(x + 2)(x + 1)}$

68. $\dfrac{(x + 3)(x - 4)}{(x - 4)(x + 2)} \cdot \dfrac{(x + 5)(x - 6)}{(x + 3)(x - 6)}$

69. $\dfrac{(2x + 3)(x - 4)}{(x + 8)(x - 4)} \div \dfrac{(x - 4)(x + 2)}{(x - 4)(x + 8)}$

70. $\dfrac{(6x + 5)(x - 3)}{(x + 9)(x - 1)} \div \dfrac{(x - 3)(2x + 7)}{(x - 1)(x + 9)}$

71. $\dfrac{4x}{8x + 4} \cdot \dfrac{14x + 7}{6}$

72. $\dfrac{12x - 20}{5x} \cdot \dfrac{6}{9x - 15}$

73. $\dfrac{p^2 - 25}{4p} \cdot \dfrac{2}{5 - p}$

74. $\dfrac{a^2 - 1}{4a} \cdot \dfrac{2}{1 - a}$

75. $(7k + 7) \div \dfrac{4k + 4}{5}$

76. $(8y - 16) \div \dfrac{3y - 6}{10}$

77. $(z^2 - 1) \cdot \dfrac{1}{1 - z}$

78. $(y^2 - 4) \div \dfrac{2 - y}{8y}$

79. $\dfrac{m^2 - 49}{m + 1} \div \dfrac{7 - m}{m}$

80. $\dfrac{k^2 - 4}{3k^2} \div \dfrac{2 - k}{11k}$

81. $\dfrac{12x - 10y}{3x + 2y} \cdot \dfrac{6x + 4y}{10y - 12x}$

82. $\dfrac{9s - 12t}{2s + 2t} \cdot \dfrac{3s + 3t}{4t - 3s}$

83. $\dfrac{x^2 - 25}{x^2 + x - 20} \cdot \dfrac{x^2 + 7x + 12}{x^2 - 2x - 15}$

84. $\dfrac{t^2 - 49}{t^2 + 4t - 21} \cdot \dfrac{t^2 + 8t + 15}{t^2 - 2t - 35}$

85. $\dfrac{6x^2 + 5xy - 6y^2}{12x^2 - 11xy + 2y^2} \div \dfrac{4x^2 - 12xy + 9y^2}{8x^2 - 14xy + 3y^2}$

86. $\dfrac{8a^2 - 6ab - 9b^2}{6a^2 - 5ab - 6b^2} \div \dfrac{4a^2 + 11ab + 6b^2}{9a^2 + 12ab + 4b^2}$

87. $\dfrac{3k^2 + 17kp + 10p^2}{6k^2 + 13kp - 5p^2} \div \dfrac{6k^2 + kp - 2p^2}{6k^2 - 5kp + p^2}$

88. $\dfrac{16c^2 + 24cd + 9d^2}{16c^2 - 16cd + 3d^2} \div \dfrac{16c^2 - 9d^2}{16c^2 - 24cd + 9d^2}$

89. $\left(\dfrac{6k^2 - 13k - 5}{k^2 + 7k} \div \dfrac{2k - 5}{k^3 + 6k^2 - 7k} \right) \cdot \dfrac{k^2 - 5k + 6}{3k^2 - 8k - 3}$

90. $\left(\dfrac{2x^3 + 3x^2 - 2x}{3x - 15} \div \dfrac{2x^3 - x^2}{x^2 - 3x - 10} \right) \cdot \dfrac{5x^2 - 10x}{3x^2 + 12x + 12}$

91. $\dfrac{a^2(2a + b) + 6a(2a + b) + 5(2a + b)}{3a^2(a + 2b) - 2a(a + 2b) - (a + 2b)} \div \dfrac{a + 1}{a - 1}$

92. $\dfrac{2x^2(x - 3z) - 5x(x - 3z) + 2(x - 3z)}{4x^2(3z - x) - 11x(3z - x) + 6(3z - x)} \div \dfrac{4x + 1}{4x - 3}$

7.2 Adding and Subtracting Rational Expressions

OBJECTIVES

1 Add and subtract rational expressions with the same denominator.

2 Find a least common denominator.

3 Add and subtract rational expressions with different denominators.

OBJECTIVE 1 Add and subtract rational expressions with the same denominator. The following steps, used to add or subtract rational numbers, are also used to add or subtract rational expressions.

Adding or Subtracting Rational Expressions

Step 1 **If the denominators are the same,** add or subtract the numerators. Place the result over the common denominator.

If the denominators are different, first find the least common denominator. Write all rational expressions with this least common denominator, and then add or subtract the numerators. Place the result over the common denominator.

Step 2 **Simplify.** Write all answers in lowest terms.

EXAMPLE 1 Adding and Subtracting Rational Expressions with the Same Denominators

Add or subtract as indicated.

(a) $\dfrac{3y}{5} + \dfrac{x}{5} = \dfrac{3y + x}{5}$ ← Add numerators.
    ← Keep the common denominator.

(b) $\dfrac{7}{2r^2} - \dfrac{11}{2r^2} = \dfrac{7 - 11}{2r^2}$ Subtract numerators; keep the common denominator.

$= \dfrac{-4}{2r^2}$

$= -\dfrac{2}{r^2}$ Lowest terms

(c) $\dfrac{m}{m^2 - p^2} + \dfrac{p}{m^2 - p^2} = \dfrac{m + p}{m^2 - p^2}$ Add numerators; keep the common denominator.

$= \dfrac{m + p}{(m + p)(m - p)}$ Factor.

$= \dfrac{1}{m - p}$ Lowest terms

(d) $\dfrac{4}{x^2 + 2x - 8} + \dfrac{x}{x^2 + 2x - 8} = \dfrac{4 + x}{x^2 + 2x - 8}$

$= \dfrac{4 + x}{(x - 2)(x + 4)}$

$= \dfrac{1}{x - 2}$

Now Try Exercises 7, 9, and 15.

OBJECTIVE 2 Find a least common denominator.   We add or subtract rational expressions with different denominators by first writing them with a common denominator, usually the **least common denominator (LCD).**

Finding the Least Common Denominator

Step 1   **Factor** each denominator.

Step 2   **Find the least common denominator.** The LCD is the product of all different factors from each denominator, with each factor raised to the *greatest* power that occurs in any denominator.

EXAMPLE 2   Finding Least Common Denominators

Assume that the given expressions are denominators of two fractions. Find the LCD for each group.

(a) $5xy^2$, $2x^3y$

Each denominator is already factored.

$$5xy^2 = 5 \cdot x \cdot y^2$$
$$2x^3y = 2 \cdot x^3 \cdot y$$

Greatest exponent on x is 3.

$$\text{LCD} = 5 \cdot 2 \cdot x^3 \cdot y^2 \;\leftarrow\; \text{Greatest exponent on } y \text{ is 2.}$$
$$= 10x^3y^2$$

(b) $k - 3, \quad k$

Each denominator is already factored. The LCD, an expression divisible by *both* $k - 3$ and k, is

$$k(k - 3).$$

It is usually best to leave a least common denominator in factored form.

(c) $y^2 - 2y - 8, \quad y^2 + 3y + 2$

Factor the denominators.

$$\left.\begin{array}{l} y^2 - 2y - 8 = (y - 4)(y + 2) \\ y^2 + 3y + 2 = (y + 2)(y + 1) \end{array}\right\} \text{Factor.}$$

The LCD, divisible by both polynomials, is

$$(y - 4)(y + 2)(y + 1).$$

(d) $8z - 24, \quad 5z^2 - 15z$

$$\left.\begin{array}{l} 8z - 24 = 8(z - 3) \\ 5z^2 - 15z = 5z(z - 3) \end{array}\right\} \text{Factor.}$$

The LCD is $8 \cdot 5z \cdot (z - 3) = 40z(z - 3)$.

(e) $m^2 + 5m + 6, \quad m^2 + 4m + 4, \quad 2(m^2 + 2m - 3)$

$$\left.\begin{array}{l} m^2 + 5m + 6 = (m + 3)(m + 2) \\ m^2 + 4m + 4 = (m + 2)^2 \\ 2(m^2 + 2m - 3) = 2(m + 3)(m - 1) \end{array}\right\} \text{Factor.}$$

The LCD is $2(m + 3)(m + 2)^2(m - 1)$.

> **Now Try Exercises 21, 23, 27, and 35.**

OBJECTIVE 3 Add and subtract rational expressions with different denominators. Before adding or subtracting two rational expressions, we write each expression with the least common denominator by multiplying its numerator and denominator by the factors needed to get the LCD. This procedure is valid because we are multiplying each rational expression by a form of 1, the identity element for multiplication.

Adding or subtracting rational expressions follows the same procedure as that used for rational numbers. Consider the sum $\frac{7}{15} + \frac{5}{12}$. The LCD for 15 and 12 is 60. Multiply $\frac{7}{15}$ by $\frac{4}{4}$ (a form of 1) and multiply $\frac{5}{12}$ by $\frac{5}{5}$ so that each fraction has denominator 60, and then add the numerators.

$$\frac{7}{15} + \frac{5}{12} = \frac{7 \cdot 4}{15 \cdot 4} + \frac{5 \cdot 5}{12 \cdot 5} \qquad \text{Fundamental property}$$

$$= \frac{28}{60} + \frac{25}{60}$$

$$= \frac{28 + 25}{60} \qquad \text{Add the numerators.}$$

$$= \frac{53}{60}$$

▌**EXAMPLE 3** Adding and Subtracting Rational Expressions with Different Denominators

Add or subtract as indicated.

(a) $\dfrac{5}{2p} + \dfrac{3}{8p}$

The LCD for $2p$ and $8p$ is $8p$. To write the first rational expression with a denominator of $8p$, multiply by $\frac{4}{4}$.

$$\frac{5}{2p} + \frac{3}{8p} = \frac{5 \cdot 4}{2p \cdot 4} + \frac{3}{8p} \qquad \text{Fundamental principle}$$

$$= \frac{20}{8p} + \frac{3}{8p}$$

$$= \frac{20 + 3}{8p} \qquad \text{Add numerators.}$$

$$= \frac{23}{8p}$$

(b) $\dfrac{6}{r} - \dfrac{5}{r - 3}$

The LCD is $r(r - 3)$. Rewrite each rational expression with this denominator.

$$\frac{6}{r} - \frac{5}{r - 3} = \frac{6(r - 3)}{r(r - 3)} - \frac{r \cdot 5}{r(r - 3)} \qquad \text{Fundamental principle}$$

$$= \frac{6r - 18}{r(r - 3)} - \frac{5r}{r(r - 3)} \qquad \text{Distributive and commutative properties}$$

$$= \frac{6r - 18 - 5r}{r(r - 3)} \qquad \text{Subtract numerators.}$$

$$= \frac{r - 18}{r(r - 3)} \qquad \text{Combine terms in the numerator.}$$

Now Try Exercises 39 and 43.

CAUTION One of the most common sign errors in algebra occurs when subtracting a rational expression with two or more terms in the numerator. Remember that in this situation, the subtraction sign must be distributed to *every* term in the numerator of the fraction that follows it. Study Example 4 carefully to see how this is done.

▌**EXAMPLE 4** Using the Distributive Property When Subtracting Rational Expressions

Subtract.

(a) $\dfrac{7x}{3x + 1} - \dfrac{x - 2}{3x + 1}$

The denominators are the same for both rational expressions. The subtraction sign must be applied to *both* terms in the numerator of the second rational expression.

Notice the careful use of the distributive property here.

$$\frac{7x}{3x+1} - \frac{x-2}{3x+1} = \frac{7x - (x-2)}{3x+1}$$ Write as a single rational expression.

$$= \frac{7x - x + 2}{3x+1}$$ Distributive property; be careful with signs.

$$= \frac{6x+2}{3x+1}$$ Combine terms in the numerator.

$$= \frac{2(3x+1)}{3x+1}$$ Factor the numerator.

$$= 2$$ Lowest terms

(b) $\dfrac{1}{q-1} - \dfrac{1}{q+1}$

$$= \frac{1(q+1)}{(q-1)(q+1)} - \frac{1(q-1)}{(q+1)(q-1)}$$ Fundamental property

$$= \frac{(q+1) - (q-1)}{(q-1)(q+1)}$$ Subtract.

$$= \frac{q + 1 - q + 1}{(q-1)(q+1)}$$ Distributive property

$$= \frac{2}{(q-1)(q+1)}$$ Combine terms in the numerator.

Now Try Exercises 47 and 53.

In some problems, rational expressions to be added or subtracted have denominators that are opposites of each other. The next example illustrates how to proceed in such a problem.

EXAMPLE 5   Adding Rational Expressions with Denominators That Are Opposites

Add.

$$\frac{y}{y-2} + \frac{8}{2-y}$$

To get a common denominator of $y - 2$, multiply the second expression by -1 in both the numerator and the denominator.

$$\frac{y}{y-2} + \frac{8}{2-y} = \frac{y}{y-2} + \frac{8(-1)}{(2-y)(-1)}$$

$$= \frac{y}{y-2} + \frac{-8}{y-2}$$

$$= \frac{y-8}{y-2}$$ Add the numerators.

Now Try Exercise 49.

The next example illustrates addition and subtraction involving more than two rational expressions.

EXAMPLE 6   Adding and Subtracting Three Rational Expressions

Add and subtract as indicated.

$$\frac{3}{x-2} + \frac{5}{x} - \frac{6}{x^2-2x}$$

The denominator of the third rational expression factors as $x(x-2)$, which is the LCD for the three rational expressions.

$$\frac{3}{x-2} + \frac{5}{x} - \frac{6}{x^2-2x}$$

$$= \frac{3x}{x(x-2)} + \frac{5(x-2)}{x(x-2)} - \frac{6}{x(x-2)} \qquad \text{Fundamental property}$$

$$= \frac{3x + 5(x-2) - 6}{x(x-2)} \qquad \text{Add and subtract the numerators.}$$

$$= \frac{3x + 5x - 10 - 6}{x(x-2)} \qquad \text{Distributive property}$$

$$= \frac{8x - 16}{x(x-2)} \qquad \text{Combine terms in the numerator.}$$

$$= \frac{8(x-2)}{x(x-2)} \qquad \text{Factor the numerator.}$$

$$= \frac{8}{x} \qquad \text{Lowest terms}$$

Now Try Exercise 55.

EXAMPLE 7   Subtracting Rational Expressions

Subtract.

$$\frac{m+4}{m^2-2m-3} - \frac{2m-3}{m^2-5m+6}$$

$$= \frac{m+4}{(m-3)(m+1)} - \frac{2m-3}{(m-3)(m-2)} \qquad \text{Factor each denominator.}$$

The LCD is $(m-3)(m+1)(m-2)$.

$$= \frac{(m+4)(m-2)}{(m-3)(m+1)(m-2)} - \frac{(2m-3)(m+1)}{(m-3)(m-2)(m+1)} \qquad \text{Fundamental property}$$

$$= \frac{(m+4)(m-2) - (2m-3)(m+1)}{(m-3)(m+1)(m-2)} \qquad \text{Subtract.}$$

$$= \frac{m^2 + 2m - 8 - (2m^2 - m - 3)}{(m-3)(m+1)(m-2)} \qquad \text{Multiply in the numerator.}$$

$$= \frac{m^2 + 2m - 8 - 2m^2 + m + 3}{(m - 3)(m + 1)(m - 2)} \qquad \text{Distributive property; be careful with signs.}$$

$$= \frac{-m^2 + 3m - 5}{(m - 3)(m + 1)(m - 2)} \qquad \text{Combine terms in the numerator.}$$

If we try to factor the numerator, we find that this rational expression is in lowest terms.

Now Try Exercise 69.

EXAMPLE 8 Adding Rational Expressions

Add.

$$\frac{5}{x^2 + 10x + 25} + \frac{2}{x^2 + 7x + 10}$$

$$= \frac{5}{(x + 5)^2} + \frac{2}{(x + 5)(x + 2)} \qquad \text{Factor each denominator.}$$

The LCD is $(x + 5)^2(x + 2)$.

$$= \frac{5(x + 2)}{(x + 5)^2(x + 2)} + \frac{2(x + 5)}{(x + 5)^2(x + 2)} \qquad \text{Fundamental property}$$

$$= \frac{5(x + 2) + 2(x + 5)}{(x + 5)^2(x + 2)} \qquad \text{Add.}$$

$$= \frac{5x + 10 + 2x + 10}{(x + 5)^2(x + 2)} \qquad \text{Distributive property}$$

$$= \frac{7x + 20}{(x + 5)^2(x + 2)} \qquad \text{Combine terms in the numerator.}$$

Now Try Exercise 77.

7.2 EXERCISES

For Extra Help

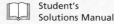
Student's
Solutions Manual

MyMathLab

InterAct Math
Tutorial Software

AW Math
Tutor Center

*Math*XP MathXL

Digital Video Tutor
CD 7/Videotape 8

RELATING CONCEPTS (EXERCISES 1–6)

For Individual or Group Work

Work Exercises 1–6 in order.

1. Let $x = 4$ and $y = 2$. Evaluate $\frac{1}{x} + \frac{1}{y}$.

2. Let $x = 4$ and $y = 2$. Evaluate $\frac{1}{x + y}$.

3. Are the answers for Exercises 1 and 2 the same? What can you conclude?

4. Let $x = 3$ and $y = 5$. Evaluate $\frac{1}{x} - \frac{1}{y}$.

5. Let $x = 3$ and $y = 5$. Evaluate $\frac{1}{x - y}$.

6. Are the answers for Exercises 4 and 5 the same? What can you conclude?

Add or subtract as indicated. Write all answers in lowest terms. See Example 1.

7. $\dfrac{7}{t} + \dfrac{2}{t}$

8. $\dfrac{5}{r} + \dfrac{9}{r}$

9. $\dfrac{11}{5x} - \dfrac{1}{5x}$

10. $\dfrac{7}{4y} - \dfrac{3}{4y}$

11. $\dfrac{5x+4}{6x+5} + \dfrac{x+1}{6x+5}$

12. $\dfrac{6y+12}{4y+3} + \dfrac{2y-6}{4y+3}$

13. $\dfrac{x^2}{x+5} - \dfrac{25}{x+5}$

14. $\dfrac{y^2}{y+6} - \dfrac{36}{y+6}$

15. $\dfrac{-3p+7}{p^2+7p+12} + \dfrac{8p+13}{p^2+7p+12}$

16. $\dfrac{5x+6}{x^2+x-20} + \dfrac{4-3x}{x^2+x-20}$

17. $\dfrac{a^3}{a^2+ab+b^2} - \dfrac{b^3}{a^2+ab+b^2}$

18. $\dfrac{p^3}{p^2-pq+q^2} + \dfrac{q^3}{p^2-pq+q^2}$

✍ 19. Write a step-by-step method for adding or subtracting rational expressions that have a common denominator. Illustrate with an example.

✍ 20. Write a step-by-step method for adding or subtracting rational expressions that have different denominators. Give an example.

Assume that the expressions given are denominators of fractions. Find the least common denominator (LCD) for each group. See Example 2.

21. $18x^2y^3, \quad 24x^4y^5$

22. $24a^3b^4, \quad 18a^5b^2$

23. $z - 2, \quad z$

24. $k + 3, \quad k$

25. $2y + 8, \quad y + 4$

26. $3r - 21, \quad r - 7$

27. $x^2 - 81, \quad x^2 + 18x + 81$

28. $y^2 - 16, \quad y^2 - 8y + 16$

29. $m + n, \quad m - n, \quad m^2 - n^2$

30. $r + s, \quad r - s, \quad r^2 - s^2$

31. $x^2 - 3x - 4, \quad x + x^2$

32. $y^2 - 8y + 12, \quad y^2 - 6y$

33. $2t^2 + 7t - 15, \quad t^2 + 3t - 10$

34. $s^2 - 3s - 4, \quad 3s^2 + s - 2$

35. $2y + 6, \quad y^2 - 9, \quad y$

36. $9x + 18, \quad x^2 - 4, \quad x$

✍ 37. One student added two rational expressions and obtained the answer $\dfrac{3}{5-y}$. Another student obtained the answer $\dfrac{-3}{y-5}$ for the same problem. Is it possible that both answers are correct? Explain.

✍ 38. What is *wrong* with the following work?

$$\frac{x}{x+2} - \frac{4x-1}{x+2} = \frac{x-4x-1}{x+2} = \frac{-3x-1}{x+2}$$

Add or subtract as indicated. Write all answers in lowest terms. See Examples 3–6.

39. $\dfrac{8}{t} + \dfrac{7}{3t}$

40. $\dfrac{5}{x} + \dfrac{9}{4x}$

41. $\dfrac{5}{12x^2y} - \dfrac{11}{6xy}$

42. $\dfrac{7}{18a^3b^2} - \dfrac{2}{9ab}$

43. $\dfrac{1}{x-1} - \dfrac{1}{x}$

44. $\dfrac{3}{x-3} - \dfrac{1}{x}$

45. $\dfrac{3a}{a+1} + \dfrac{2a}{a-3}$

46. $\dfrac{2x}{x+4} + \dfrac{3x}{x-7}$

47. $\dfrac{17y+3}{9y+7} - \dfrac{-10y-18}{9y+7}$

48. $\dfrac{7x+8}{3x+2} - \dfrac{x+4}{3x+2}$

49. $\dfrac{2}{4-x} + \dfrac{5}{x-4}$

50. $\dfrac{3}{2-t} + \dfrac{1}{t-2}$

51. $\dfrac{w}{w-z} - \dfrac{z}{z-w}$

52. $\dfrac{a}{a-b} - \dfrac{b}{b-a}$

53. $\dfrac{5}{12 + 4x} - \dfrac{7}{9 + 3x}$

54. $\dfrac{3}{10x + 15} - \dfrac{8}{12x + 18}$

55. $\dfrac{4x}{x - 1} - \dfrac{2}{x + 1} - \dfrac{4}{x^2 - 1}$

56. $\dfrac{4}{x + 3} - \dfrac{x}{x - 3} - \dfrac{18}{x^2 - 9}$

57. $\dfrac{15}{y^2 + 3y} + \dfrac{2}{y} + \dfrac{5}{y + 3}$

58. $\dfrac{7}{t - 2} - \dfrac{6}{t^2 - 2t} - \dfrac{3}{t}$

59. $\dfrac{5}{x - 2} + \dfrac{1}{x} + \dfrac{2}{x^2 - 2x}$

60. $\dfrac{5x}{x - 3} + \dfrac{2}{x} + \dfrac{6}{x^2 - 3x}$

61. $\dfrac{3x}{x + 1} + \dfrac{4}{x - 1} - \dfrac{6}{x^2 - 1}$

62. $\dfrac{5x}{x + 3} + \dfrac{x + 2}{x} - \dfrac{6}{x^2 + 3x}$

63. $\dfrac{4}{x + 1} + \dfrac{1}{x^2 - x + 1} - \dfrac{12}{x^3 + 1}$

64. $\dfrac{5}{x + 2} + \dfrac{2}{x^2 - 2x + 4} - \dfrac{60}{x^3 + 8}$

65. $\dfrac{2x + 4}{x + 3} + \dfrac{3}{x} - \dfrac{6}{x^2 + 3x}$

66. $\dfrac{4x + 1}{x + 5} - \dfrac{2}{x} + \dfrac{10}{x^2 + 5x}$

67. $\dfrac{3}{(p - 2)^2} - \dfrac{5}{p - 2} + 4$

68. $\dfrac{8}{(3r - 1)^2} + \dfrac{2}{3r - 1} - 6$

*Add or subtract as indicated. Write all answers in lowest terms. See Examples 7 and 8.**

69. $\dfrac{3}{x^2 - 5x + 6} - \dfrac{2}{x^2 - 4x + 4}$

70. $\dfrac{2}{m^2 - 4m + 4} + \dfrac{3}{m^2 + m - 6}$

71. $\dfrac{5x}{x^2 + xy - 2y^2} - \dfrac{3x}{x^2 + 5xy - 6y^2}$

72. $\dfrac{6x}{6x^2 + 5xy - 4y^2} - \dfrac{2y}{9x^2 - 16y^2}$

73. $\dfrac{5x - y}{x^2 + xy - 2y^2} - \dfrac{3x + 2y}{x^2 + 5xy - 6y^2}$

74. $\dfrac{6x + 5y}{6x^2 + 5xy - 4y^2} - \dfrac{x + 2y}{9x^2 - 16y^2}$

75. $\dfrac{r + s}{3r^2 + 2rs - s^2} - \dfrac{s - r}{6r^2 - 5rs + s^2}$

76. $\dfrac{3y}{y^2 + yz - 2z^2} + \dfrac{4y - 1}{y^2 - z^2}$

77. $\dfrac{3}{x^2 + 4x + 4} + \dfrac{7}{x^2 + 5x + 6}$

78. $\dfrac{5}{x^2 + 6x + 9} - \dfrac{2}{x^2 + 4x + 3}$

Work each problem.

79. A *concours d'elegance* is a competition in which a maximum of 100 points is awarded to a car based on its general attractiveness. The function defined by the rational expression

$$c(x) = \dfrac{1010}{49(101 - x)} - \dfrac{10}{49}$$

approximates the cost, in thousands of dollars, of restoring a car so that it will win x points.

(a) Simplify the expression for $c(x)$ by performing the indicated subtraction.

(b) Use the simplified expression to determine how much it would cost to win 95 points.

---

* The authors wish to thank Joyce Nemeth of Broward Community College for her suggestions regarding some of these exercises.

80. A *cost-benefit model* expresses the cost of an undertaking in terms of the benefits received. One cost-benefit model gives the cost in thousands of dollars to remove x percent of a certain pollutant as

$$c(x) = \frac{6.7x}{100 - x}.$$

Another model produces the relationship

$$c(x) = \frac{6.5x}{102 - x}.$$

(a) What is the cost found by averaging the two models? (*Hint:* The average of two quantities is half their sum.)

(b) Using the two given models and your answer to part (a), find the cost to the nearest dollar to remove 95% ($x = 95$) of the pollutant.

(c) Average the two costs in part (b) from the given models. What do you notice about this result compared to the cost using the average of the two models?

RELATING CONCEPTS (EXERCISES 81–86)

For Individual or Group Work

In Example 6 we showed that

$$\frac{3}{x - 2} + \frac{5}{x} - \frac{6}{x^2 - 2x}$$

is equal to $\frac{8}{x}$. Algebra is, in a sense, a generalized form of arithmetic. **Work Exercises 81–86 in order,** to see how the algebra in this example is related to the arithmetic of common fractions.

81. Perform the following operations, and express your answer in lowest terms.

$$\frac{3}{7} + \frac{5}{9} - \frac{6}{63}$$

82. Substitute 9 for x in the given problem from Example 6. Compare this problem to the one given in Exercise 81. What do you notice?

83. Now substitute 9 for x in the answer given in Example 6. Do your results agree with the result you obtained in Exercise 81?

84. Replace x in the problem from Example 6 with the number of letters in your last name, assuming that this number is not 2. If your last name has two letters, let $x = 3$. Now predict the answer to your problem. Verify that your prediction is correct.

85. Why will $x = 2$ not work for the problem from Example 6?

86. What other value of x is not allowed in the problem from Example 6?

7.3 Complex Fractions

A **complex fraction** is an expression having a fraction in the numerator, denominator, or both. Examples of complex fractions include

$$\frac{1 + \dfrac{1}{x}}{2}, \qquad \frac{\dfrac{4}{y}}{6 - \dfrac{3}{y}}, \qquad \text{and} \qquad \frac{\dfrac{m^2 - 9}{m + 1}}{\dfrac{m + 3}{m^2 - 1}}.$$

OBJECTIVE 1   Simplify complex fractions by simplifying the numerator and denominator (Method 1).   There are two different methods for simplifying complex fractions.

Simplifying a Complex Fraction: Method 1

Step 1   Simplify the numerator and denominator separately.

Step 2   Divide by multiplying the numerator by the reciprocal of the denominator.

Step 3   Simplify the resulting fraction, if possible.

In Step 2, we are treating the complex fraction as a quotient of two rational expressions and dividing. Before performing this step, be sure that both the numerator and denominator are single fractions.

EXAMPLE 1   Simplifying Complex Fractions by Method 1

Use Method 1 to simplify each complex fraction.

(a) $\dfrac{\dfrac{x + 1}{x}}{\dfrac{x - 1}{2x}}$

Both the numerator and the denominator are already simplified, so divide by multiplying the numerator by the reciprocal of the denominator.

$$\frac{\dfrac{x + 1}{x}}{\dfrac{x - 1}{2x}} = \frac{x + 1}{x} \div \frac{x - 1}{2x} \qquad \text{Write as a division problem.}$$

$$= \frac{x + 1}{x} \cdot \frac{2x}{x - 1} \qquad \text{Multiply by the reciprocal of } \tfrac{x - 1}{2x}.$$

$$= \frac{2(x + 1)}{x - 1} \qquad \text{Multiply and simplify.}$$

(b) $\dfrac{2 + \dfrac{1}{y}}{3 - \dfrac{2}{y}} = \dfrac{\dfrac{2y}{y} + \dfrac{1}{y}}{\dfrac{3y}{y} - \dfrac{2}{y}}$

$= \dfrac{\dfrac{2y + 1}{y}}{\dfrac{3y - 2}{y}}$ Simplify the numerator and denominator.

$= \dfrac{2y + 1}{y} \div \dfrac{3y - 2}{y}$ Write as a division problem.

$= \dfrac{2y + 1}{y} \cdot \dfrac{y}{3y - 2}$ Multiply by the reciprocal of $\frac{3y - 2}{y}$.

$= \dfrac{2y + 1}{3y - 2}$ Multiply and simplify.

Now Try Exercises 5 and 9.

OBJECTIVE 2 Simplify complex fractions by multiplying by a common denominator (Method 2). The second method for simplifying complex fractions uses the identity property for multiplication.

> **Simplifying a Complex Fraction: Method 2**
>
> *Step 1* Multiply the numerator and denominator of the complex fraction by the least common denominator of the fractions in the numerator and the fractions in the denominator of the complex fraction.
>
> *Step 2* Simplify the resulting fraction, if possible.

EXAMPLE 2   Simplifying Complex Fractions by Method 2

Use Method 2 to simplify each complex fraction.

(a) $\dfrac{2 + \dfrac{1}{y}}{3 - \dfrac{2}{y}}$

Multiply the numerator and denominator by the LCD of all the fractions in the numerator and the denominator of the complex fraction. (This is the same as multiplying by 1.) Here the LCD is y.

$$\dfrac{2 + \dfrac{1}{y}}{3 - \dfrac{2}{y}} = \dfrac{2 + \dfrac{1}{y}}{3 - \dfrac{2}{y}} \cdot 1$$

$$= \frac{\left(2 + \dfrac{1}{y}\right) \cdot y}{\left(3 - \dfrac{2}{y}\right) \cdot y}$$

Multiply the numerator and denominator by y, since $\frac{y}{y} = 1$.

$$= \frac{2 \cdot y + \dfrac{1}{y} \cdot y}{3 \cdot y - \dfrac{2}{y} \cdot y}$$

Distributive property

$$= \frac{2y + 1}{3y - 2}$$

Compare this method of solution with that used in Example 1(b).

(b) $\dfrac{2p + \dfrac{5}{p-1}}{3p - \dfrac{2}{p}}$

The LCD is $p(p-1)$.

$$\frac{2p + \dfrac{5}{p-1}}{3p - \dfrac{2}{p}} = \frac{\left(2p + \dfrac{5}{p-1}\right) \cdot p(p-1)}{\left(3p - \dfrac{2}{p}\right) \cdot p(p-1)}$$

Multiply the numerator and denominator by the LCD.

$$= \frac{2p[p(p-1)] + \dfrac{5}{p-1} \cdot p(p-1)}{3p[p(p-1)] - \dfrac{2}{p} \cdot p(p-1)}$$

Distributive property

$$= \frac{2p[p(p-1)] + 5p}{3p[p(p-1)] - 2(p-1)}$$

$$= \frac{2p^3 - 2p^2 + 5p}{3p^3 - 3p^2 - 2p + 2}$$

This rational expression is in lowest terms.

Now Try Exercises 9 (using Method 2) and 11.

OBJECTIVE 3 Compare the two methods of simplifying complex fractions. Choosing whether to use Method 1 or Method 2 to simplify a complex fraction is usually a matter of preference. Some students prefer one method over the other, while other students feel comfortable with both methods and rely on practice with many examples to determine which method they will use on a particular problem. In the next example, we illustrate how to simplify a complex fraction using both methods so that you can observe the processes and decide for yourself the pros and cons of each method.

EXAMPLE 3 Simplifying Complex Fractions Using Both Methods

Use both Method 1 and Method 2 to simplify each complex fraction.

Method 1 | **Method 2**

Method 1

(a) $\dfrac{\dfrac{2}{x-3}}{\dfrac{5}{x^2-9}}$

$= \dfrac{\dfrac{2}{x-3}}{\dfrac{5}{(x-3)(x+3)}}$

$= \dfrac{2}{x-3} \div \dfrac{5}{(x-3)(x+3)}$

$= \dfrac{2}{x-3} \cdot \dfrac{(x-3)(x+3)}{5}$

$= \dfrac{2(x+3)}{5}$

Method 2

(a) $\dfrac{\dfrac{2}{x-3}}{\dfrac{5}{x^2-9}}$

$= \dfrac{\dfrac{2}{x-3} \cdot (x-3)(x+3)}{\dfrac{5}{(x-3)(x+3)} \cdot (x-3)(x+3)}$

$= \dfrac{2(x+3)}{5}$

Method 1

(b) $\dfrac{\dfrac{1}{x}+\dfrac{1}{y}}{\dfrac{1}{x^2}-\dfrac{1}{y^2}}$

$= \dfrac{\dfrac{y}{xy}+\dfrac{x}{xy}}{\dfrac{y^2}{x^2y^2}-\dfrac{x^2}{x^2y^2}}$

$= \dfrac{\dfrac{y+x}{xy}}{\dfrac{y^2-x^2}{x^2y^2}}$

$= \dfrac{y+x}{xy} \div \dfrac{y^2-x^2}{x^2y^2}$

$= \dfrac{y+x}{xy} \cdot \dfrac{x^2y^2}{(y-x)(y+x)}$

$= \dfrac{xy}{y-x}$

Method 2

(b) $\dfrac{\dfrac{1}{x}+\dfrac{1}{y}}{\dfrac{1}{x^2}-\dfrac{1}{y^2}}$

$= \dfrac{\left(\dfrac{1}{x}+\dfrac{1}{y}\right) \cdot x^2y^2}{\left(\dfrac{1}{x^2}-\dfrac{1}{y^2}\right) \cdot x^2y^2}$

$= \dfrac{xy^2+x^2y}{y^2-x^2}$

$= \dfrac{xy(y+x)}{(y+x)(y-x)}$

$= \dfrac{xy}{y-x}$

Now Try Exercises 13 and 17.

OBJECTIVE 4 Simplify rational expressions with negative exponents. Rational expressions and complex fractions sometimes involve negative exponents. To simplify such expressions, we begin by rewriting the expressions with only positive exponents.

EXAMPLE 4 Simplifying Rational Expressions with Negative Exponents

Simplify each expression, using only positive exponents in the answer.

(a) $\dfrac{m^{-1} + p^{-2}}{2m^{-2} - p^{-1}}$

First write the expression with only positive exponents using the definition of a negative exponent.

$$\frac{m^{-1} + p^{-2}}{2m^{-2} - p^{-1}} = \frac{\dfrac{1}{m} + \dfrac{1}{p^2}}{\dfrac{2}{m^2} - \dfrac{1}{p}}$$

Note that the 2 in $2m^{-2}$ is *not* raised to the -2 power (since m is the base for the exponent -2), so $2m^{-2} = \dfrac{2}{m^2}$. Simplify the complex fraction using Method 2, multiplying the numerator and denominator by the LCD, $m^2 p^2$.

$$\frac{\dfrac{1}{m} + \dfrac{1}{p^2}}{\dfrac{2}{m^2} - \dfrac{1}{p}} = \frac{m^2 p^2 \left(\dfrac{1}{m} + \dfrac{1}{p^2} \right)}{m^2 p^2 \left(\dfrac{2}{m^2} - \dfrac{1}{p} \right)}$$

$$= \frac{m^2 p^2 \cdot \dfrac{1}{m} + m^2 p^2 \cdot \dfrac{1}{p^2}}{m^2 p^2 \cdot \dfrac{2}{m^2} - m^2 p^2 \cdot \dfrac{1}{p}} \qquad \text{Distributive property}$$

$$= \frac{m p^2 + m^2}{2 p^2 - m^2 p} \qquad \text{Lowest terms}$$

(b) $\dfrac{k^{-1}}{k^{-1} + 1} = \dfrac{\dfrac{1}{k}}{\dfrac{1}{k} + 1}$ Write with positive exponents.

$$= \frac{k \cdot \dfrac{1}{k}}{k \left(\dfrac{1}{k} + 1 \right)} \qquad \text{Use Method 2.}$$

$$= \frac{k \cdot \dfrac{1}{k}}{k \cdot \dfrac{1}{k} + k \cdot 1}$$

$$= \frac{1}{1 + k}$$

Now Try Exercises 29 and 31.

7.3 EXERCISES

1. Explain in your own words the two methods of simplifying complex fractions.

2. Method 2 of simplifying complex fractions says that we can multiply both the numerator and the denominator of the complex fraction by the same nonzero expression. What property of real numbers from Section 1.4 justifies this method?

Use either method to simplify each complex fraction. See Examples 1–3.

3. $\dfrac{\dfrac{12}{x-1}}{\dfrac{6}{x}}$

4. $\dfrac{\dfrac{24}{t+4}}{\dfrac{6}{t}}$

5. $\dfrac{\dfrac{k+1}{2k}}{\dfrac{3k-1}{4k}}$

6. $\dfrac{\dfrac{1-r}{4r}}{\dfrac{-1-r}{8r}}$

7. $\dfrac{\dfrac{4z^2x^4}{9}}{\dfrac{12x^2z^5}{15}}$

8. $\dfrac{\dfrac{3y^2x^3}{8}}{\dfrac{9y^3x^4}{16}}$

9. $\dfrac{\dfrac{1}{x}+1}{-\dfrac{1}{x}+1}$

10. $\dfrac{\dfrac{2}{k}-1}{\dfrac{2}{k}+1}$

11. $\dfrac{\dfrac{3}{x}+\dfrac{3}{y}}{\dfrac{3}{x}-\dfrac{3}{y}}$

12. $\dfrac{\dfrac{4}{t}-\dfrac{4}{s}}{\dfrac{4}{t}+\dfrac{4}{s}}$

13. $\dfrac{\dfrac{8x-24y}{10}}{\dfrac{x-3y}{5x}}$

14. $\dfrac{\dfrac{10x-5y}{12}}{\dfrac{2x-y}{6y}}$

15. $\dfrac{\dfrac{x^2-16y^2}{xy}}{\dfrac{1}{y}-\dfrac{4}{x}}$

16. $\dfrac{\dfrac{2}{s}-\dfrac{3}{t}}{\dfrac{4t^2-9s^2}{st}}$

17. $\dfrac{y-\dfrac{y-3}{3}}{\dfrac{4}{9}+\dfrac{2}{3y}}$

18. $\dfrac{p-\dfrac{p+2}{4}}{\dfrac{3}{4}-\dfrac{5}{2p}}$

19. $\dfrac{\dfrac{x+2}{x}+\dfrac{1}{x+2}}{\dfrac{5}{x}+\dfrac{x}{x+2}}$

20. $\dfrac{\dfrac{y+3}{y}-\dfrac{4}{y-1}}{\dfrac{y}{y-1}+\dfrac{1}{y}}$

RELATING CONCEPTS (EXERCISES 21–26)

For Individual or Group Work

Simplifying a complex fraction by Method 1 is a good way to review the methods of adding, subtracting, multiplying, and dividing rational expressions. Method 2 gives a good review of the fundamental principle of rational expressions. Refer to the following complex fraction and **work Exercises 21–26 in order.**

$$\dfrac{\dfrac{4}{m}+\dfrac{m+2}{m-1}}{\dfrac{m+2}{m}-\dfrac{2}{m-1}}$$

21. Add the fractions in the numerator.

22. Subtract as indicated in the denominator.

23. Divide your answer from Exercise 21 by your answer from Exercise 22.

24. Go back to the original complex fraction and find the LCD of all denominators.

25. Multiply the numerator and denominator of the complex fraction by your answer from Exercise 24.

26. Your answers for Exercises 23 and 25 should be the same. Write a paragraph comparing the two methods. Which method do you prefer? Explain why.

Simplify each expression, using only positive exponents in your answer. See Example 4.

27. $\dfrac{1}{x^{-2} + y^{-2}}$

28. $\dfrac{1}{p^{-2} - q^{-2}}$

29. $\dfrac{x^{-2} + y^{-2}}{x^{-1} + y^{-1}}$

30. $\dfrac{x^{-1} - y^{-1}}{x^{-2} - y^{-2}}$

31. $\dfrac{x^{-1} + 2y^{-1}}{2y + 4x}$

32. $\dfrac{a^{-2} - 4b^{-2}}{3b - 6a}$

33. (a) Start with the complex fraction $\dfrac{\frac{3}{mp} - \frac{4}{p} + \frac{8}{m}}{2m^{-1} - 3p^{-1}}$ and write it so that there are no negative exponents in your expression.

(b) Explain why $\dfrac{\frac{3}{mp} - \frac{4}{p} + \frac{8}{m}}{\frac{1}{2m} - \frac{1}{3p}}$ would *not* be a correct response in part (a).

(c) Simplify the complex fraction in part (a).

34. Is $\dfrac{m^{-1} + n^{-1}}{m^{-2} + n^{-2}} = \dfrac{m^2 + n^2}{m + n}$ a true statement? Explain why or why not.

7.4 Equations with Rational Expressions and Graphs

OBJECTIVES

1 Determine the domain of a rational equation.

2 Solve rational equations.

3 Recognize the graph of a rational function.

4 Solve rational equations using a graphing calculator.

At the beginning of this chapter, we defined the domain of a rational expression as the set of all possible values of the variable. Any value that makes the denominator 0 is excluded.

OBJECTIVE 1 Determine the domain of a rational equation. The **domain of a rational equation** is the intersection (overlap) of the domains of the rational expressions in the equation.

EXAMPLE 1 Determining the Domains of Rational Equations

Find the domain of each equation.

(a) $\dfrac{2}{x} - \dfrac{3}{2} = \dfrac{7}{2x}$

The domains of the three rational terms of the equation $\frac{2}{x} - \frac{3}{2} = \frac{7}{2x}$ are, in order, $\{x \mid x \neq 0\}$, $(-\infty, \infty)$, and $\{x \mid x \neq 0\}$. The intersection of these three domains is all real numbers except 0, which may be written $\{x \mid x \neq 0\}$.

(b) $\dfrac{2}{x-3} - \dfrac{3}{x+3} = \dfrac{12}{x^2 - 9}$

The domains of the three terms are, respectively, $\{x \mid x \neq 3\}$, $\{x \mid x \neq -3\}$, and $\{x \mid x \neq \pm 3\}$. The domain of the equation is the intersection of the three domains, all real numbers except 3 and -3, written $\{x \mid x \neq \pm 3\}$.

Now Try Exercises 5 and 9.

OBJECTIVE 2 Solve rational equations.   The easiest way to solve most equations with rational expressions is to multiply all terms in the equation by the least common denominator. This step will clear the equation of all denominators, as the next examples show. *We can do this only with equations, not expressions.*

Because the first step in solving a rational equation is to multiply both sides of the equation by a common denominator, it is *necessary* to either check the potential solutions or verify that they are in the domain.

CAUTION   When both sides of an equation are multiplied by a *variable* expression, the resulting "solutions" may not satisfy the original equation. You *must* either determine and observe the domain or check all potential solutions in the original equation. *It is wise to do both.*

EXAMPLE 2 Solving an Equation with Rational Expressions

Solve $\dfrac{2}{x} - \dfrac{3}{2} = \dfrac{7}{2x}$.

The domain, which excludes 0, was found in Example 1(a).

$$2x\left(\frac{2}{x} - \frac{3}{2}\right) = 2x\left(\frac{7}{2x}\right) \qquad \text{Multiply by the LCD, 2x.}$$

$$2x\left(\frac{2}{x}\right) - 2x\left(\frac{3}{2}\right) = 2x\left(\frac{7}{2x}\right) \qquad \text{Distributive property}$$

$$4 - 3x = 7 \qquad \text{Multiply.}$$

$$-3x = 3 \qquad \text{Subtract 4.}$$

$$x = -1 \qquad \text{Divide by } -3.$$

Check: $\qquad \dfrac{2}{x} - \dfrac{3}{2} = \dfrac{7}{2x} \qquad$ Original equation

$$\dfrac{2}{-1} - \dfrac{3}{2} = \dfrac{7}{2(-1)} \quad ? \qquad \text{Let } x = -1.$$

$$-\dfrac{7}{2} = -\dfrac{7}{2} \qquad \text{True}$$

The solution set is $\{-1\}$.

Now Try Exercise 11.

EXAMPLE 3 Solving an Equation with No Solution

Solve $\dfrac{2}{x - 3} - \dfrac{3}{x + 3} = \dfrac{12}{x^2 - 9}$.

Using the result from Example 1(b), we know that the domain excludes 3 and -3. We multiply each side by the LCD, $(x + 3)(x - 3)$.

$$(x + 3)(x - 3)\left(\frac{2}{x - 3} - \frac{3}{x + 3}\right) = (x + 3)(x - 3)\left(\frac{12}{x^2 - 9}\right)$$

$2(x + 3) - 3(x - 3) = 12$	Distributive property
$2x + 6 - 3x + 9 = 12$	Distributive property
$-x + 15 = 12$	Combine terms.
$-x = -3$	Subtract 15.
$x = 3$	Divide by -1.

Since 3 is not in the domain, it cannot be a solution of the equation. Substituting 3 in the original equation shows why.

$$\frac{2}{x - 3} - \frac{3}{x + 3} = \frac{12}{x^2 - 9}$$

$$\frac{2}{3 - 3} - \frac{3}{3 + 3} = \frac{12}{3^2 - 9} \qquad ? \qquad \text{Let } x = 3.$$

$$\frac{2}{0} - \frac{3}{6} = \frac{12}{0} \qquad ?$$

Since division by 0 is undefined, the given equation has no solution and the solution set is $\emptyset$.

Now Try Exercise 25.

EXAMPLE 4 Solving an Equation with Rational Expressions

Solve $\dfrac{3}{p^2 + p - 2} - \dfrac{1}{p^2 - 1} = \dfrac{7}{2(p^2 + 3p + 2)}$.

Factor each denominator to find the LCD, $2(p - 1)(p + 2)(p + 1)$. The domain excludes 1, -2, and -1. Multiply each side by the LCD.

$$2(p - 1)(p + 2)(p + 1)\left[\frac{3}{(p + 2)(p - 1)} - \frac{1}{(p + 1)(p - 1)}\right]$$

$$= 2(p - 1)(p + 2)(p + 1)\left[\frac{7}{2(p + 2)(p + 1)}\right]$$

$2 \cdot 3(p + 1) - 2(p + 2) = 7(p - 1)$	Distributive property
$6p + 6 - 2p - 4 = 7p - 7$	Distributive property
$4p + 2 = 7p - 7$	Combine terms.
$9 = 3p$	
$3 = p$	

Note that 3 is in the domain; substitute 3 for p in the original equation to check that the solution set is $\{3\}$.

Now Try Exercise 37.

▇ **EXAMPLE 5** Solving an Equation That Leads to a Quadratic Equation

Solve $\dfrac{2}{3x+1} = \dfrac{1}{x} - \dfrac{6x}{3x+1}$.

Since the denominator $3x + 1$ cannot equal 0, $-\frac{1}{3}$ is excluded from the domain, as is 0. Multiply each side by the LCD, $x(3x + 1)$.

$$x(3x+1)\left(\frac{2}{3x+1}\right) = x(3x+1)\left(\frac{1}{x} - \frac{6x}{3x+1}\right)$$

$$2x = 3x + 1 - 6x^2$$

Since this equation is quadratic, write it in standard form with 0 on the right side.

$$6x^2 - 3x + 2x - 1 = 0$$
$$6x^2 - x - 1 = 0 \qquad \text{Standard form}$$
$$(3x+1)(2x-1) = 0 \qquad \text{Factor.}$$
$$3x + 1 = 0 \quad \text{or} \quad 2x - 1 = 0 \qquad \text{Zero-factor property}$$
$$x = -\frac{1}{3} \quad \text{or} \quad x = \frac{1}{2}$$

Because $-\frac{1}{3}$ is not in the domain of the equation, it is not a solution. Check that the solution set is $\left\{\frac{1}{2}\right\}$.

Now Try Exercise 31.

OBJECTIVE 3 Recognize the graph of a rational function. As mentioned in Section 7.1, a function defined by a rational expression is a *rational function*. Because one or more values of x are excluded from the domain of most rational functions, their graphs are usually *discontinuous*. That is, there will be one or more breaks in the graph. For example, we use point plotting and observing the domain to graph the simple rational function defined by

$$f(x) = \frac{1}{x}.$$

The domain of this function includes all real numbers except 0. Thus, there will be no point on the graph with $x = 0$. The vertical line with equation $x = 0$ is called a **vertical asymptote** of the graph. We show some typical ordered pairs in the table for both negative and positive x-values.

x	-3	-2	-1	$-.5$	$-.25$	$-.1$	$.1$	$.25$	$.5$	1	2	3
y	$-\frac{1}{3}$	$-\frac{1}{2}$	-1	-2	-4	-10	10	4	2	1	$\frac{1}{2}$	$\frac{1}{3}$

Notice that the closer positive values of x are to 0, the larger y is. Similarly, the closer negative values of x are to 0, the smaller (more negative) y is. Using this observation, the fact that the domain excludes 0, and plotting the points found above produces the graph in Figure 2.

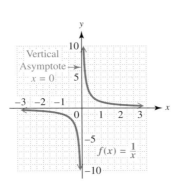

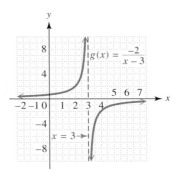

FIGURE **2**   FIGURE **3**

The graph of

$$g(x) = \frac{-2}{x - 3},$$

is shown in Figure 3. Some ordered pairs that belong to the function are shown in the table.

x	-2	-1	0	1	2	2.5	2.75	3.25	3.5	4	5	6	7
y	$\frac{2}{5}$	$\frac{1}{2}$	$\frac{2}{3}$	1	2	4	8	-8	-4	-2	-1	$-\frac{2}{3}$	$-\frac{1}{2}$

There is no point on the graph for $x = 3$, because 3 is excluded from the domain. The dashed line $x = 3$ represents the asymptote and is not part of the graph. As suggested by the points from the table, the graph gets closer to the vertical asymptote $x = 3$ as the x-values get closer to 3.

OBJECTIVE 4 Solve rational equations using a graphing calculator. Earlier, we solved linear and quadratic equations using a graphing calculator. The procedure is similar with rational equations. Because rational functions usually have values of x that are excluded from the domain, a calculator in *connected mode* may show a vertical line on the screen where an asymptote occurs. Using *dot mode* will usually give a more realistic picture.

In Figure 4, we show the graph of

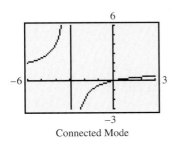

Connected Mode

$$g(x) = \frac{x}{x + 3}$$

generated in connected mode and dot mode. If dot mode is used, we must remember that, theoretically, the function is continuous (unbroken) on its domain, in this case, $(-\infty, -3) \cup (-3, \infty)$. As before, the x-intercepts of the graph give the solutions of the equation

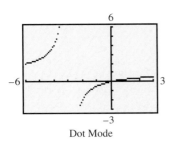

Dot Mode

FIGURE **4**

$$\frac{x}{x + 3} = 0.$$

From the graph, since the x-intercept is $(0, 0)$, we see that the only solution of this equation is $\{0\}$.

■ **EXAMPLE 6** Finding Solutions of a Rational Equation from a Graphing Calculator Screen

Two views of the graph of $f(x) = x^{-2} + x^{-1} - 1.5$ are shown in Figure 5. Use the graph to determine the solution set of $x^{-2} + x^{-1} - 1.5 = 0$.

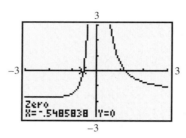 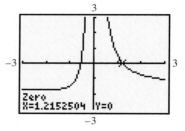

FIGURE 5

Look at the bottom of each calculator screen for the x-values for which $y = 0$, $-.5485838$ and 1.2152504. These are the solutions of the equation. The solution set is $\{-.5485838, 1.2152504\}$.

Now Try Exercise 49.

7.4 EXERCISES

As explained in this section, any values that would cause a denominator to equal 0 *must be excluded from the domain and consequently as solutions of an equation that has variable expressions in the denominators.* **(a)** *Without actually solving the equation, list all possible numbers that would have to be rejected if they appeared as potential solutions.* **(b)** *Then give the domain using set notation. See Example 1.*

1. $\dfrac{1}{x+1} - \dfrac{1}{x-2} = 0$

2. $\dfrac{3}{x+4} - \dfrac{2}{x-9} = 0$

3. $\dfrac{5}{3x+5} - \dfrac{1}{x} = \dfrac{1}{2x+3}$

4. $\dfrac{6}{4x+7} - \dfrac{3}{x} = \dfrac{5}{6x-13}$

5. $\dfrac{1}{3x} + \dfrac{1}{2x} = \dfrac{x}{3}$

6. $\dfrac{5}{6x} - \dfrac{8}{2x} = \dfrac{x}{4}$

7. $\dfrac{3x+1}{x-4} = \dfrac{6x+5}{2x-7}$

8. $\dfrac{4x-1}{2x+3} = \dfrac{12x-25}{6x-2}$

9. $\dfrac{2}{x^2-x} + \dfrac{1}{x+3} = \dfrac{4}{x-2}$

✍ **10.** Is it possible that any potential solutions to the equation

$$\dfrac{x+7}{4} - \dfrac{x+3}{3} = \dfrac{x}{12}$$

would have to be rejected? Explain.

Solve each equation. See Examples 2–5.

11. $\dfrac{-5}{2x} + \dfrac{3}{4x} = \dfrac{-7}{4}$

12. $\dfrac{6}{5x} - \dfrac{2}{3x} = \dfrac{-8}{45}$

13. $x - \dfrac{24}{x} = -2$

14. $p + \dfrac{15}{p} = -8$

15. $\dfrac{x-4}{x+6} = \dfrac{2x+3}{2x-1}$

16. $\dfrac{5x-8}{x+2} = \dfrac{5x-1}{x+3}$

17. $\dfrac{3x+1}{x-4} = \dfrac{6x+5}{2x-7}$

18. $\dfrac{4x-1}{2x+3} = \dfrac{12x-25}{6x-2}$

19. $\dfrac{1}{y-1} + \dfrac{5}{12} = \dfrac{-2}{3y-3}$

20. $\dfrac{4}{m+2} - \dfrac{11}{9} = \dfrac{1}{3m+6}$

21. $\dfrac{-2}{3t-6} - \dfrac{1}{36} = \dfrac{-3}{4t-8}$

22. $\dfrac{3}{4m+2} = \dfrac{17}{2} - \dfrac{7}{2m+1}$

23. $\dfrac{3}{k+2} - \dfrac{2}{k^2-4} = \dfrac{1}{k-2}$

24. $\dfrac{3}{x-2} + \dfrac{21}{x^2-4} = \dfrac{14}{x+2}$

25. $\dfrac{1}{y+2} + \dfrac{3}{y+7} = \dfrac{5}{y^2+9y+14}$

26. $\dfrac{1}{t+3} + \dfrac{4}{t+5} = \dfrac{2}{t^2+8t+15}$

27. $\dfrac{9}{x} + \dfrac{4}{6x-3} = \dfrac{2}{6x-3}$

28. $\dfrac{5}{n} + \dfrac{4}{6-3n} = \dfrac{2n}{6-3n}$

29. $\dfrac{6}{w+3} + \dfrac{-7}{w-5} = \dfrac{-48}{w^2-2w-15}$

30. $\dfrac{2}{r-5} + \dfrac{3}{2r+1} = \dfrac{22}{2r^2-9r-5}$

31. $\dfrac{x}{x-3} + \dfrac{4}{x+3} = \dfrac{18}{x^2-9}$

32. $\dfrac{2x}{x-3} + \dfrac{4}{x+3} = \dfrac{-24}{x^2-9}$

33. $\dfrac{1}{x+4} + \dfrac{x}{x-4} = \dfrac{-8}{x^2-16}$

34. $\dfrac{5}{x-4} - \dfrac{3}{x-1} = \dfrac{x^2-1}{x^2-5x+4}$

35. $\dfrac{2}{4x+7} + \dfrac{x}{3} = \dfrac{6}{12x+21}$

36. $\dfrac{2}{k^2+k-6} + \dfrac{1}{k^2-k-2} = \dfrac{4}{k^2+4k+3}$

37. $\dfrac{5}{p^2+3p+2} - \dfrac{3}{p^2-4} = \dfrac{1}{p^2-p-2}$

38. $\dfrac{5x+14}{x^2-9} = \dfrac{-2x^2-5x+2}{x^2-9} + \dfrac{2x+4}{x-3}$

39. $\dfrac{4x-7}{4x^2-9} = \dfrac{-2x^2+5x-4}{4x^2-9} + \dfrac{x+1}{2x+3}$

40. Professor Dan Abbey asked the following question on a test: What is the solution set of $\dfrac{x+3}{x+3} = 1$? Only one student answered it correctly.

  (a) What is the solution set?
  (b) Many students answered {all real numbers}. Why is this not correct?

Graph each rational function. Give the equation of the vertical asymptote. See Figures 2 and 3.

41. $f(x) = \dfrac{2}{x}$

42. $f(x) = \dfrac{3}{x}$

43. $f(x) = \dfrac{1}{x-2}$

44. $f(x) = \dfrac{1}{x+2}$

Solve each problem.

45. The average number of vehicles waiting in line to enter a sports arena parking area is modeled by the function defined by

$$w(x) = \dfrac{x^2}{2(1-x)},$$

where x is a quantity between 0 and 1 known as the *traffic intensity*. (*Source:* Mannering, F., and W. Kilareski, *Principles of Highway Engineering and Traffic Control,* John Wiley and Sons, 1990.) To the nearest tenth, find the average number of vehicles waiting for each traffic intensity.

  (a) .1 **(b)** .8 **(c)** .9
  (d) What happens to waiting time as traffic intensity increases?

46. The percent of deaths caused by smoking is modeled by the rational function defined by

$$p(x) = \frac{x - 1}{x},$$

where x is the number of times a smoker is more likely to die of lung cancer than a nonsmoker. This is called the *incidence rate*. (*Source:* Walker, A., *Observation and Inference: An Introduction to the Methods of Epidemiology,* Epidemiology Resources Inc., 1991.) For example, $x = 10$ means that a smoker is 10 times more likely than a nonsmoker to die of lung cancer.

(a) Find $p(x)$ if x is 10.
(b) For what values of x is $p(x) = 80\%$? (*Hint:* Change 80% to a decimal.)
(c) Can the incidence rate equal 0? Explain.

47. The force required to keep a 2000-lb car, going 30 mph, from skidding on a curve is given by

$$F(r) = \frac{225,000}{r},$$

where r is the radius of the curve in feet.

(a) What radius must a curve have if a force of 450 lb is needed to keep the car from skidding?
(b) As the radius of the curve is lengthened, how is the force affected?

48. The amount of heating oil produced (in gallons per day) by an oil refinery is modeled by the rational function defined by

$$f(x) = \frac{125,000 - 25x}{125 + 2x},$$

where x is the amount of gasoline produced (in hundreds of gallons per day). Suppose the refinery must produce 300 gal of heating oil per day to meet the needs of its customers.

(a) How much gasoline will be produced per day?

(b) The graph of f is shown in the figure. Use it to decide what happens to the amount of gasoline (x) produced as the amount of heating oil (y) produced increases.

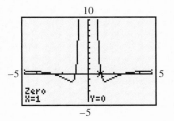

TECHNOLOGY INSIGHTS (EXERCISES 49–52)

Two views of the graph of $f(x) = 7x^{-4} - 8x^{-2} + 1$ are shown. Use the graphs to respond to Exercises 49 and 50. See Example 6.

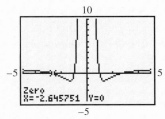

49. How many solutions does the equation $f(x) = 0$ have?

50. Give the solutions of $f(x) = 0$ (to the nearest hundredth if necessary).

In Exercises 51 and 52, use the graph to determine the solution set of the equation $f(x) = 0$. All solutions are integers.

51. $f(x) = \dfrac{x^3 - x^2 - 6x}{x^2 - 1}$

52. $f(x) = \dfrac{-x^3 - x^2 + 2x}{x^2}$

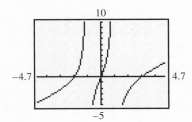

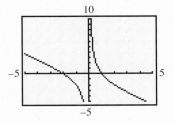

SUMMARY EXERCISES ON OPERATIONS AND EQUATIONS WITH RATIONAL EXPRESSIONS

A common student error is to confuse an equation, such as $\frac{x}{2} + \frac{x}{3} = -5$, with an operation, such as $\frac{x}{2} + \frac{x}{3}$. Look for the equals sign to distinguish between them. Equations are solved for a numerical answer, while problems involving operations result in simplified expressions.

<table>
<tr><td>

Solving an Equation

Solve: $\dfrac{x}{2} + \dfrac{x}{3} = -5$.

Multiply each side by the LCD, 6.

$$6\left(\dfrac{x}{2} + \dfrac{x}{3}\right) = 6(-5)$$

$$6\left(\dfrac{x}{2}\right) + 6\left(\dfrac{x}{3}\right) = 6(-5)$$

$$3x + 2x = -30$$
$$5x = -30$$
$$x = -6$$

Check that the solution set is $\{-6\}$.

</td><td>

Performing an Operation

Add: $\dfrac{x}{2} + \dfrac{x}{3}$.

Write both fractions with the LCD, 6.

$$\dfrac{x}{2} + \dfrac{x}{3} = \dfrac{x \cdot 3}{2 \cdot 3} + \dfrac{x \cdot 2}{3 \cdot 2}$$

$$= \dfrac{3x}{6} + \dfrac{2x}{6}$$

$$= \dfrac{3x + 2x}{6}$$

$$= \dfrac{5x}{6}$$

</td></tr>
</table>

In each exercise, identify as an equation *or an* operation. *Then perform the indicated operation or solve the given equation, as appropriate.*

1. $\dfrac{x}{2} - \dfrac{x}{4} = 5$

2. $\dfrac{4x - 20}{x^2 - 25} \cdot \dfrac{(x + 5)^2}{10}$

3. $\dfrac{6}{7x} - \dfrac{4}{x}$

4. $\dfrac{\dfrac{1}{x} + \dfrac{1}{y}}{\dfrac{1}{x} - \dfrac{1}{y}}$

5. $\dfrac{5}{7t} = \dfrac{52}{7} - \dfrac{3}{t}$

6. $\dfrac{x - 5}{3} + \dfrac{1}{3} = \dfrac{x - 2}{5}$

7. $\dfrac{7}{6x} + \dfrac{5}{8x}$

8. $\dfrac{4}{x} - \dfrac{8}{x + 1} = 0$

9. $\dfrac{\dfrac{6}{x + 1} - \dfrac{1}{x}}{\dfrac{2}{x} - \dfrac{4}{x + 1}}$

10. $\dfrac{8}{r + 2} - \dfrac{7}{4r + 8}$

11. $\dfrac{x}{x + y} + \dfrac{2y}{x - y}$

12. $\dfrac{3p^2 - 6p}{p + 5} \div \dfrac{p^2 - 4}{8p + 40}$

13. $\dfrac{x - 2}{9} \cdot \dfrac{5}{8 - 4x}$

14. $\dfrac{a - 4}{3} + \dfrac{11}{6} = \dfrac{a + 1}{2}$

15. $\dfrac{b^2 + b - 6}{b^2 + 2b - 8} \cdot \dfrac{b^2 + 8b + 16}{3b + 12}$

16. $\dfrac{10z^2 - 5z}{3z^3 - 6z^2} \div \dfrac{2z^2 + 5z - 3}{z^2 + z - 6}$

17. $\dfrac{5}{x^2 - 2x} - \dfrac{3}{x^2 - 4}$

18. $\dfrac{6}{t + 1} + \dfrac{4}{5t + 5} = \dfrac{34}{15}$

19. $\dfrac{\dfrac{5}{x} - \dfrac{3}{y}}{\dfrac{9x^2 - 25y^2}{x^2 y}}$

20. $\dfrac{-2}{a^2 + 2a - 3} - \dfrac{5}{3 - 3a} = \dfrac{4}{3a + 9}$

21. $\dfrac{4y^2 - 13y + 3}{2y^2 - 9y + 9} \div \dfrac{4y^2 + 11y - 3}{6y^2 - 5y - 6}$

22. $\dfrac{8}{3k + 9} - \dfrac{8}{15} = \dfrac{2}{5k + 15}$

23. $\dfrac{3r}{r - 2} = 1 + \dfrac{6}{r - 2}$

24. $\dfrac{6z^2 - 5z - 6}{6z^2 + 5z - 6} \cdot \dfrac{12z^2 - 17z + 6}{12z^2 - z - 6}$

25. $\dfrac{-1}{3 - x} - \dfrac{2}{x - 3}$

26. $\dfrac{\dfrac{t}{4} - \dfrac{1}{t}}{1 + \dfrac{t + 4}{t}}$

27. $\dfrac{2}{y + 1} - \dfrac{3}{y^2 - y - 2} = \dfrac{3}{y - 2}$

28. $\dfrac{7}{2x^2 - 8x} + \dfrac{3}{x^2 - 16}$

29. $\dfrac{3}{y - 3} - \dfrac{3}{y^2 - 5y + 6} = \dfrac{2}{y - 2}$

30. $\dfrac{2k + \dfrac{5}{k - 1}}{3k - \dfrac{2}{k}}$

7.5 Applications of Rational Expressions

OBJECTIVES

1 Find the value of an unknown variable in a formula.

2 Solve a formula for a specified variable.

3 Solve applications using proportions.

4 Solve applications about distance, rate, and time.

5 Solve applications about work rates.

OBJECTIVE 1 Find the value of an unknown variable in a formula. Formulas may contain rational expressions, as does $t = \frac{d}{r}$. We now show how to work with formulas of this type.

EXAMPLE 1 Finding the Value of a Variable in a Formula

In physics the focal length, f, of a lens is given by the formula

$$\frac{1}{f} = \frac{1}{p} + \frac{1}{q}.$$

In the formula, p is the distance from the object to the lens and q is the distance from the lens to the image. See Figure 6. Find q if $p = 20$ cm and $f = 10$ cm.

Replace f with 10 and p with 20.

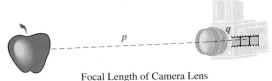

Focal Length of Camera Lens

FIGURE 6

$$\frac{1}{f} = \frac{1}{p} + \frac{1}{q}$$

$$\frac{1}{10} = \frac{1}{20} + \frac{1}{q} \qquad \text{Let } f = 10, p = 20.$$

$$20q \cdot \frac{1}{10} = 20q\left(\frac{1}{20} + \frac{1}{q}\right) \qquad \text{Multiply by the LCD, } 20q.$$

$$2q = q + 20$$

$$q = 20$$

The distance from the lens to the image is 20 cm.

Now Try Exercise 5.

OBJECTIVE 2 Solve a formula for a specified variable. Recall that the goal in solving for a specified variable is to isolate it on one side of the equals sign.

■ EXAMPLE 2 Solving a Formula for a Specified Variable

Solve $\dfrac{1}{f} = \dfrac{1}{p} + \dfrac{1}{q}$ for p.

$$fpq \cdot \frac{1}{f} = fpq\left(\frac{1}{p} + \frac{1}{q}\right) \qquad \text{Multiply by the LCD, } fpq.$$

$$pq = fq + fp \qquad \text{Distributive property}$$

Transform the equation so that the terms with p (the specified variable) are on the same side. One way to do this is to subtract fp from each side.

$$pq - fp = fq \qquad \text{Subtract } fp.$$

$$p(q - f) = fq \qquad \text{Factor out } p.$$

$$p = \frac{fq}{q - f} \qquad \text{Divide by } q - f.$$

Now Try Exercise 11.

■ EXAMPLE 3 Solving a Formula for a Specified Variable

Solve $I = \dfrac{nE}{R + nr}$ for n.

$$(R + nr)I = (R + nr)\frac{nE}{R + nr} \qquad \text{Multiply by } R + nr.$$

$$RI + nrI = nE$$

$$RI = nE - nrI \qquad \text{Subtract } nrI.$$

$$RI = n(E - rI) \qquad \text{Factor out } n.$$

$$\frac{RI}{E - rI} = n \qquad \text{Divide by } E - rI.$$

Now Try Exercise 15.

CAUTION Refer to the steps in Examples 2 and 3 that factor out the desired variable. This is a step that often gives students difficulty. Remember that the variable for which you are solving *must* be a factor on only one side of the equation, so each side can be divided by the remaining factor in the last step.

We now solve problems that translate into equations with rational expressions, using the six-step problem-solving method from Chapter 2.

OBJECTIVE 3 Solve applications using proportions. A **ratio** is a comparison of two quantities. The ratio of a to b may be written in any of the following ways:

$$a \text{ to } b, \qquad a:b, \qquad \text{or} \qquad \frac{a}{b}.$$

Ratios are usually written as quotients in algebra. A **proportion** is a statement that two ratios are equal. Proportions are a useful and important type of rational equation.

EXAMPLE 4 Solving a Proportion

In 1999, 16 of every 100 Americans had no health insurance coverage. The population at that time was about 274 million. How many million had no health insurance? (*Source:* U.S. Bureau of the Census.)

Step 1 **Read** the problem.

Step 2 **Assign a variable.** Let $x =$ the number (in millions) who had no health insurance.

Step 3 **Write an equation.** To get an equation, set up a proportion. The ratio x to 274 should equal the ratio 16 to 100. Write the proportion and solve the equation.

$$\frac{16}{100} = \frac{x}{274}$$

Step 4 **Solve.** $27{,}400\left(\dfrac{16}{100}\right) = 27{,}400\left(\dfrac{x}{274}\right)$ Multiply by a common denominator.

$$4384 = 100x \qquad \text{Simplify.}$$
$$x = 43.84$$

Step 5 **State the answer.** There were 43.84 million Americans with no health insurance in 1999.

Step 6 **Check** that the ratio of this number to 274 million is equivalent to $\frac{16}{100}$.

Now Try Exercise 31.

EXAMPLE 5 Solving a Proportion Involving Rates

Marissa's car uses 10 gal of gas to travel 210 mi. She has 5 gal of gas in the car, and she wants to know how much more gas she will need to drive 640 mi. If we assume the car continues to use gas at the same rate, how many more gallons will she need?

Step 1 **Read** the problem.

Step 2 **Assign a variable.** Let $x =$ the additional number of gallons of gas needed.

Step 3 **Write an equation.** To get an equation, set up a proportion.

$$\begin{matrix} \text{gallons} \rightarrow \\ \text{miles} \rightarrow \end{matrix} \;\; \frac{10}{210} = \frac{5+x}{640} \;\; \begin{matrix} \leftarrow \text{gallons} \\ \leftarrow \text{miles} \end{matrix}$$

Step 4 **Solve.** The LCD is $10 \cdot 21 \cdot 64$.

$$10 \cdot 21 \cdot 64\left(\frac{10}{210}\right) = 10 \cdot 21 \cdot 64\left(\frac{5+x}{640}\right)$$
$$64 \cdot 10 = 21(5 + x)$$
$$640 = 105 + 21x \qquad \text{Distributive property}$$
$$535 = 21x \qquad \text{Subtract 105.}$$
$$25.5 \approx x \qquad \text{Divide by 21; round to the nearest tenth.}$$

Step 5 **State the answer.** Marissa will need about 25.5 more gallons of gas.

Step 6 **Check** the answer in the words of the problem. The 25.5 gal plus the 5 gal equals 30.5 gal.

$$\frac{30.5}{640} \approx .048 \quad \text{and} \quad \frac{10}{210} \approx .048$$

Since the rates are equal, the solution is correct.

Now Try Exercise 35.

OBJECTIVE 4 **Solve applications about distance, rate, and time.** A familiar example of a rate is speed, which is the ratio of distance to time. The next examples use the distance formula $d = rt$ introduced in Chapter 2.

EXAMPLE 6 Solving a Problem about Distance, Rate, and Time

A tour boat goes 10 mi against the current in a small river in the same time that it goes 15 mi with the current. If the speed of the current is 3 mph, find the speed of the boat in still water.

Step 1 **Read** the problem. We must find the speed of the boat in still water.

Step 2 **Assign a variable.**

Let $x =$ the speed of the boat in still water.

When the boat is traveling *against* the current, the current slows the boat down, and the speed of the boat is the difference between its speed in still water and the speed of the current, that is, $x - 3$ mph.

When the boat is traveling *with* the current, the current speeds the boat up, and the speed of the boat is the sum of its speed in still water and the speed of the current, that is, $x + 3$ mph.

Thus, $x - 3 =$ the speed of the boat *against* the current,

and $x + 3 =$ the speed of the boat *with* the current.

Because the time is the same going against the current as with the current, find time in terms of distance and rate (speed) for each situation. Start with the distance formula, $d = rt$, and divide each side by r to get

$$t = \frac{d}{r}.$$

Going against the current, the distance is 10 mi and the rate is $x - 3$ mph, giving

$$t = \frac{d}{r} = \frac{10}{x - 3}.$$

Going with the current, the distance is 15 mi and the rate is $x + 3$ mph, so

$$t = \frac{d}{r} = \frac{15}{x + 3}.$$

This information is summarized in the following table.

	Distance	Rate	Time
Against Current	10	$x - 3$	$\dfrac{10}{x - 3}$
With Current	15	$x + 3$	$\dfrac{15}{x + 3}$

Times are equal.

Step 3   **Write an equation.** Because the times are equal,

$$\frac{10}{x - 3} = \frac{15}{x + 3}.$$

Step 4   **Solve** this equation. The LCD is $(x + 3)(x - 3)$.

$$(x + 3)(x - 3)\left(\frac{10}{x - 3}\right) = (x + 3)(x - 3)\left(\frac{15}{x + 3}\right) \qquad \text{Multiply by the LCD.}$$

$$10(x + 3) = 15(x - 3)$$

$$10x + 30 = 15x - 45 \qquad \text{Distributive property}$$

$$30 = 5x - 45 \qquad \text{Subtract 10x.}$$

$$75 = 5x \qquad \text{Add 45.}$$

$$15 = x \qquad \text{Divide by 5.}$$

Step 5   **State the answer.** The speed of the boat in still water is 15 mph.

Step 6   **Check** the answer: $\dfrac{10}{15 - 3} = \dfrac{15}{15 + 3}$ is true.

Now Try Exercise 39.

EXAMPLE 7   Solving a Problem about Distance, Rate, and Time

At O'Hare International Airport in Chicago, Cheryl and Bill are walking to the gate (at the same speed) to catch their flight to Akron, Ohio. Since Bill wants a window seat, he steps onto the moving sidewalk and continues to walk while Cheryl uses the stationary sidewalk. If the sidewalk moves at 1 m per sec and Bill saves 50 sec covering the 300-m distance, what is their walking speed?

Step 1   **Read** the problem. We must find their walking speed.

Step 2   **Assign a variable.** Let x represent their walking speed in meters per second. Thus Cheryl travels at x m per sec and Bill travels at $x + 1$ m per sec. Since Bill's time is 50 sec less than Cheryl's time, express their times in terms of the known distances and the variable rates. As in Example 6, start with $d = rt$ and divide each side by r to get $t = \frac{d}{r}$. For Cheryl, the distance is 300 m and the rate is x mph. Cheryl's time is

$$t = \frac{d}{r} = \frac{300}{x}.$$

Bill travels 300 m at a rate of $x + 1$ mph, so his time is

$$t = \frac{d}{r} = \frac{300}{x + 1}.$$

This information is summarized in the following table.

	Distance	Rate	Time
Cheryl	300	x	$\dfrac{300}{x}$
Bill	300	$x + 1$	$\dfrac{300}{x + 1}$

Step 3 **Write an equation** using the times from the table.

Bill's time	is	Cheryl's time	less 50 seconds.
$\dfrac{300}{x + 1}$	$=$	$\dfrac{300}{x}$	$-\quad 50$

Step 4 **Solve.**

$$x(x + 1)\left(\frac{300}{x + 1}\right) = x(x + 1)\left(\frac{300}{x} - 50\right) \qquad \text{Multiply by the LCD, } x(x + 1).$$

$$300x = 300(x + 1) - 50x(x + 1)$$

$$300x = 300x + 300 - 50x^2 - 50x \qquad \text{Distributive property}$$

$$0 = 50x^2 + 50x - 300 \qquad \text{Standard form}$$

$$0 = x^2 + x - 6 \qquad \text{Divide by 50.}$$

$$0 = (x + 3)(x - 2) \qquad \text{Factor.}$$

$$x + 3 = 0 \quad \text{or} \quad x - 2 = 0 \qquad \text{Zero-factor property}$$

$$x = -3 \quad \text{or} \qquad x = 2$$

Discard the negative answer, since speed cannot be negative.

Step 5 **State the answer.** Their walking speed is 2 m per sec.

Step 6 **Check** the solution in the words of the original problem.

Now Try Exercise 45.

OBJECTIVE 5 Solve applications about work rates. Problems about work are closely related to distance problems.

PROBLEM SOLVING

People work at different rates. If the letters r, t, and A represent the rate at which the work is done, the time required, and the amount of work accomplished, respectively, then $A = rt$. Notice the similarity to the distance formula, $d = rt$.

Amount of work can be measured in terms of jobs accomplished. Thus, if 1 job is completed, $A = 1$, and the formula gives the rate as

$$1 = rt$$

$$r = \frac{1}{t}.$$

Rate of Work

If a job can be accomplished in t units of time, then the rate of work is

$$\frac{1}{t} \text{ job per unit of time.}$$

To solve a work problem, we begin by using this fact to express all rates of work. See if you can identify the six steps used in the following example.

■ EXAMPLE 8   Solving a Problem about Work

Letitia and Kareem are working on a neighborhood cleanup. Kareem can clean up all the trash in the area in 7 hr, while Letitia can do the same job in 5 hr. How long will it take them if they work together?

Let $x =$ the number of hours it will take the two people working together. Just as we made a table for the distance formula, $d = rt$, make a table here for $A = rt$, with $A = 1$. Since $A = 1$, the rate for each person will be $\frac{1}{t}$, where t is the time it takes the person to complete the job alone. For example, since Kareem can clean up all the trash in 7 hr, his rate is $\frac{1}{7}$ of the job per hour. Similarly, Letitia's rate is $\frac{1}{5}$ of the job per hour. Fill in the table as shown.

	Rate	Time Working Together	Fractional Part of the Job Done
Kareem	$\frac{1}{7}$	x	$\frac{1}{7}x$
Letitia	$\frac{1}{5}$	x	$\frac{1}{5}x$

Since together they complete 1 job, the sum of the fractional parts accomplished by them should equal 1.

$$\underset{\substack{\text{Part done} \\ \text{by Kareem}}}{\frac{1}{7}x} \; \underset{+}{+} \; \underset{\substack{\text{part done} \\ \text{by Letitia}}}{\frac{1}{5}x} \; \underset{\text{is}}{=} \; \underset{\substack{1 \text{ whole} \\ \text{job.}}}{1}$$

Solve this equation. The LCD is 35.

$$35\left(\frac{1}{7}x + \frac{1}{5}x\right) = 35 \cdot 1$$

$$5x + 7x = 35$$

$$12x = 35$$

$$x = \frac{35}{12}$$

Working together, Kareem and Letitia can do the job in $\frac{35}{12}$ hr, or 2 hr 55 min. Check this result in the original problem.   ■

Now Try Exercise 47.

There is another way to approach problems about work. For instance, in Example 8, x represents the number of hours it will take the two people working together to complete the entire job. In one hour, $\frac{1}{x}$ of the entire job will be completed. In one hour, Kareem completes $\frac{1}{7}$ of the job and Letitia completes $\frac{1}{5}$ of the job. The sum of their rates should equal $\frac{1}{x}$. This gives the equation

$$\frac{1}{7} + \frac{1}{5} = \frac{1}{x}.$$

When each side of this equation is multiplied by $35x$, the result is $5x + 7x = 35$. Notice that this is the same equation we got in Example 8 in the third line from the bottom. Thus the solution of the equation is the same using either approach.

7.5 EXERCISES

For Extra Help

Student's
Solutions Manual

MyMathLab

InterAct Math
Tutorial Software

AW Math
Tutor Center

MathXL

Digital Video Tutor
CD 7/Videotape 8

In Exercises 1–4, a familiar formula is given. Give the letter of the choice that is an equivalent form of the given formula.

1. $p = br$ (percent)

   A. $b = \dfrac{p}{r}$ **B.** $r = \dfrac{b}{p}$ **C.** $b = \dfrac{r}{p}$ **D.** $p = \dfrac{r}{b}$

2. $V = LWH$ (geometry)

   A. $H = \dfrac{LW}{V}$ **B.** $L = \dfrac{V}{WH}$ **C.** $L = \dfrac{WH}{V}$ **D.** $W = \dfrac{H}{VL}$

3. $m = \dfrac{F}{a}$ (physics)

   A. $a = mF$ **B.** $F = \dfrac{m}{a}$ **C.** $F = \dfrac{a}{m}$ **D.** $F = ma$

4. $I = \dfrac{E}{R}$ (electricity)

   A. $R = \dfrac{I}{E}$ **B.** $R = IE$ **C.** $E = \dfrac{I}{R}$ **D.** $E = RI$

Solve each problem. See Example 1.

5. A gas law in chemistry says that

$$\frac{PV}{T} = \frac{pv}{t}.$$

Suppose that $T = 300$, $t = 350$, $V = 9$, $P = 50$, and $v = 8$. Find p.

6. In work with electric circuits, the formula

$$\frac{1}{a} = \frac{1}{b} + \frac{1}{c}$$

occurs. Find b if $a = 8$ and $c = 12$.

7. A formula from anthropology says that

$$c = \frac{100b}{L}.$$

Find L if $c = 80$ and $b = 5$.

8. The gravitational force between two masses is given by

$$F = \frac{GMm}{d^2}.$$

Find M if $F = 10$, $G = 6.67 \times 10^{-11}$, $m = 1$, and $d = 3 \times 10^{-6}$.

Solve each formula for the specified variable. See Examples 2 and 3.

9. $F = \dfrac{GMm}{d^2}$ for G (physics)

10. $F = \dfrac{GMm}{d^2}$ for M (physics)

11. $\dfrac{1}{a} = \dfrac{1}{b} + \dfrac{1}{c}$ for a (electricity)

12. $\dfrac{1}{a} = \dfrac{1}{b} + \dfrac{1}{c}$ for b (electricity)

13. $\dfrac{PV}{T} = \dfrac{pv}{t}$ for v (chemistry)

14. $\dfrac{PV}{T} = \dfrac{pv}{t}$ for T (chemistry)

15. $I = \dfrac{nE}{R + nr}$ for r (engineering)

16. $a = \dfrac{V - v}{t}$ for V (physics)

17. $A = \dfrac{1}{2} h(B + b)$ for b (mathematics)

18. $S = \dfrac{n}{2}(a + \ell)d$ for n (mathematics)

19. $\dfrac{E}{e} = \dfrac{R + r}{r}$ for r (engineering)

20. $y = \dfrac{x + z}{a - x}$ for x

21. To solve the equation $m = \dfrac{ab}{a - b}$ for a, what is the first step?

22. Suppose you are asked to solve the equation

$$rp - rq = p + q$$

for r. What is the first step?

Solve each problem mentally. Use proportions in Exercises 23 and 24.

23. In a mathematics class, 3 of every 4 students are girls. If there are 20 students in the class, how many are girls? How many are boys?

24. In a certain southern state, sales tax on a purchase of $1.50 is $.12. What is the sales tax on a purchase of $6.00?

25. If Marin can mow her yard in 2 hr, what is her rate (in job per hour)?

26. A van traveling from Atlanta to Detroit averages 50 mph and takes 14 hr to make the trip. How far is it from Atlanta to Detroit?

Use the bar graph to answer the questions in Exercises 27–30.

27. In which year was the ratio of truck accidents to car accidents the least?

28. In which year was the ratio of truck accidents to car accidents the greatest?

29. In which year was the ratio of car accidents to truck accidents closest to 3 to 1?

30. In which year was the ratio of car accidents to truck accidents closest to 2 to 1?

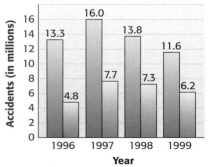

MOTOR VEHICLE ACCIDENTS INVOLVING CARS AND TRUCKS

Source: National Safety Council.

Use a proportion to solve each problem. See Examples 4 and 5.

31. During the 1997–1998 academic year, the ratio of teachers to students in private high schools was approximately 1 to 24. If a private high school had 554 students, how many teachers would be at the school if this ratio was valid for that school? Round your answer to the nearest whole number. (*Source:* U.S. National Center for Education Statistics, *Private School Universe Survey,* 1997–98.)

32. During the 1998–1999 National Basketball Association season, Shaquille O'Neal of the Los Angeles Lakers played in 49 games for a total of 1705 min. If he had played in all 50 of the team's games, how many minutes would he have played, assuming that the ratio of games to minutes stayed the same? Round your answer to the nearest whole number. (*Source: Sports Illustrated 2000 Sports Almanac.*)

33. Biologists tagged 500 fish in a lake on January 1. On February 1 they returned and collected a random sample of 400 fish, 8 of which had been previously tagged. Approximately how many fish does the lake have based on this experiment?

34. Suppose that in the experiment of Exercise 33, 10 of the previously tagged fish were collected on February 1. What would be the estimate of the fish population?

Nurses use proportions to determine the amount of a drug to administer when the dose of the drug is measured in milligrams but the drug is packaged in a diluted form in milliliters. (Source: Hoyles, Celia, Richard Noss, and Stefano Pozzi, "Proportional Reasoning in Nursing Practice," Journal for Research in Mathematics Education, January 2001.) For example, to find the number of milliliters of fluid needed to administer 300 mg of a drug that comes packaged as 120 mg in 2 mL of fluid, a nurse sets up the proportion

$$\frac{120 \text{ mg}}{2 \text{ mL}} = \frac{300 \text{ mg}}{x \text{ mL}},$$

where x represents the amount to administer in milliliters. Use this method to find the correct dose for each prescription.

35. 120 mg of Amakacine packaged as 100 mg in 2-mL vials

36. 1.5 mg of morphine packaged as 20 mg ampules diluted in 10 mL of fluid

In geometry, it is shown that two triangles with corresponding angle measures equal, called **similar triangles,** *have corresponding sides proportional. For example, in the figure at the top of the next page, angle A = angle D, angle B = angle E, and angle C = angle F, so the triangles are similar. Then the following ratios of corresponding sides are equal.*

$$\frac{4}{6} = \frac{6}{9} = \frac{2x + 1}{2x + 5}$$

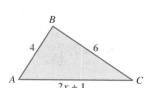
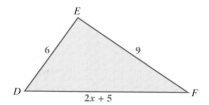

37. Solve for x using the given proportion to find the lengths of the third sides of the triangles.

38. Suppose the following triangles are similar. Find y and the lengths of the two longest sides of each triangle.

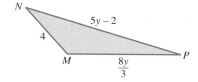
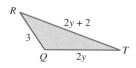

Solve each problem. See Examples 6 and 7.

39. Kellen's boat goes 12 mph. Find the rate of the current of the river if she can go 6 mi upstream in the same amount of time she can go 10 mi downstream.

	Distance	Rate	Time
Downstream	10	$12 + x$	
Upstream	6	$12 - x$	

40. Kasey can travel 8 mi upstream in the same time it takes her to go 12 mi downstream. Her boat goes 15 mph in still water. What is the rate of the current?

	Distance	Rate	Time
Downstream			
Upstream			

41. Driving from Tulsa to Detroit, Jeff averaged 50 mph. He figured that if he had averaged 60 mph, his driving time would have decreased 3 hr. How far is it from Tulsa to Detroit?

42. If Dr. Dawson rides his bike to his office, he averages 12 mph. If he drives his car, he averages 36 mph. His time driving is $\frac{1}{4}$ hr less than his time riding his bike. How far is his office from home?

43. A private plane traveled from San Francisco to a secret rendezvous. It averaged 200 mph. On the return trip, the average speed was 300 mph. If the total traveling time was 4 hr, how far from San Francisco was the secret rendezvous?

44. Johnny averages 30 mph when he drives on the old highway to his favorite fishing hole, and he averages 50 mph when most of his route is on the interstate. If both routes are the same length, and he saves 2 hr by traveling on the interstate, how far away is the fishing hole?

45. On the first part of a trip to Carmel traveling on the freeway, Marge averaged 60 mph. On the rest of the trip, which was 10 mi longer than the first part, she averaged 50 mph. Find the total distance to Carmel if the second part of the trip took 30 min more than the first part.

46. While on vacation, Jim and Annie decided to drive all day. During the first part of their trip on the highway, they averaged 60 mph. When they got to Houston, traffic caused them to average only 30 mph. The distance they drove in Houston was 100 mi less than their distance on the highway. What was their total driving distance if they spent 50 min more on the highway than they did in Houston?

Solve each problem. See Example 8.

47. Butch and Peggy want to pick up the mess that their grandson, Grant, has made in his playroom. Butch could do it in 15 min working alone. Peggy, working alone, could clean it in 12 min. How long will it take them if they work together?

	Rate	Time Working Together	Fractional Part of the Job Done
Butch	$\frac{1}{15}$	x	
Peggy	$\frac{1}{12}$	x	

48. Lou can groom Jay Beckenstein's dogs in 8 hr, but it takes his business partner, Janet, only 5 hr to groom the same dogs. How long will it take them to groom Jay's dogs if they work together?

	Rate	Time Working Together	Fractional Part of the Job Done
Lou	$\frac{1}{8}$	x	
Janet	$\frac{1}{5}$	x	

49. Ron Wood can paint a room in 6 hr working alone. If his son Jason helps him, the job takes 4 hr. How long would it take Jason to do the job if he worked alone?

50. Sandi and Cary Goldstein are refinishing a table. Working alone, Cary could do the job in 7 hr. If the two work together, the job takes 5 hr. How long will it take Sandi to refinish the table working alone?

51. If a vat of acid can be filled by an inlet pipe in 10 hr and emptied by an outlet pipe in 20 hr, how long will it take to fill the vat if both pipes are open?

52. A winery has a vat to hold chardonnay. An inlet pipe can fill the vat in 9 hr, while an outlet pipe can empty it in 12 hr. How long will it take to fill the vat if both the outlet and the inlet pipes are open?

53. Suppose that Hortense and Mort can clean their entire house in 7 hr, while their toddler, Mimi, just by being around, can completely mess it up in only 2 hr. If Hortense and Mort clean the house while Mimi is at her grandma's, and then start cleaning up after Mimi the minute she gets home, how long does it take from the time Mimi gets home until the whole place is a shambles?

54. An inlet pipe can fill an artificial lily pond in 60 min, while an outlet pipe can empty it in 80 min. Through an error, both pipes are left open. How long will it take for the pond to fill?

It Depends on What You Mean by "Average"

OBJECTIVE Learn when and how to calculate a harmonic mean.

Finding an average seems to be a simple process. Don't we just add the values and divide by the number of values? Well, for rational expressions, it all depends on what you mean by "average."

- To find the average of two fractions, say $\frac{1}{3}$ and $\frac{3}{4}$, add the two fractions and divide by 2.

$$\frac{\frac{1}{3} + \frac{3}{4}}{2} = \frac{\left(\frac{1}{3} + \frac{3}{4}\right) \cdot 12}{2 \cdot 12} = \frac{4 + 9}{24} = \frac{13}{24}$$

On a number line, the fraction $\frac{13}{24}$ is the **arithmetic mean,** which is exactly halfway between the fractions $\frac{1}{3}$ and $\frac{3}{4}$.

- Suppose you travel one direction at 60 mph and return at 30 mph.

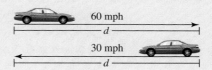

To find your average rate, you have to calculate the total distance divided by the total time. Recall that $d = rt$, so if the distance *each* way is d, then the total distance is $2d$, the time going is $\frac{d}{60}$, and the time returning is $\frac{d}{30}$. Since $r = \frac{d}{t}$,

$$r = \frac{2d}{\frac{d}{60} + \frac{d}{30}} = \frac{2d \cdot 60}{\left(\frac{d}{60} + \frac{d}{30}\right) \cdot 60} = \frac{120d}{d + 2d} = \frac{120d}{3d} = 40 \text{ mph.}$$

The average rate is 40 mph. This is the *harmonic mean* of 60 and 30. The **harmonic mean** of two numbers a and b is defined as

$$\frac{2ab}{a + b}.$$

Note that

$$\frac{2 \cdot 60 \cdot 30}{60 + 30} = \frac{3600}{90} = 40.$$

- To calculate a batting average, you find the **ratio** of the number of hits to the number of "at bats." Suppose a baseball player has 72 hits in 364 "at bats." His batting average would be $\frac{72}{364} \approx .198$. If the same player gets an additional 3 hits from 8 more "at bats" during the next week, then his revised batting average would be

$$\frac{72 + 3}{364 + 8} = \frac{75}{372} \approx .202.$$

A carpenter builds wine racks. As a group, find the appropriate "average" quantity for each situation.

A. The carpenter told his helper to cut $\frac{1}{2}$-ft pieces from a dowel. The helper could not find a measuring tape, but he did recall that the distance from the tip of his middle finger to the tip of his thumb was approximately 6 in., so he estimated the lengths. When the carpenter checked his work, he found that the helper had actually cut two pieces that were $\frac{5}{12}$ and $\frac{1}{2}$ ft long. What was the average length of the two pieces?

B. Once the pieces are cut, the carpenter can assemble and finish a wine rack in 2 hr, working alone. His helper takes 4 hr to accomplish the same task, working alone. If the carpenter and the helper work together, what is their average time to assemble and finish a wine rack?

C. Of 115 wine racks built, 112 passed a quality control check. What was the acceptance rate? During the next week, the carpenter built 35 additional wine racks, of which 28 were acceptable. What was the revised acceptance rate? Round answers to the nearest thousandth.

CHAPTER 7 SUMMARY

KEY TERMS

7.1 rational expression
rational function

7.2 least common
denominator (LCD)
7.3 complex fraction

7.4 vertical asymptote
domain of a rational
equation

7.5 ratio
proportion

TEST YOUR WORD POWER

See how well you have learned the vocabulary in this chapter. Answers, with examples, follow the Quick Review.

1. A **rational expression** is
   A. an algebraic expression made up of a term or the sum of a finite number of terms with real coefficients and integer exponents
   B. a polynomial equation of degree 2
   C. an expression with one or more fractions in the numerator, denominator, or both
   D. the quotient of two polynomials with denominator not 0.

2. In a given set of fractions, the **least common denominator** is
   A. the smallest denominator of all the denominators
   B. the smallest expression that is divisible by all the denominators

C. the largest integer that evenly divides the numerator and denominator of all the fractions
D. the largest denominator of all the denominators.

3. A **complex fraction** is
   A. an algebraic expression made up of a term or the sum of a finite number of terms with real coefficients and integer exponents
   B. a polynomial equation of degree 2
   C. an expression with one or more fractions in the numerator, denominator, or both
   D. the quotient of two polynomials with denominator not 0.

4. A **ratio**
   A. compares two quantities using a quotient
   B. says that two quotients are equal
   C. is a product of two quantities
   D. is a difference between two quantities.

5. A **proportion**
   A. compares two quantities using a quotient
   B. says that two quotients are equal
   C. is a product of two quantities
   D. is a difference between two quantities.

QUICK REVIEW

CONCEPTS	EXAMPLES

7.1 RATIONAL EXPRESSIONS AND FUNCTIONS; MULTIPLYING AND DIVIDING

Fundamental Property of Rational Numbers
If $\frac{a}{b}$ is a rational number and if c is any nonzero real number, then

$$\frac{a}{b} = \frac{ac}{bc}.$$

$$\frac{3}{4} = \frac{3 \cdot 5}{4 \cdot 5} = \frac{15}{20}$$

Writing a Rational Expression in Lowest Terms
Step 1 Factor the numerator and the denominator completely.

Step 2 Apply the fundamental property.

Write in lowest terms.

$$\frac{2x + 8}{x^2 - 16} = \frac{2(x + 4)}{(x - 4)(x + 4)}$$

$$= \frac{2}{x - 4}$$

(continued)

CONCEPTS	EXAMPLES

Multiplying Rational Expressions

Step 1 Factor numerators and denominators.

Step 2 Apply the fundamental property.

Step 3 Multiply the remaining factors in the numerator and in the denominator.

Step 4 Check that the product is in lowest terms.

Multiply.

$$\frac{x^2 + 2x + 1}{x^2 - 1} \cdot \frac{5}{3x + 3} = \frac{(x + 1)^2}{(x - 1)(x + 1)} \cdot \frac{5}{3(x + 1)}$$

$$= \frac{5}{3(x - 1)}$$

Dividing Rational Expressions

Multiply the first rational expression by the reciprocal of the second.

Divide.

$$\frac{2x + 5}{x - 3} \div \frac{2x^2 + 3x - 5}{x^2 - 9} = \frac{2x + 5}{x - 3} \cdot \frac{(x + 3)(x - 3)}{(2x + 5)(x - 1)}$$

$$= \frac{x + 3}{x - 1}$$

7.2 ADDING AND SUBTRACTING RATIONAL EXPRESSIONS

Adding or Subtracting Rational Expressions

Step 1 If the denominators are the same, add or subtract the numerators. Place the result over the common denominator. If the denominators are different, write all rational expressions with the LCD. Then add or subtract the numerators, and place the result over the common denominator.

Step 2 Be sure the answer is in lowest terms.

Subtract.

$$\frac{1}{x + 6} - \frac{3}{x + 2} = \frac{x + 2}{(x + 6)(x + 2)} - \frac{3(x + 6)}{(x + 6)(x + 2)}$$

$$= \frac{x + 2 - 3(x + 6)}{(x + 6)(x + 2)}$$

$$= \frac{x + 2 - 3x - 18}{(x + 6)(x + 2)}$$

$$= \frac{-2x - 16}{(x + 6)(x + 2)}$$

7.3 COMPLEX FRACTIONS

Simplifying a Complex Fraction

Method 1 Simplify the numerator and denominator separately, as much as possible. Then multiply the numerator by the reciprocal of the denominator. Write the answer in lowest terms.

Simplify the complex fraction.

Method 1

$$\frac{\dfrac{1}{x^2} - \dfrac{1}{y^2}}{\dfrac{1}{x} + \dfrac{1}{y}} = \frac{\dfrac{y^2}{x^2y^2} - \dfrac{x^2}{x^2y^2}}{\dfrac{y}{xy} + \dfrac{x}{xy}}$$

$$= \frac{\dfrac{y^2 - x^2}{x^2y^2}}{\dfrac{y + x}{xy}}$$

$$= \frac{y^2 - x^2}{x^2y^2} \div \frac{y + x}{xy}$$

$$= \frac{(y + x)(y - x)}{x^2y^2} \cdot \frac{xy}{y + x}$$

$$= \frac{y - x}{xy}$$

CONCEPTS	EXAMPLES

Method 2 Multiply the numerator and denominator of the complex fraction by the least common denominator of all fractions appearing in the complex fraction. Then simplify the result.

Method 2

$$\frac{\dfrac{1}{x^2} - \dfrac{1}{y^2}}{\dfrac{1}{x} + \dfrac{1}{y}} = \frac{x^2 y^2 \left(\dfrac{1}{x^2} - \dfrac{1}{y^2} \right)}{x^2 y^2 \left(\dfrac{1}{x} + \dfrac{1}{y} \right)}$$

$$= \frac{y^2 - x^2}{xy^2 + x^2 y}$$

$$= \frac{(y - x)(y + x)}{xy(y + x)}$$

$$= \frac{y - x}{xy}$$

7.4 EQUATIONS WITH RATIONAL EXPRESSIONS AND GRAPHS

Solving an Equation with Rational Expressions
To solve an equation involving rational expressions, first determine the domain. Then multiply all the terms in the equation by the least common denominator. Solve the resulting equation. Each potential solution *must* be checked to see that it is in the domain of the equation.

Solve.

$$\frac{1}{x} + x = \frac{26}{5}$$

Note that 0 is excluded from the domain.

$$5 + 5x^2 = 26x \qquad \text{Multiply by } 5x.$$
$$5x^2 - 26x + 5 = 0$$
$$(5x - 1)(x - 5) = 0$$
$$x = \frac{1}{5} \quad \text{or} \quad x = 5$$

Both check. The solution set is $\left\{ \frac{1}{5}, 5 \right\}$.

The graph of a rational function may have one or more breaks. At such points, the graph will approach an asymptote.

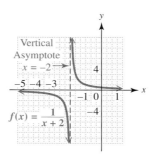

$$f(x) = \frac{1}{x + 2}$$

Vertical Asymptote $x = -2$

7.5 APPLICATIONS OF RATIONAL EXPRESSIONS

To solve a formula for a particular variable, isolate that variable on one side.

Solve for L.

$$c = \frac{100b}{L}$$
$$cL = 100b \qquad \text{Multiply by } L.$$
$$L = \frac{100b}{c} \qquad \text{Divide by } c.$$

(continued)

CONCEPTS	EXAMPLES

CONCEPTS

To solve a motion problem, use the formula

$$d = rt$$

or one of its equivalents,

$$t = \frac{d}{r} \quad \text{or} \quad r = \frac{d}{t}.$$

To solve a work problem, use the fact that if a complete job is done in t units of time, the rate of work is $\frac{1}{t}$ job per unit of time. The amount of work accomplished is $A = rt$, so if one job is accomplished in time t, use the formula

$$1 = rt.$$

EXAMPLES

Solve.

A canal has a current of 2 mph. Find the speed of Amy's boat in still water if it goes 11 mi downstream in the same time that it goes 8 mi upstream.

Let x represent the speed of the boat in still water.

	Distance	Rate	Time
Downstream	11	$x + 2$	$\dfrac{11}{x + 2}$
Upstream	8	$x - 2$	$\dfrac{8}{x - 2}$

Because the times are the same, the equation is

$$\frac{11}{x + 2} = \frac{8}{x - 2}. \qquad \text{Use } t = \tfrac{d}{r}.$$

The LCD is $(x + 2)(x - 2)$.

$11(x - 2) = 8(x + 2)$	Multiply by the LCD.
$11x - 22 = 8x + 16$	Distributive property
$3x = 38$	Subtract 8x and add 22.
$x = 12\dfrac{2}{3}$	Divide by 3.

The speed in still water is $12\frac{2}{3}$ mph.

Answers to Test Your Word Power

1. D; *Examples:* $-\dfrac{3}{4y^2}, \dfrac{5x^3}{x + 2}, \dfrac{a + 3}{a^2 - 4a - 5}$ **2.** B; *Example:* The LCD of $\dfrac{1}{x}, \dfrac{2}{3}$, and $\dfrac{5}{x + 1}$ is $3x(x + 1)$.

3. C; *Examples:* $\dfrac{\frac{2}{3}}{\frac{4}{7}}, \dfrac{x - \frac{1}{x}}{x + \frac{1}{y}}, \dfrac{\frac{2}{a + 1}}{a^2 - 1}$ **4.** A; *Example:* $\dfrac{7 \text{ in.}}{12 \text{ in.}}$ compares two quantities. **5.** B; *Example:* The proportion $\dfrac{2}{3} = \dfrac{8}{12}$ states that the two ratios are equal.

CHAPTER **7** REVIEW EXERCISES

[7.1] (a) *Find all real numbers that are excluded from the domain.* (b) *Give the domain using set notation.*

1. $f(x) = \dfrac{-7}{3x + 18}$ **2.** $f(x) = \dfrac{5x + 17}{x^2 - 7x + 10}$ **3.** $f(x) = \dfrac{9}{x^2 - 18x + 81}$

Write in lowest terms.

4. $\dfrac{12x^2 + 6x}{24x + 12}$ **5.** $\dfrac{25m^2 - n^2}{25m^2 - 10mn + n^2}$ **6.** $\dfrac{r - 2}{4 - r^2}$

✎ **7.** What is meant by the reciprocal of a rational expression?

Multiply or divide. Write the answer in lowest terms.

8. $\dfrac{(2y + 3)^2}{5y} \cdot \dfrac{15y^3}{4y^2 - 9}$ **9.** $\dfrac{w^2 - 16}{w} \cdot \dfrac{3}{4 - w}$

10. $\dfrac{z^2 - z - 6}{z - 6} \cdot \dfrac{z^2 - 6z}{z^2 + 2z - 15}$ **11.** $\dfrac{m^3 - n^3}{m^2 - n^2} \div \dfrac{m^2 + mn + n^2}{m + n}$

[7.2] *Assume that each expression is the denominator of a rational expression. Find the least common denominator for each group.*

12. $32b^3, \quad 24b^5$ **13.** $9r^2, \quad 3r + 1, \quad 9$

14. $6x^2 + 13x - 5, \quad 9x^2 + 9x - 4$

Add or subtract as indicated.

15. $\dfrac{8}{z} - \dfrac{3}{2z^2}$ **16.** $\dfrac{5y + 13}{y + 1} - \dfrac{1 - 7y}{y + 1}$

17. $\dfrac{6}{5a + 10} + \dfrac{7}{6a + 12}$ **18.** $\dfrac{3r}{10r^2 - 3rs - s^2} + \dfrac{2r}{2r^2 + rs - s^2}$

[7.3] *Simplify each complex fraction.*

19. $\dfrac{\dfrac{3}{t} + 2}{\dfrac{4}{t} - 7}$ **20.** $\dfrac{\dfrac{2}{m - 3n}}{\dfrac{1}{3n - m}}$ **21.** $\dfrac{\dfrac{3}{p} - \dfrac{2}{q}}{\dfrac{9q^2 - 4p^2}{qp}}$ **22.** $\dfrac{x^{-2} - y^{-2}}{x^{-1} - y^{-1}}$

[7.4] *Solve each equation.*

23. $\dfrac{1}{t + 4} + \dfrac{1}{2} = \dfrac{3}{2t + 8}$ **24.** $\dfrac{-5m}{m + 1} + \dfrac{m}{3m + 3} = \dfrac{56}{6m + 6}$

25. $\dfrac{2}{k - 1} - \dfrac{4k + 1}{k^2 - 1} = \dfrac{-1}{k + 1}$ **26.** $\dfrac{5}{x + 2} + \dfrac{3}{x + 3} = \dfrac{x}{x^2 + 5x + 6}$

27. After solving the equation

$$\frac{3}{x - 3} - \frac{2}{x - 2} = \frac{3}{x^2 - 5x + 6},$$

a student got $x = 3$ as her final step. She could not understand why the answer in the back of the book was "∅," because she checked her algebra several times and was sure that all her algebraic work was correct. Was she wrong or was the answer in the back of the book wrong? Explain.

28. Explain the difference between simplifying the expression

$$\frac{4}{x} + \frac{1}{2} - \frac{1}{3}$$

and solving the equation

$$\frac{4}{x} + \frac{1}{2} = \frac{1}{3}.$$

29. Which graph has a vertical asymptote, and what is its equation?

A. 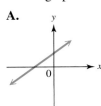 **B.** 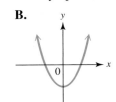 **C.** 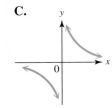 **D.**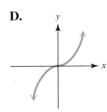

[7.5]

30. According to a law from physics, $\frac{1}{A} = \frac{1}{B} + \frac{1}{C}$. Find A if $B = 30$ and $C = 10$.

Solve each formula for the specified variable.

31. $F = \dfrac{GMm}{d^2}$ for m (physics)

32. $\mu = \dfrac{Mv}{M + m}$ for M (electronics)

Solve each problem.

33. An article in *Scientific American* predicts that, in the year 2050, about 23,200 of the 58,000 passenger-km per day in North America will be provided by high-speed trains. If the traffic volume in a typical region of North America is 15,000, how many passenger-kilometers per day will high-speed trains provide there? (*Source:* Schafer, Andreas and David Victor, "The Past and Future of Global Mobility," *Scientific American,* October 1997.)

34. A river has a current of 4 km per hr. Find the speed of Lynn McTernan's boat in still water if it goes 40 km downstream in the same time that it takes to go 24 km upstream.

	d	r	t
Upstream	24	x − 4	
Downstream	40		

35. A sink can be filled by a cold-water tap in 8 min and filled by the hot-water tap in 12 min. How long would it take to fill the sink with both taps open?

36. Amber Allen and Steven Pusztai need to sort a pile of bottles at the recycling center. Working alone, Amber could do the entire job in 9 hr, while Steven could do the entire job in 6 hr. How long will it take them if they work together?

▨ MIXED REVIEW EXERCISES

Write in lowest terms.

37. $\dfrac{x + 2y}{x^2 - 4y^2}$

38. $\dfrac{x^2 + 2x - 15}{x^2 - x - 6}$

Perform the indicated operations.

39. $\dfrac{2}{m} + \dfrac{5}{3m^2}$

40. $\dfrac{k^2 - 6k + 9}{1 - 216k^3} \cdot \dfrac{6k^2 + 17k - 3}{9 - k^2}$

41. $\dfrac{\dfrac{-3}{x} + \dfrac{x}{2}}{1 + \dfrac{x + 1}{x}}$

42. $\dfrac{9x^2 + 46x + 5}{3x^2 - 2x - 1} \div \dfrac{x^2 + 11x + 30}{x^3 + 5x^2 - 6x}$

43. $\dfrac{\dfrac{3}{x} - 5}{6 + \dfrac{1}{x}}$

44. $\dfrac{9}{3 - x} - \dfrac{2}{x - 3}$

45. $\dfrac{4y + 16}{30} \div \dfrac{2y + 8}{5}$

46. $\dfrac{t^{-2} + s^{-2}}{t^{-1} - s^{-1}}$

47. $\dfrac{4a}{a^2 - ab - 2b^2} - \dfrac{6b - a}{a^2 + 4ab + 3b^2}$

48. $\dfrac{a}{b} + \dfrac{b}{c} + \dfrac{c}{d}$

Solve.

49. $\dfrac{x + 3}{x^2 - 5x + 4} - \dfrac{1}{x} = \dfrac{2}{x^2 - 4x}$

50. $A = \dfrac{Rr}{R + r}$ for r

51. $1 - \dfrac{5}{r} = \dfrac{-4}{r^2}$

52. $\dfrac{3x}{x - 4} + \dfrac{2}{x} = \dfrac{48}{x^2 - 4x}$

53. The strength of a contact lens is given in units called diopters and also in millimeters of arc. As the diopters increase, the millimeters of arc decrease. The rational function defined by

$$a = \dfrac{337}{d}$$

relates the arc measurement a to the diopter measurement d. (*Source:* Bausch and Lomb.)

 (a) What arc measurement will correspond to 40.5-diopter lenses?
 (b) A lens with an arc measurement of 7.51 mm will provide what diopter strength?

54. The hot-water tap can fill a tub in 20 min. The cold-water tap takes 15 min to fill the tub. How long would it take to fill the tub with both taps open?

55. At a certain gasoline station, 3 gal of unleaded gasoline cost $4.86. How much would 13 gal of the same gasoline cost?

56. Three-fourths of a number is subtracted from seven-sixths of the number, giving 10. Find the number.

CHAPTER 7 TEST

1. Find all real numbers excluded from the domain of $f(x) = \dfrac{x + 3}{3x^2 + 2x - 8}$. Then give the domain using set notation.

2. Write $\dfrac{6x^2 - 13x - 5}{9x^3 - x}$ in lowest terms.

Multiply or divide.

3. $\dfrac{(x + 3)^2}{4} \cdot \dfrac{6}{2x + 6}$

4. $\dfrac{y^2 - 16}{y^2 - 25} \cdot \dfrac{y^2 + 2y - 15}{y^2 - 7y + 12}$

5. $\dfrac{x^2 - 9}{x^3 + 3x^2} \div \dfrac{x^2 + x - 12}{x^3 + 9x^2 + 20x}$

6. Find the least common denominator for the following group of denominators:
   $t^2 + t - 6, \quad t^2 + 3t, \quad t^2.$

Add or subtract as indicated.

7. $\dfrac{7}{6t^2} - \dfrac{1}{3t}$

8. $\dfrac{9}{(x - 3)^2} + \dfrac{2}{(x - 3)(x + 3)}$

9. $\dfrac{6}{x + 4} + \dfrac{1}{x + 2} - \dfrac{3x}{x^2 + 6x + 8}$

Simplify each complex fraction.

10. $\dfrac{\dfrac{12}{r + 4}}{\dfrac{11}{6r + 24}}$

11. $\dfrac{\dfrac{1}{a} - \dfrac{1}{b}}{\dfrac{a}{b} - \dfrac{b}{a}}$

12. $\dfrac{2x^{-2} + y^{-2}}{x^{-1} - y^{-1}}$

13. One of the following is an expression to be simplified by algebraic operations, and the other is an equation to be solved. Identify each, then simplify the one that requires operations and solve the one that is an equation.

   (a) $\dfrac{2x}{3} + \dfrac{x}{4} - \dfrac{11}{2}$

   (b) $\dfrac{2x}{3} + \dfrac{x}{4} = \dfrac{11}{2}$

Solve each equation.

14. $\dfrac{1}{x} - \dfrac{4}{3x} = \dfrac{1}{x - 2}$

15. $\dfrac{y}{y + 2} - \dfrac{1}{y - 2} = \dfrac{8}{y^2 - 4}$

16. Checking the solution(s) of an equation in earlier chapters verified that the algebraic steps were performed correctly. When an equation includes a term with a variable denominator, what additional reason *requires* that the solutions be checked?

17. Solve for the variable ℓ in this formula from mathematics:

$$S = \frac{n}{2}(a + \ell).$$

18. Sketch the graph of the function defined by $f(x) = \dfrac{-2}{x + 1}$. Give the equation of its vertical asymptote.

Solve each problem.

19. Wayne can do a job in 9 hr, while Susan can do the same job in 5 hr. How long would it take them to do the job if they worked together?

20. The rate of the current in a stream is 3 mph. Nana's boat can go 36 mi downstream in the same time that it takes to go 24 mi upstream. Find the rate of her boat in still water.

21. Biologists collected a sample of 600 fish from Lake Linda on May 1 and tagged each of them. When they returned on June 1, a new sample of 800 fish was collected and 10 of these had been previously tagged. Use this experiment to determine the approximate fish population of Lake Linda.

22. In biology, the function defined by

$$g(x) = \frac{5x}{2 + x}$$

gives the growth rate of a population for x units of available food. (*Source:* Smith, J. Maynard, *Models in Ecology,* Cambridge University Press, 1974.)

(a) What amount of food (in appropriate units) would produce a growth rate of 3 units of growth per unit of food?

(b) What is the growth rate if no food is available?

CUMULATIVE REVIEW EXERCISES CHAPTERS **1–7**

Evaluate if $x = -4$, $y = 3$, and $z = 6$.

1. $|2x| + 3y - z^3$

2. $\dfrac{x(2x - 1)}{3y - z}$

Solve each equation.

3. $7(2x + 3) - 4(2x + 1) = 2(x + 1)$

4. $|6x - 8| - 4 = 0$

5. $ax + by = cx + d$ for x

Solve each inequality.

6. $\dfrac{2}{3}y + \dfrac{5}{12}y \leq 20$

7. $|3x + 2| \geq 4$

Solve each problem.

8. Otis Taylor invested some money at 4% interest and twice as much at 3% interest. His interest for the first year was $400. How much did he invest at each rate?

9. A triangle has area 42 m². The base is 14 m long. Find the height of the triangle.

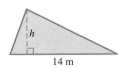

10. Graph $-4x + 2y = 8$ and give the intercepts.

Find the slope of each line described in Exercises 11 and 12.

11. Through $(-5, 8)$ and $(-1, 2)$ **12.** Perpendicular to $4x - 3y = 12$

13. Write an equation of the line in Exercise 11. Give the equation in the form $y = mx + b$.

Graph the solution set of each inequality.

14. $2x + 5y > 10$ **15.** $x - y \geq 3$ and $3x + 4y \leq 12$

Decide whether each relation defined in Exercises 16–18 is a function, and give its domain and range.

16. Average Hourly Wages in Mexico

Year	Wage (in dollars)
1990	1.25
1992	1.61
1994	1.80
1996	1.21
1998	1.94
2000	2.26

Source: John Christman, CIEMEX-WEFA.

17.

18. $y = -\sqrt{x + 2}$

19. Given the equation $5x - 3y = 8$,

   (a) write it with function notation $f(x)$;
   (b) find $f(1)$.

20. If $f(x) = 3x + 6$, what is $f(x + 3)$?

Solve each system.

21. $4x - y = -7$
    $5x + 2y = 1$

22. $x + y - 2z = -1$
    $2x - y + z = -6$
    $3x + 2y - 3z = -3$

23. $x + 2y + z = 5$
    $x - y + z = 3$
    $2x + 4y + 2z = 11$

24. Taking traffic into account, an automobile can travel on the average 7 km in the same time that an airplane can travel 100 km. The average speed of an airplane is 558 km per hr greater than that of an automobile. Find both speeds. (*Source:* Schafer, Andreas and David Victor, "The Past and Future of Global Mobility," *Scientific American,* October 1997.)

25. Evaluate $\begin{vmatrix} 2 & 4 & 1 \\ 1 & 3 & 6 \\ 2 & 3 & -1 \end{vmatrix}$.

Simplify. Write each answer with only positive exponents. Assume all variables represent nonzero real numbers.

26. $\left(\dfrac{a^{-3}b^4}{a^2 b^{-1}} \right)^{-2}$

27. $\left(\dfrac{m^{-4}n^2}{m^2 n^{-3}} \right) \cdot \left(\dfrac{m^5 n^{-1}}{m^{-2} n^5} \right)$

Perform the indicated operations.

28. $(3y^2 - 2y + 6) - (-y^2 + 5y + 12)$ **29.** $(4f + 3)(3f - 1)$

30. $(7t^3 + 8)(7t^3 - 8)$

31. $\left(\dfrac{1}{4}x + 5\right)^2$

32. $(3x^3 + 13x^2 - 17x - 7) \div (3x + 1)$

33. For the polynomial functions defined by
$$f(x) = x^2 + 2x - 3 \quad \text{and} \quad g(x) = 2x^3 - 3x^2 + 4x - 1,$$
find **(a)** $(f + g)(x)$, **(b)** $(g - f)(x)$, and **(c)** $(f + g)(-1)$.

Factor each polynomial completely.

34. $2x^2 - 13x - 45$ **35.** $100t^4 - 25$ **36.** $8p^3 + 125$

37. Solve the equation $3x^2 + 4x = 7$.

Write each rational expression in lowest terms.

38. $\dfrac{y^2 - 16}{y^2 - 8y + 16}$

39. $\dfrac{8x^2 - 18}{8x^2 + 4x - 12}$

Perform the indicated operations. Express the answer in lowest terms.

40. $\dfrac{2a^2}{a + b} \cdot \dfrac{a - b}{4a}$

41. $\dfrac{x + 4}{x - 2} + \dfrac{2x - 10}{x - 2}$

42. $\dfrac{2x}{2x - 1} + \dfrac{4}{2x + 1} + \dfrac{8}{4x^2 - 1}$

43. Solve the equation $\dfrac{-3x}{x + 1} + \dfrac{4x + 1}{x} = \dfrac{-3}{x^2 + x}$.

44. Solve the formula for q: $\dfrac{1}{f} = \dfrac{1}{p} + \dfrac{1}{q}$.

Solve each problem.

45. Erika Suco can fly her plane 200 mi against the wind in the same time it takes her to fly 300 mi with the wind. The wind blows at 30 mph. Find the speed of her plane in still air.

46. Machine A can complete a certain job in 2 hr. To speed up the work, Machine B, which could complete the job alone in 3 hr, is brought in to help. How long will it take the two machines to complete the job working together?

Roots, Radicals, and Root Functions

8

Tom Skilling is the chief meteorologist for the *Chicago Tribune*. He writes a column titled "Ask Tom Why," where readers question him on a variety of topics. In the Saturday, August 17, 2002 issue, reader Ted Fleischaker wrote: "I cannot remember the formula to calculate the distance to the horizon. I have a stunning view from my 14th floor condo, 150 feet above the ground. How far can I see?" (See Exercise 118 in Section 8.3.)

In Skilling's answer, he explained the formula for finding the distance d to the horizon in miles,

$$d = 1.224\sqrt{h},$$

where h is the height in feet. Square roots such as this one are often found in formulas. This chapter deals with roots and radicals.

8.1 Radical Expressions and Graphs

OBJECTIVE 1 Find roots of numbers. In Chapter 1 we found square roots of positive numbers such as

$$\sqrt{36} = 6 \text{ because } 6 \cdot 6 = 36 \quad \text{and} \quad \sqrt{144} = 12 \text{ because } 12 \cdot 12 = 144.$$

In this section we extend our discussion of roots to cube roots, fourth roots, and higher roots. In general, $\sqrt[n]{a}$ is a number whose nth power equals a. That is,

$$\sqrt[n]{a} = b \quad \text{means} \quad b^n = a.$$

The number a is the **radicand,** n is the **index** or **order,** and the expression $\sqrt[n]{a}$ is a **radical.**

EXAMPLE 1 Simplifying Higher Roots

Simplify.

(a) $\sqrt[3]{27} = 3$ because $3^3 = 27$.   **(b)** $\sqrt[3]{125} = 5$ because $5^3 = 125$.

(c) $\sqrt[4]{16} = 2$ because $2^4 = 16$.   **(d)** $\sqrt[5]{32} = 2$ because $2^5 = 32$.

Now Try Exercises 5, 19, 21, and 35.

OBJECTIVE 2 Find principal roots. If n is even, positive numbers have two nth roots. For example, both 4 and -4 are square roots of 16, and 2 and -2 are fourth roots of 16. In such cases, the notation $\sqrt[n]{a}$ represents the positive root, called the **principal root.**

---

nth Root

If n is *even* and a is *positive* or 0, then

$$\sqrt[n]{a} \text{ represents the principal } n\text{th root of } a,$$

and $-\sqrt[n]{a}$ represents the negative nth root of a.

If n is *even* and a is *negative,* then

$$\sqrt[n]{a} \text{ is not a real number.}$$

If n is *odd,* then

$$\text{there is exactly one } n\text{th root of } a, \text{ written } \sqrt[n]{a}.$$

If n is even, the two nth roots of a are often written together as $\pm\sqrt[n]{a}$, with $\pm$ read "positive or negative."

EXAMPLE 2 Finding Roots

Find each root.

(a) $\sqrt{100} = 10$

Because the radicand is positive, there are two square roots, 10 and -10. We want the principal root, which is 10.

(b) $-\sqrt{100} = -10$

Here, we want the negative square root, -10.

(c) $\sqrt[4]{81} = 3$

(d) $\sqrt[6]{-64}$

The index is even and the radicand is negative, so this is not a real number.

(e) $\sqrt[3]{-8} = -2$ because $(-2)^3 = -8$.

Now Try Exercises 1, 15, 17, and 25.

OBJECTIVE 3 Graph functions defined by radical expressions. A **radical expression** is an algebraic expression that contains radicals. For example,

$$3 - \sqrt{x}, \qquad \sqrt[3]{x}, \qquad \text{and} \qquad \sqrt{2x - 1}$$

are radical expressions.

In earlier chapters we graphed functions defined by polynomial and rational expressions. Now we examine the graphs of functions defined by the basic radical expressions $f(x) = \sqrt{x}$ and $f(x) = \sqrt[3]{x}$.

Figure 1 shows the graph of the **square root function** with a table of selected points. Only nonnegative values can be used for x, so the domain is $[0, \infty)$. Because $\sqrt{x}$ is the principal square root of x, it always has a nonnegative value, so the range is also $[0, \infty)$.

x	$f(x) = \sqrt{x}$
0	0
1	1
4	2
9	3

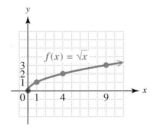

x	$f(x) = \sqrt[3]{x}$
-8	-2
-1	-1
0	0
1	1
8	2

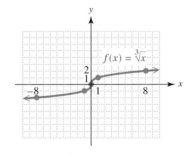

FIGURE 1   **FIGURE 2**

Figure 2 shows the graph of the **cube root function** and a table of selected points. Since any real number (positive, negative, or 0) can be used for x in the cube root function, $\sqrt[3]{x}$ can be positive, negative, or 0. Thus both the domain and the range of the cube root function are $(-\infty, \infty)$.

EXAMPLE 3 Graphing Functions Defined with Radicals

Graph each function by creating a table of values. Give the domain and range.

(a) $f(x) = \sqrt{x - 3}$

A table of values is given on the next page. The x-values were chosen in such a way that the function values are all integers. For the radicand to be nonnegative, we must have $x - 3 \geq 0$, or $x \geq 3$. Therefore, the domain is $[3, \infty)$. Function values are positive or 0, so the range is $[0, \infty)$. The graph is shown in Figure 3 on the next page.

x	$f(x) = \sqrt{x-3}$
3	$\sqrt{3-3} = 0$
4	$\sqrt{4-3} = 1$
7	$\sqrt{7-3} = 2$

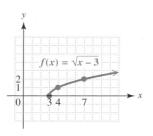

FIGURE 3

(b) $f(x) = \sqrt[3]{x} + 2$

See the table and Figure 4. Both the domain and range are $(-\infty, \infty)$.

x	$f(x) = \sqrt[3]{x} + 2$
-8	$\sqrt[3]{-8} + 2 = 0$
-1	$\sqrt[3]{-1} + 2 = 1$
0	$\sqrt[3]{0} + 2 = 2$
1	$\sqrt[3]{1} + 2 = 3$
8	$\sqrt[3]{8} + 2 = 4$

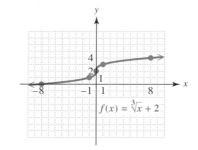

FIGURE 4

Now Try Exercises 37 and 41.

OBJECTIVE 4 Find *n*th roots of *n*th powers. A square root of a^2 (where $a \neq 0$) is a number that can be squared to give a^2. This number is either a or $-a$. Since the symbol $\sqrt{a^2}$ represents the *nonnegative* square root, we express $\sqrt{a^2}$ with absolute value bars as $|a|$, because a may be a negative number.

> **$\sqrt{a^2}$**
>
> For any real number a,
> $$\sqrt{a^2} = |a|.$$

EXAMPLE 4 Simplifying Square Roots Using Absolute Value

Find each square root.

(a) $\sqrt{7^2} = |7| = 7$

(b) $\sqrt{(-7)^2} = |-7| = 7$

(c) $\sqrt{k^2} = |k|$

(d) $\sqrt{(-k)^2} = |-k| = |k|$

Now Try Exercises 45, 47, and 53.

We can generalize this idea to any *n*th root.

$\sqrt[n]{a^n}$

If n is an *even* positive integer, then $\sqrt[n]{a^n} = |a|$,

and if n is an *odd* positive integer, then $\sqrt[n]{a^n} = a$.

That is, use absolute value when n is even; absolute value is not necessary when n is odd.

EXAMPLE 5   Simplifying Higher Roots Using Absolute Value

Simplify each root.

(a) $\sqrt[6]{(-3)^6} = |-3| = 3$ n is even; use absolute value.

(b) $\sqrt[5]{(-4)^5} = -4$ n is odd.

(c) $-\sqrt[4]{(-9)^4} = -|-9| = -9$ **(d)** $\sqrt[3]{\dfrac{8}{27}} = \sqrt[3]{\left(\dfrac{2}{3}\right)^3} = \dfrac{2}{3}$

(e) $-\sqrt{m^4} = -|m^2| = -m^2$

No absolute value bars are needed here because m^2 is nonnegative for any real number value of m.

(f) $\sqrt[3]{a^{12}} = a^4$ because $a^{12} = (a^4)^3$.

(g) $\sqrt[4]{x^{12}} = |x^3|$

We use absolute value bars to guarantee that the result is not negative (because x^3 can be either positive or negative, depending on x). Also, $|x^3|$ can be written as $x^2 \cdot |x|$.

Now Try Exercises 49, 51, 55, and 57.

OBJECTIVE 5 Use a calculator to find roots.   While numbers such as $\sqrt{9}$ and $\sqrt[3]{-8}$ are rational, radicals are often irrational numbers. To find approximations of roots such as $\sqrt{15}$, $\sqrt[3]{10}$, and $\sqrt[4]{2}$, we usually use scientific or graphing calculators. Using a calculator, we find

$$\sqrt{15} \approx 3.872983346, \quad \sqrt[3]{10} \approx 2.15443469, \quad \text{and} \quad \sqrt[4]{2} \approx 1.189207115,$$

where the symbol $\approx$ means "is approximately equal to." In this book we usually show approximations rounded to three decimal places. Thus, we would write

$$\sqrt{15} \approx 3.873, \quad \sqrt[3]{10} \approx 2.154, \quad \text{and} \quad \sqrt[4]{2} \approx 1.189.$$

Figure 5 shows how the preceding approximations are displayed on a TI-83 Plus graphing calculator. In Figure 5(a), eight or nine decimal places are shown, while in Figure 5(b), the number of decimal places is fixed at three.

There is a simple way to check that a calculator approximation is "in the ball-park." Because 16 is a little larger than 15, $\sqrt{16} = 4$ should be a little larger than $\sqrt{15}$. Thus, 3.873 is a reasonable approximation for $\sqrt{15}$.

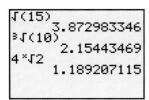

(a)

```
√(15)
           3.873
³√(10)
           2.154
4×√2
           1.189
```

(b)

FIGURE 5

NOTE   The methods for finding approximations differ among makes and models of calculators. You should always consult your owner's manual for keystroke instructions. Be aware that graphing calculators often differ from scientific calculators in the order in which keystrokes are made.

EXAMPLE 6 Finding Approximations for Roots

Use a calculator to verify that each approximation is correct.

(a) $\sqrt{39} \approx 6.245$

(b) $-\sqrt{72} \approx -8.485$

(c) $\sqrt[3]{93} \approx 4.531$

(d) $\sqrt[4]{39} \approx 2.499$

Now Try Exercises 59, 65, 67, and 69.

EXAMPLE 7 Using Roots to Calculate Resonant Frequency

In electronics, the resonant frequency f of a circuit may be found by the formula

$$f = \frac{1}{2\pi\sqrt{LC}},$$

where f is in cycles per second, L is in henrys, and C is in farads.* Find the resonant frequency f if $L = 5 \times 10^{-4}$ henrys and $C = 3 \times 10^{-10}$ farads. Give your answer to the nearest thousand.

Find the value of f when $L = 5 \times 10^{-4}$ and $C = 3 \times 10^{-10}$.

$$f = \frac{1}{2\pi\sqrt{LC}} \qquad \text{Given formula}$$

$$= \frac{1}{2\pi\sqrt{(5 \times 10^{-4})(3 \times 10^{-10})}} \qquad \text{Substitute for } L \text{ and } C.$$

$$\approx 411,000 \qquad \text{Use a calculator.}$$

The resonant frequency f is approximately 411,000 cycles per sec.

Now Try Exercise 73.

8.1 EXERCISES

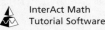
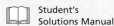
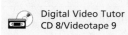
Match each expression from Column I with the equivalent choice from Column II. Answers may be used more than once. See Example 2.

I	II
1. $-\sqrt{16}$	**A.** 3
2. $\sqrt{-16}$	**B.** -2
3. $\sqrt[3]{-27}$	**C.** 2
4. $\sqrt[5]{-32}$	**D.** -3
5. $\sqrt[4]{81}$	**E.** -4
6. $\sqrt[3]{8}$	**F.** Not a real number

---

*Henrys and farads are units of measure in electronics.

Choose the closest approximation of each square root.

7. $\sqrt{123.5}$

    A. 9   **B.** 10   **C.** 11   **D.** 12

8. $\sqrt{67.8}$

    A. 7   **B.** 8   **C.** 9   **D.** 10

Refer to the rectangle to answer the questions in Exercises 9 and 10.

9. Which one of the following is the best estimate of its area?

    A. 2500   **B.** 250   **C.** 50   **D.** 100

10. Which one of the following is the best estimate of its perimeter?

    A. 15   **B.** 250   **C.** 100   **D.** 30

11. Consider the expression $-\sqrt{-a}$. Decide whether it is *positive*, *negative*, 0, or *not a real number* if

    (a) $a > 0$,   **(b)** $a < 0$,   **(c)** $a = 0$.

12. If n is odd, under what conditions is $\sqrt[n]{a}$

    (a) positive,   **(b)** negative,   **(c)** 0?

Find each root that is a real number. Use a calculator as necessary. See Examples 1 and 2.

13. $\sqrt{81}$ **14.** $\sqrt{121}$ **15.** $-\sqrt[3]{216}$ **16.** $-\sqrt[3]{343}$

17. $\sqrt[3]{-64}$ **18.** $\sqrt[3]{-125}$ **19.** $\sqrt[3]{512}$ **20.** $\sqrt[3]{1000}$

21. $\sqrt[4]{1296}$ **22.** $\sqrt[4]{625}$ **23.** $-\sqrt[4]{81}$ **24.** $-\sqrt[4]{256}$

25. $\sqrt[4]{-16}$ **26.** $\sqrt[4]{-81}$ **27.** $\sqrt[6]{64}$ **28.** $\sqrt[4]{256}$

29. $\sqrt[6]{-32}$ **30.** $\sqrt[8]{-1}$ **31.** $\sqrt{\dfrac{64}{81}}$ **32.** $\sqrt{\dfrac{100}{9}}$

33. $\sqrt[3]{\dfrac{8}{27}}$ **34.** $\sqrt[4]{\dfrac{81}{16}}$ **35.** $\sqrt[6]{\dfrac{1}{64}}$ **36.** $\sqrt[5]{\dfrac{1}{32}}$

Graph each function and give its domain and range. See Example 3.

37. $f(x) = \sqrt{x + 3}$ **38.** $f(x) = \sqrt{x - 5}$ **39.** $f(x) = \sqrt{x} - 2$

40. $f(x) = \sqrt{x} + 4$ **41.** $f(x) = \sqrt[3]{x} - 3$ **42.** $f(x) = \sqrt[3]{x} + 1$

43. $f(x) = \sqrt[3]{x - 3}$ **44.** $f(x) = \sqrt[3]{x + 1}$

Simplify each root. See Examples 4 and 5.

45. $\sqrt{12^2}$ **46.** $\sqrt{19^2}$ **47.** $\sqrt{(-10)^2}$ **48.** $-\sqrt{13^2}$

49. $\sqrt[6]{(-2)^6}$ **50.** $\sqrt[6]{(-4)^6}$ **51.** $\sqrt[5]{(-9)^5}$ **52.** $\sqrt[5]{(-8)^5}$

53. $\sqrt{x^2}$ **54.** $-\sqrt{x^2}$ **55.** $\sqrt[3]{x^3}$ **56.** $-\sqrt[3]{x^3}$

57. $\sqrt[3]{x^{15}}$ **58.** $\sqrt[4]{k^{20}}$

Find a decimal approximation for each radical. Round the answer to three decimal places. See Example 6.

59. $\sqrt{9483}$ **60.** $\sqrt{6825}$ **61.** $\sqrt{284.361}$ **62.** $\sqrt{846.104}$

63. $\sqrt{7}$ **64.** $\sqrt{11}$ **65.** $-\sqrt{82}$ **66.** $-\sqrt{91}$

67. $\sqrt[3]{423}$ **68.** $\sqrt[3]{555}$ **69.** $\sqrt[4]{100}$ **70.** $\sqrt[4]{250}$

71. $\sqrt[5]{23.8}$ **72.** $\sqrt[5]{98.4}$

Solve each problem. See Example 7.

73. Use the formula in Example 7 to calculate the resonant frequency of a circuit to the nearest thousand if $L = 7.237 \times 10^{-5}$ henrys and $C = 2.5 \times 10^{-10}$ farads.

74. The threshold weight T for a person is the weight above which the risk of death increases greatly. The threshold weight in pounds for men aged 40–49 is related to height in inches by the formula

$$h = 12.3\sqrt[3]{T}.$$

What height corresponds to a threshold weight of 216 lb for a 43-year-old man? Round your answer to the nearest inch, and then to the nearest tenth of a foot.

75. According to an article in *The World Scanner Report* (August 1991), the distance D, in miles, to the horizon from an observer's point of view over water or "flat" earth is given by

$$D = \sqrt{2H},$$

where H is the height of the point of view, in feet. If a person whose eyes are 6 ft above ground level is standing at the top of a hill 44 ft above "flat" earth, approximately how far to the horizon will she be able to see?

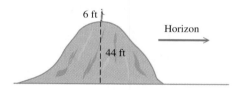

76. The time for one complete swing of a simple pendulum is

$$t = 2\pi\sqrt{\frac{L}{g}},$$

where t is time in seconds, L is the length of the pendulum in feet, and g, the force due to gravity, is about 32 ft per sec². Find the time of a complete swing of a 2-ft pendulum to the nearest tenth of a second.

77. Heron's formula gives a method of finding the area of a triangle if the lengths of its sides are known. Suppose that a, b, and c are the lengths of the sides. Let s denote one-half of the perimeter of the triangle (called the *semiperimeter*); that is,

$$s = \frac{1}{2}(a + b + c).$$

Then the area of the triangle is

$$A = \sqrt{s(s - a)(s - b)(s - c)}.$$

Find the area of the Bermuda Triangle, if the "sides" of this triangle measure approximately 850 mi, 925 mi, and 1300 mi. Give your answer to the nearest thousand square miles.

78. The Vietnam Veterans' Memorial in Washington, D.C., is in the shape of an unenclosed isosceles triangle with equal sides of length 246.75 ft. If the triangle were enclosed, the third side would have length 438.14 ft. Use Heron's formula from the previous exercise to find the area of this enclosure to the nearest hundred square feet. (*Source:* Information pamphlet obtained at the Vietnam Veterans' Memorial.)

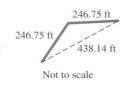

246.75 ft

246.75 ft

438.14 ft

Not to scale

79. The formula

$$I = \sqrt{\frac{2P}{L}}$$

relates the coefficient of self-induction L (in henrys), the energy P stored in an electronic circuit (in joules), and the current I (in amps). Find I if $P = 120$ and $L = 80$.

80. When the wind blows, the air feels much colder than the actual temperature. The **windchill factor** measures the cooling effect that the wind has on one's skin. Until recently, the formula that the National Weather Service used to compute windchill was

$$T_{wc} = .0817\left(3.71\sqrt{V} + 5.81 - .25V\right)(T - 91.4) + 91.4,$$

where T_{wc} is windchill, V is wind speed in miles per hour (mph), and T is air temperature in degrees Fahrenheit. The windchill for various wind speeds and temperatures is shown in the table.

Windchill Factor

| | Air Temperature (°Fahrenheit) | | | | | | | | | | | | | | |
	35	**30**	**25**	**20**	**15**	**10**	**5**	**0**	**−5**	**−10**	**−15**	**−20**	**−25**	**−30**	**−35**
4	35	30	25	20	15	10	5	0	−5	−10	−15	−20	−25	−30	−35
5	32	27	22	16	11	6	0	−5	−10	−15	−21	−26	−31	−36	−42
10	22	16	10	3	−3	−9	−15	−22	−27	−34	−40	−46	−52	−58	−64
15	16	9	2	−5	−11	−18	−25	−31	−38	−45	−51	−58	−65	−72	−78
20	12	4	−3	−10	−17	−24	−31	−39	−46	−53	−60	−67	−74	−81	−88
25	8	1	−7	−15	−22	−29	−36	−44	−51	−59	−66	−74	−81	−88	−96
30	6	−2	−10	−18	−25	−33	−41	−49	−56	−64	−71	−79	−86	−93	−101
35	4	−4	−12	−20	−27	−35	−43	−52	−58	−67	−74	−82	−89	−97	−105
40	3	−5	−13	−21	−29	−37	−45	−53	−60	−69	−76	−84	−92	−100	−107
45	2	−6	−14	−22	−30	−38	−46	−54	−62	−70	−78	−85	−93	−102	−109

Wind Speed (mph)

Source: USA Today.

Choose a temperature of 10°F. Use the formula to calculate the windchill for wind speeds of 4, 10, 25, and 40 mph. Round the results to the nearest degree. Do your results match those in the tables?

8.2 Rational Exponents

OBJECTIVE 1 Use exponential notation for *n*th roots. In mathematics we often formulate definitions so that previous rules remain valid. In Chapter 5 we defined 0 as an exponent in such a way that the rules for products, quotients, and powers would still be valid. Now we look at exponents that are rational numbers of the form $\frac{1}{n}$, where *n* is a natural number.

For the rules of exponents to remain valid, the product $(3^{1/2})^2 = 3^{1/2} \cdot 3^{1/2}$ should be found by adding exponents.

$$(3^{1/2})^2 = 3^{1/2} \cdot 3^{1/2}$$
$$= 3^{1/2+1/2}$$
$$= 3^1$$
$$= 3$$

However, by definition $(\sqrt{3})^2 = \sqrt{3} \cdot \sqrt{3} = 3$. Since both $(3^{1/2})^2$ and $(\sqrt{3})^2$ are equal to 3, it is reasonable to have

$$3^{1/2} = \sqrt{3}.$$

This suggests the following generalization.

$a^{1/n}$

If $\sqrt[n]{a}$ is a real number, then $\qquad a^{1/n} = \sqrt[n]{a}.$

EXAMPLE 1 Evaluating Exponentials of the Form $a^{1/n}$

Evaluate each expression.

(a) $64^{1/3} = \sqrt[3]{64} = 4$ **(b)** $100^{1/2} = \sqrt{100} = 10$

(c) $-256^{1/4} = -\sqrt[4]{256} = -4$

(d) $(-256)^{1/4} = \sqrt[4]{-256}$ is not a real number because the radicand, -256, is negative and the index is even.

(e) $(-32)^{1/5} = \sqrt[5]{-32} = -2$ **(f)** $\left(\dfrac{1}{8}\right)^{1/3} = \sqrt[3]{\dfrac{1}{8}} = \dfrac{1}{2}$

Now Try Exercises 11, 13, 19, and 25.

CAUTION Notice the difference between parts (c) and (d) in Example 1. The radical in part (c) is the *negative fourth root* of a positive number, while the radical in part (d) is the *principal fourth root of a negative number,* which is not a real number.

OBJECTIVE 2 Define and use expressions of the form $a^{m/n}$. We know that $8^{1/3} = \sqrt[3]{8}$. How should we define a number like $8^{2/3}$? For past rules of exponents to be valid,

$$8^{2/3} = 8^{(1/3)2} = (8^{1/3})^2.$$

Since $8^{1/3} = \sqrt[3]{8}$,

$$8^{2/3} = \left(\sqrt[3]{8}\right)^2 = 2^2 = 4.$$

Generalizing from this example, we define $a^{m/n}$ as follows.

$a^{m/n}$

If m and n are positive integers with m/n in lowest terms, then

$$a^{m/n} = (a^{1/n})^m,$$

provided that $a^{1/n}$ is a real number. If $a^{1/n}$ is not a real number, then $a^{m/n}$ is not a real number.

▇ **EXAMPLE 2** Evaluating Exponentials of the Form $a^{m/n}$

Evaluate each exponential.

(a) $36^{3/2} = (36^{1/2})^3 = 6^3 = 216$ (b) $125^{2/3} = (125^{1/3})^2 = 5^2 = 25$

(c) $-4^{5/2} = -(4^{5/2}) = -(4^{1/2})^5 = -(2)^5 = -32$

(d) $(-27)^{2/3} = [(-27)^{1/3}]^2 = (-3)^2 = 9$

Notice how the $-$ sign is used in parts (c) and (d). In part (c), we first evaluate the exponential and then find its negative. In part (d), the $-$ sign is part of the base, -27.

(e) $(-100)^{3/2}$ is not a real number since $(-100)^{1/2}$ is not a real number.   ▇

Now Try Exercises 21 and 23.

When a rational exponent is negative, the earlier interpretation of negative exponents is applied.

$a^{-m/n}$

If $a^{m/n}$ is a real number, then

$$a^{-m/n} = \frac{1}{a^{m/n}} \quad (a \neq 0).$$

▇ **EXAMPLE 3** Evaluating Exponentials with Negative Rational Exponents

Evaluate each exponential.

(a) $16^{-3/4}$

By the definition of a negative exponent,

$$16^{-3/4} = \frac{1}{16^{3/4}}.$$

Since $16^{3/4} = \left(\sqrt[4]{16}\right)^3 = 2^3 = 8$,

$$16^{-3/4} = \frac{1}{16^{3/4}} = \frac{1}{8}.$$

(b) $25^{-3/2} = \dfrac{1}{25^{3/2}} = \dfrac{1}{(\sqrt{25})^3} = \dfrac{1}{5^3} = \dfrac{1}{125}$

(c) $\left(\dfrac{8}{27}\right)^{-2/3} = \dfrac{1}{\left(\dfrac{8}{27}\right)^{2/3}} = \dfrac{1}{\left(\sqrt[3]{\dfrac{8}{27}}\right)^2} = \dfrac{1}{\left(\dfrac{2}{3}\right)^2} = \dfrac{1}{\dfrac{4}{9}} = \dfrac{9}{4}$

We could also use the rule $\left(\dfrac{b}{a}\right)^{-m} = \left(\dfrac{a}{b}\right)^{m}$ here, as follows.

$$\left(\dfrac{8}{27}\right)^{-2/3} = \left(\dfrac{27}{8}\right)^{2/3} = \left(\sqrt[3]{\dfrac{27}{8}}\right)^2 = \left(\dfrac{3}{2}\right)^2 = \dfrac{9}{4}$$

Now Try Exercises 27 and 29.

CAUTION When using the rule in Example 3(c), we take the reciprocal only of the base, *not* the exponent. Also, be careful to distinguish between exponential expressions like $-16^{1/4}$, $16^{-1/4}$, and $-16^{-1/4}$.

$$-16^{1/4} = -2, \qquad 16^{-1/4} = \dfrac{1}{2}, \qquad \text{and} \qquad -16^{-1/4} = -\dfrac{1}{2}.$$

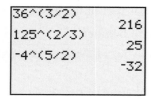

FIGURE 6

```
(-27)^(2/3)
                 9
16^(-3/4)▸Frac
               1/8
```

FIGURE 7

The screens in Figures 6 and 7 illustrate how a graphing calculator performs some of the evaluations seen in Examples 2 and 3. (All results on the screens are rational numbers.)

We obtain an alternative definition of $a^{m/n}$ by using the power rule for exponents a little differently than in the earlier definition. If all indicated roots are real numbers, then

$$a^{m/n} = a^{m(1/n)} = (a^m)^{1/n},$$

so

$$a^{m/n} = (a^m)^{1/n}.$$

$a^{m/n}$

If all indicated roots are real numbers, then

$$a^{m/n} = (a^{1/n})^m = (a^m)^{1/n}.$$

We can now evaluate an expression such as $27^{2/3}$ in two ways:

$$27^{2/3} = (27^{1/3})^2 = 3^2 = 9$$

or

$$27^{2/3} = (27^2)^{1/3} = 729^{1/3} = 9.$$

In most cases, it is easier to use $(a^{1/n})^m$.

This rule can also be expressed with radicals as follows.

Radical Form of $a^{m/n}$

If all indicated roots are real numbers, then

$$a^{m/n} = \sqrt[n]{a^m} = \left(\sqrt[n]{a}\right)^m.$$

That is, raise a to the power and then take the root, or take the root and then raise a to the power.

For example,
$$8^{2/3} = \sqrt[3]{8^2} = \sqrt[3]{64} = 4, \quad \text{and} \quad 8^{2/3} = \left(\sqrt[3]{8}\right)^2 = 2^2 = 4,$$
so
$$8^{2/3} = \sqrt[3]{8^2} = \left(\sqrt[3]{8}\right)^2.$$

OBJECTIVE 3 Convert between radicals and rational exponents. Using the definition of rational exponents, we can simplify many problems involving radicals by converting the radicals to numbers with rational exponents. After simplifying, we convert the answer back to radical form.

■ **EXAMPLE 4 Converting between Rational Exponents and Radicals**

Write each exponential as a radical. Assume all variables represent positive real numbers. Use the definition that takes the root first.

(a) $13^{1/2} = \sqrt{13}$

(b) $6^{3/4} = \left(\sqrt[4]{6}\right)^3$

(c) $9m^{5/8} = 9\left(\sqrt[8]{m}\right)^5$

(d) $6x^{2/3} - (4x)^{3/5} = 6\left(\sqrt[3]{x}\right)^2 - \left(\sqrt[5]{4x}\right)^3$

(e) $r^{-2/3} = \dfrac{1}{r^{2/3}} = \dfrac{1}{\left(\sqrt[3]{r}\right)^2}$

(f) $(a^2 + b^2)^{1/2} = \sqrt{a^2 + b^2}$ Note that $\sqrt{a^2 + b^2} \neq a + b$.

In (g)–(i), write each radical as an exponential. Simplify. Assume all variables represent positive real numbers.

(g) $\sqrt{10} = 10^{1/2}$

(h) $\sqrt[4]{3^8} = 3^{8/4} = 3^2 = 9$

(i) $\sqrt[6]{z^6} = z$ since z is positive.

■

Now Try Exercises 33, 35, 37, 49, and 51.

NOTE In Example 4(i), it was not necessary to use absolute value bars since the directions specifically stated that the variable represents a positive real number. Since the absolute value of the positive real number z is z itself, the answer is simply z. When working exercises with radicals, we often assume that variables represent positive real numbers, which will eliminate the need for absolute value.

OBJECTIVE 4 Use the rules for exponents with rational exponents. The definition of rational exponents allows us to apply the rules for exponents first introduced in Chapter 5.

Rules for Rational Exponents

Let r and s be rational numbers. For all real numbers a and b for which the indicated expressions exist:

$$a^r \cdot a^s = a^{r+s} \qquad a^{-r} = \frac{1}{a^r} \qquad \frac{a^r}{a^s} = a^{r-s} \qquad \left(\frac{a}{b}\right)^{-r} = \frac{b^r}{a^r}$$

$$(a^r)^s = a^{rs} \qquad (ab)^r = a^r b^r \qquad \left(\frac{a}{b}\right)^r = \frac{a^r}{b^r} \qquad a^{-r} = \left(\frac{1}{a}\right)^r.$$

■ **EXAMPLE 5** Applying Rules for Rational Exponents

Write with only positive exponents. Assume all variables represent positive real numbers.

(a) $2^{1/2} \cdot 2^{1/4} = 2^{1/2+1/4} = 2^{3/4}$ \quad Product rule

(b) $\dfrac{5^{2/3}}{5^{7/3}} = 5^{2/3-7/3} = 5^{-5/3} = \dfrac{1}{5^{5/3}}$ \quad Quotient rule

(c) $\dfrac{(x^{1/2}y^{2/3})^4}{y} = \dfrac{(x^{1/2})^4(y^{2/3})^4}{y}$ \quad Power rule

$\qquad = \dfrac{x^2 y^{8/3}}{y^1}$ \quad Power rule

$\qquad = x^2 y^{8/3-1}$ \quad Quotient rule

$\qquad = x^2 y^{5/3}$

(d) $\left(\dfrac{x^4 y^{-6}}{x^{-2}y^{1/3}}\right)^{-2/3} = \dfrac{(x^4)^{-2/3}(y^{-6})^{-2/3}}{(x^{-2})^{-2/3}(y^{1/3})^{-2/3}}$

$\qquad = \dfrac{x^{-8/3}y^4}{x^{4/3}y^{-2/9}}$ \quad Power rule

$\qquad = x^{-8/3-4/3}y^{4-(-2/9)}$ \quad Quotient rule

$\qquad = x^{-4}y^{38/9}$

$\qquad = \dfrac{y^{38/9}}{x^4}$ \quad Definition of negative exponent

The same result is obtained if we simplify within the parentheses first, leading to $(x^6 y^{-19/3})^{-2/3}$.

Then, apply the power rule. (Show that the result is the same.)

(e) $m^{3/4}(m^{5/4} - m^{1/4}) = m^{3/4} \cdot m^{5/4} - m^{3/4} \cdot m^{1/4}$ \quad Distributive property

$\qquad = m^{3/4+5/4} - m^{3/4+1/4}$ \quad Product rule

$\qquad = m^{8/4} - m^{4/4}$

$\qquad = m^2 - m$

Do not make the common mistake of multiplying exponents in the first step. ■

Now Try Exercises 57, 59, 65, 75, and 77.

CAUTION Use the rules of exponents in problems like those in Example 5. Do not convert the expressions to radical form.

■ **EXAMPLE 6** Applying Rules for Rational Exponents

Rewrite all radicals as exponentials, and then apply the rules for rational exponents. Leave answers in exponential form. Assume all variables represent positive real numbers.

(a) $\sqrt[3]{x^2} \cdot \sqrt[4]{x} = x^{2/3} \cdot x^{1/4}$ Convert to rational exponents.

$= x^{2/3+1/4}$ Product rule

$= x^{8/12+3/12}$ Write exponents with a common denominator.

$= x^{11/12}$

(b) $\dfrac{\sqrt{x^3}}{\sqrt[3]{x^2}} = \dfrac{x^{3/2}}{x^{2/3}} = x^{3/2-2/3} = x^{5/6}$

(c) $\sqrt{\sqrt[4]{z}} = \sqrt{z^{1/4}} = (z^{1/4})^{1/2} = z^{1/8}$

Now Try Exercises 83, 85, and 89.

8.2 EXERCISES

For Extra Help

 Student's Solutions Manual

 MyMathLab

 InterAct Math Tutorial Software

 AW Math Tutor Center

*Math*XP MathXL

 Digital Video Tutor CD 8/Videotape 9

Match each expression from Column I with the equivalent choice from Column II.

I		II	
1. $2^{1/2}$		**A.** -4	
2. $(-27)^{1/3}$		**B.** 8	
3. $-16^{1/2}$		**C.** $\sqrt{2}$	
4. $(-16)^{1/2}$		**D.** $-\sqrt{6}$	
5. $(-32)^{1/5}$		**E.** -3	
6. $(-32)^{2/5}$		**F.** $\sqrt{6}$	
7. $4^{3/2}$		**G.** 4	
8. $6^{2/4}$		**H.** -2	
9. $-6^{2/4}$		**I.** 6	
10. $36^{.5}$		**J.** Not a real number	

Simplify each expression involving rational exponents. See Examples 1–3.

11. $169^{1/2}$ **12.** $121^{1/2}$ **13.** $729^{1/3}$ **14.** $512^{1/3}$

15. $16^{1/4}$ **16.** $625^{1/4}$ **17.** $\left(\dfrac{64}{81}\right)^{1/2}$ **18.** $\left(\dfrac{8}{27}\right)^{1/3}$

19. $(-27)^{1/3}$ **20.** $(-32)^{1/5}$ **21.** $100^{3/2}$ **22.** $64^{3/2}$

23. $-16^{5/2}$ **24.** $-32^{3/5}$ **25.** $(-144)^{1/2}$ **26.** $(-36)^{1/2}$

27. $64^{-3/2}$ **28.** $81^{-3/2}$ **29.** $\left(-\dfrac{8}{27}\right)^{-2/3}$ **30.** $\left(-\dfrac{64}{125}\right)^{-2/3}$

31. Explain why $(-64)^{1/2}$ is not a real number, while $-64^{1/2}$ is a real number.

32. Explain why $a^{1/n}$ is defined to be equal to $\sqrt[n]{a}$ when $\sqrt[n]{a}$ is real.

Write with radicals. Assume all variables represent positive real numbers. See Example 4.

33. $12^{1/2}$ **34.** $3^{1/2}$ **35.** $8^{3/4}$

36. $7^{2/3}$ **37.** $(9q)^{5/8} - (2x)^{2/3}$ **38.** $(3p)^{3/4} + (4x)^{1/3}$

39. $(2m)^{-3/2}$ **40.** $(5y)^{-3/5}$ **41.** $(2y + x)^{2/3}$

42. $(r + 2z)^{3/2}$ **43.** $(3m^4 + 2k^2)^{-2/3}$ **44.** $(5x^2 + 3z^3)^{-5/6}$

45. Show that, in general, $\sqrt{a^2 + b^2} \neq a + b$ by replacing a with 3 and b with 4.

✎ **46.** Suppose someone claims that $\sqrt[n]{a^n + b^n}$ must equal $a + b$, since when $a = 1$ and $b = 0$, a true statement results:

$$\sqrt[n]{a^n + b^n} = \sqrt[n]{1^n + 0^n} = \sqrt[n]{1^n} = 1 = 1 + 0 = a + b.$$

Explain why this is faulty reasoning.

Simplify by first converting to rational exponents. Assume all variables represent positive real numbers. See Example 4.

47. $\sqrt{2^{12}}$ **48.** $\sqrt{5^{10}}$ **49.** $\sqrt[3]{4^9}$ **50.** $\sqrt[4]{6^8}$ **51.** $\sqrt{x^{20}}$

52. $\sqrt{r^{50}}$ **53.** $\sqrt[3]{x} \cdot \sqrt{x}$ **54.** $\sqrt[4]{y} \cdot \sqrt[5]{y^2}$ **55.** $\dfrac{\sqrt[3]{t^4}}{\sqrt[5]{t^4}}$ **56.** $\dfrac{\sqrt[4]{w^3}}{\sqrt[6]{w}}$

Use the rules of exponents to simplify each expression. Write all answers with positive exponents. Assume all variables represent positive real numbers. See Example 5.

57. $3^{1/2} \cdot 3^{3/2}$ **58.** $6^{4/3} \cdot 6^{2/3}$ **59.** $\dfrac{64^{5/3}}{64^{4/3}}$

60. $\dfrac{125^{7/3}}{125^{5/3}}$ **61.** $y^{7/3} \cdot y^{-4/3}$ **62.** $r^{-8/9} \cdot r^{17/9}$

63. $\dfrac{k^{1/3}}{k^{2/3} \cdot k^{-1}}$ **64.** $\dfrac{z^{3/4}}{z^{5/4} \cdot z^{-2}}$ **65.** $\dfrac{(x^{1/4}y^{2/5})^{20}}{x^2}$

66. $\dfrac{(r^{1/5}s^{2/3})^{15}}{r^2}$ **67.** $\dfrac{(x^{2/3})^2}{(x^2)^{7/3}}$ **68.** $\dfrac{(p^3)^{1/4}}{(p^{5/4})^2}$

69. $\dfrac{m^{3/4}n^{-1/4}}{(m^2n)^{1/2}}$ **70.** $\dfrac{(a^2b^5)^{-1/4}}{(a^{-3}b^2)^{1/6}}$ **71.** $\dfrac{p^{1/5}p^{7/10}p^{1/2}}{(p^3)^{-1/5}}$

72. $\dfrac{z^{1/3}z^{-2/3}z^{1/6}}{(z^{-1/6})^3}$ **73.** $\left(\dfrac{b^{-3/2}}{c^{-5/3}}\right)^2 (b^{-1/4}c^{-1/3})^{-1}$ **74.** $\left(\dfrac{m^{-2/3}}{a^{-3/4}}\right)^4 (m^{-3/8}a^{1/4})^{-2}$

75. $\left(\dfrac{p^{-1/4}q^{-3/2}}{3^{-1}p^{-2}q^{-2/3}}\right)^{-2}$ **76.** $\left(\dfrac{2^{-2}w^{-3/4}x^{-5/8}}{w^{3/4}x^{-1/2}}\right)^{-3}$ **77.** $p^{2/3}(p^{1/3} + 2p^{4/3})$

78. $z^{5/8}(3z^{5/8} + 5z^{11/8})$ **79.** $k^{1/4}(k^{3/2} - k^{1/2})$ **80.** $r^{3/5}(r^{1/2} + r^{3/4})$

81. $6a^{7/4}(a^{-7/4} + 3a^{-3/4})$ **82.** $4m^{5/3}(m^{-2/3} - 4m^{-5/3})$

Write with rational exponents, and then apply the properties of exponents. Assume all radicands represent positive real numbers. Give answers in exponential form. See Example 6.

83. $\sqrt[5]{x^3} \cdot \sqrt[4]{x}$ **84.** $\sqrt[6]{y^5} \cdot \sqrt[3]{y^2}$ **85.** $\dfrac{\sqrt{x^5}}{\sqrt{x^8}}$ **86.** $\dfrac{\sqrt[3]{k^5}}{\sqrt[3]{k^7}}$

87. $\sqrt{y} \cdot \sqrt[3]{yz}$ **88.** $\sqrt[3]{xz} \cdot \sqrt{z}$ **89.** $\sqrt[4]{\sqrt[3]{m}}$ **90.** $\sqrt[3]{\sqrt{k}}$

91. $\sqrt{\sqrt[3]{\sqrt[4]{x}}}$ **92.** $\sqrt[3]{\sqrt[5]{\sqrt{y}}}$

Solve each problem.

93. Meteorologists can determine the duration of a storm by using the function defined by

$$T(D) = .07D^{3/2},$$

where D is the diameter of the storm in miles and T is the time in hours. Find the duration of a storm with a diameter of 16 mi. Round your answer to the nearest tenth of an hour.

94. The threshold weight T, in pounds, for a person is the weight above which the risk of death increases greatly. The threshold weight in pounds for men aged 40–49 is related to height in inches by the function defined by

$$h(T) = (1860.867T)^{1/3}.$$

What height corresponds to a threshold weight of 200 lb for a 46-yr-old man? Round your answer to the nearest inch, and then to the nearest tenth of a foot.

RELATING CONCEPTS (EXERCISES 95–102)

For Individual or Group Work

*Earlier, we factored expressions like $x^4 - x^5$ by factoring out the greatest common factor to get $x^4 - x^5 = x^4(1 - x)$. We can adapt this approach to factor expressions with rational exponents. When one or more of the exponents is negative or a fraction, we use order on the number line discussed in Chapter 1 to decide on the common factor. In this type of factoring, we want the binomial factor to have only positive exponents, so we always factor out the variable with the *least* exponent. A positive exponent is greater than a negative exponent, so in $7z^{5/8} + z^{-3/4}$, we factor out $z^{-3/4}$, because $-\frac{3}{4}$ is less than $\frac{5}{8}$.*

Factor out the given common factor from each expression. Assume all variables represent positive real numbers.

95. $3x^{-1/2} - 4x^{1/2}$; $x^{-1/2}$

96. $m^3 - 3m^{5/2}$; $m^{5/2}$

97. $4t^{-1/2} + 7t^{3/2}$; $t^{-1/2}$

98. $8x^{2/3} + 5x^{-1/3}$; $x^{-1/3}$

99. $4p - p^{3/4}$; $p^{3/4}$

100. $2m^{1/8} - m^{5/8}$; $m^{1/8}$

101. $9k^{-3/4} - 2k^{-1/4}$; $k^{-3/4}$

102. $7z^{-5/8} - z^{-3/4}$; $z^{-3/4}$

8.3 Simplifying Radical Expressions

OBJECTIVE 1 Use the product rule for radicals. We now develop rules for multiplying and dividing radicals that have the same index. For example, is the product of two nth-root radicals equal to the nth root of the product of the radicands? Are the expressions $\sqrt{36 \cdot 4}$ and $\sqrt{36} \cdot \sqrt{4}$ equal? To find out, we do the computations:

$$\sqrt{36 \cdot 4} = \sqrt{144} = 12$$
$$\sqrt{36} \cdot \sqrt{4} = 6 \cdot 2 = 12.$$

Notice that in both cases the result is the same. This is an example of the **product rule for radicals.**

Product Rule for Radicals

If $\sqrt[n]{a}$ and $\sqrt[n]{b}$ are real numbers and n is a natural number, then

$$\sqrt[n]{a} \cdot \sqrt[n]{b} = \sqrt[n]{ab}.$$

That is, the product of two radicals is the radical of the product.

We justify the product rule using the rules for rational exponents. Since $\sqrt[n]{a} = a^{1/n}$ and $\sqrt[n]{b} = b^{1/n}$,

$$\sqrt[n]{a} \cdot \sqrt[n]{b} = a^{1/n} \cdot b^{1/n} = (ab)^{1/n} = \sqrt[n]{ab}.$$

CAUTION Use the product rule only when the radicals have the *same* index.

EXAMPLE 1 Using the Product Rule

Multiply. Assume all variables represent positive real numbers.

(a) $\sqrt{5} \cdot \sqrt{7} = \sqrt{5 \cdot 7} = \sqrt{35}$ **(b)** $\sqrt{2} \cdot \sqrt{19} = \sqrt{2 \cdot 19} = \sqrt{38}$

(c) $\sqrt{11} \cdot \sqrt{p} = \sqrt{11p}$ **(d)** $\sqrt{7} \cdot \sqrt{11xyz} = \sqrt{77xyz}$

Now Try Exercises 7, 9, and 11.

EXAMPLE 2 Using the Product Rule

Multiply. Assume all variables represent positive real numbers.

(a) $\sqrt[3]{3} \cdot \sqrt[3]{12} = \sqrt[3]{3 \cdot 12} = \sqrt[3]{36}$ **(b)** $\sqrt[4]{8y} \cdot \sqrt[4]{3r^2} = \sqrt[4]{24yr^2}$

(c) $\sqrt[6]{10m^4} \cdot \sqrt[6]{5m} = \sqrt[6]{50m^5}$

(d) $\sqrt[4]{2} \cdot \sqrt[5]{2}$ cannot be simplified using the product rule for radicals because the indexes (4 and 5) are different.

Now Try Exercises 13, 15, 17, and 19.

OBJECTIVE 2 Use the quotient rule for radicals. The **quotient rule for radicals** is similar to the product rule.

Quotient Rule for Radicals

If $\sqrt[n]{a}$ and $\sqrt[n]{b}$ are real numbers, $b \neq 0$, and n is a natural number, then

$$\sqrt[n]{\frac{a}{b}} = \frac{\sqrt[n]{a}}{\sqrt[n]{b}}.$$

That is, the radical of a quotient is the quotient of the radicals.

EXAMPLE 3 Using the Quotient Rule

Simplify. Assume all variables represent positive real numbers.

(a) $\sqrt{\dfrac{16}{25}} = \dfrac{\sqrt{16}}{\sqrt{25}} = \dfrac{4}{5}$

(b) $\sqrt{\dfrac{7}{36}} = \dfrac{\sqrt{7}}{\sqrt{36}} = \dfrac{\sqrt{7}}{6}$

(c) $\sqrt[3]{-\dfrac{8}{125}} = \sqrt[3]{\dfrac{-8}{125}} = \dfrac{\sqrt[3]{-8}}{\sqrt[3]{125}} = \dfrac{-2}{5} = -\dfrac{2}{5}$

(d) $\sqrt[3]{\dfrac{7}{216}} = \dfrac{\sqrt[3]{7}}{\sqrt[3]{216}} = \dfrac{\sqrt[3]{7}}{6}$

(e) $\sqrt[5]{\dfrac{x}{32}} = \dfrac{\sqrt[5]{x}}{\sqrt[5]{32}} = \dfrac{\sqrt[5]{x}}{2}$

(f) $-\sqrt[3]{\dfrac{m^6}{125}} = -\dfrac{\sqrt[3]{m^6}}{\sqrt[3]{125}} = -\dfrac{m^2}{5}$

Now Try Exercises 23, 25, 31, 33, and 35.

OBJECTIVE 3 Simplify radicals.   We use the product and quotient rules to simplify radicals. A radical is **simplified** if the following four conditions are met.

Simplified Radical

1. The radicand has no factor raised to a power greater than or equal to the index.

2. The radicand has no fractions.

3. No denominator contains a radical.

4. Exponents in the radicand and the index of the radical have no common factor (except 1).

EXAMPLE 4 Simplifying Roots of Numbers

Simplify.

(a) $\sqrt{24}$

   Check to see whether 24 is divisible by a perfect square (the square of a natural number) such as 4, 9, ... Choose the largest perfect square that divides into 24. The largest such number is 4. Write 24 as the product of 4 and 6, and then use the product rule.

$$\sqrt{24} = \sqrt{4 \cdot 6} = \sqrt{4} \cdot \sqrt{6} = 2\sqrt{6}$$

(b) $\sqrt{108}$

The number 108 is divisible by the perfect square 36: $\sqrt{108} = \sqrt{36 \cdot 3}$. If this is not obvious, try factoring 108 into its prime factors.

$$\sqrt{108} = \sqrt{2^2 \cdot 3^3}$$
$$= \sqrt{2^2 \cdot 3^2 \cdot 3}$$
$$= 2 \cdot 3 \cdot \sqrt{3} \qquad \text{Product rule}$$
$$= 6\sqrt{3}$$

(c) $\sqrt{10}$

No perfect square (other than 1) divides into 10, so $\sqrt{10}$ cannot be simplified further.

(d) $\sqrt[3]{16}$

Look for the largest perfect *cube* that divides into 16. The number 8 satisfies this condition, so write 16 as $8 \cdot 2$ (or factor 16 into prime factors).

$$\sqrt[3]{16} = \sqrt[3]{8 \cdot 2} = \sqrt[3]{8} \cdot \sqrt[3]{2} = 2\sqrt[3]{2}$$

(e) $-\sqrt[4]{162} = -\sqrt[4]{81 \cdot 2}$ 81 is a perfect 4th power.

$$= -\sqrt[4]{81} \cdot \sqrt[4]{2} \qquad \text{Product rule}$$
$$= -3\sqrt[4]{2}$$

Now Try Exercises 39, 41, 49, and 55.

CAUTION In simplifying an expression like that in Example 4(b), be careful with which factors belong *outside* the radical sign and which belong *inside*. Note how $2 \cdot 3$ is written outside because $\sqrt{2^2} = 2$ and $\sqrt{3^2} = 3$, while the remaining 3 is left inside the radical.

EXAMPLE 5   Simplifying Radicals Involving Variables

Simplify. Assume all variables represent positive real numbers.

(a) $\sqrt{16m^3} = \sqrt{16m^2 \cdot m}$

$$= \sqrt{16m^2} \cdot \sqrt{m}$$
$$= 4m\sqrt{m}$$

No absolute value bars are needed around the m in color because of the assumption that all the variables represent *positive* real numbers.

(b) $\sqrt{200k^7q^8} = \sqrt{10^2 \cdot 2 \cdot (k^3)^2 \cdot k \cdot (q^4)^2}$ Factor.

$$= 10k^3q^4\sqrt{2k} \qquad \text{Remove perfect square factors.}$$

(c) $\sqrt[3]{8x^4y^5} = \sqrt[3]{(8x^3y^3)(xy^2)}$ $8x^3y^3$ is the largest perfect cube that divides $8x^4y^5$.

$$= \sqrt[3]{8x^3y^3} \cdot \sqrt[3]{xy^2}$$
$$= 2xy\sqrt[3]{xy^2}$$

(d) $-\sqrt[4]{32y^9} = -\sqrt[4]{(16y^8)(2y)}$ $16y^8$ is the largest 4th power that divides $32y^9$.

$\qquad\qquad = -\sqrt[4]{16y^8} \cdot \sqrt[4]{2y}$

$\qquad\qquad = -2y^2\sqrt[4]{2y}$

Now Try Exercises 75, 79, 83, and 87.

N O T E From Example 5 we see that if a variable is raised to a power with an exponent divisible by 2, it is a perfect square. If it is raised to a power with an exponent divisible by 3, it is a perfect cube. In general, if it is raised to a power with an exponent divisible by n, it is a perfect nth power.

The conditions for a simplified radical given earlier state that an exponent in the radicand and the index of the radical should have no common factor (except 1). The next example shows how to simplify radicals with such common factors.

EXAMPLE 6   Simplifying Radicals by Using Smaller Indexes

Simplify. Assume all variables represent positive real numbers.

(a) $\sqrt[9]{5^6}$

We can write this radical using rational exponents and then write the exponent in lowest terms. We then express the answer as a radical.

$$\sqrt[9]{5^6} = 5^{6/9} = 5^{2/3} = \sqrt[3]{5^2} \quad \text{or} \quad \sqrt[3]{25}$$

(b) $\sqrt[4]{p^2} = p^{2/4} = p^{1/2} = \sqrt{p}$ (Recall the assumption that $p > 0$.)

Now Try Exercises 93 and 97.

These examples suggest the following rule.

If m is an integer, n and k are natural numbers, and all indicated roots exist, then

$$\sqrt[kn]{a^{km}} = \sqrt[n]{a^m}.$$

OBJECTIVE 4   Simplify products and quotients of radicals with different indexes.   Since the product and quotient rules for radicals apply only when they have the same index, we multiply and divide radicals with different indexes by using rational exponents.

EXAMPLE 7   Multiplying Radicals with Different Indexes

Simplify $\sqrt{7} \cdot \sqrt[3]{2}$.

Because the different indexes, 2 and 3, have a least common index of 6, use rational exponents to write each radical as a sixth root.

$$\sqrt{7} = 7^{1/2} = 7^{3/6} = \sqrt[6]{7^3} = \sqrt[6]{343}$$
$$\sqrt[3]{2} = 2^{1/3} = 2^{2/6} = \sqrt[6]{2^2} = \sqrt[6]{4}$$

Therefore,

$$\sqrt{7} \cdot \sqrt[3]{2} = \sqrt[6]{343} \cdot \sqrt[6]{4} = \sqrt[6]{1372}. \qquad \text{Product rule}$$

Now Try Exercise 99.

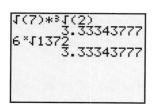

FIGURE 8

Results such as the one in Example 7 can be supported using a calculator, as shown in Figure 8. Notice that the calculator gives the same approximation for the initial product and the final radical that we obtained.

CAUTION The computation in Figure 8 is not *proof* that the two expressions are equal. The algebra in Example 7, however, is valid proof of their equality.

OBJECTIVE 5 Use the Pythagorean formula. The **Pythagorean formula** relates the lengths of the three sides of a right triangle.

Pythagorean Formula

If c is the length of the longest side of a right triangle and a and b are the lengths of the shorter sides, then

$$c^2 = a^2 + b^2.$$

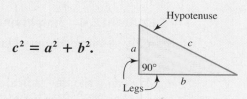

The longest side is the **hypotenuse** and the two shorter sides are the **legs** of the triangle. The hypotenuse is the side opposite the right angle.

■ **EXAMPLE 8 Using the Pythagorean Formula**

Use the Pythagorean formula to find the length of the hypotenuse in the triangle in Figure 9.

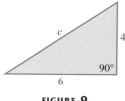

FIGURE 9

To find the length of the hypotenuse c, let $a = 4$ and $b = 6$. Then use the formula.

$$c^2 = a^2 + b^2$$
$$c^2 = 4^2 + 6^2 \qquad \text{Let } a = 4 \text{ and } b = 6.$$
$$c^2 = 52$$
$$c = \sqrt{52} \qquad \text{Choose the principal root.}$$
$$c = \sqrt{4 \cdot 13} \qquad \text{Factor.}$$
$$c = \sqrt{4} \cdot \sqrt{13} \qquad \text{Product rule}$$
$$c = 2\sqrt{13}$$

The length of the hypotenuse is $2\sqrt{13}$.

Now Try Exercise 109.

CAUTION When using the equation $c^2 = a^2 + b^2$, be sure that the length of the hypotenuse is substituted for c, and that the lengths of the legs are substituted for a and b. Errors often occur because values are substituted incorrectly.

| CONNECTIONS |

The Pythagorean formula is undoubtedly one of the most widely used and oldest formulas we have. It is very important in trigonometry, which is used in surveying, drafting, engineering, navigation, and many other fields. There is evidence that the Babylonians knew the concept quite well. Although attributed to Pythagoras, it was known to every surveyor from Egypt to China for a thousand years before Pythagoras. In the 1939 movie *The Wizard of Oz*, the Scarecrow asks the Wizard for a brain. When the Wizard presents him with a diploma granting him a Th.D. (Doctor of Thinkology), the Scarecrow recites the following:

> The sum of the square roots of any two sides of an isosceles triangle is equal to the square root of the remaining side. . . . Oh joy! Rapture! I've got a brain.

For Discussion or Writing

Did the Scarecrow recite the Pythagorean formula? (An *isosceles triangle* is a triangle with two equal sides.) Is his statement true? Explain.

■ **EXAMPLE 9** Using a Formula from Electronics

The impedance Z of an alternating series circuit is given by the formula

$$Z = \sqrt{R^2 + X^2},$$

where R is the resistance and X is the reactance, both in ohms. Find the value of the impedance if $R = 40$ ohms and $X = 30$ ohms.

Substitute 40 for R and 30 for X in the formula.

$$Z = \sqrt{R^2 + X^2} \qquad \text{Given formula}$$
$$= \sqrt{40^2 + 30^2} \qquad \text{Let } R = 40 \text{ and } X = 30.$$
$$= \sqrt{1600 + 900}$$
$$= \sqrt{2500}$$
$$= 50$$

The impedance is 50 ohms.

Now Try Exercise 113.

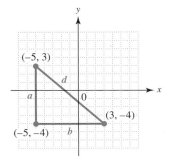

FIGURE 10

OBJECTIVE 6 Use the distance formula. An important result in algebra is derived by using the Pythagorean formula. The *distance formula* allows us to find the distance between two points in the coordinate plane, or the length of the line segment joining those two points. Figure 10 shows the points $(3, -4)$ and $(-5, 3)$. The vertical line through $(-5, 3)$ and the horizontal line through $(3, -4)$ intersect at the point $(-5, -4)$. Thus, the point $(-5, -4)$ becomes the vertex of the right angle in a right triangle. By the Pythagorean formula, the square of the length of the hypotenuse, d, of the right triangle in Figure 10 is equal to the sum of the squares of the lengths of the two legs a and b:

$$d^2 = a^2 + b^2.$$

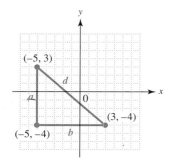

FIGURE 10
(repeated)

The length a is the difference between the y-coordinates of the endpoints. Since the x-coordinate of both points in Figure 10 is -5, the side is vertical, and we can find a by finding the difference between the y-coordinates. We subtract -4 from 3 to get a positive value for a.

$$a = 3 - (-4) = 7$$

Similarly, we find b by subtracting -5 from 3.

$$b = 3 - (-5) = 8$$

Substituting these values into the formula, we obtain

$$d^2 = a^2 + b^2$$
$$d^2 = 7^2 + 8^2 \qquad \text{Let } a = 7 \text{ and } b = 8.$$
$$d^2 = 49 + 64$$
$$d^2 = 113$$
$$d = \sqrt{113}.$$

We choose the principal root since distance cannot be negative. Therefore, the distance between $(-5, 3)$ and $(3, -4)$ is $\sqrt{113}$.

NOTE It is customary to leave the distance in radical form. Do not use a calculator to get an approximation unless you are specifically directed to do so.

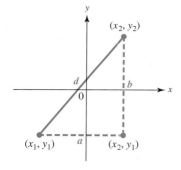

FIGURE 11

This result can be generalized. Figure 11 shows the two points (x_1, y_1) and (x_2, y_2). To find a formula for the distance d between these two points, notice that the distance between (x_1, y_1) and (x_2, y_1) is given by

$$a = |x_2 - x_1|,$$

and the distance between (x_2, y_2) and (x_2, y_1) is given by

$$b = |y_2 - y_1|.$$

From the Pythagorean formula,

$$d^2 = a^2 + b^2$$
$$d^2 = (x_2 - x_1)^2 + (y_2 - y_1)^2. \qquad |p - q|^2 = (p - q)^2$$

Choosing the principal square root gives the **distance formula.**

Distance Formula

The distance between the points (x_1, y_1) and (x_2, y_2) is

$$d = \sqrt{(x_2 - x_1)^2 + (y_2 - y_1)^2}.$$

EXAMPLE 10 Using the Distance Formula

Find the distance between the points $(-3, 5)$ and $(6, 4)$.

When using the distance formula to find the distance between two points, designating the points as (x_1, y_1) and (x_2, y_2) is arbitrary. We choose $(x_1, y_1) = (-3, 5)$ and $(x_2, y_2) = (6, 4)$.

$$d = \sqrt{(x_2 - x_1)^2 + (y_2 - y_1)^2}$$
$$= \sqrt{[6 - (-3)]^2 + (4 - 5)^2} \quad x_2 = 6,\ y_2 = 4,\ x_1 = -3,\ y_1 = 5$$
$$= \sqrt{9^2 + (-1)^2}$$
$$= \sqrt{82}$$

Now Try Exercise 121.

8.3 EXERCISES

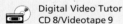
Decide whether each statement is true *or* false *by using the product rule explained in this section. Then support your answer by finding a calculator approximation for each expression.*

1. $2\sqrt{12} = \sqrt{48}$ **2.** $\sqrt{72} = 2\sqrt{18}$

3. $3\sqrt{8} = 2\sqrt{18}$ **4.** $5\sqrt{72} = 6\sqrt{50}$

5. Which one of the following is *not* equal to $\sqrt{\frac{1}{2}}$? (Do not use calculator approximations.)

     A. $\sqrt{.5}$ **B.** $\sqrt{\frac{2}{4}}$ **C.** $\sqrt{\frac{3}{6}}$ **D.** $\frac{\sqrt{4}}{\sqrt{16}}$

6. Use the π key on your calculator to get a value for π. Now find an approximation for $\sqrt[4]{\frac{2143}{22}}$. Does the result mean that π is actually equal to $\sqrt[4]{\frac{2143}{22}}$? Why or why not?

Multiply using the product rule. Assume all variables represent positive real numbers. See Examples 1 and 2.

7. $\sqrt{5} \cdot \sqrt{6}$ **8.** $\sqrt{10} \cdot \sqrt{3}$ **9.** $\sqrt{14} \cdot \sqrt{x}$ **10.** $\sqrt{23} \cdot \sqrt{t}$

11. $\sqrt{14} \cdot \sqrt{3pqr}$ **12.** $\sqrt{7} \cdot \sqrt{5xt}$ **13.** $\sqrt[3]{7x} \cdot \sqrt[3]{2y}$ **14.** $\sqrt[3]{9x} \cdot \sqrt[3]{4y}$

15. $\sqrt[4]{11} \cdot \sqrt[4]{3}$ **16.** $\sqrt[4]{6} \cdot \sqrt[4]{9}$ **17.** $\sqrt[4]{2x} \cdot \sqrt[4]{3y^2}$ **18.** $\sqrt[4]{3y^2} \cdot \sqrt[4]{6yz}$

19. $\sqrt[3]{7} \cdot \sqrt[4]{3}$ **20.** $\sqrt[5]{8} \cdot \sqrt[6]{12}$

21. Explain the product rule for radicals in your own words. Give examples.

22. Explain the quotient rule for radicals in your own words. Give examples.

Simplify each radical. Assume all variables represent positive real numbers. See Example 3.

23. $\sqrt{\frac{64}{121}}$ **24.** $\sqrt{\frac{16}{49}}$ **25.** $\sqrt{\frac{3}{25}}$ **26.** $\sqrt{\frac{13}{49}}$

27. $\sqrt{\frac{x}{25}}$ **28.** $\sqrt{\frac{k}{100}}$ **29.** $\sqrt{\frac{p^6}{81}}$ **30.** $\sqrt{\frac{w^{10}}{36}}$

31. $\sqrt[3]{\frac{27}{64}}$ **32.** $\sqrt[3]{\frac{216}{125}}$ **33.** $\sqrt[3]{-\frac{r^2}{8}}$ **34.** $\sqrt[3]{-\frac{t}{125}}$

35. $-\sqrt[4]{\frac{81}{x^4}}$ **36.** $-\sqrt[4]{\frac{625}{y^4}}$ **37.** $\sqrt[5]{\frac{1}{x^{15}}}$ **38.** $\sqrt[5]{\frac{32}{y^{20}}}$

Express each radical in simplified form. See Example 4.

39. $\sqrt{12}$ **40.** $\sqrt{18}$ **41.** $\sqrt{288}$ **42.** $\sqrt{72}$ **43.** $-\sqrt{32}$

44. $-\sqrt{48}$ **45.** $-\sqrt{28}$ **46.** $-\sqrt{24}$ **47.** $\sqrt{-300}$ **48.** $\sqrt{-150}$

49. $\sqrt[3]{128}$ **50.** $\sqrt[3]{24}$ **51.** $\sqrt[3]{-16}$ **52.** $\sqrt[3]{-250}$ **53.** $\sqrt[3]{40}$

54. $\sqrt[3]{375}$ **55.** $-\sqrt[4]{512}$ **56.** $-\sqrt[4]{1250}$ **57.** $\sqrt[5]{64}$ **58.** $\sqrt[5]{128}$

59. A student claimed that $\sqrt[3]{14}$ is not in simplified form, since $14 = 8 + 6$, and 8 is a perfect cube. Was his reasoning correct? Why or why not?

60. Explain in your own words why $\sqrt[3]{k^4}$ is not a simplified radical.

Express each radical in simplified form. Assume all variables represent positive real numbers. See Example 5.

61. $\sqrt{72k^2}$ **62.** $\sqrt{18m^2}$ **63.** $\sqrt{144x^3y^9}$

64. $\sqrt{169s^5t^{10}}$ **65.** $\sqrt{121x^6}$ **66.** $\sqrt{256z^{12}}$

67. $-\sqrt[3]{27t^{12}}$ **68.** $-\sqrt[3]{64y^{18}}$ **69.** $-\sqrt{100m^8z^4}$

70. $-\sqrt{25t^6s^{20}}$ **71.** $-\sqrt[3]{-125a^6b^9c^{12}}$ **72.** $-\sqrt[3]{-216y^{15}x^6z^3}$

73. $\sqrt[4]{\dfrac{1}{16}r^8t^{20}}$ **74.** $\sqrt[4]{\dfrac{81}{256}t^{12}u^8}$ **75.** $\sqrt{50x^3}$ **76.** $\sqrt{300z^3}$

77. $-\sqrt{500r^{11}}$ **78.** $-\sqrt{200p^{13}}$ **79.** $\sqrt{13x^7y^8}$ **80.** $\sqrt{23k^9p^{14}}$

81. $\sqrt[3]{8z^6w^9}$ **82.** $\sqrt[3]{64a^{15}b^{12}}$ **83.** $\sqrt[3]{-16z^5t^7}$ **84.** $\sqrt[3]{-81m^4n^{10}}$

85. $\sqrt[4]{81x^{12}y^{16}}$ **86.** $\sqrt[4]{81t^8u^{28}}$ **87.** $-\sqrt[4]{162r^{15}s^{10}}$ **88.** $-\sqrt[4]{32k^5m^{10}}$

89. $\sqrt{\dfrac{y^{11}}{36}}$ **90.** $\sqrt{\dfrac{v^{13}}{49}}$ **91.** $\sqrt[3]{\dfrac{x^{16}}{27}}$ **92.** $\sqrt[3]{\dfrac{y^{17}}{125}}$

Simplify each radical. Assume that $x \geq 0$. See Example 6.

93. $\sqrt[4]{48^2}$ **94.** $\sqrt[4]{50^2}$ **95.** $\sqrt[4]{25}$

96. $\sqrt[6]{8}$ **97.** $\sqrt[10]{x^{25}}$ **98.** $\sqrt[12]{x^{44}}$

Simplify by first writing the radicals as radicals with the same index. Then multiply. Assume all variables represent positive real numbers. See Example 7.

99. $\sqrt[3]{4} \cdot \sqrt{3}$ **100.** $\sqrt[3]{5} \cdot \sqrt{6}$ **101.** $\sqrt[4]{3} \cdot \sqrt[3]{4}$

102. $\sqrt[5]{7} \cdot \sqrt[4]{5}$ **103.** $\sqrt{x} \cdot \sqrt[3]{x}$ **104.** $\sqrt[3]{y} \cdot \sqrt[4]{y}$

TECHNOLOGY INSIGHTS (EXERCISES 105–108)

A graphing calculator can be used to test whether two quantities are equal. In the screen shown here, the first two lines of entries both represent true statements, and thus the calculator returns a 1 to indicate true. *The third entry is* false, *and the calculator returns a 0. These can be verified algebraically using the rules for radicals found in this section.*

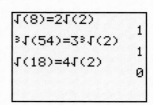

Determine whether the calculator should return a 1 or a 0 for each screen.

105.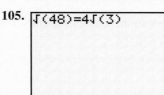

106.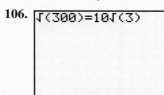

107.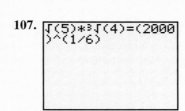

108.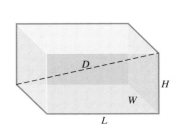

Find the unknown length in each right triangle. Simplify the answer if necessary. See Example 8.

109.

110.

111.
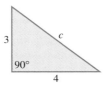

112.

Solve each problem. See Example 9.

113. The length of the diagonal of a box is given by

$$D = \sqrt{L^2 + W^2 + H^2},$$

where L, W, and H are the length, width, and height of the box. Find the length of the diagonal, D, of a box that is 4 ft long, 3 ft high, and 2 ft wide. Give the exact value, then round to the nearest tenth of a foot.

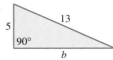

114. A Sanyo color television, model AVM-2755, has a rectangular screen with a 21.7-in. width. Its height is 16 in. What is the diagonal of the screen to the nearest tenth of an inch? (*Source:* Actual measurements of the author's television.)

16 in.

21.7 in.

115. A formula from electronics dealing with impedance of parallel resonant circuits is

$$I = \frac{E}{\sqrt{R^2 + \omega^2 L^2}},$$

where the variables are in appropriate units. Find I if $E = 282$, $R = 100$, $L = 264$, and $\omega = 120\pi$. Give your answer to the nearest thousandth.

116. In the study of sound, one version of the law of tensions is

$$f_1 = f_2 \sqrt{\frac{F_1}{F_2}}.$$

If $F_1 = 300$, $F_2 = 60$, and $f_2 = 260$, find f_1 to the nearest unit.

117. The illumination I, in foot-candles, produced by a light source is related to the distance d, in feet, from the light source by the equation

$$d = \sqrt{\frac{k}{I}},$$

where k is a constant. If $k = 640$, how far from the light source will the illumination be 2 foot-candles? Give the exact value, and then round to the nearest tenth of a foot.

118. The following letter appeared in the column "Ask Tom Why," written by Tom Skilling of the *Chicago Tribune*.

> *Dear Tom,*
> *I cannot remember the formula to calculate the distance to the horizon. I have a stunning view from my 14th floor condo, 150 feet above the ground. How far can I see?*
> *Ted Fleischaker; Indianapolis, Ind.*

Skilling's answer was as follows.

> To find the distance to the horizon in miles, take the square root of the height of your view in feet and multiply that result by 1.224. Your answer will be the number of miles to the horizon. (*Source: Chicago Tribune,* August 17, 2002.)

Assuming Ted's eyes are 6 ft above the ground, the total height from the ground is $150 + 6 = 156$ ft. To the nearest tenth of a mile, how far can he see to the horizon?

Find the distance between each pair of points. See Example 10.

119. (6, 13) and (1, 1)   **120.** (8, 13) and (2, 5)

121. (−6, 5) and (3, −4)   **122.** (−1, 5) and (−7, 7)

123. (−8, 2) and (−4, 1)   **124.** (−1, 2) and (5, 3)

125. $(4.7, 2.3)$ and $(1.7, -1.7)$

126. $(-2.9, 18.2)$ and $(2.1, 6.2)$

127. $(\sqrt{2}, \sqrt{6})$ and $(-2\sqrt{2}, 4\sqrt{6})$

128. $(\sqrt{7}, 9\sqrt{3})$ and $(-\sqrt{7}, 4\sqrt{3})$

129. $(x + y, y)$ and $(x - y, x)$

130. $(c, c - d)$ and $(d, c + d)$

131. As given in the text, the distance formula is expressed with a radical. Write the distance formula using rational exponents.

132. An alternative form of the distance formula is

$$d = \sqrt{(x_1 - x_2)^2 + (y_1 - y_2)^2}.$$

Compare this to the form given in this section, and explain why the two forms are equivalent.

Find the perimeter of each triangle. ($Hint: For Exercise 133, \sqrt{k} + \sqrt{k} = 2\sqrt{k}.$)

133.

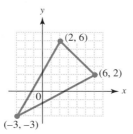

134.

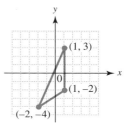

8.4 Adding and Subtracting Radical Expressions

OBJECTIVES

1 Define a radical expression.

2 Simplify radical expressions involving addition and subtraction.

OBJECTIVE 1 Define a radical expression. Recall from Section 8.1 that a **radical expression** is an algebraic expression that contains radicals. For example,

$$\sqrt[4]{3} + \sqrt{6}, \qquad \sqrt{x + 2y} - 1, \qquad \text{and} \qquad \sqrt{8} - \sqrt{2r}$$

are radical expressions. The examples in the previous section discussed simplifying radical expressions that involve multiplication and division. Now we show how to simplify radical expressions that involve addition and subtraction.

OBJECTIVE 2 Simplify radical expressions involving addition and subtraction. An expression such as $4\sqrt{2} + 3\sqrt{2}$ can be simplified by using the distributive property.

$$4\sqrt{2} + 3\sqrt{2} = (4 + 3)\sqrt{2} = 7\sqrt{2}$$

As another example, $2\sqrt{3} - 5\sqrt{3} = (2 - 5)\sqrt{3} = -3\sqrt{3}$. This is similar to simplifying $4x + 3x$ to $7x$ or $2y - 5y$ to $-3y$.

CAUTION Only radical expressions with the *same index* and the *same radicand* may be combined. Expressions such as $5\sqrt{3} + 2\sqrt{2}$ or $3\sqrt{3} + 2\sqrt[3]{3}$ cannot be simplified by combining terms.

EXAMPLE 1 Adding and Subtracting Radicals

Add or subtract to simplify each radical expression.

(a) $3\sqrt{24} + \sqrt{54}$

Begin by simplifying each radical; then use the distributive property to combine terms.

$$
\begin{aligned}
3\sqrt{24} + \sqrt{54} &= 3\sqrt{4} \cdot \sqrt{6} + \sqrt{9} \cdot \sqrt{6} && \text{Product rule} \\
&= 3 \cdot 2\sqrt{6} + 3\sqrt{6} \\
&= 6\sqrt{6} + 3\sqrt{6} \\
&= 9\sqrt{6} && \text{Combine terms.}
\end{aligned}
$$

(b)
$$
\begin{aligned}
2\sqrt{20x} - \sqrt{45x} &= 2\sqrt{4} \cdot \sqrt{5x} - \sqrt{9} \cdot \sqrt{5x} && \text{Product rule} \\
&= 2 \cdot 2\sqrt{5x} - 3\sqrt{5x} \\
&= 4\sqrt{5x} - 3\sqrt{5x} \\
&= \sqrt{5x}, \quad x \geq 0 && \text{Combine terms.}
\end{aligned}
$$

(c) $2\sqrt{3} - 4\sqrt{5}$

Here the radicands differ and are already simplified, so $2\sqrt{3} - 4\sqrt{5}$ cannot be simplified further.

Now Try Exercises 7, 15, and 19.

CAUTION Do not confuse the product rule with combining like terms. The root of a sum *does not equal* the sum of the roots. For example,

$$\sqrt{9 + 16} \neq \sqrt{9} + \sqrt{16}$$

since $\sqrt{9 + 16} = \sqrt{25} = 5$, but $\sqrt{9} + \sqrt{16} = 3 + 4 = 7$.

EXAMPLE 2 Adding and Subtracting Radicals with Higher Indexes

Add or subtract to simplify each radical expression. Assume all variables represent positive real numbers.

(a)
$$
\begin{aligned}
2\sqrt[3]{16} - 5\sqrt[3]{54} &= 2\sqrt[3]{8 \cdot 2} - 5\sqrt[3]{27 \cdot 2} && \text{Factor.} \\
&= 2\sqrt[3]{8} \cdot \sqrt[3]{2} - 5\sqrt[3]{27} \cdot \sqrt[3]{2} && \text{Product rule} \\
&= 2 \cdot 2 \cdot \sqrt[3]{2} - 5 \cdot 3 \cdot \sqrt[3]{2} \\
&= 4\sqrt[3]{2} - 15\sqrt[3]{2} \\
&= -11\sqrt[3]{2} && \text{Combine terms.}
\end{aligned}
$$

(b)
$$
\begin{aligned}
2\sqrt[3]{x^2 y} + \sqrt[3]{8x^5 y^4} &= 2\sqrt[3]{x^2 y} + \sqrt[3]{(8x^3 y^3)x^2 y} && \text{Factor.} \\
&= 2\sqrt[3]{x^2 y} + 2xy\sqrt[3]{x^2 y} && \text{Product rule} \\
&= (2 + 2xy)\sqrt[3]{x^2 y} && \text{Distributive property}
\end{aligned}
$$

Now Try Exercises 23 and 29.

CAUTION Remember to write the index when working with cube roots, fourth roots, and so on.

▓ **EXAMPLE 3 Adding and Subtracting Radicals with Fractions**

Perform the indicated operations. Assume all variables represent positive real numbers.

(a) $2\sqrt{\dfrac{75}{16}} + 4\dfrac{\sqrt{8}}{\sqrt{32}} = 2\dfrac{\sqrt{25 \cdot 3}}{\sqrt{16}} + 4\dfrac{\sqrt{4 \cdot 2}}{\sqrt{16 \cdot 2}}$ Quotient rule

$$= 2\left(\dfrac{5\sqrt{3}}{4}\right) + 4\left(\dfrac{2\sqrt{2}}{4\sqrt{2}}\right)$$ Product rule

$$= \dfrac{5\sqrt{3}}{2} + 2$$ Multiply; $\dfrac{\sqrt{2}}{\sqrt{2}} = 1$.

$$= \dfrac{5\sqrt{3}}{2} + \dfrac{4}{2}$$ Write with a common denominator.

$$= \dfrac{5\sqrt{3} + 4}{2}$$

(b) $10\sqrt[3]{\dfrac{5}{x^6}} - 3\sqrt[3]{\dfrac{4}{x^9}} = 10\dfrac{\sqrt[3]{5}}{\sqrt[3]{x^6}} - 3\dfrac{\sqrt[3]{4}}{\sqrt[3]{x^9}}$ Quotient rule

$$= \dfrac{10\sqrt[3]{5}}{x^2} - \dfrac{3\sqrt[3]{4}}{x^3}$$

$$= \dfrac{10x\sqrt[3]{5}}{x^3} - \dfrac{3\sqrt[3]{4}}{x^3}$$ Write with a common denominator.

$$= \dfrac{10x\sqrt[3]{5} - 3\sqrt[3]{4}}{x^3}$$

Now Try Exercises 47 and 53.

A calculator can support some of the results obtained in the examples of this section. In Example 1(a), we simplified $3\sqrt{24} + \sqrt{54}$ to obtain $9\sqrt{6}$. The screen in Figure 12(a) shows that the approximations are the same, suggesting that our simplification was correct. Figure 12(b) shows support for the result of Example 2(a): $2\sqrt[3]{16} - 5\sqrt[3]{54} = -11\sqrt[3]{2}$. Figure 12(c) supports the result of Example 3(a).

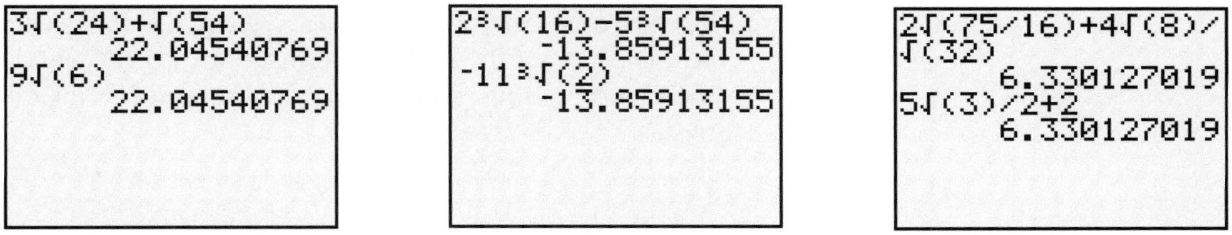

(a)   (b)   (c)

FIGURE 12

┌─ **CONNECTIONS** ─┐

A triangle that has whole number measures for the lengths of two sides may have an irrational number as the measure of the third side. For example, a right triangle with the two shorter sides measuring 1 and 2 units will have a longest side measuring $\sqrt{5}$ units. The ratio of the dimensions of the *golden rectangle,* considered to have the most pleasing dimensions of any rectangle, is irrational. To sketch a golden rectangle, begin with the square *ONRS*. Divide it into two equal parts by segment *MK*, as shown in the figure. Let *M* be the center of a circle with radius *MN*. Sketch the rectangle *PQRS*. This is a golden rectangle, with the property that if the original square is taken away, *PQNO* is still a golden rectangle. If the square with side *OP* is taken away, another golden rectangle results, and so on.

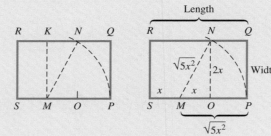

If the sides of the generating square have measure $2x$, then by the Pythagorean formula,

$$MN = \sqrt{x^2 + (2x)^2} = \sqrt{x^2 + 4x^2} = \sqrt{5x^2}.$$

Since *NP* is an arc of the circle with radius *MN*,

$$MP = MN = \sqrt{5x^2}.$$

The ratio of length to width is

$$\frac{\text{length}}{\text{width}} = \frac{x + \sqrt{5x^2}}{2x} = \frac{x + x\sqrt{5}}{2x} = \frac{x\left(1 + \sqrt{5}\right)}{2x} = \frac{1 + \sqrt{5}}{2},$$

which is an irrational number.

For Discussion or Writing

1. The golden rectangle has been widely used in art and architecture. See whether you can find some examples of its use. Use a calculator to approximate the ratio found above, called the *golden ratio.*

2. The sequence 1, 1, 2, 3, 5, 8, 13, 21, 34, 55, . . . is called the *Fibonacci sequence.* After the first two terms, both 1, every term is found by adding the two preceding terms. Form a sequence of ratios of the successive terms:

$$\frac{1}{1}, \frac{2}{1}, \frac{3}{2}, \frac{5}{3}, \frac{8}{5}, \frac{13}{8}, \frac{21}{13}, \frac{34}{21}, \frac{55}{34}, \dots$$

Now use a calculator to find approximations of these ratios. What seems to be happening?

8.4 EXERCISES

1. Which one of the following sums could be simplified without first simplifying the individual radical expressions?

   A. $\sqrt{50} + \sqrt{32}$ **B.** $3\sqrt{6} + 9\sqrt{6}$ **C.** $\sqrt[3]{32} - \sqrt[3]{108}$ **D.** $\sqrt[5]{6} - \sqrt[5]{192}$

2. Let $a = 1$ and let $b = 64$.
   (a) Evaluate $\sqrt{a} + \sqrt{b}$. Then find $\sqrt{a+b}$. Are they equal?
   (b) Evaluate $\sqrt[3]{a} + \sqrt[3]{b}$. Then find $\sqrt[3]{a+b}$. Are they equal?
   (c) Complete the following: In general, $\sqrt[n]{a} + \sqrt[n]{b} \neq$ _____,
   based on the observations in parts (a) and (b) of this exercise.

3. Even though the root indexes of the terms are not equal, the sum $\sqrt{64} + \sqrt[3]{125} + \sqrt[4]{16}$ can be simplified quite easily. What is this sum? Why can we add these terms so easily?

4. Explain why $28 - 4\sqrt{2}$ is not equal to $24\sqrt{2}$. (This is a common error among algebra students.)

Simplify. Assume all variables represent positive real numbers. See Examples 1 and 2.

5. $\sqrt{36} - \sqrt{100}$ **6.** $\sqrt{25} - \sqrt{81}$ **7.** $-2\sqrt{48} + 3\sqrt{75}$

8. $4\sqrt{32} - 2\sqrt{8}$ **9.** $\sqrt[3]{16} + 4\sqrt[3]{54}$ **10.** $3\sqrt[3]{24} - 2\sqrt[3]{192}$

11. $\sqrt[4]{32} + 3\sqrt[4]{2}$ **12.** $\sqrt[4]{405} - 2\sqrt[4]{5}$

13. $6\sqrt{18} - \sqrt{32} + 2\sqrt{50}$ **14.** $5\sqrt{8} + 3\sqrt{72} - 3\sqrt{50}$

15. $5\sqrt{6} + 2\sqrt{10}$ **16.** $3\sqrt{11} - 5\sqrt{13}$

17. $2\sqrt{5} + 3\sqrt{20} + 4\sqrt{45}$ **18.** $5\sqrt{54} - 2\sqrt{24} - 2\sqrt{96}$

19. $8\sqrt{2x} - \sqrt{8x} + \sqrt{72x}$ **20.** $4\sqrt{18k} - \sqrt{72k} + \sqrt{50k}$

21. $3\sqrt{72m^2} - 5\sqrt{32m^2} - 3\sqrt{18m^2}$ **22.** $9\sqrt{27p^2} - 14\sqrt{108p^2} + 2\sqrt{48p^2}$

23. $-\sqrt[3]{54} + 2\sqrt[3]{16}$ **24.** $15\sqrt[3]{81} - 4\sqrt[3]{24}$

25. $2\sqrt[3]{27x} - 2\sqrt[3]{8x}$ **26.** $6\sqrt[3]{128m} + 3\sqrt[3]{16m}$

27. $\sqrt[3]{x^2y} - \sqrt[3]{8x^2y}$ **28.** $3\sqrt[3]{x^2y^2} - 2\sqrt[3]{64x^2y^2}$

29. $3x\sqrt[3]{xy^2} - 2\sqrt[3]{8x^4y^2}$ **30.** $6q^2\sqrt[3]{5q} - 2q\sqrt[3]{40q^4}$

31. $5\sqrt[4]{32} + 3\sqrt[4]{162}$ **32.** $2\sqrt[4]{512} + 4\sqrt[4]{32}$

33. $3\sqrt[4]{x^5y} - 2x\sqrt[4]{xy}$ **34.** $2\sqrt[4]{m^9p^6} - 3m^2p\sqrt[4]{mp^2}$

35. $2\sqrt[4]{32a^3} + 5\sqrt[4]{2a^3}$ **36.** $-\sqrt[4]{16r} + 5\sqrt[4]{r}$

37. $\sqrt[3]{64xy^2} + \sqrt[3]{27x^4y^5}$ **38.** $\sqrt[4]{625s^3t} - \sqrt[4]{81s^7t^5}$

Simplify. Assume all variables represent positive real numbers. See Example 3.

39. $\sqrt{8} - \dfrac{\sqrt{64}}{\sqrt{16}}$ **40.** $\sqrt{48} - \dfrac{\sqrt{81}}{\sqrt{9}}$ **41.** $\dfrac{2\sqrt{5}}{3} + \dfrac{\sqrt{5}}{6}$

42. $\dfrac{4\sqrt{3}}{3} + \dfrac{2\sqrt{3}}{9}$

43. $\sqrt{\dfrac{8}{9}} + \sqrt{\dfrac{18}{36}}$

44. $\sqrt{\dfrac{12}{16}} + \sqrt{\dfrac{48}{64}}$

45. $\dfrac{\sqrt{32}}{3} + \dfrac{2\sqrt{2}}{3} - \dfrac{\sqrt{2}}{\sqrt{9}}$

46. $\dfrac{\sqrt{27}}{2} - \dfrac{3\sqrt{3}}{2} + \dfrac{\sqrt{3}}{\sqrt{4}}$

47. $3\sqrt{\dfrac{50}{9}} + 8\dfrac{\sqrt{2}}{\sqrt{8}}$

48. $9\sqrt{\dfrac{48}{25}} - 2\dfrac{\sqrt{2}}{\sqrt{98}}$

49. $\sqrt{\dfrac{25}{x^8}} - \sqrt{\dfrac{9}{x^6}}$

50. $\sqrt{\dfrac{100}{y^4}} + \sqrt{\dfrac{81}{y^{10}}}$

51. $3\sqrt[3]{\dfrac{m^5}{27}} - 2m\sqrt[3]{\dfrac{m^2}{64}}$

52. $2a\sqrt[4]{\dfrac{a}{16}} - 5a\sqrt[4]{\dfrac{a}{81}}$

53. $3\sqrt[3]{\dfrac{2}{x^6}} - 4\sqrt[3]{\dfrac{5}{x^9}}$

54. $-4\sqrt[3]{\dfrac{4}{t^9}} + 3\sqrt[3]{\dfrac{9}{t^{12}}}$

In Example 1(a) we showed that $3\sqrt{24} + \sqrt{54} = 9\sqrt{6}$. *To support this result, we can find a calculator approximation of* $3\sqrt{24}$, *then find a calculator approximation of* $\sqrt{54}$, *and add these two approximations. Then, we find a calculator approximation of* $9\sqrt{6}$. *It should correspond to the sum that we just found. (For this example, both approximations are 22.04540769. Due to rounding procedures, there may be a discrepancy in the final digit if you try to duplicate this work.) Follow this procedure to support the statements in Exercises 55 and 56.*

55. $3\sqrt{32} - 2\sqrt{8} = 8\sqrt{2}$

56. $2\sqrt{40} + 6\sqrt{90} - 3\sqrt{160} = 10\sqrt{10}$

57. A rectangular yard has a length of $\sqrt{192}$ m and a width of $\sqrt{48}$ m. Choose the best estimate of its dimensions. Then estimate the perimeter.

A. 14 m by 7 m **B.** 5 m by 7 m **C.** 14 m by 8 m **D.** 15 m by 8 m

58. If the sides of a triangle are $\sqrt{65}$ in., $\sqrt{35}$ in., and $\sqrt{26}$ in., which one of the following is the best estimate of its perimeter?

A. 20 in. **B.** 26 in. **C.** 19 in. **D.** 24 in.

Solve each problem. Give answers as simplified radical expressions.

59. Find the perimeter of the triangle.

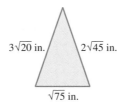

$3\sqrt{20}$ in. $2\sqrt{45}$ in.

$\sqrt{75}$ in.

60. Find the perimeter of the rectangle.

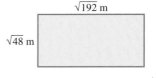

$\sqrt{192}$ m

$\sqrt{48}$ m

61. What is the perimeter of the computer graphic?

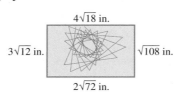

$4\sqrt{18}$ in.

$3\sqrt{12}$ in. $\sqrt{108}$ in.

$2\sqrt{72}$ in.

62. Find the area of the trapezoid.

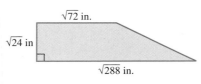

$\sqrt{72}$ in.

$\sqrt{24}$ in

$\sqrt{288}$ in.

8.5 Multiplying and Dividing Radical Expressions

OBJECTIVE 1 Multiply radical expressions. We multiply binomial expressions involving radicals by using the FOIL (First, Outer, Inner, Last) method. For example, we find the product of the binomials $\sqrt{5} + 3$ and $\sqrt{6} + 1$ as follows.

$$\left(\sqrt{5} + 3\right)\left(\sqrt{6} + 1\right) = \overbrace{\sqrt{5} \cdot \sqrt{6}}^{\text{First}} + \overbrace{\sqrt{5} \cdot 1}^{\text{Outer}} + \overbrace{3 \cdot \sqrt{6}}^{\text{Inner}} + \overbrace{3 \cdot 1}^{\text{Last}}$$

$$= \sqrt{30} + \sqrt{5} + 3\sqrt{6} + 3$$

This result cannot be simplified further.

EXAMPLE 1 Multiplying Binomials Involving Radical Expressions

Multiply using FOIL.

$$\text{(a)} \ \left(7 - \sqrt{3}\right)\left(\sqrt{5} + \sqrt{2}\right) = \overset{\text{F}}{7\sqrt{5}} + \overset{\text{O}}{7\sqrt{2}} - \overset{\text{I}}{\sqrt{3} \cdot \sqrt{5}} - \overset{\text{L}}{\sqrt{3} \cdot \sqrt{2}}$$

$$= 7\sqrt{5} + 7\sqrt{2} - \sqrt{15} - \sqrt{6}$$

(b) $\left(\sqrt{10} + \sqrt{3}\right)\left(\sqrt{10} - \sqrt{3}\right)$

$$= \sqrt{10} \cdot \sqrt{10} - \sqrt{10} \cdot \sqrt{3} + \sqrt{10} \cdot \sqrt{3} - \sqrt{3} \cdot \sqrt{3}$$

$$= 10 - 3$$

$$= 7$$

Notice that this is the kind of product that results in the difference of squares:

$$(x + y)(x - y) = x^2 - y^2.$$

Here, $x = \sqrt{10}$ and $y = \sqrt{3}$.

(c) $\left(\sqrt{7} - 3\right)^2 = \left(\sqrt{7} - 3\right)\left(\sqrt{7} - 3\right)$

$$= \sqrt{7} \cdot \sqrt{7} - 3\sqrt{7} - 3\sqrt{7} + 3 \cdot 3$$

$$= 7 - 6\sqrt{7} + 9$$

$$= 16 - 6\sqrt{7}$$

(d) $\left(5 - \sqrt[3]{3}\right)\left(5 + \sqrt[3]{3}\right) = 5 \cdot 5 + 5\sqrt[3]{3} - 5\sqrt[3]{3} - \sqrt[3]{3} \cdot \sqrt[3]{3}$

$$= 25 - \sqrt[3]{3^2}$$

$$= 25 - \sqrt[3]{9}$$

(e) $\left(\sqrt{k} + \sqrt{y}\right)\left(\sqrt{k} - \sqrt{y}\right) = \left(\sqrt{k}\right)^2 - \left(\sqrt{y}\right)^2$

$$= k - y, \quad k \geq 0 \text{ and } y \geq 0$$

Now Try Exercises 13, 17, 23, 27, and 39.

NOTE   In Example 1(c) we could have used the formula for the square of a binomial,

$$(x - y)^2 = x^2 - 2xy + y^2,$$

to obtain the same result:

$$\left(\sqrt{7} - 3\right)^2 = \left(\sqrt{7}\right)^2 - 2\left(\sqrt{7}\right)(3) + 3^2$$
$$= 7 - 6\sqrt{7} + 9$$
$$= 16 - 6\sqrt{7}.$$

OBJECTIVE 2 Rationalize denominators with one radical term.   As defined earlier, a simplified radical expression will have no radical in the denominator. The origin of this agreement no doubt occurred before the days of high-speed calculation, when computation was a tedious process performed by hand. To see this, consider the radical expression $\frac{1}{\sqrt{2}}$. To find a decimal approximation by hand, it would be necessary to divide 1 by a decimal approximation for $\sqrt{2}$, such as 1.414. It would be much easier if the divisor were a whole number. This can be accomplished by multiplying $\frac{1}{\sqrt{2}}$ by 1 in the form $\frac{\sqrt{2}}{\sqrt{2}}$:

$$\frac{1}{\sqrt{2}} \cdot \frac{\sqrt{2}}{\sqrt{2}} = \frac{\sqrt{2}}{2}.$$

Now the computation would require dividing 1.414 by 2 to obtain .707, a much easier task.

With current technology, either form of this fraction can be approximated with the same number of keystrokes. See Figure 13, which shows how a calculator gives the same approximation for both forms of the expression.

A common way of "standardizing" the form of a radical expression is to have the denominator contain no radicals. The process of removing radicals from a denominator so that the denominator contains only rational numbers is called **rationalizing the denominator.**

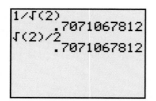

FIGURE 13

■ **EXAMPLE 2   Rationalizing Denominators with Square Roots**

Rationalize each denominator.

(a) $\dfrac{3}{\sqrt{7}}$

Multiply the numerator and denominator by $\sqrt{7}$. This is, in effect, multiplying by 1.

$$\frac{3}{\sqrt{7}} = \frac{3 \cdot \sqrt{7}}{\sqrt{7} \cdot \sqrt{7}}$$

In the denominator, since $\sqrt{7} \cdot \sqrt{7} = \sqrt{7 \cdot 7} = \sqrt{49} = 7$,

$$\frac{3}{\sqrt{7}} = \frac{3\sqrt{7}}{7}.$$

The denominator is now a rational number.

(b) $\dfrac{5\sqrt{2}}{\sqrt{5}} = \dfrac{5\sqrt{2} \cdot \sqrt{5}}{\sqrt{5} \cdot \sqrt{5}} = \dfrac{5\sqrt{10}}{5} = \sqrt{10}$

(c) $\dfrac{6}{\sqrt{12}}$

Less work is involved if we simplify the radical in the denominator first.

$$\frac{6}{\sqrt{12}} = \frac{6}{\sqrt{4 \cdot 3}} = \frac{6}{2\sqrt{3}} = \frac{3}{\sqrt{3}}$$

Now we rationalize the denominator by multiplying the numerator and denominator by $\sqrt{3}$.

$$\frac{3 \cdot \sqrt{3}}{\sqrt{3} \cdot \sqrt{3}} = \frac{3\sqrt{3}}{3} = \sqrt{3}$$

Now Try Exercises 43, 49, and 51.

EXAMPLE 3 Rationalizing Denominators in Roots of Fractions

Simplify each radical. In part (b), $p > 0$.

(a) $\sqrt{\dfrac{18}{125}} = \dfrac{\sqrt{18}}{\sqrt{125}}$ Quotient rule

$= \dfrac{\sqrt{9 \cdot 2}}{\sqrt{25 \cdot 5}}$ Factor.

$= \dfrac{3\sqrt{2}}{5\sqrt{5}}$ Product rule

$= \dfrac{3\sqrt{2} \cdot \sqrt{5}}{5\sqrt{5} \cdot \sqrt{5}}$ Multiply by $\dfrac{\sqrt{5}}{\sqrt{5}}$.

$= \dfrac{3\sqrt{10}}{5 \cdot 5}$ Product rule

$= \dfrac{3\sqrt{10}}{25}$

(b) $\sqrt{\dfrac{50m^4}{p^5}} = \dfrac{\sqrt{50m^4}}{\sqrt{p^5}}$ Quotient rule

$= \dfrac{5m^2\sqrt{2}}{p^2\sqrt{p}}$ Product rule

$= \dfrac{5m^2\sqrt{2} \cdot \sqrt{p}}{p^2\sqrt{p} \cdot \sqrt{p}}$ Multiply by $\dfrac{\sqrt{p}}{\sqrt{p}}$.

$= \dfrac{5m^2\sqrt{2p}}{p^2 \cdot p}$ Product rule

$= \dfrac{5m^2\sqrt{2p}}{p^3}$

Now Try Exercises 55 and 63.

■ **EXAMPLE 4** Rationalizing Denominators with Cube and Fourth Roots

Simplify.

(a) $\sqrt[3]{\dfrac{27}{16}}$

Use the quotient rule and simplify the numerator and denominator.

$$\sqrt[3]{\frac{27}{16}} = \frac{\sqrt[3]{27}}{\sqrt[3]{16}} = \frac{3}{\sqrt[3]{8} \cdot \sqrt[3]{2}} = \frac{3}{2\sqrt[3]{2}}$$

To get a rational denominator, multiply the numerator and denominator by a number that will result in a perfect cube in the radicand in the denominator. Since $2 \cdot 4 = 8$, a perfect cube, multiply the numerator and denominator by $\sqrt[3]{4}$.

$$\sqrt[3]{\frac{27}{16}} = \frac{3}{2\sqrt[3]{2}} = \frac{3 \cdot \sqrt[3]{4}}{2\sqrt[3]{2} \cdot \sqrt[3]{4}} = \frac{3\sqrt[3]{4}}{2\sqrt[3]{8}} = \frac{3\sqrt[3]{4}}{2 \cdot 2} = \frac{3\sqrt[3]{4}}{4}$$

(b) $\sqrt[4]{\dfrac{5x}{z}} = \dfrac{\sqrt[4]{5x}}{\sqrt[4]{z}} \cdot \dfrac{\sqrt[4]{z^3}}{\sqrt[4]{z^3}} = \dfrac{\sqrt[4]{5xz^3}}{\sqrt[4]{z^4}} = \dfrac{\sqrt[4]{5xz^3}}{z}, \quad x \geq 0, z > 0$

Now Try Exercises 71 and 81.

CAUTION It is easy to make mistakes in problems like the one in Example 4(a). A typical error is to multiply the numerator and denominator by $\sqrt[3]{2}$, forgetting that

$$\sqrt[3]{2} \cdot \sqrt[3]{2} \neq 2.$$

You need *three* factors of 2 to obtain 2^3 under the radical. As implied in Example 4(a),

$$\sqrt[3]{2} \cdot \sqrt[3]{2} \cdot \sqrt[3]{2} = 2.$$

OBJECTIVE 3 Rationalize denominators with binomials involving radicals. Recall the special product

$$(x + y)(x - y) = x^2 - y^2.$$

To rationalize a denominator that contains a binomial expression (one that contains exactly two terms) involving radicals, such as

$$\frac{3}{1 + \sqrt{2}},$$

we must use *conjugates*. The conjugate of $1 + \sqrt{2}$ is $1 - \sqrt{2}$. In general, $x + y$ and $x - y$ are **conjugates.**

Rationalizing a Binomial Denominator

Whenever a radical expression has a sum or difference with square root radicals in the denominator, rationalize the denominator by multiplying both the numerator and denominator by the conjugate of the denominator.

For the expression $\dfrac{3}{1 + \sqrt{2}}$, we rationalize the denominator by multiplying both the numerator and denominator by $1 - \sqrt{2}$, the conjugate of the denominator.

$$\frac{3}{1 + \sqrt{2}} = \frac{3\left(1 - \sqrt{2}\right)}{\left(1 + \sqrt{2}\right)\left(1 - \sqrt{2}\right)}$$

Then $\left(1 + \sqrt{2}\right)\left(1 - \sqrt{2}\right) = 1^2 - \left(\sqrt{2}\right)^2 = 1 - 2 = -1$. Placing -1 in the denominator gives

$$= \frac{3\left(1 - \sqrt{2}\right)}{-1}$$

$$= \frac{3}{-1}\left(1 - \sqrt{2}\right)$$

$$= -3\left(1 - \sqrt{2}\right) \quad \text{or} \quad -3 + 3\sqrt{2}.$$

▇ EXAMPLE 5 Rationalizing Binomial Denominators

Rationalize each denominator.

(a) $\dfrac{5}{4 - \sqrt{3}}$

To rationalize the denominator, multiply both the numerator and denominator by the conjugate of the denominator, $4 + \sqrt{3}$.

$$\frac{5}{4 - \sqrt{3}} = \frac{5\left(4 + \sqrt{3}\right)}{\left(4 - \sqrt{3}\right)\left(4 + \sqrt{3}\right)}$$

$$= \frac{5\left(4 + \sqrt{3}\right)}{16 - 3}$$

$$= \frac{5\left(4 + \sqrt{3}\right)}{13}$$

Notice that the numerator is left in factored form. This makes it easier to determine whether the expression is written in lowest terms.

(b) $\dfrac{\sqrt{2} - \sqrt{3}}{\sqrt{5} + \sqrt{3}}$

Multiply the numerator and denominator by $\sqrt{5} - \sqrt{3}$ to rationalize the denominator.

$$\frac{\sqrt{2} - \sqrt{3}}{\sqrt{5} + \sqrt{3}} = \frac{\left(\sqrt{2} - \sqrt{3}\right)\left(\sqrt{5} - \sqrt{3}\right)}{\left(\sqrt{5} + \sqrt{3}\right)\left(\sqrt{5} - \sqrt{3}\right)}$$

$$= \frac{\sqrt{10} - \sqrt{6} - \sqrt{15} + 3}{5 - 3}$$

$$= \frac{\sqrt{10} - \sqrt{6} - \sqrt{15} + 3}{2}$$

(c) $\dfrac{3}{\sqrt{5m} - \sqrt{p}} = \dfrac{3\left(\sqrt{5m} + \sqrt{p}\right)}{\left(\sqrt{5m} - \sqrt{p}\right)\left(\sqrt{5m} + \sqrt{p}\right)}$

$\qquad\qquad = \dfrac{3\left(\sqrt{5m} + \sqrt{p}\right)}{5m - p}, \quad 5m \ne p, m > 0, p > 0$

Now Try Exercises 85, 91, and 95.

OBJECTIVE 4 Write radical quotients in lowest terms.

EXAMPLE 6 Writing Radical Quotients in Lowest Terms

Write each quotient in lowest terms.

(a) $\dfrac{6 + 2\sqrt{5}}{4}$

Factor the numerator and denominator, then write in lowest terms.

$$\dfrac{6 + 2\sqrt{5}}{4} = \dfrac{2\left(3 + \sqrt{5}\right)}{2 \cdot 2} = \dfrac{3 + \sqrt{5}}{2}$$

Here is an alternative method for writing this expression in lowest terms.

$$\dfrac{6 + 2\sqrt{5}}{4} = \dfrac{6}{4} + \dfrac{2\sqrt{5}}{4} = \dfrac{3}{2} + \dfrac{\sqrt{5}}{2} = \dfrac{3 + \sqrt{5}}{2}$$

(b) $\dfrac{5y - \sqrt{8y^2}}{6y} = \dfrac{5y - 2y\sqrt{2}}{6y}, \quad y > 0 \qquad$ Product rule

$\qquad\qquad = \dfrac{y\left(5 - 2\sqrt{2}\right)}{6y} \qquad\qquad$ Factor the numerator.

$\qquad\qquad = \dfrac{5 - 2\sqrt{2}}{6}$

Note that the final fraction cannot be simplified further because there is no common factor of 2 in the numerator.

Now Try Exercises 107 and 109.

CAUTION Be careful to factor *before* writing a quotient in lowest terms.

| **CONNECTIONS** |

In calculus, it is sometimes desirable to rationalize the *numerator* in an expression. The procedure is similar to rationalizing the denominator. For example, to rationalize the numerator of

$$\dfrac{6 - \sqrt{2}}{4},$$

we multiply the numerator and the denominator by the conjugate of the numerator.

$$\frac{6 - \sqrt{2}}{4} = \frac{(6 - \sqrt{2})(6 + \sqrt{2})}{4(6 + \sqrt{2})} = \frac{36 - 2}{4(6 + \sqrt{2})} = \frac{34}{4(6 + \sqrt{2})} = \frac{17}{2(6 + \sqrt{2})}$$

In the final expression, the numerator is rationalized and is in lowest terms.

For Discussion or Writing

Rationalize the numerators of the following expressions, assuming a and b are nonnegative real numbers.

1. $\dfrac{8\sqrt{5} - 1}{6}$ **2.** $\dfrac{3\sqrt{a} + \sqrt{b}}{b}$ **3.** $\dfrac{3\sqrt{a} + \sqrt{b}}{\sqrt{b} - \sqrt{a}}$

4. Rationalize the denominator of the expression in Exercise 3, and then describe the difference in the procedure you used from what you did in Exercise 3.

8.5 EXERCISES

For Extra Help

Student's Solutions Manual

MyMathLab

InterAct Math Tutorial Software

Tutor Center AW Math Tutor Center

*Math*XL MathXL

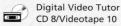
Digital Video Tutor CD 8/Videotape 10

Match each part of a rule for a special product in Column I with the other part in Column II.

        I   **II**

1. $(x + \sqrt{y})(x - \sqrt{y})$ **A.** $x - y$

2. $(\sqrt{x} + y)(\sqrt{x} - y)$ **B.** $x + 2y\sqrt{x} + y^2$

3. $(\sqrt{x} + \sqrt{y})(\sqrt{x} - \sqrt{y})$ **C.** $x - y^2$

4. $(\sqrt{x} + \sqrt{y})^2$ **D.** $x - 2\sqrt{xy} + y$

5. $(\sqrt{x} - \sqrt{y})^2$ **E.** $x^2 - y$

6. $(\sqrt{x} + y)^2$ **F.** $x + 2\sqrt{xy} + y$

Multiply, then simplify each product. Assume all variables represent positive real numbers. See Example 1.

7. $\sqrt{6}(3 + \sqrt{2})$ **8.** $\sqrt{2}(\sqrt{32} - \sqrt{9})$ **9.** $5(\sqrt{72} - \sqrt{8})$

10. $\sqrt{3}(\sqrt{12} + 2)$ **11.** $(\sqrt{7} + 3)(\sqrt{7} - 3)$ **12.** $(\sqrt{3} - 5)(\sqrt{3} + 5)$

13. $(\sqrt{2} - \sqrt{3})(\sqrt{2} + \sqrt{3})$ **14.** $(\sqrt{7} + \sqrt{3})(\sqrt{7} - \sqrt{3})$

15. $(\sqrt{8} - \sqrt{2})(\sqrt{8} + \sqrt{2})$ **16.** $(\sqrt{20} - \sqrt{5})(\sqrt{20} + \sqrt{5})$

17. $(\sqrt{2} + 1)(\sqrt{3} - 1)$ **18.** $(\sqrt{3} + 3)(\sqrt{5} - 2)$

19. $(\sqrt{11} - \sqrt{7})(\sqrt{2} + \sqrt{5})$ **20.** $(\sqrt{6} + \sqrt{2})(\sqrt{3} + \sqrt{2})$

21. $(2\sqrt{3} + \sqrt{5})(3\sqrt{3} - 2\sqrt{5})$ **22.** $(\sqrt{7} - \sqrt{11})(2\sqrt{7} + 3\sqrt{11})$

23. $(\sqrt{5} + 2)^2$ **24.** $(\sqrt{11} - 1)^2$

25. $(\sqrt{21} - \sqrt{5})^2$ **26.** $(\sqrt{6} - \sqrt{2})^2$

27. $(2 + \sqrt[3]{6})(2 - \sqrt[3]{6})$ **28.** $(\sqrt[3]{3} + 6)(\sqrt[3]{3} - 6)$

29. $\left(2 + \sqrt[3]{2}\right)\left(4 - 2\sqrt[3]{2} + \sqrt[3]{4}\right)$

30. $\left(\sqrt[3]{3} - 1\right)\left(\sqrt[3]{9} + \sqrt[3]{3} + 1\right)$

31. $\left(3\sqrt{x} - \sqrt{5}\right)\left(2\sqrt{x} + 1\right)$

32. $\left(4\sqrt{p} + \sqrt{7}\right)\left(\sqrt{p} - 9\right)$

33. $\left(3\sqrt{r} - \sqrt{s}\right)\left(3\sqrt{r} + \sqrt{s}\right)$

34. $\left(\sqrt{k} + 4\sqrt{m}\right)\left(\sqrt{k} - 4\sqrt{m}\right)$

35. $\left(\sqrt[3]{2y} - 5\right)\left(4\sqrt[3]{2y} + 1\right)$

36. $\left(\sqrt[3]{9z} - 2\right)\left(5\sqrt[3]{9z} + 7\right)$

37. $\left(\sqrt{3x} + 2\right)\left(\sqrt{3x} - 2\right)$

38. $\left(\sqrt{6y} - 4\right)\left(\sqrt{6y} + 4\right)$

39. $\left(2\sqrt{x} + \sqrt{y}\right)\left(2\sqrt{x} - \sqrt{y}\right)$

40. $\left(\sqrt{p} + 5\sqrt{s}\right)\left(\sqrt{p} - 5\sqrt{s}\right)$

41. $\left[\left(\sqrt{2} + \sqrt{3}\right) - \sqrt{6}\right]\left[\left(\sqrt{2} + \sqrt{3}\right) + \sqrt{6}\right]$

42. $\left[\left(\sqrt{5} - \sqrt{2}\right) - \sqrt{3}\right]\left[\left(\sqrt{5} - \sqrt{2}\right) + \sqrt{3}\right]$

Rationalize the denominator in each expression. Assume all variables represent positive real numbers. See Examples 2 and 3.

43. $\dfrac{7}{\sqrt{7}}$

44. $\dfrac{11}{\sqrt{11}}$

45. $\dfrac{15}{\sqrt{3}}$

46. $\dfrac{12}{\sqrt{6}}$

47. $\dfrac{\sqrt{3}}{\sqrt{2}}$

48. $\dfrac{\sqrt{7}}{\sqrt{6}}$

49. $\dfrac{9\sqrt{3}}{\sqrt{5}}$

50. $\dfrac{3\sqrt{2}}{\sqrt{11}}$

51. $\dfrac{-6}{\sqrt{18}}$

52. $\dfrac{-5}{\sqrt{24}}$

53. $\sqrt{\dfrac{7}{2}}$

54. $\sqrt{\dfrac{10}{3}}$

55. $-\sqrt{\dfrac{7}{50}}$

56. $-\sqrt{\dfrac{13}{75}}$

57. $\sqrt{\dfrac{24}{x}}$

58. $\sqrt{\dfrac{52}{y}}$

59. $\dfrac{-8\sqrt{3}}{\sqrt{k}}$

60. $\dfrac{-4\sqrt{13}}{\sqrt{m}}$

61. $-\sqrt{\dfrac{150m^5}{n^3}}$

62. $-\sqrt{\dfrac{98r^3}{s^5}}$

63. $\sqrt{\dfrac{288x^7}{y^9}}$

64. $\sqrt{\dfrac{242t^9}{u^{11}}}$

65. $\dfrac{5\sqrt{2m}}{\sqrt{y^3}}$

66. $\dfrac{2\sqrt{5r}}{\sqrt{m^3}}$

67. $-\sqrt{\dfrac{48k^2}{z}}$

68. $-\sqrt{\dfrac{75m^3}{p}}$

Simplify. Assume all variables represent positive real numbers. See Example 4.

69. $\sqrt[3]{\dfrac{2}{3}}$

70. $\sqrt[3]{\dfrac{4}{5}}$

71. $\sqrt[3]{\dfrac{4}{9}}$

72. $\sqrt[3]{\dfrac{5}{16}}$

73. $\sqrt[3]{\dfrac{9}{32}}$

74. $\sqrt[3]{\dfrac{10}{9}}$

75. $-\sqrt[3]{\dfrac{2p}{r^2}}$

76. $-\sqrt[3]{\dfrac{6x}{y^2}}$

77. $\sqrt[3]{\dfrac{x^6}{y}}$

78. $\sqrt[3]{\dfrac{m^9}{q}}$

79. $\sqrt[4]{\dfrac{16}{x}}$

80. $\sqrt[4]{\dfrac{81}{y}}$

81. $\sqrt[4]{\dfrac{2y}{z}}$

82. $\sqrt[4]{\dfrac{7t}{s^2}}$

83. Explain the procedure you will use to rationalize the denominator of the expression in Exercise 85: $\dfrac{3}{4 + \sqrt{5}}$. Would multiplying both the numerator and the denominator of this fraction by $4 + \sqrt{5}$ lead to a rationalized denominator? Why or why not?

84. Show, in two ways, that the reciprocal of $\sqrt{6} - \sqrt{5}$ is $\sqrt{6} + \sqrt{5}$. (In general, however, the conjugate is not equal to the reciprocal.)

Rationalize the denominator in each expression. Assume all variables represent positive real numbers and no denominators are 0. See Example 5.

85. $\dfrac{3}{4 + \sqrt{5}}$

86. $\dfrac{4}{3 - \sqrt{7}}$

87. $\dfrac{\sqrt{8}}{3 - \sqrt{2}}$

88. $\dfrac{\sqrt{27}}{2 + \sqrt{3}}$

89. $\dfrac{2}{3\sqrt{5} + 2\sqrt{3}}$

90. $\dfrac{-1}{3\sqrt{2} - 2\sqrt{7}}$

91. $\dfrac{\sqrt{2} - \sqrt{3}}{\sqrt{6} - \sqrt{5}}$

92. $\dfrac{\sqrt{5} + \sqrt{6}}{\sqrt{3} - \sqrt{2}}$

93. $\dfrac{m - 4}{\sqrt{m} + 2}$

94. $\dfrac{r - 9}{\sqrt{r} - 3}$

95. $\dfrac{4\sqrt{x}}{\sqrt{x} - 2\sqrt{y}}$

96. $\dfrac{5\sqrt{r}}{3\sqrt{r} + \sqrt{s}}$

97. $\dfrac{\sqrt{x} - \sqrt{y}}{\sqrt{x} + \sqrt{y}}$

98. $\dfrac{\sqrt{a} + \sqrt{b}}{\sqrt{a} - \sqrt{b}}$

99. $\dfrac{5\sqrt{k}}{2\sqrt{k} + \sqrt{q}}$

100. $\dfrac{3\sqrt{x}}{\sqrt{x} - 2\sqrt{y}}$

101. If a and b are both positive numbers and $a^2 = b^2$, then $a = b$. Use this fact to show that $\dfrac{\sqrt{6} - \sqrt{2}}{4} = \dfrac{\sqrt{2 - \sqrt{3}}}{2}$.

102. Use a calculator approximation to support your result in Exercise 101.

Write each expression in lowest terms. Assume all variables represent positive real numbers. See Example 6.

103. $\dfrac{30 - 20\sqrt{6}}{10}$

104. $\dfrac{24 + 12\sqrt{5}}{12}$

105. $\dfrac{3 - 3\sqrt{5}}{3}$

106. $\dfrac{-5 + 5\sqrt{2}}{5}$

107. $\dfrac{16 - 4\sqrt{8}}{12}$

108. $\dfrac{12 - 9\sqrt{72}}{18}$

109. $\dfrac{6p + \sqrt{24p^3}}{3p}$

110. $\dfrac{11y - \sqrt{242y^5}}{22y}$

Rationalize each denominator. Assume all radicals represent real numbers and no denominators are 0.

111. $\dfrac{1}{\sqrt{x + y}}$

112. $\dfrac{5}{\sqrt{m - n}}$

113. $\dfrac{p}{\sqrt{p + 2}}$

114. $\dfrac{3q}{\sqrt{5 + q}}$

115. The following expression occurs in a certain standard problem in trigonometry:

$$\frac{1}{\sqrt{2}} \cdot \frac{\sqrt{3}}{2} - \frac{1}{\sqrt{2}} \cdot \frac{1}{2}.$$

Show that it simplifies to $\frac{\sqrt{6} - \sqrt{2}}{4}$. Then verify using a calculator approximation.

116. The following expression occurs in a certain standard problem in trigonometry:

$$\frac{\sqrt{3} + 1}{1 - \sqrt{3}}.$$

Show that it simplifies to $-2 - \sqrt{3}$. Then verify using a calculator approximation.

RELATING CONCEPTS (EXERCISES 117–124)

For Individual or Group Work

In Chapter 6 we presented methods of factoring, where the terms in the factors were integers. For example, the binomial $x^2 - 9$ is a difference of squares and factors as $(x + 3)(x - 3)$. However, we can also use this pattern to factor any binomial if we allow square root radicals in the terms of the factors. For example, $t - 5$ can be factored as $(\sqrt{t} + \sqrt{5})(\sqrt{t} - \sqrt{5})$.

Similarly, we can factor any binomial as the sum or difference of cubes, using the patterns $x^3 + y^3 = (x + y)(x^2 - xy + y^2)$ and $x^3 - y^3 = (x - y)(x^2 + xy + y^2)$. For example, we can factor $y + 2$ and $y - 2$ as follows:

$$y + 2 = \left(\sqrt[3]{y} + \sqrt[3]{2}\right)\left(\sqrt[3]{y^2} - \sqrt[3]{2y} + \sqrt[3]{4}\right)$$

$$y - 2 = \left(\sqrt[3]{y} - \sqrt[3]{2}\right)\left(\sqrt[3]{y^2} + \sqrt[3]{2y} + \sqrt[3]{4}\right).$$

Use these ideas to **work Exercises 117–124 in order.**

117. Factor $x - 7$ as the difference of squares.

118. Factor $x - 7$ as the difference of cubes.

119. Factor $x + 7$ as the sum of cubes.

120. Use the result of Exercise 117 to rationalize the denominator of $\dfrac{x + 3}{\sqrt{x} - \sqrt{7}}$.

121. Use the result of Exercise 118 to rationalize the denominator of $\dfrac{x + 3}{\sqrt[3]{x} - \sqrt[3]{7}}$.

122. Use the result of Exercise 119 to rationalize the denominator of $\dfrac{x + 3}{\sqrt[3]{x^2} - \sqrt[3]{7x} + \sqrt[3]{49}}$.

123. Factor the integer 2 as a difference of cubes by first writing it as $5 - 3$.

124. Use the result of Exercise 123 to rationalize the denominator of $\dfrac{2}{\sqrt[3]{5} - \sqrt[3]{3}}$.

Rationalize the numerator in each expression. Assume all variables represent positive real numbers. (Hint: See the Connections box following Example 6.)

125. $\dfrac{6 - \sqrt{2}}{4}$

126. $\dfrac{8\sqrt{5} - 1}{6}$

127. $\dfrac{3\sqrt{a} + \sqrt{b}}{b}$

128. $\dfrac{\sqrt{p} - 3\sqrt{q}}{4q}$

SUMMARY EXERCISES ON OPERATIONS WITH RADICALS

Recall that a simplified radical satisfies the following conditions.

1. The radicand has no factor raised to a power greater than or equal to the index.
2. The radicand has no fractions.
3. No denominator contains a radical.
4. Exponents in the radicand and the index of the radical have no common factor (except 1).

Perform all indicated operations and express each answer in simplest form. Assume all variables represent positive real numbers.

1. $6\sqrt{10} - 12\sqrt{10}$

2. $\sqrt{7}(\sqrt{7} - \sqrt{2})$

3. $(1 - \sqrt{3})(2 + \sqrt{6})$

4. $\sqrt{50} - \sqrt{98} + \sqrt{72}$

5. $(3\sqrt{5} + 2\sqrt{7})^2$

6. $\dfrac{-3}{\sqrt{6}}$

7. $\dfrac{8}{\sqrt{7} + \sqrt{5}}$

8. $\sqrt[3]{16x^2} - \sqrt[3]{54x^2} + \sqrt[3]{128x^2}$

9. $\dfrac{1 - \sqrt{2}}{1 + \sqrt{2}}$

10. $(1 - \sqrt[3]{3})(1 + \sqrt[3]{3} + \sqrt[3]{9})$

11. $(\sqrt{5} + 7)(\sqrt{5} - 7)$

12. $\dfrac{1}{\sqrt{x} - \sqrt{5}}, \quad x \neq 5$

13. $\sqrt[3]{8a^3b^5c^9}$

14. $\dfrac{15}{\sqrt[3]{9}}$

15. $\dfrac{3}{\sqrt{5} + 2}$

16. $\sqrt{\dfrac{3}{5x}}$

17. $\dfrac{16\sqrt{3}}{5\sqrt{12}}$

18. $\dfrac{2\sqrt{25}}{8\sqrt{50}}$

19. $\dfrac{-10}{\sqrt[3]{10}}$

20. $\dfrac{\sqrt{6} + \sqrt{5}}{\sqrt{6} - \sqrt{5}}$

21. $\sqrt{12x} - \sqrt{75x}$

22. $(5 - 3\sqrt{3})^2$

23. $(\sqrt{74} - \sqrt{73})(\sqrt{74} + \sqrt{73})$

24. $\sqrt[3]{\dfrac{13}{81}}$

25. $-t^2\sqrt[4]{t} + 3\sqrt[4]{t^9} - t\sqrt[4]{t^5}$

8.6 Solving Equations with Radicals

OBJECTIVES

1 Solve radical equations using the power rule.

2 Solve radical equations that require additional steps.

3 Solve radical equations with indexes greater than 2.

4 Solve radical equations using a graphing calculator.

5 Use the power rule to solve a formula for a specified variable.

OBJECTIVE 1 Solve radical equations using the power rule. An equation that includes one or more radical expressions with a variable is called a **radical equation.** Some examples of radical equations are

$$\sqrt{x - 4} = 8, \qquad \sqrt{5x + 12} = 3\sqrt{2x - 1}, \qquad \text{and} \qquad \sqrt[3]{6 + x} = 27.$$

The equation $x = 1$ has only one solution. Its solution set is $\{1\}$. If we square both sides of this equation, we get $x^2 = 1$. This new equation has two solutions: -1 and 1. Notice that the solution of the original equation is also a solution of the squared equation. However, the squared equation has another solution, -1, that is *not* a solution of the original equation. When solving equations with radicals, we use this idea of raising both sides to a power. It is an application of the **power rule.**

Power Rule for Solving an Equation with Radicals

If both sides of an equation are raised to the same power, all solutions of the original equation are also solutions of the new equation.

Read the power rule carefully; it does *not* say that all solutions of the new equation are solutions of the original equation. They may or may not be. Solutions that do not satisfy the original equation are called **extraneous solutions;** they must be discarded.

CAUTION When the power rule is used to solve an equation, *every solution of the new equation* must *be checked in the original equation.*

EXAMPLE 1 Using the Power Rule

Solve $\sqrt{3x + 4} = 8$.

Use the power rule and square both sides to obtain

$$\left(\sqrt{3x + 4}\right)^2 = 8^2$$
$$3x + 4 = 64$$
$$3x = 60$$
$$x = 20.$$

To check, substitute the potential solution in the *original* equation.

Check:
$$\sqrt{3x + 4} = 8$$
$$\sqrt{3 \cdot 20 + 4} = 8 \quad ? \qquad \text{Let } x = 20.$$
$$\sqrt{64} = 8 \quad ?$$
$$8 = 8 \qquad \text{True}$$

Since 20 satisfies the *original* equation, the solution set is {20}.

Now Try Exercise 9.

The solution of the equation in Example 1 can be generalized to give a method for solving equations with radicals.

Solving an Equation with Radicals

Step 1 **Isolate the radical.** Make sure that one radical term is alone on one side of the equation.

Step 2 **Apply the power rule.** Raise both sides of the equation to a power that is the same as the index of the radical.

Step 3 **Solve** the resulting equation; if it still contains a radical, repeat Steps 1 and 2.

Step 4 **Check** all potential solutions in the original equation.

CAUTION Remember Step 4 or you may get an incorrect solution set.

EXAMPLE 2 Using the Power Rule

Solve $\sqrt{5q - 1} + 3 = 0$.

Step 1 To get the radical alone on one side, subtract 3 from each side.

$$\sqrt{5q - 1} = -3$$

Step 2   Now square both sides.

$$\left(\sqrt{5q - 1}\right)^2 = (-3)^2$$

Step 3
$$5q - 1 = 9$$
$$5q = 10$$
$$q = 2$$

Step 4   Check the potential solution, 2, by substituting it in the original equation.

Check:
$$\sqrt{5q - 1} + 3 = 0$$
$$\sqrt{5 \cdot 2 - 1} + 3 = 0 \qquad ? \qquad \text{Let } q = 2.$$
$$3 + 3 = 0 \qquad \qquad \text{False}$$

This false result shows that 2 is *not* a solution of the original equation; it is extraneous. The solution set is $\emptyset$.

Now Try Exercise 11.

NOTE   We could have determined after Step 1 that the equation in Example 2 has no solution because the expression on the left cannot be negative.

OBJECTIVE 2   Solve radical equations that require additional steps.   The next examples involve finding the square of a binomial. Recall that

$$(x + y)^2 = x^2 + 2xy + y^2.$$

EXAMPLE 3   Using the Power Rule; Squaring a Binomial

Solve $\sqrt{4 - x} = x + 2$.

Step 1   The radical is alone on the left side of the equation.

Step 2   Square both sides; the square of $x + 2$ is $(x + 2)^2 = x^2 + 4x + 4$.

$$\left(\sqrt{4 - x}\right)^2 = (x + 2)^2$$
$$4 - x = x^2 + 4x + 4$$

⎣—Twice the product of 2 and x

Step 3   The new equation is quadratic, so get 0 on one side.

$$0 = x^2 + 5x \qquad \text{Subtract 4 and add } x.$$
$$0 = x(x + 5) \qquad \text{Factor.}$$
$$x = 0 \quad \text{or} \quad x + 5 = 0 \qquad \text{Zero-factor property}$$
$$x = -5$$

Step 4   Check each potential solution in the original equation.

Check: If $x = 0$, then

$$\sqrt{4 - x} = x + 2$$
$$\sqrt{4 - 0} = 0 + 2 \qquad ?$$
$$\sqrt{4} = 2 \qquad ?$$
$$2 = 2. \qquad \text{True}$$

If $x = -5$, then

$$\sqrt{4 - x} = x + 2$$
$$\sqrt{4 - (-5)} = -5 + 2 \qquad ?$$
$$\sqrt{9} = -3 \qquad ?$$
$$3 = -3. \qquad \text{False}$$

The solution set is $\{0\}$. The other potential solution, -5, is extraneous.

Now Try Exercise 27.

CAUTION When a radical equation requires squaring a binomial as in Example 3, remember to include the middle term.

$$(x + 2)^2 \neq x^2 + 4 \qquad\qquad (x + 2)^2 = x^2 + 4x + 4$$
$$\textbf{INCORRECT} \qquad\qquad\qquad \textbf{CORRECT}$$

EXAMPLE 4 Using the Power Rule; Squaring a Binomial

Solve $\sqrt{x^2 - 4x + 9} = x - 1$.

Squaring both sides gives $(x - 1)^2 = x^2 - 2(x)(1) + 1^2$ on the right.

$$\left(\sqrt{x^2 - 4x + 9}\right)^2 = (x - 1)^2$$
$$x^2 - 4x + 9 = x^2 - 2x + 1$$

⤷ Twice the product of x and -1

Subtract x^2 and 1 from each side; then add $4x$ to each side to obtain

$$8 = 2x$$
$$4 = x.$$

Check this potential solution in the original equation.

Check:
$$\sqrt{x^2 - 4x + 9} = x - 1$$
$$\sqrt{4^2 - 4 \cdot 4 + 9} = 4 - 1 \qquad ? \qquad \text{Let } x = 4.$$
$$3 = 3 \qquad\qquad\qquad \text{True}$$

The solution set of the original equation is $\{4\}$.

Now Try Exercise 29.

EXAMPLE 5 Using the Power Rule; Squaring Twice

Solve $\sqrt{5x + 6} + \sqrt{3x + 4} = 2$.

Start by getting one radical alone on one side of the equation by subtracting $\sqrt{3x + 4}$ from each side.

$$\sqrt{5x + 6} = 2 - \sqrt{3x + 4}$$
$$\left(\sqrt{5x + 6}\right)^2 = \left(2 - \sqrt{3x + 4}\right)^2 \qquad\qquad \text{Square both sides.}$$
$$5x + 6 = 4 - 4\sqrt{3x + 4} + (3x + 4)$$

⤷ Twice the product of 2 and $-\sqrt{3x + 4}$

This equation still contains a radical, so square both sides again. Before doing this, isolate the radical term on the right.

$$5x + 6 = 8 + 3x - 4\sqrt{3x + 4}$$
$$2x - 2 = -4\sqrt{3x + 4} \qquad\qquad \text{Subtract 8 and } 3x.$$
$$x - 1 = -2\sqrt{3x + 4} \qquad\qquad \text{Divide by 2.}$$
$$(x - 1)^2 = \left(-2\sqrt{3x + 4}\right)^2 \qquad\qquad \text{Square both sides again.}$$
$$x^2 - 2x + 1 = (-2)^2\left(\sqrt{3x + 4}\right)^2 \qquad\qquad (ab)^2 = a^2 b^2$$
$$x^2 - 2x + 1 = 4(3x + 4)$$
$$x^2 - 2x + 1 = 12x + 16 \qquad\qquad \text{Distributive property}$$

$$x^2 - 14x - 15 = 0 \qquad \text{Standard form}$$
$$(x - 15)(x + 1) = 0 \qquad \text{Factor.}$$
$$x - 15 = 0 \quad \text{or} \quad x + 1 = 0 \qquad \text{Zero-factor property}$$
$$x = 15 \quad \text{or} \qquad x = -1$$

Check each of these potential solutions in the original equation. Only -1 checks, so the solution set, $\{-1\}$, has only one element.

Now Try Exercise 51.

OBJECTIVE 3 Solve radical equations with indexes greater than 2. The power rule also works for powers greater than 2.

EXAMPLE 6 Using the Power Rule for a Power Greater than 2

Solve $\sqrt[3]{z + 5} = \sqrt[3]{2z - 6}$.

Raise both sides to the third power.

$$\left(\sqrt[3]{z + 5}\right)^3 = \left(\sqrt[3]{2z - 6}\right)^3$$
$$z + 5 = 2z - 6$$
$$11 = z$$

Check: $\qquad \sqrt[3]{z + 5} = \sqrt[3]{2z - 6} \qquad$ Original equation

$\qquad\qquad\quad \sqrt[3]{11 + 5} = \sqrt[3]{2 \cdot 11 - 6} \quad\; ? \qquad$ Let $z = 11$.

$\qquad\qquad\qquad\quad \sqrt[3]{16} = \sqrt[3]{16} \qquad\qquad$ True

The solution set is $\{11\}$.

Now Try Exercise 37.

 OBJECTIVE 4 Solve radical equations using a graphing calculator. In Example 4 we solved the equation $\sqrt{x^2 - 4x + 9} = x - 1$ using algebraic methods. If we write this equation with one side equal to 0, we get

$$\sqrt{x^2 - 4x + 9} - x + 1 = 0.$$

Using a graphing calculator to graph the function defined by

$$f(x) = \sqrt{x^2 - 4x + 9} - x + 1,$$

we obtain the graph shown in Figure 14. Notice that its zero (x-value of the x-intercept) is 4, which is the solution we found in Example 4.

In Example 3, we found that the single solution of $\sqrt{4 - x} = x + 2$ is 0, with an extraneous value of -5. If we graph $f(x) = \sqrt{4 - x}$ and $g(x) = x + 2$ in the same window, we find that the x-coordinate of the point of intersection of the two graphs is 0, which is the solution of the equation. See Figure 15.

We solved the equation in Example 3 by squaring both sides, obtaining $4 - x = x^2 + 4x + 4$. In Figure 16 on the next page, we show that the two functions defined by $f(x) = 4 - x$ and $g(x) = x^2 + 4x + 4$ have two points of intersection. The extraneous value -5 that we found in Example 3 shows up as an x-value of one of these points of intersection. However, our check showed that -5 was not a solution of the *original* equation (before the squaring step). Here we see a graphical interpretation of the extraneous value.

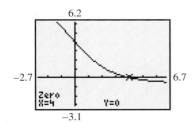

FIGURE 14

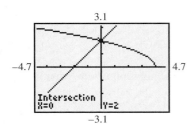

FIGURE 15

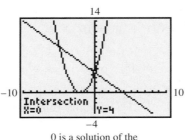
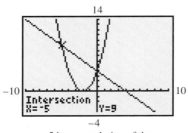

0 is a solution of the
original equation.

−5 is *not* a solution of the
original equation.

FIGURE 16

OBJECTIVE 5 Use the power rule to solve a formula for a specified variable.

■ **EXAMPLE 7** Solving a Formula from Electronics for a Variable

An important property of a radio frequency transmission line is its *characteristic impedance,* represented by Z and measured in ohms. If L and C are the inductance and capacitance, respectively, per unit of length of the line, then these quantities are related by the formula $Z = \sqrt{\dfrac{L}{C}}$. Solve this formula for C.

$$Z = \sqrt{\frac{L}{C}} \qquad \text{Given formula}$$

$$Z^2 = \frac{L}{C} \qquad \text{Square both sides.}$$

$$CZ^2 = L \qquad \text{Multiply by } C.$$

$$C = \frac{L}{Z^2} \qquad \text{Divide by } Z^2.$$

Now Try Exercise 67.

8.6 EXERCISES

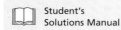

Check each equation to see if the given value for x is a solution.

1. $\sqrt{3x + 18} = x$

    (a) 6   **(b)** −3

2. $\sqrt{3x - 3} = x - 1$

    (a) 1   **(b)** 4

3. $\sqrt{x + 2} = \sqrt{9x - 2} - 2\sqrt{x - 1}$

    (a) 2   **(b)** 7

4. $\sqrt{8x - 3} = 2x$

    (a) $\dfrac{3}{2}$ **(b)** $\dfrac{1}{2}$

✎ **5.** Is 9 a solution of the equation $\sqrt{x} = -3$? If not, what is the solution of this equation? Explain.

✎ **6.** Before even attempting to solve $\sqrt{3x + 18} = x$, how can you be sure that the equation cannot have a negative solution?

Solve each equation. See Examples 1–4.

7. $\sqrt{r-2} = 3$

8. $\sqrt{q+1} = 7$

9. $\sqrt{6k-1} = 1$

10. $\sqrt{7m-3} = 5$

11. $\sqrt{4r+3} + 1 = 0$

12. $\sqrt{5k-3} + 2 = 0$

13. $\sqrt{3k+1} - 4 = 0$

14. $\sqrt{5z+1} - 11 = 0$

15. $4 - \sqrt{x-2} = 0$

16. $9 - \sqrt{4k+1} = 0$

17. $\sqrt{9a-4} = \sqrt{8a+1}$

18. $\sqrt{4p-2} = \sqrt{3p+5}$

19. $2\sqrt{x} = \sqrt{3x+4}$

20. $2\sqrt{m} = \sqrt{5m-16}$

21. $3\sqrt{z-1} = 2\sqrt{2z+2}$

22. $5\sqrt{4a+1} = 3\sqrt{10a+25}$

23. $k = \sqrt{k^2+4k-20}$

24. $p = \sqrt{p^2-3p+18}$

25. $a = \sqrt{a^2+3a+9}$

26. $z = \sqrt{z^2-4z-8}$

27. $\sqrt{9-x} = x+3$

28. $\sqrt{5-x} = x+1$

29. $\sqrt{k^2+2k+9} = k+3$

30. $\sqrt{a^2-3a+3} = a-1$

31. $\sqrt{r^2+9r+3} = -r$

32. $\sqrt{p^2-15p+15} = p-5$

33. $\sqrt{z^2+12z-4} + 4 - z = 0$

34. $\sqrt{m^2+3m+12} - m - 2 = 0$

🖉 **35.** What is *wrong* with this first step in the solution process for $\sqrt{3x+4} = 8 - x$. Solve it correctly.

$$3x + 4 = 64 + x^2$$

🖉 **36.** Explain what is *wrong* with this first step in the solution process for the equation $\sqrt{5x+6} - \sqrt{x+3} = 3$. Then solve it correctly.

$$(5x+6) + (x+3) = 9$$

Solve each equation. See Examples 5 and 6.

37. $\sqrt[3]{2x+5} = \sqrt[3]{6x+1}$

38. $\sqrt[3]{p-1} = 2$

39. $\sqrt[3]{a^2+5a+1} = \sqrt[3]{a^2+4a}$

40. $\sqrt[3]{r^2+2r+8} = \sqrt[3]{r^2}$

41. $\sqrt[3]{2m-1} = \sqrt[3]{m+13}$

42. $\sqrt[3]{2k-11} - \sqrt[3]{5k+1} = 0$

43. $\sqrt[4]{a+8} = \sqrt[4]{2a}$

44. $\sqrt[4]{z+11} = \sqrt[4]{2z+6}$

45. $\sqrt[3]{x-8} + 2 = 0$

46. $\sqrt[3]{r+1} + 1 = 0$

47. $\sqrt[4]{2k-5} + 4 = 0$

48. $\sqrt[4]{8z-3} + 2 = 0$

49. $\sqrt{k+2} - \sqrt{k-3} = 1$

50. $\sqrt{r+6} - \sqrt{r-2} = 2$

51. $\sqrt{2r+11} - \sqrt{5r+1} = -1$

52. $\sqrt{3x-2} - \sqrt{x+3} = 1$

53. $\sqrt{3p+4} - \sqrt{2p-4} = 2$

54. $\sqrt{4x+5} - \sqrt{2x+2} = 1$

55. $\sqrt{3-3p} - 3 = \sqrt{3p+2}$

56. $\sqrt{4x+7} - 4 = \sqrt{4x-1}$

57. $\sqrt{2\sqrt{x+11}} = \sqrt{4x+2}$

58. $\sqrt{1+\sqrt{24-10x}} = \sqrt{3x+5}$

59. What is the smallest power to which you can raise both sides of the radical equation $\sqrt{x+3} = \sqrt[3]{5+4x}$ so that the radicals are eliminated?

60. What is the smallest power to which you can raise both sides of the radical equation $\sqrt[4]{x+3} = \sqrt[3]{10x+14}$ so that the radicals are eliminated?

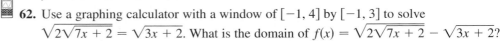

61. Use a graphing calculator to solve $\sqrt{3 - 3x} = 3 + \sqrt{3x + 2}$. What is the domain of $y = \sqrt{3 - 3x} - 3 - \sqrt{3x + 2}$?

62. Use a graphing calculator with a window of $[-1, 4]$ by $[-1, 3]$ to solve $\sqrt{2\sqrt{7x + 2}} = \sqrt{3x + 2}$. What is the domain of $f(x) = \sqrt{2\sqrt{7x + 2}} - \sqrt{3x + 2}$?

For each equation, rewrite the expressions with rational exponents as radical expressions, and then solve using the procedures explained in this section.

63. $(2x - 9)^{1/2} = 2 + (x - 8)^{1/2}$ **64.** $(3w + 7)^{1/2} = 1 + (w + 2)^{1/2}$

65. $(2w - 1)^{2/3} - w^{1/3} = 0$ **66.** $(x^2 - 2x)^{1/3} - x^{1/3} = 0$

Solve each formula from electricity and radio for the indicated variable. See Example 7. (Source: Cooke, Nelson M., and Joseph B. Orleans, Mathematics Essential to Electricity and Radio, *McGraw-Hill, 1943.)*

67. $V = \sqrt{\dfrac{2K}{m}}$ for K **68.** $V = \sqrt{\dfrac{2K}{m}}$ for m

69. $f = \dfrac{1}{2\pi\sqrt{LC}}$ for L **70.** $r = \sqrt{\dfrac{Mm}{F}}$ for F

71. A number of useful formulas involve radicals or radical expressions. Many occur in the mathematics needed for working with objects in space. The formula

$$N = \frac{1}{2\pi}\sqrt{\frac{a}{r}}$$

is used to find the rotational rate N of a space station. Here a is the acceleration and r represents the radius of the space station in meters. To find the value of r that will make N simulate the effect of gravity on Earth, the equation must be solved for r, using the required value of N. (*Source:* Kastner, Bernice, *Space Mathematics*, NASA, 1972.)

(a) Solve the equation for r.
(b) Find the value of r that makes $N = .063$ rotation per sec, if $a = 9.8$ m per sec^2.
(c) Find the value of r that makes $N = .04$ rotation per sec, if $a = 9.8$ m per sec^2.

If x is the number of years since 1900, the equation $y = x^{.7}$ approximates the timber grown in the United States in billions of cubic feet. Let $x = 20$ represent 1920, $x = 52$ represent 1952, and so on.

72. Replace x in the equation for each year shown in the graph and use a calculator to find the value of y. (Round answers to the nearest billion.)

73. Use the graph to estimate the amount of timber grown for each year shown.

74. Compare the values found from the equation with your estimates from the graph. Does the equation give a good approximation of the data from the graph? In which year is the approximation best?

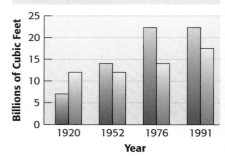

U.S. TIMBER GROWTH AND HARVEST

■ Timber grown
□ Timber harvested

Source: Figures from U.S. Forest Service. Adapted from "*The Truth about America's Forests,*" Evergreen, 4025 Crater Lake Hwy., Medford, Ore. 97504.

75. From the graph, estimate the amount of timber harvested in each year shown.

76. Use the equation $y = x^{.62}$ and a calculator to approximate the amount of timber harvested in each of the given years. (Round answers to the nearest billion.)

77. Compare your answers from Exercises 75 and 76. Does the equation give a good approximation? For which year is it poorest?

8.7 Complex Numbers

OBJECTIVES

1 Simplify numbers of the form $\sqrt{-b}$, where $b > 0$.

2 Recognize imaginary complex numbers.

3 Add and subtract complex numbers.

4 Multiply complex numbers.

5 Divide complex numbers.

6 Find powers of i.

As we saw in Chapter 1, the set of real numbers includes many other number sets (the rational numbers, integers, and natural numbers, for example). In this section a new set of numbers is introduced that includes the set of real numbers, as well as numbers that are even roots of negative numbers, like $\sqrt{-2}$.

OBJECTIVE 1 Simplify numbers of the form $\sqrt{-b}$, where $b > 0$. The equation $x^2 + 1 = 0$ has no real number solution since any solution must be a number whose square is -1. In the set of real numbers, all squares are nonnegative numbers because the product of two positive numbers or two negative numbers is positive and $0^2 = 0$. To provide a solution for the equation $x^2 + 1 = 0$, a new number i is defined so that

$$i^2 = -1.$$

That is, i is a number whose square is -1, so $i = \sqrt{-1}$. This definition of i makes it possible to define any square root of a negative real number as follows.

$\sqrt{-b}$

For any positive real number b,

$$\sqrt{-b} = i\sqrt{b}.$$

▓ EXAMPLE 1 Simplifying Square Roots of Negative Numbers

Write each number as a product of a real number and i.

(a) $\sqrt{-100} = i\sqrt{100} = 10i$ **(b)** $-\sqrt{-36} = -i\sqrt{36} = -6i$

(c) $\sqrt{-2} = i\sqrt{2}$

Now Try Exercises 7, 9, and 11.

CAUTION It is easy to mistake $\sqrt{2}i$ for $\sqrt{2i}$, with the i under the radical. For this reason, we usually write $\sqrt{2}i$ as $i\sqrt{2}$, as in the definition of $\sqrt{-b}$.

When finding a product such as $\sqrt{-4} \cdot \sqrt{-9}$, we cannot use the product rule for radicals because it applies only to nonnegative radicands. For this reason, we change $\sqrt{-b}$ to the form $i\sqrt{b}$ before performing any multiplications or divisions. For example,

$$\sqrt{-4} \cdot \sqrt{-9} = i\sqrt{4} \cdot i\sqrt{9}$$
$$= i \cdot 2 \cdot i \cdot 3$$
$$= 6i^2$$
$$= 6(-1) \qquad \text{Substitute: } i^2 = -1.$$
$$= -6.$$

CAUTION Using the product rule for radicals *before* using the definition of $\sqrt{-b}$ gives a *wrong* answer. The preceding example shows that

$$\sqrt{-4} \cdot \sqrt{-9} = -6,$$

but

$$\sqrt{-4(-9)} = \sqrt{36} = 6,$$

so

$$\sqrt{-4} \cdot \sqrt{-9} \neq \sqrt{-4(-9)}.$$

■ **EXAMPLE 2** Multiplying Square Roots of Negative Numbers

Multiply.

(a) $\sqrt{-3} \cdot \sqrt{-7} = i\sqrt{3} \cdot i\sqrt{7}$
$$= i^2\sqrt{3 \cdot 7}$$
$$= (-1)\sqrt{21} \qquad \text{Substitute: } i^2 = -1.$$
$$= -\sqrt{21}$$

(b) $\sqrt{-2} \cdot \sqrt{-8} = i\sqrt{2} \cdot i\sqrt{8}$
$$= i^2\sqrt{2 \cdot 8}$$
$$= (-1)\sqrt{16}$$
$$= (-1)4$$
$$= -4$$

(c) $\sqrt{-5} \cdot \sqrt{6} = i\sqrt{5} \cdot \sqrt{6} = i\sqrt{30}$

Now Try Exercises 15, 17, and 19.

The methods used to find products also apply to quotients.

■ **EXAMPLE 3** Dividing Square Roots of Negative Numbers

Divide.

(a) $\dfrac{\sqrt{-75}}{\sqrt{-3}} = \dfrac{i\sqrt{75}}{i\sqrt{3}} = \sqrt{\dfrac{75}{3}} = \sqrt{25} = 5$

(b) $\dfrac{\sqrt{-32}}{\sqrt{8}} = \dfrac{i\sqrt{32}}{\sqrt{8}} = i\sqrt{\dfrac{32}{8}} = i\sqrt{4} = 2i$

Now Try Exercises 21 and 23.

OBJECTIVE 2 Recognize imaginary complex numbers. With the imaginary number i and the real numbers, a new set of numbers can be formed that includes the real numbers as a subset. The *complex numbers* are defined as follows.

Complex Number

If a and b are real numbers, then any number of the form $a + bi$ is called a **complex number.**

In the complex number $a + bi$, the number a is called the **real part** and b is called the **imaginary part.** When $b = 0$, $a + bi$ is a real number, so the real numbers are a subset of the complex numbers. Complex numbers with $b \neq 0$ are called **imaginary numbers.*** In spite of their name, imaginary numbers are very useful in applications, particularly in work with electricity.

The relationships among the various sets of numbers discussed in this book are shown in Figure 17.

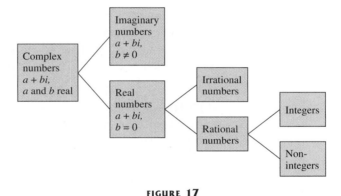

FIGURE 17

OBJECTIVE 3 Add and subtract complex numbers. The commutative, associative, and distributive properties for real numbers are also valid for complex numbers. Thus, to add complex numbers, we add their real parts and add their imaginary parts.

▨ EXAMPLE 4 Adding Complex Numbers

Add.

(a) $(2 + 3i) + (6 + 4i)$
  $= (2 + 6) + (3 + 4)i$ Commutative, associative, and distributive properties
  $= 8 + 7i$

(b) $(4 + 2i) + (3 - i) + (-6 + 3i) = [4 + 3 + (-6)] + [2 + (-1) + 3]i$
                 $= 1 + 4i$

Now Try Exercises 27 and 35.

---

*Some texts define bi as the imaginary part of the complex number $a + bi$. Also, imaginary numbers are sometimes defined as complex numbers with $a = 0$ and $b \neq 0$.

We subtract complex numbers by subtracting their real parts and subtracting their imaginary parts.

EXAMPLE 5   Subtracting Complex Numbers

Subtract.

(a) $(6 + 5i) - (3 + 2i) = (6 - 3) + (5 - 2)i$
$$= 3 + 3i$$

(b) $(7 - 3i) - (8 - 6i) = (7 - 8) + [-3 - (-6)]i$
$$= -1 + 3i$$

(c) $(-9 + 4i) - (-9 + 8i) = (-9 + 9) + (4 - 8)i$
$$= 0 - 4i$$
$$= -4i$$

Now Try Exercises 31 and 33.

In Example 5(c), the answer was written as $0 - 4i$ and then as just $-4i$. A complex number written in the form $a + bi$, like $0 - 4i$, is in **standard form.** In this section, most answers will be given in standard form, but if a or b is 0, we consider answers such as a or bi to be in standard form.

OBJECTIVE 4 Multiply complex numbers.   We multiply complex numbers as we multiply polynomials. Complex numbers of the form $a + bi$ have the same form as binomials, so we multiply two complex numbers in standard form by using the FOIL method for multiplying binomials. (Recall that FOIL stands for *First, Outer, Inner, Last.*)

EXAMPLE 6   Multiplying Complex Numbers

Multiply.

(a) $4i(2 + 3i) = 4i(2) + 4i(3i)$ Distributive property
$$= 8i + 12i^2$$
$$= 8i + 12(-1)$$ Substitute: $i^2 = -1$.
$$= -12 + 8i$$

(b) $(3 + 5i)(4 - 2i)$
Use the FOIL method.
$$(3 + 5i)(4 - 2i) = \underbrace{3(4)}_{\text{First}} + \underbrace{3(-2i)}_{\text{Outer}} + \underbrace{5i(4)}_{\text{Inner}} + \underbrace{5i(-2i)}_{\text{Last}}$$
$$= 12 - 6i + 20i - 10i^2$$
$$= 12 + 14i - 10(-1)$$ Substitute: $i^2 = -1$.
$$= 12 + 14i + 10$$
$$= 22 + 14i$$

(c) $(2 + 3i)(1 - 5i) = 2(1) + 2(-5i) + 3i(1) + 3i(-5i)$ FOIL

$$= 2 - 10i + 3i - 15i^2$$

$$= 2 - 7i - 15(-1)$$

$$= 2 - 7i + 15$$

$$= 17 - 7i$$

Now Try Exercises 45 and 47.

The two complex numbers $a + bi$ and $a - bi$ are called **complex conjugates,** or simply *conjugates,* of each other. The product of a complex number and its conjugate is always a real number, as shown here.

$$(a + bi)(a - bi) = a^2 - abi + abi - b^2i^2$$

$$= a^2 - b^2(-1)$$

$$(a + bi)(a - bi) = a^2 + b^2$$

For example, $(3 + 7i)(3 - 7i) = 3^2 + 7^2 = 9 + 49 = 58$.

OBJECTIVE 5 Divide complex numbers.   The quotient of two complex numbers should be a complex number. To write the quotient as a complex number, we need to eliminate i in the denominator. We use conjugates to do this.

EXAMPLE 7   Dividing Complex Numbers

Find each quotient.

(a) $\dfrac{8 + 9i}{5 + 2i}$

Multiply both the numerator and denominator by the conjugate of the denominator. The conjugate of $5 + 2i$ is $5 - 2i$.

$$\frac{8 + 9i}{5 + 2i} = \frac{(8 + 9i)(5 - 2i)}{(5 + 2i)(5 - 2i)} \qquad \frac{5 - 2i}{5 - 2i} = 1$$

$$= \frac{40 - 16i + 45i - 18i^2}{5^2 + 2^2}$$

$$= \frac{58 + 29i}{29} \qquad \text{Substitute: } i^2 = -1; \text{ combine terms.}$$

$$= \frac{29(2 + i)}{29} \qquad \text{Factor the numerator.}$$

$$= 2 + i \qquad \text{Lowest terms}$$

Notice that this is just like rationalizing a denominator. The final result is in standard form.

(b) $\dfrac{1 + i}{i}$

The conjugate of i is $-i$. Multiply both the numerator and denominator by $-i$.

$$\dfrac{1 + i}{i} = \dfrac{(1 + i)(-i)}{i(-i)}$$

$$= \dfrac{-i - i^2}{-i^2}$$

$$= \dfrac{-i - (-1)}{-(-1)} \qquad \text{Substitute: } i^2 = -1.$$

$$= \dfrac{-i + 1}{1}$$

$$= 1 - i$$

Now Try Exercises 61 and 67.

In Examples 4–7, we showed how complex numbers can be added, subtracted, multiplied, and divided using algebraic methods. Many current models of graphing calculators can perform these operations. Figure 18 shows how the computations in parts of Examples 4–7 are carried out by a TI-83 Plus calculator. It is important to use parentheses as shown.

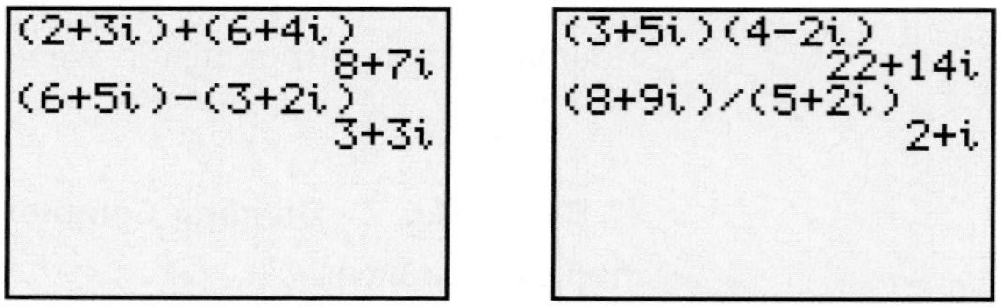

```
(2+3i)+(6+4i)
            8+7i
(6+5i)-(3+2i)
            3+3i
```

```
(3+5i)(4-2i)
          22+14i
(8+9i)/(5+2i)
             2+i
```

FIGURE 18

OBJECTIVE 6 Find powers of i. Because i^2 is defined to be -1, we can find higher powers of i as shown in the following examples.

$$i^3 = i \cdot i^2 = i(-1) = -i \qquad\qquad i^6 = i^2 \cdot i^4 = (-1) \cdot 1 = -1$$
$$i^4 = i^2 \cdot i^2 = (-1)(-1) = 1 \qquad i^7 = i^3 \cdot i^4 = (-i) \cdot 1 = -i$$
$$i^5 = i \cdot i^4 = i \cdot 1 = i \qquad\qquad\quad i^8 = i^4 \cdot i^4 = 1 \cdot 1 = 1$$

As these examples suggest, the powers of i rotate through the four numbers i, -1, $-i$, and 1. Larger powers of i can be simplified by using the fact that $i^4 = 1$. For example,

$$i^{75} = (i^4)^{18} \cdot i^3 = 1^{18} \cdot i^3 = 1 \cdot i^3 = i^3 = -i.$$

This example suggests a quick method for simplifying larger powers of i.

EXAMPLE 8 Simplifying Powers of i

Find each power of i.

(a) $i^{12} = (i^4)^3 = 1^3 = 1$

(b) $i^{39} = i^{36} \cdot i^3 = (i^4)^9 \cdot i^3 = 1^9 \cdot (-i) = -i$

(c) $i^{-2} = \dfrac{1}{i^2} = \dfrac{1}{-1} = -1$

(d) $i^{-1} = \dfrac{1}{i}$

To simplify this quotient, multiply both the numerator and denominator by $-i$, the conjugate of i.

$$\frac{1}{i} = \frac{1(-i)}{i(-i)} = \frac{-i}{-i^2} = \frac{-i}{-(-1)} = \frac{-i}{1} = -i$$

Now Try Exercises 73 and 81.

8.7 EXERCISES

Decide whether each expression is equal to 1, -1, i, *or* $-i$.

1. $\sqrt{-1}$ **2.** $-\sqrt{-1}$ **3.** i^2 **4.** $-i^2$ **5.** $\dfrac{1}{i}$ **6.** $(-i)^2$

Write each number as a product of a real number and i. Simplify all radical expressions. See Example 1.

7. $\sqrt{-169}$ **8.** $\sqrt{-225}$ **9.** $-\sqrt{-144}$ **10.** $-\sqrt{-196}$

11. $\sqrt{-5}$ **12.** $\sqrt{-21}$ **13.** $\sqrt{-48}$ **14.** $\sqrt{-96}$

Multiply or divide as indicated. See Examples 2 and 3.

15. $\sqrt{-7} \cdot \sqrt{-15}$ **16.** $\sqrt{-3} \cdot \sqrt{-19}$ **17.** $\sqrt{-4} \cdot \sqrt{-25}$ **18.** $\sqrt{-9} \cdot \sqrt{-81}$

19. $\sqrt{-3} \cdot \sqrt{11}$ **20.** $\sqrt{-10} \cdot \sqrt{2}$ **21.** $\dfrac{\sqrt{-300}}{\sqrt{-100}}$ **22.** $\dfrac{\sqrt{-40}}{\sqrt{-10}}$

23. $\dfrac{\sqrt{-75}}{\sqrt{3}}$ **24.** $\dfrac{\sqrt{-160}}{\sqrt{10}}$

25. (a) Every real number is a complex number. Explain why this is so.
  (b) Not every complex number is a real number. Give an example of this and explain why this statement is true.

26. Explain how to add, subtract, multiply, and divide complex numbers. Give examples.

Add or subtract as indicated. Write your answers in the form a + bi. See Examples 4 and 5.

27. $(3 + 2i) + (-4 + 5i)$

28. $(7 + 15i) + (-11 + 14i)$

29. $(5 - i) + (-5 + i)$

30. $(-2 + 6i) + (2 - 6i)$

31. $(4 + i) - (-3 - 2i)$

32. $(9 + i) - (3 + 2i)$

33. $(-3 - 4i) - (-1 - 4i)$

34. $(-2 - 3i) - (-5 - 3i)$

35. $(-4 + 11i) + (-2 - 4i) + (7 + 6i)$

36. $(-1 + i) + (2 + 5i) + (3 + 2i)$

37. $[(7 + 3i) - (4 - 2i)] + (3 + i)$

38. $[(7 + 2i) + (-4 - i)] - (2 + 5i)$

39. Fill in the blank with the correct response:

Because $(4 + 2i) - (3 + i) = 1 + i$, using the definition of subtraction, we can check this to find that $(1 + i) + (3 + i) = $ _____.

40. Fill in the blank with the correct response:

Because $\dfrac{-5}{2 - i} = -2 - i$, using the definition of division, we can check this to find that $(-2 - i)(2 - i) = $ _____.

Multiply. See Example 6.

41. $(3i)(27i)$

42. $(5i)(125i)$

43. $(-8i)(-2i)$

44. $(-32i)(-2i)$

45. $5i(-6 + 2i)$

46. $3i(4 + 9i)$

47. $(4 + 3i)(1 - 2i)$

48. $(7 - 2i)(3 + i)$

49. $(4 + 5i)^2$

50. $(3 + 2i)^2$

51. $2i(-4 - i)^2$

52. $3i(-3 - i)^2$

53. $(12 + 3i)(12 - 3i)$

54. $(6 + 7i)(6 - 7i)$

55. $(4 + 9i)(4 - 9i)$

56. $(7 + 2i)(7 - 2i)$

57. What is the conjugate of $a + bi$?

58. If we multiply $a + bi$ by its conjugate, we get _____, which is always a real number.

Write each expression in standard form $a + bi$. See Example 7.

59. $\dfrac{2}{1 - i}$

60. $\dfrac{29}{5 + 2i}$

61. $\dfrac{-7 + 4i}{3 + 2i}$

62. $\dfrac{-38 - 8i}{7 + 3i}$

63. $\dfrac{8i}{2 + 2i}$

64. $\dfrac{-8i}{1 + i}$

65. $\dfrac{2 - 3i}{2 + 3i}$

66. $\dfrac{-1 + 5i}{3 + 2i}$

67. $\dfrac{3 + i}{i}$

68. $\dfrac{5 - i}{-i}$

TECHNOLOGY INSIGHTS (EXERCISES 69–70)

Predict the answer that the calculator screen will provide for the given complex number operation entry.

69.

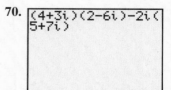

70.

71. Recall that if $a \neq 0$, then $\frac{1}{a}$ is called the reciprocal of a. Use this definition to express the reciprocal of $5 - 4i$ in the form $a + bi$.

72. Recall that if $a \neq 0$, then a^{-1} is defined to be $\frac{1}{a}$. Use this definition to express $(4 - 3i)^{-1}$ in the form $a + bi$.

Find each power of i. See Example 8.

73. i^{18} **74.** i^{26} **75.** i^{89} **76.** i^{48} **77.** i^{38}

78. i^{102} **79.** i^{43} **80.** i^{83} **81.** i^{-5} **82.** i^{-17}

83. A student simplified i^{-18} as follows:

$$i^{-18} = i^{-18} \cdot i^{20} = i^{-18+20} = i^2 = -1.$$

Explain the mathematical justification for this correct work.

84. Explain why

$$(46 + 25i)(3 - 6i) \quad \text{and} \quad (46 + 25i)(3 - 6i)i^{12}$$

must be equal. (Do not actually perform the computation.)

Ohm's law for the current I in a circuit with voltage E, resistance R, capacitance reactance X_c, and inductive reactance X_L is

$$I = \frac{E}{R + (X_L - X_c)i}.$$

Use this law to work Exercises 85 and 86.

85. Find I if $E = 2 + 3i$, $R = 5$, $X_L = 4$, and $X_c = 3$.

86. Find E if $I = 1 - i$, $R = 2$, $X_L = 3$, and $X_c = 1$.

Complex numbers will appear again in this book in Chapter 9, when we study quadratic equations. The following exercises examine how a complex number can be a solution of a quadratic equation.

87. Show that $1 + 5i$ is a solution of $x^2 - 2x + 26 = 0$. Then show that its conjugate is also a solution.

88. Show that $3 + 2i$ is a solution of $x^2 - 6x + 13 = 0$. Then show that its conjugate is also a solution.

RELATING CONCEPTS (EXERCISES 89–94)

For Individual or Group Work

Consider the following expressions:

Binomials	**Complex Numbers**
$x + 2$, $3x - 1$	$1 + 2i$, $3 - i$

*When we add, subtract, or multiply complex numbers in standard form, the rules are the same as those for the corresponding operations on binomials. That is, we add or subtract like terms, and we use FOIL to multiply. Division, however, is comparable to division by the sum or difference of radicals, where we multiply by the conjugate to get a rational denominator. To express the quotient of two complex numbers in standard form, we also multiply by the conjugate of the denominator. **Work Exercises 89–94 in order,** to better understand these ideas.*

89. (a) Add the two binomials. **(b)** Add the two complex numbers.

(continued)

90. (a) Subtract the second binomial from the first.
    (b) Subtract the second complex number from the first.

91. (a) Multiply the two binomials.
    (b) Multiply the two complex numbers.

92. (a) Rationalize the denominator: $\dfrac{\sqrt{3}-1}{1+\sqrt{2}}$.

    (b) Write in standard form: $\dfrac{3-i}{1+2i}$.

93. Explain why the answers for (a) and (b) in Exercise 91 do not correspond as the answers in Exercises 89–90 do.

94. Explain why the answers for (a) and (b) in Exercise 92 do not correspond as the answers in Exercises 89–90 do.

Perform the indicated operations. Give answers in standard form.

95. $\dfrac{3}{2-i} + \dfrac{5}{1+i}$

96. $\dfrac{2}{3+4i} + \dfrac{4}{1-i}$

97. $\left(\dfrac{2+i}{2-i} + \dfrac{i}{1+i}\right)i$

98. $\left(\dfrac{4-i}{1+i} - \dfrac{2i}{2+i}\right)4i$

<table>
<tr><td>Chapter</td><td>8</td><td>**Group Activity**</td></tr>
</table>

Solar Electricity

OBJECTIVE Apply the Pythagorean formula.

In this activity you will determine the sizes of frames needed to support solar electric panels on a flat roof.

A. The following table gives three different solar modules by Solarex. Have each member of the group choose one of the solar panels.

Model	Watts	Volts	Amps	Size (in inches)	Cost (in dollars)
MSX-77	77	16.9	4.56	44 × 26	475
MSX-83	83	17.1	4.85	44 × 24	490
MSX-60	60	17.1	3.5	44 × 20	382

Source: Solarex table in *Jade Mountain* catalog.

B. To use your solar panel, you must make a wooden frame to support it. The sides of this frame will form a right triangle. The hypotenuse of the triangle will be the width of the solar panel you chose. Make a sketch and use the Pythagorean formula to find the dimensions of the legs for each frame given the following conditions. Round answers to the nearest tenth.

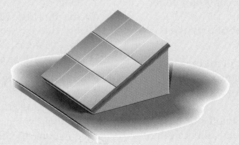

1. The legs have equal length.
2. One leg is twice the length of the other.
3. One leg is 3 times the length of the other.

C. Compare the different frame sizes for each panel. What factors might determine which of the triangles you would use in your frame?

CHAPTER 8 SUMMARY

KEY TERMS

8.1 radicand	square root function	conjugates	imaginary part
index (order)	cube root function	8.6 radical equation	imaginary numbers
radical	8.3 simplified radical	extraneous solution	standard form
principal root	8.5 rationalizing the	8.7 complex number	complex conjugates
radical expression	denominator	real part	

NEW SYMBOLS

$\sqrt{}$ radical sign

$\sqrt[n]{a}$ radical; principal nth root of a

$\pm$ positive or negative

$\approx$ is approximately equal to

$a^{1/n}$ a to the power $\dfrac{1}{n}$

$a^{m/n}$ a to the power $\dfrac{m}{n}$

i a number whose square is -1

TEST YOUR WORD POWER

See how well you have learned the vocabulary in this chapter. Answers, with examples, follow the Quick Review.

1. A **radicand** is
   A. the index of a radical
   B. the number or expression under the radical sign
   C. the positive root of a number
   D. the radical sign.

2. The **Pythagorean formula** states that, in a right triangle,
   A. the sum of the measures of the angles is 180°
   B. the sum of the lengths of the two shorter sides equals the length of the longest side
   C. the longest side is opposite the right angle
   D. the square of the length of the longest side equals the sum of the squares of the lengths of the two shorter sides.

3. A **hypotenuse** is
   A. either of the two shorter sides of a triangle
   B. the shortest side of a triangle
   C. the side opposite the right angle in a triangle
   D. the longest side in any triangle.

4. **Rationalizing the denominator** is the process of
   A. eliminating fractions from a radical expression
   B. changing the denominator of a fraction from a radical to a rational number
   C. clearing a radical expression of radicals
   D. multiplying radical expressions.

5. An **extraneous solution** is a solution
   A. that does not satisfy the original equation
   B. that makes an equation true
   C. that makes an expression equal 0
   D. that checks in the original equation.

6. A **complex number** is
   A. a real number that includes a complex fraction
   B. a zero multiple of i
   C. a number of the form $a + bi$, where a and b are real numbers
   D. the square root of -1.

QUICK REVIEW

CONCEPTS	EXAMPLES

8.1 RADICAL EXPRESSIONS AND GRAPHS

$\sqrt[n]{a} = b$ means $b^n = a$.

$\sqrt[n]{a}$ is the principal nth root of a.

$\sqrt[n]{a^n} = |a|$ if n is even.

$\sqrt[n]{a^n} = a$ if n is odd.

The two square roots of 64 are $\sqrt{64} = 8$, the principal square root, and $-\sqrt{64} = -8$.

$$\sqrt[4]{(-2)^4} = |-2| = 2$$
$$\sqrt[3]{-27} = -3$$

CONCEPTS	**EXAMPLES**

Functions Defined by Radical Expressions

The square root function with $f(x) = \sqrt{x}$ and the cube root function with $f(x) = \sqrt[3]{x}$ are two important functions defined by radical expressions.

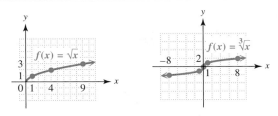

8.2 RATIONAL EXPONENTS

$a^{1/n} = \sqrt[n]{a}$ whenever $\sqrt[n]{a}$ exists.

If m and n are positive integers with $\frac{m}{n}$ in lowest terms, then $a^{m/n} = (a^{1/n})^m$, provided that $a^{1/n}$ is a real number.

All of the usual definitions and rules for exponents are valid for rational exponents.

$$81^{1/2} = \sqrt{81} = 9 \qquad -64^{1/3} = -\sqrt[3]{64} = -4$$
$$8^{5/3} = (8^{1/3})^5 = 2^5 = 32$$

$$5^{-1/2} \cdot 5^{1/4} = 5^{-1/2+1/4} = 5^{-1/4} = \frac{1}{5^{1/4}} \qquad (y^{2/5})^{10} = y^4$$

$$\frac{x^{-1/3}}{x^{-1/2}} = x^{-1/3-(-1/2)} = x^{-1/3+1/2} = x^{1/6}, \quad x > 0$$

8.3 SIMPLIFYING RADICAL EXPRESSIONS

Product and Quotient Rules for Radicals

If $\sqrt[n]{a}$ and $\sqrt[n]{b}$ are real numbers and n is a natural number, then

$$\sqrt[n]{a} \cdot \sqrt[n]{b} = \sqrt[n]{ab}$$

and

$$\sqrt[n]{\frac{a}{b}} = \frac{\sqrt[n]{a}}{\sqrt[n]{b}}, \qquad b \neq 0.$$

$$\sqrt{3} \cdot \sqrt{7} = \sqrt{21}$$
$$\sqrt[5]{x^3 y} \cdot \sqrt[5]{xy^2} = \sqrt[5]{x^4 y^3}$$

$$\frac{\sqrt{x^5}}{\sqrt{x^4}} = \sqrt{\frac{x^5}{x^4}} = \sqrt{x}, \quad x > 0$$

Simplified Radical

1. The radicand has no factor raised to a power greater than or equal to the index.
2. The radicand has no fractions.
3. No denominator contains a radical.
4. Exponents in the radicand and the index of the radical have no common factor (except 1).

$$\sqrt{18} = \sqrt{9 \cdot 2} = 3\sqrt{2}$$
$$\sqrt[3]{54x^5 y^3} = \sqrt[3]{27x^3 y^3 \cdot 2x^2} = 3xy\sqrt[3]{2x^2}$$
$$\sqrt{\frac{7}{4}} = \frac{\sqrt{7}}{\sqrt{4}} = \frac{\sqrt{7}}{2}$$
$$\sqrt[9]{x^3} = x^{3/9} = x^{1/3} \quad \text{or} \quad \sqrt[3]{x}$$

Pythagorean Formula

If c is the length of the longest side of a right triangle and a and b are the lengths of the shorter sides, then $c^2 = a^2 + b^2$. The longest side is the hypotenuse and the two shorter sides are the legs of the triangle. The hypotenuse is opposite the right angle.

Find b for the triangle in the figure.

$$10^2 + b^2 = \left(2\sqrt{61}\right)^2$$
$$b^2 = 4(61) - 100$$
$$b^2 = 144$$
$$b = 12$$

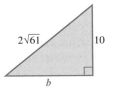

Distance Formula

The distance between (x_1, y_1) and (x_2, y_2) is

$$d = \sqrt{(x_2 - x_1)^2 + (y_2 - y_1)^2}.$$

The distance between $(3, -2)$ and $(-1, 1)$ is

$$\sqrt{(-1 - 3)^2 + [1 - (-2)]^2} = \sqrt{(-4)^2 + 3^2}$$
$$= \sqrt{16 + 9} = \sqrt{25} = 5.$$

(continued)

CONCEPTS	EXAMPLES

8.4 ADDING AND SUBTRACTING RADICAL EXPRESSIONS

Only radical expressions with the same index and the same radicand may be combined.

$$3\sqrt{17} + 2\sqrt{17} - 8\sqrt{17} = (3 + 2 - 8)\sqrt{17}$$
$$= -3\sqrt{17}$$

$$\sqrt[3]{2} - \sqrt[3]{250} = \sqrt[3]{2} - 5\sqrt[3]{2}$$
$$= -4\sqrt[3]{2}$$

$$\left.\begin{array}{l} \sqrt{15} + \sqrt{30} \\ \sqrt{3} + \sqrt[3]{9} \end{array}\right\} \quad \begin{array}{l} \text{cannot be} \\ \text{simplified further} \end{array}$$

8.5 MULTIPLYING AND DIVIDING RADICAL EXPRESSIONS

Multiply binomial radical expressions by using the FOIL method. Special products from Section 5.4 may apply.

$$\left(\sqrt{2} + \sqrt{7}\right)\left(\sqrt{3} - \sqrt{6}\right)$$
$$= \sqrt{6} - 2\sqrt{3} + \sqrt{21} - \sqrt{42} \qquad \sqrt{12} = 2\sqrt{3}$$

$$\left(\sqrt{5} - \sqrt{10}\right)\left(\sqrt{5} + \sqrt{10}\right) = 5 - 10 = -5$$

$$\left(\sqrt{3} - \sqrt{2}\right)^2 = 3 - 2\sqrt{3} \cdot \sqrt{2} + 2 = 5 - 2\sqrt{6}$$

Rationalize the denominator by multiplying both the numerator and denominator by the same expression.

$$\frac{\sqrt{7}}{\sqrt{5}} = \frac{\sqrt{7}}{\sqrt{5}} \cdot \frac{\sqrt{5}}{\sqrt{5}} = \frac{\sqrt{35}}{5}$$

$$\frac{\sqrt[3]{2}}{\sqrt[3]{4}} = \frac{\sqrt[3]{2}}{\sqrt[3]{4}} \cdot \frac{\sqrt[3]{2}}{\sqrt[3]{2}} = \frac{\sqrt[3]{4}}{\sqrt[3]{8}} = \frac{\sqrt[3]{4}}{2}$$

$$\frac{4}{\sqrt{5} - \sqrt{2}} = \frac{4}{\sqrt{5} - \sqrt{2}} \cdot \frac{\sqrt{5} + \sqrt{2}}{\sqrt{5} + \sqrt{2}}$$
$$= \frac{4\left(\sqrt{5} + \sqrt{2}\right)}{5 - 2} = \frac{4\left(\sqrt{5} + \sqrt{2}\right)}{3}$$

$$\frac{5 + 15\sqrt{6}}{10} = \frac{5\left(1 + 3\sqrt{6}\right)}{5 \cdot 2} = \frac{1 + 3\sqrt{6}}{2}$$

8.6 SOLVING EQUATIONS WITH RADICALS

Solving an Equation with Radicals

Step 1 Isolate one radical on one side of the equation.

Step 2 Raise both sides of the equation to a power that is the same as the index of the radical.

Step 3 Solve the resulting equation; if it still contains a radical, repeat Steps 1 and 2.

Step 4 Check all potential solutions in the *original* equation.

Potential solutions that do not check are extraneous; they are not part of the solution set.

Solve $\sqrt{2x + 3} - x = 0$.

$$\sqrt{2x + 3} = x$$
$$\left(\sqrt{2x + 3}\right)^2 = x^2$$
$$2x + 3 = x^2$$
$$x^2 - 2x - 3 = 0$$
$$(x - 3)(x + 1) = 0$$
$$x - 3 = 0 \quad \text{or} \quad x + 1 = 0$$
$$x = 3 \quad \text{or} \qquad x = -1$$

A check shows that 3 is a solution, but -1 is extraneous. The solution set is $\{3\}$.

CONCEPTS	EXAMPLES

8.7 COMPLEX NUMBERS

$i^2 = -1$, so $i = \sqrt{-1}$.

For any positive number b, $\sqrt{-b} = i\sqrt{b}$.

To multiply radicals with negative radicands, first change each factor to the form $i\sqrt{b}$, then multiply. The same procedure applies to quotients.

$$\sqrt{-25} = i\sqrt{25} = 5i$$

$$\sqrt{-3} \cdot \sqrt{-27} = i\sqrt{3} \cdot i\sqrt{27}$$
$$= i^2\sqrt{81}$$
$$= -1 \cdot 9$$
$$= -9$$

$$\frac{\sqrt{-18}}{\sqrt{-2}} = \frac{i\sqrt{18}}{i\sqrt{2}} = \sqrt{\frac{18}{2}} = \sqrt{9} = 3$$

Adding and Subtracting Complex Numbers
Add (or subtract) the real parts and add (or subtract) the imaginary parts.

$$(5 + 3i) + (8 - 7i) = 13 - 4i$$
$$(5 + 3i) - (8 - 7i) = -3 + 10i$$

Multiplying and Dividing Complex Numbers
Multiply complex numbers by using the FOIL method.

$$(2 + i)(5 - 3i) = 10 - 6i + 5i - 3i^2$$
$$= 10 - i - 3(-1)$$
$$= 10 - i + 3$$
$$= 13 - i$$

Divide complex numbers by multiplying the numerator and the denominator by the conjugate of the denominator.

$$\frac{20}{3 + i} = \frac{20(3 - i)}{(3 + i)(3 - i)} = \frac{20(3 - i)}{9 - i^2}$$
$$= \frac{20(3 - i)}{10} = 2(3 - i) = 6 - 2i$$

Answers to Test Your Word Power

1. B; *Example:* In $\sqrt{3xy}$, $3xy$ is the radicand.   **2.** D; *Example:* In a right triangle where $a = 6$, $b = 8$, and $c = 10$, $6^2 + 8^2 = 10^2$.
3. C; *Example:* In a right triangle where the sides measure 9, 12, and 15 units, the hypotenuse is the side opposite the right angle, with measure 15 units.   **4.** B; *Example:* To rationalize the denominator of $\dfrac{5}{\sqrt{3} + 1}$, multiply both the numerator and denominator by $\sqrt{3} - 1$ to get $\dfrac{5(\sqrt{3} - 1)}{2}$.   **5.** A; *Example:* The potential solution 2 is extraneous in $\sqrt{5q - 1} + 3 = 0$.   **6.** C; *Examples:* -5 (or $-5 + 0i$), $7i$ (or $0 + 7i$), $\sqrt{2} - 4i$

CHAPTER 8 REVIEW EXERCISES

[8.1] *Find each root.*

1. $\sqrt{1764}$ **2.** $-\sqrt{289}$ **3.** $\sqrt[3]{216}$

4. $\sqrt[3]{-125}$ **5.** $-\sqrt[3]{27}$ **6.** $\sqrt[5]{-32}$

7. Under what conditions is $\sqrt[n]{a}$ not a real number?

8. Simplify each radical so that no radicals appear. Assume x represents any real number.

(a) $\sqrt{x^2}$ (b) $-\sqrt{x^2}$ (c) $\sqrt[3]{x^3}$

Use a calculator to find a decimal approximation for each number. Give the answer to the nearest thousandth.

9. $-\sqrt{47}$

10. $\sqrt[3]{-129}$

11. $\sqrt[4]{605}$

12. $500^{-3/4}$

13. $-500^{4/3}$

14. $-28^{-1/2}$

Graph each function. Give the domain and the range.

15. $f(x) = \sqrt{x-1}$

16. $f(x) = \sqrt[3]{x} + 4$

17. What is the best estimate of the area of the triangle shown here?

    A. 3600   **B.** 30   **C.** 60   **D.** 360

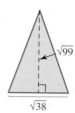

$\sqrt{99}$

$\sqrt{38}$

[8.2]

18. Fill in the blanks with the correct responses: One way to evaluate $8^{2/3}$ is to first find the _____ root of _____, which is _____. Then raise that result to the _____ power, to get an answer of _____. Therefore, $8^{2/3} =$ _____.

19. Which one of the following is a positive number?

    A. $(-27)^{2/3}$ **B.** $(-64)^{5/3}$ **C.** $(-100)^{1/2}$ **D.** $(-32)^{1/5}$

20. If a is a negative number and n is odd, then what must be true about m for $a^{m/n}$ to be

    (a) positive   **(b)** negative?

21. If a is negative and n is even, then what can be said about $a^{1/n}$?

Simplify. If the expression does not represent a real number, say so.

22. $49^{1/2}$

23. $-121^{1/2}$

24. $16^{5/4}$

25. $-8^{2/3}$

26. $-\left(\dfrac{36}{25}\right)^{3/2}$

27. $\left(-\dfrac{1}{8}\right)^{-5/3}$

28. $\left(\dfrac{81}{10,000}\right)^{-3/4}$

29. $(-16)^{3/4}$

30. Solve the Pythagorean formula $a^2 + b^2 = c^2$ for b, where $b > 0$.

31. Explain the relationship between the expressions $a^{m/n}$ and $\sqrt[n]{a^m}$. Give an example.

Write each expression as a radical.

32. $(m + 3n)^{1/2}$

33. $(3a + b)^{-5/3}$

Write each expression with a rational exponent.

34. $\sqrt{7^9}$

35. $\sqrt[5]{p^4}$

Use the rules for exponents to simplify each expression. Write the answer with only positive exponents. Assume all variables represent positive real numbers.

36. $5^{1/4} \cdot 5^{7/4}$

37. $\dfrac{96^{2/3}}{96^{-1/3}}$

38. $\dfrac{(a^{1/3})^4}{a^{2/3}}$

39. $\dfrac{y^{-1/3} \cdot y^{5/6}}{y}$

40. $\left(\dfrac{z^{-1}x^{-3/5}}{2^{-2}z^{-1/2}x}\right)^{-1}$

41. $r^{-1/2}(r + r^{3/2})$

Simplify by first writing each radical in exponential form. Leave the answer in exponential form. Assume all variables represent positive real numbers.

42. $\sqrt[8]{s^4}$

43. $\sqrt[6]{r^9}$

44. $\dfrac{\sqrt{p^5}}{p^2}$

45. $\sqrt[4]{k^3} \cdot \sqrt{k^3}$

46. $\sqrt[3]{m^5} \cdot \sqrt[3]{m^8}$

47. $\sqrt[4]{\sqrt[3]{z}}$

48. $\sqrt{\sqrt{\sqrt{x}}}$

49. $\sqrt[3]{\sqrt[5]{x}}$

50. $\sqrt{\sqrt[6]{\sqrt[3]{x}}}$

✍ **51.** By the product rule for exponents, we know that $2^{1/4} \cdot 2^{1/5} = 2^{9/20}$. However, there is no exponent rule to simplify $3^{1/4} \cdot 2^{1/5}$. Why?

[8.3]   *Simplify each radical. Assume all variables represent positive real numbers.*

52. $\sqrt{6} \cdot \sqrt{11}$

53. $\sqrt{5} \cdot \sqrt{r}$

54. $\sqrt[3]{6} \cdot \sqrt[3]{5}$

55. $\sqrt[4]{7} \cdot \sqrt[4]{3}$

56. $\sqrt{20}$

57. $\sqrt{75}$

58. $-\sqrt{125}$

59. $\sqrt[3]{-108}$

60. $\sqrt{100y^7}$

61. $\sqrt[3]{64p^4q^6}$

62. $\sqrt[3]{108a^8b^5}$

63. $\sqrt[3]{632r^8t^4}$

64. $\sqrt{\dfrac{y^3}{144}}$

65. $\sqrt[3]{\dfrac{m^{15}}{27}}$

66. $\sqrt[3]{\dfrac{r^2}{8}}$

67. $\sqrt[4]{\dfrac{a^9}{81}}$

Simplify each radical expression.

68. $\sqrt[6]{15^3}$

69. $\sqrt[4]{p^6}$

70. $\sqrt[3]{2} \cdot \sqrt[4]{5}$

71. $\sqrt{x} \cdot \sqrt[5]{x}$

72. Find the missing length in the right triangle. Simplify the answer if applicable.

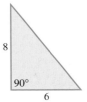

73. Find the distance between the points $(-4, 7)$ and $(10, 6)$.

[8.4]   *Perform the indicated operations. Assume all variables represent positive real numbers.*

74. $2\sqrt{8} - 3\sqrt{50}$

75. $8\sqrt{80} - 3\sqrt{45}$

76. $-\sqrt{27y} + 2\sqrt{75y}$

77. $2\sqrt{54m^3} + 5\sqrt{96m^3}$

78. $3\sqrt[3]{54} + 5\sqrt[3]{16}$

79. $-6\sqrt[4]{32} + \sqrt[4]{512}$

80. $\dfrac{3}{\sqrt{16}} - \dfrac{\sqrt{5}}{2}$

81. $\dfrac{4}{\sqrt{25}} + \dfrac{\sqrt{5}}{4}$

In Exercises 82 and 83, leave answers as simplified radicals.

82. Find the perimeter of a rectangular electronic billboard having sides of lengths shown in the figure.

83. Find the perimeter of a triangular electronic highway road sign having the dimensions shown in the figure.

[8.5] *Multiply.*

84. $\left(\sqrt{3} + 1\right)\left(\sqrt{3} - 2\right)$ **85.** $\left(\sqrt{7} + \sqrt{5}\right)\left(\sqrt{7} - \sqrt{5}\right)$

86. $\left(3\sqrt{2} + 1\right)\left(2\sqrt{2} - 3\right)$ **87.** $\left(\sqrt{13} - \sqrt{2}\right)^2$

88. $\left(\sqrt[3]{2} + 3\right)\left(\sqrt[3]{4} - 3\sqrt[3]{2} + 9\right)$ **89.** $\left(\sqrt[3]{4y} - 1\right)\left(\sqrt[3]{4y} + 3\right)$

90. Use a calculator to show that the answer to Exercise 87, $15 - 2\sqrt{26}$, is not equal to $13\sqrt{26}$.

91. A friend wants to rationalize the denominator of the fraction $\dfrac{5}{\sqrt[3]{6}}$, and she decides to multiply the numerator and denominator by $\sqrt[3]{6}$. Why will her plan *not* work?

Rationalize each denominator. Assume all variables represent positive real numbers.

92. $\dfrac{\sqrt{6}}{\sqrt{5}}$ **93.** $\dfrac{-6\sqrt{3}}{\sqrt{2}}$ **94.** $\dfrac{3\sqrt{7p}}{\sqrt{y}}$ **95.** $\sqrt{\dfrac{11}{8}}$

96. $-\sqrt[3]{\dfrac{9}{25}}$ **97.** $\sqrt[3]{\dfrac{108m^3}{n^5}}$ **98.** $\dfrac{1}{\sqrt{2} + \sqrt{7}}$ **99.** $\dfrac{-5}{\sqrt{6} - 3}$

Write in lowest terms.

100. $\dfrac{2 - 2\sqrt{5}}{8}$ **101.** $\dfrac{4 - 8\sqrt{8}}{12}$ **102.** $\dfrac{-18 + \sqrt{27}}{6}$

[8.6] *Solve each equation.*

103. $\sqrt{8x + 9} = 5$ **104.** $\sqrt{2z - 3} - 3 = 0$

105. $\sqrt{3m + 1} - 2 = -3$ **106.** $\sqrt{7z + 1} = z + 1$

107. $3\sqrt{m} = \sqrt{10m - 9}$ **108.** $\sqrt{p^2 + 3p + 7} = p + 2$

109. $\sqrt{a + 2} - \sqrt{a - 3} = 1$ **110.** $\sqrt[3]{5m - 1} = \sqrt[3]{3m - 2}$

111. $\sqrt[3]{2x^2 + 3x - 7} = \sqrt[3]{2x^2 + 4x + 6}$ **112.** $\sqrt[3]{3y^2 - 4y + 6} = \sqrt[3]{3y^2 - 2y + 8}$

113. $\sqrt[3]{1 - 2k} - \sqrt[3]{-k - 13} = 0$ **114.** $\sqrt[3]{11 - 2t} - \sqrt[3]{-1 - 5t} = 0$

115. $\sqrt[4]{x - 1} + 2 = 0$ **116.** $\sqrt[4]{2k + 3} + 1 = 0$

117. $\sqrt[4]{x + 7} = \sqrt[4]{2x}$ **118.** $\sqrt[4]{x + 8} = \sqrt[4]{3x}$

RELATING CONCEPTS (EXERCISES 119–125)

For Individual or Group Work

Solve the equations in Exercises 119–124 in order, and then use a generalization to fill in the blanks in Exercise 125.

119. $x = 3$ **120.** $x = -3$ **121.** $x^2 = 9$

122. $x^3 = 27$ **123.** $x^4 = 81$ **124.** $x^5 = -243$

125. Suppose both sides of $x = k$ are raised to the nth power.

  (a) If n is even, the number of solutions of the new equation is
  _____ the number of solutions of the original
  (more than/the same as/fewer than)
  equation.

  (b) If n is odd, the number of solutions of the new equation is
  _____ the number of solutions of the original
  (more than/the same as/fewer than)
  equation.

126. Carpenters stabilize wall frames with a diagonal brace, as shown in the figure. The length of the brace is given by $L = \sqrt{H^2 + W^2}$.

(a) Solve this formula for H.

(b) If the bottom of the brace is attached 9 ft from the corner and the brace is 12 ft long, how far up the corner post should it be nailed? Give your answer to the nearest tenth of a foot.

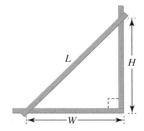

[8.7] *Write each expression as a product of a real number and i.*

127. $\sqrt{-25}$ **128.** $\sqrt{-200}$

129. If a is a positive real number, is $-\sqrt{-a}$ a real number?

Perform the indicated operations. Write each imaginary number answer in standard form $a + bi$.

130. $(-2 + 5i) + (-8 - 7i)$ **131.** $(5 + 4i) - (-9 - 3i)$ **132.** $\sqrt{-5} \cdot \sqrt{-7}$

133. $\sqrt{-25} \cdot \sqrt{-81}$ **134.** $\dfrac{\sqrt{-72}}{\sqrt{-8}}$ **135.** $(2 + 3i)(1 - i)$

136. $(6 - 2i)^2$ **137.** $\dfrac{3 - i}{2 + i}$ **138.** $\dfrac{5 + 14i}{2 + 3i}$

Find each power of i.

139. i^{11} **140.** i^{36} **141.** i^{-10} **142.** i^{-8}

MIXED REVIEW EXERCISES

Simplify. Assume all variables represent positive real numbers.

143. $-\sqrt[4]{256}$ **144.** $1000^{-2/3}$ **145.** $\dfrac{z^{-1/5} \cdot z^{3/10}}{z^{7/10}}$

146. $\sqrt[4]{k^{24}}$ **147.** $\sqrt[3]{54z^9 t^8}$ **148.** $-5\sqrt{18} + 12\sqrt{72}$

149. $8\sqrt[3]{x^3 y^2} - 2x\sqrt[3]{y^2}$ **150.** $\left(\sqrt{5} - \sqrt{3}\right)\left(\sqrt{7} + \sqrt{3}\right)$

151. $\dfrac{-1}{\sqrt{12}}$ **152.** $\sqrt[3]{\dfrac{12}{25}}$ **153.** i^{-1000}

154. $\sqrt{-49}$ **155.** $(4 - 9i) + (-1 + 2i)$ **156.** $\dfrac{\sqrt{50}}{\sqrt{-2}}$

157. $\dfrac{3 + \sqrt{54}}{6}$ **158.** $(3 + 2i)^2$

Solve each equation.

159. $\sqrt{x + 4} = x - 2$ **160.** $\sqrt[3]{2x - 9} = \sqrt[3]{5x + 3}$

161. $\sqrt{6 + 2x} - 1 = \sqrt{7 - 2x}$ **162.** $\sqrt{7x + 11} - 5 = 0$

163. $\sqrt{6x + 2} - \sqrt{5x + 3} = 0$ **164.** $\sqrt{3 + 5x} - \sqrt{x + 11} = 0$

165. $3\sqrt{x} = \sqrt{8x + 9}$ **166.** $6\sqrt{p} = \sqrt{30p + 24}$

167. $\sqrt{11 + 2x} + 1 = \sqrt{5x + 1}$ **168.** $\sqrt{5x + 6} - \sqrt{x + 3} = 3$

CHAPTER 8 TEST

Evaluate.

1. $-\sqrt{841}$

2. $\sqrt[3]{-512}$

3. $125^{1/3}$

4. For $\sqrt{146.25}$, which choice gives the best estimate?
   A. 10   **B.** 11   **C.** 12   **D.** 13

Use a calculator to approximate each root to the nearest thousandth.

5. $\sqrt{478}$

6. $\sqrt[3]{-832}$

7. Graph the function defined by $f(x) = \sqrt{x + 6}$, and give the domain and range.

Simplify each expression. Assume all variables represent positive real numbers.

8. $\left(\dfrac{16}{25}\right)^{-3/2}$

9. $(-64)^{-4/3}$

10. $\dfrac{3^{2/5}x^{-1/4}y^{2/5}}{3^{-8/5}x^{7/4}y^{1/10}}$

11. $\left(\dfrac{x^{-4}y^{-6}}{x^{-2}y^{3}}\right)^{-2/3}$

12. $7^{3/4} \cdot 7^{-1/4}$

13. $\sqrt[3]{a^4} \cdot \sqrt[3]{a^7}$

14. Use the Pythagorean formula to find the exact length of side b in the figure.

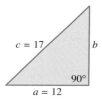

15. Find the distance between the points $(-4, 2)$ and $(2, 10)$.

Simplify each expression. Assume all variables represent positive real numbers.

16. $\sqrt{54x^5y^6}$

17. $\sqrt[4]{32a^7b^{13}}$

18. $\sqrt{2} \cdot \sqrt[3]{5}$ (Express as a radical.)

19. $3\sqrt{20} - 5\sqrt{80} + 4\sqrt{500}$

20. $\sqrt[3]{16t^3s^5} - \sqrt[3]{54t^6s^2}$

21. $\left(7\sqrt{5} + 4\right)\left(2\sqrt{5} - 1\right)$

22. $\left(\sqrt{3} - 2\sqrt{5}\right)^2$

23. $\dfrac{-5}{\sqrt{40}}$

24. $\dfrac{2}{\sqrt[3]{5}}$

25. $\dfrac{-4}{\sqrt{7} + \sqrt{5}}$

26. Write $\dfrac{6 + \sqrt{24}}{2}$ in lowest terms.

27. The following formula is used in physics, relating the velocity of sound V to the temperature T.

$$V = \dfrac{V_0}{\sqrt{1 - kT}}$$

   (a) Find an approximation of V to the nearest tenth if $V_0 = 50$, $k = .01$, and $T = 30$. Use a calculator.

   (b) Solve the formula for T.

Solve each equation.

28. $\sqrt[3]{5x} = \sqrt[3]{2x - 3}$

29. $x + \sqrt{x + 6} = 9 - x$

30. $\sqrt{x + 4} - \sqrt{1 - x} = -1$

Perform the indicated operations. Express the answers in standard form $a + bi$.

31. $(-2 + 5i) - (3 + 6i) - 7i$

32. $(1 + 5i)(3 + i)$

33. $\dfrac{7 + i}{1 - i}$

34. Simplify i^{37}.

35. Answer *true* or *false* to each of the following.

    (a) $i^2 = -1$ **(b)** $i = \sqrt{-1}$ **(c)** $i = -1$ **(d)** $\sqrt{-3} = i\sqrt{3}$

CUMULATIVE REVIEW EXERCISES CHAPTERS **1–8**

Evaluate each expression if $a = -3$, $b = 5$, and $c = -4$.

1. $|2a^2 - 3b + c|$

2. $\dfrac{(a + b)(a + c)}{3b - 6}$

Solve each equation.

3. $3(x + 2) - 4(2x + 3) = -3x + 2$

4. $\dfrac{1}{3}x + \dfrac{1}{4}(x + 8) = x + 7$

5. $.04x + .06(100 - x) = 5.88$

6. $|6x + 7| = 13$

7. $|-2x + 4| = |-2x - 3|$

8. Find the solution set of $-5 - 3(m - 2) < 11 - 2(m + 2)$. Give it using interval notation.

9. Find the measures of the marked angles.

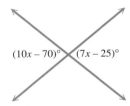

$(10x - 70)°$ $(7x - 25)°$

Solve each problem.

10. A piggy bank has 50 coins, all of which are nickels and quarters. The total value of the money is $8.90. How many of each denomination are there in the bank?

11. How many liters of pure alcohol must be mixed with 40 L of 18% alcohol to obtain a 22% alcohol solution?

12. Graph the equation $4x - 3y = 12$.

13. Find the slope of the line passing through the points $(-4, 6)$ and $(2, -3)$. Then find the equation of the line and write it in the form $y = mx + b$.

14. If $f(x) = 3x - 7$, find $f(-10)$.

15. Solve the system by elimination or substitution.

$$3x - y = 23$$
$$2x + 3y = 8$$

16. Solve the system by matrix methods.

$$x + y + z = 1$$
$$x - y - z = -3$$
$$x + y - z = -1$$

Solve the problem by using a system of equations.

17. In 1997, if you had sent five 2-oz letters and three 3-oz letters using first-class mail, it would have cost you $5.09. Sending three 2-oz letters and five 3-oz letters would have cost $5.55. What was the 1997 postage rate for one 2-oz letter and for one 3-oz letter? (*Source:* U.S. Postal Service.)

Perform the indicated operations.

18. $(3k^3 - 5k^2 + 8k - 2) - (4k^3 + 11k + 7) + (2k^2 - 5k)$

19. $(8x - 7)(x + 3)$

20. $\dfrac{8z^3 - 16z^2 + 24z}{8z^2}$

21. $\dfrac{6y^4 - 3y^3 + 5y^2 + 6y - 9}{2y + 1}$

Factor each polynomial completely.

22. $2p^2 - 5pq + 3q^2$

23. $3k^4 + k^2 - 4$

24. $x^3 + 512$

Solve by factoring.

25. $2x^2 + 11x + 15 = 0$

26. $5t(t - 1) = 2(1 - t)$

27. What is the domain of $f(x) = \dfrac{2}{x^2 - 9}$?

Perform each operation and express the answer in lowest terms.

28. $\dfrac{y^2 + y - 12}{y^3 + 9y^2 + 20y} \div \dfrac{y^2 - 9}{y^3 + 3y^2}$

29. $\dfrac{1}{x + y} + \dfrac{3}{x - y}$

Simplify each complex fraction.

30. $\dfrac{\dfrac{-6}{x - 2}}{\dfrac{8}{3x - 6}}$

31. $\dfrac{\dfrac{1}{a} - \dfrac{1}{b}}{\dfrac{a}{b} - \dfrac{b}{a}}$

32. $\dfrac{x^{-1}}{y - x^{-1}}$

33. Natalie can ride her bike 4 mph faster than her husband, Chuck. If Natalie can ride 48 mi in the same time that Chuck can ride 24 mi, what are their speeds?

34. Solve the equation $\dfrac{p+1}{p-3} = \dfrac{4}{p-3} + 6$.

Write each expression in simplest form, using only positive exponents. Assume all variables represent positive real numbers.

35. $27^{-2/3}$

36. $\sqrt{200x^4}$

37. $\sqrt[3]{16x^2y} \cdot \sqrt[3]{3x^3y}$

38. $\sqrt{50} + \sqrt{8}$

39. $\dfrac{1}{\sqrt{10} - \sqrt{8}}$

40. $\left(2\sqrt{x} + \sqrt{y}\right)\left(-3\sqrt{x} - 4\sqrt{y}\right)$

41. Find the distance between the points $(-4, 4)$ and $(-2, 9)$.

42. Solve the equation $\sqrt{3r - 8} = r - 2$.

43. The *fall speed,* in miles per hour, of a vehicle running off the road into a ditch is given by

$$S = \frac{2.74D}{\sqrt{h}},$$

where D is the horizontal distance traveled from the level surface to the bottom of the ditch and h is the height (or depth) of the ditch. What is the fall speed of a vehicle that traveled 32 ft horizontally into a 5-ft-deep ditch?

Write in standard form $a + bi$.

44. $(5 + 7i) - (3 - 2i)$

45. $\dfrac{6 - 2i}{1 - i}$

Quadratic Equations and Inequalities

In recent years, the number of publicly traded U.S. companies filing for bankruptcy has been at its highest level since the recession of the early 1990s. One casualty of this trend was retailer Montgomery Ward. Started in 1872 as a mail-order catalog business, the company grew to include 250 stores in 30 states. After filing for Chapter 11 bankruptcy protection, the retailer closed for good in 2001. (*Source: USA Today,* December 29, 2000.)

Since then, other high-profile U.S. companies including Enron Corp., Adelphia Communications Corp., Kmart Corp., WorldCom Inc., US Airways, and United Airlines have become victims of this trend. In Example 6 of Section 9.4, we use a *quadratic function* to model the number of company bankruptcy filings.

9.1 The Square Root Property and Completing the Square

We introduced quadratic equations in Section 6.5. Recall that a *quadratic equation* is defined as follows.

Quadratic Equation

An equation that can be written in the form
$$ax^2 + bx + c = 0,$$
where a, b, and c are real numbers, with $a \neq 0$, is a **quadratic equation.** The given form is called **standard form.**

A quadratic equation is a *second-degree equation,* that is, an equation with a squared term and no terms of higher degree. For example,
$$4m^2 + 4m - 5 = 0 \qquad \text{and} \qquad 3x^2 = 4x - 8$$
are quadratic equations, with the first equation in standard form.

OBJECTIVE 1   Review the zero-factor property.   In Section 6.5 we used factoring and the zero-factor property to solve quadratic equations.

Zero-Factor Property

If two numbers have a product of 0, then at least one of the numbers must be 0. That is, if $ab = 0$, then $a = 0$ or $b = 0$.

We solved a quadratic equation such as $3x^2 - 5x - 28 = 0$ using the zero-factor property as follows.

$$3x^2 - 5x - 28 = 0$$
$$(3x + 7)(x - 4) = 0 \qquad \text{Factor.}$$
$$3x + 7 = 0 \quad \text{or} \quad x - 4 = 0 \qquad \text{Zero-factor property}$$
$$3x = -7 \quad \text{or} \qquad x = 4 \qquad \text{Solve each equation.}$$
$$x = -\frac{7}{3}$$

The solution set is $\left\{-\frac{7}{3}, 4\right\}$.

OBJECTIVE 2   Learn the square root property.   Although factoring is the simplest way to solve quadratic equations, not every quadratic equation can be solved easily by factoring. In this section and the next, we develop three other methods of solving quadratic equations based on the following property.

Square Root Property

If x and k are complex numbers and $x^2 = k$, then
$$x = \sqrt{k} \quad \text{or} \quad x = -\sqrt{k}.$$

The following steps justify the square root property.

$$x^2 = k$$
$$x^2 - k = 0 \qquad \text{Subtract } k.$$
$$\left(x - \sqrt{k}\right)\left(x + \sqrt{k}\right) = 0 \qquad \text{Factor.}$$
$$x - \sqrt{k} = 0 \quad \text{or} \quad x + \sqrt{k} = 0 \qquad \text{Zero-factor property}$$
$$x = \sqrt{k} \quad \text{or} \quad x = -\sqrt{k} \qquad \text{Solve each equation.}$$

CAUTION Remember that if $k \neq 0$, using the square root property always produces *two* square roots, one positive and one negative.

EXAMPLE 1 Using the Square Root Property

Solve each equation.

(a) $x^2 = 5$

By the square root property,
$$x = \sqrt{5} \quad \text{or} \quad x = -\sqrt{5},$$
and the solution set is $\left\{\sqrt{5}, -\sqrt{5}\right\}$.

(b) $4x^2 - 48 = 0$

Solve for x^2.

$$4x^2 - 48 = 0$$
$$4x^2 = 48 \qquad \text{Add 48.}$$
$$x^2 = 12 \qquad \text{Divide by 4.}$$
$$x = \sqrt{12} \quad \text{or} \quad x = -\sqrt{12} \qquad \text{Square root property}$$
$$x = 2\sqrt{3} \quad \text{or} \quad x = -2\sqrt{3} \qquad \sqrt{12} = \sqrt{4} \cdot \sqrt{3} = 2\sqrt{3}$$

The solutions are $2\sqrt{3}$ and $-2\sqrt{3}$. Check each in the original equation.

Check: $\qquad\qquad 4x^2 - 48 = 0 \qquad$ Original equation

$$4\left(2\sqrt{3}\right)^2 - 48 = 0 \quad ? \qquad\qquad 4\left(-2\sqrt{3}\right)^2 - 48 = 0 \quad ?$$
$$4(12) - 48 = 0 \quad ? \qquad\qquad 4(12) - 48 = 0 \quad ?$$
$$48 - 48 = 0 \quad ? \qquad\qquad 48 - 48 = 0 \quad ?$$
$$0 = 0 \quad \text{True} \qquad\qquad 0 = 0 \quad \text{True}$$

The solution set is $\left\{2\sqrt{3}, -2\sqrt{3}\right\}$.

Now Try Exercises 7 and 13.

NOTE Recall that solutions such as those in Example 1 are sometimes abbreviated with the symbol $\pm$ (read "positive or negative"); with this symbol the solutions in Example 1 would be written $\pm\sqrt{5}$ and $\pm 2\sqrt{3}$.

EXAMPLE 2 Using the Square Root Property in an Application

Galileo Galilei (1564–1642) developed a formula for freely falling objects described by

$$d = 16t^2,$$

where d is the distance in feet that an object falls (disregarding air resistance) in t sec, regardless of weight. Galileo dropped objects from the Leaning Tower of Pisa to develop this formula. If the Leaning Tower is about 180 ft tall, use Galileo's formula to determine how long it would take an object dropped from the tower to fall to the ground. (*Source: Microsoft Encarta Encyclopedia 2002.*)

We substitute 180 for d in Galileo's formula.

$$d = 16t^2$$
$$180 = 16t^2 \qquad \text{Let } d = 180.$$
$$11.25 = t^2 \qquad \text{Divide by 16.}$$
$$t = \sqrt{11.25} \quad \text{or} \quad t = -\sqrt{11.25} \qquad \text{Square root property}$$

Since time cannot be negative, we discard the negative solution. In applied problems, we usually prefer approximations to exact values. Using a calculator, $\sqrt{11.25} \approx 3.4$ so $t \approx 3.4$. The object would fall to the ground in about 3.4 sec.

Now Try Exercise 25.

OBJECTIVE 3 Solve quadratic equations of the form $(ax + b)^2 = c$ by using the square root property. To solve more complicated equations using the square root property, such as

$$(x - 5)^2 = 36,$$

substitute $(x - 5)^2$ for x^2 and 36 for k in the square root property to obtain

$$x - 5 = \sqrt{36} \quad \text{or} \quad x - 5 = -\sqrt{36}$$
$$x - 5 = 6 \quad \text{or} \quad x - 5 = -6$$
$$x = 11 \quad \text{or} \quad x = -1.$$

Check: $(x - 5)^2 = 36$ \qquad Original equation

$$(11 - 5)^2 = 36 \quad ? \qquad\qquad (-1 - 5)^2 = 36 \quad ?$$
$$6^2 = 36 \quad ? \qquad\qquad (-6)^2 = 36 \quad ?$$
$$36 = 36 \quad \text{True} \qquad\qquad 36 = 36 \quad \text{True}$$

Since both solutions satisfy the original equation, the solution set is $\{-1, 11\}$.

EXAMPLE 3 Using the Square Root Property

Solve $(2x - 3)^2 = 18$.

$$2x - 3 = \sqrt{18} \quad \text{or} \quad 2x - 3 = -\sqrt{18} \qquad \text{Square root property}$$
$$2x = 3 + \sqrt{18} \quad \text{or} \quad 2x = 3 - \sqrt{18}$$
$$x = \frac{3 + \sqrt{18}}{2} \quad \text{or} \quad x = \frac{3 - \sqrt{18}}{2}$$
$$x = \frac{3 + 3\sqrt{2}}{2} \quad \text{or} \quad x = \frac{3 - 3\sqrt{2}}{2} \qquad \sqrt{18} = \sqrt{9 \cdot 2} = 3\sqrt{2}$$

We show the check for the first solution. The check for the second solution is similar.

Check:

$$(2x - 3)^2 = 18 \qquad \text{Original equation}$$

$$\left[2\left(\frac{3 + 3\sqrt{2}}{2}\right) - 3\right]^2 = 18 \quad ?$$

$$\left(3 + 3\sqrt{2} - 3\right)^2 = 18 \quad ?$$

$$\left(3\sqrt{2}\right)^2 = 18 \quad ?$$

$$18 = 18 \qquad \text{True}$$

The solution set is $\left\{\dfrac{3 + 3\sqrt{2}}{2}, \dfrac{3 - 3\sqrt{2}}{2}\right\}$.

Now Try Exercise 21.

OBJECTIVE 4 Solve quadratic equations by completing the square. We can use the square root property to solve *any* quadratic equation by writing it in the form $(x + k)^2 = n$. That is, we must write the left side of the equation as a perfect square trinomial that can be factored as $(x + k)^2$, the square of a binomial, and the right side must be a constant. Rewriting a quadratic equation in this form is called **completing the square.**

Recall that the perfect square trinomial

$$x^2 + 10x + 25$$

can be factored as $(x + 5)^2$. In the trinomial, the coefficient of x (the first-degree term) is 10 and the constant term is 25. Notice that if we take half of 10 and square it, we get the constant term, 25.

$$\underset{\text{Coefficient of } x}{} \qquad \underset{\text{Constant}}{}$$

$$\left[\frac{1}{2}(10)\right]^2 = 5^2 = 25$$

Similarly, in

$$x^2 + 12x + 36, \qquad \left[\frac{1}{2}(12)\right]^2 = 6^2 = 36,$$

and in

$$m^2 - 6m + 9, \qquad \left[\frac{1}{2}(-6)\right]^2 = (-3)^2 = 9.$$

This relationship is true in general and is the idea behind completing the square.

EXAMPLE 4 Solving a Quadratic Equation by Completing the Square

Solve $x^2 + 8x + 10 = 0$.

This quadratic equation cannot be solved by factoring, and it is not in the correct form to solve using the square root property. To solve it by completing the square, we need a perfect square trinomial on the left side of the equation. To get this form, we first subtract 10 from each side.

$$x^2 + 8x + 10 = 0$$

$$x^2 + 8x = -10 \qquad \text{Subtract 10.}$$

We must add a constant to get a perfect square trinomial on the left.

$$x^2 + 8x + \underbrace{\quad ? \quad}$$

Needs to be a perfect
square trinomial

To find this constant, we apply the ideas preceding this example—we take half the coefficient of the first-degree term and square the result.

$$\left[\frac{1}{2}(8)\right]^2 = 4^2 = 16 \leftarrow \text{Desired constant}$$

We add this constant, 16, to *each* side of the equation. (Why?)

$$x^2 + 8x + 16 = -10 + 16$$

Now we factor the perfect square trinomial on the left and add on the right.

$$(x + 4)^2 = 6$$

We can solve this equation using the square root property.

$$x + 4 = \sqrt{6} \qquad \text{or} \quad x + 4 = -\sqrt{6}$$

$$x = -4 + \sqrt{6} \quad \text{or} \qquad x = -4 - \sqrt{6}$$

Check:
$$x^2 + 8x + 10 = 0 \qquad \text{Original equation}$$

$$\left(-4 + \sqrt{6}\right)^2 + 8\left(-4 + \sqrt{6}\right) + 10 = 0 \qquad ? \qquad \text{Let } x = -4 + \sqrt{6}.$$

$$16 - 8\sqrt{6} + 6 - 32 + 8\sqrt{6} + 10 = 0 \qquad ?$$

$$0 = 0 \qquad \text{True}$$

The check for the second solution is similar. The solution set is

$$\left\{-4 + \sqrt{6}, -4 - \sqrt{6}\right\}.$$

Now Try Exercise 39.

The procedure from Example 4 can be generalized.

Completing the Square

To solve $ax^2 + bx + c = 0$ $(a \neq 0)$ by completing the square, use these steps.

Step 1 **Be sure the squared term has coefficient 1.** If the coefficient of the squared term is 1, proceed to Step 2. If the coefficient of the squared term is not 1 but some other nonzero number a, divide each side of the equation by a.

Step 2 **Put the equation in correct form.** Rewrite so that terms with variables are on one side of the equals sign and the constant is on the other side.

Step 3 **Square half the coefficient of the first-degree term.**

Step 4   **Add the square to each side.**

Step 5   **Factor the perfect square trinomial.** One side should now be a perfect square trinomial. Factor it as the square of a binomial. Simplify the other side.

Step 6   **Solve the equation.** Apply the square root property to complete the solution.

NOTE   Steps 1 and 2 can be done in either order. With some equations, it is more convenient to do Step 2 first.

▉ **EXAMPLE 5   Solving a Quadratic Equation with *a* = 1 by Completing the Square**

Solve $k^2 + 5k - 1 = 0$.

Follow the steps in the box. Since the coefficient of the squared term is 1, begin with Step 2.

Step 2 $\qquad\qquad k^2 + 5k = 1$ $\qquad$ Add 1 to each side.

Step 3   Take half the coefficient of the first-degree term and square the result.

$$\left[\frac{1}{2}(5)\right]^2 = \left(\frac{5}{2}\right)^2 = \frac{25}{4}$$

Step 4   Add the square to each side of the equation to get

$$k^2 + 5k + \frac{25}{4} = 1 + \frac{25}{4}.$$

Step 5 $\qquad\qquad \left(k + \frac{5}{2}\right)^2 = \frac{29}{4}$ $\qquad$ Factor on the left; add on the right.

Step 6 $k + \dfrac{5}{2} = \sqrt{\dfrac{29}{4}}$ $\qquad$ or $\qquad$ $k + \dfrac{5}{2} = -\sqrt{\dfrac{29}{4}}$ $\qquad$ Square root property

$\qquad k + \dfrac{5}{2} = \dfrac{\sqrt{29}}{2}$ $\qquad$ or $\qquad$ $k + \dfrac{5}{2} = -\dfrac{\sqrt{29}}{2}$

$\qquad\qquad k = -\dfrac{5}{2} + \dfrac{\sqrt{29}}{2}$ $\qquad$ or $\qquad$ $k = -\dfrac{5}{2} - \dfrac{\sqrt{29}}{2}$

$\qquad\qquad k = \dfrac{-5 + \sqrt{29}}{2}$ $\qquad$ or $\qquad$ $k = \dfrac{-5 - \sqrt{29}}{2}$

Check that the solution set is $\left\{\dfrac{-5 + \sqrt{29}}{2}, \dfrac{-5 - \sqrt{29}}{2}\right\}$.

Now Try Exercise 41.

■ **EXAMPLE 6** Solving a Quadratic Equation with $a \neq 1$ by Completing the Square

Solve $2x^2 - 4x - 5 = 0$.

First divide each side of the equation by 2 to get 1 as the coefficient of the squared term.

$$x^2 - 2x - \frac{5}{2} = 0 \qquad \text{Step 1}$$

$$x^2 - 2x = \frac{5}{2} \qquad \text{Step 2}$$

$$\left[\frac{1}{2}(-2) \right]^2 = (-1)^2 = 1 \qquad \text{Step 3}$$

$$x^2 - 2x + 1 = \frac{5}{2} + 1 \qquad \text{Step 4}$$

$$(x - 1)^2 = \frac{7}{2} \qquad \text{Step 5}$$

$$x - 1 = \sqrt{\frac{7}{2}} \quad \text{or} \quad x - 1 = -\sqrt{\frac{7}{2}} \qquad \text{Step 6}$$

$$x = 1 + \sqrt{\frac{7}{2}} \quad \text{or} \quad x = 1 - \sqrt{\frac{7}{2}}$$

$$x = 1 + \frac{\sqrt{14}}{2} \quad \text{or} \quad x = 1 - \frac{\sqrt{14}}{2} \qquad \text{Rationalize denominators.}$$

Add the two terms in each solution as follows.

$$1 + \frac{\sqrt{14}}{2} = \frac{2}{2} + \frac{\sqrt{14}}{2} = \frac{2 + \sqrt{14}}{2}$$

$$1 - \frac{\sqrt{14}}{2} = \frac{2}{2} - \frac{\sqrt{14}}{2} = \frac{2 - \sqrt{14}}{2}.$$

Check that the solution set is $\left\{ \dfrac{2 + \sqrt{14}}{2}, \dfrac{2 - \sqrt{14}}{2} \right\}$.

Now Try Exercise 47.

OBJECTIVE 5 Solve quadratic equations with solutions that are not real numbers. So far, all the equations we have solved using the square root property have had two real solutions. In the equation $x^2 = k$, if $k < 0$, there will be two imaginary solutions.

■ **EXAMPLE 7** Solving Quadratic Equations with Imaginary Solutions

Solve each equation.

(a) $x^2 = -15$

$$x = \sqrt{-15} \quad \text{or} \quad x = -\sqrt{-15} \qquad \text{Square root property}$$

$$x = i\sqrt{15} \quad \text{or} \quad x = -i\sqrt{15} \qquad \sqrt{-1} = i$$

The solution set is $\left\{ i\sqrt{15}, -i\sqrt{15} \right\}$.

(b) $(t + 2)^2 = -16$

$$t + 2 = \sqrt{-16} \quad \text{or} \quad t + 2 = -\sqrt{-16} \qquad \text{Square root property}$$
$$t + 2 = 4i \quad \text{or} \quad t + 2 = -4i \qquad \sqrt{-16} = 4i$$
$$t = -2 + 4i \quad \text{or} \quad t = -2 - 4i$$

The solution set is $\{-2 + 4i, -2 - 4i\}$.

(c) $x^2 + 2x + 7 = 0$

Solve by completing the square.

$$x^2 + 2x = -7 \qquad \text{Subtract 7.}$$
$$x^2 + 2x + 1 = -7 + 1 \qquad \left[\tfrac{1}{2}(2)\right]^2 = 1; \text{ add 1 to each side.}$$
$$(x + 1)^2 = -6 \qquad \text{Factor on the left; add on the right.}$$
$$x + 1 = \pm i\sqrt{6} \qquad \text{Square root property}$$
$$x = -1 \pm i\sqrt{6} \qquad \text{Subtract 1.}$$

The solution set is $\left\{-1 + i\sqrt{6}, -1 - i\sqrt{6}\right\}$.

Now Try Exercises 55, 57, and 61.

NOTE The procedure for completing the square is also used in other areas of mathematics. For example, we use it in Section 10.3 when we graph quadratic equations and again in Chapter 13 when we work with circles.

9.1 EXERCISES

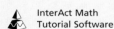

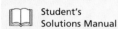
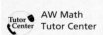

1. A student was asked to solve the quadratic equation $x^2 = 16$ and did not get full credit for the solution set $\{4\}$. Why?

2. Why can't the zero-factor property be used to solve every quadratic equation?

3. Give a one-sentence description or explanation of each phrase.

    (a) Quadratic equation in standard form   **(b)** Zero-factor property
    (c) Square root property

4. What is wrong with the following "solution"?

$$x^2 - x - 2 = 5$$
$$(x - 2)(x + 1) = 5$$
$$x - 2 = 5 \quad \text{or} \quad x + 1 = 5 \qquad \text{Zero-factor property}$$
$$x = 7 \quad \text{or} \quad x = 4$$

Use the square root property to solve each equation. See Examples 1 and 3.

5. $x^2 = 81$ **6.** $z^2 = 225$ **7.** $t^2 = 17$

8. $x^2 = 19$ **9.** $m^2 = 32$ **10.** $x^2 = 54$

11. $t^2 - 20 = 0$ **12.** $p^2 - 50 = 0$ **13.** $3n^2 - 72 = 0$

14. $5z^2 - 200 = 0$ **15.** $(x + 2)^2 = 25$ **16.** $(t + 8)^2 = 9$

17. $(x - 4)^2 = 3$ **18.** $(x + 3)^2 = 11$ **19.** $(t + 5)^2 = 48$

20. $(m - 6)^2 = 27$ **21.** $(3x - 1)^2 = 7$ **22.** $(2x + 4)^2 = 10$

23. $(4p + 1)^2 = 24$ **24.** $(5t - 2)^2 = 12$

Solve Exercises 25 and 26 using Galileo's formula, $d = 16t^2$. Round answers to the nearest tenth. See Example 2.

25. The Gateway Arch in St. Louis, Missouri, is 630 ft tall. How long would it take an object dropped from the top of it to fall to the ground? (*Source: Home & Away,* November/December 2000.)

26. Mount Rushmore National Memorial in South Dakota features a sculpture of four of America's favorite presidents carved into the rim of the mountain, 500 ft above the valley floor. How long would it take a rock dropped from the top of the sculpture to fall to the ground? (*Source: Microsoft Encarta Encyclopedia 2002.*)

27. Of the two equations

$$(2x + 1)^2 = 5 \qquad \text{and} \qquad x^2 + 4x = 12,$$

one is more suitable for solving by the square root property, and the other is more suitable for solving by completing the square. Which method do you think most students would use for each equation?

28. Why would most students find the equation $x^2 + 4x = 20$ easier to solve by completing the square than the equation $5x^2 + 2x = 3$?

29. Decide what number must be added to make each expression a perfect square trinomial.

     (a) $x^2 + 6x + \underline{\hspace{1cm}}$ **(b)** $x^2 + 14x + \underline{\hspace{1cm}}$ **(c)** $p^2 - 12p + \underline{\hspace{1cm}}$

     (d) $x^2 + 3x + \underline{\hspace{1cm}}$ **(e)** $q^2 - 9q + \underline{\hspace{1cm}}$ **(f)** $t^2 - \dfrac{1}{2}t + \underline{\hspace{1cm}}$

30. What would be the first step in solving $2x^2 + 8x = 9$ by completing the square?

Determine the number that will complete the square to solve each equation after the constant term has been written on the right side. Do not actually solve. See Examples 4–6.

31. $x^2 + 4x - 2 = 0$ **32.** $t^2 + 2t - 1 = 0$ **33.** $x^2 + 10x + 18 = 0$

34. $x^2 + 8x + 11 = 0$ **35.** $3w^2 - w - 24 = 0$ **36.** $4z^2 - z - 39 = 0$

Solve each equation by completing the square. Use the results of Exercises 31–36 to solve Exercises 39–44. See Examples 4–6.

37. $x^2 - 2x - 24 = 0$ **38.** $m^2 - 4m - 32 = 0$ **39.** $x^2 + 4x - 2 = 0$

40. $t^2 + 2t - 1 = 0$ **41.** $x^2 + 10x + 18 = 0$ **42.** $x^2 + 8x + 11 = 0$

43. $3w^2 - w = 24$

44. $4z^2 - z = 39$

45. $2k^2 + 5k - 2 = 0$

46. $3r^2 + 2r - 2 = 0$

47. $5x^2 - 10x + 2 = 0$

48. $2x^2 - 16x + 25 = 0$

49. $9x^2 - 24x = -13$

50. $25n^2 - 20n = 1$

51. $z^2 - \dfrac{4}{3}z = -\dfrac{1}{9}$

52. $p^2 - \dfrac{8}{3}p = -1$

53. $.1x^2 - .2x - .1 = 0$

54. $.1p^2 - .4p + .1 = 0$

Find the imaginary solutions of each equation. See Example 7.

55. $x^2 = -12$

56. $x^2 = -18$

57. $(r - 5)^2 = -3$

58. $(t + 6)^2 = -5$

59. $(6k - 1)^2 = -8$

60. $(4m - 7)^2 = -27$

61. $m^2 + 4m + 13 = 0$

62. $t^2 + 6t + 10 = 0$

63. $3r^2 + 4r + 4 = 0$

64. $4x^2 + 5x + 5 = 0$

65. $-m^2 - 6m - 12 = 0$

66. $-k^2 - 5k - 10 = 0$

Solve for x. Assume that a and b represent positive real numbers.

67. $x^2 - b = 0$

68. $x^2 = 4b$

69. $4x^2 = b^2 + 16$

70. $9x^2 - 25a = 0$

71. $(5x - 2b)^2 = 3a$

72. $x^2 - a^2 - 36 = 0$

RELATING CONCEPTS (EXERCISES 73–78)

For Individual or Group Work

The Greeks had a method of completing the square geometrically in which they literally changed a figure into a square. For example, to complete the square for $x^2 + 6x$, we begin with a square of side x, as in the figure on the left. We add three rectangles of width 1 to the right side and the bottom to get a region with area $x^2 + 6x$. To fill in the corner (complete the square), we must add 9 1-by-1 squares as shown in the figure on the right.

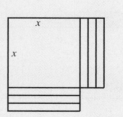 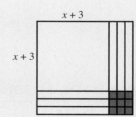

Work Exercises 73–78 in order.

73. What is the area of the original square?

74. What is the area of each strip?

75. What is the total area of the six strips?

76. What is the area of each small square in the corner of the second figure?

77. What is the total area of the small squares?

78. What is the area of the new, larger square?

TECHNOLOGY INSIGHTS (EXERCISES 79–80)

Two of the following calculator screens show the intersection points of the graph of $y = x^2$ and the graph of the horizontal line $y = 5$. The other screen shows that $\sqrt{5} \approx 2.236068$, so the graphs intersect at $x = -\sqrt{5}$ and $x = \sqrt{5}$. This supports our solution $\pm\sqrt{5}$ for $x^2 = 5$ using the square root property in Example 1(a).

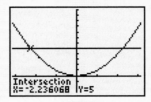 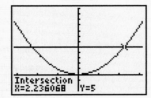

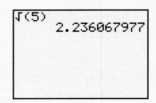

Use the screens in Exercises 79 and 80 to give the exact value of the solutions of $x^2 = k$.

79.

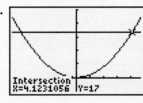 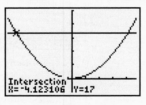

80.

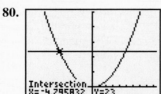 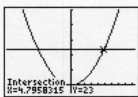

9.2 The Quadratic Formula

OBJECTIVES

1 Derive the quadratic formula.

2 Solve quadratic equations using the quadratic formula.

3 Use the discriminant to determine the number and type of solutions.

The examples in the previous section showed that any quadratic equation can be solved by completing the square; however, completing the square can be tedious and time consuming. In this section, we complete the square to solve the general quadratic equation

$$ax^2 + bx + c = 0,$$

where a, b, and c are complex numbers and $a \neq 0$. The solution of this general equation gives a formula for finding the solution of any specific quadratic equation.

OBJECTIVE 1 Derive the quadratic formula. To solve $ax^2 + bx + c = 0$ by completing the square (assuming $a > 0$), we follow the steps given in Section 9.1.

$$ax^2 + bx + c = 0$$

$$x^2 + \frac{b}{a}x + \frac{c}{a} = 0 \qquad \text{Divide by } a. \text{ (Step 1)}$$

$$x^2 + \frac{b}{a}x = -\frac{c}{a} \qquad \text{Subtract } \tfrac{c}{a}. \text{ (Step 2)}$$

$$\left[\frac{1}{2}\left(\frac{b}{a}\right)\right]^2 = \left(\frac{b}{2a}\right)^2 = \frac{b^2}{4a^2} \qquad \text{(Step 3)}$$

$$x^2 + \frac{b}{a}x + \frac{b^2}{4a^2} = -\frac{c}{a} + \frac{b^2}{4a^2} \qquad \text{Add } \tfrac{b^2}{4a^2} \text{ to each side. (Step 4)}$$

Write the left side as a perfect square and rearrange the right side.

$$\left(x + \frac{b}{2a}\right)^2 = \frac{b^2}{4a^2} + \frac{-c}{a} \qquad \text{(Step 5)}$$

$$\left(x + \frac{b}{2a}\right)^2 = \frac{b^2}{4a^2} + \frac{-4ac}{4a^2} \qquad \text{Write with a common denominator.}$$

$$\left(x + \frac{b}{2a}\right)^2 = \frac{b^2 - 4ac}{4a^2} \qquad \text{Add fractions.}$$

$$x + \frac{b}{2a} = \sqrt{\frac{b^2 - 4ac}{4a^2}} \quad \text{or} \quad x + \frac{b}{2a} = -\sqrt{\frac{b^2 - 4ac}{4a^2}} \qquad \begin{array}{l}\text{Square root}\\ \text{property}\\ \text{(Step 6)}\end{array}$$

Since $\quad \sqrt{\dfrac{b^2 - 4ac}{4a^2}} = \dfrac{\sqrt{b^2 - 4ac}}{\sqrt{4a^2}} = \dfrac{\sqrt{b^2 - 4ac}}{2a},$

the right side of each equation can be expressed as

$$x + \frac{b}{2a} = \frac{\sqrt{b^2 - 4ac}}{2a} \qquad \text{or} \quad x + \frac{b}{2a} = \frac{-\sqrt{b^2 - 4ac}}{2a}$$

$$x = \frac{-b}{2a} + \frac{\sqrt{b^2 - 4ac}}{2a} \quad \text{or} \qquad x = \frac{-b}{2a} - \frac{\sqrt{b^2 - 4ac}}{2a}$$

$$x = \frac{-b + \sqrt{b^2 - 4ac}}{2a} \quad \text{or} \qquad x = \frac{-b - \sqrt{b^2 - 4ac}}{2a}.$$

If $a < 0$, the same two solutions are obtained. The result is the **quadratic formula,** which is abbreviated as follows.

Quadratic Formula

The solutions of $ax^2 + bx + c = 0$ $(a \neq 0)$ are given by

$$x = \frac{-b \pm \sqrt{b^2 - 4ac}}{2a}.$$

CAUTION In the quadratic formula, $x = \dfrac{-b \pm \sqrt{b^2 - 4ac}}{2a}$, the square root is added to or subtracted from the value of $-b$ *before* dividing by $2a$.

OBJECTIVE 2 Solve quadratic equations using the quadratic formula. To use the quadratic formula, first write the given equation in standard form $ax^2 + bx + c = 0$; then identify the values of a, b, and c and substitute them into the formula.

EXAMPLE 1 Using the Quadratic Formula (Rational Solutions)

Solve $6x^2 - 5x - 4 = 0$.

First, identify the values of a, b, and c of the general quadratic equation, $ax^2 + bx + c = 0$. Here a, the coefficient of the second-degree term, is 6, while b, the coefficient of the first-degree term, is -5, and the constant c is -4. Substitute these values into the quadratic formula.

$$x = \frac{-b \pm \sqrt{b^2 - 4ac}}{2a}$$

$$x = \frac{-(-5) \pm \sqrt{(-5)^2 - 4(6)(-4)}}{2(6)} \qquad a = 6,\ b = -5,\ c = -4$$

$$x = \frac{5 \pm \sqrt{25 + 96}}{12}$$

$$x = \frac{5 \pm \sqrt{121}}{12}$$

$$x = \frac{5 \pm 11}{12}$$

This last statement leads to two solutions, one from $+$ and one from $-$.

$$x = \frac{5 + 11}{12} = \frac{16}{12} = \frac{4}{3} \qquad \text{or} \qquad x = \frac{5 - 11}{12} = \frac{-6}{12} = -\frac{1}{2}$$

Check each solution by substituting it in the original equation. The solution set is $\left\{ -\frac{1}{2}, \frac{4}{3} \right\}$.

Now Try Exercise 5.

We could have used factoring to solve the equation in Example 1.

$$6x^2 - 5x - 4 = 0$$

$(3x - 4)(2x + 1) = 0$		Factor.
$3x - 4 = 0$ or $2x + 1 = 0$		Zero-factor property
$3x = 4$ or $2x = -1$		Solve each equation.
$x = \dfrac{4}{3}$ or $x = -\dfrac{1}{2}$		Same solutions as in Example 1

When solving quadratic equations, it is a good idea to try factoring first. If the equation cannot be factored or if factoring is difficult, then use the quadratic formula.

Later in this section, we will show a way to determine whether factoring can be used to solve a quadratic equation.

EXAMPLE 2   Using the Quadratic Formula (Irrational Solutions)

Solve $4r^2 = 8r - 1$.

Write the equation in standard form as

$$4r^2 - 8r + 1 = 0,$$

and identify $a = 4$, $b = -8$, and $c = 1$. Now use the quadratic formula.

$$r = \frac{-b \pm \sqrt{b^2 - 4ac}}{2a}$$

$$r = \frac{-(-8) \pm \sqrt{(-8)^2 - 4(4)(1)}}{2(4)} \qquad a = 4, b = -8, c = 1$$

$$= \frac{8 \pm \sqrt{64 - 16}}{8}$$

$$= \frac{8 \pm \sqrt{48}}{8}$$

$$= \frac{8 \pm 4\sqrt{3}}{8} \qquad \sqrt{48} = \sqrt{16} \cdot \sqrt{3} = 4\sqrt{3}$$

$$= \frac{4(2 \pm \sqrt{3})}{4(2)} \qquad \text{Factor.}$$

$$= \frac{2 \pm \sqrt{3}}{2} \qquad \text{Lowest terms}$$

The solution set is $\left\{ \dfrac{2 + \sqrt{3}}{2}, \dfrac{2 - \sqrt{3}}{2} \right\}$.

Now Try Exercise 9.

CAUTION   *Every* quadratic equation must be expressed in standard form $ax^2 + bx + c = 0$ before we begin to solve it, whether we use factoring or the quadratic formula. Also, when writing solutions in lowest terms, be sure to *factor first;* then divide out the common factor, as shown in the last two steps in Example 2.

EXAMPLE 3   Using the Quadratic Formula (Imaginary Solutions)

Solve $(9q + 3)(q - 1) = -8$.

To write this equation in standard form, we first multiply and collect all nonzero terms on the left.

$$(9q + 3)(q - 1) = -8$$
$$9q^2 - 6q - 3 = -8$$
$$9q^2 - 6q + 5 = 0 \qquad \text{Standard form}$$

From the equation $9q^2 - 6q + 5 = 0$, we identify $a = 9$, $b = -6$, and $c = 5$, and use the quadratic formula.

$$q = \frac{-(-6) \pm \sqrt{(-6)^2 - 4(9)(5)}}{2(9)}$$

$$= \frac{6 \pm \sqrt{-144}}{18}$$

$$= \frac{6 \pm 12i}{18} \qquad \sqrt{-144} = 12i$$

$$= \frac{6(1 \pm 2i)}{6(3)} \qquad \text{Factor.}$$

$$= \frac{1 \pm 2i}{3} \qquad \text{Lowest terms}$$

The solution set, written in standard form $a + bi$ for complex numbers, is $\left\{ \frac{1}{3} + \frac{2}{3}i, \frac{1}{3} - \frac{2}{3}i \right\}$.

Now Try Exercise 33.

OBJECTIVE 3 Use the discriminant to determine the number and type of solutions. The solutions of the quadratic equation $ax^2 + bx + c = 0$ are given by

$$x = \frac{-b \pm \sqrt{b^2 - 4ac}}{2a}. \quad \longleftarrow \text{Discriminant}$$

If a, b, and c are integers, the type of solutions of a quadratic equation—that is, rational, irrational, or imaginary—is determined by the expression under the radical sign, $b^2 - 4ac$. Because it distinguishes among the three types of solutions, $b^2 - 4ac$ is called the *discriminant*. By calculating the discriminant before solving a quadratic equation, we can predict whether the solutions will be rational numbers, irrational numbers, or imaginary numbers. (This can be useful in an applied problem, for example, where irrational or imaginary solutions are not acceptable.)

Discriminant

The **discriminant** of $ax^2 + bx + c = 0$ is $b^2 - 4ac$. If a, b, and c are integers, then the number and type of solutions are determined as follows.

Discriminant	Number and Type of Solutions
Positive, and the square of an integer	Two rational solutions
Positive, but not the square of an integer	Two irrational solutions
Zero	One rational solution
Negative	Two imaginary solutions

Calculating the discriminant can also help you decide whether to solve a quadratic equation by factoring or by using the quadratic formula. If the discriminant is a perfect square (including 0), then the equation can be solved by factoring. Otherwise, the quadratic formula should be used.

EXAMPLE 4   Using the Discriminant

Find the discriminant. Use it to predict the number and type of solutions for each equation. Tell whether the equation can be solved by factoring or whether the quadratic formula should be used.

(a) $6x^2 - x - 15 = 0$

We find the discriminant by evaluating $b^2 - 4ac$.

$$b^2 - 4ac = (-1)^2 - 4(6)(-15) \qquad a = 6, \, b = -1, \, c = -15$$
$$= 1 + 360$$
$$= 361$$

A calculator shows that $361 = 19^2$, a perfect square. Since a, b, and c are integers and the discriminant is a perfect square, there will be two rational solutions and the equation can be solved by factoring.

(b) $3m^2 - 4m = 5$

Write the equation in standard form as $3m^2 - 4m - 5 = 0$ to find that $a = 3$, $b = -4$, and $c = -5$.

$$b^2 - 4ac = (-4)^2 - 4(3)(-5)$$
$$= 16 + 60$$
$$= 76$$

Because 76 is positive but not the square of an integer and a, b, and c are integers, the equation will have two irrational solutions and is best solved using the quadratic formula.

(c) $4x^2 + x + 1 = 0$

Since $a = 4$, $b = 1$, and $c = 1$, the discriminant is

$$1^2 - 4(4)(1) = -15.$$

Since the discriminant is negative and a, b, and c are integers, this quadratic equation will have two imaginary solutions. The quadratic formula should be used to solve it.

(d) $4t^2 + 9 = 12t$

Write the equation as $4t^2 - 12t + 9 = 0$ to find $a = 4$, $b = -12$, and $c = 9$. The discriminant is

$$b^2 - 4ac = (-12)^2 - 4(4)(9)$$
$$= 144 - 144$$
$$= 0.$$

Because the discriminant is 0, the quantity under the radical in the quadratic formula is 0, and there is only one rational solution. Again, the equation can be solved by factoring.

Now Try Exercises 37 and 39.

EXAMPLE 5   Using the Discriminant

Find k so that $9x^2 + kx + 4 = 0$ will have only one rational solution.

The equation will have only one rational solution if the discriminant is 0. Since $a = 9$, $b = k$, and $c = 4$, the discriminant is

$$b^2 - 4ac = k^2 - 4(9)(4) = k^2 - 144.$$

Set the discriminant equal to 0 and solve for k.

$$k^2 - 144 = 0$$
$$k^2 = 144 \qquad \text{Subtract 144.}$$
$$k = 12 \quad \text{or} \quad k = -12 \qquad \text{Square root property}$$

The equation will have only one rational solution if $k = 12$ or $k = -12$.

Now Try Exercise 53.

9.2 EXERCISES

✐ *Answer each question in Exercises 1–4.*

1. An early version of Microsoft *Word* for Windows included the 1.0 edition of *Equation Editor*. The documentation used the following for the quadratic formula.

$$x = -b \pm \frac{\sqrt{b^2 - 4ac}}{2a}$$

Was this correct? Explain.

2. The Cadillac Bar in Houston, Texas, encourages patrons to write (tasteful) messages on the walls. One person attempted to write the quadratic formula, as shown here.

$$x = \frac{-b\sqrt{b^2 - 4ac}}{2a}$$

Was this correct? Explain.

3. What is wrong with the following "solution" of $5x^2 - 5x + 1 = 0$?

$$x = \frac{5 \pm \sqrt{25 - 4(5)(1)}}{2(5)} \qquad a = 5, b = -5, c = 1$$
$$x = \frac{5 \pm \sqrt{5}}{10}$$
$$x = \frac{1}{2} \pm \sqrt{5}$$

4. A student claimed that the equation $2x^2 - 5 = 0$ cannot be solved using the quadratic formula because there is no first-degree x-term. Was the student correct? Explain.

Use the quadratic formula to solve each equation. (All solutions for these equations are real numbers.) See Examples 1 and 2.

5. $m^2 - 8m + 15 = 0$

6. $x^2 + 3x - 28 = 0$

7. $2k^2 + 4k + 1 = 0$

8. $2w^2 + 3w - 1 = 0$

9. $2x^2 - 2x = 1$

10. $9t^2 + 6t = 1$

11. $x^2 + 18 = 10x$

12. $x^2 - 4 = 2x$

13. $4k^2 + 4k - 1 = 0$

14. $4r^2 - 4r - 19 = 0$

15. $2 - 2x = 3x^2$

16. $26r - 2 = 3r^2$

17. $\dfrac{x^2}{4} - \dfrac{x}{2} = 1$

18. $p^2 + \dfrac{p}{3} = \dfrac{1}{6}$

19. $-2t(t + 2) = -3$

20. $-3x(x + 2) = -4$

21. $(r - 3)(r + 5) = 2$

22. $(k + 1)(k - 7) = 1$

23. $(g + 2)(g - 3) = 1$

24. $(x - 5)(x + 2) = 6$

25. $p = \dfrac{5(5 - p)}{3(p + 1)}$

26. $k = \dfrac{k + 15}{3(k - 1)}$

Use the quadratic formula to solve each equation. (All solutions for these equations are imaginary numbers.) See Example 3.

27. $x^2 - 3x + 6 = 0$ **28.** $x^2 - 5x + 20 = 0$ **29.** $r^2 - 6r + 14 = 0$

30. $t^2 + 4t + 11 = 0$ **31.** $4x^2 - 4x = -7$ **32.** $9x^2 - 6x = -7$

33. $x(3x + 4) = -2$ **34.** $z(2z + 3) = -2$

35. $(x + 5)(x - 6) = (2x - 1)(x - 4)$ **36.** $(3x - 4)(x + 2) = (2x - 5)(x + 5)$

Use the discriminant to determine whether the solutions for each equation are

    A. *two rational numbers;*   **B.** *one rational number;*
    C. *two irrational numbers;*   **D.** *two imaginary numbers.*

Do not actually solve. See Example 4.

37. $25x^2 + 70x + 49 = 0$ **38.** $4k^2 - 28k + 49 = 0$ **39.** $x^2 + 4x + 2 = 0$

40. $9x^2 - 12x - 1 = 0$ **41.** $3x^2 = 5x + 2$ **42.** $4x^2 = 4x + 3$

43. $3m^2 - 10m + 15 = 0$ **44.** $18x^2 + 60x + 82 = 0$

45. Using the discriminant, which equations in Exercises 37–44 can be solved by factoring?

Based on your answer in Exercise 45, solve the equation given in each exercise.

46. Exercise 37   **47.** Exercise 38   **48.** Exercise 41   **49.** Exercise 42

50. Find the discriminant for each quadratic equation. Use it to tell whether the equation can be solved by factoring or whether the quadratic formula should be used. Then solve each equation.

    (a) $3k^2 + 13k = -12$ **(b)** $2x^2 + 19 = 14x$

51. Is it possible for the solution of a quadratic equation with integer coefficients to include just one irrational number? Why or why not?

52. Can the solution of a quadratic equation with integer coefficients include one real and one imaginary number? Why or why not?

Find the value of a, b, or c so that each equation will have exactly one rational solution. See Example 5.

53. $p^2 + bp + 25 = 0$ **54.** $r^2 - br + 49 = 0$ **55.** $am^2 + 8m + 1 = 0$

56. $at^2 + 24t + 16 = 0$ **57.** $9x^2 - 30x + c = 0$ **58.** $4m^2 + 12m + c = 0$

59. One solution of $4x^2 + bx - 3 = 0$ is $-\frac{5}{2}$. Find b and the other solution.

60. One solution of $3x^2 - 7x + c = 0$ is $\frac{1}{3}$. Find c and the other solution.

9.3 Equations Quadratic in Form

OBJECTIVES

1 Solve an equation with fractions by writing it in quadratic form.

2 Use quadratic equations to solve applied problems.

3 Solve an equation with radicals by writing it in quadratic form.

4 Solve an equation that is quadratic in form by substitution.

We have introduced four methods for solving quadratic equations written in standard form $ax^2 + bx + c = 0$. The following table lists some advantages and disadvantages of each method.

Methods for Solving Quadratic Equations

Method	Advantages	Disadvantages
Factoring	This is usually the fastest method.	Not all polynomials are factorable; some factorable polynomials are hard to factor.
Square root property	This is the simplest method for solving equations of the form $(ax + b)^2 = c$.	Few equations are given in this form.
Completing the square	This method can always be used, although most people prefer the quadratic formula.	It requires more steps than other methods.
Quadratic formula	This method can always be used.	It is more difficult than factoring because of the square root, although calculators can simplify its use.

OBJECTIVE 1 Solve an equation with fractions by writing it in quadratic form. A variety of nonquadratic equations can be written in the form of a quadratic equation and solved by using one of the methods in the table. As you solve the equations in this section, try to decide which method is best for each equation.

EXAMPLE 1 Solving an Equation with Fractions that Leads to a Quadratic Equation

Solve $\dfrac{1}{x} + \dfrac{1}{x - 1} = \dfrac{7}{12}$.

Clear fractions by multiplying each term by the least common denominator, $12x(x - 1)$. (Note that the domain must be restricted to $x \neq 0$ and $x \neq 1$.)

$$12x(x - 1)\frac{1}{x} + 12x(x - 1)\frac{1}{x - 1} = 12x(x - 1)\frac{7}{12}$$

$$12(x - 1) + 12x = 7x(x - 1)$$

$$12x - 12 + 12x = 7x^2 - 7x \qquad \text{Distributive property}$$

$$24x - 12 = 7x^2 - 7x \qquad \text{Combine terms.}$$

Recall that a quadratic equation must be in standard form before it can be solved by factoring or the quadratic formula. Combine and rearrange terms so that one side is 0. Then factor to solve the resulting equation.

$$7x^2 - 31x + 12 = 0 \qquad \text{Standard form}$$
$$(7x - 3)(x - 4) = 0 \qquad \text{Factor.}$$

Using the zero-factor property gives the solutions $\frac{3}{7}$ and 4. Check by substituting these solutions in the original equation. The solution set is $\left\{\frac{3}{7}, 4\right\}$.

Now Try Exercise 19.

OBJECTIVE 2 Use quadratic equations to solve applied problems. Earlier we solved distance-rate-time (or motion) problems that led to linear equations or rational equations. Now we solve motion problems that lead to quadratic equations. We continue to use the six-step problem-solving method from Chapter 2.

EXAMPLE 2 Solving a Motion Problem

A riverboat for tourists averages 12 mph in still water. It takes the boat 1 hr 4 min to go 6 mi upstream and return. Find the speed of the current. See Figure 1.

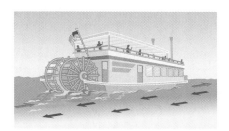

FIGURE 1

Step 1 **Read** the problem carefully.

Step 2 **Assign a variable.** Let $x =$ the speed of the current. The current slows down the boat when it is going upstream, so the rate (or speed) of the boat going upstream is its speed in still water less the speed of the current, or $12 - x$. Similarly, the current speeds up the boat as it travels downstream, so its speed downstream is $12 + x$. Thus,

$$12 - x = \text{the rate upstream;}$$
$$12 + x = \text{the rate downstream.}$$

Use the distance formula, $d = rt$, solved for time t.

$$t = \frac{d}{r}$$

This information can be used to complete a table.

	d	r	t	
Upstream	6	$12 - x$	$\dfrac{6}{12 - x}$	Times in hours
Downstream	6	$12 + x$	$\dfrac{6}{12 + x}$	

Step 3 **Write an equation.** The total time of 1 hr 4 min can be written as

$$1 + \frac{4}{60} = 1 + \frac{1}{15} = \frac{16}{15} \text{ hr.}$$

Because the time upstream plus the time downstream equals $\frac{16}{15}$ hr,

Time upstream	+	Time downstream	=	Total time
↓		↓		↓
$\dfrac{6}{12 - x}$	$+$	$\dfrac{6}{12 + x}$	$=$	$\dfrac{16}{15}.$

Step 4 **Solve** the equation. Multiply each side by $15(12 - x)(12 + x)$, the LCD, and solve the resulting quadratic equation.

$$15(12 + x)6 + 15(12 - x)6 = 16(12 - x)(12 + x)$$
$$90(12 + x) + 90(12 - x) = 16(144 - x^2)$$

$1080 + 90x + 1080 - 90x = 2304 - 16x^2$ Distributive property

$\qquad\qquad\qquad 2160 = 2304 - 16x^2$ Combine terms.

$\qquad\qquad\qquad 16x^2 = 144$

$\qquad\qquad\qquad\quad x^2 = 9$ Divide by 16.

$\qquad\qquad x = 3 \quad \text{or} \quad x = -3$ Square root property

Step 5 **State the answer.** The speed of the current cannot be -3, so the answer is 3 mph.

Step 6 **Check** that this value satisfies the original problem.

Now Try Exercise 31.

CAUTION As shown in Example 2, when a quadratic equation is used to solve an applied problem, sometimes only *one* answer satisfies the application. *Always* check each answer in the words of the original problem.

In Chapter 7 we solved problems about work rates. Recall that a person's work rate is $\frac{1}{t}$ part of the job per hour, where t is the time in hours required to do the complete job. Thus, the part of the job the person will do in x hr is $\frac{1}{t}x$.

EXAMPLE 3 Solving a Work Problem

In takes two carpet layers 4 hr to carpet a room. If each worked alone, one of them could do the job in 1 hr less time than the other. How long would it take each carpet layer to complete the job alone?

Step 1 **Read** the problem again. There will be two answers.

Step 2 **Assign a variable.** Let x represent the number of hours for the slower carpet layer to complete the job alone. Then the faster carpet layer could do the entire job in $(x - 1)$ hr. The slower person's rate is $\frac{1}{x}$, and the faster person's rate is $\frac{1}{x - 1}$. Together, they do the job in 4 hr. Complete a table as shown.

	Rate	Time Working Together	Fractional Part of the Job Done	
Slower Worker	$\dfrac{1}{x}$	4	$\dfrac{1}{x}(4)$	⟵ Sum is 1
Faster Worker	$\dfrac{1}{x-1}$	4	$\dfrac{1}{x-1}(4)$	⟵ whole job.

Step 3 **Write an equation.** The sum of the fractional parts done by the workers should equal 1 (the whole job).

Part done by slower worker $+$ Part done by faster worker $=$ 1 whole job
$$\downarrow \qquad\qquad \downarrow \qquad\qquad \downarrow$$
$$\frac{4}{x} \qquad + \qquad \frac{4}{x-1} \qquad = \qquad 1$$

Step 4 **Solve** the equation. Multiply each side by the LCD, $x(x-1)$.

$$4(x-1) + 4x = x(x-1)$$
$$4x - 4 + 4x = x^2 - x \qquad \text{Distributive property}$$
$$x^2 - 9x + 4 = 0 \qquad \text{Standard form}$$

This equation cannot be solved by factoring, so use the quadratic formula.

$$x = \frac{9 \pm \sqrt{81 - 16}}{2} = \frac{9 \pm \sqrt{65}}{2} \qquad a = 1,\, b = -9,\, c = 4$$

To the nearest tenth,

$$x = \frac{9 + \sqrt{65}}{2} \approx 8.5 \quad \text{or} \quad x = \frac{9 - \sqrt{65}}{2} \approx .5. \qquad \text{Use a calculator.}$$

Step 5 **State the answer.** Only the solution 8.5 makes sense in the original problem. (Why?) Thus, the slower worker could do the job in about 8.5 hr and the faster in about $8.5 - 1 = 7.5$ hr.

Step 6 **Check** that these results satisfy the original problem. ▨

Now Try Exercise 37.

OBJECTIVE 3 Solve an equation with radicals by writing it in quadratic form.

▨ **EXAMPLE 4** Solving Radical Equations That Lead to Quadratic Equations

Solve each equation.

(a) $k = \sqrt{6k - 8}$

This equation is not quadratic. However, squaring both sides of the equation gives a quadratic equation that can be solved by factoring.

$$k^2 = 6k - 8 \qquad \text{Square both sides.}$$
$$k^2 - 6k + 8 = 0 \qquad \text{Standard form}$$
$$(k - 4)(k - 2) = 0 \qquad \text{Factor.}$$
$$k - 4 = 0 \quad \text{or} \quad k - 2 = 0 \qquad \text{Zero-factor property}$$
$$k = 4 \quad \text{or} \qquad k = 2 \qquad \text{Potential solutions}$$

Recall from our work with radical equations in Section 8.6 that squaring both sides of an equation can introduce extraneous solutions that do not satisfy the original equation. Therefore, *all potential solutions must be checked in the original (not the squared) equation.*

Check: If $k = 4$, then

$$k = \sqrt{6k - 8}$$
$$4 = \sqrt{6(4) - 8} \quad \text{?}$$
$$4 = \sqrt{16} \quad \text{?}$$
$$4 = 4. \qquad \text{True}$$

If $k = 2$, then

$$k = \sqrt{6k - 8}$$
$$2 = \sqrt{6(2) - 8} \quad \text{?}$$
$$2 = \sqrt{4} \quad \text{?}$$
$$2 = 2. \qquad \text{True}$$

Both solutions check, so the solution set is $\{2, 4\}$.

(b) $x + \sqrt{x} = 6$

$$\sqrt{x} = 6 - x \qquad \text{Isolate the radical on one side.}$$
$$x = 36 - 12x + x^2 \qquad \text{Square both sides.}$$
$$0 = x^2 - 13x + 36 \qquad \text{Standard form}$$
$$0 = (x - 4)(x - 9) \qquad \text{Factor.}$$
$$x - 4 = 0 \quad \text{or} \quad x - 9 = 0 \qquad \text{Zero-factor property}$$
$$x = 4 \quad \text{or} \quad x = 9 \qquad \text{Potential solutions}$$

Check both potential solutions in the *original* equation.

If $x = 4$, then

$$x + \sqrt{x} = 6$$
$$4 + \sqrt{4} = 6 \quad \text{?}$$
$$6 = 6. \qquad \text{True}$$

If $x = 9$, then

$$x + \sqrt{x} = 6$$
$$9 + \sqrt{9} = 6 \quad \text{?}$$
$$12 = 6. \qquad \text{False}$$

Only the solution 4 checks, so the solution set is $\{4\}$.

Now Try Exercises 41 and 47.

OBJECTIVE 4   Solve an equation that is quadratic in form by substitution.   A nonquadratic equation that can be written in the form $au^2 + bu + c = 0$, for $a \neq 0$ and an algebraic expression u, is called **quadratic in form.**

EXAMPLE 5   Solving Equations That Are Quadratic in Form

Solve each equation.

(a) $x^4 - 13x^2 + 36 = 0$

Because $x^4 = (x^2)^2$, we can write this equation in quadratic form with $u = x^2$ and $u^2 = x^4$. (Any letter except x could be used instead of u.)

$$x^4 - 13x^2 + 36 = 0$$
$$(x^2)^2 - 13x^2 + 36 = 0 \qquad x^4 = (x^2)^2$$
$$u^2 - 13u + 36 = 0 \qquad \text{Let } u = x^2.$$
$$(u - 4)(u - 9) = 0 \qquad \text{Factor.}$$

SECTION 9.3   Equations Quadratic in Form   **557**

$$u - 4 = 0 \quad \text{or} \quad u - 9 = 0 \qquad \text{Zero-factor property}$$
$$u = 4 \quad \text{or} \quad u = 9 \qquad \text{Solve.}$$

To find x, we substitute x^2 for u.

$$x^2 = 4 \quad \text{or} \quad x^2 = 9$$
$$x = \pm 2 \quad \text{or} \quad x = \pm 3 \qquad \text{Square root property}$$

The equation $x^4 - 13x^2 + 36 = 0$, a fourth-degree equation, has four solutions.* The solution set is $\{-3, -2, 2, 3\}$, which can be verified by substituting into the original equation for x.

(b) $4x^6 + 1 = 5x^3$

This equation is quadratic in form with $u = x^3$ and $u^2 = x^6$.

$$4x^6 + 1 = 5x^3$$
$$4(x^3)^2 + 1 = 5x^3$$
$$4u^2 + 1 = 5u \qquad \text{Let } u = x^3.$$
$$4u^2 - 5u + 1 = 0 \qquad \text{Standard form}$$
$$(4u - 1)(u - 1) = 0 \qquad \text{Factor.}$$
$$4u - 1 = 0 \quad \text{or} \quad u - 1 = 0 \qquad \text{Zero-factor property}$$
$$u = \frac{1}{4} \quad \text{or} \quad u = 1 \qquad \text{Solve.}$$
$$x^3 = \frac{1}{4} \quad \text{or} \quad x^3 = 1 \qquad u = x^3$$

From these equations,

$$x = \sqrt[3]{\frac{1}{4}} = \frac{\sqrt[3]{1}}{\sqrt[3]{4}} = \frac{1}{\sqrt[3]{4}} \cdot \frac{\sqrt[3]{2}}{\sqrt[3]{2}} = \frac{\sqrt[3]{2}}{2} \qquad \text{or} \qquad x = \sqrt[3]{1} = 1.$$

There are other complex solutions for this equation, but finding them involves trigonometry. The real number solution set of $4x^6 + 1 = 5x^3$ is $\left\{ \frac{\sqrt[3]{2}}{2}, 1 \right\}$.

(c) $x^4 = 6x^2 - 3$

First write the equation as

$$x^4 - 6x^2 + 3 = 0 \qquad \text{or} \qquad (x^2)^2 - 6(x^2) + 3 = 0,$$

which is quadratic in form with $u = x^2$. Substitute u for x^2 and u^2 for x^4 to get

$$u^2 - 6u + 3 = 0.$$

---

*In general, an equation in which an nth-degree polynomial equals 0 has n complex solutions, although some of them may be repeated.

Since this equation cannot be solved by factoring, use the quadratic formula.

$$u = \frac{6 \pm \sqrt{36 - 12}}{2} \qquad a = 1, b = -6, c = 3$$

$$u = \frac{6 \pm \sqrt{24}}{2}$$

$$u = \frac{6 \pm 2\sqrt{6}}{2} \qquad \sqrt{24} = \sqrt{4} \cdot \sqrt{6} = 2\sqrt{6}$$

$$u = \frac{2(3 \pm \sqrt{6})}{2} \qquad \text{Factor.}$$

$$u = 3 \pm \sqrt{6} \qquad \text{Lowest terms}$$

Since $u = x^2$, find x by using the square root property.

$$x^2 = 3 + \sqrt{6} \qquad \text{or} \quad x^2 = 3 - \sqrt{6}$$

$$x = \pm\sqrt{3 + \sqrt{6}} \quad \text{or} \quad x = \pm\sqrt{3 - \sqrt{6}}$$

The solution set contains four numbers:

$$\left\{ \sqrt{3 + \sqrt{6}}, -\sqrt{3 + \sqrt{6}}, \sqrt{3 - \sqrt{6}}, -\sqrt{3 - \sqrt{6}} \right\}.$$

Now Try Exercises 55, 79, and 83.

NOTE Some students prefer to solve equations like those in Examples 5(a) and (b) by factoring directly. For example,

$$x^4 - 13x^2 + 36 = 0 \qquad \text{Example 5(a) equation}$$
$$(x^2 - 9)(x^2 - 4) = 0 \qquad \text{Factor.}$$
$$(x + 3)(x - 3)(x + 2)(x - 2) = 0. \qquad \text{Factor again.}$$

Using the zero-factor property gives the same solutions obtained in Example 5(a). Equations that cannot be solved by factoring, like that in Example 5(c), must be solved using the method of substitution and the quadratic formula.

EXAMPLE 6 Solving Equations That Are Quadratic in Form

Solve each equation.

(a) $2(4m - 3)^2 + 7(4m - 3) + 5 = 0$

Because of the repeated quantity $4m - 3$, this equation is quadratic in form with $u = 4m - 3$.

$$2(4m - 3)^2 + 7(4m - 3) + 5 = 0$$
$$2u^2 + 7u + 5 = 0 \qquad \text{Let } 4m - 3 = u.$$
$$(2u + 5)(u + 1) = 0 \qquad \text{Factor.}$$
$$2u + 5 = 0 \qquad \text{or} \qquad u + 1 = 0 \qquad \text{Zero-factor property}$$
$$u = -\frac{5}{2} \qquad \text{or} \qquad u = -1$$
$$4m - 3 = -\frac{5}{2} \qquad \text{or} \quad 4m - 3 = -1 \qquad \text{Substitute } 4m - 3 \text{ for } u.$$

$$4m = \frac{1}{2} \qquad \text{or} \qquad 4m = 2 \qquad \text{Solve for } m.$$

$$m = \frac{1}{8} \qquad \text{or} \qquad m = \frac{1}{2}$$

Check that the solution set of the original equation is $\left\{\frac{1}{8}, \frac{1}{2}\right\}$.

(b) $2a^{2/3} - 11a^{1/3} + 12 = 0$

Let $a^{1/3} = u$; then $a^{2/3} = (a^{1/3})^2 = u^2$. Substitute into the given equation.

$$2u^2 - 11u + 12 = 0 \qquad \text{Let } a^{1/3} = u; a^{2/3} = u^2.$$

$$(2u - 3)(u - 4) = 0 \qquad \text{Factor.}$$

$$2u - 3 = 0 \qquad \text{or} \qquad u - 4 = 0 \qquad \text{Zero-factor property}$$

$$u = \frac{3}{2} \qquad \text{or} \qquad u = 4$$

$$a^{1/3} = \frac{3}{2} \qquad \text{or} \qquad a^{1/3} = 4 \qquad u = a^{1/3}$$

$$(a^{1/3})^3 = \left(\frac{3}{2}\right)^3 \qquad \text{or} \qquad (a^{1/3})^3 = 4^3 \qquad \text{Cube each side.}$$

$$a = \frac{27}{8} \qquad \text{or} \qquad a = 64$$

Check that the solution set is $\left\{\frac{27}{8}, 64\right\}$.

Now Try Exercises 59 and 65.

CAUTION A common error when solving problems like those in Examples 5 and 6 is to stop too soon. Once you have solved for u, remember to substitute and solve for the values of the *original* variable.

9.3 EXERCISES

For Extra Help

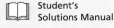
Student's
Solutions Manual

MyMathLab

InterAct Math
Tutorial Software

AW Math
Tutor Center

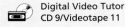
MathXL

Digital Video Tutor
CD 9/Videotape 11

Refer to the box at the beginning of this section. Decide whether factoring, the square root property, or the quadratic formula is most appropriate for solving each quadratic equation. Do not actually solve the equations.

1. $(2x + 3)^2 = 4$
2. $4x^2 - 3x = 1$
3. $z^2 + 5z - 8 = 0$

4. $2k^2 + 3k = 1$
5. $3m^2 = 2 - 5m$
6. $p^2 = 5$

Write a sentence describing the first step you would take to solve each equation. Do not actually solve.

7. $\dfrac{14}{x} = x - 5$

8. $\sqrt{1 + x} + x = 5$

9. $(r^2 + r)^2 - 8(r^2 + r) + 12 = 0$

10. $3t = \sqrt{16 - 10t}$

11. What is wrong with the following "solution"?

$$x = \sqrt{3x + 4}$$
$$x^2 = 3x + 4 \quad \text{Square both sides.}$$
$$x^2 - 3x - 4 = 0$$
$$(x - 4)(x + 1) = 0$$
$$x - 4 = 0 \quad \text{or} \quad x + 1 = 0$$
$$x = 4 \quad \text{or} \quad x = -1$$

Solution set: $\{4, -1\}$

12. What is wrong with the following "solution"?

$$2(m - 1)^2 - 3(m - 1) + 1 = 0$$
$$2u^2 - 3u + 1 = 0$$
$$\text{Let } u = m - 1.$$
$$(2u - 1)(u - 1) = 0$$
$$2u - 1 = 0 \quad \text{or} \quad u - 1 = 0$$
$$u = \frac{1}{2} \quad \text{or} \quad u = 1$$

Solution set: $\left\{\dfrac{1}{2}, 1\right\}$

Solve each equation. Check your solutions. See Example 1.

13. $\dfrac{14}{x} = x - 5$

14. $\dfrac{-12}{x} = x + 8$

15. $1 - \dfrac{3}{x} - \dfrac{28}{x^2} = 0$

16. $4 - \dfrac{7}{r} - \dfrac{2}{r^2} = 0$

17. $3 - \dfrac{1}{t} = \dfrac{2}{t^2}$

18. $1 + \dfrac{2}{k} = \dfrac{3}{k^2}$

19. $\dfrac{1}{x} + \dfrac{2}{x + 2} = \dfrac{17}{35}$

20. $\dfrac{2}{m} + \dfrac{3}{m + 9} = \dfrac{11}{4}$

21. $\dfrac{2}{x + 1} + \dfrac{3}{x + 2} = \dfrac{7}{2}$

22. $\dfrac{4}{3 - p} + \dfrac{2}{5 - p} = \dfrac{26}{15}$

23. $\dfrac{3}{2x} - \dfrac{1}{2(x + 2)} = 1$

24. $\dfrac{4}{3x} - \dfrac{1}{2(x + 1)} = 1$

25. $3 = \dfrac{1}{t + 2} + \dfrac{2}{(t + 2)^2}$

26. $1 + \dfrac{2}{3z + 2} = \dfrac{15}{(3z + 2)^2}$

27. $\dfrac{6}{p} = 2 + \dfrac{p}{p + 1}$

28. $\dfrac{k}{2 - k} + \dfrac{2}{k} = 5$

Use the concepts of this section to answer each question.

29. A boat goes 20 mph in still water, and the rate of the current is t mph.

　　(a) What is the rate of the boat when it travels upstream?

　　(b) What is the rate of the boat when it travels downstream?

30. If it takes m hr to grade a set of papers, what is the grader's rate (in job per hour)?

Solve each problem. See Examples 2 and 3.

31. On a windy day Yoshiaki found that he could go 16 mi downstream and then 4 mi back upstream at top speed in a total of 48 min. What was the top speed of Yoshiaki's boat if the current was 15 mph?

	d	r	t
Upstream	4	$x - 15$	
Downstream	16		

32. Lekesha flew her plane for 6 hr at a constant speed. She traveled 810 mi with the wind, then turned around and traveled 720 mi against the wind. The wind speed was a constant 15 mph. Find the speed of the plane.

	d	r	t
With Wind	810		
Against Wind	720		

33. In Canada, Medicine Hat and Cranbrook are 300 km apart. Harry rides his Honda 20 km per hr faster than Yoshi rides his Yamaha. Find Harry's average speed if he travels from Cranbrook to Medicine Hat in $1\frac{1}{4}$ hr less time than Yoshi. (*Source: State Farm Road Atlas.*)

ALBERTA

BRITISH COLUMBIA

Medicine Hat

300 km

Cranbrook

34. In California, the distance from Jackson to Lodi is about 40 mi, as is the distance from Lodi to Manteca. Rico drove from Jackson to Lodi during the rush hour, stopped in Lodi for a root beer, and then drove on to Manteca at 10 mph faster. Driving time for the entire trip was 88 min. Find his speed from Jackson to Lodi. (*Source: State Farm Road Atlas.*)

CALIFORNIA

40 mi — Jackson

Lodi

40 mi

Manteca

35. Working together, two people can cut a large lawn in 2 hr. One person can do the job alone in 1 hr less time than the other. How long (to the nearest tenth) would it take the faster person to do the job? (*Hint:* x is the time of the faster person.)

	Rate	Time Working Together	Fractional Part of the Job Done
Faster Worker	$\dfrac{1}{x}$	2	
Slower Worker		2	

36. A janitorial service provides two people to clean an office building. Working together, the two can clean the building in 5 hr. One person is new to the job and would take 2 hr longer than the other person to clean the building alone. How long (to the nearest tenth) would it take the new worker to clean the building alone?

	Rate	Time Working Together	Fractional Part of the Job Done
Faster Worker			
Slower Worker			

37. Rusty and Nancy Brauner are planting flats of spring flowers. Working alone, Rusty would take 2 hr longer than Nancy to plant the flowers. Working together, they do the job in 12 hr. How long would it have taken each person working alone?

38. Jay Beckenstein can work through a stack of invoices in 1 hr less time than Colleen Manley Jones can. Working together they take $1\frac{1}{2}$ hr. How long would it take each person working alone?

39. A washing machine can be filled in 6 min if both the hot and cold water taps are fully opened. Filling the washer with hot water alone takes 9 min longer than filling it with cold water alone. How long does it take to fill the washer with cold water?

40. Two pipes together can fill a large tank in 2 hr. One of the pipes, used alone, takes 3 hr longer than the other to fill the tank. How long would each pipe take to fill the tank alone?

Solve each equation. Check your solutions. See Example 4.

41. $x = \sqrt{7x - 10}$ **42.** $z = \sqrt{5z - 4}$ **43.** $2x = \sqrt{11x + 3}$ **44.** $4x = \sqrt{6x + 1}$

45. $3x = \sqrt{16 - 10x}$ **46.** $4t = \sqrt{8t + 3}$ **47.** $p - 2\sqrt{p} = 8$ **48.** $k + \sqrt{k} = 12$

49. $m = \sqrt{\dfrac{6 - 13m}{5}}$ **50.** $r = \sqrt{\dfrac{20 - 19r}{6}}$

Solve each equation. Check your solutions. See Examples 5 and 6.

51. $t^4 - 18t^2 + 81 = 0$ **52.** $x^4 - 8x^2 + 16 = 0$ **53.** $4k^4 - 13k^2 + 9 = 0$

54. $9x^4 - 25x^2 + 16 = 0$ **55.** $x^4 + 48 = 16x^2$ **56.** $z^4 = 17z^2 - 72$

57. $(x + 3)^2 + 5(x + 3) + 6 = 0$ **58.** $(k - 4)^2 + (k - 4) - 20 = 0$

59. $3(m + 4)^2 - 8 = 2(m + 4)$ **60.** $(t + 5)^2 + 6 = 7(t + 5)$

61. $2 + \dfrac{5}{3k - 1} = \dfrac{-2}{(3k - 1)^2}$ **62.** $3 - \dfrac{7}{2p + 2} = \dfrac{6}{(2p + 2)^2}$

63. $2 - 6(m - 1)^{-2} = (m - 1)^{-1}$ **64.** $3 - 2(x - 1)^{-1} = (x - 1)^{-2}$

65. $x^{2/3} + x^{1/3} - 2 = 0$ **66.** $x^{2/3} - 2x^{1/3} - 3 = 0$

67. $r^{2/3} + r^{1/3} - 12 = 0$ **68.** $3x^{2/3} - x^{1/3} - 24 = 0$

69. $4k^{4/3} - 13k^{2/3} + 9 = 0$ **70.** $9m^{2/5} = 16 - 10m^{1/5}$

71. $2(1 + \sqrt{r})^2 = 13(1 + \sqrt{r}) - 6$ **72.** $(k^2 + k)^2 + 12 = 8(k^2 + k)$

73. $2x^4 + x^2 - 3 = 0$ **74.** $4k^4 + 5k^2 + 1 = 0$

The equations in Exercises 75–84 are not grouped by type. Decide which method of solution applies, and then solve each equation. Give only real solutions. See Examples 1 and 4–6.

75. $12x^4 - 11x^2 + 2 = 0$ **76.** $\left(x - \dfrac{1}{2}\right)^2 + 5\left(x - \dfrac{1}{2}\right) - 4 = 0$

77. $\sqrt{2x + 3} = 2 + \sqrt{x - 2}$ **78.** $\sqrt{m + 1} = -1 + \sqrt{2m}$

79. $2m^6 + 11m^3 + 5 = 0$ **80.** $8x^6 + 513x^3 + 64 = 0$

81. $6 = 7(2w - 3)^{-1} + 3(2w - 3)^{-2}$ **82.** $m^6 - 10m^3 = -9$

83. $2x^4 - 9x^2 = -2$ **84.** $8x^4 + 1 = 11x^2$

RELATING CONCEPTS (EXERCISES 85–90)

For Individual or Group Work

*Consider the following equation, and **work Exercises 85–90 in order.***

$$\frac{x^2}{(x - 3)^2} + \frac{3x}{x - 3} - 4 = 0.$$

85. Why must 3 be excluded from the domain of this equation?

86. Multiply both sides of the equation by the LCD, $(x - 3)^2$, and solve. There is only one solution—what is it?

87. Write the equation so that it is quadratic in form using the rational expression $\frac{x}{x-3}$.

88. Explain why the expression $\frac{x}{x-3}$ cannot equal 1.

89. Solve the equation from Exercise 87 by making the substitution $t = \frac{x}{x-3}$. You should get two values for t. Why is one of them impossible for this equation?

90. Solve the equation $x^2(x-3)^{-2} + 3x(x-3)^{-1} - 4 = 0$ by letting $s = (x-3)^{-1}$. You should get two values for s. Why is this impossible for this equation?

SUMMARY EXERCISES ON SOLVING QUADRATIC EQUATIONS

*Exercises marked * require knowledge of imaginary numbers.*

We have introduced four algebraic methods for solving quadratic equations written in the form $ax^2 + bx + c = 0$: **factoring, the square root property, completing the square, and the quadratic formula.** *Refer to the summary box at the beginning of Section 9.3 to review some of the advantages and disadvantages of each method. Then solve each quadratic equation by the method of your choice.*

1. $p^2 = 7$

2. $6x^2 - x - 15 = 0$

3. $n^2 + 6n + 4 = 0$

4. $(x - 3)^2 = 25$

5. $\dfrac{5}{m} + \dfrac{12}{m^2} = 2$

6. $3m^2 = 3 - 8m$

7. $2r^2 - 4r + 1 = 0$

*8. $x^2 = -12$

9. $x\sqrt{2} = \sqrt{5x - 2}$

10. $m^4 - 10m^2 + 9 = 0$

11. $(2k + 3)^2 = 8$

12. $\dfrac{2}{x} + \dfrac{1}{x-2} = \dfrac{5}{3}$

13. $t^4 + 14 = 9t^2$

14. $8x^2 - 4x = 2$

*15. $z^2 + z + 1 = 0$

16. $5x^6 + 2x^3 - 7 = 0$

17. $4t^2 - 12t + 9 = 0$

18. $x\sqrt{3} = \sqrt{2 - x}$

19. $r^2 - 72 = 0$

20. $-3x^2 + 4x = -4$

21. $x^2 - 5x - 36 = 0$

22. $w^2 = 169$

*23. $3p^2 = 6p - 4$

24. $z = \sqrt{\dfrac{5z + 3}{2}}$

25. $2(3k - 1)^2 + 5(3k - 1) = -2$

*26. $\dfrac{4}{r^2} + 3 = \dfrac{1}{r}$

9.4 Formulas and Further Applications

OBJECTIVE 1 Solve formulas for variables involving squares and square roots. The methods presented earlier can be used to solve such formulas.

EXAMPLE 1 Solving for Variables Involving Squares or Square Roots

Solve each formula for the given variable.

(a) $w = \dfrac{kFr}{v^2}$ for v

$$w = \frac{kFr}{v^2}$$ ⟵ Get v alone on one side.

$$v^2 w = kFr$$ Multiply by v^2.

$$v^2 = \frac{kFr}{w}$$ Divide by w.

$$v = \pm\sqrt{\frac{kFr}{w}}$$ Square root property

$$v = \frac{\pm\sqrt{kFr}}{\sqrt{w}} \cdot \frac{\sqrt{w}}{\sqrt{w}} = \frac{\pm\sqrt{kFrw}}{w}$$ Rationalize the denominator.

(b) $d = \sqrt{\dfrac{4A}{\pi}}$ for A

$$d = \sqrt{\frac{4A}{\pi}}$$ ⟵ Get A alone on one side.

$$d^2 = \frac{4A}{\pi}$$ Square both sides.

$$\pi d^2 = 4A$$ Multiply by π.

$$\frac{\pi d^2}{4} = A$$ Divide by 4.

Now Try Exercises 9 and 19.

NOTE In many formulas like $v = \dfrac{\pm\sqrt{kFrw}}{w}$ in Example 1(a), we choose the positive value. In our work here, we will include both positive and negative values.

EXAMPLE 2 Solving for a Variable that Appears in First- and Second-Degree Terms

Solve $s = 2t^2 + kt$ for t.

Since the given equation has terms with t^2 and t, write it in standard form $ax^2 + bx + c = 0$, with t as the variable instead of x.

$$2t^2 + kt - s = 0$$

Now use the quadratic formula with $a = 2$, $b = k$, and $c = -s$.

$$t = \frac{-k \pm \sqrt{k^2 - 4(2)(-s)}}{2(2)}$$ Solve for t.

$$t = \frac{-k \pm \sqrt{k^2 + 8s}}{4}$$

The solutions are $t = \dfrac{-k + \sqrt{k^2 + 8s}}{4}$ and $t = \dfrac{-k - \sqrt{k^2 + 8s}}{4}$.

Now Try Exercise 15.

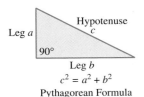

Leg a

Hypotenuse c

$90°$

Leg b

$c^2 = a^2 + b^2$

Pythagorean Formula

OBJECTIVE 2 Solve applied problems using the Pythagorean formula. The Pythagorean formula $a^2 + b^2 = c^2$, illustrated by the figure in the margin, was introduced in Chapter 8 and is used to solve applications involving right triangles. Such problems often require solving quadratic equations.

EXAMPLE 3 Using the Pythagorean Formula

Two cars left an intersection at the same time, one heading due north, the other due west. Some time later, they were exactly 100 mi apart. The car headed north had gone 20 mi farther than the car headed west. How far had each car traveled?

Step 1 **Read** the problem carefully.

Step 2 **Assign a variable.** Let x be the distance traveled by the car headed west. Then $(x + 20)$ is the distance traveled by the car headed north. See Figure 2. The cars are 100 mi apart, so the hypotenuse of the right triangle equals 100.

Step 3 **Write an equation.** Use the Pythagorean formula.

$$c^2 = a^2 + b^2$$
$$100^2 = x^2 + (x + 20)^2$$

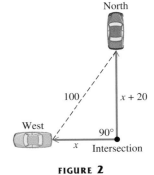

North

100 $x + 20$

West $90°$

x Intersection

FIGURE 2

Step 4 **Solve.**

$10,000 = x^2 + x^2 + 40x + 400$	Square the binomial.
$0 = 2x^2 + 40x - 9600$	Standard form
$0 = x^2 + 20x - 4800$	Divide by 2.
$0 = (x + 80)(x - 60)$	Factor.
$x + 80 = 0$ or $x - 60 = 0$	Zero-factor property
$x = -80$ or $x = 60$	

Step 5 **State the answer.** Since distance cannot be negative, discard the negative solution. The required distances are 60 mi and $60 + 20 = 80$ mi.

Step 6 **Check.** Since $60^2 + 80^2 = 100^2$, the answer is correct.

Now Try Exercise 31.

OBJECTIVE 3 Solve applied problems using area formulas.

EXAMPLE 4 Solving an Area Problem

A rectangular reflecting pool in a park is 20 ft wide and 30 ft long. The park gardener wants to plant a strip of grass of uniform width around the edge of the pool. She has enough seed to cover 336 ft². How wide will the strip be?

Step 1 **Read** the problem carefully.

Step 2 **Assign a variable.** The pool is shown in Figure 3. If x represents the unknown width of the grass strip, the width of the large rectangle is given by $20 + 2x$ (the width of the pool plus two grass strips), and the length is given by $30 + 2x$.

FIGURE 3

Step 3 **Write an equation.** The area of the large rectangle is given by the product of its length and width, $(30 + 2x)(20 + 2x)$. The area of the pool is $30 \cdot 20 = 600$ ft². The area of the large rectangle minus the area of the pool should equal the area of the grass strip. Since the area of the grass strip is to be 336 ft², the equation is

$$\underset{\substack{\text{Area}\\\text{of}\\\text{rectangle}}}{\underbrace{(30 + 2x)(20 + 2x)}} - \underset{\substack{\text{Area}\\\text{of}\\\text{pool}}}{\underbrace{600}} = \underset{\substack{\text{Area}\\\text{of}\\\text{grass}}}{\underbrace{336}}.$$

Step 4 **Solve.**

$$600 + 100x + 4x^2 - 600 = 336 \qquad \text{Multiply.}$$
$$4x^2 + 100x - 336 = 0 \qquad \text{Standard form}$$
$$x^2 + 25x - 84 = 0 \qquad \text{Divide by 4.}$$
$$(x + 28)(x - 3) = 0 \qquad \text{Factor.}$$
$$x = -28 \quad \text{or} \quad x = 3 \qquad \text{Zero-factor property}$$

Step 5 **State the answer.** The width cannot be -28 ft, so the grass strip should be 3 ft wide.

Step 6 **Check.** If $x = 3$, then the area of the large rectangle (which includes the grass strip) is

$$(30 + 2 \cdot 3)(20 + 2 \cdot 3) = 36 \cdot 26 = 936 \text{ ft}^2. \qquad \text{Area of pool and strip}$$

The area of the pool is $30 \cdot 20 = 600$ ft². So, the area of the grass strip is $936 - 600 = 336$ ft², which is the area the gardener had enough seed to cover. The answer is correct.

Now Try Exercise 37.

OBJECTIVE 4 Solve applied problems using quadratic functions as models. Some applied problems can be modeled by *quadratic functions*, which can be written in the form

$$f(x) = ax^2 + bx + c,$$

for real numbers a, b, and c, $a \neq 0$.

EXAMPLE 5 Solving an Applied Problem Using a Quadratic Function

If an object is propelled upward from the top of a 144-ft building at 112 ft per sec, its position (in feet above the ground) is given by

$$s(t) = -16t^2 + 112t + 144,$$

where t is time in seconds after it was propelled. When does it hit the ground?

When the object hits the ground, its distance above the ground is 0. We must find the value of t that makes $s(t) = 0$.

$$0 = -16t^2 + 112t + 144 \qquad \text{Let } s(t) = 0.$$
$$0 = t^2 - 7t - 9 \qquad \text{Divide by } -16.$$
$$t = \frac{7 \pm \sqrt{49 + 36}}{2} \qquad \text{Quadratic formula}$$
$$t = \frac{7 \pm \sqrt{85}}{2} \approx \frac{7 \pm 9.2}{2} \qquad \text{Use a calculator.}$$

The solutions are $t \approx 8.1$ or $t \approx -1.1$. Time cannot be negative, so we discard the negative solution. The object hits the ground about 8.1 sec after it is propelled.

Now Try Exercise 43.

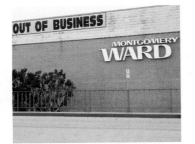

EXAMPLE 6 Using a Quadratic Function to Model Company Bankruptcy Filings

The number of companies filing for bankruptcy was high in the early 1990s due to an economic recession. The number then declined during the middle 1990s, and in recent years has increased again. The quadratic function defined by

$$f(x) = 3.37x^2 - 28.6x + 133$$

approximates the number of company bankruptcy filings during the years 1990–2001, where x is the number of years since 1990. (*Source:* www.BankruptcyData.com)

(a) Use the model to approximate the number of company bankruptcy filings in 1995.
For 1995, $x = 5$, so find $f(5)$.

$$f(5) = 3.37(5)^2 - 28.6(5) + 133 \qquad \text{Let } x = 5.$$
$$= 74.25$$

There were about 74 company bankruptcy filings in 1995.

(b) In what year did company bankruptcy filings reach 150?
Find the value of x that makes $f(x) = 150$.

$$f(x) = 3.37x^2 - 28.6x + 133$$
$$150 = 3.37x^2 - 28.6x + 133 \qquad \text{Let } f(x) = 150.$$
$$0 = 3.37x^2 - 28.6x - 17 \qquad \text{Standard form}$$

Now use $a = 3.37$, $b = -28.6$, and $c = -17$ in the quadratic formula.

$$x = \frac{28.6 \pm \sqrt{(-28.6)^2 - 4(3.37)(-17)}}{2(3.37)}$$

$$x \approx 9.0 \quad \text{or} \quad x \approx -.56 \qquad \text{Use a calculator.}$$

The positive solution is $x \approx 9$, so company bankruptcy filings reached 150 in the year $1990 + 9 = 1999$. (Reject the negative solution since the model is not valid for negative values of x.) Note that company bankruptcy filings doubled from about 74 in 1995 to 150 in 1999.

Now Try Exercises 55 and 57.

9.4 EXERCISES

✎ *Answer each question in Exercises 1–4.*

1. What is the first step in solving a formula that has the specified variable in the denominator?

2. What is the first step in solving a formula like $gw^2 = 2r$ for w?

3. What is the first step in solving a formula like $gw^2 = kw + 24$ for w?

4. Why is it particularly important to check all proposed solutions to an applied problem against the information in the original problem?

In Exercises 5 and 6, solve for m in terms of the other variables ($m > 0$).

5.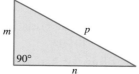

6.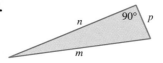

Solve each equation for the indicated variable. (Leave $\pm$ in your answers.) See Examples 1 and 2.

7. $d = kt^2$ for t

8. $s = kwd^2$ for d

9. $I = \dfrac{ks}{d^2}$ for d

10. $R = \dfrac{k}{d^2}$ for d

11. $F = \dfrac{kA}{v^2}$ for v

12. $L = \dfrac{kd^4}{h^2}$ for h

13. $V = \dfrac{1}{3}\pi r^2 h$ for r

14. $V = \pi(r^2 + R^2)h$ for r

15. $At^2 + Bt = -C$ for t

16. $S = 2\pi rh + \pi r^2$ for r

17. $D = \sqrt{kh}$ for h

18. $F = \dfrac{k}{\sqrt{d}}$ for d

19. $p = \sqrt{\dfrac{k\ell}{g}}$ for ℓ

20. $p = \sqrt{\dfrac{k\ell}{g}}$ for g

✎ **21.** If g is a positive number in the formula of Exercise 19, explain why k and ℓ must have the same sign in order for p to be a real number.

✐ **22.** Refer to Example 2 of this section. Suppose that k and s both represent positive numbers.

(a) Which one of the two solutions given is positive?

(b) Which one is negative?　　(c) How can you tell?

Solve each equation for the indicated variable.

23. $p = \dfrac{E^2 R}{(r + R)^2}$ for R　$(E > 0)$

24. $S(6S - t) = t^2$ for S

25. $10p^2 c^2 + 7pcr = 12r^2$ for r

26. $S = vt + \dfrac{1}{2} gt^2$ for t

27. $LI^2 + RI + \dfrac{1}{c} = 0$ for I

28. $P = EI - RI^2$ for I

Solve each problem. When appropriate, round answers to the nearest tenth. See Example 3.

29. Find the lengths of the sides of the triangle.

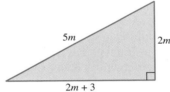

30. Find the lengths of the sides of the triangle.

31. Two ships leave port at the same time, one heading due south and the other heading due east. Several hours later, they are 170 mi apart. If the ship traveling south traveled 70 mi farther than the other ship, how many miles did they each travel?

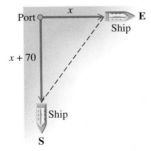

32. Allyson Pellissier is flying a kite that is 30 ft farther above her hand than its horizontal distance from her. The string from her hand to the kite is 150 ft long. How high is the kite?

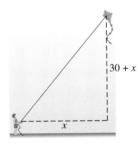

33. A toy manufacturer needs a piece of plastic in the shape of a right triangle with the longer leg 2 cm more than twice as long as the shorter leg, and the hypotenuse 1 cm more than the longer leg. How long should the three sides of the triangular piece be?

34. Michael Fuentes, a developer, owns a piece of land enclosed on three sides by streets, giving it the shape of a right triangle. The hypotenuse is 8 m longer than the longer leg, and the shorter leg is 9 m shorter than the hypotenuse. Find the lengths of the three sides of the property.

35. Two pieces of a large wooden puzzle fit together to form a rectangle with length 1 cm less than twice the width. The diagonal, where the two pieces meet, is 2.5 cm in length. Find the length and width of the rectangle.

36. A 13-ft ladder is leaning against a house. The distance from the bottom of the ladder to the house is 7 ft less than the distance from the top of the ladder to the ground. How far is the bottom of the ladder from the house?

Solve each problem. See Example 4.

37. A couple wants to buy a rug for a room that is 20 ft long and 15 ft wide. They want to leave an even strip of flooring uncovered around the edges of the room. How wide a strip will they have if they buy a rug with an area of 234 ft²?

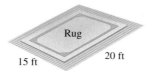

38. A club swimming pool is 30 ft wide and 40 ft long. The club members want an exposed aggregate border in a strip of uniform width around the pool. They have enough material for 296 ft². How wide can the strip be?

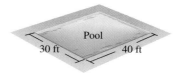

39. Arif's backyard is 20 m by 30 m. He wants to put a flower garden in the middle of the backyard, leaving a strip of grass of uniform width around the flower garden. Arif must have 184 m² of grass. Under these conditions, what will the length and width of the garden be?

40. A rectangle has a length 2 m less than twice its width. When 5 m are added to the width, the resulting figure is a square with an area of 144 m². Find the dimensions of the original rectangle.

41. A rectangular piece of sheet metal has a length that is 4 in. less than twice the width. A square piece 2 in. on a side is cut from each corner. The sides are then turned up to form an uncovered box of volume 256 in.³. Find the length and width of the original piece of metal.

42. Another rectangular piece of sheet metal is 2 in. longer than it is wide. A square piece 3 in. on a side is cut from each corner. The sides are then turned up to form an uncovered box of volume 765 in.³. Find the dimensions of the original piece of metal.

Solve each problem. When appropriate, round answers to the nearest tenth. See Example 5.

43. An object is projected directly upward from the ground. After t sec its distance in feet above the ground is

$$s(t) = 144t - 16t^2.$$

After how many seconds will the object be 128 ft above the ground? (*Hint:* Look for a common factor before solving the equation.)

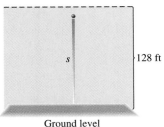

128 ft

Ground level

44. When does the object in Exercise 43 strike the ground?

45. A ball is projected upward from the ground. Its distance in feet from the ground in t sec is given by

$$s(t) = -16t^2 + 128t.$$

At what times will the ball be 213 ft from the ground?

46. A toy rocket is launched from ground level. Its distance in feet from the ground in t sec is given by

$$s(t) = -16t^2 + 208t.$$

At what times will the rocket be 550 ft from the ground?

213 ft

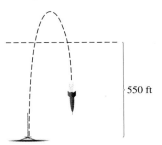

550 ft

47. The function defined by

$$D(t) = 13t^2 - 100t$$

gives the distance in feet a car going approximately 68 mph will skid in t sec. Find the time it would take for the car to skid 180 ft.

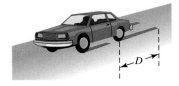

48. The function given in Exercise 47 becomes

$$D(t) = 13t^2 - 73t$$

for a car going 50 mph. Find the time for this car to skid 218 ft.

✍ *A rock is projected upward from ground level, and its distance in feet from the ground in t sec is given by $s(t) = -16t^2 + 160t$. Use algebra and a short explanation to answer Exercises 49 and 50.*

49. After how many seconds does it reach a height of 400 ft? How would you describe in words its position at this height?

50. After how many seconds does it reach a height of 425 ft? How would you interpret the mathematical result here?

Solve each problem using a quadratic equation.

51. A certain bakery has found that the daily demand for bran muffins is $\frac{3200}{p}$, where p is the price of a muffin in cents. The daily supply is $3p - 200$. Find the price at which supply and demand are equal.

52. In one area the demand for compact discs is $\frac{700}{P}$ per day, where P is the price in dollars per disc. The supply is $5P - 1$ per day. At what price does supply equal demand?

53. The formula $A = P(1 + r)^2$ gives the amount A in dollars that P dollars will grow to in 2 yr at interest rate r (where r is given as a decimal), using compound interest. What interest rate will cause $2000 to grow to $2142.25 in 2 yr?

54. If a square piece of cardboard has 3-in. squares cut from its corners and then has the flaps folded up to form an open-top box, the volume of the box is given by the formula $V = 3(x - 6)^2$, where x is the length of each side of the original piece of cardboard in inches. What original length would yield a box with volume 432 in.^{3}?

Sales of SUVs (sport utility vehicles) in the United States (in millions) for the years 1990 through 1999 are shown in the bar graph and can be modeled by the quadratic function defined by

$$f(x) = .016x^2 + .124x + .787.$$

Here, $x = 0$ represents 1990, $x = 1$ represents 1991, and so on. Use the graph and the model to work Exercises 55–58. See Example 6.

SALES OF SUVS IN THE UNITED STATES (IN MILLIONS)

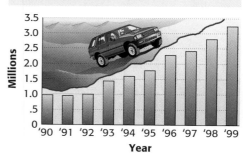

Source: CNW Marketing Research of Bandon, OR, based on automakers' reported sales.

55. **(a)** Use the graph to estimate sales in 1997 to the nearest tenth.

(b) Use the model to approximate sales in 1997 to the nearest tenth. How does this result compare to your estimate from part (a)?

56. **(a)** Use the model to estimate sales in 2000 to the nearest tenth.

(b) Sales through October 2000 were about 2.9 million. Based on this, is the sales estimate for 2000 from part (a) reasonable? Explain.

57. Based on the model, in what year did sales reach 2 million? (Round down to the nearest year.) How does this result compare to the sales shown in the graph?

58. Based on the model, in what year did sales reach 3 million? (Round down to the nearest year.) How does this result compare to the sales shown in the graph?

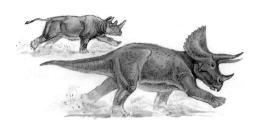

William Froude was a 19th century naval architect who used the expression

$$\frac{v^2}{g\ell}$$

in shipbuilding. This expression, known as the Froude number, was also used by R. McNeill Alexander in his research on dinosaurs. (Source: *"How Dinosaurs Ran,"* Scientific American, *April 1991.) In Exercises 59 and 60, find the value of v (in meters per second), given g = 9.8 m per sec².*

59. Rhinoceros: $\ell = 1.2$;
Froude number $= 2.57$

60. Triceratops: $\ell = 2.8$;
Froude number $= .16$

Recall that corresponding sides of similar triangles are proportional. Use this fact to find the lengths of the indicated sides of each pair of similar triangles. Check all possible solutions in both triangles. Sides of a triangle cannot be negative (and are not drawn to scale here).

61. Side *AC*

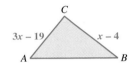

62. Side *RQ*

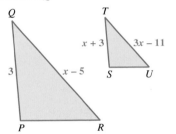

9.5 Quadratic and Rational Inequalities

OBJECTIVES

1 Solve quadratic inequalities.

2 Solve polynomial inequalities of degree 3 or more.

3 Solve rational inequalities.

We discussed methods of solving linear inequalities in Chapter 3 and methods of solving quadratic equations in this chapter. Now we combine these ideas to solve *quadratic inequalities.*

Quadratic Inequality

A **quadratic inequality** can be written in the form

$$ax^2 + bx + c < 0 \qquad \text{or} \qquad ax^2 + bx + c > 0,$$

where a, b, and c are real numbers, with $a \neq 0$.

As before, the symbols $<$ and $>$ may be replaced with $\leq$ and $\geq$.

OBJECTIVE 1 Solve quadratic inequalities.

■ **EXAMPLE 1 Solving a Quadratic Inequality Using Test Numbers**

Solve $x^2 - x - 12 > 0$.

First solve the related quadratic *equation* $x^2 - x - 12 = 0$ by factoring.

$$(x - 4)(x + 3) = 0 \qquad \text{Factor.}$$

$$x - 4 = 0 \quad \text{or} \quad x + 3 = 0 \qquad \text{Zero-factor property}$$

$$x = 4 \quad \text{or} \qquad x = -3$$

The numbers 4 and -3 divide a number line into the three intervals shown in Figure 4. Be careful to put the smaller number on the left.

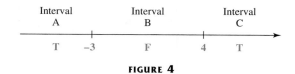

FIGURE 4

The numbers 4 and -3 are the only numbers that make the expression $x^2 - x - 12$ equal to 0. All other numbers make the expression either positive or negative. The sign of the expression can change from positive to negative or from negative to positive only at a number that makes it 0. Therefore, if one number in an interval satisfies the inequality, then all the numbers in that interval will satisfy the inequality.

To see if the numbers in Interval A satisfy the inequality, choose any number from Interval A in Figure 4 (that is, any number less than -3). Substitute this test number for x in the original inequality $x^2 - x - 12 > 0$. If the result is *true,* then all numbers in Interval A satisfy the inequality.

Try -5 from Interval A. Substitute -5 for x.

$$x^2 - x - 12 > 0 \qquad \text{Original inequality}$$

$$(-5)^2 - (-5) - 12 > 0 \qquad ?$$

$$25 + 5 - 12 > 0 \qquad ?$$

$$18 > 0 \qquad \text{True}$$

Because -5 from Interval A satisfies the inequality, all numbers from Interval A are solutions.

Now try 0 from Interval B. If $x = 0$, then

$$0^2 - 0 - 12 > 0 \qquad ?$$

$$-12 > 0. \qquad \text{False}$$

The numbers in Interval B are *not* solutions. Now try 5 from Interval C. If $x = 5$, then

$$5^2 - 5 - 12 > 0 \qquad ?$$

$$8 > 0. \qquad \text{True}$$

Since the test number 5 satisfies the inequality, the numbers in Interval C are solutions.

Based on these results (shown by the colored letters in Figure 4), the solution set includes the numbers in Intervals A and C, as shown on the graph in Figure 5. The solution set is written in interval notation as

$$(-\infty, -3) \cup (4, \infty).$$

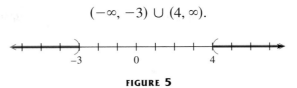

FIGURE 5

Now Try Exercise 7.

In summary, follow these steps to solve a quadratic inequality.

> **Solving a Quadratic Inequality**
>
> *Step 1*   **Write the inequality as an equation and solve it.**
>
> *Step 2*   **Use the solutions from Step 1 to determine intervals.** Graph the numbers found in Step 1 on a number line. These numbers divide the number line into intervals.
>
> *Step 3*   **Find the intervals that satisfy the inequality.** Substitute a test number from each interval into the original inequality to determine the intervals that satisfy the inequality. All numbers in those intervals are in the solution set. A graph of the solution set will usually look like one of these. (Square brackets might be used instead of parentheses.)
>
>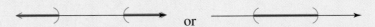
>
> *Step 4*   **Consider the endpoints separately.** The numbers from Step 1 are included in the solution set if the inequality symbol is $\le$ or $\ge$; they are not included if it is $<$ or $>$.

Special cases of quadratic inequalities may occur, as in the next example.

EXAMPLE 2   Solving Special Cases

Solve $(2t - 3)^2 > -1$. Then solve $(2t - 3)^2 < -1$.

Because $(2t - 3)^2$ is never negative, it is always greater than -1. Thus, the solution set for $(2t - 3)^2 > -1$ is the set of all real numbers, $(-\infty, \infty)$. In the same way, there is no solution for $(2t - 3)^2 < -1$ and the solution set is $\emptyset$.

Now Try Exercises 21 and 23.

OBJECTIVE 2 Solve polynomial inequalities of degree 3 or more.   Higher-degree polynomial inequalities that can be factored are solved in the same way as quadratic inequalities.

EXAMPLE 3 Solving a Third-Degree Polynomial Inequality

Solve $(x - 1)(x + 2)(x - 4) \leq 0$.

This is a *cubic* (third-degree) inequality rather than a quadratic inequality, but it can be solved using the method shown in the box by extending the zero-factor property to more than two factors. Begin by setting the factored polynomial *equal* to 0 and solving the equation. (Step 1)

$$(x - 1)(x + 2)(x - 4) = 0$$
$$x - 1 = 0 \quad \text{or} \quad x + 2 = 0 \quad \text{or} \quad x - 4 = 0$$
$$x = 1 \quad \text{or} \quad x = -2 \quad \text{or} \quad x = 4$$

Locate the numbers -2, 1, and 4 on a number line, as in Figure 6, to determine the Intervals A, B, C, and D. (Step 2)

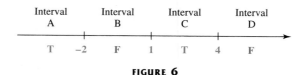

FIGURE 6

Substitute a test number from each interval in the *original* inequality to determine which intervals satisfy the inequality. (Step 3) Use a table to organize this information.

Interval	Test Number	Test of Inequality	True or False?
A	-3	$-28 \leq 0$	T
B	0	$8 \leq 0$	F
C	2	$-8 \leq 0$	T
D	5	$28 \leq 0$	F

Verify the information given in the table and graphed in Figure 7. The numbers in Intervals A and C are in the solution set, which is written in interval notation as

$$(-\infty, -2] \cup [1, 4].$$

The three endpoints are included since the inequality symbol is $\leq$. (Step 4)

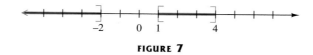

FIGURE 7

Now Try Exercise 25.

OBJECTIVE 3 Solve rational inequalities. Inequalities that involve rational expressions, called **rational inequalities,** are solved similarly using the following steps.

Solving a Rational Inequality

Step 1   **Write the inequality** so that 0 is on one side and there is a single fraction on the other side.

Step 2   **Determine the numbers that make the numerator or denominator equal to 0.**

Step 3   **Divide a number line into intervals.** Use the numbers from Step 2.

Step 4   **Find the intervals that satisfy the inequality.** Test a number from each interval by substituting it into the *original* inequality.

Step 5   **Consider the endpoints separately.** Exclude any values that make the denominator 0.

CAUTION   As indicated in Step 5, any number that makes the denominator 0 *must* be excluded from the solution set.

■ **EXAMPLE 4**   Solving a Rational Inequality

Solve $\dfrac{-1}{p-3} > 1$.

Write the inequality so that 0 is on one side. (Step 1)

$$\frac{-1}{p-3} - 1 > 0 \qquad \text{Subtract 1.}$$

$$\frac{-1}{p-3} - \frac{p-3}{p-3} > 0 \qquad \text{Use } p-3 \text{ as the common denominator.}$$

$$\frac{-1 - p + 3}{p-3} > 0 \qquad \begin{array}{l}\text{Write the left side as a single fraction;}\\ \text{be careful with signs in the numerator.}\end{array}$$

$$\frac{-p + 2}{p-3} > 0 \qquad \text{Combine terms.}$$

The sign of the rational expression $\frac{-p+2}{p-3}$ will change from positive to negative or negative to positive only at those numbers that make the numerator or denominator 0. The number 2 makes the numerator 0, and 3 makes the denominator 0. (Step 2) These two numbers, 2 and 3, divide a number line into three intervals. See Figure 8. (Step 3)

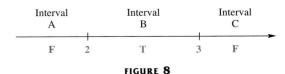

FIGURE 8

Testing a number from each interval in the *original* inequality, $\dfrac{-1}{p-3} > 1$, gives the results shown in the table on the next page. (Step 4)

Interval	Test Number	Test of Inequality	True or False?
A	0	$\frac{1}{3} > 1$	F
B	2.5	$2 > 1$	T
C	4	$-1 > 1$	F

The solution set is the interval $(2, 3)$. This interval does not include 3 since it would make the denominator of the original equality 0; 2 is not included either since the inequality symbol is $>$. (Step 5) A graph of the solution set is given in Figure 9.

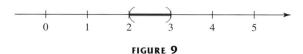

FIGURE 9

Now Try Exercise 33.

EXAMPLE 5 Solving a Rational Inequality

Solve $\dfrac{m - 2}{m + 2} \le 2$.

Write the inequality so that 0 is on one side. (Step 1)

$$\frac{m - 2}{m + 2} - 2 \le 0 \qquad \text{Subtract 2.}$$

$$\frac{m - 2}{m + 2} - \frac{2(m + 2)}{m + 2} \le 0 \qquad \text{Use } m + 2 \text{ as the common denominator.}$$

$$\frac{m - 2 - 2m - 4}{m + 2} \le 0 \qquad \text{Write as a single fraction.}$$

$$\frac{-m - 6}{m + 2} \le 0 \qquad \text{Combine terms.}$$

The number -6 makes the numerator 0, and -2 makes the denominator 0. (Step 2) These two numbers determine three intervals. See Figure 10. (Step 3)

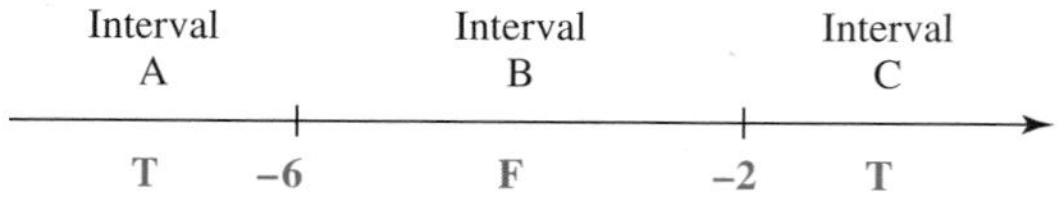

	Interval A		Interval B		Interval C	
	T	-6	F	-2	T	

FIGURE 10

Test one number from each interval in the original inequality. (Step 4)

Interval	Test Number	Test of Inequality	True or False?
A	-8	$\frac{5}{3} \le 2$	T
B	-4	$3 \le 2$	F
C	0	$-1 \le 2$	T

The solution set is

$$(-\infty, -6] \cup (-2, \infty).$$

The number -6 satisfies the original inequality, but -2 cannot be used as a solution since it makes the denominator 0. (Step 5) A graph of the solution set is shown in Figure 11.

FIGURE 11

Now Try Exercise 37.

9.5 EXERCISES

1. Explain how to determine whether to include or exclude endpoints when solving a quadratic or higher-degree inequality.

2. The solution set of the inequality $x^2 + x - 12 < 0$ is the interval $(-4, 3)$. Without actually performing any work, give the solution set of the inequality $x^2 + x - 12 \geq 0$.

Solve each inequality and graph the solution set. See Example 1. (Hint: In Exercises 19 and 20, use the quadratic formula.)

3. $(x + 1)(x - 5) > 0$ **4.** $(m + 6)(m - 2) > 0$ **5.** $(r + 4)(r - 6) < 0$

6. $(x + 4)(x - 8) < 0$ **7.** $x^2 - 4x + 3 \geq 0$ **8.** $m^2 - 3m - 10 \geq 0$

9. $10t^2 + 9t \geq 9$ **10.** $3r^2 + 10r \geq 8$ **11.** $4x^2 - 9 \leq 0$

12. $9x^2 - 25 \leq 0$ **13.** $6x^2 + x \geq 1$ **14.** $4p^2 + 7p \geq -3$

15. $z^2 - 4z \geq 0$ **16.** $x^2 + 2x < 0$ **17.** $3k^2 - 5k \leq 0$

18. $2z^2 + 3z > 0$ **19.** $x^2 - 6x + 6 \geq 0$ **20.** $3k^2 - 6k + 2 \leq 0$

Solve each inequality. See Example 2.

21. $(4 - 3x)^2 \geq -2$ **22.** $(6z + 7)^2 \geq -1$

23. $(3x + 5)^2 \leq -4$ **24.** $(8t + 5)^2 \leq -5$

Solve each inequality and graph the solution set. See Example 3.

25. $(p - 1)(p - 2)(p - 4) < 0$ **26.** $(2r + 1)(3r - 2)(4r + 7) < 0$

27. $(x - 4)(2x + 3)(3x - 1) \geq 0$ **28.** $(z + 2)(4z - 3)(2z + 7) \geq 0$

Solve each inequality and graph the solution set. See Examples 4 and 5.

29. $\dfrac{x - 1}{x - 4} > 0$ **30.** $\dfrac{x + 1}{x - 5} > 0$

31. $\dfrac{2n + 3}{n - 5} \leq 0$ **32.** $\dfrac{3t + 7}{t - 3} \leq 0$

33. $\dfrac{8}{x - 2} \geq 2$ **34.** $\dfrac{20}{x - 1} \geq 1$

35. $\dfrac{3}{2t - 1} < 2$ **36.** $\dfrac{6}{m - 1} < 1$

37. $\dfrac{g - 3}{g + 2} \geq 2$

38. $\dfrac{m + 4}{m + 5} \geq 2$

39. $\dfrac{x - 8}{x - 4} < 3$

40. $\dfrac{2t - 3}{t + 1} > 4$

41. $\dfrac{4k}{2k - 1} < k$

42. $\dfrac{r}{r + 2} < 2r$

43. $\dfrac{2x - 3}{x^2 + 1} \geq 0$

44. $\dfrac{9x - 8}{4x^2 + 25} < 0$

45. $\dfrac{(3x - 5)^2}{x + 2} > 0$

46. $\dfrac{(5x - 3)^2}{2x + 1} \leq 0$

RELATING CONCEPTS (EXERCISES 47–50)

For Individual or Group Work

A rock is projected vertically upward from the ground. Its distance s in feet above the ground after t sec is given by the quadratic function defined by

$$s(t) = -16t^2 + 256t.$$

Work Exercises 47–50 in order, *to see how quadratic equations and inequalities are related.*

47. At what times will the rock be 624 ft above the ground? (*Hint:* Let $s(t) = 624$ and solve the quadratic *equation.*)

48. At what times will the rock be more than 624 ft above the ground? (*Hint:* Set $s(t) > 624$ and solve the quadratic *inequality.*)

49. At what times will the rock be at ground level? (*Hint:* Let $s(t) = 0$ and solve the quadratic *equation.*)

50. At what times will the rock be less than 624 ft above the ground? (*Hint:* Set $s(t) < 624$, solve the quadratic *inequality,* and observe the solutions in Exercises 48 and 49 to determine the smallest and largest possible values of t.)

Chapter 9   Group Activity

Kepler's Three Laws

OBJECTIVE   Investigate Kepler's three laws of planetary motion.

Johannes Kepler's three laws of planetary motion are landmarks in the history of astronomy and mathematics. Kepler's deep interest in astronomy led him to accept a lectureship at the University of Grätz in Austria at the age of 23. In 1599, Kepler became an assistant to the famous Danish-Swedish astronomer Tycho Brahe, court astronomer to Kaiser Rudolph II. Two years later Brahe suddenly died leaving a mass of data on planetary motion and his position as court astronomer to Kepler. Kepler formulated his first two laws in 1609 and ten years later his third law of planetary motion.*

The three laws are:

A. *All planets travel around the sun in elliptical-shaped orbits with the sun at one focus.* See the figure.

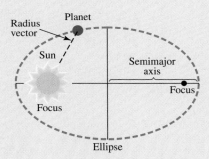

1. Construct an ellipse using a string approximately 11 in. long, a pencil, and a sheet of graph paper. Begin by choosing two points to represent the foci (plural of focus) of the ellipse. One group member should hold the ends of the string to the foci with his or her index fingers, making sure to leave the string slack between the foci. The other group member then slips the pencil tip under the string, pulling it tight and tracing around the foci completely.

2. Practice this activity several times. Try adjusting the distance between the foci. What do you notice about the shape of the ellipse at these different positions?

B. *The radius vector, an imaginary straight line joining a planet to the sun, sweeps over equal areas in equal intervals of time.* This means that a planet moves fastest in its orbit when it is at *perihelion,* the point at which a body in orbit around the sun comes closest to the sun, and slowest when it is at *aphelion,* the point at which a body orbiting the sun is farthest from the sun.

---

*Sources: Howard Eves, *An Introduction to the History of Mathematics,* Holt, Rinehart, and Winston, 1990. *Longman Illustrated Dictionary of Astronomy and Astronautics, The Terminology of Space,* Longman York Press, 1988.

Use one of the ellipses created in part A1 to demonstrate Kepler's second law. Mark two points P_1 and P_2 near the perihelion and points P_3 and P_4 near the aphelion. Choose the points carefully using squares on the graph paper to approximate the areas swept between points. Shade the areas created between the points.

C. *For each planet in the solar system, the square of its orbital period, P, in years, equals the cube of its semimajor axis, a, in astronomical units, or $P^2 = a^3$. This means the greater the average distance of a planet from the sun, the longer it takes to complete one orbit.*

Complete the table by using the square root property to find the orbital period in years given the average distance a planet lies from the sun in astronomical units.

Planet	a (AU)	P, orbital period in years
Mercury	.387	
Venus	.723	.615
Earth	1	
Mars	1.524	
Jupiter	5.20	11.858
Saturn	9.54	
Neptune	30.1	
Uranus	19.18	
Pluto	39.53	

Source: James B. Kaler, *Astronomy! A Brief Edition,* Addison Wesley, 1997.

Note: One astronomical unit (AU) is the average distance of Earth from the sun, i.e., 149,600,000 km. The astronomical unit is the basic unit of distance in the solar system.

CHAPTER 9 SUMMARY

KEY TERMS

9.1 quadratic equation
9.2 quadratic formula
     discriminant

9.3 quadratic in form

9.5 quadratic inequality
     rational inequality

TEST YOUR WORD POWER

See how well you have learned the vocabulary in this chapter. Answers, with examples, follow the Quick Review.

1. The **quadratic formula** is
  A. a formula to find the number of
    solutions of a quadratic equation
  B. a formula to find the type of
    solutions of a quadratic equation
  C. the standard form of a quadratic
    equation
  D. a general formula for solving any
    quadratic equation.

2. The **discriminant** is
  A. the quantity under the radical in
    the quadratic formula
  B. the quantity in the denominator in
    the quadratic formula
  C. the solution set of a quadratic
    equation
  D. the result of using the quadratic
    formula.

3. A **quadratic inequality** is a
    polynomial inequality of
  A. degree one
  B. degree two
  C. degree three
  D. degree four.

4. A **rational inequality** is an
    inequality that
  A. has a second-degree term
  B. has three factors
  C. involves a fraction
  D. involves a radical.

QUICK REVIEW

CONCEPTS	EXAMPLES

9.1 THE SQUARE ROOT PROPERTY AND COMPLETING THE SQUARE

Square Root Property
If x and k are complex numbers and $x^2 = k$, then
$$x = \sqrt{k} \quad \text{or} \quad x = -\sqrt{k}.$$

Solve $(x - 1)^2 = 8$.

$$x - 1 = \sqrt{8} \qquad \text{or} \quad x - 1 = -\sqrt{8}$$
$$x = 1 + 2\sqrt{2} \quad \text{or} \qquad x = 1 - 2\sqrt{2}$$

Solution set: $\left\{1 + 2\sqrt{2},\, 1 - 2\sqrt{2}\right\}$

Completing the Square
To solve $ax^2 + bx + c = 0 \; (a \neq 0)$:

Step 1 If $a \neq 1$, divide each side by a.

Step 2 Write the equation with the variable
    terms on one side and the constant on the
    other.

Step 3 Take half the coefficient of x and square it.

Solve $2x^2 - 4x - 18 = 0$.

$$x^2 - 2x - 9 = 0 \qquad \text{Divide by 2.}$$
$$x^2 - 2x = 9 \qquad \text{Add 9.}$$

$$\left[\frac{1}{2}(-2)\right]^2 = (-1)^2 = 1$$

(continued)

CONCEPTS	EXAMPLES

Step 4 Add the square to each side.

Step 5 Factor the perfect square trinomial, and write it as the square of a binomial. Simplify the other side.

Step 6 Use the square root property to complete the solution.

$$x^2 - 2x + 1 = 9 + 1$$
$$(x - 1)^2 = 10$$

$$x - 1 = \sqrt{10} \quad \text{or} \quad x - 1 = -\sqrt{10}$$
$$x = 1 + \sqrt{10} \quad \text{or} \quad x = 1 - \sqrt{10}$$

Solution set: $\left\{ 1 + \sqrt{10}, 1 - \sqrt{10} \right\}$

9.2 THE QUADRATIC FORMULA

Quadratic Formula

The solutions of $ax^2 + bx + c = 0 \ (a \neq 0)$ are given by

$$x = \frac{-b \pm \sqrt{b^2 - 4ac}}{2a}.$$

Solve $3x^2 + 5x + 2 = 0$.

$$x = \frac{-5 \pm \sqrt{5^2 - 4(3)(2)}}{2(3)} = \frac{-5 \pm 1}{6}$$

$$x = -1 \quad \text{or} \quad x = -\frac{2}{3}$$

Solution set: $\left\{ -1, -\frac{2}{3} \right\}$

The Discriminant

If a, b, and c are integers, then the discriminant, $b^2 - 4ac$, of $ax^2 + bx + c = 0$ determines the number and type of solutions as follows.

Discriminant	Number and Type of Solutions
Positive, the square of an integer	Two rational solutions
Positive, not the square of an integer	Two irrational solutions
Zero	One rational solution
Negative	Two imaginary solutions

For $x^2 + 3x - 10 = 0$, the discriminant is

$$3^2 - 4(1)(-10) = 49. \quad \text{Two rational solutions}$$

For $4x^2 + x + 1 = 0$, the discriminant is

$$1^2 - 4(4)(1) = -15. \quad \text{Two imaginary solutions}$$

9.3 EQUATIONS QUADRATIC IN FORM

A nonquadratic equation that can be written in the form

$$au^2 + bu + c = 0,$$

for $a \neq 0$ and an algebraic expression u, is called quadratic in form. Substitute u for the expression, solve for u, and then solve for the variable in the expression.

Solve $3(x + 5)^2 + 7(x + 5) + 2 = 0$.

$$3u^2 + 7u + 2 = 0 \quad \text{Let } u = x + 5.$$
$$(3u + 1)(u + 2) = 0$$

$$u = -\frac{1}{3} \quad \text{or} \quad u = -2$$

$$x + 5 = -\frac{1}{3} \quad \text{or} \quad x + 5 = -2 \quad x + 5 = u$$

$$x = -\frac{16}{3} \quad \text{or} \quad x = -7$$

Solution set: $\left\{ -7, -\frac{16}{3} \right\}$

CONCEPTS	EXAMPLES

9.4 FORMULAS AND FURTHER APPLICATIONS

To solve a formula for a squared variable, proceed as follows.

(a) If the variable appears only to the second power: Isolate the squared variable on one side of the equation, and then use the square root property.

Solve $A = \dfrac{2mp}{r^2}$ for r.

$$r^2 A = 2mp \qquad \text{Multiply by } r^2.$$

$$r^2 = \frac{2mp}{A} \qquad \text{Divide by } A.$$

$$r = \pm \sqrt{\frac{2mp}{A}} \qquad \text{Square root property}$$

$$r = \frac{\pm\sqrt{2mpA}}{A} \qquad \text{Rationalize the denominator.}$$

(b) If the variable appears to the first and second powers: Write the equation in standard form, and then use the quadratic formula.

Solve $m^2 + rm = t$ for m.

$$m^2 + rm - t = 0 \qquad \text{Standard form}$$

$$m = \frac{-r \pm \sqrt{r^2 - 4(1)(-t)}}{2(1)} \qquad a = 1,\ b = r,\ c = -t$$

$$m = \frac{-r \pm \sqrt{r^2 + 4t}}{2}$$

9.5 QUADRATIC AND RATIONAL INEQUALITIES

Solving a Quadratic (or Higher-Degree Polynomial) Inequality

Step 1 Write the inequality as an equation and solve.

Solve $2x^2 + 5x + 2 < 0$.

$$2x^2 + 5x + 2 = 0$$

$$x = -\frac{1}{2} \quad \text{or} \quad x = -2$$

Step 2 Use the numbers found in Step 1 to divide a number line into intervals.

Step 3 Substitute a test number from each interval into the original inequality to determine the intervals that belong to the solution set.

$x = -3$ makes the original inequality false; $x = -1$ makes it true; $x = 0$ makes it false.

Step 4 Consider the endpoints separately.

The solution set is the interval $\left(-2, -\frac{1}{2}\right)$.

Solving a Rational Inequality

Step 1 Write the inequality so that 0 is on one side and there is a single fraction on the other side.

Solve $\dfrac{x}{x+2} \geq 4$.

$$\frac{x}{x+2} - 4 \geq 0$$

$$\frac{x}{x+2} - \frac{4(x+2)}{x+2} \geq 0$$

$$\frac{-3x - 8}{x+2} \geq 0$$

(continued)

CONCEPTS	EXAMPLES
Step 2 Determine the numbers that make the numerator or denominator 0.	$-\frac{8}{3}$ makes the numerator 0; -2 makes the denominator 0.
Step 3 Use the numbers from Step 2 to divide a number line into intervals.	
Step 4 Substitute a test number from each interval into the original inequality to determine the intervals that belong to the solution set.	-4 makes the original inequality false; $-\frac{7}{3}$ makes it true; 0 makes it false.
Step 5 Consider the endpoints separately.	The solution set is the interval $\left[-\frac{8}{3}, -2\right)$. The endpoint -2 is not included since it makes the denominator 0.

Answers to Test Your Word Power

1. D; *Example:* The solutions of $ax^2 + bx + c = 0 \ (a \neq 0)$ are given by $x = \dfrac{-b \pm \sqrt{b^2 - 4ac}}{2a}$. **2.** A; *Example:* In the quadratic formula,

the discriminant is $b^2 - 4ac$. **3.** B; *Examples:* $(x + 1)(2x - 5) > 4$, $2x^2 + 5x < 0$, $x^2 - 4x \leq 4$ **4.** C; *Examples:* $\dfrac{x - 1}{x + 3} < 0$,

$\dfrac{5}{y - 2} > 3$, $\dfrac{x}{2x - 1} \geq 3x$

CHAPTER 9 REVIEW EXERCISES

Exercises marked * require knowledge of imaginary numbers.

[9.1] *Solve each equation by using the square root property or completing the square.*

1. $t^2 = 121$ **2.** $p^2 = 3$ **3.** $(2x + 5)^2 = 100$

*4. $(3k - 2)^2 = -25$ **5.** $x^2 + 4x = 15$ **6.** $2m^2 - 3m = -1$

7. A student gave the following "solution" to the equation $x^2 = 12$.

$$x^2 = 12$$
$$x = \sqrt{12} \qquad \text{Square root property}$$
$$x = 2\sqrt{3}$$

What is wrong with this solution?

8. Navy Pier Center in Chicago, Illinois, features a 150-ft tall Ferris wheel. Use Galileo's formula $d = 16t^2$ to find how long it would take a wallet dropped from the top of the Ferris wheel to fall to the ground. Round your answer to the nearest tenth of a second. (*Source: Microsoft Encarta Encyclopedia 2002.*)

[9.2] *Solve each equation using the quadratic formula.*

9. $2x^2 + x - 21 = 0$ **10.** $k^2 + 5k = 7$ **11.** $(t + 3)(t - 4) = -2$

*12. $2x^2 + 3x + 4 = 0$ ***13.** $3p^2 = 2(2p - 1)$ **14.** $m(2m - 7) = 3m^2 + 3$

Use the discriminant to predict whether the solutions to each equation are

A. *two rational numbers;*   **B.** *one rational number;*
C. *two irrational numbers;*   **D.** *two imaginary numbers.*

15. $x^2 + 5x + 2 = 0$ **16.** $4t^2 = 3 - 4t$

17. $4x^2 = 6x - 8$ **18.** $9z^2 + 30z + 25 = 0$

[9.3] *Solve each equation.*

19. $\dfrac{15}{x} = 2x - 1$ **20.** $\dfrac{1}{n} + \dfrac{2}{n + 1} = 2$

21. $-2r = \sqrt{\dfrac{48 - 20r}{2}}$ **22.** $8(3x + 5)^2 + 2(3x + 5) - 1 = 0$

23. $2x^{2/3} - x^{1/3} - 28 = 0$ **24.** $p^4 - 10p^2 + 9 = 0$

Solve each problem. Round answers to the nearest tenth, as necessary.

25. Phong paddled his canoe 20 mi upstream, then paddled back. If the speed of the current was 3 mph and the total trip took 7 hr, what was Phong's speed?

26. Maureen O'Connor drove 8 mi to pick up her friend Laurie, and then drove 11 mi to a mall at a speed 15 mph faster. If Maureen's total travel time was 24 min, what was her speed on the trip to pick up Laurie?

27. An old machine processes a batch of checks in 1 hr more time than a new one. How long would it take the old machine to process a batch of checks that the two machines together process in 2 hr?

28. Greg Tobin can process a stack of invoices 1 hr faster than Carter Fenton can. Working together, they take 1.5 hr. How long would it take each person working alone?

[9.4] *Solve each formula for the indicated variable. (Give answers with ±.)*

29. $k = \dfrac{rF}{wv^2}$ for v **30.** $p = \sqrt{\dfrac{yz}{6}}$ for y **31.** $mt^2 = 3mt + 6$ for t

Solve each problem. Round answers to the nearest tenth, as necessary.

32. A large machine requires a part in the shape of a right triangle with a hypotenuse 9 ft less than twice the length of the longer leg. The shorter leg must be $\frac{3}{4}$ the length of the longer leg. Find the lengths of the three sides of the part.

33. A square has an area of 256 cm². If the same amount is removed from one dimension and added to the other, the resulting rectangle has an area 16 cm² less. Find the dimensions of the rectangle.

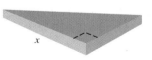

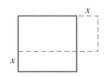

34. Nancy wants to buy a mat for a photograph that measures 14 in. by 20 in. She wants to have an even border around the picture when it is mounted on the mat. If the area of the mat she chooses is 352 in.², how wide will the border be?

35. A searchlight moves horizontally back and forth along a wall with the distance of the light from a starting point at t min given by the quadratic function defined by

$$f(t) = 100t^2 - 300t.$$

How long will it take before the light returns to the starting point?

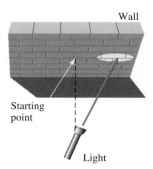

36. The Mart Hotel in Dallas, Texas, is 400 ft high. Suppose that a ball is projected upward from the top of the Mart, and its position in feet above the ground is given by the quadratic function defined by

$$f(t) = -16t^2 + 45t + 400,$$

where t is the number of seconds elapsed. How long will it take for the ball to reach a height of 200 ft above the ground? (*Source: World Almanac and Book of Facts,* 2002.)

37. The Toronto Dominion Center in Winnipeg, Manitoba, is 407 ft high. Suppose that a ball is projected upward from the top of the center, and its position in feet above the ground is given by the quadratic function defined by

$$s(t) = -16t^2 + 75t + 407,$$

where t is the number of seconds elapsed. How long will it take for the ball to reach a height of 450 ft above the ground? (*Source: World Almanac and Book of Facts,* 2002.)

38. The manager of a restaurant has determined that the demand for frozen yogurt is $\frac{25}{p}$ units per day, where p is the price (in dollars) per unit. The supply is $70p + 15$ units per day. Find the price at which supply and demand are equal.

39. Use the formula $A = P(1 + r)^2$ to find the interest rate r at which a principal P of $10,000 will increase to $10,920.25 in 2 yr.

40. The number of e-mail boxes in North America (in millions) for the years 1995 through 2001 are shown in the graph and can be modeled by the quadratic function defined by

$$f(x) = 3.29x^2 - 10.4x + 21.6.$$

In the model, $x = 5$ represents 1995, $x = 10$ represents 2000, and so on.

(a) Use the model to approximate the number of e-mail boxes in 2001 to the nearest whole number. How does this result compare to the number shown in the graph?

GROWTH OF E-MAIL BOXES IN NORTH AMERICA

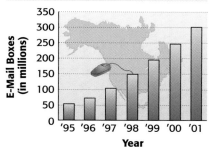

Source: IDC research.

(b) Based on the model, in what year did the number of e-mail boxes reach 200 million? (Round down to the nearest year.) How does this result compare to the number shown in the graph?

[9.5]   *Solve each inequality and graph the solution set.*

41. $x^2 + x \le 12$

42. $(x - 4)(2x + 3) > 0$

43. $(4m + 3)^2 \le -4$

44. $(x + 2)(x - 3)(x + 5) \le 0$

45. $\dfrac{3t + 4}{t - 2} \le 1$

46. $\dfrac{6}{2z - 1} < 2$

![] **MIXED REVIEW EXERCISES**

Solve.

47. $3t^2 - 6t = -4$

48. $V = r^2 + R^2h$ for R

49. $(x^2 - 2x)^2 = 11(x^2 - 2x) - 24$

50. $x^4 - 1 = 0$

51. $2x - \sqrt{x} = 6$

52. $(r - 1)(2r + 3)(r + 6) < 0$

53. $S = \dfrac{Id^2}{k}$ for d

54. $(3k + 11)^2 = 7$

55. $6 + \dfrac{15}{s^2} = -\dfrac{19}{s}$

56. $(8k - 7)^2 \ge -1$

57. $\dfrac{-2}{x + 5} \le -5$

58. $x^4 - 8x^2 = -1$

Solve each problem.

59. Natural gas use in the United States in trillions of cubic feet (ft³) from 1970 through 1999 can be modeled by the quadratic function defined by

$$f(x) = .014x^2 - .396x + 21.2,$$

where $x = 0$ represents 1970, $x = 5$ represents 1975, and so on. (*Source:* Energy Information Administration.)

(a) Use the model to approximate natural gas use in 2000.

(b) Based on the model, in what year will natural gas use reach 25 trillion ft³? (Round down to the nearest year.)

60. In 4 hr, Kerrie can go 15 mi upriver and come back. The speed of the current is 5 mph. Find the speed of the boat in still water.

61. Refer to Exercise 36. Suppose that a wire is attached to the top of the Mart and pulled tight. It is attached to the ground 100 ft from the base of the building. How long is the wire?

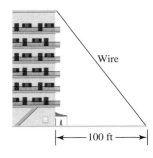

RELATING CONCEPTS (EXERCISES 62–66)

For Individual or Group Work

Work Exercises 62–66 in order, to see the connections between equations and inequalities.

62. Use the methods of Chapter 2 to solve the equation or inequality, and graph the solution set.

    (a) $3x - (4x + 2) = 0$ **(b)** $3x - (4x + 2) > 0$ **(c)** $3x - (4x + 2) < 0$

63. Use the methods of this chapter to solve the equation or inequality, and graph the solution set.

    (a) $x^2 - 6x + 5 = 0$ **(b)** $x^2 - 6x + 5 > 0$ **(c)** $x^2 - 6x + 5 < 0$

64. Use the methods of Sections 7.4 and 9.5 to solve the equation or inequality, and graph the solution set.

    (a) $\dfrac{-5x + 20}{x - 2} = 0$ **(b)** $\dfrac{-5x + 20}{x - 2} > 0$ **(c)** $\dfrac{-5x + 20}{x - 2} < 0$

65. Fill in the blanks in the following statement: If we solve a linear, quadratic, or rational equation and the two inequalities associated with it, the union of the three solution sets will be _____ ; the only exception will be in the case of the rational equation and inequalities, where the number or numbers that cause the _____ to be 0 will be excluded.

66. Suppose that the solution set of a quadratic equation is $\{-5, 3\}$ and the solution set of one of the associated inequalities is $(-\infty, -5) \cup (3, \infty)$. What is the solution set of the other associated inequality?

CHAPTER **9** TEST

*Problems marked * require knowledge of imaginary numbers.*

Solve each equation by using the square root property or completing the square.

    1. $t^2 = 54$ **2.** $(7x + 3)^2 = 25$ **3.** $2x^2 + 4x = 8$

Solve using the quadratic formula.

4. $x^2 - 4x + 2 = 0$ **5.** $2x^2 - 3x - 1 = 0$ *****6.** $3t^2 - 4t = -5$

*****7.** If k is a negative number, then which one of the following equations will have two imaginary solutions?

     A. $x^2 = 4k$ **B.** $x^2 = -4k$ **C.** $(x + 2)^2 = -k$ **D.** $x^2 + k = 0$

8. What is the discriminant for $2x^2 - 8x - 3 = 0$? How many and what type of solutions does this equation have? (Do not actually solve.)

Solve by any method.

9. $3 - \dfrac{16}{x} - \dfrac{12}{x^2} = 0$ **10.** $4x^2 + 7x - 3 = 0$ **11.** $3x = \sqrt{\dfrac{9x + 2}{2}}$

12. $9x^4 + 4 = 37x^2$ **13.** $12 = (2n + 1)^2 + (2n + 1)$

14. Solve $S = 4\pi r^2$ for r. (Leave $\pm$ in your answer.)

Solve each problem.

15. Andrew and Kent do word processing. For a certain prospectus, Kent can prepare it 2 hr faster than Andrew can. If they work together, they can do the entire prospectus in 5 hr. How long will it take each of them working alone to prepare the prospectus? Round your answers to the nearest tenth of an hour.

16. Abby Tanenbaum paddled her canoe 10 mi upstream and then paddled back to her starting point. If the rate of the current was 3 mph and the entire trip took $3\frac{1}{2}$ hr, what was Abby's rate?

17. Tyler McGinnis has a pool 24 ft long and 10 ft wide. He wants to construct a concrete walk around the pool. If he plans for the walk to be of uniform width and cover 152 ft^2, what will the width of the walk be?

18. At a point 30 m from the base of a tower, the distance to the top of the tower is 2 m more than twice the height of the tower. Find the height of the tower.

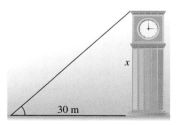

Solve, and graph each solution set.

19. $2x^2 + 7x > 15$ **20.** $\dfrac{5}{t - 4} \le 1$

CUMULATIVE REVIEW EXERCISES CHAPTERS **1–9**

1. Let $S = \left\{ -\frac{7}{3}, -2, -\sqrt{3}, 0, .7, \sqrt{12}, \sqrt{-8}, 7, \frac{32}{3} \right\}$. List the elements of S that are elements of each set.

    (a) Integers   **(b)** Rational numbers   **(c)** Real numbers   **(d)** Complex numbers

Simplify each expression.

2. $|-3| + 8 - |-9| - (-7 + 3)$ **3.** $2(-3)^2 + (-8)(-5) + (-17)$

Solve each equation.

4. $7 - (4 + 3t) + 2t = -6(t - 2) - 5$ **5.** $|6x - 9| = |-4x + 2|$

6. $2x = \sqrt{\dfrac{5x + 2}{3}}$ **7.** $\dfrac{3}{x - 3} - \dfrac{2}{x - 2} = \dfrac{3}{x^2 - 5x + 6}$

8. $(r - 5)(2r + 3) = 1$ **9.** $x^4 - 5x^2 + 4 = 0$

Solve each inequality.

10. $-2x + 4 \le -x + 3$ **11.** $|3x - 7| \le 1$

12. $x^2 - 4x + 3 < 0$ **13.** $\dfrac{3}{p + 2} > 1$

Graph each relation. Tell whether or not each is a function, and if it is, give its domain and range.

14. $4x - 5y = 15$ **15.** $4x - 5y < 15$

16. Find the slope and intercepts of the line with equation $-2x + 7y = 16$.

Write an equation for the specified line. Express each equation in slope-intercept form.

17. Through $(2, -3)$ and parallel to the line with equation $5x + 2y = 6$

18. Through $(-4, 1)$ and perpendicular to the line with equation $5x + 2y = 6$

19. The record track-qualifying speeds at North Carolina Motor Speedway since Richard Petty captured the first pole in 1965 are given in the table. Let $x = 0$ represent 1965, $x = 10$ represent 1975, and so on.

    (a) Use the ordered pairs (year, speed) to make a scatter diagram of the data.

    (b) A linear equation can be used to model the data. Will its slope be positive or negative?

    (c) Use the ordered pairs (0, 116.26) and (20, 141.85) to write a linear equation that models the data.

    (d) Use your model to approximate the record speed for 1998 to the nearest hundredth. How does it compare to the actual value from the table?

QUALIFYING RECORDS

Year	Speed (in mph)
1965	116.26
1975	132.02
1985	141.85
1995	155.38
1998	156.36

Source: NASCAR.

20. Does the relation $x = 5$ define a function? Explain.

21. For the function defined by $f(x) = 2(x - 1)^2 - 5$, find $f(-2)$.

Solve each system of equations.

22. $2x - 4y = 10$
$\quad\ 9x + 3y = 3$

23. $\quad x + \ y + 2z = 3$
$\quad -x + \ y + \ z = -5$
$\quad 2x + 3y - \ z = -8$

24. The merger in 2000 of America Online and Time Warner was the largest in U.S. history. The two companies had combined sales of \$34.2 billion. Sales for AOL were \$.3 billion less than 4 times the sales for Time Warner. What were sales for each company? (*Source:* Company reports.)

(a) Write a system of equations to solve the problem.
(b) Solve the problem.

Write with positive exponents only. Assume variables represent positive real numbers.

25. $\left(\dfrac{x^{-3}y^2}{x^5 y^{-2}}\right)^{-1}$

26. $\dfrac{(4x^{-2})^2(2y^3)}{8x^{-3}y^5}$

Perform the indicated operations.

27. $(7x + 4)(2x - 3)$

28. $\left(\dfrac{2}{3}t + 9\right)^2$

29. $(3t^3 + 5t^2 - 8t + 7) - (6t^3 + 4t - 8)$

30. Divide $4x^3 + 2x^2 - x + 26$ by $x + 2$.

Factor completely.

31. $16x - x^3$

32. $24m^2 + 2m - 15$

33. $8x^3 + 27y^3$

34. $9x^2 - 30xy + 25y^2$

Perform the indicated operations and express each answer in lowest terms. Assume denominators are nonzero.

35. $\dfrac{x^2 - 3x - 10}{x^2 + 3x + 2} \cdot \dfrac{x^2 - 2x - 3}{x^2 + 2x - 15}$

36. $\dfrac{3}{2 - k} - \dfrac{5}{k} + \dfrac{6}{k^2 - 2k}$

37. $\dfrac{\dfrac{r}{s} - \dfrac{s}{r}}{\dfrac{r}{s} + 1}$

Simplify each radical expression.

38. $\sqrt[3]{\dfrac{27}{16}}$

39. $\dfrac{2}{\sqrt{7} - \sqrt{5}}$

Solve each problem.

40. Tri rode his bicycle for 12 mi and then walked an additional 8 mi. The total time for the trip was 5 hr. If his rate while walking was 10 mph less than his rate while riding, what was each rate?

41. Two cars left an intersection at the same time, one heading due south and the other due east. Later they were exactly 95 mi apart. The car heading east had gone 38 mi less than twice as far as the car heading south. How far had each car traveled?

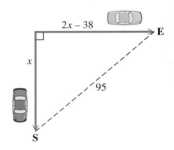

In this day of Automated Teller Machines (ATMs), people often find themselves doing what they have done for years when faced with a soft drink machine that won't respond: They talk to it. According to one report, the following are percentages of people in the United States, the United Kingdom (UK), and Germany who talk to ATMs and what they say.

	United States	UK	Germany
Thanking the ATM	22%	24%	14%
Cursing the ATM	31%	41%	53%
Telling the ATM to Hurry Up	47%	36%	33%

Source: BMRB International for NCR.

In a random sample of 3000 people, how many would there be in each category?

42. People in the United States who curse the ATM

43. People in the UK who thank the ATM

44. People in Germany who tell the ATM to hurry up

45. How many more German cursers would there be than United States thankers?

Additional Graphs of Functions and Relations

10

10.1 Operations and Composition

10.2 Graphs of Quadratic Functions

10.3 More About Parabolas; Applications

10.4 Symmetry; Increasing and Decreasing Functions

10.5 Piecewise Linear Functions

During the past two decades, the number of multiple births in the United States has increased 59%, primarily due to greater use of fertility drugs and greater numbers of births to women over age 40. The number of higher-order multiple births, that is, births involving triplets or more, has increased even more dramatically, some 423%. One of the most publicized higher-order multiple births occurred November 19, 1997, with the birth of the McCaughey septuplets in Des Moines, IA. All seven premature babies survived, a first in medical history. (*Source:* American College of Obstetricians and Gynecologists; *The Gazette,* November 19, 2003.)

In Example 6 of Section 10.2, we determine a quadratic function that models the number of higher-order multiple births in the United States.

10.1 Operations and Composition

OBJECTIVE 1 Review how functions are formed using operations on functions. In Chapter 5 we defined the sum, difference, product, and quotient of functions. We review these definitions here.

Operations on Functions

If f and g are functions, then for all values of x for which both $f(x)$ and $g(x)$ exist,

$$(f + g)(x) = f(x) + g(x), \qquad \text{Sum}$$

$$(f - g)(x) = f(x) - g(x), \qquad \text{Difference}$$

$$(fg)(x) = f(x) \cdot g(x), \qquad \text{Product}$$

and

$$\left(\frac{f}{g}\right)(x) = \frac{f(x)}{g(x)}, \quad \text{where } g(x) \neq 0. \qquad \text{Quotient}$$

NOTE The condition $g(x) \neq 0$ in the definition of the quotient means that the domain of $\left(\frac{f}{g}\right)(x)$ consists of all values of x for which $f(x)$ is defined and $g(x)$ is not 0.

EXAMPLE 1 Using the Operations on Functions

Let $f(x) = x^2 + 1$ and $g(x) = 3x + 5$. Find each of the following.

(a) $(f + g)(1)$

Since $f(1) = 1^2 + 1 = 2$ and $g(1) = 3(1) + 5 = 8$,

$$(f + g)(1) = f(1) + g(1) = 2 + 8 = 10.$$

(b) $(f - g)(-3) = f(-3) - g(-3)$

$$= [(-3)^2 + 1] - [3(-3) + 5]$$

$$= 10 - (-4) = 14$$

(c) $(fg)(5) = f(5) \cdot g(5)$

$$= [5^2 + 1] \cdot [3(5) + 5]$$

$$= 26 \cdot 20 = 520$$

(d) $\left(\frac{f}{g}\right)(0) = \frac{f(0)}{g(0)} = \frac{0^2 + 1}{3 \cdot 0 + 5} = \frac{1}{5}$

Now Try Exercises 1, 3, and 5.

EXAMPLE 2 Using the Operations on Functions

Let $f(x) = 8x - 9$ and $g(x) = \sqrt{2x - 1}$. Find each of the following.

(a) $(f + g)(x) = f(x) + g(x) = 8x - 9 + \sqrt{2x - 1}$

(b) $(f - g)(x) = f(x) - g(x) = 8x - 9 - \sqrt{2x - 1}$

(c) $(fg)(x) = f(x) \cdot g(x) = (8x - 9)\sqrt{2x - 1}$

(d) $\left(\dfrac{f}{g}\right)(x) = \dfrac{f(x)}{g(x)} = \dfrac{8x - 9}{\sqrt{2x - 1}}$

In parts (a)–(d), the domain of f is the set of all real numbers, while the domain of g, where $g(x) = \sqrt{2x - 1}$, includes just those real numbers that make $2x - 1 \geq 0$; the domain of g is the interval $\left[\frac{1}{2}, \infty\right)$. The domain of $f + g$, $f - g$, and fg is thus $\left[\frac{1}{2}, \infty\right)$. With $\frac{f}{g}$, the denominator cannot be 0, so the value $\frac{1}{2}$ is excluded from the domain. The domain of $\frac{f}{g}$ of is $\left(\frac{1}{2}, \infty\right)$. ▇

Now Try Exercise 13.

The domains of $f + g$, $f - g$, fg, and $\frac{f}{g}$ are summarized below. (Recall that the intersection of two sets is the set of all elements belonging to *both* sets.)

Domains of $f + g$, $f - g$, fg, $\dfrac{f}{g}$

For functions f and g, the domains of $f + g$, $f - g$, and fg include all real numbers in the intersection of the domains of f and g, while the domain of $\frac{f}{g}$ includes those real numbers in the intersection of the domains of f and g for which $g(x) \neq 0$.

OBJECTIVE 2 Find a difference quotient. If the coordinates of point P are $(x, f(x))$ and the coordinates of point Q are $(x + h, f(x + h))$, then the expression

$$\frac{f(x + h) - f(x)}{h}$$

gives the slope of the line PQ. This expression is called the **difference quotient** and is important in the study of calculus.

EXAMPLE 3 Finding Difference Quotients

For $f(x) = x^2 - 2x$, find the following and simplify each expression.

(a) $f(x + h) - f(x)$

$$f(x + h) = (x + h)^2 - 2(x + h)$$
$$= x^2 + 2xh + h^2 - 2x - 2h$$
$$f(x + h) - f(x) = (x^2 + 2xh + h^2 - 2x - 2h) - (x^2 - 2x)$$
$$= 2xh + h^2 - 2h$$

(b) $\dfrac{f(x + h) - f(x)}{h}$

Use the result from part (a).

$$\frac{f(x + h) - f(x)}{h} = \frac{2xh + h^2 - 2h}{h} = 2x + h - 2$$

Now Try Exercise 21.

OBJECTIVE 3 **Form composite functions and find their domains.** The diagram in Figure 1 shows a function f that assigns to each element x of set X some element y of set Y. Suppose that a function g takes each element of set Y and assigns a value z of set Z. Using both f and g, then, an element x in X is assigned to an element z in Z. The result of this process is a new function h, which takes an element x in X and assigns it an element z in Z.

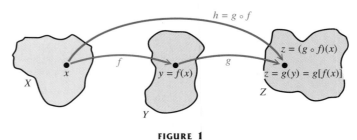

FIGURE 1

This function h is called the *composition* of functions g and f, written $g \circ f$, and is defined as follows.

Composition of Functions

If f and g are functions, then the **composite function,** or **composition,** of g and f is defined by

$$(g \circ f)(x) = g[f(x)]$$

for all x in the domain of f such that $f(x)$ is in the domain of g.

Read $g \circ f$ as "g of f."

As a real-life example of how composite functions occur, suppose an oil well off the California coast is leaking, with the leak spreading oil in a circular layer over the surface. See Figure 2. At any time t, in minutes, after the beginning of the leak, the radius of the circular oil slick is given by $r(t) = 5t$ ft. Since $A(r) = \pi r^2$ gives the area of a circle of radius r, the area can be expressed as a function of time by substituting $5t$ for r in $A(r) = \pi r^2$ to get

FIGURE 2

$$A(r) = \pi r^2$$
$$A[r(t)] = \pi(5t)^2 = 25\pi t^2.$$

The function $A[r(t)]$ is a composite function of the functions A and r.

EXAMPLE 4 Evaluating Composite Functions

Given $f(x) = 2x - 1$ and $g(x) = \dfrac{4}{x - 1}$, find each of the following.

(a) $(f \circ g)(2)$

First find $g(2)$. Since

$$g(x) = \frac{4}{x - 1},$$

$$g(2) = \frac{4}{2 - 1} = \frac{4}{1} = 4.$$

Now find $(f \circ g)(2) = f[g(2)] = f(4)$. Since
$$f(x) = 2x - 1,$$
$$f[g(2)] = f(4) = 2(4) - 1 = 7.$$

(b) $(g \circ f)(-3)$
$$f(-3) = 2(-3) - 1 = -7$$
$$(g \circ f)(-3) = g[f(-3)] = g(-7) = \frac{4}{-7-1} = \frac{4}{-8} = -\frac{1}{2}$$

Now Try Exercises 7 and 9.

EXAMPLE 5 Finding Composite Functions

Let $f(x) = 4x + 1$ and $g(x) = 2x^2 + 5x$. Find each of the following.

(a) $(g \circ f)(x)$

By definition, $(g \circ f)(x) = g[f(x)]$. Using the given functions,

$$\begin{aligned}
(g \circ f)(x) &= g[f(x)] \\
&= g(4x + 1) & f(x) = 4x + 1 \\
&= 2(4x + 1)^2 + 5(4x + 1) & g(x) = 2x^2 + 5x \\
&= 2(16x^2 + 8x + 1) + 20x + 5 & \text{Multiply.} \\
&= 32x^2 + 16x + 2 + 20x + 5 & \text{Distributive property} \\
&= 32x^2 + 36x + 7. & \text{Combine terms.}
\end{aligned}$$

(b) $(f \circ g)(x)$

If we use the preceding definition with f and g interchanged, $(f \circ g)(x)$ becomes $f[g(x)]$.

$$\begin{aligned}
(f \circ g)(x) &= f[g(x)] \\
&= f(2x^2 + 5x) & g(x) = 2x^2 + 5x \\
&= 4(2x^2 + 5x) + 1 & f(x) = 4x + 1 \\
&= 8x^2 + 20x + 1 & \text{Distributive property}
\end{aligned}$$

The domain of both composite functions is the set of all real numbers.

Now Try Exercise 25.

As Example 5 shows, it is not always true that $f \circ g = g \circ f$. In fact, two composite functions are equal only for a special class of functions discussed in Chapter 11.

CAUTION In general, the composite function $f \circ g$ is not the same as the product fg. For example, with f and g defined as in Example 5,
$$(f \circ g)(x) = 8x^2 + 20x + 1$$
but
$$(fg)(x) = (4x + 1)(2x^2 + 5x) = 8x^3 + 22x^2 + 5x.$$

EXAMPLE 6 Finding Composite Functions and Their Domains

Let $f(x) = \frac{1}{x}$ and $g(x) = \sqrt{3 - x}$. Find $f \circ g$ and $g \circ f$. Give the domain of each.

First find $f \circ g$.

$$(f \circ g)(x) = f[g(x)]$$
$$= f(\sqrt{3 - x})$$
$$= \frac{1}{\sqrt{3 - x}}$$

The radical expression $\sqrt{3 - x}$ is a positive real number only when $3 - x > 0$ or $x < 3$, so the domain of $f \circ g$ is the interval $(-\infty, 3)$.

Use the same functions to find $g \circ f$, as follows.

$$(g \circ f)(x) = g[f(x)]$$
$$= g\left(\frac{1}{x}\right) \qquad f(x) = \frac{1}{x}$$
$$= \sqrt{3 - \frac{1}{x}} \qquad g(x) = \sqrt{3 - x}$$
$$= \sqrt{\frac{3x - 1}{x}} \qquad \text{Write as a single fraction.}$$

The domain of $g \circ f$ is the set of all real numbers x such that $x \neq 0$ and $3 - f(x) \geq 0$. By the methods of Section 3.5, the domain of $g \circ f$ is the set $(-\infty, 0) \cup \left[\frac{1}{3}, \infty\right)$.

Now Try Exercise 27.

EXAMPLE 7 Finding the Functions That Form a Given Composite

Find functions f and g such that

$$(f \circ g)(x) = (x^2 - 5)^3 - 4(x^2 - 5) + 3.$$

Note the repeated quantity $x^2 - 5$. If $g(x) = x^2 - 5$ and $f(x) = x^3 - 4x + 3$, then

$$(f \circ g)(x) = f[g(x)]$$
$$= f(x^2 - 5)$$
$$= (x^2 - 5)^3 - 4(x^2 - 5) + 3.$$

There are other pairs of functions f and g that also work. For instance,

$$f(x) = (x - 5)^3 - 4(x - 5) + 3 \qquad \text{and} \qquad g(x) = x^2.$$

Now Try Exercise 51.

CONNECTIONS

Without mentioning it, we have used composite functions in several instances in this book. In Chapter 6, we used substitution to factor expressions like

$$2(3m + 1)^2 - (3m + 1) + 1,$$

by substituting $x = 3m + 1$ before factoring. If we choose $f(x) = 2x^2 - x + 1$ and $g(m) = 3m + 1$, then

$$f[g(m)] = 2(3m + 1)^2 - (3m + 1) + 1.$$

We also used substitution in Chapter 7 to rewrite the expression

$$5 + \frac{7}{p^2 - 4} - \frac{6}{(p^2 - 4)^2} \qquad \text{as} \qquad 5 + \frac{7}{x} - \frac{6}{x^2}.$$

Here, we chose

$$f(x) = 5 + \frac{7}{x} - \frac{6}{x^2} \qquad \text{and} \qquad g(p) = p^2 - 4$$

to obtain the composite function

$$f[g(p)] = 5 + \frac{7}{p^2 - 4} - \frac{6}{(p^2 - 4)^2}.$$

For Discussion or Writing

Give one other example where we have used composite functions in this book before this section.

10.1 EXERCISES

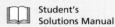

Let $f(x) = 4x^2 - 2x$ and let $g(x) = 8x + 1$. Find each given value. See Examples 1 and 4.

1. $(f + g)(3)$　　　　**2.** $(f - g)(-5)$　　　　**3.** $(fg)(4)$

4. $(gf)(-3)$　　　　**5.** $\left(\dfrac{g}{f}\right)(-1)$　　　　**6.** $\left(\dfrac{f}{g}\right)(4)$

7. $(f \circ g)(2)$　　　　**8.** $(f \circ g)(-5)$　　　　**9.** $(g \circ f)(2)$

10. $(g \circ f)(-5)$　　　　**11.** $(f \circ g)(-2)$　　　　**12.** $(g \circ f)(-2)$

*For each pair of functions f and g, find **(a)** $f + g$, **(b)** $f - g$, **(c)** fg, and **(d)** $\frac{f}{g}$. Give the domain for each of these in Exercises 13–16. See Example 2.*

13. $f(x) = 4x - 1, \quad g(x) = 6x + 3$

14. $f(x) = 9 - 2x, \quad g(x) = -5x + 2$

15. $f(x) = 3x^2 - 2x, \quad g(x) = x^2 - 2x + 1$

16. $f(x) = 6x^2 - 11x, \quad g(x) = x^2 - 4x - 5$

17. $f(x) = \sqrt{2x + 5}, \quad g(x) = \sqrt{4x + 9}$

18. $f(x) = \sqrt{11x - 3}, \quad g(x) = \sqrt{2x - 15}$

19. If f and g are functions, explain in your own words how to find the function values for $f + g$. Give an example.

20. If f and g are functions, explain in your own words how to find the function values for fg. Give an example.

*For each function, find and simplify **(a)** $f(x + h) - f(x)$ and **(b)** $\dfrac{f(x + h) - f(x)}{h}$. See Example 3.*

21. $f(x) = 2x^2 - 1$　　　　　　　**22.** $f(x) = 3x^2 + 2$

23. $f(x) = x^2 + 4x$　　　　　　　**24.** $f(x) = x^2 - 5x$

Find f ∘ g and g ∘ f and their domains for each pair of functions. See Examples 5 and 6.

25. $f(x) = 5x + 3$, $g(x) = -x^2 + 4x + 3$ **26.** $f(x) = 4x^2 + 2x + 8$, $g(x) = x + 5$

27. $f(x) = \dfrac{1}{x}$, $g(x) = x^2$ **28.** $f(x) = \dfrac{2}{x^4}$, $g(x) = 2 - x$

29. $f(x) = \sqrt{x + 2}$, $g(x) = 8x - 6$ **30.** $f(x) = 9x - 11$, $g(x) = 2\sqrt{x + 2}$

31. $f(x) = \dfrac{1}{x - 5}$, $g(x) = \dfrac{2}{x}$ **32.** $f(x) = \dfrac{8}{x - 6}$, $g(x) = \dfrac{4}{3x}$

33. Describe the steps required to find the composite function $f \circ g$. Give an example.

34. Composition is an operation that is unique to functions. Is composition of functions commutative? That is, does $f \circ g = g \circ f$ for all functions f and g? Explain, using an example.

The graphs of functions f and g are shown. Use these graphs to find each indicated value.

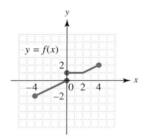
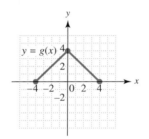

35. $f(1) + g(1)$ **36.** $f(4) - g(3)$ **37.** $f(-2) \cdot g(4)$ **38.** $\dfrac{f(4)}{g(2)}$

39. $(f \circ g)(2)$ **40.** $(g \circ f)(2)$ **41.** $(g \circ f)(-4)$ **42.** $(f \circ g)(-2)$

The tables give some selected ordered pairs for functions f and g.

x	3	4	6
$f(x)$	1	3	9

x	2	7	1	9
$g(x)$	3	6	9	12

Find each of the following.

43. $(f \circ g)(2)$ **44.** $(f \circ g)(7)$ **45.** $(g \circ f)(3)$

46. $(g \circ f)(6)$ **47.** $(f \circ f)(4)$ **48.** $(g \circ g)(1)$

49. Why can you not determine $(f \circ g)(1)$ given the information in the tables for Exercises 43–48?

50. Extend the concept of composition of functions to evaluate $(g \circ (f \circ g))(7)$ using the tables for Exercises 43–48.

A function h is given. Find functions f and g such that $h(x) = (f \circ g)(x)$. Many such pairs of functions exist. See Example 7.

51. $h(x) = (6x - 2)^2$ **52.** $h(x) = \sqrt{x^2 - 1}$

53. $h(x) = \dfrac{1}{x^2 + 2}$ **54.** $h(x) = (x + 2)^3 - 3(x + 2)^2$

RELATING CONCEPTS (EXERCISES 55–62)

For Individual or Group Work

Work Exercises 55–62 in order, to see how important properties of operations with real numbers are related to similar properties of composition of functions. Fill in the blanks when appropriate.*

55. Because _____ is the identity element for addition, $a +$ _____ $=$ _____ $+ a$ for all real numbers a.

56. Because _____ is the identity element for multiplication, $a \cdot$ _____ $=$ _____ $\cdot a$ for all real numbers a.

57. Consider the function defined by $f(x) = x$. Choose any function g that you wish, and find $(f \circ g)(x)$. Then find $(g \circ f)(x)$. How do your two results compare?

58. Based on your observation in Exercise 57, what do you think the function defined by $f(x) = x$ is called (with respect to composition of functions)?

59. The inverse property of addition says that for every real number a, there exists a unique real number _____ such that $a +$ _____ $=$ _____ $+ a = 0$.

60. The inverse property of multiplication says that for every nonzero real number a, there exists a unique real number _____ such that $a \cdot$ _____ $=$ _____ $\cdot a = 1$.

61. Consider the functions defined by $f(x) = x^3 + 2$ and $g(x) = \sqrt[3]{x - 2}$. Find $(f \circ g)(x)$ and then find $(g \circ f)(x)$. How do your results compare?

62. Based on your observation in Exercise 61, what do you think the functions f and g are called with respect to each other (and regarding composition of functions)?

Solve each problem.

63. The function defined by $f(x) = 12x$ computes the number of inches in x ft and the function defined by $g(x) = 5280x$ computes the number of feet in x mi. What is $(f \circ g)(x)$ and what does it compute?

64. The perimeter x of a square with sides of length s is given by the formula $x = 4s$.

   (a) Solve for s in terms of x.

   (b) If y represents the area of this square, write y as a function of the perimeter x.

   (c) Use the composite function of part (b) to find the area of a square with perimeter 6.

65. Suppose the demand for a certain brand of vacuum cleaner is given by

$$D(p) = \frac{-p^2}{100} + 500,$$

where p is the price in dollars. If the price in terms of the cost, c, is expressed as

$$p(c) = 2c - 10,$$

find $D(c)$, the demand in terms of the cost.

---

*The ideas developed in these Relating Concepts exercises are more fully developed in Section 11.1.

66. Suppose the population P of a certain species of fish depends on the number x (in hundreds) of a smaller kind of fish that serves as its food supply, where
$$P(x) = 2x^2 + 1.$$
Suppose, also, that the number x (in hundreds) of the smaller species of fish depends on the amount a (in appropriate units) of its food supply, a kind of plankton, where
$$x = f(a) = 3a + 2.$$
Find $(P \circ f)(a)$, the relationship between the population P of the large fish and the amount a of plankton available.

67. When a thermal inversion layer is over a city (as happens often in Los Angeles), pollutants cannot rise vertically but are trapped below the layer and must disperse horizontally. Assume that a factory smokestack begins emitting a pollutant at 8 A.M. Assume that the pollutant disperses horizontally over a circular area. Suppose that t represents the time, in hours, since the factory began emitting pollutants ($t = 0$ represents 8 A.M.), and assume that the radius of the circle of pollution is $r(t) = 2t$ mi. Let $A(r) = \pi r^2$ represent the area of a circle of radius r. Find and interpret $(A \circ r)(t)$.

68. An oil well off the Gulf Coast is leaking, with the leak spreading oil over the surface as a circle. At any time t, in minutes, after the beginning of the leak, the radius of the circular oil slick on the surface is $r(t) = 4t$ ft. Let $A(r) = \pi r^2$ represent the area of a circle of radius r. Find and interpret $(A \circ r)(t)$.

10.2 Graphs of Quadratic Functions

OBJECTIVES

1 Graph a quadratic function.

2 Graph parabolas with horizontal and vertical shifts.

3 Predict the shape and direction of a parabola from the coefficient of x^2.

4 Find a quadratic function to model data.

5 Use the geometric definition of a parabola.

OBJECTIVE 1 Graph a quadratic function. Polynomial functions were defined in Chapter 5, where we graphed a few simple second-degree polynomial functions by point-plotting. In Figure 3, we repeat a table of ordered pairs for the simplest quadratic function, defined by $y = x^2$, and the resulting graph.

x	y
-2	4
-1	1
0	0
1	1
2	4

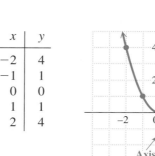

FIGURE 3

As mentioned in Chapter 5, this graph is called a **parabola.** The point $(0, 0)$, the lowest point on the curve, is the **vertex** of this parabola. The vertical line through the vertex is the **axis** of the parabola, here $x = 0$. A parabola is **symmetric about its axis;** that is, if the graph were folded along the axis, the two portions of the curve would coincide. As Figure 3 suggests, x can be any real number, so the domain of the function defined by $y = x^2$ is $(-\infty, \infty)$. Since y is always nonnegative, the range is $[0, \infty)$.

In Section 9.4, we solved applications modeled by quadratic functions. In this section and the next, we consider graphs of more general quadratic functions as defined here.

Quadratic Function

A function that can be written in the form

$$f(x) = ax^2 + bx + c$$

for real numbers a, b, and c, with $a \neq 0$, is a **quadratic function.**

The graph of any quadratic function is a parabola with a vertical axis.

N O T E   We use the variable y and function notation $f(x)$ interchangeably when discussing parabolas. Although we use the letter f most often to name quadratic functions, other letters can be used. We use the capital letter F to distinguish between different parabolas graphed on the same coordinate axes.

Parabolas, which are a type of *conic section* (Chapter 13), have many applications. The large dishes seen on the sidelines of televised football games, which are used by television crews to pick up the shouted signals of players on the field, have cross sections that are parabolas. Cross sections of satellite dishes and automobile headlights also form parabolas. The cables that are used to support suspension bridges are shaped like parabolas.

The reflectors of solar ovens and flashlights are made by revolving a parabola about its axis. The *focus* of a parabola is a point on its axis that determines the curvature. See Figure 4. When the parabolic reflector of a solar oven is aimed at the sun, the light rays bounce off the reflector and collect at the focus, creating an intense temperature at that point. On the other hand, when a light bulb is placed at the focus of a parabolic reflector, light rays reflect out parallel to the axis.

There are several ways to use operations on functions to get variations of $f(x) = x^2$.

1. $f(x) = ax^2$ Multiply $g(x) = a$ and $h(x) = x^2$ to get $f(x)$.

2. $f(x) = x^2 + k$ Add $h(x) = x^2$ and $g(x) = k$ to get $f(x)$.

3. $f(x) = (x - h)^2$ Form the composite function $H[g(x)]$ with $H(x) = x^2$ and $g(x) = x - h$ to get $f(x)$.

4. $f(x) = a(x - h)^2 + k$ Form $H[g(x)]$ with $H(x) = ax^2 + k$ and $g(x) = x - h$ to get $f(x)$.

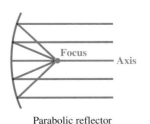

Parabolic reflector

FIGURE 4

Each of these functions has a graph that is a parabola, but in each case the graph is modified from that of $f(x) = x^2$.

OBJECTIVE 2 Graph parabolas with horizontal and vertical shifts. Parabolas need not have their vertices at the origin, as does the graph of $f(x) = x^2$. For example, to graph a parabola of the form $F(x) = x^2 + k$, start by selecting sample values of x like those that were used to graph $f(x) = x^2$. The corresponding values of $F(x)$ in $F(x) = x^2 + k$ differ by k from those of $f(x) = x^2$. For this reason, the graph of $F(x) = x^2 + k$ is *shifted*, or *translated*, k units vertically compared with that of $f(x) = x^2$.

EXAMPLE 1 Graphing a Parabola with a Vertical Shift

Graph $F(x) = x^2 - 2$.

This graph has the same shape as that of $f(x) = x^2$, but since k here is -2, the graph is shifted 2 units down, with vertex $(0, -2)$. Every function value is 2 less than the corresponding function value of $f(x) = x^2$. Plotting points on both sides of the vertex gives the graph in Figure 5. Notice that since the parabola is symmetric about its axis $x = 0$, the plotted points are "mirror images" of each other. Since x can be any real number, the domain is still $(-\infty, \infty)$; the value of y (or $F(x)$) is always greater than or equal to -2, so the range is $[-2, \infty)$. The graph of $f(x) = x^2$ is shown in Figure 5 for comparison.

x	$f(x) = x^2$	$F(x) = x^2 - 2$
-2	4	2
-1	1	-1
0	0	-2
1	1	-1
2	4	2

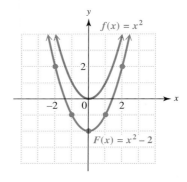

FIGURE 5

Now Try Exercise 23.

Vertical Shift

The graph of $F(x) = x^2 + k$ is a parabola with the same shape as the graph of $f(x) = x^2$. The parabola is shifted k units up if $k > 0$, and $|k|$ units down if $k < 0$. The vertex is $(0, k)$.

The graph of $F(x) = (x - h)^2$ is also a parabola with the same shape as that of $f(x) = x^2$. Because $(x - h)^2 \geq 0$ for all x, the vertex of $F(x) = (x - h)^2$ is the lowest point on the parabola. The lowest point occurs here when $F(x)$ is 0. To get $F(x)$ equal to 0, let $x = h$ so the vertex of $F(x) = (x - h)^2$ is $(h, 0)$. Based on this, the graph of $F(x) = (x - h)^2$ is shifted h units horizontally compared with that of $f(x) = x^2$.

▨ EXAMPLE 2   Graphing a Parabola with a Horizontal Shift

Graph $F(x) = (x - 2)^2$.

If $x = 2$, then $F(x) = 0$, giving the vertex $(2, 0)$. The graph of $F(x) = (x - 2)^2$ has the same shape as that of $f(x) = x^2$ but is shifted 2 units to the right. Plotting several points on one side of the vertex and using symmetry about the axis $x = 2$ to find corresponding points on the other side of the vertex gives the graph in Figure 6. Again, the domain is $(-\infty, \infty)$; the range is $[0, \infty)$.

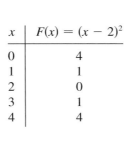

x	$F(x) = (x - 2)^2$
0	4
1	1
2	0
3	1
4	4

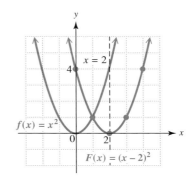

FIGURE 6

Now Try Exercise 27.

Horizontal Shift

The graph of $\boldsymbol{F(x) = (x - h)^2}$ is a parabola with the same shape as the graph of $f(x) = x^2$. The parabola is shifted h units horizontally: h units to the right if $h > 0$, and $|h|$ units to the left if $h < 0$. The vertex is $(h, 0)$.

CAUTION   Errors frequently occur when horizontal shifts are involved. To determine the direction and magnitude of a horizontal shift, find the value that would cause the expression $x - h$ to equal 0. For example, the graph of $F(x) = (x - 5)^2$ would be shifted 5 units to the *right,* because $+5$ would cause $x - 5$ to equal 0. On the other hand, the graph of $F(x) = (x + 5)^2$ would be shifted 5 units to the *left,* because -5 would cause $x + 5$ to equal 0.

A parabola can have both horizontal and vertical shifts.

▨ EXAMPLE 3   Graphing a Parabola with Horizontal and Vertical Shifts

Graph $F(x) = (x + 3)^2 - 2$.

This graph has the same shape as that of $f(x) = x^2$, but is shifted 3 units to the left (since $x + 3 = 0$ if $x = -3$) and 2 units down (because of the -2). As shown in

Figure 7, the vertex is $(-3, -2)$, with axis $x = -3$. This function has domain $(-\infty, \infty)$ and range $[-2, \infty)$.

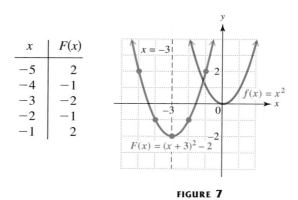

x	$F(x)$
-5	2
-4	-1
-3	-2
-2	-1
-1	2

FIGURE 7

Now Try Exercise 29.

The characteristics of the graph of a parabola of the form $F(x) = (x - h)^2 + k$ are summarized as follows.

Vertex and Axis of a Parabola

The graph of $F(x) = (x - h)^2 + k$ is a parabola with the same shape as the graph of $f(x) = x^2$, but with vertex (h, k). The axis is the vertical line $x = h$.

NOTE   The vertical and horizontal translations discussed in this section also occur with other types of graphs. For example, the graph of $F(x) = \sqrt{x - h} + k$, with $h > 0$ and $k > 0$, would have the same shape as the graph of $f(x) = \sqrt{x}$, but shifted h units to the right and k units up.

OBJECTIVE 3   Predict the shape and direction of a parabola from the coefficient of x^2. Not all parabolas open up, and not all parabolas have the same shape as the graph of $f(x) = x^2$.

EXAMPLE 4   Graphing a Parabola That Opens Down and Is Wider Than the Graph of $f(x) = x^2$

Graph $f(x) = -\dfrac{1}{2}x^2$.

This parabola is shown in Figure 8. The coefficient $-\frac{1}{2}$ affects the shape of the graph; the $\frac{1}{2}$ makes the parabola wider $\left(\text{since the values of } \frac{1}{2}x^2 \text{ increase more slowly}\right.$ than those of $x^2\big)$, and the negative sign makes the parabola open down. The graph is not shifted in any direction; the vertex is still $(0, 0)$ and the axis is $x = 0$. Unlike the parabolas graphed in Examples 1–3, the vertex here has the *largest* function value of any point on the graph. The domain is $(-\infty, \infty)$; the range is $(-\infty, 0]$.

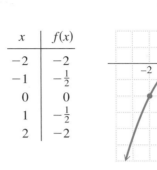

x	$f(x)$
-2	-2
-1	$-\frac{1}{2}$
0	0
1	$-\frac{1}{2}$
2	-2

FIGURE 8

Now Try Exercise 21.

Some general principles concerning the graph of $F(x) = a(x - h)^2 + k$ are summarized as follows.

General Principles

1. The graph of the quadratic function defined by

$$F(x) = a(x - h)^2 + k, \quad a \neq 0$$

is a parabola with vertex (h, k) and the vertical line $x = h$ as axis.

2. The graph opens up if a is positive and down if a is negative.

3. The graph is wider than that of $f(x) = x^2$ if $0 < |a| < 1$. The graph is narrower than that of $f(x) = x^2$ if $|a| > 1$.

EXAMPLE 5 Using the General Principles to Graph a Parabola

Graph $F(x) = -2(x + 3)^2 + 4$.

The parabola opens down (because $a < 0$) and is narrower than the graph of $f(x) = x^2$, since $|-2| = 2 > 1$, causing values of $F(x)$ to decrease more quickly than those of $f(x) = -x^2$. This parabola has vertex $(-3, 4)$, as shown in Figure 9. To complete the graph, we plotted the ordered pairs $(-4, 2)$ and, by symmetry, $(-2, 2)$. Symmetry can be used to find additional ordered pairs that satisfy the equation, if desired.

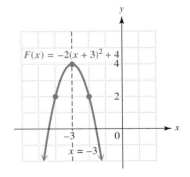

FIGURE 9

Now Try Exercise 33.

OBJECTIVE 4 Find a quadratic function to model data.

EXAMPLE 6 Finding a Quadratic Function to Model the Rise in Multiple Births

The number of higher-order multiple births in the United States is rising. Let x represent the number of years since 1970 and y represent the rate of higher-order multiples born per 100,000 births since 1971. The data are shown in the following table.

Year	x	y
1971	1	29.1
1976	6	35.0
1981	11	40.0
1986	16	47.0
1991	21	100.0
1996	26	152.6

Source: National Center for Health Statistics.

Find a quadratic function that models the data.

A scatter diagram of the ordered pairs (x, y) is shown in Figure 10. Notice that the graphed points do not follow a linear pattern, so a linear function would not model the data very well. Instead, the general shape suggested by the scatter diagram indicates that a parabola should approximate these points, as shown by the dashed curve in Figure 11. The equation for such a parabola would have a positive coefficient for x^2 since the graph opens up. To find a quadratic function of the form

$$y = ax^2 + bx + c$$

that models, or *fits*, these data, we choose three representative ordered pairs and use them to write a system of three equations. Using $(1, 29.1)$, $(11, 40)$, and $(21, 100)$, we substitute the x- and y-values from the ordered pairs into the quadratic form $y = ax^2 + bx + c$ to get the following three equations.

$$a(1)^2 + b(1) + c = 29.1 \quad \text{or} \quad a + b + c = 29.1 \quad (1)$$
$$a(11)^2 + b(11) + c = 40 \quad \text{or} \quad 121a + 11b + c = 40 \quad (2)$$
$$a(21)^2 + b(21) + c = 100 \quad \text{or} \quad 441a + 21b + c = 100 \quad (3)$$

We can find the values of a, b, and c by solving this system of three equations in three variables using the methods of Section 4.2. Multiplying equation (1) by -1 and adding the result to equation (2) gives

$$120a + 10b = 10.9. \quad (4)$$

Multiplying equation (2) by -1 and adding the result to equation (3) gives

$$320a + 10b = 60. \quad (5)$$

We eliminate b from this system of two equations in two variables by multiplying equation (4) by -1 and adding the result to equation (5) to obtain

$$200a = 49.1$$
$$a = .2455. \quad \text{Use a calculator.}$$

U.S. HIGHER-ORDER MULTIPLE BIRTHS

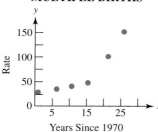

Years Since 1970

FIGURE 10

U.S. HIGHER-ORDER MULTIPLE BIRTHS

Years Since 1970

FIGURE 11

We substitute .2455 for a in equation (4) or (5) to find that $b = -1.856$. Substituting the values of a and b into equation (1) gives $c = 30.7105$. Using these values of a, b, and c, our model is defined by

$$y = .2455x^2 - 1.856x + 30.7105.$$

Now Try Exercise 49.

NOTE In Example 6, if we had chosen three different ordered pairs of data, a slightly different model would result. The *quadratic regression* feature on a graphing calculator can also be used to generate the quadratic model that best fits given data. See your owner's manual for details.

OBJECTIVE 5 Use the geometric definition of a parabola. Geometrically, a parabola is defined as the set of all points in a plane that are equally distant from a fixed point and a fixed line not containing the point. The point is the **focus** and the line is the **directrix.** The line through the focus and perpendicular to the directrix is the axis of the parabola. The point on the axis that is equally distant from the focus and the directrix is the vertex of the parabola. See Figure 12.

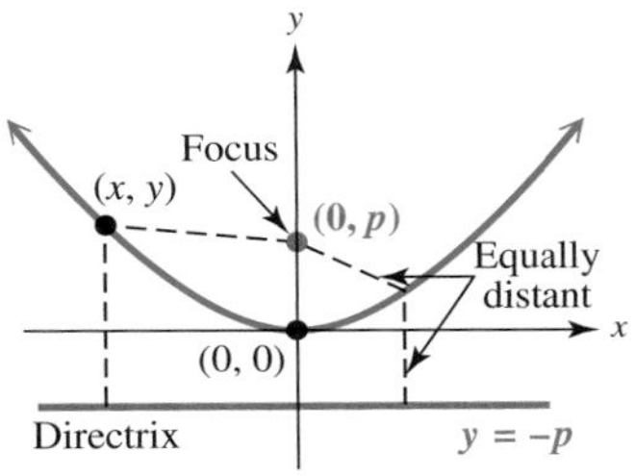

The parabola in Figure 12 has the point $(0, p)$ as focus and the line $y = -p$ as directrix. The vertex is $(0, 0)$. Let (x, y) be any point on the parabola. The distance from (x, y) to the directrix is $|y - (-p)|$, while the distance from (x, y) to $(0, p)$ is $\sqrt{(x - 0)^2 + (y - p)^2}$. Since (x, y) is equally distant from the directrix and the focus,

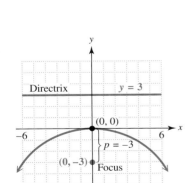

FIGURE 12

$$|y - (-p)| = \sqrt{(x - 0)^2 + (y - p)^2}$$
$$(y + p)^2 = x^2 + (y - p)^2 \qquad \text{Square both sides.}$$
$$y^2 + 2py + p^2 = x^2 + y^2 - 2py + p^2 \qquad \text{Square of a binomial}$$
$$4py = x^2, \qquad \text{Combine terms.}$$

which is the equation of the parabola with focus $(0, p)$ and directrix $y = -p$. Solving $4py = x^2$ for y gives

$$y = \frac{1}{4p}x^2,$$

so $\frac{1}{4p} = a$ when the equation is written as $y = ax^2 + bx + c$.

This result could be extended to a parabola with vertex at (h, k), focus $|p|$ units above (h, k), and directrix $|p|$ units below (h, k), or to a parabola with vertex at (h, k), focus $|p|$ units below (h, k), and directrix $|p|$ units above (h, k).

EXAMPLE 7 Using the Geometric Definition of a Parabola

Use the geometric definition to find the equation of the parabola with focus at $(0, -3)$ and directrix $y = 3$.

Since the vertex is halfway between the focus and the directrix, the vertex is $(0, 0)$, as shown in Figure 13. Here $p = -3$, so the distance between the focus and the vertex is $|p| = 3$. The equation is

$$y = \frac{1}{4p}x^2 = \frac{1}{4(-3)}x^2 = \frac{1}{-12}x^2 \qquad \text{or} \qquad -12y = x^2.$$

Now Try Exercise 53.

FIGURE 13

10.2 EXERCISES

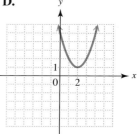
1. Match each quadratic function with its graph from choices A–D.

(a) $f(x) = (x + 2)^2 - 1$

(b) $f(x) = (x + 2)^2 + 1$

(c) $f(x) = (x - 2)^2 - 1$

(d) $f(x) = (x - 2)^2 + 1$

A.
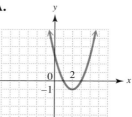

B.
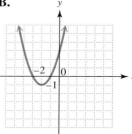

C.
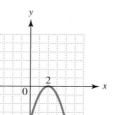

D.
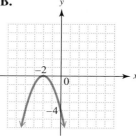

2. Match each quadratic function with its graph from choices A–D.

(a) $f(x) = -x^2 + 2$

(b) $f(x) = -x^2 - 2$

(c) $f(x) = -(x + 2)^2$

(d) $f(x) = -(x - 2)^2$

A.
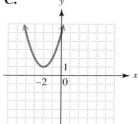

B.
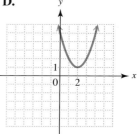

C.
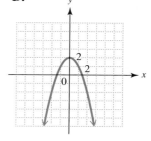

D.

Identify the vertex of each parabola. See Examples 1–4.

3. $f(x) = -3x^2$

4. $f(x) = \dfrac{1}{2}x^2$

5. $f(x) = x^2 + 4$

6. $f(x) = x^2 - 4$

7. $f(x) = (x - 1)^2$

8. $f(x) = (x + 3)^2$

9. $f(x) = (x + 3)^2 - 4$

10. $f(x) = (x - 5)^2 - 8$

11. Describe how each of the parabolas in Exercises 9 and 10 is shifted compared to the graph of $f(x) = x^2$.

12. What does the value of a in $F(x) = a(x - h)^2 + k$ tell you about the graph of the function compared to the graph of $f(x) = x^2$?

For each quadratic function, tell whether the graph opens up or down and whether the graph is wider, narrower, or the same shape as the graph of $f(x) = x^2$. See Examples 4 and 5.

13. $f(x) = -\dfrac{2}{5}x^2$

14. $f(x) = -2x^2$

15. $f(x) = 3x^2 + 1$

16. $f(x) = \dfrac{2}{3}x^2 - 4$

17. For $f(x) = a(x - h)^2 + k$, in what quadrant is the vertex if

(a) $h > 0, k > 0$; (b) $h > 0, k < 0$; (c) $h < 0, k > 0$; (d) $h < 0, k < 0$?

18. (a) What is the value of h if the graph of $f(x) = a(x - h)^2 + k$ has vertex on the y-axis?

(b) What is the value of k if the graph of $f(x) = a(x - h)^2 + k$ has vertex on the x-axis?

19. Match each quadratic function with the description of the parabola that is its graph.

(a) $f(x) = (x - 4)^2 - 2$ **A.** Vertex $(2, -4)$, opens down
(b) $f(x) = (x - 2)^2 - 4$ **B.** Vertex $(2, -4)$, opens up
(c) $f(x) = -(x - 4)^2 - 2$ **C.** Vertex $(4, -2)$, opens down
(d) $f(x) = -(x - 2)^2 - 4$ **D.** Vertex $(4, -2)$, opens up

20. Explain in your own words the meaning of each term.

(a) Vertex of a parabola   (b) Axis of a parabola

Graph each parabola. Plot at least two points in addition to the vertex. Give the vertex, axis, domain, and range in Exercises 27–36. See Examples 1–5.

21. $f(x) = -2x^2$

22. $f(x) = \dfrac{1}{3}x^2$

23. $f(x) = x^2 - 1$

24. $f(x) = x^2 + 3$

25. $f(x) = -x^2 + 2$

26. $f(x) = 2x^2 - 2$

27. $f(x) = (x - 4)^2$

28. $f(x) = -2(x + 1)^2$

29. $f(x) = (x + 2)^2 - 1$

30. $f(x) = (x - 1)^2 + 2$

31. $f(x) = 2(x - 2)^2 - 4$

32. $f(x) = -3(x - 2)^2 + 1$

33. $f(x) = -\dfrac{1}{2}(x + 1)^2 + 2$

34. $f(x) = -\dfrac{2}{3}(x + 2)^2 + 1$

35. $f(x) = 2(x - 2)^2 - 3$

36. $f(x) = \dfrac{4}{3}(x - 3)^2 - 2$

RELATING CONCEPTS (EXERCISES 37–42)

For Individual or Group Work

The procedures that allow the graph of $y = x^2$ to be shifted vertically and horizontally apply to other types of functions. In Section 3.5 we introduced linear functions of the form $g(x) = ax + b$. Consider the graph of the simplest linear function defined by $g(x) = x$, shown here. **Work Exercises 37–42 in order.**

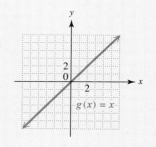

37. Based on the concepts of this section, how does the graph of $F(x) = x^2 + 6$ compare to the graph of $f(x) = x^2$ if a *vertical* shift is considered?

38. Graph the linear function defined by $G(x) = x + 6$.

39. Based on the concepts of Chapter 3, how does the graph of $G(x) = x + 6$ compare to the graph of $g(x) = x$ if a vertical shift is considered? (*Hint:* Look at the y-intercept.)

40. Based on the concepts of this section, how does the graph of $F(x) = (x - 6)^2$ compare to the graph of $f(x) = x^2$ if a *horizontal* shift is considered?

41. Graph the linear function $G(x) = x - 6$.

42. Based on the concepts of Chapter 3, how does the graph of $G(x) = x - 6$ compare to the graph of $g(x) = x$ if a horizontal shift is considered? (*Hint:* Look at the x-intercept.)

In Exercises 43–48, tell whether a linear or quadratic function would be a more appropriate model for each set of graphed data. If linear, tell whether the slope should be positive or negative. If quadratic, tell whether the coefficient a of x^2 should be positive or negative. See Example 6.

43.

U.S. TRADE DEFICIT

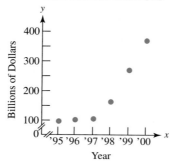

Year

Source: U.S. Department
of Commerce.

44. **AVERAGE DAILY VOLUME OF FIRST-CLASS MAIL***

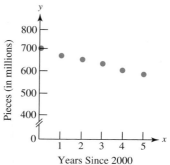

Years Since 2000

*Projected
Source: General Accounting Office.

45.
SOCIAL SECURITY ASSETS*

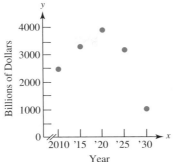

Year

*Projected
Source: Social Security Administration.

**46. CEDAR RAPIDS SCHOOLS—
GENERAL RESERVE FUND**

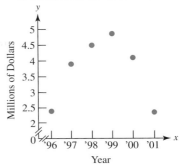

Year

Source: Cedar Rapids School District.

**47. CONSUMER DEMAND FOR
ELECTRICITY**

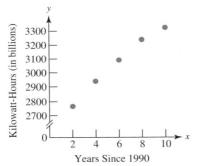

Years Since 1990

Source: U.S. Department of Energy.

**48. U.S. COMMERCIAL
BANK FAILURES**

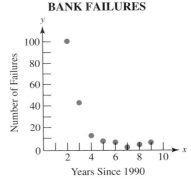

Years Since 1990

Source: www.ABA.com

Solve each problem. See Example 6.

49. The number of publicly traded companies filing for bankruptcy for selected years between 1990 and 2000 are shown in the table. In the year column, 0 represents 1990, 2 represents 1992, and so on.

Year	Number of Bankruptcies
0	115
2	91
4	70
6	84
8	120
10	176

Source: www.BankruptcyData.com

(a) Use the ordered pairs (year, number of bankruptcies) to make a scatter diagram of the data.

(b) Use the scatter diagram to decide whether a linear or quadratic function would better model the data. If quadratic, should the coefficient a of x^2 be positive or negative?

(c) Use the ordered pairs (0, 115), (4, 70), and (8, 120) to find a quadratic function that models the data. Round the values of a, b, and c in your model to three decimal places, as necessary.

(d) Use your model from part (c) to approximate the number of company bankruptcy filings in 2002. Round your answer to the nearest whole number.

✐ **(e)** The number of company bankruptcy filings through August 16, 2002 was 129. Based on this, is your estimate from part (d) reasonable? Explain.

50. In a study, the number of new AIDS patients who survived the first year for the years from 1991 through 1997 are shown in the table. In the year column, 1 represents 1991, 2 represents 1992, and so on.

Year	Number of Patients
1	55
2	130
3	155
4	160
5	155
6	150
7	115

Source: HIV Health Services Planning Council.

(a) Use the ordered pairs (year, number of patients) to make a scatter diagram of the data.

(b) Would a linear or quadratic function better model the data?

(c) Should the coefficient a of x^2 in a quadratic model be positive or negative?

(d) Use the ordered pairs (2, 130), (3, 155), and (7, 115) to find a quadratic function that models the data.

(e) Use your model from part (d) to approximate the number of AIDS patients who survived the first year in 1994 and 1996. How well does the model approximate the actual data from the table?

51. In Example 6, we determined that the quadratic function defined by

$$y = .2455x^2 - 1.856x + 30.7105$$

modeled the rate of higher-order multiple births, where x represents the number of years since 1970.

(a) Use this model to approximate the rate of higher-order births in 1999 to the nearest tenth.

(b) The actual rate of higher-order births in 1999 was 184.9. (*Source:* National Center for Health Statistics.) How does the approximation using the model compare to the actual rate for 1999?

Use the geometric definition of a parabola to find the equation of each parabola described. See Example 7.

52. Focus (0, 2), directrix $y = -2$

53. Focus $\left(0, -\dfrac{1}{2}\right)$, directrix $y = \dfrac{1}{2}$

54. Focus (0, −1), directrix $y = 1$

55. Focus $\left(0, \dfrac{1}{4}\right)$, directrix $y = -\dfrac{1}{4}$

56. Focus (0, 4), directrix $y = 2$

57. Focus (0, −1), directrix $y = -2$

TECHNOLOGY INSIGHTS (EXERCISES 58–62)

Recall from Chapter 3 that the x-value of the x-intercept of the graph of the line $y = mx + b$ is the solution of the linear equation $mx + b = 0$. In the same way, the x-values of the x-intercepts of the graph of the parabola $y = ax^2 + bx + c$ are the real solutions of the quadratic equation $ax^2 + bx + c = 0$.

In Exercises 58–61, the calculator graphs show the x-values of the x-intercepts of the graph of the polynomial in the equation. Use the graphs to solve each equation.

58. $x^2 - x - 20 = 0$

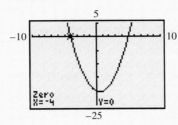 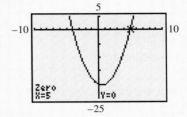

59. $x^2 + 9x + 14 = 0$

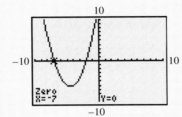 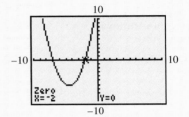

60. $-2x^2 + 5x + 3 = 0$

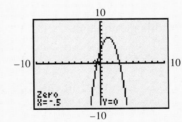 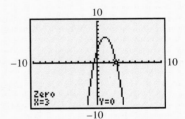

61. $-8x^2 + 6x + 5 = 0$

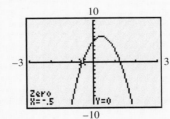 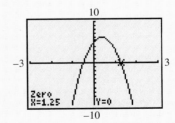

62. The graph of a quadratic function defined by $y = f(x)$ is shown in the standard viewing window, without *x*-axis tick marks. Which one of the following choices would be the only possible solution set for the equation $f(x) = 0$?

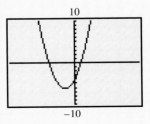

A. $\{-4, 1\}$ **B.** $\{1, 4\}$ **C.** $\{-1, -4\}$ **D.** $\{4, -1\}$

Explain your answer.

10.3 More About Parabolas; Applications

OBJECTIVE 1 Find the vertex of a vertical parabola.   When the equation of a parabola is given in the form $f(x) = ax^2 + bx + c$, we need to locate the vertex to sketch an accurate graph. There are two ways to do this: complete the square as shown in Examples 1 and 2, or use a formula derived by completing the square.

■ **EXAMPLE 1 Completing the Square to Find the Vertex**

Find the vertex of the graph of $f(x) = x^2 - 4x + 5$.

To find the vertex, we need to express $x^2 - 4x + 5$ in the form $(x - h)^2 + k$. We do this by completing the square on $x^2 - 4x$, as in Section 9.1. The process is slightly different here because we want to keep $f(x)$ alone on one side of the equation. Instead of adding the appropriate number to each side, we *add and subtract* it on the right. This is equivalent to adding 0.

$$f(x) = x^2 - 4x + 5$$
$$= (x^2 - 4x \qquad) + 5 \qquad \text{Group the variable terms.}$$
$$\left[\frac{1}{2}(-4)\right]^2 = (-2)^2 = 4$$
$$= (x^2 - 4x + 4 - 4) + 5 \qquad \text{Add and subtract 4.}$$
$$= (x^2 - 4x + 4) - 4 + 5 \qquad \text{Bring } -4 \text{ outside the parentheses.}$$
$$f(x) = (x - 2)^2 + 1 \qquad \text{Factor; combine terms.}$$

The vertex of this parabola is (2, 1).

Now Try Exercise 5.

■ **EXAMPLE 2 Completing the Square to Find the Vertex When $a \neq 1$**

Find the vertex of the graph of $f(x) = -3x^2 + 6x - 1$.

We must complete the square on $-3x^2 + 6x$. Because the x^2-term has a coefficient other than 1, we factor that coefficient out of the first two terms and then proceed as in Example 1.

$$f(x) = -3x^2 + 6x - 1$$
$$= -3(x^2 - 2x) - 1 \qquad \text{Factor out } -3.$$
$$\left[\frac{1}{2}(-2)\right]^2 = (-1)^2 = 1$$
$$= -3(x^2 - 2x + 1 - 1) - 1 \qquad \text{Add and subtract 1 within the parentheses.}$$

Bring -1 outside the parentheses; be sure to multiply it by -3.

$$= -3(x^2 - 2x + 1) + (-3)(-1) - 1 \qquad \text{Distributive property}$$
$$= -3(x^2 - 2x + 1) + 3 - 1$$
$$f(x) = -3(x - 1)^2 + 2 \qquad \text{Factor; combine terms.}$$

The vertex is (1, 2).

Now Try Exercise 7.

To derive a formula for the vertex of the graph of the quadratic function defined by $f(x) = ax^2 + bx + c$ $(a \neq 0)$, complete the square.

$$f(x) = ax^2 + bx + c \qquad \text{Standard form}$$

$$= a\left(x^2 + \frac{b}{a}x\right) + c \qquad \text{Factor } a \text{ from the first two terms.}$$

$$\left[\frac{1}{2}\left(\frac{b}{a}\right)\right]^2 = \left(\frac{b}{2a}\right)^2 = \frac{b^2}{4a^2}$$

$$= a\left(x^2 + \frac{b}{a}x + \frac{b^2}{4a^2} - \frac{b^2}{4a^2}\right) + c \qquad \text{Add and subtract } \frac{b^2}{4a^2}.$$

$$= a\left(x^2 + \frac{b}{a}x + \frac{b^2}{4a^2}\right) + a\left(-\frac{b^2}{4a^2}\right) + c \qquad \text{Distributive property}$$

$$= a\left(x^2 + \frac{b}{a}x + \frac{b^2}{4a^2}\right) - \frac{b^2}{4a} + c$$

$$= a\left(x + \frac{b}{2a}\right)^2 + \frac{4ac - b^2}{4a} \qquad \text{Factor; combine terms.}$$

$$f(x) = a\left[x - \left(\frac{-b}{2a}\right)\right]^2 + \frac{4ac - b^2}{4a} \qquad f(x) = (x - h)^2 + k$$

$$\underbrace{\qquad}_{h} \qquad \underbrace{\qquad}_{k}$$

Thus, the vertex (h, k) can be expressed in terms of a, b, and c. However, it is not necessary to remember this expression for k, since it can be found by replacing x with $\frac{-b}{2a}$. Using function notation, if $y = f(x)$, then the y-value of the vertex is $f\left(\frac{-b}{2a}\right)$.

Vertex Formula

The graph of the quadratic function defined by $f(x) = ax^2 + bx + c$ $(a \neq 0)$ has vertex

$$\left(\frac{-b}{2a}, f\left(\frac{-b}{2a}\right)\right),$$

and the axis of the parabola is the line

$$x = \frac{-b}{2a}.$$

EXAMPLE 3 Using the Formula to Find the Vertex

Use the vertex formula to find the vertex of the graph of
$$f(x) = x^2 - x - 6.$$

For this function, $a = 1$, $b = -1$, and $c = -6$. The x-coordinate of the vertex of the parabola is given by

$$\frac{-b}{2a} = \frac{-(-1)}{2(1)} = \frac{1}{2}.$$

The y-coordinate is $f\left(\frac{-b}{2a}\right) = f\left(\frac{1}{2}\right)$.

$$f\left(\frac{1}{2}\right) = \left(\frac{1}{2}\right)^2 - \frac{1}{2} - 6 = \frac{1}{4} - \frac{1}{2} - 6 = -\frac{25}{4}$$

The vertex is $\left(\frac{1}{2}, -\frac{25}{4}\right)$.

Now Try Exercise 9.

OBJECTIVE 2 Graph a quadratic function. We give a general approach for graphing any quadratic function here.

Graphing a Quadratic Function f

Step 1 **Determine whether the graph opens up or down.** If $a > 0$, the parabola opens up; if $a < 0$, it opens down.

Step 2 **Find the vertex.** Use either the vertex formula or completing the square.

Step 3 **Find any intercepts.** To find the x-intercepts (if any), solve $f(x) = 0$. To find the y-intercept, evaluate $f(0)$.

Step 4 **Complete the graph.** Plot the points found so far. Find and plot additional points as needed, using symmetry about the axis.

EXAMPLE 4 Using the Steps to Graph a Quadratic Function

Graph the quadratic function defined by

$$f(x) = x^2 - x - 6.$$

Step 1 From the equation, $a = 1$, so the graph of the function opens up.

Step 2 The vertex, $\left(\frac{1}{2}, -\frac{25}{4}\right)$, was found in Example 3 using the vertex formula.

Step 3 Find any intercepts. Since the vertex, $\left(\frac{1}{2}, -\frac{25}{4}\right)$, is in quadrant IV and the graph opens up, there will be two x-intercepts. To find them, let $f(x) = 0$ and solve.

$$f(x) = x^2 - x - 6$$

$\quad 0 = x^2 - x - 6 \qquad\qquad$ Let $f(x) = 0$.

$\quad 0 = (x - 3)(x + 2) \qquad$ Factor.

$x - 3 = 0 \quad$ or $\quad x + 2 = 0 \qquad$ Zero-factor property

$\quad\; x = 3 \quad$ or $\qquad\quad x = -2$

The x-intercepts are $(3, 0)$ and $(-2, 0)$. To find the y-intercept, evaluate $f(0)$.

$$f(x) = x^2 - x - 6$$
$$f(0) = 0^2 - 0 - 6 \qquad \text{Let } x = 0.$$
$$f(0) = -6$$

The y-intercept is $(0, -6)$.

Step 4 Plot the points found so far and additional points as needed using symmetry about the axis, $x = \frac{1}{2}$. The graph is shown in Figure 14. The domain is $(-\infty, \infty)$, and the range is $\left[-\frac{25}{4}, \infty\right)$.

x	y
-2	0
-1	-4
0	-6
$\frac{1}{2}$	$-\frac{25}{4}$
2	-4
3	0

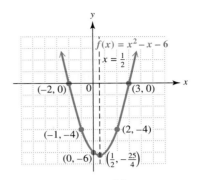

FIGURE 14

Now Try Exercise 17.

OBJECTIVE 3 Use the discriminant to find the number of x-intercepts of a vertical parabola. The graph of a quadratic function may have two x-intercepts, one x-intercept, or no x-intercepts, as shown in Figure 15. Recall from Section 9.2 that $b^2 - 4ac$ is called the *discriminant* of the quadratic equation $ax^2 + bx + c = 0$ and that we can use it to determine the number of real solutions of a quadratic equation.

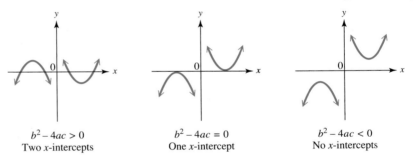

FIGURE 15

In a similar way, we can use the discriminant of a quadratic *function* to determine the number of x-intercepts of its graph. If the discriminant is positive, the parabola will have two x-intercepts. If the discriminant is 0, there will be only one x-intercept, and it will be the vertex of the parabola. If the discriminant is negative, the graph will have no x-intercepts.

EXAMPLE 5 Using the Discriminant to Determine the Number of x-Intercepts

Find the discriminant and use it to determine the number of x-intercepts of the graph of each quadratic function.

(a) $f(x) = 2x^2 + 3x - 5$

The discriminant is $b^2 - 4ac$. Here $a = 2$, $b = 3$, and $c = -5$, so

$$b^2 - 4ac = 9 - 4(2)(-5) = 49.$$

Since the discriminant is positive, the parabola has two x-intercepts.

(b) $f(x) = -3x^2 - 1$

Here, $a = -3$, $b = 0$, and $c = -1$. The discriminant is

$$b^2 - 4ac = 0 - 4(-3)(-1) = -12.$$

The discriminant is negative, so the graph has no x-intercepts.

(c) $f(x) = 9x^2 + 6x + 1$

Here, $a = 9$, $b = 6$, and $c = 1$. The discriminant is

$$b^2 - 4ac = 36 - 4(9)(1) = 0.$$

The parabola has only one x-intercept (its vertex) because the value of the discriminant is 0.

Now Try Exercises 11 and 13.

OBJECTIVE 4 **Use quadratic functions to solve problems involving maximum or minimum value.** The vertex of the graph of a quadratic function is either the highest or the lowest point on the parabola. The y-value of the vertex gives the maximum or minimum value of y, while the x-value tells where that maximum or minimum occurs.

PROBLEM SOLVING

In many applied problems we must find the largest or smallest value of some quantity. When we can express that quantity in terms of a quadratic function, the value of k in the vertex (h, k) gives that optimum value.

EXAMPLE 6 **Finding the Maximum Area of a Rectangular Region**

A farmer has 120 ft of fencing. He wants to put a fence around a rectangular field next to a building. Find the maximum area he can enclose.

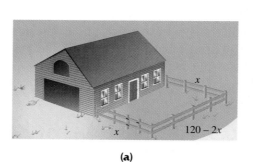

(a)

(Area) $A(x)$

When the width is 30 ft,...
area is a maximum, 1800 ft^2.

$(30, 1800)$

1800

(Width) x

0   30   60

When the width is between 0 and 30 ft, the area is less than 1800 ft^2.

When the width is between 30 and 60 ft, the area is less than 1800 ft^2.

(b)

FIGURE 16

Figure 16(a) shows the field. Let x represent the width of the field. Since he has 120 ft of fencing,

$x + x + \text{length} = 120$	Sum of the sides is 120 ft.
$2x + \text{length} = 120$	Combine terms.
$\text{length} = 120 - 2x.$	Subtract $2x$.

The area is given by the product of the width and length, so

$$A(x) = x(120 - 2x)$$
$$= 120x - 2x^2.$$

To determine the maximum area, find the vertex of the parabola given by $A(x) = 120x - 2x^2$ using the vertex formula. Writing the equation in standard form as $A(x) = -2x^2 + 120x$ gives $a = -2$, $b = 120$, and $c = 0$, so

$$h = \frac{-b}{2a} = \frac{-120}{2(-2)} = \frac{-120}{-4} = 30;$$

$$A(30) = -2(30)^2 + 120(30) = -2(900) + 3600 = 1800.$$

The graphical interpretation in Figure 16(b) shows that the graph is a parabola that opens down, and its vertex is $(30, 1800)$. Thus, the maximum area will be 1800 ft^2. This area will occur if x, the width of the field, is 30 ft.

Now Try Exercise 35.

CAUTION Be careful when interpreting the meanings of the coordinates of the vertex. The first coordinate, x, gives the value for which the *function value* is a maximum or a minimum. Be sure to read the problem carefully to determine whether you are asked to find the value of the independent variable, the function value, or both.

■ **EXAMPLE 7** Finding the Maximum Height Attained by a Projectile

If air resistance is neglected, a projectile on Earth shot straight upward with an initial velocity of 40 m per sec will be at a height s in meters given by

$$s(t) = -4.9t^2 + 40t,$$

where t is the number of seconds elapsed after projection. After how many seconds will it reach its maximum height, and what is this maximum height?

For this function, $a = -4.9$, $b = 40$, and $c = 0$. Use the vertex formula.

$$h = \frac{-b}{2a} = \frac{-40}{2(-4.9)} \approx 4.1 \qquad \text{Use a calculator.}$$

This indicates that the maximum height is attained at 4.1 sec. To find this maximum height, calculate $s(4.1)$.

$$s(4.1) = -4.9(4.1)^2 + 40(4.1)$$
$$\approx 81.6 \qquad\qquad \text{Use a calculator.}$$

The projectile will attain a maximum height of approximately 81.6 m.

Now Try Exercise 37.

OBJECTIVE 5 Graph horizontal parabolas. The directrix of a parabola could be the *vertical* line $x = -p$, where $p > 0$, with focus on the x-axis at $(p, 0)$, producing a parabola opening to the right. This parabola is the graph of the relation

$$y^2 = 4px \qquad \text{or} \qquad x = \frac{1}{4p}y^2.$$

For example, we obtain the equation $x = y^2$ from $y = x^2$ by interchanging x and y. Choosing values of y and finding the corresponding values of x gives the parabola in Figure 17. The graph of $x = y^2$, shown in red, is symmetric with respect to the line $y = 0$ and has vertex $(0, 0)$. For comparison, the graph of $y = x^2$ is shown in blue.

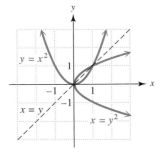

FIGURE 17

These graphs are mirror images of each other with respect to the line $y = x$. From the graph, the domain of $x = y^2$ is $[0, \infty)$, and the range is $(-\infty, \infty)$.

If x and y are interchanged in the equation $y = ax^2 + bx + c$, the equation becomes $x = ay^2 + by + c$.

Graph of a Horizontal Parabola

The graph of

$$x = ay^2 + by + c \qquad \text{or} \qquad x = a(y - k)^2 + h$$

is a parabola with vertex (h, k) and the horizontal line $y = k$ as axis. The graph opens to the right if $a > 0$ and to the left if $a < 0$.

EXAMPLE 8 Graphing a Horizontal Parabola

Graph $x = (y - 2)^2 - 3$.

This graph has its vertex at $(-3, 2)$, since the roles of x and y are reversed. It opens to the right because $a = 1 > 0$, and has the same shape as $y = x^2$. Plotting a few additional points gives the graph shown in Figure 18. Note that the graph is symmetric about its axis, $y = 2$. The domain is $[-3, \infty)$, and the range is $(-\infty, \infty)$.

x	y
-3	2
-2	3
-2	1
1	4
1	0

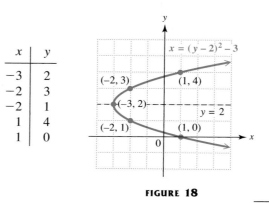

FIGURE 18

Now Try Exercise 21.

When a quadratic equation is given in the form $x = ay^2 + by + c$, completing the square on y allows us to find the vertex.

EXAMPLE 9 Completing the Square to Graph a Horizontal Parabola

Graph $x = -2y^2 + 4y - 3$. Give the domain and range of the relation.

$$\begin{aligned}
x &= -2y^2 + 4y - 3 \\
&= -2(y^2 - 2y) - 3 & &\text{Factor out } -2. \\
&= -2(y^2 - 2y + 1 - 1) - 3 & &\text{Complete the square within the} \\
& & &\text{parentheses; add and subtract 1.} \\
&= -2(y^2 - 2y + 1) + (-2)(-1) - 3 & &\text{Distributive property} \\
x &= -2(y - 1)^2 - 1 & &\text{Factor; simplify.}
\end{aligned}$$

Because of the negative coefficient (-2) in $x = -2(y - 1)^2 - 1$, the graph opens to the left (the negative x-direction) and is narrower than the graph of $y = x^2$ because

$|-2| > 1$. As shown in Figure 19, the vertex is $(-1, 1)$ and the axis is $y = 1$. The domain is $(-\infty, -1]$, and the range is $(-\infty, \infty)$.

x	y
-3	2
-3	0
-1	1

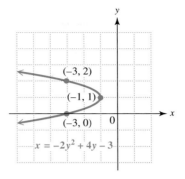

FIGURE 19

Now Try Exercise 25.

CAUTION Only quadratic equations solved for y (whose graphs are vertical parabolas) are examples of functions. *The horizontal parabolas in Examples 8 and 9 are not graphs of functions,* because they do not satisfy the vertical line test. Furthermore, the vertex formula given earlier does not apply to parabolas with horizontal axes.

In summary, the graphs of parabolas studied in this section and the previous one fall into the following categories.

Graphs of Parabolas

Equation	Graph
$y = ax^2 + bx + c$ **or** $y = a(x - h)^2 + k$	 (h, k) These graphs $a > 0$ represent functions. $a < 0$
$x = ay^2 + by + c$ **or** $x = a(y - k)^2 + h$	 These graphs are not $a > 0$ graphs of functions. $a < 0$

10.3 EXERCISES

1. How can you determine just by looking at the equation of a parabola whether it has a vertical or a horizontal axis?

2. Why can't the graph of a quadratic function be a parabola with a horizontal axis?

3. How can you determine the number of x-intercepts of the graph of a quadratic function without graphing the function?

4. If the vertex of the graph of a quadratic function is $(1, -3)$, and the graph opens down, how many x-intercepts does the graph have?

Find the vertex of each parabola. See Examples 1–3.

5. $f(x) = x^2 + 8x + 10$ **6.** $f(x) = x^2 + 10x + 23$ **7.** $f(x) = -2x^2 + 4x - 5$

8. $f(x) = -3x^2 + 12x - 8$ **9.** $f(x) = -\dfrac{1}{2}x^2 + 2x - 3$ **10.** $f(x) = 4x^2 - x + 5$

Find the vertex of each parabola. For each equation, decide whether the graph opens up, down, to the left, or to the right, and whether it is wider, narrower, or the same shape as the graph of $y = x^2$. If it is a vertical parabola, find the discriminant and use it to determine the number of x-intercepts. See Examples 1–3, 5, 8, and 9.

11. $f(x) = 2x^2 + 4x + 5$ **12.** $f(x) = 3x^2 - 6x + 4$ **13.** $f(x) = -x^2 + 5x + 3$

14. $x = -y^2 + 7y + 2$ **15.** $x = \dfrac{1}{3}y^2 + 6y + 24$ **16.** $x = \dfrac{1}{2}y^2 + 10y - 5$

Use the concepts of this section to match each equation in Exercises 17–22 with its graph in A–F.

17. $y = 2x^2 + 4x - 3$ **18.** $y = -x^2 + 3x + 5$ **19.** $y = -\dfrac{1}{2}x^2 - x + 1$

20. $x = y^2 + 6y + 3$ **21.** $x = -y^2 - 2y + 4$ **22.** $x = 3y^2 + 6y + 5$

A. 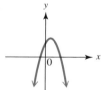   **B.**    **C.**

D. 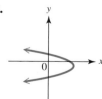   **E.** 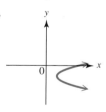   **F.**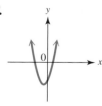

Graph each parabola. (Use the results of Exercises 5–8 to help graph the parabolas in Exercises 23–26.) Give the vertex, axis, domain, and range. See Examples 4, 8, and 9.

23. $f(x) = x^2 + 8x + 10$ **24.** $f(x) = x^2 + 10x + 23$ **25.** $f(x) = -2x^2 + 4x - 5$

26. $f(x) = -3x^2 + 12x - 8$ **27.** $x = (y + 2)^2 + 1$ **28.** $x = (y + 3)^2 - 2$

29. $x = -\dfrac{1}{5}y^2 + 2y - 4$ **30.** $x = -\dfrac{1}{2}y^2 - 4y - 6$ **31.** $x = 3y^2 + 12y + 5$

32. $x = 4y^2 + 16y + 11$

Solve each problem. See Examples 6 and 7.

33. Find the pair of numbers whose sum is 60 and whose product is a maximum. (*Hint:* Let x and $60 - x$ represent the two numbers.)

34. Find the pair of numbers whose sum is 40 and whose product is a maximum.

35. Morgan's Department Store wants to construct a rectangular parking lot on land bordered on one side by a highway. It has 280 ft of fencing that is to be used to fence off the other three sides. What should be the dimensions of the lot if the enclosed area is to be a maximum? What is the maximum area?

36. Keisha Hughes has 100 m of fencing material to enclose a rectangular exercise run for her dog. What width will give the enclosure the maximum area?

37. If an object on Earth is propelled upward with an initial velocity of 32 ft per sec, then its height after t sec is given by

$$h(t) = 32t - 16t^2.$$

Find the maximum height attained by the object and the number of seconds it takes to hit the ground.

38. A projectile on Earth is fired straight upward so that its distance (in feet) above the ground t sec after firing is given by

$$s(t) = -16t^2 + 400t.$$

Find the maximum height it reaches and the number of seconds it takes to reach that height.

39. After experimentation, two University of Miami physics students find that when a bottle of wine is shaken several times, held upright, and uncorked, its cork travels according to the function defined by

$$s(t) = -16t^2 + 64t + 3,$$

where s is its height in feet above the ground t sec after being released. After how many seconds will it reach its maximum height? What is the maximum height?

40. Professor Levy has found that the number of students attending his intermediate algebra class is approximated by

$$S(x) = -x^2 + 20x + 80,$$

where x is the number of hours that the Campus Center is open daily. Find the number of hours that the center should be open so that the number of students attending class is a maximum. What is this maximum number of students?

41. The annual percent increase in the amount pharmacies paid wholesalers for drugs in the years 1990 through 1999 can be modeled by the quadratic function defined by

$$f(x) = .228x^2 - 2.57x + 8.97,$$

where $x = 0$ represents 1990, $x = 1$ represents 1991, and so on. (*Source: IMS Health, Retail and Provider Perspective.*)

(a) Since the coefficient of x^2 in the model is positive, the graph of this quadratic function is a parabola that opens up. Will the y-value of the vertex of this graph be a maximum or minimum?

(b) In what year was the minimum percent increase? (Round down to the nearest year.) Use the actual x-value of the vertex, to the nearest tenth, to find this increase.

42. The U.S. domestic oyster catch (in millions) for the years 1990 through 1998 can be approximated by the quadratic function defined by

$$f(x) = -.566x^2 + 5.08x + 29.2,$$

where $x = 0$ represents 1990, $x = 1$ represents 1991, and so on. (*Source:* National Marine Fisheries Service.)

(a) Since the coefficient of x^2 in the model is negative, the graph of this quadratic function is a parabola that opens down. Will the y-value of the vertex of this graph be a maximum or minimum?

(b) In what year was the maximum domestic oyster catch? (Round down to the nearest year.) Use the actual x-value of the vertex, to the nearest tenth, to find this catch.

43. The graph shows how Social Security assets are expected to change as the number of retirees receiving benefits increases. The graph suggests that a quadratic function would be a good fit to the data. The data are approximated by the function defined by

$$f(x) = -20.57x^2 + 758.9x - 3140.$$

In the model, $x = 10$ represents 2010, $x = 15$ represents 2015, and so on, and $f(x)$ is in billions of dollars.

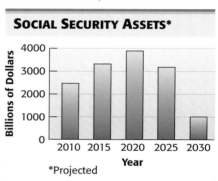

SOCIAL SECURITY ASSETS*

*Projected

Source: Social Security Administration.

(a) Explain why the coefficient of x^2 in the model is negative, based on the graph.

(b) Algebraically determine the vertex of the graph, with coordinates to four significant digits.

✐ **(c)** Interpret the answer to part (b) as it applies to the application.

44. The graph shows the performance of investment portfolios with different mixtures of U.S. and foreign investments for the period January 1, 1971, to December 31, 1996.

✐ **(a)** Is this the graph of a function? Explain.

(b) What investment mixture shown on the graph appears to represent the vertex? What relative amount of risk does this point represent? What return on investment does it provide?

(c) Which point on the graph represents the riskiest investment mixture? What return on investment does it provide?

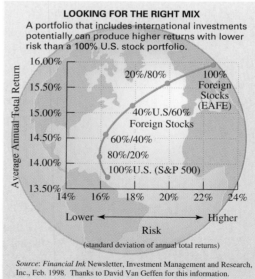

LOOKING FOR THE RIGHT MIX

A portfolio that includes international investments potentially can produce higher returns with lower risk than a 100% U.S. stock portfolio.

Source: Financial Ink Newsletter, Investment Management and Research, Inc., Feb. 1998. Thanks to David Van Geffen for this information.

45. A charter flight charges a fare of $200 per person, plus $4 per person for each unsold seat on the plane. If the plane holds 100 passengers and if x represents the number of unsold seats, find the following.

(a) A function defined by $R(x)$ that describes the total revenue received for the flight (*Hint:* Multiply the number of people flying, $100 - x$, by the price per ticket, $200 + 4x$.)

(b) The graph of the function from part (a)

(c) The number of unsold seats that will produce the maximum revenue

(d) The maximum revenue

46. For a trip to a resort, a charter bus company charges a fare of $48 per person, plus $2 per person for each unsold seat on the bus. If the bus has 42 seats and x represents the number of unsold seats, find the following.

(a) A function defined by $R(x)$ that describes the total revenue from the trip (*Hint:* Multiply the total number riding, $42 - x$, by the price per ticket, $48 + 2x$.)

(b) The graph of the function from part (a)

(c) The number of unsold seats that produces the maximum revenue

(d) The maximum revenue

TECHNOLOGY INSIGHTS (EXERCISES 47–50)

Graphing calculators are capable of determining the coordinates of "peaks" and "valleys" of graphs. In the case of quadratic functions, these peaks and valleys are the vertices and are called maximum and minimum points. For example, the vertex of the graph of $f(x) = -x^2 - 6x - 13$ is $(-3, -4)$, as indicated in the display at the bottom of the screen. In this case, the vertex is a maximum point.

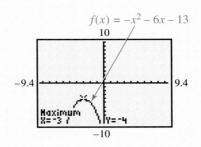

In Exercises 47–50, match the function with its calculator graph in A–D by determining the vertex and using the display at the bottom of the screen.

47. $f(x) = x^2 - 8x + 18$

48. $f(x) = x^2 + 8x + 18$

49. $f(x) = x^2 - 8x + 14$

50. $f(x) = x^2 + 8x + 14$

A.

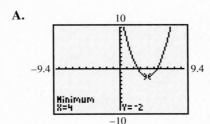

B.

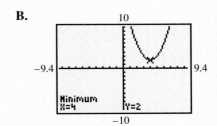

C.

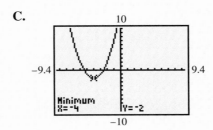

D.

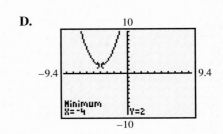

RELATING CONCEPTS (EXERCISES 51–54)

For Individual or Group Work

In Example 1 of Section 9.5, we determined the solution set of the quadratic inequality $x^2 - x - 12 > 0$ by using regions on a number line and testing values in the inequality. If we graph $f(x) = x^2 - x - 12$, the x-intercepts will determine the solutions of the quadratic equation $x^2 - x - 12 = 0$. The solution set is $\{-3, 4\}$. The x-values of the points on the graph that are above *the x-axis form the solution set of $x^2 - x - 12 > 0$. As seen in the figure, this solution set is $(-\infty, -3) \cup (4, \infty)$, which supports the result found in Section 9.5.*

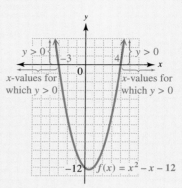

The graph is *above* the x-axis for $(-\infty, -3) \cup (4, \infty)$.

Similarly, the solution set of the quadratic inequality $x^2 - x - 12 < 0$ is found by locating the points on the graph that lie below *the x-axis. Those x-values belong to the interval $(-3, 4)$.*

In Exercises 51–54, the graph of a quadratic function f is given. Use the graph to find the solution set of each equation or inequality.

51. (a) $x^2 - 4x + 3 = 0$
   (b) $x^2 - 4x + 3 > 0$
   (c) $x^2 - 4x + 3 < 0$

52. (a) $3x^2 + 10x - 8 = 0$
   (b) $3x^2 + 10x - 8 \geq 0$
   (c) $3x^2 + 10x - 8 < 0$

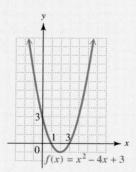

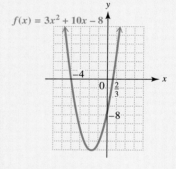

53. (a) $-2x^2 - x + 15 = 0$
   (b) $-2x^2 - x + 15 \geq 0$
   (c) $-2x^2 - x + 15 \leq 0$

54. (a) $-x^2 + 3x + 10 = 0$
   (b) $-x^2 + 3x + 10 \geq 0$
   (c) $-x^2 + 3x + 10 \leq 0$

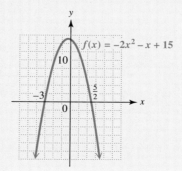

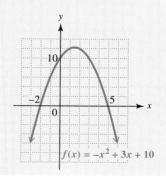

10.4 Symmetry; Increasing and Decreasing Functions

OBJECTIVE 1 Understand how multiplying a function by a real number *a* affects its graph. In Section 10.2, we saw that the value of *a* affects the graph of $g(x) = ax^2$ in several ways. If *a* is positive, then the graph opens up; if *a* is negative, then the graph opens down—that is, the graph of $g(x) = -x^2$ is the same as the graph of $f(x) = x^2$ reflected about the *x*-axis. Also, if $0 < |a| < 1$, then the graph is wider than the graph of $f(x) = x^2$; if $|a| > 1$, then the graph is narrower than that of $f(x) = x^2$. Figure 8 in Section 10.2 demonstrates these effects with the graph of $f(x) = -\frac{1}{2}x^2$. The same effects are true with the graphs of any function.

Reflection, Stretching, and Shrinking

The graph of $g(x) = a \cdot f(x)$ has the same general shape as the graph of $f(x)$.

It is reflected about the *x*-axis if *a* is negative;

It is stretched vertically compared to the graph of $f(x)$ if $|a| > 1$;

It is shrunken vertically compared to the graph of $f(x)$ if $0 < |a| < 1$.

EXAMPLE 1 Comparing the Graph of $g(x) = a \cdot f(x)$ with the Graph of $f(x)$

Graph each function.

(a) $g(x) = 2\sqrt{x}$

We graphed $f(x) = \sqrt{x}$ in Section 8.1. Because the coefficient 2 is positive and greater than 1, the graph of $g(x) = 2\sqrt{x}$ will have the same general shape as the graph of $f(x)$ but is stretched vertically. See Figure 20.

(b) $h(x) = -\sqrt{x}$

Because of the negative sign, the graph of $h(x) = -\sqrt{x}$ will have the same general shape as the graph of $f(x) = \sqrt{x}$ but is reflected about the *x*-axis. See Figure 21.

x	$\sqrt{x}$	$2\sqrt{x}$
0	0	0
1	1	2
4	2	4
9	3	6

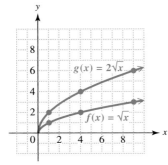

FIGURE 20

x	$\sqrt{x}$	$-\sqrt{x}$
0	0	0
1	1	-1
4	2	-2
9	3	-3

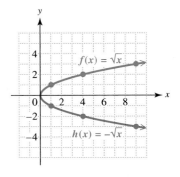

FIGURE 21

Now Try Exercise 1.

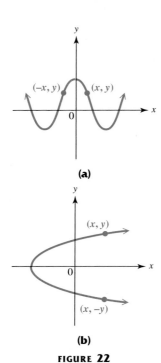

(a)

(b)

FIGURE 22

The parabolas graphed in the previous two sections were symmetric with respect to the axis of the parabola, a vertical or horizontal line through the vertex. The graphs of many other relations also are symmetric with respect to a line or a point. As we saw when graphing parabolas, symmetry is helpful in drawing graphs.

OBJECTIVE 2 **Test for symmetry with respect to an axis.** The graph in Figure 22(a) is cut in half by the y-axis with each half the mirror image of the other half. Such a graph is *symmetric with respect to the y-axis.* As suggested by Figure 22(a), for a graph to be symmetric with respect to the y-axis, the point $(-x, y)$ must be on the graph whenever the point (x, y) is on the graph.

Similarly, if the graph in Figure 22(b) were folded in half along the x-axis, the portion from the top would exactly match the portion from the bottom. Such a graph is *symmetric with respect to the x-axis.* As the graph suggests, symmetry with respect to the x-axis means that the point $(x, -y)$ must be on the graph whenever the point (x, y) is on the graph.

The following tests tell when a graph is symmetric with respect to the x-axis or y-axis.

> **Symmetry with Respect to an Axis**
>
> The graph of a relation is **symmetric with respect to the y-axis** if the replacement of x with $-x$ results in an equivalent equation.
>
> The graph of a relation is **symmetric with respect to the x-axis** if the replacement of y with $-y$ results in an equivalent equation.

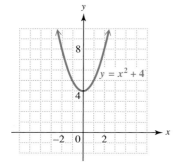

FIGURE 23

EXAMPLE 2 **Testing for Symmetry with Respect to an Axis**

Test for symmetry with respect to the x-axis and the y-axis.

(a) $y = x^2 + 4$

Replace x with $-x$.

$$y = x^2 + 4 \quad \text{becomes} \quad y = (-x)^2 + 4 = x^2 + 4.$$

The result is equivalent to the original equation, so the graph (shown in Figure 23) is symmetric with respect to the y-axis. The graph is *not* symmetric with respect to the x-axis, since replacing y with $-y$ gives

$$-y = x^2 + 4 \quad \text{or} \quad y = -x^2 - 4,$$

which is not equivalent to the original equation.

(b) $2x + y = 4$

Replace x with $-x$ and then replace y with $-y$; in neither case does an equivalent equation result. This graph is neither symmetric with respect to the x-axis nor to the y-axis.

(c) $x = |y|$

Replacing x with $-x$ gives $-x = |y|$, which is not equivalent to the original equation. The graph is not symmetric with respect to the y-axis. Replacing y with $-y$ gives $x = |-y| = |y|$. Thus, the graph is symmetric with respect to the x-axis. ∎

OBJECTIVE 3 Test for symmetry with respect to the origin.   Another kind of symmetry occurs when a graph can be rotated 180° about the origin and have the result coincide exactly with the original graph. Symmetry of this type is called *symmetry with respect to the origin*. It can be shown that rotating a graph 180° is equivalent to saying that the point $(-x, -y)$ is on the graph whenever the point (x, y) is on the graph. Figure 24 shows two graphs that are symmetric with respect to the origin.

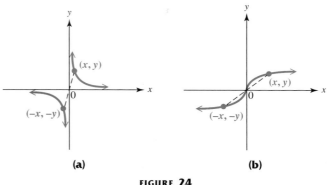

(a)   **(b)**

FIGURE 24

A test for this type of symmetry follows.

Symmetry with Respect to the Origin

The graph of a relation is **symmetric with respect to the origin** if the replacement of both x with $-x$ and y with $-y$ results in an equivalent equation.

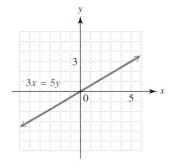

FIGURE 25

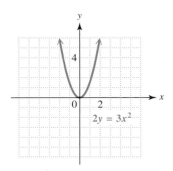

FIGURE 26

EXAMPLE 3   Testing for Symmetry with Respect to the Origin

For each equation, decide if the graph is symmetric with respect to the origin.

(a) $3x = 5y$

  Replace x with $-x$ and y with $-y$ in the equation.

$$3(-x) = 5(-y) \qquad \text{Substitute } -x \text{ for } x \text{ and } -y \text{ for } y.$$
$$-3x = -5y \qquad \text{Multiply.}$$
$$3x = 5y \qquad \text{Multiply each side by } -1.$$

Since this equation is equivalent to the original equation, the graph has symmetry with respect to the origin. See Figure 25.

(b) $2y = 3x^2$

  Substituting $-x$ for x and $-y$ for y gives

$$2(-y) = 3(-x)^2 \qquad \text{Replace } y \text{ with } -y \text{ and } x \text{ with } -x.$$
$$-2y = 3x^2, \qquad \text{Multiply.}$$

which is not equivalent to the original equation. As Figure 26 shows, the graph is not symmetric with respect to the origin.

A summary of the tests for symmetry follows.

Tests for Symmetry

	Symmetric with Respect to		
	x-axis	*y-axis*	*Origin*
Test	Replace y with $-y$.	Replace x with $-x$.	Replace x with $-x$ and replace y with $-y$.
Example			

Now Try Exercises 9, 11, and 13.

 CONNECTIONS

A figure has *rotational symmetry* around an axis *I* if it coincides with itself by all rotations about *I*. Because of their complete rotational symmetry, the circle in the plane and the sphere in space were considered by the early Greeks to be the most perfect geometric figures. Aristotle assumed a spherical shape for the celestial bodies because any other would detract from their heavenly perfection.

Symmetry has been an important characteristic of art from the earliest times. The art of M. C. Escher (1898–1972) is composed of symmetries and translations, and Leonardo da Vinci's sketches indicate a superior understanding of symmetry. Almost all nature exhibits symmetry—from the hexagons of snowflakes to the diatom, a microscopic sea plant. Perhaps the most striking examples of symmetry in nature are crystals.

A diatom   A cross section of tourmaline

Source: Mathematics, Life Science Library, Time Inc., New York, 1963.

For Discussion or Writing

Discuss other examples of symmetry in art and nature.

OBJECTIVE 4 Decide if a function is increasing or decreasing on an interval. Intuitively, a function is said to be *increasing* if its graph goes up from left to right. The function graphed in Figure 27(a) is an increasing function. On the other hand, a function is *decreasing* if its graph goes down from left to right, like the function in Figure 27(b). The function graphed in Figure 27(c) is neither an increasing function nor a decreasing function. However, it is increasing on the interval $(-\infty, -1)$ and decreasing on the interval $(-1, \infty)$.

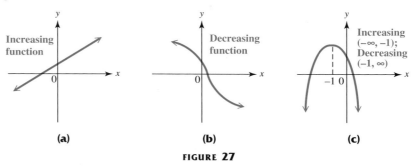

FIGURE 27

In the following definition of a function increasing or decreasing on an interval, I represents any interval of real numbers.

Increasing and Decreasing Functions

Let f be a function, with x_1 and x_2 in an interval I in the domain of f. Then

f is **increasing** on I if $f(x_1) < f(x_2)$ whenever $x_1 < x_2$;

f is **decreasing** on I if $f(x_1) > f(x_2)$ whenever $x_1 < x_2$.

EXAMPLE 4 Determining Increasing or Decreasing Intervals

Give the intervals where each function is increasing or decreasing.

(a) The function graphed in Figure 28(a) is decreasing on $(-\infty, -2)$ and $(1, \infty)$. The function is increasing on $(-2, 1)$.

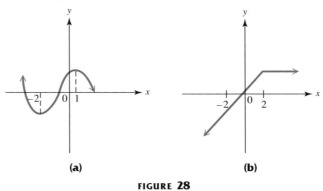

FIGURE 28

(b) The function graphed in Figure 28(b) is increasing on $(-\infty, 2)$. On the interval $(2, \infty)$ the function is neither increasing nor decreasing.

Now Try Exercises 25 and 27.

A function that is neither increasing nor decreasing on an interval is said to be **constant** on that interval. The function graphed in Figure 28(a) is constant on the interval $(2, \infty)$.

CAUTION When identifying intervals over which a function is increasing, decreasing, or constant, remember that we are interested in identifying *domain* intervals. Range values do not appear in these stated intervals.

10.4 EXERCISES

 For Exercises 1 and 2, see Example 1.

1. Use the graph of $y = f(x)$ in the figure to obtain the graph of each equation. Explain how each graph is related to the graph of $y = f(x)$.

   (a) $y = -f(x)$ **(b)** $y = 2f(x)$

2. Use the graph of $y = g(x)$ in the figure to obtain the graph of each equation. Explain how each graph is related to the graph of $y = g(x)$.

   (a) $y = \dfrac{1}{2} g(x)$ **(b)** $y = -g(x)$

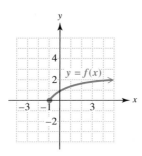

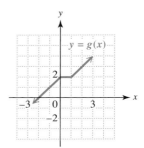

  3. In your own words, state how you would go about testing an equation in x and y to see if its graph is symmetric with respect to **(a)** the x-axis, **(b)** the y-axis, and **(c)** the origin.

  4. Explain why the graph of a function cannot be symmetric with respect to the x-axis.

*Plot each point, and then use the same axes to plot the points that are symmetric to the given point with respect to the following: **(a)** x-axis, **(b)** y-axis, **(c)** origin.*

  5. $(-4, -2)$ **6.** $(-8, 3)$ **7.** $(-8, 0)$ **8.** $(0, -3)$

Use the tests for symmetry to decide whether the graph of each relation is symmetric with respect to the x-axis, the y-axis, or the origin. Remember that more than one of these symmetries may apply, and that perhaps none apply. See Examples 2 and 3. (Do not graph.)

  9. $x^2 + y^2 = 5$ **10.** $y^2 = 4 - x^2$ **11.** $y = x^2 - 8x$

 12. $y = 4x - x^2$ **13.** $y = |x|$ **14.** $y = |x| + 1$

 15. $y = x^3$ **16.** $y = -x^3$ **17.** $f(x) = \dfrac{1}{1 + x^2}$

 18. $f(x) = \dfrac{-1}{x^2 + 9}$ **19.** $xy = 2$ **20.** $xy = -6$

RELATING CONCEPTS (EXERCISES 21–23)

For Individual or Group Work

*A function f is an **even function** if $f(-x) = f(x)$ for all x in the domain of f.*
*A function f is an **odd function** if $f(-x) = -f(x)$ for all x in the domain of f.*

To see how this relates to another mathematical concept, **work Exercises 21–23 in order.**

21. Use the preceding definition to determine whether the function defined by $f(x) = x^n$ is an even function or an odd function for $n = 2$, $n = 4$, and $n = 6$.

22. Use the preceding definition to determine whether the function defined by $f(x) = x^n$ is an even function or an odd function for $n = 1$, $n = 3$, and $n = 5$.

23. (a) Describe how the words *even* and *odd* are appropriate choices for the definitions.
   (b) If a function is even, what do we know about its symmetry?
   (c) If a function is odd, what do we know about its symmetry?

For each function, give the interval where f is decreasing and the interval where f is increasing. In Exercises 28 and 29, first graph the function. See Example 4.

24.

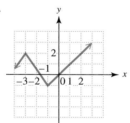

25.

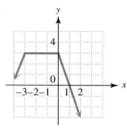

26.

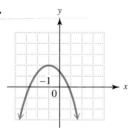

27.

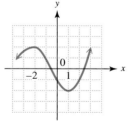

28. $f(x) = 3 - x^2$

29. $f(x) = 2x^2 + 1$

30. The graph shows the number of successful announced payloads launched into space for the years 1993 through 1997. Refer to the graph and answer the following questions.

   (a) Over what interval, in years, is the function increasing?
   (b) Over what interval, in years, is the function decreasing?

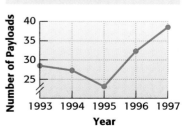

SUCCESSFUL ANNOUNCED SPACE PAYLOADS

Source: National Aeronautics and Space Administration.

*Decide whether each figure is symmetric with respect to (**a**) the given line, and (**b**) the given point.*

31.

32.

33.

34.

Assume that for $y = f(x)$, $f(2) = 3$. For each given statement, find another value for the function.

35. The graph of $y = f(x)$ is symmetric with respect to the origin.

36. The graph of $y = f(x)$ is symmetric with respect to the y-axis.

37. The graph of $y = f(x)$ is symmetric with respect to the line $x = 3$.

38. A graph that is symmetric with respect to both the x-axis and the y-axis is also symmetric with respect to the origin. Explain why.

Complete the left half of the graph of $y = f(x)$ based on the given assumption.

39. For all x, $f(-x) = f(x)$.

40. For all x, $f(-x) = -f(x)$.

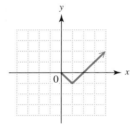

TECHNOLOGY INSIGHTS (EXERCISES 41–46)

Shown on the left is the graph of $Y_1 = (X - 2)^2 + 1$ in the standard viewing window of a graphing calculator. Six other functions, Y_2 through Y_7, are graphed according to the rules shown in the screen on the right.

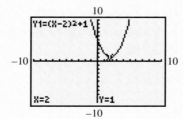

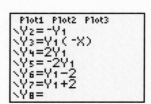

Match each function with its calculator graph from choices A–F without using a calculator, by applying the techniques of this section.

41. Y_2 **42.** Y_3 **43.** Y_4 **44.** Y_5 **45.** Y_6 **46.** Y_7

A.

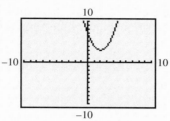

B.

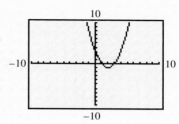

C.

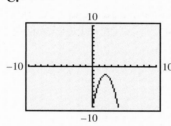

D.

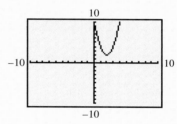

E.

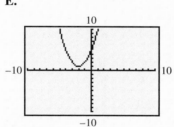

F.

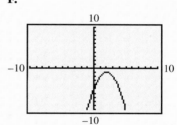

10.5 Piecewise Linear Functions

OBJECTIVES

1 Graph absolute value functions.

2 Graph other piecewise linear functions.

3 Graph step functions.

A function defined by different linear equations over different intervals of its domain is called a **piecewise linear function.**

OBJECTIVE 1 Graph absolute value functions.   An example of a function with a graph that includes portions of two lines is the **absolute value function,** defined by $f(x) = |x|$. By the definition of absolute value, this function is composed of portions of two lines, written as follows.

$$f(x) = \begin{cases} x & \text{if } x \ge 0 \\ -x & \text{if } x < 0 \end{cases}$$

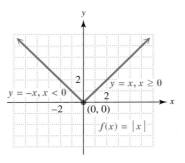

FIGURE 29

Since $|x|$ can be found for any real number x, the domain of $f(x) = |x|$ is $(-\infty, \infty)$. Also, $|x| \geq 0$ for any real number x, so the range is $[0, \infty)$. If $x \geq 0$, then $y = |x| = x$, so for nonnegative values of x we graph $y = x$. On the other hand, if $x < 0$, then $y = |x| = -x$, and we graph $y = -x$ for negative values of x. The final graph is shown in Figure 29. By the vertical line test, the graph is that of a function. Notice that the graph of $f(x) = |x|$ is symmetric with respect to the y-axis. Like parabolas, the graphs of absolute value functions always have an axis of symmetry.

EXAMPLE 1   Graphing Absolute Value Functions

Graph each function.

(a) $f(x) = -|x|$

As shown in Section 10.4, the negative sign indicates that the graph of $f(x)$ is the reflection of the graph of $y = |x|$ about the x-axis. The domain is $(-\infty, \infty)$, and the range is $(-\infty, 0]$. As shown in Figure 30, on the interval $(-\infty, 0]$, the graph is the same as the graph of $y = x$; on the interval $(0, \infty)$, it is the graph of $y = -x$.

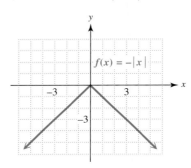

FIGURE 30

(b) $f(x) = |3x + 4| + 1$

In Section 10.2, we saw that the graph of $y = a(x - h)^2 + k$ is a parabola that, depending on the absolute value of a, is stretched or shrunken compared with the graph of $y = x^2$, and that is shifted h units horizontally and k units vertically. The same idea applies here. If we write

$$y = |3x + 4| + 1 \quad \text{as} \quad y = a|x - h| + k,$$

its graph will compare similarly with the graph of $y = |x|$.

$$y = |3x + 4| + 1$$

$$y = \left|3\left(x + \frac{4}{3}\right)\right| + 1 \qquad \text{Factor 3 from the absolute value expression.}$$

$$y = |3|\left|x + \frac{4}{3}\right| + 1 \qquad |ab| = |a| \cdot |b|; \text{ Write each factor in absolute value bars.}$$

$$y = 3\left|x + \frac{4}{3}\right| + 1 \qquad |3| = 3$$

In this form, we see that the graph is narrower than the graph of $y = |x|$ (that is, stretched vertically), with the "vertex" at the point $\left(-\frac{4}{3}, 1\right)$. See Figure 31. The axis

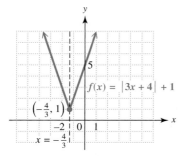

FIGURE 31

of symmetry is $x = -\frac{4}{3}$. The coefficient of x, 3, determines the slopes of the two partial lines that form the graph. One has slope 3, and the other has slope -3. Because the absolute value of the slopes is greater than 1, the lines are steeper than the lines that form the graph of $y = |x|$.

Now Try Exercises 5 and 11.

OBJECTIVE 2 Graph other piecewise linear functions. The different parts of piecewise linear functions may have completely different equations.

▨ **EXAMPLE 2** Graphing Piecewise Linear Functions

Graph each function.

(a) $f(x) = \begin{cases} x + 1 & \text{if } x \le 2 \\ -2x + 7 & \text{if } x > 2 \end{cases}$

Graph the function over each interval of the domain separately. If $x \le 2$, this portion of the graph has an endpoint at $x = 2$. Find the y-value by substituting 2 for x in $y = x + 1$ to get $y = 3$. Another point is needed to complete this portion of the graph. Choose an x-value less than 2. Choosing $x = -1$ gives $y = -1 + 1 = 0$. Draw the graph through $(2, 3)$ and $(-1, 0)$ as a partial line with an endpoint at $(2, 3)$.

Graph the function over interval $x > 2$ similarly. This line will have an open endpoint when $x = 2$ and $y = -2(2) + 7 = 3$. Choosing $x = 4$ gives $y = -2(4) + 7 = -1$. The partial line through $(2, 3)$ and $(4, -1)$ completes the graphs. The two parts meet at $(2, 3)$. See Figure 32.

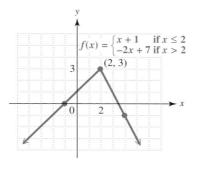

FIGURE 32   **FIGURE 33**

(b) $f(x) = \begin{cases} 2x + 3 & \text{if } x \le 1 \\ -x + 4 & \text{if } x > 1 \end{cases}$

Graph the function over each interval of the domain separately. If $x \le 1$, the graph has an endpoint at $x = 1$. Substitute 1 for x in $y = 2x + 3$ to get the ordered pair $(1, 5)$. For another point on this portion of the graph, choose a number less than 1, say $x = -2$. This gives the ordered pair $(-2, -1)$. Draw the partial line through these points with an endpoint at $(1, 5)$. Graph the function over the interval $x > 1$ similarly. This line has an open endpoint at $(1, 3)$ and goes through $(4, 0)$. The completed graph is shown in Figure 33.

Now Try Exercises 15 and 17.

CAUTION In Example 2, we did not graph the entire lines but only those portions with domain intervals as given. Graphs of these functions should not be two intersecting lines.

▧ **EXAMPLE 3** Applying a Piecewise Linear Function

The table and the graph in Figure 34 show the number of cable TV stations from 1984 through 1999. Write equations for each part of the graph and use them to define a function that models the number of cable TV stations from 1984 through 1999. Let $x = 0$ represent 1984, $x = 1$ represent 1985, and so on.

Year	Number
1984	48
1992	87
1999	181

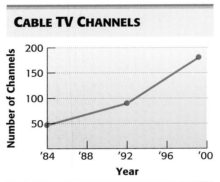

CABLE TV CHANNELS

Source: National Cable Television Association.

FIGURE 34

The data in the table can be used to write the ordered pairs $(0, 48)$ and $(8, 87)$. The slope of the line through these points is

$$m = \frac{87 - 48}{8 - 0} = \frac{39}{8} = 4.875.$$

Using the ordered pair $(0, 48)$ and $m = 4.875$ in the point-slope form of the equation of a line gives an equation of the first line.

$$y - y_1 = m(x - x_1)$$
$$y - 48 = 4.875(x - 0)$$
$$y = 4.875x + 48$$

Similarly, the equation of the other line is $y = 13.429x - 20.432$ (with the slope rounded to 3 decimal places). Verify this. Thus, the number of cable TV channels can be modeled by the function defined by

$$f(x) = \begin{cases} 4.875x + 48 & \text{if } 0 \leq x \leq 8 \\ 13.429x - 20.432 & \text{if } 8 < x \leq 15. \end{cases}$$

Now Try Exercise 43.

OBJECTIVE 3 Graph step functions. The symbol $[\![x]\!]$ is used to represent the greatest integer less than or equal to x. For example, $[\![8.4]\!] = 8$, $[\![-5]\!] = -5$, $[\![-6.9]\!] = -7$, and $[\![\pi]\!] = 3$. The function defined by $f(x) = [\![x]\!]$ is called the **greatest integer function.** It is an example of a **step function,** with a graph that looks like a series of steps.

▨ **EXAMPLE 4** Graphing the Greatest Integer Function $f(x) = [\![x]\!]$

Graph $f(x) = [\![x]\!]$.

For any value of x in the interval $[0, 1)$, $[\![x]\!] = 0$. Also, for x in $[1, 2)$, $[\![x]\!] = 1$. This process continues; for x in $[2, 3)$, the value of $[\![x]\!]$ is 2. The values of y are constant between integers, but they jump at integer values of x. Thus the graph, shown in Figure 35 over the interval $[-3, 4)$, is a series of line segments. In each case, the left endpoint of the segment is included, and the right endpoint is excluded. The domain of the function is $(-\infty, \infty)$, while the range is the set of integers, $\{\ldots, -2, -1, 0, 1, 2, \ldots\}$.

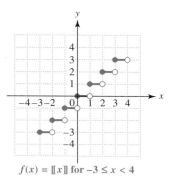

$f(x) = [\![x]\!]$ for $-3 \le x < 4$

FIGURE 35

▨

Now Try Exercise 29.

▨ **EXAMPLE 5** Graphing the Greatest Integer Function $f(x) = [\![ax + b]\!]$

Graph $f(x) = \left[\!\!\left[\dfrac{1}{2}x + 1\right]\!\!\right]$.

If x is in the interval $[0, 2)$, then $y = 1$. For x in $[2, 4)$, $y = 2$, and so on. The graph is shown in Figure 36. Again, the domain of the function is $(-\infty, \infty)$. The range is the set of integers, $\{\ldots, -2, -1, 0, 1, 2, \ldots\}$. (Again, we show only a portion of the graph.)

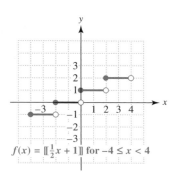

$f(x) = [\![\frac{1}{2}x + 1]\!]$ for $-4 \le x < 4$

FIGURE 36

▨

Now Try Exercise 31.

EXAMPLE 6 Applying the Greatest Integer Function

An express mail company charges $10 for a package weighing 1 lb or less. Each additional pound or part of a pound costs $3 more. Find the cost to send a package weighing 2 lb; 2.5 lb; 5.8 lb. Graph the ordered pairs (pounds, cost). Is this the graph of a function?

The cost for a package weighing 2 lb is $10 for the first pound and $3 for the second pound, for a total of $13. For a 2.5-lb package, the cost will be the same as for 3 lb:

$$10 + 2(3) = 16, \quad \text{or} \quad \$16.$$

A 5.8-lb package will cost the same as a 6-lb package:

$$10 + 5(3) = 25, \quad \text{or} \quad \$25.$$

The graph of this step function is shown in Figure 37. Notice that the right endpoints are included in this case, instead of the left endpoints.

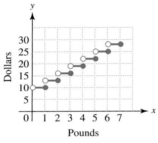

FIGURE 37

Now Try Exercise 45.

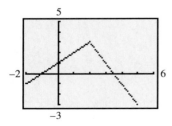

FIGURE 38

CONNECTIONS

The absolute value function is often a built-in function in a graphing calculator. It is designated by $|x|$ or abs(x). If neither of these forms is available, use the fact that $|x| = \sqrt{x^2}$.

Piecewise functions are graphed with some graphing calculators by entering each function separately with its domain and graphing them simultaneously. See Figure 38, which is the graph of the function in Example 2(a). The graph does not distinguish between open and closed endpoints, so it is still important for you to understand what the graph should look like.

Some graphing calculators have a built-in function for the greatest integer function. If your calculator does not have this feature, these functions can be graphed like other piecewise functions.

10.5 EXERCISES

Without actually plotting points, match each function with its graph from choices A–D.

1. $f(x) = |x - 2| + 2$

2. $f(x) = |x + 2| + 2$

3. $f(x) = |x - 2| - 2$

4. $f(x) = |x + 2| - 2$

A.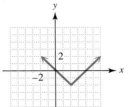

B.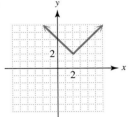

C.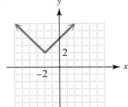

D.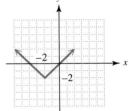

Sketch the graph of each function defined by the absolute value expression. See Example 1.

5. $f(x) = |x + 1|$

6. $f(x) = |x - 1|$

7. $f(x) = |2 - x|$

8. $f(x) = |-3 - x|$

9. $y = |x| + 4$

10. $y = 2|x| - 1$

11. $y = 3|x - 2| - 1$

12. $y = \dfrac{1}{2}|x + 3| + 1$

*For each piecewise linear function, find **(a)** $f(-5)$, **(b)** $f(-1)$, **(c)** $f(0)$, **(d)** $f(3)$, and **(e)** $f(5)$. See Example 2.*

13. $f(x) = \begin{cases} 2x & \text{if } x \le -1 \\ x - 1 & \text{if } x > -1 \end{cases}$

14. $f(x) = \begin{cases} 3x + 5 & \text{if } x \le 0 \\ 4 - 2x & \text{if } 0 < x < 2 \\ x & \text{if } x \ge 2 \end{cases}$

Graph each piecewise linear function. See Example 2.

15. $f(x) = \begin{cases} x - 1 & \text{if } x \le 3 \\ 2 & \text{if } x > 3 \end{cases}$

16. $f(x) = \begin{cases} 6 - x & \text{if } x \le 3 \\ 3x - 6 & \text{if } x > 3 \end{cases}$

17. $f(x) = \begin{cases} 4 - x & \text{if } x < 2 \\ 1 + 2x & \text{if } x \ge 2 \end{cases}$

18. $f(x) = \begin{cases} -2 & \text{if } x \ge 1 \\ 2 & \text{if } x < 1 \end{cases}$

19. $f(x) = \begin{cases} 2x + 1 & \text{if } x \ge 0 \\ x & \text{if } x < 0 \end{cases}$

20. $f(x) = \begin{cases} 5x - 4 & \text{if } x \ge 1 \\ x & \text{if } x < 1 \end{cases}$

Graph each piecewise function. (Hint: At least one part is not linear.)

21. $f(x) = \begin{cases} 2 + x & \text{if } x < -4 \\ -x^2 & \text{if } x \geq -4 \end{cases}$

22. $f(x) = \begin{cases} -2x & \text{if } x \leq 2 \\ -x^2 & \text{if } x > 2 \end{cases}$

23. $f(x) = \begin{cases} |x| & \text{if } x > -2 \\ x^2 - 2 & \text{if } x \leq -2 \end{cases}$

24. $f(x) = \begin{cases} |x| - 1 & \text{if } x > -1 \\ x^2 - 1 & \text{if } x \leq -1 \end{cases}$

RELATING CONCEPTS (EXERCISES 25–28)

For Individual or Group Work

Graphs describing trends often are examples of piecewise linear functions. The graph shows the number of successful announced payloads launched into space each year from 1993 through 1997. (See also Section 10.4, Exercise 30.)

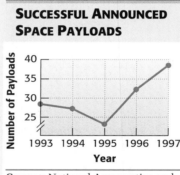

SUCCESSFUL ANNOUNCED SPACE PAYLOADS

Source: National Aeronautics and Space Administration.

Work Exercises 25–28 in order, to see how the slopes of these segments are interpreted.

25. In 1995, there were 24 successful announced payloads; in 1996, there were 32. What is the slope of the segment connecting these data points?

26. Based on your result in Exercise 25, fill in the blanks: The slope of the segment indicates that between the years _____ and _____ , the number of successful announced payloads _____ by _____ .
  (increased/decreased)

27. From 1996 to 1997, the number of payloads rose from 32 to 37. What is the slope of the segment connecting these data points?

28. In your own words, interpret the magnitude of the slopes.

Graph each step function defined by the greatest integer expressions. See Examples 4 and 5.

29. $f(x) = [\![-x]\!]$

30. $f(x) = [\![2x]\!]$

31. $f(x) = [\![2x - 1]\!]$

32. $f(x) = [\![3x + 1]\!]$

33. $f(x) = [\![3x]\!]$

34. $f(x) = [\![3x]\!] + 1$

| **TECHNOLOGY INSIGHTS** (EXERCISES 35–42) |

Absolute value equations and inequalities were introduced in Section 2.7. We saw there that the solution set of $|2x + 1| = 7$ is $\{-4, 3\}$, the solution set of $|2x + 1| > 7$ is $(-\infty, -4) \cup (3, \infty)$, and the solution set of $|2x + 1| < 7$ is $(-4, 3)$. If we graph $y_1 = |2x + 1|$ and $y_2 = 7$ in the standard viewing window of a graphing calculator, and use the feature that allows us to find the points of intersection of the two graphs, we obtain the two figures that follow.

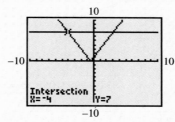 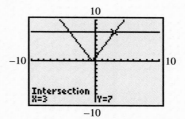

Notice that the x-coordinates of the points of intersection are the solutions of the equation. Because the graph of y_1 lies *above* the graph of y_2 for values of x less than -4 or greater than 3, we have provided graphical support for the solution set of $y_1 > y_2$ given above. Similarly, the graph of y_1 lies *below* the graph of y_2 for values of x between -4 and 3, the solution set of $y_1 < y_2$.

For the given equations and inequalities, where y_1 represents the expression on the left and y_2 represents the expression on the right, use the graphing calculator screens to give the solution sets.

35. (a) $|x + 4| = 6$ **(b)** $|x + 4| < 6$ **(c)** $|x + 4| > 6$

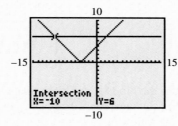 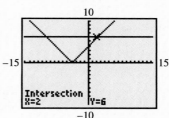

36. (a) $|2x + 3| = 5$ **(b)** $|2x + 3| > 5$ **(c)** $|2x + 3| < 5$

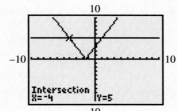 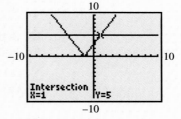

Use the algebraic methods described in Section 2.7 to solve each absolute value equation or inequality. Then use the method described in the discussion preceding Exercise 35 to give graphical support for your answer.

37. $|x - 5| = 4$ **38.** $|x + 2| = 7$ **39.** $|7 - 4x| > 1$

40. $|3 - 2x| > 3$ **41.** $|.5x - 2| < 1$ **42.** $|.4x + 2| < 2$

Work each problem. See Examples 3 and 6.

43. The light vehicle market share (in percent) in the U.S. for domestic cars is shown in the graph.

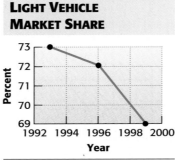

LIGHT VEHICLE MARKET SHARE

Source: J.D. Power & Associates.

Let $x = 3$ represent 1993, $x = 6$ represent 1996, and so on. Use the points on the graph to write equations for the line segments in the intervals $[3, 6]$ and $(6, 9]$. Then define $f(x)$ for the piecewise linear function.

44. When a diabetic takes long-acting insulin, the insulin reaches its peak effect on the blood sugar level in about 3 hr. This effect remains fairly constant for 5 hr, then declines, and is very low until the next injection. In a typical patient, the level of insulin might be given by the following function.

$$i(t) = \begin{cases} 40t + 100 & \text{if } 0 \le t \le 3 \\ 220 & \text{if } 3 < t \le 8 \\ -80t + 860 & \text{if } 8 < t \le 10 \\ 60 & \text{if } 10 < t \le 24 \end{cases}$$

Here $i(t)$ is the blood sugar level, in appropriate units, at time t measured in hours from the time of the injection. Chuck takes his insulin at 6 A.M. Find the blood sugar level at each of the following times.

(a) 7 A.M.　　**(b)** 9 A.M.　　**(c)** 10 A.M.　　**(d)** noon
(e) 3 P.M.　　**(f)** 5 P.M.　　**(g)** midnight　　**(h)** Graph $y = i(t)$.

45. Suppose a chain-saw rental firm charges a fixed $4 sharpening fee plus $7 per day or fraction of a day. Let $S(x)$ represent the cost of renting a saw for x days. Find the value of

(a) $S(1)$　　**(b)** $S(1.25)$　　**(c)** $S(3.5)$.
(d) Graph $y = S(x)$.　　**(e)** Give the domain and range of S.

46. To rent a midsized car from Avis costs $30 per day or fraction of a day. If you pick up the car in Lansing and drop it in West Lafayette, there is a fixed $50 dropoff charge. Let $C(x)$ represent the cost of renting the car for x days, taking it from Lansing to West Lafayette. Find each of the following.

(a) $C\left(\dfrac{3}{4}\right)$　　**(b)** $C\left(\dfrac{9}{10}\right)$　　**(c)** $C(1)$　　**(d)** $C\left(1\dfrac{5}{8}\right)$　　**(e)** $C(2.4)$
(f) Graph $y = C(x)$.

■ Chapter **10** **Group Activity** ■

Finding the Path of a Comet

OBJECTIVE Find and graph an equation of a parabola with a given focus.

The orbit that a comet takes as it approaches the sun depends on its velocity, as well as other factors. If the velocity of a comet equals escape velocity—that is, it is going just fast enough to get away from the sun—then its orbit will be parabolic. (Other possible orbits are *hyperbolic* or *elliptical*. These figures are discussed in Chapter 13.) The sun is at the focus of the parabola. The vertex is the point where the comet is closest to the sun.

A. Refer to Section 10.2, Objective 5, and make a sketch of the comet and the sun.

    1. Place the vertex of the parabola at the origin, with the focus on the *y*-axis and a horizontal directrix.

    2. Assume that the comet is .75 astronomical unit* from the sun at its closest point.

B. What are the coordinates of the focus?

C. What is the equation of the directrix?

D. Using the information from Section 10.2, Objective 5, find an equation of the parabola.

E. Graph the parabola (as a vertical parabola). Include the focus and directrix on your graph.

---

*One astronomical unit (AU) is the average distance between Earth and the sun.

CHAPTER 10 SUMMARY

KEY TERMS

10.1 difference quotient
composite function
(composition)
10.2 parabola
vertex
axis
quadratic function

focus
directrix
10.4 symmetric with
respect to the y-axis
symmetric with
respect to the x-axis

symmetric with
respect to the
origin
increasing function
decreasing function
10.5 piecewise linear
function

absolute value
function
greatest integer
function
step function

NEW SYMBOLS

$f \circ g$ composite function

$[\![x]\!]$ greatest integer
function

TEST YOUR WORD POWER

See how well you have learned the vocabulary in this chapter. Answers, with examples, follow the Quick Review.

1. A **quadratic function** is a function
that can be written in the form
   A. $f(x) = mx + b$ for real numbers
   m and b
   B. $f(x) = \frac{P(x)}{Q(x)}$, where $Q(x) \neq 0$
   C. $f(x) = ax^2 + bx + c$ for real
   numbers a, b, and c ($a \neq 0$)
   D. $f(x) = \sqrt{x}$ for $x \geq 0$.

2. A **parabola** is the graph of
   A. any equation in two variables
   B. a linear equation
   C. an equation of degree 3
   D. an equation of the form
   $y = ax^2 + bx + c$ or
   $x = ay^2 + by + c$.

3. The **vertex** of a vertical parabola is
   A. the point where the graph
   intersects the y-axis

   B. the point where the graph
   intersects the x-axis
   C. the lowest point on a parabola
   that opens up or the highest point
   on a parabola that opens down
   D. the origin.

4. The **axis** of a parabola is
   A. either the x-axis or the y-axis
   B. the vertical line (of a vertical
   parabola) or the horizontal line
   (of a horizontal parabola) through
   the vertex
   C. the lowest or highest point on the
   graph of a parabola
   D. a line through the origin.

5. A parabola is **symmetric about its
axis** since
   A. its graph is near the axis
   B. its graph is identical on each side

   of the axis
   C. its graph looks different on each
   side of the axis
   D. its graph intersects the axis.

6. An **absolute value function** is a
function that can be written in
the form
   A. $f(x) = ax^2 + bx + c$, where
   $a \neq 0$
   B. $f(x) = a(x - h)^2 + k$, where
   $a \neq 0$
   C. $f(x) = a|x - h| + k$, where
   $a \neq 0$
   D. $f(x) = [\![x - h]\!] + k$.

QUICK REVIEW

CONCEPTS	EXAMPLES

10.1 OPERATIONS AND COMPOSITION

Operations on Functions

$$(f + g)(x) = f(x) + g(x) \quad \text{Sum}$$
$$(f - g)(x) = f(x) - g(x) \quad \text{Difference}$$
$$(fg)(x) = f(x) \cdot g(x) \quad \text{Product}$$
$$\left(\frac{f}{g}\right)(x) = \frac{f(x)}{g(x)}, \quad g(x) \neq 0 \quad \text{Quotient}$$

If $f(x) = 3x^2 + 2$ and $g(x) = \sqrt{x}$, then

$$(f + g)(x) = 3x^2 + 2 + \sqrt{x}$$
$$(f - g)(x) = 3x^2 + 2 - \sqrt{x}$$
$$(fg)(x) = (3x^2 + 2)(\sqrt{x})$$
$$\left(\frac{f}{g}\right)(x) = \frac{3x^2 + 2}{\sqrt{x}}, \quad x > 0.$$

Composition of Functions
If f and g are functions, then the composite function of g and f is

$$(g \circ f)(x) = g[f(x)].$$

If $g(x) = \sqrt{x}$ and $f(x) = x^2 - 1$, then the composite function of g and f is

$$(g \circ f)(x) = g[f(x)] = \sqrt{x^2 - 1}.$$

10.2 GRAPHS OF QUADRATIC FUNCTIONS

1. The graph of the quadratic function defined by $F(x) = a(x - h)^2 + k, a \neq 0$, is a parabola with vertex at (h, k) and the vertical line $x = h$ as axis.

2. The graph opens up if a is positive and down if a is negative.

3. The graph is wider than the graph of $f(x) = x^2$ if $0 < |a| < 1$ and narrower if $|a| > 1$.

The parabola with focus at $(0, p)$ and directrix $y = -p$ has equation

$$y = \frac{1}{4p}x^2.$$

Graph $f(x) = -(x + 3)^2 + 1$.

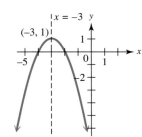

The graph opens down since $a < 0$. It is shifted 3 units left and 1 unit up, so the vertex is $(-3, 1)$, with axis $x = -3$. The domain is $(-\infty, \infty)$; the range is $(-\infty, 1]$.

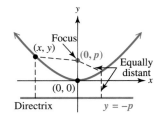

10.3 MORE ABOUT PARABOLAS; APPLICATIONS

The vertex of the graph of $f(x) = ax^2 + bx + c$, $a \neq 0$, may be found by completing the square. The vertex has coordinates

$$\left(\frac{-b}{2a}, f\left(\frac{-b}{2a}\right)\right).$$

Graphing a Quadratic Function
Step 1 Determine whether the graph opens up or down.
Step 2 Find the vertex.
Step 3 Find the x-intercepts (if any). Find the y-intercept.
Step 4 Find and plot additional points as needed.

Graph $f(x) = x^2 + 4x + 3$.

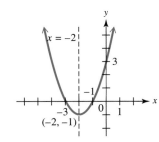

The graph opens up since $a > 0$. The vertex is $(-2, -1)$. The solutions of $x^2 + 4x + 3 = 0$ are -1 and -3, so the x-intercepts are $(-1, 0)$ and $(-3, 0)$. Since $f(0) = 3$, the y-intercept is $(0, 3)$. The domain is $(-\infty, \infty)$; the range is $[-1, \infty)$.

(continued)

CONCEPTS	EXAMPLES

Horizontal Parabolas

The graph of $x = ay^2 + by + c$ is a horizontal parabola, opening to the right if $a > 0$ or to the left if $a < 0$. Horizontal parabolas do not represent functions.

Graph $x = 2y^2 + 6y + 5$.

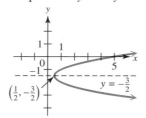

The graph opens to the right since $a > 0$. The vertex is $\left(\frac{1}{2}, -\frac{3}{2}\right)$. The axis is $y = -\frac{3}{2}$. The domain is $\left[\frac{1}{2}, \infty\right)$; the range is $(-\infty, \infty)$.

10.4 SYMMETRY; INCREASING AND DECREASING FUNCTIONS

The graph of $g(x) = a \cdot f(x)$ has the same general shape as the graph of $f(x)$.

It is reflected about the x-axis if a is negative.

It is stretched vertically compared to the graph of $f(x)$ if $|a| > 1$.

It is shrunken vertically compared to the graph of $f(x)$ if $0 < |a| < 1$.

Let $g(x) = -\frac{1}{2}\sqrt{x}$; the graph of $g(x)$ is shown with the graph of $f(x) = \sqrt{x}$.

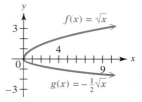

Symmetry

To decide whether the graph of a relation is symmetric with respect to the following, perform the indicated test.

Test each relation for symmetry.

(a) The x-axis
Replace y with $-y$.

(a) $x = y^2 - 5$
$x = (-y)^2 - 5 = y^2 - 5$
The graph is symmetric with respect to the x-axis.

(b) The y-axis
Replace x with $-x$.

(b) $y = -2x^2 + 1$
$y = -2(-x)^2 + 1 = -2x^2 + 1$
The graph is symmetric with respect to the y-axis.

(c) The origin
Replace x with $-x$ and y with $-y$.

(c) $\qquad x^2 + y^2 = 4$
$(-x)^2 + (-y)^2 = 4$
$x^2 + y^2 = 4$

The symmetry holds if the resulting equation is equivalent to the original equation.

The graph is symmetric with respect to the origin (and to the x-axis and y-axis).

Increasing and Decreasing Functions

A function f is increasing on an interval if $f(x_1) < f(x_2)$ whenever $x_1 < x_2$.

A function f is decreasing on an interval if $f(x_1) > f(x_2)$ whenever $x_1 < x_2$.

If f is neither increasing nor decreasing on an interval, it is constant there.

f is increasing on $(-\infty, a)$.

f is decreasing on (a, b).

f is constant on (b, ∞).

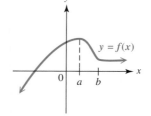

CONCEPTS	EXAMPLES

10.5 PIECEWISE LINEAR FUNCTIONS

Absolute Value Function

$$f(x) = |x|$$

The graph of $f(x) = |ax + b| + c$ has "vertex" at $\left(-\frac{b}{a}, c\right)$ and it is symmetric with respect to a vertical axis through the "vertex."

Graph $f(x) = 2|x - 1| + 3$.

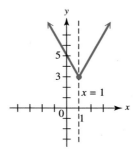

Other Piecewise Linear Functions

Graph each portion with an open or solid endpoint as appropriate.

Graph $f(x) = \begin{cases} x - 2 & \text{if } x \geq 1 \\ 3x & \text{if } x < 1. \end{cases}$

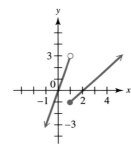

Greatest Integer Function

$$f(x) = [\![x]\!]$$

$[\![x]\!]$ is the greatest integer less than or equal to x.

Graph $f(x) = [\![2x - 1]\!]$.

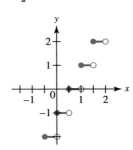

Answers to Test Your Word Power

1. C; *Examples:* $f(x) = x^2 - 2$, $f(x) = (x + 4)^2 + 1$, $f(x) = x^2 - 4x + 5$ **2.** D; *Examples:* See the figures in the Quick Review for Sections 10.2 and 10.3.   **3.** C; *Example:* The graph of $y = (x + 3)^2$ has vertex $(-3, 0)$, which is the lowest point on the graph.
4. B; *Example:* The axis of $y = (x + 3)^2$ is the vertical line $x = -3$.   **5.** B; *Example:* Since the graph of $y = (x + 3)^2$ is symmetric about its axis $x = -3$, the points $(-2, 1)$ and $(-4, 1)$ are on the graph.   **6.** C; *Examples:* $f(x) = |x - 1|$, $f(x) = 2|x| + 5$, $f(x) = -|x|$

CHAPTER **10** REVIEW EXERCISES

[10.1] *Given $f(x) = x^2 - 2x$ and $g(x) = 5x + 3$, find the following. Give the domain of each.*

1. $(f + g)(x)$ **2.** $(f - g)(x)$ **3.** $(fg)(x)$

4. $\left(\dfrac{f}{g}\right)(x)$ **5.** $(g \circ f)(x)$ **6.** $(f \circ g)(x)$

For $f(x) = 2x - 3$ and $g(x) = \sqrt{x}$, find the following.

7. $(f - g)(4)$ **8.** $\left(\dfrac{f}{g}\right)(9)$ **9.** $(fg)(5)$

10. $(f + g)(5)$ **11.** $(fg)(2b)$ **12.** $(g \circ f)(2)$

13. After working Exercise 12, find $(f \circ g)(2)$. Are your answers equal? Is composition of functions a commutative operation?

14. Explain in your own words why 5 is not in the domain of $f(x) = \sqrt{9 - 2x}$.

[10.2, 10.3] *Identify the vertex of each parabola.*

15. $y = 6 - 2x^2$ **16.** $f(x) = -(x - 1)^2$ **17.** $f(x) = (x - 3)^2 + 7$

18. $y = -3x^2 + 4x - 2$ **19.** $x = (y - 3)^2 - 4$

Graph each parabola. Give the vertex, axis, domain, and range.

20. $f(x) = -5x^2$ **21.** $f(x) = 3x^2 - 2$ **22.** $y = (x + 2)^2$

23. $y = 2(x - 2)^2 - 3$ **24.** $f(x) = -2x^2 + 8x - 5$ **25.** $y = x^2 + 3x + 2$

26. $x = (y - 1)^2 + 2$ **27.** $x = 2(y + 3)^2 - 4$ **28.** $x = -\dfrac{1}{2}y^2 + 6y - 14$

Solve each problem.

29. Consumer spending for home video games in dollars per person per year is given in the table. Let $x = 0$ represent 1990, $x = 2$ represent 1992, and so on.

CONSUMER SPENDING FOR HOME VIDEO GAMES

Year	Dollars
1990	12.39
1992	13.08
1994	15.78
1996	19.43
1997	22.71
1998	24.14
1999	25.08

Source: Statistical Abstract of the United States.

(a) Use the data for 1990, 1994, and 1997 in the quadratic form $ax^2 + bx + c = y$ to write a system of three equations.

(b) Solve the system from part (a) to find a quadratic function f that models the data.

(c) Use the model found in part (b) to approximate consumer spending for home video games in 1998 to the nearest cent. How does your answer compare to the actual data from the table?

30. The height (in feet) of a projectile t sec after being fired from Earth into the air is given by

$$f(t) = -16t^2 + 160t.$$

Find the number of seconds required for the projectile to reach maximum height. What is the maximum height?

31. Find the length and width of a rectangle having a perimeter of 200 m if the area is to be a maximum.

32. Find two numbers whose sum is 10 and whose product is a maximum.

[10.4]   *Use the tests for symmetry to determine any symmetries of the graph of each relation. Do not graph.*

33. $2x^2 - y^2 = 4$ **34.** $3x^2 + 4y^2 = 12$ **35.** $2x - y^2 = 8$

36. $y = 2x^2 + 3$ **37.** $y = 2\sqrt{x} - 4$ **38.** $y = \dfrac{1}{x^2}$

39. What is wrong with this statement? "A function whose graph is a circle centered at the origin has symmetry with respect to both axes and the origin."

40. Suppose that a circle has its center at the origin. Is it symmetric with respect to **(a)** the x-axis, **(b)** the y-axis, **(c)** the origin?

41. Suppose that a linear function in the form $f(x) = mx + b$ has $m < 0$. Is it increasing or decreasing over all real numbers?

Give the intervals where each function is increasing, decreasing, or constant.

42.

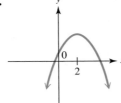

43.

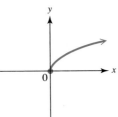

44.

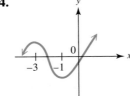

45.

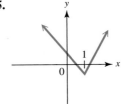

[10.5] *Graph each function.*

46. $f(x) = 2|x| + 3$ **47.** $f(x) = |x - 2|$ **48.** $f(x) = -|x - 1|$

49. $f(x) = \begin{cases} 2x + 1 & \text{if } x \le -1 \\ x + 3 & \text{if } x > -1 \end{cases}$ **50.** $f(x) = \begin{cases} -x & \text{if } x \le 0 \\ x^2 & \text{if } x > 0 \end{cases}$

51. $f(x) = -[\![x]\!]$ **52.** $f(x) = [\![x + 1]\!]$

✏ **53.** Describe how the graph of $y = 2|x + 4| - 3$ relates to the graph of $y = |x|$.

54. Montreal taxi rates in a recent year were 90¢ for the first $\frac{1}{9}$ mi and 10¢ for each additional $\frac{1}{9}$ mi or fraction of $\frac{1}{9}$ mi. Let $C(x)$ be the cost for a taxi ride of $\frac{x}{9}$ mi. Find
(a) $C(1)$ **(b)** $C(2.3)$ **(c)** $C(8)$.
(d) Graph $y = C(x)$. **(e)** Give the domain and range of C.

■ MIXED REVIEW EXERCISES

Match each equation with the figure that most closely resembles its graph.

55. $g(x) = x^2 - 5$ **A.** 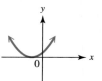 **B.**

56. $h(x) = -x^2 + 4$

57. $F(x) = (x - 1)^2$

58. $G(x) = (x + 1)^2$

59. $H(x) = (x - 1)^2 + 1$ **C.** 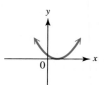 **D.**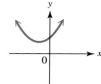

60. $K(x) = (x + 1)^2 + 1$

E. 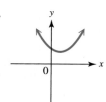 **F.**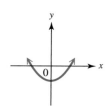

61. Graph $f(x) = 4x^2 + 4x - 2$. Give the vertex, axis, domain, and range.

Graph each relation.

62. $f(x) = -\sqrt{2 - x}$ **63.** $f(x) = |2x + 1|$

64. $y^2 = x - 1$ **65.** $f(x) = [\![x]\!] - 2$

66. A car rental costs $37 for one day, which includes 50 free miles. Each additional 25 mi or portion of 25 mi costs $10. Graph the function $F = \{(\text{miles, cost})\}$. Give the domain and range of F.

RELATING CONCEPTS (EXERCISES 67–70)

For Individual or Group Work

If you compare the graph of $y = x^2 + 3$ to the graph of $y = x^2$, you will notice that the first is simply a vertical translation of the second 3 units up. **Work Exercises 67–70 in order,** *to see how other transformations are accomplished.*

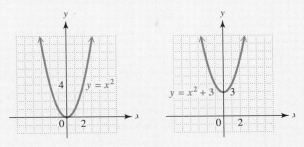

67. Graph $y = x^2$ and $y = x^2 - 4$. The second is obtained by translating the first
_____ units _____ .
                  (up/down)

68. Graph $y = x^2$ and $y = (x + 3)^2$. The second is obtained by translating the first
_____ units to the _____ .
                        (left/right)

69. Graph $y = x^2$ and $y = (x - 4)^2$. The second is obtained by translating the first
_____ units to the _____ .
                        (left/right)

70. Based on your observations in Exercises 67–69, describe how the graph of $y = f(x - h)^2 + k$ compares to the graph of $y = f(x)$ for any function f.

CHAPTER **10** TEST

Let f and g be functions defined by $f(x) = 4x + 2$ and $g(x) = -x^2 + 3$. Find each function value.

1. $g(1)$ **2.** $(f + g)(-2)$ **3.** $\left(\dfrac{f}{g}\right)(3)$ **4.** $(f \circ g)(2)$

5. For the functions defined in the directions for Exercises 1–4, find and simplify $(f - g)(x)$ and give its domain.

6. Which one of the following most closely resembles the graph of $f(x) = a(x - h)^2 + k$ if $a < 0$, $h > 0$, and $k < 0$?

A. 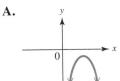 **B.** 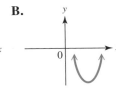 **C.** 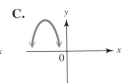 **D.**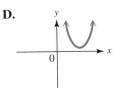

Graph each parabola. Identify the vertex, axis, domain, and range.

7. $f(x) = \dfrac{1}{2}x^2 - 2$ **8.** $f(x) = -x^2 + 4x - 1$ **9.** $x = -(y - 2)^2 + 2$

10. The percent increase for in-state tuition at Iowa public universities during the years 1992 through 2002 can be modeled by the quadratic function defined by

$$f(x) = .156x^2 - 2.05x + 10.2,$$

where $x = 2$ represents 1992, $x = 3$ represents 1993, and so on. (*Source:* Iowa Board of Regents.)

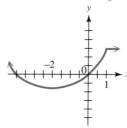

  (a) Based on this model, by what percent (to the nearest tenth) did tuition increase in 2001?

  (b) In what year was the minimum tuition increase? (Round down to the nearest year.) To the nearest tenth, by what percent did tuition increase that year?

11. Palo Alto College is planning to construct a rectangular parking lot on land bordered on one side by a highway. The plan is to use 640 ft of fencing to fence off the other three sides. What should the dimensions of the lot be if the enclosed area is to be a maximum?

Use the tests for symmetry to determine the symmetries, if any, for each relation.

12. $f(x) = -x^2 + 1$ **13.** $x = y^2 + 7$ **14.** $x^2 + y^2 = 4$

15. Give the intervals where the function is increasing, decreasing, and constant.

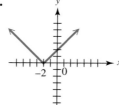

16. Match each function with its graph from choices A–D.

  (a) $f(x) = |x - 2|$ **(b)** $f(x) = |x + 2|$ **(c)** $f(x) = |x| + 2$ **(d)** $f(x) = |x| - 2$

A.

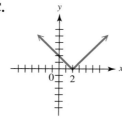

B.

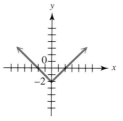

C.

D.

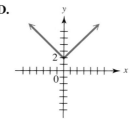

Graph each relation.

17. $f(x) = |x - 3| + 4$

18. $f(x) = [\![2x]\!]$

19. $f(x) = \begin{cases} -x & \text{if } x \le 2 \\ x - 4 & \text{if } x > 2 \end{cases}$

20. Assume that postage rates are 37¢ for the first ounce, plus 23¢ for each additional ounce, and that each letter carries one 37¢ stamp and as many 23¢ stamps as necessary. Graph the function defined by $p(x) =$ the number of stamps on a letter weighing x oz. Use the interval $(0, 5]$.

CUMULATIVE REVIEW EXERCISES CHAPTERS **1–10**

Find the slope of each line.

1. Through $(-2, 1)$ and $(5, -4)$

2. With equation $2x - 3y = 6$

3. Find $f(-3)$ if $f(x) = x^2 - 1$.

4. Does $\dfrac{x^2}{4} - \dfrac{y^2}{9} = 1$ define a function? Explain.

Perform the indicated operations. Assume all variables represent positive real numbers.

5. $(7y - 6)(4y + 3)$

6. $(4x^3 - 3x^2 + 2x - 3) \div (x - 1)$

7. $\dfrac{3p}{p + 1} + \dfrac{4}{p - 1} - \dfrac{6}{p^2 - 1}$

8. $\dfrac{6x - 10}{3x} \cdot \dfrac{6}{18x - 30}$

9. $4\sqrt{2z} - \sqrt{72z} + \sqrt{32z}$

10. $\left(\sqrt{5y} - 3\right)\left(\sqrt{5y} + 3\right)$

11. Write $\dfrac{2(m^{-1})^{-4}}{(3m^{-3})^2}$ with only positive exponents.

12. Write $\sqrt[3]{k^2} \cdot \sqrt{k}$ with rational exponents and simplify.

13. Factor $2k^4 + k^2 - 3$ completely.

Solve.

14. $2(3x - 1) = -(4 - x) - 28$

15. $3p = q(p + q)$ for p

16. $2x + 3 \le 5 - (x - 4)$

17. $2x + 1 > 5$ or $2 - x \ge 2$

18. $|5 - 3m| = 12$

19. $|12t + 7| \ge 0$

20. $\begin{aligned} x + y + z &= 2 \\ 2x + y - z &= 5 \\ x - y + z &= -2 \end{aligned}$
(elimination)

21. $\begin{aligned} x + 2y &= 10 \\ 3x - y &= 9 \end{aligned}$
(row operations)

22. $2x^3 - x^2 - 28x = 0$

23. $\dfrac{4}{a + 2} - \dfrac{11}{9} = \dfrac{5}{3a + 6}$

24. $\sqrt{m + 1} - \sqrt{m - 2} = 1$

25. $5m^2 + 3m = 3$

26. $S = 2\pi rh + \pi r^2$ for r

27. $3t^2 - 8 > 10t$

28. Find the value of the determinant. $\begin{vmatrix} 3 & 1 & 0 \\ -2 & 0 & 2 \\ 0 & -1 & 1 \end{vmatrix}$

29. Use Cramer's rule to solve the system with $D = 28$, $D_x = 35$, $D_y = 56$, and $D_z = 42$.

Solve each problem.

30. A toy rocket is propelled directly upward. Its height in feet after t sec is defined by $h(t) = -16t^2 + 256t$ (neglecting air resistance).

   (a) When will it return to the ground?
   (b) After how many seconds will it be 768 ft above the ground?

31. A plane delivering supplies to an isolated community flew there at an average speed of 200 mph. The average speed on the return trip was 300 mph. Find the distance to the community if total flight time was 4 hr.

Graph.

32. $2x - 3y = 6$

33. $x + 2y \le 4$

34. $f(x) = -2x^2 + 5x + 3$

35. $f(x) = |x + 1|$

36. Over what interval is the function defined by $f(x) = x^2 + 6$

   (a) increasing, **(b)** decreasing?

Tell whether the graph of each relation exhibits any symmetries.

37. $y^2 = x + 3$

38. $x^2 - 6y^2 = 18$

39. If $f(x) = [\![x + 2]\!]$, find **(a)** $f(3)$ and **(b)** $f(-3.1)$.

40. Does the relation $x = y^2 - 2$ define a function? Explain.

Inverse, Exponential, and Logarithmic Functions

11

The exponential and logarithmic functions introduced in this chapter are used to model a wide variety of situations, including environmental issues, compound interest, earthquake intensity, fossil dating, and sound levels.

Recently, there has been concern about the level of sound Americans are subjected to daily. For example, action sequences in *Pearl Harbor, The Movie* reached 107 decibels, while the sound levels in *Lethal Weapon 4* often reached 100 decibels or more, compared to an average of 95 decibels for a motorcycle. In Section 11.5, Exercise 41, we give a logarithmic function to measure sound levels and to find the decibel levels of other recent movies. (*Source: World Almanac and Book of Facts, 2001;* www.lhh.org/noise/)

11.1 Inverse Functions

In this chapter we study two important types of functions, *exponential* and *logarithmic*. These functions are related in a special way: They are *inverses* of one another. We begin by discussing inverse functions in general.

OBJECTIVE 1 Decide whether a function is one-to-one and, if it is, find its inverse. Suppose we define the function

$$G = \{(-2, 2), (-1, 1), (0, 0), (1, 3), (2, 5)\}.$$

We can form another set of ordered pairs from G by interchanging the x- and y-values of each pair in G. We can call this set F, so

$$F = \{(2, -2), (1, -1), (0, 0), (3, 1), (5, 2)\}.$$

To show that these two sets are related, F is called the *inverse* of G. For a function f to have an inverse, f must be a *one-to-one function*.

One-to-One Function

In a **one-to-one function,** each x-value corresponds to only one y-value, and each y-value corresponds to only one x-value.

The function shown in Figure 1(a) is not one-to-one because the y-value 7 corresponds to *two* x-values, 2 and 3. That is, the ordered pairs (2, 7) and (3, 7) both belong to the function. The function in Figure 1(b) is one-to-one.

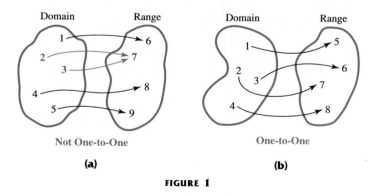

FIGURE 1

The *inverse* of any one-to-one function f is found by interchanging the components of the ordered pairs of f. The inverse of f is written f^{-1}. Read f^{-1} as "the inverse of f" or "f-inverse."

CAUTION The symbol $f^{-1}(x)$ does not represent $\dfrac{1}{f(x)}$.

The definition of the inverse of a function follows.

Inverse of a Function

The **inverse** of a one-to-one function f, written f^{-1}, is the set of all ordered pairs of the form (y, x), where (x, y) belongs to f. Since the inverse is formed by interchanging x and y, the domain of f becomes the range of f^{-1} and the range of f becomes the domain of f^{-1}.

For inverses f and f^{-1}, it follows that

$$f(f^{-1}(x)) = x \quad \text{and} \quad f^{-1}(f(x)) = x.$$

■ EXAMPLE 1 Finding the Inverses of One-to-One Functions

Find the inverse of each one-to-one function.

(a) $F = \{(-2, 1), (-1, 0), (0, 1), (1, 2), (2, 2)\}$

Each x-value in F corresponds to just one y-value. However, the y-value 2 corresponds to two x-values, 1 and 2. Also, the y-value 1 corresponds to both -2 and 0. Because some y-values correspond to more than one x-value, F is not one-to-one and does not have an inverse.

(b) $G = \{(3, 1), (0, 2), (2, 3), (4, 0)\}$

Every x-value in G corresponds to only one y-value, and every y-value corresponds to only one x-value, so G is a one-to-one function. The inverse function is found by interchanging the x- and y-values in each ordered pair.

$$G^{-1} = \{(1, 3), (2, 0), (3, 2), (0, 4)\}$$

Notice how the domain and range of G become the range and domain, respectively, of G^{-1}.

(c) The U.S. Environmental Protection Agency has developed an indicator of air quality called the Pollutant Standard Index (PSI). If the PSI exceeds 100 on a particular day, then that day is classified as unhealthy. The table shows the number of unhealthy days in Chicago for the years 1991 through 1997.

Year	Number of Unhealthy Days
1991	21
1992	4
1993	3
1994	8
1995	21
1996	6
1997	9

Source: U.S. Environmental Protection Agency.

Let f be the function defined in the table, with the years forming the domain and the numbers of unhealthy days forming the range. Then f is not one-to-one, because in two different years (1991 and 1995), the number of unhealthy days was the same, 21.

Now Try Exercises 1, 9, and 11.

OBJECTIVE 2 **Use the horizontal line test to determine whether a function is one-to-one.** It may be difficult to decide whether a function is one-to-one just by looking at the equation that defines the function. However, by graphing the function and observing the graph, we can use the *horizontal line test* to tell whether the function is one-to-one.

> **Horizontal Line Test**
>
> If any horizontal line intersects the graph of a function in no more than one point, then the function is one-to-one.

The horizontal line test follows from the definition of a one-to-one function. Any two points that lie on the same horizontal line have the same y-coordinate. No two ordered pairs that belong to a one-to-one function may have the same y-coordinate, and therefore no horizontal line will intersect the graph of a one-to-one function more than once.

EXAMPLE 2 **Using the Horizontal Line Test**

Use the horizontal line test to determine whether the graphs in Figures 2 and 3 are graphs of one-to-one functions.

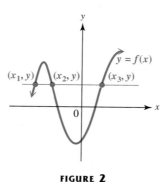

FIGURE 2

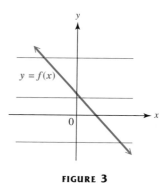

FIGURE 3

Because the horizontal line shown in Figure 2 intersects the graph in more than one point (actually three points), the function is not one-to-one.

Every horizontal line will intersect the graph in Figure 3 in exactly one point. This function is one-to-one.

Now Try Exercise 7.

OBJECTIVE 3 **Find the equation of the inverse of a function.** By definition, the inverse of a function is found by interchanging the x- and y-values of each of its ordered pairs. The equation of the inverse of a function defined by $y = f(x)$ is found in the same way.

Finding the Equation of the Inverse of $y = f(x)$

For a one-to-one function f defined by an equation $y = f(x)$, find the defining equation of the inverse as follows.

Step 1   **Interchange** x and y.

Step 2   **Solve** for y.

Step 3   **Replace** y with $f^{-1}(x)$.

■ EXAMPLE 3   Finding Equations of Inverses

Decide whether each equation defines a one-to-one function. If so, find the equation of the inverse.

(a) $f(x) = 2x + 5$

The graph of $y = 2x + 5$ is a nonvertical line, so by the horizontal line test, f is a one-to-one function. To find the inverse, let $y = f(x)$ so that

$$y = 2x + 5$$
$$x = 2y + 5 \qquad \text{Interchange } x \text{ and } y. \text{ (Step 1)}$$
$$2y = x - 5 \qquad \text{Solve for } y. \text{ (Step 2)}$$
$$y = \frac{x - 5}{2}$$
$$f^{-1}(x) = \frac{x - 5}{2}. \qquad \text{(Step 3)}$$

Thus, f^{-1} is a linear function. In the function defined by $y = 2x + 5$, the value of y is found by starting with a value of x, multiplying by 2, and adding 5. The equation for the inverse has us *subtract* 5, and then *divide* by 2. This shows how an inverse is used to "undo" what a function does to the variable x.

(b) $y = x^2 + 2$

This equation has a vertical parabola as its graph, so some horizontal lines will intersect the graph at two points. For example, both $x = 3$ and $x = -3$ correspond to $y = 11$. Because of the x^2-term, there are many pairs of x-values that correspond to the same y-value. This means that the function defined by $y = x^2 + 2$ is not one-to-one and does not have an inverse.

If this is not noticed, then following the steps for finding the equation of an inverse leads to

$$y = x^2 + 2$$
$$x = y^2 + 2 \qquad \text{Interchange } x \text{ and } y.$$
$$x - 2 = y^2 \qquad \text{Solve for } y.$$
$$\pm\sqrt{x - 2} = y. \qquad \text{Square root property}$$

The last step shows that there are two y-values for each choice of $x > 2$, so the given function is not one-to-one and cannot have an inverse.

(c) $f(x) = (x - 2)^3$

Refer to Section 5.3 to see that the graphs of cubing functions are one-to-one.

$$y = (x - 2)^3 \qquad \text{Replace } f(x) \text{ with } y.$$

$$x = (y - 2)^3 \qquad \text{Interchange } x \text{ and } y.$$

$$\sqrt[3]{x} = \sqrt[3]{(y - 2)^3} \qquad \text{Take the cube root on each side.}$$

$$\sqrt[3]{x} = y - 2$$

$$\sqrt[3]{x} + 2 = y \qquad \text{Solve for } y.$$

$$f^{-1}(x) = \sqrt[3]{x} + 2 \qquad \text{Replace } y \text{ with } f^{-1}(x).$$

> **Now Try Exercises 13, 17, and 19.**

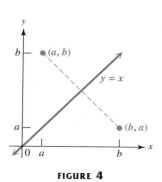

FIGURE 4

OBJECTIVE 4 Graph f^{-1} from the graph of f. One way to graph the inverse of a function f whose equation is known is to find some ordered pairs that belong to f, interchange x and y to get ordered pairs that belong to f^{-1}, plot those points, and sketch the graph of f^{-1} through the points. A simpler way is to select points on the graph of f and use symmetry to find corresponding points on the graph of f^{-1}.

For example, suppose the point (a, b) shown in Figure 4 belongs to a one-to-one function f. Then the point (b, a) belongs to f^{-1}. The line segment connecting (a, b) and (b, a) is perpendicular to, and cut in half by, the line $y = x$. The points (a, b) and (b, a) are "mirror images" of each other with respect to $y = x$. For this reason we can find the graph of f^{-1} from the graph of f by locating the mirror image of each point in f with respect to the line $y = x$.

EXAMPLE 4 Graphing the Inverse

Graph the inverses of the functions f (shown in blue) in Figure 5.

In Figure 5 the graphs of two functions f are shown in blue. Their inverses are shown in red. In each case, the graph of f^{-1} is a reflection of the graph of f with respect to the line $y = x$.

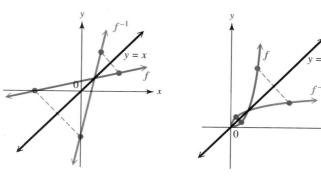

FIGURE 5

> **Now Try Exercises 25 and 29.**

OBJECTIVE 5 Use a graphing calculator to graph inverse functions. We have described how inverses of one-to-one functions may be determined algebraically. We also explained how the graph of a one-to-one function f compares to the graph of its inverse f^{-1}: It is a reflection of the graph of f^{-1} across the line $y = x$. In Example 3 we showed that the inverse of the one-to-one function defined by $f(x) = 2x + 5$ is given by $f^{-1}(x) = \frac{x-5}{2}$. If we use a square viewing window of a graphing calculator and graph $y_1 = f(x) = 2x + 5$, $y_2 = f^{-1}(x) = \frac{x-5}{2}$, and $y_3 = x$, we can see how this reflection appears on the screen. See Figure 6.

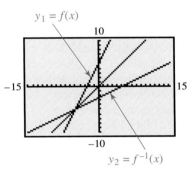

FIGURE 6

FIGURE 7

Some graphing calculators have the capability to "draw" the inverse of a function. Figure 7 shows the graphs of $f(x) = x^3 + 2$ and its inverse in a square viewing window.

11.1 EXERCISES

In Exercises 1–4, write a few sentences of explanation. See Example 1.

1. The table shows the number of uncontrolled hazardous waste sites that require further investigation to determine whether remedies are needed under the Superfund program. The seven states listed are ranked in the top ten in the United States.

If this correspondence is considered to be a function that pairs each state with its number of uncontrolled waste sites, is it one-to-one? If not, explain why. (See Example 1(c).)

State	Number of Sites
New Jersey	108
Pennsylvania	101
California	94
New York	79
Florida	53
Illinois	40
Wisconsin	40

Source: U.S. Environmental Protection Agency.

2. The table shows emissions of a major air pollutant, carbon monoxide, in the United States for the years 1992 through 1998.

   If this correspondence is considered to be a function that pairs each year with its emissions amount, is it one-to-one? If not, explain why.

Year	Amount of Emissions (in thousands of tons)
1992	97,630
1993	98,160
1994	102,643
1995	93,353
1996	95,479
1997	94,410
1998	89,454

Source: U.S. Environmental Protection Agency.

3. The road mileage between Denver, Colorado, and several selected U.S. cities is shown in the table. If we consider this as a function that pairs each city with a distance, is it a one-to-one function? How could we change the answer to this question by adding 1 mile to one of the distances shown?

City	Distance to Denver (in miles)
Atlanta	1398
Dallas	781
Indianapolis	1058
Kansas City, MO	600
Los Angeles	1059
San Francisco	1235

4. Suppose you consider the set of ordered pairs (x, y) such that x represents a person in your mathematics class and y represents that person's mother. Explain how this function might not be a one-to-one function.

In Exercises 5–8, choose the correct response from the given list.

5. If a function is made up of ordered pairs in such a way that the same y-value appears in a correspondence with two different x-values, then

   A. the function is one-to-one
   B. the function is not one-to-one
   C. its graph does not pass the vertical line test
   D. it has an inverse function associated with it.

6. Which equation defines a one-to-one function? Explain why the others are not, using specific examples.

   A. $f(x) = x$ **B.** $f(x) = x^2$ **C.** $f(x) = |x|$ **D.** $f(x) = -x^2 + 2x - 1$

7. Only one of the graphs illustrates a one-to-one function. Which one is it? (See Example 2.)

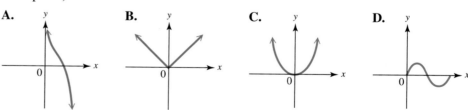

8. If a function f is one-to-one and the point (p, q) lies on the graph of f, then which point *must* lie on the graph of f^{-1}?

   A. $(-p, q)$ **B.** $(-q, -p)$ **C.** $(p, -q)$ **D.** (q, p)

If the function is one-to-one, find its inverse. See Examples 1–3.

9. $\{(3, 6), (2, 10), (5, 12)\}$

10. $\left\{(-1, 3), (0, 5), (5, 0), \left(7, -\dfrac{1}{2}\right)\right\}$

11. $\{(-1, 3), (2, 7), (4, 3), (5, 8)\}$

12. $\{(-8, 6), (-4, 3), (0, 6), (5, 10)\}$

13. $f(x) = 2x + 4$

14. $f(x) = 3x + 1$

15. $g(x) = \sqrt{x - 3}, \quad x \geq 3$

16. $g(x) = \sqrt{x + 2}, \quad x \geq -2$

17. $f(x) = 3x^2 + 2$

18. $f(x) = -4x^2 - 1$

19. $f(x) = x^3 - 4$

20. $f(x) = x^3 - 3$

Let $f(x) = 2^x$. We will see in the next section that this function is one-to-one. Find each value, always working part (a) before part (b).

21. (a) $f(3)$ **(b)** $f^{-1}(8)$

22. (a) $f(4)$ **(b)** $f^{-1}(16)$

23. (a) $f(0)$ **(b)** $f^{-1}(1)$

24. (a) $f(-2)$ **(b)** $f^{-1}\left(\dfrac{1}{4}\right)$

The graphs of some functions are given in Exercises 25–30. (a) Use the horizontal line test to determine whether the function is one-to-one. (b) If the function is one-to-one, then graph the inverse of the function. (Remember that if f is one-to-one and (a, b) is on the graph of f, then (b, a) is on the graph of f^{-1}.) See Example 4.

25.

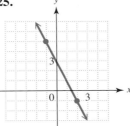

26.

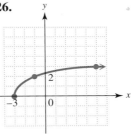

27.

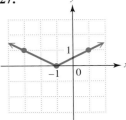

28.

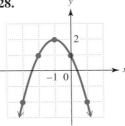

29.

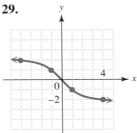

30.

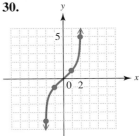

Each function defined in Exercises 31–38 is a one-to-one function. Graph the function as a solid line (or curve) and then graph its inverse on the same set of axes as a dashed line (or curve). In Exercises 35–38 you are given a table to complete so that graphing the function will be a bit easier. See Example 4.

31. $f(x) = 2x - 1$ **32.** $f(x) = 2x + 3$ **33.** $g(x) = -4x$ **34.** $g(x) = -2x$

35. $f(x) = \sqrt{x}, \quad x \geq 0$

36. $f(x) = -\sqrt{x}, \quad x \geq 0$

x	$f(x)$
0	
1	
4	

x	$f(x)$
0	
1	
4	

37. $f(x) = x^3 - 2$

x	$f(x)$
-1	
0	
1	
2	

38. $f(x) = x^3 + 3$

x	$f(x)$
-2	
-1	
0	
1	

RELATING CONCEPTS (EXERCISES 39–42)

For Individual or Group Work

Inverse functions are used by government agencies and other businesses to send and receive coded information. The functions they use are usually very complicated. A simple example might use the function defined by $f(x) = 2x + 5$. (Note that it is one-to-one.) Suppose that each letter of the alphabet is assigned a numerical value according to its position, as follows:

A	1	G	7	L	12	Q	17	V	22
B	2	H	8	M	13	R	18	W	23
C	3	I	9	N	14	S	19	X	24
D	4	J	10	O	15	T	20	Y	25
E	5	K	11	P	16	U	21	Z	26
F	6								

This is an Enigma machine, used by the Germans in World War II to send coded messages.

Using the function, the word ALGEBRA *would be encoded as*

$$7 \quad 29 \quad 19 \quad 15 \quad 9 \quad 41 \quad 7,$$

because

$$f(\text{A}) = f(1) = 2(1) + 5 = 7, \quad f(\text{L}) = f(12) = 2(12) + 5 = 29,$$

and so on. The message would then be decoded by using the inverse of f, defined by $f^{-1}(x) = \frac{x-5}{2}$. For example,

$$f^{-1}(7) = \frac{7-5}{2} = 1 = \text{A}, \quad f^{-1}(29) = \frac{29-5}{2} = 12 = \text{L},$$

and so on. **Work Exercises 39–42 in order.**

39. Suppose that you are an agent for a detective agency and you know that today's function for your code is defined by $f(x) = 4x - 5$. Find the rule for f^{-1} algebraically.

40. You receive the following coded message today. (Read across from left to right.)

47 95 23 67 −1 59 27 31 51 23 7 −1 43 7 79 43 −1 75 55 67

31 71 75 27 15 23 67 15 −1 75 15 71 75 75 27 31 51

23 71 31 51 7 15 71 43 31 7 15 11 3 67 15 −1 11

Use the letter/number assignment described earlier to decode the message.

41. Why is a one-to-one function essential in this encoding/decoding process?

42. Use $f(x) = x^3 + 4$ to encode your name, using the letter/number assignment described earlier.

Each function defined is one-to-one. Find the inverse algebraically, and then graph both the function and its inverse on the same graphing calculator screen. Use a square viewing window. See Objective 5.

43. $f(x) = 2x - 7$

44. $f(x) = -3x + 2$

45. $f(x) = x^3 + 5$

46. $f(x) = \sqrt[3]{x + 2}$

Some graphing calculators have the capability to draw the "inverse" of a function even if the function is not one-to-one; therefore, the inverse is not technically a function, but it is a relation. For example, the graphs of $y = x^2$ and $x = y^2$ are shown in the screen using a square viewing window.

Read your instruction manual to see if your model has this capability. If so, draw both Y_1 and its inverse in the same square window.

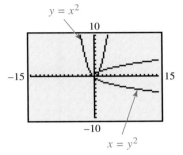

47. $Y_1 = X^2 + 3X + 4$

48. $Y_1 = X^3 - 9X$

49. Explain why the "inverse" of the function in Exercise 47 does not actually satisfy the definition of inverse as given in this section.

50. At what points do the graphs of $y = x^2$ and $x = y^2$ intersect? (See the graph above.) Verify this using algebraic methods.

11.2 Exponential Functions

OBJECTIVES

1 Define an exponential function.

2 Graph an exponential function.

3 Solve exponential equations of the form $a^x = a^k$ for x.

4 Use exponential functions in applications involving growth or decay.

OBJECTIVE 1 Define an exponential function. In Section 8.2 we showed how to evaluate 2^x for rational values of x. For example,

$$2^3 = 8, \qquad 2^{-1} = \frac{1}{2}, \qquad 2^{1/2} = \sqrt{2}, \qquad \text{and} \qquad 2^{3/4} = \sqrt[4]{2^3} = \sqrt[4]{8}.$$

In more advanced courses it is shown that 2^x exists for all real number values of x, both rational and irrational. (Later in this chapter, we will see how to approximate the value of 2^x for irrational x.) The following definition of an exponential function assumes that a^x exists for all real numbers x.

> **Exponential Function**
>
> For $a > 0$, $a \neq 1$, and all real numbers x,
>
> $$f(x) = a^x$$
>
> defines the **exponential function with base a.**

> **NOTE** The two restrictions on a in the definition of an exponential function are important. The restriction that a must be positive is necessary so that the function can be defined for all real numbers x. For example, letting a be negative ($a = -2$, for instance) and letting $x = \frac{1}{2}$ would give the expression $(-2)^{1/2}$, which is not real. The other restriction, $a \neq 1$, is necessary because 1 raised to any power is equal to 1, and the function would then be the linear function defined by $f(x) = 1$.

OBJECTIVE 2 Graph an exponential function. We graph an exponential function by finding several ordered pairs that belong to the function, plotting these points, and connecting them with a smooth curve.

> **CAUTION** Be sure to plot enough points to see how rapidly the graph rises.

▨ EXAMPLE 1   Graphing an Exponential Function with $a > 1$

Graph $f(x) = 2^x$.

Choose some values of x, and find the corresponding values of $f(x)$.

x	-3	-2	-1	0	1	2	3	4
$f(x) = 2^x$	$\frac{1}{8}$	$\frac{1}{4}$	$\frac{1}{2}$	1	2	4	8	16

Plotting these points and drawing a smooth curve through them gives the blue graph shown in Figure 8. This graph is typical of the graphs of exponential functions of the form $F(x) = a^x$, where $a > 1$. The larger the value of a, the faster the graph rises. To see this, compare the red graph of $F(x) = 5^x$ with the graph of $f(x) = 2^x$ in Figure 8.

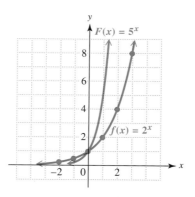

FIGURE 8

By the vertical line test, the graphs in Figure 8 represent functions. As these graphs suggest, the domain of an exponential function includes all real numbers. Because y is always positive, the range is $(0, \infty)$. Figure 8 also shows an important characteristic of exponential functions where $a > 1$: As x gets larger, y increases at a faster and faster rate.

Now Try Exercise 5.

EXAMPLE 2   Graphing an Exponential Function with $a < 1$

Graph $g(x) = \left(\dfrac{1}{2}\right)^x$.

Again, find some points on the graph.

x	-3	-2	-1	0	1	2	3
$g(x) = \left(\frac{1}{2}\right)^x$	8	4	2	1	$\frac{1}{2}$	$\frac{1}{4}$	$\frac{1}{8}$

The graph, shown in Figure 9, is very similar to that of $f(x) = 2^x$ (Figure 8) with the same domain and range, except that here as x gets larger, y *decreases*. This graph is typical of the graph of a function of the form $F(x) = a^x$, where $0 < a < 1$.

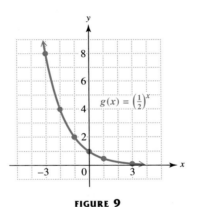

FIGURE 9

Now Try Exercise 7.

Based on Examples 1 and 2, we make the following generalizations about the graphs of exponential functions of the form $F(x) = a^x$.

Graph of $F(x) = a^x$

1. The graph contains the point $(0, 1)$.

2. When $a > 1$, the graph will *rise* from left to right. When $0 < a < 1$, the graph will *fall* from left to right. In both cases, the graph goes from the second quadrant to the first.

3. The graph will approach the x-axis, but never touch it. (Recall from Chapter 7 that such a line is called an *asymptote*.)

4. The domain is $(-\infty, \infty)$, and the range is $(0, \infty)$.

▨ **EXAMPLE 3** **Graphing a More Complicated Exponential Function**

Graph $f(x) = 3^{2x-4}$.

Find some ordered pairs.

$$\text{If } x = 0, \text{ then } y = 3^{2(0)-4} = 3^{-4} = \frac{1}{81}.$$

$$\text{If } x = 2, \text{ then } y = 3^{2(2)-4} = 3^0 = 1.$$

These ordered pairs, $\left(0, \frac{1}{81}\right)$ and $(2, 1)$, along with the other ordered pairs shown in the table, lead to the graph in Figure 10. The graph is similar to the graph of $f(x) = 3^x$ except that it is shifted to the right and rises more rapidly.

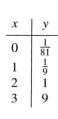

x	y
0	$\frac{1}{81}$
1	$\frac{1}{9}$
2	1
3	9

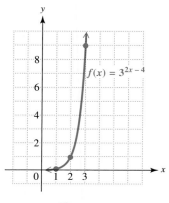

FIGURE 10

▨

Now Try Exercise 11.

OBJECTIVE 3 **Solve exponential equations of the form $a^x = a^k$ for x.** Until this chapter, we have solved only equations that had the variable as a base, like $x^2 = 8$; all exponents have been constants. An **exponential equation** is an equation that has a variable in an exponent, such as

$$9^x = 27.$$

By the horizontal line test, the exponential function defined by $F(x) = a^x$ is a one-to-one function, so we can use the following property to solve many exponential equations.

Property for Solving an Exponential Equation

For $a > 0$ and $a \neq 1$,   if $a^x = a^y$ then $x = y.$

This property would not necessarily be true if $a = 1$.

To solve an exponential equation using this property, follow these steps.

Solving an Exponential Equation

Step 1 **Each side must have the same base.** If the two sides of the equation do not have the same base, express each as a power of the same base if possible.

Step 2 **Simplify exponents** if necessary, using the rules of exponents.

Step 3 **Set exponents equal** using the property given in this section.

Step 4 **Solve** the equation obtained in Step 3.

NOTE These steps cannot be applied to an exponential equation like

$$3^x = 12$$

because Step 1 cannot easily be done. A method for solving such equations is given in Section 11.6.

EXAMPLE 4 Solving an Exponential Equation

Solve the equation $9^x = 27$.

We can use the property given at the bottom of the previous page if both sides are written with the same base.

$$9^x = 27$$

$$(3^2)^x = 3^3 \qquad \text{Write with the same base;}$$
$$\qquad\qquad 9 = 3^2 \text{ and } 27 = 3^3. \text{ (Step 1)}$$

$$3^{2x} = 3^3 \qquad \text{Power rule for exponents (Step 2)}$$

$$2x = 3 \qquad \text{If } a^x = a^y, \text{ then } x = y. \text{ (Step 3)}$$

$$x = \frac{3}{2} \qquad \text{(Step 4)}$$

Check that the solution set is $\left\{\frac{3}{2}\right\}$ by substituting $\frac{3}{2}$ for x in the original equation.

Now Try Exercise 17.

EXAMPLE 5 Solving Exponential Equations

Solve each equation.

(a) $\quad 4^{3x-1} = 16^{x+2}$

$$(2^2)^{3x-1} = (2^4)^{x+2} \qquad \text{Write with the same base;}$$
$$\qquad\qquad 4 = 2^2 \text{ and } 16 = 2^4.$$

$$2^{6x-2} = 2^{4x+8} \qquad \text{Power rule for exponents}$$

$$6x - 2 = 4x + 8 \qquad \text{Set exponents equal.}$$

$$2x = 10 \qquad \text{Subtract } 4x; \text{ add 2.}$$

$$x = 5 \qquad \text{Divide by 2.}$$

Verify that the solution set is $\{5\}$.

(b) $6^x = \dfrac{1}{216}$

$\qquad 6^x = \dfrac{1}{6^3} \qquad$ $216 = 6^3$

$\qquad 6^x = 6^{-3} \qquad$ Write with the same base; $\dfrac{1}{6^3} = 6^{-3}$.

$\qquad x = -3 \qquad$ Set exponents equal.

Verify that the solution set is $\{-3\}$.

(c) $\left(\dfrac{2}{3}\right)^x = \dfrac{9}{4}$

$\qquad \left(\dfrac{2}{3}\right)^x = \left(\dfrac{4}{9}\right)^{-1} \qquad$ $\dfrac{9}{4} = \left(\dfrac{4}{9}\right)^{-1}$

$\qquad \left(\dfrac{2}{3}\right)^x = \left[\left(\dfrac{2}{3}\right)^2\right]^{-1} \qquad$ Write with the same base.

$\qquad \left(\dfrac{2}{3}\right)^x = \left(\dfrac{2}{3}\right)^{-2} \qquad$ Power rule for exponents

$\qquad x = -2 \qquad$ Set exponents equal.

Check that the solution set is $\{-2\}$.

Now Try Exercises 19, 21, and 25.

OBJECTIVE 4 Use exponential functions in applications involving growth or decay.

EXAMPLE 6 Solving an Application Involving Exponential Growth

One result of the rapidly increasing world population is an increase of carbon dioxide in the air, which scientists believe may be contributing to global warming. Both population and amounts of carbon dioxide in the air are increasing exponentially. This means that the growth rate is continually increasing. The graph in Figure 11 shows the concentration of carbon dioxide (in parts per million) in the air.

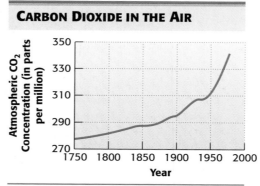

CARBON DIOXIDE IN THE AIR

Source: Sacramento Bee, Monday, September 13, 1993.

FIGURE 11

The data are approximated by the function defined by

$$f(x) = 278(1.00084)^x,$$

where x is the number of years since 1750. Use this function and a calculator to approximate the concentration of carbon dioxide in parts per million for each year.

(a) 1900

Because x represents the number of years since 1750, in this case

$$x = 1900 - 1750 = 150.$$

Thus, evaluate $f(150)$.

$$f(150) = 278(1.00084)^{150} \qquad \text{Let } x = 150.$$
$$\approx 315 \text{ parts per million} \qquad \text{Use a calculator.}$$

(b) 1950

Use $x = 1950 - 1750 = 200$.

$$f(200) = 278(1.00084)^{200}$$
$$\approx 329 \text{ parts per million}$$

Now Try Exercise 39.

EXAMPLE 7   Applying an Exponential Decay Function

The atmospheric pressure (in millibars) at a given altitude x, in meters, can be approximated by the function defined by

$$f(x) = 1038(1.000134)^{-x},$$

for values of x between 0 and 10,000. Because the base is greater than 1 and the coefficient of x in the exponent is negative, the function values decrease as x increases. This means that as the altitude increases, the atmospheric pressure decreases. (*Source:* Miller, A. and J. Thompson, *Elements of Meteorology,* Fourth Edition, Charles E. Merrill Publishing Company, 1993.)

(a) According to this function, what is the pressure at ground level?

At ground level, $x = 0$, so

$$f(0) = 1038(1.000134)^{-0} = 1038(1) = 1038.$$

The pressure is 1038 millibars.

(b) What is the pressure at 5000 m?

Use a calculator to find $f(5000)$.

$$f(5000) = 1038(1.000134)^{-5000}$$
$$\approx 531$$

The pressure is approximately 531 millibars.

Now Try Exercise 41.

11.2 EXERCISES

Choose the correct response in Exercises 1–4.

1. Which point lies on the graph of $f(x) = 2^x$?

A. $(1, 0)$ **B.** $(2, 1)$ **C.** $(0, 1)$ **D.** $\left(\sqrt{2}, \dfrac{1}{2}\right)$

2. Which statement is true?

A. The y-intercept of the graph of $f(x) = 10^x$ is $(0, 10)$.
B. For any $a > 1$, the graph of $f(x) = a^x$ falls from left to right.
C. The point $\left(\frac{1}{2}, \sqrt{5}\right)$ lies on the graph of $f(x) = 5^x$.
D. The graph of $y = 4^x$ rises at a faster rate than the graph of $y = 10^x$.

3. The asymptote of the graph of $F(x) = a^x$

A. is the x-axis   **B.** is the y-axis
C. has equation $x = 1$ **D.** has equation $y = 1$.

4. Which equation is graphed here?

A. $y = 1000\left(\dfrac{1}{2}\right)^{.3x}$ **B.** $y = 1000\left(\dfrac{1}{2}\right)^{x}$

C. $y = 1000(2)^{.3x}$ **D.** $y = 1000^x$

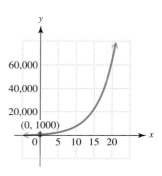

Graph each exponential function. See Examples 1–3.

5. $f(x) = 3^x$ **6.** $f(x) = 5^x$ **7.** $g(x) = \left(\dfrac{1}{3}\right)^x$ **8.** $g(x) = \left(\dfrac{1}{5}\right)^x$

9. $y = 4^{-x}$ **10.** $y = 6^{-x}$ **11.** $y = 2^{2x-2}$ **12.** $y = 2^{2x+1}$

13. **(a)** For an exponential function defined by $f(x) = a^x$, if $a > 1$, the graph _____
(rises/falls)
from left to right. If $0 < a < 1$, the graph _____ from left to right.
(rises/falls)

(b) Based on your answers in part (a), make a conjecture (an educated guess)
concerning whether an exponential function defined by $f(x) = a^x$ is one-to-one.
Then decide whether it has an inverse based on the concepts of Section 11.1.

 14. In your own words, describe the characteristics of the graph of an exponential function.
Use the exponential function defined by $f(x) = 3^x$ (Exercise 5) and the words
asymptote, domain, and *range* in your explanation.

Solve each equation. See Examples 4 and 5.

15. $6^x = 36$ **16.** $8^x = 64$ **17.** $100^x = 1000$ **18.** $8^x = 4$

19. $16^{2x+1} = 64^{x+3}$ **20.** $9^{2x-8} = 27^{x-4}$ **21.** $5^x = \dfrac{1}{125}$ **22.** $3^x = \dfrac{1}{81}$

23. $5^x = .2$ **24.** $10^x = .1$ **25.** $\left(\dfrac{3}{2}\right)^x = \dfrac{8}{27}$ **26.** $\left(\dfrac{4}{3}\right)^x = \dfrac{27}{64}$

Use the exponential key of a calculator to find an approximation to the nearest thousandth.

27. $12^{2.6}$ **28.** $13^{1.8}$ **29.** $.5^{3.921}$ **30.** $.6^{4.917}$ **31.** $2.718^{2.5}$ **32.** $2.718^{-3.1}$

33. Try to evaluate $(-2)^4$ on a scientific calculator. You may get an error message, since the exponential function key on many calculators does not allow negative bases. Discuss the concept introduced in this section that is closely related to this "peculiarity" of many scientific calculators.

34. Explain why the exponential equation $4^x = 6$ cannot be solved using the method explained in this section. Change 6 to another number that *will* allow the method of this section to be used, and then solve the equation.

The graph shown here accompanied the article "Is Our World Warming?" which appeared in the October 1990 issue of National Geographic. *It shows projected temperature increases using two graphs: one an exponential-type curve, and the other linear. From the graph, approximate the increase (a) for the exponential curve and (b) for the linear graph for each year.*

35. 2000

36. 2010

37. 2020

38. 2040

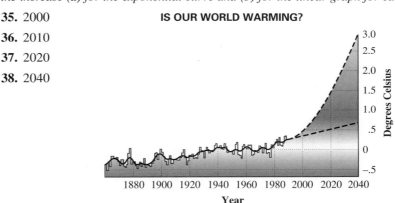

IS OUR WORLD WARMING?

Graph, "Zero Equals Average Global Temperature for the Period 1950–1979." Dale D. Glasgow, © National Geographic Society. Reprinted by permission.

Solve each problem. See Examples 6 and 7.

39. Based on figures from 1970 through 1998, the worldwide carbon monoxide emissions in thousands of tons are approximated by the exponential function defined by

$$f(x) = 132,359(1.0124)^{-x},$$

where $x = 0$ corresponds to 1970, $x = 5$ corresponds to 1975, and so on. (*Source:* U.S. Environmental Protection Agency.)

(a) Use this model to approximate the emissions in 1970.

(b) Use this model to approximate the emissions in 1995.

(c) In 1998, the actual amount of emissions was 89,454 million tons. How does this compare to the number that the model provides?

40. Based on figures from 1980 through 1999, the municipal solid waste generated in millions of tons can be approximated by the exponential function defined by

$$f(x) = 157.28(1.0204)^x,$$

where $x = 0$ corresponds to 1980, $x = 5$ corresponds to 1985, and so on. (*Source:* U.S. Environmental Protection Agency.)

(a) Use the model to approximate the number of tons of this waste in 1980.

(b) Use the model to approximate the number of tons of this waste in 1995.

(c) In 1999, the actual number of millions of tons of this waste was 229.9. How does this compare to the number that the model provides?

41. A small business estimates that the value $V(t)$ of a copy machine is decreasing according to the function defined by

$$V(t) = 5000(2)^{-.15t},$$

where t is the number of years that have elapsed since the machine was purchased, and $V(t)$ is in dollars.

(a) What was the original value of the machine?

(b) What is the value of the machine 5 yr after purchase? Give your answer to the nearest dollar.

(c) What is the value of the machine 10 yr after purchase? Give your answer to the nearest dollar.

(d) Graph the function.

42. The amount of radioactive material in an ore sample is given by the function defined by

$$A(t) = 100(3.2)^{-.5t},$$

where $A(t)$ is the amount present, in grams, of the sample t months after the initial measurement.

(a) How much was present at the initial measurement? (*Hint:* $t = 0$.)

(b) How much was present 2 months later?

(c) How much was present 10 months later?

(d) Graph the function.

43. Refer to the function in Exercise 41. When will the value of the machine be $2500? (*Hint:* Let $V(t) = 2500$, divide both sides by 5000, and use the method of Example 4.)

44. Refer to the function in Exercise 41. When will the value of the machine be $1250?

RELATING CONCEPTS (EXERCISES 45–50)

For Individual or Group Work

In these exercises we examine several methods of simplifying the expression $16^{3/4}$. *Work Exercises 45–50 in order.*

45. Write $16^{3/4}$ as a radical expression with the exponent outside the radical. Then simplify the expression.

46. Write $16^{3/4}$ as a radical expression with the exponent under the radical. Then simplify the expression.

47. Use a calculator to find the square root of 16^3. Now find the square root of that result.

48. Explain why the result in Exercise 47 is equal to $16^{3/4}$.

49. Predict the result a calculator will give when 16 is raised to the .75 power. Then check your answer by actually performing the operation on your calculator.

50. Write $\sqrt[100]{16^{75}}$ as an exponential expression. Then write the exponent in lowest terms, rewrite as a radical, and evaluate this radical expression.

11.3 Logarithmic Functions

The graph of $y = 2^x$ is the curve shown in blue in Figure 12. Because $y = 2^x$ defines a one-to-one function, it has an inverse. Interchanging x and y gives $x = 2^y$, the inverse of $y = 2^x$. As we saw in Section 11.1, the graph of the inverse is found by reflecting the graph of $y = 2^x$ about the line $y = x$. The graph of $x = 2^y$ is shown as a red curve in Figure 12.

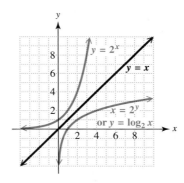

FIGURE 12

OBJECTIVE 1   Define a logarithm.   We cannot solve the equation $x = 2^y$ for the dependent variable y with the methods presented up to now. The following definition is used to solve $x = 2^y$ for y.

Logarithm

For all positive numbers a, $a \neq 1$, and all positive numbers x,

$$y = \log_a x \quad \text{means the same as} \quad x = a^y.$$

This key statement should be memorized. The abbreviation **log** is used for the word **logarithm**. Read $\log_a x$ as "the logarithm of x to the base a" or "the base a logarithm of x." To remember the location of the base and the exponent in each form, refer to the following diagrams.

$$\overset{\text{Exponent}}{\underset{\text{Base}}{\text{Logarithmic form: } y = \log_a x}}$$ $$\overset{\text{Exponent}}{\underset{\text{Base}}{\text{Exponential form: } x = a^y}}$$

In working with logarithmic form and exponential form, remember the following.

Meaning of $\log_a x$

A logarithm is an exponent; $\log_a x$ is the exponent to which the base a must be raised to obtain x.

OBJECTIVE 2 Convert between exponential and logarithmic forms. We can use the definition of logarithm to write exponential statements in logarithmic form and logarithmic statements in exponential form. The following table shows several pairs of equivalent statements.

Exponential Form	Logarithmic Form
$3^2 = 9$	$\log_3 9 = 2$
$\left(\frac{1}{5}\right)^{-2} = 25$	$\log_{1/5} 25 = -2$
$10^5 = 100{,}000$	$\log_{10} 100{,}000 = 5$
$4^{-3} = \frac{1}{64}$	$\log_4 \frac{1}{64} = -3$

OBJECTIVE 3 Solve logarithmic equations of the form $\log_a b = k$ for a, b, or k. A **logarithmic equation** is an equation with a logarithm in at least one term. We solve logarithmic equations of the form $\log_a b = k$ for any of the three variables by first writing the equation in exponential form.

▬ EXAMPLE 1   Solving Logarithmic Equations

Solve each equation.

(a) $\log_4 x = -2$

By the definition of logarithm, $\log_4 x = -2$ is equivalent to $x = 4^{-2}$. Solve this exponential equation.

$$x = 4^{-2} = \frac{1}{16}$$

The solution set is $\left\{\frac{1}{16}\right\}$.

(b) $\log_{1/2}(3x + 1) = 2$

$$3x + 1 = \left(\frac{1}{2}\right)^2 \qquad \text{Write in exponential form.}$$

$$3x + 1 = \frac{1}{4}$$

$$12x + 4 = 1 \qquad \text{Multiply by 4.}$$

$$12x = -3 \qquad \text{Subtract 4.}$$

$$x = -\frac{1}{4} \qquad \text{Divide by 12.}$$

The solution set is $\left\{-\frac{1}{4}\right\}$.

(c) $\log_x 3 = 2$

$$x^2 = 3 \qquad \text{Write in exponential form.}$$

$$x = \pm\sqrt{3} \qquad \text{Take square roots.}$$

Notice that only the principal square root satisfies the equation since the base must be a positive number. The solution set is $\left\{\sqrt{3}\right\}$.

(d) $\log_{49} \sqrt[3]{7} = x$

$$49^x = \sqrt[3]{7} \qquad \text{Write in exponential form.}$$
$$(7^2)^x = 7^{1/3} \qquad \text{Write with the same base.}$$
$$7^{2x} = 7^{1/3} \qquad \text{Power rule for exponents}$$
$$2x = \frac{1}{3} \qquad \text{Set exponents equal.}$$
$$x = \frac{1}{6} \qquad \text{Divide by 2.}$$

The solution set is $\left\{\frac{1}{6}\right\}$.

Now Try Exercises 21, 25, 37, and 39.

For any real number b, we know that $b^1 = b$ and for $b \neq 0$, $b^0 = 1$. Writing these two statements in logarithmic form gives the following two properties of logarithms.

Properties of Logarithms

For any positive real number b, $b \neq 1$,

$$\log_b b = 1 \qquad \text{and} \qquad \log_b 1 = 0.$$

EXAMPLE 2   Using Properties of Logarithms

Use the preceding two properties of logarithms to evaluate each logarithm.

(a) $\log_7 7 = 1$ **(b)** $\log_{\sqrt{2}} \sqrt{2} = 1$

(c) $\log_9 1 = 0$ **(d)** $\log_{.2} 1 = 0$

Now Try Exercise 1.

OBJECTIVE 4 Define and graph logarithmic functions. Now we define the logarithmic function with base a.

Logarithmic Function

If a and x are positive numbers, with $a \neq 1$, then

$$G(x) = \log_a x$$

defines the **logarithmic function with base a.**

The graph of $x = 2^y$ in Figure 12, which is equivalent to $y = g(x) = \log_2 x$, is typical of graphs of logarithmic functions with base $a > 1$. To graph a logarithmic function, it is helpful to write it in exponential form first. Then plot selected ordered pairs to determine the graph.

EXAMPLE 3   Graphing a Logarithmic Function

Graph $f(x) = \log_{1/2} x$.

By writing $y = f(x) = \log_{1/2} x$ in exponential form as $x = \left(\frac{1}{2}\right)^y$, we can identify ordered pairs that satisfy the equation. Here it is easier to choose values for y and find the corresponding values of x. See the table of ordered pairs.

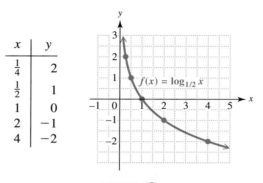

x	y
$\frac{1}{4}$	2
$\frac{1}{2}$	1
1	0
2	-1
4	-2

$f(x) = \log_{1/2} x$

FIGURE 13

Plotting these points (be careful to get the x- and y-values in the right order) and connecting them with a smooth curve gives the graph in Figure 13. This graph is typical of logarithmic functions with $0 < a < 1$.

Now Try Exercise 43.

Based on the graphs of the functions defined by $y = \log_2 x$ in Figure 12 and $y = \log_{1/2} x$ in Figure 13, we make the following generalizations about the graphs of logarithmic functions of the form $G(x) = \log_a x$.

Graph of $G(x) = \log_a x$

1. The graph contains the point $(1, 0)$.

2. When $a > 1$, the graph will *rise* from left to right, from the fourth quadrant to the first. When $0 < a < 1$, the graph will *fall* from left to right, from the first quadrant to the fourth.

3. The graph will approach the y-axis, but never touch it. (The y-axis is an asymptote.)

4. The domain is $(0, \infty)$, and the range is $(-\infty, \infty)$.

Compare these generalizations to the similar ones for exponential functions found in Section 11.2.

OBJECTIVE 5 Use logarithmic functions in applications involving growth or decay.
Logarithmic functions, like exponential functions, can be applied to growth or decay of real-world phenomena.

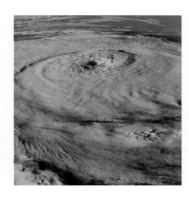

■ EXAMPLE 4 Solving an Application of a Logarithmic Function

The function defined by

$$f(x) = 27 + 1.105 \log_{10}(x + 1)$$

approximates the barometric pressure in inches of mercury at a distance of x mi from the eye of a typical hurricane. (*Source:* Miller, A. and R. Anthes, *Meteorology,* Fifth Edition, Charles E. Merrill Publishing Company, 1985.)

(a) Approximate the pressure 9 mi from the eye of the hurricane.

Let $x = 9$, and find $f(9)$.

$$
\begin{aligned}
f(9) &= 27 + 1.105 \log_{10}(9 + 1) & \text{Let } x = 9. \\
&= 27 + 1.105 \log_{10} 10 & \text{Add inside parentheses.} \\
&= 27 + 1.105(1) & \log_{10} 10 = 1 \\
&= 28.105
\end{aligned}
$$

The pressure 9 mi from the eye of the hurricane is 28.105 in.

(b) Approximate the pressure 99 mi from the eye of the hurricane.

$$
\begin{aligned}
f(99) &= 27 + 1.105 \log_{10}(99 + 1) & \text{Let } x = 99. \\
&= 27 + 1.105 \log_{10} 100 & \text{Add inside parentheses.} \\
&= 27 + 1.105(2) & \log_{10} 100 = 2 \\
&= 29.21
\end{aligned}
$$

The pressure 99 mi from the eye of the hurricane is 29.21 in.

Now Try Exercise 55.

CONNECTIONS

In the United States, the intensity of an earthquake is rated using the *Richter scale.* The Richter scale rating of an earthquake of intensity x is given by

$$R = \log_{10} \frac{x}{x_0},$$

where x_0 is the intensity of an earthquake of a certain (small) size. The graph here shows Richter scale ratings for major Southern California earthquakes since 1920. As the graph indicates, earthquakes "come in bunches," and the 1990s were an especially busy time.

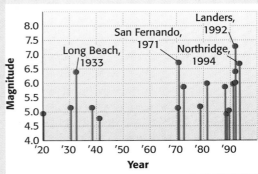

MAJOR SOUTHERN CALIFORNIA EARTHQUAKES

Earthquakes with magnitudes greater than 4.8

Source: Caltech; U.S. Geological Survey.

(continued)

For Discussion or Writing

Writing the given logarithmic equation in exponential form, we obtain

$$10^R = \frac{x}{x_0} \quad \text{or} \quad x = 10^R x_0.$$

1. The 1994 Northridge earthquake had a Richter scale rating of 6.7; the Landers earthquake had a rating of 7.3. How much more powerful was the Landers earthquake than the Northridge earthquake?

2. Compare the smallest rated earthquake in the figure (at 4.8) with the Landers quake. How much more powerful was the Landers quake?

11.3 EXERCISES

For Extra Help

- Student's Solutions Manual
- MyMathLab
- InterAct Math Tutorial Software
- AW Math Tutor Center
- MathXL
- Digital Video Tutor CD 11/Videotape 13

1. By definition, $\log_a x$ is the exponent to which the base a must be raised in order to obtain x. Use this definition to match the logarithm in Column I with its value in Column II. (*Example:* $\log_3 9$ is equal to 2 because 2 is the exponent to which 3 must be raised in order to obtain 9.)

I	II
(a) $\log_4 16$	A. -2
(b) $\log_3 81$	B. -1
(c) $\log_3\left(\frac{1}{3}\right)$	C. 2
(d) $\log_{10} .01$	D. 0
(e) $\log_5 \sqrt{5}$	E. $\frac{1}{2}$
(f) $\log_{13} 1$	F. 4

2. Match the logarithmic equation in Column I with the corresponding exponential equation from Column II.

I	II
(a) $\log_{1/3} 3 = -1$	A. $8^{1/3} = \sqrt[3]{8}$
(b) $\log_5 1 = 0$	B. $\left(\frac{1}{3}\right)^{-1} = 3$
(c) $\log_2 \sqrt{2} = \frac{1}{2}$	C. $4^1 = 4$
(d) $\log_{10} 1000 = 3$	D. $2^{1/2} = \sqrt{2}$
(e) $\log_8 \sqrt[3]{8} = \frac{1}{3}$	E. $5^0 = 1$
(f) $\log_4 4 = 1$	F. $10^3 = 1000$

Write in logarithmic form. See the table in Objective 2.

3. $4^5 = 1024$

4. $3^6 = 729$

5. $\left(\frac{1}{2}\right)^{-3} = 8$

6. $\left(\frac{1}{6}\right)^{-3} = 216$

7. $10^{-3} = .001$

8. $36^{1/2} = 6$

9. $\sqrt[4]{625} = 5$

10. $\sqrt[3]{343} = 7$

Write in exponential form. See the table in Objective 2.

11. $\log_4 64 = 3$

12. $\log_2 512 = 9$

13. $\log_{10} \frac{1}{10,000} = -4$

14. $\log_{100} 100 = 1$

15. $\log_6 1 = 0$

16. $\log_\pi 1 = 0$

17. $\log_9 3 = \frac{1}{2}$

18. $\log_{64} 2 = \frac{1}{6}$

19. When a student asked his teacher to explain how to evaluate $\log_9 3$ without showing any work, his teacher told him, "Think radically." Explain what the teacher meant by this hint.

✍ **20.** A student told her teacher, "I know that $\log_2 1$ is the exponent to which 2 must be raised in order to obtain 1, but I can't think of any such number." How would you explain to the student that the value of $\log_2 1$ is 0?

Solve each equation for x. See Examples 1 and 2.

21. $x = \log_{27} 3$ **22.** $x = \log_{125} 5$ **23.** $\log_x 9 = \dfrac{1}{2}$ **24.** $\log_x 5 = \dfrac{1}{2}$

25. $\log_x 125 = -3$ **26.** $\log_x 64 = -6$ **27.** $\log_{12} x = 0$ **28.** $\log_4 x = 0$

29. $\log_x x = 1$ **30.** $\log_x 1 = 0$ **31.** $\log_x \dfrac{1}{25} = -2$ **32.** $\log_x \dfrac{1}{10} = -1$

33. $\log_8 32 = x$ **34.** $\log_{81} 27 = x$ **35.** $\log_\pi \pi^4 = x$ **36.** $\log_{\sqrt{2}} \sqrt{2^9} = x$

37. $\log_6 \sqrt{216} = x$ **38.** $\log_4 \sqrt{64} = x$

39. $\log_4(2x + 4) = 3$ **40.** $\log_3(2x + 7) = 4$

If the point (p, q) is on the graph of $f(x) = a^x$ (for $a > 0$ and $a \neq 1$), then the point (q, p) is on the graph of $f^{-1}(x) = \log_a x$. Use this fact, and refer to the graphs required in Exercises 5–8 in Section 11.2 to graph each logarithmic function. See Example 3.

41. $y = \log_3 x$ **42.** $y = \log_5 x$ **43.** $y = \log_{1/3} x$ **44.** $y = \log_{1/5} x$

✍ **45.** Explain why 1 is not allowed as a base for a logarithmic function.

✍ **46.** Compare the summary of facts about the graph of $F(x) = a^x$ in Section 11.2 with the similar summary of facts about the graph of $G(x) = \log_a x$ in this section. Make a list of the facts that reinforce the concept that F and G are inverse functions.

47. The domain of $F(x) = a^x$ is $(-\infty, \infty)$, while the range is $(0, \infty)$. Therefore, since $G(x) = \log_a x$ defines the inverse of F, the domain of G is _____, while the range of G is _____.

48. The graphs of both $F(x) = 3^x$ and $G(x) = \log_3 x$ rise from left to right. Which one rises at a faster rate?

Use the graph at the right to predict the value of $f(t)$ for the given value of t.

49. $t = 0$

50. $t = 10$

51. $t = 60$

52. Show that the points determined in Exercises 49–51 lie on the graph of $f(t) = 8 \log_5(2t + 5)$.

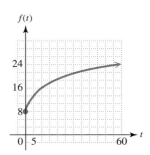

Solve each application of a logarithmic function. See Example 4.

53. According to selected figures from 1981 through 1995, the number of Superfund hazardous waste sites in the United States can be approximated by the function defined by

$$f(x) = 11.34 + 317.01 \log_2 x,$$

where $x = 1$ corresponds to 1981, $x = 2$ to 1982, and so on. (*Source:* U.S. Environmental Protection Agency.) Use the function to approximate the number of sites in each year.

(a) 1984   **(b)** 1988   **(c)** 1996

54. According to selected figures from 1980 through 1993, the number of trillion cubic feet of dry natural gas consumed worldwide can be approximated by the function defined by

$$f(x) = 51.47 + 6.044 \log_2 x,$$

where $x = 1$ corresponds to 1980, $x = 2$ to 1981, and so on. (*Source:* Energy Information Administration.) Use the function to approximate consumption in each year.

(a) 1980   **(b)** 1987   **(c)** 1995

55. Sales (in thousands of units) of a new product are approximated by the function defined by

$$S(t) = 100 + 30 \log_3(2t + 1),$$

where t is the number of years after the product is introduced.

(a) What were the sales after 1 yr?
(b) What were the sales after 13 yr?
(c) Graph $y = S(t)$.

56. A study showed that the number of mice in an old abandoned house was approximated by the function defined by

$$M(t) = 6 \log_4(2t + 4),$$

where t is measured in months and $t = 0$ corresponds to January 1998. Find the number of mice in the house in

(a) January 1998   **(b)** July 1998   **(c)** July 2000.
(d) Graph the function.

57. A supply of hybrid striped bass were introduced into a lake in January 1990. Biologists researching the bass population found that the number of bass in the lake was approximated by the function defined by

$$B(t) = 500 \log_3(2t + 3),$$

where $t = 0$ corresponds to January 1990, $t = 1$ to January 1991, $t = 2$ to January 1992, and so on. Use this function to find the bass population in

(a) January 1990   **(b)** January 1993   **(c)** January 2002.
(d) Graph the function for $0 \leq t \leq 12$.

58. Use the exponential key of your calculator to find approximations for the expression $\left(1 + \frac{1}{x}\right)^x$, using x values of 1, 10, 100, 1000, and 10,000. Explain what seems to be happening as x gets larger and larger.

As mentioned in Section 11.1, some graphing calculators have the capability of drawing the inverse of a function. For example, the two screens that follow show the graphs of $f(x) = 2^x$ and $g(x) = \log_2 x$. The graph of g was obtained by drawing the graph of f^{-1}, since $g(x) = f^{-1}(x)$. (Compare to Figure 12 in this section.)

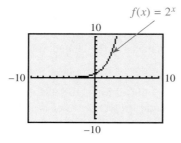

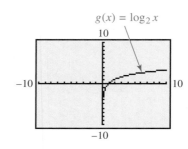

Use a graphing calculator with the capability of drawing the inverse of a function to draw the graph of each logarithmic function. Use the standard viewing window.

59. $g(x) = \log_3 x$ (Compare to Exercise 41.)

60. $g(x) = \log_5 x$ (Compare to Exercise 42.)

61. $g(x) = \log_{1/3} x$ (Compare to Exercise 43.)

62. $g(x) = \log_{1/5} x$ (Compare to Exercise 44.)

11.4 Properties of Logarithms

OBJECTIVES

1 Use the product rule for logarithms.

2 Use the quotient rule for logarithms.

3 Use the power rule for logarithms.

4 Use properties to write alternative forms of logarithmic expressions.

Logarithms have been used as an aid to numerical calculation for several hundred years. Today the widespread use of calculators has made the use of logarithms for calculation obsolete. However, logarithms are still very important in applications and in further work in mathematics.

OBJECTIVE 1 Use the product rule for logarithms. One way in which logarithms simplify problems is by changing a problem of multiplication into one of addition. We know that $\log_2 4 = 2$, $\log_2 8 = 3$, and $\log_2 32 = 5$. Since $2 + 3 = 5$,

$$\log_2 32 = \log_2 4 + \log_2 8$$
$$\log_2 (4 \cdot 8) = \log_2 4 + \log_2 8.$$

This is true in general.

Product Rule for Logarithms

If x, y, and b are positive real numbers, where $b \neq 1$, then
$$\log_b xy = \log_b x + \log_b y.$$

That is, the logarithm of a product is the sum of the logarithms of the factors.

NOTE The word statement of the product rule can be restated by replacing "logarithm" with "exponent." The rule then becomes the familiar rule for multiplying exponential expressions: The *exponent* of a product is the sum of the *exponents* of the factors.

To prove this rule, let $m = \log_b x$ and $n = \log_b y$, and recall that
$$\log_b x = m \quad \text{means} \quad b^m = x.$$
$$\log_b y = n \quad \text{means} \quad b^n = y.$$

Now consider the product xy.

$$xy = b^m \cdot b^n \qquad \text{Substitute.}$$
$$xy = b^{m+n} \qquad \text{Product rule for exponents}$$
$$\log_b xy = m + n \qquad \text{Convert to logarithmic form.}$$
$$\log_b xy = \log_b x + \log_b y \qquad \text{Substitute.}$$

The last statement is the result we wished to prove.

EXAMPLE 1 Using the Product Rule

Use the product rule to rewrite each expression. Assume $x > 0$.

(a) $\log_5(6 \cdot 9)$

By the product rule,

$$\log_5(6 \cdot 9) = \log_5 6 + \log_5 9.$$

(b) $\log_7 8 + \log_7 12 = \log_7(8 \cdot 12) = \log_7 96$

(c) $\log_3(3x) = \log_3 3 + \log_3 x$

$\qquad\qquad = 1 + \log_3 x \qquad\qquad$ $\log_3 3 = 1$

(d) $\log_4 x^3 = \log_4(x \cdot x \cdot x) \qquad\qquad$ $x^3 = x \cdot x \cdot x$

$\qquad\qquad = \log_4 x + \log_4 x + \log_4 x \qquad$ Product rule

$\qquad\qquad = 3 \log_4 x$

Now Try Exercises 7 and 21.

OBJECTIVE 2 Use the quotient rule for logarithms. The rule for division is similar to the rule for multiplication.

> **Quotient Rule for Logarithms**
>
> If x, y, and b are positive real numbers, where $b \neq 1$, then
>
> $$\log_b \frac{x}{y} = \log_b x - \log_b y.$$
>
> That is, the logarithm of a quotient is the difference between the logarithm of the numerator and the logarithm of the denominator.

The proof of this rule is very similar to the proof of the product rule.

EXAMPLE 2 Using the Quotient Rule

Use the quotient rule to rewrite each logarithm.

(a) $\log_4 \frac{7}{9} = \log_4 7 - \log_4 9$

(b) $\log_5 6 - \log_5 x = \log_5 \frac{6}{x}, \quad x > 0$

(c) $\log_3 \frac{27}{5} = \log_3 27 - \log_3 5$

$\qquad\qquad = 3 - \log_3 5 \qquad\qquad$ $\log_3 27 = 3$

Now Try Exercises 9 and 23.

CAUTION Remember that there is no property of logarithms to rewrite the logarithm of a *sum* or *difference*. For example, we *cannot* write $\log_b(x + y)$ in terms of $\log_b x$ and $\log_b y$. Also,

$$\log_b \frac{x}{y} \neq \frac{\log_b x}{\log_b y}.$$

OBJECTIVE 3 Use the power rule for logarithms. An exponential expression such as 2^3 means $2 \cdot 2 \cdot 2$; the base is used as a factor 3 times. Thus, it seems reasonable that the product rule can be extended to rewrite the logarithm of a power as the product of the exponent and the logarithm of the base. For example, by the product rule for logarithms,

$$\begin{aligned} \log_5 2^3 &= \log_5(2 \cdot 2 \cdot 2) \\ &= \log_5 2 + \log_5 2 + \log_5 2 \\ &= 3 \log_5 2. \end{aligned}$$

Also,

$$\begin{aligned} \log_2 7^4 &= \log_2(7 \cdot 7 \cdot 7 \cdot 7) \\ &= \log_2 7 + \log_2 7 + \log_2 7 + \log_2 7 \\ &= 4 \log_2 7. \end{aligned}$$

Furthermore, we saw in Example 1(d) that $\log_4 x^3 = 3 \log_4 x$. These examples suggest the following rule.

Power Rule for Logarithms

If x and b are positive real numbers, where $b \neq 1$, and if r is any real number, then

$$\log_b x^r = r \log_b x.$$

That is, the logarithm of a number to a power equals the exponent times the logarithm of the number.

As further examples of this result,

$$\log_b m^5 = 5 \log_b m \qquad \text{and} \qquad \log_3 5^4 = 4 \log_3 5.$$

To prove the power rule, let

$$\begin{aligned} \log_b x &= m. \\ b^m &= x && \text{Convert to exponential form.} \\ (b^m)^r &= x^r && \text{Raise to the power } r. \\ b^{mr} &= x^r && \text{Power rule for exponents} \\ \log_b x^r &= mr && \text{Convert to logarithmic form.} \\ \log_b x^r &= rm \\ \log_b x^r &= r \log_b x && m = \log_b x \end{aligned}$$

This is the statement to be proved.

As a special case of the power rule, let $r = \frac{1}{p}$, so

$$\log_b \sqrt[p]{x} = \log_b x^{1/p} = \frac{1}{p} \log_b x.$$

For example, using this result, with $x > 0$,

$$\log_b \sqrt[5]{x} = \log_b x^{1/5} = \frac{1}{5} \log_b x \qquad \text{and} \qquad \log_b \sqrt[3]{x^4} = \log_b x^{4/3} = \frac{4}{3} \log_b x.$$

Another special case is

$$\log_b \frac{1}{x} = \log_b x^{-1} = -\log_b x.$$

NOTE For a review of rational exponents, refer to Section 8.2.

EXAMPLE 3 Using the Power Rule

Use the power rule to rewrite each logarithm. Assume $b > 0$, $x > 0$, and $b \neq 1$.

(a) $\log_5 4^2 = 2 \log_5 4$ 　　　　　　　　**(b)** $\log_b x^5 = 5 \log_b x$

(c) $\log_b \sqrt{7}$

Begin by rewriting the radical expression with a rational exponent.

$$\log_b \sqrt{7} = \log_b 7^{1/2} \qquad \sqrt{x} = x^{1/2}$$

$$= \frac{1}{2} \log_b 7 \qquad \text{Power rule}$$

(d) $\log_2 \sqrt[5]{x^2} = \log_2 x^{2/5} \qquad \sqrt[5]{x^2} = x^{2/5}$

$$= \frac{2}{5} \log_2 x \qquad \text{Power rule}$$

Now Try Exercise 11.

Two special properties involving both exponential and logarithmic expressions come directly from the fact that logarithmic and exponential functions are inverses of each other.

Special Properties

If $b > 0$ and $b \neq 1$, then

$$b^{\log_b x} = x, \quad x > 0 \qquad \text{and} \qquad \log_b b^x = x.$$

To prove the first statement, let

$$y = \log_b x.$$

$$b^y = x \qquad \qquad \text{Convert to exponential form.}$$

$$b^{\log_b x} = x \qquad \qquad \text{Replace } y \text{ with } \log_b x.$$

The proof of the second statement is similar.

▓ **EXAMPLE 4** Using the Special Properties

Find the value of each logarithmic expression.

(a) $\log_5 5^4 = 4$, since $\log_b b^x = x$.   **(b)** $\log_3 9 = \log_3 3^2 = 2$

(c) $4^{\log_4 10} = 10$

Now Try Exercises 3 and 5.

Here is a summary of the properties of logarithms.

Properties of Logarithms

If x, y, and b are positive real numbers, where $b \neq 1$, and r is any real number, then

Product Rule $\log_b xy = \log_b x + \log_b y$

Quotient Rule $\log_b \dfrac{x}{y} = \log_b x - \log_b y$

Power Rule $\log_b x^r = r \log_b x$

Special Properties $b^{\log_b x} = x$ and $\log_b b^x = x.$

OBJECTIVE 4 Use properties to write alternative forms of logarithmic expressions. Applying the properties of logarithms is important for solving equations with logarithms and in calculus.

▓ **EXAMPLE 5** Writing Logarithms in Alternative Forms

Use the properties of logarithms to rewrite each expression if possible. Assume all variables represent positive real numbers.

(a) $\log_4 4x^3 = \log_4 4 + \log_4 x^3$ Product rule

$\qquad\qquad = 1 + 3 \log_4 x$ $\log_4 4 = 1$; power rule

(b) $\log_7 \sqrt{\dfrac{m}{n}} = \log_7 \left(\dfrac{m}{n}\right)^{1/2}$

$\qquad\qquad = \dfrac{1}{2} \log_7 \dfrac{m}{n}$ Power rule

$\qquad\qquad = \dfrac{1}{2}(\log_7 m - \log_7 n)$ Quotient rule

(c) $\log_5 \dfrac{a^2}{bc} = \log_5 a^2 - \log_5 bc$ Quotient rule

$\qquad\qquad = 2 \log_5 a - \log_5 bc$ Power rule

$\qquad\qquad = 2 \log_5 a - (\log_5 b + \log_5 c)$ Product rule

$\qquad\qquad = 2 \log_5 a - \log_5 b - \log_5 c$

Notice the careful use of parentheses in the third step. Since we are subtracting the logarithm of a product and rewriting it as a sum of two terms, we must place parentheses around the sum.

(d) $4 \log_b m - \log_b n = \log_b m^4 - \log_b n$ Power rule

$$= \log_b \frac{m^4}{n}$$ Quotient rule

(e) $\log_b(x + 1) + \log_b(2x - 1) - \dfrac{2}{3} \log_b x$

$$= \log_b(x + 1) + \log_b(2x - 1) - \log_b x^{2/3}$$ Power rule

$$= \log_b \frac{(x + 1)(2x - 1)}{x^{2/3}}$$ Product and quotient rules

$$= \log_b \frac{2x^2 + x - 1}{x^{2/3}}$$

(f) $\log_8(2p + 3r)$ cannot be rewritten using the properties of logarithms.   ◼

Now Try Exercises 13, 15, 27, and 31.

In the next example, we use numerical values for $\log_2 5$ and $\log_2 3$. While we use the equals sign to give these values, they are actually just approximations since most logarithms of this type are irrational numbers. We use $=$ with the understanding that the values are correct to four decimal places.

◼ **EXAMPLE 6**   **Using the Properties of Logarithms with Numerical Values**

Given that $\log_2 5 = 2.3219$ and $\log_2 3 = 1.5850$, evaluate the following.

(a) $\log_2 15 = \log_2(3 \cdot 5)$

$$= \log_2 3 + \log_2 5$$ Product rule

$$= 1.5850 + 2.3219$$

$$= 3.9069$$

(b) $\log_2 .6 = \log_2 \dfrac{3}{5}$ $.6 = \frac{6}{10} = \frac{3}{5}$

$$= \log_2 3 - \log_2 5$$ Quotient rule

$$= 1.5850 - 2.3219$$

$$= -.7369$$

(c) $\log_2 27 = \log_2 3^3$

$$= 3 \log_2 3$$ Power rule

$$= 3(1.5850)$$

$$= 4.7550$$

◼

Now Try Exercises 33, 35, and 43.

■ EXAMPLE 7 Deciding Whether Statements about Logarithms Are True

Decide whether each statement is *true* or *false*.

(a) $\log_2 8 - \log_2 4 = \log_2 4$

Evaluate both sides.

Left side: $\log_2 8 - \log_2 4 = \log_2 2^3 - \log_2 2^2 = 3 - 2 = 1$

Right side: $\log_2 4 = \log_2 2^2 = 2$

The statement is false because $1 \neq 2$.

(b) $\log_3(\log_2 8) = \dfrac{\log_7 49}{\log_8 64}$

Evaluate both sides.

Left side: $\log_3(\log_2 8) = \log_3 3 = 1$

Right side: $\dfrac{\log_7 49}{\log_8 64} = \dfrac{\log_7 7^2}{\log_8 8^2} = \dfrac{2}{2} = 1$

The statement is true because $1 = 1$.

Now Try Exercises 45 and 51.

Napier's Rods

CONNECTIONS

Long before the days of calculators and computers, the search for making calculations easier was an ongoing process. Machines built by Charles Babbage and Blaise Pascal, a system of "rods" used by John Napier, and slide rules were the forerunners of today's electronic marvels. The invention of logarithms by John Napier in the sixteenth century was a great breakthrough in the search for easier methods of calculation.

Since logarithms are exponents, their properties allowed users of tables of common logarithms to multiply by adding, divide by subtracting, raise to powers by multiplying, and take roots by dividing. Although logarithms are no longer used for computations, they play an important part in higher mathematics.

For Discussion or Writing

1. To multiply 458.3 by 294.6 using logarithms, we add $\log_{10} 458.3$ and $\log_{10} 294.6$, then find 10 to the sum. Perform this multiplication using the $\boxed{\log x}$ key* and the $\boxed{10^x}$ key on your calculator. Check your answer by multiplying directly with your calculator.

2. Try division, raising to a power, and taking a root by this method.

---

*In this text, the notation $\log x$ is used to mean $\log_{10} x$. This is also the meaning of the log key on calculators.

11.4 EXERCISES

Use the indicated rule of logarithms to complete each equation in Exercises 1–5.

1. $\log_{10}(3 \cdot 4) = $ _____ (product rule)

2. $\log_{10} \dfrac{3}{4} = $ _____ (quotient rule)

3. $3^{\log_3 4} = $ _____ (special property)

4. $\log_{10} 3^4 = $ _____ (power rule)

5. $\log_3 3^4 = $ _____ (special property)

6. Evaluate $\log_2(8 + 8)$. Then evaluate $\log_2 8 + \log_2 8$. Are the results the same? How could you change the operation in the first expression to make the two expressions equal?

Use the properties of logarithms to express each logarithm as a sum or difference of logarithms, or as a single number if possible. Assume all variables represent positive real numbers. See Examples 1–5.

7. $\log_7(4 \cdot 5)$ **8.** $\log_8(9 \cdot 11)$ **9.** $\log_5 \dfrac{8}{3}$

10. $\log_3 \dfrac{7}{5}$ **11.** $\log_4 6^2$ **12.** $\log_5 7^4$

13. $\log_3 \dfrac{\sqrt[3]{4}}{x^2 y}$ **14.** $\log_7 \dfrac{\sqrt[3]{13}}{pq^2}$ **15.** $\log_3 \sqrt{\dfrac{xy}{5}}$

16. $\log_6 \sqrt{\dfrac{pq}{7}}$ **17.** $\log_2 \dfrac{\sqrt[3]{x} \cdot \sqrt[5]{y}}{r^2}$ **18.** $\log_4 \dfrac{\sqrt[4]{z} \cdot \sqrt[5]{w}}{s^2}$

19. A student erroneously wrote $\log_a(x + y) = \log_a x + \log_a y$. When his teacher explained that this was indeed wrong, the student claimed that he had used the distributive property. Write a few sentences explaining why the distributive property does not apply in this case.

20. Write a few sentences explaining how the rules for multiplying and dividing powers of the same base are similar to the rules for finding logarithms of products and quotients.

Use the properties of logarithms to write each expression as a single logarithm. Assume all variables are defined in such a way that the variable expressions are positive, and bases are positive numbers not equal to 1. See Examples 1–5.

21. $\log_b x + \log_b y$ **22.** $\log_b 2 + \log_b z$

23. $\log_a m - \log_a n$ **24.** $\log_b x - \log_b y$

25. $(\log_a r - \log_a s) + 3 \log_a t$ **26.** $(\log_a p - \log_a q) + 2 \log_a r$

27. $3 \log_a 5 - 4 \log_a 3$ **28.** $3 \log_a 5 + \dfrac{1}{2} \log_a 9$

29. $\log_{10}(x + 3) + \log_{10}(x - 3)$ **30.** $\log_{10}(y + 4) + \log_{10}(y - 4)$

31. $3 \log_p x + \dfrac{1}{2} \log_p y - \dfrac{3}{2} \log_p z - 3 \log_p a$

32. $\dfrac{1}{3} \log_b x + \dfrac{2}{3} \log_b y - \dfrac{3}{4} \log_b s - \dfrac{2}{3} \log_b t$

To four decimal places, the values of $\log_{10} 2$ *and* $\log_{10} 9$ *are*

$$\log_{10} 2 = .3010 \qquad \log_{10} 9 = .9542.$$

Evaluate each logarithm by applying the appropriate rule or rules from this section. **DO NOT USE A CALCULATOR.** *See Example 6.*

33. $\log_{10} 18$ **34.** $\log_{10} \dfrac{9}{2}$ **35.** $\log_{10} \dfrac{2}{9}$ **36.** $\log_{10} 4$

37. $\log_{10} 36$ **38.** $\log_{10} 162$ **39.** $\log_{10} 3$ **40.** $\log_{10} \sqrt[5]{2}$

41. $\log_{10} \sqrt[4]{9}$ **42.** $\log_{10} \dfrac{1}{9}$ **43.** $\log_{10} 9^5$ **44.** $\log_{10} 2^{19}$

Decide whether each statement is true *or* false. *See Example 7.*

45. $\log_2(8 + 32) = \log_2 8 + \log_2 32$ **46.** $\log_2(64 - 16) = \log_2 64 - \log_2 16$

47. $\log_3 7 + \log_3 7^{-1} = 0$ **48.** $\log_9 14 - \log_{14} 9 = 0$

49. $\log_6 60 - \log_6 10 = 1$ **50.** $\log_3 8 + \log_3 \dfrac{1}{8} = 0$

51. $\dfrac{\log_{10} 7}{\log_{10} 14} = \dfrac{1}{2}$ **52.** $\dfrac{\log_{10} 10}{\log_{10} 100} = \dfrac{1}{10}$

53. Refer to the Note following the word statement of the product rule for logarithms in this section. Now, state the quotient rule in words, replacing "logarithm" with "exponent."

54. Explain why the statement for the power rule for logarithms requires that x be a positive real number.

55. Refer to Example 7(a). Change the left side of the equation using the quotient rule so that the statement becomes true, and simplify.

56. What is wrong with the following "proof" that $\log_2 16$ does not exist? Explain.

$$\log_2 16 = \log_2(-4)(-4)$$
$$= \log_2(-4) + \log_2(-4)$$

Since the logarithm of a negative number is not defined, the final step cannot be evaluated, and so $\log_2 16$ does not exist.

RELATING CONCEPTS (EXERCISES 57–62)

For Individual or Group Work

Work Exercises 57–62 in order.

57. Evaluate $\log_3 81$.

58. Write the *meaning* of the expression $\log_3 81$.

59. Evaluate $3^{\log_3 81}$.

60. Write the *meaning* of the expression $\log_2 19$.

61. Evaluate $2^{\log_2 19}$.

62. Keeping in mind that a logarithm is an exponent and using the results from Exercises 57–61, what is the simplest form of the expression $k^{\log_k m}$?

11.5 Common and Natural Logarithms

As mentioned earlier, logarithms are important in many applications of mathematics to everyday problems, particularly in biology, engineering, economics, and social science. In this section we find numerical approximations for logarithms. Traditionally, base 10 logarithms were used most often because our number system is base 10. Logarithms to base 10 are called **common logarithms,** and $\log_{10} x$ is abbreviated as simply $\log x$, where the base is understood to be 10.

OBJECTIVE 1 Evaluate common logarithms using a calculator. We use calculators to evaluate common logarithms. In the next example we give the results of evaluating some common logarithms using a calculator with a $\boxed{\text{LOG}}$ key. (This may be a second function key on some calculators.) For simple scientific calculators, just enter the number, then press the $\boxed{\text{LOG}}$ key. For graphing calculators, these steps are reversed. We give all logarithms to four decimal places.

EXAMPLE 1 Evaluating Common Logarithms

Evaluate each logarithm using a calculator.

(a) $\log 327.1 \approx 2.5147$ **(b)** $\log 437{,}000 \approx 5.6405$

(c) $\log .0615 \approx -1.2111$

> **Now Try Exercises 7, 9, and 11.**

Figure 14 shows how a graphing calculator displays the common logarithms in Example 1. The calculator is set to give four decimal places.

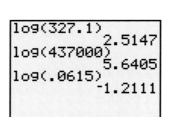

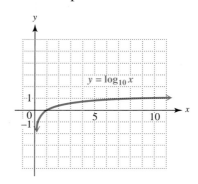

| FIGURE 14 | FIGURE 15 |

Notice that $\log .0615 \approx -1.2111$, a negative result. The common logarithm of a number between 0 and 1 is always negative because the logarithm is the exponent on 10 that produces the number. For example,

$$10^{-1.2111} \approx .0615.$$

If the exponent (the logarithm) were positive, the result would be greater than 1 because $10^0 = 1$. See Figure 15.

OBJECTIVE 2 Use common logarithms in applications. In chemistry, pH is a measure of the acidity or alkalinity of a solution; pure water, for example, has pH 7. In general, acids have pH numbers less than 7, and alkaline solutions have pH values greater than 7. The **pH** of a solution is defined as

$$pH = -\log[H_3O^+],$$

where $[H_3O^+]$ is the hydronium ion concentration in moles per liter. It is customary to round pH values to the nearest tenth.

EXAMPLE 2 Using pH in an Application

Wetlands are classified as *bogs, fens, marshes,* and *swamps.* These classifications are based on pH values. A pH value between 6.0 and 7.5, such as that of Summerby Swamp in Michigan's Hiawatha National Forest, indicates that the wetland is a "rich fen." When the pH is between 4.0 and 6.0, the wetland is a "poor fen," and if the pH falls to 3.0 or less, it is a "bog." (*Source:* Mohlenbrock, R., "Summerby Swamp, Michigan," *Natural History,* March 1994.) Suppose that the hydronium ion concentration of a sample of water from a wetland is 6.3×10^{-3}. How would this wetland be classified?

Use the definition of pH.

$$
\begin{aligned}
pH &= -\log(6.3 \times 10^{-3}) \\
&= -(\log 6.3 + \log 10^{-3}) && \text{Product rule} \\
&= -[.7993 - 3(1)] && \text{Use a calculator to find log 6.3.} \\
&= -.7993 + 3 \\
&\approx 2.2
\end{aligned}
$$

Since the pH is less than 3.0, the wetland is a bog.

Now Try Exercise 29.

EXAMPLE 3 Finding Hydronium Ion Concentration

Find the hydronium ion concentration of drinking water with pH 6.5.

$$
\begin{aligned}
pH &= -\log[H_3O^+] \\
6.5 &= -\log[H_3O^+] && \text{Let pH = 6.5.} \\
\log[H_3O^+] &= -6.5 && \text{Multiply by } -1.
\end{aligned}
$$

Solve for $[H_3O^+]$ by writing the equation in exponential form, remembering that the base is 10.

$$
\begin{aligned}
[H_3O^+] &= 10^{-6.5} \\
[H_3O^+] &\approx 3.2 \times 10^{-7} && \text{Use a calculator.}
\end{aligned}
$$

Now Try Exercise 35.

OBJECTIVE 3 Evaluate natural logarithms using a calculator. The most important logarithms used in applications are **natural logarithms,** which have as base the number e. The number e is a fundamental number in our universe. For this reason e, like π, is called a *universal constant.* The letter e is used to honor Leonhard Euler,

Leonhard Euler (1707–1783)

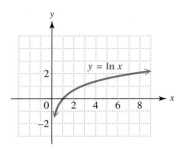

FIGURE 16

who published extensive results on the number in 1748. Since it is an irrational number, its decimal expansion never terminates and never repeats. The first few digits of the decimal value of e are 2.7182818285. A calculator key $\boxed{e^x}$ or the two keys $\boxed{\text{INV}}$ and $\boxed{\ln x}$ are used to approximate powers of e. For example, a calculator gives

$$e^2 \approx 7.389056099, \quad e^3 \approx 20.08553692, \quad \text{and} \quad e^{.6} \approx 1.8221188.$$

Logarithms to base e are called **natural logarithms** because they occur in biology and the social sciences in natural situations that involve growth or decay. The base e logarithm of x is written $\ln x$ (read "el en x"). A graph of $y = \ln x$, the equation that defines the natural logarithmic function, is given in Figure 16.

A calculator key labeled $\boxed{\ln x}$ is used to evaluate natural logarithms. If your calculator has an $\boxed{e^x}$ key, but not a key labeled $\boxed{\ln x}$, find natural logarithms by entering the number, pressing the $\boxed{\text{INV}}$ key, and then pressing the $\boxed{e^x}$ key. This works because $y = e^x$ defines the inverse function of $y = \ln x$ (or $y = \log_e x$).

```
ln(.5841)
          -.5377
ln(192.7)
          5.2611
ln(10.84)
          2.3832
```

FIGURE 17

EXAMPLE 4 Finding Natural Logarithms

Evaluate each logarithm with a calculator.

(a) $\ln .5841 \approx -.5377$

As with common logarithms, a number between 0 and 1 has a negative natural logarithm.

(b) $\ln 192.7 \approx 5.2611$ **(c)** $\ln 10.84 \approx 2.3832$

Figure 17 shows how a graphing calculator displays these natural logarithms to four decimal places.

> **Now Try Exercises 15, 17, and 19.**

OBJECTIVE 4 Use natural logarithms in applications. Some applications involve functions that use natural logarithms, as seen in the next example.

EXAMPLE 5 Applying a Natural Logarithmic Function

The altitude in meters that corresponds to an atmospheric pressure of x millibars is given by the logarithmic function defined by

$$f(x) = 51{,}600 - 7457 \ln x.$$

(*Source:* Miller, A. and J. Thompson, *Elements of Meteorology,* Fourth Edition, Charles E. Merrill Publishing Company, 1993.) Use this function to find the altitude when atmospheric pressure is 400 millibars.

Let $x = 400$ and substitute in the expression for $f(x)$.

$$f(400) = 51{,}600 - 7457 \ln 400$$
$$\approx 6900$$

Atmospheric pressure is 400 millibars at approximately 6900 m.

> **Now Try Exercise 39.**

NOTE In Example 5, the final answer was obtained using a calculator *without* rounding the intermediate values. In general, it is best to wait until the final step to round the answer; otherwise, a buildup of round-off error may cause the final answer to have an incorrect final decimal place digit.

OBJECTIVE 5 Use the change-of-base rule. We have used a calculator to approximate the values of common logarithms (base 10) and natural logarithms (base e). However, some applications involve logarithms to other bases. For example, for the years 1980–1996, the percentage of women who had a baby in the last year and returned to work is given by

$$f(x) = 38.83 + 4.208 \log_2 x,$$

for year x. (*Source:* U.S. Bureau of the Census.) To use this function, we need to find a base 2 logarithm. The following rule is used to convert logarithms from one base to another.

Change-of-Base Rule

If $a > 0$, $a \neq 1$, $b > 0$, $b \neq 1$, and $x > 0$, then

$$\log_a x = \frac{\log_b x}{\log_b a}.$$

NOTE Any positive number other than 1 can be used for base b in the change-of-base rule, but usually the only practical bases are e and 10 because calculators give logarithms only for these two bases.

To derive the change-of-base rule, let $\log_a x = m$.

$$\log_a x = m$$
$$a^m = x \qquad \text{Change to exponential form.}$$

Since logarithmic functions are one-to-one, if all variables are positive and if $x = y$, then $\log_b x = \log_b y$.

$$\log_b(a^m) = \log_b x$$
$$m \log_b a = \log_b x \qquad \text{Power rule}$$
$$(\log_a x)(\log_b a) = \log_b x \qquad \text{Substitute for } m.$$
$$\log_a x = \frac{\log_b x}{\log_b a} \qquad \text{Divide by } \log_b a.$$

The last step gives the change-of-base rule.

EXAMPLE 6 Using the Change-of-Base Rule

Find $\log_5 12$.

Use common logarithms and the change-of-base rule.

$$\log_5 12 = \frac{\log 12}{\log 5}$$

$$\approx 1.5440 \qquad \text{Use a calculator.}$$

Verify that the same value is found when using natural logarithms.

Now Try Exercise 47.

EXAMPLE 7 Using the Change-of-Base Rule in an Application

Use natural logarithms in the change-of-base rule and the function

$$f(x) = 38.83 + 4.208 \log_2 x$$

(given earlier) to find the percent of women who returned to work after having a baby in 1995. In the equation, $x = 0$ represents 1980.

Substitute $1995 - 1980 = 15$ for x in the equation.

$$f(15) = 38.83 + 4.208 \log_2 15$$

$$= 38.83 + 4.208 \left(\frac{\ln 15}{\ln 2} \right) \qquad \text{Change-of-base rule}$$

$$\approx 55.3\% \qquad \text{Use a calculator.}$$

This is very close to the actual value of 55%.

Now Try Exercise 59.

| CONNECTIONS |

As previously mentioned, the number $e \approx 2.718281828$ is a fundamental number in our universe. If there are intelligent beings elsewhere, they too will have to use e to do higher mathematics.

The properties of e are used extensively in calculus and in higher mathematics. In Section 11.6 we see how it applies to growth and decay in the physical world.

For Discussion or Writing

The value of e can be expressed as

$$e = 1 + \frac{1}{1} + \frac{1}{1 \cdot 2} + \frac{1}{1 \cdot 2 \cdot 3} + \frac{1}{1 \cdot 2 \cdot 3 \cdot 4} + \cdots.$$

Approximate e using two terms of this expression, then three terms, four terms, five terms, and six terms. How close is the approximation to the value of e given above with six terms? Does this infinite sum approach the value of e very quickly?

11.5 EXERCISES

Choose the correct response in Exercises 1–4.

1. What is the base in the expression log x?

   A. e **B.** 1   **C.** 10   **D.** x

2. What is the base in the expression ln x?

   A. e **B.** 1   **C.** 10   **D.** x

3. Since $10^0 = 1$ and $10^1 = 10$, between what two consecutive integers is the value of log 5.6?

   A. 5 and 6   **B.** 10 and 11   **C.** 0 and 1   **D.** -1 and 0

4. Since $e^1 \approx 2.718$ and $e^2 \approx 7.389$, between what two consecutive integers is the value of ln 5.6?

   A. 5 and 6   **B.** 2 and 3   **C.** 1 and 2   **D.** 0 and 1

5. Without using a calculator, give the value of log $10^{19.2}$.

6. Without using a calculator, give the value of ln $e^{\sqrt{2}}$.

You will need a calculator for the remaining exercises in this set.

Find each logarithm. Give an approximation to four decimal places. See Examples 1 and 4.

7. log 43	**8.** log 98	**9.** log 328.4
10. log 457.2	**11.** log .0326	**12.** log .1741
13. log(4.76×10^9)	**14.** log(2.13×10^4)	**15.** ln 7.84
16. ln 8.32	**17.** ln .0556	**18.** ln .0217
19. ln 388.1	**20.** ln 942.6	**21.** ln($8.59 \times e^2$)
22. ln($7.46 \times e^3$)	**23.** ln 10	**24.** log e

25. Use your calculator to find approximations of the following logarithms:

   (a) log 356.8   **(b)** log 35.68   **(c)** log 3.568.

   ✐ **(d)** Observe your answers and make a conjecture concerning the decimal values of the common logarithms of numbers greater than 1 that have the same digits.

26. Let k represent the number of letters in your last name.

   (a) Use your calculator to find log k.

   (b) Raise 10 to the power indicated by the number you found in part (a). What is your result?

   ✐ **(c)** Use the concepts of Section 11.1 to explain why you obtained the answer you found in part (b). Would it matter what number you used for k to observe the same result?

✐ **27.** Try to find log(-1) using a calculator. (If you have a graphing calculator, it should be in real number mode.) What happens? Explain.

Refer to Example 2. In Exercises 28–30, suppose that water from a wetland area is sampled and found to have the given hydronium ion concentration. Determine whether the wetland is a rich fen, *a* poor fen, *or a* bog.

28. 2.5×10^{-5} **29.** 2.5×10^{-2} **30.** 2.5×10^{-7}

Find the pH *of the substance with the given hydronium ion concentration. See Example 2.*

31. Ammonia, 2.5×10^{-12} **32.** Sodium bicarbonate, 4.0×10^{-9}

33. Grapes, 5.0×10^{-5} **34.** Tuna, 1.3×10^{-6}

Use the formula for pH *to find the hydronium ion concentration of the substance with the given pH. See Example 3.*

35. Human blood plasma, 7.4 **36.** Human gastric contents, 2.0

37. Spinach, 5.4 **38.** Bananas, 4.6

Solve each problem. See Example 5.

39. The number of years, $N(r)$, since two independently evolving languages split off from a common ancestral language is approximated by

$$N(r) = -5000 \ln r,$$

where r is the percent of words (in decimal form) from the ancestral language common to both languages now. Find the number of years since the split for each percent of common words.

(a) 85% (or .85)   **(b)** 35% (or .35)   **(c)** 10% (or .10)

40. The time t in years for an amount increasing at a rate of r (in decimal form) to double is given by

$$t(r) = \frac{\ln 2}{\ln(1 + r)}.$$

This is called *doubling time.* Find the doubling time to the nearest tenth for an investment at each interest rate.

(a) 2% (or .02)   **(b)** 5% (or .05)   **(c)** 8% (or .08)

41. The loudness of sounds is measured in a unit called a *decibel,* abbreviated dB. A very faint sound, called the *threshold sound,* is assigned an intensity I_0. If a particular sound has intensity I, then the decibel level of this louder sound is

$$D = 10 \log\left(\frac{I}{I_0}\right).$$

Find the average decibel level for each popular movie with the given intensity I. For comparison, a motorcycle or power saw has a decibel level of about 95 dB, and the sound of a jackhammer or helicopter is about 105 dB. (*Source: World Almanac and Book of Facts,* 2001; www.lhh.org/noise/)

(a) *Armageddon;* $5.012 \times 10^{10} I_0$

(b) *Godzilla;* $10^{10} I_0$

(c) *Saving Private Ryan;* $6,310,000,000 \, I_0$

42. The concentration of a drug injected into the bloodstream decreases with time. The intervals of time T when the drug should be administered are given by

$$T = \frac{1}{k} \ln \frac{C_2}{C_1},$$

where k is a constant determined by the drug in use, C_2 is the concentration at which the drug is harmful, and C_1 is the concentration below which the drug is ineffective. (*Source:* Horelick, Brindell and Sinan Koont, "Applications of Calculus to Medicine: Prescribing Safe and Effective Dosage," *UMAP Module 202,* 1977.) Thus, if $T = 4$, the drug should be administered every 4 hr. For a certain drug, $k = \frac{1}{3}$, $C_2 = 5$, and $C_1 = 2$. How often should the drug be administered? (*Hint:* Round down.)

43. The growth of outpatient surgeries as a percent of total surgeries at hospitals is approximated by

$$f(x) = -1317 + 304 \ln x,$$

where x represents the number of years since 1900. (*Source:* American Hospital Association.)

(a) What does this function predict for the percent of outpatient surgeries in 1998?
(b) When did outpatient surgeries reach 50%? (*Hint:* Substitute for y, then write the equation in exponential form to solve it.)

44. In the central Sierra Nevada of California, the percent of moisture p that falls as snow rather than rain is approximated reasonably well by

$$f(x) = 86.3 \ln x - 680,$$

where x is the altitude in feet.

(a) What percent of the moisture at 5000 ft falls as snow?
(b) What percent at 7500 ft falls as snow?

45. The *cost-benefit equation*

$$T = -.642 - 189 \ln(1 - p)$$

describes the approximate tax T, in dollars per ton, that would result in a $p\%$ (in decimal form) reduction in carbon dioxide emissions.

(a) What tax will reduce emissions 25%?
(b) Explain why the equation is not valid for $p = 0$ or $p = 1$.

46. The age in years of a female blue whale is approximated by

$$t = -2.57 \ln\left(\frac{87 - L}{63}\right),$$

where L is its length in feet.

(a) How old is a female blue whale that measures 80 ft?
(b) The equation that defines t has domain $24 < L < 87$. Explain why.

Use the change-of-base rule (with either common or natural logarithms) to find each logarithm to four decimal places. See Example 6.

47. $\log_3 12$ **48.** $\log_4 18$ **49.** $\log_5 3$

50. $\log_7 4$ **51.** $\log_3 \sqrt{2}$ **52.** $\log_6 \sqrt[3]{5}$

53. $\log_\pi e$ **54.** $\log_\pi 10$ **55.** $\log_e 12$

📝 **56.** Explain why the answer to Exercise 55 is the same one that you get when you use a calculator to approximate ln 12.

57. Let m be the number of letters in your first name, and let n be the number of letters in your last name.

📝 **(a)** In your own words, explain what $\log_m n$ means.
(b) Use your calculator to find $\log_m n$.
(c) Raise m to the power indicated by the number you found in part (b). What is your result?

58. The equation $5^x = 7$ cannot be solved using the methods described in Section 11.2. However, in solving this equation, we must find the exponent to which 5 must be raised in order to obtain 7: this is $\log_5 7$.

(a) Use the change-of-base rule and your calculator to find $\log_5 7$.
(b) Raise 5 to the number you found in part (a). What is your result?
(c) Using as many decimal places as your calculator gives, write the solution set of $5^x = 7$. (Equations of this type will be studied in more detail in Section 11.6.)

Solve each application of a logarithmic function. See Example 7.

59. Refer to Exercise 53 in Section 11.3. Determine the number of waste sites in 1998.

60. Refer to Exercise 54 in Section 11.3. Determine the approximate consumption in 1998.

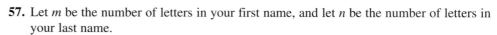
TECHNOLOGY INSIGHTS (EXERCISES 61–64)

Because graphing calculators are equipped with log *x and* ln *x keys, it is possible to graph the functions defined by* $f(x) = \log x$ *and* $g(x) = \ln x$ *directly, as shown in the figures that follow.*

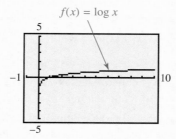

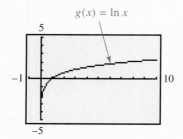

To graph functions defined by logarithms to bases other than 10 or e, however, we must use the change-of-base rule. For example, to graph $y = \log_2 x$, *we may enter* Y_1 *as* $\dfrac{\log X}{\log 2}$ *or* $\dfrac{\ln X}{\ln 2}$. *This is shown in the figure at the right. (Compare it to the figure in Exercises 59–62 of Section 11.3, where it was drawn using the fact that* $y = \log_2 x$ *is the inverse of* $y = 2^x$.*)*

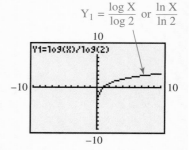

$$Y_1 = \frac{\log X}{\log 2} \text{ or } \frac{\ln X}{\ln 2}$$

📊 *Use the change-of-base rule to graph each logarithmic function with a graphing calculator. Use a viewing window with* Xmin $= -1$, Xmax $= 10$, Ymin $= -5$, *and* Ymax $= 5$.

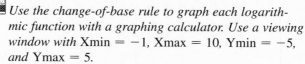

61. $g(x) = \log_3 x$ **62.** $g(x) = \log_5 x$ **63.** $g(x) = \log_{1/3} x$ **64.** $g(x) = \log_{1/5} x$

11.6 Exponential and Logarithmic Equations; Further Applications

As mentioned earlier, exponential and logarithmic functions are important in many applications of mathematics. Using these functions in applications requires solving exponential and logarithmic equations. Some simple equations were solved in Sections 11.2 and 11.3. More general methods for solving these equations depend on the following properties.

Properties for Solving Exponential and Logarithmic Equations

For all real numbers $b > 0$, $b \neq 1$, and any real numbers x and y:

1. If $x = y$, then $b^x = b^y$.

2. If $b^x = b^y$, then $x = y$.

3. If $x = y$, and $x > 0$, $y > 0$, then $\log_b x = \log_b y$.

4. If $x > 0$, $y > 0$, and $\log_b x = \log_b y$, then $x = y$.

We used Property 2 to solve exponential equations in Section 11.2.

OBJECTIVE 1 Solve equations involving variables in the exponents.   The first two examples illustrate the method for solving exponential equations using Property 3.

EXAMPLE 1 Solving an Exponential Equation

Solve $3^x = 12$.

$$3^x = 12$$
$$\log 3^x = \log 12 \qquad \text{Property 3}$$
$$x \log 3 = \log 12 \qquad \text{Power rule}$$
$$x = \frac{\log 12}{\log 3} \qquad \text{Divide by log 3.}$$

This quotient is the exact solution. To get a decimal approximation for the solution, use a calculator.

$$x \approx 2.262$$

The solution set is $\{2.262\}$. Check that $3^{2.262} \approx 12$.

Now Try Exercise 5.

CAUTION   Be careful: $\frac{\log 12}{\log 3}$ is *not* equal to log 4 because log 4 $\approx$ .6021, but $\frac{\log 12}{\log 3} \approx 2.262$.

When an exponential equation has e as the base, it is easiest to use base e logarithms.

EXAMPLE 2 Solving an Exponential Equation with Base e

Solve $e^{.003x} = 40$.

Take base e logarithms on both sides.

$$\ln e^{.003x} = \ln 40$$

$$.003x \ln e = \ln 40 \qquad \text{Power rule}$$

$$.003x = \ln 40 \qquad \ln e = \ln e^1 = 1$$

$$x = \frac{\ln 40}{.003} \qquad \text{Divide by .003.}$$

$$x \approx 1230 \qquad \text{Use a calculator.}$$

The solution set is $\{1230\}$. Check that $e^{.003(1230)} \approx 40$.

Now Try Exercise 15.

General Method for Solving an Exponential Equation

Take logarithms to the same base on both sides and then use the power rule of logarithms or the special property $\log_b b^x = x$. (See Examples 1 and 2.)

As a special case, if both sides can be written as exponentials with the same base, do so, and set the exponents equal. (See Section 11.2.)

OBJECTIVE 2 Solve equations involving logarithms. The properties of logarithms from Section 11.4 are useful here, as is using the definition of a logarithm to change the equation to exponential form.

EXAMPLE 3 Solving a Logarithmic Equation

Solve $\log_2(x + 5)^3 = 4$. Give the exact solution.

$$(x + 5)^3 = 2^4 \qquad \text{Convert to exponential form.}$$

$$(x + 5)^3 = 16$$

$$x + 5 = \sqrt[3]{16} \qquad \text{Take the cube root on each side.}$$

$$x = -5 + \sqrt[3]{16} \qquad \text{Subtract 5.}$$

$$x = -5 + 2\sqrt[3]{2} \qquad \text{Simplify the radical.}$$

Verify that the solution satisfies the equation, so the solution set is $\left\{-5 + 2\sqrt[3]{2}\right\}$.

Now Try Exercise 29.

CAUTION Recall that the domain of $y = \log_b x$ is $(0, \infty)$. For this reason, it is always necessary to check that the solution of an equation with logarithms yields only logarithms of positive numbers in the original equation.

EXAMPLE 4   Solving a Logarithmic Equation

Solve $\log_2(x + 1) - \log_2 x = \log_2 7$.

$$\log_2(x + 1) - \log_2 x = \log_2 7$$

$$\log_2 \frac{x + 1}{x} = \log_2 7 \qquad \text{Quotient rule}$$

$$\frac{x + 1}{x} = 7 \qquad \text{Property 4}$$

$$x + 1 = 7x \qquad \text{Multiply by } x.$$

$$1 = 6x$$

$$\frac{1}{6} = x$$

Check this solution by substituting in the original equation. Here, both $x + 1$ and x must be positive. If $x = \frac{1}{6}$, this condition is satisfied, so the solution set is $\left\{\frac{1}{6}\right\}$.

Now Try Exercise 35.

EXAMPLE 5   Solving a Logarithmic Equation

Solve $\log x + \log(x - 21) = 2$.

Write the left side as a single logarithm, write in exponential form, and solve the equation.

$$\log x + \log(x - 21) = 2$$

$$\log x(x - 21) = 2 \qquad \text{Product rule}$$

$$x(x - 21) = 10^2 \qquad \begin{array}{l}\log x = \log_{10} x\text{; write in}\\ \text{exponential form.}\end{array}$$

$$x^2 - 21x = 100$$

$$x^2 - 21x - 100 = 0 \qquad \text{Standard form}$$

$$(x - 25)(x + 4) = 0 \qquad \text{Factor.}$$

$$x - 25 = 0 \quad \text{or} \quad x + 4 = 0 \qquad \text{Zero-factor property}$$

$$x = 25 \quad \text{or} \qquad x = -4$$

The value -4 must be rejected as a solution since it leads to the logarithm of a negative number in the original equation:

$$\log(-4) + \log(-4 - 21) = 2. \qquad \text{The left side is undefined.}$$

The only solution, therefore, is 25, and the solution set is $\{25\}$.

Now Try Exercise 39.

CAUTION   Do not reject a potential solution just because it is nonpositive. Reject any value that *leads to* the logarithm of a nonpositive number.

In summary, we use the following steps to solve a logarithmic equation.

Solving a Logarithmic Equation

Step 1 **Transform the equation so that a single logarithm appears on one side.** Use the product rule or quotient rule of logarithms to do this.

Step 2 **(a) Use Property 4.** If $\log_b x = \log_b y$, then $x = y$. (See Example 4.)

(b) Write the equation in exponential form. If $\log_b x = k$, then $x = b^k$. (See Examples 3 and 5.)

OBJECTIVE 3 Solve applications of compound interest. So far in this book, problems involving applications of interest have been limited to simple interest using the formula $I = prt$. In most cases, interest paid or charged is *compound interest* (interest paid on both principal and interest). The formula for compound interest is an important application of exponential functions.

Compound Interest Formula (for a Finite Number of Periods)

If a principal of P dollars is deposited at an annual rate of interest r compounded (paid) n times per year, the account will contain

$$A = P\left(1 + \frac{r}{n}\right)^{nt}$$

dollars after t years. (In this formula, r is expressed as a decimal.)

EXAMPLE 6 Solving a Compound Interest Problem for *A*

How much money will there be in an account at the end of 5 yr if $1000 is deposited at 6% compounded quarterly? (Assume no withdrawals are made.)

Because interest is compounded quarterly, $n = 4$. The other values given in the problem are $P = 1000$, $r = .06$ (because $6\% = .06$), and $t = 5$. Substitute into the compound interest formula to find the value of A.

$$A = 1000\left(1 + \frac{.06}{4}\right)^{4 \cdot 5}$$

$$A = 1000(1.015)^{20}$$

Now use the $\boxed{y^x}$ key on a calculator and round the answer to the nearest cent.

$$A = 1346.86$$

The account will contain $1346.86. (The actual amount of interest earned is $1346.86 - $1000 = $346.86. Why?)

Now Try Exercise 45(a).

Interest can be compounded annually, semiannually, quarterly, daily, and so on. The number of compounding periods can get larger and larger. If the value of n is allowed to approach infinity, we have an example of *continuous compounding*. However, the compound interest formula above cannot be used for continuous compounding since there is no finite value for n. The formula for continuous compounding is an example of exponential growth involving the number e.

Continuous Compound Interest Formula

If a principal of P dollars is deposited at an annual rate of interest r compounded continuously for t years, the final amount on deposit is

$$A = Pe^{rt}.$$

EXAMPLE 7 Solving a Continuous Compound Interest Problem

In Example 6 we found that $1000 invested for 5 yr at 6% interest compounded quarterly would grow to $1346.86.

(a) How much would this same investment grow to if interest were compounded continuously?

Use the formula for continuous compounding with $P = 1000$, $r = .06$, and $t = 5$.

$$
\begin{aligned}
A &= Pe^{rt} && \text{Formula} \\
&= 1000e^{.06(5)} && \text{Substitute.} \\
&= 1000e^{.30} \\
&= 1349.86 && \text{Use a calculator and round to the nearest cent.}
\end{aligned}
$$

Continuous compounding would cause the investment to grow to $1349.86. Notice that this is $3.00 more than the amount in Example 6, when interest was compounded quarterly.

(b) How long would it take for the initial investment to double its original amount? (This is called the *doubling time.*)

We must find the value of t that will cause A to be $2(\$1000) = \$2000$.

$$
\begin{aligned}
A &= Pe^{rt} \\
2000 &= 1000e^{.06t} && \text{Let } A = 2P = 2000. \\
2 &= e^{.06t} && \text{Divide by 1000.} \\
\ln 2 &= .06t && \text{Take natural logarithms; } \ln e^k = k. \\
t &= \frac{\ln 2}{.06} && \text{Divide by .06.} \\
t &\approx 11.55 && \text{Use a calculator.}
\end{aligned}
$$

It would take about 11.55 yr for the original investment to double.

Now Try Exercise 47.

OBJECTIVE 4 Solve applications involving exponential growth and decay. One of the most common applications of exponential functions depends on the fact that in many situations involving growth or decay of a population, the amount or number of some quantity present at time t can be closely approximated by

$$y = y_0 e^{kt},$$

where y_0 is the amount or number present at time $t = 0$, k is a constant, and e is the base of natural logarithms.

EXAMPLE 8   Applying an Exponential Function

The *greenhouse effect* refers to the phenomenon whereby emissions of gases such as carbon dioxide, methane, and chlorofluorocarbons (CFCs) have the potential to alter the climate of the earth and destroy the ozone layer. Concentrations of CFC-12, used in refrigeration technology, in parts per billion (ppb) can be modeled by the exponential function defined by

$$f(x) = .48e^{.04x},$$

where $x = 0$ represents 1990. Use this function to approximate the concentration in 1998.

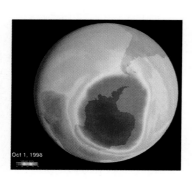

Oct 1, 1998

Since $x = 0$ represents 1990, $x = 8$ represents 1998. Evaluate $f(8)$ using a calculator.

$$f(8) = .48e^{.04(8)} = .48e^{.32} \approx .66$$

In 1998, the concentration of CFC-12 was about .66 ppb.

Now Try Exercise 53.

You have probably heard of the carbon 14 dating process used to determine the age of fossils. The method used involves a base e exponential decay function.

EXAMPLE 9   Solving an Exponential Decay Problem

Carbon 14 is a radioactive form of carbon that is found in all living plants and animals. After a plant or animal dies, the radioactive carbon 14 disintegrates according to the function defined by

$$y = y_0 e^{-.000121t},$$

where t is time in years, y is the amount of the sample at time t, and y_0 is the initial amount present at $t = 0$.

(a) If an initial sample contains $y_0 = 10$ g of carbon 14, how many grams will be present after 3000 yr?

Let $y_0 = 10$ and $t = 3000$ in the formula, and use a calculator.

$$y = 10e^{-.000121(3000)} \approx 6.96 \text{ g}$$

(b) How long would it take for the initial sample to decay to half of its original amount? (This is called the *half-life.*)

Let $y = \frac{1}{2}(10) = 5$, and solve for t.

$$5 = 10e^{-.000121t} \qquad \text{Substitute.}$$

$$\frac{1}{2} = e^{-.000121t} \qquad \text{Divide by 10.}$$

$$\ln \frac{1}{2} = -.000121t \qquad \text{Take natural logarithms; } \ln e^k = k.$$

$$t = \frac{\ln \frac{1}{2}}{-.000121} \qquad \text{Divide by } -.000121.$$

$$t \approx 5728 \qquad \text{Use a calculator.}$$

The half-life is just over 5700 yr.

Now Try Exercise 59.

OBJECTIVE 5 **Use a graphing calculator to solve exponential and logarithmic equations.** Earlier we saw that the *x*-intercepts of the graph of a function *f* correspond to the real solutions of the equation $f(x) = 0$. This idea was applied to linear and quadratic equations and can be extended to exponential and logarithmic equations as well. In Example 1, we solved the equation $3^x = 12$ algebraically using rules for logarithms and found the solution set to be $\{2.262\}$. This can be supported graphically by showing that the *x*-intercept of the graph of the function defined by $y = 3^x - 12$ corresponds to this solution. See Figure 18.

In Example 5, we solved $\log x + \log(x - 21) = 2$ and found the solution set to be $\{25\}$. (We rejected the apparent solution -4 since it led to the logarithm of a negative number.) Figure 19 shows that the *x*-intercept of the graph of the function defined by $y = \log x + \log(x - 21) - 2$ supports this result.

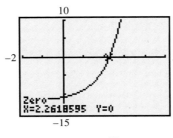

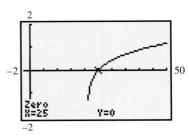

FIGURE 18   **FIGURE 19**

11.6 EXERCISES

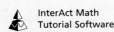

RELATING CONCEPTS (EXERCISES 1–4)

For Individual or Group Work

In Section 11.2 we solved an equation such as $5^x = 125$ by writing each side as a power of the same base, setting exponents equal, and then solving the resulting equation as follows.

$$5^x = 125 \qquad \text{Original equation}$$
$$5^x = 5^3 \qquad 125 = 5^3$$
$$x = 3 \qquad \text{Set exponents equal.}$$

Solution set: $\{3\}$

The method described in this section can also be used to solve this equation.
Work Exercises 1–4 in order, *to see how this is done.*

1. Take common logarithms on both sides, and write this equation.

2. Apply the power rule for logarithms on the left.

3. Write the equation so that *x* is alone on the left.

4. Use a calculator to find the decimal form of the solution. What is the solution set?

Many of the problems in the remaining exercises require a scientific calculator.

Solve each equation. Give solutions to three decimal places. See Example 1.

5. $7^x = 5$ **6.** $4^x = 3$ **7.** $9^{-x+2} = 13$

8. $6^{-t+1} = 22$ **9.** $3^{2x} = 14$ **10.** $5^{\cdot 3x} = 11$

11. $2^{x+3} = 5^x$ **12.** $6^{m+3} = 4^m$ **13.** $2^{x+3} = 3^{x-4}$

Solve each equation. Use natural logarithms. When appropriate, give solutions to three decimal places. See Example 2.

14. $e^{.006x} = 30$ **15.** $e^{.012x} = 23$ **16.** $e^{-.103x} = 7$

17. $e^{-.205x} = 9$ **18.** $\ln e^x = 4$ **19.** $\ln e^{3x} = 9$

20. $\ln e^{.04x} = \sqrt{3}$ **21.** $\ln e^{.45x} = \sqrt{7}$ **22.** $\ln e^{2x} = \pi$

23. Try solving one of the equations in Exercises 14–17 using common logarithms rather than natural logarithms. (You should get the same solution.) Explain why using natural logarithms is a better choice.

24. If you were asked to solve $10^{.0025x} = 75$, would natural or common logarithms be a better choice? Explain.

Solve each equation. Give the exact solution. See Example 3.

25. $\log_3(6x + 5) = 2$ **26.** $\log_5(12x - 8) = 3$ **27.** $\log_2(2x - 1) = 5$

28. $\log_6(4x + 2) = 2$ **29.** $\log_7(x + 1)^3 = 2$ **30.** $\log_4(x - 3)^3 = 4$

31. Suppose that in solving a logarithmic equation having the term $\log(x - 3)$ you obtain an apparent solution of 2. All algebraic work is correct. Explain why you must reject 2 as a solution of the equation.

32. Suppose that in solving a logarithmic equation having the term $\log(3 - x)$ you obtain an apparent solution of -4. All algebraic work is correct. Should you reject -4 as a solution of the equation? Explain why or why not.

Solve each equation. Give exact solutions. See Examples 4 and 5.

33. $\log(6x + 1) = \log 3$

34. $\log(7 - x) = \log 12$

35. $\log_5(3t + 2) - \log_5 t = \log_5 4$

36. $\log_2(x + 5) - \log_2(x - 1) = \log_2 3$

37. $\log 4x - \log(x - 3) = \log 2$

38. $\log(-x) + \log 3 = \log(2x - 15)$

39. $\log_2 x + \log_2(x - 7) = 3$

40. $\log(2x - 1) + \log 10x = \log 10$

41. $\log 5x - \log(2x - 1) = \log 4$

42. $\log_3 x + \log_3(2x + 5) = 1$

43. $\log_2 x + \log_2(x - 6) = 4$

44. $\log_2 x + \log_2(x + 4) = 5$

Solve each problem. See Examples 6 and 7.

45. (a) How much money will there be in an account at the end of 6 yr if $2000 is deposited at 4% compounded quarterly? (Assume no withdrawals are made.)
(b) To one decimal place, how long will it take for the account to grow to $3000?

46. (a) How much money will there be in an account at the end of 7 yr if $3000 is deposited at 3.5% compounded quarterly? (Assume no withdrawals are made.)
(b) To one decimal place, when will the account grow to $5000?

47. (a) What will be the amount A in an account with initial principal $4000 if interest is compounded continuously at an annual rate of 3.5% for 6 yr?
(b) How long will it take for the initial amount to double?

48. Refer to Exercise 46. Does the money grow to a larger value under those conditions, or when invested for 7 yr at 3% compounded continuously?

49. Find the amount of money in an account after 12 yr if $5000 is deposited at 7% annual interest compounded as follows.

(a) Annually   (b) Semiannually   (c) Quarterly
(d) Daily (Use $n = 365$.)   (e) Continuously

50. How much money will be in an account at the end of 8 yr if $4500 is deposited at 6% annual interest compounded as follows?

(a) Annually   (b) Semiannually   (c) Quarterly
(d) Daily (Use $n = 365$.)   (e) Continuously

51. How much money must be deposited today to amount to $1850 in 40 yr at 6.5% compounded continuously?

52. How much money must be deposited today to amount to $1000 in 10 yr at 5% compounded continuously?

Solve each problem. See Examples 8 and 9.

53. The total expenditures in millions of current dollars for pollution abatement and control during the period from 1985 through 1993 can be approximated by the function defined by

$$P(x) = 70,967e^{.0526x},$$

where $x = 0$ corresponds to 1985, $x = 1$ to 1986, and so on. Approximate the expenditures for each year. (*Source:* U.S. Bureau of Economic Analysis, *Survey of Current Business, May 1995.*)

(a) 1987   (b) 1990   (c) 1993
(d) What were the approximate expenditures for 1985?

54. The emission of the greenhouse gas nitrous oxide increased yearly during the first half of the 1990s. Based on figures during the period from 1990 through 1994, the emissions in thousands of metric tons can be modeled by the function defined by

$$N(x) = 446.5e^{.0118x},$$

where $x = 0$ corresponds to 1990, $x = 1$ to 1991, and so on. Approximate the emissions for each year. (*Source:* U.S. Energy Information Administration, *Emission of Greenhouse Gases in the United States, annual.*)

(a) 1991   (b) 1992   (c) 1994
(d) What were the approximate emissions in 1990?

55. Based on selected figures obtained during the 1980s and 1990s, consumer expenditures on all types of books in the United States can be modeled by the function defined by

$$B(x) = 8768e^{.072x},$$

where $x = 0$ represents 1980, $x = 1$ represents 1981, and so on, and $B(x)$ is in millions of dollars. Approximate consumer expenditures for 1998. (*Source:* Book Industry Study Group.)

56. Based on selected figures obtained during the 1970s, 1980s, and 1990s, the total number of bachelor's degrees earned in the United States can be modeled by the function defined by

$$D(x) = 815,427e^{.0137x},$$

where $x = 1$ corresponds to 1971, $x = 10$ corresponds to 1980, and so on. Approximate the number of bachelor's degrees earned in 1994. (*Source:* U.S. National Center for Education Statistics.)

57. Suppose that the amount, in grams, of plutonium 241 present in a given sample is determined by the function defined by

$$A(t) = 2.00e^{-.053t},$$

where t is measured in years. Find the amount present in the sample after the given number of years.

(a) 4   (b) 10   (c) 20

(d) What was the initial amount present?

58. Suppose that the amount, in grams, of radium 226 present in a given sample is determined by the function defined by

$$A(t) = 3.25e^{-.00043t},$$

where t is measured in years. Find the amount present in the sample after the given number of years.

(a) 20   (b) 100   (c) 500

(d) What was the initial amount present?

59. A sample of 400 g of lead 210 decays to polonium 210 according to the function defined by

$$A(t) = 400e^{-.032t},$$

where t is time in years.

(a) How much lead will be left in the sample after 25 yr?

(b) How long will it take the initial sample to decay to half of its original amount?

60. The concentration of a drug in a person's system decreases according to the function defined by

$$C(t) = 2e^{-.125t},$$

where $C(t)$ is in appropriate units, and t is in hours.

(a) How much of the drug will be in the system after 1 hr?

(b) Find the time that it will take for the concentration to be half of its original amount.

61. Refer to Exercise 53. Assuming that the function continued to apply past 1993, in what year could we have expected total expenditures to have been 133,500 million dollars? (*Source:* U.S. Bureau of Economic Analysis, *Survey of Current Business, May 1995.*)

62. Refer to Exercise 54. Assuming that the function continued to apply past 1994, in what year could we have expected nitrous oxide emissions to have been 485 thousand metric tons? (*Source:* U.S. Energy Information Administration, *Emission of Greenhouse Gases in the United States, annual.*)

63. The number of ants in an anthill grows according to the function defined by

$$f(t) = 300e^{.4t},$$

where t is time measured in days. Find the time it will take for the number of ants to double.

TECHNOLOGY INSIGHTS (EXERCISES 64–67)

64. The function defined by $P(x) = 70,967e^{.0526x}$, described in Exercise 53, is graphed in the screen at the right. Interpret the meanings of X and Y in the display at the bottom of the screen in the context of Exercise 53.

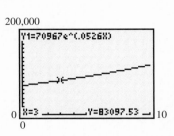

65. The function defined by $A(x) = 3.25e^{-.00043x}$, with $x = t$, described in Exercise 58, is graphed in the following figure. Interpret the meanings of X and Y in the display at the bottom of the screen in the context of Exercise 58.

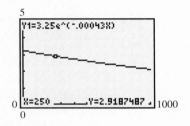

66. The screen shows a table of selected values for the function defined by $Y_1 = \left(1 + \frac{1}{X}\right)^X$.

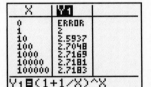

(a) Why is there an error message for X = 0?
(b) What number does the function value seem to approach as X takes on larger and larger values?
(c) Use a calculator to evaluate this function for X = 1,000,000. What value do you get? Now evaluate $e = e^1$. How close are these two values?
(d) Make a conjecture: As the values of x approach infinity, the value of $\left(1 + \frac{1}{x}\right)^x$ approaches _____.

67. Here is another property of logarithms: For $b > 0$, $x > 0$, $b \neq 1$, $x \neq 1$,

$$\log_b x = \frac{1}{\log_x b}.$$

Now observe the following calculator screen.

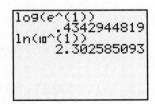

(a) Without using a calculator, give a decimal representation for $\frac{1}{.4342944819}$. Then support your answer using the reciprocal key of your calculator.
(b) Without using a calculator, give a decimal representation for $\frac{1}{2.302585093}$. Then support your answer using the reciprocal key of your calculator.

Chapter **11** **Group Activity**

How Much Space Do We Need?

OBJECTIVE Use natural logarithms and exponential equations to calculate how long it will take to fully populate Earth with people.

Applications of exponential growth and decay were introduced in this chapter. Using function notation, the formula for exponential population growth is

$$P(t) = P_0 e^{kt},$$

where $P(t)$ is population after t yr, P_0 is initial population, k is annual growth rate, and t is number of years elapsed.

A. If Earth's population will double in 30 yr at the current growth rate, what is this growth rate? (Express your answer as a percent.)

B. Earth has a total surface area of approximately 5.1×10^{14} m^2. Seventy percent of this surface area is rock, ice, sand, and open ocean. Another 8% of the total surface area, made up of tundra, lakes and streams, continental shelves, algae beds and reefs, and estuaries, is unfit for living space. The remaining area is suitable for growing food and for living space.

   1. Determine the surface area available for growing food.

   2. Determine the surface area available for living space. Notice that the surface area available for living space is also considered available for growing food.

C. Suppose that each person needs 100 m^2 of Earth's surface for living space. Use 5.5×10^9 for Earth's population.

   1. If none of the surface area available for living space is used for food, how long will it take for the livable surface of Earth to be covered with people? (Use the growth rate you found in part A and the surface area you found in part B2.)

   2. How much surface area would be left to grow food?

D. Measure a space that is 1 m^2 in area. Discuss with your partner whether or not you would want to be packed this closely together on Earth. Take into account that many people live in high-rise apartment buildings and how that translates into surface area used per person.

E. Now suppose that for each person 100 m^2 of Earth's surface is needed for living space and growing food.

1. Using the same population and growth rate as in part C, determine how long it will take to fill Earth with people. (Use the surface area from part B2.)

2. Does 100 m^2 per person for living space and growing food seem reasonable? Consider the following questions in your discussion.

- How much space do you think it takes to raise animals for food? To grow grains, nuts, fruits, and vegetables?
- Would food grow as well in desert areas, mountainous areas, or jungle areas?
- Would there be any space left for wild animals or natural plant life?
- Would there be any space left for shopping malls, movie theaters, concert halls, factories, office buildings, or parking lots?

3. Write a paragraph summarizing your results and your discussion.

CHAPTER **11** SUMMARY

KEY TERMS

11.1 one-to-one function
inverse of a function

11.2 exponential function
asymptote
exponential equation

11.3 logarithm
logarithmic equation
logarithmic function
with base a

11.5 common logarithm
natural logarithm

NEW SYMBOLS

$f^{-1}(x)$ the inverse of $f(x)$

$\log_a x$ the logarithm of x to the base a

$\log x$ common (base 10) logarithm of x

$\ln x$ natural (base e) logarithm of x

e a constant, approximately 2.7182818

TEST YOUR WORD POWER

See how well you have learned the vocabulary in this chapter. Answers, with examples, follow the Quick Review.

1. In a **one-to-one function**
   A. each x-value corresponds to only one y-value
   B. each x-value corresponds to one or more y-values
   C. each x-value is the same as each y-value
   D. each x-value corresponds to only one y-value and each y-value corresponds to only one x-value.

2. If f is a one-to-one function, then the **inverse** of f is
   A. the set of all solutions of f
   B. the set of all ordered pairs formed by interchanging the coordinates of the ordered pairs of f
   C. the set of all ordered pairs that are the opposite (negative) of the coordinates of the ordered pairs of f

   D. an equation involving an exponential expression.

3. An **exponential function** is a function defined by an expression of the form
   A. $f(x) = ax^2 + bx + c$ for real numbers a, b, c $(a \neq 0)$
   B. $f(x) = \log_a x$ for positive numbers a and x $(a \neq 1)$
   C. $f(x) = a^x$ for all real numbers x $(a > 0, a \neq 1)$
   D. $f(x) = \sqrt{x}$ for $x \geq 0$.

4. An **asymptote** is
   A. a line that a graph intersects just once
   B. a line that the graph of a function more and more closely approaches as the x-values increase or decrease

   C. the x-axis or y-axis
   D. a line about which a graph is symmetric.

5. A **logarithm** is
   A. an exponent
   B. a base
   C. an equation
   D. a polynomial.

6. A **logarithmic function** is a function that is defined by an expression of the form
   A. $f(x) = ax^2 + bx + c$ for real numbers a, b, c $(a \neq 0)$
   B. $f(x) = \log_a x$ for positive numbers a and x $(a \neq 1)$
   C. $f(x) = a^x$ for all real numbers x $(a > 0, a \neq 1)$
   D. $f(x) = \sqrt{x}$ for $x \geq 0$.

QUICK REVIEW

CONCEPTS	EXAMPLES

11.1 INVERSE FUNCTIONS

Horizontal Line Test

If any horizontal line intersects the graph of a function in no more than one point, then the function is one-to-one.

Find f^{-1} if $f(x) = 2x - 3$.
The graph of f is a straight line, so f is one-to-one by the horizontal line test.

CONCEPTS	EXAMPLES

Inverse Functions

For a one-to-one function f defined by an equation $y = f(x)$, the equation that defines the inverse function f^{-1} is found by interchanging x and y, solving for y, and replacing y with $f^{-1}(x)$.

To find $f^{-1}(x)$, interchange x and y in the equation $y = 2x - 3$.

$$x = 2y - 3$$

Solve for y to get $\qquad y = \dfrac{x + 3}{2}.$

Therefore, $\qquad f^{-1}(x) = \dfrac{x + 3}{2}.$

In general, the graph of f^{-1} is the mirror image of the graph of f with respect to the line $y = x$.

The graphs of a function f and its inverse f^{-1} are shown here.

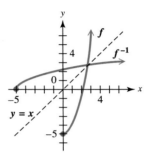

11.2　EXPONENTIAL FUNCTIONS

For $a > 0$, $a \neq 1$, $f(x) = a^x$ defines the exponential function with base a.

Graph of $F(x) = a^x$

1. The graph contains the point $(0, 1)$.
2. When $a > 1$, the graph rises from left to right. When $0 < a < 1$, the graph falls from left to right.
3. The x-axis is an asymptote.
4. The domain is $(-\infty, \infty)$; the range is $(0, \infty)$.

$F(x) = 3^x$ defines the exponential function with base 3.

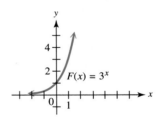

11.3　LOGARITHMIC FUNCTIONS

$y = \log_a x$ 　means　 $x = a^y.$

For $b > 0$, $b \neq 1$, $\log_b b = 1$ and $\log_b 1 = 0$.

For $a > 0$, $a \neq 1$, $x > 0$, $G(x) = \log_a x$ defines the logarithmic function with base a.

Graph of $G(x) = \log_a x$

1. The graph contains the point $(1, 0)$.
2. When $a > 1$, the graph rises from left to right. When $0 < a < 1$, the graph falls from left to right.
3. The y-axis is an asymptote.
4. The domain is $(0, \infty)$; the range is $(-\infty, \infty)$.

$y = \log_2 x$ means $x = 2^y$.

$$\log_3 3 = 1 \qquad \log_5 1 = 0$$

$G(x) = \log_3 x$ defines the logarithmic function with base 3.

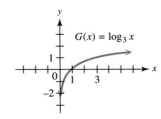

(continued)

CONCEPTS	EXAMPLES

11.4 PROPERTIES OF LOGARITHMS

Product Rule

$$\log_a xy = \log_a x + \log_a y$$

$$\log_2 3m = \log_2 3 + \log_2 m$$

Quotient Rule

$$\log_a \frac{x}{y} = \log_a x - \log_a y$$

$$\log_5 \frac{9}{4} = \log_5 9 - \log_5 4$$

Power Rule

$$\log_a x^r = r \log_a x$$

$$\log_{10} 2^3 = 3 \log_{10} 2$$

Special Properties

$$b^{\log_b x} = x \quad \text{and} \quad \log_b b^x = x$$

$$6^{\log_6 10} = 10 \qquad \log_3 3^4 = 4$$

11.5 COMMON AND NATURAL LOGARITHMS

Common logarithms (base 10) are used in applications such as pH, sound level, and intensity of an earthquake. Use the (LOG) key of a calculator to evaluate common logarithms.

Use the formula $\text{pH} = -\log[H_3O^+]$ to find the pH (to one decimal place) of grapes with hydronium ion concentration 5.0×10^{-5}.

$$\begin{aligned} \text{pH} &= -\log(5.0 \times 10^{-5}) && \text{Substitute.} \\ &= -(\log 5.0 + \log 10^{-5}) && \text{Property of logarithms} \\ &\approx 4.3 && \text{Evaluate.} \end{aligned}$$

Natural logarithms (base e) are often used in applications of growth and decay, such as time for money invested to double, decay of chemical compounds, and biological growth. Use the (ln x) key or both the (INV) and (e^x) keys to evaluate natural logarithms.

Use the formula for doubling time (in years) $t = \frac{\ln 2}{\ln(1 + r)}$ to find the doubling time to the nearest tenth at an interest rate of 4%.

$$\begin{aligned} t &= \frac{\ln 2}{\ln(1 + .04)} && \text{Substitute.} \\ &\approx 17.7 && \text{Evaluate.} \end{aligned}$$

The doubling time is about 17.7 yr.

Change-of-Base Rule

If $a > 0$, $a \neq 1$, $b > 0$, $b \neq 1$, $x > 0$, then

$$\log_a x = \frac{\log_b x}{\log_b a}.$$

$$\log_3 17 = \frac{\ln 17}{\ln 3} = \frac{\log 17}{\log 3} \approx 2.5789$$

11.6 EXPONENTIAL AND LOGARITHMIC EQUATIONS; FURTHER APPLICATIONS

To solve exponential equations, use these properties ($b > 0$, $b \neq 1$).

1. If $b^x = b^y$, then $x = y$.

Solve

$$2^{3x} = 2^5.$$
$$3x = 5$$
$$x = \frac{5}{3}$$

The solution set is $\left\{\frac{5}{3}\right\}$.

CONCEPTS	EXAMPLES

2. If $x = y$, $x > 0$, $y > 0$, then $\log_b x = \log_b y$.

Solve $5^m = 8$.

$$\log 5^m = \log 8$$
$$m \log 5 = \log 8$$
$$m = \frac{\log 8}{\log 5} \approx 1.2920$$

The solution set is $\{1.2920\}$.

To solve logarithmic equations, use these properties, where $b > 0$, $b \neq 1$, $x > 0$, $y > 0$. First use the properties of Section 11.4, if necessary, to write the equation in the proper form.

1. If $\log_b x = \log_b y$, then $x = y$.

Solve $\log_3 2x = \log_3(x + 1)$.

$$2x = x + 1$$
$$x = 1$$

The solution set is $\{1\}$.

2. If $\log_b x = y$, then $b^y = x$.

Solve $\log_2(3a - 1) = 4$.

$$3a - 1 = 2^4$$
$$3a - 1 = 16$$
$$3a = 17$$
$$a = \frac{17}{3}$$

The solution set is $\left\{\frac{17}{3}\right\}$.

Answers to Test Your Word Power

1. D; *Example:* The function $f = \{(0, 2), (1, -1), (3, 5), (-2, 3)\}$ is one-to-one. **2.** B; *Example:* The inverse of the one-to-one function f defined in Answer 1 is $f^{-1} = \{(2, 0), (-1, 1), (5, 3), (3, -2)\}$. **3.** C; *Examples:* $f(x) = 4^x$, $g(x) = \left(\frac{1}{2}\right)^x$, $h(x) = 2^{-x+3}$
4. B; *Example:* The graph of $F(x) = 2^x$ has the x-axis ($y = 0$) as an asymptote. **5.** A; *Example:* $\log_a x$ is the exponent to which a must be raised to obtain x; $\log_3 9 = 2$ since $3^2 = 9$. **6.** B; *Examples:* $y = \log_3 x$, $y = \log_{1/3} x$

CHAPTER **11** REVIEW EXERCISES

[11.1]   *Determine whether each graph is the graph of a one-to-one function.*

1.

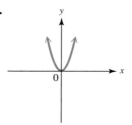

2.

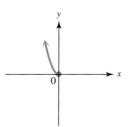

3. The table lists caffeine amounts in several popular 12-oz sodas. If the set of sodas is the domain and the set of caffeine amounts is the range of the function consisting of the six pairs listed, is it a one-to-one function? Why or why not?

Soda	Caffeine (mg)
Mountain Dew	55
Diet Coke	45
Dr. Pepper	41
Sunkist Orange Soda	41
Diet Pepsi-Cola	36
Coca-Cola Classic	34

Source: National Soft Drink Association.

Determine whether each function is one-to-one. If it is, find its inverse.

4. $f(x) = -3x + 7$ **5.** $f(x) = \sqrt[3]{6x - 4}$ **6.** $f(x) = -x^2 + 3$

Each function graphed is one-to-one. Graph its inverse.

7.

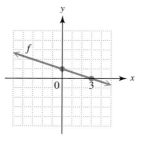

8.

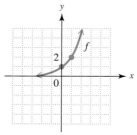

[11.2]   *Graph each function.*

9. $f(x) = 3^x$ **10.** $f(x) = \left(\dfrac{1}{3}\right)^x$ **11.** $y = 3^{x+1}$ **12.** $y = 2^{2x+3}$

Solve each equation.

13. $4^{3x} = 8^{x+4}$ **14.** $\left(\dfrac{1}{27}\right)^{x-1} = 9^{2x}$

15. The gross wastes generated in plastics, in millions of tons, from 1960 through 1990 can be approximated by the exponential function defined by

$$W(x) = .67(1.123)^x,$$

where $x = 0$ corresponds to 1960, $x = 5$ to 1965, and so on. Use this function to approximate the plastic waste amounts for each year. (*Source:* U.S. Environmental Protection Agency, *Characterization of Municipal Solid Waste in the United States: 1994 Update,* 1995.)

(a) 1965   **(b)** 1975   **(c)** 1990

[11.3]   *Graph each function.*

16. $g(x) = \log_3 x$ (*Hint:* See Exercise 9.)   **17.** $g(x) = \log_{1/3} x$ (*Hint:* See Exercise 10.)

Solve each equation.

18. $\log_8 64 = x$ **19.** $\log_2 \sqrt{8} = x$ **20.** $\log_x \left(\dfrac{1}{49} \right) = -2$

21. $\log_4 x = \dfrac{3}{2}$ **22.** $\log_k 4 = 1$ **23.** $\log_b b^2 = 2$

24. In your own words, explain the meaning of $\log_b a$.

25. Based on the meaning of $\log_b a$, what is the simplest form of $b^{\log_b a}$?

26. A company has found that total sales, in thousands of dollars, are given by the function defined by

$$S(x) = 100 \log_2(x + 2),$$

where x is the number of weeks after a major advertising campaign was introduced.

(a) What were the total sales 6 weeks after the campaign was introduced?

(b) Graph the function.

[11.4]   *Apply the properties of logarithms to express each logarithm as a sum or difference of logarithms. Assume all variables represent positive real numbers.*

27. $\log_2 3xy^2$ **28.** $\log_4 \dfrac{\sqrt{x} \cdot w^2}{z}$

Apply the properties of logarithms to write each expression as a single logarithm. Assume all variables represent positive real numbers, $b \neq 1$.

29. $\log_b 3 + \log_b x - 2 \log_b y$ **30.** $\log_3(x + 7) - \log_3(4x + 6)$

[11.5]   *Evaluate each logarithm. Give approximations to four decimal places.*

31. $\log 28.9$ **32.** $\log .257$ **33.** $\ln 28.9$ **34.** $\ln .257$

Use the change-of-base rule (with either common or natural logarithms) to find each logarithm. Give approximations to four decimal places.

35. $\log_{16} 13$ **36.** $\log_4 12$

Use the formula $pH = -\log[H_3O^+]$ *to find the* pH *of each substance with the given hydronium ion concentration.*

37. Milk, 4.0×10^{-7}

38. Crackers, 3.8×10^{-9}

39. If orange juice has pH 4.6, what is its hydronium ion concentration?

40. Suppose the quantity, measured in grams, of a radioactive substance present at time t is given by

$$Q(t) = 500e^{-.05t},$$

where t is measured in days. Find the quantity present at the following times.

  (a) $t = 0$ **(b)** $t = 4$

41. Section 11.5, Exercise 40 introduced the *doubling function* defined by

$$t(r) = \frac{\ln 2}{\ln(1 + r)},$$

that gives the number of years required to double your money when it is invested at interest rate r (in decimal form) compounded annually. How long does it take to double your money at each rate? Round answers to the nearest year.

  (a) 4%   **(b)** 6%   **(c)** 10%   **(d)** 12%

  ✍ **(e)** Compare each answer in parts (a)–(d) with these numbers:

$$\frac{72}{4}, \frac{72}{6}, \frac{72}{10}, \frac{72}{12}.$$

  What do you find?

42. The graph shows the percent change in commercial rents in California from 1992 through 1999. The percent change in rents is approximated by the logarithmic function defined by

$$g(x) = -650 + 143 \ln x,$$

where x represents the number of years since 1900.

  (a) Find $g(92)$ and $g(99)$.
  (b) Compare your results with the corresponding values from the graph.

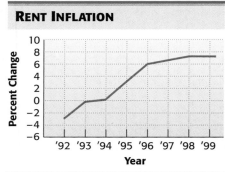

RENT INFLATION

Source: CB Commercial/Torto Wheaton Research.

[11.6]   *Solve each equation. Give solutions to three decimal places.*

43. $3^x = 9.42$

44. $2^{x-1} = 15$

45. $e^{.06x} = 3$

Solve each equation. Give exact solutions.

46. $\log_3(9x + 8) = 2$

47. $\log_5(y + 6)^3 = 2$

48. $\log_3(p + 2) - \log_3 p = \log_3 2$

49. $\log(2x + 3) = 1 + \log x$

50. $\log_4 x + \log_4(8 - x) = 2$

51. $\log_2 x + \log_2(x + 15) = \log_2 16$

📝 **52.** Explain the error in the following "solution" of the equation $\log x^2 = 2$.

$\log x^2 = 2$	Original equation
$2 \log x = 2$	Power rule for logarithms
$\log x = 1$	Divide both sides by 2.
$x = 10^1$	Write in exponential form.
$x = 10$	$10^1 = 10$

Solution set: $\{10\}$

Solve each problem. Use a calculator as necessary.

53. If $20,000 is deposited at 7% annual interest compounded quarterly, how much will be in the account after 5 yr, assuming no withdrawals are made?

54. How much will $10,000 compounded continuously at 6% annual interest amount to in 3 yr?

55. Which is a better plan?

> Plan A: Invest $1000 at 4% compounded quarterly for 3 yr
>
> Plan B: Invest $1000 at 3.9% compounded monthly for 3 yr

56. What is the half-life of the radioactive substance described in Exercise 40?

57. A machine purchased for business use *depreciates,* or loses value, over a period of years. The value of the machine at the end of its useful life is called its *scrap value.* By one method of depreciation (where it is assumed a constant percentage of the value depreciates annually), the scrap value, S, is given by

$$S = C(1 - r)^n,$$

where C is the original cost, n is the useful life in years, and r is the constant percent of depreciation.

(a) Find the scrap value of a machine costing $30,000, having a useful life of 12 yr and a constant annual rate of depreciation of 15%.

(b) A machine has a "half-life" of 6 yr. Find the constant annual rate of depreciation.

58. Recall from Exercise 39 in Section 11.5 that the number of years, $N(r)$, since two independently evolving languages split off from a common ancestral language is approximated by

$$N(r) = -5000 \ln r,$$

where r is the percent of words from the ancestral language common to both languages now. Find r if the split occurred 2000 yr ago.

59. Which one of the following is *not* equal to the solution of $7^x = 23$?

A. $\dfrac{\log 23}{\log 7}$ **B.** $\dfrac{\ln 23}{\ln 7}$ **C.** $\log_7 23$ **D.** $\log_{23} 7$

60. Consider the logarithmic equation

$$\log(2x + 3) = \log x + 1.$$

(a) Solve the equation using properties of logarithms.

📝 **(b)** If $Y_1 = \log(2X + 3)$ and $Y_2 = \log X + 1$, then the graph of $Y_1 - Y_2$ looks like that shown. Explain how the display at the bottom of the screen confirms the solution set found in part (a).

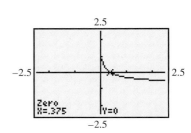

RELATING CONCEPTS (EXERCISES 61–72)

For Individual or Group Work

Work Exercises 61–72 in order, to see some of the relationships between exponential and logarithmic properties and functions.

61. Complete the table, and graph the function defined by $f(x) = 2^x$.

x	$f(x)$
-2	
-1	
0	
1	
2	
3	

62. Complete the table, and graph the function defined by $g(x) = \log_2 x$.

x	$g(x)$
$\frac{1}{4}$	
$\frac{1}{2}$	
1	
2	
4	
8	

63. What do you notice about the ordered pairs found in Exercises 61 and 62? What do we call the functions f and g in relationship to each other?

64. Fill in the blanks with the word *vertical* or *horizontal:* The graph of f in Exercise 61 has a _____ asymptote, while the graph of g in Exercise 62 has a _____ asymptote.

65. Using properties of exponents, $2^2 \cdot 2^3 = 2^{\frac{?}{}}$, because __?__ + __?__ = __?__ .

66. It is a fact that $32 = 4 \cdot 8$. Therefore, using properties of logarithms,
$$\log_2 32 = \log_2 \underline{} + \log_2 \underline{}.$$

67. Use the change-of-base rule to find an approximation for $\log_2 13$. Give as many digits as your calculator displays, and store this approximation in memory.

68. In your own words, explain what $\log_2 13$ means.

69. Simplify without using a calculator: $2^{\log_2 13}$.

70. Use the exponential key of your calculator to raise 2 to the power obtained in Exercise 67. What is the result? Why is this so?

71. Based on your result in Exercise 67, the point $(13, \underline{})$ lies on the graph of $g(x) = \log_2 x$.

72. Use the method of Section 11.2 to solve the equation $2^{x+1} = 8^{2x+3}$.

MIXED REVIEW EXERCISES

Evaluate.

73. $\log_2 128$

74. $5^{\log_5 36}$

75. $e^{\ln 4}$

76. $10^{\log e}$

77. $\log_3 3^{-5}$

78. $\ln e^{5.4}$

Solve.

79. $\log_3(x + 9) = 4$

80. $\ln e^x = 3$

81. $\log_x \dfrac{1}{81} = 2$

82. $27^x = 81$

83. $2^{2x-3} = 8$

84. $5^{x+2} = 25^{2x+1}$

85. $\log_3(x + 1) - \log_3 x = 2$ **86.** $\log(3x - 1) = \log 10$ **87.** $\ln(x^2 + 3x + 4) = \ln 2$

88. A small business estimates that the value of a copy machine is decreasing according to the function defined by

$$f(t) = 5000(2)^{-.15t},$$

where t is the number of years that have elapsed since the machine was purchased and $f(t)$ is in dollars.

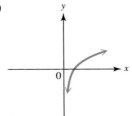

(a) What was the original value of the machine? (*Hint:* Find $f(0)$.)

(b) What is the value of the machine 5 yr after purchase? Give your answer to the nearest dollar.

(c) What is the value of the machine 10 yr after purchase? Give your answer to the nearest dollar.

89. Find the useful life of the machine in Exercise 57 if the scrap value is $10,000, the cost is $30,000, and the depreciation rate is 15%.

90. Based on selected figures from 1970 through 1995, the fractional part of the generation of municipal solid waste recovered can be approximated by the function defined by

$$R(x) = .0597e^{.0553x},$$

where $x = 0$ corresponds to 1970, $x = 10$ to 1980, and so on. Based on this model, what *percent* of municipal solid waste was recovered in 1990? (*Source:* Franklin Associates, Ltd., Prairie Village, KS, *Characterization of Municipal Solid Waste in the United States: 1995.*)

One measure of the diversity of the species in an ecological community is the index of diversity, *the logarithmic expression*

$$-(p_1 \ln p_1 + p_2 \ln p_2 + \cdots + p_n \ln p_n),$$

where $p_1, p_2, \ldots, p_n$ are the proportions of a sample belonging to each of n species in the sample. (Source: Ludwig, John and James Reynolds, Statistical Ecology: A Primer on Methods and Computing, New York, John Wiley and Sons, 1988.) Find the index of diversity to three decimal places if a sample of 100 from a community produces the following numbers.

91. 90 of one species, 10 of another

92. 60 of one species, 40 of another

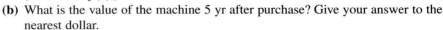

CHAPTER **11** TEST

1. Decide whether each function is one-to-one.

(a) $f(x) = x^2 + 9$ (b)

2. Find $f^{-1}(x)$ for the one-to-one function defined by $f(x) = \sqrt[3]{x + 7}$.

3. Graph the inverse of f, given the graph of f at the right.

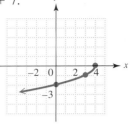

Graph each function.

4. $f(x) = 6^x$

5. $g(x) = \log_6 x$

6. Explain how the graph of the function in Exercise 5 can be obtained from the graph of the function in Exercise 4.

Solve each equation. Give the exact solution.

7. $5^x = \dfrac{1}{625}$

8. $2^{3x-7} = 8^{2x+2}$

9. A recent report predicts that the U.S. Hispanic population will increase from 26.7 million in 1995 to 96.5 million in 2050. (*Source:* U.S. Bureau of the Census.) Assuming an exponential growth pattern, the population is approximated by

$$f(x) = 26.7e^{.023x},$$

where x represents the number of years since 1995. Use this function to estimate the Hispanic population in each year.

(a) 2000   **(b)** 2010

10. Write in logarithmic form: $4^{-2} = .0625$.   **11.** Write in exponential form: $\log_7 49 = 2$.

Solve each equation.

12. $\log_{1/2} x = -5$

13. $x = \log_9 3$

14. $\log_x 16 = 4$

15. Fill in the blanks with the correct responses: The value of $\log_2 32$ is _____. This means that if we raise _____ to the _____ power, the result is _____.

Use properties of logarithms to write each expression as a sum or difference of logarithms. Assume variables represent positive real numbers.

16. $\log_3 x^2 y$

17. $\log_5 \left(\dfrac{\sqrt{x}}{yz} \right)$

Use properties of logarithms to write each expression as a single logarithm. Assume variables represent positive real numbers, $b \neq 1$.

18. $3 \log_b s - \log_b t$

19. $\dfrac{1}{4} \log_b r + 2 \log_b s - \dfrac{2}{3} \log_b t$

20. Use a calculator to approximate each logarithm to four decimal places.

(a) $\log 23.1$ **(b)** $\ln .82$

21. Use the change-of-base rule to express $\log_3 19$

   (a) in terms of common logarithms;   **(b)** in terms of natural logarithms;
   (c) correct to four decimal places.

22. Solve, giving the correct solution to four decimal places.

$$3^x = 78$$

23. Solve $\log_8(x + 5) + \log_8(x - 2) = 1$.

24. Suppose that $10,000 is invested at 4.5% annual interest, compounded quarterly. How much will be in the account in 5 yr if no money is withdrawn?

25. Suppose that $15,000 is invested at 5% annual interest, compounded continuously.

   (a) How much will be in the account in 5 yr if no money is withdrawn?
   (b) How long will it take for the initial principal to double?

CUMULATIVE REVIEW EXERCISES CHAPTERS 1–11

Let $S = \left\{-\frac{9}{4}, -2, -\sqrt{2}, 0, .6, \sqrt{11}, \sqrt{-8}, 6, \frac{30}{3}\right\}$. List the elements of S that are members of each set.

   1. Integers   **2.** Rational numbers

   3. Irrational numbers   **4.** Real numbers

Simplify each expression.

   5. $|-8| + 6 - |-2| - (-6 + 2)$ **6.** $-12 - |-3| - 7 - |-5|$

   7. $2(-5) + (-8)(4) - (-3)$

Solve each equation or inequality.

   8. $7 - (3 + 4a) + 2a = -5(a - 1) - 3$ **9.** $2m + 2 \le 5m - 1$

   10. $|2x - 5| = 9$ **11.** $|3p| - 4 = 12$

   12. $|3k - 8| \le 1$ **13.** $|4m + 2| > 10$

Graph.

   14. $5x + 2y = 10$ **15.** $-4x + y \le 5$

16. The graph indicates that timber harvests by Sierra Pacific Industries dropped from 17,716 acres in 1997 to 9733 acres in 1999.

   (a) Is this the graph of a function?
   (b) What is the slope of the line in the graph? Interpret the slope in the context of the timber harvests.

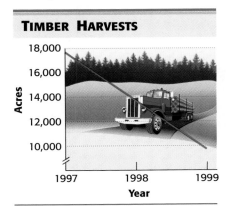

TIMBER HARVESTS

Source: Department of Forestry and Fire Protection.

17. Find an equation of the line through $(5, -1)$ and parallel to the line with equation $3x - 4y = 12$. Write the equation in slope-intercept form.

Solve each system.

18. $5x - 3y = 14$
$2x + 5y = 18$

19. $2x - 7y = 8$
$4x - 14y = 3$

20. $x + 2y + 3z = 11$
$3x - y + z = 8$
$2x + 2y - 3x = -12$

21. Candy worth $1.00 per lb is to be mixed with 10 lb of candy worth $1.96 per lb to get a mixture that will be sold for $1.60 per lb. How many pounds of the $1.00 candy should be used?

Number of Pounds	Price per Pound	Value
x	$1.00	$1x$
	$1.60	

Perform the indicated operations.

22. $(2p + 3)(3p - 1)$

23. $(4k - 3)^2$

24. $(3m^3 + 2m^2 - 5m) - (8m^3 + 2m - 4)$

25. Divide $6t^4 + 17t^3 - 4t^2 + 9t + 4$ by $3t + 1$.

Factor.

26. $8x + x^3$

27. $24y^2 - 7y - 6$

28. $5z^3 - 19z^2 - 4z$

29. $16a^2 - 25b^4$

30. $8c^3 + d^3$

31. $16r^2 + 56rq + 49q^2$

Perform the indicated operations.

32. $\dfrac{(5p^3)^4(-3p^7)}{2p^2(4p^4)}$

33. $\dfrac{x^2 - 9}{x^2 + 7x + 12} \div \dfrac{x - 3}{x + 5}$

34. $\dfrac{2}{k + 3} - \dfrac{5}{k - 2}$

35. $\dfrac{3}{p^2 - 4p} - \dfrac{4}{p^2 + 2p}$

Simplify.

36. $\sqrt{288}$

37. $2\sqrt{32} - 5\sqrt{98}$

38. Solve $\sqrt{2x + 1} - \sqrt{x} = 1$.

39. Multiply $(5 + 4i)(5 - 4i)$.

Solve each equation or inequality.

40. $3x^2 - x - 1 = 0$

41. $k^2 + 2k - 8 > 0$

42. $x^4 - 5x^2 + 4 = 0$

43. Find two numbers whose sum is 300 and whose product is a maximum.

44. Graph $f(x) = \dfrac{1}{3}(x - 1)^2 + 2$.

45. Graph $f(x) = 2^x$.

46. Solve $5^{x+3} = \left(\dfrac{1}{25}\right)^{3x+2}$.

47. Graph $f(x) = \log_3 x$.

48. Given that $\log_2 9 = 3.1699$, what is the value of $\log_2 81$?

49. Rewrite the following using the product, quotient, and power properties of logarithms.

$$\log \frac{x^3 \sqrt{y}}{z}$$

50. Let the number of bacteria present in a certain culture be given by

$$B(t) = 25{,}000e^{.2t},$$

where t is time measured in hours, and $t = 0$ corresponds to noon. Find, to the nearest hundred, the number of bacteria present at

(a) noon;   **(b)** 1 P.M.;   **(c)** 2 P.M.;
(d) When will the population double?

Polynomial and Rational Functions 12

12.1 Synthetic Division

12.2 Zeros of Polynomial Functions

12.3 Graphs and Applications of Polynomial Functions

Summary Exercises on Polynomial Functions and Graphs

12.4 Graphs and Applications of Rational Functions

The emergence of new diseases and drug-resistant strains of old ones has eliminated modern medicine's hope of eradicating infectious diseases. One such flu-like illness that scientists believe is a new form of the common cold virus is severe acute respiratory syndrome (SARS). SARS first appeared in 2003 in Asia and spread rapidly around the globe. To understand how these diseases spread, mathematicians and scientists analyze data reported to health officials and create mathematical models.

AIDS is a disease that has caused an estimated 21.8 million deaths worldwide since it first appeared in the late 1970s. (*Source:* UNAIDS, Joint United Nations Program on HIV/AIDS.) In Sections 12.3 and 12.4, we use polynomial and rational functions and their graphs to model the number of AIDS cases diagnosed and deaths from AIDS in the United States.

12.1 Synthetic Division

To prepare for later work in this chapter, we now examine a shortcut method for division of polynomials, called *synthetic division*. Polynomial division was first covered in Section 5.5, and you may wish to review it at this time.

OBJECTIVE 1 Know the meaning of the division algorithm. The following algorithm will be used in this chapter.

Division Algorithm

Let $f(x)$ and $g(x)$ be polynomials with $g(x)$ of lower degree than $f(x)$ and $g(x)$ of degree 1 or more. There exist unique polynomials $q(x)$ and $r(x)$ such that

$$f(x) = g(x) \cdot q(x) + r(x),$$

where either $r(x) = 0$ or the degree of $r(x)$ is less than the degree of $g(x)$.

In Example 4 of Section 5.5, we saw how to perform long division of two polynomials. Using x as the variable (rather than r), we found that the following is true:

$$\underbrace{6x^4 + 9x^3 + 2x^2 - 8x + 7}_{\substack{f(x) \\ \text{Dividend} \\ \text{(Original polynomial)}}} = \underbrace{(3x^2 - 2)}_{\substack{g(x) \\ \text{Divisor} \cdot}} \underbrace{(2x^2 + 3x + 2)}_{\substack{q(x) \\ \text{Quotient}}} + \underbrace{(-2x + 11)}_{\substack{r(x) \\ + \text{ Remainder}}}.$$

OBJECTIVE 2 Use synthetic division to divide by a polynomial of the form $x - k$. Often, when one polynomial is divided by a second, the second polynomial is of the form $x - k$, where the coefficient of the x-term is 1. To see how a shortcut for these divisions works, look first below left, where the division of $3x^3 - 2x + 5$ by $x - 3$ is shown. Notice that 0 was inserted for the missing x^2-term.

$$
\begin{array}{r}
3x^2 + 9x + 25 \\
x - 3\overline{)3x^3 + 0x^2 - 2x + 5} \\
\underline{3x^3 - 9x^2} \\
9x^2 - 2x \\
\underline{9x^2 - 27x} \\
25x + 5 \\
\underline{25x - 75} \\
80
\end{array}
\qquad
\begin{array}{r}
3 \quad 9 \quad 25 \\
1 - 3\overline{)3 \quad 0 \quad -2 \quad 5} \\
\underline{3 \quad -9} \\
9 \quad -2 \\
\underline{9 \quad -27} \\
25 \quad 5 \\
\underline{25 \quad -75} \\
80
\end{array}
$$

On the right, exactly the same division is shown written without the variables. This is why it is *essential* to use 0 as a placeholder in synthetic division. All the numbers in color on the right are repetitions of the numbers directly above them, so we omit them, as shown on the left at the top of the next page.

$$
\begin{array}{r}
3 \quad\;\; 9 \quad\; 25 \\
1-3\overline{)3 \quad\; 0 \;\; -2 \quad\; 5} \\
-9 \\
\overline{\quad 9 \;\; -2} \\
-27 \\
\overline{\quad\; 25 \quad\; 5} \\
-75 \\
\overline{\quad\quad 80}
\end{array}
\qquad
\begin{array}{r}
3 \quad\;\; 9 \quad\; 25 \\
1-3\overline{)3 \quad\; 0 \;\; -2 \quad\; 5} \\
-9 \\
\overline{\quad 9} \\
-27 \\
\overline{\quad\; 25} \\
-75 \\
\overline{\quad\quad 80}
\end{array}
$$

The numbers in color on the left are again repetitions of the numbers directly above them; they too are omitted, as shown on the right above.

Now we can condense the problem. If we bring the 3 in the dividend down to the beginning of the bottom row, the top row can be omitted, since it duplicates the bottom row.

$$
\begin{array}{r}
1-3\overline{)3 \quad\;\; 0 \quad\; -2 \quad\;\; 5} \\
-9 \;\; -27 \;\; -75 \\
\overline{3 \quad\; 9 \quad\;\; 25 \quad\;\; 80}
\end{array}
$$

Finally, we omit the 1 at the upper left. Also, to simplify the arithmetic, we replace subtraction in the second row by addition. To compensate for this, we change the -3 at the upper left to its additive inverse, 3. The final result is now shown.

Additive inverse $\rightarrow$ $3\overline{)3 \quad\;\; 0 \quad\; -2 \quad\;\; 5}$
$\qquad\qquad\qquad\quad\underline{\quad\;\; 9 \quad\;\; 27 \quad\;\; 75} \leftarrow$ Signs changed
$\qquad\qquad\qquad\;\; 3 \quad\; 9 \quad\;\; 25 \quad\;\; 80 \leftarrow$ Remainder
$\qquad\qquad\qquad\;\; \downarrow \quad\;\; \downarrow \quad\;\;\; \downarrow \qquad\; \downarrow$

The quotient is read $\qquad 3x^2 + 9x + 25 + \dfrac{80}{x-3}$
from the bottom row.

The first three numbers in the bottom row are the coefficients of the quotient polynomial with degree 1 less than the degree of the dividend. The last number gives the remainder.

Synthetic Division

This shortcut procedure is called **synthetic division.** It is used only when dividing a polynomial by a binomial of the form $x - k$.

▮ EXAMPLE 1 Using Synthetic Division

Use synthetic division to divide $5x^2 + 16x + 15$ by $x + 2$.

As mentioned, we use synthetic division only when dividing by a polynomial of the form $x - k$. Change $x + 2$ into this form by writing it as

$$x + 2 = x - (-2),$$

where $k = -2$. Now write the coefficients of $5x^2 + 16x + 15$, placing -2 to the left.

$x + 2$ leads to -2. $\rightarrow$ $-2\overline{)5 \quad 16 \quad 15}$ $\leftarrow$ Coefficients

Bring down the 5, and multiply: $-2 \cdot 5 = -10$.

$$-2\overline{)5 \quad\; 16 \quad 15}$$
$$\downarrow \quad -10$$
$$5$$

Add 16 and -10, getting 6, and multiply 6 by -2 to get -12.

$$-2\overline{)5 \quad\; 16 \quad\;\; 15}$$
$$-10 \quad -12$$
$$5 \quad\;\; 6$$

Add 15 and -12, getting 3.

$$-2\overline{)5 \quad\; 16 \quad\;\; 15}$$
$$-10 \quad -12$$
$$5 \quad\;\; 6 \quad\;\; 3$$

The result is read from the bottom row.

$$\frac{5x^2 + 16x + 15}{x + 2} = 5x + 6 + \frac{3}{x + 2}$$

Now Try Exercise 7.

The result of the division in Example 1 can be written as

$$5x^2 + 16x + 15 = (x + 2)(5x + 6) + 3$$

by multiplying both sides by the denominator $x + 2$. The following theorem is a generalization of the division process illustrated above.

For any polynomial $f(x)$ and any complex number k, there exists a unique polynomial $q(x)$ and number r such that

$$f(x) = (x - k)q(x) + r.$$

For example, in the synthetic division above,

$$\underbrace{5x^2 + 16x + 15}_{f(x)} = \underbrace{(x + 2)}_{(x - k)\,\cdot} \underbrace{(5x + 6)}_{q(x)} + \underbrace{3}_{+\; r}.$$

This theorem is a special case of the division algorithm given earlier. Here $g(x)$ is the first-degree polynomial $x - k$.

EXAMPLE 2 Using Synthetic Division with a Missing Term

Use synthetic division to find $(-4x^5 + x^4 + 6x^3 + 2x^2 + 50) \div (x - 2)$.

Use the steps given above, inserting 0 as coefficient for the missing x-term.

$$
\begin{array}{r|rrrrrr}
2) & -4 & 1 & 6 & 2 & 0 & 50 \\
   &   & -8 & -14 & -16 & -28 & -56 \\
\hline
   & -4 & -7 & -8 & -14 & -28 & -6
\end{array}
$$

Read the result from the bottom row.

$$
\frac{-4x^5 + x^4 + 6x^3 + 2x^2 + 50}{x - 2} = -4x^4 - 7x^3 - 8x^2 - 14x - 28 + \frac{-6}{x - 2}
$$

Now Try Exercise 13.

CAUTION To avoid errors, use 0 as coefficient for any missing terms, including a missing constant, when setting up the division.

OBJECTIVE 3 Use the remainder theorem to evaluate a polynomial. By the division algorithm, $f(x) = (x - k)q(x) + r$. This equality is true for all complex values of x, so it is true for $x = k$. Replacing x with k gives

$$
\begin{aligned}
f(k) &= (k - k)q(k) + r \\
f(k) &= r.
\end{aligned}
$$

This proves the following **remainder theorem,** which gives a new method of evaluating functions defined by polynomials.

Remainder Theorem

If the polynomial $f(x)$ is divided by $x - k$, then the remainder is equal to $f(k)$.

For example, in the synthetic division of Example 2, where the polynomial was divided by $x - 2$, the remainder was -6. Replacing x in the polynomial with 2 gives

$$
\begin{aligned}
-4x^5 + x^4 + 6x^3 + 2x^2 + 50 &= -4 \cdot 2^5 + 2^4 + 6 \cdot 2^3 + 2 \cdot 2^2 + 50 \\
&= -4 \cdot 32 + 16 + 6 \cdot 8 + 2 \cdot 4 + 50 \\
&= -128 + 16 + 48 + 8 + 50 \\
&= -6,
\end{aligned}
$$

the same number as the remainder; that is, dividing by $x - 2$ produced a remainder equal to the result when x is replaced with 2. Thus, a simpler way to find the value of a polynomial is often by using synthetic division. By the remainder theorem, instead of replacing x by 2 to find $f(2)$, divide $f(x)$ by $x - 2$ using synthetic division as in Example 2. Then $f(2)$ is the remainder, -6.

$$
\begin{array}{r|rrrrrr}
2) & -4 & 1 & 6 & 2 & 0 & 50 \\
   &   & -8 & -14 & -16 & -28 & -56 \\
\hline
   & -4 & -7 & -8 & -14 & -28 & -6 \quad \leftarrow f(2)
\end{array}
$$

EXAMPLE 3 Using the Remainder Theorem

Let $f(x) = 2x^3 - 5x^2 - 3x + 11$. Find $f(-2)$.

Use the remainder theorem; divide $f(x)$ by $x - (-2)$.

$$
\begin{array}{r}
\text{Value of } x \rightarrow \quad -2)\overline{2 \quad -5 \quad -3 \quad 11} \\
\underline{-4 \quad 18 \quad -30} \\
2 \quad -9 \quad 15 \quad -19 \leftarrow \text{Remainder}
\end{array}
$$

By this result, $f(-2) = -19$.

Now Try Exercise 27.

OBJECTIVE 4 Decide whether a given number is a zero of a polynomial function. The function defined by $f(x) = 2x^3 - 5x^2 - 3x + 11$ in Example 3 is an example of a *polynomial function.* We extend the definition given in Section 5.3 to include complex numbers as coefficients.

Polynomial Function

A **polynomial function of degree n** is a function defined by

$$f(x) = a_n x^n + a_{n-1} x^{n-1} + \cdots + a_1 x + a_0,$$

for complex numbers $a_n, a_{n-1}, \ldots, a_1$, and a_0, where $a_n \neq 0$.

We are often required to find values of x that make $f(x)$ equal 0. A **zero** of a polynomial function f is a value of k such that $f(k) = 0$.

The remainder theorem gives a quick way to decide if a number k is a zero of a polynomial function defined by $f(x)$. Use synthetic division to find $f(k)$; if the remainder is 0, then $f(k) = 0$ and k is a zero of $f(x)$. A zero of $f(x)$ is called a **root** or **solution** of the equation $f(x) = 0$.

EXAMPLE 4 Deciding Whether a Number is a Zero

Decide whether the given number is a zero of the given polynomial function.

(a) -5; $\quad f(x) = 2x^4 + 12x^3 + 6x^2 - 5x + 75$

Use synthetic division and the remainder theorem.

$$
\begin{array}{r}
\text{Proposed zero} \rightarrow \quad -5)\overline{2 \quad 12 \quad 6 \quad -5 \quad 75} \\
\underline{-10 \quad -10 \quad 20 \quad -75} \\
2 \quad 2 \quad -4 \quad 15 \quad 0 \leftarrow \text{Remainder}
\end{array}
$$

Since the remainder is 0, $f(-5) = 0$, and -5 is a zero of $f(x) = 2x^4 + 12x^3 + 6x^2 - 5x + 75$.

(b) -4; $\quad f(x) = x^4 + x^2 - 3x + 1$

Remember to use 0 as coefficient for the missing x^3-term in the synthetic division.

$$
\begin{array}{r}
-4)\overline{1 \quad 0 \quad 1 \quad -3 \quad 1} \\
\underline{-4 \quad 16 \quad -68 \quad 284} \\
1 \quad -4 \quad 17 \quad -71 \quad 285
\end{array}
$$

The remainder is not 0, so -4 is not a zero of $f(x) = x^4 + x^2 - 3x + 1$. In fact, $f(-4) = 285$.

(c) $1 + 2i$; $\quad f(x) = x^4 - 2x^3 + 4x^2 + 2x - 5$

Use synthetic division and operations with complex numbers.

$$
\begin{array}{r|rrrrr}
1 + 2i) & 1 & -2 & 4 & 2 & -5 \\
 & & 1 + 2i & -5 & -1 - 2i & 5 \\
\hline
 & 1 & -1 + 2i & -1 & 1 - 2i & 0
\end{array}
$$

Since the remainder is 0, $1 + 2i$ is a zero of the given polynomial function.

Now Try Exercises 41, 43, and 47.

The synthetic division in Example 4(a) shows that $x - (-5)$ divides the polynomial with 0 remainder. Thus $x - (-5) = x + 5$ is a *factor* of the polynomial and
$$2x^4 + 12x^3 + 6x^2 - 5x + 75 = (x + 5)(2x^3 + 2x^2 - 4x + 15).$$

The second factor is the quotient polynomial of degree $4 - 1 = 3$, whose coefficients are found in the last row of the synthetic division.

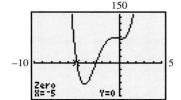

CONNECTIONS

The procedure in Example 4 is exactly how we use a graphing calculator to find zeros of a function by determining the x-intercepts of the graph. The screen shows the graph of $f(x) = 2x^4 + 12x^3 + 6x^2 - 5x + 75$ and shows that one value of x that makes $f(x) = 0$ is -5. This agrees with our result in Example 4(a).

For Discussion or Writing

Estimate the other x-intercept to the nearest tenth. Verify your answer using synthetic division.

12.1 EXERCISES

For Extra Help

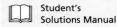
Student's
Solutions Manual

MyMathLab

InterAct Math
Tutorial Software

AW Math
Tutor Center

*Math*XP MathXL

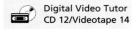
Digital Video Tutor
CD 12/Videotape 14

✐ **1.** What is the purpose of synthetic division?

2. What type of polynomial divisors may be used with synthetic division?

Use synthetic division to find each quotient. See Examples 1 and 2.

3. $\dfrac{x^2 - 6x + 5}{x - 1}$

4. $\dfrac{x^2 - 4x - 21}{x + 3}$

5. $\dfrac{4m^2 + 19m - 5}{m + 5}$

6. $\dfrac{3k^2 - 5k - 12}{k - 3}$

7. $\dfrac{2a^2 + 8a + 13}{a + 2}$

8. $\dfrac{4y^2 - 5y - 20}{y - 4}$

9. $(p^2 - 3p + 5) \div (p + 1)$

10. $(z^2 + 4z - 6) \div (z - 5)$

11. $\dfrac{4a^3 - 3a^2 + 2a - 3}{a - 1}$

12. $\dfrac{5p^3 - 6p^2 + 3p + 14}{p + 1}$

13. $(x^5 - 2x^3 + 3x^2 - 4x - 2) \div (x - 2)$

14. $(2y^5 - 5y^4 - 3y^2 - 6y - 23) \div (y - 3)$

15. $(-4r^6 - 3r^5 - 3r^4 + 5r^3 - 6r^2 + 3r + 3) \div (r - 1)$

16. $(2t^6 - 3t^5 + 2t^4 - 5t^3 + 6t^2 - 3t - 2) \div (t - 2)$

17. $(-3y^5 + 2y^4 - 5y^3 - 6y^2 - 1) \div (y + 2)$

18. $(m^6 + 2m^4 - 5m + 11) \div (m - 2)$

19. $\dfrac{y^3 + 1}{y - 1}$

20. $\dfrac{z^4 + 81}{z - 3}$

Express each polynomial function in the form $f(x) = (x - k)q(x) + r$ for the given value of k.

21. $f(x) = 2x^3 + x^2 + x - 8; \quad k = -1$

22. $f(x) = 2x^3 + 3x^2 - 16x + 10; \quad k = -4$

23. $f(x) = -x^3 + 2x^2 + 4; \quad k = -2$

24. $f(x) = -4x^3 + 2x^2 - 3x - 10; \quad k = 2$

25. $f(x) = 4x^4 - 3x^3 - 20x^2 - x; \quad k = 3$

26. $f(x) = 2x^4 + x^3 - 15x^2 + 3x; \quad k = -3$

For each polynomial function, use the remainder theorem and synthetic division to find $f(k)$. See Example 3.

27. $k = 3; \quad f(x) = x^2 - 4x + 5$ 28. $k = -2; \quad f(x) = x^2 + 5x + 6$

29. $k = 2; \quad f(x) = 2x^2 - 3x - 3$ 30. $k = 4; \quad f(x) = -x^3 + 8x^2 + 63$

31. $k = -1; \quad f(x) = x^3 - 4x^2 + 2x + 1$ 32. $k = 2; \quad f(x) = 2x^3 - 3x^2 - 5x + 4$

33. $k = 3; \quad f(x) = 2x^5 - 10x^3 - 19x^2 - 45$

34. $k = 4; \quad f(x) = x^4 + 6x^3 + 9x^2 + 3x - 3$

35. $k = -8; \quad f(x) = x^6 + 7x^5 - 5x^4 + 22x^3 - 16x^2 + x + 19$

36. $k = -\dfrac{1}{2}; \quad f(x) = 6x^3 - 31x^2 - 15x$

37. $k = 2 + i; \quad f(x) = x^2 - 5x + 1$

38. $k = 3 - 2i; \quad f(x) = x^2 - x + 3$

39. Explain why a 0 remainder in synthetic division of $f(x)$ by $x - k$ indicates that k is a solution of the equation $f(x) = 0$.

40. Explain why it is important to insert 0s as placeholders for missing terms before performing synthetic division.

Use synthetic division to decide whether the given number is a zero of the given polynomial function. See Example 4.

41. $3; \quad f(x) = 2x^3 - 6x^2 - 9x + 4$ 42. $-6; \quad f(x) = 2x^3 + 9x^2 - 16x + 12$

43. $-5; \quad f(x) = x^3 + 7x^2 + 10x$ 44. $-2; \quad f(x) = 2x^3 - 3x^2 - 5x$

45. $\dfrac{2}{5}; \quad f(x) = 5x^4 + 2x^3 - x + 15$ 46. $\dfrac{1}{2}; \quad f(x) = 2x^4 - 3x^2 + 4$

47. $2 - i; \quad f(x) = x^2 + 3x + 4$ 48. $1 - 2i; \quad f(x) = x^2 - 3x + 5$

RELATING CONCEPTS (EXERCISES 49–53)

For Individual or Group Work

In Section 5.5 we saw the close connection between polynomial division and writing a quotient of polynomials in lowest terms after factoring the numerator. Now we can show a connection between dividing one polynomial by another and factoring the first polynomial. Let

$$f(x) = 2x^2 + 5x - 12.$$

Work Exercises 49–53 in order.

49. Factor $f(x)$.

50. Solve $f(x) = 0$.

51. Find $f(-4)$ and $f\left(\frac{3}{2}\right)$.

52. Complete the following sentence:

If $f(a) = 0$, then $x -$ _____ is a factor of $f(x)$.

53. Use the conclusion in Exercise 52 to decide whether $x - 3$ is a factor of

$$f(x) = 3x^3 - 4x^2 - 17x + 6.$$

Factor $f(x)$ completely.

TECHNOLOGY INSIGHTS (EXERCISES 54–57)

Use the graph to determine a solution of each equation.

54. $2x^3 + 12x^2 + 24x + 16 = 0$

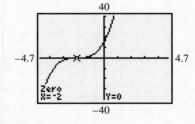

55. $x^3 - x^2 - 21x + 45 = 0$

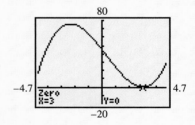

56. $x^3 + 3x^2 - 10x - 24 = 0$

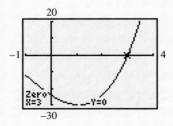

57. $x^3 + 3x^2 - 13x - 15 = 0$

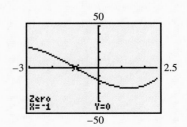

12.2 Zeros of Polynomial Functions

OBJECTIVES

1 Use the factor theorem.

2 Use the rational zeros theorem.

3 Find polynomial functions that satisfy given conditions.

4 Apply the conjugate zeros theorem.

In this section we build upon some of the ideas presented in the previous section to learn about finding zeros of polynomial functions.

OBJECTIVE 1 Use the factor theorem.   By the remainder theorem, if $f(k) = 0$, then the remainder when $f(x)$ is divided by $x - k$ is 0. This means that $x - k$ is a factor of $f(x)$. Conversely, if $x - k$ is a factor of $f(x)$, then $f(k)$ must equal 0. This is summarized in the following **factor theorem.**

> **Factor Theorem**
>
> The polynomial $x - k$ is a factor of the polynomial $f(x)$ if and only if $f(k) = 0$.

EXAMPLE 1   Deciding Whether $x - k$ Is a Factor of $f(x)$

Is $x - 1$ a factor of $f(x) = 2x^4 + 3x^2 - 5x + 7$?

   By the factor theorem, $x - 1$ will be a factor of $f(x)$ only if $f(1) = 0$. Use synthetic division and the remainder theorem to decide.

$$
\begin{array}{r|rrrrr}
1) & 2 & 0 & 3 & -5 & 7 \\
   &   & 2 & 2 & 5 & 0 \\
\hline
   & 2 & 2 & 5 & 0 & 7 \leftarrow \text{Remainder}
\end{array}
$$

Since the remainder is 7, $f(1) = 7$ and not 0, so $x - 1$ is not a factor of $f(x)$.

Now Try Exercise 9.

The factor theorem can be used to factor a polynomial of higher degree into linear factors. Linear factors are factors of the form $ax - b$.

EXAMPLE 2   Factoring a Polynomial Given a Zero

Factor $f(x) = 6x^3 + 19x^2 + 2x - 3$ into linear factors given that -3 is a zero of f.

   Since -3 is a zero of f, $x - (-3) = x + 3$ is a factor. Use synthetic division to divide $f(x)$ by $x + 3$.

$$
\begin{array}{r|rrrr}
-3) & 6 & 19 & 2 & -3 \\
    &   & -18 & -3 & 3 \\
\hline
    & 6 & 1 & -1 & 0
\end{array}
$$

The quotient is $6x^2 + x - 1$, so

$$
\begin{aligned}
f(x) &= (x + 3)(6x^2 + x - 1) \\
     &= (x + 3)(2x + 1)(3x - 1), \qquad \text{Factor } 6x^2 + x - 1.
\end{aligned}
$$

where all factors are linear.

Now Try Exercise 11.

OBJECTIVE 2 Use the rational zeros theorem.   The **rational zeros theorem** gives a method to determine all possible candidates for rational zeros of a polynomial function with integer coefficients.

Rational Zeros Theorem

Let $f(x) = a_n x^n + a_{n-1} x^{n-1} + \cdots + a_1 x + a_0$, where $a_n \neq 0$, define a polynomial function with integer coefficients. If $\frac{p}{q}$ is a rational number written in lowest terms, and if $\frac{p}{q}$ is a zero of f, then p is a factor of the constant term a_0, and q is a factor of the leading coefficient a_n.

PROOF $f\left(\frac{p}{q}\right) = 0$ since $\frac{p}{q}$ is a zero of $f(x)$, so

$$a_n\left(\frac{p}{q}\right)^n + a_{n-1}\left(\frac{p}{q}\right)^{n-1} + \cdots + a_1\left(\frac{p}{q}\right) + a_0 = 0.$$

This also can be written as

$$a_n\left(\frac{p^n}{q^n}\right) + a_{n-1}\left(\frac{p^{n-1}}{q^{n-1}}\right) + \cdots + a_1\left(\frac{p}{q}\right) + a_0 = 0.$$

Multiply both sides of this last result by q^n and add $-a_0 q^n$ to both sides.

$$a_n p^n + a_{n-1} p^{n-1} q + \cdots + a_1 p q^{n-1} = -a_0 q^n$$

Factoring out p gives

$$p(a_n p^{n-1} + a_{n-1} p^{n-2} q + \cdots + a_1 q^{n-1}) = -a_0 q^n.$$

This result shows that $-a_0 q^n$ equals the product of the two factors, p and $(a_n p^{n-1} + \cdots + a_1 q^{n-1})$. For this reason, p must be a factor of $-a_0 q^n$. Since it was assumed that $\frac{p}{q}$ is written in lowest terms, p and q have no common factor other than 1, so p is not a factor of q^n. Thus, p must be a factor of a_0. In a similar way, it can be shown that q is a factor of a_n.

▨ EXAMPLE 3   Using the Rational Zeros Theorem

Do each of the following for the polynomial function defined by

$$f(x) = 6x^4 + 7x^3 - 12x^2 - 3x + 2.$$

(a) List all possible rational zeros.

For a rational number $\frac{p}{q}$ to be zero, p must be a factor of $a_0 = 2$ and q must be a factor of $a_4 = 6$. Thus, p can be ± 1 or ± 2, and q can be ± 1, ± 2, ± 3, or ± 6. The possible rational zeros, $\frac{p}{q}$, are

$$\pm 1, \qquad \pm 2, \qquad \pm\frac{1}{2}, \qquad \pm\frac{1}{3}, \qquad \pm\frac{1}{6}, \qquad \pm\frac{2}{3}.$$

(b) Find all rational zeros and factor $f(x)$.

Use the remainder theorem to show that 1 and -2 are zeros.

```
1)6    7    -12    -3     2
       6     13     1    -2
   ─────────────────────────
   6   13      1    -2     0
```

The 0 remainder shows that 1 is a zero. Now, use the quotient polynomial

$6x^3 + 13x^2 + x - 2$ and synthetic division to find that -2 is also a zero.

$$
\begin{array}{r}
-2\,\overline{)\,6 \quad\ 13 \quad\ 1 \quad -2\ } \\
-12 \quad -2 \quad\ 2 \\
\hline
6 \quad\ 1 \quad -1 \quad\ 0
\end{array}
$$

The new quotient polynomial is $6x^2 + x - 1$. Factor to solve $6x^2 + x - 1 = 0$, obtaining $(3x - 1)(2x + 1) = 0$. The remaining two zeros are $\frac{1}{3}$ and $-\frac{1}{2}$.

Since the four zeros of $f(x) = 6x^4 + 7x^3 - 12x^2 - 3x + 2$ are 1, -2, $\frac{1}{3}$, and $-\frac{1}{2}$, the factors of $f(x)$ are $x - 1$, $x + 2$, $x - \frac{1}{3}$, and $x + \frac{1}{2}$, and

$$f(x) = a(x - 1)(x + 2)\left(x - \frac{1}{3}\right)\left(x + \frac{1}{2}\right)$$

$$f(x) = 6(x - 1)(x + 2)\left(x - \frac{1}{3}\right)\left(x + \frac{1}{2}\right) \qquad \text{The leading coefficient of } f(x) \text{ is } 6.$$

$$f(x) = (x - 1)(x + 2)(3)\left(x - \frac{1}{3}\right)(2)\left(x + \frac{1}{2}\right)$$

$$f(x) = (x - 1)(x + 2)(3x - 1)(2x + 1).$$

Now Try Exercise 25.

CAUTION The rational zeros theorem gives only *possible* rational zeros; it does not tell us whether these rational numbers are *actual* zeros. We must rely on other methods to determine whether they are indeed zeros. Furthermore, the function must have integer coefficients. To apply the rational zeros theorem to a polynomial with fractional coefficients, multiply through by the least common denominator of all the fractions. For example, any rational zeros of

$$p(x) = x^4 - \frac{1}{6}x^3 + \frac{2}{3}x^2 - \frac{1}{6}x - \frac{1}{3}$$

will also be rational zeros of

$$q(x) = 6x^4 - x^3 + 4x^2 - x - 2.$$

The function q was obtained by multiplying the terms of p by 6.

OBJECTIVE 3 **Find polynomial functions that satisfy given conditions.** The next theorem says that every function defined by polynomial of degree 1 or more has a zero, which means that every such polynomial can be factored. The theorem was first proved by Carl Friedrich Gauss (1777–1855) as part of his doctoral dissertation completed in 1799. Gauss's proof used advanced mathematical concepts outside the field of algebra.

Fundamental Theorem of Algebra

Every function defined by a polynomial of degree 1 or more has at least one complex zero.

From the fundamental theorem, if $f(x)$ is of degree 1 or more, then there is some number k_1 such that $f(k_1) = 0$. By the factor theorem,

$$f(x) = (x - k_1) \cdot q_1(x)$$

for some polynomial $q_1(x)$. If $q_1(x)$ is of degree 1 or more, the fundamental theorem and the factor theorem can be used to factor $q_1(x)$ in the same way. There is some number k_2 such that $q_1(k_2) = 0$, so that

$$q_1(x) = (x - k_2)q_2(x)$$

and
$$f(x) = (x - k_1)(x - k_2)q_2(x).$$

Assuming that $f(x)$ has degree n and repeating this process n times gives

$$f(x) = a(x - k_1)(x - k_2) \cdots (x - k_n),$$

where a is the leading coefficient of $f(x)$. Each of these factors leads to a zero of $f(x)$, so $f(x)$ has the n zeros $k_1, k_2, \ldots, k_n$. This result suggests the next theorem.

Number of Zeros Theorem

A function defined by a polynomial of degree n has at most n distinct zeros.

The theorem says that there exist *at most n* distinct zeros. For example, the polynomial function defined by $f(x) = x^3 + 3x^2 + 3x + 1 = (x + 1)^3$ is of degree 3 but has only one distinct zero, -1. Actually, the zero -1 occurs three times, since there are three factors of $x + 1$; this zero is called a **zero of multiplicity** 3.

■ EXAMPLE 4 Finding a Polynomial Function That Satisfies Given Conditions (Real Zeros)

Find a function f defined by a polynomial of degree 3 that satisfies the following conditions.

(a) Zeros of -1, 2, and 4; $f(1) = 3$

These three zeros give $x - (-1) = x + 1$, $x - 2$, and $x - 4$ as factors of $f(x)$. Since $f(x)$ is to be of degree 3, these are the only possible factors by the theorem just stated. Therefore, $f(x)$ has the form

$$f(x) = a(x + 1)(x - 2)(x - 4)$$

for some real number a. To find a, use the fact that $f(1) = 3$.

$$f(1) = a(1 + 1)(1 - 2)(1 - 4) = 3$$
$$a(2)(-1)(-3) = 3$$
$$6a = 3$$
$$a = \frac{1}{2}$$

Thus,
$$f(x) = \frac{1}{2}(x + 1)(x - 2)(x - 4)$$

or
$$f(x) = \frac{1}{2}x^3 - \frac{5}{2}x^2 + x + 4. \qquad \text{Multiply.}$$

(b) -2 is a zero of multiplicity 3; $f(-1) = 4$

The polynomial function defined by $f(x)$ has the form

$$f(x) = a(x + 2)(x + 2)(x + 2)$$
$$= a(x + 2)^3.$$

Since $f(-1) = 4$,

$$f(-1) = a(-1 + 2)^3 = 4$$
$$a(1)^3 = 4$$
$$a = 4,$$

and $f(x) = 4(x + 2)^3 = 4x^3 + 24x^2 + 48x + 32.$

Now Try Exercise 45.

NOTE   In Example 4(a), we cannot clear the denominators in $f(x)$ by multiplying through by 2, because the result would equal $2 \cdot f(x)$, not $f(x)$.

OBJECTIVE 4 Apply the conjugate zeros theorem.   The remainder theorem can be used to show that both $2 + i$ and $2 - i$ are zeros of $f(x) = x^3 - x^2 - 7x + 15$. In general, if $a + bi$ is a zero of a polynomial function with *real* coefficients, then so is $a - bi$. To prove this requires the following properties of complex conjugates. Let $z = a + bi$, and write $\bar{z}$ for the conjugate of z, so $\bar{z} = a - bi$. For example, if $z = -5 + 2i$, then $\bar{z} = -5 - 2i$. The proofs of these properties are left for the exercises. (See Exercises 57–60.)

Properties of Complex Conjugates

For any complex numbers c and d,

$$\overline{c + d} = \bar{c} + \bar{d}, \qquad \overline{c \cdot d} = \bar{c} \cdot \bar{d}, \qquad \text{and} \qquad \overline{c^n} = (\bar{c})^n.$$

We now show that if the complex number z is a zero of $f(x)$, then the conjugate of z is also a zero of $f(x)$.

PROOF   Start with the polynomial function defined by

$$f(x) = a_n x^n + a_{n-1} x^{n-1} + \cdots + a_1 x + a_0,$$

where all coefficients are real numbers. If $z = a + bi$ is a zero of $f(x)$, then

$$f(z) = a_n z^n + a_{n-1} z^{n-1} + \cdots + a_1 z + a_0 = 0.$$

Taking the conjugate of both sides of this last equation gives

$$\overline{a_n z^n + a_{n-1} z^{n-1} + \cdots + a_1 z + a_0} = \bar{0}.$$

Using generalizations of the properties $\overline{c + d} = \bar{c} + \bar{d}$, and $\overline{c \cdot d} = \bar{c} \cdot \bar{d}$ gives

$$\overline{a_n z^n} + \overline{a_{n-1} z^{n-1}} + \cdots + \overline{a_1 z} + \overline{a_0} = \bar{0}$$

or $$\overline{a_n}\, \overline{z^n} + \overline{a_{n-1}}\, \overline{z^{n-1}} + \cdots + \overline{a_1}\, \bar{z} + \overline{a_0} = \bar{0}.$$

Now use the third property above and the fact that for any real number a, $\bar{a} = a$, to obtain

$$a_n(\bar{z})^n + a_{n-1}(\bar{z})^{n-1} + \cdots + a_1(\bar{z}) + a_0 = 0.$$

Here $\bar{z}$ is also a zero of $f(x)$, completing the proof of the **conjugate zeros theorem.**

Conjugate Zeros Theorem

If $f(x)$ is a polynomial function *having only real coefficients* and if $a + bi$ is a zero of $f(x)$, where a and b are real numbers, then $a - bi$ is also a zero of $f(x)$.

CAUTION It is *essential* that the polynomial function have only real coefficients. For example, $f(x) = x - (1 + i)$ has $1 + i$ as a zero, but the conjugate $1 - i$ is not a zero.

EXAMPLE 5 **Finding a Polynomial Function That Satisfies Given Conditions (Complex Zeros)**

Find a polynomial function of lowest degree having only real coefficients and zeros 3 and $2 + i$.

The complex number $2 - i$ also must be a zero, so the polynomial function has at least three zeros, 3, $2 + i$, and $2 - i$. For the polynomial to be of lowest degree, these must be the only zeros. By the factor theorem there must be three factors, $x - 3$, $x - (2 + i)$, and $x - (2 - i)$. A polynomial function of lowest degree is

$$
\begin{aligned}
f(x) &= (x - 3)[x - (2 + i)][x - (2 - i)] \\
&= (x - 3)(x - 2 - i)(x - 2 + i) \\
&= x^3 - 7x^2 + 17x - 15.
\end{aligned}
$$

Polynomials such as $2(x^3 - 7x^2 + 17x - 15)$ or $\sqrt{5}(x^3 - 7x^2 + 17x - 15)$ also satisfy the given conditions on zeros. The information on zeros given in the problem is not enough to give a specific value for the leading coefficient.

Now Try Exercise 41.

The theorem on conjugate zeros can help predict the number of real zeros of polynomial functions with real coefficients. A polynomial function with real coefficients of odd degree n, where $n \geq 1$, must have at least one real zero (since zeros of the form $a + bi$, where $b \neq 0$, occur in conjugate pairs). On the other hand, a polynomial function with real coefficients of even degree n may have no real zeros.

EXAMPLE 6 **Finding All Zeros of a Polynomial Function Given One Zero**

Find all zeros of $f(x) = x^4 - 7x^3 + 18x^2 - 22x + 12$, given that $1 - i$ is a zero.

Since the polynomial function has only real coefficients and since $1 - i$ is a zero, by the conjugate zeros theorem $1 + i$ is also a zero. To find the remaining zeros, we first divide the original polynomial by $x - (1 - i)$.

$$
\begin{array}{r|rrrrr}
1 - i) & 1 & -7 & 18 & -22 & 12 \\
 & & 1 - i & -7 + 5i & 16 - 6i & -12 \\
\hline
 & 1 & -6 - i & 11 + 5i & -6 - 6i & 0
\end{array}
$$

By the factor theorem, since $x = 1 - i$ is a zero of $f(x)$, $x - (1 - i)$ is a factor, and

$f(x)$ can be written as

$$f(x) = [x - (1 - i)][x^3 + (-6 - i)x^2 + (11 + 5i)x + (-6 - 6i)].$$

We know that $x = 1 + i$ is also a zero of $f(x)$, so

$$f(x) = [x - (1 - i)][x - (1 + i)]q(x)$$

for some polynomial $q(x)$. Thus,

$$x^3 + (-6 - i)x^2 + (11 + 5i)x + (-6 - 6i) = [x - (1 + i)]q(x).$$

We use synthetic division to find $q(x)$.

$$
\begin{array}{r|rrrr}
1 + i & 1 & -6 - i & 11 + 5i & -6 - 6i \\
      &   & 1 + i & -5 - 5i & 6 + 6i \\
\hline
      & 1 & -5   & 6   & 0
\end{array}
$$

Since $q(x) = x^2 - 5x + 6$, $f(x)$ can be written as

$$f(x) = [x - (1 - i)][x - (1 + i)](x^2 - 5x + 6).$$

Factoring the quadratic polynomial $x^2 - 5x + 6$ shows that the factors are $x - 2$ and $x - 3$, which leads to the zeros 2 and 3. Thus, the four zeros of $f(x)$ are $1 - i$, $1 + i$, 2, and 3.

Now Try Exercise 17.

12.2 EXERCISES

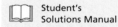

Decide whether each statement is true *or* false.

1. Given that $x - 1$ is a factor of $f(x) = x^6 - x^4 + 2x^2 - 2$, we are assured that $f(1) = 0$.

2. Given that $f(1) = 0$ for $f(x) = x^6 - x^4 + 2x^2 - 2$, we are assured that $x - 1$ is a factor of $f(x)$.

3. For the function defined by $f(x) = (x + 2)^4(x - 3)$, 2 is a zero of multiplicity 4.

4. Given that $2 + 3i$ is a zero of $f(x) = x^2 - 4x + 13$, we are assured that $2 - 3i$ is also a zero.

Use the factor theorem to decide whether the second polynomial is a factor of the first. See Example 1.

5. $4x^2 + 2x + 54$; $x - 4$

6. $5x^2 - 14x + 10$; $x + 2$

7. $x^3 + 2x^2 - 3$; $x - 1$

8. $2x^3 + x + 2$; $x + 1$

9. $2x^4 + 5x^3 - 2x^2 + 5x + 6$; $x + 3$

10. $5x^4 + 16x^3 - 15x^2 + 8x + 16$; $x + 4$

Factor $f(x)$ into linear factors given that k is a zero of $f(x)$. See Example 2.

11. $f(x) = 2x^3 - 3x^2 - 17x + 30$; $k = 2$

12. $f(x) = 2x^3 - 3x^2 - 5x + 6$; $k = 1$

13. $f(x) = 6x^3 + 13x^2 - 14x + 3$; $k = -3$

14. $f(x) = 6x^3 + 17x^2 - 63x + 10$; $k = -5$

For each polynomial function, one zero is given. Find all others. See Examples 2 and 6.

15. $f(x) = x^3 - x^2 - 4x - 6$;   3

16. $f(x) = x^3 + 4x^2 - 5$;   1

17. $f(x) = x^3 - 7x^2 + 17x - 15$; $2 - i$

18. $f(x) = 4x^3 + 6x^2 - 2x - 1$; $\dfrac{1}{2}$

19. $f(x) = x^4 + 5x^2 + 4$; $-i$

20. $f(x) = x^4 + 10x^3 + 27x^2 + 10x + 26$; i

*For each polynomial function (**a**) list all possible rational zeros, (**b**) find all rational zeros, and (**c**) factor $f(x)$. See Example 3.*

21. $f(x) = x^3 - 2x^2 - 13x - 10$

22. $f(x) = x^3 + 5x^2 + 2x - 8$

23. $f(x) = x^3 + 6x^2 - x - 30$

24. $f(x) = x^3 - x^2 - 10x - 8$

25. $f(x) = 6x^3 + 17x^2 - 31x - 12$

26. $f(x) = 15x^3 + 61x^2 + 2x - 8$

27. $f(x) = 12x^3 + 20x^2 - x - 6$

28. $f(x) = 12x^3 + 40x^2 + 41x + 12$

For each polynomial function, find all zeros and their multiplicities.

29. $f(x) = 7x^3 + x$

30. $f(x) = (x + 1)^2(x - 1)^3(x^2 - 10)$

31. $f(x) = 3(x - 2)(x + 3)(x^2 - 1)$

32. $f(x) = 5x^2(x + 1 - \sqrt{2})(2x + 5)$

33. $f(x) = (x^2 + x - 2)^5(x - 1 + \sqrt{3})^2$

34. $f(x) = (7x - 2)^3(x^2 + 9)^2$

For each of the following, find a polynomial function of lowest degree with only real coefficients and having the given zeros. See Examples 4 and 5.

35. $3 + i$ and $3 - i$

36. $7 - 2i$ and $7 + 2i$

37. $1 + \sqrt{2}, 1 - \sqrt{2},$ and 3

38. $1 - \sqrt{3}, 1 + \sqrt{3},$ and 1

39. $-2 + i, -2 - i, 3,$ and -3

40. $3 + 2i, -1,$ and 2

41. 2 and $3i$

42. -1 and $6 - 3i$

43. $1 + 2i, 2$ (multiplicity 2)

44. $2 + i, -3$ (multiplicity 2)

Find a polynomial function of degree 3 with only real coefficients that satisfies the given conditions. See Examples 4 and 5.

45. Zeros of $-3, 1,$ and 4; $f(2) = 30$

46. Zeros of $1, -1,$ and 0; $f(2) = 3$

47. Zeros of $-2, 1,$ and 0; $f(-1) = -1$

48. Zeros of $2, -3,$ and 5; $f(3) = 6$

49. Zeros of $5, i,$ and $-i;$ $f(2) = 5$

50. Zeros of $-2, i,$ and $-i;$ $f(-3) = 30$

RELATING CONCEPTS (EXERCISES 51–54)

For Individual or Group Work

 A calculator graph of the polynomial function defined by

$$f(x) = x^3 - 21x - 20$$

is shown here. Notice that there are three x-intercepts and thus three real zeros of the function. **Work Exercises 51–54 in order.**

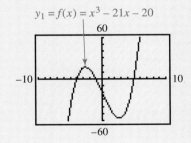

$y_1 = f(x) = x^3 - 21x - 20$

51. Given that $x + 4$ is a factor of $f(x)$, find the other factor and call it $g(x)$. Use synthetic division and the factor theorem.

52. Using the same window as shown above, graph both f and g on the same screen. What kind of function is g? What do you notice about the x-intercepts of the two functions in comparison to each other?

53. Given that $x - 5$ is a factor of $g(x)$, find the other factor and call it $h(x)$. Use either synthetic division and the factor theorem, or direct factorization.

54. Using the same window as shown above, graph both g and h on the same screen. What kind of function is h? What do you notice about the x-intercepts in comparison to each other?

Use the concepts of this section to work Exercises 55 and 56.

55. Show that -2 is a zero of multiplicity 2 of $f(x) = x^4 + 2x^3 - 7x^2 - 20x - 12$ and find all other complex zeros. Then write $f(x)$ in factored form.

56. Show that -1 is a zero of multiplicity 3 of $f(x) = x^5 - 4x^3 - 2x^2 + 3x + 2$ and find all other complex zeros. Then write $f(x)$ in factored form.

If c and d are complex numbers, prove each statement. (Hint: Let $c = a + bi$ and $d = m + ni$ and form all the conjugates, the sums, and the products.)

57. $\overline{c + d} = \overline{c} + \overline{d}$
58. $\overline{cd} = \overline{c} \cdot \overline{d}$

59. $\overline{a} = a$ for any real number a
60. $\overline{c^n} = (\overline{c})^n$

 Graphing calculators have the capability of finding (or approximating) real zeros using a "solver" feature. For example, the screen shown here indicates the three zeros of

$$Y_1 = f(X) = .5X^3 - 2.5X^2 + X + 4.$$

Since different makes and models of calculators require different keystrokes, consult your owner's manual to see how your particular model accomplishes equation solving.

```
solve(Y1,X,-5)
              -1
solve(Y1,X,1)
               2
solve(Y1,X,6)
               4
```

Shown here are –5, 1, and 6 as guesses that yield the three zeros, –1, 2, and 4.

Use a graphing calculator to find the real zeros of each function defined by $f(x)$. Use the "solver" feature. Express decimal approximations to the nearest hundredth.

61. $f(x) = .86x^3 - 5.24x^2 + 3.55x + 7.84$

62. $f(x) = -2.47x^3 - 6.58x^2 - 3.33x + .14$

63. $f(x) = 2.45x^4 - 3.22x^3 + .47x^2 - 6.54x + 3$

64. $f(x) = 4x^4 + 8x^3 - 4x^2 + 4x + 1$

65. $f(x) = -\sqrt{7}x^3 + \sqrt{5}x + \sqrt{17}$

66. $f(x) = \sqrt{10}x^3 - \sqrt{11}x - \sqrt{8}$

Descartes' rule of signs can help determine the number of positive and the number of negative real zeros of a polynomial function.

Descartes' Rule of Signs

Let $f(x)$ define a polynomial function with real coefficients and a nonzero constant term, with terms in descending powers of x.

(a) The number of positive real zeros of f either equals the number of variations in sign occurring in the coefficients of $f(x)$, or is less than the number of variations by a positive even integer.

(b) The number of negative real zeros of f either equals the number of variations in sign occurring in the coefficients of $f(-x)$, or is less than the number of variations by a positive even integer.

In the theorem, a *variation in sign* is a change from positive to negative or negative to positive in successive terms of the polynomial. Missing terms (those with 0 coefficients) are counted as no change in sign and can be ignored. For example, in the polynomial function defined by $f(x) = x^4 - 6x^3 + 8x^2 + 2x - 1$, $f(x)$ has three variations in sign:

$$+x^4 - 6x^3 + 8x^2 + 2x - 1.$$

Thus, by Descartes' rule of signs, f has either 3 or $3 - 2 = 1$ positive real zeros. Since

$$f(-x) = (-x)^4 - 6(-x)^3 + 8(-x)^2 + 2(-x) - 1$$
$$= x^4 + 6x^3 + 8x^2 - 2x - 1$$

has only one variation in sign, f has only one negative real zero.

Use Descartes' rule of signs to determine the possible number of positive real zeros and the possible number of negative real zeros for each function.

67. $f(x) = 2x^3 - 4x^2 + 2x + 7$

68. $f(x) = x^3 + 2x^2 + x - 10$

69. $f(x) = 5x^4 + 3x^2 + 2x - 9$

70. $f(x) = 3x^4 + 2x^3 - 8x^2 - 10x - 1$

71. $f(x) = x^5 + 3x^4 - x^3 + 2x + 3$

72. $f(x) = 2x^5 - x^4 + x^3 - x^2 + x + 5$

73. The polynomial function defined by $f(x) = x^3 - 6x^2 + 11x - 6$ has three integer zeros, one of which is 1. Explain how you would use synthetic division to find the complete factored form. Give the factored form.

74. Explain why $f(x) = x^3 + x^2 + x + 2$ cannot have any positive zeros.

12.3 | Graphs and Applications of Polynomial Functions

OBJECTIVES

1. Graph functions of the form $f(x) = a(x - h)^n + k$.
2. Graph general polynomial functions.
3. Use the intermediate value and boundedness theorems. (continued)

We have already discussed the graphs of polynomial functions of degree 0 to 2. In this section we show how to graph polynomial functions of degree 3 or more. The domains will be restricted to real numbers, since we will be graphing on the real number plane.

The concepts presented here allow you to understand the ideas of finding real zeros of polynomial functions. Once these ideas are mastered, you may wish to investigate the use of computers and graphing calculators to find real zeros of polynomial functions. Learning the methods of this section first will underscore the power of technology.

OBJECTIVES (continued)

4 Approximate real zeros of polynomial functions using a graphing calculator.

5 Solve applications using polynomial functions as models.

OBJECTIVE 1 Graph functions of the form $f(x) = a(x - h)^n + k$.

EXAMPLE 1 Graphing Functions of the Form $f(x) = ax^n$

Graph each function f defined as follows.

(a) $f(x) = x^3$

Choose several values for x, and find the corresponding values of $f(x)$, or y, as shown in the left table beside Figure 1. Plot the resulting ordered pairs and connect the points with a smooth curve. The graph of $f(x) = x^3$ is shown in blue in Figure 1.

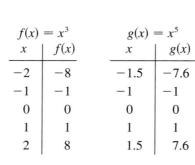

$f(x) = x^3$		$g(x) = x^5$	
x	$f(x)$	x	$g(x)$
-2	-8	-1.5	-7.6
-1	-1	-1	-1
0	0	0	0
1	1	1	1
2	8	1.5	7.6

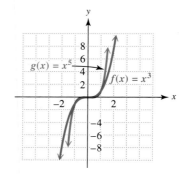

FIGURE 1

(b) $g(x) = x^5$

Work as in part (a) of this example to obtain the graph shown in red in Figure 1. Notice that the graphs of $f(x) = x^3$ and $g(x) = x^5$ are both symmetric with respect to the origin.

(c) $f(x) = x^4$, $g(x) = x^6$

Some typical ordered pairs for the graphs of $f(x) = x^4$ and $g(x) = x^6$ are given in the tables in Figure 2. These graphs are symmetric with respect to the y-axis, as is the graph of $f(x) = ax^2$ for a nonzero real number a.

$f(x) = x^4$		$g(x) = x^6$	
x	$f(x)$	x	$g(x)$
-2	16	-1.5	11.4
-1	1	-1	1
0	0	0	0
1	1	1	1
2	16	1.5	11.4

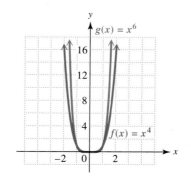

FIGURE 2

The ZOOM feature of a graphing calculator is useful with graphs like those in Example 1 to show the difference between the graphs of $y = x^3$ and $y = x^5$ and

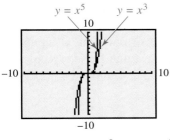

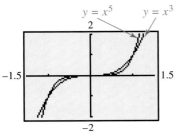

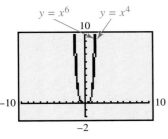

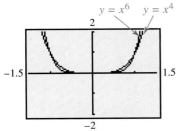

FIGURE 3

between $y = x^4$ and $y = x^6$ for values of x in the interval $[-1.5, 1.5]$. See Figure 3. In each case the first window shows x in $[-10, 10]$ and the second window shows x in $[-1.5, 1.5]$.

The value of a in $f(x) = ax^n$ determines the width of the graph. When $|a| > 1$, the graph is stretched vertically, making it narrower, while when $0 < |a| < 1$, the graph is shrunk or compressed vertically, so the graph is broader. The graph of $f(x) = -ax^n$ is reflected across the x-axis compared to the graph of $f(x) = ax^n$. ∎

Now Try Exercises 1 and 3.

Compared with the graph $f(x) = ax^n$, the graph of $f(x) = ax^n + k$ is translated (shifted) k units up if $k > 0$ and $|k|$ units down if $k < 0$. Also, the graph of $f(x) = a(x - h)^n$ is translated h units to the right if $h > 0$ and $|h|$ units to the left is $h < 0$, when compared with the graph of $f(x) = ax^n$.

The graph of $f(x) = a(x - h)^n + k$ shows a combination of these translations. The effects here are the same as those we saw earlier with quadratic functions in Chapter 10.

EXAMPLE 2 Examining Vertical and Horizontal Translations

Graph each function.

(a) $f(x) = x^5 - 2$

The graph will be the same as that of $f(x) = x^5$, but translated down 2 units. See Figure 4.

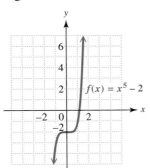

FIGURE 4

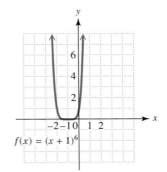

FIGURE 5

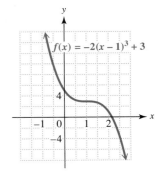

FIGURE 6

(b) $f(x) = (x + 1)^6$

This function f has a graph like that of $f(x) = x^6$, but since $x + 1 = x - (-1)$, it is translated 1 unit to the left as shown in Figure 5.

(c) $f(x) = -2(x - 1)^3 + 3$

The negative sign in -2 causes the graph to be reflected about the x-axis when compared with the graph of $f(x) = x^3$. Because $|-2| > 1$, the graph is stretched vertically as compared to the graph of $f(x) = x^3$. As shown in Figure 6, the graph is also translated 1 unit to the right and 3 units up. ∎

Now Try Exercises 5, 7, and 9.

OBJECTIVE 2 Graph general polynomial functions. The domain of every polynomial function is the set of all real numbers. The range of a polynomial function of odd degree is also the set of all real numbers. Some typical graphs of polynomial functions of odd degree are shown in Figure 7(a). These graphs suggest that for every polynomial function f of odd degree there is at least one real value of x that makes $f(x) = 0$. The zeros are the x-intercepts of the graph.

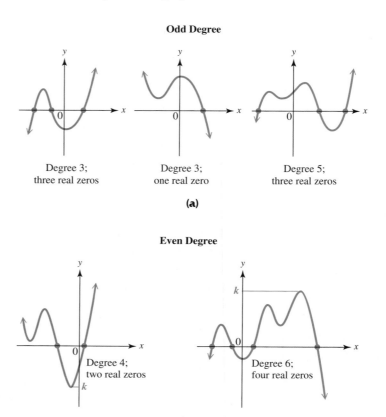

Odd Degree

Degree 3;
three real zeros

Degree 3;
one real zero

Degree 5;
three real zeros

(a)

Even Degree

Degree 4;
two real zeros

Degree 6;
four real zeros

(b)

FIGURE 7

A polynomial function of even degree will have a range of the form $(-\infty, k]$ or $[k, \infty)$ for some real number k. Figure 7(b) shows two typical graphs of polynomial functions of even degree.

As mentioned in Section 12.2, a zero k of a polynomial function has *multiplicity* that corresponds to the exponent of the factor $x - k$. For example, in

$$f(x) = (x - 1)^2(x + 4)^3,$$

the zero 1 has multiplicity 2 and the zero -4 has multiplicity 3. Determining whether a zero has even or odd multiplicity aids in sketching the graph near that zero. If the zero has odd multiplicity, the graph will cross the x-axis at the corresponding x-intercept. If the zero has even multiplicity, the graph will be tangent to the x-axis at the corresponding x-intercept (that is, it will touch but not cross the x-axis). See Figure 8.

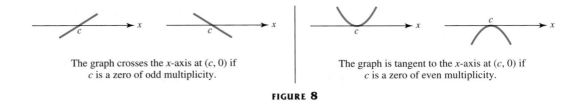

The graph crosses the x-axis at $(c, 0)$ if c is a zero of odd multiplicity.

The graph is tangent to the x-axis at $(c, 0)$ if c is a zero of even multiplicity.

FIGURE 8

The graphs in Figures 7(a) and 7(b) show that polynomial functions often have **turning points** where the function changes from increasing to decreasing or from decreasing to increasing.

Turning Points

A polynomial function of degree n has at most $n - 1$ turning points with at least one turning point between each pair of successive zeros.

The end behavior of a polynomial graph is determined by the term of highest degree. That is, a polynomial of the form $f(x) = a_n x^n + a_{n-1} x^{n-1} + \cdots + a_0$ has the same end behavior as the polynomial $f(x) = a_n x^n$. For instance, the polynomial $f(x) = 2x^3 - 8x^2 + 9$ has the same end behavior as $f(x) = 2x^3$. It is large and positive for large positive values of x and large and negative for negative values of x with large absolute value. The arrowheads at the ends of the graph look like those of the first graph in Figure 7(a); the right one points up and the left one points down. The end behavior of polynomial graphs is summarized in the following table.

End Behavior of the Graph of $f(x) = a_n x^n + a_{n-1} x^{n-1} + \cdots + a_0$

Degree	Sign of a_n	Left Arrow	Right Arrow	Example
Odd	Positive	Down	Up	First graph of Figure 7(a)
Odd	Negative	Up	Down	Second graph of Figure 7(a)
Even	Positive	Up	Up	First graph of Figure 7(b)
Even	Negative	Down	Down	Second graph of Figure 7(b)

We have discussed several characteristics of the graphs of polynomial functions that are useful in graphing the function. We now define what we mean by a *comprehensive graph* of a polynomial function.

Comprehensive Graph of a Polynomial Function

A **comprehensive graph** of a polynomial function will show the following characteristics.

1. all x-intercepts (zeros)
2. the y-intercept
3. all turning points
4. enough of the domain to show the end behavior

If the zeros of a polynomial function are known, its graph can be approximated without plotting very many points, as shown in the next example.

■ EXAMPLE 3 Graphing a Polynomial Function

Graph the polynomial function defined by

$$f(x) = 2x^3 + 5x^2 - x - 6,$$

given that 1 is a zero.

Since 1 is a zero, we know that $x - 1$ is a factor of $f(x)$. To find the remaining quadratic factor, use synthetic division.

$$
\begin{array}{r|rrrr}
1) & 2 & 5 & -1 & -6 \\
   &   & 2 & 7 & 6 \\
\hline
   & 2 & 7 & 6 & 0
\end{array}
$$

The final line indicates that $2x^2 + 7x + 6$ is a factor. This factors further as $(2x + 3)(x + 2)$, so the factored form of $f(x)$ is

$$f(x) = (x - 1)(2x + 3)(x + 2).$$

The three zeros are 1, $-\frac{3}{2}$, and -2. To sketch the graph, we note that these zeros divide the x-axis into four intervals:

$$(-\infty, -2), \qquad \left(-2, -\frac{3}{2}\right), \qquad \left(-\frac{3}{2}, 1\right), \qquad \text{and} \qquad (1, \infty).$$

These intervals are shown in Figure 9.

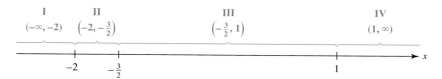

FIGURE 9

In any of these regions, the values of $f(x)$ are either always positive or always negative. To find the sign of $f(x)$ in each region, select an x-value in each region and substitute it into the expression for $f(x)$ to determine if the values of the function are positive or negative in that region. A typical selection of test points and the results of the tests are shown below.

Region	Test point	Value of f(x)	Sign of f(x)
I $(-\infty, -2)$	-3	-12	Negative
II $\left(-2, -\frac{3}{2}\right)$	$-\frac{7}{4}$	$\frac{11}{32}$	Positive
III $\left(-\frac{3}{2}, 1\right)$	0	-6	Negative
IV $(1, \infty)$	2	28	Positive

Plot the three zeros and the test points and join them with a smooth curve to get the graph. As expected, because each zero has odd multiplicity (that is, 1), the graph

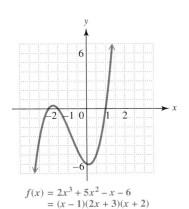

$f(x) = 2x^3 + 5x^2 - x - 6$
$= (x - 1)(2x + 3)(x + 2)$

(a)

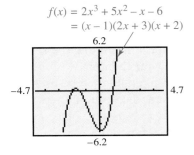

$f(x) = 2x^3 + 5x^2 - x - 6$
$= (x - 1)(2x + 3)(x + 2)$

(b)

FIGURE 10

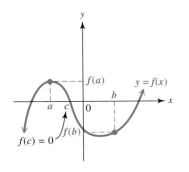

FIGURE 11

crosses the x-axis each time. The graph in Figure 10(a) shows that this function has two turning points, the maximum number for a third-degree polynomial function. When the values of $f(x)$ are negative, the graph is below the x-axis, and when $f(x)$ takes on positive values, the graph is above the x-axis, as shown in Figure 10(a). The sketch could be improved by plotting additional points in each region. Notice that the left arrowhead points down and the right one points up. This end behavior is correct since the first term is $2x^3$. Figure 10(b) shows how a graphing calculator graphs this function.

Now Try Exercise 33.

OBJECTIVE 3 Use the intermediate value and boundedness theorems. As Example 3 shows, a key to graphing a polynomial function is locating its zeros. In the special case where the zeros are rational numbers, the zeros can be found by the rational zeros theorem of Section 12.2. Occasionally, irrational zeros can be found by inspection. For instance, $f(x) = x^3 - 2$ has the irrational zero $\sqrt[3]{2}$. Two theorems presented in this section apply to the zeros of every polynomial function with real coefficients. The first theorem uses the fact that graphs of polynomial functions are unbroken curves, with no gaps or sudden jumps. The proof requires advanced methods, so it is not given here. Figure 11 illustrates the theorem.

Intermediate Value Theorem (as applied to locating zeros)

If $f(x)$ defines a polynomial function with only real coefficients, and if for real numbers a and b the values $f(a)$ and $f(b)$ are opposite in sign, then there exists at least one real zero between a and b.

This theorem helps to identify intervals where zeros of polynomial functions are located. If $f(a)$ and $f(b)$ are opposite in sign, then 0 is between $f(a)$ and $f(b)$, and there must be a number c between a and b where $f(c) = 0$.

CAUTION Be careful how you interpret the intermediate value theorem. If $f(a)$ and $f(b)$ are *not* opposite in sign, it does not necessarily mean that there is no zero between a and b. For example, in Figure 12, $f(a)$ and $f(b)$ are both negative, but -3 and -1, which are between a and b, are zeros of $f(x)$.

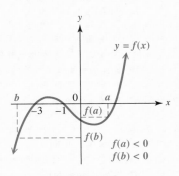

FIGURE 12

EXAMPLE 4 Using the Intermediate Value Theorem to Locate a Zero

Show that $f(x) = x^3 - 2x^2 - x + 1$ has a real zero between 2 and 3.

Use synthetic division to find $f(2)$ and $f(3)$.

$$\begin{array}{r|rrrr} 2) & 1 & -2 & -1 & 1 \\ & & 2 & 0 & -2 \\ \hline & 1 & 0 & -1 & -1 = f(2) \end{array} \qquad \begin{array}{r|rrrr} 3) & 1 & -2 & -1 & 1 \\ & & 3 & 3 & 6 \\ \hline & 1 & 1 & 2 & 7 = f(3) \end{array}$$

The results show that $f(2) = -1$ and $f(3) = 7$. Since $f(2)$ is negative but $f(3)$ is positive, by the intermediate value theorem there must be a real zero between 2 and 3.

Now Try Exercise 41(a).

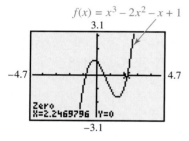

FIGURE 13

A graphing calculator can locate the zero established in Example 4. See Figure 13, which indicates that this zero is approximately 2.2469796. (Notice that there are two other zeros as well.)

The intermediate value theorem for polynomials is helpful in limiting the search for real zeros to smaller and smaller intervals. In Example 4 the theorem was used to verify that there is a real zero between 2 and 3. The theorem could then be used repeatedly to locate the zero more accurately. The next theorem, the **boundedness theorem,** shows how the bottom row of a synthetic division can be used to place upper and lower bounds on the possible real zeros of a polynomial function.

Boundedness Theorem

Let $f(x)$ be a polynomial function of degree $n \geq 1$ with real coefficients and with a positive leading coefficient. If $f(x)$ is divided synthetically by $x - c$ and

(a) if $c > 0$ and all numbers in the bottom row of the synthetic division are nonnegative, then $f(x)$ has no zero greater than c;

(b) if $c < 0$ and the numbers in the bottom row of the synthetic division alternate in sign (with 0 considered positive or negative, as needed), then $f(x)$ has no zero less than c.

EXAMPLE 5 Using the Boundedness Theorem

Show that the real zeros of $f(x) = 2x^4 - 5x^3 + 3x + 1$ satisfy the following conditions.

(a) No real zero is greater than 3.

Since $f(x)$ has real coefficients and the leading coefficient, 2, is positive, the boundedness theorem can be used. Divide $f(x)$ synthetically by $x - 3$.

$$\begin{array}{r|rrrrr} 3) & 2 & -5 & 0 & 3 & 1 \\ & & 6 & 3 & 9 & 36 \\ \hline & 2 & 1 & 3 & 12 & 37 \end{array}$$

Since $3 > 0$ and all numbers in the last row of the synthetic division are nonnegative, $f(x)$ has no real zero greater than 3.

(b) No real zero is less than -1.

Divide $f(x)$ by $x + 1$.

$$
\begin{array}{r}
-1\,\overline{)\,2 \quad -5 \quad 0 \quad 3 \quad 1\,} \\
-2 \quad 7 \quad -7 \quad 4 \\
\hline
2 \quad -7 \quad 7 \quad -4 \quad 5
\end{array}
$$

Here $-1 < 0$ and the numbers in the last row alternate in sign, so $f(x)$ has no zero less than -1.

Now Try Exercises 45 and 47.

OBJECTIVE 4 Approximate real zeros of polynomial functions using a graphing calculator.

EXAMPLE 6 Approximating Real Zeros of a Polynomial Function

Use a graphing calculator to approximate the real zeros of the function defined by

$$f(x) = x^4 - 6x^3 + 8x^2 + 2x - 1.$$

The greatest degree term is x^4, so the graph will have end behavior similar to the graph of $f(x) = x^4$, which is positive for all values of x with large absolute values. That is, the end behavior is upward at the left and the right. There are at most four real zeros, since the polynomial is fourth-degree.

Since $f(0) = -1$, the y-intercept is $(0, -1)$. Because the end behavior is positive on the left and the right and the y-value of the y-intercept is negative, by the intermediate value theorem f has at least one zero on either side of $x = 0$. The calculator graph in Figure 14 supports these facts. We can see that there are four zeros, and the table in Figure 14 indicates that they are between -1 and 0, 0 and 1, 2 and 3, and 3 and 4, because there is a change of sign in $f(x)$ in each case.

Using the capability of the calculator, we can find the zeros to a great degree of accuracy. Figure 15 shows that the negative zero is approximately $-.4142136$. To the nearest hundredth, the four zeros are $-.41$, $.27$, 2.24, and 3.73.

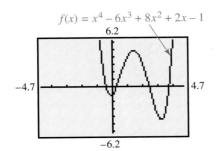

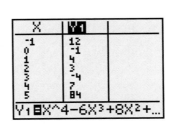

FIGURE 14

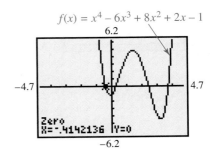

FIGURE 15

Now Try Exercise 53.

OBJECTIVE 5 Solve applications using polynomial functions as models. In Section 10.3, we saw that a quadratic function of the form $f(x) = ax^2 + bx + c$ can be written in the form $f(x) = a(x - h)^2 + k$ by completing the square. This latter form allows us to identify the vertex of the parabola, (h, k). We use this fact in the next example.

■ EXAMPLE 7   Finding a Polynomial Model

The table lists the total (cumulative) number of AIDS cases diagnosed in the United States up to 1993. For example, a total of 22,620 AIDS cases were diagnosed between 1981 and 1985.

Year	AIDS Cases	Year	AIDS Cases
1982	1563	1988	105,489
1983	4647	1989	147,170
1984	10,845	1990	193,245
1985	22,620	1991	248,023
1986	41,662	1992	315,329
1987	70,222	1993	361,508

Source: U.S. Dept. of Health and Human Services, Centers for Disease Control and Prevention, *HIV/AIDS Surveillance*, March 1994.

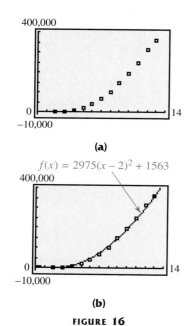

(a)

$$f(x) = 2975(x - 2)^2 + 1563$$

(b)

FIGURE 16

Figure 16(a) shows these data plotted using a graphing calculator, with $x = 2$ corresponding to 1982, $x = 3$ corresponding to 1983, and so on.

(a) Find a quadratic function of the form $f(x) = a(x - h)^2 + k$ that models these data, using (2, 1563) as the vertex and the point (13, 361,508) to determine a.

$$f(x) = a(x - h)^2 + k \qquad \text{Given form}$$
$$f(x) = a(x - 2)^2 + 1563 \qquad (h, k) = (2, 1563)$$

Let $x = 13$ and $f(13) = 361{,}508$ to find a.

$$361{,}508 = a(13 - 2)^2 + 1563$$
$$361{,}508 = 121a + 1563$$
$$359{,}945 = 121a$$
$$a = 2975 \qquad \text{Nearest whole number}$$

The desired function is $f(x) = 2975(x - 2)^2 + 1563$. It is graphed with the data points in Figure 16(b).

(b) Use the model from part (a) to approximate the number of cases for 1989, and compare the result to the actual data from the table.

$$f(x) = 2975(x - 2)^2 + 1563$$
$$f(9) = 2975(9 - 2)^2 + 1563 \qquad \text{For 1989, } x = 9.$$
$$= 147{,}338$$

The number 147,338 compares very closely with the actual figure 147,170 given in the table.

Now Try Exercise 75.

In conclusion, *we emphasize that there are important relationships among the following ideas:*

1. the x-intercepts of the graph of $y = f(x)$;
2. the zeros of the function f;
3. the solutions of the equation $f(x) = 0$.

For example, look back at the function from Example 3,

$$f(x) = 2x^3 + 5x^2 - x - 6 = (x - 1)(2x + 3)(x + 2).$$

Its graph, shown in Figure 10, has x-intercepts $(1, 0)$, $\left(-\frac{3}{2}, 0\right)$, and $(-2, 0)$. Since 1, $-\frac{3}{2}$, and -2 are the x-values for which the function is 0, they are the zeros of f. Furthermore, 1, $-\frac{3}{2}$, and -2 are solutions of the polynomial equation $2x^3 + 5x^2 - x - 6 = 0$. This discussion is summarized as follows.

> **x-Intercepts, Zeros, and Solutions**
>
> If the point $(a, 0)$ is an x-intercept of the graph of $y = f(x)$, then a is a zero of f and a is a solution of the equation $f(x) = 0$.

12.3 EXERCISES

Sketch the graph of each function. See Examples 1 and 2.

1. $f(x) = \frac{1}{4}x^6$ **2.** $f(x) = 2x^4$ **3.** $f(x) = -\frac{5}{4}x^5$ **4.** $f(x) = -\frac{2}{3}x^5$

5. $f(x) = \frac{1}{2}x^3 + 1$ **6.** $f(x) = -x^4 + 2$ **7.** $f(x) = -(x + 1)^3$

8. $f(x) = \frac{1}{3}(x + 3)^4$ **9.** $f(x) = (x - 1)^4 + 2$ **10.** $f(x) = (x + 2)^3 - 1$

The graphs of four polynomial functions are shown in A–D. They represent the graphs of functions defined by these four equations, but not necessarily in the order listed.

$$y = x^3 - 3x^2 - 6x + 8 \qquad y = x^4 + 7x^3 - 5x^2 - 75x$$
$$y = -x^3 + 9x^2 - 27x + 17 \qquad y = -x^5 + 36x^3 - 22x^2 - 147x - 90.$$

Apply the concepts of this section to answer each question in Exercises 11–18.

11. Which one of the graphs is that of $y = x^3 - 3x^2 - 6x + 8$?

12. Which one of the graphs is that of $y = x^4 + 7x^3 - 5x^2 - 75x$?

13. How many real zeros does the graph in C have?

14. Which one of C and D is the graph of $y = -x^3 + 9x^2 - 27x + 17$? (*Hint:* Look at the y-intercept.)

15. Which of the graphs cannot be that of a cubic polynomial function?

16. How many positive real zeros does the function graphed in D have?

17. How many negative real zeros does the function graphed in A have?

18. Which one of the graphs is that of a function whose range is *not* $(-\infty, \infty)$?

A.
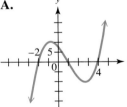

B.
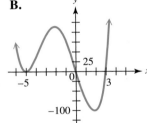

C.
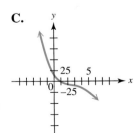

D.
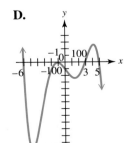

Determine the maximum possible number of turning points of the graph of each function.

19. $f(x) = x^3 - 3x^2 - 6x + 8$

20. $f(x) = x^3 + 4x^2 - 11x - 30$

21. $f(x) = 2x^4 - 9x^3 + 5x^2 + 57x - 45$

22. $f(x) = 4x^4 + 27x^3 - 42x^2 - 445x - 300$

23. $f(x) = -x^4 - 4x^3 + 3x^2 + 18x$

24. $f(x) = -x^4 + 2x^3 + 8x^2$

The polynomial functions in Exercises 25–32 are given in factored form. Graph each function. See Example 3.

25. $f(x) = (x - 4)(x + 2)(x - 1)$

26. $f(x) = (x + 2)(x - 1)(x + 1)$

27. $f(x) = 2x(x - 3)(x + 2)$

28. $f(x) = x(x - 2)(x + 1)$

29. $f(x) = x^2(x + 1)(x - 1)$

30. $f(x) = x^2(x - 2)(x + 3)^2$

31. $f(x) = -x(x + 3)(x + 1)$

32. $f(x) = (3x - 1)(x + 2)^2$

For each polynomial function, one zero is given. Find all rational zeros and factor the polynomial. Then graph the function. See Example 3.

33. $f(x) = 2x^3 - 5x^2 - x + 6$; zero: -1

34. $f(x) = 3x^3 + x^2 - 10x - 8$; zero: 2

35. $f(x) = x^3 + x^2 - 8x - 12$; zero: 3

36. $f(x) = x^3 + 6x^2 - 32$; zero: -4

37. $f(x) = -x^3 - x^2 + 8x + 12$; zero: -2

38. $f(x) = -x^3 + 10x^2 - 33x + 36$; zero: 3

39. $f(x) = x^4 - 18x^2 + 81$; zero: -3 (multiplicity 2)

40. $f(x) = x^4 - 8x^2 + 16$; zero: 2 (multiplicity 2)

 *For each of the following, **(a)** show that the polynomial function has a zero between the two given integers; **(b)** use a graphing calculator to evaluate all zeros to the nearest thousandth. See Examples 4 and 6.*

41. $f(x) = x^4 + x^3 - 6x^2 - 20x - 16$; between -2 and -1

42. $f(x) = x^4 - 2x^3 - 2x^2 - 18x + 5$; between 0 and 1

43. $f(x) = x^4 - 4x^3 - 20x^2 + 32x + 12$; between -4 and -3

44. $f(x) = x^4 - 4x^3 - 44x^2 + 160x - 80$; between 2 and 3

Use the boundedness theorem to show that the real zeros of each polynomial function satisfy the given conditions. See Example 5.

45. $f(x) = x^4 - x^3 + 3x^2 - 8x + 8$; no real zero greater than 2

46. $f(x) = 2x^5 - x^4 + 2x^3 - 2x^2 + 4x - 4$; no real zero greater than 1

47. $f(x) = x^4 + x^3 - x^2 + 3$; no real zero less than -2

48. $f(x) = x^5 + 2x^3 - 2x^2 + 5x + 5$; no real zero less than -1

49. $f(x) = 3x^4 + 2x^3 - 4x^2 + x - 1$; no real zero greater than 1

50. $f(x) = 3x^4 + 2x^3 - 4x^2 + x - 1$; no real zero less than -2

In Exercises 51 and 52, find a cubic polynomial having the graph shown.

51.

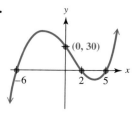

52.

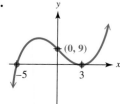

Use a graphing calculator to approximate all real zeros of each function to the nearest hundredth. See Example 6.

53. $f(x) = .86x^3 - 5.24x^2 + 3.55x + 7.84$ **54.** $f(x) = -2.47x^3 - 6.58x^2 - 3.33x + .14$

55. $f(x) = \sqrt{7}x^3 + \sqrt{5}x^2 + \sqrt{17}$ **56.** $f(x) = \sqrt{10}x^3 - \sqrt{11}x - \sqrt{8}$

57. $f(x) = 2.45x^4 - 3.22x^3 + .47x^2 - 6.54x - 3$

58. $f(x) = \sqrt{17}x^4 - \sqrt{22}x^2 + 1$

A graphing calculator can be used to find the coordinates of the turning points of the graph of a polynomial function. Use this capability of your calculator to find the coordinates of the turning point in the given interval. Give your answers to the nearest hundredth.

59. $f(x) = x^3 + 4x^2 - 8x - 8$, $[-3.8, -3]$

60. $f(x) = x^3 + 4x^2 - 8x - 8$, $[.3, 1]$

61. $f(x) = 2x^3 - 5x^2 - x + 1$, $[-1, 0]$

62. $f(x) = 2x^3 - 5x^2 - x + 1$, $[1.4, 2]$

63. $f(x) = x^4 - 7x^3 + 13x^2 + 6x - 28$, $[-1, 0]$

64. $f(x) = x^3 - x + 3$, $[-1, 0]$

RELATING CONCEPTS (EXERCISES 65–74)

For Individual or Group Work

In Chapter 10 we briefly discussed even and odd functions. Some polynomial functions are examples of even or odd functions.

A function $y = f(x)$ is an **even function** if $f(-x) = f(x)$ for all x in the domain of f.

A function $y = f(x)$ is an **odd function** if $f(-x) = -f(x)$ for all x in the domain of f.

For example, $f(x) = x^2$ is an even function, because

$$f(-x) = (-x)^2 = x^2 = f(x).$$

Also, $f(x) = x^3$ is an odd function, because

$$f(-x) = (-x)^3 = -x^3 = -f(x).$$

A function may be neither even nor odd. Decide whether each polynomial function is even, odd, or neither.

65. $f(x) = 2x^3$ **66.** $f(x) = -4x^5$

67. $f(x) = .2x^4$ **68.** $f(x) = -x^6$

69. $f(x) = -x^5$ **70.** $f(x) = 3x^2 + 2x$

71. $f(x) = 2x^3 + 3x^2$ **72.** $f(x) = 4x^3 - x$

73. $f(x) = x^4 + 3x^2 + 5$

74. Refer to the discussion of symmetry in Section 10.4 and fill in the blanks with the correct responses. By the definition of an even function, if (a, b) lies on the graph of an even function, then so does $(-a, b)$. Therefore, the graph of an even function is symmetric with respect to the _____. If (a, b) lies on the graph of an odd function, then by definition, so does $(-a, -b)$. Therefore, the graph of an odd function is symmetric with respect to the _____.

Solve each problem. See Example 7.

75. The table lists the total (cumulative) number of known deaths caused by AIDS in the United States up to 1993.

Year	Deaths	Year	Deaths
1982	620	1988	61,723
1983	2122	1989	89,172
1984	5600	1990	119,821
1985	12,529	1991	154,567
1986	24,550	1992	191,508
1987	40,820	1993	220,592

Source: U.S. Dept. of Health and Human Services, Centers for Disease Control and Prevention, *HIV/AIDS Surveillance*, March 1994.

(a) Plot the data. Let $x = 2$ correspond to the year 1982.

(b) Find a quadratic function of the form $g(x) = a(x - h)^2 + k$ that models the data. Use $(2, 620)$ as the vertex and the ordered pair $(13, 220{,}592)$ to approximate the value of a. Graph this function over the data points.

(c) Use the function from part (b) to approximate the number of deaths for 1987. Compare the result to the actual figure given in the table.

76. Repeat Example 7, but use the ordered pair $(10, 193{,}245)$ to approximate a.

(a) Find the function $f(x) = a(x - h)^2 + k$ that models the data.

(b) Use this new function to approximate the number of cases for 1989. Compare this result to the one found in part (b) of Example 7. Which model is more accurate for 1989?

77. A piece of rectangular sheet metal is 20 in. wide. It is to be made into a rain gutter by turning up the edges to form parallel sides. Let x represent the length of each of the parallel sides.

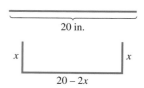

(a) Give the restrictions on x.

(b) Describe a function A that gives the area of a cross section of the gutter.

(c) For what value of x will A be a maximum (and thus maximize the amount of water that the gutter will hold)? What is this maximum area?

(d) For what values of x will the area of a cross section be less than 40 in.^{2}?

78. A certain right triangle has area 84 in.^{2} One leg of the triangle measures 1 in. less than the hypotenuse. Let x represent the length of the hypotenuse.

(a) Express the length of the leg mentioned above in terms of x.

(b) Express the length of the other leg in terms of x.

(c) Write an equation based on the information determined thus far. Square both sides and then write the equation with one side as a polynomial with integer coefficients, in descending powers, and the other side equal to 0.

(d) Solve the equation in part (c) graphically. Find the lengths of the three sides of the triangle.

79. Find the value of x in the figure that will maximize the area of rectangle $ABCD$.

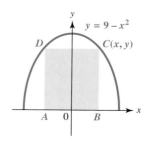

80. A technique for measuring cardiac output depends on the concentration of a dye in the bloodstream after a known amount is injected into a vein near the heart. For a normal heart, the concentration of dye in appropriate units in the bloodstream at time x (in seconds) is given by the function defined as follows.

$$g(x) = -.006x^4 + .140x^3 - .053x^2 + 1.79x$$

What is the concentration after 20 sec?

81. From 1930 to 1990 the rate of breast cancer was nearly constant at 30 cases per 100,000 females, whereas the rate of lung cancer in females over the same period increased. The rate of lung cancer cases per 100,000 females in the year t (where $t = 0$ corresponds to 1930) can be modeled using the function defined by

$$f(t) = .00028t^3 - .011t^2 + .23t + .93.$$

(*Source:* Valanis, B., *Epidemiology in Nursing and Health Care,* Appleton & Lange, 1992.)

(a) Use a graphing calculator to graph the rates of breast and lung cancer for $0 \le t \le 60$ in the window $[0, 60]$ by $[-10, 40]$.

(b) Determine the year when rates for lung cancer first exceeded those for breast cancer.

(c) Discuss reasons for the rapid increase of lung cancer in females.

82. The linear function defined by

$$f(x) = 1.5457x - 3067.7,$$

where x is the year, can be used to estimate the percentage of gonorrhea cases with antibiotic resistance diagnosed from 1985 to 1990. (*Source:* Teutsch, S., and R. Churchill, *Principles and Practice of Public Health Surveillance,* Oxford University Press, 1994.)

(a) Determine this percentage in 1988.

(b) Interpret the slope of the graph of f.

83. The time delay between an individual's initial infection with HIV and when that individual develops symptoms of AIDS is an important issue. In one study of HIV patients who were infected by intravenous drug use, it was found that after 4 yr 17% of the patients had AIDS and after 7 yr 33% had developed the disease. The relationship between the time interval and the percent of patients with AIDS can accurately be modeled with a linear function defined by

$$f(x) = .05\overline{3}x - .04\overline{3},$$

where x represents the time interval in years. (*Source:* Alcabes, P., A. Munoz, D. Vlahov, and G. Friedland, "Incubation Period of Human Immunodeficiency Virus," *Epidemiologic Review,* vol. 15, no. 2, The Johns Hopkins University School of Hygiene and Public Health, 1993.)

(a) Assuming the function continues to model the situation, determine the percent of patients with AIDS after 10 yr.

(b) Predict the number of years before half of these patients will have AIDS.

84. The number of military personnel on active duty in the United States during the period 1985 to 1990 can be determined by the cubic model defined by

$$f(x) = -7.66x^3 + 52.71x^2 - 93.43x + 2151,$$

where $x = 0$ corresponds to 1985, and $f(x)$ is in thousands. Based on this model, how many military personnel were on active duty in 1990? (*Source:* U.S. Department of Defense.)

85. The table lists the total annual amount (in millions of dollars) of government-guaranteed student loans from 1986 to 1994.
  (a) Graph the data with the following three function definitions, where x represents the year.
   (i) $f(x) = .4(x - 1986)^2 + 8.6$
   (ii) $g(x) = 1.088(x - 1986) + 8.6$
   (iii) $h(x) = 1.455\sqrt{x - 1986} + 8.6$
  (b) Discuss which function definition best models the data.

Year	Amount	Year	Amount
1986	8.6	1991	13.5
1987	9.8	1992	14.7
1988	11.8	1993	16.5
1989	12.5	1994	18.2
1990	12.3		

Source: USA Today.

86. One result of improved health care is that people are living longer. The table lists the number of Americans (in thousands) who are expected to be over 100 years old for selected years.
  (a) Use graphing to determine which polynomial best models the number of Americans over 100 yr old, where $x = 0$ corresponds to 1994.
   (i) $f(x) = 6.057x + 44.714$
   (ii) $g(x) = .4018x^2 + 2.039x + 50.071$
   (iii) $h(x) = -.06x^3 + .506x^2 + 1.659x + 50.238$
  (b) Use your choice from part (a) to predict the number of Americans who will be over 100 yr old in the year 2008.

Year	Number
1994	50
1996	56
1998	65
2000	75
2002	94
2004	110

Source: U.S. Bureau of the Census.

87. A simple pendulum will swing back and forth in regular time intervals. Grandfather clocks use pendulums to keep accurate time. The relationship between the length of a pendulum L and the time T for one complete oscillation can be expressed by the equation $L = kT^n$, where k is a constant and n is a positive integer to be determined. The data in the table were taken for different lengths of pendulums.

L (ft)	T (sec)
1.0	1.11
1.5	1.36
2.0	1.57
2.5	1.76
3.0	1.92
3.5	2.08
4.0	2.22

  (a) If the length of the pendulum increases, what happens to T?
  (b) Discuss how n and k can be found.
  (c) Use the data to approximate k and determine the best value for n.
  (d) Using the values of k and n from part (c), predict T for a pendulum having length 5 ft.
  (e) If the length L of a pendulum doubles, what happens to the period T?

88. A storage tank for butane gas is to be built in the shape of a right circular cylinder of altitude 12 ft, with a half sphere attached to each end. If x represents the radius of each half sphere, what radius should be used to cause the volume of the tank to be 144π ft^3?

SUMMARY EXERCISES ON POLYNOMIAL FUNCTIONS AND GRAPHS

In Sections 12.2 and 12.3 we introduced many characteristics of polynomial functions and their graphs. This set of review exercises is designed to put these ideas together in examining typical polynomial functions.

For each polynomial function, do the following in order.

(a) (Optional) Use Descartes' rule of signs (see Section 12.2 Exercises) to find the possible number of positive and negative real zeros.

(b) Use the rational zeros theorem to determine the possible rational zeros of the function.

(c) Find the rational zeros, if any.

(d) Find all other real zeros, if any.

(e) Find any other complex zeros (that is, zeros that are not real), if any.

(f) Find the x-intercepts of the graph, if any.

(g) Find the y-intercept of the graph.

(h) Use synthetic division to find $f(4)$, and give the coordinates of the corresponding point on the graph.

(i) Sketch the graph of the function.

1. $f(x) = x^4 + 3x^3 - 3x^2 - 11x - 6$

2. $f(x) = -2x^5 + 5x^4 + 34x^3 - 30x^2 - 84x + 45$

3. $f(x) = 2x^5 - 10x^4 + x^3 - 5x^2 - x + 5$

4. $f(x) = 3x^4 - 4x^3 - 22x^2 + 15x + 18$

5. $f(x) = -2x^4 - x^3 + x + 2$

6. $f(x) = 4x^5 + 8x^4 + 9x^3 + 27x^2 + 27x$
(*Hint:* Factor out x first.)

7. $f(x) = 3x^4 - 14x^2 - 5$
(*Hint:* Factor the polynomial.)

8. $f(x) = -x^5 - x^4 + 10x^3 + 10x^2 - 9x - 9$

9. $f(x) = -3x^4 + 22x^3 - 55x^2 + 52x - 12$

10. For the polynomial functions in Exercises 1–9 that have irrational zeros, use a graphing calculator to find their approximations to the nearest thousandth.

12.4 Graphs and Applications of Rational Functions

OBJECTIVES

1 Graph rational functions using reflection and translation.

2 Find asymptotes of the graph of a rational function.

3 Graph rational functions where the degree of the numerator is less than the degree of the denominator.

4 Graph rational functions where the degrees of the numerator and the denominator are equal.

Rational functions were introduced in Section 7.1. We begin by reviewing the definition given there.

> **Rational Function**
>
> A function of the form
> $$f(x) = \frac{P(x)}{Q(x)}$$
> where $P(x)$ and $Q(x)$ are polynomial functions, $Q(x) \neq 0$, is called a **rational function.**

Since any values of x for which $Q(x) = 0$ are excluded from the domain of a rational function, this type of function often has a *discontinuous graph,* that is, a graph with one or more breaks in it.

OBJECTIVE 1 Graph rational functions using reflection and translation. The simplest rational function with a variable denominator is
$$f(x) = \frac{1}{x}.$$

5 Graph rational functions where the degree of the numerator is greater than the degree of the denominator.

6 Graph a rational function that is not in lowest terms.

7 Graph rational functions and solve applications using a graphing calculator.

As we saw in Section 7.4, the domain of this function is the set of all real numbers except 0. The number 0 cannot be used as a value of x, but for graphing it is helpful to find the values of $f(x)$ for some values of x close to 0. The table shows what happens to $f(x)$ as x gets closer and closer to 0 from either side.

x approaches 0.

x	-1	$-.1$	$-.01$	$-.001$	.001	.01	.1	1
$f(x)$	-1	-10	-100	-1000	1000	100	10	1

$|f(x)|$ gets larger and larger.

The table suggests that $|f(x)|$ gets larger and larger as x gets closer and closer to 0, which is written in symbols as

$$|f(x)| \to \infty \text{ as } x \to 0.$$

(The symbol $x \to 0$ means that x approaches as close as desired to 0, without necessarily ever being equal to 0.) Since x cannot equal 0, the graph of $f(x) = \frac{1}{x}$ will never intersect the vertical line $x = 0$. This line is called a *vertical asymptote*.

On the other hand, as $|x|$ gets larger and larger, the values of $f(x) = \frac{1}{x}$ get closer and closer to 0, as shown in the table.

x	$-10{,}000$	-1000	-100	-10	10	100	1000	10,000
$f(x)$	$-.0001$	$-.001$	$-.01$	$-.1$	.1	.01	.001	.0001

Letting $|x|$ get larger and larger without bound (written $|x| \to \infty$) causes the graph of $y = f(x) = \frac{1}{x}$ to move closer and closer to the horizontal line $y = 0$. This line is called a *horizontal asymptote*.

If the point (a, b) lies on the graph of $f(x) = \frac{1}{x}$, then so does the point $(-a, -b)$. Therefore, the graph of f is symmetric with respect to the origin. Choosing some positive values of x and finding the corresponding values of $f(x)$ gives the first-quadrant part of the graph shown in Figure 17. The other part of the graph (in the third quadrant) can be found by symmetry.

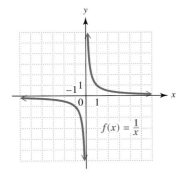

FIGURE 17

EXAMPLE 1 Graphing a Rational Function Using Reflection

Graph $f(x) = -\dfrac{2}{x}$.

The expression on the right side of the equation can be rewritten so that

$$f(x) = -2 \cdot \frac{1}{x}.$$

Compared to $f(x) = \frac{1}{x}$, the graph will be reflected about the x-axis (because of the negative sign), and each point will be twice as far from the x-axis. See Figure 18.

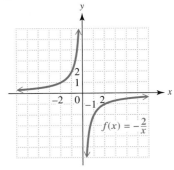

FIGURE 18

Now Try Exercise 7.

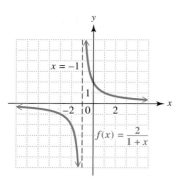

FIGURE 19

EXAMPLE 2   Graphing a Rational Function Using Translation

Graph $f(x) = \dfrac{2}{1+x}$.

The domain of this function is the set of all real numbers except -1. As shown in Figure 19, the graph is that of $f(x) = \frac{1}{x}$, translated 1 unit to the left, with each y-value doubled. This can be seen by writing the expression as

$$f(x) = 2 \cdot \frac{1}{x - (-1)}.$$

Now Try Exercise 9.

OBJECTIVE 2 Find asymptotes of the graph of a rational function.   The preceding examples suggest the following definitions of vertical and horizontal asymptotes.

Vertical and Horizontal Asymptotes

Given the rational function defined by $f(x) = \frac{P(x)}{Q(x)}$, written in lowest terms, the following statements apply:

1. If $|f(x)| \to \infty$ as $x \to a$, then the line $x = a$ is a **vertical asymptote.**
2. If $f(x) \to a$ as $|x| \to \infty$, then the line $y = a$ is a **horizontal asymptote.**

Locating asymptotes is important when graphing rational functions. Vertical asymptotes are found by setting the denominator of a rational function equal to 0 and then solving. Horizontal asymptotes (and, in some cases, *oblique asymptotes*) are found by considering what happens to $f(x)$ as $|x| \to \infty$.

EXAMPLE 3   Finding the Equations of Asymptotes

For each rational function, find all asymptotes.

(a) $f(x) = \dfrac{x+1}{(2x-1)(x+3)}$

To find the vertical asymptotes, set the denominator equal to 0 and solve.

$$(2x - 1)(x + 3) = 0$$

$$2x - 1 = 0 \quad \text{or} \quad x + 3 = 0 \qquad \text{Zero-factor property}$$

$$x = \frac{1}{2} \quad \text{or} \qquad x = -3$$

The equations of the vertical asymptotes are $x = \frac{1}{2}$ and $x = -3$.

To find the equation of the horizontal asymptote, divide each term by the largest power of x in the expression. Begin by multiplying the factors in the denominator to get

$$f(x) = \frac{x+1}{(2x-1)(x+3)} = \frac{x+1}{2x^2 + 5x - 3}.$$

Now divide each term in the numerator and denominator by x^2, since 2 is the largest exponent on x. This gives

$$f(x) = \frac{\dfrac{x}{x^2} + \dfrac{1}{x^2}}{\dfrac{2x^2}{x^2} + \dfrac{5x}{x^2} - \dfrac{3}{x^2}} = \frac{\dfrac{1}{x} + \dfrac{1}{x^2}}{2 + \dfrac{5}{x} - \dfrac{3}{x^2}}.$$

As $|x|$ gets larger and larger, the quotients $\frac{1}{x}, \frac{1}{x^2}, \frac{5}{x}$, and $\frac{3}{x^2}$ all approach 0, and the value of $f(x)$ approaches

$$\frac{0 + 0}{2 + 0 - 0} = \frac{0}{2} = 0.$$

The line $y = 0$ (that is, the x-axis) is therefore the horizontal asymptote.

(b) $f(x) = \dfrac{2x + 1}{x - 3}$

Set the denominator equal to 0 to find that the vertical asymptote has the equation $x = 3$. To find the horizontal asymptote, divide each term in the rational expression by x, since the greatest power of x in the expression is 1.

$$f(x) = \frac{2x + 1}{x - 3} = \frac{\dfrac{2x}{x} + \dfrac{1}{x}}{\dfrac{x}{x} - \dfrac{3}{x}} = \frac{2 + \dfrac{1}{x}}{1 - \dfrac{3}{x}}$$

As $|x|$ gets larger and larger, both $\frac{1}{x}$ and $\frac{3}{x}$ approach 0, and $f(x)$ approaches

$$\frac{2 + 0}{1 - 0} = \frac{2}{1} = 2,$$

so the line $y = 2$ is the horizontal asymptote.

(c) $f(x) = \dfrac{x^2 + 1}{x - 2}$

Setting the denominator equal to 0 shows that the vertical asymptote has the equation $x = 2$. If we divide by the largest power of x as before (x^2 in this case), we see that there is no horizontal asymptote because

$$f(x) = \frac{\dfrac{x^2}{x^2} + \dfrac{1}{x^2}}{\dfrac{x}{x^2} - \dfrac{2}{x^2}} = \frac{1 + \dfrac{1}{x^2}}{\dfrac{1}{x} - \dfrac{2}{x^2}}$$

does not approach any real number as $|x| \to \infty$, since $\frac{1}{0}$ is undefined. This happens whenever the degree of the numerator is greater than the degree of the denominator. In such cases we divide the denominator into the numerator to write the expression in another form. Using synthetic division gives

$$\begin{array}{r} 2)\overline{1 \quad 0 \quad 1} \\ \quad 2 \quad 4 \\ \hline 1 \quad 2 \quad 5. \end{array}$$

The function can now be written as

$$f(x) = \frac{x^2 + 1}{x - 2} = x + 2 + \frac{5}{x - 2}.$$

For very large values of $|x|$, $\frac{5}{x-2}$ is close to 0, and the graph approaches the line $y = x + 2$. This line is an **oblique asymptote** (neither vertical nor horizontal) for the function.

In general, if the degree of its numerator is exactly one more than the degree of its denominator, a rational function may have an oblique asymptote. We find the equation of this asymptote by dividing the numerator by the denominator and disregarding any remainder.

Now Try Exercises 15, 21, and 25.

The results of Example 3 can be summarized as follows.

Determining Asymptotes

To find asymptotes of a rational function written *in lowest terms,* use the following procedures.

1. **Vertical Asymptotes**

   Find any vertical asymptotes by setting the denominator equal to 0 and solving for x. If a is a zero of the denominator, then the line $x = a$ is a vertical asymptote.

2. **Other Asymptotes**

   Determine any other asymptotes. There are three possibilities:

   (a) If the numerator has lesser degree than the denominator, there is a horizontal asymptote, $y = 0$ (the x-axis).

   (b) If the numerator and denominator have the same degree, and the function is of the form

   $$f(x) = \frac{a_n x^n + \cdots + a_0}{b_n x^n + \cdots + b_0}, \quad \text{where } b_n \neq 0,$$

   dividing by x^n in the numerator and denominator produces the horizontal asymptote

   $$y = \frac{a_n}{b_n}.$$

   (c) If the numerator is of degree exactly one more than the denominator, there may be an oblique asymptote. To find it, divide the numerator by the denominator and disregard any remainder. Set the rest of the quotient equal to y to obtain the equation of the asymptote.

NOTE The graph of a rational function may have more than one vertical asymptote, or it may have none at all. The graph cannot intersect any vertical asymptote. There can be only one other (nonvertical) asymptote, and the graph *may* intersect that asymptote. See Example 6.

The following procedure can be used to graph rational functions written in lowest terms.

Graphing Rational Functions

Let $f(x) = \frac{P(x)}{Q(x)}$ be a rational function written in lowest terms. To sketch its graph, follow the steps below.

Step 1   **Find any vertical asymptotes.**

Step 2   **Find any horizontal or oblique asymptote.**

Step 3   **Find the y-intercept** by evaluating $f(0)$.

Step 4   **Find the x-intercepts,** if any, by solving $f(x) = 0$. (These will be the zeros of the numerator, $P(x)$.)

Step 5   **Determine whether the graph will intersect its nonvertical asymptote** by solving $f(x) = k$, where k is the y-value of the nonvertical asymptote.

Step 6   **Plot a few selected points, as necessary.** Choose an x-value in each interval of the domain as determined by the vertical asymptotes and x-intercepts.

Step 7   **Complete the sketch.**

OBJECTIVE 3 **Graph rational functions where the degree of the numerator is less than the degree of the denominator.** The next example shows how the preceding guidelines are used to graph a rational function.

▧ **EXAMPLE 4** **Graphing a Rational Function; Degree of Numerator < Degree of Denominator**

Graph $f(x) = \dfrac{x + 1}{(2x - 1)(x + 3)}$.

Step 1   As shown in Example 3(a), the vertical asymptotes have equations $x = \frac{1}{2}$ and $x = -3$.

Step 2   Again, as shown in Example 3(a), the horizontal asymptote is the x-axis.

Step 3   The y-intercept is $\left(0, -\frac{1}{3}\right)$, since

$$f(0) = \frac{0 + 1}{(2(0) - 1)(0 + 3)} = -\frac{1}{3}.$$

Step 4   The x-intercept is found by solving $f(x) = 0$.

$$\frac{x + 1}{(2x - 1)(x + 3)} = 0$$

$$x + 1 = 0 \qquad \text{Multiply by } (2x - 1)(x + 3).$$

$$x = -1$$

The x-intercept is $(-1, 0)$.

Step 5   To determine whether the graph intersects its horizontal asymptote, solve

$$f(x) = 0. \quad \leftarrow y\text{-value of horizontal asymptote}$$

Since the horizontal asymptote is the x-axis, the solution of this equation was found in Step 4. The graph intersects its horizontal asymptote at $(-1, 0)$.

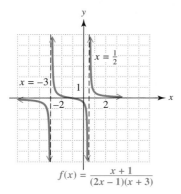

$$f(x) = \frac{x+1}{(2x-1)(x+3)}$$

FIGURE 20

Step 6 Plot a point in each of the intervals determined by the x-intercepts and vertical asymptotes, $(-\infty, -3)$, $(-3, -1)$, $\left(-1, \frac{1}{2}\right)$, and $\left(\frac{1}{2}, \infty\right)$, to get an idea of how the graph behaves in each region.

Step 7 Complete the sketch. The graph is shown in Figure 20.

Now Try Exercise 35.

OBJECTIVE 4 Graph rational functions where the degrees of the numerator and the denominator are equal. In the remaining examples, we will not specifically number the steps.

EXAMPLE 5 Graphing a Rational Function; Degree of Numerator = Degree of Denominator

Graph $f(x) = \dfrac{2x+1}{x-3}$.

As shown in Example 3(b), the equation of the vertical asymptote is $x = 3$ and the equation of the horizontal asymptote is $y = 2$. Since $f(0) = -\frac{1}{3}$, the y-intercept is $\left(0, -\frac{1}{3}\right)$. The solution of $f(x) = 0$ is $-\frac{1}{2}$, so the only x-intercept is $\left(-\frac{1}{2}, 0\right)$. The graph does not intersect its horizontal asymptote, since $f(x) = 2$ has no solution. (Verify this.) The points $(-4, 1)$ and $\left(6, \frac{13}{3}\right)$ are on the graph and can be used to complete the sketch, as shown in Figure 21.

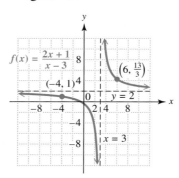

FIGURE 21

Now Try Exercise 39.

EXAMPLE 6 Graphing a Rational Function; Graph Intersects the Asymptote

Graph $f(x) = \dfrac{3(x+1)(x-2)}{(x+4)^2}$.

The only vertical asymptote is the line $x = -4$. To find any horizontal asymptotes, we multiply the factors in the numerator and denominator.

$$f(x) = \frac{3x^2 - 3x - 6}{x^2 + 8x + 16}$$

As explained in the guidelines for determining asymptotes, the equation of the horizontal asymptote can be shown to be

$$y = \frac{3}{1} \quad \begin{matrix} \leftarrow \text{ Leading coefficient of numerator} \\ \leftarrow \text{ Leading coefficient of denominator} \end{matrix}$$

or $y = 3$. The y-intercept is $\left(0, -\frac{3}{8}\right)$, and the x-intercepts are $(-1, 0)$ and $(2, 0)$. By setting $f(x) = 3$ and solving, we find the point where the graph intersects the horizontal asymptote.

$$f(x) = \frac{3x^2 - 3x - 6}{x^2 + 8x + 16}$$

$$3 = \frac{3x^2 - 3x - 6}{x^2 + 8x + 16} \qquad \text{Let } f(x) = 3.$$

$$3x^2 + 24x + 48 = 3x^2 - 3x - 6 \qquad \text{Multiply by } x^2 + 8x + 16.$$

$$24x + 48 = -3x - 6 \qquad \text{Subtract } 3x^2.$$

$$27x = -54 \qquad \text{Add } 3x; \text{ subtract } 48.$$

$$x = -2 \qquad \text{Divide by } 27.$$

The graph intersects its horizontal asymptote at $(-2, 3)$.

Some other points that lie on the graph are $(-10, 9)$, $(-3, 30)$, and $\left(5, \frac{2}{3}\right)$. These can be used to complete the graph, shown in Figure 22.

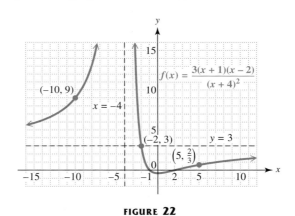

FIGURE 22

Now Try Exercise 47.

OBJECTIVE 5 Graph rational functions where the degree of the numerator is greater than the degree of the denominator.

EXAMPLE 7 Graphing a Rational Function; Degree of Numerator > Degree of Denominator

Graph $f(x) = \dfrac{x^2 + 1}{x - 2}$.

As shown in Example 3(c), the vertical asymptote has the equation $x = 2$, and the graph has an oblique asymptote with equation $y = x + 2$. The y-intercept is $\left(0, -\frac{1}{2}\right)$, and the graph has no x-intercepts, since the numerator, $x^2 + 1$, has no real zeros. Using the intercepts, asymptotes, the points $\left(4, \frac{17}{2}\right)$ and $\left(-1, -\frac{2}{3}\right)$, and the general behavior of the graph near its asymptotes, we obtain the graph shown in Figure 23.

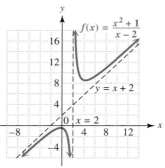

FIGURE 23

Now Try Exercise 45.

OBJECTIVE 6 **Graph a rational function that is not in lowest terms.** As mentioned earlier, a rational function must be in lowest terms before we can use the methods discussed in this section to determine its graph. A rational function that is not in lowest terms will have a "hole," or *point of discontinuity,* in its graph.

EXAMPLE 8 **Graphing a Rational Function That is Not in Lowest Terms**

Graph $f(x) = \dfrac{x^2 - 4}{x - 2}$.

Start by noticing that the domain of this function does not include 2. The rational expression $\dfrac{x^2 - 4}{x - 2}$ can be written in lowest terms by factoring the numerator, and using the fundamental property.

$$\frac{x^2 - 4}{x - 2} = \frac{(x + 2)(x - 2)}{x - 2} = x + 2 \quad (x \neq 2)$$

Therefore, the graph of this function will be the same as the graph of $y = x + 2$ (a straight line), with the exception of the point with x-value 2. Instead of an asymptote, there is a hole in the graph at $(2, 4)$. See Figure 24.

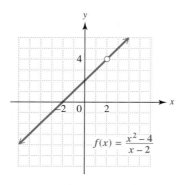

FIGURE 24

Now Try Exercise 51.

OBJECTIVE 7 Graph rational functions and solve applications using a graphing calculator. Special care must be taken in interpreting rational function graphs generated by a graphing calculator. If the calculator is in connected mode, it may show a vertical line for x-values that produce vertical asymptotes. While this may be interpreted as a graph of the asymptote, a more realistic graph can be obtained by using dot mode. In Figure 25, there are three "views": the first shows the graph of $f(x) = \frac{1}{x+3}$ in connected mode, the second shows the same window but in dot mode, and the third shows a carefully chosen window, where the calculator is in connected mode but does not show the vertical line.

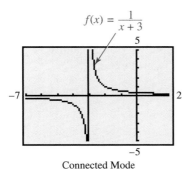
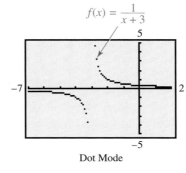
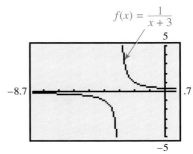

Connected Mode	Dot Mode	Carefully Chosen Window, Connected Mode

FIGURE 25

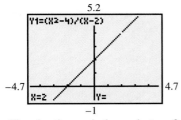

There is a tiny gap in the graph at $x = 2$.

FIGURE 26

In Example 8, if the window of a graphing calculator is set so that 2 is an x-value for the location of the cursor, then the display will show an unlit pixel at 2. To see this, look carefully in Figure 26 at the point on the screen where $x = 2$. However, such points will often *not* be evident from calculator graphs—once again showing us a reason for studying the concepts along with the technology.

Rational functions have a variety of applications. The next example discusses one, and several others are given in the exercises for this section.

EXAMPLE 9 Applying Rational Functions to Traffic Intensity

Traffic intensity can be modeled by rational functions. For example, suppose that vehicles arrive randomly at a parking ramp with an average rate of 2.6 vehicles per minute. The parking attendant can admit 3.2 vehicles per minute. However, since arrivals are random, lines form at various times. (*Source:* Mannering, F., and W. Kilareski, *Principles of Highway Engineering and Traffic Control,* John Wiley & Sons, 1990.)

(a) The traffic intensity x is defined as the ratio of the average arrival rate to the average admittance rate. Determine x for this parking ramp.

The average arrival rate is 2.6 vehicles and the average admittance rate is 3.2 vehicles, so

$$x = \frac{2.6}{3.2} = .8125.$$

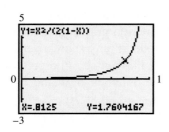

FIGURE 27

(b) The average number of vehicles waiting in line to enter the ramp is given by the rational function defined by

$$f(x) = \frac{x^2}{2(1 - x)},$$

where $0 \le x < 1$ is the traffic intensity. Graph this function using a graphing calculator, and compute $f(.8125)$ for this parking ramp.

Figure 27 shows the graph. Find $f(.8125)$ by substitution.

$$f(.8125) = \frac{.8125^2}{2(1 - .8125)} \approx 1.76 \text{ vehicles}$$

(c) What happens to the number of vehicles waiting as the traffic intensity approaches 1?

From the graph in Figure 27, we see that as x approaches 1, $y = f(x)$ gets very large. This is not surprising; it is what we would expect.

Now Try Exercise 65.

12.4 EXERCISES

For Extra Help

Student's Solutions Manual

MyMathLab

InterAct Math Tutorial Software

Tutor Center — AW Math Tutor Center

MathXL — MathXL

Digital Video Tutor CD 12/Videotape 14

Use the graphs of the rational functions in A–D to answer the questions in Exercises 1–6. Give all possible answers; there may be more than one correct choice.

A.

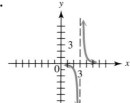

B.

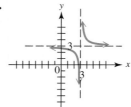

C.

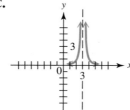

D.

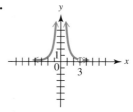

1. Which choices have domain $(-\infty, 3) \cup (3, \infty)$?

2. Which choices have range $(-\infty, 3) \cup (3, \infty)$?

3. Which choices have range $(-\infty, 0) \cup (0, \infty)$?

4. If f represents the function, only one choice has a single solution to the equation $f(x) = 3$. Which one is it?

5. Which choices have the x-axis as a horizontal asymptote?

6. Which choices are symmetric with respect to a vertical line?

Use reflections and/or translations to graph each rational function. See Examples 1 and 2.

7. $f(x) = -\dfrac{3}{x}$

8. $f(x) = \dfrac{2}{x}$

9. $f(x) = \dfrac{1}{x + 2}$

10. $f(x) = \dfrac{1}{x - 3}$

11. $f(x) = \dfrac{1}{x} + 1$

12. $f(x) = \dfrac{1}{x} - 2$

13. Sketch the following graphs and compare them with the graph of $f(x) = \dfrac{1}{x^2}$.

  (a) $f(x) = \dfrac{1}{(x - 3)^2}$ **(b)** $f(x) = -\dfrac{2}{x^2}$ **(c)** $f(x) = \dfrac{-2}{(x - 3)^2}$

14. Describe in your own words what is meant by an asymptote of the graph of a rational function.

Give the equations of any vertical, horizontal, or oblique asymptotes for the graph of each rational function. See Example 3.

15. $f(x) = \dfrac{2}{x - 5}$

16. $f(x) = \dfrac{-1}{x + 2}$

17. $f(x) = \dfrac{-8}{3x - 7}$

18. $f(x) = \dfrac{5}{4x - 9}$

19. $f(x) = \dfrac{2 - x}{x + 2}$

20. $f(x) = \dfrac{x - 4}{5 - x}$

21. $f(x) = \dfrac{3x - 5}{2x + 9}$

22. $f(x) = \dfrac{4x + 3}{3x - 7}$

23. $f(x) = \dfrac{2}{x^2 - 4x + 3}$

24. $f(x) = \dfrac{-5}{x^2 - 3x - 10}$

25. $f(x) = \dfrac{x^2 - 1}{x + 3}$

26. $f(x) = \dfrac{x^2 + 4}{x - 1}$

27. $f(x) = \dfrac{(x - 3)(x + 1)}{(x + 2)(2x - 5)}$

28. $f(x) = \dfrac{3(x + 2)(x - 4)}{(5x - 1)(x - 5)}$

29. Which choice has a graph that does not have a vertical asymptote?

  A. $f(x) = \dfrac{1}{x^2 + 2}$ **B.** $f(x) = \dfrac{1}{x^2 - 2}$

  C. $f(x) = \dfrac{3}{x^2}$ **D.** $f(x) = \dfrac{2x + 1}{x - 8}$

30. Which choice has a graph that does not have a horizontal asymptote?

A. $f(x) = \dfrac{2x - 7}{x + 3}$ **B.** $f(x) = \dfrac{3x}{x^2 - 9}$

C. $f(x) = \dfrac{x^2 - 9}{x + 3}$ **D.** $f(x) = \dfrac{x + 5}{(x + 2)(x - 3)}$

Graph each rational function. See Examples 4–8.

31. $f(x) = \dfrac{4}{5 + 3x}$

32. $f(x) = \dfrac{1}{(x - 2)(x + 4)}$

33. $f(x) = \dfrac{3}{(x + 4)^2}$

34. $f(x) = \dfrac{3x}{(x + 1)(x - 2)}$

35. $f(x) = \dfrac{2x + 1}{(x + 2)(x + 4)}$

36. $f(x) = \dfrac{5x}{x^2 - 1}$

37. $f(x) = \dfrac{-x}{x^2 - 4}$

38. $f(x) = \dfrac{3x}{x - 1}$

39. $f(x) = \dfrac{4x}{1 - 3x}$

40. $f(x) = \dfrac{x + 1}{x - 4}$

41. $f(x) = \dfrac{x - 5}{x + 3}$

42. $f(x) = \dfrac{x}{x^2 - 9}$

43. $f(x) = \dfrac{3x}{x^2 - 16}$

44. $f(x) = \dfrac{2x^2 + 3}{x - 4}$

45. $f(x) = \dfrac{x^2 + 1}{x + 3}$

46. $f(x) = \dfrac{x^2 + 2x}{2x - 1}$

47. $f(x) = \dfrac{(x - 3)(x + 1)}{(x - 1)^2}$

48. $f(x) = \dfrac{x^2 - x}{x + 2}$

49. $f(x) = \dfrac{x(x - 2)}{(x + 3)^2}$

50. $f(x) = \dfrac{(x - 5)(x - 2)}{x^2 + 9}$

51. $f(x) = \dfrac{x^2 - 9}{x + 3}$

52. $f(x) = \dfrac{1}{x^2 + 1}$

RELATING CONCEPTS (EXERCISES 53–59)

For Individual or Group Work

Consider the rational function defined by

$$f(x) = \frac{x^2 - 16}{x + 4}.$$

Work Exercises 53–59 in order. Use a graphing calculator as necessary.

53. Using an integer viewing window, graph this rational function in connected mode. Graph it as y_1.

54. Look closely at your calculator screen. What do you notice about the graph at the x-value -4?

55. Try to evaluate the function for $x = -4$ using your calculator. What happens? Why?

56. Write the expression for $f(x)$ in lowest terms, and call the new function $g(x)$. How does the domain of f compare to the domain of g?

57. Using an integer viewing window, graph $y = g(x)$ as y_2 in connected mode. Look closely at your calculator screen. How does this graph compare to that of y_1 you found in Exercises 53 and 54?

58. Try to evaluate the function g for $x = -4$. What happens? Why?

59. Write a short discussion of how Exercises 53–58 reinforce the concept of the domain of a function.

60. The graphs in A–D show the four ways that a rational function can approach the vertical line $x = 2$ as an asymptote. Match each graph in A–D with the appropriate rational function in (a)–(d).

A.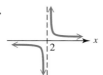

B.

C.

D.

(a) $f(x) = \dfrac{1}{(x - 2)^2}$ (b) $f(x) = \dfrac{1}{x - 2}$

(c) $f(x) = \dfrac{-1}{x - 2}$ (d) $f(x) = \dfrac{-1}{(x - 2)^2}$

61. Suppose that a friend tells you that the graph of

$$f(x) = \frac{x^2 - 25}{x + 5}$$

has a vertical asymptote with equation $x = -5$. Is this correct? If not, describe the behavior of the graph at $x = -5$.

62. Let $f(x) = \dfrac{P(x)}{Q(x)}$ define a rational function where the expression is written in lowest terms. Suppose that the degree of $P(x)$ is m and the degree of $Q(x)$ is n. Write an explanation of how you would determine the nonvertical asymptote in each situation. Give examples.

(a) $m < n$ (b) $m = n$ (c) $m = n + 1$

Solve each application of rational functions. See Example 9.

63. (Refer to Example 7 and Exercise 75 in Section 12.3.) The following data were first seen in Section 12.3.

Year	AIDS Cases	Deaths	Year	AIDS Cases	Deaths
1982	1563	620	1988	105,489	61,723
1983	4647	2122	1989	147,170	89,172
1984	10,845	5600	1990	193,245	119,821
1985	22,620	12,529	1991	248,023	154,567
1986	41,662	24,550	1992	315,329	191,508
1987	70,222	40,820	1993	361,509	220,592

Source: U.S. Dept. of Health and Human Services, Centers for Disease Control and Prevention, *HIV/AIDS Surveillance*, March 1994.

(a) Make a table listing the ratio of total deaths caused by AIDS to total cases of AIDS in the United States for each year from 1982 to 1993. For example, in 1982 there were 620 deaths and 1563 cases so the ratio is

$$\frac{620}{1563} \approx .397.$$

(b) As time progresses, what happens to the values of the ratio?

(c) The models found in Section 12.3 for these data are as follows:

For AIDS cases: $f(x) = 2975(x - 2)^2 + 1563$
SPXACEor AIDS deaths: $g(x) = 1818(x - 2)^2 + 620$.

Define the rational function h, where $h(x) = \frac{g(x)}{f(x)}$. Graph $h(x)$ in the window [2, 20] by [0, 1]. Compare $h(x)$ to the values for the ratio found in your table.

(d) Use $h(x)$ to write an equation that approximates the relationship between the functions $f(x)$ and $g(x)$ as x increases.

(e) The ratio of AIDS deaths to AIDS cases can be used to estimate the total number of AIDS deaths. According to the World Health Organization, in 1994 there had been 4 million AIDS cases diagnosed worldwide since the disease began. Predict the total number of deaths caused by AIDS.

64. The table contains incidence ratios by age for deaths due to coronary heart disease (CHD) and lung cancer (LC) when comparing smokers (21–39 cigarettes per day) to nonsmokers.

Age	CHD	LC
55–64	1.9	10
65–74	1.7	9

The incidence ratio of 10 means that smokers are 10 times more likely than nonsmokers to die of lung cancer between the ages of 55 and 64. If the incidence ratio is x, then the percent P (in decimal form) of deaths caused by smoking can be calculated using the rational function defined by

$$P(x) = \frac{x - 1}{x}.$$

(*Source:* Walker, A., *Observation and Inference: An Introduction to the Methods of Epidemiology*, Epidemiology Resources Inc., 1991.)

(a) As x increases, what value does $P(x)$ approach?

(b) Why do you suppose the incidence ratios are slightly smaller for ages 65–74 than for ages 55–64?

65. Refer to Example 9. Let the average number of vehicles arriving at the gate of an amusement park per minute equal k, and let the average number of vehicles admitted by the park attendants equal r. Then the average waiting time T (in minutes) for each vehicle arriving at the park is given by the rational function defined by

$$T(r) = \frac{2r - k}{2r^2 - 2kr}, \quad \text{where } r > k.$$

(*Source:* Mannering, F., and W. Kilareski, *Principles of Highway Engineering and Traffic Control*, John Wiley and Sons, 1990.)

(a) It is known from experience that on Saturday afternoon $k = 25$. Use graphing to estimate the admittance rate r that is necessary to keep the average waiting time T for each vehicle to 30 sec.

(b) If one park attendant can serve 5.3 vehicles per minute, how many park attendants will be needed to keep the average wait to 30 sec?

66. The rational function defined by

$$d(x) = \frac{8710x^2 - 69{,}400x + 470{,}000}{1.08x^2 - 324x + 82{,}200}$$

can be used to accurately model the braking distance for automobiles traveling at x mph, where $20 \le x \le 70$. (*Source:* Mannering, F., and W. Kilareski, *Principles of Highway Engineering and Traffic Control*, John Wiley and Sons, 1990.)

(a) Use graphing to estimate x when $d(x) = 300$.

(b) Complete the table for each value of x.

x	20	25	30	35	40	45	50	55	60	65	70
$d(x)$											

(c) If a car doubles its speed, does the stopping distance double or more than double? Explain.

(d) Suppose the stopping distance doubled whenever the speed doubled. What type of relationship would exist between the stopping distance and the speed?

Chapter **12** **Group Activity**

Using a Function to Model AIDS Deaths

OBJECTIVE Create polynomial functions to model data from a graph; use the model to make predictions.

Throughout the 1980s and 1990s, AIDS became a major health issue in the United States and the rest of the world. In this activity you will create a simple polynomial function to model the death rate of male AIDS patients, ages 25–44. The following graph from the Centers for Disease Control and Prevention shows different causes of death over a 10-yr period for men aged 25–44. The graph showing deaths from AIDS approximates a straight line and can be modeled by a linear function.

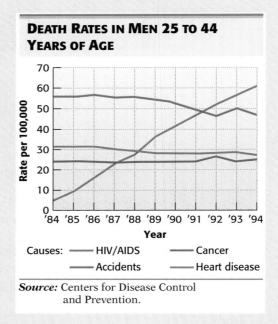

DEATH RATES IN MEN 25 TO 44 YEARS OF AGE

Causes: —— HIV/AIDS —— Cancer
—— Accidents —— Heart disease

Source: Centers for Disease Control and Prevention.

A. Create polynomial models.

   1. Look at the graph for HIV/AIDS deaths. Have each member of the group select two points on the graph. (*Note:* Pick points that you can estimate easily.) Write your points as ordered pairs, letting $x = 0$ represent 1984, $x = 1$ represent 1985, and so on.

   2. Use the methods of Chapter 3 to write an equation of the line through the two points you selected.

   3. Write your equation as a function.

B. Use your models to make predictions.

   1. Compare functions within your group. Are they the same? Why or why not?

2. Select another point on the graph, one not used by anyone in your group to create his or her model. Find the value of each function at this point. Which function more accurately predicted the value shown on the graph?

3. Suppose the death rate from AIDS continued to increase. Use your model to predict the death rate from AIDS for this age group of men in 1997.

4. In 1997 the total number of deaths from AIDS was 14,185. (*Source:* National Center for Health Statistics, U.S. Dept. of Health and Human Services.) The total population of the United States was 267,636,061. Compare the actual rate of AIDS deaths per 100,000 for 1997 with your model. What does this imply about your prediction? What happened to the death rate from AIDS between 1994 and 1997?

CHAPTER **12** SUMMARY

KEY TERMS

12.1 division algorithm
synthetic division
remainder theorem

polynomial function
of degree n
zero (root)

12.2 multiplicity (of a
zero)

12.3 turning points
comprehensive graph

12.4 rational function
vertical asymptote
horizontal asymptote
oblique asymptote

NEW SYMBOLS

$\bar{c}$ the conjugate of
the complex
number c

$x \rightarrow a$ x approaches a

TEST YOUR WORD POWER

See how well you have learned the vocabulary in this chapter. Answers, with examples, follow the Quick Review.

1. **Synthetic division** is
   A. a method for dividing any two
   polynomials
   B. a method for multiplying two
   binomials
   C. a shortcut method for dividing a
   polynomial by a binomial of the
   form $x - k$
   D. the process of adding to a
   binomial the number that makes
   it a perfect square trinomial.

2. A **polynomial function** is a function
   defined by
   A. $f(x) = \sqrt{x}$, for $x \geq 0$
   B. $f(x) = |mx + b|$
   C. $f(x) = a_n x^n + a_{n-1} x^{n-1} + \cdots + a_1 x + a_0$ for complex numbers a_n,
   $a_{n-1}, \ldots, a_1$, and a_0, where
   $a_n \neq 0$
   D. $f(x) = \frac{P(x)}{Q(x)}$, where $Q(x) \neq 0$.

3. A **zero of a function** f is
   A. a point where the function
   changes from increasing to
   decreasing or from decreasing to
   increasing
   B. a value of x that satisfies $f(x) = 0$
   C. a value for which the function is
   undefined
   D. a point where the graph of the
   function intersects the x-axis or
   the y-axis.

4. A **turning point** is
   A. a point where the function
   changes from increasing to
   decreasing or from decreasing to
   increasing
   B. a value of x that satisfies $f(x) = 0$
   C. a value for which the function is
   undefined
   D. a point where the graph of the
   function intersects the x-axis or
   the y-axis.

5. A **rational function** is a function
   defined by
   A. $f(x) = \sqrt{x}$, for $x \geq 0$
   B. $f(x) = |mx + b|$
   C. $f(x) = a_n x^n + a_{n-1} x^{n-1} + \cdots + a_1 x + a_0$ for complex numbers a_n,
   $a_{n-1}, \ldots, a_1$, and a_0, where
   $a_n \neq 0$
   D. $f(x) = \frac{P(x)}{Q(x)}$, where $Q(x) \neq 0$.

6. An **asymptote** is
   A. a line that a graph intersects just
   once
   B. the x-axis or y-axis
   C. a line about which a graph is
   symmetric
   D. a line that a graph approaches
   more closely as x approaches a
   certain value or as $|x|$ gets larger
   without bound.

▰ QUICK REVIEW

CONCEPTS	EXAMPLES

12.1 SYNTHETIC DIVISION

Synthetic division provides a shortcut method for dividing a polynomial by a binomial of the form $x - k$.

Use synthetic division to divide
$$f(x) = 2x^3 - 3x + 2$$
by $x - 1$, and write the result in the form $f(x) = g(x) \cdot q(x) + r(x)$.

$$
\begin{array}{r|rrrr}
1) & 2 & 0 & -3 & 2 \\
   &   & 2 & 2 & -1 \\
\hline
   & 2 & 2 & -1 & 1 \\
\end{array}
$$

$\underbrace{\text{Coefficients of}}_{\text{the quotient}}$ $\underbrace{\text{Remainder}}$

$$2x^3 - 3x + 2 = (x - 1) \cdot (2x^2 + 2x - 1) + 1$$

12.2 ZEROS OF POLYNOMIAL FUNCTIONS

Factor Theorem
The polynomial $x - k$ is a factor of the polynomial $f(x)$ if and only if $f(k) = 0$.

For the polynomial function defined by
$$f(x) = x^3 + x + 2,$$
$f(-1) = 0$. Therefore, $x - (-1)$, or $x + 1$, is a factor of $f(x)$.
Also, since $x - 1$ is a factor of $g(x) = x^3 - 1$, $g(1) = 0$.

Rational Zeros Theorem
Let $f(x) = a_n x^n + a_{n-1} x^{n-1} + \cdots + a_1 x + a_0$, where $a_n \neq 0$, define a polynomial function with integer coefficients. If $\frac{p}{q}$ is a rational number written in lowest terms and if $\frac{p}{q}$ is a zero of f, then p is a factor of the constant term a_0 and q is a factor of the leading coefficient a_n.

The only rational numbers that can possibly be zeros of
$$f(x) = 2x^3 - 9x^2 - 4x - 5$$
are ± 1, $\pm \frac{1}{2}$, ± 5, and $\pm \frac{5}{2}$.
By synthetic division, it can be shown that the only rational zero of $f(x)$ is 5.

$$
\begin{array}{r|rrrr}
5) & 2 & -9 & -4 & -5 \\
   &   & 10 & 5 & 5 \\
\hline
   & 2 & 1 & 1 & 0 \leftarrow f(5) \\
\end{array}
$$

Fundamental Theorem of Algebra
Every function defined by a polynomial of degree 1 or more has at least one complex zero.

$f(x) = x^3 + x + 2$ has at least 1 and at most 3 zeros.

Number of Zeros Theorem
A function defined by a polynomial of degree n has at most n distinct zeros.

Conjugate Zeros Theorem
If $f(x)$ is a polynomial function having only real coefficients and if $a + bi$ is a zero of $f(x)$, then $a - bi$ is also a zero of $f(x)$.

Since $1 + 2i$ is a zero of
$$f(x) = x^3 - 5x^2 + 11x - 15,$$
its conjugate $1 - 2i$ is a zero as well.   (continued)

CONCEPTS	EXAMPLES

12.3 GRAPHS AND APPLICATIONS OF POLYNOMIAL FUNCTIONS

Graphing Using Translations

The graph of the function defined by

$$f(x) = a(x - h)^n + k$$

can be found by considering the effects of the constants a, h, and k on the graph of

$$y = x^n.$$

Sketch the graph of $f(x) = -(x + 2)^4 + 1$.

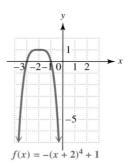

$f(x) = -(x + 2)^4 + 1$

Turning Points

A polynomial function of degree n has at most $n - 1$ turning points.

The graph of

$$f(x) = 4x^5 - 2x^3 + 3x^2 + x - 10$$

has at most $5 - 1 = 4$ turning points.

Graphing Polynomial Functions

To graph a polynomial function f, where $f(x)$ is factorable, find x-intercepts and y-intercepts. Choose a value in each region determined by the x-intercepts to decide whether the graph is above or below the x-axis.

If f is not factorable, use the procedure in the Summary Exercises that follow Section 12.3 to graph $f(x)$.

Sketch the graph of $f(x) = (x + 2)(x - 1)(x + 3)$.

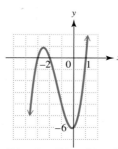

$f(x) = (x + 2)(x - 1)(x + 3)$

Intermediate Value Theorem for Polynomials

If $f(x)$ defines a polynomial function with *real coefficients*, and if for real numbers a and b the values of $f(a)$ and $f(b)$ are opposite in sign, then there exists at least one real zero between a and b.

For the polynomial function defined by

$$f(x) = -x^4 + 2x^3 + 3x^2 + 6,$$
$$f(3.1) = 2.0599 \quad \text{and} \quad f(3.2) = -2.6016.$$

Since $f(3.1) > 0$ and $f(3.2) < 0$, there exists at least one real zero between 3.1 and 3.2.

CONCEPTS	EXAMPLES

Boundedness Theorem

Let $f(x)$ be a polynomial function with *real coefficients* and with a *positive* leading coefficient. If $f(x)$ is divided synthetically by $x - c$, and

a. if $c > 0$ and all numbers in the bottom row of the synthetic division are nonnegative, then $f(x)$ has no zero greater than c;

b. if $c < 0$ and the numbers in the bottom row of the synthetic division alternate in sign (with 0 considered positive or negative, as needed), then $f(x)$ has no zero less than c.

Show that $f(x) = x^3 - x^2 - 8x + 12$ has no zero greater than 4 and no zero less than -4.

$$
\begin{array}{r|rrrr}
4) & 1 & -1 & -8 & 12 \\
  &   & 4 & 12 & 16 \\
\hline
  & 1 & 3 & 4 & 28 \leftarrow \text{All positive}
\end{array}
$$

$$
\begin{array}{r|rrrr}
-4) & 1 & -1 & -8 & 12 \\
   &   & -4 & 20 & -48 \\
\hline
   & 1 & -5 & 12 & -36 \leftarrow \text{Alternating signs}
\end{array}
$$

12.4 GRAPHS AND APPLICATIONS OF RATIONAL FUNCTIONS

Graphing Rational Functions

To graph a rational function in lowest terms, find asymptotes and intercepts. Determine whether the graph intersects the nonvertical asymptote. Plot a few points, as necessary, to complete the sketch.

Graph the rational function defined by $f(x) = \dfrac{x^2 - 1}{(x + 3)(x - 2)}$.

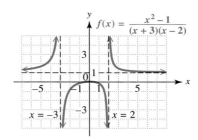

Point of Discontinuity

If a rational function is not written in lowest terms, there may be a "hole" in the graph instead of an asymptote.

Graph the rational function defined by $f(x) = \dfrac{x^2 - 1}{x + 1}$.

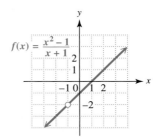

Answers to Test Your Word Power

1. C; *Example:* See Section 12.1 of the Quick Review.   **2.** C; *Examples:* $f(x) = x^2 - 4x + 5$, $f(x) = 6x^3 - 31x^2 - 15x$, $f(x) = x^4 + 2$

3. B; *Example:* 3 is a zero of the polynomial function defined by $f(x) = x^2 - 3x$ since $f(3) = 0$.   **4.** A; *Example:* The graph of

$f(x) = 2x^3 + 5x^2 - x - 6$ has two turning points. See Figure 10 of Section 12.3.   **5.** D; *Examples:* $f(x) = \dfrac{4}{x}$, $f(x) = \dfrac{1}{x - 5}$,

$f(x) = \dfrac{x^2 - 9}{x + 1}$ **6.** D; *Example:* The graph of $f(x) = \dfrac{2}{x + 4}$ has $x = -4$ as a vertical asymptote.

CHAPTER **12** REVIEW EXERCISES

[12.1] *Use synthetic division to perform each division.*

1. $\dfrac{3x^2 - x - 2}{x - 1}$

2. $\dfrac{10x^2 - 3x - 15}{x + 2}$

3. $(2x^3 - 5x^2 + 12) \div (x - 3)$

4. $(-x^4 + 19x^2 + 18x + 15) \div (x + 4)$

Use synthetic division to decide whether -5 is a solution of each equation.

5. $2x^3 + 8x^2 - 14x - 20 = 0$

6. $-3x^4 + 2x^3 + 5x^2 - 9x + 1 = 0$

Use synthetic division to evaluate $f(k)$ for the given value of k.

7. $f(x) = 3x^3 - 5x^2 + 4x - 1; \quad k = -1$

8. $f(x) = x^4 - 2x^3 - 9x - 5; \quad k = 3$

9. Use synthetic division to write the quotient

$$\frac{2x^3 + x - 6}{x + 2}$$

in the form $f(x) = g(x) \cdot q(x) + r(x)$.

[12.2]

✎ **10.** Explain how the rational zeros theorem can also be used to determine the answer in Exercise 6.

Find all rational zeros of each polynomial function in Exercises 11–14.

11. $f(x) = 2x^3 - 9x^2 - 6x + 5$

12. $f(x) = 3x^3 - 10x^2 - 27x + 10$

13. $f(x) = x^3 - \dfrac{17}{6}x^2 - \dfrac{13}{3}x - \dfrac{4}{3}$

14. $f(x) = 8x^4 - 14x^3 - 29x^2 - 4x + 3$

15. Is -1 a zero of f, if $f(x) = 2x^4 + x^3 - 4x^2 + 3x + 1$?

16. Is -2 a zero of f, if $f(x) = 2x^4 + x^3 - 4x^2 + 3x + 1$?

17. Is $x + 1$ a factor of $f(x) = x^3 + 2x^2 + 3x - 1$?

18. Is $x + 1$ a factor of $f(x) = 2x^3 - x^2 + x + 4$?

19. Find a function f defined by a polynomial of degree 3 with real coefficients, having -2, 1, and 4 as zeros, and such that $f(2) = 16$.

20. Find a function f defined by a polynomial of degree 4 with real coefficients, having 1, -1, and $3i$ as zeros, and such that $f(2) = 39$.

21. Find a lowest-degree polynomial $f(x)$ with real coefficients defining a function having zeros 2, -2, and $-i$.

22. Find a lowest-degree polynomial $f(x)$ with real coefficients defining a function having zeros 2, -3, and $5i$.

23. Find a polynomial $f(x)$ of lowest degree with real coefficients defining a function having -3 and $1 - i$ as zeros.

24. Find all zeros of f, if $f(x) = x^4 - 3x^3 - 8x^2 + 22x - 24$, given $1 - i$ is a zero, and factor $f(x)$.

25. Is it possible for a polynomial function of degree 4 with real coefficients to have three zeros that are not real and one zero that is real? Explain.

26. Suppose that a polynomial function has six real zeros. Is it possible for the function to be of degree 5? Explain.

[12.3]

27. For the polynomial function defined by $f(x) = x^3 - 3x^2 - 7x + 12$, $f(4) = 0$. Therefore, we can say that 4 is a(n) _____ of the function, 4 is a(n) _____ of the equation $x^3 - 3x^2 - 7x + 12 = 0$, and that $(4, 0)$ is a(n) _____ of the graph of the function.

28. Which one of the following is not a polynomial function?

A. $f(x) = x^2$ **B.** $f(x) = 2x + 5$ **C.** $f(x) = \dfrac{1}{x}$ **D.** $f(x) = x^{100}$

Give the maximum possible number of turning points of the graph of each polynomial function.

29. $f(x) = x^3 - 9x$

30. $f(x) = 4x^4 - 6x^2 + 2$

Graph each polynomial function.

31. $f(x) = x^3 + 5$

32. $f(x) = 1 - x^4$

33. $f(x) = x^2(2x + 1)(x - 2)$

34. $f(x) = 2x^3 + 13x^2 + 15x$

35. $f(x) = 12x^3 - 13x^2 - 5x + 6$, given that 1 is a zero

36. $f(x) = x^4 - 2x^3 - 5x^2 + 6x$, given that 0 and 1 are zeros

Show that each polynomial function has real zeros satisfying the given conditions.

37. $f(x) = 3x^3 - 8x^2 + x + 2$, zeros in $[-1, 0]$ and $[2, 3]$

38. $f(x) = 4x^3 - 37x^2 + 50x + 60$, zeros in $[2, 3]$ and $[7, 8]$

39. $f(x) = x^3 + 2x^2 - 22x - 8$, zeros in $[-1, 0]$ and $[-6, -5]$

40. $f(x) = 2x^4 - x^3 - 21x^2 + 51x - 36$, no zero greater than 4

41. $f(x) = 6x^4 + 13x^3 - 11x^2 - 3x + 5$, no zero greater than 1 or less than -3

42. Use a graphing calculator to approximate the real zeros for each polynomial function. Round to the nearest thousandth for irrational zeros.

(a) $f(x) = 2x^3 - 11x^2 - 2x + 2$ **(b)** $f(x) = x^4 - 4x^3 - 5x^2 + 14x - 15$
(c) $f(x) = x^3 + 3x^2 - 4x - 2$

Graph each polynomial function.

43. $f(x) = 2x^3 - 11x^2 - 2x + 2$ (See Exercise 42(a).)

44. $f(x) = x^4 - 4x^3 - 5x^2 + 14x - 15$ (See Exercise 42(b).)

45. $f(x) = x^3 + 3x^2 - 4x - 2$ (See Exercise 42(c).)

46. $f(x) = 2x^4 - 3x^3 + 4x^2 + 5x - 1$

[12.4] *Graph each rational function.*

47. $f(x) = \dfrac{8}{x}$

48. $f(x) = \dfrac{2}{3x - 1}$

49. $f(x) = \dfrac{4x - 2}{3x + 1}$

50. $f(x) = \dfrac{6x}{(x - 1)(x + 2)}$

51. $f(x) = \dfrac{2x}{x^2 - 1}$

52. $f(x) = \dfrac{x^2 + 4}{x + 2}$

53. $f(x) = \dfrac{x^2 - 1}{x}$

54. $f(x) = \dfrac{x^2 + 6x + 5}{x - 3}$

55. $f(x) = \dfrac{4x^2 - 9}{2x + 3}$

56. Economist Arthur Laffer has been a center of controversy because of his **Laffer curve,** an idealized version of which is shown here. According to this curve, increasing a tax rate, say from x_1 percent to x_2 percent on the graph here, can actually lead to a decrease in government revenue. All economists agree on the endpoints, 0 revenue at tax rates of both 0% and 100%, but there is much disagreement on the location of the rate x_1 that produces maximum revenue.

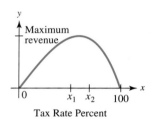

(a) Suppose an economist studying the Laffer curve produces the rational function defined by

$$R(x) = \frac{80x - 8000}{x - 110},$$

with $R(x)$ giving government revenue in tens of millions of dollars for a tax rate of x percent, with the function valid for $55 \leq x \leq 100$. Find the revenue for each tax rate.

(i) 55%   **(ii)** 60%   **(iii)** 70%   **(iv)** 90%   **(v)** 100%
(vi) Graph R in the window [0, 100] by [0, 80].

(b) Suppose an economist studies a different tax, this time producing

$$R(x) = \frac{60x - 6000}{x - 120},$$

where $R(x)$ is government revenue in millions of dollars from a tax rate of x percent, with $R(x)$ valid for $50 \leq x \leq 100$. Find the revenue for each tax rate.

(i) 50%   **(ii)** 60%   **(iii)** 80%   **(iv)** 100%
(v) Graph R in the window [0, 100] by [0, 50].

MIXED REVIEW EXERCISES

Graph each function.

57. $f(x) = -\dfrac{1}{x^3}$

58. $f(x) = (x - 2)^4$

59. $f(x) = \dfrac{3 - 4x}{2x + 1}$

60. $f(x) = (x + 3)^2(x - 1)$

61. $f(x) = 2x^5 - 3x^4 + x^2 - 2$

62. $f(x) = \dfrac{(x + 4)(2x + 5)}{x - 1}$

63. $f(x) = \dfrac{x^3 + 1}{x + 1}$

64. $f(x) = 3x^3 + 2x^2 - 27x - 18$

CHAPTER **12** TEST

Use synthetic division in Exercises 1–5.

1. Find the quotient when $2x^3 - 3x - 10$ is divided by $x - 2$.

2. Write the quotient $\dfrac{x^4 + 2x^3 - x^2 + 3x - 5}{x + 1}$ in the form $f(x) = g(x) \cdot q(x) + r(x)$.

3. Determine whether $x + 3$ is a factor of $x^4 - 2x^3 - 15x^2 - 4x - 12$.

4. Evaluate $f(-4)$ if $f(x) = 3x^3 - 4x^2 - 5x + 9$.

5. Is 3 a zero of $f(x) = 6x^4 - 11x^3 - 35x^2 + 34x + 24$? Why or why not?

6. Find a function f defined by a polynomial having real coefficients, with degree 4, zeros 2, -1, and i, and such that $f(3) = 80$.

7. For the polynomial function defined by
$$f(x) = 6x^3 - 25x^2 + 12x + 7,$$
  (a) list all possible rational zeros;
  (b) find all rational zeros.

8. Show that the polynomial function defined by
$$f(x) = 2x^4 - 3x^3 + 4x^2 - 5x - 1$$
  has no real zeros greater than 2 or less than -1.

9. For the polynomial function defined by
$$f(x) = 2x^3 - x + 3,$$
  (a) use the intermediate value theorem to show that there is a zero between -2 and -1;
  (b) use a graphing calculator to find this zero to the nearest thousandth.

10. For the graph of the polynomial function defined by
$$f(x) = 2x^3 - 9x^2 + 4x + 8,$$
  without actually graphing,
  (a) determine the maximum possible number of x-intercepts;
  (b) determine the maximum possible number of turning points.

Graph each polynomial function.

11. $f(x) = (x - 1)^4$

12. $f(x) = x(x + 1)(x - 2)$

13. $f(x) = 2x^3 - 7x^2 + 2x + 3$, given that 3 is a zero

14. $f(x) = x^4 - 5x^2 + 6$

15. The polynomial function defined by

$$f(x) = -.184x^3 + 1.45x^2 + 10.7x - 27.9$$

models the average temperature at Trout Lake, Canada, in degrees Fahrenheit, where $x = 1$ corresponds to January and $x = 12$ to December. What is the average temperature in June?

Graph each rational function.

16. $f(x) = \dfrac{-2}{x + 3}$

17. $f(x) = \dfrac{3x - 1}{x - 2}$

18. $f(x) = \dfrac{x^2 - 1}{x^2 - 9}$

19. What is the equation of the oblique asymptote of the graph of

$$f(x) = \frac{2x^2 + x - 6}{x - 1}?$$

20. Which one of the functions defined below has a graph with no x-intercepts?

A. $f(x) = (x - 2)(x + 3)^2$ **B.** $f(x) = \dfrac{x + 7}{x - 2}$

C. $f(x) = x^3 - x$ **D.** $f(x) = \dfrac{1}{x^2 + 4}$

CUMULATIVE REVIEW EXERCISES CHAPTERS 1–12

Simplify.

1. $-5 + [-3 - (-2 - 3)] - 6$

2. $5(3x + 2) - 4(8x - 6)$

Solve each equation or inequality.

3. $4x + 8x = 17x - 10$

4. $.10(x - 6) + .05x = .06(50)$

5. $\dfrac{x + 1}{3} + \dfrac{2x}{3} = x + \dfrac{1}{3}$

6. $4(x + 2) \geq 6x - 8$

7. $5 < 3x - 4 < 9$

8. $|m + 2| - 3 > 2$

9. A jar of 38 coins contains only nickels and dimes. The amount of money in the jar is \$2.50. How many nickels and how many dimes are in the jar?

10. Find the slope of the line through the points $(-2, 7)$ and $(4, -3)$.

11. Find an equation of the line passing through the point $(5, -3)$ and perpendicular to the graph of $y = \dfrac{1}{3}x - 6$. Give it in slope-intercept form.

Sketch the graph of each equation or inequality.

12. $-3x + 5y = -15$ **13.** $y \le -2x + 7$

14. If $f(x) = x^3 - 2x^2 + 3x - 6$, find $f(-2)$.

15. If y varies directly as x and $y = 4$ when $x = 6$, find y when $x = 20$.

Solve each system using the method indicated.

16. $3x + 2y = -4$
     $y = 2x + 5$ (substitution)

17. $-2x + \ y - 4z = 2$
     $3x + 2y - \ z = -3$
     $-x - 4y - 2z = -17$ (row operations)

18. $\ 2x + \ y + 3z = 1$
     $\ x - 2y + \ z = -3$
     $-3x + \ y - 2z = -4$ (Cramer's rule)

19. Sketch the graph of $f(x) = x^2$.

Perform the indicated operations.

20. $(r^4 - 2r^3 + 6)(3r - 1)$ **21.** $[(k - 5h) + 2]^2$

Factor.

22. $6x^2 - 15x - 9$ **23.** $729 + 8y^6$

24. Solve $x^3 + 3x^2 - x - 3 = 0$ by factoring. (*Hint:* Factor by grouping.)

25. Consider the rational expression $\dfrac{x^2 - 36}{4x^2 - 21x - 18}$.

  (a) For what values of x is this expression undefined?
  (b) Express it in lowest terms.

Perform each operation and express the answer in lowest terms.

26. $\dfrac{2r + 4}{5r} \cdot \dfrac{3r}{5r + 10}$ **27.** $\dfrac{y^2 - 2y - 3}{y^2 + 4y + 4} \div \dfrac{y^2 - 1}{y^2 + y - 2}$

28. $\dfrac{3x + 12}{2x + 7} + \dfrac{-7x - 26}{2x + 7}$ **29.** $\dfrac{2}{r - 2} - \dfrac{r + 3}{r - 1}$

30. Simplify the complex fraction $\dfrac{\dfrac{1}{y} + \dfrac{1}{y - 1}}{\dfrac{1}{y} - \dfrac{2}{y - 1}}$.

Solve each equation.

31. $\dfrac{10}{m^2} - \dfrac{3}{m} = 1$ **32.** $\dfrac{3}{a + 1} = \dfrac{1}{a - 1} - \dfrac{2}{a^2 - 1}$

Simplify each radical expression.

33. $\sqrt{32} - \sqrt{128} + \sqrt{162}$ **34.** $\dfrac{3 - 4\sqrt{2}}{1 - \sqrt{2}}$

35. Solve $x = \sqrt{x + 2}$.   **36.** Express in $a + bi$ form: $\dfrac{6 + 2i}{1 + i}$.

Consider the quadratic equation $x^2 + 4x + 2 = 0$ in Exercises 37–39.

37. What is the value of the discriminant?

38. Based on your answer to Exercise 37, are the solutions rational, irrational, or not real?

39. Find the solution set of the equation.

40. Solve the quadratic inequality $3x^2 - 13x - 10 \leq 0$.

Decide whether the graph of the relation is symmetric with respect to the x-axis, the y-axis, the origin, or none of these.

41. $x = y^2 + 3$ **42.** $y = x^2 + 3$ **43.** $y = 5x^3$

44. Give the interval over which the graph of $f(x) = -x^2 + 3$ is increasing.

45. Find $f(3)$ if $f(x) = \begin{cases} x^2 - 6 & \text{if } x \leq 3 \\ 2x & \text{if } x > 3. \end{cases}$

46. Find $f^{-1}(x)$ if $f(x) = \sqrt[3]{3x + 5}$.

47. Evaluate $\log_{16} \dfrac{1}{8}$.

48. Use a calculator to find an approximation to four decimal places.

  (a) $\ln 4.6$ **(b)** $\log .0035$

49. Solve $\log_2 x + \log_2(x + 2) - 3 = 0$.

50. Find the quotient by using synthetic division:
$$(x^4 + 7x^3 - 5x^2 + 2x + 13) \div (x + 1).$$

51. Consider the polynomial function defined by
$$f(x) = x^3 + 7x^2 + 7x - 15.$$

  (a) Given that -3 is a zero of f, find all other zeros.
  (b) Factor $f(x)$ into linear factors.
  (c) What are the x-intercepts of the graph of f? What is the y-intercept?
  (d) Sketch the graph.

52. Consider the rational function defined by
$$f(x) = \frac{x^2 - 4}{x^2 - 9}.$$

  (a) Give the equations of the vertical asymptotes of the graph of f.
  (b) What are the x-intercepts of the graph of f?
  (c) What is the equation of the horizontal asymptote of the graph of f?
  (d) Sketch the graph.

53. Antique-car owners often enter their cars in a *concours d'elegance* in which a maximum of 100 points can be awarded to a particular car. Points are awarded for the general attractiveness of the car. The function defined by

$$C(x) = \frac{10x}{49(101 - x)}$$

expresses the cost, in thousands of dollars, of restoring a car so that it will win x points. Graph the function for $0 \leq x < 101$.

Conic Sections

When a plane intersects an infinite cone at different angles, it produces curves called *conic sections*. In Chapter 10 we studied one conic section, the *parabola*. In 1609, Johannes Kepler (1571–1630) established the importance of another conic section, the *ellipse,* when he discovered that the orbits of the planets around the sun are elliptical, not circular. Exercises 51 and 52 of Section 13.1 involve the equations of the elliptical orbits formed by the planets Mars and Venus.

13.1 The Circle and the Ellipse

OBJECTIVES

1 Find the equation of a circle given the center and radius.

2 Determine the center and radius of a circle given its equation.

3 Recognize the equation of an ellipse.

4 Graph ellipses.

5 Graph circles and ellipses using a graphing calculator.

When an infinite cone is intersected by a plane, the resulting figure is called a **conic section.** The parabola is one example of a conic section; circles, ellipses, and hyperbolas may also result. See Figure 1.

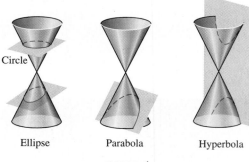

Circle

Ellipse   Parabola   Hyperbola

FIGURE 1

OBJECTIVE 1 Find the equation of a circle given the center and radius. A **circle** is the set of all points in a plane that lie a fixed distance from a fixed point. The fixed point is called the **center,** and the fixed distance is called the **radius.** We use the distance formula to find an equation of a circle.

▧ **EXAMPLE 1 Finding the Equation of a Circle and Graphing It**

Find an equation of the circle with radius 3 and center at $(0, 0)$, and graph it.

If the point (x, y) is on the circle, then the distance from (x, y) to the center $(0, 0)$ is 3. By the distance formula,

$$\sqrt{(x_2 - x_1)^2 + (y_2 - y_1)^2} = d$$
$$\sqrt{(x - 0)^2 + (y - 0)^2} = 3$$
$$x^2 + y^2 = 9. \qquad \text{Square both sides.}$$

An equation of this circle is $x^2 + y^2 = 9$. The graph is shown in Figure 2.

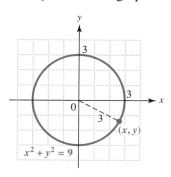

FIGURE 2

Now Try Exercise 1.

A circle may not be centered at the origin, as seen in the next example.

EXAMPLE 2   Finding an Equation of a Circle and Graphing It

Find an equation of the circle with center at $(4, -3)$ and radius 5, and graph it.

Use the distance formula again.

$$\sqrt{(x-4)^2 + [y - (-3)]^2} = 5$$

$$(x-4)^2 + (y+3)^2 = 25 \qquad \text{Square both sides.}$$

To graph the circle, plot the center $(4, -3)$, then move 5 units right, left, up, and down from the center. Draw a smooth curve through these four points, sketching one quarter of the circle at a time. The graph of this circle is shown in Figure 3.

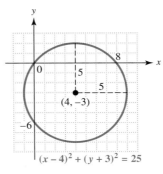

FIGURE 3

Now Try Exercises 7 and 23.

Examples 1 and 2 suggest the form of an equation of a circle with radius r and center at (h, k). If (x, y) is a point on the circle, then the distance from the center (h, k) to the point (x, y) is r. By the distance formula,

$$\sqrt{(x-h)^2 + (y-k)^2} = r.$$

Squaring both sides gives the **center-radius form** of the equation of a circle.

Equation of a Circle (Center-Radius Form)

$$(x - h)^2 + (y - k)^2 = r^2$$

is an equation of the circle with radius r and center at (h, k).

EXAMPLE 3   Using the Center-Radius Form of the Equation of a Circle

Find an equation of the circle with center at $(-1, 2)$ and radius 4.

Use the center-radius form, with $h = -1$, $k = 2$, and $r = 4$.

$$(x - h)^2 + (y - k)^2 = r^2$$

$$[x - (-1)]^2 + (y - 2)^2 = 4^2$$

$$(x + 1)^2 + (y - 2)^2 = 16$$

Now Try Exercise 9.

OBJECTIVE 2 Determine the center and radius of a circle given its equation. In the equation found in Example 2, multiplying out $(x - 4)^2$ and $(y + 3)^2$ gives

$$(x - 4)^2 + (y + 3)^2 = 25$$
$$x^2 - 8x + 16 + y^2 + 6y + 9 = 25$$
$$x^2 + y^2 - 8x + 6y = 0.$$

This general form suggests that an equation with both x^2- and y^2-terms with equal coefficients may represent a circle. The next example shows how to tell, by completing the square. This procedure was introduced in Chapter 9.

EXAMPLE 4 Completing the Square to Find the Center and Radius

Find the center and radius of the circle $x^2 + y^2 + 2x + 6y - 15 = 0$, and graph it.

Since the equation has x^2- and y^2-terms with equal coefficients, its graph might be that of a circle. To find the center and radius, complete the squares on x and y.

$$x^2 + y^2 + 2x + 6y = 15 \qquad \text{Get the constant on the right.}$$

$$(x^2 + 2x \quad) + (y^2 + 6y \quad) = 15 \qquad \text{Rewrite in anticipation of completing the square.}$$

$$\left[\frac{1}{2}(2)\right]^2 = 1 \qquad \left[\frac{1}{2}(6)\right]^2 = 9 \qquad \text{Square half the coefficient of each middle term.}$$

$$(x^2 + 2x + 1) + (y^2 + 6y + 9) = 15 + 1 + 9 \qquad \text{Complete the squares on both } x \text{ and } y.$$

$$(x + 1)^2 + (y + 3)^2 = 25 \qquad \text{Factor on the left; add on the right.}$$

$$[x - (-1)]^2 + [y - (-3)]^2 = 5^2 \qquad \text{Center-radius form}$$

The final equation

$$[x - (-1)]^2 + [y - (-3)]^2 = 5^2$$

or

$$(x + 1)^2 + (y + 3)^2 = 5^2$$

indicates that the graph is a circle with center at $(-1, -3)$ and radius 5. The graph is shown in Figure 4.

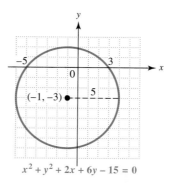

$$x^2 + y^2 + 2x + 6y - 15 = 0$$

FIGURE 4

Now Try Exercise 11.

NOTE If the procedure of Example 4 leads to an equation of the form $(x - h)^2 + (y - k)^2 = 0$, then the graph is the single point (h, k). If the constant on the right side is negative, then the equation has no graph.

OBJECTIVE 3 Recognize the equation of an ellipse. An **ellipse** is the set of all points in a plane the *sum* of whose distances from two fixed points is constant. These fixed points are called **foci** (singular: *focus*). Figure 5 shows an ellipse whose foci are $(c, 0)$ and $(-c, 0)$, with x-intercepts $(a, 0)$ and $(-a, 0)$ and y-intercepts $(0, b)$ and $(0, -b)$. It is shown in more advanced courses that $c^2 = a^2 - b^2$ for an ellipse of this type. The origin is the **center** of the ellipse.

An ellipse has the following equation.

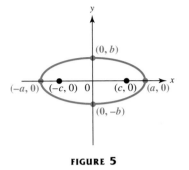

FIGURE 5

> **Equation of an Ellipse**
>
> The ellipse whose x-intercepts are $(a, 0)$ and $(-a, 0)$ and whose y-intercepts are $(0, b)$ and $(0, -b)$ has an equation of the form
> $$\frac{x^2}{a^2} + \frac{y^2}{b^2} = 1.$$

NOTE A circle is a special case of an ellipse, where $a^2 = b^2$.

The paths of Earth and other planets around the sun are approximately ellipses; the sun is at one focus and a point in space is at the other. The orbits of communication satellites and other space vehicles are elliptical. Elliptical bicycle gears are designed to respond to the legs' natural strengths and weaknesses. At the top and bottom of the powerstroke, where the legs have the least leverage, the gear offers little resistance, but as the gear rotates, the resistance increases. This allows the legs to apply more power where it is most naturally available. See Figure 6.

FIGURE 6

OBJECTIVE 4 Graph ellipses. To graph an ellipse centered at the origin, we plot the four intercepts and then sketch the ellipse through those points.

EXAMPLE 5 Graphing Ellipses

Graph each ellipse.

(a) $\dfrac{x^2}{49} + \dfrac{y^2}{36} = 1$

Here, $a^2 = 49$, so $a = 7$, and the x-intercepts for this ellipse are $(7, 0)$ and $(-7, 0)$. Similarly, $b^2 = 36$, so $b = 6$, and the y-intercepts are $(0, 6)$ and $(0, -6)$. Plotting the intercepts and sketching the ellipse through them gives the graph in Figure 7 on the next page.

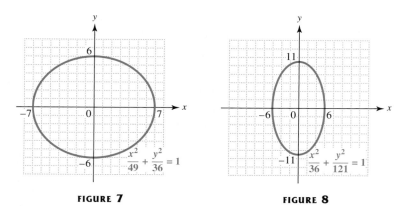

FIGURE 7　　　　　　**FIGURE 8**

(b) $\dfrac{x^2}{36} + \dfrac{y^2}{121} = 1$

The x-intercepts for this ellipse are $(6, 0)$ and $(-6, 0)$, and the y-intercepts are $(0, 11)$ and $(0, -11)$. Join these with the smooth curve of an ellipse. The graph has been sketched in Figure 8.

Now Try Exercises 27 and 31.

As with the graphs of functions and circles, the graph of an ellipse may be shifted horizontally and vertically, as in the next example.

EXAMPLE 6 **Graphing an Ellipse Shifted Horizontally and Vertically**

Graph $\dfrac{(x - 2)^2}{25} + \dfrac{(y + 3)^2}{49} = 1$.

Just as $(x - 2)^2$ and $(y + 3)^2$ would indicate that the center of a circle would be $(2, -3)$, so it is with this ellipse. Figure 9 shows that the graph goes through the four points $(2, 4)$, $(7, -3)$, $(2, -10)$, and $(-3, -3)$. The x-values of these points are found by adding $\pm a = \pm 5$ to 2, and the y-values come from adding $\pm b = \pm 7$ to -3.

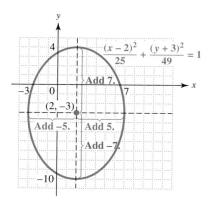

FIGURE 9

Now Try Exercise 35.

OBJECTIVE 5 Graph circles and ellipses using a graphing calculator. The only conic section whose graph is a function is the vertical parabola with equation $f(x) = ax^2 + bx + c$. Therefore, a graphing calculator in function mode cannot directly graph a circle or an ellipse. We must first solve the equation for y, getting two functions y_1 and y_2. The union of these two graphs is the graph of the entire figure. For example, to graph $(x + 3)^2 + (y + 2)^2 = 25$, begin by solving for y.

$$(x + 3)^2 + (y + 2)^2 = 25$$
$$(y + 2)^2 = 25 - (x + 3)^2 \qquad \text{Subtract } (x + 3)^2.$$
$$y + 2 = \pm\sqrt{25 - (x + 3)^2} \qquad \text{Take square roots.}$$
$$y = -2 \pm \sqrt{25 - (x + 3)^2} \qquad \text{Subtract 2.}$$

The two functions to be graphed are

$$y_1 = -2 + \sqrt{25 - (x + 3)^2} \qquad \text{and} \qquad y_2 = -2 - \sqrt{25 - (x + 3)^2}.$$

To get an undistorted screen, a *square viewing window* must be used. (Refer to your instruction manual for details.) See Figure 10. The two semicircles seem to be disconnected. This is because the graphs are nearly vertical at those points, and the calculator cannot show a true picture of the behavior there.

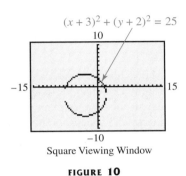

$(x + 3)^2 + (y + 2)^2 = 25$

Square Viewing Window

FIGURE 10

13.1 EXERCISES

For Extra Help

 Student's Solutions Manual

 MyMathLab

 InterAct Math Tutorial Software

 AW Math Tutor Center

 MathXL

 Digital Video Tutor CD 13/Videotape 15

1. See Example 1. Consider the circle whose equation is $x^2 + y^2 = 25$.

  (a) What are the coordinates of its center?   **(b)** What is its radius?
  (c) Sketch its graph.

2. Explain why a set of points defined by a circle does not satisfy the definition of a function.

Match each equation with the correct graph. See Examples 1–3.

3. $(x - 3)^2 + (y - 2)^2 = 25$ **A.**

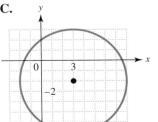

B.

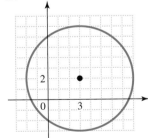

4. $(x - 3)^2 + (y + 2)^2 = 25$

5. $(x + 3)^2 + (y - 2)^2 = 25$ **C.**

D.

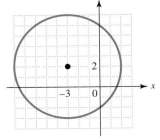

6. $(x + 3)^2 + (y + 2)^2 = 25$

Find the equation of a circle satisfying the given conditions. See Examples 2 and 3.

7. Center: $(-4, 3)$; radius: 2

8. Center: $(5, -2)$; radius: 4

9. Center: $(-8, -5)$; radius: $\sqrt{5}$

10. Center: $(-12, 13)$; radius: $\sqrt{7}$

Find the center and radius of each circle. (Hint: In Exercises 15 and 16, divide each side by a common factor.) See Example 4.

11. $x^2 + y^2 + 4x + 6y + 9 = 0$

12. $x^2 + y^2 - 8x - 12y + 3 = 0$

13. $x^2 + y^2 + 10x - 14y - 7 = 0$

14. $x^2 + y^2 - 2x + 4y - 4 = 0$

15. $3x^2 + 3y^2 - 12x - 24y + 12 = 0$

16. $2x^2 + 2y^2 + 20x + 16y + 10 = 0$

17. A circle can be drawn on a piece of posterboard by fastening one end of a string with a thumbtack, pulling the string taut with a pencil, and tracing a curve, as shown in the figure. Explain why this method works.

18. This figure shows how the crawfish race is held at the Crawfish Festival in Breaux Bridge, Louisiana. Explain why a circular "racetrack" is appropriate for such a race.

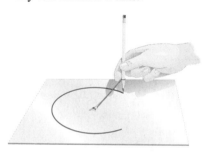

Graph each circle. Identify the center if it is not at the origin. See Examples 1, 2, and 4.

19. $x^2 + y^2 = 9$

20. $x^2 + y^2 = 4$

21. $2y^2 = 10 - 2x^2$

22. $3x^2 = 48 - 3y^2$

23. $(x + 3)^2 + (y - 2)^2 = 9$

24. $(x - 1)^2 + (y + 3)^2 = 16$

25. $x^2 + y^2 - 4x - 6y + 9 = 0$

26. $x^2 + y^2 + 8x + 2y - 8 = 0$

Graph each ellipse. See Examples 5 and 6.

27. $\dfrac{x^2}{9} + \dfrac{y^2}{25} = 1$

28. $\dfrac{x^2}{9} + \dfrac{y^2}{16} = 1$

29. $\dfrac{x^2}{36} = 1 - \dfrac{y^2}{16}$

30. $\dfrac{x^2}{9} = 1 - \dfrac{y^2}{4}$

31. $\dfrac{y^2}{25} = 1 - \dfrac{x^2}{49}$

32. $\dfrac{y^2}{9} = 1 - \dfrac{x^2}{16}$

33. $\dfrac{x^2}{16} + \dfrac{y^2}{4} = 1$

34. $\dfrac{x^2}{49} + \dfrac{y^2}{81} = 1$

35. $\dfrac{(x + 1)^2}{64} + \dfrac{(y - 2)^2}{49} = 1$

36. $\dfrac{(x - 4)^2}{9} + \dfrac{(y + 2)^2}{4} = 1$

37. $\dfrac{(x - 2)^2}{16} + \dfrac{(y - 1)^2}{9} = 1$

38. $\dfrac{(x + 3)^2}{25} + \dfrac{(y + 2)^2}{36} = 1$

39. It is possible to sketch an ellipse on a piece of posterboard by fastening two ends of a length of string, pulling the string taut with a pencil, and tracing a curve, as shown in the figure. Explain why this method works.

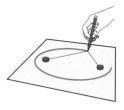

40. Discuss the similarities and differences between the equations of a circle and an ellipse.

41. Explain why a set of ordered pairs whose graph forms an ellipse does not satisfy the definition of a function.

42. (a) How many points are there on the graph of $(x - 4)^2 + (y - 1)^2 = 0$? Explain.
(b) How many points are there on the graph of $(x - 4)^2 + (y - 1)^2 = -1$? Explain.

TECHNOLOGY INSIGHTS (EXERCISES 43 AND 44)

---

43. The circle shown in the calculator graph was created using function mode, with a square viewing window. It is the graph of $(x + 2)^2 + (y - 4)^2 = 16$. What are the two functions y_1 and y_2 that were used to obtain this graph?

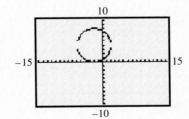

44. The ellipse shown in the calculator graph was graphed using function mode, with a square viewing window. It is the graph of $\frac{x^2}{4} + \frac{y^2}{9} = 1$. What are the two functions y_1 and y_2 that were used to obtain this graph?

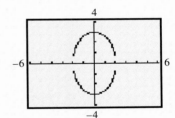

Use a graphing calculator in function mode to graph each circle or ellipse. Use a square viewing window. See Objective 5.

45. $x^2 + y^2 = 36$

46. $(x - 2)^2 + y^2 = 49$

47. $\frac{x^2}{16} + \frac{y^2}{4} = 1$

48. $\frac{(x - 3)^2}{25} + \frac{y^2}{9} = 1$

Solve each problem.

49. An arch has the shape of half an ellipse. The equation of the ellipse is $100x^2 + 324y^2 = 32,400$, where x and y are in meters.

(a) How high is the center of the arch?
(b) How wide is the arch across the bottom?

NOT TO SCALE

50. A one-way street passes under an overpass, which is in the form of the top half of an ellipse, as shown in the figure. Suppose that a truck 12 ft wide passes directly under the overpass. What is the maximum possible height of this truck?

In Exercises 51 and 52, see Figure 15 and use the fact that $c^2 = a^2 - b^2$ where $a^2 > b^2$.

51. The orbit of Mars is an ellipse with the sun at one focus. For x and y in millions of miles, the equation of the orbit is

$$\frac{x^2}{141.7^2} + \frac{y^2}{141.1^2} = 1.$$

(*Source:* Kaler, James B., *Astronomy!*, Addison-Wesley, 1997.)

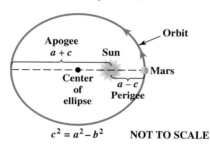

c² = a² – b²   NOT TO SCALE

(a) Find the greatest distance (the *apogee*) from Mars to the sun.
(b) Find the smallest distance (the *perigee*) from Mars to the sun.

52. The orbit of Venus around the sun (one of the foci) is an ellipse with equation

$$\frac{x^2}{5013} + \frac{y^2}{4970} = 1,$$

where x and y are measured in millions of miles. (*Source:* Kaler, James B., *Astronomy!*, Addison-Wesley, 1997.)

(a) Find the greatest distance between Venus and the sun.
(b) Find the smallest distance between Venus and the sun.

A lithotripter is a machine used to crush kidney stones using shock waves. The patient is placed in an elliptical tub with the kidney stone at one focus of the ellipse. A beam is projected from the other focus to the tub, so that it reflects to hit the kidney stone.

53. Suppose a lithotripter is based on the ellipse with equation

$$\frac{x^2}{36} + \frac{y^2}{9} = 1.$$

How far from the center of the ellipse must the kidney stone and the source of the beam be placed?

54. Rework Exercise 53 if the equation of the ellipse is $9x^2 + 4y^2 = 36$. (*Hint:* Write the equation in fraction form by dividing each term by 36, and use $c^2 = b^2 - a^2$, since $b > a$ here.)

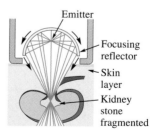

The top of an ellipse is illustrated in this depiction of how a lithotripter crushes kidney stones.

Source: Adapted drawing of an ellipse in illustration of a lithotripter. The American Medical Association, *Encyclopedia of Medicine*, 1989.

55. (a) Suppose that $(c, 0)$ and $(-c, 0)$ are the foci of an ellipse and that the sum of the distances from any point (x, y) on the ellipse to the two foci is $2a$. See the figure. Show that the equation of the resulting ellipse is

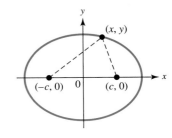

$$\frac{x^2}{a^2} + \frac{y^2}{a^2 - c^2} = 1.$$

   (b) Show that in the equation in part (a), the x-intercepts are $(a, 0)$ and $(-a, 0)$.

   (c) Let $b^2 = a^2 - c^2$, and show that $(0, b)$ and $(0, -b)$ are the y-intercepts in the equation in part (a).

56. Use the result of Exercise 55(a) to find an equation of an ellipse with foci $(3, 0)$ and $(-3, 0)$, where the sum of the distances from any point of the ellipse to the two foci is 10.

13.2   The Hyperbola and Functions Defined by Radicals

OBJECTIVES

1 Recognize the equation of a hyperbola.

2 Graph hyperbolas by using asymptotes.

3 Identify conic sections by their equations.

4 Graph certain square root functions.

OBJECTIVE 1   Recognize the equation of a hyperbola.   A **hyperbola** is the set of all points in a plane such that the absolute value of the *difference* of the distances from two fixed points (called *foci*) is constant. Figure 11 shows a hyperbola; using the distance formula and the definition above, we can show that this hyperbola has equation

$$\frac{x^2}{16} - \frac{y^2}{12} = 1.$$

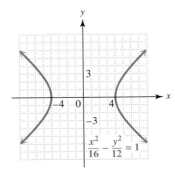

FIGURE 11

   To graph hyperbolas centered at the origin, we need to find their intercepts. For the hyperbola in Figure 11, we proceed as follows.

x-Intercepts	**y-Intercepts**
Let $y = 0$.	Let $x = 0$.
$\dfrac{x^2}{16} - \dfrac{0^2}{12} = 1$ Let $y = 0$.	$\dfrac{0^2}{16} - \dfrac{y^2}{12} = 1$ Let $x = 0$.
$\dfrac{x^2}{16} = 1$	$-\dfrac{y^2}{12} = 1$
$x^2 = 16$ Multiply by 16.	$y^2 = -12$ Multiply by -12.
$x = \pm 4$	
The x-intercepts are $(4, 0)$ and $(-4, 0)$.	Because there are no *real* solutions to $y^2 = -12$, the graph has no y-intercepts.

The graph of $\dfrac{x^2}{16} - \dfrac{y^2}{12} = 1$ has no y-intercepts. On the other hand, the hyperbola in Figure 12 has no x-intercepts. Its equation is

$$\frac{y^2}{25} - \frac{x^2}{9} = 1,$$

with y-intercepts $(0, 5)$ and $(0, -5)$.

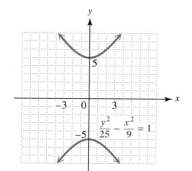

FIGURE 12

Equations of Hyperbolas

A hyperbola with x-intercepts $(a, 0)$ and $(-a, 0)$ has an equation of the form

$$\frac{x^2}{a^2} - \frac{y^2}{b^2} = 1,$$

and a hyperbola with y-intercepts $(0, b)$ and $(0, -b)$ has an equation of the form

$$\frac{y^2}{b^2} - \frac{x^2}{a^2} = 1.$$

OBJECTIVE 2 Graph hyperbolas by using asymptotes. The two branches of the graph of a hyperbola approach a pair of intersecting straight lines, which are its asymptotes. See Figure 13 on the next page. The asymptotes are useful for sketching the graph of the hyperbola.

Asymptotes of Hyperbolas

The extended diagonals of the rectangle with vertices (corners) at the points (a, b), $(-a, b)$, $(-a, -b)$, and $(a, -b)$ are the **asymptotes** of the hyperbolas

$$\frac{x^2}{a^2} - \frac{y^2}{b^2} = 1 \quad \text{and} \quad \frac{y^2}{b^2} - \frac{x^2}{a^2} = 1.$$

This rectangle is called the **fundamental rectangle.** Using the methods of Chapter 3, we could show that the equations of these asymptotes are

$$y = \frac{b}{a}x \quad \text{and} \quad y = -\frac{b}{a}x.$$

To graph hyperbolas, follow these steps.

Graphing a Hyperbola

Step 1   **Find the intercepts.** Locate the intercepts at $(a, 0)$ and $(-a, 0)$ if the x^2-term has a positive coefficient, or at $(0, b)$ and $(0, -b)$ if the y^2- term has a positive coefficient.

Step 2   **Find the fundamental rectangle.** Locate the vertices of the fundamental rectangle at (a, b), $(-a, b)$, $(-a, -b)$, and $(a, -b)$.

Step 3   **Sketch the asymptotes.** The extended diagonals of the rectangle are the asymptotes of the hyperbola, and they have equations $y = \pm\frac{b}{a}x$.

Step 4   **Draw the graph.** Sketch each branch of the hyperbola through an intercept and approaching (but not touching) the asymptotes.

■ **EXAMPLE 1   Graphing a Horizontal Hyperbola**

Graph $\dfrac{x^2}{16} - \dfrac{y^2}{25} = 1$.

Step 1   Here $a = 4$ and $b = 5$. The x-intercepts are $(4, 0)$ and $(-4, 0)$.

Step 2   The four points $(4, 5)$, $(-4, 5)$, $(-4, -5)$, and $(4, -5)$ are the vertices of the fundamental rectangle, as shown in Figure 13.

Steps 3   The equations of the asymptotes are $y = \pm\frac{5}{4}x$, and the hyperbola approaches
and 4   these lines as x and y get larger and larger in absolute value.

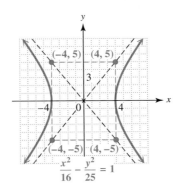

FIGURE 13

Now Try Exercise 7.

CAUTION   When sketching the graph of a hyperbola, be sure that the branches do not touch the asymptotes.

EXAMPLE 2   Graphing a Vertical Hyperbola

Graph $\dfrac{y^2}{49} - \dfrac{x^2}{16} = 1$.

This hyperbola has y-intercepts $(0, 7)$ and $(0, -7)$. The asymptotes are the extended diagonals of the rectangle with vertices at $(4, 7)$, $(-4, 7)$, $(-4, -7)$, and $(4, -7)$. Their equations are $y = \pm\frac{7}{4}x$. See Figure 14.

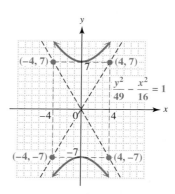

FIGURE 14

Now Try Exercise 9.

Hyperbolas are graphed with a graphing calculator in much the same way as circles and ellipses, by first writing the equations of two root functions whose union is equivalent to the equation of the hyperbola. A square window gives a truer shape for hyperbolas, too.

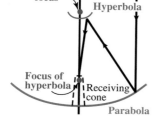

| CONNECTIONS |

A hyperbola and a parabola are used together in one kind of microwave antenna system. The cross sections of the system consist of a parabola and a hyperbola with the focus of the parabola coinciding with one focus of the hyperbola. See the figure.

The incoming microwaves that are parallel to the axis of the parabola are reflected from the parabola up toward the hyperbola and back to the other focus of the hyperbola, where the cone of the antenna is located to capture the signal.

For Discussion or Writing

The property of the parabola and the hyperbola that is used here is a "reflection property" of the foci. Explain why this name is appropriate.

OBJECTIVE 3 Identify conic sections by their equations.   Rewriting a second-degree equation in one of the forms given for ellipses, hyperbolas, circles, or parabolas makes it possible to determine when the graph is one of these.

Summary of Conic Sections

Equation	Graph	Description	Identification
$y = ax^2 + bx + c$ or $y = a(x - h)^2 + k$	 Parabola	It opens up if $a > 0$, down if $a < 0$. The vertex is (h, k).	It has an x^2-term. y is not squared.
$x = ay^2 + by + c$ or $x = a(y - k)^2 + h$	 Parabola	It opens to the right if $a > 0$, to the left if $a < 0$. The vertex is (h, k).	It has a y^2-term. x is not squared.
$(x - h)^2 +$ $(y - k)^2 = r^2$	 Circle	The center is (h, k), and the radius is r.	x^2- and y^2-terms have the same positive coefficient.
$\dfrac{x^2}{a^2} + \dfrac{y^2}{b^2} = 1$	 Ellipse	The x-intercepts are $(a, 0)$ and $(-a, 0)$. The y-intercepts are $(0, b)$ and $(0, -b)$.	x^2- and y^2-terms have different positive coefficients.
$\dfrac{x^2}{a^2} - \dfrac{y^2}{b^2} = 1$	 Hyperbola	The x-intercepts are $(a, 0)$ and $(-a, 0)$. The asymptotes are found from (a, b), $(a, -b)$, $(-a, -b)$, and $(-a, b)$.	x^2 has a positive coefficient. y^2 has a negative coefficient.
$\dfrac{y^2}{b^2} - \dfrac{x^2}{a^2} = 1$	 Hyperbola	The y-intercepts are $(0, b)$ and $(0, -b)$. The asymptotes are found from (a, b), $(a, -b)$, $(-a, -b)$, and $(-a, b)$.	y^2 has a positive coefficient. x^2 has a negative coefficient.

EXAMPLE 3 Identifying the Graphs of Equations

Identify the graph of each equation.

(a) $9x^2 = 108 + 12y^2$

Both variables are squared, so the graph is either an ellipse or a hyperbola. (This situation also occurs for a circle, which is a special case of an ellipse.) To see whether the graph is an ellipse or a hyperbola, rewrite the equation so that the x^2- and y^2-terms are on one side of the equation and 1 is on the other.

$$9x^2 - 12y^2 = 108 \qquad \text{Subtract } 12y^2.$$

$$\frac{x^2}{12} - \frac{y^2}{9} = 1 \qquad \text{Divide by 108.}$$

Because of the minus sign, the graph of this equation is a hyperbola.

(b) $x^2 = y - 3$

Only one of the two variables, x, is squared, so this is the vertical parabola $y = x^2 + 3$.

(c) $x^2 = 9 - y^2$

Get the variable terms on the same side of the equation.

$$x^2 + y^2 = 9 \qquad \text{Add } y^2.$$

The graph of this equation is a circle with center at the origin and radius 3.

Now Try Exercises 17 and 21.

OBJECTIVE 4 Graph certain square root functions. Recall that no vertical line will intersect the graph of a function in more than one point. Thus, horizontal parabolas, all circles and ellipses, and most hyperbolas discussed in this chapter are examples of graphs that do not satisfy the conditions of a function. However, by considering only a part of the graph of each of these we have the graph of a function, as seen in Figure 15.

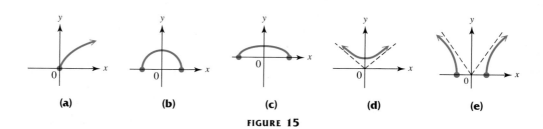

FIGURE 15

In parts (a), (b), (c), and (d) of Figure 15, the top portion of a conic section is shown (parabola, circle, ellipse, and hyperbola, respectively). In part (e), the top two portions of a hyperbola are shown. In each case, the graph is that of a function since the graph satisfies the conditions of the vertical line test.

In Sections 8.1 and 10.4 we observed the square root function defined by $f(x) = \sqrt{x}$. To find equations for the types of graphs shown in Figure 15, we extend its definition.

> **Square Root Function**
>
> A function of the form
>
> $$f(x) = \sqrt{u}$$
>
> for an algebraic expression u, with $u \geq 0$, is called a **square root function.**

EXAMPLE 4   Graphing a Semicircle

Graph $f(x) = \sqrt{25 - x^2}$. Give the domain and range.

Replace $f(x)$ with y and square both sides to get the equation

$$y^2 = 25 - x^2 \quad \text{or} \quad x^2 + y^2 = 25.$$

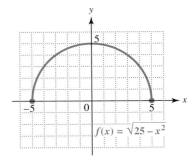

FIGURE 16

This is the graph of a circle with center at $(0, 0)$ and radius 5. Since $f(x)$, or y, represents a principal square root in the original equation, $f(x)$ must be nonnegative. This restricts the graph to the upper half of the circle, as shown in Figure 16. Use the graph and the vertical line test to verify that it is indeed a function. The domain is $[-5, 5]$, and the range is $[0, 5]$.

Now Try Exercise 25.

EXAMPLE 5   Graphing a Portion of an Ellipse

Graph $\dfrac{y}{6} = -\sqrt{1 - \dfrac{x^2}{16}}$. Give the domain and range.

Square both sides to get an equation whose form is known.

$$\frac{y^2}{36} = 1 - \frac{x^2}{16}$$

$$\frac{x^2}{16} + \frac{y^2}{36} = 1 \qquad \text{Add } \tfrac{x^2}{16}.$$

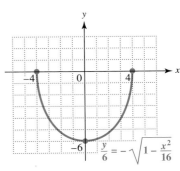

FIGURE 17

This is the equation of an ellipse with x-intercepts $(4, 0)$ and $(-4, 0)$ and y-intercepts $(0, 6)$ and $(0, -6)$. Since $\frac{y}{6}$ equals a negative square root in the original equation, y must be nonpositive, restricting the graph to the lower half of the ellipse, as shown in Figure 17. Verify that this is the graph of a function, using the vertical line test. The domain is $[-4, 4]$, and the range is $[-6, 0]$.

Now Try Exercise 27.

Root functions, since they are functions, can be entered and graphed directly with a graphing calculator.

13.2 EXERCISES

Based on the discussions of ellipses in the previous section and of hyperbolas in this section, match each equation with its graph.

1. $\dfrac{x^2}{25} + \dfrac{y^2}{9} = 1$

2. $\dfrac{x^2}{9} + \dfrac{y^2}{25} = 1$

3. $\dfrac{x^2}{9} - \dfrac{y^2}{25} = 1$

4. $\dfrac{x^2}{25} - \dfrac{y^2}{9} = 1$

A.

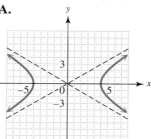

B.

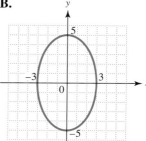

C.

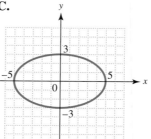

D.

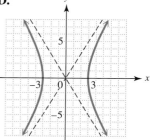

 5. Write an explanation of how you can tell from the equation whether the branches of a hyperbola open up and down or left and right.

 6. Describe how the fundamental rectangle is used to sketch a hyperbola.

Graph each hyperbola. See Examples 1 and 2.

7. $\dfrac{x^2}{16} - \dfrac{y^2}{9} = 1$

8. $\dfrac{y^2}{4} - \dfrac{x^2}{25} = 1$

9. $\dfrac{y^2}{9} - \dfrac{x^2}{9} = 1$

10. $\dfrac{x^2}{49} - \dfrac{y^2}{16} = 1$

11. $\dfrac{x^2}{25} - \dfrac{y^2}{36} = 1$

12. $\dfrac{y^2}{9} - \dfrac{x^2}{4} = 1$

13. $\dfrac{y^2}{16} - \dfrac{x^2}{16} = 1$

14. $\dfrac{x^2}{25} - \dfrac{y^2}{9} = 1$

Identify the graph of each equation as a parabola, circle, ellipse, *or* hyperbola, *and then sketch. See Example 3.*

15. $x^2 - y^2 = 16$

16. $x^2 + y^2 = 16$

17. $4x^2 + y^2 = 16$

18. $x^2 - 2y = 0$

19. $y^2 = 36 - x^2$

20. $9x^2 + 25y^2 = 225$

21. $9x^2 = 144 + 16y^2$

22. $x^2 + 9y^2 = 9$

23. $y^2 = 4 + x^2$

 24. State in your own words the major difference between the definitions of *ellipse* and *hyperbola*.

Graph each function defined by a radical expression. Give the domain and range. See Examples 4 and 5.

25. $f(x) = \sqrt{16 - x^2}$

26. $f(x) = \sqrt{9 - x^2}$

27. $f(x) = -\sqrt{36 - x^2}$

28. $f(x) = -\sqrt{25 - x^2}$

29. $\dfrac{y}{3} = \sqrt{1 + \dfrac{x^2}{9}}$

30. $y = \sqrt{\dfrac{x + 4}{2}}$

In Section 13.1, Example 6, we saw that the center of an ellipse may be shifted away from the origin. The same process applies to hyperbolas. For example, the hyperbola

$$\frac{(x + 5)^2}{4} - \frac{(y - 2)^2}{9} = 1,$$

shown at the right, has the same graph as $\dfrac{x^2}{4} - \dfrac{y^2}{9} = 1$, *but it is centered at* $(-5, 2)$. *Graph each hyperbola with center shifted away from the origin.*

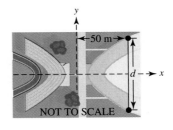
Shift 2 units up $\{$ $(-5, 2)$
$(0, 0)$
Shift 5 units left

31. $\dfrac{(x - 2)^2}{4} - \dfrac{(y + 1)^2}{9} = 1$

32. $\dfrac{(x + 3)^2}{16} - \dfrac{(y - 2)^2}{25} = 1$

33. $\dfrac{y^2}{36} - \dfrac{(x - 2)^2}{49} = 1$

34. $\dfrac{(y - 5)^2}{9} - \dfrac{x^2}{25} = 1$

Solve each problem.

35. Two buildings in a sports complex are shaped and positioned like a portion of the branches of the hyperbola with equation

$$400x^2 - 625y^2 = 250,000,$$

where x and y are in meters.

(a) How far apart are the buildings at their closest point?

(b) Find the distance d in the figure.

$\leftarrow$50 m$\rightarrow$
$d \rightarrow x$
NOT TO SCALE

36. In rugby, after a *try* (similar to a touchdown in American football) the scoring team attempts a kick for extra points. The ball must be kicked from directly behind the point where the try was scored. The kicker can choose the distance but cannot move the ball sideways. It can be shown that the kicker's best choice is on the hyperbola with equation

$$\frac{x^2}{g^2} - \frac{y^2}{g^2} = 1,$$

where $2g$ is the distance between the goal posts. Since the hyperbola approaches its asymptotes, it is easier for the kicker to estimate points on the asymptotes instead of on the hyperbola. What are the asymptotes of this hyperbola? Why is it relatively easy to estimate them? (*Source:* Isaksen, Daniel C., "How to Kick a Field Goal," *The College Mathematics Journal*, September 1996.)

37. When a satellite is launched into orbit, the shape of its trajectory is determined by its velocity. The trajectory will be hyperbolic if the velocity V, in meters per second, satisfies the inequality

$$V > \frac{2.82 \times 10^7}{\sqrt{D}},$$

where D is the distance, in meters, from the center of Earth. For what values of V will the trajectory be hyperbolic if $D = 4.25 \times 10^7$ m? (*Source:* Kaler, James B., *Astronomy!*, Addison-Wesley, 1997.)

38. The percent of women in the work force has increased steadily for many years. The line graph shows the change for the period from 1975 through 1999, where $x = 75$ represents 1975, $x = 80$ represents 1980, and so on. The graph resembles the upper branch of a horizontal hyperbola. Using statistical methods, we found the corresponding square root equation

$$y = .607\sqrt{383.9 + x^2},$$

which closely approximates the line graph.

(a) According to the graph, what percent of women were in the work force in 1985?

(b) According to the equation, what percent of women worked in 1985? (Round to the nearest percent.)

WOMEN IN THE WORK FORCE

Source: U.S. Bureau of Labor Statistics.

TECHNOLOGY INSIGHTS (EXERCISES 39 AND 40)

39. The hyperbola shown in the figure was graphed in function mode, with a square viewing window. It is the graph of $\frac{x^2}{9} - y^2 = 1$. What are the two functions y_1 and y_2 that were used to obtain this graph?

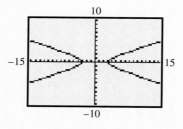

40. Repeat Exercise 39 for the graph of $\frac{y^2}{9} - x^2 = 1$, shown in the figure.

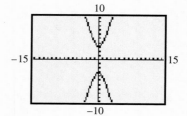

Use a graphing calculator in function mode to graph each hyperbola. Use a square viewing window.

41. $\dfrac{x^2}{25} - \dfrac{y^2}{49} = 1$ **42.** $\dfrac{x^2}{4} - \dfrac{y^2}{16} = 1$ **43.** $\dfrac{y^2}{9} - x^2 = 1$ **44.** $\dfrac{y^2}{36} - \dfrac{x^2}{4} = 1$

45. Suppose that a hyperbola has center at the origin, foci at $(-c, 0)$ and $(c, 0)$, and the absolute value of the difference between the distances from any point (x, y) of the hyperbola to the two foci is $2a$. See the figure. Let $b^2 = c^2 - a^2$, and show that an equation of the hyperbola is

$$\frac{x^2}{a^2} - \frac{y^2}{b^2} = 1.$$

46. Use the result of Exercise 45 to find an equation of a hyperbola with center at the origin, foci at $(-2, 0)$ and $(2, 0)$, and the absolute value of the difference between the distances from any point of the hyperbola to the two foci equal to 2.

13.3   Nonlinear Systems of Equations

An equation in which some terms have more than one variable or a variable of degree 2 or greater is called a **nonlinear equation**. A **nonlinear system of equations** includes at least one nonlinear equation.

When solving a nonlinear system, it helps to visualize the types of graphs of the equations of the system to determine the possible number of points of intersection. For example, if a system includes two equations where the graph of one is a parabola and the graph of the other is a line, then there may be zero, one, or two points of intersection, as illustrated in Figure 18.

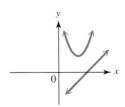
No points of intersection

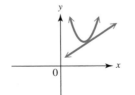
One point of intersection

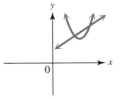
Two points of intersection

FIGURE 18

OBJECTIVE 1 Solve a nonlinear system by substitution. We solve nonlinear systems by the elimination method, the substitution method, or a combination of the two. The substitution method is usually best when one of the equations is linear.

◾ **EXAMPLE 1** Solving a Nonlinear System by Substitution

Solve the system

$$x^2 + y^2 = 9 \qquad (1)$$
$$2x - y = 3. \qquad (2)$$

The graph of (1) is a circle and the graph of (2) is a line. Visualizing the possible ways the graphs could intersect indicates that there may be zero, one, or two points of intersection. It is best to solve the linear equation first for one of the two variables; then substitute the resulting expression into the nonlinear equation to obtain an equation in one variable.

$$2x - y = 3 \qquad (2)$$
$$y = 2x - 3 \qquad (3)$$

Substitute $2x - 3$ for y in equation (1).

$$x^2 + (2x - 3)^2 = 9$$
$$x^2 + 4x^2 - 12x + 9 = 9$$
$$5x^2 - 12x = 0$$
$$x(5x - 12) = 0 \qquad \text{GCF is } x.$$
$$x = 0 \quad \text{or} \quad x = \frac{12}{5} \qquad \text{Zero-factor property}$$

Let $x = 0$ in equation (3) to get $y = -3$. If $x = \frac{12}{5}$, then $y = \frac{9}{5}$. The solution set of the system is $\left\{ (0, -3), \left(\frac{12}{5}, \frac{9}{5} \right) \right\}$. The graph in Figure 19 confirms the two points of intersection.

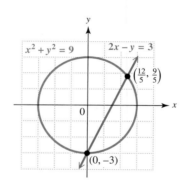

FIGURE 19

◾
Now Try Exercise 19.

◾ **EXAMPLE 2** Solving a Nonlinear System by Substitution

Solve the system

$$6x - y = 5 \qquad (1)$$
$$xy = 4. \qquad (2)$$

The graph of (1) is a line. We have not specifically mentioned equations like (2); however, it can be shown by plotting points that its graph is a hyperbola. Visualizing a line and a hyperbola indicates that there may be zero, one, or two points of intersection. Since neither equation has a squared term, we can solve either equation for one of the variables and then substitute the result into the other equation. Solving $xy = 4$ for x gives $x = \frac{4}{y}$. We substitute $\frac{4}{y}$ for x in equation (1).

$$6\left(\frac{4}{y}\right) - y = 5 \qquad \text{Let } x = \tfrac{4}{y} \text{ in Equation (1).}$$

$$\frac{24}{y} - y = 5$$

$$24 - y^2 = 5y \qquad \text{Multiply by } y, \ y \neq 0.$$

$$0 = y^2 + 5y - 24 \qquad \text{Standard form}$$

$$0 = (y - 3)(y + 8) \qquad \text{Factor.}$$

$$y = 3 \quad \text{or} \quad y = -8 \qquad \text{Zero-factor property}$$

We substitute these results into $x = \frac{4}{y}$ to obtain the corresponding values of x.

$$\text{If } y = 3, \text{ then } x = \frac{4}{3}. \qquad \text{If } y = -8, \text{ then } x = -\frac{1}{2}.$$

The solution set of the system is $\left\{\left(\frac{4}{3}, 3\right), \left(-\frac{1}{2}, -8\right)\right\}$. The graph in Figure 20 shows that there are two points of intersection.

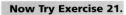

Now Try Exercise 21.

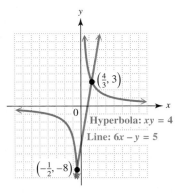

FIGURE 20

OBJECTIVE 2 Use the elimination method to solve a system with two second-degree equations.   The elimination method is often used when both equations are second degree.

EXAMPLE 3   Solving a Nonlinear System by Elimination

Solve the system

$$x^2 + y^2 = 9 \qquad (1)$$
$$2x^2 - y^2 = -6. \qquad (2)$$

The graph of (1) is a circle, while the graph of (2) is a hyperbola. By analyzing the possibilities we conclude that there may be zero, one, two, three, or four points of intersection. Adding the two equations will eliminate y, leaving an equation that can be solved for x.

$$
\begin{aligned}
x^2 + y^2 &= 9 \\
2x^2 - y^2 &= -6 \\
\hline
3x^2 &= 3 \\
x^2 &= 1 \\
x = 1 \quad \text{or} \quad x &= -1
\end{aligned}
$$

Each value of x gives corresponding values for y when substituted into one of the original equations. Using equation (1) gives the following.

If $x = 1$, then

$$1^2 + y^2 = 9$$
$$y^2 = 8$$
$$y = \sqrt{8} \quad \text{or} \quad y = -\sqrt{8}$$
$$y = 2\sqrt{2} \quad \text{or} \quad y = -2\sqrt{2}.$$

If $x = -1$, then

$$(-1)^2 + y^2 = 9$$
$$y^2 = 8$$
$$y = 2\sqrt{2} \quad \text{or} \quad y = -2\sqrt{2}.$$

The solution set is $\{(1, 2\sqrt{2}), (1, -2\sqrt{2}), (-1, 2\sqrt{2}), (-1, -2\sqrt{2})\}$. Figure 21 shows the four points of intersection.

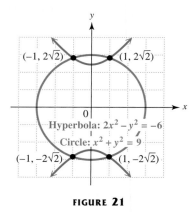

FIGURE 21

Now Try Exercise 35.

OBJECTIVE 3 Solve a system that requires a combination of methods. Solving a system of second-degree equations may require a combination of methods.

EXAMPLE 4 Solving a Nonlinear System by a Combination of Methods

Solve the system

$$x^2 + 2xy - y^2 = 7 \qquad (1)$$
$$x^2 - y^2 = 3. \qquad (2)$$

While we have not graphed equations like (1), its graph is a hyperbola. The graph of (2) is also a hyperbola. Two hyperbolas may have zero, one, two, three, or four points of intersection. We use the elimination method here in combination with the substitution method. We begin by eliminating the squared terms by multiplying each side of equation (2) by -1 and then adding the result to equation (1).

$$
\begin{array}{rcr}
x^2 + 2xy - y^2 = & 7 \\
-x^2 \qquad\;\; + y^2 = & -3 \\
\hline
2xy \qquad\quad = & 4
\end{array}
$$

Next, we solve $2xy = 4$ for y. (Either variable would do.)

$$2xy = 4$$
$$y = \frac{2}{x} \qquad (3)$$

Now, we substitute $y = \frac{2}{x}$ into one of the original equations. It is easier to do this with equation (2).

$$x^2 - y^2 = 3 \qquad (2)$$
$$x^2 - \left(\frac{2}{x}\right)^2 = 3$$
$$x^2 - \frac{4}{x^2} = 3$$

$$x^4 - 4 = 3x^2 \qquad \text{Multiply by } x^2, x \neq 0.$$
$$x^4 - 3x^2 - 4 = 0 \qquad \text{Subtract } 3x^2.$$
$$(x^2 - 4)(x^2 + 1) = 0 \qquad \text{Factor.}$$
$$x^2 - 4 = 0 \quad \text{or} \quad x^2 + 1 = 0$$
$$x^2 = 4 \quad \text{or} \qquad x^2 = -1$$
$$x = 2 \quad \text{or} \quad x = -2 \quad \text{or} \quad x = i \quad \text{or} \quad x = -i$$

Substituting these four values of x into equation (3) gives the corresponding values for y.

If $x = 2$, then $y = 1$.   If $x = i$, then $y = -2i$.

If $x = -2$, then $y = -1$.   If $x = -i$, then $y = 2i$.

Note that if we substitute the x-values we found into equation (1) or (2) instead of into equation (3), we get extraneous solutions. It is always wise to check all solutions in both of the given equations. There are four ordered pairs in the solution set, two with real values and two with imaginary values. The solution set is

$$\{(2, 1), (-2, -1), (i, -2i), (-i, 2i)\}.$$

The graph of the system, shown in Figure 22, shows only the two real intersection points because the graph is in the real number plane. The two ordered pairs with imaginary components are solutions of the system, but do not appear on the graph.

Now Try Exercise 39.

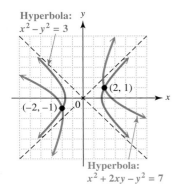

Hyperbola:
$x^2 - y^2 = 3$

(2, 1)

(−2, −1)

Hyperbola:
$x^2 + 2xy - y^2 = 7$

FIGURE 22

NOTE   In the examples of this section, we analyzed the possible number of points of intersection of the graphs in each system. However, in Examples 2 and 4, we worked with equations whose graphs had not been studied. Keep in mind that it is not absolutely essential to visualize the number of points of intersection in order to solve the system. Furthermore, as in Example 4, there are sometimes imaginary solutions to nonlinear systems that do not appear as points of intersection in the real plane. Visualizing the geometry of the graphs is only an aid to solving these systems.

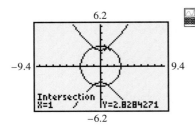

FIGURE 23

OBJECTIVE 4   Use a graphing calculator to solve a nonlinear system.   If the equations in a nonlinear system can be solved for y, then we can graph the equations of the system with a graphing calculator and use the capabilities of the calculator to identify all intersection points. For instance, the two equations in Example 3 would require graphing the four separate functions

$$Y_1 = \sqrt{9 - X^2}, \quad Y_2 = -\sqrt{9 - X^2}, \quad Y_3 = \sqrt{2X^2 + 6}, \quad \text{and} \quad Y_4 = -\sqrt{2X^2 + 6}.$$

Figure 23 indicates the coordinates of one of the points of intersection.

13.3 EXERCISES

1. Write an explanation of the steps you would use to solve the system

$$x^2 + y^2 = 25$$
$$y = x - 1$$

by the substitution method. Why would the elimination method not be appropriate for this system?

2. Write an explanation of the steps you would use to solve the system

$$x^2 + y^2 = 12$$
$$x^2 - y^2 = 13$$

by the elimination method.

Each sketch represents the graphs of a pair of equations in a system. How many points are in each solution set?

3.

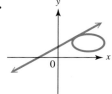

4.

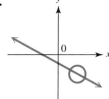

5.

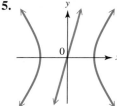

6.

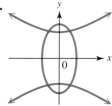

Suppose that a nonlinear system is composed of equations whose graphs are those described, and the number of points of intersection of the two graphs is as given. Make a sketch satisfying these conditions. (There may be more than one way to do this.)

7. A line and a circle; no points

8. A line and a circle; one point

9. A line and a hyperbola; one point

10. A line and an ellipse; no points

11. A circle and an ellipse; four points

12. A parabola and an ellipse; one point

13. A parabola and an ellipse; four points

14. A parabola and a hyperbola; two points

Solve each system by the substitution method. See Examples 1 and 2.

15. $y = 4x^2 - x$
$y = x$

16. $y = x^2 + 6x$
$3y = 12x$

17. $y = x^2 + 6x + 9$
$x + y = 3$

18. $y = x^2 + 8x + 16$
$x - y = -4$

19. $x^2 + y^2 = 2$
$2x + y = 1$

20. $2x^2 + 4y^2 = 4$
$x = 4y$

21. $xy = 4$
$3x + 2y = -10$

22. $xy = -5$
$2x + y = 3$

23. $xy = -3$
$x + y = -2$

24. $xy = 12$
$x + y = 8$

25. $y = 3x^2 + 6x$
$y = x^2 - x - 6$

26. $y = 2x^2 + 1$
$y = 5x^2 + 2x - 7$

27. $2x^2 - y^2 = 6$
$y = x^2 - 3$

28. $x^2 + y^2 = 4$
$y = x^2 - 2$

29. $x^2 - xy + y^2 = 0$
$x - 2y = 1$

30. $x^2 - 3x + y^2 = 4$
$2x - y = 3$

Solve each system by the elimination method or a combination of the elimination and substitution methods. See Examples 3 and 4.

31. $3x^2 + 2y^2 = 12$
$x^2 + 2y^2 = 4$

32. $2x^2 + y^2 = 28$
$4x^2 - 5y^2 = 28$

33. $2x^2 + 3y^2 = 6$
$x^2 + 3y^2 = 3$

34. $6x^2 + y^2 = 9$
$3x^2 + 4y^2 = 36$

35. $5x^2 - 2y^2 = -13$
$3x^2 + 4y^2 = 39$

36. $x^2 + 6y^2 = 9$
$4x^2 + 3y^2 = 36$

37. $2x^2 = 8 - 2y^2$
$3x^2 = 24 - 4y^2$

38. $5x^2 = 20 - 5y^2$
$2y^2 = 2 - x^2$

39. $x^2 + xy + y^2 = 15$
$x^2 + y^2 = 10$

40. $2x^2 + 3xy + 2y^2 = 21$
$x^2 + y^2 = 6$

41. $3x^2 + 2xy - 3y^2 = 5$
$-x^2 - 3xy + y^2 = 3$

42. $-2x^2 + 7xy - 3y^2 = 4$
$2x^2 - 3xy + 3y^2 = 4$

 Use a graphing calculator to solve each system. Then confirm your answer algebraically.

43. $xy = -6$
$x + y = -1$

44. $y = 2x^2 + 4x$
$y = -x^2 - 1$

Solve each problem by using a nonlinear system.

45. The area of a rectangular rug is 84 ft^2 and its perimeter is 38 ft. Find the length and width of the rug.

46. Find the length and width of a rectangular room whose perimeter is 50 m and whose area is 100 m^2.

47. A company has found that the price p (in dollars) of its scientific calculator is related to the supply x (in thousands) by the equation

$$px = 16.$$

The price is related to the demand x (in thousands) for the calculator by the equation

$$p = 10x + 12.$$

The *equilibrium price* is the value of p where demand equals supply. Find the equilibrium price and the supply/demand at that price by solving a system of equations. (*Hint:* Demand, price, and supply must all be positive.)

48. The calculator company in Exercise 47 has also determined that the cost y to make x (thousand) calculators is

$$y = 4x^2 + 36x + 20,$$

while the revenue y from the sale of x (thousand) calculators is

$$36x^2 - 3y = 0.$$

Find the *break-even point*, where cost equals revenue, by solving a system of equations.

49. In the 1970s, the number of bachelor's degrees earned by men began to decrease. It stayed fairly constant in the 1980s, and then in the 1990s slowly began to increase again. Meanwhile, the number of bachelor's degrees earned by women continued to rise steadily throughout this period. Functions that model the situation are defined by the following equations, where y is the number of degrees (in thousands) granted in year x, with $x = 0$ corresponding to 1970.

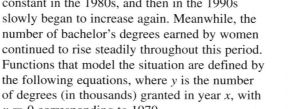

Men: $\quad y = .138x^2 + .064x + 451$

Women: $\quad y = 12.1x + 334$

Solve this system of equations to find the year when the same number of bachelor's degrees were awarded to men and women. How many bachelor's degrees were awarded to each sex in that year? Give the answer to the nearest ten thousand. (*Source:* U.S. National Center for Education Statistics, *Digest of Education Statistics,* annual.)

50. Andy Grove, chairman of chip maker Intel Corp., once noted that decreasing prices for computers and stable prices for Internet access implied that the trend lines for these costs either have crossed or soon will. He predicted that the time is not far away when computers, like cell phones, may be given away to sell on-line time. To see this, assume a price of $1000 for a computer, and let x represent the number of months it will be used. (*Source:* Corcoran, Elizabeth, "Can Free Computers Be That Far Away?", *Washington Post,* from *Sacramento Bee,* February 3, 1999.)

(a) Write an equation for the monthly cost y of the computer over this period.

(b) The average monthly on-line cost is about $20. Assume this will remain constant and write an equation to express this cost.

(c) Solve the system of equations from parts (a) and (b). Interpret your answer in relation to the situation.

13.4 Second-Degree Inequalities, Systems of Inequalities, and Linear Programming

OBJECTIVES

1 Graph second-degree inequalities.

2 Graph the solution set of a system of inequalities.

3 Use a graphing calculator to graph a system of inequalities.

4 Solve linear programming problems by graphing.

OBJECTIVE 1 Graph second-degree inequalities. The linear inequality $3x + 2y \leq 5$ is graphed by first graphing the boundary line $3x + 2y = 5$. *Second-degree inequalities* are graphed in the same way. A **second-degree inequality** is an inequality with at least one variable of degree 2 and no variable with degree greater than 2. An example is $x^2 + y^2 \leq 36$. The boundary of the inequality $x^2 + y^2 \leq 36$ is the graph of the equation $x^2 + y^2 = 36$, a circle with radius 6 and center at the origin, as shown in Figure 24.

The inequality $x^2 + y^2 \leq 36$ will include either the points outside the circle or the points inside the circle, as well as the boundary. We decide which region to shade by substituting any test point not on the circle, such as $(0, 0)$, into the original inequality. Since $0^2 + 0^2 \leq 36$ is a true statement, the original inequality includes the points inside the circle, the shaded region in Figure 24, and the boundary.

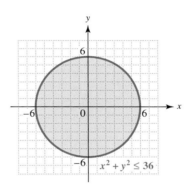

$x^2 + y^2 \leq 36$

FIGURE 24

EXAMPLE 1 Graphing a Second-Degree Inequality

Graph $y < -2(x - 4)^2 - 3$.

The boundary, $y = -2(x - 4)^2 - 3$, is a parabola that opens down with vertex at $(4, -3)$. Using $(0, 0)$ as a test point gives

$$0 < -2(0 - 4)^2 - 3 \qquad ?$$
$$0 < -32 - 3 \qquad ?$$
$$0 < -35. \qquad \text{False}$$

Because the final inequality is a false statement, the points in the region containing $(0, 0)$ do not satisfy the inequality. Figure 25 shows the final graph; the parabola is drawn as a dashed curve since the points of the parabola itself do not satisfy the inequality, and the region inside (or below) the parabola is shaded.

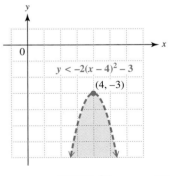

$y < -2(x - 4)^2 - 3$

$(4, -3)$

FIGURE 25

Now Try Exercise 11.

> **NOTE** Since the substitution is easy, the origin is the test point of choice unless the graph actually passes through $(0, 0)$.

EXAMPLE 2 Graphing a Second-Degree Inequality

Graph $16y^2 \leq 144 + 9x^2$.

First rewrite the inequality as follows.

$$16y^2 - 9x^2 \leq 144 \qquad \text{Subtract } 9x^2.$$

$$\frac{y^2}{9} - \frac{x^2}{16} \leq 1 \qquad \text{Divide by 144.}$$

This form shows that the boundary is the hyperbola given by

$$\frac{y^2}{9} - \frac{x^2}{16} = 1.$$

Since the graph is a vertical hyperbola, the desired region will be either the region between the branches or the regions above the top branch and below the bottom branch. Choose $(0, 0)$ as a test point. Substituting into the original inequality leads to $0 \leq 144$, a true statement, so the region between the branches containing $(0, 0)$ is shaded, as shown in Figure 26.

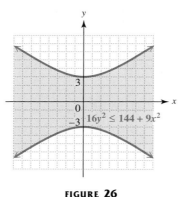

FIGURE 26

Now Try Exercise 17.

OBJECTIVE 2 Graph the solution set of a system of inequalities. If two or more inequalities are considered at the same time, we have a **system of inequalities.** To find the solution set of the system, we find the intersection of the graphs (solution sets) of the inequalities in the system.

EXAMPLE 3 Graphing a System of Two Inequalities

Graph the solution set of the system

$$2x + 3y > 6$$
$$x^2 + y^2 < 16.$$

Begin by graphing the solution set of $2x + 3y > 6$. The boundary line is the graph of $2x + 3y = 6$ and is a dashed line because of the symbol $>$. The test point $(0, 0)$ leads to a false statement in the inequality $2x + 3y > 6$, so shade the region above the line, as shown in Figure 27. The graph of $x^2 + y^2 < 16$ is the interior of a dashed circle centered at the origin with radius 4. This is shown in Figure 28.

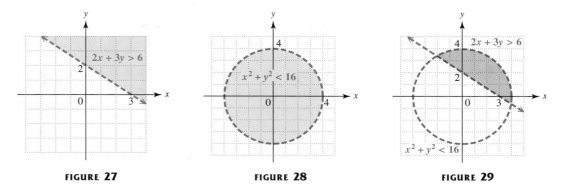

FIGURE 27   **FIGURE 28**   **FIGURE 29**

Finally, to get the graph of the solution set of the system, determine the intersection of the graphs of the two inequalities. The overlapping region in Figure 29 is the solution set.

Now Try Exercise 29.

EXAMPLE 4   **Graphing a Linear System with Three Inequalities**

Graph the solution set of the system

$$x + y < 1$$
$$y \le 2x + 3$$
$$y \ge -2.$$

Graph each inequality separately, on the same axes. The graph of $x + y < 1$ consists of all points below the dashed line $x + y = 1$. The graph of $y \le 2x + 3$ is the region that lies below the solid line $y = 2x + 3$. Finally, the graph of $y \ge -2$ is the region above the solid horizontal line $y = -2$.

The graph of the system, the intersection of these three graphs, is the triangular region enclosed by the three boundary lines in Figure 30, including two of its boundaries.

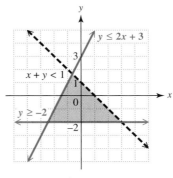

FIGURE 30

Now Try Exercise 31.

EXAMPLE 5 Graphing a System with Three Inequalities

Graph the solution set of the system

$$y \geq x^2 - 2x + 1$$
$$2x^2 + y^2 > 4$$
$$y < 4.$$

The graph of $y = x^2 - 2x + 1$ is a parabola with vertex at $(1, 0)$. Those points above (or in the interior of) the parabola satisfy the condition $y > x^2 - 2x + 1$. Thus, points on the parabola or in the interior are the solution set of $y \geq x^2 - 2x + 1$. The graph of the equation $2x^2 + y^2 = 4$ is an ellipse. We draw it as a dashed curve. To satisfy the inequality $2x^2 + y^2 > 4$, a point must lie outside the ellipse. The graph of $y < 4$ includes all points below the dashed line $y = 4$. Finally, the graph of the system is the shaded region in Figure 31, which lies outside the ellipse, inside or on the boundary of the parabola, and below the line $y = 4$.

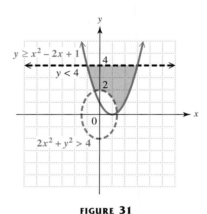

FIGURE 31

Now Try Exercise 33.

OBJECTIVE 3 Use a graphing calculator to graph a system of inequalities. Graphing calculators can show the solution set of a system of inequalities. We graph the boundary curves and then use the appropriate commands to shade the region. For example, to view the solution set of the system in Example 3,

$$2x + 3y > 6$$
$$x^2 + y^2 < 16,$$

we first solve the corresponding equations for y. Doing so gives

$$y = -\frac{2}{3}x + 2, \qquad y = \sqrt{16 - x^2}, \qquad \text{and} \qquad y = -\sqrt{16 - x^2}.$$

Now we direct the calculator to graph these equations and shade above the first and third graphs and below the second graph. Compare the result, shown in Figure 32, with Figure 29. Notice that the calculator graph does not distinguish between solid boundary lines and dashed boundary lines. We still need to understand the mathematics to correctly interpret a calculator graph.

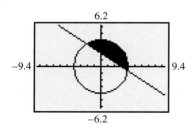

FIGURE 32

OBJECTIVE 4 Solve linear programming problems by graphing.   An important application of mathematics to business and social science is called *linear programming.* We use **linear programming** to find an optimum value, for example, minimum cost or maximum profit. It was first developed during World War II to solve supply allocation problems for the U.S. Air Force.

Suppose that the Audiophone Company makes two products, tape decks and amplifiers. Each tape deck produces a profit of $3, while each amplifier produces $7 profit. The company must manufacture at least one tape deck per day to satisfy a customer, but no more than five, because of production problems. The number of amplifiers produced cannot exceed six per day, and the number of tape decks cannot exceed the number of amplifiers. How many of each should the company manufacture to obtain the maximum profit?

Begin by translating the statement of the problem into symbols.

Let $x =$ number of tape decks to be produced daily

and $y =$ number of amplifiers to be produced daily.

The company must produce at least one tape deck (one or more), so

$$x \geq 1.$$

Since no more than 5 tape decks may be produced,

$$x \leq 5.$$

No more than 6 amplifiers may be made in one day, so

$$y \leq 6.$$

The number of tape decks may not exceed the number of amplifiers translates as

$$x \leq y.$$

The numbers of tape decks and amplifiers cannot be negative, so

$$x \geq 0 \quad \text{and} \quad y \geq 0.$$

These restrictions, or **constraints,** that are placed on production form the system of inequalities

$$x \geq 1, \quad x \leq 5, \quad y \leq 6, \quad x \leq y, \quad x \geq 0, \quad y \geq 0.$$

To find the maximum possible profit that the company can make, subject to these constraints, we sketch the graph of the solution set of the system. See Figure 33. The only feasible values of x and y are those that satisfy all constraints. These values correspond to points that lie on the boundary or in the shaded region, called the **region of feasible solutions.**

Each tape deck gives a profit of $3, so the daily profit from production of x tape decks is $3x$ dollars. Also, the profit from production of y amplifiers will be $7y$ dollars per day. Total daily profit is thus

$$\text{profit} = 3x + 7y.$$

This equation defines the function to be maximized, called the **objective function.**

We can now state the problem of the Audiophone Company as follows: Find values of x and y in the shaded region of Figure 33 that will produce the maximum possible value of $3x + 7y$.

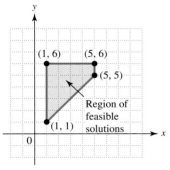

FIGURE 33

It can be shown that any optimum value (maximum or minimum) will always occur at a **vertex** (or **corner**) **point** of the region of feasible solutions. Locate the point (x, y) that gives the maximum profit by checking the coordinates of the vertex points, shown in Figure 33. Find the profit that corresponds to each coordinate pair and choose the one that gives maximum profit.

Point	Profit = $3x + 7y$
(1, 1)	$3(1) + 7(1) = 10$
(1, 6)	$3(1) + 7(6) = 45$
(5, 6)	$3(5) + 7(6) = 57$ ← Maximum
(5, 5)	$3(5) + 7(5) = 50$

The maximum profit of $57 is obtained when 5 tape decks and 6 amplifiers are produced each day.

Now Try Exercise 47.

EXAMPLE 6 Solving a Problem Using Linear Programming

Packages of food and clothing are being sent to assist hurricane victims in Hawaii. Commercial carriers have volunteered to transport the packages, provided they fit in the available cargo space. Each 20-cubic-foot box of food weighs 40 lb, and each 30-cubic-foot box of clothing weighs 10 lb. The total weight cannot exceed 16,000 lb, and the total volume must not exceed 18,000 ft³. Each carton of food will feed 10 people, while each carton of clothing will help clothe 8 people. How many packages of food and how many packages of clothing should be sent in order to maximize the number of people assisted? How many people will be assisted?

Let $\qquad$ $F =$ the number of cartons of food to send

and $\qquad$ $C =$ the number of cartons of clothing to send.

Because the total weight cannot exceed 16,000 lb, one constraint is

$$40F + 10C \le 16{,}000.$$

The total volume cannot exceed 18,000 ft³, so another constraint is

$$20F + 30C \le 18{,}000.$$

Since C and F cannot be negative, two further constraints are

$$C \ge 0 \quad \text{and} \quad F \ge 0.$$

The system we consider for this problem is

$$40F + 10C \le 16{,}000$$
$$20F + 30C \le 18{,}000$$
$$C \ge 0$$
$$F \ge 0.$$

Its graph is shown in Figure 34, with C on the vertical axis and F on the horizontal axis.

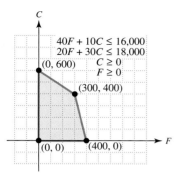

$$40F + 10C \leq 16{,}000$$
$$20F + 30C \leq 18{,}000$$
$$(0, 600) \quad C \geq 0$$
$$F \geq 0$$
$$(300, 400)$$
$$(0, 0) \quad (400, 0)$$

FIGURE 34

Since each food carton will feed 10 people and each clothing carton will help clothe 8 people, the objective function is

$$\text{people assisted} = 10F + 8C.$$

We want to *maximize* $10F + 8C$. The vertices of the feasible region are $(0, 0)$, $(0, 600)$, $(300, 400)$, and $(400, 0)$. The following table shows that $(300, 400)$ maximizes the objective function.

Point	People Assisted $= 10F + 8C$	
$(0, 0)$	$10(0) + 8(0) = 0$	
$(0, 600)$	$10(0) + 8(600) = 4800$	
$(300, 400)$	$10(300) + 8(400) = 6200$	← Maximum
$(400, 0)$	$10(400) + 8(0) = 4000$	

Therefore, 300 cartons of food and 400 cartons of clothing should be sent to assist the maximum number of victims, 6200.

Now Try Exercise 55.

13.4 EXERCISES

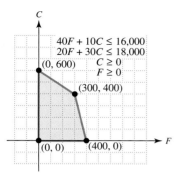

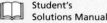

1. Which one of the following is a description of the graph of the solution set of the following system?

$$x^2 + y^2 < 25$$
$$y > -2$$

A. All points outside the circle $x^2 + y^2 = 25$ and above the line $y = -2$
B. All points outside the circle $x^2 + y^2 = 25$ and below the line $y = -2$
C. All points inside the circle $x^2 + y^2 = 25$ and above the line $y = -2$
D. All points inside the circle $x^2 + y^2 = 25$ and below the line $y = -2$

2. Fill in each blank with the appropriate response. The graph of the system

$$y > x^2 + 1$$
$$\frac{x^2}{9} + \frac{y^2}{4} > 1$$
$$y < 5$$

consists of all points _____ the parabola $y = x^2 + 1$, _____ the
　　　　　　　　　(above/below)　　　　　　　　　　　　(inside/outside)

ellipse $\frac{x^2}{9} + \frac{y^2}{4} = 1$, and _____ the line $y = 5$.
　　　　　　　　　　　　(above/below)

3. Explain how to graph the solution set of a nonlinear inequality.

4. Explain how to graph the solution set of a system of nonlinear inequalities.

Match each nonlinear inequality with its graph.

5. $y \geq x^2 + 4$ **6.** $y \leq x^2 + 4$ **7.** $y < x^2 + 4$ **8.** $y > x^2 + 4$

A. 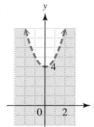   **B.** 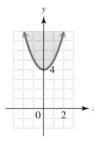   **C.** 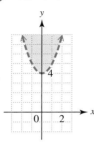   **D.**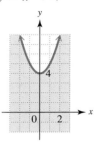

Graph each nonlinear inequality. See Examples 1 and 2.

9. $y^2 > 4 + x^2$

10. $y^2 \leq 4 - 2x^2$

11. $y + 2 \geq x^2$

12. $x^2 \leq 16 - y^2$

13. $2y^2 \geq 8 - x^2$

14. $x^2 \leq 16 + 4y^2$

15. $y \leq x^2 + 4x + 2$

16. $9x^2 < 16y^2 - 144$

17. $9x^2 > 16y^2 + 144$

18. $4y^2 \leq 36 - 9x^2$

19. $x^2 - 4 \geq -4y^2$

20. $x \geq y^2 - 8y + 14$

21. $x \leq -y^2 + 6y - 7$

22. $y^2 - 16x^2 \leq 16$

Graph each system of inequalities. See Examples 3–5.

23. $2x + 5y < 10$
  $x - 2y < 4$

24. $3x - y > -6$
  $4x + 3y > 12$

25. $5x - 3y \leq 15$
  $4x + y \geq 4$

26. $4x - 3y \leq 0$
  $x + y \leq 5$

27. $x \leq 5$
  $y \leq 4$

28. $x \geq -2$
  $y \leq 4$

29. $y > x^2 - 4$
  $y < -x^2 + 3$

30. $x^2 - y^2 \geq 9$
  $\dfrac{x^2}{16} + \dfrac{y^2}{9} \leq 1$

31. $x^2 + y^2 \geq 4$
  $x + y \leq 5$
  $x \geq 0$
  $y \geq 0$

32. $y^2 - x^2 \geq 4$
  $-5 \leq y \leq 5$

33. $y \leq -x^2$
  $y \geq x - 3$
  $y \leq -1$
  $x < 1$

34. $y < x^2$
  $y > -2$
  $x + y < 3$
  $3x - 2y > -6$

For each nonlinear inequality in Exercises 35–42, a restriction is placed on one or both variables. For example, the graph of

$$x^2 + y^2 \leq 4, \quad x \geq 0$$

would be as shown in the figure. Only the right half of the interior of the circle and its boundary is shaded, because of the restriction that x must be nonnegative. Graph each nonlinear inequality with the given restrictions.

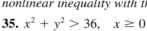

35. $x^2 + y^2 > 36, \quad x \geq 0$

36. $4x^2 + 25y^2 < 100, \quad y < 0$

37. $x < y^2 - 3, \quad x < 0$

38. $x^2 - y^2 < 4, \quad x < 0$

39. $4x^2 - y^2 > 16, \quad x < 0$

40. $x^2 + y^2 > 4, \quad y < 0$

41. $x^2 + 4y^2 \geq 1, \quad x \geq 0, y \geq 0$

42. $2x^2 - 32y^2 \leq 8, \quad x \leq 0, y \geq 0$

 Use the shading feature of a graphing calculator to graph each system. See Objective 3.

43. $y \geq x - 3$
  $y \leq -x + 4$

44. $y \geq -x^2 + 5$
  $y \leq x^2 - 3$

45. $y < x^2 + 4x + 4$
$y > -3$

46. $y > (x - 4)^2 - 3$
$y < 5$

The graphs in Exercises 47–50 show regions of feasible solutions. Find the maximum and minimum values of each expression.

47. $3x + 5y$

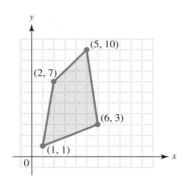

48. $6x + y$

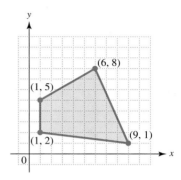

49. $40x + 75y$

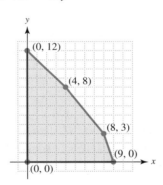

50. $35x + 125y$

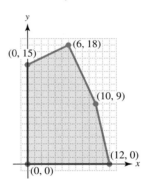

In Exercises 51–54 use graphical methods to solve each problem. See Example 6.

51. Find $x \geq 0$ and $y \geq 0$ such that
$$2x + 3y \leq 6$$
$$4x + y \leq 6$$
and $5x + 2y$ is maximized.

52. Find $x \geq 0$ and $y \geq 0$ such that
$$x + y \leq 10$$
$$5x + 2y \geq 20$$
$$2y \geq x$$
and $x + 3y$ is minimized.

53. Find $x \geq 2$ and $y \geq 5$ such that
$$3x - y \geq 12$$
$$x + y \leq 15$$
and $2x + y$ is minimized.

54. Find $x \geq 10$ and $y \geq 20$ such that
$$2x + 3y \leq 100$$
$$5x + 4y \leq 200$$
and $x + 3y$ is maximized.

Solve each problem. See Example 6.

55. Farmer Jones raises only pigs and geese. He wants to raise no more than 16 animals with no more than 12 geese. He spends $50 to raise a pig and $20 to raise a goose. He has $500 available for this purpose. Find the maximum profit he can make if he makes a profit of $80 per goose and $40 per pig.

56. A wholesaler of sporting goods wishes to display her products at a convention of dealers in such a way that she gets the maximum number of inquiries about her uniforms and hats. Her booth at the convention has 12 m² of floor space to be used for display purposes. A display unit for hats requires 2 m²; and for uniforms, 4 m². Experience tells the wholesaler that she should never have more than a total of 5 units of uniforms and hats on display at one time. If she receives three inquiries for each unit of hats and two inquiries for each unit of uniforms on display, how many of each should she display in order to get the maximum number of inquiries?

57. An office manager wants to buy some filing cabinets. She knows that cabinet #1 costs $20 each, requires 6 ft² of floor space, and holds 8 ft³ of files. Cabinet #2 costs $40 each, requires 8 ft² of floor space, and holds 12 ft³. She can spend no more than $280 due to budget limitations, while her office has room for no more than 72 ft² of cabinets. She wants the maximum storage capacity within the limits imposed by funds and space. How many of each type of cabinet should she buy?

58. In a small town in South Carolina, zoning rules require that the window space (in square feet) in a house be at least one-sixth of the space used up by solid walls. The cost to heat the house is 10¢ for each square foot of solid walls and 40¢ for each square foot of windows. Find the maximum total area (windows plus walls) if $80 is available to pay for heat.

59. The manufacturing process requires that oil refineries manufacture at least 2 gal of gasoline for each gallon of fuel oil. To meet the winter demand for fuel oil, at least 3 million gal per day must be produced. The demand for gasoline is no more than 6.4 million gal per day. If the price of gasoline is $1.90 per gal and the price of fuel oil is $1.50 per gal, how much of each should be produced to maximize revenue?

60. Seall Manufacturing Company makes color television sets. It produces a bargain set that sells for $100 profit and a deluxe set that sells for $150 profit. On the assembly line the bargain set requires 3 hr, while the deluxe set takes 5 hr. The cabinet shop spends 1 hr on the cabinet for the bargain set and 3 hr on the cabinet for the deluxe set. Both sets require 2 hr of time for testing and packing. On a particular production run the Seall Company has available 3900 work hr on the assembly line, 2100 work hr in the cabinet shop, and 2200 work hr in the testing and packing department. How many sets of each type should it produce to make maximum profit? What is the maximum profit?

Chapter 13 Group Activity

Finding the Paths of Natural Satellites

OBJECTIVE Write and graph equations of ellipses from given data.

The moon, which orbits Earth, and Halley's comet, which orbits the sun, are both natural satellites. In Section 13.1, you solved problems where you were given equations of ellipses for the orbits of planets and were asked to find apogees (greatest distance from the sun) and perigees (smallest distance from the sun). This activity reverses the process; that is, given apogees and perigees you must find equations of ellipses.

A. Have each student choose a natural satellite from the table. Predict the shape of the orbital ellipse for your satellite.

Natural Satellite	Apogee	Perigee
Moon	406.7 thousand km from Earth	356.4 thousand km from Earth
Halley's comet	35 astronomical units* from the sun	.6 astronomical unit* from the sun

Source: World Book Encyclopedia.
*One astronomical unit (AU) is the distance from Earth to the sun.

B. For your satellite, do the following.

1. Find values for a, b, and c. Note that apogee $= a + c$, perigee $= a - c$, and $c^2 = a^2 - b^2$.

2. Write the equation of the ellipse in the form $\dfrac{x^2}{a^2} + \dfrac{y^2}{b^2} = 1$.

3. Rewrite the equation so it can be graphed on a graphing calculator. (See Section 13.1, Objective 5.)

4. Graph your equation on a graphing calculator. Adjust the window setting in order to see the entire graph. Once the window is set correctly, get a square window to see the true shape of the ellipse.

C. Compare your graph with your partner's graph.

1. Do the graphs reflect the shapes you predicted in part A?

2. What window was used to graph each ellipse?

CHAPTER **13** SUMMARY

■ KEY TERMS

13.1 conic section
circle
center
radius
center-radius form
ellipse
foci

13.2 hyperbola
asymptotes of a
hyperbola
fundamental
rectangle
square root function

13.3 nonlinear equation
nonlinear system of
equations
13.4 second-degree
inequality
system of
inequalities

linear programming
constraints
region of feasible
solutions
objective function
vertex (corner) point

■ TEST YOUR WORD POWER

See how well you have learned the vocabulary in this chapter. Answers, with examples, follow the Quick Review.

1. **Conic sections** are
   A. graphs of first-degree equations
   B. the result of two or more intersecting planes
   C. graphs of first-degree inequalities
   D. figures that result from the intersection of an infinite cone with a plane.

2. A **circle** is the set of all points in a plane
   A. such that the absolute value of the difference of the distances from two fixed points is constant
   B. that lie a fixed distance from a fixed point
   C. the sum of whose distances from two fixed points is constant
   D. that make up the graph of any second-degree equation.

3. An **ellipse** is the set of all points in a plane
   A. such that the absolute value of the difference of the distances from two fixed points is constant

B. that lie a fixed distance from a fixed point
   C. the sum of whose distances from two fixed points is constant
   D. that make up the graph of any second-degree equation.

4. A **hyperbola** is the set of all points in a plane
   A. such that the absolute value of the difference of the distances from two fixed points is constant
   B. that lie a fixed distance from a fixed point
   C. the sum of whose distances from two fixed points is constant
   D. that make up the graph of any second-degree equation.

5. A **nonlinear equation** is an equation
   A. in which some terms have more than one variable or a variable of degree 2 or greater
   B. in which the terms have only one variable

C. of degree 1
   D. of a linear function.

6. A **nonlinear system of equations** is a system
   A. with at least one linear equation
   B. with two or more inequalities
   C. with at least one nonlinear equation
   D. with at least two linear equations.

7. **Linear programming** is
   A. a way to program a graphing calculator
   B. a systematic method of solving a system of inequalities
   C. a systematic method of solving a system of equations
   D. a method for finding the solution of a system of inequalities that optimizes a function.

QUICK REVIEW

CONCEPTS	EXAMPLES

13.1 THE CIRCLE AND THE ELLIPSE

Circle

The circle with radius r and center at (h, k) has an equation of the form

$$(x - h)^2 + (y - k)^2 = r^2.$$

The circle with equation $(x + 2)^2 + (y - 3)^2 = 25$, which can be written $[x - (-2)]^2 + (y - 3)^2 = 5^2$, has center $(-2, 3)$ and radius 5.

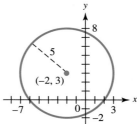

Ellipse

The ellipse whose x-intercepts are $(a, 0)$ and $(-a, 0)$ and whose y-intercepts are $(0, b)$ and $(0, -b)$ has an equation of the form

$$\frac{x^2}{a^2} + \frac{y^2}{b^2} = 1.$$

Graph $\dfrac{x^2}{9} + \dfrac{y^2}{4} = 1$.

13.2 THE HYPERBOLA AND FUNCTIONS DEFINED BY RADICALS

Hyperbola

A hyperbola with x-intercepts $(a, 0)$ and $(-a, 0)$ has an equation of the form

$$\frac{x^2}{a^2} - \frac{y^2}{b^2} = 1,$$

and a hyperbola with y-intercepts $(0, b)$ and $(0, -b)$ has an equation of the form

$$\frac{y^2}{b^2} - \frac{x^2}{a^2} = 1.$$

Graph $\dfrac{x^2}{4} - \dfrac{y^2}{4} = 1$.

The graph has x-intercepts $(2, 0)$ and $(-2, 0)$.

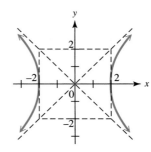

The extended diagonals of the fundamental rectangle with vertices at the points (a, b), $(-a, b)$, $(-a, -b)$, and $(a, -b)$ are the asymptotes of these hyperbolas.

The fundamental rectangle has vertices at $(2, 2)$, $(-2, 2)$, $(-2, -2)$, and $(2, -2)$.

(continued)

CONCEPTS	EXAMPLES

Graphing a Square Root Function

To graph a square root function defined by $f(x) = u$ for an algebraic expression u, with $u \geq 0$, square both sides so that the equation can be easily recognized. Then graph only the part indicated by the original equation.

Graph $y = -\sqrt{4 - x^2}$.

Square both sides and rearrange terms to get

$$x^2 + y^2 = 4.$$

This equation has a circle as its graph. However, graph only the lower half of the circle, since the original equation indicates that y cannot be positive.

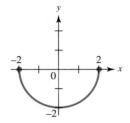

13.3 NONLINEAR SYSTEMS OF EQUATIONS

Solving a Nonlinear System

A nonlinear system can be solved by the substitution method, the elimination method, or a combination of the two.

Solve the system

$$x^2 + 2xy - y^2 = 14 \quad (1)$$
$$x^2 - y^2 = -16. \quad (2)$$

Multiply equation (2) by -1 and use elimination.

$$
\begin{array}{r}
x^2 + 2xy - y^2 = 14 \\
-x^2 \qquad\; + y^2 = 16 \\
\hline
2xy \qquad\quad = 30 \\
xy = 15
\end{array}
$$

Solve for y to obtain $y = \frac{15}{x}$, and substitute into equation (2).

$$x^2 - \left(\frac{15}{x}\right)^2 = -16$$

$$x^2 - \frac{225}{x^2} = -16$$

$$x^4 + 16x^2 - 225 = 0 \qquad \text{Multiply by } x^2; \text{ add } 16x^2.$$
$$(x^2 - 9)(x^2 + 25) = 0 \qquad \text{Factor.}$$
$$x = \pm 3 \quad \text{or} \quad x = \pm 5i \qquad \text{Zero-factor property}$$

Find corresponding y-values to get the solution set

$$\{(3, 5), (-3, -5), (5i, -3i), (-5i, 3i)\}.$$

CONCEPTS

EXAMPLES

13.4 SECOND-DEGREE INEQUALITIES, SYSTEMS OF INEQUALITIES, AND LINEAR PROGRAMMING

Graphing a Second-Degree Inequality
To graph a second-degree inequality, graph the corresponding equation as a boundary and use test points to determine which region(s) form the solution set. Shade the appropriate region(s).

Graphing a System of Inequalities
The solution set of a system of inequalities is the intersection of the solution sets of the individual inequalities.

Graph $y \geq x^2 - 2x + 3$.   Graph the solution set of the system

$$3x - 5y > -15$$
$$x^2 + y^2 \leq 25.$$

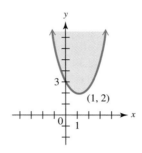

Solving a Linear Programming Problem
Write the objective function and all constraints, graph the feasible region, identify all vertex points (corner points), and find the value of the objective function at each vertex point. Choose the required maximum or minimum value accordingly.

The feasible region for

$$x + 2y \leq 14$$
$$3x + 4y \leq 36$$
$$x \geq 0$$
$$y \geq 0$$

is given in the figure. Maximize the objective function $8x + 12y$.

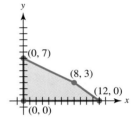

Vertex Point	Value of $8x + 12y$
(0, 0)	0
(0, 7)	84
(12, 0)	96
(8, 3)	100 ← Maximum

The objective function is maximized for $x = 8$ and $y = 3$.

CHAPTER 13 REVIEW EXERCISES

[13.1] *Write an equation for each circle.*

1. Center $(-2, 4)$, $r = 3$ **2.** Center $(-1, -3)$, $r = 5$ **3.** Center $(4, 2)$, $r = 6$

Find the center and radius of each circle.

4. $x^2 + y^2 + 6x - 4y - 3 = 0$ **5.** $x^2 + y^2 - 8x - 2y + 13 = 0$

6. $2x^2 + 2y^2 + 4x + 20y = -34$ **7.** $4x^2 + 4y^2 - 24x + 16y = 48$

Graph each equation.

8. $x^2 + y^2 = 16$ **9.** $\dfrac{x^2}{16} + \dfrac{y^2}{9} = 1$ **10.** $\dfrac{x^2}{49} + \dfrac{y^2}{25} = 1$

11. A satellite is in an elliptical orbit around Earth with perigee altitude of 160 km and apogee altitude of 16,000 km. See the figure. (*Source:* Kastner, Bernice, *Space Mathematics,* NASA.) Find the equation of the ellipse.

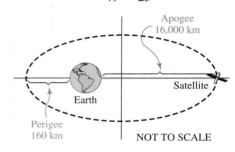

12. (a) The Roman Colosseum is an ellipse with $a = 310$ ft and $b = \frac{513}{2}$ ft. Find the distance between the foci of this ellipse.

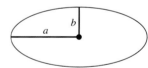

  (b) A formula for the approximate circumference of an ellipse is

$$C \approx 2\pi \sqrt{\dfrac{a^2 + b^2}{2}},$$

  where a and b are the lengths given in part (a). Use this formula to find the approximate circumference of the Roman Colosseum.

[13.2] *Graph each equation.*

13. $\dfrac{x^2}{16} - \dfrac{y^2}{25} = 1$ **14.** $\dfrac{y^2}{25} - \dfrac{x^2}{4} = 1$ **15.** $f(x) = -\sqrt{16 - x^2}$

Identify the graph of each equation as a parabola, circle, ellipse, *or* hyperbola.

16. $x^2 + y^2 = 64$ **17.** $y = 2x^2 - 3$ **18.** $y^2 = 2x^2 - 8$

19. $y^2 = 8 - 2x^2$ **20.** $x = y^2 + 4$ **21.** $x^2 - y^2 = 64$

22. Ships and planes often use a location-finding system called LORAN. With this system, a radio transmitter at M sends out a series of pulses. (See the figure.) When each pulse is received at transmitter S, it then sends out a pulse. A ship at P receives pulses from both M and S. A receiver on the ship measures the difference in the arrival times of the pulses. A special map gives hyperbolas that correspond to the differences in arrival times (which give the distances d_1 and d_2 in the figure.) The ship can then be located as lying on a branch of a particular hyperbola.

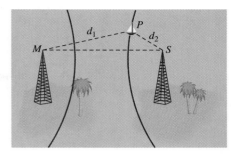

   Suppose $d_1 = 80$ mi and $d_2 = 30$ mi, and the distance between transmitters M and S is 100 mi. Use the definition to find an equation of the hyperbola on which the ship is located.

[13.3] *Solve each system.*

23. $2y = 3x - x^2$
    $x + 2y = -12$

24. $y + 1 = x^2 + 2x$
    $y + 2x = 4$

25. $x^2 + 3y^2 = 28$
    $y - x = -2$

26. $xy = 8$
    $x - 2y = 6$

27. $x^2 + y^2 = 6$
    $x^2 - 2y^2 = -6$

28. $3x^2 - 2y^2 = 12$
    $x^2 + 4y^2 = 18$

29. How many solutions are possible for a system of two equations whose graphs are a circle and a line?

30. How many solutions are possible for a system of two equations whose graphs are a parabola and a hyperbola?

[13.4] *Graph each inequality.*

31. $9x^2 \geq 16y^2 + 144$

32. $4x^2 + y^2 \geq 16$

33. $y < -(x + 2)^2 + 1$

Graph each system of inequalities.

34. $2x + 5y \leq 10$
    $3x - y \leq 6$

35. $|x| \leq 2$
    $|y| > 1$
    $4x^2 + 9y^2 \leq 36$

36. $9x^2 \leq 4y^2 + 36$
    $x^2 + y^2 \leq 16$

Set up a system of inequalities for Exercises 37 and 38 and then graph the region of feasible solutions for each system.

37. A bakery makes both cakes and cookies. Each batch of cakes requires 2 hr in the oven and 3 hr in the decorating room. Each batch of cookies needs $1\frac{1}{2}$ hr in the oven and $\frac{2}{3}$ hr in the decorating room. The oven is available no more than 15 hr per day, while the decorating room can be used no more than 13 hr per day.

38. A company makes two kinds of pizza, basic and plain. Basic contains cheese and beef, while plain contains onions and beef. The company sells at least 3 units per day of basic, and at least 2 units of plain. The beef costs $5 per unit for basic and $4 per unit for plain. The company can spend no more than $50 per day on beef. Dough for basic is $2 per unit, while dough for plain is $1 per unit. The company can spend no more than $16 per day on dough.

39. How many batches of cakes and cookies should the bakery of Exercise 37 make in order to maximize profits if cookies produce a profit of $20 per batch and cakes produce a profit of $30 per batch?

40. How many units of each kind of pizza should the company of Exercise 38 make in order to maximize revenue if basic sells for $20 per unit and plain for $15 per unit?

RELATING CONCEPTS (EXERCISES 41–45)

For Individual or Group Work

In Chapter 4 we discussed several methods of solving systems of linear equations in three variables. These methods can be used to find an equation of a circle through three points in a plane that are not on the same line. The equation of a circle can be written in the form $x^2 + y^2 + ax + by + c = 0$ for some values of a, b, and c. **Work Exercises 41–45 in order,** *to find the equation of the circle through the points* (2, 4), (5, 1), *and* (−1, 1).

41. Determine one equation in a, b, and c by letting $x = 2$ and $y = 4$ in the general form given above. Write it with a, b, and c on the left and the constant on the right.

42. Repeat Exercise 41 for the point (5, 1).

43. Repeat Exercise 41 for the point (−1, 1).

44. Solve the system formed by the equations found in Exercises 41–43, and give the equation of the circle that satisfies these conditions.

45. Use the methods of this chapter to find the center and the radius of the circle in Exercise 44.

MIXED REVIEW EXERCISES

Graph.

46. $\dfrac{x^2}{64} + \dfrac{y^2}{25} = 1$ **47.** $\dfrac{y^2}{4} - 1 = \dfrac{x^2}{9}$ **48.** $x^2 + y^2 = 25$ **49.** $x^2 + 9y^2 = 9$

50. $x^2 - 9y^2 = 9$ **51.** $f(x) = \sqrt{4 - x}$ **52.** $3x + 2y \geq 0$ **53.** $4y > 3x - 12$
$\qquad\qquad\qquad\qquad\qquad\qquad\qquad\qquad\qquad\qquad\qquad\;\; y \leq 4 \qquad\qquad\; x^2 < 16 - y^2$
$\qquad\qquad\qquad\qquad\qquad\qquad\qquad\qquad\qquad\qquad\qquad\;\; x \leq 4$

54. Explain why a set of points that form an ellipse does not satisfy the definition of a function.

55. The orbit of Mercury around the sun (a focus) is an ellipse with equation

$$\frac{x^2}{3352} + \frac{y^2}{3211} = 1,$$

where x and y are measured in million kilometers.

(a) Find its apogee, its greatest distance from the sun. (*Hint:* Refer to Section 13.1, Exercise 51.)

(b) Find its perigee, its smallest distance from the sun.

56. Find $x \geq 0$ and $y \geq 0$ such that

$$3x + 2y \leq 6$$
$$-2x + 4y \leq 8$$

and $2x + 5y$ is maximized.

CHAPTER **13** TEST

1. Find the center and radius of the circle whose equation is $(x - 2)^2 + (y + 3)^2 = 16$. Sketch the graph.

2. Find the center and radius of the circle whose equation is $x^2 + y^2 + 8x - 2y = 8$.

Graph.

3. $f(x) = \sqrt{9 - x^2}$

4. $4x^2 + 9y^2 = 36$

5. $16y^2 - 4x^2 = 64$

6. $\dfrac{y}{2} = -\sqrt{1 - \dfrac{x^2}{9}}$

Identify the graph of each equation as a parabola, hyperbola, ellipse, *or* circle.

7. $6x^2 + 4y^2 = 12$

8. $16x^2 = 144 + 9y^2$

9. $4y^2 + 4x = 9$

Solve each nonlinear system.

10. $2x - y = 9$
$xy = 5$

11. $x - 4 = 3y$
$x^2 + y^2 = 8$

12. $x^2 + y^2 = 25$
$x^2 - 2y^2 = 16$

13. Graph the inequality $y < x^2 - 2$.

14. Graph the system $\begin{aligned} x^2 + 25y^2 &\leq 25 \\ x^2 + y^2 &\leq 9. \end{aligned}$

15. Use the given region to find the maximum values of the objective function $z = 2x + 4y$.

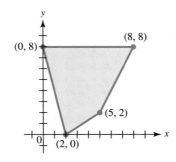

CUMULATIVE REVIEW EXERCISES CHAPTERS **1–13**

1. Simplify $-10 + |-5| - |3| + 4$.

Solve.

2. $4 - (2x + 3) + x = 5x - 3$

3. $-4k + 7 \geq 6k + 1$

4. $|5m| - 6 = 14$

5. $|2p - 5| > 15$

6. Find the slope of the line through $(2, 5)$ and $(-4, 1)$.

7. Find the equation of the line through $(-3, -2)$ and perpendicular to the graph of $2x - 3y = 7$.

Solve each system.

8. $3x - y = 12$
$2x + 3y = -3$

9. $x + y - 2z = 9$
$2x + y + z = 7$
$3x - y - z = 13$

10. $xy = -5$
$2x + y = 3$

Solve each problem.

11. Al and Bev traveled from their apartment to a picnic 20 mi away. Al traveled on his bike while Bev, who left later, took her car. Al's average speed was half of Bev's average speed. The trip took Al $\frac{1}{2}$ hr longer than Bev. What was Bev's average speed?

12. The president of InstaTune, a chain of franchised automobile tune-up shops, reports that people who buy a franchise and open a shop pay a weekly fee (in dollars) to company headquarters, according to the linear function defined by

$$f(x) = .07x + 135,$$

where $f(x)$ is the fee and x is the total amount of money taken in during the week by the shop. Find the weekly fee if \$2000 is taken in for the week. (*Source: Business Week.*)

Perform the indicated operations.

13. $(5y - 3)^2$

14. $(2r + 7)(6r - 1)$

15. $\dfrac{8x^4 - 4x^3 + 2x^2 + 13x + 8}{2x + 1}$

Factor.

16. $12x^2 - 7x - 10$

17. $2y^4 + 5y^2 - 3$

18. $z^4 - 1$

19. $a^3 - 27b^3$

Perform the indicated operations.

20. $\dfrac{5x - 15}{24} \cdot \dfrac{64}{3x - 9}$

21. $\dfrac{y^2 - 4}{y^2 - y - 6} \div \dfrac{y^2 - 2y}{y - 1}$

22. $\dfrac{5}{c + 5} - \dfrac{2}{c + 3}$

23. $\dfrac{p}{p^2 + p} + \dfrac{1}{p^2 + p}$

Solve.

24. Kareem and Jamal want to clean their office. Kareem can do the job alone in 3 hr, while Jamal can do it alone in 2 hr. How long will it take them if they work together?

Simplify. Assume all variables represent positive real numbers.

25. $\left(\dfrac{4}{3}\right)^{-1}$

26. $\dfrac{(2a)^{-2}a^4}{a^{-3}}$

27. $4\sqrt[3]{16} - 2\sqrt[3]{54}$

28. $\dfrac{3\sqrt{5x}}{\sqrt{2x}}$

29. $\dfrac{5 + 3i}{2 - i}$

Solve.

30. $2\sqrt{k} = \sqrt{5k + 3}$ **31.** $10q^2 + 13q = 3$ **32.** $(4x - 1)^2 = 8$

33. $3k^2 - 3k - 2 = 0$ **34.** $2(x^2 - 3)^2 - 5(x^2 - 3) = 12$

35. $F = \dfrac{kwv^2}{r}$ for v

36. If $f(x) = x^2 + 2x - 4$ and $g(x) = 3x + 2$, find

   (a) $(g \circ f)(1)$;   **(b)** $(f \circ g)(x)$.

37. If $f(x) = x^3 + 4$, find $f^{-1}(x)$.

38. Evaluate each expression.

   (a) $3^{\log_3 4}$ **(b)** $e^{\ln 7}$

39. Use properties of logarithms to write $2 \log(3x + 7) - \log 4$ as a single logarithm.

40. Solve $\log(x + 2) + \log(x - 1) = 1$.

41. If $10,000 is invested at 5% for 4 yr, how much will there be in the account if interest is compounded

   (a) quarterly;   **(b)** continuously?

42. The bar graph shows on-line retail sales (in billions of dollars) over the Internet. A reasonable model for sales y in billions of dollars is the exponential function defined by

$$y = 1.38(1.65)^x.$$

The years are coded such that x is the number of years since 1995.

   (a) Use the model to estimate sales in 2000. (*Hint:* Let $x = 5$.)

   (b) Use the model to estimate sales in 2003.

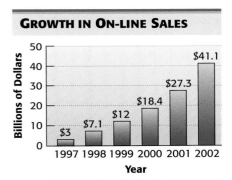

GROWTH IN ON-LINE SALES

Source: Jupiter Communications.

43. Use synthetic division to determine $f(3)$, if $f(x) = 2x^3 - 4x^2 + 5x - 10$.

44. Use the factor theorem to determine whether $x + 2$ is a factor of

$$5x^4 + 10x^3 + 6x^2 + 8x - 8.$$

If it is, what is the other factor? If it is not, explain why.

45. Find all zeros of $f(x) = 3x^3 + x^2 - 22x - 24$, given that one zero is -2.

Graph.

46. $f(x) = -3x + 5$ **47.** $f(x) = -2(x - 1)^2 + 3$ **48.** $\dfrac{x^2}{25} + \dfrac{y^2}{16} \le 1$

49. $\dfrac{x^2}{4} - \dfrac{y^2}{16} = 1$ **50.** $f(x) = 3^x$

Further Topics in Algebra

14

Amazing as it may seem, the male honeybee hatches from an unfertilized egg, while the female hatches from a fertilized one. The "family tree" of a male honeybee is shown here, where M represents male and F represents female. If we start with the male honeybee at the top, and count the number of bees in each generation, we obtain the following numbers in the order shown.

$$1, 1, 2, 3, 5, 8$$

Do you see the pattern here? After the first two terms (1 and 1), each successive term is obtained by adding the two previous terms. Thus, the term following 8 is $5 + 8 = 13$. The sequence of numbers described here is called the *Fibonacci sequence,* named after the 13th century Italian mathematician Leonardo of Pisa, who was also known as Fibonacci. This fascinating sequence has countless interesting properties and appears in many places in nature.

   In this chapter we study *sequences* and sums of terms of sequences, known as *series.*

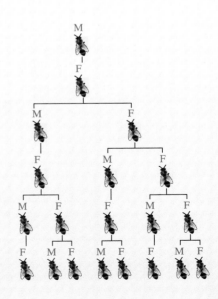

849

14.1 Sequences and Series

OBJECTIVES

1 Find the terms of a sequence given the general term.

2 Find the general term of a sequence.

3 Use sequences to solve applied problems.

4 Use summation notation to evaluate a series.

5 Write a series using summation notation.

6 Find the arithmetic mean (average) of a group of numbers.

A **sequence** is a function whose domain is the set of natural numbers. Intuitively, a sequence is a list of numbers in which the order of their appearance is important. Sequences appear in many places in daily life. For instance, the interest portions of monthly loan payments made to pay off an automobile or home loan form a sequence.

In the Palace of the Alhambra, residence of the Moorish rulers of Granada, Spain, the Sultana's quarters feature an interesting architectural pattern. There are 2 matched marble slabs inlaid in the floor, 4 walls, an octagon (8-sided) ceiling, 16 windows, 32 arches, and so on. If this pattern is continued indefinitely, the set of numbers forms an *infinite sequence.*

Infinite Sequence

An **infinite sequence** is a function with the set of positive integers as the domain.

OBJECTIVE 1 Find the terms of a sequence given the general term. For any positive integer n, the function value (y-value) of a sequence is written as a_n (read "a sub-n") instead of $a(n)$ or $f(n)$. The function values $a_1, a_2, a_3, \ldots$, written in order, are the **terms** of the sequence, with a_1 the first term, a_2 the second term, and so on. The expression a_n, which defines the sequence, is called the **general term** of the sequence.

In the Palace of the Alhambra example, the first five terms of the sequence are

$$a_1 = 2, \quad a_2 = 4, \quad a_3 = 8, \quad a_4 = 16, \quad \text{and} \quad a_5 = 32.$$

The general term for this sequence is $a_n = 2^n$.

EXAMPLE 1 Writing the Terms of Sequences from the General Term

Given an infinite sequence with $a_n = n + \dfrac{1}{n}$, find the following.

(a) The second term of the sequence
To get a_2, the second term, replace n with 2.

$$a_2 = 2 + \frac{1}{2} = \frac{5}{2}$$

(b) $a_{10} = 10 + \dfrac{1}{10} = \dfrac{101}{10}$ **(c)** $a_{12} = 12 + \dfrac{1}{12} = \dfrac{145}{12}$

Now Try Exercises 1 and 11.

As mentioned earlier, a sequence is a special kind of function. Graphing calculators can be used to generate and graph sequences, as shown in Figure 1. The calculator must be in graphing dot mode, so the discrete points on the graph are not connected. Remember that the domain of a sequence consists only of natural numbers.

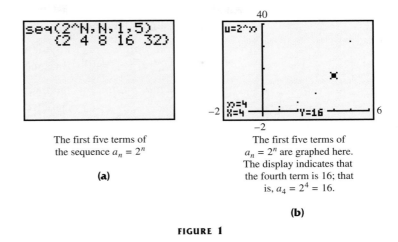

The first five terms of
the sequence $a_n = 2^n$

(a)

The first five terms of
$a_n = 2^n$ are graphed here.
The display indicates that
the fourth term is 16; that
is, $a_4 = 2^4 = 16$.

(b)

FIGURE 1

OBJECTIVE 2 Find the general term of a sequence.   Sometimes we need to find a general term to fit the first few terms of a given sequence. There are no rules for finding the general term of a sequence from the first few terms. In fact, it is possible to give more than one general term that produce the same first three or four terms. However, in many examples, the terms may suggest a general term.

EXAMPLE 2 Finding the General Term of a Sequence

Find an expression for the general term a_n of the sequence.

$$5, 10, 15, 20, 25, \ldots$$

The first term is $5(1)$, the second is $5(2)$, and so on. By inspection, $a_n = 5n$ will produce the given first five terms.

Now Try Exercise 17.

CAUTION   One problem with using just a few terms to suggest a general term, as in Example 2, is that there may be more than one general term that gives the same first few terms.

OBJECTIVE 3 Use sequences to solve applied problems.   Practical problems often involve *finite sequences.*

Finite Sequence

A **finite sequence** has a domain that includes only the first n positive integers.

For example, if n is 5, the domain is $\{1, 2, 3, 4, 5\}$, and the sequence has five terms.

▦ EXAMPLE 3 Using a Sequence in an Application

Keshon borrows $5000 and agrees to pay $500 monthly, plus interest of 1% on the unpaid balance from the beginning of that month. Find the payments for the first 4 months and the remaining debt at the end of this period.

The payments and remaining balances are calculated as follows.

First month	Payment:	$500 + .01(5000) = 550$ dollars
	Balance:	$5000 - 500 = 4500$ dollars
Second month	Payment:	$500 + .01(4500) = 545$ dollars
	Balance:	$5000 - 2 \cdot 500 = 4000$ dollars
Third month	Payment:	$500 + .01(4000) = 540$ dollars
	Balance:	$5000 - 3 \cdot 500 = 3500$ dollars
Fourth month	Payment:	$500 + .01(3500) = 535$ dollars
	Balance:	$5000 - 4 \cdot 500 = 3000$ dollars

The payments for the first four months, in dollars, are

$$550, \ 545, \ 540, \ 535$$

and the remaining debt at the end of this period is 3000 dollars.

Now Try Exercise 21.

OBJECTIVE 4 Use summation notation to evaluate a series. By adding the terms of a sequence, we obtain a *series*.

Series

The indicated sum of the terms of a sequence is called a **series.**

For example, if we consider the sum of the payments listed in Example 3,

$$550 + 545 + 540 + 535,$$

we obtain a series that represents the total amount of payments for the first four months.

Since a sequence can be finite or infinite, there are finite or infinite series. One type of infinite series is discussed in Section 14.3, and the binomial theorem discussed in Section 14.4 defines an important finite series. In this section we discuss only finite series.

We use a compact notation, called **summation notation,** to write a series from the general term of the corresponding sequence. For example, the sum of the first six terms of the sequence with general term $a_n = 3n + 2$ is written with the Greek letter Σ (sigma) as

$$\sum_{i=1}^{6} (3i + 2).$$

We read this as "the sum from $i = 1$ to 6 of $3i + 2$." To find this sum, we replace the letter i in $3i + 2$ with 1, 2, 3, 4, 5, and 6, as follows.

$$\sum_{i=1}^{6} (3i + 2) = (3 \cdot 1 + 2) + (3 \cdot 2 + 2) + (3 \cdot 3 + 2)$$
$$+ (3 \cdot 4 + 2) + (3 \cdot 5 + 2) + (3 \cdot 6 + 2)$$
$$= 5 + 8 + 11 + 14 + 17 + 20$$
$$= 75$$

The letter i is called the **index of summation.**

CAUTION This use of i has no connection with the complex number i.

■ **EXAMPLE 4** Evaluating Series Written in Summation Notation

Write out the terms and evaluate each series.

(a) $\displaystyle\sum_{i=1}^{5} (i - 4) = (1 - 4) + (2 - 4) + (3 - 4) + (4 - 4) + (5 - 4)$
$$= -3 - 2 - 1 + 0 + 1$$
$$= -5$$

(b) $\displaystyle\sum_{i=3}^{7} 3i^2 = 3(3)^2 + 3(4)^2 + 3(5)^2 + 3(6)^2 + 3(7)^2$
$$= 27 + 48 + 75 + 108 + 147$$
$$= 405$$

Now Try Exercises 25 and 27.

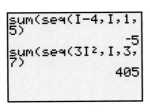

FIGURE 2

Figure 2 shows how a graphing calculator can be used to obtain the results found in Example 4.

OBJECTIVE 5 Write a series using summation notation. In Example 4, we started with summation notation and wrote each series using + signs. It is possible to go in the other direction; that is, given a series, we can write it using summation notation. To do this, we observe a pattern in the terms and write the general term accordingly.

■ **EXAMPLE 5** Writing Series with Summation Notation

Write each sum with summation notation.

(a) $2 + 5 + 8 + 11$

First, find a general term a_n that will give these four terms for a_1, a_2, a_3, and a_4. Inspection (and trial and error) shows that $3i - 1$ will work for these four terms, since

$$3(1) - 1 = 2$$
$$3(2) - 1 = 5$$
$$3(3) - 1 = 8$$
$$3(4) - 1 = 11.$$

(Remember, there may be other expressions that also work. These four terms may be the first terms of more than one sequence.) Since i ranges from 1 to 4, write the sum as

$$2 + 5 + 8 + 11 = \sum_{i=1}^{4} (3i - 1).$$

(b) $8 + 27 + 64 + 125 + 216$

Since these numbers are the cubes of 2, 3, 4, 5, and 6,

$$8 + 27 + 64 + 125 + 216 = \sum_{i=2}^{6} i^3.$$

Now Try Exercises 37 and 41.

OBJECTIVE 6 Find the arithmetic mean (average) of a group of numbers.

Arithmetic Mean or Average

The **arithmetic mean,** or **average,** of a group of numbers is symbolized $\bar{x}$ and is found by dividing the sum of the numbers by the number of numbers. That is,

$$\bar{x} = \frac{\sum_{i=1}^{n} x_i}{n}.$$

Here the values of x_i represent the individual numbers in the group, and n represents the number of numbers.

EXAMPLE 6 Finding the Arithmetic Mean or Average

The following table shows the number of companies listed on the New York Stock Exchange for each year during the period 1994 through 2000. What was the average number of listings per year for this 7-yr period?

Year	Number of Listings
1994	2570
1995	2675
1996	2907
1997	3047
1998	3114
1999	3025
2000	2862

Source: New York Stock Exchange.

Let $x_1 = 2570$, $x_2 = 2675$, and so on. Since there are 7 numbers in the group, $n = 7$. Therefore,

$$\bar{x} = \frac{\sum_{i=1}^{7} x_i}{7}$$

$$= \frac{2570 + 2675 + 2907 + 3047 + 3114 + 3025 + 2862}{7}$$

$$= 2886 \quad \text{(rounded to the nearest unit)}.$$

The average number of listings per year for this 7-yr period was 2886.

Now Try Exercise 51.

14.1 EXERCISES

Write out the first five terms of each sequence. See Example 1.

1. $a_n = n + 1$

2. $a_n = n - 4$

3. $a_n = \dfrac{n + 3}{n}$

4. $a_n = \dfrac{n + 2}{n + 1}$

5. $a_n = 3^n$

6. $a_n = 1^{n-1}$

7. $a_n = \dfrac{1}{n^2}$

8. $a_n = \dfrac{n^2}{n + 1}$

9. $a_n = (-1)^n$

10. $a_n = (-1)^{2n-1}$

Find the indicated term for each sequence. See Example 1.

11. $a_n = -9n + 2;\ a_8$

12. $a_n = 3n - 7;\ a_{12}$

13. $a_n = \dfrac{3n + 7}{2n - 5};\ a_{14}$

14. $a_n = \dfrac{5n - 9}{3n + 8};\ a_{16}$

15. $a_n = (n + 1)(2n + 3);\ a_8$

16. $a_n = (5n - 2)(3n + 1);\ a_{10}$

Find a general term, a_n, for the given terms of each sequence. See Example 2.

17. 4, 8, 12, 16, . . .

18. $-10, -20, -30, -40, . . .$

19. $\dfrac{1}{3}, \dfrac{1}{9}, \dfrac{1}{27}, \dfrac{1}{81}, . . .$

20. $\dfrac{1}{2}, \dfrac{2}{3}, \dfrac{3}{4}, \dfrac{4}{5}, . . .$

Solve each applied problem by writing the first few terms of a sequence. See Example 3.

21. Anne borrows $1000 and agrees to pay $100 plus interest of 1% on the unpaid balance each month. Find the payments for the first 6 months and the remaining debt at the end of this period.

22. Larissa Perez is offered a new modeling job with a salary of $20{,}000 + 2500n$ dollars per year at the end of the nth year. Write a sequence showing her salary at the end of each of the first 5 yr. If she continues in this way, what will her salary be at the end of the tenth year?

23. Suppose that an automobile loses $\frac{1}{5}$ of its value each year; that is, at the end of any given year, the value is $\frac{4}{5}$ of the value at the beginning of that year. If a car costs $20,000 new, what is its value at the end of 5 yr?

24. A certain car loses $\frac{1}{2}$ of its value each year. If this car cost $40,000 new, what is its value at the end of 6 yr?

Write out each series and evaluate it. See Example 4.

25. $\displaystyle\sum_{i=1}^{5} (i + 3)$

26. $\displaystyle\sum_{i=1}^{6} (i + 9)$

27. $\displaystyle\sum_{i=1}^{3} (i^2 + 2)$

28. $\displaystyle\sum_{i=1}^{4} i(i + 3)$

29. $\displaystyle\sum_{i=1}^{6} (-1)^i$

30. $\displaystyle\sum_{i=1}^{5} (-1)^i \cdot i$

31. $\displaystyle\sum_{i=3}^{7} (i - 3)(i + 2)$

32. $\displaystyle\sum_{i=2}^{6} \dfrac{i^2 + 1}{2}$

Write out the terms of each series.

33. $\displaystyle\sum_{i=1}^{5} 2x \cdot i$

34. $\displaystyle\sum_{i=1}^{6} x^i$

35. $\displaystyle\sum_{i=1}^{5} i \cdot x^i$

36. $\displaystyle\sum_{i=2}^{6} \dfrac{x + i}{x - i}$

Write each series using summation notation. See Example 5.

37. $3 + 4 + 5 + 6 + 7$

38. $1 + 4 + 9 + 16$

39. $\dfrac{1}{2} + \dfrac{1}{3} + \dfrac{1}{4} + \dfrac{1}{5} + \dfrac{1}{6}$

40. $-1 + 2 - 3 + 4 - 5 + 6$

41. $1 + 4 + 9 + 16 + 25$

42. $1 + 16 + 81 + 256$

43. Suppose that f is a function with domain all real numbers, where $f(x) = 2x + 4$. Suppose that an infinite sequence is defined by $a_n = 2n + 4$. Discuss the similarities and differences between the function and the sequence. Give examples using each.

44. What is wrong with the following?
For the sequence defined by $a_n = 2n + 4$, find $a_{1/2}$.

45. Explain the basic difference between a sequence and a series.

46. Evaluate $\displaystyle\sum_{i=1}^{3} 5i$ and $5 \displaystyle\sum_{i=1}^{3} i$. Notice that the sums are the same. Explain how the distributive property plays a role in assuring us that the two sums are equal.

Find the arithmetic mean for each collection of numbers. See Example 6.

47. 8, 11, 14, 9, 3, 6, 8

48. 10, 12, 8, 19, 23

49. 5, 9, 8, 2, 4, 7, 3, 2

50. 2, 1, 4, 8, 3, 7

Solve each problem. See Example 6.

51. The number of mutual funds available to investors for each year during the period 1996 through 2000 is given in the table.

Year	Number of Funds Available
1996	6254
1997	6684
1998	7314
1999	7791
2000	8171

Source: Investment Company Institute.

To the nearest whole number, what was the average number of funds available per year during this period?

52. The total assets of mutual funds, in billions of dollars, for each year during the period 1992 through 1996 are shown in the table. To the nearest tenth (in billions of dollars), what were the average assets per year during this period?

Year	Assets (in billions of dollars)
1992	1646.3
1993	2075.4
1994	2161.5
1995	2820.4
1996	3539.2

Source: Investment Company Institute.

RELATING CONCEPTS (EXERCISES 53–60)

For Individual or Group Work

The following properties of series provide useful shortcuts for evaluating series.

If $a_1, a_2, a_3, \ldots, a_n$ and $b_1, b_2, b_3, \ldots, b_n$ are two sequences, and c is a constant, then for every positive integer n,

(a) $\displaystyle\sum_{i=1}^{n} c = nc$ **(b)** $\displaystyle\sum_{i=1}^{n} ca_i = c\sum_{i=1}^{n} a_i$

(c) $\displaystyle\sum_{i=1}^{n} (a_i + b_i) = \sum_{i=1}^{n} a_i + \sum_{i=1}^{n} b_i$ **(d)** $\displaystyle\sum_{i=1}^{n} (a_i - b_i) = \sum_{i=1}^{n} a_i - \sum_{i=1}^{n} b_i.$

Work Exercises 53–60 in order, *to see how these shortcuts can make work easier.*

53. Use property (c) to write $\displaystyle\sum_{i=1}^{6} (i^2 + 3i + 5)$ as the sum of three summations.

54. Use property (b) to rewrite the second summation from Exercise 53.

55. Use property (a) to rewrite the third summation from Exercise 53.

56. Rewrite $1 + 2 + 3 + 4 + \cdots + n = \dfrac{n(n+1)}{2}$ using summation notation.

57. Rewrite $1^2 + 2^2 + 3^2 + 4^2 + \cdots + n^2 = \dfrac{n(n+1)(2n+1)}{6}$ using summation notation.

58. Use the summations you wrote in Exercises 56 and 57 and the given properties to evaluate the three summations from Exercises 53–55. This gives the value of $\displaystyle\sum_{i=1}^{6} (i^2 + 3i + 5)$ without writing out all six terms.

59. Use the given properties and summations to evaluate $\displaystyle\sum_{i=1}^{12} (i^2 - i).$

60. Use the given properties and summations to evaluate $\displaystyle\sum_{i=1}^{20} (2 + i - i^2).$

14.2 Arithmetic Sequences

OBJECTIVES

1 Find the common difference of an arithmetic sequence.

2 Find the general term of an arithmetic sequence.

3 Use an arithmetic sequence in an application.

4 Find any specified term or the number of terms of an arithmetic sequence. (continued)

OBJECTIVE 1 Find the common difference of an arithmetic sequence. In this section we introduce a special type of sequence that has many applications.

Arithmetic Sequence

A sequence in which each term after the first differs from the preceding term by a constant amount is called an **arithmetic sequence** or **arithmetic progression.**

For example, the sequence

$$6, 11, 16, 21, 26, \ldots$$

is an arithmetic sequence, since the difference between any two adjacent terms is always 5. The number 5 is called the **common difference** of the arithmetic sequence.

OBJECTIVES (continued)

5 Find the sum of a specified number of terms of an arithmetic sequence.

The common difference, d, is found by subtracting any pair of terms a_n and a_{n+1}. That is,

$$d = a_{n+1} - a_n.$$

EXAMPLE 1 Finding the Common Difference

Find d for the arithmetic sequence.

$$-11, -4, 3, 10, 17, 24, \ldots$$

Since the sequence is arithmetic, d is the difference between any two adjacent terms. Choosing the terms 10 and 17 gives

$$d = 17 - 10$$
$$= 7.$$

The terms -11 and -4 would give $d = -4 - (-11) = 7$, the same result.

Now Try Exercise 7.

EXAMPLE 2 Writing the Terms of a Sequence from the First Term and Common Difference

Write the first five terms of the arithmetic sequence with first term 3 and common difference -2.

The second term is found by adding -2 to the first term 3, getting 1. For the next term, add -2 to 1, and so on. The first five terms are

$$3, 1, -1, -3, -5.$$

Now Try Exercise 9.

OBJECTIVE 2 Find the general term of an arithmetic sequence. Generalizing from Example 2, if we know the first term a_1 and the common difference d of an arithmetic sequence, then the sequence is completely defined as

$$a_1, \quad a_2 = a_1 + d, \quad a_3 = a_1 + 2d, \quad a_4 = a_1 + 3d, \ldots.$$

Writing the terms of the sequence in this way suggests the following rule.

General Term of an Arithmetic Sequence

The general term of an arithmetic sequence with first term a_1 and common difference d is

$$a_n = a_1 + (n - 1)d.$$

Since $a_n = a_1 + (n - 1)d = dn + (a_1 - d)$ is a linear function in n, any linear expression of the form $kn + c$, where k and c are real numbers, defines an arithmetic sequence.

EXAMPLE 3   Finding the General Term of an Arithmetic Sequence

Find the general term for the arithmetic sequence.

$$-9, -6, -3, 0, 3, 6, \ldots$$

Then use the general term to find a_{20}.

Here the first term is $a_1 = -9$. To find d, subtract any two adjacent terms. For example,

$$d = -3 - (-6) = 3.$$

Now find a_n.

$$
\begin{array}{ll}
a_n = a_1 + (n-1)d & \text{Formula for } a_n \\
 = -9 + (n-1)(3) & \text{Let } a_1 = -9, d = 3. \\
 = -9 + 3n - 3 & \text{Distributive property} \\
a_n = 3n - 12 & \text{Combine terms.}
\end{array}
$$

Thus, the general term is $a_n = 3n - 12$. To find a_{20}, let $n = 20$.

$$a_{20} = 3(20) - 12 = 60 - 12 = 48$$

Now Try Exercise 13.

OBJECTIVE 3 Use an arithmetic sequence in an application.

EXAMPLE 4   Applying an Arithmetic Sequence

Howie Sorkin's uncle decides to start a fund for Howie's education. He makes an initial contribution of $3000 and each month deposits an additional $500. Thus, after one month there will be $3000 + $500 = $3500. How much will there be after 24 months? (Disregard any interest.)

The contributions can be described using an arithmetic sequence. After n months, the fund will contain

$$a_n = 3000 + 500n \text{ dollars.}$$

To find the amount in the fund after 24 months, find a_{24}.

$$
\begin{array}{ll}
a_{24} = 3000 + 500(24) & \text{Let } n = 24. \\
\phantom{a_{24}} = 3000 + 12{,}000 & \text{Multiply.} \\
\phantom{a_{24}} = 15{,}000 & \text{Add.}
\end{array}
$$

The account will contain $15,000 (disregarding interest) after 24 months.

Now Try Exercise 47.

OBJECTIVE 4 Find any specified term or the number of terms of an arithmetic sequence.
The formula for the general term has four variables: a_n, a_1, n, and d. If we know any three of these, the formula can be used to find the value of the fourth variable. The next example shows how to find a particular term.

EXAMPLE 5 Finding Specified Terms

Find the indicated term for each arithmetic sequence.

(a) $a_1 = -6$, $d = 12$; a_{15}

We use the formula $a_n = a_1 + (n - 1)d$. Since we want $a_n = a_{15}$, $n = 15$.

$$
\begin{aligned}
a_{15} &= a_1 + (15 - 1)d & \text{Let } n = 15. \\
&= -6 + 14(12) & \text{Let } a_1 = -6, d = 12. \\
&= 162
\end{aligned}
$$

(b) $a_5 = 2$ and $a_{11} = -10$; a_{17}

Any term can be found if a_1 and d are known. Use the formula for a_n with the two given terms.

$$
\begin{aligned}
a_5 &= a_1 + (5 - 1)d \\
a_5 &= a_1 + 4d \\
2 &= a_1 + 4d \qquad \text{Let } a_5 = 2.
\end{aligned}
\qquad
\begin{aligned}
a_{11} &= a_1 + (11 - 1)d \\
a_{11} &= a_1 + 10d \\
-10 &= a_1 + 10d \qquad \text{Let } a_{11} = -10.
\end{aligned}
$$

This gives a system of two equations with two variables, a_1 and d. Find d by adding -1 times one equation to the other to eliminate a_1.

$$
\begin{aligned}
-10 &= a_1 + 10d \\
\underline{-2} &= \underline{-a_1 - 4d} \qquad \text{Multiply } 2 = a_1 + 4d \text{ by } -1. \\
-12 &= 6d \qquad \text{Add.} \\
-2 &= d \qquad \text{Divide by 6.}
\end{aligned}
$$

Now find a_1 by substituting -2 for d into either equation.

$$
\begin{aligned}
-10 &= a_1 + 10(-2) & \text{Let } d = -2. \\
-10 &= a_1 - 20 \\
10 &= a_1
\end{aligned}
$$

Use the formula for a_n to find a_{17}.

$$
\begin{aligned}
a_{17} &= a_1 + (17 - 1)d & \text{Let } n = 17. \\
&= a_1 + 16d \\
&= 10 + 16(-2) & \text{Let } a_1 = 10, d = -2. \\
&= -22
\end{aligned}
$$

Now Try Exercises 19 and 23.

Sometimes we need to find out how many terms are in a sequence, as shown in the following example.

EXAMPLE 6 Finding the Number of Terms in a Sequence

Find the number of terms in the arithmetic sequence.

$$-8, -2, 4, 10, \ldots, 52$$

Let n represent the number of terms in the sequence. Since $a_n = 52$, $a_1 = -8$, and $d = -2 - (-8) = 6$, use the formula $a_n = a_1 + (n - 1)d$ to find n. Substituting the known values into the formula gives

$$a_n = a_1 + (n - 1)d$$

$52 = -8 + (n - 1)6$ Let $a_n = 52$, $a_1 = -8$, $d = 6$.

$52 = -8 + 6n - 6$ Distributive property

$66 = 6n$ Combine terms.

$n = 11$.   Divide by 6.

The sequence has 11 terms.

Now Try Exercise 25.

OBJECTIVE 5 **Find the sum of a specified number of terms of an arithmetic sequence.** To find a formula for the sum, S_n, of the first n terms of an arithmetic sequence, we can write out the terms as

$$S_n = a_1 + (a_1 + d) + (a_1 + 2d) + \cdots + [a_1 + (n - 1)d].$$

This same sum can be written in reverse as

$$S_n = a_n + (a_n - d) + (a_n - 2d) + \cdots + [a_n - (n - 1)d].$$

Now add the corresponding terms of these two expressions for S_n to get

$$2S_n = (a_1 + a_n) + (a_1 + a_n) + (a_1 + a_n) + \cdots + (a_1 + a_n).$$

The right-hand side of this expression contains n terms, each equal to $a_1 + a_n$, so

$$2S_n = n(a_1 + a_n)$$

$$S_n = \frac{n}{2}(a_1 + a_n).$$

EXAMPLE 7 **Finding the Sum of the First n Terms**

Find the sum of the first five terms of the arithmetic sequence in which $a_n = 2n - 5$.
We can use the formula $S_n = \frac{n}{2}(a_1 + a_n)$ to find the sum of the first five terms. Here $n = 5$, $a_1 = 2(1) - 5 = -3$, and $a_5 = 2(5) - 5 = 5$. From the formula,

$$S_5 = \frac{5}{2}(-3 + 5) = \frac{5}{2}(2) = 5.$$

Now Try Exercise 39.

It is sometimes useful to express the sum of an arithmetic sequence, S_n, in terms of a_1 and d, the quantities that define the sequence. We can do this as follows. Since

$$S_n = \frac{n}{2}(a_1 + a_n) \quad \text{and} \quad a_n = a_1 + (n - 1)d,$$

by substituting the expression for a_n into the expression for S_n we obtain

$$S_n = \frac{n}{2}(a_1 + [a_1 + (n - 1)d])$$

$$S_n = \frac{n}{2}[2a_1 + (n - 1)d].$$

The summary on the next page gives both of the alternative forms that may be used to find the sum of the first n terms of an arithmetic sequence.

> ### Sum of the First n Terms of an Arithmetic Sequence
>
> The sum of the first n terms of the arithmetic sequence with first term a_1, nth term a_n, and common difference d is
>
> $$S_n = \frac{n}{2}(a_1 + a_n) \quad \text{or} \quad S_n = \frac{n}{2}[2a_1 + (n-1)d].$$

■ EXAMPLE 8   Finding the Sum of the First n Terms

Find the sum of the first eight terms of the arithmetic sequence having first term 3 and common difference -2.

   Since the known values, $a_1 = 3$, $d = -2$, and $n = 8$, appear in the second formula for S_n, we use it.

$$S_n = \frac{n}{2}[2a_1 + (n-1)d]$$

$$S_8 = \frac{8}{2}[2(3) + (8-1)(-2)] \qquad \text{Let } a_1 = 3, d = -2, n = 8.$$

$$= 4[6 - 14]$$

$$= -32$$

Now Try Exercise 35.

   As mentioned earlier, linear expressions of the form $kn + c$, where k and c are real numbers, define an arithmetic sequence. For example, the sequences defined by $a_n = 2n + 5$ and $a_n = n - 3$ are arithmetic sequences. For this reason,

$$\sum_{i=1}^{n} (ki + c)$$

represents the sum of the first n terms of an arithmetic sequence having first term $a_1 = k(1) + c = k + c$ and general term $a_n = k(n) + c = kn + c$. We can find this sum with the first formula for S_n, as shown in the next example.

■ EXAMPLE 9   Using S_n to Evaluate a Summation

Find $\displaystyle\sum_{i=1}^{12} (2i - 1)$.

   This is the sum of the first 12 terms of the arithmetic sequence having $a_n = 2n - 1$. This sum, S_{12}, is found with the first formula for S_n,

$$S_n = \frac{n}{2}(a_1 + a_n).$$

Here $n = 12$, $a_1 = 2(1) - 1 = 1$, and $a_{12} = 2(12) - 1 = 23$. Substitute these values into the formula to find

$$S_{12} = \frac{12}{2}(1 + 23) = 6(24) = 144.$$

Now Try Exercise 41.

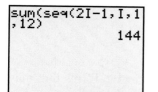

FIGURE 3

Figure 3 shows how a graphing calculator supports the result of Example 9.

14.2 EXERCISES

1. Using several examples, explain the meaning of *arithmetic sequence*.

2. Can any two terms of an arithmetic sequence be used to find the common difference? Explain.

If the given sequence is arithmetic, find the common difference, d. If the sequence is not arithmetic, say so. See Example 1.

3. 1, 2, 3, 4, 5, . . .

4. 2, 5, 8, 11, . . .

5. 2, −4, 6, −8, 10, −12, . . .

6. −6, −10, −14, −18, . . .

7. −10, −5, 0, 5, 10, . . .

8. 1, 2, 4, 7, 11, 16, . . .

Write the first five terms of each arithmetic sequence. See Example 2.

9. $a_1 = 5, d = 4$

10. $a_1 = 6, d = 7$

11. $a_1 = -2, d = -4$

12. $a_1 = -3, d = -5$

Use the formula for a_n to find the general term for each arithmetic sequence. See Example 3.

13. $a_1 = 2, d = 5$

14. $a_1 = 5, d = -3$

15. $3, \dfrac{15}{4}, \dfrac{9}{2}, \dfrac{21}{4}, \ldots$

16. 4, 14, 24, . . .

17. −3, 0, 3, . . .

18. −10, −5, 0, 5, 10, . . .

Find the indicated term for each arithmetic sequence. See Examples 2 and 5.

19. $a_1 = 4, d = 3; a_{25}$

20. $a_1 = 1, d = -\dfrac{1}{2}; a_{12}$

21. 2, 4, 6, . . . ; a_{24}

22. 1, 5, 9, . . . ; a_{50}

23. $a_{12} = -45, a_{10} = -37; a_1$

24. $a_{10} = -2, a_{15} = -8; a_3$

Find the number of terms in each arithmetic sequence. See Example 6.

25. 3, 5, 7, . . . , 33

26. $2, \dfrac{3}{2}, 1, \dfrac{1}{2}, \ldots, -5$

27. $\dfrac{3}{4}, 3, \dfrac{21}{4}, \ldots, 12$

28. 4, 1, −2, . . . , −32

29. In the formula for S_n, what does n represent?

30. Explain when you would use each of the two formulas for S_n.

RELATING CONCEPTS (EXERCISES 31–34)

For Individual or Group Work

Exercises 31–34 show how to find the sum $1 + 2 + 3 + \cdots + 99 + 100$ in an ingenious way. **Work them in order.**

31. Consider the following:

$$S = 1 + 2 + 3 + \cdots + 99 + 100$$
$$\underline{S = 100 + 99 + 98 + \cdots + 2 + 1.}$$

Add the left sides of this equation. The result is _____. Add the columns on the right side. The sum _____ appears _____ times, so by multiplication, the sum of the right sides of the equations is _____.

(continued)

32. Form an equation by setting the sum of the left sides equal to the sum of the right sides.

33. Solve the equation from Exercise 32 to find that the desired sum, S, is _____.

34. Find the sum $S = 1 + 2 + 3 + \cdots + 199 + 200$ using the procedure described in Exercises 31–33.

Find S_6 for each arithmetic sequence. See Examples 7 and 8.

35. $a_1 = 6, d = 3$ **36.** $a_1 = 5, d = 4$ **37.** $a_1 = 7, d = -3$

38. $a_1 = -5, d = -4$ **39.** $a_n = 4 + 3n$ **40.** $a_n = 9 + 5n$

Use a formula for S_n to evaluate each series. See Example 9.

41. $\displaystyle\sum_{i=1}^{10} (8i - 5)$ **42.** $\displaystyle\sum_{i=1}^{17} (i - 1)$ **43.** $\displaystyle\sum_{i=1}^{20} (2i - 5)$

44. $\displaystyle\sum_{i=1}^{10} \left(\frac{1}{2}i - 1\right)$ **45.** $\displaystyle\sum_{i=1}^{250} i$ **46.** $\displaystyle\sum_{i=1}^{2000} i$

Solve each applied problem. (Hint: Determine whether you need to find a specific term of a sequence or the sum of the terms of a sequence immediately after reading the problem.) See Example 4.

47. Nancy Bondy's aunt has promised to deposit $1 in her account on the first day of her birthday month, $2 on the second day, $3 on the third day, and so on for 30 days. How much will this amount to over the entire month?

48. Repeat Exercise 47, but assume that the deposits are $2, $4, $6, and so on, and that the month is February of a leap year.

49. Suppose that Randy Morgan is offered a job at $1600 per month with a guaranteed increase of $50 every 6 months for 5 yr. What will his salary be at the end of this period of time?

50. Repeat Exercise 49, but assume that the starting salary is $2000 per month, and the guaranteed increase is $100 every 4 months for 3 yr.

51. A seating section in a theater-in-the-round has 20 seats in the first row, 22 in the second row, 24 in the third row, and so on for 25 rows. How many seats are there in the last row? How many seats are there in the section?

52. José Valdevielso has started on a fitness program. He plans to jog 10 min per day for the first week, and then add 10 min per day each week until he is jogging an hour each day. In which week will this occur? What is the total number of minutes he will run during the first 4 weeks?

53. A child builds with blocks, placing 35 blocks in the first row, 31 in the second row, 27 in the third row, and so on. Continuing this pattern, can she end with a row containing exactly 1 block? If not, how many blocks will the last row contain? How many rows can she build this way?

54. A stack of firewood has 28 pieces on the bottom, 24 on top of those, then 20, and so on. If there are 108 pieces of wood, how many rows are there? (*Hint:* $n \leq 7$.)

14.3 Geometric Sequences

In an arithmetic sequence, each term after the first is found by *adding* a fixed number to the previous term. A *geometric sequence* is defined as follows.

Geometric Sequence

A **geometric sequence** or **geometric progression** is a sequence in which each term after the first is a constant multiple of the preceding term.

OBJECTIVE 1 Find the common ratio of a geometric sequence. We find the constant multiplier, called the **common ratio,** by dividing any term after the first by the preceding term. That is, the common ratio is

$$r = \frac{a_{n+1}}{a_n}.$$

For example,

$$2, 6, 18, 54, 162, \ldots$$

is a geometric sequence in which the first term, a_1, is 2 and the common ratio is

$$r = \frac{6}{2} = \frac{18}{6} = \frac{54}{18} = \frac{162}{54} = 3.$$

EXAMPLE 1 Finding the Common Ratio

Find r for the geometric sequence.

$$15, \frac{15}{2}, \frac{15}{4}, \frac{15}{8}, \ldots$$

To find r, choose any two successive terms and divide the second one by the first. Choosing the second and third terms of the sequence gives

$$r = \frac{a_3}{a_2} = \frac{15}{4} \div \frac{15}{2} = \frac{1}{2}.$$

Any other two successive terms could have been used to find r. Additional terms of the sequence can be found by multiplying each successive term by $\frac{1}{2}$.

Now Try Exercise 3.

OBJECTIVE 2 Find the general term of a geometric sequence. The general term a_n of a geometric sequence $a_1, a_2, a_3, \ldots$ is expressed in terms of a_1 and r by writing the first few terms as

$$a_1, \quad a_2 = a_1 r, \quad a_3 = a_1 r^2, \quad a_4 = a_1 r^3, \ldots,$$

which suggests the rule on the next page.

> **General Term of a Geometric Sequence**
>
> The general term of the geometric sequence with first term a_1 and common ratio r is
>
> $$a_n = a_1 r^{n-1}.$$

CAUTION Be careful to use the correct order of operations when finding $a_1 r^{n-1}$. The value of r^{n-1} must be found first. Then multiply the result by a_1.

EXAMPLE 2 Finding the General Term

Find the general term of the sequence in Example 1.

The first term is $a_1 = 15$ and the common ratio is $r = \frac{1}{2}$. Substituting into the formula for the general term gives

$$a_n = a_1 r^{n-1} = 15\left(\frac{1}{2}\right)^{n-1},$$

the required general term. Notice that it is not possible to simplify further, because the exponent must be applied before the multiplication can be done.

Now Try Exercise 11.

OBJECTIVE 3 Find any specified term of a geometric sequence. We can use the formula for the general term to find any particular term.

EXAMPLE 3 Finding Specified Terms

Find the indicated term for each geometric sequence.

(a) $a_1 = 4$, $r = -3$; a_6

Let $n = 6$. From the general term $a_n = a_1 r^{n-1}$,

$$
\begin{aligned}
a_6 &= a_1 \cdot r^{6-1} \qquad &\text{Let } n = 6.\\
&= 4 \cdot (-3)^5 \qquad &\text{Let } a_1 = 4, r = -3.\\
&= -972. \qquad &\text{Evaluate } (-3)^5 \text{ first.}
\end{aligned}
$$

(b) $\dfrac{3}{4}, \dfrac{3}{8}, \dfrac{3}{16}, \ldots$; a_7

Here, $r = \frac{1}{2}$, $a_1 = \frac{3}{4}$, and $n = 7$.

$$a_7 = \frac{3}{4} \cdot \left(\frac{1}{2}\right)^6 = \frac{3}{4} \cdot \frac{1}{64} = \frac{3}{256}$$

Now Try Exercises 17 and 19.

EXAMPLE 4 Writing the Terms of a Sequence

Write the first five terms of the geometric sequence whose first term is 5 and whose common ratio is $\frac{1}{2}$.

Using the formula $a_n = a_1 r^{n-1}$,

$$a_1 = 5, \quad a_2 = 5\left(\frac{1}{2}\right) = \frac{5}{2}, \quad a_3 = 5\left(\frac{1}{2}\right)^2 = \frac{5}{4},$$

$$a_4 = 5\left(\frac{1}{2}\right)^3 = \frac{5}{8}, \quad a_5 = 5\left(\frac{1}{2}\right)^4 = \frac{5}{16}.$$

Now Try Exercise 23.

OBJECTIVE 4 **Find the sum of a specified number of terms of a geometric sequence.** It is convenient to have a formula for the sum of the first n terms of a geometric sequence, S_n. We can develop a formula by first writing out S_n.

$$S_n = a_1 + a_1 r + a_1 r^2 + a_1 r^3 + \cdots + a_1 r^{n-1}$$

Next, we multiply both sides by r.

$$r S_n = a_1 r + a_1 r^2 + a_1 r^3 + a_1 r^4 + \cdots + a_1 r^n$$

We subtract the first result from the second.

$$r S_n - S_n = (a_1 r - a_1) + (a_1 r^2 - a_1 r) + (a_1 r^3 - a_1 r^2)$$
$$+ (a_1 r^4 - a_1 r^3) + \cdots + (a_1 r^n - a_1 r^{n-1})$$

Using the commutative, associative, and distributive properties, we obtain

$$r S_n - S_n = (a_1 r - a_1 r) + (a_1 r^2 - a_1 r^2)$$
$$+ (a_1 r^3 - a_1 r^3) + \cdots + (a_1 r^n - a_1)$$
$$S_n(r - 1) = a_1 r^n - a_1.$$

If $r \neq 1$, then

$$S_n = \frac{a_1 r^n - a_1}{r - 1} = \frac{a_1(r^n - 1)}{r - 1}. \qquad \text{Divide by } r - 1.$$

A summary of this discussion follows.

Sum of the First n Terms of a Geometric Sequence

The sum of the first n terms of the geometric sequence with first term a_1 and common ratio r is

$$S_n = \frac{a_1(r^n - 1)}{r - 1} \quad (r \neq 1).$$

If $r = 1$, then $S_n = a_1 + a_1 + a_1 + \cdots + a_1 = n a_1$.

Multiplying the formula for S_n by $\frac{-1}{-1}$ gives an alternative form that is sometimes preferable.

$$S_n = \frac{a_1(r^n - 1)}{r - 1} \cdot \frac{-1}{-1} = \frac{a_1(1 - r^n)}{1 - r}$$

▨ **EXAMPLE 5** Finding the Sum of the First n Terms

Find the sum of the first six terms of the geometric sequence with first term -2 and common ratio 3.

Substitute $n = 6$, $a_1 = -2$, and $r = 3$ into the formula for S_n.

$$S_n = \frac{a_1(r^n - 1)}{r - 1}$$

$$S_6 = \frac{-2(3^6 - 1)}{3 - 1} \qquad \text{Let } n = 6, a_1 = -2, r = 3.$$

$$= \frac{-2(729 - 1)}{2} \qquad \text{Evaluate } 3^6.$$

$$= -728$$

Now Try Exercise 27.

A series of the form

$$\sum_{i=1}^{n} a \cdot b^i$$

represents the sum of the first n terms of a geometric sequence having first term $a_1 = a \cdot b^1 = ab$ and common ratio b. The next example illustrates this form.

▨ **EXAMPLE 6** Using the Formula for S_n to Find a Summation

Find $\displaystyle\sum_{i=1}^{4} 3 \cdot 2^i$.

Since the series is in the form

$$\sum_{i=1}^{n} a \cdot b^i,$$

it represents the sum of the first n terms of the geometric sequence with $a_1 = a \cdot b^1$ and $r = b$. The sum is found by using the formula

$$S_n = \frac{a_1(r^n - 1)}{r - 1}.$$

Here $n = 4$. Also, $a_1 = 6$ and $r = 2$. Now substitute into the formula for S_n.

$$S_4 = \frac{6(2^4 - 1)}{2 - 1} \qquad \text{Let } n = 4, a_1 = 6, r = 2.$$

$$= \frac{6(16 - 1)}{1} \qquad \text{Evaluate } 2^4.$$

$$= 90$$

Now Try Exercise 31.

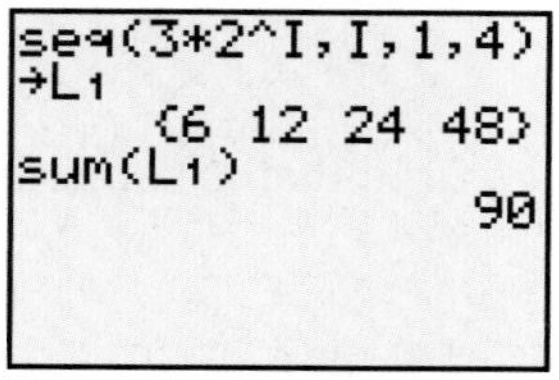

```
seq(3*2^I,I,1,4)
→L₁
        {6 12 24 48}
sum(L₁)
                  90
```

FIGURE 4

Figure 4 shows how a graphing calculator can store the terms in a list, and then find the sum of these terms. This supports the result of Example 6.

OBJECTIVE 5 Apply the formula for the future value of an ordinary annuity. A sequence of equal payments made at equal periods of time is called an **annuity.** If the payments are made at the end of the time period, and if the frequency of payments is the same as the frequency of compounding, the annuity is called an **ordinary annuity.** The time between payments is the **payment period,** and the time from the beginning of the first payment period to the end of the last period is called the **term of the annuity.** The **future value of the annuity,** the final sum on deposit, is defined as the sum of the compound amounts of all the payments, compounded to the end of the term.

For example, suppose $1500 is deposited at the end of the year for the next 6 yr in an account paying 8% per yr compounded annually. To find the future value of this annuity, look separately at each of the $1500 payments. The first of these payments will produce a compound amount of

$$1500(1 + .08)^5 = 1500(1.08)^5.$$

Use 5 as the exponent instead of 6 since the money is deposited at the *end* of the first year and earns interest for only 5 yr. The second payment of $1500 will produce a compound amount of $1500(1.08)^4$. Continuing in this way and finding the sum of all the terms gives

$$1500(1.08)^5 + 1500(1.08)^4 + 1500(1.08)^3 + 1500(1.08)^2 + 1500(1.08)^1 + 1500.$$

(The last payment earns no interest at all.) Reading in reverse order, we see that this expression is the sum of the first six terms of a geometric sequence with $a_1 = 1500$, $r = 1.08$, and $n = 6$. Therefore, the sum is

$$\frac{a_1(r^n - 1)}{r - 1} = \frac{1500[(1.08)^6 - 1]}{1.08 - 1} = 11{,}003.89$$

or $11,003.89.

We state the following formula without proof.

Future Value of an Ordinary Annuity

$$S = R\left[\frac{(1 + i)^n - 1}{i}\right]$$

where

S is future value,

R is the payment at the end of each period,

i is the interest rate per period, and

n is the number of periods.

EXAMPLE 7   Applying the Formula for the Future Value of an Annuity

(a) Rocky Rhodes is an athlete who feels that his playing career will last 7 yr. To prepare for his future, he deposits $22,000 at the end of each year for 7 yr in an account paying 6% compounded annually. How much will he have on deposit after 7 yr?

His payments form an ordinary annuity with $R = 22{,}000$, $n = 7$, and $i = .06$. The future value of this annuity (using the formula) is

$$S = 22{,}000\left[\frac{(1.06)^7 - 1}{.06}\right] = 184{,}664.43, \qquad \text{Use a calculator.}$$

or $184,664.43.

(b) Experts say that the baby boom generation (born between 1946 and 1960) cannot count on a company pension or Social Security to provide a comfortable retirement, as their parents did. It is recommended that they start to save early and regularly. Judy Zahrndt, a baby boomer, has decided to deposit $200 at the end of each month in an account that pays interest of 7.2% compounded monthly for retirement in 20 yr. How much will be in the account at that time?

Because the interest is compounded monthly, $i = \frac{.072}{12}$. Also, $R = 200$ and $n = 12(20)$. The future value is

$$S = 200\left[\frac{\left(1 + \dfrac{.072}{12}\right)^{12(20)} - 1}{\dfrac{.072}{12}}\right] = 106{,}752.47,$$

or $106,752.47.

Now Try Exercise 35.

OBJECTIVE 6 Find the sum of an infinite number of terms of certain geometric sequences. Consider an infinite geometric sequence such as

$$\frac{1}{3}, \frac{1}{6}, \frac{1}{12}, \frac{1}{24}, \frac{1}{48}, \ldots .$$

Can the sum of the terms of such a sequence be found somehow? The sum of the first two terms is

$$S_2 = \frac{1}{3} + \frac{1}{6} = \frac{1}{2} = .5.$$

In a similar manner,

$$S_3 = S_2 + \frac{1}{12} = \frac{1}{2} + \frac{1}{12} = \frac{7}{12} \approx .583, \quad S_4 = S_3 + \frac{1}{24} = \frac{7}{12} + \frac{1}{24} = \frac{15}{24} = .625,$$

$$S_5 = \frac{31}{48} \approx .64583, \quad S_6 = \frac{21}{32} = .65625, \quad S_7 = \frac{127}{192} \approx .6614583.$$

Each term of the geometric sequence is smaller than the preceding one, so each additional term is contributing less and less to the sum. In decimal form (to the nearest thousandth) the first seven terms and the tenth term are given in the table.

Term	a_1	a_2	a_3	a_4	a_5	a_6	a_7	a_{10}
Value	.333	.167	.083	.042	.021	.010	.005	.001

As the table suggests, the value of a term gets closer and closer to 0 as the number of the term increases. To express this idea, we say that as n increases without bound (written $n \to \infty$), the limit of the term a_n is 0, written

$$\lim_{n \to \infty} a_n = 0.$$

A number that can be defined as the sum of an infinite number of terms of a geometric sequence can be found by starting with the expression for the sum of a finite number of terms:

$$S_n = \frac{a_1(r^n - 1)}{r - 1}.$$

If $|r| < 1$, then as n increases without bound the value of r^n gets closer and closer to 0. For example, in the infinite sequence just discussed, $r = \frac{1}{2} = .5$. The following table shows how $r^n = (.5)^n$, given to the nearest thousandth, gets smaller as n increases.

n	1	2	3	4	5	6	7	10
r^n	.5	.25	.125	.063	.031	.016	.008	.001

As r^n approaches 0, $r^n - 1$ approaches $0 - 1 = -1$, and S_n approaches the quotient $\dfrac{-a_1}{r - 1}$. Thus,

$$\lim_{r^n \to 0} S_n = \lim_{r^n \to 0} \frac{a_1(r^n - 1)}{r - 1} = \frac{a_1(0 - 1)}{r - 1} = \frac{-a_1}{r - 1} = \frac{a_1}{1 - r}.$$

This limit is defined to be the sum of the infinite geometric sequence:

$$a_1 + a_1 r + a_1 r^2 + a_1 r^3 + \cdots = \frac{a_1}{1 - r}, \quad \text{if } |r| < 1.$$

What happens if $|r| > 1$? For example, suppose the sequence is

$$6, 12, 24, \ldots, 3(2)^n, \ldots.$$

In this kind of sequence, as n increases, the value of r^n also increases and so does the sum S_n. Since each new term adds a larger and larger amount to the sum, there is no limit to the value of S_n, and the sum S_n does not exist. A similar situation exists if $r = 1$.

In summary, the sum of the terms of an infinite geometric sequence is defined as follows.

Sum of the Terms of an Infinite Geometric Sequence

The sum S of the terms of an infinite geometric sequence with first term a_1 and common ratio r, where $|r| < 1$, is

$$S = \frac{a_1}{1 - r}.$$

If $|r| \geq 1$, then the sum does not exist.

▓ **EXAMPLE 8** **Finding the Sum of the Terms of an Infinite Geometric Sequence**

Find the sum of the terms of the infinite geometric sequence with $a_1 = 3$ and $r = -\frac{1}{3}$.

From the preceding rule, the sum is

$$S = \frac{a_1}{1-r} = \frac{3}{1-\left(-\frac{1}{3}\right)} = \frac{3}{\frac{4}{3}} = \frac{9}{4}.$$

Now Try Exercise 39.

Using summation notation, the sum of an infinite geometric sequence is written as

$$\sum_{i=1}^{\infty} a_i.$$

For instance, the sum in Example 8 would be written

$$\sum_{i=1}^{\infty} 3\left(-\frac{1}{3}\right)^{i-1}.$$

▓ **EXAMPLE 9** **Finding the Sum of the Terms of an Infinite Geometric Series**

Find $\displaystyle\sum_{i=1}^{\infty} \left(\frac{1}{2}\right)^i$.

This is the infinite geometric series

$$\frac{1}{2} + \frac{1}{4} + \frac{1}{8} + \cdots,$$

with $a_1 = \frac{1}{2}$ and $r = \frac{1}{2}$. Since $|r| < 1$, we find the sum as follows.

$$S = \frac{a_1}{1-r} = \frac{\frac{1}{2}}{1-\frac{1}{2}} = \frac{\frac{1}{2}}{\frac{1}{2}} = 1$$

Now Try Exercise 43.

14.3 EXERCISES

For Extra Help

 Student's
Solutions Manual

 MyMathLab

 InterAct Math
Tutorial Software

Tutor Center AW Math
Tutor Center

Math*XP* MathXL

 Digital Video Tutor
CD 14/Videotape 16

✎ **1.** Using several examples, explain the meaning of *geometric sequence*.

✎ **2.** Explain why the sequence 5, 5, 5, 5, . . . can be considered either arithmetic or geometric.

If the given sequence is geometric, find the common ratio, r. If the sequence is not geometric, say so. See Example 1.

3. 4, 8, 16, 32, . . .

4. 5, 15, 45, 135, . . .

5. $\frac{1}{3}, \frac{2}{3}, \frac{3}{3}, \frac{4}{3}, \frac{5}{3}, \ldots$

6. $\frac{1}{3}, \frac{2}{3}, \frac{4}{3}, \frac{8}{3}, \ldots$

7. 1, −3, 9, −27, 81, . . .

8. 1, −3, 7, −11, . . .

9. 1, $-\frac{1}{2}, \frac{1}{4}, -\frac{1}{8}, \frac{1}{16}, \ldots$

10. $\frac{2}{3}, \frac{2}{15}, \frac{2}{75}, \frac{2}{375}, \ldots$

Find a general term for each geometric sequence. See Example 2.

11. $5, 10, \ldots$

12. $-2, -6, \ldots$

13. $\dfrac{1}{9}, \dfrac{1}{3}, \ldots$

14. $-3, \dfrac{3}{2}, \ldots$

15. $10, -2, \ldots$

16. $-4, 8, \ldots$

Find the indicated term for each geometric sequence. See Example 3.

17. $a_1 = 2, r = 5; a_{10}$

18. $a_1 = -1, r = 3; a_{15}$

19. $\dfrac{1}{2}, \dfrac{1}{6}, \dfrac{1}{18}, \ldots; a_{12}$

20. $\dfrac{2}{3}, -\dfrac{1}{3}, \dfrac{1}{6}, \ldots; a_{18}$

21. $a_3 = \dfrac{1}{2}, a_7 = \dfrac{1}{32}; a_{25}$

22. $a_5 = 48, a_8 = -384; a_{10}$

Write the first five terms of each geometric sequence. See Example 4.

23. $a_1 = 2, r = 3$

24. $a_1 = 4, r = 2$

25. $a_1 = 5, r = -\dfrac{1}{5}$

26. $a_1 = 6, r = -\dfrac{1}{3}$

Use the formula for S_n to find the sum for each geometric sequence. See Examples 5 and 6. In Exercises 29–34, give the answer to the nearest thousandth.

27. $\dfrac{1}{3}, \dfrac{1}{9}, \dfrac{1}{27}, \dfrac{1}{81}, \dfrac{1}{243}$

28. $\dfrac{4}{3}, \dfrac{8}{3}, \dfrac{16}{3}, \dfrac{32}{3}, \dfrac{64}{3}, \dfrac{128}{3}$

29. $-\dfrac{4}{3}, -\dfrac{4}{9}, -\dfrac{4}{27}, -\dfrac{4}{81}, -\dfrac{4}{243}, -\dfrac{4}{729}$

30. $\dfrac{5}{16}, -\dfrac{5}{32}, \dfrac{5}{64}, -\dfrac{5}{128}, \dfrac{5}{256}$

31. $\displaystyle\sum_{i=1}^{7} 4\left(\dfrac{2}{5}\right)^i$

32. $\displaystyle\sum_{i=1}^{8} 5\left(\dfrac{2}{3}\right)^i$

33. $\displaystyle\sum_{i=1}^{10} (-2)\left(\dfrac{3}{5}\right)^i$

34. $\displaystyle\sum_{i=1}^{6} (-2)\left(-\dfrac{1}{2}\right)^i$

Solve each problem involving an ordinary annuity. See Example 7.

35. A father opened a savings account for his daughter on the day she was born, depositing $1000. Each year on her birthday he deposits another $1000, making the last deposit on her twenty-first birthday. If the account pays 9.5% interest compounded annually, how much is in the account at the end of the day on the daughter's twenty-first birthday?

36. A 45-year-old man puts $1000 in a retirement account at the end of each quarter $\left(\tfrac{1}{4} \text{ of a year}\right)$ until he reaches age 60. If the account pays 11% annual interest compounded quarterly, how much will be in the account at that time?

37. At the end of each quarter a 50-year-old woman puts $1200 in a retirement account that pays 7% interest compounded quarterly. When she reaches age 60, she withdraws the entire amount and places it in a mutual fund that pays 9% interest compounded monthly. From then on she deposits $300 in the mutual fund at the end of each month. How much is in the account when she reaches age 65?

38. John Bray deposits $10,000 at the beginning of each year for 12 yr in an account paying 5% compounded annually. He then puts the total amount on deposit in another account paying 6% compounded semiannually for another 9 yr. Find the final amount on deposit after the entire 21-yr period.

Find the sum, if it exists, of the terms of each infinite geometric sequence. See Examples 8 and 9.

39. $a_1 = 6, r = \dfrac{1}{3}$

40. $a_1 = 10, r = \dfrac{1}{5}$

41. $a_1 = 1000, r = -\dfrac{1}{10}$

42. $a_1 = 8500, r = \dfrac{3}{5}$

43. $\displaystyle\sum_{i=1}^{\infty} \dfrac{9}{8}\left(-\dfrac{2}{3}\right)^i$

44. $\displaystyle\sum_{i=1}^{\infty} \dfrac{3}{5}\left(\dfrac{5}{6}\right)^i$

45. $\displaystyle\sum_{i=1}^{\infty} \dfrac{12}{5}\left(\dfrac{5}{4}\right)^i$

46. $\displaystyle\sum_{i=1}^{\infty} \left(-\dfrac{16}{3}\right)\left(-\dfrac{9}{8}\right)^i$

*Solve each application. (*Hint: *Determine whether you need to find a specific term of a sequence or the sum of the terms of a sequence immediately after reading the problem.)*

47. A certain ball when dropped from a height rebounds $\frac{3}{5}$ of the original height. How high will the ball rebound after the fourth bounce if it was dropped from a height of 10 ft?

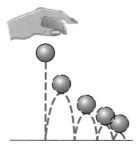

48. A fully wound yo-yo has a string 40 in. long. It is allowed to drop and on its first rebound, it returns to a height 15 in. lower than its original height. Assuming this "rebound ratio" remains constant until the yo-yo comes to rest, how far does it travel on its third trip up the string?

49. A particular substance decays in such a way that it loses half its weight each day. In how many days will 256 g of the substance be reduced to 32 g? How much of the substance is left after 10 days?

50. A tracer dye is injected into a system with an input and an excretion. After one hour $\frac{2}{3}$ of the dye is left. At the end of the second hour $\frac{2}{3}$ of the remaining dye is left, and so on. If one unit of the dye is injected, how much is left after 6 hr?

51. In a certain community the consumption of electricity has increased about 6% per yr.

(a) If a community uses 1.1 billion units of electricity now, how much will it use 5 yr from now?

(b) Find the number of years it will take for the consumption to double.

52. Suppose the community in Exercise 51 reduces its increase in consumption to 2% per yr.

(a) How much will it use 5 yr from now?

(b) Find the number of years it will take for the consumption to double.

53. A machine depreciates by $\frac{1}{4}$ of its value each year. If it cost $50,000 new, what is its value after 8 yr?

54. Refer to Exercise 48. Theoretically, how far does the yo-yo travel before coming to rest?

RELATING CONCEPTS (EXERCISES 55–60)

For Individual or Group Work

In Chapter 1 we learned that any repeating decimal is a rational number; that is, it can be expressed as a quotient of integers. Thus, the repeating decimal

$$.99999\ldots,$$

an endless string of 9s, must be a rational number. **Work Exercises 55–60 in order, to discover the surprising simplest form of this rational number.**

55. Use long division or your previous experience to write a repeating decimal representation for $\frac{1}{3}$.

56. Use long division or your previous experience to write a repeating decimal representation for $\frac{2}{3}$.

57. Because $\frac{1}{3} + \frac{2}{3} = 1$, the sum of the decimal representations in Exercises 55 and 56 must also equal 1. Line up the decimals in the usual vertical method for addition, and obtain the repeating decimal result. The value of this decimal is exactly 1.

58. The repeating decimal $.99999\ldots$ can be written as the sum of the terms of a geometric sequence with $a_1 = .9$ and $r = .1$:

$$.99999\ldots = .9 + .9(.1) + .9(.1)^2 + .9(.1)^3 + .9(.1)^4 + .9(.1)^5 + \cdots.$$

Since $|.1| < 1$, this sum can be found using the formula $S = \dfrac{a_1}{1 - r}$. Use this formula to support the result you found another way in Exercises 55–57.

59. Which one of the following is true, based on your results in Exercises 57 and 58?

A. $.99999\ldots < 1$ **B.** $.99999\ldots = 1$ **C.** $.99999\ldots \approx 1$

60. Show that $.49999\ldots = \frac{1}{2}$.

14.4 The Binomial Theorem

OBJECTIVE 1 Expand a binomial raised to a power. Writing out the binomial expression $(x + y)^n$ for nonnegative integer values of n gives a family of expressions that is important in the study of mathematics and its applications. For example,

$$(x + y)^0 = 1,$$
$$(x + y)^1 = x + y,$$
$$(x + y)^2 = x^2 + 2xy + y^2,$$
$$(x + y)^3 = x^3 + 3x^2y + 3xy^2 + y^3,$$
$$(x + y)^4 = x^4 + 4x^3y + 6x^2y^2 + 4xy^3 + y^4,$$
$$(x + y)^5 = x^5 + 5x^4y + 10x^3y^2 + 10x^2y^3 + 5xy^4 + y^5.$$

Inspection shows that these expansions follow a pattern. By identifying the pattern, we can write a general expression for $(x + y)^n$.

First, if n is a positive integer, each expansion after $(x + y)^0$ begins with x raised to the same power to which the binomial is raised. That is, the expansion of $(x + y)^1$ has a first term of x^1, the expansion of $(x + y)^2$ has a first term of x^2, the expansion of $(x + y)^3$ has a first term of x^3, and so on. Also, the last term in each expansion is y to this same power, so the expansion of $(x + y)^n$ should begin with the term x^n and end with the term y^n.

The exponents on x decrease by 1 in each term after the first, while the exponents on y, beginning with y in the second term, increase by 1 in each succeeding term. Thus, the *variables* in the expansion of $(x + y)^n$ have the following pattern.

$$x^n, \quad x^{n-1}y, \quad x^{n-2}y^2, \quad x^{n-3}y^3, \dots, xy^{n-1}, \quad y^n$$

This pattern suggests that the sum of the exponents on x and y in each term is n. For example, in the third term above, the variable part is $x^{n-2}y^2$ and the sum of the exponents, $n - 2$ and 2, is n.

Now examine the pattern for the *coefficients* in the terms of the preceding expansions. Writing the coefficients alone in a triangular pattern gives **Pascal's triangle,** named in honor of the 17th century mathematician Blaise Pascal, one of the first to use it extensively.

Blaise Pascal (1623–1662)

Pascal's Triangle
1
1   1
1   2   1
1   3   3   1
1   4   6   4   1
1   5   10   10   5   1   and so on

Arranging the coefficients in this way shows that each number in the triangle is the sum of the two numbers just above it (one to the right and one to the left). For example, in the fifth row from the top, 1 is the sum of 1 (the only number above it), 4 is the sum of 1 and 3, 6 is the sum of 3 and 3, and so on.

To obtain the coefficients for $(x + y)^6$, we attach the seventh row to the table by adding pairs of numbers from the sixth row.

$$1 \quad 6 \quad 15 \quad 20 \quad 15 \quad 6 \quad 1$$

We then use these coefficients to expand $(x + y)^6$ as

$$(x + y)^6 = x^6 + 6x^5y + 15x^4y^2 + 20x^3y^3 + 15x^2y^4 + 6xy^5 + y^6.$$

CONNECTIONS

Over the years, many interesting patterns have been discovered in Pascal's triangle. In the following figure, the triangular array is written in a different form. The indicated sums along the diagonals shown are the terms of the *Fibonacci sequence,* mentioned in the chapter introduction. The presence of this sequence in the triangle apparently was not recognized by Pascal.

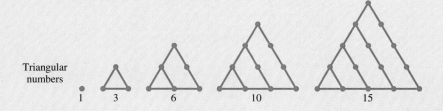

Triangular numbers are found by counting the number of points in triangular arrangements of points. The first few triangular numbers are shown in the figure below.

Triangular numbers

1   3   6   10   15

The number of points in these figures form the sequence 1, 3, 6, 10, . . . , a sequence that is found in Pascal's triangle, as shown in the next figure.

```
              1
           1     1
        1     2     1
     1     3     3     1
  1     4     6     4     1
1     5    10    10     5     1
```

For Discussion or Writing

1. Predict the next two numbers in the sequence of sums of the diagonals of Pascal's triangle.

2. Predict the next five numbers in the list of triangular numbers.

3. Describe other sequences that can be found in Pascal's triangle.

Although it is possible to use Pascal's triangle to find the coefficients of $(x + y)^n$ for any positive integer value of n, it is impractical for large values of n. A more efficient way to determine these coefficients uses a notational shorthand with the symbol $n!$ (read "n factorial") defined as follows.

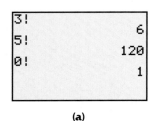

(a)

13! 6227020800
25! 1.551121004E25
69! 1.711224524E98

A graphing calculator with a 10-digit display will give the exact value of $n!$ for $n \leq 13$ and approximate values of $n!$ for $14 \leq n \leq 69$.

(b)

FIGURE 5

---

n Factorial ($n!$)

For any positive integer n,

$$n! = n(n - 1)(n - 2) \cdots (3)(2)(1).$$

For example,

$$3! = 3 \cdot 2 \cdot 1 = 6 \qquad \text{and} \qquad 5! = 5 \cdot 4 \cdot 3 \cdot 2 \cdot 1 = 120.$$

From the definition of n factorial, $n[(n - 1)!] = n!$. If $n = 1$, then $1(0!) = 1! = 1$. Because of this, $0!$ is defined as

$$0! = 1.$$

Now Try Exercise 1.

Scientific and graphing calculators can compute factorials. The three example factorial expressions above are shown in Figure 5(a). Figure 5(b) shows some larger factorials.

EXAMPLE 1 Evaluating Expressions with $n!$

Find the value of each expression.

(a) $\dfrac{5!}{4!1!} = \dfrac{5 \cdot 4 \cdot 3 \cdot 2 \cdot 1}{(4 \cdot 3 \cdot 2 \cdot 1)(1)} = 5$

(b) $\dfrac{5!}{3!2!} = \dfrac{5 \cdot 4 \cdot 3 \cdot 2 \cdot 1}{(3 \cdot 2 \cdot 1)(2 \cdot 1)} = \dfrac{5 \cdot 4}{2 \cdot 1} = 10$

(c) $\dfrac{6!}{3!3!} = \dfrac{6 \cdot 5 \cdot 4 \cdot 3 \cdot 2 \cdot 1}{(3 \cdot 2 \cdot 1)(3 \cdot 2 \cdot 1)} = \dfrac{6 \cdot 5 \cdot 4}{3 \cdot 2 \cdot 1} = 20$

(d) $\dfrac{4!}{4!0!} = \dfrac{4 \cdot 3 \cdot 2 \cdot 1}{(4 \cdot 3 \cdot 2 \cdot 1)(1)} = 1$

Now Try Exercises 3 and 7.

Now look again at the coefficients of the expansion

$$(x + y)^5 = x^5 + 5x^4y + 10x^3y^2 + 10x^2y^3 + 5xy^4 + y^5.$$

The coefficient of the second term is 5 and the exponents on the variables in that term are 4 and 1. From Example 1(a), $\frac{5!}{4!1!} = 5$. The coefficient of the third term is 10, and the exponents are 3 and 2. From Example 1(b), $\frac{5!}{3!2!} = 10$. Similar results hold true for the remaining terms. The first term can be written as $1x^5y^0$, and the last term can be written as $1x^0y^5$. Then the coefficient of the first term should be $\frac{5!}{5!0!} = 1$, and the coefficient of the last term would be $\frac{5!}{0!5!} = 1$. Generalizing, the coefficient for a term of $(x + y)^n$ in which the variable part is $x^r y^{n-r}$ will be

$$\frac{n!}{r!(n - r)!}.$$

NOTE The denominator factorials in the coefficient of a term are the same as the exponents on the variables in that term.

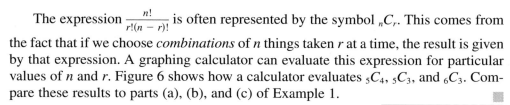

FIGURE 6

The expression $\frac{n!}{r!(n - r)!}$ is often represented by the symbol $_nC_r$. This comes from the fact that if we choose *combinations* of n things taken r at a time, the result is given by that expression. A graphing calculator can evaluate this expression for particular values of n and r. Figure 6 shows how a calculator evaluates $_5C_4$, $_5C_3$, and $_6C_3$. Compare these results to parts (a), (b), and (c) of Example 1.

Now Try Exercise 5.

Summarizing this work gives the **binomial theorem,** or the **general binomial expansion.**

> **Binomial Theorem**
>
> For any positive integer n,
>
> $$(x + y)^n = x^n + \frac{n!}{(n - 1)!1!}x^{n-1}y + \frac{n!}{(n - 2)!2!}x^{n-2}y^2$$
>
> $$+ \frac{n!}{(n - 3)!3!}x^{n-3}y^3 + \cdots + \frac{n!}{1!(n - 1)!}xy^{n-1} + y^n.$$

The binomial theorem can be written in summation notation as

$$(x + y)^n = \sum_{i=0}^{n} \frac{n!}{(n - i)!i!}x^{n-i}y^i.$$

NOTE The letter i is used here instead of r because we are using summation notation. It is not the imaginary number i.

EXAMPLE 2 Using the Binomial Theorem

Expand $(2m + 3)^4$.

$$(2m + 3)^4 = (2m)^4 + \frac{4!}{3!1!}(2m)^3(3) + \frac{4!}{2!2!}(2m)^2(3)^2 + \frac{4!}{1!3!}(2m)(3)^3 + 3^4$$

$$= 16m^4 + 4(8m^3)(3) + 6(4m^2)(9) + 4(2m)(27) + 81$$

$$= 16m^4 + 96m^3 + 216m^2 + 216m + 81$$

Now Try Exercise 17.

EXAMPLE 3 Using the Binomial Theorem

Expand $\left(a - \dfrac{b}{2} \right)^5$.

$$\left(a - \frac{b}{2} \right)^5 = a^5 + \frac{5!}{4!1!}a^4\left(-\frac{b}{2} \right) + \frac{5!}{3!2!}a^3\left(-\frac{b}{2} \right)^2 + \frac{5!}{2!3!}a^2\left(-\frac{b}{2} \right)^3$$

$$+ \frac{5!}{1!4!}a\left(-\frac{b}{2} \right)^4 + \left(-\frac{b}{2} \right)^5$$

$$= a^5 + 5a^4\left(-\frac{b}{2} \right) + 10a^3\left(\frac{b^2}{4} \right) + 10a^2\left(-\frac{b^3}{8} \right)$$

$$+ 5a\left(\frac{b^4}{16} \right) + \left(-\frac{b^5}{32} \right)$$

$$= a^5 - \frac{5}{2}a^4b + \frac{5}{2}a^3b^2 - \frac{5}{4}a^2b^3 + \frac{5}{16}ab^4 - \frac{1}{32}b^5$$

Now Try Exercise 19.

CAUTION When the binomial is the *difference* of two terms as in Example 3, the signs of the terms in the expansion will alternate. Those terms with odd exponents on the second variable expression $\left(-\frac{b}{2} \text{ in Example 3} \right)$ will be negative, while those with even exponents on the second variable expression will be positive.

OBJECTIVE 2 Find any specified term of the expansion of a binomial. Any single term of a binomial expansion can be determined without writing out the whole expansion. For example, if $n \geq 10$, then the tenth term of $(x + y)^n$ has y raised to the ninth power (since y has the power of 1 in the second term, the power of 2 in the third term, and so on). Since the exponents on x and y in any term must have a sum of n, the exponent on x in the tenth term is $n - 9$. These quantities, 9 and $n - 9$, determine the factorials in the denominator of the coefficient. Thus,

$$\frac{n!}{(n-9)!9!}x^{n-9}y^9$$

is the tenth term of $(x + y)^n$. A generalization of this idea follows.

---

rth Term of the Binomial Expansion

If $n \geq r - 1$, then the rth term of the expansion of $(x + y)^n$ is

$$\frac{n!}{[n-(r-1)]!(r-1)!}x^{n-(r-1)}y^{r-1}.$$

In this general expression, remember to start with the exponent on y, which is 1 less than the term number r. Then subtract that exponent from n to get the exponent on x: $n - (r - 1)$. The two exponents are then used as the factorials in the denominator of the coefficient.

EXAMPLE 4 Finding a Single Term of a Binomial Expansion

Find the fourth term of $(a + 2b)^{10}$.

In the fourth term, $2b$ has an exponent of $4 - 1 = 3$ and a has an exponent of $10 - 3 = 7$. The fourth term is

$$\frac{10!}{7!3!}(a^7)(2b)^3 = \frac{10 \cdot 9 \cdot 8}{3 \cdot 2 \cdot 1}(a^7)(8b^3)$$

$$= 120a^7(8b^3)$$

$$= 960a^7b^3.$$

Now Try Exercise 29.

14.4 EXERCISES

For Extra Help

Student's
Solutions Manual

MyMathLab

InterAct Math
Tutorial Software

AW Math
Tutor Center

MathXL

Digital Video Tutor
CD 14/Videotape 16

Evaluate each expression. See Example 1.

1. $6!$ **2.** $4!$ **3.** $\dfrac{6!}{4!2!}$ **4.** $\dfrac{7!}{3!4!}$

5. $_6C_2$ **6.** $_7C_4$ **7.** $\dfrac{4!}{0!4!}$ **8.** $\dfrac{5!}{5!0!}$

9. $4! \cdot 5$ **10.** $6! \cdot 7$ **11.** $_{13}C_{11}$ **12.** $_{13}C_2$

Use the binomial theorem to expand each expression. See Examples 2 and 3.

13. $(m + n)^4$ **14.** $(x + r)^5$ **15.** $(a - b)^5$ **16.** $(p - q)^4$

17. $(2x + 3)^3$ **18.** $\left(\dfrac{x}{3} + 2y\right)^5$ **19.** $\left(\dfrac{x}{2} - y\right)^4$ **20.** $(x^2 + 1)^4$

21. $(mx - n^2)^3$ **22.** $(2p^2 - q^2)^3$

Write the first four terms of each binomial expansion.

23. $(r + 2s)^{12}$ **24.** $(m - n)^{20}$ **25.** $(3x - y)^{14}$

26. $(2p + 3q)^{11}$ **27.** $(t^2 + u^2)^{10}$ **28.** $(x^2 - y^2)^{15}$

Find the indicated term of each binomial expansion. See Example 4.

29. $(2m + n)^{10}$; fourth term   **30.** $(a - 3b)^{12}$; fifth term

31. $\left(x + \dfrac{y}{2}\right)^8$; seventh term   **32.** $(3p - 2q)^{15}$; eighth term

33. $(k - 1)^9$; third term   **34.** $(-4 - s)^{11}$; fourth term

35. The middle term of $(x^2 - 2y)^6$ **36.** The middle term of $(m^3 + 3)^8$

37. The term with x^9y^4 in $(3x^3 - 4y^2)^5$ **38.** The term with x^{10} in $\left(x^3 - \dfrac{2}{x}\right)^6$

14.5 Mathematical Induction

OBJECTIVES

1 Learn the principle of mathematical induction.

2 Use the principle of mathematical induction to prove a statement.

OBJECTIVE 1 Learn the principle of mathematical induction. Many results in mathematics are claimed true for any positive integer. Any of these results can be checked for $n = 1$, $n = 2$, $n = 3$, and so on, but since the set of positive integers is infinite it is impossible to check every possible case. For example, let S_n represent the statement that the sum of the first n positive integers is $\frac{n(n + 1)}{2}$, that is,

$$S_n: 1 + 2 + 3 + \cdots + n = \frac{n(n + 1)}{2}.$$

The truth of this statement can be checked quickly for the first few values of n.

If $n = 1$, S_1 is $\quad\quad 1 = \dfrac{1(1 + 1)}{2}$, a true statement, since $1 = 1$.

If $n = 2$, S_2 is $\quad 1 + 2 = \dfrac{2(2 + 1)}{2}$, a true statement, since $3 = 3$.

If $n = 3$, S_3 is $\quad 1 + 2 + 3 = \dfrac{3(3 + 1)}{2}$, a true statement, since $6 = 6$.

If $n = 4$, S_4 is $1 + 2 + 3 + 4 = \dfrac{4(4 + 1)}{2}$, a true statement, since $10 = 10$.

Since the statement is true for $n = 1, 2, 3,$ and 4, can we conclude that the statement is true for all positive integers by observing this finite number of examples? The answer is no. To prove that such a statement is true for every positive integer, we use the following principle.

> **Principle of Mathematical Induction**
>
> Let S_n be a statement concerning the positive integer n. Suppose that
> 1. S_1 is true;
> 2. for any positive integer k, $k \le n$, if S_k is true, then S_{k+1} is also true.
>
> Then S_n is true for every positive integer value of n.

A proof by mathematical induction can be explained as follows. By assumption (1) above, the statement is true when $n = 1$. If (2) has been proven, the fact that the statement is true for $n = 1$ implies that it is true for $n = 1 + 1 = 2$. Using (2) again, the statement is thus true for $2 + 1 = 3$, for $3 + 1 = 4$, for $4 + 1 = 5$, and so on. Continuing in this way shows that the statement must be true for *every* positive integer, no matter how large.

The situation is similar to that of an infinite number of dominoes lined up as suggested in Figure 7. If the first domino is pushed over, it pushes the next, which pushes the next, and so on, continuing indefinitely.

FIGURE 7

Another example of the principle of mathematical induction might be an infinite ladder. Suppose the rungs are spaced so that, whenever you are on a rung, you know you can move to the next rung. Thus *if* you can get to the first rung, then you can go as high up the ladder as you wish.

Two separate steps are required for a proof by mathematical induction.

Proof by Mathematical Induction

Step 1 Prove that the statement is true for $n = 1$.

Step 2 Show that for any positive integer k, $k \leq n$, if S_k is true, then S_{k+1} is also true.

OBJECTIVE 2 Use the principle of mathematical induction to prove a statement. Mathematical induction is used in the next example to prove the statement discussed earlier.

▧ **EXAMPLE 1** Proving an Equality Statement by Mathematical Induction

Let S_n represent the statement

$$1 + 2 + 3 + \cdots + n = \frac{n(n + 1)}{2}.$$

Prove that S_n is true for every positive integer n.

PROOF   The proof by mathematical induction is as follows.

Step 1 Show that the statement is true when $n = 1$. If $n = 1$, S_1 becomes

$$1 = \frac{1(1 + 1)}{2},$$

which is true.

Step 2 Show that S_k implies S_{k+1}, where S_k is the statement

$$1 + 2 + 3 + \cdots + k = \frac{k(k + 1)}{2},$$

and S_{k+1} is the statement

$$1 + 2 + 3 + \cdots + k + (k + 1) = \frac{(k + 1)[(k + 1) + 1]}{2}.$$

Start with S_k and assume it is a true statement.

$$1 + 2 + 3 + \cdots + k = \frac{k(k + 1)}{2}$$

Add $k + 1$ to both sides of this equation to obtain S_{k+1}.

$$1 + 2 + 3 + \cdots + k + (k + 1) = \frac{k(k + 1)}{2} + (k + 1)$$

Now, factor out the common factor $k + 1$ on the right to obtain

$$= (k + 1)\left(\frac{k}{2} + 1\right)$$

$$= (k + 1)\left(\frac{k + 2}{2}\right)$$

$$1 + 2 + 3 + \cdots + k + (k + 1) = \frac{(k + 1)[(k + 1) + 1]}{2}.$$

This final result is the statement for $n = k + 1$; it has been shown that if S_k is true, then S_{k+1} is also true. The two steps required for a proof by mathematical induction are now complete, so the statement S_n is true for every positive integer value of n.

Now Try Exercise 5.

CAUTION Notice that the left side of the statement always includes *all* the terms up to the *n*th term, as well as the *n*th term.

EXAMPLE 2 **Proving an Equality Statement by Mathematical Induction**

Prove that $4 + 7 + 10 + \cdots + (3n + 1) = \dfrac{n(3n + 5)}{2}$ for all positive integers n.

PROOF

Step 1 Show that the statement is true for S_1. S_1 is

$$4 = \frac{1(3 \cdot 1 + 5)}{2}.$$

Since the right side equals 4, S_1 is a true statement.

Step 2 Show that if S_k is true, then S_{k+1} is true, where S_k is

$$4 + 7 + 10 + \cdots + (3k + 1) = \frac{k(3k + 5)}{2},$$

and S_{k+1} is

$$4 + 7 + 10 + \cdots + (3k + 1) + [3(k + 1) + 1]$$
$$= \frac{(k + 1)[3(k + 1) + 5]}{2}.$$

Start with S_k:

$$4 + 7 + 10 + \cdots + (3k + 1) = \frac{k(3k + 5)}{2}.$$

To transform the left side of the equation S_k to become the left side of the equation S_{k+1}, we must add the $(k + 1)$st term. Adding $[3(k + 1) + 1]$ to both sides of S_k gives

$$4 + 7 + 10 + \cdots + (3k + 1) + [3(k + 1) + 1]$$
$$= \frac{k(3k + 5)}{2} + [3(k + 1) + 1].$$

Clear the parentheses in the new term on the right side of the equals sign and simplify.

$$= \frac{k(3k + 5)}{2} + 3k + 3 + 1$$

$$= \frac{k(3k + 5)}{2} + 3k + 4$$

Now combine the two terms on the right.

$$= \frac{k(3k + 5)}{2} + \frac{2(3k + 4)}{2}$$

$$= \frac{k(3k + 5) + 2(3k + 4)}{2}$$

$$= \frac{3k^2 + 5k + 6k + 8}{2}$$

$$= \frac{3k^2 + 11k + 8}{2}$$

$$= \frac{(k + 1)(3k + 8)}{2}$$

Since $3k + 8$ can be written as $3(k + 1) + 5$,

$$4 + 7 + 10 + \cdots + (3k + 1) + [3(k + 1) + 1]$$
$$= \frac{(k + 1)[3(k + 1) + 5]}{2}.$$

The final result is the statement for S_{k+1}. Therefore, if S_k is true, then S_{k+1} is true. The two steps required for a proof by mathematical induction are completed, so the general statement S_n is true for every positive integer value of n.

Now Try Exercise 7.

■ EXAMPLE 3 Proving an Inequality Statement by Mathematical Induction

Prove that if x is a real number between 0 and 1, then for every positive integer n,

$$0 < x^n < 1.$$

PROOF

Step 1 Here S_1 is the statement

$$\text{if} \quad 0 < x < 1, \quad \text{then} \quad 0 < x^1 < 1,$$

which is true.

Step 2 S_k is the statement

$$\text{if} \quad 0 < x < 1, \quad \text{then} \quad 0 < x^k < 1.$$

To show that this implies that S_{k+1} is true, multiply the three parts of $0 < x^k < 1$ by x.

$$x \cdot 0 < x \cdot x^k < x \cdot 1$$

(Here the fact that $0 < x$ is used.) Simplify to obtain

$$0 < x^{k+1} < x.$$

Since $x < 1$,

$$x^{k+1} < x < 1$$

and

$$0 < x^{k+1} < 1.$$

By this work, if S_k is true, then S_{k+1} is true, so the given statement is true for every positive integer n.

Now Try Exercise 25.

14.5 EXERCISES

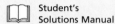

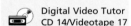
1. A proof by mathematical induction allows us to prove that a statement is true for all
_____.

2. Suppose that Step 2 in a proof by mathematical induction can be satisfied, but Step 1 cannot. May we conclude that the proof is complete? Explain.

3. What is *wrong* with the following proof by mathematical induction?

Prove: Any natural number equals the next natural number; that is, $n = n + 1$.

PROOF
To begin, we assume that the statement is true for some natural number $n = k$:

$$k = k + 1.$$

We must now show that the statement is true for $n = k + 1$. If we add 1 to both sides, we have

$$k + 1 = k + 1 + 1$$
$$k + 1 = k + 2.$$

Hence, if the statement is true for $n = k$, then it is also true for $n = k + 1$. Thus, the theorem is proved.

4. Write out in full and verify the statements S_1, S_2, S_3, S_4, and S_5 for the following statement.

$$2 + 4 + 6 + \cdots + 2n = n(n + 1)$$

Then use mathematical induction to prove that the statement is true for every positive integer n.

Use the method of mathematical induction to prove each statement. Assume that n is a positive integer. See Examples 1–3.

5. $1 + 3 + 5 + \cdots + (2n - 1) = n^2$

6. $5 + 10 + 15 + \cdots + 5n = \dfrac{5n(n + 1)}{2}$

7. $3 + 6 + 9 + \cdots + 3n = \dfrac{3n(n + 1)}{2}$

8. $2 + 4 + 8 + \cdots + 2^n = 2^{n+1} - 2$

9. $3 + 3^2 + 3^3 + \cdots + 3^n = \dfrac{3(3^n - 1)}{2}$

10. $1^2 + 2^2 + 3^2 + \cdots + n^2 = \dfrac{n(n + 1)(2n + 1)}{6}$

11. $1^3 + 2^3 + 3^3 + \cdots + n^3 = \dfrac{n^2(n + 1)^2}{4}$

12. $5 \cdot 6 + 5 \cdot 6^2 + 5 \cdot 6^3 + \cdots + 5 \cdot 6^n = 6(6^n - 1)$

13. $7 \cdot 8 + 7 \cdot 8^2 + 7 \cdot 8^3 + \cdots + 7 \cdot 8^n = 8(8^n - 1)$

14. $\dfrac{1}{1 \cdot 4} + \dfrac{1}{4 \cdot 7} + \dfrac{1}{7 \cdot 10} + \cdots + \dfrac{1}{(3n - 2)(3n + 1)} = \dfrac{n}{3n + 1}$

15. $\dfrac{1}{1 \cdot 2} + \dfrac{1}{2 \cdot 3} + \dfrac{1}{3 \cdot 4} + \cdots + \dfrac{1}{n(n + 1)} = \dfrac{n}{n + 1}$

16. $\dfrac{1}{2} + \dfrac{1}{2^2} + \dfrac{1}{2^3} + \cdots + \dfrac{1}{2^n} = 1 - \dfrac{1}{2^n}$

17. $\dfrac{4}{5} + \dfrac{4}{5^2} + \dfrac{4}{5^3} + \cdots + \dfrac{4}{5^n} = 1 - \dfrac{1}{5^n}$

18. $2 + 2^2 + 2^3 + \cdots + 2^n = 2^{n+1} - 2$

19. $x^{2n} + x^{2n-1}y + \cdots + xy^{2n-1} + y^{2n} = \dfrac{x^{2n+1} - y^{2n+1}}{x - y}$

20. $x^{2n-1} + x^{2n-2}y + \cdots + xy^{2n-2} + y^{2n-1} = \dfrac{x^{2n} - y^{2n}}{x - y}$

21. $(a^m)^n = a^{mn}$ (Assume that a and m are constant.)

22. $(ab)^n = a^n b^n$ (Assume that a and b are constant.)

23. If $a > 1$, then $a^n > 1$.

24. If $a > 1$, then $a^n > a^{n-1}$.

25. If $0 < a < 1$, then $a^n < a^{n-1}$.

RELATING CONCEPTS (EXERCISES 26–30)

For Individual or Group Work

Many of the statements you are asked to prove in these exercises are generalizations of properties of algebra used earlier in the book. ***Work Exercises 26–30 in order.***

26. In the statement given in Exercise 19, replace n with 1.

27. What factorization is this a rearrangement of?

28. Write a statement similar to the one in Exercise 19 for

$$\dfrac{x^{2n+1} + y^{2n+1}}{x + y}.$$

29. The statement in Exercise 21 is a _____ rule for _____.

30. The statement in Exercise 22 is another one of the _____ rules for _____.

31. Suppose that n straight lines (with $n \geq 2$) are drawn in a plane, where no two lines are parallel and no three lines pass through the same point. Show that the number of points of intersection of the lines is

$$\frac{n^2 - n}{2}.$$

32. The series of sketches at the right starts with an equilateral triangle having sides of length 1. In the following steps, equilateral triangles are constructed on each side of the preceding figure. The lengths of the sides of these new triangles are $\frac{1}{3}$ the length of the sides of the preceding triangles. Develop a formula for the number of sides of the nth figure. Use mathematical induction to prove your answer.

33. Find the perimeter of the nth figure in Exercise 32.

34. Show that the area of the nth figure in Exercise 32 is

$$\sqrt{3}\left[\frac{2}{5} - \frac{3}{20}\left(\frac{4}{9}\right)^{n-1}\right].$$

14.6   Counting Theory

OBJECTIVES

1 Use the fundamental principle of counting.

2 Learn the formula $P(n, r)$ for permutations.

3 Use the permutations formula to solve counting problems.

4 Learn the formula $\binom{n}{r}$ for combinations.

5 Use the combinations formula to solve counting problems.

6 Distinguish between permutations and combinations.

If there are 3 roads from Albany to Baker and 2 roads from Baker to Creswich, in how many ways can one travel from Albany to Creswich by way of Baker? For each of the 3 roads from Albany to Baker, there are 2 different roads from Baker to Creswich, so there are $3 \cdot 2 = 6$ different ways to make the trip, as shown in the **tree diagram** in Figure 8.

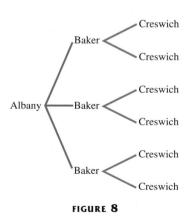

FIGURE 8

This example illustrates the following property.

Fundamental Principle of Counting

If one event can occur in m ways and a second event can occur in n ways, then both events can occur in mn ways, provided the outcome of the first event does not influence the outcome of the second.

OBJECTIVE 1 Use the fundamental principle of counting. The fundamental principle of counting can be extended to any number of events, provided the outcome of no one event influences the outcome of another. Such events are called **independent events.**

EXAMPLE 1 Using the Fundamental Principle of Counting

A restaurant offers a choice of 3 salads, 5 main dishes, and 2 desserts. Use the fundamental principle of counting to find the number of different 3-course meals that can be selected.

Three independent events are involved: selecting a salad, selecting a main dish, and selecting a dessert. The first event can occur in 3 ways, the second event can occur in 5 ways, and the third event can occur in 2 ways; thus there are

$$3 \cdot 5 \cdot 2 = 30 \text{ possible meals.}$$

Now Try Exercise 11.

EXAMPLE 2 Using the Fundamental Principle of Counting

Gioran Lucas has 5 different books that he wishes to arrange on his desk. How many different arrangements are possible?

Five events are involved: selecting a book for the first spot, selecting a book for the second spot, and so on. Here the outcome of the first event *does* influence the outcome of the other events (since one book has already been chosen). For the first spot Gioran has 5 choices, for the second spot 4 choices, for the third spot 3 choices, and so on. Now use the fundamental principle of counting to find that there are

$$5 \cdot 4 \cdot 3 \cdot 2 \cdot 1 = 120 \text{ different arrangements.}$$

Now Try Exercise 13.

When using the fundamental principle of counting, we often encounter products such as $5 \cdot 4 \cdot 3 \cdot 2 \cdot 1$ from Example 2. For convenience in writing these products, we use the symbol $n!$ (read "n factorial"), which was first defined in Section 14.4.

n Factorial ($n!$)

For any positive integer n,

$$n! = n(n - 1)(n - 2) \cdots (3)(2)(1)$$

and

$$0! = 1.$$

By the definition,

$$5 \cdot 4 \cdot 3 \cdot 2 \cdot 1 \quad \text{is written as} \quad 5!.$$

Also, $3! = 3 \cdot 2 \cdot 1 = 6$. The definition of $n!$ means that $n[(n-1)!] = n!$ for all natural numbers $n \geq 2$. It is a natural extension to have this relationship also hold for $n = 1$, so, by definition, $0! = 1$.

■ EXAMPLE 3 Arranging *r* of *n* Items (*r < n*)

Suppose Gioran (from Example 2) wishes to place only 3 of the 5 books on his desk. How many arrangements of 3 books are possible?

He still has 5 ways to fill the first spot, 4 ways to fill the second spot, and 3 ways to fill the third. Since he wants to use 3 books, there are only 3 spots to be filled (3 events) instead of 5, so there are

$$5 \cdot 4 \cdot 3 = 60 \text{ arrangements.}$$

Now Try Exercise 19.

OBJECTIVE 2 Learn the formula *P(n, r)* for permutations. The number 60 in Example 3 is called the number of *permutations* of 5 things taken 3 at a time, written $P(5, 3) = 60$. Example 2 showed that the number of ways of arranging 5 elements from a set of 5 elements, written $P(5, 5)$, is 120.

A **permutation** of n elements taken r at a time is one of the ways of arranging r elements taken from a set of n elements ($r \leq n$). Generalizing from the examples above, the number of permutations of n elements taken r at a time, denoted by $P(n, r)$, is

$$
\begin{aligned}
P(n, r) &= n(n-1)(n-2) \cdots (n-r+1) \\
&= \frac{n(n-1)(n-2) \cdots (n-r+1)(n-r)(n-r-1) \cdots (2)(1)}{(n-r)(n-r-1) \cdots (2)(1)} \\
&= \frac{n!}{(n-r)!}.
\end{aligned}
$$

This proves the following result.

Permutations of *n* Elements Taken *r* at a Time

If $P(n, r)$ denotes the number of **permutations** of n elements taken r at a time, $r \leq n$, then

$$P(n, r) = \frac{n!}{(n-r)!}.$$

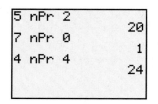

FIGURE 9

Some other symbols used for permutations of n things taken r at a time are $_nP_r$ and P_r^n.

Graphing calculators use the notation $_nP_r$ for evaluating permutation expressions. The screen in Figure 9 shows how a calculator evaluates $P(5, 2)$, $P(7, 0)$, and $P(4, 4)$.

NOTE Some scientific calculators have a key for permutations. Consult your owner's manual for instructions on how to use it.

OBJECTIVE 3 Use the permutations formula to solve counting problems.

EXAMPLE 4 Using the Permutations Formula

Suppose 8 people enter an event in a swim meet. Assuming there are no ties, in how many ways could the gold, silver, and bronze medals be awarded?

Using the fundamental principle of counting, there are 3 choices to be made, giving $8 \cdot 7 \cdot 6 = 336$. However, we can also use the formula for $P(n, r)$ and obtain the same result.

$$P(8, 3) = \frac{8!}{(8 - 3)!} = \frac{8!}{5!}$$

$$= \frac{8 \cdot 7 \cdot 6 \cdot 5 \cdot 4 \cdot 3 \cdot 2 \cdot 1}{5 \cdot 4 \cdot 3 \cdot 2 \cdot 1}$$

$$= 8 \cdot 7 \cdot 6 = 336$$

Now Try Exercise 15.

EXAMPLE 5 Using the Permutations Formula

In how many ways can 6 students be seated in a row of 6 desks?

$$P(6, 6) = \frac{6!}{(6 - 6)!} = \frac{6!}{0!} = 6! = 6 \cdot 5 \cdot 4 \cdot 3 \cdot 2 \cdot 1 = 720$$

Now Try Exercise 21.

The next example involves using the fundamental counting principle with some restrictions.

EXAMPLE 6 Using the Fundamental Counting Principle with Restrictions

In how many ways can three letters of the alphabet be arranged if a vowel cannot be used in the middle position, and repetitions of the letters are allowed?

We cannot use $P(26, 3)$ here, because of the restriction on the middle position, and because repetition is allowed. In the first and third positions, we can use any of the 26 letters of the alphabet, but in the middle position, we can use only one of $26 - 5 = 21$ letters (since there are 5 vowels). Now, using the fundamental counting principle, there are

$$26 \cdot 21 \cdot 26 = 14,196$$

ways to arrange the letters according to the problem.

Now Try Exercise 17.

OBJECTIVE 4 Learn the formula $\binom{n}{r}$ for combinations. We have discussed a method for finding the number of ways to arrange r elements taken from a set of n elements. Sometimes, however, the arrangement (or order) of the elements is not important.

For example, suppose three people (Ms. Opelka, Mr. Adams, and Ms. Jacobs) apply for 2 identical jobs. Ignoring all other factors, in how many ways can the personnel officer select 2 people from the 3 applicants? Here the arrangement or order of the people is unimportant. Selecting Ms. Opelka and Mr. Adams is the same as selecting Mr. Adams and Ms. Opelka. Therefore, there are only 3 ways to select 2 of the 3 applicants:

Ms. Opelka and Mr. Adams;

Ms. Opelka and Ms. Jacobs;

Mr. Adams and Ms. Jacobs.

These three choices are called the *combinations* of 3 elements taken 2 at a time. A **combination** of n elements taken r at a time is one of the ways in which r elements can be chosen from n elements.

Each combination of r elements forms $r!$ permutations. Therefore, the number of combinations of n elements taken r at a time is found by dividing the number of permutations, $P(n, r)$, by $r!$ to obtain

$$\frac{P(n, r)}{r!}$$

combinations. This expression can be rewritten as follows.

$$\frac{P(n, r)}{r!} = \frac{\dfrac{n!}{(n - r)!}}{r!} = \frac{n!}{(n - r)!} \cdot \frac{1}{r!} = \frac{n!}{(n - r)!\, r!}$$

The symbol $\binom{n}{r}$ is used to represent the number of combinations of n things taken r at a time. With this symbol the preceding results are stated as follows.

Combinations of n Elements Taken r at a Time

If $\binom{n}{r}$ represents the number of **combinations** of n elements taken r at a time, for $r \le n$, then

$$\binom{n}{r} = \frac{n!}{(n - r)!\, r!}.$$

Other symbols used for $\binom{n}{r}$ are $C(n, r)$, $_nC_r$, and C_r^n.

NOTE The combinations notation $\binom{n}{r}$ also represents the binomial coefficient defined in Section 14.4. That is, binomial coefficients are the combinations of n elements chosen r at a time.

The $_nC_r$ notation is used by graphing calculators to find combinations. Figure 10 shows a calculator screen with $_6C_2$ and $_7C_5$.

```
6 nCr 2
              15
7 nCr 5
              21
```

OBJECTIVE 5 Use the combinations formula to solve counting problems. In the preceding discussion, it was shown that $\binom{3}{2} = 3$. The same result can be found using the formula.

$$\binom{3}{2} = \frac{3!}{(3-2)!\,2!} = \frac{3 \cdot 2 \cdot 1}{1 \cdot 2 \cdot 1} = 3$$

The combinations formula is used in the next examples.

EXAMPLE 7 Using the Combinations Formula

How many different committees of 3 people can be chosen from a group of 8 people?

Since the order in which the members of the committee are chosen does not affect the result, use combinations.

$$\binom{8}{3} = \frac{8!}{5!\,3!} = \frac{8 \cdot 7 \cdot 6 \cdot 5 \cdot 4 \cdot 3 \cdot 2 \cdot 1}{5 \cdot 4 \cdot 3 \cdot 2 \cdot 1 \cdot 3 \cdot 2 \cdot 1} = 56$$

Now Try Exercise 23.

EXAMPLE 8 Using the Combinations Formula

From a group of 30 bank employees, 3 are to be selected to work on a special project.

(a) In how many different ways can the employees be selected?

The number of 3-element combinations from a set of 30 elements must be found. (Use combinations, not permutations, because order within the group of 3 does not affect the result.) Using the formula gives

$$\binom{30}{3} = \frac{30!}{27!\,3!} = 4060.$$

There are 4060 ways to select the project group.

(b) In how many different ways can the group of 3 be selected if it has already been decided that a certain employee must work on the project?

Since one employee has already been selected to work on the project, the problem is reduced to selecting 2 more employees from the 29 employees that are left.

$$\binom{29}{2} = \frac{29!}{27!\,2!} = 406$$

In this case, the project group can be selected in 406 different ways.

Now Try Exercise 43.

EXAMPLE 9 Using Combinations to Choose a Delegation

A congressional committee consists of four senators and six representatives. A delegation of five members is to be chosen. In how many ways could this delegation include exactly three senators?

"Exactly three senators" implies that there must be $5 - 3 = 2$ representatives as well. The three senators could be chosen in $\binom{4}{3} = 4$ ways. The two representatives could be chosen in $\binom{6}{2} = 15$ ways. Now apply the fundamental principle of counting to find that there are $4 \cdot 15 = 60$ ways to form the delegation.

Now Try Exercise 37.

OBJECTIVE 6 Distinguish between permutations and combinations. Students often have difficulty determining whether to use permutations or combinations in solving problems. The following chart lists some of the similarities and differences between these two concepts.

Permutations	Combinations
Number of ways of selecting r items out of n items	
Repetitions are not allowed.	
Order is important.	Order is not important.
Arrangements of r items from a set of n items	Subsets of r items from a set of n items
$P(n, r) = \dfrac{n!}{(n-r)!}$	$\dbinom{n}{r} = \dfrac{n!}{(n-r)!\,r!}$
Clue words: arrangement, schedule, order	Clue words: group, committee, sample, selection

EXAMPLE 10   Distinguishing between Permutations and Combinations

Tell whether permutations or combinations should be used to solve each problem.

(a) How many 4-digit codes are possible if no digits are repeated?

Since changing the order of the 4 digits results in a different code, permutations should be used.

(b) A sample of 3 light bulbs is randomly selected from a batch of 15 bulbs. How many different samples are possible?

The order in which the 3 light bulbs are selected is not important. The sample is unchanged if the items are rearranged, so combinations should be used.

(c) In a basketball tournament with 8 teams, how many games must be played so that each team plays every other team exactly once?

Selection of 2 teams for a game is an *unordered* subset of 2 from the set of 8 teams. Use combinations.

(d) In how many ways can 4 stockbrokers be assigned to 6 offices so that each broker has a private office?

The office assignments are an *ordered* selection of 4 offices from the 6 offices. Exchanging the offices of any 2 brokers within a selection of 4 offices gives a different assignment, so permutations should be used.

Now Try Exercise 33.

14.6 EXERCISES

Evaluate each expression.

1. $P(6, 4)$ **2.** $P(7, 5)$ **3.** $P(9, 2)$ **4.** $P(6, 5)$

5. $P(5, 1)$ **6.** $P(6, 0)$ **7.** $\binom{4}{2}$ **8.** $\binom{9}{3}$

9. $\binom{6}{0}$ **10.** $\binom{8}{1}$

Use the fundamental counting principle or permutations to solve each problem. See Examples 1–6.

11. How many different types of homes are available if a builder offers a choice of 6 basic plans, 4 roof styles, and 2 exterior finishes?

12. A menu offers a choice of 4 salads, 8 main dishes, and 5 desserts. How many different 3-course meals (salad, main dish, dessert) are possible?

13. In an experiment on social interaction, 8 people will sit in 8 seats in a row. In how many different ways can the 8 people be seated?

14. For many years, the state of California used 3 letters followed by 3 digits on its automobile license plates.

(a) How many different license plates are possible with this arrangement?
(b) When the state ran out of new plates, the order was reversed to 3 digits followed by 3 letters. How many additional plates were then possible?
(c) Several years ago, the plates described in (b) were also used up. The state then issued plates with 1 letter followed by 3 digits and then 3 letters. How many plates does this scheme provide?

15. In how many ways can 7 of 10 mice be arranged in a row for a genetics experiment?

16. How many 7-digit telephone numbers are possible if the first digit cannot be 0, and

(a) only odd digits may be used?
(b) the telephone number must be a multiple of 10 (that is, it must end in 0)?
(c) the first three digits must be 456?

17. If your college offers 400 courses, 20 of which are in mathematics, and your counselor arranges your schedule of 4 courses by random selection, how many schedules are possible that do not include a math course?

18. In how many ways can 5 players be assigned to the 5 positions on a basketball team, assuming that any player can play any position? In how many ways can 10 players be assigned to the 5 positions?

19. In a sales force of 35 people, how many ways can 3 salespeople be selected for 3 different leadership jobs?

20. A softball team has 20 players. How many 9-player batting orders are possible?

21. In how many ways can 6 bank tellers be assigned to 6 different windows? In how many ways can 10 tellers be assigned to the 6 windows?

Use combinations to solve each problem. See Examples 7–9.

22. How many different samples of 4 light bulbs can be selected from a carton of 2 dozen bulbs?

23. A professional stockbrokers' association has 50 members. If a committee of 6 is to be selected at random, how many different committees are possible?

24. A group of 5 financial planners is to be selected at random from a professional organization with 30 members to participate in a seminar. In how many ways can this be done? In how many ways can the group that will not participate be selected?

25. Harry's Hamburger Heaven sells hamburgers with cheese, relish, lettuce, tomato, onion, mustard, or ketchup. How many different hamburgers can be concocted using any 4 of the extras?

26. How many different 5-card poker hands can be dealt from a deck of 52 playing cards?

27. Seven cards are marked with the numbers 1 through 7 and are shuffled, and then 3 cards are drawn. How many different 3-card combinations are possible?

28. A bag contains 18 marbles. How many samples of 3 can be drawn from it? How many samples of 5 marbles?

29. In Exercise 28, if the bag contains 5 purple, 4 green, and 9 black marbles, how many samples of 3 can be drawn in which all the marbles are black? How many samples of 3 can be drawn in which exactly 2 marbles are black?

30. In Exercise 22, assume it is known that there are 5 defective light bulbs in the carton. How many samples of 4 can be drawn in which all are defective? How many samples of 4 can be drawn in which there are 2 good bulbs and 2 defective bulbs?

31. Explain the difference between a permutation and a combination. What should you look for in a problem to decide which of these is an appropriate method of solution?

32. Is the choice of 2 kittens from a litter of 6 kittens an example of a permutation or a combination? Explain.

33. Determine whether each of the following is a permutation or a combination.

(a) Your 5-digit postal zip code
(b) A particular 5-card hand in a game of poker
(c) A committee of school board members

34. Padlocks with digit dials are often referred to as "combination locks." According to the mathematical definition of combination, is this an accurate description? Why or why not?

Solve each problem using any method. See Examples 1–10.

35. From a pool of 7 secretaries, 3 are selected to be assigned to 3 managers, with 1 secretary for each manager. In how many ways can this be done?

36. In a game of musical chairs, 12 children, staying in the same order, circle around 11 chairs. Each child who is next to a chair must sit down when the music stops. (One will be left out.) How many seatings are possible?

37. In an office with 8 men and 11 women, how many 5-member groups with the following compositions can be chosen for a training session?

(a) All men
(b) All women
(c) 3 men and 2 women
(d) No more than 3 women

38. In an experiment on plant hardiness, a researcher gathers 6 wheat plants, 3 barley plants, and 2 rye plants. Four plants are to be selected at random.

(a) In how many ways can this be done?
(b) In how many ways can this be done if exactly 2 wheat plants must be included?

39. From 10 names on a ballot, 4 will be elected to a political party committee. How many different committees are possible? In how many ways can the committee of 4 be formed if each person will have a different responsibility?

40. In how many ways can 5 of 9 plants be arranged in a row on a windowsill?

41. Johanna Lucas specializes in making different vegetable soups with carrots, celery, onions, beans, peas, tomatoes, and potatoes. How many different soups can she make using any 4 ingredients?

42. How many 4-letter radio-station call letters can be made if the first letter must be K or W and no letter may be repeated? How many if repeats are allowed? How many of the call letters with no repeats can end in K?

43. A group of 12 workers decides to send a delegation of 3 to their supervisor to discuss their work assignments.

(a) How many delegations of 3 are possible?
(b) How many are possible if one of the 12, the foreman, must be in the delegation?
(c) If there are 5 women and 7 men in the group, how many possible delegations would include exactly 1 woman?

44. The Riverdale board of supervisors is composed of 2 liberals and 5 conservatives. Three members are to be selected randomly as delegates to a convention.

(a) How many delegations are possible?
(b) How many delegations could have all liberals?
(c) How many delegations could have 2 conservatives and 1 liberal?
(d) If the supervisor who serves as chairman of the board must be included, how many delegations are possible?

Prove each statement for every positive integer n.

45. $P(n, n - 1) = P(n, n)$

46. $P(n, 1) = n$

47. $P(n, 0) = 1$

48. $\dbinom{n}{n} = 1$

49. $\dbinom{n}{0} = 1$

50. $\dbinom{n}{n - 1} = n$

14.7 Basics of Probability

OBJECTIVES

1 Learn the terminology of probability theory.

2 Find the probability of an event.

3 Find the probability of the complement of E, given the probability of E.

4 Find the odds in favor of an event.

5 Find the probability of a compound event.

The study of probability has become increasingly popular because it has a wide range of practical applications. The basic ideas of probability are introduced in this section.

OBJECTIVE 1 Learn the terminology of probability theory. In probability, each repetition of an experiment is called a **trial.** The possible results of each trial are called **outcomes** of the experiment. In this section, we are concerned with outcomes that are equally likely to occur. For example, the experiment of tossing a coin has two equally likely possible outcomes: landing heads up (H) or landing tails up (T). Also, the experiment of rolling a fair die has 6 equally likely outcomes: landing so the face that is up shows 1, 2, 3, 4, 5, or 6 dots.

The set S of all possible outcomes of a given experiment is called the **sample space** for the experiment. (In this text all sample spaces are finite.) The sample space for the experiment of tossing a coin consists of the outcomes H and T. This sample space can be written in set notation as

$$S = \{H, T\}.$$

Similarly, the sample space for the experiment of rolling a single die is

$$S = \{1, 2, 3, 4, 5, 6\}.$$

Any subset of the sample space is called an **event.** In the experiment with the die, for example, "the number showing is a three" is an event, say E_1, such that $E_1 = \{3\}$. "The number showing is greater than three" is also an event, say E_2, such that $E_2 = \{4, 5, 6\}$. To represent the number of outcomes that belong to event E, the notation $n(E)$ is used. Then $n(E_1) = 1$ and $n(E_2) = 3$.

OBJECTIVE 2 Find the probability of an event. $P(E)$ is used to designate the *probability* of event E.

Probability of Event *E*

In a sample space with equally likely outcomes, the **probability** of an event E, written $P(E)$, is the ratio of the number of outcomes in sample space S that belong to event E, $n(E)$, to the total number of outcomes in sample space S, $n(S)$. That is

$$P(E) = \frac{n(E)}{n(S)}.$$

This definition is used to find the probability of event E_1 given above, by starting with the sample space for the experiment, $S = \{1, 2, 3, 4, 5, 6\}$, and the desired event $E_1 = \{3\}$. Since $n(E_1) = 1$ and since there are 6 outcomes in the sample space,

$$P(E_1) = \frac{n(E_1)}{n(S)} = \frac{1}{6}.$$

EXAMPLE 1 Finding Probabilities of Events

A single die is rolled. Write the following events in set notation and give the probability for each event.

(a) E_3: the number showing is even

Use the definition above. Since $E_3 = \{2, 4, 6\}$, $n(E_3) = 3$. Also shown above, $n(S) = 6$, so

$$P(E_3) = \frac{3}{6} = \frac{1}{2}.$$

(b) E_4: the number showing is greater than 4

Again, $n(S) = 6$. Event $E_4 = \{5, 6\}$, with $n(E_4) = 2$. By the definition,

$$P(E_4) = \frac{2}{6} = \frac{1}{3}.$$

(c) E_5: the number showing is less than 7

$$E_5 = \{1, 2, 3, 4, 5, 6\} \quad \text{and} \quad P(E_5) = \frac{6}{6} = 1$$

(d) E_6: the number showing is 7

$$E_6 = \emptyset \quad \text{and} \quad P(E_6) = \frac{0}{6} = 0$$

Now Try Exercises 5 and 9.

In Example 1(c), $E_5 = S$. Therefore, the event E_5 is certain to occur every time the experiment is performed. An event that is certain to occur, such as E_5, always has a probability of 1. On the other hand, in Example 1(d), $E_6 = \emptyset$ and $P(E_6)$ is 0.

The probability of an impossible event, such as E_6, is always 0, since none of the outcomes in the sample space satisfy the event. For any event E, $P(E)$ is between 0 and 1 inclusive.

OBJECTIVE 3 **Find the probability of the complement of E, given the probability of E.** The set of all outcomes in the sample space that do *not* belong to event E is called the **complement** of E, written E'. For example, in the experiment of drawing a single card from a standard deck of 52 cards, let E be the event "the card is an ace." Then E' is the event "the card is not an ace." From the definition of E', for any event E,

$$E \cup E' = S \qquad \text{and} \qquad E \cap E' = \emptyset.$$

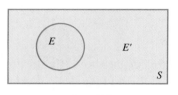

FIGURE 11

Probability concepts can be illustrated using **Venn diagrams,** as shown in Figure 11. The rectangle in Figure 11 represents the sample space in an experiment. The area inside the circle represents event E, while the area inside the rectangle, but outside the circle, represents event E'.

A standard deck of 52 cards has four suits: hearts ♥, clubs ♣, diamonds ♦, and spades ♠, with 13 cards of each suit. Each suit has a jack, a queen, and a king (sometimes called the "face cards"), an ace, and cards numbered from 2 to 10. The hearts and diamonds are red, and the spades and clubs are black. We refer to this standard deck of cards in this section.

EXAMPLE 2   **Using the Complement in a Probability Problem**

In the experiment of drawing a card from a well-shuffled deck, find the probability of events E, the card is an ace, and E'.

Since there are four aces in the deck of 52 cards, $n(E) = 4$ and $n(S) = 52$. Therefore, $P(E) = \frac{n(E)}{n(S)} = \frac{4}{52} = \frac{1}{13}$. Of the 52 cards, 48 are not aces, so

$$P(E') = \frac{n(E')}{n(S)} = \frac{48}{52} = \frac{12}{13}.$$

Now Try Exercise 19.

In Example 2, $P(E) + P(E') = \frac{1}{13} + \frac{12}{13} = 1$. This is always true for any event E and its complement E'. That is,

$$P(E) + P(E') = 1.$$

This can be restated as

$$P(E) = 1 - P(E') \qquad \text{or} \qquad P(E') = 1 - P(E).$$

These two equations suggest an alternative way to compute the probability of an event. For example, if it is known that $P(E) = \frac{1}{10}$, then

$$P(E') = 1 - \frac{1}{10} = \frac{9}{10}.$$

OBJECTIVE 4 **Find the odds in favor of an event.**   Sometimes probability statements are expressed in terms of odds, a comparison $P(E)$ with $P(E')$. The **odds** in favor of an event E are expressed as the ratio of $P(E)$ to $P(E')$ or as the fraction $\frac{P(E)}{P(E')}$. For example, if the probability of rain can be established as $\frac{1}{3}$, the odds that it will rain are

$$P(\text{rain}) \text{ to } P(\text{no rain}) = \frac{1}{3} \text{ to } \frac{2}{3} = \frac{\frac{1}{3}}{\frac{2}{3}} = \frac{1}{2} \quad \text{or} \quad 1 \text{ to } 2.$$

On the other hand, the odds that it will not rain are 2 to 1 $\left(\text{or } \frac{2}{3} \text{ to } \frac{1}{3}\right)$. If the odds in favor of an event are, say, 3 to 5, then the probability of the event is $\frac{3}{8}$, while the probability of the complement of the event is $\frac{5}{8}$. If the odds favoring event E are m to n, then

$$P(E) = \frac{m}{m+n} \quad \text{and} \quad P(E') = \frac{n}{m+n}.$$

▨ EXAMPLE 3 Finding Odds of an Event

(a) A manager is to be selected at random from 6 sales managers and 4 office managers. Find the odds in favor of a sales manager being selected.

Let E represent the event "a sales manager is selected." Then

$$P(E) = \frac{6}{10} = \frac{3}{5} \quad \text{and} \quad P(E') = 1 - \frac{3}{5} = \frac{2}{5}.$$

Therefore, the odds in favor of a sales manager being selected are

$$P(E) \text{ to } P(E') = \frac{3}{5} \text{ to } \frac{2}{5} = \frac{\frac{3}{5}}{\frac{2}{5}} = \frac{3}{2} \quad \text{or} \quad 3 \text{ to } 2.$$

(b) In a recent year, the probability that corporate stock was owned by a pension fund was .227. (*Source:* Federal Reserve Board and New York Stock Exchange.) Find the odds that year against a corporate stock being owned by a pension fund.

Let E represent the event "corporate stock is owned by a pension fund." Then $P(E) = .227$ and $P(E') = 1 - .227 = .773$. Since

$$\frac{P(E')}{P(E)} = \frac{.773}{.227} \approx 3.4,$$

the odds against a corporate stock being owned by a pension fund were about 3.4 to 1.

Now Try Exercise 21.

OBJECTIVE 5 Find the probability of a compound event.   A **compound event** involves an *alternative,* such as E or F, where E and F are events. For example, in the experiment of rolling a die, suppose H is the event "the result is a 3," and K is the event "the result is an even number." What is the probability of the compound event "the result is a 3 or an even number"? From the information stated above,

$$H = \{3\} \qquad K = \{2, 4, 6\} \qquad H \text{ or } K = \{2, 3, 4, 6\}$$

$$P(H) = \frac{1}{6} \qquad P(K) = \frac{3}{6} = \frac{1}{2} \qquad P(H \text{ or } K) = \frac{4}{6} = \frac{2}{3}.$$

Notice that $P(H) + P(K) = P(H \text{ or } K)$.

Before assuming that this relationship is true in general, consider another event for this experiment, "the result is a 2," event G. Now

$$G = \{2\} \qquad K = \{2, 4, 6\} \qquad K \text{ or } G = \{2, 4, 6\}$$

$$P(G) = \frac{1}{6} \qquad P(K) = \frac{3}{6} = \frac{1}{2} \qquad P(K \text{ or } G) = \frac{3}{6} = \frac{1}{2}.$$

In this case $P(K) + P(G) \neq P(K \text{ or } G)$.

As Figure 12 shows, the difference in the two examples above comes from the fact that events H and K cannot occur simultaneously. Such events are called **mutually exclusive events.** In fact, $H \cap K = \emptyset$, which is true for any two mutually exclusive events. Events K and G, however, can occur simultaneously. Both are satisfied if the result of the roll is a 2, the element in their intersection ($K \cap G = \{2\}$). This example suggests the following property.

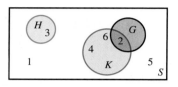

FIGURE 12

Probability of Alternative Events

For any events E and F,
$$P(E \text{ or } F) = P(E \cup F) = P(E) + P(F) - P(E \cap F).$$

EXAMPLE 4 Finding the Probability of Alternative Events

One card is drawn from a well-shuffled deck of 52 cards. What is the probability of each event?

(a) The card is an ace or a spade.

The events "drawing an ace" and "drawing a spade" are not mutually exclusive since it is possible to draw the ace of spades, an outcome satisfying both events. The probability is

$$P(\text{ace or spade}) = P(\text{ace}) + P(\text{spade}) - P(\text{ace and spade})$$
$$= \frac{4}{52} + \frac{13}{52} - \frac{1}{52}$$
$$= \frac{16}{52} = \frac{4}{13}.$$

(b) The card is a 3 or a king.

"Drawing a 3" and "drawing a king" are mutually exclusive events because it is impossible to draw one card that is both a 3 and a king. Using the rule given above,

$$P(3 \text{ or } K) = P(3) + P(K) - P(3 \text{ and } K)$$
$$= \frac{4}{52} + \frac{4}{52} - 0$$
$$= \frac{8}{52} = \frac{2}{13}.$$

Now Try Exercise 25.

EXAMPLE 5   Finding the Probability of Alternative Events

For the experiment consisting of one roll of a pair of dice, find the probability that the sum of the dots showing is at most 4.

The description "at most 4" can be rewritten as "2 or 3 or 4." (A sum of 1 is meaningless here.) Then

$$P(\text{at most 4}) = P(2 \text{ or } 3 \text{ or } 4)$$
$$= P(2) + P(3) + P(4), \qquad (1)$$

since the events represented by "2," "3," and "4" are mutually exclusive.

The sample space for this experiment includes 36 possible pairs of numbers from 1 to 6: (1, 1), (1, 2), (1, 3), (1, 4), (1, 5), (1, 6), (2, 1), (2, 2), and so on. The pair (1, 1) is the only one with a sum of 2, so $P(2) = \frac{1}{36}$. Also $P(3) = \frac{2}{36}$ since both (1, 2) and (2, 1) give a sum of 3. The pairs (1, 3), (2, 2), and (3, 1) have a sum of 4, so $P(4) = \frac{3}{36}$. Substituting into equation (1) above gives

$$P(\text{at most 4}) = \frac{1}{36} + \frac{2}{36} + \frac{3}{36} = \frac{6}{36} = \frac{1}{6}.$$

Now Try Exercise 29.

The box summarizes the properties of probability discussed in this section.

Properties of Probability

For any events E and F,

1. $0 \le P(E) \le 1$

2. $P(\text{a certain event}) = 1$

3. $P(\text{an impossible event}) = 0$

4. $P(E') = 1 - P(E)$

5. $P(E \text{ or } F) = P(E \cup F) = P(E) + P(F) - P(E \cap F)$.

| CONNECTIONS |

Games of chance and gambling enterprises (the earliest motivators for the study of probability) are a major force today.

For Discussion or Writing

One state lottery game requires you to pick 6 different numbers from 1 to 99.

1. How many ways are there to choose 6 numbers if order is not important?

(continued)

2. How many ways are there if order is important?

3. Assume order is unimportant. What is the probability of picking all 6 numbers correctly to win the big prize?

4. Discuss the probability of winning in a state lottery in your area.

14.7 EXERCISES

For Extra Help

 Student's
Solutions Manual

 MyMathLab

 InterAct Math
Tutorial Software

 AW Math
Tutor Center

*Math*XP MathXL

 Digital Video Tutor
CD 14/Videotape 17

1. State the definition of the probability of event E in a sample space with equally likely outcomes in your own words, and give examples.

2. List the properties of probability in words. Give an example of each property.

Write a sample space with equally likely outcomes for each experiment.

3. Two ordinary coins are tossed.   4. Three ordinary coins are tossed.

5. Five slips of paper marked with the numbers 1, 2, 3, 4, and 5 are placed in a box. After mixing well, two slips are drawn.

6. A die is rolled and then a coin is tossed.

Write the events in Exercises 7–10 in set notation and give the probability of each event. See Example 1.

7. In the experiment from Exercise 3:

   (a) Both coins show the same face.   (b) At least one coin turns up heads.

8. In Exercise 4:

   (a) The result of the toss is exactly 2 heads and 1 tail.
   (b) At least 2 coins show tails.

9. In Exercise 5:

   (a) Both slips are marked with even numbers.
   (b) Both slips are marked with odd numbers.
   (c) Both slips are marked with the same number.
   (d) One slip is marked with an odd number and the other with an even number.

10. In Exercise 6:

   (a) The die shows an even number.   (b) The coin shows heads.
   (c) The die shows 6.   (d) The die shows 2 and the coin shows tails.

11. A student gives the answer to a probability problem as $\frac{6}{5}$. Tell why this answer must be incorrect.

12. If the probability of an event is .857, what is the probability that the event will not occur?

13. A marble is drawn at random from a box containing 3 yellow, 4 white, and 8 blue marbles. Find the probability in (a)–(c). Then work (d) and (e). See Examples 1 and 3.

   (a) A yellow marble is drawn.   (b) A blue marble is drawn.
   (c) A black marble is drawn.
   (d) What are the odds in favor of drawing a yellow marble?
   (e) What are the odds against drawing a blue marble?

RELATING CONCEPTS (EXERCISES 14–18)

For Individual or Group Work

Many probability problems involve numbers that are too large to determine the number of outcomes easily, even with a tree diagram. In such cases we can use combinations. For example, if 3 engines are tested from a shipping container packed with 12 engines, 1 of which is defective, what is P(E), the probability that the defective engine will be found? **Work Exercises 14–18 in order.**

14. How many ways are there to choose the sample of 3 from the 12 engines?

15. How many ways are there to choose a sample of 3 with 1 defective and 2 good engines?

16. What is $n(E)$ in this experiment if E is the event, "The defective engine is in the sample"?

17. What is $n(S)$ in this experiment?

18. Find $P(E)$.

Solve each problem. See Examples 2–5.

19. In the experiment of drawing a card from a well-shuffled deck, find the probability of the events, E, the card is a face card (K, Q, J of any suit), and E'.

20. A baseball player with a batting average of .300 comes to bat. What are the odds in favor of his getting a hit?

21. The probability that a bank with assets greater than or equal to $30 billion will make a loan to a small business is .002. (*Source: Wall Street Journal* analysis of *Cal Reports* filed with federal banking authorities.) What are the odds against such a bank making a small business loan?

22. If the odds that it will rain are 4 to 5, what is the probability of rain?

23. Ms. Bezzone invites 10 relatives to a party: her mother, 2 uncles, 3 brothers, and 4 cousins. If the chances of any one guest arriving first are equally likely, find the following probabilities.

(a) The first guest is an uncle or a cousin.
(b) The first guest is a brother or a cousin.
(c) The first guest is an uncle or her mother.

24. A card is drawn from a well-shuffled deck of 52 cards. Find the probability that the card is the following.

(a) A queen   (b) Red
(c) A black 3   (d) A club or red

25. In Exercise 24, find the probability of the following.

(a) A face card (K, Q, J of any suit)   (b) Red or a 3
(c) Less than a 4 (consider aces as 1s)

26. Two dice are rolled. Find the probability of the following events.

(a) The sum of the dots is at least 10.
(b) The sum of the dots is either 7 or at least 10.
(c) The sum of the dots is 3 or the dice both show the same number.

27. If a marble is drawn from a bag containing 2 yellow, 5 red, and 3 blue marbles, what are the probabilities of the following results?

   (a) The marble is yellow or blue.   (b) The marble is yellow or red.

   (c) The marble is green.

28. The law firm of Alam, Bartolini, Chinn, Dickinson, and Ellsberg has two senior partners, Alam and Bartolini. Two of the attorneys are to be selected to attend a conference. Assuming that all are equally likely to be selected, find the following probabilities.

   (a) Chinn is selected.   (b) Alam and Dickinson are selected.

   (c) At least one senior partner is selected.

29. The management of a bank wants to survey its employees, who are classified as follows for the purpose of an interview: 30% have worked for the bank more than 5 yr; 28% are female; 65% contribute to a voluntary retirement plan; half of the female employees contribute to the retirement plan. Find the following probabilities.

   (a) A male employee is selected.

   (b) An employee is selected who has worked for the bank for 5 yr or less.

   (c) An employee is selected who contributes to the retirement plan or is female.

30. The table shows the probabilities of a person accumulating specific amounts of credit card charges over a 12-month period.

Charges	Probability
Under $100	.31
$100–$499	.18
$500–$999	.18
$1000–$1999	.13
$2000–$2999	.08
$3000–$4999	.05
$5000–$9999	.06
$10,000 or more	.01

Find the probabilities that a person's total charges during the period are the following.

   (a) $500–$999   (b) $500–$2999   (c) $5000–$9999   (d) $3000 or more

In most animals and plants, it is very unusual for the number of main parts of the organism (arms, legs, toes, flower petals, etc.) to vary from generation to generation. Some species, however, have meristic variability, *in which the number of certain body parts varies from generation to generation. One researcher studied the front feet of certain guinea pigs and produced the following probabilities.**

$$P(\text{only four toes, all perfect}) = .77$$
$$P(\text{one imperfect toe and four good ones}) = .13$$
$$P(\text{exactly five good toes}) = .10$$

Find the probability of each event.

31. No more than 4 good toes

32. 5 toes, whether perfect or not

The probabilities for the outcomes of an experiment having sample space $S = \{s_1, s_2, s_3, s_4, s_5, s_6\}$ *are shown here.*

Outcomes	s_1	s_2	s_3	s_4	s_5	s_6
Probability	.17	.03	.09	.46	.21	.04

Let $E = \{s_1, s_2, s_5\}$, *and let* $F = \{s_4, s_5\}$. *Find each probability.*

33. $P(E)$

34. $P(F)$

35. $P(E \cap F)$

36. $P(E \cup F)$

37. $P(E' \cup F')$

38. $P(E' \cap F)$

---

*Excerpt from "Analysis of Variability in Number of Digits in an Inbred Strain of Guinea Pigs," by S. Wright in *Genetics*, v. 19 (1934), 506–36. Reprinted by permission of Genetics Society of America.

■■■ Chapter **14** Group Activity ■■■

Investing for the Future

OBJECTIVE Calculate compound interest; understand the effects of monthly, quarterly, and annual compounding.

In this chapter you have seen many different types of financing and investing options including loans, mutual funds, savings accounts, and annuities. In this activity you will analyze annuities with different periods of compound interest.

Consider a family that has a 10-year-old child, for whom they want to save for college. The family is considering three different savings options.

Option 1: An annuity that compounds quarterly

Option 2: An annuity that compounds monthly

Option 3: Put off saving until high school (that is, the last 4 yr); then put money into an annuity that compounds annually

A. Have each member of your group calculate total savings for one of the three options. Use the following formula, where S is future value, R is payment amount made each *period, i* is annual interest rate divided by the number of periods per year, and n is total number of compounding periods, along with the specific information given below for each option.

$$S = R\left[\frac{(1 + i)^n - 1}{i}\right]$$

Option 1: The family plans to save $300 per quarter (3 months), the interest rate is 5%, and the annuity compounds quarterly. Savings will be for 8 yr.

Option 2: The family plans to save $100 per month, the interest rate is 5%, and the annuity compounds monthly. Again, savings will be for 8 yr.

Option 3: The family plans to wait and save only the last 4 yr. They will save $2400 a yr, the interest rate is 5%, and the annuity compounds annually.

B. Compare your answers.

1. How much is being invested using each option?

2. Which option resulted in the largest amount of savings?

3. Explain why this option produced more savings.

4. What other considerations might be involved in deciding how to save?

CHAPTER **14** SUMMARY

KEY TERMS

14.1 sequence
infinite sequence
terms of a sequence
general term
finite sequence
series
summation notation
index of summation
arithmetic mean
(average)

14.2 arithmetic sequence
(arithmetic
progression)
common difference
14.3 geometric sequence
(geometric
progression)
common ratio
annuity
ordinary annuity
payment period

future value of an
annuity
term of an annuity
14.4 Pascal's triangle
binomial theorem
(general binomial
expansion)
14.6 tree diagram
independent events
permutation
combination

14.7 trial
outcome
sample space
event
probability
complement
Venn diagram
odds
compound event
mutually exclusive
events

NEW SYMBOLS

a_n nth term of a
sequence

$\sum\limits_{i=1}^{n} a_i$ summation
notation

S_n sum of first
n terms of a
sequence

$\lim\limits_{n \to \infty} a_n$ limit of a_n as n gets
larger and larger

$\sum\limits_{i=1}^{\infty} a_i$ sum of an infinite
number of terms

$n!$ n factorial

$_nC_r$ binomial coefficient
(combinations
of n elements
taken r at a time)

$P(n, r)$ permutations of
n elements taken
r at a time

$\binom{n}{r}$ combinations of
n elements
taken r at a time

$n(E)$ number of
outcomes that
belong to
event E

$P(E)$ probability of
event E

E' complement of
event E

TEST YOUR WORD POWER

See how well you have learned the vocabulary in this chapter. Answers, with examples, follow the Quick Review.

1. An **infinite sequence** is
    A. the values of a function
    B. a function whose domain is the
 set of natural numbers
    C. the sum of the terms of a function
    D. the average of a group of
 numbers.

2. A **series** is
    A. the sum of the terms of a
 sequence
    B. the product of the terms of a
 sequence
    C. the average of the terms of a
 sequence
    D. the function values of a sequence.

3. An **arithmetic sequence** is a
 sequence in which
    A. each term after the first is a
 constant multiple of the preceding
 term
    B. the numbers are written in a
 triangular array
    C. the terms are added
    D. each term after the first differs
 from the preceding term by a
 common amount.

4. A **geometric sequence** is a sequence
 in which
    A. each term after the first is a
 constant multiple of the preceding
 term

    B. the numbers are written in a
 triangular array
    C. the terms are multiplied
    D. each term after the first differs
 from the preceding term by a
 common amount.

5. The **common difference** is
    A. the average of the terms in a
 sequence
    B. the constant multiplier in a
 geometric sequence
    C. the difference between any two
 adjacent terms in an arithmetic
 sequence
    D. the sum of the terms of an
 arithmetic sequence.

(continued)

6. The **common ratio** is
   A. the average of the terms in a sequence
   B. the constant multiplier in a geometric sequence
   C. the difference between any two adjacent terms in an arithmetic sequence
   D. the product of the terms of a geometric sequence.

7. A **permutation** is
   A. the ratio of the number of outcomes in an equally likely sample space that satisfy an event to the total number of outcomes in the sample space
   B. one of the ways r elements taken from a set of n elements can be arranged

   C. one of the (unordered) subsets of r elements taken from a set of n elements
   D. the ratio of the probability that an event will occur to the probability that it will not occur.

8. A **combination** is
   A. the ratio of the number of outcomes in an equally likely sample space that satisfy an event to the total number of outcomes in the sample space
   B. one of the ways r elements taken from a set of n elements can be arranged
   C. one of the (unordered) subsets of r elements taken from a set of n elements

   D. the ratio of the probability that an event will occur to the probability that it will not occur.

9. The **probability of an event** is
   A. the ratio of the number of outcomes in an equally likely sample space that satisfy an event to the total number of outcomes in the sample space
   B. one of the ways r elements taken from a set of n elements can be arranged
   C. one of the (unordered) subsets of r elements taken from a set of n elements
   D. the ratio of the probability that an event will occur to the probability that it will not occur.

QUICK REVIEW

CONCEPTS	EXAMPLES

14.1 SEQUENCES AND SERIES

Sequence

General Term a_n

Series

$1, \dfrac{1}{2}, \dfrac{1}{3}, \dfrac{1}{4}, \ldots, \dfrac{1}{n}$ has general term $a_n = \dfrac{1}{n}$.

The corresponding series is the *sum*

$$1 + \frac{1}{2} + \frac{1}{3} + \frac{1}{4} + \cdots + \frac{1}{n}.$$

14.2 ARITHMETIC SEQUENCES

Assume a_1 is the first term, a_n is the nth term, and d is the common difference.

The arithmetic sequence 2, 5, 8, 11, . . . has $a_1 = 2$.

Common Difference

$$d = a_{n+1} - a_n$$

$$d = 5 - 2 = 3$$

(Any two successive terms could have been used.)

nth Term

$$a_n = a_1 + (n - 1)d$$

Suppose that $n = 10$. Then the 10th term is

$$a_{10} = 2 + (10 - 1)3$$
$$= 2 + 9 \cdot 3 = 29.$$

CONCEPTS	EXAMPLES

Sum of the First n Terms

$$S_n = \frac{n}{2}(a_1 + a_n)$$

or $\quad S_n = \frac{n}{2}[2a_1 + (n-1)d]$

The sum of the first 10 terms is

$$S_{10} = \frac{10}{2}(2 + a_{10})$$

$$= 5(2 + 29) = 5(31) = 155$$

or $\quad S_{10} = \frac{10}{2}[2(2) + (10-1)3]$

$$= 5(4 + 9 \cdot 3)$$

$$= 5(4 + 27) = 5(31) = 155.$$

14.3 GEOMETRIC SEQUENCES

Assume a_1 is the first term, a_n is the nth term, and r is the common ratio.

Common Ratio

$$r = \frac{a_{n+1}}{a_n}$$

The geometric sequence 1, 2, 4, 8, . . . has $a_1 = 1$.

$$r = \frac{8}{4} = 2$$

(Any two successive terms could have been used.)

nth Term

$$a_n = a_1 r^{n-1}$$

Suppose that $n = 6$. Then the sixth term is

$$a_6 = (1)(2)^{6-1} = 1(2)^5 = 32.$$

Sum of the First n Terms

$$S_n = \frac{a_1(1 - r^n)}{1 - r} \quad \text{or} \quad S_n = \frac{a_1(r^n - 1)}{r - 1} \quad (r \neq 1)$$

The sum of the first six terms is

$$S_6 = \frac{1(2^6 - 1)}{2 - 1} = \frac{64 - 1}{1} = 63.$$

Future Value of an Ordinary Annuity

$$S = R\left[\frac{(1 + i)^n - 1}{i}\right],$$

where S is future value, R is the payment at the end of each period, i is the interest rate per period, and n is the number of periods.

If \$5800 is deposited into an ordinary annuity at the end of each quarter for 4 yr and interest is earned at 6.4% compounded quarterly, then

$$R = \$5800, \quad i = \frac{.064}{4} = .016, \quad n = 4(4) = 16,$$

and

$$S = 5800\left[\frac{(1 + .016)^{16} - 1}{.016}\right]$$

$$= \$104,812.44.$$

Sum of the Terms of an Infinite Geometric Sequence with $|r| < 1$

$$S = \frac{a_1}{1 - r}$$

The sum S of the terms of an infinite geometric sequence with $a_1 = 1$ and $r = \frac{1}{2}$ is

$$S = \frac{1}{1 - \frac{1}{2}} = \frac{1}{\frac{1}{2}} = 2.$$

(continued)

CONCEPTS	EXAMPLES

14.4 THE BINOMIAL THEOREM

For any positive integer n,

$$n! = n(n - 1)(n - 2) \cdots (3)(2)(1).$$
$$0! = 1$$
$$_nC_r = \frac{n!}{r!(n - r)!}$$

$$4! = 4 \cdot 3 \cdot 2 \cdot 1 = 24$$

$$_5C_3 = \frac{5!}{3!(5 - 3)!}$$
$$= \frac{5!}{3!2!}$$
$$= \frac{5 \cdot 4 \cdot 3 \cdot 2 \cdot 1}{3 \cdot 2 \cdot 1 \cdot 2 \cdot 1}$$
$$= 10$$

General Binomial Expansion

For any positive integer n,

$$(x + y)^n = x^n + \frac{n!}{(n - 1)!1!}x^{n-1}y$$
$$+ \frac{n!}{(n - 2)!2!}x^{n-2}y^2$$
$$+ \frac{n!}{(n - 3)!3!}x^{n-3}y^3 + \cdots$$
$$+ \frac{n!}{1!(n - 1)!}xy^{n-1} + y^n.$$

$$(2m + 3)^4 = (2m)^4 + \frac{4!}{3!1!}(2m)^3(3) + \frac{4!}{2!2!}(2m)^2(3)^2$$
$$+ \frac{4!}{1!3!}(2m)(3)^3 + 3^4$$
$$= 2^4 m^4 + 4(2)^3 m^3(3) + 6(2)^2 m^2(9)$$
$$+ 4(2m)(27) + 81$$
$$= 16m^4 + 12(8)m^3 + 54(4)m^2 + 216m + 81$$
$$= 16m^4 + 96m^3 + 216m^2 + 216m + 81$$

rth Term of the Binomial Expansion of $(x + y)^n$

$$\frac{n!}{[n - (r - 1)]!(r - 1)!}x^{n-(r-1)}y^{r-1}$$

The eighth term of $(a - 2b)^{10}$ is

$$\frac{10!}{3!7!}a^3(-2b)^7 = \frac{10 \cdot 9 \cdot 8}{3 \cdot 2 \cdot 1}a^3(-2)^7b^7$$
$$= 120(-128)a^3b^7$$
$$= -15{,}360a^3b^7.$$

14.5 MATHEMATICAL INDUCTION

Proof by Mathematical Induction

Let S_n be a statement concerning the positive integer n. Suppose that

1. S_1 is true;

2. for any positive integer k, $k \le n$, if S_k is true, then S_{k+1} is also true.

Then S_n is true for every positive integer value of n.

See Examples 1–3 in Section 14.5.

CONCEPTS	EXAMPLES

14.6 COUNTING THEORY

Fundamental Principle of Counting
If one event can occur in m ways and a second event can occur in n ways, then both events can occur in mn ways, provided the outcome of the first event does not influence the outcome of the second.

If there are 2 ways to choose a pair of socks and 5 ways to choose a pair of shoes, there are $2 \cdot 5 = 10$ ways to choose socks and shoes.

Permutations Formula
If $P(n, r)$ denotes the number of permutations of n elements taken r at a time, $r \le n$, then

$$P(n, r) = \frac{n!}{(n-r)!}.$$

How many ways are there to arrange the letters of the word *triangle* using 5 letters at a time?
Here, $n = 8$ and $r = 5$, so the number of ways is

$$P(8, 5) = \frac{8!}{(8-5)!} = \frac{8!}{3!} = 6720.$$

Combinations Formula
The number of combinations of n elements taken r at a time is

$$\binom{n}{r} = \frac{n!}{r!(n-r)!}.$$

How many committees of 4 senators can be formed from a group of 9 senators?
Since the arrangement of senators does not matter, this is a combinations problem. The number of possible committees is

$$\binom{9}{4} = \frac{9!}{4!\,5!} = 126.$$

14.7 BASICS OF PROBABILITY

Probability of an Event E
In a sample space with equally likely outcomes, the probability of an event E, written $P(E)$, is the ratio of the number of outcomes in sample space S that belong to event E, $n(E)$, to the total number of outcomes in sample space S, $n(S)$. That is,

$$P(E) = \frac{n(E)}{n(S)}.$$

A number is chosen at random from the set

$$S = \{1, 2, 3, 4, 5, 6\}.$$

What is the probability that the number is less than 3?
The event is $E = \{1, 2\}$; $n(S) = 6$ and $n(E) = 2$. Therefore,

$$P(E) = \frac{2}{6} = \frac{1}{3}.$$

Properties of Probability
For any events E and F:
1. $0 \le P(E) \le 1$
2. $P(\text{a certain event}) = 1$
3. $P(\text{an impossible event}) = 0$
4. $P(E') = 1 - P(E)$
5. $P(E \text{ or } F) = P(E \cup F)$
$= P(E) + P(F) - P(E \cap F).$

What is the probability that the number is 3 or more?
This event is E'.

$$P(E') = 1 - \frac{1}{3} = \frac{2}{3}$$

(continued)

Answers to Test Your Word Power

1. B; *Example:* The ordered list of numbers 3, 6, 9, 12, 15, . . . is an infinite sequence. **2.** A; *Example:* $3 + 6 + 9 + 12 + 15$, written in summation notation as $\sum_{i=1}^{5} 3i$, is a series. **3.** D; *Example:* The sequence $-3, 2, 7, 12, 17, \ldots$ is arithmetic. **4.** A; *Example:* The sequence 1, 4, 16, 64, 256, . . . is geometric. **5.** C; *Example:* The common difference of the arithmetic sequence in Answer 3 is 5 since $2 - (-3) = 5, 7 - 2 = 5, 12 - 7 = 5$, and so on. **6.** B; *Example:* The common ratio of the geometric sequence in Answer 4 is 4 since $\frac{4}{1} = \frac{16}{4} = \frac{64}{16} = \frac{256}{64} = 4$. **7.** B; *Example:* The permutations of the three letters m, n, and t taken two at a time are mn, mt, nt, nm, tm, and tn. **8.** C; *Example:* The combinations of the letters in Answer 7 are mn, mt, and nt. **9.** A; *Examples:* Suppose the letters m, n, and t are written on slips of paper and one slip is randomly drawn. The probability of drawing m is $\frac{1}{3}$. The probability of drawing m or n is $\frac{2}{3}$.

CHAPTER 14 REVIEW EXERCISES

[14.1] *Write out the first four terms of each sequence.*

1. $a_n = 2n - 3$

2. $a_n = \dfrac{n - 1}{n}$

3. $a_n = n^2$

4. $a_n = \left(\dfrac{1}{2}\right)^n$

5. $a_n = (n + 1)(n - 1)$

Write each series as a sum of terms.

6. $\sum_{i=1}^{5} i^2 x$

7. $\sum_{i=1}^{6} (i + 1)x^i$

Evaluate each series.

8. $\sum_{i=1}^{4} (i + 2)$

9. $\sum_{i=1}^{6} 2^i$

10. $\sum_{i=4}^{7} \dfrac{i}{i + 1}$

11. Find the arithmetic mean, or average, of the mutual fund retirement assets for the years 1996 through 2000 shown in the table. Give your answer to the nearest tenth (in billions of dollars).

Year	Assets (in billions of dollars)
1996	1166
1997	1509
1998	1899
1999	2462
2000	2408

Source: Investment Company Institute.

[14.2–14.3] *Decide whether each sequence is* arithmetic, geometric, *or* neither. *If the sequence is arithmetic, find the common difference, d. If it is geometric, find the common ratio, r.*

12. 2, 5, 8, 11, . . .

13. $-6, -2, 2, 6, 10, \ldots$

14. $\dfrac{2}{3}, -\dfrac{1}{3}, \dfrac{1}{6}, -\dfrac{1}{12}, \ldots$

15. $-1, 1, -1, 1, -1, \ldots$

16. $64, 32, 8, \dfrac{1}{2}, \ldots$

17. $64, 32, 16, 8, \ldots$

18. $10, 8, 6, 4, \ldots$

[14.2] *Find the indicated term for each arithmetic sequence.*

19. $a_1 = -2, d = 5; a_{16}$

20. $a_6 = 12, a_8 = 18; a_{25}$

Find the general term for each arithmetic sequence.

21. $a_1 = -4, d = -5$

22. $6, 3, 0, -3, \ldots$

Find the number of terms in each arithmetic sequence.

23. $7, 10, 13, \ldots, 49$

24. $5, 1, -3, \ldots, -79$

Find S_8 for each arithmetic sequence.

25. $a_1 = -2, d = 6$

26. $a_n = -2 + 5n$

[14.3] *Find the general term for each geometric sequence.*

27. $-1, -4, \ldots$

28. $\dfrac{2}{3}, \dfrac{2}{15}, \ldots$

Find the indicated term for each geometric sequence.

29. $2, -6, 18, \ldots; a_{11}$

30. $a_3 = 20, a_5 = 80; a_{10}$

Find each sum, if it exists.

31. $\displaystyle\sum_{i=1}^{5} \left(\dfrac{1}{4}\right)^i$

32. $\displaystyle\sum_{i=1}^{8} \dfrac{3}{4}(-1)^i$

33. $\displaystyle\sum_{i=1}^{\infty} 4\left(\dfrac{1}{5}\right)^i$

34. $\displaystyle\sum_{i=1}^{\infty} 2(3)^i$

[14.4] *Use the binomial theorem to expand each binomial.*

35. $(2p - q)^5$

36. $(x^2 + 3y)^4$

37. $\left(\sqrt{m} + \sqrt{n}\right)^4$

38. Write the fourth term of the expansion of $(3a + 2b)^{19}$.

39. Write the twenty-third term of the expansion of $(-2k + 3)^{25}$.

[14.5]

40. What is mathematical induction?

Use mathematical induction to prove that each statement is true for every positive integer n.

41. $2 + 6 + 10 + 14 + \cdots + (4n - 2) = 2n^2$

42. $2^2 + 4^2 + 6^2 + \cdots + (2n)^2 = \dfrac{2n(n + 1)(2n + 1)}{3}$

43. $2 + 2^2 + 2^3 + \cdots + 2^n = 2(2^n - 1)$

44. $1^3 + 3^3 + 5^3 + \cdots + (2n - 1)^3 = n^2(2n^2 - 1)$

[14.6] *Evaluate.*

45. $P(5, 5)$ **46.** $P(9, 2)$ **47.** $\binom{7}{3}$ **48.** $\binom{8}{5}$

49. Write a sentence or two telling what permutations and combinations are used for and how they differ.

Solve each problem.

50. Two people are planning their wedding. They can select from 2 different chapels, 4 soloists, 3 organists, and 2 ministers. How many different wedding arrangements will be possible?

51. John Jacobs, who is furnishing his apartment, wants to buy a new couch. He can select from 5 different styles, each available in 3 different fabrics, with 6 color choices. How many different couches are available?

52. Four students are to be assigned to 4 different summer jobs. Each student is qualified for all 4 jobs. In how many ways can the jobs be assigned?

53. How many different license plates can be formed with a letter followed by 3 digits and then 3 letters? How many such license plates have no repeats?

[14.7] *Solve each problem.*

54. A marble is drawn at random from a box containing 4 green, 5 black, and 6 white marbles. Find the probability of each event.

    (a) A green marble is drawn.
    (b) A marble that is not black is drawn.
    (c) A blue marble is drawn.

55. Refer to Exercise 54 and answer each question.

    (a) What are the odds in favor of drawing a green marble?
    (b) What are the odds against drawing a white marble?

A card is drawn from a standard deck of 52 cards. Find the probability that each card described is drawn.

56. A black king

57. A face card or an ace

58. An ace or a diamond

59. A card that is not a diamond

60. A card that is not a diamond or not black

MIXED REVIEW EXERCISES

Find the indicated term and S_{10} for each sequence.

61. a_{40}: arithmetic; $1, 7, 13, \ldots$

62. a_{10}: geometric; $-3, 6, -12, \ldots$

63. a_9: geometric; $a_1 = 1, r = -3$

64. a_{15}: arithmetic; $a_1 = -4, d = 3$

Find the general term for each arithmetic or geometric sequence.

65. $2, 7, 12, \ldots$

66. $2, 8, 32, \ldots$

67. $27, 9, 3, \ldots$

68. $12, 9, 6, \ldots$

Solve each problem.

69. When Mary's sled goes down the hill near her home, she covers 3 ft in the first second; then for each second after that she goes 4 ft more than in the preceding second. If the distance she covers going down is 210 ft, how long does it take her to reach the bottom?

70. An ordinary annuity is set up so that $672 is deposited at the end of each quarter for 7 yr. The money earns 8% annual interest compounded quarterly. What is the future value of the annuity?

71. The school population in Pfleugerville has been dropping 3% per yr. The current population is 50,000. If this trend continues, what will the population be in 6 yr?

72. A pump removes $\frac{1}{2}$ of the liquid in a container with each stroke. What fraction of the liquid is left in the container after 7 strokes?

73. Consider the repeating decimal number .55555 . . . .

   (a) Write it as the sum of the terms of an infinite geometric sequence.
   (b) What is r for this sequence?
   (c) Find this infinite sum, if it exists, and write it as a common fraction in lowest terms.

74. Can the sum of the terms of the infinite geometric sequence with $a_n = 5(2)^n$ be found? Explain.

75. Can any two terms of a geometric sequence be used to find the common ratio? Explain.

76. A student council consists of a president, vice-president, secretary/treasurer, and 3 representatives at large. Three members are to be selected to attend a conference.

   (a) How many different such delegations are possible?
   (b) How many are possible if the president must attend?

77. Nine football teams are competing for first-, second-, and third-place titles in a statewide tournament. In how many ways can the winners be determined?

78. A sample shipment of 5 swimming pool filters is chosen. The probability of exactly 0, 1, 2, 3, 4, or 5 filters being defective is given in the following table.

Number Defective	0	1	2	3	4	5
Probability	.31	.25	.18	.12	.08	.06

Find the probability that the given number of filters are defective.

(a) No more than 3   **(b)** At least 2

CHAPTER 14 TEST

Write the first five terms of each sequence described.

1. $a_n = (-1)^n + 1$

2. Arithmetic, with $a_1 = 4$ and $d = 2$

3. Geometric, with $a_4 = 6$ and $r = \frac{1}{2}$

Find a_4 for each sequence described.

4. Arithmetic, with $a_1 = 6$ and $d = -2$

5. Geometric, with $a_5 = 16$ and $a_7 = 9$

Find S_5 for each sequence described.

6. Arithmetic, with $a_2 = 12$ and $a_3 = 15$

7. Geometric, with $a_5 = 4$ and $a_7 = 1$

8. The numbers of commercial banks in the United States for the years 1996 through 2000 are given in the table. What was the average number of banks per year for this period?

Year	Number
1996	66,733
1997	69,468
1998	70,731
1999	72,265
2000	72,998

Source: U.S. Federal Deposit Insurance Corporation, Statistics on Banking, annual.

9. If $4000 is deposited in an ordinary annuity at the end of each quarter for 7 yr and earns 6% interest compounded quarterly, how much will be in the account at the end of this term?

10. Under what conditions does an infinite geometric series have a sum?

Find each sum that exists.

11. $\sum\limits_{i=1}^{5} (2i + 8)$

12. $\sum\limits_{i=1}^{6} (3i - 5)$

13. $\sum\limits_{i=1}^{500} i$

14. $\sum\limits_{i=1}^{3} \frac{1}{2}(4^i)$

15. $\sum\limits_{i=1}^{\infty} \left(\frac{1}{4}\right)^i$

16. $\sum\limits_{i=1}^{\infty} 6\left(\frac{3}{2}\right)^i$

17. Expand $(3k - 5)^4$.

18. Write the fifth term of $\left(2x - \dfrac{y}{3}\right)^{12}$.

Solve each problem.

19. Cheryl bought a new sewing machine for $300. She agreed to pay $20 per month for 15 months plus interest of 1% each month on the unpaid balance. Find the total cost of the machine.

20. During the summer months, the population of a certain insect colony triples each week. If there are 20 insects in the colony at the end of the first week in July, how many are present by the end of September? (Assume exactly 4 weeks in a month.)

21. Given

$$S_n: 8 + 14 + 20 + 26 + \cdots + (6n + 2) = 3n^2 + 5n$$

for every positive integer n. Write out S_k and S_{k+1}.

Evaluate.

22. $P(11, 3)$

23. $\dbinom{10}{2}$

24. A clothing manufacturer makes women's coats in 4 different styles. Each coat can be made from one of 3 fabrics. Each fabric comes in 5 different colors. How many different coats can be made?

25. A club with 30 members is to elect a president, secretary, and treasurer from its membership. If a member can hold at most one position, in how many ways can the offices be filled?

26. In how many ways can a committee of 3 representatives be chosen from a group of 9 representatives?

A card is drawn from a standard deck of 52 cards. Find the probability that each card described is drawn.

27. A red 3

28. A card that is not a face card

29. A king or a spade

30. In the card-drawing experiment above, what are the odds in favor of drawing a face card?

CUMULATIVE REVIEW EXERCISES CHAPTERS **1–14**

This set of exercises may be considered a final examination for the course.

Let $P = \left\{-\frac{8}{3}, 10, 0, \sqrt{13}, -\sqrt{3}, \frac{45}{15}, \sqrt{-7}, .82, -3\right\}$. List the elements of P that are members of each set.

1. Integers

2. Rational numbers

3. Irrational numbers

4. Real numbers

Simplify each expression.

5. $|-7| + 6 - |-10| - (-8 + 3)$

6. $-15 - |-4| - 10 - |-6|$

7. $4(-6) + (-8)(5) - (-9)$

Solve each equation or inequality.

8. $9 - (5 + 3a) + 5a = -4(a - 3) - 7$

9. $7m + 18 \leq 9m - 2$

10. $|4x - 3| = 21$

11. $\dfrac{x + 3}{12} - \dfrac{x - 3}{6} = 0$

12. $2x > 8$ or $-3x > 9$

13. $|2m - 5| \geq 11$

14. Find the slope of the line through $(4, -5)$ and $(-12, -17)$.

15. Find the standard form of the equation of the line through $(-2, 10)$ and parallel to the line with equation $3x + y = 7$.

Graph.

16. $x - 3y = 6$

17. $4x - y < 4$

18. Consider the set of ordered pairs

$$\{(-3, 2), (-2, 6), (0, 4), (1, 2), (2, 6)\}.$$

    (a) Is this a function?   **(b)** What is its domain?   **(c)** What is its range?

Solve each system of equations using the method indicated.

19. $2x + 5y = -19$
     $-3x + 2y = -19$ (Elimination)

20. $y = 5x + 3$
     $2x + 3y = -8$ (Substitution)

21. $x + 2y + z = 8$
     $2x - y + 3z = 15$
     $-x + 3y - 3z = -11$ (Row operations)

22. Evaluate $\begin{vmatrix} -3 & -2 \\ 6 & 9 \end{vmatrix}$.

23. Use Cramer's rule to solve the system with $D = 4$, $D_x = 24$, $D_y = 16$, and $D_z = 14$.

24. Nuts worth \$3 per lb are to be mixed with 8 lb of nuts worth \$4.25 per lb to obtain a mixture that will be sold for \$4 per lb. How many pounds of the \$3 nuts should be used?

Perform the indicated operations.

25. $(4p + 2)(5p - 3)$

26. $(3k - 7)^2$

27. $(2m^3 - 3m^2 + 8m) - (7m^3 + 5m - 8)$

28. Divide $6t^4 + 5t^3 - 18t^2 + 14t - 1$ by $3t - 2$.

Factor.

29. $6z^3 + 5z^2 - 4z$

30. $49a^4 - 9b^2$

31. $c^3 + 27d^3$

32. $64r^2 + 48rq + 9q^2$

Solve each equation or inequality.

33. $2x^2 + x = 10$

34. $k^2 - k - 6 \leq 0$

Simplify.

35. $\left(\dfrac{2}{3}\right)^{-2}$

36. $\dfrac{(3p^2)^3(-2p^6)}{4p^3(5p^7)}$

37. What is the domain of the rational function defined by $f(x) = \dfrac{2}{x^2 - 81}$?

Simplify.

38. $\dfrac{x^2 - 16}{x^2 + 2x - 8} \div \dfrac{x - 4}{x + 7}$

39. $\dfrac{5}{p^2 + 3p} - \dfrac{2}{p^2 - 4p}$

Solve.

40. $\dfrac{4}{x-3} - \dfrac{6}{x+3} = \dfrac{24}{x^2-9}$

41. $6x^2 + 5x = 8$

42. $\sqrt{3x-2} = x$

43. $3^{2x-1} = 81$

44. $\log_8 x + \log_8(x+2) = 1$

45. Simplify $5\sqrt{72} - 4\sqrt{50}$.

46. Multiply $(8+3i)(8-3i)$.

47. Find $f^{-1}(x)$, if $f(x) = 9x + 5$.

Graph.

48. $f(x) = 2(x-2)^2 - 3$

49. $g(x) = \left(\dfrac{1}{3}\right)^x$

50. $y = \log_{1/3} x$

51. $f(x) = \dfrac{2}{x-3}$

52. $\dfrac{x^2}{9} + \dfrac{y^2}{25} = 1$

53. $x^2 - y^2 = 9$

54. Factor $f(x) = 2x^3 + 9x^2 + 3x - 4$ into linear factors given that $f(-4) = 0$.

55. Solve the system $\begin{aligned} xy &= -5 \\ 2x + y &= 3. \end{aligned}$

56. Find the equation of a circle with center at $(-5, 12)$ and radius 9.

57. Write the first five terms of the sequence defined by $a_n = 5n - 12$.

58. Find each sum.

    (a) The sum of the first six terms of the arithmetic sequence with $a_1 = 8$ and $d = 2$

    (b) The sum of the geometric series $15 - 6 + \frac{12}{5} - \frac{24}{25} + \cdots$

59. Find the sum: $\displaystyle\sum_{i=1}^{4} 3i$.

60. Use the binomial theorem to expand $(2a - 1)^5$.

61. What is the fourth term in the expansion of $\left(3x^4 - \frac{1}{2}y^2\right)^5$?

62. Use mathematical induction to prove that
$$4 + 8 + 12 + 16 + \cdots + 4n = 2n(n+1).$$

63. Evaluate each expression.

    (a) $P(7, 3)$ **(b)** $\dbinom{10}{4}$

64. Find the probability of rolling a sum of 11 with two dice.

65. If the odds that it will rain are 3 to 7, what is the probability of rain?

Appendix A
An Introduction to Calculators

There is little doubt that the appearance of handheld calculators three decades ago and the later development of scientific and graphing calculators have changed the methods of learning and studying mathematics forever. For example, computations with tables of logarithms and slide rules made up an important part of mathematics courses prior to 1970. Today, with the widespread availability of calculators, these topics are studied only for their historical significance.

Calculators come in a large array of different types, sizes, and prices. *For the course for which this textbook is intended, the most appropriate type is the scientific calculator,* which costs $10–$20.

In this introduction, we explain some of the features of scientific and graphing calculators. However, remember that calculators vary among manufacturers and models, and that while the methods explained here apply to many of them, they may not apply to your specific calculator. *This introduction is only a guide and is not intended to take the place of your owner's manual.* Always refer to the manual in the event you need an explanation of how to perform a particular operation.

Scientific Calculators

Scientific calculators are capable of much more than the typical four-function calculator that you might use for balancing your checkbook. Most scientific calculators use *algebraic logic.* (Models sold by Texas Instruments, Sharp, Casio, and Radio Shack, for example, use algebraic logic.) A notable exception is Hewlett-Packard, a company whose calculators use *Reverse Polish Notation* (RPN). In this introduction, we explain the use of calculators with algebraic logic.

Arithmetic Operations To perform an operation of arithmetic, simply enter the first number, press the operation key, $+$, $-$, $\times$, or $\div$, enter the second number, and then press the $=$ key. For example, to add 4 and 3, use the following keystrokes.

$$\boxed{4} \quad \boxed{+} \quad \boxed{3} \quad \boxed{=} \qquad \boxed{\quad 7 \quad}$$

Change Sign Key The key marked $+/-$ allows you to change the sign of a display. This is particularly useful when you wish to enter a negative number. For example, to enter −3, use the following keystrokes.

$$\boxed{3} \;\; \boxed{+/-} \;\; \boxed{\qquad -3}$$

Memory Key   Scientific calculators can hold a number in memory for later use. The label of the memory key varies among models; two of these are $\boxed{M}$ and $\boxed{STO}$. The $\boxed{M+}$ and $\boxed{M-}$ keys allow you to add to or subtract from the value currently in memory. The memory recall key, labeled $\boxed{MR}$, $\boxed{RM}$, or $\boxed{RCL}$, allows you to retrieve the value stored in memory.

Suppose that you wish to store the number 5 in memory. Enter 5, then press the key for memory. You can then perform other calculations. When you need to retrieve the 5, press the key for memory recall.

If a calculator has a constant memory feature, the value in memory will be retained even after the power is turned off. Some advanced calculators have more than one memory. Read the owner's manual for your model to see exactly how memory is activated.

Clearing/Clear Entry Keys   The keys $\boxed{C}$ or $\boxed{CE}$ allow you to clear the display or clear the last entry entered into the display. In some models, pressing the $\boxed{C}$ key once will clear the last entry, while pressing it twice will clear the entire operation in progress.

Second Function Key   This key, usually marked $\boxed{2nd}$, is used in conjunction with another key to activate a function that is printed *above* an operation key (and not on the key itself). For example, suppose you wish to find the square of a number, and the squaring function (explained in more detail later) is printed above another key. You would need to press $\boxed{2nd}$ before the desired squaring function can be activated.

Square Root Key   Pressing $\boxed{\sqrt{\;}}$ or $\boxed{\sqrt{x}}$ will give the square root (or an approximation of the square root) of the number in the display. On many newer scientific calculators, the square root key is pressed *before* entering the number, while other calculators use the opposite order. Experiment with your calculator to see which method it uses. For example, to find the square root of 36, use the following keystrokes.

$$\boxed{\sqrt{\;}} \;\; \boxed{3} \;\; \boxed{6} \;\; \boxed{\qquad 6} \qquad \text{or} \qquad \boxed{3} \;\; \boxed{6} \;\; \boxed{\sqrt{\;}} \;\; \boxed{\qquad 6}$$

The square root of 2 is an example of an irrational number (Chapter 8). The calculator will give an approximation of its value, since the decimal for $\sqrt{2}$ never terminates and never repeats. The number of digits shown will vary among models. To find an approximation for $\sqrt{2}$, use the following keystrokes.

$$\boxed{\sqrt{\;}} \;\; \boxed{2} \;\; \boxed{1.4142136} \qquad \text{or} \qquad \boxed{2} \;\; \boxed{\sqrt{\;}} \;\; \boxed{1.4142136}$$
An approximation for $\sqrt{2}$

Squaring Key   The $\boxed{x^2}$ key allows you to square the entry in the display. For example, to square 35.7, use the following keystrokes.

$$\boxed{3} \;\; \boxed{5} \;\; \boxed{.} \;\; \boxed{7} \;\; \boxed{x^2} \;\; \boxed{1274.49}$$

The squaring key and the square root key are often found on the same key, with one of them being a second function (that is, activated by the second function key previously described).

Reciprocal Key   The key marked $\boxed{1/x}$ is the reciprocal key. (When two numbers have a product of 1, they are called *reciprocals.* See Chapter 1.) Suppose that you wish to find the reciprocal of 5. Use the following keystrokes.

$$\boxed{5} \quad \boxed{1/x} \quad \boxed{\qquad 0.2 \qquad}$$

Inverse Key   Some calculators have an inverse key, marked $\boxed{\text{INV}}$. Inverse operations are operations that "undo" each other. For example, the operations of squaring and taking the square root are inverse operations. The use of the $\boxed{\text{INV}}$ key varies among different models of calculators, so read your owner's manual carefully.

Exponential Key   The key marked $\boxed{x^y}$ or $\boxed{y^x}$ allows you to raise a number to a power. For example, if you wish to raise 4 to the fifth power (that is, find 4^5, as explained in Chapter 1), use the following keystrokes.

$$\boxed{4} \quad \boxed{x^y} \quad \boxed{5} \quad \boxed{=} \quad \boxed{\qquad 1024 \qquad}$$

Root Key   Some calculators have a key specifically marked $\boxed{\sqrt[x]{x}}$ or $\boxed{\sqrt[y]{y}}$; with others, the operation of taking roots is accomplished by using the inverse key in conjunction with the exponential key. Suppose, for example, your calculator is of the latter type and you wish to find the fifth root of 1024. Use the following keystrokes.

$$\boxed{1} \quad \boxed{0} \quad \boxed{2} \quad \boxed{4} \quad \boxed{\text{INV}} \quad \boxed{x^y} \quad \boxed{5} \quad \boxed{=} \quad \boxed{\qquad 4 \qquad}$$

Notice how this "undoes" the operation explained in the exponential key discussion.

Pi Key   The number π is an important number in mathematics. It occurs, for example, in the area and circumference formulas for a circle. By pressing the $\boxed{\pi}$ key, you can display the first few digits of π. (Because π is irrational, the display shows only an approximation.) One popular model gives the following display when the $\boxed{\pi}$ key is pressed.

$$\boxed{3.1415927} \qquad \text{An approximation for } \pi$$

Methods of Display   When decimal approximations are shown on scientific calculators, they are either *truncated* or *rounded.* To see how a particular model is programmed, evaluate 1/18 as an example. If the display shows .0555555 (last digit 5), it truncates the display. If it shows .0555556 (last digit 6), it rounds the display.

When very large or very small numbers are obtained as answers, scientific calculators often express these numbers in scientific notation (Chapter 5). For example, if you multiply 6,265,804 by 8,980,591, the display might look like this:

$$\boxed{5.6270623 \ 13}$$

The 13 at the far right means that the number on the left is multiplied by 10^{13}. This means that the decimal point must be moved 13 places to the right if the answer is to be expressed in its usual form. Even then, the value obtained will only be an approximation: 56,270,623,000,000.

Graphing Calculators

While you are not expected to have a graphing calculator to study from this book, we include the following as background information and reference should your course or future courses require the use of graphing calculators.

Basic Features   In addition to the typical keys found on scientific calculators, graphing calculators have keys that can be used to create graphs, make tables, analyze data, and change settings. One of the major differences between graphing and scientific calculators is that a graphing calculator has a larger viewing screen with graphing capabilities. The screens below illustrate the graphs of $Y = X$ and $Y = X^2$.

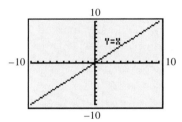 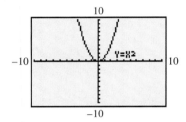

If you look closely at the screens, you will see that the graphs appear to be jagged rather than smooth, as they should be. The reason for this is that graphing calculators have much lower resolution than computer screens. Because of this, graphs generated by graphing calculators must be interpreted carefully.

Editing Input   The screen of a graphing calculator can display several lines of text at a time. This feature allows you to view both previous and current expressions. If an incorrect expression is entered, an error message is displayed. The erroneous expression can be viewed and corrected by using various editing keys, much like a word-processing program. You do not need to enter the entire expression again. Many graphing calculators can also recall past expressions for editing or updating. The screen on the left shows how two expressions are evaluated. The final line is entered incorrectly, and the resulting error message is shown in the screen on the right.

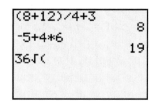 

Order of Operations   Arithmetic operations on graphing calculators are usually entered as they are written in mathematical expressions. For example, to evaluate $\sqrt{36}$ you would first press the square root key, and then enter 36. See the left screen at the top of the next page. The order of operations on a graphing calculator is also important, and current models insert parentheses when typical errors might occur. The open parenthesis that follows the square root symbol is automatically entered by the calculator so that an expression such as $\sqrt{2 \times 8}$ will not be calculated incorrectly as $\sqrt{2} \times 8$. Compare the two entries and their results in the screen on the right.

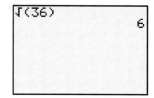

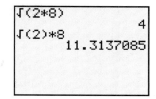

Viewing Windows   The viewing window for a graphing calculator is similar to the viewfinder in a camera. A camera usually cannot take a photograph of an entire view of a scene. The camera must be centered on some object and can capture only a portion of the available scenery. A camera with a zoom lens can photograph different views of the same scene by zooming in and out. Graphing calculators have similar capabilities. The *xy*-coordinate plane is infinite. The calculator screen can only show a finite, rectangular region in the plane, and it must be specified before the graph can be drawn. This is done by setting both minimum and maximum values for the *x*- and *y*-axes. The scale (distance between tick marks) is usually specified as well. Determining an appropriate viewing window for a graph is often a challenge, and many times it will take a few attempts before a satisfactory window is found.

The screen on the left shows a standard viewing window, and the graph of Y = 2X + 1 is shown on the right. Using a different window would give a different view of the line.

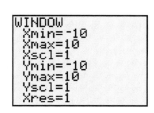

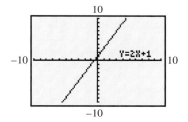

Locating Points on a Graph: Tracing and Tables   Graphing calculators allow you to trace along the graph of an equation and display the coordinates of points on the graph. See the screen on the left below, which indicates that the point (2, 5) lies on the graph of Y = 2X + 1. Tables for equations can also be displayed. The screen on the right shows a partial table for this same equation. Note the middle of the screen, which indicates that when X = 2, Y = 5.

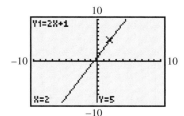

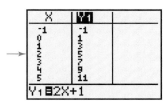

Additional Features　There are many features of graphing calculators that go far beyond the scope of this book. These calculators can be programmed, much like computers. Many of them can solve equations at the stroke of a key, analyze statistical data, and perform symbolic algebraic manipulations. Calculators also provide the opportunity to ask "What if . . . ?" more easily. Values in algebraic expressions can be altered and conjectures tested quickly.

Final Comments　Despite the power of today's calculators, they cannot replace human thought. ***In the entire problem-solving process, your brain is the most important component.*** Calculators are only tools and, like any tool, they must be used appropriately in order to enhance our ability to understand mathematics. Mathematical insight may often be the quickest and easiest way to solve a problem; a calculator may neither be needed nor appropriate. By applying mathematical concepts, you can make the decision whether or not to use a calculator.

Appendix B
Properties of Matrices

OBJECTIVES

1 Know the basic definitions for matrices.

2 Add and subtract matrices.

3 Multiply a matrix by a scalar.

4 Multiply matrices.

5 Use matrices in applications.

The use of matrices to solve systems of linear equations was shown in Section 4.4. Here we discuss algebraic properties of matrices.

OBJECTIVE 1 Know the basic definitions for matrices. We use capital letters to name matrices. Subscript notation is often used to name the elements of a matrix, as in the following matrix A.

$$A = \begin{bmatrix} a_{11} & a_{12} & a_{13} & \cdots & a_{1n} \\ a_{21} & a_{22} & a_{23} & \cdots & a_{2n} \\ a_{31} & a_{32} & a_{33} & \cdots & a_{3n} \\ \vdots & \vdots & \vdots & & \vdots \\ a_{m1} & a_{m2} & a_{m3} & \cdots & a_{mn} \end{bmatrix}$$

With this notation, the first-row, first-column element is a_{11} (read "a-sub-one-one"); the second-row, third-column element is a_{23}; and in general, the ith-row, jth-column element is a_{ij}.

Certain matrices have special names. An $n \times n$ matrix is a **square matrix of order n.** Also, a matrix with just one row is a **row matrix,** and a matrix with just one column is a **column matrix.**

Two matrices are equal if they are the same size and if corresponding elements, position by position, are equal. Using this definition, the matrices

$$\begin{bmatrix} 2 & 1 \\ 3 & -5 \end{bmatrix} \quad \text{and} \quad \begin{bmatrix} 1 & 2 \\ -5 & 3 \end{bmatrix}$$

are *not* equal (even though they contain the same elements and are the same size), since the corresponding elements differ.

■ EXAMPLE 1 Deciding Whether Two Matrices Are Equal

Find the values of the variables for which each statement is true.

(a) $\begin{bmatrix} 2 & 1 \\ p & q \end{bmatrix} = \begin{bmatrix} x & y \\ -1 & 0 \end{bmatrix}$

The only way this statement can be true is if $2 = x$, $1 = y$, $p = -1$, and $q = 0$.

(b) $\begin{bmatrix} x \\ y \end{bmatrix} = \begin{bmatrix} 1 \\ 4 \\ 0 \end{bmatrix}$

This statement can never be true, since the two matrices are different sizes. (One is 2×1 and the other is 3×1.)

Now Try Exercise 1.

OBJECTIVE 2 Add and subtract matrices. Addition of matrices is defined as follows.

Addition of Matrices

To add two matrices of the same size, add corresponding elements. Only matrices of the same size can be added.

In Exercises 51 and 52 you are asked to show that matrix addition satisfies the commutative and associative properties of addition.

EXAMPLE 2 Adding Matrices

Find each sum.

(a) $\begin{bmatrix} 5 & -6 \\ 8 & 9 \end{bmatrix} + \begin{bmatrix} -4 & 6 \\ 8 & -3 \end{bmatrix} = \begin{bmatrix} 5 + (-4) & -6 + 6 \\ 8 + 8 & 9 + (-3) \end{bmatrix} = \begin{bmatrix} 1 & 0 \\ 16 & 6 \end{bmatrix}$

(b) $\begin{bmatrix} 2 \\ 5 \\ 8 \end{bmatrix} + \begin{bmatrix} -6 \\ 3 \\ 12 \end{bmatrix} = \begin{bmatrix} -4 \\ 8 \\ 20 \end{bmatrix}$

(c) Because matrices $A = \begin{bmatrix} 5 & 8 \\ 6 & 2 \end{bmatrix}$ and $B = \begin{bmatrix} 3 & 9 & 1 \\ 4 & 2 & 5 \end{bmatrix}$ are different sizes, the sum $A + B$ cannot be found.

Now Try Exercises 15 and 21.

A matrix containing only zero elements is called a **zero matrix.** For example, $O = [0 \quad 0 \quad 0]$ is the 1×3 zero matrix, while

$$O = \begin{bmatrix} 0 & 0 & 0 \\ 0 & 0 & 0 \end{bmatrix}$$

is the 2×3 zero matrix.

By the additive inverse property in Chapter 1, each real number has an additive inverse: if a is a real number, there is a real number $-a$ such that

$$a + (-a) = 0 \quad \text{and} \quad -a + a = 0.$$

What about matrices? Given the matrix

$$A = \begin{bmatrix} -5 & 2 & -1 \\ 3 & 4 & -6 \end{bmatrix},$$

is there a matrix $-A$ such that

$$A + (-A) = O$$

where O is the 2×3 zero matrix? The answer is yes: the matrix $-A$ has as elements the additive inverses of the elements of A. (Remember, each element of A is a real number and therefore has an additive inverse.)

$$-A = \begin{bmatrix} 5 & -2 & 1 \\ -3 & -4 & 6 \end{bmatrix}$$

To check, test that $A + (-A)$ equals the zero matrix, O.

$$A + (-A) = \begin{bmatrix} -5 & 2 & -1 \\ 3 & 4 & -6 \end{bmatrix} + \begin{bmatrix} 5 & -2 & 1 \\ -3 & -4 & 6 \end{bmatrix} = \begin{bmatrix} 0 & 0 & 0 \\ 0 & 0 & 0 \end{bmatrix} = O$$

Matrix $-A$ is called the **additive inverse,** or **negative,** of matrix A. Every matrix has an additive inverse.

The real number b is subtracted from the real number a, written $a - b$, by adding a and the additive inverse of b. That is,

$$a - b = a + (-b).$$

The same definition works for subtraction of matrices.

Subtraction of Matrices

If A and B are two matrices of the same size, then
$$A - B = A + (-B).$$

In practice, the difference of two matrices of the same size is found by subtracting corresponding elements.

EXAMPLE 3   Subtracting Matrices

Find each difference.

(a) $\begin{bmatrix} -5 & 6 \\ 2 & 4 \end{bmatrix} - \begin{bmatrix} -3 & 2 \\ 5 & -8 \end{bmatrix} = \begin{bmatrix} -5 - (-3) & 6 - 2 \\ 2 - 5 & 4 - (-8) \end{bmatrix} = \begin{bmatrix} -2 & 4 \\ -3 & 12 \end{bmatrix}$

(b) $\begin{bmatrix} 8 & 6 & -4 \end{bmatrix} - \begin{bmatrix} 3 & 5 & -8 \end{bmatrix} = \begin{bmatrix} 5 & 1 & 4 \end{bmatrix}$

(c) The matrices
$$\begin{bmatrix} -2 & 5 \\ 0 & 1 \end{bmatrix} \quad \text{and} \quad \begin{bmatrix} 3 \\ 5 \end{bmatrix}$$
are different sizes and cannot be subtracted.

Now Try Exercise 17.

OBJECTIVE 3   Multiply a matrix by a scalar.   In work with matrices, a real number is called a **scalar** to distinguish it from a matrix. The product of a scalar k and a matrix X is the matrix kX, each of whose elements is k times the corresponding element of X.

EXAMPLE 4   Multiplying a Matrix by a Scalar

Find each product.

(a) $5 \begin{bmatrix} 2 & -3 \\ 0 & 4 \end{bmatrix} = \begin{bmatrix} 10 & -15 \\ 0 & 20 \end{bmatrix}$

(b) $\dfrac{3}{4} \begin{bmatrix} 20 & 36 \\ 12 & -16 \end{bmatrix} = \begin{bmatrix} 15 & 27 \\ 9 & -12 \end{bmatrix}$

Now Try Exercise 23.

The proofs of the following properties of scalar multiplication are left for Exercises 55–58.

Properties of Scalar Multiplication

If A and B are matrices of the same size and c and d are real numbers, then

$$(c + d)A = cA + dA \qquad c(A)d = cd(A)$$
$$c(A + B) = cA + cB \qquad (cd)A = c(dA).$$

OBJECTIVE 4 Multiply matrices. Multiplying two matrices is a little more complicated than scalar multiplication, but it is useful in applications. To find the product of

$$A = \begin{bmatrix} -3 & 4 & 2 \\ 5 & 0 & 4 \end{bmatrix} \qquad \text{and} \qquad B = \begin{bmatrix} -6 & 4 \\ 2 & 3 \\ 3 & -2 \end{bmatrix},$$

first locate *row* 1 of A and *column* 1 of B, shown shaded below.

$$A = \begin{bmatrix} -3 & 4 & 2 \\ 5 & 0 & 4 \end{bmatrix} \qquad B = \begin{bmatrix} -6 & 4 \\ 2 & 3 \\ 3 & -2 \end{bmatrix}$$

Multiply corresponding elements, and find the sum of the products.

$$-3(-6) + 4(2) + 2(3) = 32$$

This result is the element for row 1, column 1 of the product matrix.

Now use *row* 1 of A and *column* 2 of B (shaded below) to determine the element in row 1, column 2 of the product matrix.

$$A = \begin{bmatrix} -3 & 4 & 2 \\ 5 & 0 & 4 \end{bmatrix} \qquad B = \begin{bmatrix} -6 & 4 \\ 2 & 3 \\ 3 & -2 \end{bmatrix}$$

Multiply corresponding elements, and add the products.

$$-3(4) + 4(3) + 2(-2) = -4$$

Next, use *row* 2 of A and *column* 1 of B; this will give the row 2, column 1 entry of the product matrix.

$$\begin{bmatrix} -3 & 4 & 2 \\ 5 & 0 & 4 \end{bmatrix} \begin{bmatrix} -6 & 4 \\ 2 & 3 \\ 3 & -2 \end{bmatrix} \qquad 5(-6) + 0(2) + 4(3) = -18$$

Finally, use *row* 2 of A and *column* 2 of B to find the entry for row 2, column 2 of the product matrix.

$$\begin{bmatrix} -3 & 4 & 2 \\ 5 & 0 & 4 \end{bmatrix} \begin{bmatrix} -6 & 4 \\ 2 & 3 \\ 3 & -2 \end{bmatrix} \qquad 5(4) + 0(3) + 4(-2) = 12$$

The product matrix can now be written.

$$\begin{bmatrix} -3 & 4 & 2 \\ 5 & 0 & 4 \end{bmatrix} \begin{bmatrix} -6 & 4 \\ 2 & 3 \\ 3 & -2 \end{bmatrix} = \begin{bmatrix} 32 & -4 \\ -18 & 12 \end{bmatrix}$$

We see that the product of a 2×3 matrix and a 3×2 matrix is a 2×2 matrix.

By definition, the product AB of an $m \times n$ matrix A and an $n \times p$ matrix B is found as follows: Multiply each element of the first row of A by the corresponding element of the first column of B. The sum of these n products is the first-row, first-column element of AB. Also, the sum of the products found by multiplying the elements of the first row of A times the corresponding elements of the second column of B gives the first-row, second-column element of AB, and so on.

To find the ith-row, jth-column element of AB, multiply each element in the ith row of A by the corresponding element in the jth column of B. (Note the shaded areas in the matrices below.) The sum of these products will give the element of row i, column j of AB.

$$A = \begin{bmatrix} a_{11} & a_{12} & a_{13} & \cdots & a_{1n} \\ a_{21} & a_{22} & a_{23} & \cdots & a_{2n} \\ \vdots & & & & \\ a_{i1} & a_{i2} & a_{i3} & \cdots & a_{in} \\ \vdots & & & & \\ a_{m1} & a_{m2} & a_{m3} & \cdots & a_{mn} \end{bmatrix}$$

$$B = \begin{bmatrix} b_{11} & b_{12} & \cdots & b_{1j} & \cdots & b_{1p} \\ b_{21} & b_{22} & \cdots & b_{2j} & \cdots & b_{2p} \\ \vdots & & & \vdots & & \\ b_{n1} & b_{n2} & \cdots & b_{nj} & \cdots & b_{np} \end{bmatrix}$$

Matrix Multiplication

If the number of columns of matrix A is the same as the number of rows of matrix B, then entry c_{ij} of the product matrix $C = AB$ is found as follows:

$$c_{ij} = a_{i1}b_{1j} + a_{i2}b_{2j} + \cdots + a_{in}b_{nj}.$$

The final product will have as many rows as A and as many columns as B.

■ EXAMPLE 5 Deciding Whether Two Matrices Can Be Multiplied

Suppose matrix A is 3×2, while matrix B is 2×4. Can the product AB be calculated? What is the size of the product? Can the product BA be calculated? What is the size of BA?

The following diagram helps answer the questions about the product AB.

Matrix A 3×2 Matrix B 2×4

Must match

Size of AB

3×4

The product AB exists, since the number of columns of A equals the number of rows of B. (Both are 2.) The product is a 3×4 matrix. Make a similar diagram for BA.

Matrix B Matrix A

2×4 3×2

Different

The product BA does not exist, since B has 4 columns and A has only 3 rows.

Now Try Exercise 29.

EXAMPLE 6 Multiplying Two Matrices

Find AB and BA, if possible, where

$$A = \begin{bmatrix} 1 & -3 \\ 7 & 2 \end{bmatrix} \quad \text{and} \quad B = \begin{bmatrix} 1 & 0 & -1 & 2 \\ 3 & 1 & 4 & -1 \end{bmatrix}.$$

First decide whether AB can be found. Since A is 2×2 and B is 2×4, the product can be found and will be a 2×4 matrix. Now use the definition of matrix multiplication.

$$AB = \begin{bmatrix} 1 & -3 \\ 7 & 2 \end{bmatrix} \begin{bmatrix} 1 & 0 & -1 & 2 \\ 3 & 1 & 4 & -1 \end{bmatrix}$$

$$= \begin{bmatrix} 1(1) + (-3)3 & 1(0) + (-3)1 & 1(-1) + (-3)4 & 1(2) + (-3)(-1) \\ 7(1) + 2(3) & 7(0) + 2(1) & 7(-1) + 2(4) & 7(2) + 2(-1) \end{bmatrix}$$

$$= \begin{bmatrix} -8 & -3 & -13 & 5 \\ 13 & 2 & 1 & 12 \end{bmatrix}$$

Since B is a 2×4 matrix and A is a 2×2 matrix, the number of columns of B (4) does not equal the number of rows of A (2). Therefore, the product BA cannot be found.

Now Try Exercise 33.

EXAMPLE 7 Multiplying Square Matrices in Different Orders

If $A = \begin{bmatrix} 1 & 3 \\ -2 & 5 \end{bmatrix}$ and $B = \begin{bmatrix} -2 & 7 \\ 0 & 2 \end{bmatrix}$, then the definition of matrix multiplication can be used to show that

$$AB = \begin{bmatrix} -2 & 13 \\ 4 & -4 \end{bmatrix} \quad \text{and} \quad BA = \begin{bmatrix} -16 & 29 \\ -4 & 10 \end{bmatrix}.$$

Now Try Exercise 39.

CAUTION Examples 5 and 6 showed that the order in which two matrices are to be multiplied may determine whether their product can be found. Example 7 showed that even when both products AB and BA can be found, they may not be equal. In general, for matrices A and B, $AB \neq BA$, so *matrix multiplication is not commutative.*

Matrix multiplication does, however, satisfy the associative and distributive properties. Proofs of these results for the special cases when A, B, and C are square matrices are left for Exercises 53 and 54.

Properties of Matrix Multiplication

If A, B, and C are matrices such that all the following products and sums exist, then

$$(AB)C = A(BC)$$
$$A(B + C) = AB + AC$$
$$(B + C)A = BA + CA.$$

A graphing calculator with matrix capability will perform matrix multiplication, as well as other matrix operations. The three screens in Figure 1 show matrix multiplication using a calculator. Compare to the product AB in Example 6.

```
[A]
        [[1  -3]
         [7   2]]
```
```
[B]
        [[1  0  -1  2 ]
         [3  1   4  -1]]
```
```
[A][B]
    [[-8  -3  -13  5 ]
     [13   2    1  12]]
```

FIGURE 1

OBJECTIVE 5 Use matrices in applications.

EXAMPLE 8 Applying Matrix Multiplication

A contractor builds three kinds of houses, models A, B, and C, with a choice of two styles, colonial or ranch. Matrix P below shows the number of each kind of house the contractor is planning to build for a new 100-home subdivision. The amounts for each of the main materials used depend on the style of the house. These amounts are shown in matrix Q below, while matrix R gives the cost in dollars for each kind of material. Concrete is measured here in cubic yards, lumber in 1000 board feet, brick in 1000s, and shingles in 100 square feet.

$$\begin{array}{c} \\ \text{Model A} \\ \text{Model B} \\ \text{Model C} \end{array} \overset{\begin{array}{cc}\text{Colonial} & \text{Ranch}\end{array}}{\begin{bmatrix} 0 & 30 \\ 10 & 20 \\ 20 & 20 \end{bmatrix}} = P$$

$$\begin{array}{c} \\ \text{Colonial} \\ \text{Ranch} \end{array} \overset{\begin{array}{cccc}\text{Concrete} & \text{Lumber} & \text{Brick} & \text{Shingles}\end{array}}{\begin{bmatrix} 10 & 2 & 0 & 2 \\ 50 & 1 & 20 & 2 \end{bmatrix}} = Q$$

$$\begin{array}{c} \\ \text{Concrete} \\ \text{Lumber} \\ \text{Brick} \\ \text{Shingles} \end{array} \overset{\begin{array}{c}\text{Cost per}\\\text{unit}\end{array}}{\begin{bmatrix} 20 \\ 180 \\ 60 \\ 25 \end{bmatrix}} = R$$

(a) What is the total cost of materials for all houses of each model?

To find the materials cost of each model, first find matrix PQ, which will show the total amount of each material needed for all houses of each model.

$$PQ = \begin{bmatrix} 0 & 30 \\ 10 & 20 \\ 20 & 20 \end{bmatrix} \begin{bmatrix} 10 & 2 & 0 & 2 \\ 50 & 1 & 20 & 2 \end{bmatrix} = \begin{matrix} \text{Concrete} & \text{Lumber} & \text{Brick} & \text{Shingles} \\ \begin{bmatrix} 1500 & 30 & 600 & 60 \\ 1100 & 40 & 400 & 60 \\ 1200 & 60 & 400 & 80 \end{bmatrix} & \begin{matrix} \text{Model A} \\ \text{Model B} \\ \text{Model C} \end{matrix} \end{matrix}$$

Multiplying PQ and the cost matrix R gives the total cost of materials for each model.

$$(PQ)R = \begin{bmatrix} 1500 & 30 & 600 & 60 \\ 1100 & 40 & 400 & 60 \\ 1200 & 60 & 400 & 80 \end{bmatrix} \begin{bmatrix} 20 \\ 180 \\ 60 \\ 25 \end{bmatrix} = \begin{matrix} \text{Cost} \\ \begin{bmatrix} 72,900 \\ 54,700 \\ 60,800 \end{bmatrix} & \begin{matrix} \text{Model A} \\ \text{Model B} \\ \text{Model C} \end{matrix} \end{matrix}$$

(b) How much of each of the four kinds of material must be ordered?

The totals of the columns of matrix PQ will give a matrix whose elements represent the total amounts of each material needed for the subdivision. Call this matrix T and write it as a row matrix.

$$T = \begin{bmatrix} 3800 & 130 & 1400 & 200 \end{bmatrix}$$

(c) What is the total cost of the materials?

The total cost of all the materials is given by the product of matrix R, the cost matrix, and matrix T, the total amounts matrix. To multiply these and get a 1×1 matrix, representing the total cost, requires multiplying a 1×4 matrix and a 4×1 matrix. This is why in part (b) a row matrix was written rather than a column matrix. The total materials cost is given by TR, so

$$TR = \begin{bmatrix} 3800 & 130 & 1400 & 200 \end{bmatrix} \begin{bmatrix} 20 \\ 180 \\ 60 \\ 25 \end{bmatrix} = \begin{bmatrix} 188,400 \end{bmatrix}.$$

The total cost of the materials is $188,400.

Now Try Exercise 49.

To help keep track of the quantities a matrix represents, let matrix P, from Example 8, represent models/styles, matrix Q represent styles/materials, and matrix R represent materials/cost. In each case the meaning of the rows is written first and that of the columns second. When the product PQ was found in Example 8, the rows of the matrix represented models and the columns represented materials. Therefore, the matrix product PQ represents models/materials. The common quantity, styles, in both P and Q was eliminated in the product PQ. Do you see that the product $(PQ)R$ represents models/cost?

In practical problems this notation helps to identify the order in which two matrices should be multiplied so that the results are meaningful. In Example 8(c), either product RT or product TR could have been found. However, since T represents subdivisions/materials and R represents materials/cost, only TR gave the required matrix representing subdivisions/cost.

APPENDIX B EXERCISES

Find the values of each variable. See Example 1.

1. $\begin{bmatrix} w & x \\ y & z \end{bmatrix} = \begin{bmatrix} 3 & 2 \\ -1 & 4 \end{bmatrix}$

2. $\begin{bmatrix} 0 & 5 & x \\ -1 & 3 & y+2 \\ 4 & 1 & z \end{bmatrix} = \begin{bmatrix} 0 & w+3 & 6 \\ -1 & 3 & 0 \\ 4 & 1 & 8 \end{bmatrix}$

3. $\begin{bmatrix} 2 & 5 & 6 \\ 1 & m & n \end{bmatrix} = \begin{bmatrix} z & y & w \\ 1 & 8 & -2 \end{bmatrix}$

4. $\begin{bmatrix} -7+z & 4r & 8s \\ 6p & 2 & 5 \end{bmatrix} + \begin{bmatrix} -9 & 8r & 3 \\ 2 & 5 & 4 \end{bmatrix} = \begin{bmatrix} 2 & 36 & 27 \\ 20 & 7 & 12a \end{bmatrix}$

5. $\begin{bmatrix} a+2 & 3z+1 & 5m \\ 8k & 0 & 3 \end{bmatrix} + \begin{bmatrix} 3a & 2z & 5m \\ 2k & 5 & 6 \end{bmatrix} = \begin{bmatrix} 10 & -14 & 80 \\ 10 & 5 & 9 \end{bmatrix}$

6. A 3×8 matrix has _____ columns and _____ rows.

Find the size of each matrix. Identify any square, column, or row matrices.

7. $\begin{bmatrix} -4 & 8 \\ 2 & 3 \end{bmatrix}$

8. $\begin{bmatrix} -9 & 6 & 2 \\ 4 & 1 & 8 \end{bmatrix}$

9. $\begin{bmatrix} -6 & 8 & 0 & 0 \\ 4 & 1 & 9 & 2 \\ 3 & -5 & 7 & 1 \end{bmatrix}$

10. $\begin{bmatrix} 8 & -2 & 4 & 6 & 3 \end{bmatrix}$

11. $\begin{bmatrix} 2 \\ 4 \end{bmatrix}$

12. $\begin{bmatrix} -9 \end{bmatrix}$

13. Your friend missed the lecture on adding matrices. In your own words, explain to him how to add two matrices. Give an example.

14. Explain in your own words how to subtract two matrices. Give an example.

Perform each operation in Exercises 15–22, whenever possible. See Examples 2 and 3.

15. $\begin{bmatrix} 6 & -9 & 2 \\ 4 & 1 & 3 \end{bmatrix} + \begin{bmatrix} -8 & 2 & 5 \\ 6 & -3 & 4 \end{bmatrix}$

16. $\begin{bmatrix} 9 & 4 \\ -8 & 2 \end{bmatrix} + \begin{bmatrix} -3 & 2 \\ -4 & 7 \end{bmatrix}$

17. $\begin{bmatrix} -6 & 8 \\ 0 & 0 \end{bmatrix} - \begin{bmatrix} 0 & 0 \\ -4 & -2 \end{bmatrix}$

18. $\begin{bmatrix} 1 & -4 \\ 2 & -3 \\ -8 & 4 \end{bmatrix} - \begin{bmatrix} -6 & 9 \\ -2 & 5 \\ -7 & -12 \end{bmatrix}$

19. $\begin{bmatrix} 3x+y & x-2y & 2x \\ 5x & 3y & x+y \end{bmatrix} + \begin{bmatrix} 2x & 3y & 5x+y \\ 3x+2y & x & 2x \end{bmatrix}$

20. $\begin{bmatrix} 4k-8y \\ 6z-3x \\ 2k+5a \\ -4m+2n \end{bmatrix} - \begin{bmatrix} 5k+6y \\ 2z+5x \\ 4k+6a \\ 4m-2n \end{bmatrix}$

21. $\begin{bmatrix} 3 \\ 2 \end{bmatrix} + \begin{bmatrix} 2 & 3 \end{bmatrix}$

22. $\begin{bmatrix} 0 \\ 0 \end{bmatrix} - \begin{bmatrix} 0 & 0 & 0 \end{bmatrix}$

Let $A = \begin{bmatrix} -2 & 4 \\ 0 & 3 \end{bmatrix}$ and $B = \begin{bmatrix} -6 & 2 \\ 4 & 0 \end{bmatrix}$. *Find each of the following. See Example 4.*

23. $2A$

24. $-3B$

25. $2A - B$

26. $-2A + 4B$

27. $-A + \dfrac{1}{2}B$

28. $\dfrac{3}{4}A - B$

Decide whether each product can be found given

$$A = \begin{bmatrix} 3 & 7 & 1 \\ -2 & 4 & 0 \end{bmatrix} \quad and \quad B = \begin{bmatrix} 1 & 2 \\ 5 & 7 \end{bmatrix}.$$

Give the size of each product if it exists. See Example 5.

29. AB

30. BA

Find each matrix product, whenever possible. See Examples 6 and 7.

31. $\begin{bmatrix} 1 & 2 \\ 3 & 4 \end{bmatrix}\begin{bmatrix} -1 \\ 7 \end{bmatrix}$

32. $\begin{bmatrix} -1 & 5 \\ 7 & 0 \end{bmatrix}\begin{bmatrix} 6 \\ 2 \end{bmatrix}$

33. $\begin{bmatrix} 3 & -4 & 1 \\ 5 & 0 & 2 \end{bmatrix}\begin{bmatrix} -1 \\ 4 \\ 2 \end{bmatrix}$

34. $\begin{bmatrix} -6 & 3 & 5 \\ 2 & 9 & 1 \end{bmatrix}\begin{bmatrix} -2 \\ 0 \\ 3 \end{bmatrix}$

35. $\begin{bmatrix} 5 & 2 \\ -1 & 4 \end{bmatrix}\begin{bmatrix} 3 & -2 \\ 1 & 0 \end{bmatrix}$

36. $\begin{bmatrix} -4 & 0 \\ 1 & 3 \end{bmatrix}\begin{bmatrix} -2 & 4 \\ 0 & 1 \end{bmatrix}$

37. $\begin{bmatrix} 2 & 2 & -1 \\ 3 & 0 & 1 \end{bmatrix}\begin{bmatrix} 0 & 2 \\ -1 & 4 \\ 0 & 2 \end{bmatrix}$

38. $\begin{bmatrix} -9 & 2 & 1 \\ 3 & 0 & 0 \end{bmatrix}\begin{bmatrix} 2 \\ -1 \\ 4 \end{bmatrix}$

39. $\begin{bmatrix} -1 & 2 & 0 \\ 0 & 3 & 2 \\ 0 & 1 & 4 \end{bmatrix}\begin{bmatrix} 2 & -1 & 2 \\ 0 & 2 & 1 \\ 3 & 0 & -1 \end{bmatrix}$

40. $\begin{bmatrix} -2 & -3 & -4 \\ 2 & -1 & 0 \\ 4 & -2 & 3 \end{bmatrix}\begin{bmatrix} 0 & 1 & 4 \\ 1 & 2 & -1 \\ 3 & 2 & -2 \end{bmatrix}$

41. $\begin{bmatrix} -2 & 4 & 1 \end{bmatrix}\begin{bmatrix} 3 & -2 & 4 \\ 2 & 1 & 0 \\ 0 & -1 & 4 \end{bmatrix}$

42. $\begin{bmatrix} 0 & 3 & -4 \end{bmatrix}\begin{bmatrix} -2 & 6 & 3 \\ 0 & 4 & 2 \\ -1 & 1 & 4 \end{bmatrix}$

43. $\begin{bmatrix} -3 & 0 & 2 & 1 \\ 4 & 0 & 2 & 6 \end{bmatrix}\begin{bmatrix} -4 & 2 \\ 0 & 1 \end{bmatrix}$

44. $\begin{bmatrix} -1 & 2 & 4 & 1 \\ 0 & 2 & -3 & 5 \end{bmatrix}\begin{bmatrix} 1 & 2 & 4 \\ -2 & 5 & 1 \end{bmatrix}$

Work each problem. See Example 8.

45. Rite Aid Corporation has been buying small, pharmacist-owned stores at a rapid pace, as indicated by the data in the table. In what 1-yr period did the company's revenue increase most? Write the information in the last three columns of the table as a 5 × 3 matrix.

Rite Aid Grows Quickly

Year	Revenue (millions of dollars)	Net Income (millions of dollars)	Number of Employees
1998	11,375	316	83,000
1997	6970	115	73,000
1996	5446	159	35,700
1995	4534	141	36,700
1994	4059	9	27,364

Source: Hoover's Outline.

46. The table gives the same data for the Walgreen Company as given in Exercise 45 for Rite Aid Corporation.

Walgreen's Growth

Year	Revenue (millions of dollars)	Net Income (millions of dollars)	Number of Employees
1998	15,307	511	90,000
1997	13,363	436	85,000
1996	11,778	372	77,000
1995	10,395	321	68,000
1994	9235	282	61,900

Source: Hoover's Outline.

(a) Write the information in the last three columns of the table as a 5 × 3 matrix.

(b) Use the matrices from part (a) and Exercise 45 to write a matrix giving the total amounts for each year in each category for the two companies. What does the element in row 2, column 3 represent?

(c) Write a matrix representing the difference between the matrix in part (a) and the matrix in Exercise 45. What does the entry in row 1, column 2 represent?

47. A hardware chain does an inventory of a particular size of screw and finds that its Adelphi store has 100 flat-head and 150 round-head screws, its Beltsville store has 125 flat and 50 round, and its College Park store has 175 flat and 200 round. Write this information first as a 3 × 2 matrix and then as a 2 × 3 matrix.

48. At the grocery store, Miguel bought 4 quarts of milk, 2 loaves of bread, 4 potatoes, and an apple. Mary bought 2 quarts of milk, a loaf of bread, 5 potatoes, and 4 apples. Write this information first as a 2 × 4 matrix and then as a 4 × 2 matrix.

49. Yummy Yogurt sells three types of yogurt, nonfat, regular, and super creamy, at three locations. Location I sells 50 gal of nonfat, 100 gal of regular, and 30 gal of super creamy each day. Location II sells 10 gal of nonfat and Location III sells 60 gal of nonfat each day. Daily sales of regular yogurt are 90 gal at Location II and 120 gal at Location III. At Location II, 50 gal of super creamy are sold each day, and 40 gal of super creamy are sold each day at Location III.

(a) Write a 3 × 3 matrix that shows the sales figures for the three locations.

(b) The income per gallon for nonfat, regular, and super creamy is $12, $10, and $15, respectively. Write a 1 × 3 or 3 × 1 matrix displaying the income.

(c) Find a matrix product that gives the daily income at each of the three locations.

(d) What is Yummy Yogurt's total daily income from the three locations?

50. The Bread Box, a neighborhood bakery, sells four main items: sweet rolls, bread, cakes, and pies. The amount of each ingredient (in cups, except for eggs) required for these items is given by matrix A.

$$
\begin{array}{c}
\\
\text{Rolls} \\
\text{(dozen)} \\
\text{Bread} \\
\text{(loaves)} \\
\text{Cakes} \\
\text{Pies} \\
\text{(crust)}
\end{array}
\begin{array}{ccccc}
\text{Eggs} & \text{Flour} & \text{Sugar} & \text{Shortening} & \text{Milk} \\
\left[\begin{array}{ccccc}
1 & 4 & \frac{1}{4} & \frac{1}{4} & 1 \\
0 & 3 & 0 & \frac{1}{4} & 0 \\
4 & 3 & 2 & 1 & 1 \\
0 & 1 & 0 & \frac{1}{3} & 0
\end{array}\right] &&&& = A
\end{array}
$$

The cost (in cents) for each ingredient when purchased in large lots or small lots is given in matrix B.

$$
\begin{array}{c}
\\
\text{Eggs} \\
\text{Flour} \\
\text{Sugar} \\
\text{Shortening} \\
\text{Milk}
\end{array}
\begin{array}{cc}
\multicolumn{2}{c}{\text{Cost}} \\
\text{Large lot} & \text{Small lot} \\
\left[\begin{array}{cc}
5 & 5 \\
8 & 10 \\
10 & 12 \\
12 & 15 \\
5 & 6
\end{array}\right] & = B
\end{array}
$$

(a) Use matrix multiplication to find a matrix giving the comparative cost per item for the two purchase options.

(b) Suppose a day's orders consist of 20 dozen sweet rolls, 200 loaves of bread, 50 cakes, and 60 pies. Write the orders as a 1 × 4 matrix and, using matrix multiplication, write as a matrix the amount of each ingredient needed to fill the day's orders.

(c) Use matrix multiplication to find a matrix giving the costs under the two purchase options to fill the day's orders.

For Exercises 51–58, let

$$
A = \begin{bmatrix} a_{11} & a_{12} \\ a_{21} & a_{22} \end{bmatrix}, \qquad B = \begin{bmatrix} b_{11} & b_{12} \\ b_{21} & b_{22} \end{bmatrix}, \qquad \text{and} \qquad C = \begin{bmatrix} c_{11} & c_{12} \\ c_{21} & c_{22} \end{bmatrix},
$$

where all the elements are real numbers. Use these three matrices to show that each statement is true for 2 × 2 matrices.

51. $A + B = B + A$ (commutative property)

52. $A + (B + C) = (A + B) + C$ (associative property)

53. $(AB)C = A(BC)$ (associative property)

54. $A(B + C) = AB + AC$ (distributive property)

55. $c(A + B) = cA + cB$ for any real number c

56. $(c + d)A = cA + dA$ for any real numbers c and d

57. $c(A)d = cd(A)$

58. $(cd)A = c(dA)$

Appendix C
Matrix Inverses

In Appendix B we saw several parallels between the set of real numbers and the set of matrices. Another similarity is that both sets have identity and inverse elements for multiplication.

OBJECTIVES

1 Understand and write identity matrices.

2 Find multiplicative inverse matrices.

3 Use inverse matrices to solve systems of linear equations.

OBJECTIVE 1 Understand and write identity matrices. By the identity property for real numbers, $a \cdot 1 = a$ and $1 \cdot a = a$ for any real number a. If there is to be a multiplicative **identity matrix** I, such that

$$AI = A \qquad \text{and} \qquad IA = A,$$

for any matrix A, then A and I must be square matrices of the same size. Otherwise it would not be possible to find both products.

2 × 2 Identity Matrix

If I_2 represents the 2 × 2 identity matrix, then

$$I_2 = \begin{bmatrix} 1 & 0 \\ 0 & 1 \end{bmatrix}.$$

To verify that I_2 is the 2 × 2 identity matrix, we must show that $AI = A$ and $IA = A$ for any 2 × 2 matrix. Let

$$A = \begin{bmatrix} x & y \\ z & w \end{bmatrix}.$$

Then

$$AI = \begin{bmatrix} x & y \\ z & w \end{bmatrix}\begin{bmatrix} 1 & 0 \\ 0 & 1 \end{bmatrix} = \begin{bmatrix} x \cdot 1 + y \cdot 0 & x \cdot 0 + y \cdot 1 \\ z \cdot 1 + w \cdot 0 & z \cdot 0 + w \cdot 1 \end{bmatrix} = \begin{bmatrix} x & y \\ z & w \end{bmatrix} = A,$$

and

$$IA = \begin{bmatrix} 1 & 0 \\ 0 & 1 \end{bmatrix}\begin{bmatrix} x & y \\ z & w \end{bmatrix} = \begin{bmatrix} 1 \cdot x + 0 \cdot z & 1 \cdot y + 0 \cdot w \\ 0 \cdot x + 1 \cdot z & 0 \cdot y + 1 \cdot w \end{bmatrix} = \begin{bmatrix} x & y \\ z & w \end{bmatrix} = A.$$

Generalizing from this example, there is an $n \times n$ identity matrix having 1s on the main diagonal and 0s elsewhere.

n × n Identity Matrix

The $n \times n$ identity matrix is given by I_n, where

$$I_n = \begin{bmatrix} 1 & 0 & \cdots & 0 \\ 0 & 1 & \cdots & 0 \\ \vdots & \vdots & a_{ij} & \vdots \\ 0 & 0 & \cdots & 1 \end{bmatrix}.$$

The element $a_{ij} = 1$ when $i = j$ (the diagonal elements) and $a_{ij} = 0$ otherwise.

EXAMPLE 1 Stating and Verifying the 3 × 3 Identity Matrix

Let $A = \begin{bmatrix} -2 & 4 & 0 \\ 3 & 5 & 9 \\ 0 & 8 & -6 \end{bmatrix}$. Give the 3×3 identity matrix I and show that $AI = A$.

The 3×3 identity matrix is

$$I = \begin{bmatrix} 1 & 0 & 0 \\ 0 & 1 & 0 \\ 0 & 0 & 1 \end{bmatrix}.$$

By the definition of matrix multiplication,

$$AI = \begin{bmatrix} -2 & 4 & 0 \\ 3 & 5 & 9 \\ 0 & 8 & -6 \end{bmatrix}\begin{bmatrix} 1 & 0 & 0 \\ 0 & 1 & 0 \\ 0 & 0 & 1 \end{bmatrix} = \begin{bmatrix} -2 & 4 & 0 \\ 3 & 5 & 9 \\ 0 & 8 & -6 \end{bmatrix} = A.$$

The graphing calculator screen in Figure 1(a) shows identity matrices for $n = 2$ and $n = 3$. The screens in Figures 1(b) and 1(c) support the result in Example 1.

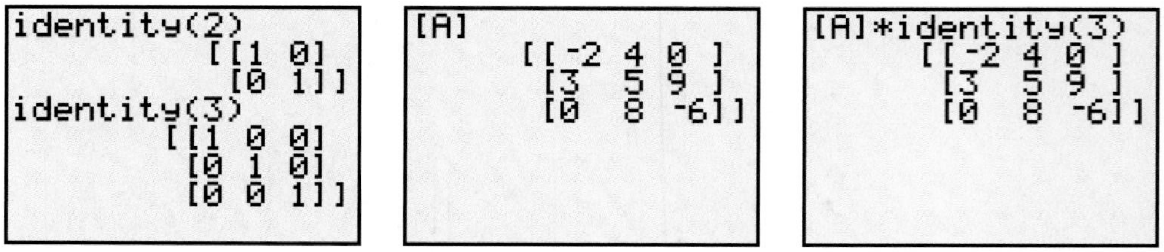

(a)   (b)   (c)

FIGURE 1

OBJECTIVE 2 Find multiplicative inverse matrices. For every nonzero real number a, there is a multiplicative inverse $\frac{1}{a}$ such that

$$a \cdot \frac{1}{a} = 1 \quad \text{and} \quad \frac{1}{a} \cdot a = 1.$$

Recall that $\frac{1}{a}$ is also written a^{-1}. In a similar way, if A is an $n \times n$ matrix, then its **multiplicative inverse,** written A^{-1}, must satisfy both

$$AA^{-1} = I_n \qquad \text{and} \qquad A^{-1}A = I_n.$$

This means that only a square matrix can have a multiplicative inverse.

Now Try Exercises 1 and 5.

CAUTION Although $a^{-1} = \frac{1}{a}$ for any nonzero real number a, if A is a matrix, then

$$A^{-1} \neq \frac{1}{A}.$$

In fact, $\frac{1}{A}$ has no meaning, since 1 is a *number* and A is a *matrix*.

We use the row operations introduced in Section 4.4 to find the matrix A^{-1}. As an example, we find the inverse of

$$A = \begin{bmatrix} 2 & 4 \\ 1 & -1 \end{bmatrix}.$$

Let the unknown inverse matrix be

$$A^{-1} = \begin{bmatrix} x & y \\ z & w \end{bmatrix}.$$

By the definition of matrix inverse, $AA^{-1} = I_2$, or

$$AA^{-1} = \begin{bmatrix} 2 & 4 \\ 1 & -1 \end{bmatrix}\begin{bmatrix} x & y \\ z & w \end{bmatrix} = \begin{bmatrix} 1 & 0 \\ 0 & 1 \end{bmatrix}.$$

By matrix multiplication,

$$\begin{bmatrix} 2x + 4z & 2y + 4w \\ x - z & y - w \end{bmatrix} = \begin{bmatrix} 1 & 0 \\ 0 & 1 \end{bmatrix}.$$

Setting corresponding elements equal gives the system of equations

$$\begin{aligned} 2x + 4z &= 1 & (1) \\ 2y + 4w &= 0 & (2) \\ x - z &= 0 & (3) \\ y - w &= 1. & (4) \end{aligned}$$

Since equations (1) and (3) involve only x and z, while equations (2) and (4) involve only y and w, these four equations lead to two systems of equations,

$$\begin{aligned} 2x + 4z &= 1 \\ x - z &= 0 \end{aligned} \qquad \text{and} \qquad \begin{aligned} 2y + 4w &= 0 \\ y - w &= 1. \end{aligned}$$

Writing the two systems as augmented matrices gives

$$\begin{bmatrix} 2 & 4 & | & 1 \\ 1 & -1 & | & 0 \end{bmatrix} \qquad \text{and} \qquad \begin{bmatrix} 2 & 4 & | & 0 \\ 1 & -1 & | & 1 \end{bmatrix}.$$

Each of these systems can be solved using row operations. However, since the elements to the left of the vertical bar are identical, the two systems can be combined into one matrix,

$$\left[\begin{array}{cc|cc} 2 & 4 & 1 & 0 \\ 1 & -1 & 0 & 1 \end{array}\right],$$

and solved simultaneously using matrix row operations. We need to change the numbers on the left of the vertical bar to the 2×2 identity matrix.

Interchange the two rows to get a 1 in the upper left corner.

$$\left[\begin{array}{cc|cc} 1 & -1 & 0 & 1 \\ 2 & 4 & 1 & 0 \end{array}\right]$$

Multiply the first row by -2, and add the results to the second row to obtain

$$\left[\begin{array}{cc|cc} 1 & -1 & 0 & 1 \\ 0 & 6 & 1 & -2 \end{array}\right]. \qquad -2R_1 + R_2$$

Now, to get a 1 in the second-row, second-column position, multiply the second row by $\frac{1}{6}$.

$$\left[\begin{array}{cc|cc} 1 & -1 & 0 & 1 \\ 0 & 1 & \frac{1}{6} & -\frac{1}{3} \end{array}\right] \qquad \frac{1}{6}R_2$$

Finally, add the second row to the first row to get a 0 in the second column above the 1.

$$\left[\begin{array}{cc|cc} 1 & 0 & \frac{1}{6} & \frac{2}{3} \\ 0 & 1 & \frac{1}{6} & -\frac{1}{3} \end{array}\right] \qquad R_2 + R_1$$

The numbers in the first column to the right of the vertical bar give the values of x and z. The second column gives the values of y and w. That is,

$$\left[\begin{array}{cc|cc} 1 & 0 & x & y \\ 0 & 1 & z & w \end{array}\right] = \left[\begin{array}{cc|cc} 1 & 0 & \frac{1}{6} & \frac{2}{3} \\ 0 & 1 & \frac{1}{6} & -\frac{1}{3} \end{array}\right]$$

so that

$$A^{-1} = \left[\begin{array}{cc} x & y \\ z & w \end{array}\right] = \left[\begin{array}{cc} \frac{1}{6} & \frac{2}{3} \\ \frac{1}{6} & -\frac{1}{3} \end{array}\right].$$

To check, multiply A by A^{-1}. The result should be I_2.

$$AA^{-1} = \left[\begin{array}{cc} 2 & 4 \\ 1 & -1 \end{array}\right]\left[\begin{array}{cc} \frac{1}{6} & \frac{2}{3} \\ \frac{1}{6} & -\frac{1}{3} \end{array}\right] = \left[\begin{array}{cc} \frac{1}{3} + \frac{2}{3} & \frac{4}{3} - \frac{4}{3} \\ \frac{1}{6} - \frac{1}{6} & \frac{2}{3} + \frac{1}{3} \end{array}\right]$$

$$= \left[\begin{array}{cc} 1 & 0 \\ 0 & 1 \end{array}\right] = I_2$$

Finally,

$$A^{-1} = \left[\begin{array}{cc} \frac{1}{6} & \frac{2}{3} \\ \frac{1}{6} & -\frac{1}{3} \end{array}\right].$$

The process for finding the multiplicative inverse A^{-1} for any $n \times n$ matrix A that has an inverse is summarized as follows.

Finding an Inverse Matrix

To obtain A^{-1} for any $n \times n$ matrix A for which A^{-1} exists, follow these steps.

Step 1 Form the augmented matrix $[A \mid I_n]$, where I_n is the $n \times n$ identity matrix.

Step 2 Perform row operations on $[A \mid I_n]$ to obtain a matrix of the form $[I_n \mid B]$.

Step 3 Matrix B is A^{-1}.

NOTE To confirm that two $n \times n$ matrices A and B are inverses of each other, it is sufficient to show that $AB = I_n$. It is not necessary to show also that $BA = I_n$.

EXAMPLE 2 **Finding the Inverse of a 3 × 3 Matrix**

Find A^{-1} if $A = \begin{bmatrix} 1 & 0 & 1 \\ 2 & -2 & -1 \\ 3 & 0 & 0 \end{bmatrix}$.

Use row operations as follows.

Step 1 Write the augmented matrix $[A \mid I_3]$.

$$\begin{bmatrix} 1 & 0 & 1 & | & 1 & 0 & 0 \\ 2 & -2 & -1 & | & 0 & 1 & 0 \\ 3 & 0 & 0 & | & 0 & 0 & 1 \end{bmatrix}$$

Step 2 Since 1 is already in the upper left-hand corner as desired, begin by using a row operation that will result in a 0 for the first element in the second row. Multiply the elements of the first row by -2, and add the results to the second row.

$$\begin{bmatrix} 1 & 0 & 1 & | & 1 & 0 & 0 \\ 0 & -2 & -3 & | & -2 & 1 & 0 \\ 3 & 0 & 0 & | & 0 & 0 & 1 \end{bmatrix} \qquad -2R_1 + R_2$$

To get 0 for the first element in the third row, multiply the elements of the first row by -3 and add to the third row.

$$\begin{bmatrix} 1 & 0 & 1 & | & 1 & 0 & 0 \\ 0 & -2 & -3 & | & -2 & 1 & 0 \\ 0 & 0 & -3 & | & -3 & 0 & 1 \end{bmatrix} \qquad -3R_1 + R_3$$

To get 1 for the second element in the second row, multiply the elements of the second row by $-\frac{1}{2}$.

$$\begin{bmatrix} 1 & 0 & 1 & | & 1 & 0 & 0 \\ 0 & 1 & \frac{3}{2} & | & 1 & -\frac{1}{2} & 0 \\ 0 & 0 & -3 & | & -3 & 0 & 1 \end{bmatrix} \qquad -\frac{1}{2}R_2$$

To get 1 for the third element in the third row, multiply the elements of the third row by $-\frac{1}{3}$.

$$\left[\begin{array}{ccc|ccc} 1 & 0 & 1 & 1 & 0 & 0 \\ 0 & 1 & \frac{3}{2} & 1 & -\frac{1}{2} & 0 \\ 0 & 0 & 1 & 1 & 0 & -\frac{1}{3} \end{array}\right] \quad -\frac{1}{3}R_3$$

To get 0 for the third element in the first row, multiply the elements of the third row by -1 and add to the first row.

$$\left[\begin{array}{ccc|ccc} 1 & 0 & 0 & 0 & 0 & \frac{1}{3} \\ 0 & 1 & \frac{3}{2} & 1 & -\frac{1}{2} & 0 \\ 0 & 0 & 1 & 1 & 0 & -\frac{1}{3} \end{array}\right] \quad -1R_3 + R_1$$

To get 0 for the third element in the second row, multiply the elements of the third row by $-\frac{3}{2}$ and add to the second row.

$$\left[\begin{array}{ccc|ccc} 1 & 0 & 0 & 0 & 0 & \frac{1}{3} \\ 0 & 1 & 0 & -\frac{1}{2} & -\frac{1}{2} & \frac{1}{2} \\ 0 & 0 & 1 & 1 & 0 & -\frac{1}{3} \end{array}\right] \quad -\frac{3}{2}R_3 + R_2$$

Step 3   The last operation shows that the inverse is

$$A^{-1} = \left[\begin{array}{ccc} 0 & 0 & \frac{1}{3} \\ -\frac{1}{2} & -\frac{1}{2} & \frac{1}{2} \\ 1 & 0 & -\frac{1}{3} \end{array}\right].$$

Confirm this by forming the product $A^{-1}A$ or AA^{-1}, each of which should equal the matrix I_3.

Now Try Exercise 15.

As the examples indicate, the most efficient order in which to perform the row operations is to make changes column by column from left to right, so for each column the required 1 is the result of the first change. Next, perform the operations that obtain the 0s in that column. Then proceed to another column.

EXAMPLE 3   Identifying a Matrix with No Inverse

Find A^{-1} given $A = \begin{bmatrix} 2 & -4 \\ 1 & -2 \end{bmatrix}$.

Using row operations to change the first column of the augmented matrix

$$\left[\begin{array}{cc|cc} 2 & -4 & 1 & 0 \\ 1 & -2 & 0 & 1 \end{array}\right]$$

results in the following matrices:

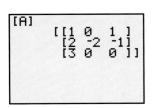

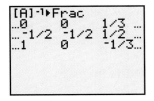

FIGURE 2

$$\begin{bmatrix} 1 & -2 & | & \frac{1}{2} & 0 \\ 1 & -2 & | & 0 & 1 \end{bmatrix} \quad \text{and} \quad \begin{bmatrix} 1 & -2 & | & \frac{1}{2} & 0 \\ 0 & 0 & | & -\frac{1}{2} & 1 \end{bmatrix}.$$

(We multiplied the elements in row one by $\frac{1}{2}$ in the first step.) At this point, the matrix should be changed so that the second-row, second-column element will be 1. Since that element is now 0, there is no way to complete the desired operations, so A^{-1} does not exist for this matrix A. Just as there is no multiplicative inverse for the real number 0, not every matrix has a multiplicative inverse. Matrix A is an example of such a matrix.

Now Try Exercise 13.

If the inverse of a matrix exists, it is unique. That is, any given square matrix has no more than one inverse.

A graphing calculator can find the inverse of a matrix. The screens in Figure 2 support the result of Example 2. The elements of the inverse are expressed as fractions.

OBJECTIVE 3 Use inverse matrices to solve systems of linear equations. Matrix inverses can be used to solve square linear systems of equations. (A **square system** has the same number of equations as variables.) For example, given the linear system

$$a_{11}x + a_{12}y + a_{13}z = b_1$$
$$a_{21}x + a_{22}y + a_{23}z = b_2$$
$$a_{31}x + a_{32}y + a_{33}z = b_3,$$

the definition of matrix multiplication can be used to rewrite the system as

$$\begin{bmatrix} a_{11} & a_{12} & a_{13} \\ a_{21} & a_{22} & a_{23} \\ a_{31} & a_{32} & a_{33} \end{bmatrix} \cdot \begin{bmatrix} x \\ y \\ z \end{bmatrix} = \begin{bmatrix} b_1 \\ b_2 \\ b_3 \end{bmatrix}. \quad (1)$$

(To see this, multiply the matrices on the left.)

$$\text{If } A = \begin{bmatrix} a_{11} & a_{12} & a_{13} \\ a_{21} & a_{22} & a_{23} \\ a_{31} & a_{32} & a_{33} \end{bmatrix}, \quad X = \begin{bmatrix} x \\ y \\ z \end{bmatrix}, \quad \text{and} \quad B = \begin{bmatrix} b_1 \\ b_2 \\ b_3 \end{bmatrix},$$

the system given in (1) becomes

$$AX = B.$$

If A^{-1} exists, then we can multiply both sides of $AX = B$ by A^{-1} on the left to obtain

$$A^{-1}(AX) = A^{-1}B$$
$$(A^{-1}A)X = A^{-1}B \qquad \text{Associative property}$$
$$I_3X = A^{-1}B \qquad \text{Inverse property}$$
$$X = A^{-1}B. \qquad \text{Identity property}$$

Matrix $A^{-1}B$ gives the solution of the system.

> **Solution of the Matrix Equation $AX = B$**
>
> If A is an $n \times n$ matrix with inverse A^{-1}, X is an $n \times 1$ matrix of variables, and B is an $n \times 1$ matrix, then the matrix equation
>
> $$AX = B$$
>
> has the solution
>
> $$X = A^{-1}B.$$

This method of using matrix inverses to solve systems of equations is useful when the inverse is already known or when many systems of the form $AX = B$ must be solved and only B changes.

EXAMPLE 4 Solving Systems of Equations Using Matrix Inverses

Use the inverse of the coefficient matrix to solve each system.

(a) $2x - 3y = 4$
$\ x + 5y = 2$

To represent the system as a matrix equation, use one matrix for the coefficients, one for the variables, and one for the constants, as follows.

$$A = \begin{bmatrix} 2 & -3 \\ 1 & 5 \end{bmatrix}, \qquad X = \begin{bmatrix} x \\ y \end{bmatrix}, \qquad \text{and} \qquad B = \begin{bmatrix} 4 \\ 2 \end{bmatrix}$$

Write the system in matrix form as the equation $AX = B$.

$$AX = \begin{bmatrix} 2 & -3 \\ 1 & 5 \end{bmatrix} \begin{bmatrix} x \\ y \end{bmatrix} = \begin{bmatrix} 2x - 3y \\ x + 5y \end{bmatrix} = \begin{bmatrix} 4 \\ 2 \end{bmatrix} = B$$

To solve the system, first find A^{-1}.

$$A^{-1} = \begin{bmatrix} \frac{5}{13} & \frac{3}{13} \\ -\frac{1}{13} & \frac{2}{13} \end{bmatrix}$$

Next, find the product $A^{-1}B$.

$$A^{-1}B = \begin{bmatrix} \frac{5}{13} & \frac{3}{13} \\ -\frac{1}{13} & \frac{2}{13} \end{bmatrix} \begin{bmatrix} 4 \\ 2 \end{bmatrix} = \begin{bmatrix} 2 \\ 0 \end{bmatrix}$$

Since $X = A^{-1}B$,

$$X = \begin{bmatrix} x \\ y \end{bmatrix} = \begin{bmatrix} 2 \\ 0 \end{bmatrix}.$$

From the final matrix, the solution set of the system is $\{(2, 0)\}$.

(b) $2x - 3y = 1$
$\ x + 5y = 20$

This system has the same matrix of coefficients as the system in part (a). Only matrix B is different. Use A^{-1} from part (a) and multiply by B to obtain

$$X = A^{-1}B = \begin{bmatrix} \frac{5}{13} & \frac{3}{13} \\ -\frac{1}{13} & \frac{2}{13} \end{bmatrix} \begin{bmatrix} 1 \\ 20 \end{bmatrix} = \begin{bmatrix} 5 \\ 3 \end{bmatrix},$$

giving the solution set $\{(5, 3)\}$.

Now Try Exercise 25.

APPENDIX C EXERCISES

Decide whether the given matrices are inverses of each other. (Check to see if their product is the identity matrix I_n.)

1. $\begin{bmatrix} 5 & 7 \\ 2 & 3 \end{bmatrix}$ and $\begin{bmatrix} 3 & -7 \\ -2 & 5 \end{bmatrix}$

2. $\begin{bmatrix} 2 & 3 \\ 1 & 1 \end{bmatrix}$ and $\begin{bmatrix} -1 & 3 \\ 1 & -2 \end{bmatrix}$

3. $\begin{bmatrix} -1 & 2 \\ 3 & -5 \end{bmatrix}$ and $\begin{bmatrix} -5 & -2 \\ -3 & -1 \end{bmatrix}$

4. $\begin{bmatrix} 2 & 1 \\ 3 & 2 \end{bmatrix}$ and $\begin{bmatrix} 2 & 1 \\ -3 & 2 \end{bmatrix}$

5. $\begin{bmatrix} 0 & 1 & 0 \\ 0 & 0 & -2 \\ 1 & -1 & 0 \end{bmatrix}$ and $\begin{bmatrix} 1 & 0 & 1 \\ 1 & 0 & 0 \\ 0 & -1 & 0 \end{bmatrix}$

6. $\begin{bmatrix} 1 & 2 & 0 \\ 0 & 1 & 0 \\ 0 & 1 & 0 \end{bmatrix}$ and $\begin{bmatrix} 1 & -2 & 0 \\ 0 & 1 & 0 \\ 0 & -1 & 1 \end{bmatrix}$

7. $\begin{bmatrix} -1 & -1 & -1 \\ 4 & 5 & 0 \\ 0 & 1 & -3 \end{bmatrix}$ and $\begin{bmatrix} 15 & 4 & -5 \\ -12 & -3 & 4 \\ -4 & -1 & 1 \end{bmatrix}$

8. $\begin{bmatrix} 1 & 3 & 3 \\ 1 & 4 & 3 \\ 1 & 3 & 4 \end{bmatrix}$ and $\begin{bmatrix} 7 & -3 & -3 \\ -1 & 1 & 0 \\ -1 & 0 & 1 \end{bmatrix}$

Find the inverse, if it exists, for each matrix. See Examples 2 and 3.

9. $\begin{bmatrix} -1 & 2 \\ -2 & -1 \end{bmatrix}$

10. $\begin{bmatrix} 1 & -1 \\ 2 & 0 \end{bmatrix}$

11. $\begin{bmatrix} -1 & -2 \\ 3 & 4 \end{bmatrix}$

12. $\begin{bmatrix} 3 & -1 \\ -5 & 2 \end{bmatrix}$

13. $\begin{bmatrix} 5 & 10 \\ -3 & -6 \end{bmatrix}$

14. $\begin{bmatrix} -6 & 4 \\ -3 & 2 \end{bmatrix}$

15. $\begin{bmatrix} 1 & 0 & 1 \\ 0 & -1 & 0 \\ 2 & 1 & 1 \end{bmatrix}$

16. $\begin{bmatrix} 1 & 0 & 0 \\ 0 & -1 & 0 \\ 1 & 0 & 1 \end{bmatrix}$

17. $\begin{bmatrix} 1 & 3 & 3 \\ 1 & 4 & 3 \\ 1 & 3 & 4 \end{bmatrix}$

18. $\begin{bmatrix} -2 & 2 & 4 \\ -3 & 4 & 5 \\ 1 & 0 & 2 \end{bmatrix}$

19. $\begin{bmatrix} 2 & 2 & -4 \\ 2 & 6 & 0 \\ -3 & -3 & 5 \end{bmatrix}$

20. $\begin{bmatrix} 2 & 4 & 6 \\ -1 & -4 & -3 \\ 0 & 1 & -1 \end{bmatrix}$

21. $\begin{bmatrix} 1 & 1 & 0 & 2 \\ 2 & -1 & 1 & -1 \\ 3 & 3 & 2 & -2 \\ 1 & 2 & 1 & 0 \end{bmatrix}$

22. $\begin{bmatrix} 1 & -2 & 3 & 0 \\ 0 & 1 & -1 & 1 \\ -2 & 2 & -2 & 4 \\ 0 & 2 & -3 & 1 \end{bmatrix}$

Solve each system by using the inverse of the coefficient matrix. See Example 4.

23. $-x + y = 1$
$2x - y = 1$

24. $x + y = 5$
$x - y = -1$

25. $2x - y = -8$
$3x + y = -2$

26. $x + 3y = -12$
$2x - y = 11$

27. $2x + 3y = -10$
$3x + 4y = -12$

28. $2x - 3y = 10$
$2x + 2y = 5$

Solve each system of equations by using the inverse of the coefficient matrix. The inverses were found in Exercises 17–20. See Example 4.

29. $x + 3y + 3z = 1$
$x + 4y + 3z = 0$
$x + 3y + 4z = -1$

30. $-2x + 2y + 4z = 3$
$-3x + 4y + 5z = 1$
$x + 2z = 2$

31. $2x + 2y - 4z = 12$
$2x + 6y = 16$
$-3x - 3y + 5z = -20$

32. $2x + 4y + 6z = 4$
$-x - 4y - 3z = 8$
$y - z = -4$

Solve each system of equations by using the inverse of the coefficient matrix. The inverses were found in Exercises 21 and 22.

33. $x + y + 2w = 3$
$2x - y + z - w = 3$
$3x + 3y + 2z - 2w = 5$
$x + 2y + z = 3$

34. $x - 2y + 3z = 1$
$y - z + w = -1$
$-2x + 2y - 2z + 4w = 2$
$2y - 3z + w = -3$

Solve each problem.

35. The amount of plate-glass sales S (in millions of dollars) can be affected by the number of new building contracts B issued (in millions) and automobiles A produced (in millions). A plate-glass company in California wants to forecast future sales by using the past three years of sales. The totals for three years are given in the table.

S	A	B
602.7	5.543	37.14
656.7	6.933	41.30
778.5	7.638	45.62

To describe the relationship between these variables, the equation

$$S = a + bA + cB$$

was used, where the coefficients a, b, and c are constants that must be determined. (*Source:* Makridakis, S., and S. Wheelwright, *Forecasting Methods for Management*, John Wiley & Sons, 1989.)

(a) Substitute the values for S, A, and B for each year from the table into the equation $S = a + bA + cB$, and obtain three linear equations involving a, b, and c.

(b) Use a graphing calculator to solve this linear system for a, b, and c. Use matrix inverse methods.

(c) Write the equation for S using these values for the coefficients.

(d) For the next year it is estimated that $A = 7.752$ and $B = 47.38$. Predict S. (The actual value for S was 877.6.)

(e) It is predicted that in six years $A = 8.9$ and $B = 66.25$. Find the value of S in this situation and discuss its validity.

36. The number of automobile tire sales is dependent on several variables. In one study the relationship between annual tire sales S (in thousands of dollars), automobile registrations R (in millions), and personal disposable income I (in millions of dollars) was investigated. The results for three years are given in the table.

S	R	I
10,170	112.9	307.5
15,305	132.9	621.63
21,289	155.2	1937.13

To describe the relationship between these variables, mathematicians often use the equation

$$S = a + bR + cI,$$

where the coefficients a, b, and c are constants that must be determined before the equation can be used. (*Source:* Jarrett, J., *Business Forecasting Methods,* Basil Blackwell, 1991.)

  (a) Substitute the values for S, R, and I for each year from the table into the equation $S = a + bR + cI$, and obtain three linear equations involving a, b, and c.
  (b) Use a graphing calculator to solve this linear system for a, b, and c. Use matrix inverse methods.
  (c) Write the equation for S using these values for the coefficients.
  (d) If $R = 117.6$ and $I = 310.73$, predict S. (The actual value for S was 11,314.)
  (e) If $R = 143.8$ and $I = 829.06$, predict S. (The actual value for S was 18,481.)

37. Give two ways to use matrices to solve a system of linear equations. Will they both work in all situations? In which situations does each method excel?

38. Discuss the similarities and differences between solving the linear equation $ax = b$ and solving the matrix equation $AX = B$.

Answers to Selected Exercises

In this section we provide the answers that we think most students will obtain when they work the exercises using the methods explained in the text. If your answer does not look exactly like the one given here, it is not necessarily wrong. In many cases there are equivalent forms of the answer that are correct. For example, if the answer section shows $\frac{3}{4}$ and your answer is .75, you have obtained the right answer but written it in a different (yet equivalent) form. Unless the directions specify otherwise, .75 is just as valid an answer as $\frac{3}{4}$.

In general, if your answer does not agree with the one given in the text, see whether it can be transformed into the other form. If it can, then it is the correct answer. If you still have doubts, talk with your instructor. You might also want to obtain a copy of the *Student's Solutions Manual* that goes with this book. Your college bookstore either has this manual or can order it for you.

CHAPTER 1 REVIEW OF THE REAL NUMBER SYSTEM

Section 1.1 (page 11)

Exercises **1.** $\{1, 2, 3, 4, 5\}$ **3.** $\{5, 6, 7, 8, \ldots\}$ **5.** $\{\ldots, -1, 0, 1, 2, 3, 4\}$ **7.** $\{10, 12, 14, 16, \ldots\}$ **9.** $\emptyset$ **11.** $\{-4, 4\}$
In Exercises 13 and 15, we give one possible answer. **13.** $\{x \mid x$ is an even natural number less than or equal to 8$\}$
15. $\{x \mid x$ is a multiple of 4 greater than 0$\}$ **17.** yes **19.** [number line from -4 to 8] **21.** $-\frac{2}{3}$ $\frac{4}{5}$ $\frac{12}{5}$ $\frac{9}{2}$ -4.8 [number line -1 to 5] **23.** The graph of a number is

a point on the number line. The coordinate of a point on the number line is the number that corresponds to the point.

25. (a) $5, 17, \frac{40}{2}$ (or 20) (b) $0, 5, 17, \frac{40}{2}$ (c) $-8, 0, 5, 17, \frac{40}{2}$ (d) $-8, -.6, 0, \frac{3}{4}, 5, \frac{13}{2}, 17, \frac{40}{2}$ (e) $-\sqrt{5}, \sqrt{3}, \pi$ (f) All are real numbers.

27. False; some are whole numbers, but negative integers are not. **29.** False; no irrational number is an integer. **31.** true **33.** true

35. true **37.** (a) -6 (b) 6 **39.** (a) 12 (b) 12 **41.** (a) $-\frac{6}{5}$ (b) $\frac{6}{5}$ **43.** 8 **45.** $\frac{3}{2}$ **47.** -5 **49.** -2 **51.** -4.5 **53.** 5

55. 6 **57.** 0 **59.** (a) Philadelphia; The population declined by 10.6%. (b) Chicago; The population increased by .6%. **61.** Pacific Ocean, Indian Ocean, Caribbean Sea, South China Sea, Gulf of California **63.** true **65.** true **67.** false **69.** true **71.** true
73. $2 < 6$ **75.** $4 > -9$ **77.** $-10 < -5$ **79.** $x > 0$ **81.** $7 > y$ **83.** $5 \geq 5$ **85.** $3t - 4 \leq 10$ **87.** $5x + 3 \neq 0$
89. $-3 < t < 5$ **91.** $-3 \leq 3x < 4$ **93.** $-6 < 10$; true **95.** $10 \geq 10$; true **97.** $-3 \geq -3$; true **99.** $-8 > -6$; false
101. $(-1, \infty)$ [number line] **103.** $(-\infty, 6]$ [number line] **105.** $(0, 3.5)$ [number line]

107. $[2, 7]$ [number line] **109.** $(-4, 3]$ [number line] **111.** $(0, 3]$ [number line] **113.** Ohio (OH), Iowa (IA),

California (CA) **115.** $x < y$ **117.** The *natural numbers* are the numbers with which we count. Some examples are 1, 2, and 3. The *whole numbers* are formed by including 0 with the natural numbers, such as 0, 10, and 100. The *integers* are formed by including the negatives of the natural numbers with the whole numbers, such as -1, -2, and -3. The *rational numbers,* such as $\frac{1}{2}$, .75, and $-\frac{3}{4}$, are formed by quotients of integers. The *irrational numbers* include positive or negative numbers that are not rational, such as π, $\sqrt{2}$, and $-\sqrt{5}$. The *real numbers* include all positive numbers, negative numbers, and 0, such as $-\pi$, 0, and $\sqrt{2}$.

Section 1.2 (page 21)

Exercises **1.** the numbers are additive inverses; $4 + (-4) = 0$ **3.** negative; $-7 + (-21) = -28$ **5.** the positive number has larger absolute value; $15 + (-2) = 13$ **7.** the number with smaller absolute value is subtracted from the one with larger absolute value; $-15 - (-3) = -12$ **9.** negative; $-5(15) = -75$ **11.** -19 **13.** 9 **15.** $-\dfrac{19}{12}$ **17.** -1.85 **19.** -11 **21.** 21 **23.** -13 **25.** -10.18 **27.** $\dfrac{67}{30}$ **29.** 14 **31.** -5 **33.** -6 **35.** -11 **37.** 16 **39.** -4 **41.** 8 **43.** 3.018 **45.** $-\dfrac{7}{4}$ **47.** $-\dfrac{7}{8}$ **49.** 1 **51.** 6 **53.** $\dfrac{13}{2}$ or $6\dfrac{1}{2}$ **55.** Answers will vary. One example is $-4 - (-9) = -4 + 9 = 5$. The sign is determined by choosing the sign of the number with the larger absolute value after subtraction has been changed to addition. **57.** It is true for multiplication (and division). It is false for addition and for subtraction when the number to be subtracted has the smaller absolute value. A more precise statement is, "The product or quotient of two negative numbers is positive." **59.** -35 **61.** 40 **63.** 2 **65.** -12 **67.** $\dfrac{6}{5}$ **69.** 1 **71.** 5.88 **73.** -10.676 **75.** -7 **77.** 6 **79.** -4 **81.** 0 **83.** undefined **85.** $\dfrac{25}{102}$ **87.** $-\dfrac{9}{13}$ **89.** -2.1 **91.** 10,000 **93.** 112°F **95.** 46.68% **97.** 2000: \$129 billion; 2010: \$206 billion; 2020: \$74 billion; 2030: $-\$501$ billion **99.** $-11,478$ **101.** C **103.** Answers will vary. **104.** less than 0 **105.** Answers will vary. **106.** less than 0; $<$

Section 1.3 (page 29)

Exercises **1.** False; $-4^6 = -(4^6)$. **3.** true **5.** true **7.** true **9.** False; the base is 3. **11. (a)** 64 **(b)** -64 **(c)** 64 **(d)** -64 **13.** 10^4 **15.** $\left(\dfrac{3}{4}\right)^5$ **17.** $(-9)^3$ **19.** z^7 **21.** 16 **23.** .021952 **25.** $\dfrac{1}{125}$ **27.** $\dfrac{256}{625}$ **29.** -125 **31.** 256 **33.** -729 **35.** -4096 **37.** 9 **39.** 13 **41.** -20 **43.** $\dfrac{10}{11}$ **45.** $-.7$ **47.** not a real number **49. (a)** B **(b)** C **(c)** A **51.** not a real number **53.** 24 **55.** 15 **57.** 55 **59.** -91 **61.** -8 **63.** -48 **65.** -2 **67.** -79 **69.** -10 **71.** 2 **73.** -2 **75.** undefined **77.** -7 **79.** -1 **81.** 17 **83.** -96 **85.** $-\dfrac{15}{238}$ **87.** \$1572 **89.** \$3296 **91.** .035 **93. (a)** 4.0% **(b)** 26.0% **(c)** 48% **(d)** The percent increased dramatically (more than ten times) from 1980 to 2000.

Section 1.4 (page 37)

Exercises **1.** B **3.** A **5.** product; 0 **7.** grouping **9.** like **11.** $2m + 2p$ **13.** $-12x + 12y$ **15.** $8k$ **17.** $-2r$ **19.** cannot be simplified **21.** $8a$ **23.** $-2d + f$ **25.** $-6y + 3$ **27.** $p + 11$ **29.** $-2k + 15$ **31.** $-3m + 2$ **33.** -1 **35.** $2p + 7$ **37.** $-6z - 39$ **39.** $(5 + 8)x = 13x$ **41.** $(5 \cdot 9)r = 45r$ **43.** $9y + 5x$ **45.** 7 **47.** 0 **49.** $8(-4) + 8x = -32 + 8x$ **51.** 0 **53.** Answers will vary. One example is washing your face and brushing your teeth; one example is putting on your socks and putting on your shoes. **55.** 1900 **57.** 75 **59.** 431 **61.** associative property **62.** associative property **63.** commutative property **64.** associative property **65.** distributive property **66.** arithmetic facts **67.** Answers will vary. **69.** No. One example is $7 + (5 \cdot 3) = (7 + 5)(7 + 3)$, which is false.

Chapter 1 Review Exercises (page 44)

1.

$\frac{9}{4}$
$-4 \quad -2 \quad 0 \quad 2 \quad 4$

3. 16 **5.** 5 **7.** $-9, -\sqrt{4}$ (or -2), $0, \frac{12}{3}$ (or 4) **9.** All are real numbers except $\sqrt{-9}$. **11.** $\{0, 1, 2, 3\}$

13. false **15.** Hyundai; 50% **17.** false **19.** $(-\infty, -5)$

$\xrightarrow{\hspace{2cm})\hspace{2cm}}$
-5

21. $\frac{41}{24}$ **23.** -3 **25.** -39 **27.** $\frac{23}{20}$

29. 11,331 ft **31.** $\frac{2}{3}$ **33.** 3.21 **35.** 10,000 **37.** -125 **39.** 20 **41.** $-.9$ **43.** -4 **45.** -2 **47.** -30 **49.** (a) 24

(b) Answers will vary. **51.** $-4z$ **53.** $4p$ **55.** $6r + 18$ **57.** $-p - 3q$ **59.** 0 **61.** $(2 + 3)x = 5x$ **63.** $(2 \cdot 4)x = 8x$

65. 0 **67.** 7 **69.** $2.32 million (in the red) **71.** $25.59 million (in the black) **73.** 25 **75.** 9 **77.** -5 **79.** -6.16 **81.** 2

83. not a real number **85.** -11.408 **87.** $-6x + 4$ **89.** -116

Chapter 1 Test (page 47)

[1.1] **1.**

$.75 \quad \frac{5}{3} \quad 6.3$
$-2 \quad 0 \quad 2 \quad 4 \quad 6$

2. $0, 3, \sqrt{25}$ (or 5), $\frac{24}{2}$ (or 12) **3.** $-1, 0, 3, \sqrt{25}$ (or 5), $\frac{24}{2}$ (or 12) **4.** $-1, -.5, 0, 3, \sqrt{25}$ (or 5), 7.5,

$\frac{24}{2}$ (or 12) **5.** All are real numbers except $\sqrt{-4}$. **6.** $(-\infty, -3)$

$\xrightarrow{\hspace{2cm})\hspace{2cm}}$
-3

7. $(-4, 2]$

$\xrightarrow{\hspace{1cm}(\hspace{1.5cm}]\hspace{1cm}}$
$-4 \qquad 2$

[1.2] **8.** 0

[1.3] **9.** -26 **10.** 19 **11.** 1 **12.** $\frac{16}{7}$ **13.** $\frac{11}{23}$ [1.2] **14.** 50,395 ft **15.** 37,486 ft **16.** 1345 ft [1.3] **17.** 14 **18.** -15

19. not a real number **20.** (a) a must be positive. (b) a must be negative. (c) a must be 0. **21.** $-\frac{6}{23}$ [1.4] **22.** $10k - 10$

23. It changes the sign of each term. The simplified form is $7r + 2$. **24.** B **25.** D **26.** A **27.** F **28.** C **29.** C **30.** E

CHAPTER 2 LINEAR EQUATIONS, INEQUALITIES, AND APPLICATIONS

Section 2.1 (page 55)

Exercises **1.** A and C **3.** Both sides are evaluated as 30, so 6 is a solution. **5.** Any number is a solution. For example, if the last name is Lincoln, $x = 7$. Both sides are evaluated as -48. **7.** (a) equation (b) expression (c) equation (d) expression **9.** $\{-1\}$ **11.** $\{-7\}$

13. $\{0\}$ **15.** $\left\{-\frac{5}{3}\right\}$ **17.** $\left\{-\frac{1}{2}\right\}$ **19.** $\{2\}$ **21.** $\{-2\}$ **23.** $\{7\}$ **25.** $\{2\}$ **27.** $\left\{\frac{3}{2}\right\}$ **29.** $\{-5\}$ **31.** 12 **33.** Yes, you will

get the correct solution. The coefficients will be larger, but in the end the solution will be the same. **35.** $\left\{-\frac{18}{5}\right\}$ **37.** $\left\{-\frac{5}{6}\right\}$ **39.** $\{6\}$

41. $\{3\}$ **43.** $\{0\}$ **45.** $\{2000\}$ **47.** $\{25\}$ **49.** $\{40\}$ **51.** Zero is a solution. He should subtract $7x$ from each side, getting $x = 0$.

53. contradiction; $\emptyset$ **55.** conditional; $\{-8\}$ **57.** conditional; $\{0\}$ **59.** identity; $\{$all real numbers$\}$ **61.** equivalent **63.** not equivalent;

The solution sets are not the same. They are $\{-3\}$ and $\emptyset$, respectively. **65.** not equivalent; The solution sets $\{4\}$ and $\{-4, 4\}$ differ.

Section 2.2 (page 62)

Exercises **1.** (a) $3x = 5x + 8$ (b) $ct = bt + k$ **2.** (a) $3x - 5x = 8$ (b) $ct - bt = k$ **3.** (a) $-2x = 8$; distributive property

(b) $t(c - b) = k$; distributive property **4.** (a) $x = -4$ (b) $t = \dfrac{k}{c - b}$ **5.** $c \neq b$; If $c = b$, the denominator is 0.

6. To solve an equation for a particular variable, such as solving the second equation for t, go through the same steps as you would in solving for x in the first equation. Treat all other variables as constants. **7.** $r = \dfrac{I}{pt}$ **9.** $L = \dfrac{P - 2W}{2}$ or $L = \dfrac{P}{2} - W$ **11.** $W = \dfrac{V}{LH}$ **13.** $r = \dfrac{C}{2\pi}$ **15.** $B = \dfrac{2A}{h} - b$ or $B = \dfrac{2A - bh}{h}$ **17.** $C = \dfrac{5}{9}(F - 32)$ **19.** D **21.** $r = \dfrac{-2k - 3y}{a - 1}$ or $r = \dfrac{2k + 3y}{1 - a}$ **23.** $y = \dfrac{-x}{w - 3}$ or $y = \dfrac{x}{3 - w}$ **25.** 4.398 hr **27.** 52 mph **29.** 104°F **31.** 230 m **33.** radius: 240 in.; diameter: 480 in. **35.** 8 ft **37.** 75% water; 25% alcohol **39.** 3% **41.** $10.51 **43.** $45.66 **45.** .600; .542; .490 **47.** 1500 **49.** 14,350 **51.** $52,846

Section 2.3 (page 76)

Connections (page 75) Our steps 1, 3, 4, and 6 correspond to Polya's steps.

Exercises 1. (a) $x + 12$ (b) $12 > x$ **3.** (a) $x - 4$ (b) $4 < x$ **5.** D **7.** $2x - 13$ **9.** $12 + 3x$ **11.** $8(x - 12)$ **13.** $\dfrac{3x}{7}$ **15.** $x + 6 = -31$; -37 **17.** $x - (-4x) = x + 9$; $\dfrac{9}{4}$ **19.** $12 - \dfrac{2}{3}x = 10$; 3 **21.** expression **23.** equation **25.** expression **27.** *Step 1:* Read the problem. *Step 2:* Assign a variable. *Step 3:* Write an equation. *Step 4:* Solve the equation. *Step 5:* State the answer. *Step 6:* Check the answer. **29.** width: 165 ft; length: 265 ft **31.** 850 mi; 925 mi; 1300 mi **33.** Wal-Mart: $220 million; Exxon Mobil: $192 million **35.** Clinton: 379 votes; Dole: 159 votes **37.** 1.9% **39.** $3765 **41.** $225 **43.** $4000 at 3%; $8000 at 4% **45.** $10,000 at 4.5%; $19,000 at 3% **47.** $58,000 **49.** 5 L **51.** 4 L **53.** 1 gal **55.** 150 lb **57.** We cannot expect the final mixture to be worth more than either of the ingredients. **59.** (a) $800 - x$ (b) $800 - y$ **60.** (a) $.05x$; $.10(800 - x)$ (b) $.05y$; $.10(800 - y)$ **61.** (a) $.05x + .10(800 - x) = 800(.0875)$ (b) $.05y + .10(800 - y) = 800(.0875)$ **62.** (a) $200 at 5%; $600 at 10% (b) 200 L of 5% acid; 600 L of 10% acid **63.** The processes are the same. The amounts of money in Problem A correspond to the amounts of solution in Problem B.

Section 2.4 (page 85)

Exercises 1. $4.50 **3.** 60 mph **5.** The problem asks for the *distance* to the workplace. To find the distance, we must multiply the rate, 10 mph, by the time, $\dfrac{3}{4}$ hr. **7.** No, the answers must be whole numbers because they represent the number of coins. **9.** 17 pennies; 17 dimes; 10 quarters **11.** 26 quarters; 21 half-dollars **13.** 28 $10 coins; 13 $20 coins **15.** floor tickets: 450; balcony tickets: 100 **17.** 7.91 m per sec **19.** 8.42 m per sec **21.** $2\dfrac{1}{2}$ hr **23.** 7:50 P.M. **25.** 15 mph **27.** $\dfrac{1}{2}$ hr **29.** 60°, 60°, 60° **31.** 40°, 45°, 95° **33.** 40°, 80° **34.** 120° **35.** The sum is equal to the measure of the angle found in Exercise 34. **36.** The sum of the measures of angles ① and ② is equal to the measure of angle ③. **37.** Both measure 122°. **39.** 64°, 26° **41.** 19, 20, 21 **43.** 61 yr old

Summary Exercises on Solving Applied Problems (page 90)

1. length: 8 in.; width: 5 in. **2.** length: 60 m; width: 30 m **3.** $425 **4.** $8.95 **5.** $800 at 4%; $1600 at 5% **6.** $12,000 at 3%; $14,000 at 4% **7.** Roosevelt: 449; Willkie: 82 **8.** Eisner: $40.1 million; Horrigan: $21.7 million **9.** 5 hr **10.** Tanui: 2.14 hr; Pippig: 2.44 hr **11.** $13\dfrac{1}{3}$ L **12.** $53\dfrac{1}{3}$ kg **13.** fives: 84; tens: 42 **14.** Great Britain: 29 million; Germany: 45 million **15.** 20°, 30°, 130° **16.** 107°, 73° **17.** 31, 32, 33 **18.** 9 and 11 **19.** 6 in., 12 in., 16 in. **20.** 23 in.

Section 2.5 (page 100)

Exercises **1.** D **3.** B **5.** F **7.** Use a parenthesis when an endpoint is not included; use a bracket when it is included.

9. $[16, \infty)$

11. $(7, \infty)$

13. $(-\infty, 4)$

15. $(-\infty, -40]$

17. $(-\infty, 4]$

19. $\left(-\infty, -\dfrac{15}{2}\right)$

21. $\left[\dfrac{1}{2}, \infty\right)$

23. $(3, \infty)$

25. $(-\infty, 4)$

27. $\left(-\infty, \dfrac{23}{6}\right]$

29. $\left(-\infty, \dfrac{76}{11}\right)$

31. $(-\infty, \infty)$

33. $\emptyset$ **35.** $\{-9\}$

36. $(-9, \infty)$

37. $(-\infty, -9)$

38. the set of all real numbers

39. $(-\infty, -3)$ **41.** $(1, 11)$

43. $[-14, 10]$

45. $[-5, 6]$

47. $\left[-\dfrac{14}{3}, 2\right]$

49. $\left[-\dfrac{1}{2}, \dfrac{35}{2}\right]$

51. $\left(-\dfrac{1}{3}, \dfrac{1}{9}\right)$

53. all numbers between -2 and 2, that is, $(-2, 2)$ **55.** all numbers greater than or equal to 3, that is, $[3, \infty)$ **57.** all numbers greater than or equal to -9, that is, $[-9, \infty)$ **59.** from about 2:30 P.M. to 6:00 P.M. **61.** about $84°F-91°F$ **63.** at least 80 **65.** 50 mi **67.** 26 tapes **69.** There is no such number x, since $4 \not< 1$. **71. (a)** 140 to 184 lb **(b)** Answers will vary.

Section 2.6 (page 109)

Exercises **1.** true **3.** false; The union is $(-\infty, 8) \cup (8, \infty)$. **5.** false; The intersection is $\emptyset$. **7.** $\{1, 3, 5\}$ or B **9.** $\{4\}$ or D **11.** $\emptyset$ **13.** $\{1, 2, 3, 4, 5, 6\}$ or A **15.** Answers will vary. One example is: The intersection of two streets is the region common to *both* streets.

17.

19.

21. $(-3, 2)$

23. $(-\infty, 2]$

25. $\emptyset$ **27.** $[5, 9]$

29. $(-3, -1)$

31. $(-\infty, 4]$

33.

35.

37. $(-\infty, 8]$

39. $[-2, \infty)$

41. $(-\infty, \infty)$

43. $(-\infty, -5) \cup (5, \infty)$

45. $(-\infty, -1) \cup (2, \infty)$

47. $[-4, -1]$ **49.** $[-9, -6]$ **51.** $(-\infty, 3)$

53. $[3, 9)$ **55.** intersection; $(-5, -1)$

57. union; $(-\infty, 4)$

59. union; $(-\infty, 0] \cup [2, \infty)$

61. intersection; $[4, 12]$

63. Maria, Joe **64.** none of them

65. none of them **66.** Luigi, Than **67.** Maria, Joe **68.** all of them **69.** none

Section 2.7 (page 118)

Connections **(page 118)** The filled carton may contain between 30.4 and 33.6 oz, inclusive.

Exercises **1.** E; C; D; B; A **3.** Use *or* for the equality statement and the $>$ statement. Use *and* for the $<$ statement. **5.** $\{-12, 12\}$

7. $\{-5, 5\}$ **9.** $\{-6, 12\}$ **11.** $\{-5, 6\}$ **13.** $\left\{-3, \dfrac{11}{2}\right\}$ **15.** $\left\{-\dfrac{19}{2}, \dfrac{9}{2}\right\}$ **17.** $\{-10, -2\}$ **19.** $\left\{-\dfrac{32}{3}, 8\right\}$

21. $(-\infty, -3) \cup (3, \infty)$

23. $(-\infty, -4] \cup [4, \infty)$

25. $(-\infty, -25] \cup [15, \infty)$

27. $(-\infty, -12) \cup (8, \infty)$

29. $(-\infty, -2) \cup (8, \infty)$

31. $\left(-\infty, -\dfrac{9}{5}\right] \cup [3, \infty)$

33. (a)   (b)

35. $[-3, 3]$

37. $(-4, 4)$

39. $[-25, 15]$

41. $[-12, 8]$

43. $[-2, 8]$

45. $\left[-\dfrac{9}{5}, 3\right]$

47. $(-\infty, -5) \cup (13, \infty)$

49. $(-\infty, -25) \cup (15, \infty)$

51. $\{-6, -1\}$

53. $\left[-\dfrac{10}{3}, 4\right]$

55. $\left[-\dfrac{7}{6}, -\dfrac{5}{6}\right]$

57. $(-\infty, -3] \cup [4, \infty)$

59. $\{-5, 5\}$ **61.** $\{1, -5\}$ **63.** $\{-5, -3\}$ **65.** $(-\infty, -3) \cup (2, \infty)$ **67.** $[-10, 0]$

69. $(-\infty, -1) \cup (5, \infty)$ **71.** $\{-1, 3\}$ **73.** $\left\{-3, \dfrac{5}{3}\right\}$ **75.** $\left\{-\dfrac{1}{3}, -\dfrac{1}{15}\right\}$ **77.** $\left\{-\dfrac{5}{4}\right\}$ **79.** $\emptyset$ **81.** $\left\{-\dfrac{1}{4}\right\}$ **83.** $\emptyset$

85. $(-\infty, \infty)$ **87.** $\left\{-\dfrac{3}{7}\right\}$ **89.** $\left\{\dfrac{2}{5}\right\}$ **91.** $\emptyset$ **93.** $|x - 1000| \le 100; 900 \le x \le 1100$ **95.** 472.9 ft   **96.** 1201 Walnut, Fidelity Bank and Trust Building, City Hall, Kansas City Power and Light, Hyatt Regency   **97.** City Center Square, Commerce Tower, Federal Office Building, 1201 Walnut, Fidelity Bank and Trust Building, City Hall, Kansas City Power and Light, Hyatt Regency **98.** (a) $|x - 472.9| \ge 75$ (b) $x \ge 547.9$ or $x \le 397.9$ (c) AT&T Town Pavilion, One Kansas City Place (d) It makes sense because it includes all buildings *not* listed earlier.

Summary Exercises on Solving Linear and Absolute Value Equations and Inequalities   (page 121)

1. $\{12\}$ **2.** $\{-5, 7\}$ **3.** $\{7\}$ **4.** $\left\{-\dfrac{2}{5}\right\}$ **5.** $\emptyset$ **6.** $(-\infty, -1]$ **7.** $\left[-\dfrac{2}{3}, \infty\right)$ **8.** $\{-1\}$ **9.** $\{-3\}$ **10.** $\left\{1, \dfrac{11}{3}\right\}$

11. $(-\infty, 5]$ **12.** $(-\infty, \infty)$ **13.** $\{2\}$ **14.** $(-\infty, -8] \cup [8, \infty)$ **15.** $\emptyset$ **16.** $(-\infty, \infty)$ **17.** $(-5.5, 5.5)$ **18.** $\left\{\dfrac{13}{3}\right\}$ **19.** $\left\{-\dfrac{96}{5}\right\}$

20. $(-\infty, 32]$ **21.** $(-\infty, -24)$ **22.** $\left\{\dfrac{3}{8}\right\}$ **23.** $\left\{\dfrac{7}{2}\right\}$ **24.** $(-6, 8)$ **25.** $(-\infty, \infty)$ **26.** $(-\infty, 5)$ **27.** $(-\infty, -4) \cup (7, \infty)$

28. $\{24\}$ **29.** $\left\{-\dfrac{1}{5}\right\}$ **30.** $\left(-\infty, -\dfrac{5}{2}\right]$ **31.** $\left[-\dfrac{1}{3}, 3\right]$ **32.** $[1, 7]$ **33.** $\left\{-\dfrac{1}{6}, 2\right\}$ **34.** $\{-3\}$ **35.** $(-\infty, -1] \cup \left[\dfrac{5}{3}, \infty\right)$

36. $\left[\dfrac{3}{4}, \dfrac{15}{8}\right]$ **37.** $\left\{-\dfrac{5}{2}\right\}$ **38.** $\{60\}$ **39.** $\left[-\dfrac{9}{2}, \dfrac{15}{2}\right]$ **40.** $(1, 9)$ **41.** $(-\infty, \infty)$ **42.** $\left\{\dfrac{1}{3}, 9\right\}$ **43.** $(-\infty, \infty)$ **44.** $\left\{-\dfrac{10}{9}\right\}$

45. $\{-2\}$ **46.** $\emptyset$ **47.** $(-\infty, -1) \cup (2, \infty)$ **48.** $[-3, -2]$

Chapter 2 Review Exercises   (page 128)

1. $\left\{-\dfrac{9}{5}\right\}$ **3.** $\left\{-\dfrac{7}{5}\right\}$ **5.** identity; $(-\infty, \infty)$ **7.** conditional; $\{0\}$ **9.** $b = \dfrac{2A - Bh}{h}$ or $b = \dfrac{2A}{h} - B$ **11.** $x = \dfrac{4}{3}(P + 12)$ or $x = \dfrac{4}{3}P + 16$ **13.** 6 ft   **15.** 6.5%   **17.** approximately 17,415,000   **19.** $9 - \dfrac{1}{3}x$ **21.** length: 13 m; width: 8 m   **23.** 12 kg

25. 10 L   **27.** 15 dimes; 8 quarters   **29.** A   **31.** 2.2 hr   **33.** 1 hr   **35.** 40°, 45°, 95°   **37.** $(-9, \infty)$ **39.** $\left(\dfrac{3}{2}, \infty\right)$ **41.** $[3, 5)$

43. 38 m or less   **45.** any score greater than or equal to 61   **47.** $\{a, c\}$ **49.** $\{a, c, e, f, g\}$ **51.** $(6, 9)$

53. $(-\infty, -3] \cup (5, \infty)$ **55.** $\emptyset$ **57.** $(-3, 4)$ **59.** $(4, \infty)$ **61.** $\{-7, 7\}$ **63.** $\left\{-\dfrac{1}{3}, 5\right\}$ **65.** $\{0, 7\}$

67. $\left\{-\dfrac{3}{4}, \dfrac{1}{2}\right\}$ **69.** $(-14, 14)$ **71.** $[-3, -2]$ **73.** $(-2, \infty)$ **75.** $[-2, 3)$ **77.** $(-\infty, \infty)$ **79.** 10 ft   **81.** $\left\{-\dfrac{7}{3}, 1\right\}$

83. $[-16, 10]$ **85.** $\left(-3, \dfrac{7}{2}\right)$ **87.** 80°   **89.** Any amount greater than or equal to $1100   **91.** $(-\infty, \infty)$ **93.** $\left\{-4, -\dfrac{2}{3}\right\}$

95. $[-4, -2]$ **97.** Gore: 266; Bush: 271   **99.**

101. **(a)** $\emptyset$ **(b)** Managerial/Professional, Technical/Sales/Administrative Support, Service, Operators/Fabricators/Laborers

Chapter 2 Test   (page 134)

[2.1] **1.** $\{-19\}$ **2.** $\{5\}$ **3.** $(-\infty, \infty)$ **4.** contradiction; $\emptyset$ [2.2] **5.** $v = \dfrac{S + 16t^2}{t}$ **6.** $r = \dfrac{-2 - 6t}{a - 3}$ or $r = \dfrac{2 + 6t}{3 - a}$

[2.3, 2.4] **7.** 3.372 hr   **8.** 6.25%   **9.** 73.3%   **10.** $8000 at 3%; $20,000 at 5%   **11.** faster car: 60 mph; slower car: 45 mph

12. 40°, 40°, 100°   [2.5] **13.** $[1, \infty)$ **14.** $(-\infty, 28)$ **15.** $[-3, 3]$

16. 82   **17.** $[500, \infty)$ [2.6] **18.** **(a)** $\{1, 5\}$ **(b)** $\{1, 2, 5, 7, 9, 12\}$ **19.** **(a)** $[2, 9)$ **(b)** $(-\infty, 3) \cup [6, \infty)$ [2.7] **20.** $\left\{-1, \dfrac{5}{2}\right\}$

21. $\left(-\infty, -\dfrac{7}{6}\right) \cup \left(\dfrac{17}{6}, \infty\right)$ **22.** $\emptyset$ **23.** $\left\{-\dfrac{5}{7}, \dfrac{11}{3}\right\}$ **24.** $\left(\dfrac{1}{3}, \dfrac{7}{3}\right)$ **25.** **(a)** $\emptyset$ **(b)** $(-\infty, \infty)$ **(c)** $\emptyset$

Cumulative Review Exercises Chapters 1–2   (page 135)

[1.1] **1.** 9, 6   **2.** 0, 9, 6   **3.** $-8, 0, 9, 6$ **4.** $-8, -\dfrac{2}{3}, 0, \dfrac{4}{5}, 9, 6$ **5.** $-\sqrt{6}$ **6.** All are real numbers.   [1.2] **7.** $-\dfrac{22}{21}$ **8.** 8

[1.3] **9.** 8   **10.** 0   **11.** -243 **12.** $\dfrac{216}{343}$ **13.** $-\dfrac{8}{27}$ **14.** -4096 **15.** $\sqrt{-36}$ is not a real number.   **16.** $\dfrac{4 + 4}{4 - 4}$ is undefined.

17. -16 **18.** 184   **19.** $\dfrac{27}{16}$ [1.4] **20.** $-20r + 17$ **21.** $13k + 42$ **22.** commutative property   **23.** distributive property

[2.1] **24.** $\{5\}$ **25.** $\{30\}$ **26.** $\{15\}$ [2.2] **27.** $b = P - a - c$ [2.1] **28.** $\emptyset$ **29.** $\{$all real numbers$\}$

[2.5] **30.** $[-14, \infty)$ **31.** $\left[\dfrac{5}{3}, 3\right)$ [2.6] **32.** $(-\infty, 0) \cup (2, \infty)$

[2.7] **33.** $\left(-\infty, -\dfrac{1}{7}\right] \cup [1, \infty)$ [2.3–2.5] **34.** $5000   **35.** $6\dfrac{1}{3}$ g   **36.** 74 or greater   **37.** 2 L

38. 9 pennies, 12 nickels, 8 quarters   [2.2] **39.** 44 mg   [2.3] **40.** **(a)** 122 **(b)** 7.6%

Section 3.1 (page 145)

Exercises **1. (a)** x represents the year; y represents the percent of women in math or computer science professions. **(b)** 1990–2000 **(c)** (1990, 36) **(d)** In 2000, the percent of women in math or computer science professions was 30%. **3.** origin **5.** $y; x; x; y$ **7.** two **9. (a)** I **(b)** III **(c)** II **(d)** IV **(e)** none **11. (a)** I or III **(b)** II or IV **(c)** II or IV **(d)** I or III **13.–21.**

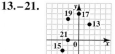

23. (a) $-3; 3; 2; -1$ **(b)** **25. (a)** $\dfrac{5}{2}; 5; \dfrac{3}{2}; 1$ **(b)** **27. (a)** $-4; 5; -\dfrac{12}{5}; \dfrac{5}{4}$ **(b)**

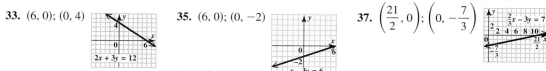

29. In quadrant III, both coordinates of the ordered pairs are negative. If $x + y = k$ and k is positive, then either x or y must be positive because the sum of two negative numbers is negative. **31.** Choose a value *other than* 0 for either x or y. For example, if $x = -5$, then $y = 4$.

33. $(6, 0); (0, 4)$ **35.** $(6, 0); (0, -2)$ **37.** $\left(\dfrac{21}{2}, 0\right); \left(0, -\dfrac{7}{3}\right)$

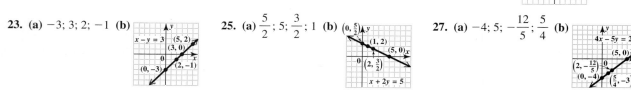

39. none; $(0, 5)$ **41.** $(2, 0)$; none **43.** $(-4, 0)$; none **45.** $(0, 0); (0, 0)$

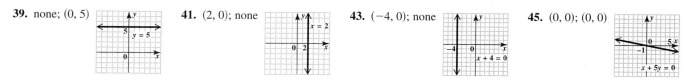

47. $(0, 0); (0, 0)$ **49.** $(0, 0); (0, 0)$ **51.** 154.6 mph **53.** B **55.** It means that when $x = 15$ (year 1995),

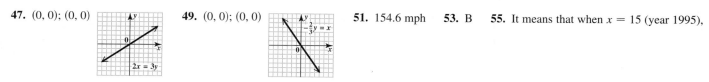

the national average family health care cost was approximately \$7483.25. **57.** $y = -\dfrac{1}{2}x$ or $y = -.5x$

59. **61.** **63.** $(6, -2)$ **64.** $(5, -2)$ **65.** $(6, 0)$ **66.** $(5, 0)$

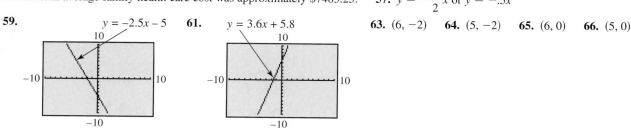

67. 5; 0 **68.** The x-coordinate of M is the average of the x-coordinates of P and Q. The y-coordinate of M is the average of the y-coordinates of P and Q. **69.** $(-5, -1)$ **71.** $\left(\dfrac{9}{2}, -\dfrac{3}{2}\right)$ **73.** $\left(0, \dfrac{11}{2}\right)$ **75.** $(2.1, .9)$

Section 3.2   (page 157)

Exercises   1. A, B, and D   **3. (a)** C **(b)** A **(c)** D **(d)** B   **5.** 0   **7.** $-\dfrac{1}{3}$ **9.** 2   **11.** $\dfrac{5}{2}$ **13.** 0   **15.** B and D are correct. Choice

A is wrong because the order of subtraction must be the same in the numerator and denominator. Choice C is wrong because slope is defined

as the change in y divided by the change in x.   **17.** 1   **19.** $\dfrac{9}{8}$ **21.** 0   **23.** 2   **25.** B   **27.** A   **29.** $-\dfrac{1}{2}$

31. $\dfrac{5}{2}$ 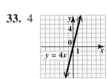   **33.** 4   **35.** undefined   **37.** 0   **39.**

41.   **43.**   **45.**   **47.**   **49.** parallel   **51.** perpendicular   **53.** neither

55. parallel   **57.** neither   **59.** perpendicular   **61.** $\dfrac{7}{10}$ **63.** $-\$4000$ per yr; The value of the machine is decreasing $\$4000$ each year

during these years.   **65.** 0% per yr (or no change); The percent of pay raise is not changing—it is 3% each year during these years.
67. (a) 1000 million per yr; 1000 million per yr; 1000 million per yr **(b)** The average rate of change is the same. When graphed, the data
points lie on a straight line.   **69. (a)** $\$200$ million per yr **(b)** The positive slope means expenditures *increased* an average of
$\$200$ million each year.   **71.** $-\$69$ per yr; The price decreased an average of $\$69$ each year from 1997 to 2002.   **73.** A is y_1 and B is y_2.

75. Since the slopes of both pairs of opposite sides are equal, the figure is a parallelogram.   **77.** $\dfrac{1}{3}$ **78.** $\dfrac{1}{3}$ **79.** $\dfrac{1}{3}$

80. $\dfrac{1}{3} = \dfrac{1}{3} = \dfrac{1}{3}$ is true.   **81.** collinear   **82.** not collinear

Section 3.3   (page 172)

Exercises   1. A   **3.** A   **5.** $3x + y = 10$ **7.** A   **9.** C   **11.** H   **13.** B   **15.** $y = 5x + 15$ **17.** $y = -\dfrac{2}{3}x + \dfrac{4}{5}$

19. $y = \dfrac{2}{5}x + 5$ **21.** $y = \dfrac{2}{3}x + 1$ **23. (a)** $y = x + 4$ **(b)** 1 **(c)** 4 **(d)**   **25. (a)** $y = -\dfrac{6}{5}x + 6$ **(b)** $-\dfrac{6}{5}$ **(c)** 6

(d)   **27. (a)** $y = \dfrac{4}{5}x - 4$ **(b)** $\dfrac{4}{5}$ **(c)** -4 **(d)**   **29. (a)** $y = -\dfrac{1}{2}x - 2$ **(b)** $-\dfrac{1}{2}$ **(c)** -2 **(d)**

31. $3x + 4y = 10$ **33.** $2x + y = 18$ **35.** $x - 2y = -13$ **37.** $4x - y = 12$ **39.** The slope-intercept form, $y = mx + b$, is used
when the slope and y-intercept are known. The point-slope form, $y - y_1 = m(x - x_1)$, is used when the slope and one point on a line or two
points on a line are known. The standard form, $Ax + By = C$, is not useful for writing the equation, but the intercepts can be found quickly

and used to graph the equation. The form $y = b$ is used for a horizontal line through the point (a, b). The form $x = a$ is used for a vertical line through the point (a, b). **41.** $y = 5$ **43.** $x = 9$ **45.** $x = .5$ **47.** $y = 8$ **49.** $2x - y = 2$ **51.** $x + 2y = 8$ **53.** $2x - 13y = -6$ **55.** $y = 5$ **57.** $x = 7$ **59.** $y = -3$ **61.** $y = 3x - 19$ **63.** $y = \frac{1}{2}x - 1$ **65.** $y = -\frac{1}{2}x + 9$ **67.** $y = 7$ **69.** $y = 45x;$ $(0, 0), (5, 225), (10, 450)$ **71.** $y = 1.50x;$ $(0, 0), (5, 7.50), (10, 15.00)$ **73. (a)** $y = 39x + 99$ **(b)** $(5, 294)$; The cost of a 5-month membership is $294. **(c)** $567 **75. (a)** $y = 50x + 25$ **(b)** $(5, 275)$; The cost of the plan for 5 months is $275. **(c)** $1225 **77. (a)** $y = .20x + 50$ **(b)** $(5, 51)$; The charge for driving 5 mi is $51. **(c)** 173 mi **79. (a)** $y = 5.6x + 9$; The percent of households accessing the Internet by broadband is increasing 5.6% per year. **(b)** 43% **81. (a)** $y = -103.2x + 28,908$ **(b)** 28,082; The result using the model is a little high. **83. (a)** $y = -3x + 9$ **(b)** 3 **(c)** $\{3\}$ **85. (a)** $y = 4x + 2$ **(b)** $-.5$ **(c)** $\{-.5\}$ **87.** D **89.** 32; 212 **90.** $(0, 32)$ and $(100, 212)$ **91.** $\frac{9}{5}$ **92.** $F = \frac{9}{5}C + 32$ **93.** $C = \frac{5}{9}(F - 32)$ **94.** When the Celsius temperature is 50°, the Fahrenheit temperature is 122°.

Section 3.4 (page 182)

Connections (page 181) **1.** $x \le 200, x \ge 100, y \ge 3000$ **2.** **3.** $C = 50x + 100y$ **4.** Some examples are $(100, 5000), (150, 3000),$ and $(150, 5000)$. The corner points are $(100, 3000)$ and $(200, 3000)$. **5.** The least cost occurs when $x = 100$ and $y = 3000$. The company should use 100 workers and manufacture 3000 units to achieve the least possible cost.

Exercises **1.** solid; below **3.** dashed; above **5.** The graph of $Ax + By = C$ divides the plane into two regions. In one of the regions, the ordered pairs satisfy $Ax + By < C$; in the other, they satisfy $Ax + By > C$. **7.** **9.**

11. 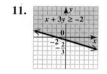 **13.** **15.** **17.** **19.** **21.**

23. **25.** $-3 < x < 3$  **27.** $-2 < x + 1 < 2$  **29.**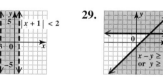

31. **33.** 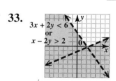 **35.** C **37.** A **39. (a)** $\{-4\}$ **(b)** $(-\infty, -4)$ **(c)** $(-4, \infty)$

41. (a) $\{3.5\}$ **(b)** $(3.5, \infty)$ **(c)** $(-\infty, 3.5)$ **We include a calculator graph and supporting explanation only with the answer to Exercise 43.**
43. (a) $\{-.6\}$ **(b)** $(-.6, \infty)$ **(c)** $(-\infty, -.6)$ The graph of $y_1 = 5x + 3$ has x-intercept $(-.6, 0)$, supporting the result of part (a). The graph of y_1 lies *above* the x-axis for values of x *greater than* $-.6$, supporting the result of part (b). The graph of y_1 lies *below* the x-axis for values of x *less*

than $-.6$, supporting the result of part (c).

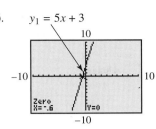

$y_1 = 5x + 3$

45. (a) $\{-1.2\}$ **(b)** $(-\infty, -1.2]$ **(c)** $[-1.2, \infty)$

Section 3.5   (page 194)

Exercises **1.** We give one of many possible answers here. A function is a set of ordered pairs in which each first element corresponds to exactly one second element. For example, $\{(0, 1), (1, 2), (2, 3), (3, 4), \ldots\}$ is a function. **3.** independent variable **5.** function **7.** not a function **9.** function **11.** not a function; domain: $\{0, 1, 2\}$; range: $\{-4, -1, 0, 1, 4\}$ **13.** function; domain: $\{2, 3, 5, 11, 17\}$; range: $\{1, 7, 20\}$ **15.** not a function; domain: $\{1\}$; range: $\{5, 2, -1, -4\}$ **17.** function; domain: $(-\infty, \infty)$; range: $(-\infty, \infty)$ **19.** function; domain: $(-\infty, \infty)$; range: $(-\infty, 4]$ **21.** not a function; domain: $[-4, 4]$; range: $[-3, 3]$ **23.** function; domain: $(-\infty, \infty)$ **25.** not a function; domain: $[0, \infty)$ **27.** function; domain: $(-\infty, \infty)$ **29.** not a function; domain: $(-\infty, \infty)$ **31.** function; domain: $[0, \infty)$

33. function; domain: $(-\infty, 0) \cup (0, \infty)$ **35.** function; domain: $\left[-\dfrac{1}{2}, \infty\right)$ **37.** function; domain: $(-\infty, 9) \cup (9, \infty)$ **39.** B **41.** 4

43. -11 **45.** $-3p + 4$ **47.** $3x + 4$ **49.** $-3x - 2$ **51.** $-6m + 13$ **53. (a)** 2 **(b)** 3 **55. (a)** 15 **(b)** 10 **57. (a)** 3 **(b)** -3

59. (a) $f(x) = \dfrac{12 - x}{3}$ **(b)** 3 **61. (a)** $f(x) = 3 - 2x^2$ **(b)** -15 **63. (a)** $f(x) = \dfrac{8 - 4x}{-3}$ **(b)** $\dfrac{4}{3}$ **65.** line; -2; $-2x + 4$; -2; 3; -2

67. domain: $(-\infty, \infty)$; range: $(-\infty, \infty)$ **69.** domain: $(-\infty, \infty)$; range: $(-\infty, \infty)$

71. domain: $(-\infty, \infty)$; range: $(-\infty, \infty)$ **73.** domain: $(-\infty, \infty)$; range: $\{-4\}$ **75. (a)** 8.25 (dollars)

(b) 3 is the value of the independent variable, which represents a package weight of 3 lb; $f(3)$ is the value of the dependent variable representing the cost to mail a 3-lb package. **(c)** $13.75; $f(5) = 13.75$ **77.** 194.53 cm **79.** 177.41 cm **81.** 1.83 m³ **83.** 4.11 m³ **85. (a)** yes **(b)** $[0, 24]$ **(c)** 1200 megawatts **(d)** at 17 hr or 5 P.M.; at 4 A.M. **(e)** $f(12) = 2100$; At 12 noon, electricity use is 2100 megawatts. **87.** $f(3) = 7$

Section 3.6   (page 205)

Exercises **1.** direct **3.** direct **5.** inverse **7.** inverse **9.** inverse **11.** direct **13.** joint **15.** combined **17.** increases; decreases **19.** 36 **21.** $\dfrac{16}{9}$ **23.** .625 **25.** $\dfrac{16}{5}$ **27.** $222\dfrac{2}{9}$ **29.** If y varies inversely as x, then x is in the denominator; however, if y varies directly as x, then x is in the numerator. If $k > 0$, then with inverse variation, as x increases, y decreases. With direct variation, y increases as x increases. **31.** $1.69\dfrac{9}{10}$ **33.** about 450 cm³ **35.** 8 lb **37.** 256 ft **39.** $106\dfrac{2}{3}$ mph **41.** 100 cycles per sec

43. $21\dfrac{1}{3}$ foot-candles **45.** $420 **47.** 448.1 lb **49.** approximately 68,600 calls **51.** 1.105 L **53.** 11.8 lb **55.** (0, 0), (1, 1.25)

56. 1.25 **57.** $y = 1.25x + 0$ or $y = 1.25x$ **58.** $a = 1.25, b = 0$ **59.** It is the price per gallon, and it is the slope of the line.

60. It can be written in the form $y = kx$ (where $k = a$). The value of a is called the constant of variation. **61.** It means that 4.6 gal cost $5.75. **62.** It means that 12 gal cost $15.00.

Chapter 3 Review Exercises (page 214)

1.

x	y
0	5
$\dfrac{10}{3}$	0
2	2
$\dfrac{14}{3}$	-2

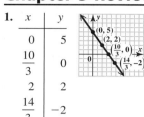

3. $(3, 0)$; $(0, -4)$

5. $(10, 0)$; $(0, 4)$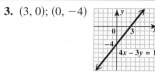

7. If both coordinates are positive, the point lies in quadrant I. If the first coordinate is negative and the second is positive, the point lies in quadrant II. To lie in quadrant III, the point must have both coordinates negative. To lie in quadrant IV, the first coordinate must be positive and the second must be negative. **9.** $-\dfrac{1}{2}$ **11.** $\dfrac{3}{4}$ **13.** $\dfrac{2}{3}$ **15.** undefined **17.** -1 **19.** negative **21.** 0 **23.** 12 ft

25. $y = -\dfrac{1}{3}x - 1$ **27.** $y = -\dfrac{4}{3}x + \dfrac{29}{3}$ **29.** $x = 2$ (Slope-intercept form is not possible.) **31.** $y = \dfrac{7}{5}x + \dfrac{16}{5}$ **33.** $y = 4x - 29$

35. (a) $y = 57x + 159$; $843 (b) $y = 47x + 159$; $723 **37.** **39.** **41.**

43. domain: $\{9, 11, 4, 17, 25\}$; range: $\{32, 47, 69, 14\}$; function **45.** domain: $(-\infty, 0]$; range: $(-\infty, \infty)$; not a function **47.** not a function; domain: $(-\infty, \infty)$ **49.** function; domain: $\left[-\dfrac{7}{4}, \infty\right)$ **51.** function; domain: $(-\infty, 6) \cup (6, \infty)$ **53.** -6 **55.** -8 **57.** (a) yes

(b) domain: $\{1994, 1995, 1996, 1997, 1998, 1999\}$; range: $\{40, 60, 80, 130, 180, 200\}$ (c) Answers will vary. Two possible answers are $(1994, 40)$ and $(1995, 60)$. (d) 40; In 1994, CNBC profits were $40 million. (e) 1998 **59.** C **61.** Because it falls from left to right, the

slope is negative. **62.** $-\dfrac{3}{2}$ **63.** $-\dfrac{3}{2}$; $\dfrac{2}{3}$ **64.** $\left(\dfrac{7}{3}, 0\right)$ **65.** $\left(0, \dfrac{7}{2}\right)$ **66.** $f(x) = -\dfrac{3}{2}x + \dfrac{7}{2}$ **67.** $f(8) = -\dfrac{17}{2}$ **68.** $x = \dfrac{23}{3}$

69. 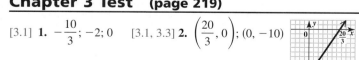 **70.** $\left\{\dfrac{7}{3}\right\}$ **71.** $\left(\dfrac{7}{3}, \infty\right)$ **72.** $\left(-\infty, \dfrac{7}{3}\right)$ **73.** C **75.** 5.59 vibrations per sec

Chapter 3 Test (page 219)

[3.1] **1.** $-\dfrac{10}{3}$; -2; 0 [3.1, 3.3] **2.** $\left(\dfrac{20}{3}, 0\right)$; $(0, -10)$ **3.** none; $(0, 5)$ **4.** $(2, 0)$; none

[3.2] **5.** $\dfrac{1}{2}$ **6.** It is a vertical line. **7.** perpendicular **8.** neither **9.** -1214 farms per yr; The number of farms decreased by about 1214 each year from 1980 to 2001. [3.3] **10.** $y = -5x + 19$ **11.** $y = 14$ **12.** $y = -\dfrac{3}{5}x - \dfrac{11}{5}$ **13.** $y = -\dfrac{1}{2}x - \dfrac{3}{2}$

14. $y = -\dfrac{1}{2}x + 2$ **15.** B   **16. (a)** $y = 1968.75x + 13{,}016.25$ **(b)** \$26,798; It is close to the actual value.   [3.4] **17.**

18.   [3.5] **19.** D   **20.** D   **21. (a)** domain: $[0, \infty)$; range: $(-\infty, \infty)$ **(b)** domain: $\{0, -2, 4\}$; range: $\{1, 3, 8\}$

22. $0; -a^2 + 2a - 1$ **23.** domain: $(-\infty, \infty)$; range: $(-\infty, \infty)$ [3.6] **24.** 200 amps   **25.** .8 lb

Cumulative Review Exercises Chapters 1–3 (page 221)

[1.1] **1.** always true   **2.** always true   **3.** never true   **4.** sometimes true; for example, $3 + (-3) = 0$, but $3 + (-1) = 2 \neq 0$ [1.2] **5.** 4

[1.3] **6.** .64   **7.** not a real number   [1.2] **8.** $\dfrac{8}{5}$ [1.4] **9.** $4m - 3$ **10.** $2x^2 + 5x + 4$ [1.3] **11.** $-\dfrac{19}{2}$ [1.1] **12.** $(-3, 5]$ **13.** no

[1.3] **14.** -24 **15.** 56   **16.** undefined   [2.1] **17.** $\left\{\dfrac{7}{6}\right\}$ **18.** $\{-1\}$ [2.2] **19.** $h = \dfrac{3V}{\pi r^2}$ **20.** $-67°F$ [2.4] **21.** 6 in.   **22.** 2 hr

[2.5] **23.** $\left(-3, \dfrac{7}{2}\right)$ **24.** $(-\infty, 1]$ [2.6] **25.** $(6, 8)$

26. $(-\infty, -2] \cup (7, \infty)$ [2.7] **27.** $\{0, 7\}$ **28.** $(-\infty, \infty)$ [2.5] **29.** The union of the three solution sets is $(-\infty, \infty)$.

[3.1] **30.** $-3; 4; -\dfrac{3}{2}$ **31.** x-intercept: $(4, 0)$; y-intercept: $\left(0, \dfrac{12}{5}\right)$ [3.2] **32. (a)** $-\dfrac{6}{5}$ **(b)** $\dfrac{5}{6}$ [3.4] **33.**

[3.3] **34.** $y = -\dfrac{3}{4}x - 1$ **35.** $y = -2$ **36.** $y = -\dfrac{4}{3}x + \dfrac{7}{3}$ [3.5] **37.** domain: $\{14, 91, 75, 23\}$; range: $\{9, 70, 56, 5\}$; not a function;

75 in the domain is paired with two different values, 70 and 56, in the range.   **38. (a)** domain: $(-\infty, \infty)$; range: $(-\infty, \infty)$ **(b)** 22

[3.2] **39.** 10.5 million per year; The number of U.S. cell phone subscribers increased by 10.5 million per year from 1992 to 2000.

[3.6] **40.** \$9.92

CHAPTER **4** SYSTEMS OF LINEAR EQUATIONS

Section 4.1 (page 232)

Exercises   **1.** $3; -6$ **3.** $\emptyset$ **5.** 0   **7.** D; The ordered pair solution must be in quadrant IV, since that is where the graphs of the equations intersect.   **9. (a)** B **(b)** C **(c)** A **(d)** D   **11.** yes   **13.** no   **15.** $\{(-2, -3)\}$ **17.** $\{(3, -1)\}$ **19.** $\{(2, -3)\}$

21. $\left\{\left(\dfrac{3}{2}, -\dfrac{3}{2}\right)\right\}$ **23.** $\{(x, y) \mid 7x + 2y = 6\}$; dependent equations **25.** $\emptyset$; inconsistent system **27.** $\{(0, -4)\}$ **29.** $\{(0, 0)\}$

31. $\{(2, -4)\}$ **33.** $y = -\dfrac{3}{7}x + \dfrac{4}{7}$; $y = -\dfrac{3}{7}x + \dfrac{3}{14}$; no solutions **35.** Both are $y = -\dfrac{2}{3}x + \dfrac{1}{3}$; infinitely many solutions

37. Answers will vary. **(a)** **(b)** **(c)** **39.** $\{(1, 2)\}$ **41.** $\left\{\left(\dfrac{22}{9}, \dfrac{22}{3}\right)\right\}$ **43.** $\{(2, 3)\}$

45. $\{(5, 4)\}$ **47.** $\left\{\left(-5, -\dfrac{10}{3}\right)\right\}$ **49.** $\{(2, 6)\}$ **51.** $\{(x, y) \mid y = 2x\}$; dependent equations **53.** $\emptyset$; inconsistent system **55.** $(3, -4)$

57. A **59. (a)** $\{(5, 5)\}$ **(b)** **61. (a)** $\{(0, -2)\}$ **(b)** **63. (a)** years 0 to 6

(b) year 6; about \$650 **65. (a)** 1989–1997 **(b)** 1997; NBC; 17% **(c)** 1989: share 20%; 1998: share 16% **(d)** NBC and ABC; (2000, 16) **(e)** Viewership has generally declined during these years. **67.** 1995–1997 **69.** $\{(2.3, 5.8)\}$ (Answers may vary slightly depending on how the system was solved.) **71.** $\{(2, 4)\}$ **73.** $\{(4, -5)\}$ **75.** $\left\{\left(\dfrac{1}{a}, \dfrac{1}{b}\right)\right\}$ **77.** $\left\{\left(-\dfrac{3}{5a}, \dfrac{7}{5}\right)\right\}$ **79.** $\{(1, 3)\}$

80. $f(x) = -3x + 6$; linear **81.** $g(x) = \dfrac{2}{3}x + \dfrac{7}{3}$; linear **82.** one; 1; 3; 1; 3; 1; 3 **83.** Begin by multiplying one or both equations by a number that will lead to opposite coefficients for one variable in the system. Then add the two equations to eliminate that variable. Solve for the remaining variable. Once that value is found, find the remaining variable value by substituting back into one of the original equations.

For example: $\begin{array}{l} 2x - y = 9 \\ x + 2y = 7. \end{array}$ Multiply the first equation by 2 and add to get $5x = 25$, or $x = 5$. Let $x = 5$ in either equation to get $y = 1$. The solution set is $\{(5, 1)\}$.

Section 4.2 (page 244)

Exercises **1.** The statement means that when -1 is substituted for x, 2 is substituted for y, and 3 is substituted for z in the three equations, the resulting three statements are true. **3.** $\{(3, 2, 1)\}$ **5.** $\{(1, 4, -3)\}$ **7.** $\{(0, 2, -5)\}$ **9.** $\{(1, 0, 3)\}$ **11.** $\{(-12, 18, 0)\}$

13. $\left\{\left(1, \dfrac{3}{10}, \dfrac{2}{5}\right)\right\}$ **15.** $\left\{\left(-\dfrac{7}{3}, \dfrac{22}{3}, 7\right)\right\}$ **17.** $\{(4, 5, 3)\}$ **19.** $\{(2, 2, 2)\}$ **21.** $\left\{\left(\dfrac{8}{3}, \dfrac{2}{3}, 3\right)\right\}$ **23.** $\{(-1, 0, 0)\}$

25. $\{(-3, 5, -6)\}$ **27.** Answers will vary. Some possible answers are **(a)** two perpendicular walls and the ceiling in a normal room, **(b)** the floors of three different levels of an office building, and **(c)** three pages of this book (since they intersect in the spine). **29.** $\emptyset$; inconsistent system **31.** $\{(x, y, z) \mid x - y + 4z = 8\}$; dependent equations **33.** $\{(x, y, z) \mid 2x + y - z = 6\}$; dependent equations **35.** $\{(0, 0, 0)\}$ **37.** $\{(2, 1, 5, 3)\}$ **39.** $\{(-2, 0, 1, 4)\}$ **41.** $128 = a + b + c$ **42.** $140 = 2.25a + 1.5b + c$ **43.** $80 = 9a + 3b + c$ **44.** $\begin{array}{l} a + b + c = 128 \\ 2.25a + 1.5b + c = 140 \\ 9a + 3b + c = 80; \{(-32, 104, 56)\} \end{array}$ **45.** $f(x) = -32x^2 + 104x + 56$ **46.** height; time **47.** 56 ft **48.** 140.5 ft

49. $a = 3, b = 1, c = -2; f(x) = 3x^2 + x - 2$ **50.** $a = 1, b = 4, c = 3; Y_1 = X^2 + 4X + 3$ **51.** If one were to eliminate *different* variables in the first two steps, the result would be two equations in three variables, and it would not be possible to solve for a single variable in the next step.

Section 4.3 (page 254)

Connections (page 250) "Mixed price" refers to the price of a mixture of the two products. The system is $9x + 7y = 107$, $7x + 9y = 101$, where x represents the price of a citron and y represents the price of a wood apple.

Exercises 1. wins: 95; losses: 67 **3.** length: 78 ft; width: 36 ft **5.** ExxonMobil: \$214 billion; General Motors: \$185 billion **7.** $x = 40$ and $y = 50$, so the angles measure 40° and 50°. **9.** NHL: \$219.74; NBA: \$203.38 **11.** single: \$2.09; double: \$3.19 **13. (a)** 6 oz **(b)** 15 oz **(c)** 24 oz **(d)** 30 oz **15.** \$.99x **17.** 6 gal of 25%; 14 gal of 35% **19.** pure acid: 6 L; 10% acid: 48 L **21.** nuts: 14 kg; cereal: 16 kg **23.** \$1000 at 2%; \$2000 at 4% **25.** 25y **27.** train: 60 km per hr; plane: 160 km per hr **29.** boat: 21 mph; current: 3 mph **31.** Turner: \$80.4 million; 'N Sync: \$76.6 million **33.** general admission: 76; with student ID: 108 **35.** 8 for a citron; 5 for a wood apple **37.** $x + y + z = 180$; angle measures: 70°, 30°, 80° **39.** first: 20°; second: 70°; third: 90° **41.** shortest: 12 cm; middle: 25 cm; longest: 33 cm **43.** Independent: 38; Democrat: 34; Republican: 28 **45.** \$10 tickets: 350; \$18 tickets: 250; \$30 tickets: 50 **47.** type A: 80; type B: 160; type C: 250 **49.** first chemical: 50 kg; second chemical: 400 kg; third chemical: 300 kg **51.** wins: 48; losses: 26; ties: 8

Section 4.4 (page 267)

Connections (page 266) 1. $\begin{bmatrix} 1 & 0 & | & -2 \\ 0 & 1 & | & -1 \end{bmatrix}$ **2.** a system of dependent equations

Exercises 1. (a) $0, 5, -3$ **(b)** $1, -3, 8$ **(c)** yes; The number of rows is the same as the number of columns (three). **(d)** $\begin{bmatrix} 1 & 4 & 8 \\ 0 & 5 & -3 \\ -2 & 3 & 1 \end{bmatrix}$

(e) $\begin{bmatrix} 1 & -\dfrac{3}{2} & -\dfrac{1}{2} \\ 0 & 5 & -3 \\ 1 & 4 & 8 \end{bmatrix}$ **(f)** $\begin{bmatrix} 1 & 15 & 25 \\ 0 & 5 & -3 \\ 1 & 4 & 8 \end{bmatrix}$ **3.** $\begin{bmatrix} 1 & 2 & | & 11 \\ 2 & -1 & | & -3 \end{bmatrix}$; $\begin{bmatrix} 1 & 2 & | & 11 \\ 0 & -5 & | & -25 \end{bmatrix}$; $\begin{bmatrix} 1 & 2 & | & 11 \\ 0 & 1 & | & 5 \end{bmatrix}$; $x + 2y = 11$, $y = 5$; $\{(1, 5)\}$

5. $\{(4, 1)\}$ **7.** $\{(1, 1)\}$ **9.** $\{(-1, 4)\}$ **11.** $\emptyset$ **13.** $\{(x, y) \mid 2x + y = 4\}$ **15.** $\begin{bmatrix} 1 & 1 & -1 & | & -3 \\ 0 & -1 & 3 & | & 10 \\ 0 & -6 & 7 & | & 38 \end{bmatrix}$; $\begin{bmatrix} 1 & 1 & -1 & | & -3 \\ 0 & 1 & -3 & | & -10 \\ 0 & -6 & 7 & | & 38 \end{bmatrix}$;

$\begin{bmatrix} 1 & 1 & -1 & | & -3 \\ 0 & 1 & -3 & | & -10 \\ 0 & 0 & -11 & | & -22 \end{bmatrix}$; $\begin{bmatrix} 1 & 1 & -1 & | & -3 \\ 0 & 1 & -3 & | & -10 \\ 0 & 0 & 1 & | & 2 \end{bmatrix}$; $x + y - z = -3$, $y - 3z = -10$, $z = 2$; $\{(3, -4, 2)\}$ **17.** $\{(4, 0, 1)\}$

19. $\{(-1, 23, 16)\}$ **21.** $\{(3, 2, -4)\}$ **23.** $\{(x, y, z) \mid x - 2y + z = 4\}$ **25.** $\emptyset$ **27.** Examples will vary. **(a)** A matrix is a rectangular array of numbers. **(b)** A horizontal arrangement of elements in a matrix is a row of a matrix. **(c)** A vertical arrangement of elements in a matrix is a column of a matrix. **(d)** A square matrix contains the same number of rows as columns. **(e)** A matrix formed by the coefficients and constants of a linear system is an augmented matrix for the system. **(f)** Row operations on a matrix allow it to be transformed into another matrix in which the solution of the associated system can be found more easily. **29.** $\{(1, 1)\}$ **31.** $\{(-1, 2, 1)\}$ **33.** $\{(1, 7, -4)\}$

Section 4.5 (page 278)

Connections (page 274) 1. -185 **2.** Every square matrix corresponds to *one and only one* real number, called its determinant.

Exercises 1. D **3.** -3 **5.** 14 **7.** 0 **9.** 59 **11.** 14 **13.** Multiply the upper left and lower right entries. Then multiply the upper right and lower left entries. Subtract the second product from the first to obtain the determinant. For example, $\begin{vmatrix} 4 & 2 \\ 7 & 1 \end{vmatrix} = 4 \cdot 1 - 2 \cdot 7 = 4 - 14 = -10$. **15.** -22 **17.** 20 **19.** -5 **21.** By choosing that row or column to expand about, all terms will have a factor of 0, and so the sum of all these terms will be 0. **23.** 6.078 **25. (a)** D **(b)** A **(c)** C **(d)** B **27.** $\{(-5, 2)\}$ **29.** $\left\{ \left(\dfrac{11}{58}, -\dfrac{5}{29} \right) \right\}$

31. $\{(-1, 2)\}$ **33.** $\{(-2, 3, 5)\}$ **35.** Cramer's rule does not apply.   **37.** $\{(20, -13, -12)\}$ **39.** $\left\{\left(\dfrac{62}{5}, -\dfrac{1}{5}, \dfrac{27}{5}\right)\right\}$ **41.** $\{(-1, 3, 5)\}$

43. $\{(1, -3, 2, -4)\}$ **45.** $\{(-a - b, a^2 + ab + b^2)\}$ **47.** $\{(1, 0)\}$ **49.** -22 **51.** 20   **53.** -5 **55.** $\{2\}$ **57.** $\{0\}$ **59.** Answers

will vary.   **61.**   **62.** $\dfrac{1}{2}\begin{vmatrix} 0 & 0 & 1 \\ -3 & -4 & 1 \\ 2 & -2 & 1 \end{vmatrix}$ **63.** 7   **64.** 8   **65.** $\dfrac{y_2 - y_1}{x_2 - x_1}$ **66.** $y - y_1 = \dfrac{y_2 - y_1}{x_2 - x_1}(x - x_1)$

67. $x_2y - x_1y - x_2y_1 - xy_2 + x_1y_2 + xy_1 = 0$ **68.** The result is the same as in Exercise 67.   **69.** 0   **71.** 0   **73.** 16
75. Rows and columns are interchanged.   **77.** Two columns are interchanged.   **79.** Each element of the second column is multiplied
by $-\dfrac{1}{2}$.   **81.** Elements of the first row are multiplied by 1; products are added to the elements of the second row.

Chapter 4 Review Exercises   (page 290)

1. (a) 1978 and 1982 **(b)** just less than 500,000   **3.** D   **5.** $\{(0, 1)\}$ **7.** $\{(-6, 3)\}$ **9.** $\emptyset$; inconsistent system   **11.** $\{(0, 4)\}$
13. (a) inconsistent **(b)** dependent equations   **15.** $\{(1, -5, 3)\}$ **17.** $\emptyset$; inconsistent system   **19. (a)** *Harry Potter and the Sorcerer's*
Stone: \$294 million; *Shrek*: \$268 million **(b)** \$241 million **(c)** \$803 million   **21.** \$2-per-lb nuts: 30 lb; \$1-per-lb candy: 70 lb
23. \$40,000 at 10%; \$100,000 at 6%; \$140,000 at 5%   **25.** Mantle: 54; Maris: 61; Blanchard: 21   **27.** $\{(-1, 5)\}$ **29.** $\{(1, 2, -1)\}$
31. -21 **33.** -139 **35.** For three unknowns, we need three equations.   **37.** $\left\{\left(\dfrac{39}{10}, \dfrac{6}{5}\right)\right\}$ **39.** $\{(-2, -2, -2)\}$ **41.** $\emptyset$
43. $\{(5, 3)\}$ **45.** $\left\{\left(\dfrac{82}{23}, -\dfrac{4}{23}\right)\right\}$ **47.** U.S.: 97; Russia: 88; China: 59   **48.** $2a + b + c = -5$ **49.** $-a + c = -1$
50. $3a + 3b + c = -18$ **51.** $a = 1, b = -7, c = 0; x^2 + y^2 + x - 7y = 0$ **52.** The relation is not a function because a vertical line
intersects its graph more than once.

Chapter 4 Test   (page 294)

[4.1] **1.** No; The graph for Ruth lies completely below the graph for Aaron.   **2.** Aaron; Ruth   **3.** $\{(6, 1)\}$ **4.** $\{(3, 3)\}$

5. $\{(0, -2)\}$ **6.** $\emptyset$; inconsistent system   [4.2] **7.** $\left\{\left(-\dfrac{2}{3}, \dfrac{4}{5}, 0\right)\right\}$ **8.** $\{(3, -2, 1)\}$ [4.1] **9.** $\{(6, -4)\}$ **10.** $\{(x, y) \mid 12x - 5y = 8\}$;

dependent equations   [4.3] **11.** *Pretty Woman*: \$178.4 million; *Runaway Bride*: \$152.3 million   **12.** 45 mph, 75 mph   **13.** 20% solution:
4 L; 50% solution: 8 L   **14.** AC adaptor: \$8; rechargeable flashlight: \$15   **15.** Orange Pekoe: 60 oz; Irish Breakfast: 30 oz; Earl Grey: 10 oz
[4.4] **16.** $\left\{\left(\dfrac{2}{5}, \dfrac{7}{5}\right)\right\}$ **17.** $\{(-1, 2, 3)\}$ [4.5] **18.** -58 **19.** -844 **20.** $\{(-6, 7)\}$ **21.** $\{(1, -2, 3)\}$
[4.2, 4.4, 4.5] **22.** $\{(-3, -2, -4)\}$

Cumulative Review Exercises Chapters 1–4   (page 295)

[1.3] **1.** 81   **2.** -81 **3.** -81 **4.** .7   **5.** $-.7$ **6.** not a real number   **7.** 4   **8.** -4 **9.** -199 **10.** 455
[1.4] **11.** commutative property   [2.1] **12.** $\left\{-\dfrac{15}{4}\right\}$ [2.7] **13.** $\left\{\dfrac{2}{3}, 2\right\}$ [2.2] **14.** $x = \dfrac{d - by}{a - c}$ or $x = \dfrac{by - d}{c - a}$ [2.1] **15.** $\{11\}$
[2.5] **16.** $\left(-\infty, \dfrac{240}{13}\right]$ [2.7] **17.** $\left[-2, \dfrac{2}{3}\right]$ **18.** $(-\infty, \infty)$ [2.2] **19.** 2010; 1813; 62.8%; 57.2%   [2.3, 2.4] **20.** 6 m

21. pennies: 35; nickels: 29; dimes: 30   **22.** 46°, 46°, 88°   [3.1] **23.** $y = 6$ **24.** $x = 4$ [3.2] **25.** $-\dfrac{4}{3}$ **26.** $\dfrac{3}{4}$

[3.3] **27.** $4x + 3y = 10$ [3.2] **28.**   [3.4] **29.**   [3.5] **30.** (a) -6 (b) $a^2 + 3a - 6$ [3.6] **31.** 17.5

[4.1, 4.4, 4.5] **32.** $\{(3, -3)\}$ **33.** $\{(x, y) \mid x - 3y = 7\}$ [4.2, 4.4, 4.5] **34.** $\{(5, 3, 2)\}$ [4.3] **35.** oranges: 5 lb; apples: 1 lb   **36.** Tickle Me Elmo: $27.63; Snacktime Kid: $36.26   **37.** small: $1.50; large: $2.50   **38.** peanuts: $2 per lb; cashews: $4 per lb   [4.1] **39.** $x = 8$ or 800 items; $3000   **40.** about $400   [4.5] **41.** 8   **42.** $\left\{\left(\dfrac{5}{4}, 2, \dfrac{3}{2}\right)\right\}$

CHAPTER 5 EXPONENTS, POLYNOMIALS, AND POLYNOMIAL FUNCTIONS

Section 5.1   (page 310)

Exercises **1.** incorrect; $(ab)^2 = a^2b^2$ **3.** incorrect; $\left(\dfrac{4}{a}\right)^3 = \dfrac{4^3}{a^3}$ **5.** The product rule says that when like bases are multiplied, the base stays the same and the exponents are added. For example, $x^5 \cdot x^6 = x^{11}$.   **7.** 13^{12} **9.** x^{17} **11.** $-27w^8$ **13.** $18x^3y^8$ **15.** The product rule does not apply.   **17.** (a) B (b) C (c) B (d) C   **19.** 1   **21.** -1 **23.** 1   **25.** 2   **27.** 0   **29.** -2 **31.** (a) B (b) D (c) B (d) D   **33.** $\dfrac{1}{5^4}$ or $\dfrac{1}{625}$ **35.** $\dfrac{1}{8}$ **37.** $\dfrac{1}{16x^2}$ **39.** $\dfrac{4}{x^2}$ **41.** $-\dfrac{1}{a^3}$ **43.** $\dfrac{1}{a^4}$ **45.** $\dfrac{11}{30}$ **47.** $-\dfrac{5}{24}$ **49.** When n is even, the expressions are opposites. When n is odd, they are equal.   **51.** 16   **53.** $\dfrac{27}{4}$ **55.** $\dfrac{27}{8}$ **57.** $\dfrac{25}{16}$ **59.** (a) B (b) D (c) D (d) B   **61.** The quotient rule says that when like bases are divided, the base stays the same and the exponents are subtracted. For example, $\dfrac{x^8}{x^5} = x^3$.   **63.** 4^2 or 16   **65.** x^4 **67.** $\dfrac{1}{r^3}$ **69.** 6^6 **71.** $\dfrac{1}{6^{10}}$ **73.** 7^2 or 49   **75.** r^3 **77.** The quotient rule does not apply.   **79.** x^{18} **81.** $\dfrac{27}{125}$ **83.** $64t^3$ **85.** $-216x^6$ **87.** $-\dfrac{64m^6}{t^3}$ **89.** $\dfrac{1}{3}$ **91.** $\dfrac{1}{a^5}$ **93.** $\dfrac{1}{k^2}$ **95.** $-4r^6$ **97.** $\dfrac{625}{a^{10}}$ **99.** $\dfrac{z^4}{x^3}$ **101.** $-14k^3$ **103.** $\dfrac{p^4}{5}$ **105.** $\dfrac{1}{2pq}$ **107.** $\dfrac{4}{a^2}$ **109.** $\dfrac{1}{6y^{13}}$ **111.** $\dfrac{4k^5}{m^2}$ **113.** $\dfrac{4k^{17}}{125}$ **115.** $\dfrac{2k^5}{3}$ **117.** $\dfrac{8}{3pq^{10}}$ **119.** $\dfrac{y^9}{8}$ **121.** $-\dfrac{3}{32m^8p^4}$ **123.** $\dfrac{2}{3y^4}$ **125.** $\dfrac{3p^8}{16q^{14}}$ **127.** 5.3×10^2 **129.** 8.3×10^{-1} **131.** 6.92×10^{-6} **133.** -3.85×10^4 **135.** 72,000   **137.** .00254   **139.** $-60,000$ **141.** .000012   **143.** .06   **145.** .0000025   **147.** 200,000   **149.** 3000   **151.** 1×10^9; 1×10^{12}; 2.128×10^{12}; 1.44419×10^5 **153.** 26,000 or 2.6×10^4 **155.** 300 sec   **157.** approximately 5.87×10^{12} mi   **159.** (a) 20,000 hr (b) 833 days   **161.** 7.5×10^9 **163.** 4×10^{17}

Section 5.2   (page 319)

Exercises **1.** $2x^3 - 3x^2 + x + 4$ **3.** $p^7 - 8p^5 + 4p^3$ **5.** $3m^4 - m^3 + 5m^2 + 10$ **7.** 7; 1   **9.** -15; 2   **11.** 1; 4   **13.** $\dfrac{1}{6}$; 1   **15.** -1; 6   **17.** monomial; 0   **19.** binomial; 1   **21.** binomial; 8   **23.** trinomial; 3   **25.** none of these; 5   **27.** A   **29.** $8z^4$ **31.** $7m^3$ **33.** $5x$ **35.** already simplified   **37.** $-3y^2 + 7y$ **39.** $8k^2 + 2k - 7$ **41.** $-2n^4 - n^3 + n^2$ **43.** $-2ab^2 + 20a^2b$ **45.** $3m + 11$ **47.** $-p - 4$ **49.** A *monomial* (or *term*) is a numeral, a variable, or a product of numerals and variables raised to positive integer powers. Some examples of monomials are 6, x, and $-4x^2y^3$. A *binomial* is a sum or difference of exactly two terms such as $x^2 + y^2$ or $x^2 - y^2$. A *trinomial* consists of exactly three terms, such as $x^2 - 3x + 8$. These are all examples of *polynomials*.   **51.** $8x^2 + x - 2$

53. $-t^4 + 2t^2 - t + 5$ **55.** $5y^3 - 3y^2 + 5y + 1$ **57.** $r + 13$ **59.** $-2a^2 - 2a - 7$ **61.** $-3z^5 + z^2 + 7z$ **63.** $12p - 4$
65. $-9p^2 + 11p - 9$ **67.** $5a + 18$ **69.** $14m^2 - 13m + 6$ **71.** $13z^2 + 10z - 3$ **73.** $10y^3 - 7y^2 + 5y + 8$
75. $-5a^4 - 6a^3 + 9a^2 - 11$ **77.** $3y^2 - 4y + 2$ **79.** $-4m^2 + 4n^2 - 7n$ **81.** $y^4 - 4y^2 - 4$ **83.** $10z^2 - 16z$

Section 5.3   (page 325)

Exercises   **1. (a)** -10 **(b)** 8   **3. (a)** 8 **(b)** 2   **5. (a)** 8 **(b)** 74   **7. (a)** -11 **(b)** 4   **9. (a)** 11,280 **(b)** 16,437 **(c)** 18,369
11. (a) 9 million **(b)** 61 million **(c)** 118 million   **13. (a)** $8x - 3$ **(b)** $2x - 17$ **15. (a)** $-x^2 + 12x - 12$ **(b)** $9x^2 + 4x + 6$
17. $x^2 + 2x - 9$ **19.** 6   **21.** $x^2 - x - 6$ **23.** 6   **25.** -33 **27.** 0   **29.** For example, let $f(x) = 2x^3 + 3x^2 + x + 4$ and
$g(x) = 2x^4 + 3x^3 - 9x^2 + 2x - 4$; $(f - g)(x) = -2x^4 - x^3 + 12x^2 - x + 8$, and $(g - f)(x) = 2x^4 + x^3 - 12x^2 + x - 8$. Because the two
differences are not equal, subtraction of polynomial functions is not commutative.   **31.** domain: $(-\infty, \infty)$; range: $(-\infty, \infty)$

33. domain: $(-\infty, \infty)$; range: $(-\infty, 0]$ **35.** domain: $(-\infty, \infty)$; range: $(-\infty, \infty)$

Section 5.4   (page 333)

Exercises   **1.** C   **3.** D   **5.** $-24m^5$ **7.** $-28x^7y^4$ **9.** $-6x^2 + 15x$ **11.** $-2q^3 - 3q^4$ **13.** $18k^4 + 12k^3 + 6k^2$
15. $6m^3 + m^2 - 14m - 3$ **17.** $m^3 - 3m^2 - 40m$ **19.** $24z^3 - 20z^2 - 16z$ **21.** $4x^5 - 4x^4 - 24x^3$ **23.** $6y^2 + y - 12$
25. $-2b^3 + 2b^2 + 18b + 12$ **27.** $25m^2 - 9n^2$ **29.** $8z^4 - 14z^3 + 17z^2 + 20z - 3$ **31.** $6p^4 + p^3 + 4p^2 - 27p - 6$
33. $m^2 - 3m - 40$ **35.** $12k^2 + k - 6$ **37.** $3z^2 + zw - 4w^2$ **39.** $12c^2 + 16cd - 3d^2$ **41.** $.1x^2 + .63x - .13$
43. $3w^2 - \dfrac{23}{4}wz - \dfrac{1}{2}z^2$ **45.** The product of two binomials is the sum of the product of the first terms, the product of the outer terms,
the product of the inner terms, and the product of the last terms.   **47.** $4p^2 - 9$ **49.** $25m^2 - 1$ **51.** $9a^2 - 4c^2$ **53.** $16x^2 - \dfrac{4}{9}$
55. $16m^2 - 49n^4$ **57.** $25y^6 - 4$ **59.** $y^2 - 10y + 25$ **61.** $4p^2 + 28p + 49$ **63.** $16n^2 + 24nm + 9m^2$ **65.** $k^2 - \dfrac{10}{7}kp + \dfrac{25}{49}p^2$
67. $(x + y)^2 = x^2 + 2xy + y^2$. The expression $x^2 + y^2$ is missing the $2xy$ term, so $(x + y)^2 \neq x^2 + y^2$ in general.
69. $25x^2 + 10x + 1 + 60xy + 12y + 36y^2$ **71.** $4a^2 + 4ab + b^2 - 12a - 6b + 9$ **73.** $4a^2 + 4ab + b^2 - 9$
75. $4h^2 - 4hk + k^2 - j^2$ **77.** $y^3 + 6y^2 + 12y + 8$ **79.** $125r^3 - 75r^2s + 15rs^2 - s^3$ **81.** $q^4 - 8q^3 + 24q^2 - 32q + 16$
83. $6a^3 + 7a^2b + 4ab^2 + b^3$ **85.** $4z^4 - 17z^3x + 12z^2x^2 - 6zx^3 + x^4$ **87.** $m^4 - 4m^2p^2 + 4mp^3 - p^4$ **89.** $a^4b - 7a^2b^3 - 6ab^4$
91. $49; 25; 49 \neq 25$ **93.** $2401; 337; 2401 \neq 337$ **95.** $\dfrac{9}{2}x^2 - 2y^2$ **97.** $15x^2 - 2x - 24$ **99.** $a - b$ **100.** $A = s^2$; $(a - b)^2$
101. $(a - b)b$ or $ab - b^2$; $2ab - 2b^2$ **102.** b^2 **103.** a^2; a **104.** $a^2 - (2ab - 2b^2) - b^2 = a^2 - 2ab + b^2$ **105. (a)** They must be
equal to each other. **(b)** $(a - b)^2 = a^2 - 2ab + b^2$; This reinforces the special product for the square of a binomial difference.
106. 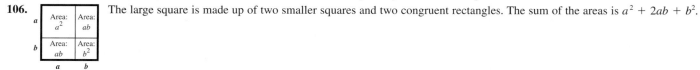 The large square is made up of two smaller squares and two congruent rectangles. The sum of the areas is $a^2 + 2ab + b^2$.

Since these expressions represent the same quantity, they must be equal. Thus $(a + b)^2 = a^2 + 2ab + b^2$.   **107.** $10x^2 - 2x$
109. $2x^2 - x - 3$ **111.** $8x^3 - 27$ **113.** $2x^3 - 18x$ **115.** -20 **117.** $2x^2 - 6x$ **119.** 36

Section 5.5 (page 340)

Exercises **1.** quotient; exponents **3.** descending powers **5.** $3x^3 - 2x^2 + 1$ **7.** $3y + 4 - \dfrac{5}{y}$ **9.** $3m + 5 + \dfrac{6}{m}$

11. $n - \dfrac{3n^2}{2m} + 2$ **13.** $\dfrac{2y}{x} + \dfrac{3}{4} + \dfrac{3w}{x}$ **15.** $r^2 - 7r + 6$ **17.** $y - 3$ **19.** $t + 5$ **21.** $p - 4 + \dfrac{44}{p + 6}$ **23.** $m^2 + 2m - 1$

25. $z^2 + 3$ **27.** $x^2 + 2x - 3 + \dfrac{6}{4x + 1}$ **29.** $2x - 5 + \dfrac{-4x + 5}{3x^2 - 2x + 4}$ **31.** $x^2 + x + 3$ **33.** $3x^2 + 6x + 11 + \dfrac{26}{x - 2}$

35. $2k^2 + 3k - 1$ **37.** $x^2 - 4x + 2 + \dfrac{9x - 4}{x^2 + 3}$ **39.** $p^2 + \dfrac{5}{2}p + 2 + \dfrac{-2}{2p + 2}$ **41.** $p^2 + p + 1$ **43.** $\dfrac{2}{3}x - 1$

45. $\dfrac{3}{4}a - 2 + \dfrac{1}{4a + 3}$ **47.** $2p + 7$ **49.** $-13; -13;$ They are the same, which suggests that when $P(x)$ is divided by $x - r$, the result is

$P(r)$. Here, $r = -1$. **51.** $5x - 1; 0$ **53.** $2x - 3; -1$ **55.** $4x^2 + 6x + 9; \dfrac{3}{2}$ **57.** $\dfrac{x^2 - 9}{2x}, \; x \neq 0$ **59.** $-\dfrac{5}{4}$ **61.** $\dfrac{x - 3}{2x}, \; x \neq 0$

63. 0

Chapter 5 Review Exercises (page 346)

1. 64 **3.** -125 **5.** $\dfrac{81}{16}$ **7.** $\dfrac{11}{30}$ **9.** 0 **11.** x^8 **13.** $\dfrac{1}{z^{15}}$ **15.** $\dfrac{r^{17}}{9}$ **17.** $\dfrac{1}{96m^7}$ **19.** $-12x^2y^8$ **21.** $\dfrac{10p^8}{q^7}$ **23.** In $(-6)^0$, the

base is -6 and the expression simplifies to 1. In -6^0, the base is 6 and the expression simplifies to -1. **25.** yes **27.** For example, let

$x = 2$ and $y = 3$. Then $(x^2 + y^2)^2 = (2^2 + 3^2)^2 = 169; x^4 + y^4 = 2^4 + 3^4 = 97 \neq 169$. **29.** 7.65×10^{-8} **31.** 2.814×10^8;

$5.0454 \times 10^4; 1 \times 10^2$ **33.** .0058 **35.** $1.5 \times 10^3; 1500$ **37.** $2.7 \times 10^{-2}; .027$ **39.** **(a)** 5.449×10^3 **(b)** 63 mi^2 **41.** -1

43. 504 **45.** **(a)** $9m^7 + 14m^6$ **(b)** binomial **(c)** 7 **47.** **(a)** $-7q^5r^3$ **(b)** monomial **(c)** 8 **49.** $-x^2 - 3x + 1$

51. $6a^3 - 4a^2 - 16a + 15$ **53.** $12x^2 + 8x + 5$ **55.** **(a)** $5x^2 - x + 5$ **(b)** $-5x^2 + 5x + 1$ **(c)** 11 **(d)** -9 **57.**

59. **61.** $15m^2 - 7m - 2$ **63.** $6w^2 - 13wt + 6t^2$ **65.** $3q^3 - 13q^2 - 14q + 20$ **67.** $36r^4 - 1$

69. $16m^2 + 24m + 9$ **71.** $y^2 - 3y + \dfrac{5}{4}$ **73.** $p^2 + 3p - 6$ **75.** $8x^2 - 10x - 3$ **77.** $\dfrac{1}{125}$ **79.** $21p^9 + 7p^8 + 14p^7$ **81.** $-\dfrac{1}{5z^9}$

83. $8x + 1 + \dfrac{5}{x - 3}$ **85.** $9m^2 - 30mn + 25n^2 - p^2$ **87.** $-3k^2 + 4k - 7$

Chapter 5 Test (page 349)

[5.1] **1.** **(a)** C **(b)** A **(c)** D **(d)** A **(e)** E **(f)** F **(g)** B **(h)** G **(i)** C **2.** $\dfrac{4x^7}{9y^{10}}$ **3.** $\dfrac{6}{r^{14}}$ **4.** $\dfrac{16}{9p^{10}q^{28}}$ **5.** $\dfrac{16}{x^6y^{16}}$ **6.** .00000091

7. $3 \times 10^{-4}; .0003$ [5.3] **8.** **(a)** -18 **(b)** $-2x^2 + 12x - 9$ **(c)** $-2x^2 - 2x - 3$ **(d)** -7 **9.** 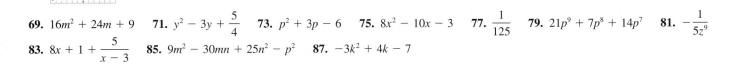 **10.** **(a)** 614 thousand

(b) 735 thousand **(c)** 800 thousand [5.2] **11.** $x^3 - 2x^2 - 10x - 13$ [5.4] **12.** $10x^2 - x - 3$ **13.** $6m^3 - 7m^2 - 30m + 25$

14. $36x^2 - y^2$ **15.** $9k^2 + 6kq + q^2$ **16.** $4y^2 - 9z^2 + 6zx - x^2$ [5.5] **17.** $4p - 8 + \dfrac{6}{p}$ **18.** $x^2 + 4x + 4$

[5.4] **19.** **(a)** $x^3 + 4x^2 + 5x + 2$ **(b)** 0 [5.5] **20.** **(a)** $x + 2, \; x \neq -1$ **(b)** 0

Cumulative Review Exercises Chapters 1–5   (page 334)

[1.1] **1.** A, B, C, D, F   **2.** B, C, D, F   **3.** D, F   **4.** C, D, F   **5.** E, F   **6.** D, F   [1.3] **7.** 32   **8.** 0   [2.1] **9.** $\{-65\}$

10. $(-\infty, \infty)$ [2.2] **11.** $t = \dfrac{A - p}{pr}$ [2.5] **12.** $(-\infty, 6)$ [2.7] **13.** $\left\{-\dfrac{1}{3}, 1\right\}$ **14.** $\left(-\infty, -\dfrac{8}{3}\right] \cup [2, \infty)$ [2.3, 2.4] **15.** 32%; 390;

270; 10%   **16.** $15°, 35°, 130°$ [3.2] **17.** $-\dfrac{4}{3}$ **18.** 0   [3.3] **19.** $y = -4x + 15$ **20.** $y = 4x$ [3.1] **21.**

[3.4] **22.**   **23.**   [3.2, 3.3] **24. (a)** 2505.1 per yr; The number of twin births increased an average of

2505.1 per yr. **(b)** $y = 2505.1x + 93{,}865$ **(c)** about 123,926   [3.5] **25.** domain: $\{-4, -1, 2, 5\}$; range: $\{-2, 0, 2\}$; function
[3.6] **26.** 800 lb   [4.1] **27.** $\{(3, 2)\}$ **28.** $\emptyset$ [4.2] **29.** $\{(1, 0, -1)\}$ [4.3] **30.** length: 42 ft; width: 30 ft   **31.** 15% solution: 6 L;
30% solution: 3 L   [5.1] **32.** $\dfrac{8m^9 n^3}{p^6}$ **33.** $\dfrac{y^7}{x^{13} z^2}$ **34.** $\dfrac{m^6}{8n^9}$ [5.2] **35.** $2x^2 - 4x + 38$ [5.4] **36.** $15x^2 + 7xy - 2y^2$ **37.** $64m^2 - 25n^2$
38. $x^3 + 8y^3$ [5.5] **39.** $m^2 - 2m + 3$ [5.1] **40. (a)** 2.814×10^8 **(b)** 1×10^{12} **(c)** \$3554

CHAPTER 6 FACTORING

Section 6.1   (page 358)

Exercises   **1.** $12(m + 5)$ **3.** $8k(k^2 + 3)$ **5.** $xy(1 - 5y)$ **7.** $-2p^2 q^4(2p + q)$ **9.** $7x^3(3x^2 + 5x + 2)$ **11.** $2t^3(5t^2 - 4t - 1)$
13. $5ac(3ac^2 - 5c + 1)$ **15.** $16zn^3(zn^3 + 4n^4 - 2z^2)$ **17.** $-9m^3 p^3(3p^2 - 4m + 8m^2 p)$ **19.** $7ab(2a^2 b + a - 3a^4 b^2 + 6b^3)$
21. $(m - 4)(2m + 5)$ **23.** $11(2z - 1)$ **25.** $(2 - x)(10 - x - x^2)$ **27.** $(3 - x)(6 + 2x - x^2)$ **29.** $20z(2z + 1)(3z + 4)$
31. $5(m + p)^2(m + p - 2 - 3m^2 - 6mp - 3p^2)$ **33.** $r(-r^2 + 3r + 5)$; $-r(r^2 - 3r - 5)$ **35.** $12s^4(-s + 4)$; $-12s^4(s - 4)$
37. $2x^2(-x^3 + 3x + 2)$; $-2x^2(x^3 - 3x - 2)$ **39.** $(m + 3q)(x + y)$ **41.** $(5m + n)(2 + k)$ **43.** $(2 - q)(2 - 3p)$
45. $(p + q)(p - 4z)$ **47.** $(x - 4)(2y + 3)$ **49.** $(m + 4)(m^2 - 6)$ **51.** $(a^2 + b^2)(-3a + 2b)$ **53.** $(y - 2)(x - 2)$
55. $(3y - 2)(3y^3 - 4)$ **57.** $(1 - a)(1 - b)$ **59.** $m^{-5}(3 + m^2)$ **61.** $p^{-3}(3 + 2p)$ **63.** The directions indicated to factor the polynomial
completely. The student's response is not the complete factored form, which is $4xy^3(xy^2 - 2)$. The teacher was justified because the directions
were not fulfilled.   **65.** C

Section 6.2   (page 365)

Exercises   **1.** D   **3.** B   **5.** $(y - 3)(y + 10)$ **7.** $(p + 8)(p + 7)$ **9.** prime   **11.** $(a + 5b)(a - 7b)$ **13.** prime
15. $(xy + 9)(xy + 2)$ **17.** $-(6m - 5)(m + 3)$ **19.** $(5x - 6)(2x + 3)$ **21.** $(4k + 3)(5k + 8)$ **23.** $(3a - 2b)(5a - 4b)$
25. $(6m - 5)^2$ **27.** prime   **29.** $(2xz - 1)(3xz + 4)$ **31.** $3(4x + 5)(2x + 1)$ **33.** $-5(a + 6)(3a - 4)$ **35.** $-11x(x - 6)(x - 4)$
37. $2xy^3(x - 12y)^2$ **39.** $6a(a - 3)(a + 5)$ **41.** $13y(y + 4)(y - 1)$ **43.** $3p(2p - 1)^2$ **45.** There is a GCF of 2. She did not factor
the polynomial *completely.* The factor $(4x + 10)$ can be factored further as $2(2x + 5)$, giving the final form as $2(2x + 5)(x - 2)$.
47. $(6p^3 - r)(2p^3 - 5r)$ **49.** $(5k + 4)(2k + 1)$ **51.** $(3m + 3p + 5)(m + p - 4)$ **53.** $(a + b)^2(a - 3b)(a + 2b)$
55. $(p + q)^2(p + 3q)$ **57.** $(z - x)^2(z + 2x)$ **59.** $(p^2 - 8)(p^2 - 2)$ **61.** $(2x^2 + 3)(x^2 - 6)$ **63.** $(4x^2 + 3)(4x^2 + 1)$

Section 6.3 (page 370)

Exercises 1. A, D **3.** B, C **5.** The sum of two squares can be factored only if the binomial has a common factor.
7. $(p + 4)(p - 4)$ **9.** $(5x + 2)(5x - 2)$ **11.** $2(3a + 7b)(3a - 7b)$ **13.** $4(4m^2 + y^2)(2m + y)(2m - y)$
15. $(y + z + 9)(y + z - 9)$ **17.** $(4 + x + 3y)(4 - x - 3y)$ **19.** $4pq$ **21.** $(k - 3)^2$ **23.** $(2z + w)^2$
25. $(4m - 1 + n)(4m - 1 - n)$ **27.** $(2r - 3 + s)(2r - 3 - s)$ **29.** $(x + y - 1)(x - y + 1)$ **31.** $2(7m + 3n)^2$ **33.** $(p + q + 1)^2$
35. $(a - b + 4)^2$ **37.** $(2x - y)(4x^2 + 2xy + y^2)$ **39.** $(4g - 3h)(16g^2 + 12gh + 9h^2)$ **41.** $3(2n + 3p)(4n^2 - 6np + 9p^2)$
43. $(y + z + 4)(y^2 + 2yz + z^2 - 4y - 4z + 16)$ **45.** $2b(3a^2 + b^2)$ **47.** $(m^2 - 5)(m^4 + 5m^2 + 25)$
49. $(10x^3 - 3)(100x^6 + 30x^3 + 9)$ **51.** $(5y^2 + z^2)(25y^4 - 5y^2z^2 + z^4)$ **52.** $(x^3 - y^3)(x^3 + y^3)$; $(x - y)(x^2 + xy + y^2) \cdot$
$(x + y)(x^2 - xy + y^2)$ **53.** $(x^2 + xy + y^2)(x^2 - xy + y^2)$ **54.** $(x^2 - y^2)(x^4 + x^2y^2 + y^4)$; $(x - y)(x + y)(x^4 + x^2y^2 + y^4)$
55. $x^4 + x^2y^2 + y^4$ **56.** The product must equal $x^4 + x^2y^2 + y^4$. Multiply $(x^2 + xy + y^2)(x^2 - xy + y^2)$ to verify this. **57.** Start by
factoring as a difference of squares. **59.** $(5p + 2q)(25p^2 - 10pq + 4q^2 + 5p - 2q)$ **61.** $(3a - 4b)(9a^2 + 12ab + 16b^2 + 5)$
63. $(t - 3)(2t + 1)(4t^2 - 2t + 1)$ **65.** $(8m - 9n)(8m + 9n - 64m^2 - 72mn - 81n^2)$

Section 6.4 (page 375)

Exercises 1. $(10a + 3b)(10a - 3b)$ **3.** $3p^2(p - 6)(p + 5)$ **5.** $3pq(a + 6b)(a - 5b)$ **7.** prime **9.** $(6b + 1)(b - 3)$
11. $(x - 10)(x^2 + 10x + 100)$ **13.** $(p + 2)(4 + m)$ **15.** $9m(m - 5 + 2m^2)$ **17.** $2(3m - 10)(9m^2 + 30m + 100)$ **19.** $(3m - 5n)^2$
21. $(k - 9)(q + r)$ **23.** $16z^2x(zx - 2)$ **25.** $(x + 7)(x - 5)$ **27.** $(x - 5)(x + 5)(x^2 + 25)$ **29.** $(p + 4)(p^2 - 4p + 16)$
31. $(8m + 25)(8m - 25)$ **33.** $6z(2z^2 - z + 3)$ **35.** $16(4b + 5c)(4b - 5c)$ **37.** $8(5z + 4)(25z^2 - 20z + 16)$ **39.** $(5r - s)(2r + 5s)$
41. $4pq(2p + q)(3p + 5q)$ **43.** $3(4k^2 + 9)(2k + 3)(2k - 3)$ **45.** $(m - n)(m^2 + mn + n^2 + m + n)$ **47.** $(x - 2m - n)(x + 2m + n)$
49. $6p^3(3p^2 - 4 + 2p^3)$ **51.** $2(x + 4)(x - 5)$ **53.** $8mn$ **55.** $2(5p + 9)(5p - 9)$ **57.** $4rx(3m^2 + mn + 10n^2)$
59. $(7a - 4b)(3a + b)$ **61.** prime **63.** $(p + 8q - 5)^2$ **65.** $(7m^2 + 1)(3m^2 - 5)$ **67.** $(2r - t)(r^2 - rt + 19t^2)$
69. $(x + 3)(x^2 + 1)(x + 1)(x - 1)$ **71.** $(m + n - 5)(m - n + 1)$

Section 6.5 (page 382)

Exercises 1. First rewrite the equation so one side is 0. Factor the other side and set each factor equal to 0. The solutions of these linear
equations are solutions of the quadratic equation. **3.** $\{5, -10\}$ **5.** $\left\{-\dfrac{8}{3}, \dfrac{5}{2}\right\}$ **7.** $\{-2, 5\}$ **9.** $\{-6, -3\}$ **11.** $\left\{-\dfrac{1}{2}, 4\right\}$

13. $\left\{-\dfrac{1}{3}, \dfrac{4}{5}\right\}$ **15.** $\{-3, 4\}$ **17.** $\left\{-5, -\dfrac{1}{5}\right\}$ **19.** $\{-4, 0\}$ **21.** $\{0, 6\}$ **23.** $\{-3, 3\}$ **25.** $\{-2, 2\}$ **27.** $\{3\}$ **29.** $\left\{-\dfrac{4}{3}\right\}$

31. $\{-4, 2\}$ **33.** $\left\{-\dfrac{1}{2}, 6\right\}$ **35.** $\{1, 6\}$ **37.** $\left\{-\dfrac{1}{2}, 0, 5\right\}$ **39.** $\left\{-\dfrac{4}{3}, 0, \dfrac{4}{3}\right\}$ **41.** $\left\{-\dfrac{5}{2}, -1, 1\right\}$ **43.** $\{-3, 3, 6\}$

45. $\{-2, 2, 3\}$ **47.** D **49.** $\left\{-\dfrac{15}{8}, -1\right\}$ **51.** $\left\{-\dfrac{5}{2}, -1\right\}$ **53.** $\left\{-1, -\dfrac{1}{4}\right\}$ **55.** mirror: 7 ft; painting: 9 ft **57.** base: 12 ft;
height: 5 ft **59.** width: 100 ft; length: 300 ft **61.** -9 and -8 or 8 and 9 **63.** Each side measures 7 in. **65.** 5 sec **67.** 6 sec
69. $\{-5, 1.5\}$ **71.** $\{-3, 4\}$

Chapter 6 Review Exercises (page 389)

1. $6p(2p - 1)$ **3.** $4qb(3q + 2b - 5q^2b)$ **5.** $(x + 3)(x - 3)$ **7.** $(m + q)(4 + n)$ **9.** $(m + 3)(2 - a)$ **11.** $(3p - 4)(p + 1)$
13. $(3r + 1)(4r - 3)$ **15.** $(2k - h)(5k - 3h)$ **17.** $2x(4 + x)(3 - x)$ **19.** $(y^2 + 4)(y^2 - 2)$ **21.** $(p + 2)^2(p + 3)(p - 2)$
23. It is not factored because there are two terms: $x^2(y^2 - 6)$ and $5(y^2 - 6)$. The correct answer is $(y^2 - 6)(x^2 + 5)$. **25.** $(4x + 5)(4x - 5)$
27. $(6m - 5n)(6m + 5n)$ **29.** $(3k - 2)^2$ **31.** $(5x - 1)(25x^2 + 5x + 1)$ **33.** $(x^4 + 1)(x^2 + 1)(x + 1)(x - 1)$ **35.** $2b(3a^2 + b^2)$

37. $\{4\}$ **39.** $\{2, 3\}$ **41.** $\left\{-\dfrac{5}{2}, \dfrac{10}{3}\right\}$ **43.** $\left\{-\dfrac{3}{2}, -\dfrac{1}{4}\right\}$ **45.** $\left\{-\dfrac{3}{2}, 0\right\}$ **47.** $\{4\}$ **49.** $\{-3, -2, 2\}$ **51.** 3 ft

53. after 16 sec   **55.** The rock reaches a height of 240 ft once on its way up and once on its way down.   **57.** $(4 + 9k)(4 - 9k)$
59. prime   **61.** $(5z - 3m)^2$ **63.** $\{0, 3\}$ **65.** 6 in.

Chapter 6 Test   (page 391)

[6.1–6.4] **1.** $11z(z - 4)$ **2.** $5x^2y^3(2y^2 - 1 - 5x^3)$ **3.** $(x + y)(3 + b)$ **4.** $-(2x + 9)(x - 4)$ **5.** $(3x - 5)(2x + 7)$
6. $(4p - q)(p + q)$ **7.** $(4a + 5b)^2$ **8.** $(x + 1 + 2z)(x + 1 - 2z)$ **9.** $(a + b)(a - b)(a + 2)$ **10.** $(3k + 11j)(3k - 11j)$
11. $(y - 6)(y^2 + 6y + 36)$ **12.** $(2k^2 - 5)(3k^2 + 7)$ **13.** $(3x^2 + 1)(9x^4 - 3x^2 + 1)$ [6.1] **14.** It is not in factored form because there are two terms: $(x^2 + 2y)p$ and $3(x^2 + 2y)$. The common factor is $x^2 + 2y$, and the factored form is $(x^2 + 2y)(p + 3)$.   [6.2] **15.** D
[6.5] **16.** $\left\{ -2, -\dfrac{2}{3} \right\}$ **17.** $\left\{ 0, \dfrac{5}{3} \right\}$ **18.** $\left\{ -\dfrac{2}{5}, 1 \right\}$ **19.** length: 8 in.; width: 5 in.   **20.** 2 sec and 4 sec

Cumulative Review Exercises Chapters 1–6   (page 392)

[1.4] **1.** $-2m + 6$ **2.** $4m - 3$ **3.** $2x^2 + 5x + 4$ [1.3] **4.** -24 **5.** 204   **6.** undefined   **7.** 10   [2.1] **8.** $\left\{ \dfrac{7}{6} \right\}$ **9.** $\{-1\}$
[2.5] **10.** $\left(-\infty, \dfrac{15}{4} \right]$ **11.** $\left(-\dfrac{1}{2}, \infty \right)$ [2.6] **12.** $(2, 3)$ **13.** $(-\infty, 2) \cup (3, \infty)$ [2.7] **14.** $\left\{ -\dfrac{16}{5}, 2 \right\}$ **15.** $(-11, 7)$
16. $(-\infty, -2] \cup [7, \infty)$ [2.2] **17.** $h = \dfrac{V}{lw}$ [2.4] **18.** 2 hr   [3.1] **19.**

   [3.2] **20.** -1 **21.** 0   [3.5] **22.** -1

[3.1] **23.** $\left(-\dfrac{7}{2}, 0 \right)$ **24.** $(0, 7)$ [4.1] **25.** $\{(1, 5)\}$ [4.2] **26.** $\{(1, 1, 0)\}$ [5.1] **27.** $\dfrac{y}{18x}$ **28.** $\dfrac{5my^4}{3}$ [5.2] **29.** $x^3 + 12x^2 - 3x - 7$
[5.4] **30.** $49x^2 + 42xy + 9y^2$ **31.** $10p^3 + 7p^2 - 28p - 24$ [6.1–6.4] **32.** $(2w + 7z)(8w - 3z)$ **33.** $(2x - 1 + y)(2x - 1 - y)$
34. $(2y - 9)^2$ **35.** $(10x^2 + 9)(10x^2 - 9)$ **36.** $(2p + 3)(4p^2 - 6p + 9)$ [6.5] **37.** $\left\{ -4, -\dfrac{3}{2}, 1 \right\}$ **38.** $\left\{ \dfrac{1}{3} \right\}$ **39.** 4 ft
40. longer sides: 18 in.; distance between: 16 in.

CHAPTER 7 RATIONAL EXPRESSIONS AND FUNCTIONS

Section 7.1   (page 402)

Exercises   1. C   **3.** D   **5.** E   **7.** Replacing x with 2 makes the denominator 0 and the value of the expression undefined. To find the values excluded from the domain, set the denominator equal to 0 and solve the equation. All solutions of the equation are excluded from the domain.   **9.** 7   **11.** $-\dfrac{1}{7}$ **13.** 0   **15.** $-2, \dfrac{3}{2}$ **17.** none   **19.** none   **21. (a)** numerator: x^2, $4x$; denominator: x, 4   **(b)** First factor the numerator, getting $x(x + 4)$, then divide the numerator and denominator by the common factor of $x + 4$ to get $\dfrac{x}{1}$ or x.   **23.** B   **25.** x
27. $\dfrac{x - 3}{x + 5}$ **29.** $\dfrac{x + 3}{2x(x - 3)}$ **31.** already in lowest terms   **33.** $\dfrac{6}{7}$ **35.** $\dfrac{z}{6}$ **37.** $\dfrac{2}{t - 3}$ **39.** $\dfrac{x - 3}{x + 1}$ **41.** $\dfrac{4x + 1}{4x + 3}$
43. $a^2 - ab + b^2$ **45.** $\dfrac{c + 6d}{c - d}$ **47.** $\dfrac{a + b}{a - b}$ **49.** -1 In Exercises 51 and 53, there are other acceptable ways to express each answer.

51. $-(x + y)$ **53.** $-\dfrac{x + y}{x - y}$ **55.** $-\dfrac{1}{2}$ **57.** already in lowest terms   **59.** Multiply the numerators, multiply the denominators, and factor each numerator and denominator. (Factoring can be performed first.) Divide the numerator and denominator by any common factors to write in lowest terms. For example, $\dfrac{6r - 5s}{3r + 2s} \cdot \dfrac{6r + 4s}{5s - 6r} = \dfrac{(6r - 5s)(6r + 4s)}{(3r + 2s)(5s - 6r)} = \dfrac{(6r - 5s)2(3r + 2s)}{(3r + 2s)(-1)(6r - 5s)} = \dfrac{2}{-1} = -2$.   **61.** $\dfrac{3y}{x^2}$

63. $\dfrac{3a^3b^2}{4}$ **65.** $\dfrac{27}{2mn^7}$ **67.** $\dfrac{x + 4}{x - 2}$ **69.** $\dfrac{2x + 3}{x + 2}$ **71.** $\dfrac{7x}{6}$ **73.** $-\dfrac{p + 5}{2p}$ (There are other ways.)   **75.** $\dfrac{35}{4}$ **77.** $-(z + 1)$ or $-z - 1$

79. $\dfrac{-m(m + 7)}{m + 1}$ (There are other ways.)   **81.** -2 **83.** $\dfrac{x + 4}{x - 4}$ **85.** $\dfrac{2x + 3y}{2x - 3y}$ **87.** $\dfrac{k + 5p}{2k + 5p}$ **89.** $(k - 1)(k - 2)$

91. $\dfrac{(a + 5)(2a + b)}{(3a + 1)(a + 2b)}$

Section 7.2   (page 411)

Exercises   **1.** $\dfrac{3}{4}$ **2.** $\dfrac{1}{6}$ **3.** no; We cannot find the sum $\dfrac{1}{x} + \dfrac{1}{y}$ by adding the denominators and keeping the common numerator.

4. $\dfrac{2}{15}$ **5.** $-\dfrac{1}{2}$ **6.** no; We cannot find the difference $\dfrac{1}{x} - \dfrac{1}{y}$ by subtracting the denominators and keeping the common numerator.

7. $\dfrac{9}{t}$ **9.** $\dfrac{2}{x}$ **11.** 1 **13.** $x - 5$ **15.** $\dfrac{5}{p + 3}$ **17.** $a - b$ **19.** First add or subtract the numerators. Then place the result over the common denominator. Write the answer in lowest terms. We give one example: $\dfrac{5}{x} - \dfrac{3x + 1}{x} = \dfrac{5 - (3x + 1)}{x} = \dfrac{5 - 3x - 1}{x} = \dfrac{4 - 3x}{x}$.

21. $72x^4y^5$ **23.** $z(z - 2)$ **25.** $2(y + 4)$ **27.** $(x + 9)^2(x - 9)$ **29.** $(m + n)(m - n)$ **31.** $x(x - 4)(x + 1)$

33. $(t + 5)(t - 2)(2t - 3)$ **35.** $2y(y + 3)(y - 3)$ **37.** Yes, they could both be correct because the expressions are equivalent.

Multiplying $\dfrac{3}{5 - y}$ by 1 in the form $\dfrac{-1}{-1}$ gives $\dfrac{-3}{y - 5}$.   **39.** $\dfrac{31}{3t}$ **41.** $\dfrac{5 - 22x}{12x^2y}$ **43.** $\dfrac{1}{x(x - 1)}$ **45.** $\dfrac{5a^2 - 7a}{(a + 1)(a - 3)}$ **47.** 3

49. $\dfrac{3}{x - 4}$ or $\dfrac{-3}{4 - x}$ **51.** $\dfrac{w + z}{w - z}$ or $\dfrac{-w - z}{z - w}$ **53.** $\dfrac{-13}{12(3 + x)}$ **55.** $\dfrac{2(2x - 1)}{x - 1}$ **57.** $\dfrac{7}{y}$ **59.** $\dfrac{6}{x - 2}$ **61.** $\dfrac{3x - 2}{x - 1}$

63. $\dfrac{4x - 7}{x^2 - x + 1}$ **65.** $\dfrac{2x + 1}{x}$ **67.** $\dfrac{4p^2 - 21p + 29}{(p - 2)^2}$ **69.** $\dfrac{x}{(x - 2)^2(x - 3)}$ **71.** $\dfrac{2x(x + 12y)}{(x + 2y)(x - y)(x + 6y)}$

73. $\dfrac{2x^2 + 21xy - 10y^2}{(x + 2y)(x - y)(x + 6y)}$ **75.** $\dfrac{3r - 2s}{(2r - s)(3r - s)}$ **77.** $\dfrac{10x + 23}{(x + 2)^2(x + 3)}$ **79. (a)** $c(x) = \dfrac{10x}{49(101 - x)}$ **(b)** approximately 3.23 thousand

dollars   **81.** $\dfrac{8}{9}$ **82.** $\dfrac{3}{7} + \dfrac{5}{9} - \dfrac{6}{63}$; They are the same.   **83.** $\dfrac{8}{9}$; yes   **84.** Answers will vary. Suppose the name is Bush, so that

$x = 4$. The problem is $\dfrac{3}{2} + \dfrac{5}{4} - \dfrac{6}{8}$. The predicted answer is $\dfrac{8}{4} = 2$, which is correct.   **85.** It causes $\dfrac{3}{x - 2}$ and $\dfrac{6}{x^2 - 2x}$ to be undefined

since 0 appears in the denominators.   **86.** 0

Section 7.3   (page 420)

Exercises   **1.** *Method 1:* Begin by simplifying the numerator to a single fraction. Then simplify the denominator to a single fraction. Write as a division problem, and multiply by the reciprocal of the denominator. Simplify the result, if possible. *Method 2:* Find the LCD of all fractions in the complex fraction. Multiply the numerator and denominator of the complex fraction by this LCD. Simplify the result, if possible.

3. $\dfrac{2x}{x - 1}$ **5.** $\dfrac{2(k + 1)}{3k - 1}$ **7.** $\dfrac{5x^2}{9z^3}$ **9.** $\dfrac{1 + x}{-1 + x}$ **11.** $\dfrac{y + x}{y - x}$ **13.** $4x$ **15.** $x + 4y$ **17.** $\dfrac{3y}{2}$ **19.** $\dfrac{x^2 + 5x + 4}{x^2 + 5x + 10}$ **21.** $\dfrac{m^2 + 6m - 4}{m(m - 1)}$

22. $\dfrac{m^2 - m - 2}{m(m - 1)}$ **23.** $\dfrac{m^2 + 6m - 4}{m^2 - m - 2}$ **24.** $m(m - 1)$ **25.** $\dfrac{m^2 + 6m - 4}{m^2 - m - 2}$ **26.** Method 1 involves simplifying the numerator and the denominator separately and then performing a division. Method 2 involves multiplying the fraction by a form of 1, the identity element for multiplication. (Preference will vary.)   **27.** $\dfrac{x^2 y^2}{y^2 + x^2}$ **29.** $\dfrac{y^2 + x^2}{xy^2 + x^2 y}$ or $\dfrac{y^2 + x^2}{xy(y + x)}$ **31.** $\dfrac{1}{2xy}$ **33.** (a) $\dfrac{\dfrac{3}{mp} - \dfrac{4}{p} + \dfrac{8}{m}}{\dfrac{2}{m} - \dfrac{3}{p}}$

(b) In the denominator, $2m^{-1} = \dfrac{2}{m}$, not $\dfrac{1}{2m}$, and $3p^{-1} = \dfrac{3}{p}$, not $\dfrac{1}{3p}$.   **(c)** $\dfrac{3 - 4m + 8p}{2p - 3m}$

Section 7.4   (page 426)

Exercises   1. (a) $-1, 2$ **(b)** $\{x \mid x \neq -1, 2\}$ **3. (a)** $-\dfrac{5}{3}, 0, -\dfrac{3}{2}$ **(b)** $\left\{ x \mid x \neq -\dfrac{5}{3}, 0, -\dfrac{3}{2} \right\}$ **5. (a)** 0 **(b)** $\{x \mid x \neq 0\}$ **7. (a)** $4, \dfrac{7}{2}$

(b) $\left\{ x \mid x \neq 4, \dfrac{7}{2} \right\}$ **9. (a)** $0, 1, -3, 2$ **(b)** $\{x \mid x \neq 0, 1, -3, 2\}$ **11.** $\{1\}$ **13.** $\{-6, 4\}$ **15.** $\left\{ -\dfrac{7}{12} \right\}$ **17.** $\emptyset$ **19.** $\{-3\}$ **21.** $\{5\}$

23. $\{5\}$ **25.** $\emptyset$ **27.** $\left\{ \dfrac{27}{56} \right\}$ **29.** $\emptyset$ **31.** $\{-10\}$ **33.** $\{-1\}$ **35.** $\{0\}$ **37.** $\{15\}$ **39.** $\left\{ x \mid x \neq -\dfrac{3}{2}, \dfrac{3}{2} \right\}$

41. $x = 0$ **43.** $x = 2$ **45. (a)** 0 **(b)** 1.6 **(c)** 4.1 **(d)** The waiting time also increases.

47. (a) 500 ft **(b)** It decreases.   **49.** four   **51.** $\{-2, 0, 3\}$

Summary Exercises on Operations and Equations with Rational Expressions
(page 430)

1. equation; $\{20\}$ **2.** operation; $\dfrac{2(x + 5)}{5}$ **3.** operation; $-\dfrac{22}{7x}$ **4.** operation; $\dfrac{y + x}{y - x}$ **5.** equation; $\left\{ \dfrac{1}{2} \right\}$ **6.** equation; $\{7\}$

7. operation; $\dfrac{43}{24x}$ **8.** equation; $\{1\}$ **9.** operation; $\dfrac{5x - 1}{-2x + 2}$ or $\dfrac{5x - 1}{-2(x - 1)}$ **10.** operation; $\dfrac{25}{4(r + 2)}$ **11.** operation; $\dfrac{x^2 + xy + 2y^2}{(x + y)(x - y)}$

12. operation; $\dfrac{24p}{p + 2}$ **13.** operation; $-\dfrac{5}{36}$ **14.** equation; $\{0\}$ **15.** operation; $\dfrac{b + 3}{3}$ **16.** operation; $\dfrac{5}{3z}$

17. operation; $\dfrac{2x + 10}{x(x - 2)(x + 2)}$ **18.** equation; $\{2\}$ **19.** operation; $\dfrac{-x}{3x + 5y}$ **20.** equation; $\{-13\}$ **21.** operation; $\dfrac{3y + 2}{y + 3}$

22. equation; $\left\{ \dfrac{5}{4} \right\}$ **23.** equation; $\emptyset$ **24.** operation; $\dfrac{2z - 3}{2z + 3}$ **25.** operation; $\dfrac{-1}{x - 3}$ or $\dfrac{1}{3 - x}$ **26.** operation; $\dfrac{t - 2}{8}$

27. equation; $\{-10\}$ **28.** operation; $\dfrac{13x + 28}{2x(x + 4)(x - 4)}$ **29.** equation; $\emptyset$ **30.** operation; $\dfrac{k(2k^2 - 2k + 5)}{(k - 1)(3k^2 - 2)}$

Section 7.5   (page 438)

Exercises   1. A   **3.** D   **5.** 65.625 **7.** $\dfrac{25}{4}$ **9.** $G = \dfrac{Fd^2}{Mm}$ **11.** $a = \dfrac{bc}{c + b}$ **13.** $v = \dfrac{PVt}{pT}$ **15.** $r = \dfrac{nE - IR}{In}$ **17.** $b = \dfrac{2A}{h} - B$

or $b = \dfrac{2A - Bh}{h}$ **19.** $r = \dfrac{eR}{E - e}$ **21.** Multiply each side by $a - b$.   **23.** 15 girls, 5 boys   **25.** $\dfrac{1}{2}$ job per hr   **27.** 1996   **29.** 1996

31. 23 teachers   **33.** 25,000 fish   **35.** 2.4 mL   **37.** $x = \dfrac{7}{2}$; $AC = 8$; $DF = 12$ **39.** 3 mph   **41.** 900 mi   **43.** 480 mi

45. 190 mi   **47.** $6\dfrac{2}{3}$ min   **49.** 12 hr   **51.** 20 hr   **53.** $2\dfrac{4}{5}$ hr

Chapter 7 Review Exercises   (page 449)

1. (a) -6 **(b)** $\{x \mid x \neq -6\}$ **3. (a)** 9 **(b)** $\{x \mid x \neq 9\}$ **5.** $\dfrac{5m + n}{5m - n}$ **7.** The reciprocal of a rational expression is another rational expression

such that the two rational expressions have a product of 1.   **9.** $\dfrac{-3(w + 4)}{w}$ **11.** 1   **13.** $9r^2(3r + 1)$ **15.** $\dfrac{16z - 3}{2z^2}$ **17.** $\dfrac{71}{30(a + 2)}$

19. $\dfrac{3 + 2t}{4 - 7t}$ **21.** $\dfrac{1}{3q + 2p}$ **23.** $\{-3\}$ **25.** $\{0\}$ **27.** Although her algebra was correct, 3 is not a solution because it is not in the domain

of the equation. Thus, $\emptyset$ is correct.   **29.** C; $x = 0$ **31.** $m = \dfrac{Fd^2}{GM}$ **33.** 6000 passenger-km per day   **35.** $4\dfrac{4}{5}$ min   **37.** $\dfrac{1}{x - 2y}$

39. $\dfrac{6m + 5}{3m^2}$ **41.** $\dfrac{x^2 - 6}{2(2x + 1)}$ **43.** $\dfrac{3 - 5x}{6x + 1}$ **45.** $\dfrac{1}{3}$ **47.** $\dfrac{5a^2 + 4ab + 12b^2}{(a + 3b)(a - 2b)(a + b)}$ **49.** $\left\{\dfrac{1}{3}\right\}$ **51.** $\{1, 4\}$ **53. (a)** 8.32 **(b)** 44.9

55. \$21.06

Chapter 7 Test   (page 452)

[7.1] 1. $-2, \dfrac{4}{3}; \left\{x \mid x \neq -2, \dfrac{4}{3}\right\}$ **2.** $\dfrac{2x - 5}{x(3x - 1)}$ **3.** $\dfrac{3(x + 3)}{4}$ **4.** $\dfrac{y + 4}{y - 5}$ **5.** $\dfrac{x + 5}{x}$ **[7.2] 6.** $t^2(t + 3)(t - 2)$ **7.** $\dfrac{7 - 2t}{6t^2}$

8. $\dfrac{11x + 21}{(x - 3)^2(x + 3)}$ **9.** $\dfrac{4}{x + 2}$ **[7.3] 10.** $\dfrac{72}{11}$ **11.** $-\dfrac{1}{a + b}$ **12.** $\dfrac{2y^2 + x^2}{xy(y - x)}$ **[7.4] 13. (a)** operation; $\dfrac{11(x - 6)}{12}$ **(b)** equation; $\{6\}$

14. $\left\{\dfrac{1}{2}\right\}$ **15.** $\{5\}$ **16.** A solution cannot make a denominator 0.   **17.** $\ell = \dfrac{2S}{n} - a$ or $\ell = \dfrac{2S - na}{n}$

18. $x = -1$ **[7.5] 19.** $3\dfrac{3}{14}$ hr   **20.** 15 mph   **21.** 48,000 fish   **22. (a)** 3 units **(b)** 0

Cumulative Review Exercises Chapters 1–7   (page 453)

[1.3] 1. -199 **2.** 12   **[2.1] 3.** $\left\{-\dfrac{15}{4}\right\}$ **[2.7] 4.** $\left\{\dfrac{2}{3}, 2\right\}$ **[2.3] 5.** $x = \dfrac{d - by}{a - c}$ or $x = \dfrac{by - d}{c - a}$ **[2.5] 6.** $\left(-\infty, \dfrac{240}{13}\right]$

[2.7] 7. $(-\infty, -2] \cup \left[\dfrac{2}{3}, \infty\right)$ **[2.3] 8.** \$4000 at 4%; \$8000 at 3%   **9.** 6 m   **[3.1] 10.** x-intercept: $(-2, 0)$;

y-intercept: $(0, 4)$ **[3.2] 11.** $-\dfrac{3}{2}$ **12.** $-\dfrac{3}{4}$ **[3.3] 13.** $y = -\dfrac{3}{2}x + \dfrac{1}{2}$ **[3.4] 14.**   **15.**

x − y ≥ 3 and
3x + 4y ≤ 12

[3.5] 16. function; domain: $\{1990, 1992, 1994, 1996, 1998, 2000\}$; range: $\{1.25, 1.61, 1.80, 1.21, 1.94, 2.26\}$ **17.** not a function; domain:

$[-2, \infty)$; range: $(-\infty, \infty)$ **18.** function; domain: $[-2, \infty)$; range: $(-\infty, 0]$ **19. (a)** $f(x) = \dfrac{5x - 8}{3}$ or $f(x) = \dfrac{5}{3}x - \dfrac{8}{3}$ **(b)** -1

20. $3x + 15$ **[4.1, 4.4, 4.5] 21.** $\{(-1, 3)\}$ **[4.2, 4.4, 4.5] 22.** $\{(-2, 3, 1)\}$ **23.** $\emptyset$ **[4.3] 24.** automobile: 42 km per hr; airplane:

600 km per hr   **[4.5] 25.** 7   **[5.1] 26.** $\dfrac{a^{10}}{b^{10}}$ **27.** $\dfrac{m}{n}$ **[5.2] 28.** $4y^2 - 7y - 6$ **[5.4] 29.** $12f^2 + 5f - 3$ **30.** $49t^6 - 64$

31. $\dfrac{1}{16}x^2 + \dfrac{5}{2}x + 25$ **[5.5] 32.** $x^2 + 4x - 7$ **[5.3] 33. (a)** $2x^3 - 2x^2 + 6x - 4$ **(b)** $2x^3 - 4x^2 + 2x + 2$ **(c)** -14

[6.1] **34.** $(2x + 5)(x - 9)$ [6.2] **35.** $25(2t^2 + 1)(2t^2 - 1)$ **36.** $(2p + 5)(4p^2 - 10p + 25)$ [6.5] **37.** $\left\{-\dfrac{7}{3}, 1\right\}$ [7.1] **38.** $\dfrac{y + 4}{y - 4}$

39. $\dfrac{2x - 3}{2(x - 1)}$ **40.** $\dfrac{a(a - b)}{2(a + b)}$ [7.2] **41.** 3   **42.** $\dfrac{2(x + 2)}{2x - 1}$ [7.4] **43.** $\{-4\}$ [7.5] **44.** $q = \dfrac{fp}{p - f}$ or $q = \dfrac{-fp}{f - p}$ **45.** 150 mph

46. $\dfrac{6}{5}$ or $1\dfrac{1}{5}$ hr

CHAPTER 8 ROOTS, RADICALS, AND ROOT FUNCTIONS

Section 8.1   (page 462)

Exercises **1.** E   **3.** D   **5.** A   **7.** C   **9.** C   **11.** **(a)** not a real number **(b)** negative **(c)** 0   **13.** 9   **15.** -6 **17.** -4 **19.** 8

21. 6   **23.** -3 **25.** not a real number   **27.** 2   **29.** not a real number   **31.** $\dfrac{8}{9}$ **33.** $\dfrac{2}{3}$ **35.** $\dfrac{1}{2}$ **In Exercises 37–43, we**

give the domain and then the range.   **37.** $[-3, \infty); [0, \infty)$ **39.** $[0, \infty); [-2, \infty)$

41. $(-\infty, \infty); (-\infty, \infty)$ **43.** $(-\infty, \infty); (-\infty, \infty)$ **45.** 12   **47.** 10   **49.** 2   **51.** -9 **53.** $|x|$ **55.** x

57. x^5 **59.** 97.381   **61.** 16.863   **63.** 2.646   **65.** -9.055 **67.** 7.507   **69.** 3.162   **71.** 1.885   **73.** 1,183,000 cycles per sec
75. 10 mi   **77.** 392,000 mi^2 **79.** 1.732 amps

Section 8.2   (page 471)

Exercises **1.** C   **3.** A   **5.** H   **7.** B   **9.** D   **11.** 13   **13.** 9   **15.** 2   **17.** $\dfrac{8}{9}$ **19.** -3 **21.** 1000   **23.** -1024 **25.** not a

real number   **27.** $\dfrac{1}{512}$ **29.** $\dfrac{9}{4}$ **31.** $(-64)^{1/2}$ is an even root of a negative number. No real number squared will give -64. On the other

hand, $-64^{1/2} = -\sqrt{64} = -8$, which is a real number. ($-64^{1/2}$ is the opposite of $64^{1/2}$.)   **33.** $\sqrt{12}$ **35.** $\left(\sqrt[4]{8}\right)^3$ **37.** $\left(\sqrt[8]{9q}\right)^5 - \left(\sqrt[3]{2x}\right)^2$

39. $\dfrac{1}{\left(\sqrt{2m}\right)^3}$ **41.** $\left(\sqrt[3]{2y + x}\right)^2$ **43.** $\dfrac{1}{\left(\sqrt[3]{3m^4 + 2k^2}\right)^2}$ **45.** $\sqrt{a^2 + b^2} = \sqrt{3^2 + 4^2} = 5; a + b = 3 + 4 = 7; 5 \neq 7$ **47.** 64

49. 64   **51.** x^{10} **53.** $\sqrt[6]{x^5}$ **55.** $\sqrt[15]{t^8}$ **57.** 9   **59.** 4   **61.** y **63.** $k^{2/3}$ **65.** $x^3 y^8$ **67.** $\dfrac{1}{x^{10/3}}$ **69.** $\dfrac{1}{m^{1/4} n^{3/4}}$ **71.** p^2 **73.** $\dfrac{c^{11/3}}{b^{11/4}}$

75. $\dfrac{q^{5/3}}{9p^{7/2}}$ **77.** $p + 2p^2$ **79.** $k^{7/4} - k^{3/4}$ **81.** $6 + 18a$ **83.** $x^{17/20}$ **85.** $\dfrac{1}{x^{3/2}}$ **87.** $y^{5/6} z^{1/3}$ **89.** $m^{1/12}$ **91.** $x^{1/24}$ **93.** 4.5 hr
95. $x^{-1/2}(3 - 4x)$ **96.** $m^{5/2}(m^{1/2} - 3)$ **97.** $t^{-1/2}(4 + 7t^2)$ **98.** $x^{-1/3}(8x + 5)$ **99.** $p^{3/4}(4p^{1/4} - 1)$ **100.** $m^{1/8}(2 - m^{1/2})$
101. $k^{-3/4}(9 - 2k^{1/2})$ **102.** $z^{-3/4}(7z^{1/8} - 1)$

Section 8.3 (page 481)

Connections (page 479) no; no; Answers will vary.

Exercises 1. true; Both are equal to $4\sqrt{3}$ and approximately 6.92820323. **3.** true; Both are equal to $6\sqrt{2}$ and approximately 8.485281374.
5. D **7.** $\sqrt{30}$ **9.** $\sqrt{14x}$ **11.** $\sqrt{42pqr}$ **13.** $\sqrt[3]{14xy}$ **15.** $\sqrt[4]{33}$ **17.** $\sqrt[4]{6xy^2}$ **19.** cannot be simplified using the product rule
21. To multiply two radical expressions with the same index, multiply the radicands and keep the same index. For example,

$\sqrt[3]{3} \cdot \sqrt[3]{5} = \sqrt[3]{15}.$ **23.** $\dfrac{8}{11}$ **25.** $\dfrac{\sqrt{3}}{5}$ **27.** $\dfrac{\sqrt{x}}{5}$ **29.** $\dfrac{p^3}{9}$ **31.** $\dfrac{3}{4}$ **33.** $-\dfrac{\sqrt[3]{r^2}}{2}$ **35.** $-\dfrac{3}{x}$ **37.** $\dfrac{1}{x^3}$ **39.** $2\sqrt{3}$ **41.** $12\sqrt{2}$
43. $-4\sqrt{2}$ **45.** $-2\sqrt{7}$ **47.** not a real number **49.** $4\sqrt[3]{2}$ **51.** $-2\sqrt[3]{2}$ **53.** $2\sqrt[3]{5}$ **55.** $-4\sqrt[4]{2}$ **57.** $2\sqrt[5]{2}$ **59.** His reasoning
was incorrect. Here 8 is a term, not a factor. **61.** $6k\sqrt{2}$ **63.** $12xy^4\sqrt{xy}$ **65.** $11x^3$ **67.** $-3t^4$ **69.** $-10m^4z^2$ **71.** $5a^2b^3c^4$
73. $\dfrac{1}{2}r^2t^5$ **75.** $5x\sqrt{2x}$ **77.** $-10r^5\sqrt{5r}$ **79.** $x^3y^4\sqrt{13x}$ **81.** $2z^2w^3$ **83.** $-2zt^2\sqrt[3]{2z^2t}$ **85.** $3x^3y^4$ **87.** $-3r^3s^2\sqrt[4]{2r^3s^2}$
89. $\dfrac{y^5\sqrt{y}}{6}$ **91.** $\dfrac{x^5\sqrt[3]{x}}{3}$ **93.** $4\sqrt{3}$ **95.** $\sqrt{5}$ **97.** $x^2\sqrt{x}$ **99.** $\sqrt[6]{432}$ **101.** $\sqrt[12]{6912}$ **103.** $\sqrt[6]{x^5}$ **105.** 1 **107.** 1 **109.** 5
111. $8\sqrt{2}$ **113.** $\sqrt{29}$ ft; 5.4 ft **115.** .003 **117.** $8\sqrt{5}$ ft; 17.9 ft **119.** 13 **121.** $9\sqrt{2}$ **123.** $\sqrt{17}$ **125.** 5 **127.** $6\sqrt{2}$
129. $\sqrt{5y^2 - 2xy + x^2}$ **131.** $d = [(x_2 - x_1)^2 + (y_2 - y_1)^2]^{1/2}$ **133.** $2\sqrt{106} + 4\sqrt{2}$

Section 8.4 (page 489)

Connections (page 488) 1. 1.618033989 **2.** As one goes farther and farther into the sequence, the successive ratios appear to become
closer and closer to the golden ratio. This is indeed the case.

Exercises 1. B **3.** 15; Each radical expression simplifies to a whole number. **5.** -4 **7.** $7\sqrt{3}$ **9.** $14\sqrt[3]{2}$ **11.** $5\sqrt[4]{2}$
13. $24\sqrt{2}$ **15.** cannot be simplified further **17.** $20\sqrt{5}$ **19.** $12\sqrt{2x}$ **21.** $-11m\sqrt{2}$ **23.** $\sqrt[3]{2}$ **25.** $2\sqrt[3]{x}$ **27.** $-\sqrt[3]{x^2y}$
29. $-x\sqrt[3]{xy^2}$ **31.** $19\sqrt[4]{2}$ **33.** $x\sqrt[4]{xy}$ **35.** $9\sqrt[4]{2a^3}$ **37.** $(4 + 3xy)\sqrt[3]{xy^2}$ **39.** $2\sqrt{2} - 2$ **41.** $\dfrac{5\sqrt{5}}{6}$ **43.** $\dfrac{7\sqrt{2}}{6}$ **45.** $\dfrac{5\sqrt{2}}{3}$
47. $5\sqrt{2} + 4$ **49.** $\dfrac{5 - 3x}{x^4}$ **51.** $\dfrac{m\sqrt[3]{m^2}}{2}$ **53.** $\dfrac{3x\sqrt[3]{2} - 4\sqrt[3]{5}}{x^3}$ **55.** Both are approximately 11.3137085. **57.** A; 42 m
59. $\left(12\sqrt{5} + 5\sqrt{3}\right)$ in. **61.** $\left(24\sqrt{2} + 12\sqrt{3}\right)$ in.

Section 8.5 (page 497)

Connections (page 497) 1. $\dfrac{319}{6(8\sqrt{5} + 1)}$ **2.** $\dfrac{9a - b}{b(3\sqrt{a} - \sqrt{b})}$ **3.** $\dfrac{9a - b}{(\sqrt{b} - \sqrt{a})(3\sqrt{a} - \sqrt{b})}$ **4.** $\dfrac{(3\sqrt{a} + \sqrt{b})(\sqrt{b} + \sqrt{a})}{b - a}$; Instead
of multiplying by the conjugate of the numerator, we use the conjugate of the denominator.

Exercises 1. E **3.** A **5.** D **7.** $3\sqrt{6} + 2\sqrt{3}$ **9.** $20\sqrt{2}$ **11.** -2 **13.** -1 **15.** 6 **17.** $\sqrt{6} - \sqrt{2} + \sqrt{3} - 1$
19. $\sqrt{22} + \sqrt{55} - \sqrt{14} - \sqrt{35}$ **21.** $8 - \sqrt{15}$ **23.** $9 + 4\sqrt{5}$ **25.** $26 - 2\sqrt{105}$ **27.** $4 - \sqrt[3]{36}$ **29.** 10
31. $6x + 3\sqrt{x} - 2\sqrt{5x} - \sqrt{5}$ **33.** $9r - s$ **35.** $4\sqrt[3]{4y^2} - 19\sqrt[3]{2y} - 5$ **37.** $3x - 4$ **39.** $4x - y$ **41.** $2\sqrt{6} - 1$ **43.** $\sqrt{7}$
45. $5\sqrt{3}$ **47.** $\dfrac{\sqrt{6}}{2}$ **49.** $\dfrac{9\sqrt{15}}{5}$ **51.** $-\sqrt{2}$ **53.** $\dfrac{\sqrt{14}}{2}$ **55.** $-\dfrac{\sqrt{14}}{10}$ **57.** $\dfrac{2\sqrt{6x}}{x}$ **59.** $\dfrac{-8\sqrt{3k}}{k}$ **61.** $\dfrac{-5m^2\sqrt{6mn}}{n^2}$
63. $\dfrac{12x^3\sqrt{2xy}}{y^5}$ **65.** $\dfrac{5\sqrt{2my}}{y^2}$ **67.** $-\dfrac{4k\sqrt{3z}}{z}$ **69.** $\dfrac{\sqrt[3]{18}}{3}$ **71.** $\dfrac{\sqrt[3]{12}}{3}$ **73.** $\dfrac{\sqrt[3]{18}}{4}$ **75.** $-\dfrac{\sqrt[3]{2pr}}{r}$ **77.** $\dfrac{x^2\sqrt[3]{y^2}}{y}$ **79.** $\dfrac{2\sqrt[4]{x^3}}{x}$
81. $\dfrac{\sqrt[4]{2yz^3}}{z}$ **83.** Multiply both the numerator and the denominator by $4 - \sqrt{5}$. No, it would not. The new denominator would be
$\left(4 + \sqrt{5}\right)^2 = 21 + 8\sqrt{5}$, which is not rational. **85.** $\dfrac{3(4 - \sqrt{5})}{11}$ **87.** $\dfrac{6\sqrt{2} + 4}{7}$ **89.** $\dfrac{2(3\sqrt{5} - 2\sqrt{3})}{33}$
91. $2\sqrt{3} + \sqrt{10} - 3\sqrt{2} - \sqrt{15}$ **93.** $\sqrt{m} - 2$ **95.** $\dfrac{4\sqrt{x}(\sqrt{x} + 2\sqrt{y})}{x - 4y}$ **97.** $\dfrac{x - 2\sqrt{xy} + y}{x - y}$ **99.** $\dfrac{5\sqrt{k}(2\sqrt{k} - \sqrt{q})}{4k - q}$

101. Square both sides to show that each is equal to $\dfrac{2 - \sqrt{3}}{4}$.   **103.** $3 - 2\sqrt{6}$ **105.** $1 - \sqrt{5}$ **107.** $\dfrac{4 - 2\sqrt{2}}{3}$ **109.** $\dfrac{6 + 2\sqrt{6p}}{3}$

111. $\dfrac{\sqrt{x + y}}{x + y}$ **113.** $\dfrac{p\sqrt{p + 2}}{p + 2}$ **115.** Each expression is approximately equal to .2588190451.   **117.** $\left(\sqrt{x} + \sqrt{7}\right)\left(\sqrt{x} - \sqrt{7}\right)$

118. $\left(\sqrt[3]{x} - \sqrt[3]{7}\right)\left(\sqrt[3]{x^2} + \sqrt[3]{7x} + \sqrt[3]{49}\right)$ **119.** $\left(\sqrt[3]{x} + \sqrt[3]{7}\right)\left(\sqrt[3]{x^2} - \sqrt[3]{7x} + \sqrt[3]{49}\right)$ **120.** $\dfrac{(x + 3)\left(\sqrt{x} + \sqrt{7}\right)}{x - 7}$

121. $\dfrac{(x + 3)\left(\sqrt[3]{x^2} + \sqrt[3]{7x} + \sqrt[3]{49}\right)}{x - 7}$ **122.** $\dfrac{(x + 3)\left(\sqrt{x} + \sqrt{7}\right)}{x + 7}$ **123.** $\left(\sqrt[3]{5} - \sqrt[3]{3}\right)\left(\sqrt[3]{25} + \sqrt[3]{15} + \sqrt[3]{9}\right)$ **124.** $\sqrt[3]{25} + \sqrt[3]{15} + \sqrt[3]{9}$

125. $\dfrac{17}{2\left(6 + \sqrt{2}\right)}$ **127.** $\dfrac{9a - b}{b\left(3\sqrt{a} - \sqrt{b}\right)}$

Summary Exercises on Operations with Radicals   (page 501)

1. $-6\sqrt{10}$ **2.** $7 - \sqrt{14}$ **3.** $2 + \sqrt{6} - 2\sqrt{3} - 3\sqrt{2}$ **4.** $4\sqrt{2}$ **5.** $73 + 12\sqrt{35}$ **6.** $\dfrac{-\sqrt{6}}{2}$ **7.** $4\left(\sqrt{7} - \sqrt{5}\right)$ **8.** $3\sqrt[3]{2x^2}$

9. $-3 + 2\sqrt{2}$ **10.** -2 **11.** -44 **12.** $\dfrac{\sqrt{x} + \sqrt{5}}{x - 5}$ **13.** $2abc^3\sqrt[3]{b^2}$ **14.** $5\sqrt[3]{3}$ **15.** $3\left(\sqrt{5} - 2\right)$ **16.** $\dfrac{\sqrt{15x}}{5x}$ **17.** $\dfrac{8}{5}$

18. $\dfrac{\sqrt{2}}{8}$ **19.** $-\sqrt[3]{100}$ **20.** $11 + 2\sqrt{30}$ **21.** $-3\sqrt{3x}$ **22.** $52 - 30\sqrt{3}$ **23.** 1 **24.** $\dfrac{\sqrt[3]{117}}{9}$ **25.** $t^2\sqrt[4]{t}$

Section 8.6   (page 506)

Exercises   1. (a) yes **(b)** no   **3. (a)** yes **(b)** no   **5.** no; There is no solution. The radical expression, which is positive, cannot equal a negative number.   **7.** $\{11\}$ **9.** $\left\{\dfrac{1}{3}\right\}$ **11.** $\varnothing$ **13.** $\{5\}$ **15.** $\{18\}$ **17.** $\{5\}$ **19.** $\{4\}$ **21.** $\{17\}$ **23.** $\{5\}$ **25.** $\varnothing$ **27.** $\{0\}$

29. $\{0\}$ **31.** $\left\{-\dfrac{1}{3}\right\}$ **33.** $\varnothing$ **35.** You cannot just square each term. The right side should be $(8 - x)^2 = 64 - 16x + x^2$. The correct first step is $3x + 4 = 64 - 16x + x^2$, and the solution set is $\{4\}$.   **37.** $\{1\}$ **39.** $\{-1\}$ **41.** $\{14\}$ **43.** $\{8\}$ **45.** $\{0\}$ **47.** $\varnothing$ **49.** $\{7\}$

51. $\{7\}$ **53.** $\{4, 20\}$ **55.** $\varnothing$ **57.** $\left\{\dfrac{5}{4}\right\}$ **59.** 6 **61.** $\varnothing$; domain: $\left[-\dfrac{2}{3}, 1\right]$ **63.** $\{9, 17\}$ **65.** $\left\{\dfrac{1}{4}, 1\right\}$ **67.** $K = \dfrac{V^2m}{2}$

69. $L = \dfrac{1}{4\pi^2f^2C}$ **71. (a)** $r = \dfrac{a}{4\pi^2N^2}$ **(b)** 62.5 m **(c)** 155.1 m   **73.** 7 billion ft³; 14 billion ft³; 22.5 billion ft³; 22.5 billion ft³

75. 12 billion ft³; 12 billion ft³; 14 billion ft³; 17.5 billion ft³   **77.** fairly good; 1920

Section 8.7   (page 515)

Exercises   1. i **3.** -1 **5.** $-i$ **7.** $13i$ **9.** $-12i$ **11.** $i\sqrt{5}$ **13.** $4i\sqrt{3}$ **15.** $-\sqrt{105}$ **17.** -10 **19.** $i\sqrt{33}$ **21.** $\sqrt{3}$

23. $5i$ **25. (a)** Any real number a can be written as $a + 0i$, and this is a complex number with imaginary part 0. **(b)** A complex number such as $2 + 3i$, with nonzero imaginary part, is not real.   **27.** $-1 + 7i$ **29.** 0 **31.** $7 + 3i$ **33.** -2 **35.** $1 + 13i$ **37.** $6 + 6i$

39. $4 + 2i$ **41.** -81 **43.** -16 **45.** $-10 - 30i$ **47.** $10 - 5i$ **49.** $-9 + 40i$ **51.** $-16 + 30i$ **53.** 153 **55.** 97

57. $a - bi$ **59.** $1 + i$ **61.** $-1 + 2i$ **63.** $2 + 2i$ **65.** $-\dfrac{5}{13} - \dfrac{12}{13}i$ **67.** $1 - 3i$ **69.** $30 + 5i$ **71.** $\dfrac{5}{41} + \dfrac{4}{41}i$ **73.** -1

75. i **77.** -1 **79.** $-i$ **81.** $-i$ **83.** Since $i^{20} = (i^4)^5 = 1^5 = 1$, the student multiplied by 1, which is justified by the identity property for multiplication.   **85.** $\dfrac{1}{2} + \dfrac{1}{2}i$ **87.** Substitute both $1 + 5i$ and $1 - 5i$ for x and show that the result is $0 = 0$ in each case.

89. (a) $4x + 1$ **(b)** $4 + i$ **90. (a)** $-2x + 3$ **(b)** $-2 + 3i$ **91. (a)** $3x^2 + 5x - 2$ **(b)** $5 + 5i$ **92. (a)** $-\sqrt{3} + \sqrt{6} + 1 - \sqrt{2}$

(b) $\dfrac{1}{5} - \dfrac{7}{5}i$ **93.** In parts (a) and (b) of Exercises 89 and 90, real and imaginary parts are added, just like coefficients of similar terms in the binomials, and the answers correspond. In Exercise 91, introducing $i^2 = -1$ when a product is found leads to answers that do not correspond.
94. In parts (a) and (b) of Exercises 89 and 90, real and imaginary parts are added, just like coefficients of similar terms in binomials, and the answers correspond. In Exercise 92, introducing $i^2 = -1$ when performing the division leads to answers that do not correspond.
95. $\dfrac{37}{10} - \dfrac{19}{10}i$ **97.** $-\dfrac{13}{10} + \dfrac{11}{10}i$

Chapter 8 Review Exercises   (page 523)

1. 42   **3.** 6   **5.** -3 **7.** $\sqrt[n]{a}$ is not a real number if n is even and a is negative.   **9.** -6.856 **11.** 4.960   **13.** -3968.503
15. domain: $[1, \infty)$; range: $[0, \infty)$ **17.** B   **19.** A   **21.** It is not a real number.   **23.** -11 **25.** -4 **27.** -32

29. It is not a real number.   **31.** The radical $\sqrt[n]{a^m}$ is equivalent to $a^{m/n}$. For example, $\sqrt[3]{8^2} = \sqrt[3]{64} = 4$, and $8^{2/3} = (8^{1/3})^2 = 2^2 = 4$.
33. $\dfrac{1}{\left(\sqrt[3]{3a+b}\right)^5}$ or $\dfrac{1}{\sqrt[3]{(3a+b)^5}}$ **35.** $p^{4/5}$ **37.** 96   **39.** $\dfrac{1}{y^{1/2}}$ **41.** $r^{1/2} + r$ **43.** $r^{3/2}$ **45.** $k^{9/4}$ **47.** $z^{1/12}$ **49.** $x^{1/15}$
51. The product rule for exponents applies only if the bases are the same.   **53.** $\sqrt{5r}$ **55.** $\sqrt[4]{21}$ **57.** $5\sqrt{3}$ **59.** $-3\sqrt[3]{4}$
61. $4pq^2\sqrt[3]{p}$ **63.** $2r^2t\sqrt[3]{79r^2t}$ **65.** $\dfrac{m^5}{3}$ **67.** $\dfrac{a^2\sqrt[4]{a}}{3}$ **69.** $p\sqrt{p}$ **71.** $\sqrt[10]{x^7}$ **73.** $\sqrt{197}$ **75.** $23\sqrt{5}$ **77.** $26m\sqrt{6m}$
79. $-8\sqrt[4]{2}$ **81.** $\dfrac{16 + 5\sqrt{5}}{20}$ **83.** $\left(12\sqrt{3} + 5\sqrt{2}\right)$ ft   **85.** 2   **87.** $15 - 2\sqrt{26}$ **89.** $2\sqrt[3]{2y^2} + 2\sqrt[3]{4y} - 3$
91. The denominator would become $\sqrt[3]{6^2} = \sqrt[3]{36}$, which is not rational.   **93.** $-3\sqrt{6}$ **95.** $\dfrac{\sqrt{22}}{4}$ **97.** $\dfrac{3m\sqrt[3]{4n}}{n^2}$ **99.** $\dfrac{5\left(\sqrt{6}+3\right)}{3}$
101. $\dfrac{1 - 4\sqrt{2}}{3}$ **103.** $\{2\}$ **105.** $\emptyset$ **107.** $\{9\}$ **109.** $\{7\}$ **111.** $\{-13\}$ **113.** $\{14\}$ **115.** $\emptyset$ **117.** $\{7\}$ **119.** $\{3\}$ **120.** $\{-3\}$
121. $\{\pm 3\}$ **122.** $\{3\}$ **123.** $\{\pm 3\}$ **124.** $\{-3\}$ **125. (a)** more than **(b)** the same as   **127.** $5i$ **129.** no   **131.** $14 + 7i$
133. -45 **135.** $5 + i$ **137.** $1 - i$ **139.** $-i$ **141.** -1 **143.** -4 **145.** $\dfrac{1}{z^{3/5}}$ **147.** $3z^3t^2\sqrt[3]{2t^2}$ **149.** $6x\sqrt[3]{y^2}$ **151.** $-\dfrac{\sqrt{3}}{6}$
153. 1   **155.** $3 - 7i$ **157.** $\dfrac{1 + \sqrt{6}}{2}$ **159.** $\{5\}$ **161.** $\left\{\dfrac{3}{2}\right\}$ **163.** $\{1\}$ **165.** $\{9\}$ **167.** $\{7\}$

Chapter 8 Test   (page 528)

[8.1] **1.** -29 **2.** -8 [8.2] **3.** 5   [8.1] **4.** C   **5.** 21.863   **6.** -9.405 **7.** domain: $[-6, \infty)$; range: $[0, \infty)$

[8.2] **8.** $\dfrac{125}{64}$ **9.** $\dfrac{1}{256}$ **10.** $\dfrac{9y^{3/10}}{x^2}$ **11.** $x^{4/3}y^6$ **12.** $7^{1/2}$ or $\sqrt{7}$ [8.3] **13.** $a^3\sqrt[3]{a^2}$ or $a^{11/3}$ **14.** $\sqrt{145}$ **15.** 10   **16.** $3x^2y^3\sqrt{6x}$

17. $2ab^3\sqrt[4]{2a^3b}$ **18.** $\sqrt[6]{200}$ [8.4] **19.** $26\sqrt{5}$ **20.** $(2ts - 3t^2)\sqrt[3]{2s^2}$ [8.5] **21.** $66 + \sqrt{5}$ **22.** $23 - 4\sqrt{15}$ **23.** $-\dfrac{\sqrt{10}}{4}$

24. $\dfrac{2\sqrt[3]{25}}{5}$ **25.** $-2\left(\sqrt{7} - \sqrt{5}\right)$ **26.** $3 + \sqrt{6}$ [8.6] **27. (a)** 59.8 **(b)** $T = \dfrac{V_0^2 - V^2}{-V^2 k}$ or $T = \dfrac{V^2 - V_0^2}{V^2 k}$ **28.** $\{-1\}$ **29.** $\{3\}$
30. $\{-3\}$ [8.7] **31.** $-5 - 8i$ **32.** $-2 + 16i$ **33.** $3 + 4i$ **34.** i **35. (a)** true **(b)** true **(c)** false **(d)** true

Cumulative Review Exercises Chapters 1–8   (page 529)

[1.3] **1.** 1   **2.** $-\dfrac{14}{9}$ [2.1] **3.** $\{-4\}$ **4.** $\{-12\}$ **5.** $\{6\}$ [2.7] **6.** $\left\{-\dfrac{10}{3}, 1\right\}$ **7.** $\left\{\dfrac{1}{4}\right\}$ [2.5] **8.** $(-6, \infty)$ [2.2] **9.** Both angles

measure 80°.   [2.3] **10.** 18 nickels; 32 quarters   **11.** $2\dfrac{2}{39}$ L   [3.1] **12.**   [3.2, 3.3] **13.** $-\dfrac{3}{2}$; $y = -\dfrac{3}{2}x$

[3.5] **14.** -37 [4.1] **15.** $\{(7, -2)\}$ [4.4] **16.** $\{(-1, 1, 1)\}$ [4.3] **17.** 2-oz letter: \$.55; 3-oz letter: \$.78

[5.2] **18.** $-k^3 - 3k^2 - 8k - 9$ [5.4] **19.** $8x^2 + 17x - 21$ [5.5] **20.** $z - 2 + \dfrac{3}{z}$ **21.** $3y^3 - 3y^2 + 4y + 1 + \dfrac{-10}{2y + 1}$

[6.2] **22.** $(2p - 3q)(p - q)$ [6.3] **23.** $(3k^2 + 4)(k - 1)(k + 1)$ **24.** $(x + 8)(x^2 - 8x + 64)$ [6.5] **25.** $\left\{-3, -\dfrac{5}{2}\right\}$ **26.** $\left\{-\dfrac{2}{5}, 1\right\}$

[7.1] **27.** $\{x \mid x \neq \pm 3\}$ **28.** $\dfrac{y}{y + 5}$ [7.2] **29.** $\dfrac{4x + 2y}{(x + y)(x - y)}$ [7.3] **30.** $-\dfrac{9}{4}$ **31.** $\dfrac{-1}{a + b}$ **32.** $\dfrac{1}{xy - 1}$ [7.6] **33.** Natalie: 8 mph;

Chuck: 4 mph   [7.5] **34.** $\emptyset$ [8.2] **35.** $\dfrac{1}{9}$ [8.3] **36.** $10x^2\sqrt{2}$ **37.** $2x\sqrt[3]{6x^2y^2}$ [8.4] **38.** $7\sqrt{2}$ [8.5] **39.** $\dfrac{\sqrt{10} + 2\sqrt{2}}{2}$

40. $-6x - 11\sqrt{xy} - 4y$ [8.3] **41.** $\sqrt{29}$ [8.6] **42.** $\{3, 4\}$ [8.1] **43.** 39.2 mph   [8.7] **44.** $2 + 9i$ **45.** $4 + 2i$

CHAPTER 9 QUADRATIC EQUATIONS AND INEQUALITIES

Section 9.1   (page 541)

Exercises   **1.** The equation is also true for $x = -4$.   **3. (a)** A quadratic equation in standard form has a second-degree polynomial in decreasing powers equal to 0.   **(b)** The zero-factor property states that if a product equals 0, then at least one of the factors equals 0.   **(c)** The square root property states that if the square of a quantity equals a number, then the quantity equals the positive or negative square root of the number.   **5.** $\{9, -9\}$ **7.** $\left\{\sqrt{17}, -\sqrt{17}\right\}$ **9.** $\left\{4\sqrt{2}, -4\sqrt{2}\right\}$ **11.** $\left\{2\sqrt{5}, -2\sqrt{5}\right\}$ **13.** $\left\{2\sqrt{6}, -2\sqrt{6}\right\}$ **15.** $\{-7, 3\}$

17. $\left\{4 + \sqrt{3}, 4 - \sqrt{3}\right\}$ **19.** $\left\{-5 + 4\sqrt{3}, -5 - 4\sqrt{3}\right\}$ **21.** $\left\{\dfrac{1 + \sqrt{7}}{3}, \dfrac{1 - \sqrt{7}}{3}\right\}$ **23.** $\left\{\dfrac{-1 + 2\sqrt{6}}{4}, \dfrac{-1 - 2\sqrt{6}}{4}\right\}$ **25.** 6.3 sec

27. square root property for $(2x + 1)^2 = 5$; completing the square for $x^2 + 4x = 12$ **29. (a)** 9 **(b)** 49 **(c)** 36 **(d)** $\dfrac{9}{4}$ **(e)** $\dfrac{81}{4}$ **(f)** $\dfrac{1}{16}$

31. 4   **33.** 25   **35.** $\dfrac{1}{36}$ **37.** $\{-4, 6\}$ **39.** $\left\{-2 + \sqrt{6}, -2 - \sqrt{6}\right\}$ **41.** $\left\{-5 + \sqrt{7}, -5 - \sqrt{7}\right\}$ **43.** $\left\{-\dfrac{8}{3}, 3\right\}$

45. $\left\{\dfrac{-5 + \sqrt{41}}{4}, \dfrac{-5 - \sqrt{41}}{4}\right\}$ **47.** $\left\{\dfrac{5 + \sqrt{15}}{5}, \dfrac{5 - \sqrt{15}}{5}\right\}$ **49.** $\left\{\dfrac{4 + \sqrt{3}}{3}, \dfrac{4 - \sqrt{3}}{3}\right\}$ **51.** $\left\{\dfrac{2 + \sqrt{3}}{3}, \dfrac{2 - \sqrt{3}}{3}\right\}$

53. $\left\{1 + \sqrt{2}, 1 - \sqrt{2}\right\}$ **55.** $\left\{2i\sqrt{3}, -2i\sqrt{3}\right\}$ **57.** $\left\{5 + i\sqrt{3}, 5 - i\sqrt{3}\right\}$ **59.** $\left\{\dfrac{1 + 2i\sqrt{2}}{6}, \dfrac{1 - 2i\sqrt{2}}{6}\right\}$ **61.** $\{-2 + 3i, -2 - 3i\}$

63. $\left\{\dfrac{-2 + 2i\sqrt{2}}{3}, \dfrac{-2 - 2i\sqrt{2}}{3}\right\}$ **65.** $\left\{-3 + i\sqrt{3}, -3 - i\sqrt{3}\right\}$ **67.** $\left\{\sqrt{b}, -\sqrt{b}\right\}$ **69.** $\left\{\dfrac{\sqrt{b^2 + 16}}{2}, -\dfrac{\sqrt{b^2 + 16}}{2}\right\}$

71. $\left\{\dfrac{2b + \sqrt{3a}}{5}, \dfrac{2b - \sqrt{3a}}{5}\right\}$ **73.** x^2 **74.** x **75.** $6x$ **76.** 1   **77.** 9   **78.** $(x + 3)^2$ or $x^2 + 6x + 9$ **79.** $\pm\sqrt{17}$

Section 9.2 (page 550)

Exercises 1. The documentation was incorrect, since the fraction bar should extend under the term $-b$. **3.** The last step is wrong. Because 5 is not a common factor in the numerator, the fraction cannot be simplified. The solutions are $\dfrac{5 \pm \sqrt{5}}{10}$. **5.** $\{3, 5\}$

7. $\left\{\dfrac{-2 + \sqrt{2}}{2}, \dfrac{-2 - \sqrt{2}}{2}\right\}$ **9.** $\left\{\dfrac{1 + \sqrt{3}}{2}, \dfrac{1 - \sqrt{3}}{2}\right\}$ **11.** $\{5 + \sqrt{7}, 5 - \sqrt{7}\}$ **13.** $\left\{\dfrac{-1 + \sqrt{2}}{2}, \dfrac{-1 - \sqrt{2}}{2}\right\}$

15. $\left\{\dfrac{-1 + \sqrt{7}}{3}, \dfrac{-1 - \sqrt{7}}{3}\right\}$ **17.** $\{1 + \sqrt{5}, 1 - \sqrt{5}\}$ **19.** $\left\{\dfrac{-2 + \sqrt{10}}{2}, \dfrac{-2 - \sqrt{10}}{2}\right\}$ **21.** $\{-1 + 3\sqrt{2}, -1 - 3\sqrt{2}\}$

23. $\left\{\dfrac{1 + \sqrt{29}}{2}, \dfrac{1 - \sqrt{29}}{2}\right\}$ **25.** $\left\{\dfrac{-4 + \sqrt{91}}{3}, \dfrac{-4 - \sqrt{91}}{3}\right\}$ **27.** $\left\{\dfrac{3}{2} + \dfrac{\sqrt{15}}{2}i, \dfrac{3}{2} - \dfrac{\sqrt{15}}{2}i\right\}$ **29.** $\{3 + i\sqrt{5}, 3 - i\sqrt{5}\}$

31. $\left\{\dfrac{1}{2} + \dfrac{\sqrt{6}}{2}i, \dfrac{1}{2} - \dfrac{\sqrt{6}}{2}i\right\}$ **33.** $\left\{-\dfrac{2}{3} + \dfrac{\sqrt{2}}{3}i, -\dfrac{2}{3} - \dfrac{\sqrt{2}}{3}i\right\}$ **35.** $\{4 + 3i\sqrt{2}, 4 - 3i\sqrt{2}\}$ **37.** B **39.** C **41.** A **43.** D

45. The equations in Exercises 37, 38, 41, and 42 can be solved by factoring. **47.** $\left\{\dfrac{7}{2}\right\}$ **49.** $\left\{-\dfrac{1}{2}, \dfrac{3}{2}\right\}$ **51.** No, because an irrational solution occurs only if the discriminant is positive, but not the square of an integer. In that case, there will be two irrational solutions.

53. -10 or 10 **55.** 16 **57.** 25 **59.** $b = \dfrac{44}{5}; \dfrac{3}{10}$

Section 9.3 (page 559)

Exercises 1. square root property **3.** quadratic formula **5.** factoring **7.** Multiply by the LCD, x. **9.** Substitute a variable for $r^2 + r$. **11.** The potential solution -1 does not check. The solution set is $\{4\}$. **13.** $\{-2, 7\}$ **15.** $\{-4, 7\}$ **17.** $\left\{-\dfrac{2}{3}, 1\right\}$

19. $\left\{-\dfrac{14}{17}, 5\right\}$ **21.** $\left\{-\dfrac{11}{7}, 0\right\}$ **23.** $\left\{\dfrac{-1 + \sqrt{13}}{2}, \dfrac{-1 - \sqrt{13}}{2}\right\}$ **25.** $\left\{-\dfrac{8}{3}, -1\right\}$ **27.** $\left\{\dfrac{2 + \sqrt{22}}{3}, \dfrac{2 - \sqrt{22}}{3}\right\}$

29. **(a)** $(20 - t)$ mph **(b)** $(20 + t)$ mph **31.** 25 mph **33.** 80 km per hr **35.** 3.6 hr **37.** Rusty: 25.0 hr; Nancy: 23.0 hr **39.** 9 min

41. $\{2, 5\}$ **43.** $\{3\}$ **45.** $\left\{\dfrac{8}{9}\right\}$ **47.** $\{16\}$ **49.** $\left\{\dfrac{2}{5}\right\}$ **51.** $\{-3, 3\}$ **53.** $\left\{-\dfrac{3}{2}, -1, 1, \dfrac{3}{2}\right\}$ **55.** $\{-2\sqrt{3}, -2, 2, 2\sqrt{3}\}$

57. $\{-6, -5\}$ **59.** $\left\{-\dfrac{16}{3}, -2\right\}$ **61.** $\left\{-\dfrac{1}{3}, \dfrac{1}{6}\right\}$ **63.** $\left\{-\dfrac{1}{2}, 3\right\}$ **65.** $\{-8, 1\}$ **67.** $\{-64, 27\}$ **69.** $\left\{-\dfrac{27}{8}, -1, 1, \dfrac{27}{8}\right\}$

71. $\{25\}$ **73.** $\left\{-1, 1, -\dfrac{\sqrt{6}}{2}i, \dfrac{\sqrt{6}}{2}i\right\}$ **75.** $\left\{-\dfrac{\sqrt{6}}{3}, -\dfrac{1}{2}, \dfrac{1}{2}, \dfrac{\sqrt{6}}{3}\right\}$ **77.** $\{3, 11\}$ **79.** $\left\{-\sqrt[3]{5}, -\dfrac{\sqrt[3]{4}}{2}\right\}$ **81.** $\left\{\dfrac{4}{3}, \dfrac{9}{4}\right\}$

83. $\left\{\dfrac{\sqrt{9 + \sqrt{65}}}{2}, -\dfrac{\sqrt{9 + \sqrt{65}}}{2}, \dfrac{\sqrt{9 - \sqrt{65}}}{2}, -\dfrac{\sqrt{9 - \sqrt{65}}}{2}\right\}$ **85.** It would cause both denominators to be 0, and division by 0 is

undefined. **86.** $\dfrac{12}{5}$ **87.** $\left(\dfrac{x}{x - 3}\right)^2 + 3\left(\dfrac{x}{x - 3}\right) - 4 = 0$ **88.** The numerator can never equal the denominator, since the denominator

is 3 less than the numerator. **89.** $\left\{\dfrac{12}{5}\right\}$; The values for t are -4 and 1. The value 1 is impossible because it leads to a contradiction

$\left(\text{since } \dfrac{x}{x - 3} \text{ is never equal to 1}\right)$. **90.** $\left\{\dfrac{12}{5}\right\}$; The values for s are $\dfrac{1}{x}$ and $\dfrac{-4}{x}$. The value $\dfrac{1}{x}$ is impossible, since $\dfrac{1}{x} \neq \dfrac{1}{x - 3}$ for all x.

Summary Exercises on Solving Quadratic Equations (page 563)

1. $\left\{\sqrt{7}, -\sqrt{7}\right\}$ **2.** $\left\{-\dfrac{3}{2}, \dfrac{5}{3}\right\}$ **3.** $\left\{-3 + \sqrt{5}, -3 - \sqrt{5}\right\}$ **4.** $\{-2, 8\}$ **5.** $\left\{-\dfrac{3}{2}, 4\right\}$ **6.** $\left\{-3, \dfrac{1}{3}\right\}$ **7.** $\left\{\dfrac{2 + \sqrt{2}}{2}, \dfrac{2 - \sqrt{2}}{2}\right\}$

8. $\left\{2i\sqrt{3}, -2i\sqrt{3}\right\}$ **9.** $\left\{\dfrac{1}{2}, 2\right\}$ **10.** $\{-3, -1, 1, 3\}$ **11.** $\left\{\dfrac{-3 + 2\sqrt{2}}{2}, \dfrac{-3 - 2\sqrt{2}}{2}\right\}$ **12.** $\left\{\dfrac{4}{5}, 3\right\}$ **13.** $\left\{-\sqrt{7}, -\sqrt{2}, \sqrt{2}, \sqrt{7}\right\}$

14. $\left\{\dfrac{1 + \sqrt{5}}{4}, \dfrac{1 - \sqrt{5}}{4}\right\}$ **15.** $\left\{-\dfrac{1}{2} + \dfrac{\sqrt{3}}{2}i, -\dfrac{1}{2} - \dfrac{\sqrt{3}}{2}i\right\}$ **16.** $\left\{-\dfrac{\sqrt[3]{175}}{5}, 1\right\}$ **17.** $\left\{\dfrac{3}{2}\right\}$ **18.** $\left\{\dfrac{2}{3}\right\}$ **19.** $\left\{6\sqrt{2}, -6\sqrt{2}\right\}$

20. $\left\{-\dfrac{2}{3}, 2\right\}$ **21.** $\{-4, 9\}$ **22.** $\{13, -13\}$ **23.** $\left\{1 + \dfrac{\sqrt{3}}{3}i, 1 - \dfrac{\sqrt{3}}{3}i\right\}$ **24.** $\{3\}$ **25.** $\left\{-\dfrac{1}{3}, \dfrac{1}{6}\right\}$

26. $\left\{\dfrac{1}{6} + \dfrac{\sqrt{47}}{6}i, \dfrac{1}{6} - \dfrac{\sqrt{47}}{6}i\right\}$

Section 9.4 (page 568)

Exercises 1. Find a common denominator, and then multiply both sides by the common denominator. **3.** Write it in standard form (with 0 on one side, in decreasing powers of w). **5.** $m = \sqrt{p^2 - n^2}$ **7.** $t = \dfrac{\pm\sqrt{dk}}{k}$ **9.** $d = \dfrac{\pm\sqrt{skI}}{I}$ **11.** $v = \dfrac{\pm\sqrt{kAF}}{F}$ **13.** $r = \dfrac{\pm\sqrt{3\pi Vh}}{\pi h}$

15. $t = \dfrac{-B \pm \sqrt{B^2 - 4AC}}{2A}$ **17.** $h = \dfrac{D^2}{k}$ **19.** $\ell = \dfrac{p^2 g}{k}$ **21.** If g is positive, the only way to have a real value for p is to have $k\ell$ positive, since the quotient of two positive numbers is positive. If k and ℓ have different signs, their product is negative, leading to a negative radicand.

23. $R = \dfrac{E^2 - 2pr \pm E\sqrt{E^2 - 4pr}}{2p}$ **25.** $r = \dfrac{5pc}{4}$ or $r = -\dfrac{2pc}{3}$ **27.** $I = \dfrac{-cR \pm \sqrt{c^2R^2 - 4cL}}{2cL}$. **29.** 2.3, 5.3, 5.8

31. eastbound ship: 80 mi; southbound ship: 150 mi **33.** 5 cm, 12 cm, 13 cm **35.** length: 2 cm; width: 1.5 cm **37.** 1 ft **39.** length: 26 m; width: 16 m **41.** 20 in. by 12 in. **43.** 1 sec and 8 sec **45.** 2.4 sec and 5.6 sec **47.** 9.2 sec **49.** It reaches its *maximum* height at 5 sec because this is the only time it reaches 400 ft. **51.** $.80 **53.** .035 or 3.5% **55. (a)** 2.4 million **(b)** 2.4 million; They are the same. **57.** 1995; The graph indicates that sales reached 2 million in 1996. **59.** 5.5 m per sec **61.** 5 or 14

Section 9.5 (page 579)

Exercises 1. Include the endpoints if the symbol is $\geq$ or $\leq$. Exclude the endpoints if the symbol is $>$ or $<$.

3. $(-\infty, -1) \cup (5, \infty)$

5. $(-4, 6)$

7. $(-\infty, 1] \cup [3, \infty)$

9. $\left(-\infty, -\dfrac{3}{2}\right] \cup \left[\dfrac{3}{5}, \infty\right)$

11. $\left[-\dfrac{3}{2}, \dfrac{3}{2}\right]$

13. $\left(-\infty, -\dfrac{1}{2}\right] \cup \left[\dfrac{1}{3}, \infty\right)$

15. $(-\infty, 0] \cup [4, \infty)$

17. $\left[0, \dfrac{5}{3}\right]$

19. $\left(-\infty, 3 - \sqrt{3}\right] \cup \left[3 + \sqrt{3}, \infty\right)$

21. $(-\infty, \infty)$ **23.** $\emptyset$ **25.** $(-\infty, 1) \cup (2, 4)$

27. $\left[-\dfrac{3}{2}, \dfrac{1}{3}\right] \cup [4, \infty)$

29. $(-\infty, 1) \cup (4, \infty)$

31. $\left[-\dfrac{3}{2}, 5\right)$

33. $(2, 6]$

35. $\left(-\infty, \dfrac{1}{2}\right) \cup \left(\dfrac{5}{4}, \infty\right)$

37. $[-7, -2)$

39. $(-\infty, 2) \cup (4, \infty)$ **41.** $\left(0, \dfrac{1}{2}\right) \cup \left(\dfrac{5}{2}, \infty\right)$ **43.** $\left[\dfrac{3}{2}, \infty\right)$

45. $\left(-2, \dfrac{5}{3}\right) \cup \left(\dfrac{5}{3}, \infty\right)$ **47.** 3 sec and 13 sec **48.** between 3 sec and 13 sec **49.** at 0 sec (the time when it is

initially projected) and at 16 sec (the time when it hits the ground) **50.** between 0 and 3 sec and between 13 and 16 sec

Chapter 9 Review Exercises (page 586)

1. $\{11, -11\}$ **3.** $\left\{-\dfrac{15}{2}, \dfrac{5}{2}\right\}$ **5.** $\left\{-2 + \sqrt{19}, -2 - \sqrt{19}\right\}$ **7.** By the square root property, the first step should be $x = \sqrt{12}$ or

$x = -\sqrt{12}$. **9.** $\left\{-\dfrac{7}{2}, 3\right\}$ **11.** $\left\{\dfrac{1 + \sqrt{41}}{2}, \dfrac{1 - \sqrt{41}}{2}\right\}$ **13.** $\left\{\dfrac{2 + i\sqrt{2}}{3}, \dfrac{2 - i\sqrt{2}}{3}\right\}$ **15.** C **17.** D **19.** $\left\{-\dfrac{5}{2}, 3\right\}$ **21.** $\{-4\}$

23. $\left\{-\dfrac{343}{8}, 64\right\}$ **25.** 7 mph **27.** 4.6 hr **29.** $v = \dfrac{\pm\sqrt{rFkw}}{kw}$ **31.** $t = \dfrac{3m \pm \sqrt{9m^2 + 24m}}{2m}$ **33.** 12 cm by 20 cm **35.** 3 min

37. .7 sec and 4.0 sec **39.** 4.5% **41.** $[-4, 3]$ **43.** $\emptyset$ **45.** $[-3, 2)$

47. $\left\{\dfrac{3 + i\sqrt{3}}{3}, \dfrac{3 - i\sqrt{3}}{3}\right\}$ **49.** $\{-2, -1, 3, 4\}$ **51.** $\{4\}$ **53.** $d = \dfrac{\pm\sqrt{SkI}}{I}$ **55.** $\left\{-\dfrac{5}{3}, -\dfrac{3}{2}\right\}$ **57.** $\left(-5, -\dfrac{23}{5}\right]$

59. (a) 21.92 trillion ft^3 (b) 2005 **61.** 412.3 ft **62.** (a) $\{-2\}$ (b) $(-\infty, -2)$

(c) $(-2, \infty)$ **63.** (a) $\{1, 5\}$ (b) $(-\infty, 1) \cup (5, \infty)$ (c) $(1, 5)$

64. (a) $\{4\}$ (b) $(2, 4)$ (c) $(-\infty, 2) \cup (4, \infty)$ **65.** $(-\infty, \infty)$; denominator

66. $(-5, 3)$

Chapter 9 Test (page 590)

[9.1] **1.** $\left\{3\sqrt{6}, -3\sqrt{6}\right\}$ **2.** $\left\{-\dfrac{8}{7}, \dfrac{2}{7}\right\}$ **3.** $\left\{-1 + \sqrt{5}, -1 - \sqrt{5}\right\}$ [9.2] **4.** $\left\{2 + \sqrt{2}, 2 - \sqrt{2}\right\}$ **5.** $\left\{\dfrac{3 + \sqrt{17}}{4}, \dfrac{3 - \sqrt{17}}{4}\right\}$

6. $\left\{\dfrac{2 + i\sqrt{11}}{3}, \dfrac{2 - i\sqrt{11}}{3}\right\}$ [9.1] **7.** A [9.2] **8.** discriminant: 88; There are two irrational solutions. [9.1–9.3] **9.** $\left\{-\dfrac{2}{3}, 6\right\}$

10. $\left\{\dfrac{-7 + \sqrt{97}}{8}, \dfrac{-7 - \sqrt{97}}{8}\right\}$ **11.** $\left\{\dfrac{2}{3}\right\}$ **12.** $\left\{-2, -\dfrac{1}{3}, \dfrac{1}{3}, 2\right\}$ **13.** $\left\{-\dfrac{5}{2}, 1\right\}$ [9.4] **14.** $r = \dfrac{\pm\sqrt{\pi S}}{2\pi}$

[9.3] **15.** Andrew: 11.1 hr; Kent: 9.1 hr **16.** 7 mph [9.4] **17.** 2 ft **18.** 16 m [9.5] **19.** $(-\infty, -5) \cup \left(\dfrac{3}{2}, \infty\right)$

20. $(-\infty, 4) \cup [9, \infty)$

Cumulative Review Exercises Chapters 1–9 (page 592)

[1.1, 8.7] **1.** (a) $-2, 0, 7$ (b) $-\dfrac{7}{3}, -2, 0, .7, 7, \dfrac{32}{3}$ (c) all except $\sqrt{-8}$ (d) All are complex numbers. [1.1–1.3] **2.** 6 **3.** 41

[2.1] **4.** $\left\{\dfrac{4}{5}\right\}$ [2.7] **5.** $\left\{\dfrac{11}{10}, \dfrac{7}{2}\right\}$ [8.6] **6.** $\left\{\dfrac{2}{3}\right\}$ [7.4] **7.** $\emptyset$ [9.1, 9.2] **8.** $\left\{\dfrac{7 + \sqrt{177}}{4}, \dfrac{7 - \sqrt{177}}{4}\right\}$ [9.3] **9.** $\{-2, -1, 1, 2\}$

[2.5] **10.** $[1, \infty)$ [2.7] **11.** $\left[2, \dfrac{8}{3}\right]$ [9.5] **12.** $(1, 3)$ **13.** $(-2, 1)$ [3.1, 3.5] **14.** function; domain: $(-\infty, \infty)$; range: $(-\infty, \infty)$

[3.4, 3.5] **15.** not a function [3.2] **16.** $m = \dfrac{2}{7}$; x-intercept: $(-8, 0)$; y-intercept: $\left(0, \dfrac{16}{7}\right)$ [3.3] **17.** $y = -\dfrac{5}{2}x + 2$

18. $y = \dfrac{2}{5}x + \dfrac{13}{5}$ **19. (a)** 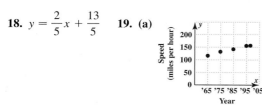 **(b)** positive **(c)** $y = 1.279x + 116.26$ **(d)** 158.47; It is a little too high.

[3.5] **20.** No, because the graph is a vertical line, which is not the graph of a function by the vertical line test. **21.** 13 [4.1] **22.** $\{(1, -2)\}$

[4.2] **23.** $\{(3, -4, 2)\}$ [4.3] **24. (a)** $x + y = 34.2$; $x = 4y - .3$ **(b)** AOL: \$27.3 billion; Time Warner: \$6.9 billion [5.1] **25.** $\dfrac{x^8}{y^4}$

26. $\dfrac{4}{xy^2}$ [5.4] **27.** $14x^2 - 13x - 12$ **28.** $\dfrac{4}{9}t^2 + 12t + 81$ [5.2] **29.** $-3t^3 + 5t^2 - 12t + 15$ [5.5] **30.** $4x^2 - 6x + 11 + \dfrac{4}{x + 2}$

[6.1–6.3] **31.** $x(4 + x)(4 - x)$ **32.** $(4m - 3)(6m + 5)$ **33.** $(2x + 3y)(4x^2 - 6xy + 9y^2)$ **34.** $(3x - 5y)^2$ [7.1] **35.** $\dfrac{x - 5}{x + 5}$

[7.2] **36.** $-\dfrac{8}{k}$ [7.3] **37.** $\dfrac{r - s}{r}$ [8.1] **38.** $\dfrac{3\sqrt[3]{4}}{4}$ [8.4] **39.** $\sqrt{7} + \sqrt{5}$ [7.5] **40.** biking: 12 mph; walking: 2 mph

[9.4] **41.** southbound car: 57 mi; eastbound car: 76 mi [2.3] **42.** 930 **43.** 720 **44.** 990 **45.** 930

CHAPTER 10 ADDITIONAL GRAPHS OF FUNCTIONS AND RELATIONS

Section 10.1 (page 601)

Connections (page 601) There are several in Sections 6.2 and 9.3; for instance, 6.2 Example 10 and 9.3 Example 6.

Exercises **1.** 55 **3.** 1848 **5.** $-\dfrac{7}{6}$ **7.** 1122 **9.** 97 **11.** 930 **13. (a)** $10x + 2$ **(b)** $-2x - 4$ **(c)** $24x^2 + 6x - 3$ **(d)** $\dfrac{4x - 1}{6x + 3}$;

All domains are $(-\infty, \infty)$, except for $\dfrac{f}{g}$, which is $\left(-\infty, -\dfrac{1}{2}\right) \cup \left(-\dfrac{1}{2}, \infty\right)$. **15. (a)** $4x^2 - 4x + 1$ **(b)** $2x^2 - 1$ **(c)** $(3x^2 - 2x)(x^2 - 2x + 1)$

(d) $\dfrac{3x^2 - 2x}{x^2 - 2x + 1}$; All domains are $(-\infty, \infty)$, except for $\dfrac{f}{g}$, which is $(-\infty, 1) \cup (1, \infty)$. **17. (a)** $\sqrt{2x + 5} + \sqrt{4x + 9}$

(b) $\sqrt{2x + 5} - \sqrt{4x + 9}$ **(c)** $\sqrt{(2x + 5)(4x + 9)}$ **(d)** $\sqrt{\dfrac{2x + 5}{4x + 9}}$ **19.** The function values for $f + g$ are found by adding $f(x) + g(x)$.

For example, if $f(x) = 2x + 3$ and $g(x) = x^2$, then $(f + g)(x) = x^2 + 2x + 3$.   **21. (a)** $4xh + 2h^2$ **(b)** $4x + 2h$ **23. (a)** $2xh + h^2 + 4h$
(b) $2x + h + 4$ **In Exercises 25–31, we give $(f \circ g)(x)$, $(g \circ f)(x)$, and the domains.**   **25.** $-5x^2 + 20x + 18$; $-25x^2 - 10x + 6$;
both domains are $(-\infty, \infty)$.   **27.** $\dfrac{1}{x^2}$; $\dfrac{1}{x^2}$; both domains are $(-\infty, 0) \cup (0, \infty)$.   **29.** $2\sqrt{2x - 1}$; $8\sqrt{x + 2} - 6$; domain of $f \circ g$: $\left[\dfrac{1}{2}, \infty\right)$;

domain of $g \circ f$: $[-2, \infty)$ **31.** $\dfrac{x}{2 - 5x}$; $2(x - 5)$; domain of $f \circ g$: $(-\infty, 0) \cup \left(0, \dfrac{2}{5}\right) \cup \left(\dfrac{2}{5}, \infty\right)$; domain of $g \circ f$: $(-\infty, 5) \cup (5, \infty)$

33. To find values of $(f \circ g)(x)$, replace x in f with $g(x)$. For example, if $f(x) = 2x - 5$ and $g(x) = x^2 + 3$, then
$(f \circ g)(x) = 2(x^2 + 3) - 5 = 2x^2 + 6 - 5 = 2x^2 + 1$.   **35.** 4   **37.** 0   **39.** 1   **41.** 2   **43.** 1   **45.** 9   **47.** 1
49. $g(1) = 9$ and $f(9)$ cannot be determined from the table.   **Other correct answers are possible in Exercises 51 and 53.**

51. $f(x) = x^2$; $g(x) = 6x - 2$ **53.** $f(x) = \dfrac{1}{x + 2}$; $g(x) = x^2$ **55.** 0; 0; 0   **56.** 1; 1; 1   **57.** $(f \circ g)(x) = g(x)$ and $(g \circ f)(x) = g(x)$.

In each case, we get $g(x)$.   **58.** $f(x) = x$ is called the identity function.   **59.** $-a$; $-a$; $-a$ **60.** $\dfrac{1}{a}$; $\dfrac{1}{a}$; $\dfrac{1}{a}$ **61.** $(f \circ g)(x) = x$ and

$(g \circ f)(x) = x$. In each case, we get x.   **62.** f and g are inverses.   **63.** $(f \circ g)(x) = 63,360x$; It computes the number of inches in x mi.

65. $D(c) = \dfrac{-c^2 + 10c - 25}{25} + 500$ **67.** $(A \circ r)(t) = 4\pi t^2$; This is the area of the circular layer as a function of time.

Section 10.2   (page 612)

Exercises   1. (a) B **(b)** C **(c)** A **(d)** D   **3.** $(0, 0)$ **5.** $(0, 4)$ **7.** $(1, 0)$ **9.** $(-3, -4)$ **11.** In Exercise 9, the parabola is shifted 3 units
to the left and 4 units down. The parabola in Exercise 10 is shifted 5 units to the right and 8 units down.   **13.** down; wider   **15.** up; narrower
17. (a) I **(b)** IV **(c)** II **(d)** III   **19. (a)** D **(b)** B **(c)** C **(d)** A   **21.** 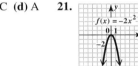   **23.**   **25.**

27. vertex: $(4, 0)$; axis: $x = 4$; domain: $(-\infty, \infty)$; range: $[0, \infty)$ **29.** vertex: $(-2, -1)$; axis: $x = -2$; domain: $(-\infty, \infty)$;

range: $[-1, \infty)$ **31.** vertex: $(2, -4)$; axis: $x = 2$; domain: $(-\infty, \infty)$; range: $[-4, \infty)$ **33.** vertex: $(-1, 2)$;

axis: $x = -1$; domain: $(-\infty, \infty)$; range: $(-\infty, 2]$ $f(x) = -\frac{1}{2}(x + 1)^2 + 2$ **35.** vertex: $(2, -3)$; axis: $x = 2$; domain: $(-\infty, \infty)$; range: $[-3, \infty)$

   37. It is shifted 6 units up.   **38.**   **39.** It is shifted 6 units up.   **40.** It is shifted 6 units to the right.

41.   **42.** It is shifted 6 units to the right.   **43.** quadratic; positive   **45.** quadratic; negative   **47.** linear; positive

49. (a) 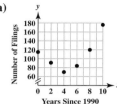   **(b)** quadratic; positive **(c)** $y = 2.969x^2 - 23.125x + 115$ **(d)** 265 **(e)** No. About 16 companies filed for

bankruptcy each month, so at this rate, filings for 2002 would be about 192. The approximation from the model seems high.   **51. (a)** 183.4

(b) The approximation using the model is quite close.   **53.** $y = -\dfrac{1}{2}x^2$ **55.** $y = x^2$ **57.** $y = \dfrac{1}{2}x^2 - \dfrac{3}{2}$ **59.** $\{-7, -2\}$

61. $\{-.5, 1.25\}$

Section 10.3   (page 626)

Exercises   1. If x is squared, it has a vertical axis; if y is squared, it has a horizontal axis.   **3.** Use the discriminant of the function. If it is positive, there are two x-intercepts. If it is 0, there is one x-intercept (at the vertex), and if it is negative, there is no x-intercept.   **5.** $(-4, -6)$

7. $(1, -3)$ **9.** $(2, -1)$ **11.** $(-1, 3)$; up; narrower; no x-intercepts   **13.** $\left(\dfrac{5}{2}, \dfrac{37}{4}\right)$; down; same; two x-intercepts   **15.** $(-3, -9)$; to the

right; wider   **17.** F   **19.** C   **21.** D   **23.** vertex: $(-4, -6)$; axis: $x = -4$; domain: $(-\infty, \infty)$; range: $[-6, \infty)$

25. vertex: $(1, -3)$; axis: $x = 1$; domain: $(-\infty, \infty)$; range: $(-\infty, -3]$ **27.** vertex: $(1, -2)$; axis: $y = -2$; domain: $[1, \infty)$;

range: $(-\infty, \infty)$ **29.** vertex: $(1, 5)$; axis: $y = 5$; domain: $(-\infty, 1]$; range: $(-\infty, \infty)$ **31.** vertex: $(-7, -2)$;

axis: $y = -2$; domain: $[-7, \infty)$; range: $(-\infty, \infty)$ **33.** 30 and 30   **35.** 140 ft by 70 ft; 9800 ft^2 **37.** 16 ft; 2 sec

39. 2 sec; 67 ft   **41. (a)** minimum **(b)** 1995; 1.7%   **43. (a)** The coefficient of x^2 is negative because the parabola opens down.
(b) $(18.45, 3860)$ **(c)** In 2018 Social Security assets will reach their maximum value of $3860 billion.

45. (a) $R(x) = (100 - x)(200 + 4x) = 20,000 + 200x - 4x^2$ (b) 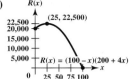 (c) 25 (d) $22,500 **47.** B **49.** A

51. (a) $\{1, 3\}$ (b) $(-\infty, 1) \cup (3, \infty)$ (c) $(1, 3)$ **52.** (a) $\left\{-4, \frac{2}{3}\right\}$ (b) $(-\infty, -4] \cup \left[\frac{2}{3}, \infty\right)$ (c) $\left(-4, \frac{2}{3}\right)$ **53.** (a) $\left\{-3, \frac{5}{2}\right\}$

(b) $\left[-3, \frac{5}{2}\right]$ (c) $(-\infty, -3] \cup \left[\frac{5}{2}, \infty\right)$ **54.** (a) $\{-2, 5\}$ (b) $[-2, 5]$ (c) $(-\infty, -2] \cup [5, \infty)$

Section 10.4 (page 636)

Exercises **1.** (a) The graph of $f(x)$ is reflected about the x-axis. (b) The graph is the same shape as that of $f(x)$,

but stretched vertically by a factor of 2. **3.** (a) Replace y with $-y$. If the equation is equivalent to the given equation, its graph is symmetric with respect to the x-axis. (b) Replace x with $-x$. If the equation is equivalent to the given equation, its graph is symmetric with respect to the y-axis. (c) If the equation is equivalent to the given equation when both $-x$ replaces x and $-y$ replaces y, its graph is symmetric with respect to the origin. **5.** **7.** **9.** x-axis, y-axis, origin **11.** None of the symmetries apply. **13.** y-axis

15. origin **17.** y-axis **19.** origin **21.** In all cases, f is an even function. **22.** In all cases, f is an odd function. **23.** (a) An even polynomial function has an even exponent on the variable. An odd polynomial function has an odd exponent on the variable. (b) An even function has its graph symmetric with respect to the y-axis. (c) An odd function has its graph symmetric with respect to the origin. **25.** f is increasing on $(-\infty, -3)$; f is decreasing on $(0, \infty)$. **27.** f is increasing on $(-\infty, -2)$ and $(1, \infty)$; f is decreasing on $(-2, 1)$. **29.** f is increasing on $(0, \infty)$; f is decreasing on $(-\infty, 0)$. **31.** (a) symmetric (b) symmetric **33.** (a) not symmetric (b) symmetric **35.** $f(-2) = -3$ **37.** $f(4) = 3$ **39.** **41.** F **43.** D **45.** B

Section 10.5 (page 645)

Exercises **1.** B **3.** A **5.** **7.** **9.** **11.** **13.** (a) -10 (b) -2 (c) -1

(d) 2 (e) 4 **15.**

$$f(x) = \begin{cases} x - 1 & \text{if } x \le 3 \\ 2 & \text{if } x > 3 \end{cases}$$

17.

$$f(x) = \begin{cases} 4 - x & \text{if } x < 2 \\ 1 + 2x & \text{if } x \ge 2 \end{cases}$$

19.

$$f(x) = \begin{cases} 2x + 1 & \text{if } x \ge 0 \\ x & \text{if } x < 0 \end{cases}$$

21.

$$f(x) = \begin{cases} 2 + x & \text{if } x < -4 \\ -x^2 & \text{if } x \ge -4 \end{cases}$$

23.

$$f(x) = \begin{cases} |x| & \text{if } x > -2 \\ x^2 - 2 & \text{if } x \le -2 \end{cases}$$

25. 8   **26.** 1995; 1996; increased; 8   **27.** 5   **28.** The larger the slope, the greater the increase in payloads.   **29.**

$f(x) = [\![-x]\!]$

31.

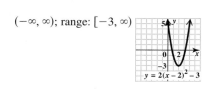
$f(x) = [\![2x - 1]\!]$

33.

$f(x) = [\![3x]\!]$

35. (a) $\{-10, 2\}$ **(b)** $(-10, 2)$ **(c)** $(-\infty, -10) \cup (2, \infty)$ **37.** $\{1, 9\}$; Show that the graph of

$y_1 = |x - 5|$ intersects the graph of $y_2 = 4$ at points with x-values equal to 1 and 9.   **39.** $(-\infty, 1.5) \cup (2, \infty)$; Show that the graph of

$y_1 = |7 - 4x|$ lies above the graph of $y_2 = 1$ for x-values less than 1.5 or greater than 2.   **41.** $(2, 6)$; Show that the graph of $y_1 = |.5x - 2|$ lies

below the graph of $y_2 = 1$ for x-values between 2 and 6.   **43.** for $[3, 6]$: $y = -\dfrac{1}{3}x + 74$; for $(6, 9]$: $y = -x + 78$;

$f(x) = \begin{cases} -\dfrac{1}{3}x + 74 & \text{if } 3 \le x \le 6 \\ -x + 78 & \text{if } 6 < x \le 9 \end{cases}$ **45. (a)** 11 **(b)** 18 **(c)** 32 **(d)**

Days

(e) domain: $(0, \infty)$; range: $\{11, 18, 25, \ldots\}$

Chapter 10 Review Exercises   (page 654)

1. $x^2 + 3x + 3$; $(-\infty, \infty)$ **3.** $(x^2 - 2x)(5x + 3)$; $(-\infty, \infty)$ **5.** $5x^2 - 10x + 3$; $(-\infty, \infty)$ **7.** 3   **9.** $7\sqrt{5}$ **11.** $(4b - 3)(\sqrt{2b})$, $b \ge 0$

13. $(f \circ g)(2) = 2\sqrt{2} - 3$; The answers are not equal, so composition of functions is not commutative.   **15.** $(0, 6)$ **17.** $(3, 7)$

19. $(-4, 3)$ **21.** vertex: $(0, -2)$; axis: $x = 0$; domain: $(-\infty, \infty)$; range: $[-2, \infty)$

$f(x) = 3x^2 - 2$

23. vertex: $(2, -3)$; axis: $x = 2$; domain:

$(-\infty, \infty)$; range: $[-3, \infty)$

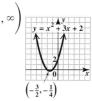
$y = 2(x - 2)^2 - 3$

25. vertex: $\left(-\dfrac{3}{2}, -\dfrac{1}{4}\right)$; axis: $x = -\dfrac{3}{2}$; domain: $(-\infty, \infty)$; range: $\left[-\dfrac{1}{4}, \infty\right)$

$y = x^2 + 3x + 2$
$\left(-\frac{3}{2}, -\frac{1}{4}\right)$

27. vertex: $(-4, -3)$; axis: $y = -3$; domain: $[-4, \infty)$; range: $(-\infty, \infty)$

$x = 2(y + 3)^2 - 4$

29. (a) $c = 12.39$, $16a + 4b + c = 15.78$,

$49a + 7b + c = 22.71$ **(b)** $f(x) = .2089x^2 + .0118x + 12.39$ **(c)** 25.85; The result using the model is a little high.

31. length: 50 m; width: 50 m   **33.** x-axis, y-axis, origin   **35.** x-axis   **37.** no symmetries   **39.** The vertical line test shows that a circle

does not represent a function.   **41.** decreasing   **43.** increasing on $(0, \infty)$ **45.** increasing on $(1, \infty)$; decreasing on $(-\infty, 1)$

47.   **49.** **51.** 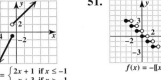 **53.** The graph is narrower than the graph of $y = |x|$, and it is shifted

$$f(x) = \begin{cases} 2x + 1 & \text{if } x \le -1 \\ x + 3 & \text{if } x > -1 \end{cases}$$

$f(x) = -[\![x]\!]$

(translated) 4 units to the left and 3 units down. **55.** F **57.** C **59.** E **61.** vertex: $\left(-\dfrac{1}{2}, -3\right)$; axis: $x = -\dfrac{1}{2}$; domain: $(-\infty, \infty)$;

range: $[-3, \infty)$ 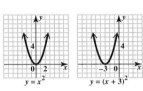 **63.** **65.** **67.** 4; down

$f(x) = 4x^2 + 4x - 2$

$f(x) = |2x + 1|$

$f(x) = [\![x]\!] - 2$

$y = x^2$ $y = x^2 - 4$

68. 3; left **69.** 4; right **70.** It is obtained by translating the graph of $y = f(x)$

$y = x^2$ $y = (x + 3)^2$

$y = x^2$ $y = (x - 4)^2$

h units to the right if $h > 0$, $|h|$ units to the left if $h < 0$, k units up if $k > 0$, $|k|$ units down if $k < 0$.

Chapter 10 Test   (page 657)

[10.1] **1.** 2   **2.** -7 **3.** $-\dfrac{7}{3}$ **4.** -2 **5.** $x^2 + 4x - 1$; $(-\infty, \infty)$ [10.2] **6.** A   **7.** vertex: $(0, -2)$; axis: $x = 0$; domain: $(-\infty, \infty)$;

range: $[-2, \infty)$ [10.3] **8.** vertex: $(2, 3)$; axis: $x = 2$; domain: $(-\infty, \infty)$; range: $(-\infty, 3]$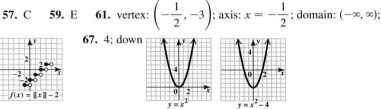

$f(x) = \frac{1}{2}x^2 - 2$

$f(x) = -x^2 + 4x - 1$

9. vertex: $(2, 2)$; axis: $y = 2$; domain: $(-\infty, 2]$; range: $(-\infty, \infty)$ **10. (a)** 6.5% **(b)** 1996; 3.5%   **11.** 160 ft by 320 ft

$x = -(y - 2)^2 + 2$

[10.4] **12.** y-axis   **13.** x-axis   **14.** x-axis, y-axis, origin   **15.** increasing: $(-2, 1)$; decreasing: $(-\infty, -2)$; constant: $(1, \infty)$
[10.5] **16. (a)** C **(b)** A **(c)** D **(d)** B   **17.** 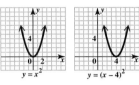 **18.** **19.** **20.**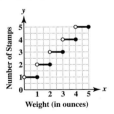

$f(x) = |x - 3| + 4$

$f(x) = [\![2x]\!]$

$f(x) = \begin{cases} -x & \text{if } x \le 2 \\ x - 4 & \text{if } x > 2 \end{cases}$

Number of Stamps

Weight (in ounces)

Cumulative Review Exercises Chapters 1–10   (page 659)

[3.2] **1.** $-\dfrac{5}{7}$ **2.** $\dfrac{2}{3}$ [3.5] **3.** 8   **4.** No, because there are x-values that correspond to two y-values. To see this, solve the equation for y.

[5.4] **5.** $28y^2 - 3y - 18$ [5.5] **6.** $4x^2 + x + 3$ [7.2] **7.** $\dfrac{3p - 2}{p - 1}$ [7.1] **8.** $\dfrac{2}{3x}$ [8.4] **9.** $2\sqrt{2z}$ [8.5] **10.** $5y - 9$ [5.1] **11.** $\dfrac{2}{9}m^{10}$

[8.2] **12.** $k\sqrt[6]{k}$ [6.4] **13.** $(2k^2 + 3)(k + 1)(k - 1)$ [2.1] **14.** $\{-6\}$ [2.2] **15.** $p = \dfrac{q^2}{3 - q}$ [2.5] **16.** $(-\infty, 2]$

[2.6] **17.** $(-\infty, 0] \cup (2, \infty)$ [2.7] **18.** $\left\{-\dfrac{7}{3}, \dfrac{17}{3}\right\}$ **19.** $(-\infty, \infty)$ [4.2] **20.** $\{(1, 2, -1)\}$ [4.4] **21.** $\{(4, 3)\}$ [6.5] **22.** $\left\{-\dfrac{7}{2}, 0, 4\right\}$

[7.4] **23.** $\left\{-\dfrac{1}{11}\right\}$ [8.6] **24.** $\{3\}$ [9.2] **25.** $\left\{\dfrac{-3 - \sqrt{69}}{10}, \dfrac{-3 + \sqrt{69}}{10}\right\}$ [9.4] **26.** $r = \dfrac{-\pi h \pm \sqrt{\pi^2 h^2 + \pi S}}{\pi}$

[9.5] **27.** $\left(-\infty, -\dfrac{2}{3}\right) \cup (4, \infty)$ [4.5] **28.** 8 **29.** $\left\{\left(\dfrac{5}{4}, 2, \dfrac{3}{2}\right)\right\}$ [6.5] **30.** (a) after 16 sec (b) 4 sec and 12 sec [7.5] **31.** 480 mi

[3.1] **32.**
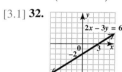
$2x - 3y = 6$

[3.4] **33.**
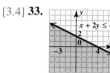
$x + 2y \le 4$

[10.3] **34.**
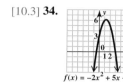
$f(x) = -2x^2 + 5x + 3$

[10.5] **35.**

$f(x) = |x + 1|$

[10.4] **36. (a)** $(0, \infty)$

(b) $(-\infty, 0)$ **37.** x-axis **38.** x-axis, y-axis, origin [10.5] **39. (a)** 5 **(b)** -2 [10.3] **40.** No. The graph is a horizontal parabola, so it fails the vertical line test.

CHAPTER **11** INVERSE, EXPONENTIAL, AND LOGARITHMIC FUNCTIONS

Section 11.1 (page 667)

Exercises **1.** It is not one-to-one because both Illinois and Wisconsin are paired with the same range element, 40. **3.** Yes. By adding 1 to 1058 two distances would be the same, so the function would not be one-to-one. **5.** B **7.** A **9.** $\{(6, 3), (10, 2), (12, 5)\}$ **11.** not one-to-one **13.** $f^{-1}(x) = \dfrac{x - 4}{2}$ **15.** $g^{-1}(x) = x^2 + 3, \; x \ge 0$ **17.** not one-to-one **19.** $f^{-1}(x) = \sqrt[3]{x + 4}$ **21. (a)** 8 **(b)** 3

23. (a) 1 **(b)** 0 **25. (a)** one-to one **(b)**
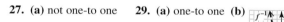
27. (a) not one to one **29. (a)** one-to one **(b)**

31.
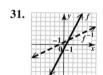
33.

35.

x	$f(x)$
0	0
1	1
4	2

37.

x	$f(x)$
-1	-3
0	-2
1	-1
2	6

39. $f^{-1}(x) = \dfrac{x + 5}{4}$

40. My graphing calculator is the greatest thing since sliced bread. **41.** If the function were not one-to-one, there would be ambiguity in some of the characters, as they could represent more than one letter. **42.** Answers will vary. For example, Jane Doe is 1004 5 2748 129 68 3379 129. **43.** $f^{-1}(x) = \dfrac{x + 7}{2}$

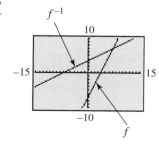

45. $f^{-1}(x) = \sqrt[3]{x} - 5$ **47.**   **49.** It is not a one-to-one function.

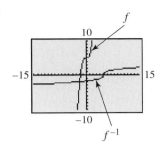

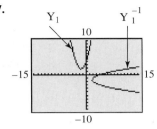

Section 11.2   (page 678)

Exercises   **1.** C   **3.** A   **5.**   **7.**   **9.**   **11.**   **13. (a)** rises; falls

(b) It is one-to-one and thus has an inverse.   **15.** $\{2\}$ **17.** $\left\{\dfrac{3}{2}\right\}$ **19.** $\{7\}$ **21.** $\{-3\}$ **23.** $\{-1\}$ **25.** $\{-3\}$ **27.** 639.545

29. .066   **31.** 12.179   **33.** In the definition of the exponential function defined by $F(x) = a^x$, a must be a positive number. This corresponds to the peculiarity that some scientific calculators do not allow negative bases.   **35. (a)** .5°C **(b)** .35°C   **37. (a)** 1.6°C **(b)** .5°C

39. (a) 132,359 thousand tons **(b)** 97,264 thousand tons **(c)** It is slightly less than what the model provides (93,733 thousand tons).

41. (a) $5000 **(b)** $2973 **(c)** $1768 **(d)** 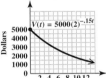   **43.** 6.67 yr after it was purchased   **45.** $\left(\sqrt[4]{16}\right)^3$; 8   **46.** $\sqrt[4]{16^3}$; 8

47. 64; 8   **48.** Because $\sqrt{\sqrt{x}} = (x^{1/2})^{1/2} = x^{1/4} = \sqrt[4]{x}$, the fourth root of 16^3 can be found by taking the square root twice.   **49.** 8

50. $16^{75/100}$; $16^{3/4} = \left(\sqrt[4]{16}\right)^3 = 2^3 = 8$

Section 11.3   (page 686)

Connections   **(page 686)**   **1.** almost 4 times as powerful   **2.** about 300 times as powerful

Exercises   **1. (a)** C **(b)** F **(c)** B **(d)** A **(e)** E **(f)** D   **3.** $\log_4 1024 = 5$ **5.** $\log_{1/2} 8 = -3$ **7.** $\log_{10} .001 = -3$ **9.** $\log_{625} 5 = \dfrac{1}{4}$

11. $4^3 = 64$ **13.** $10^{-4} = \dfrac{1}{10,000}$ **15.** $6^0 = 1$ **17.** $9^{1/2} = 3$ **19.** By using the word "radically," the teacher meant for him to consider roots. Because 3 is the square (2nd) root of 9, $\log_9 3 = \dfrac{1}{2}$.   **21.** $\left\{\dfrac{1}{3}\right\}$ **23.** $\{81\}$ **25.** $\left\{\dfrac{1}{5}\right\}$ **27.** $\{1\}$ **29.** $\{x \mid x > 0, x \neq 1\}$

31. $\{5\}$ **33.** $\left\{\dfrac{5}{3}\right\}$ **35.** $\{4\}$ **37.** $\left\{\dfrac{3}{2}\right\}$ **39.** $\{30\}$ **41.**   **43.**   **45.** Every power of 1 is equal to 1,

and thus it cannot be used as a base.   **47.** $(0, \infty)$; $(-\infty, \infty)$ **49.** 8   **51.** 24   **53. (a)** 645 sites **(b)** 962 sites **(c)** 1279 sites

55. (a) 130 thousand units **(b)** 190 thousand units **(c)**

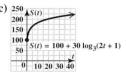

57. (a) 500 **(b)** 1000 **(c)** 1500 **(d)**

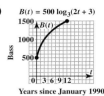

59.

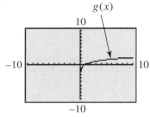

61.

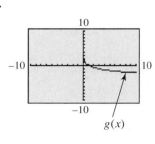

Section 11.4 (page 696)

Connections (page 695) 1.
$$\log_{10} 458.3 \approx 2.661149857$$
$$+ \log_{10} 294.6 \approx 2.469232743$$
$$\approx 5.130382600$$
$$10^{5.130382600} \approx 135,015.18$$
A calculator gives $(458.3)(294.6) = 135,015.18$.   **2.** Answers will vary.

Exercises 1. $\log_{10} 3 + \log_{10} 4$ **3.** 4   **5.** 4   **7.** $\log_7 4 + \log_7 5$ **9.** $\log_5 8 - \log_5 3$ **11.** $2 \log_4 6$ **13.** $\frac{1}{3} \log_3 4 - 2 \log_3 x -$
$\log_3 y$ **15.** $\frac{1}{2} \log_3 x + \frac{1}{2} \log_3 y - \frac{1}{2} \log_3 5$ **17.** $\frac{1}{3} \log_2 x + \frac{1}{5} \log_2 y - 2 \log_2 r$ **19.** The distributive property tells us that the *product*
$a(x + y)$ equals the sum $ax + ay$. In the notation $\log_a(x + y)$, the parentheses do not indicate multiplication. They indicate that $x + y$ is the result
of raising a to some power.   **21.** $\log_b xy$ **23.** $\log_a \frac{m}{n}$ **25.** $\log_a \frac{rt^3}{s}$ **27.** $\log_a \frac{125}{81}$ **29.** $\log_{10}(x^2 - 9)$ **31.** $\log_p \frac{x^3 y^{1/2}}{z^{3/2} a^3}$ **33.** 1.2552
35. $-.6532$ **37.** 1.5562   **39.** .4771   **41.** .2386   **43.** 4.7710   **45.** false   **47.** true   **49.** true   **51.** false   **53.** The exponent of
a quotient is the difference between the exponent of the numerator and the exponent of the denominator.   **55.** $\log_2 8 - \log_2 4 = \log_2 \frac{8}{4} =$
$\log_2 2 = 1$ **57.** 4   **58.** It is the exponent to which 3 must be raised to obtain 81.   **59.** 81   **60.** It is the exponent to which 2 must be
raised to obtain 19.   **61.** 19   **62.** m

Section 11.5 (page 703)

Connections (page 702)   2; 2.5; $2.\overline{6}$; $2.708\overline{3}$; $2.71\overline{6}$; The difference is .0016151618. It approaches e fairly quickly.

Exercises 1. C   **3.** C   **5.** 19.2   **7.** 1.6335   **9.** 2.5164   **11.** -1.4868 **13.** 9.6776   **15.** 2.0592   **17.** -2.8896 **19.** 5.9613
21. 4.1506   **23.** 2.3026   **25. (a)** 2.552424846 **(b)** 1.552424846 **(c)** .552424846 **(d)** The whole number parts will vary but the decimal
parts are the same.   **27.** An error message appears, because we cannot find the common logarithm of a negative number.   **29.** bog
31. 11.6   **33.** 4.3   **35.** 4.0×10^{-8} **37.** 4.0×10^{-6} **39. (a)** 800 yr **(b)** 5200 yr **(c)** 11,500 yr   **41. (a)** 107 dB **(b)** 100 dB
(c) 98 dB   **43. (a)** 77% **(b)** 1989   **45. (a)** \$54 per ton **(b)** If $p = 0$, then $\ln(1 - p) = \ln 1 = 0$, so T would be negative. If $p = 1$, then
$\ln(1 - p) = \ln 0$, but the domain of $\ln x$ is $(0, \infty)$.   **47.** 2.2619   **49.** .6826   **51.** .3155   **53.** .8736   **55.** 2.4849   **57.** Answers will
vary. Suppose the name is Jeffery Cole, with $m = 7$ and $n = 4$. **(a)** $\log_7 4$ is the exponent to which 7 must be raised to obtain 4.

(b) .7124143742 **(c)** 4   **59.** 1333 sites   **61.**

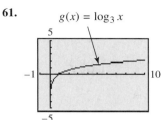

$g(x) = \log_3 x$

63.

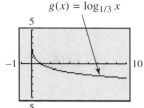

$g(x) = \log_{1/3} x$

Section 11.6   (page 713)

Exercises   **1.** $\log 5^x = \log 125$ **2.** $x \log 5 = \log 125$ **3.** $x = \dfrac{\log 125}{\log 5}$ **4.** $\dfrac{\log 125}{\log 5} = 3; \{3\}$ **5.** $\{.827\}$ **7.** $\{.833\}$ **9.** $\{1.201\}$

11. $\{2.269\}$ **13.** $\{15.967\}$ **15.** $\{261.291\}$ **17.** $\{-10.718\}$ **19.** $\{3\}$ **21.** $\{5.879\}$ **23.** Natural logarithms are a better choice

because e is the base.   **25.** $\left\{\dfrac{2}{3}\right\}$ **27.** $\left\{\dfrac{33}{2}\right\}$ **29.** $\{-1 + \sqrt[3]{49}\}$ **31.** 2 cannot be a solution because $\log(2 - 3) = \log(-1)$, and -1

is not in the domain of $\log x$.   **33.** $\left\{\dfrac{1}{3}\right\}$ **35.** $\{2\}$ **37.** $\emptyset$ **39.** $\{8\}$ **41.** $\left\{\dfrac{4}{3}\right\}$ **43.** $\{8\}$ **45. (a)** \$2539.47 **(b)** 10.2 yr

47. (a) \$4934.71 **(b)** 19.8 yr   **49. (a)** \$11,260.96 **(b)** \$11,416.64 **(c)** \$11,497.99 **(d)** \$11,580.90 **(e)** \$11,581.83   **51.** \$137.41

53. (a) 78,840 million dollars **(b)** 92,316 million dollars **(c)** 108,095 million dollars **(d)** 70,967 million dollars   **55.** 32,044 million dollars

57. (a) 1.62 g **(b)** 1.18 g **(c)** .69 g **(d)** 2.00 g   **59. (a)** 179.73 g **(b)** 21.66 yr   **61.** 1997   **63.** 1.733 days   **65.** It means that after

250 yr, approximately 2.9 g of the original sample remain.   **67. (a)** 2.302585093 **(b)** .4342944819

Chapter 11 Review Exercises   (page 724)

1. not one-to-one   **3.** This function is not one-to-one because two sodas in the list have 41 mg of caffeine.   **5.** $f^{-1}(x) = \dfrac{x^3 + 4}{6}$

7.   **9.**   **11.**   **13.** $\{4\}$ **15. (a)** 1.2 million tons **(b)** 3.8 million tons **(c)** 21.8 million tons

17.   **19.** $\left\{\dfrac{3}{2}\right\}$ **21.** $\{8\}$ **23.** $\{b \mid b > 0, b \neq 1\}$ **25.** a **27.** $\log_2 3 + \log_2 x + 2 \log_2 y$ **29.** $\log_b \dfrac{3x}{y^2}$ **31.** 1.4609

33. 3.3638   **35.** .9251   **37.** 6.4   **39.** 2.5×10^{-5} **41. (a)** 18 yr **(b)** 12 yr **(c)** 7 yr **(d)** 6 yr **(e)** Each comparison shows

approximately the same number. For example, in part (a) the doubling time is 18 yr (rounded) and $\dfrac{72}{4} = 18$. Thus, the formula $t = \dfrac{72}{100r}$ (called

the *rule of 72*) is an excellent approximation of the doubling time formula. (It is used by bankers for that purpose.)   **43.** $\{2.042\}$

45. $\{18.310\}$ **47.** $\{-6 + \sqrt[3]{25}\}$ **49.** $\left\{\dfrac{3}{8}\right\}$ **51.** $\{1\}$ **53.** \$28,295.56   **55.** Plan A is better, since it would pay \$2.92 more.

57. (a) about \$4267 **(b)** about 11%   **59.** D   **61.** $\dfrac{1}{4}, \dfrac{1}{2}, 1, 2, 4, 8$ 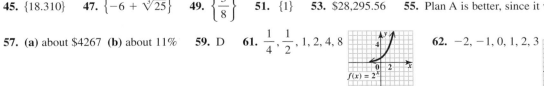   **62.** $-2, -1, 0, 1, 2, 3$

$f(x) = 2^x$

$g(x) = \log_2 x$

63. The roles of x and y are interchanged. They are inverses. **64.** horizontal; vertical **65.** 5; 2; 3; 5 **66.** 4; 8 (or 8; 4)

67. 3.700439718 (The number of displayed digits may vary.) **68.** $\log_2 13$ is the exponent to which 2 must be raised to obtain 13. **69.** 13

70. 13; The number in Exercise 67 is the exponent to which 2 must be raised to obtain 13. **71.** 3.700439718 **72.** $\left\{-\dfrac{8}{5}\right\}$ **73.** 7

75. 4 **77.** −5 **79.** {72} **81.** $\left\{\dfrac{1}{9}\right\}$ **83.** {3} **85.** $\left\{\dfrac{1}{8}\right\}$ **87.** {−2, −1} **89.** 6.8 yr **91.** .325

Chapter 11 Test (page 729)

[11.1] **1. (a)** not one-to-one **(b)** one-to-one **2.** $f^{-1}(x) = x^3 - 7$ **3.** [11.2] **4.**

[11.3] **5.** 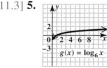 [11.1–11.3] **6.** Once the graph of $f(x) = 6^x$ is sketched, interchange the x- and y-values of its ordered pairs. The

resulting points will be on the graph of $g(x) = \log_6 x$ since f and g are inverses. [11.2] **7.** {−4} **8.** $\left\{-\dfrac{13}{3}\right\}$ [11.5] **9. (a)** 30.0 million

(b) 37.7 million [11.3] **10.** $\log_4 .0625 = -2$ **11.** $7^2 = 49$ **12.** {32} **13.** $\left\{\dfrac{1}{2}\right\}$ **14.** {2} **15.** 5; 2; 5th; 32

[11.4] **16.** $2 \log_3 x + \log_3 y$ **17.** $\dfrac{1}{2} \log_5 x - \log_5 y - \log_5 z$ **18.** $\log_b \dfrac{s^3}{t}$ **19.** $\log_b \dfrac{r^{1/4} s^2}{t^{2/3}}$ [11.5] **20. (a)** 1.3636 **(b)** −.1985

21. (a) $\dfrac{\log 19}{\log 3}$ **(b)** $\dfrac{\ln 19}{\ln 3}$ **(c)** 2.6801 [11.6] **22.** {3.9656} **23.** {3} **24.** $12{,}507.51 **25. (a)** $19{,}260.38 **(b)** approximately 13.9 yr

Cumulative Review Exercises Chapters 1–11 (page 731)

[1.1] **1.** $-2, 0, 6, \dfrac{30}{3}$ (or 10) **2.** $-\dfrac{9}{4}, -2, 0, .6, 6, \dfrac{30}{3}$ (or 10) **3.** $-\sqrt{2}, \sqrt{11}$ **4.** $-\dfrac{9}{4}, -2, -\sqrt{2}, 0, .6, \sqrt{11}, 6, \dfrac{30}{3}$ (or 10)

[1.2, 1.3] **5.** 16 **6.** −27 **7.** −39 [2.1] **8.** $\left\{-\dfrac{2}{3}\right\}$ [2.5] **9.** $[1, \infty)$ [2.7] **10.** {−2, 7} **11.** $\left\{\pm\dfrac{16}{3}\right\}$ **12.** $\left[\dfrac{7}{3}, 3\right]$

13. $(-\infty, -3) \cup (2, \infty)$ [3.1] **14.** [3.4] **15.** [3.2, 3.5] **16. (a)** yes **(b)** approximately −4000; The

number of acres harvested decreased by approximately 4000 acres per year during 1997–1999. [3.3] **17.** $y = \dfrac{3}{4}x - \dfrac{19}{4}$

[4.1, 4.4, 4.5] **18.** {(4, 2)} **19.** ∅ [4.2, 4.4, 4.5] **20.** {(1, −1, 4)} [4.3] **21.** 6 lb [5.4] **22.** $6p^2 + 7p - 3$ **23.** $16k^2 - 24k + 9$
[5.2] **24.** $-5m^3 + 2m^2 - 7m + 4$ [5.5] **25.** $2t^3 + 5t^2 - 3t + 4$ [6.1] **26.** $x(8 + x^2)$ [6.2] **27.** $(3y - 2)(8y + 3)$

28. $z(5z + 1)(z - 4)$ [6.3] **29.** $(4a + 5b^2)(4a - 5b^2)$ **30.** $(2c + d)(4c^2 - 2cd + d^2)$ **31.** $(4r + 7q)^2$ [5.1] **32.** $-\dfrac{1875p^{13}}{8}$

[7.1] **33.** $\dfrac{x + 5}{x + 4}$ [7.2] **34.** $\dfrac{-3k - 19}{(k + 3)(k - 2)}$ **35.** $\dfrac{22 - p}{p(p - 4)(p + 2)}$ [8.3] **36.** $12\sqrt{2}$ [8.4] **37.** $-27\sqrt{2}$ [8.6] **38.** {0, 4}

[8.7] 39. 41   **[9.1, 9.2] 40.** $\left\{\dfrac{1 \pm \sqrt{13}}{6}\right\}$ **[9.5] 41.** $(-\infty, -4) \cup (2, \infty)$ **[9.3] 42.** $\{\pm 1, \pm 2\}$ **[10.3] 43.** 150 and 150

[10.2] 44. $f(x) = \frac{1}{3}(x-1)^2 + 2$

[11.2] 45.

$f(x) = 2^x$

46. $\{-1\}$ **[11.3] 47.**

$f(x) = \log_3 x$

[11.4] 48. 6.3398

49. $3 \log x + \dfrac{1}{2} \log y - \log z$ **[11.6] 50. (a)** 25,000 **(b)** 30,500 **(c)** 37,300 **(d)** in about 3.5 hr, or at about 3:30 P.M.

CHAPTER **12** POLYNOMIAL AND RATIONAL FUNCTIONS

Section 12.1   (page 741)

Connections   (page 741) -2.7

Exercises   1. Synthetic division provides a quick, easy way to divide a polynomial by a binomial of the form $x - k$.   **3.** $x - 5$

5. $4m - 1$ **7.** $2a + 4 + \dfrac{5}{a+2}$ **9.** $p - 4 + \dfrac{9}{p+1}$ **11.** $4a^2 + a + 3$ **13.** $x^4 + 2x^3 + 2x^2 + 7x + 10 + \dfrac{18}{x-2}$

15. $-4r^5 - 7r^4 - 10r^3 - 5r^2 - 11r - 8 + \dfrac{-5}{r-1}$ **17.** $-3y^4 + 8y^3 - 21y^2 + 36y - 72 + \dfrac{143}{y+2}$ **19.** $y^2 + y + 1 + \dfrac{2}{y-1}$

21. $f(x) = (x+1)(2x^2 - x + 2) + (-10)$ **23.** $f(x) = (x+2)(-x^2 + 4x - 8) + 20$ **25.** $f(x) = (x-3)(4x^3 + 9x^2 + 7x + 20) + 60$

27. 2   **29.** -1 **31.** -6 **33.** 0   **35.** 11   **37.** $-6 - i$ **39.** By the remainder theorem, a 0 remainder means that $f(k) = 0$; that is,

k is a number that makes $f(x) = 0$.   **41.** no   **43.** yes   **45.** no   **47.** no   **49.** $(2x - 3)(x + 4)$ **50.** $\left\{\dfrac{3}{2}, -4\right\}$

51. $f(-4) = 0, f\left(\dfrac{3}{2}\right) = 0$ **52.** a **53.** Yes, $x - 3$ is a factor. $f(x) = (x - 3)(3x - 1)(x + 2)$ **55.** 3   **57.** -1

Section 12.2   (page 750)

Exercises   1. true   **3.** false   **5.** no   **7.** yes   **9.** yes   **11.** $f(x) = (x - 2)(2x - 5)(x + 3)$ **13.** $f(x) = (x + 3)(3x - 1)(2x - 1)$

15. $-1 \pm i$ **17.** $3, 2 + i$ **19.** $i, \pm 2i$ **21. (a)** $\pm 1, \pm 2, \pm 5, \pm 10$ **(b)** $-1, -2, 5$ **(c)** $f(x) = (x + 1)(x + 2)(x - 5)$

23. (a) $\pm 1, \pm 2, \pm 3, \pm 5, \pm 6, \pm 10, \pm 15, \pm 30$ **(b)** $-5, -3, 2$ **(c)** $f(x) = (x + 5)(x + 3)(x - 2)$ **25. (a)** $\pm 1, \pm 2, \pm 3, \pm 4, \pm 6, \pm 12, \pm\dfrac{1}{2},$

$\pm\dfrac{3}{2}, \pm\dfrac{1}{3}, \pm\dfrac{2}{3}, \pm\dfrac{4}{3}, \pm\dfrac{1}{6}$ **(b)** $-4, -\dfrac{1}{3}, \dfrac{3}{2}$ **(c)** $f(x) = (x + 4)(3x + 1)(2x - 3)$ **27. (a)** $\pm 1, \pm 2, \pm 3, \pm 6, \pm\dfrac{1}{2}, \pm\dfrac{3}{2}, \pm\dfrac{1}{3}, \pm\dfrac{2}{3},$

$\pm\dfrac{1}{6}, \pm\dfrac{1}{12}, \pm\dfrac{1}{4}, \pm\dfrac{3}{4}$ **(b)** $-\dfrac{3}{2}, -\dfrac{2}{3}, \dfrac{1}{2}$ **(c)** $f(x) = (3x + 2)(2x + 3)(2x - 1)$ **29.** $0, \pm\dfrac{\sqrt{7}}{7}i$ **31.** $2, -3, 1, -1$ **33.** -2 (mult. 5),

1 (mult. 5), $1 - \sqrt{3}$ (mult. 2)   **35.** $f(x) = x^2 - 6x + 10$ **37.** $f(x) = x^3 - 5x^2 + 5x + 3$ **39.** $f(x) = x^4 + 4x^3 - 4x^2 - 36x - 45$

41. $f(x) = x^3 - 2x^2 + 9x - 18$ **43.** $f(x) = x^4 - 6x^3 + 17x^2 - 28x + 20$ **45.** $f(x) = -3x^3 + 6x^2 + 33x - 36$

47. $f(x) = -\dfrac{1}{2}x^3 - \dfrac{1}{2}x^2 + x$ **49.** $f(x) = -\dfrac{1}{3}x^3 + \dfrac{5}{3}x^2 - \dfrac{1}{3}x + \dfrac{5}{3}$ **51.** $g(x) = x^2 - 4x - 5$ **52.** The function g is quadratic.

The x-intercepts of g are also x-intercepts of f.   **53.** $h(x) = x + 1$ **54.** The function h is linear. The x-intercept of h is also an x-intercept

of g.   **55.** $-1; 3; f(x) = (x + 2)^2(x + 1)(x - 3)$ **61.** $-.88, 2.12, 4.86$ **63.** $.44, 1.81$ **65.** 1.40 **67.** 2 or 0 positive; 1 negative

69. 1 positive; 1 negative   **71.** 2 or 0 positive; 3 or 1 negative   **73.** Since 1 is a zero of f, $x - 1$ is a factor. Use synthetic division to divide $f(x)$ by $x - 1$ to find the remaining factors.

$$\begin{array}{r|rrr} 1 & 1 & -6 & 11 & -6 \\ & & 1 & -5 & 6 \\ \hline & 1 & -5 & 6 & 0 \end{array}$$

The quotient is $x^2 - 5x + 6$, so $f(x) = (x - 1)(x^2 - 5x + 6)$. Factor again to get the complete factored form, $f(x) = (x - 1)(x - 2)(x - 3)$.

Section 12.3 (page 763)

Exercises **1.**
$f(x) = \frac{1}{4}x^6$

3.
$f(x) = -\frac{5}{4}x^5$

5.
$f(x) = \frac{1}{2}x^3 + 1$

7.
$f(x) = -(x + 1)^3$

9.
$f(x) = (x - 1)^4 + 2$

11. A **13.** one

15. B and D **17.** one **19.** 2 **21.** 3 **23.** 3 **25.**
$f(x) = (x - 4)(x + 2)$
$\cdot (x - 1)$

27.
$f(x) = 2x(x - 3)(x + 2)$

29.
$f(x) = x^2(x + 1)(x - 1)$

31.
$f(x) = -x(x + 3)(x + 1)$

33. $-1, \dfrac{3}{2}, 2; f(x) = (x + 1)(2x - 3)(x - 2)$

$f(x) = 2x^3 - 5x^2 - x + 6$

35. -2 (multiplicity 2), $3; f(x) = (x + 2)^2(x - 3)$
$f(x) = x^3 + x^2 - 8x - 12$

37. -2 (multiplicity 2), $3; f(x) = (x + 2)^2(-x + 3)$
$f(x) = -x^3 - x^2 + 8x + 12$

39. 3 (multiplicity 2), -3 (multiplicity 2); $f(x) = (x - 3)^2(x + 3)^2$
$f(x) = x^4 - 18x^2 + 81$

41. (a) $f(-2) = 8 > 0$ and $f(-1) = -2 < 0$

(b) $-1.236, 3.236$ **43. (a)** $f(-4) = 76 > 0$ and $f(-3) = -75 < 0$ **(b)** $-3.646, -.317, 1.646, 6.317$
51. $f(x) = .5(x + 6)(x - 2)(x - 5) = .5x^3 - .5x^2 - 16x + 30$ **53.** $-.88, 2.12, 4.86$ **55.** -1.52 **57.** $-.40, 2.02$
59. $(-3.44, 26.15)$ **61.** $(-.09, 1.05)$ **63.** $(-.20, -28.62)$ **65.** odd **66.** odd **67.** even **68.** even **69.** odd **70.** neither
71. neither **72.** odd **73.** even **74.** y-axis; origin **75. (a)** See part (b).
(b) $g(x) = 1818(x - 2)^2 + 620$ **(c)** 46,070; This figure is a bit higher than the figure 40,820 given in the table.

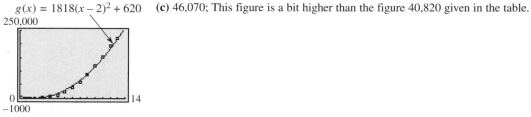
$g(x) = 1818(x - 2)^2 + 620$
250,000

0 ⌐ 14
-1000

77. **(a)** $0 < x < 10$ **(b)** $A(x) = x(20 - 2x)$ or $A(x) = -2x^2 + 20x$ **(c)** $x = 5$; maximum cross section area: 50 in.^{2} **(d)** between 0 and 2.76 in. or between 7.24 and 10 in. **79.** 1.732 **81.** **(a)**

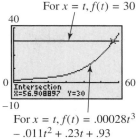

For $x = t$, $f(t) = 30$ **(b)** The graphs intersect at $x = t \approx 56.9$. Since $t = 0$

For $x = t$, $f(t) = .00028t^3 - .011t^2 + .23t + .93$

corresponds to 1930, this would be during 1986. **(c)** An increasing percentage of females have smoked during this time period. Smoking has been shown to increase the likelihood of lung cancer. **83.** **(a)** about 49% **(b)** approximately 10.2 yr

85. **(a)**

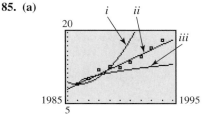

(b) All three approximate the data near 1986, but only the linear function

(ii) $g(x) = 1.088(x - 1986) + 8.6$ approximates the data near 1994. **87.** **(a)** If the length of the pendulum increases, so does the period of oscillation T. **(b)** There are a number of ways. One way is to realize that $k = \dfrac{L}{T^n}$ is true for some integer n. The ratio should be the constant k for each data point when the correct n is found. **(c)** $k \approx .81$; $n = 2$ **(d)** 2.48 sec **(e)** T increases by a factor of $\sqrt{2} \approx 1.414$.

Summary Exercises on Polynomial Functions and Graphs (page 769)

1. **(a)** positive zeros: 1; negative zeros: 3 or 1 **(b)** $\pm 1, \pm 2, \pm 3, \pm 6$ **(c)** $-3, -1$ (multiplicity 2), 2 **(d)** no other real zeros **(e)** no other complex zeros **(f)** $(-3, 0), (-1, 0), (2, 0)$ **(g)** $(0, -6)$ **(h)** $f(4) = 350$; $(4, 350)$ **(i)**

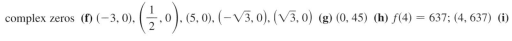

$f(x) = x^4 + 3x^3 - 3x^2 - 11x - 6$

2. **(a)** positive zeros: 3 or 1; negative zeros: 2 or 0 **(b)** $\pm 1, \pm 3, \pm 5, \pm 9, \pm 15, \pm 45, \pm \dfrac{1}{2}, \pm \dfrac{3}{2}, \pm \dfrac{5}{2}, \pm \dfrac{9}{2}, \pm \dfrac{15}{2}, \pm \dfrac{45}{2}$ **(c)** $-3, \dfrac{1}{2}, 5$ **(d)** $-\sqrt{3}, \sqrt{3}$ **(e)** no other complex zeros **(f)** $(-3, 0), \left(\dfrac{1}{2}, 0\right), (5, 0), \left(-\sqrt{3}, 0\right), \left(\sqrt{3}, 0\right)$ **(g)** $(0, 45)$ **(h)** $f(4) = 637$; $(4, 637)$ **(i)**

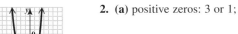

$f(x) = -2x^5 + 5x^4 + 34x^3 - 30x^2 - 84x + 45$

3. **(a)** positive zeros: 4, 2, or 0; negative zeros: 1 **(b)** $\pm 1, \pm 5, \pm \dfrac{1}{2}, \pm \dfrac{5}{2}$ **(c)** 5 **(d)** $-\dfrac{\sqrt{2}}{2}, \dfrac{\sqrt{2}}{2}$ **(e)** $-i, i$ **(f)** $\left(-\dfrac{\sqrt{2}}{2}, 0\right), \left(\dfrac{\sqrt{2}}{2}, 0\right), (5, 0)$ **(g)** $(0, 5)$ **(h)** $f(4) = -527$; $(4, -527)$ **(i)**

$f(x) = 2x^5 - 10x^4 + x^3 - 5x^2 - x + 5$

4. **(a)** positive zeros: 2 or 0; negative zeros: 2 or 0

(b) $\pm 1, \pm 2, \pm 3, \pm 6, \pm 9, \pm 18, \pm \dfrac{1}{3}, \pm \dfrac{2}{3}$ **(c)** $-\dfrac{2}{3}, 3$ **(d)** $\dfrac{-1 + \sqrt{13}}{2}, \dfrac{-1 - \sqrt{13}}{2}$ **(e)** no other complex zeros **(f)** $\left(-\dfrac{2}{3}, 0\right)$, $(3, 0)$,

$\left(\dfrac{-1 + \sqrt{13}}{2}, 0\right), \left(\dfrac{-1 - \sqrt{13}}{2}, 0\right)$ **(g)** $(0, 18)$ **(h)** $f(4) = 238$; $(4, 238)$ **(i)** **5. (a)** positive zeros: 1; negative zeros: 3 or 1

$f(x) = 3x^4 - 4x^3 - 22x^2 + 15x + 18$

(b) $\pm 1, \pm 2, \pm \dfrac{1}{2}$ **(c)** $-1, 1$ **(d)** no other real zeros **(e)** $-\dfrac{1}{4} + \dfrac{\sqrt{15}}{4}i, -\dfrac{1}{4} - \dfrac{\sqrt{15}}{4}i$ **(f)** $(-1, 0), (1, 0)$ **(g)** $(0, 2)$ **(h)** $f(4) = -570$;

$(4, -570)$ **(i)** **6. (a)** positive zeros: 0; negative zeros: 4, 2, or 0

$f(x) = -2x^4 - x^3 + x + 2$

(b) $0, \pm 1, \pm 3, \pm 9, \pm 27, \pm \dfrac{1}{2}, \pm \dfrac{3}{2}, \pm \dfrac{9}{2}, \pm \dfrac{27}{2}, \pm \dfrac{1}{4}, \pm \dfrac{3}{4}, \pm \dfrac{9}{4}, \pm \dfrac{27}{4}$ **(c)** $0, -\dfrac{3}{2}$ (multiplicity 2) **(d)** no other real zeros

(e) $\dfrac{1}{2} + \dfrac{\sqrt{11}}{2}i, \dfrac{1}{2} - \dfrac{\sqrt{11}}{2}i$ **(f)** $(0, 0), \left(-\dfrac{3}{2}, 0\right)$ **(g)** $(0, 0)$ **(h)** $f(4) = 7260$; $(4, 7260)$ **(i)**

$f(x) = 4x^5 + 8x^4 + 9x^3 + 27x^2 + 27x$

7. (a) positive zeros: 1; negative zeros: 1 **(b)** $\pm 1, \pm 5, \pm \dfrac{1}{3}, \pm \dfrac{5}{3}$ **(c)** no rational zeros **(d)** $-\sqrt{5}, \sqrt{5}$ **(e)** $-\dfrac{\sqrt{3}}{3}i, \dfrac{\sqrt{3}}{3}i$

(f) $(-\sqrt{5}, 0), (\sqrt{5}, 0)$ **(g)** $(0, -5)$ **(h)** $f(4) = 539$; $(4, 539)$ **(i)** **8. (a)** positive zeros: 2 or 0; negative zeros: 3 or 1

$f(x) = 3x^4 - 14x^2 - 5$

(b) $\pm 1, \pm 3, \pm 9$ **(c)** $-3, -1$ (multiplicity 2), 1, 3 **(d)** no other real zeros **(e)** no other complex zeros **(f)** $(-3, 0), (-1, 0), (1, 0), (3, 0)$
(g) $(0, -9)$ **(h)** $f(4) = -525$; $(4, -525)$ **(i)** **9. (a)** positive zeros: 4, 2, or 0; negative zeros: 0 **(b)** $\pm 1, \pm 2, \pm 3,$

$f(x) = -x^5 - x^4 + 10x^3 + 10x^2 - 9x - 9$

$\pm 4, \pm 6, \pm 12, \pm \dfrac{1}{3}, \pm \dfrac{2}{3}, \pm \dfrac{4}{3}$ **(c)** $\dfrac{1}{3}, 2$ (multiplicity 2), 3 **(d)** no other real zeros **(e)** no other complex zeros **(f)** $\left(\dfrac{1}{3}, 0\right), (2, 0), (3, 0)$

(g) $(0, -12)$ **(h)** $f(4) = -44$; $(4, -44)$ **(i)** **10.** For the function in Exercise 2: ± 1.732; for the function in Exercise 3:

$f(x) = -3x^4 + 22x^3 - 55x^2 + 52x - 12$

$\pm .707$; for the function in Exercise 4: $-2.303, 1.303$; for the function in Exercise 7: ± 2.236

Section 12.4 (page 780)

Exercises **1.** A, B, C **3.** A **5.** A, C, D **7.** **9.** 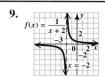 **11.**

13. (a) 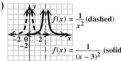 (b) 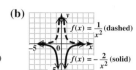 (c) 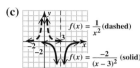 **15.** vertical asymptote: $x = 5$; horizontal

asymptote: $y = 0$ **17.** vertical asymptote: $x = \dfrac{7}{3}$; horizontal asymptote: $y = 0$ **19.** vertical asymptote: $x = -2$; horizontal asymptote:

$y = -1$ **21.** vertical asymptote: $x = -\dfrac{9}{2}$; horizontal asymptote: $y = \dfrac{3}{2}$ **23.** vertical asymptotes: $x = 3$, $x = 1$; horizontal asymptote:

$y = 0$ **25.** vertical asymptote: $x = -3$; oblique asymptote: $y = x - 3$ **27.** vertical asymptotes: $x = -2$, $x = \dfrac{5}{2}$; horizontal asymptote:

$y = \dfrac{1}{2}$ **29.** A **31.** $f(x) = \dfrac{4}{5 + 3x}$ **33.** $f(x) = \dfrac{3}{(x + 4)^2}$ **35.** $f(x) = \dfrac{2x + 1}{(x + 2)(x + 4)}$ **37.** $f(x) = \dfrac{-x}{x^2 - 4}$ **39.** $f(x) = \dfrac{4x}{1 - 3x}$

41. $f(x) = \dfrac{x - 5}{x + 3}$ **43.** $f(x) = \dfrac{3x}{x^2 - 16}$ **45.** $f(x) = \dfrac{x^2 + 1}{x + 3}$ **47.** $f(x) = \dfrac{(x - 3)(x + 1)}{(x - 1)^2}$ **49.** $f(x) = \dfrac{x(x - 2)}{(x + 3)^2}$ **51.** $f(x) = \dfrac{x^2 - 9}{x + 3}$

53.
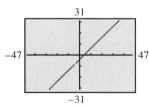
54. There is an unlit portion of the screen. (This portion is called a pixel.) **55.** There is an error

message, because -4 is not in the domain of f. **56.** $g(x) = x - 4$; The domain of f is $(-\infty, -4) \cup (-4, \infty)$, while that of g is $(-\infty, \infty)$.
57. It is the same, except the pixel at $(-4, -8)$ is lit. **58.** $g(-4) = -8$; We get a value because -4 is in the

domain of g. **59.** In this case, the graph of a rational function not in lowest terms differs by one point from the graph of the corresponding
function in lowest terms because their domains are different. **61.** It is not correct. There is a hole in the graph at $(-5, 10)$.

63. (a)

Year	1982	1983	1984	1985	1986	1987	1988	1989	1990	1991	1992	1993
Deaths/ Cases	.397	.457	.516	.554	.589	.581	.585	.606	.620	.623	.607	.610

(b) After 1985 the ratio becomes fairly constant and is equal to .6, rounded to the nearest tenth.

(c) $h(x) = \dfrac{g(x)}{f(x)}$ $\quad h(x) = \dfrac{1818(x-2)^2 + 620}{2975(x-2)^2 + 1563}$; The graph of h becomes horizontal with a value of approximately .61. The

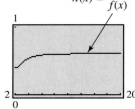

model predicts the ratios in the table quite well. Both are .6, rounded to the nearest tenth. **(d)** $g(x) \approx .6f(x)$ **(e)** 2,400,000 deaths

65. (a) For $r = x$, $\qquad$ 26 per min **(b)** 5

$$y = \frac{2x - 25}{2x^2 - 50x} \qquad y = .5$$

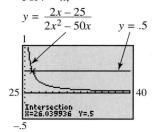

Chapter 12 Review Exercises (page 792)

1. $3x + 2$ $\quad$ **3.** $2x^2 + x + 3 + \dfrac{21}{x-3}$ $\quad$ **5.** yes $\quad$ **7.** -13 $\quad$ **9.** $2x^3 + x - 6 = (x+2) \cdot (2x^2 - 4x + 9) + (-24)$ $\quad$ **11.** $\dfrac{1}{2}, -1, 5$

13. $4, -\dfrac{1}{2}, -\dfrac{2}{3}$ $\quad$ **15.** no $\quad$ **17.** no $\quad$ **19.** $f(x) = -2x^3 + 6x^2 + 12x - 16$ $\quad$ **21.** $f(x) = x^4 - 3x^2 - 4$ (There are others.)

23. $f(x) = x^3 + x^2 - 4x + 6$ (There are others.) $\quad$ **25.** No. Because zeros that are not real come in conjugate *pairs*, there must be an even number of them. 3 is odd. $\quad$ **27.** zero; solution; x-intercept $\quad$ **29.** two $\quad$ **31.** **33.** **35.**

$f(x) = x^3 + 5$ $\qquad$ $f(x) = x^2(2x+1)(x-2)$ $\qquad$ $f(x) = 12x^3 - 13x^2 - 5x + 6$

37. $f(-1) = -10$ and $f(0) = 2$; $f(2) = -4$ and $f(3) = 14$ $\quad$ **39.** $f(-1) = 15$ and $f(0) = -8$; $f(-6) = -20$ and $f(-5) = 27$

43. **45.** **47.** **49.** **51.** **53.**

$f(x) = 2x^3 - 11x^2 - 2x + 2$ $\quad$ $f(x) = x^3 + 3x^2 - 4x - 2$ $\quad$ $f(x) = \dfrac{8}{x}$ $\quad$ $f(x) = \dfrac{4x-2}{3x+1}$ $\quad$ $f(x) = \dfrac{2x}{x^2-1}$ $\quad$ $f(x) = \dfrac{x^2-1}{x}$

55.
$f(x) = \dfrac{4x^2 - 9}{2x + 3}$

57.
$f(x) = -\dfrac{1}{x^3}$

59.
$f(x) = \dfrac{3 - 4x}{2x + 1}$

61.
$f(x) = 2x^5 - 3x^4 + x^2 - 2$

63.
$f(x) = \dfrac{x^3 + 1}{x + 1}$

Chapter 12 Test   (page 795)

[12.1] **1.** $2x^2 + 4x + 5$ **2.** $x^4 + 2x^3 - x^2 + 3x - 5 = (x + 1) \cdot (x^3 + x^2 - 2x + 5) + (-10)$ **3.** yes   **4.** -227 **5.** Yes, 3 is a zero because the last term in the bottom row of the synthetic division is 0.   **6.** $f(x) = 2x^4 - 2x^3 - 2x^2 - 2x - 4$

[12.2] **7. (a)** $\pm 1, \pm\dfrac{1}{2}, \pm\dfrac{1}{3}, \pm\dfrac{1}{6}, \pm 7, \pm\dfrac{7}{2}, \pm\dfrac{7}{3}, \pm\dfrac{7}{6}$ **(b)** $-\dfrac{1}{3}, 1, \dfrac{7}{2}$ [12.3] **9. (a)** $f(-2) = -11 < 0$ and $f(-1) = 2 > 0$

(b) -1.290 **10. (a)** 3 **(b)** 2   **11.** 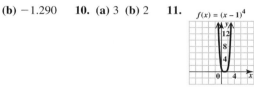 $f(x) = (x - 1)^4$ **12.** $f(x) = x(x + 1)(x - 2)$ **13.** $f(x) = 2x^3 - 7x^2 + 2x + 3$ **14.** $f(x) = x^4 - 5x^2 + 6$ **15.** $49°F$

[12.4] **16.** $f(x) = \dfrac{-2}{x + 3}$ **17.** $f(x) = \dfrac{3x - 1}{x - 2}$ **18.** $f(x) = \dfrac{x^2 - 1}{x^2 - 9}$ **19.** $y = 2x + 3$ **20.** D

Cumulative Review Exercises Chapters 1–12   (page 796)

[1.2] **1.** -9 [1.4] **2.** $-17x + 34$ [2.1] **3.** $\{2\}$ **4.** $\{24\}$ **5.** $(-\infty, \infty)$ [2.5] **6.** $(-\infty, 8]$ **7.** $\left(3, \dfrac{13}{3}\right)$

[2.7] **8.** $(-\infty, -7) \cup (3, \infty)$ [2.3] **9.** 26 nickels and 12 dimes   [3.2] **10.** $-\dfrac{5}{3}$ [3.3] **11.** $y = -3x + 12$

[3.1] **12.** $-3x + 5y = -15$ [3.4] **13.** $y \le -2x + 7$ [3.5] **14.** -28 [3.6] **15.** $\dfrac{40}{3}$ [4.1] **16.** $\{(-2, 1)\}$ [4.4] **17.** $\{(-3, 4, 2)\}$

[4.5] **18.** $\{(4, 2, -3)\}$ [5.3] **19.** $f(x) = x^2$ [5.4] **20.** $3r^5 - 7r^4 + 2r^3 + 18r - 6$ **21.** $k^2 - 10kh + 25h^2 + 4k - 20h + 4$

[6.2] **22.** $3(x - 3)(2x + 1)$ [6.3] **23.** $(9 + 2y^2)(81 - 18y^2 + 4y^4)$ [6.1, 6.5] **24.** $\{-3, -1, 1\}$ [7.1] **25. (a)** $-\dfrac{3}{4}, 6$ **(b)** $\dfrac{x + 6}{4x + 3}$

26. $\dfrac{6}{25}$ **27.** $\dfrac{y-3}{y+2}$ [7.2] **28.** -2 **29.** $\dfrac{-r^2+r+4}{(r-2)(r-1)}$ [7.3] **30.** $\dfrac{2y-1}{-y-1}$ or $\dfrac{1-2y}{y+1}$ [7.4] **31.** $\{2, -5\}$ **32.** $\emptyset$

[8.3] **33.** $5\sqrt{2}$ [8.5] **34.** $5+\sqrt{2}$ [8.6] **35.** $\{2\}$ [8.7] **36.** $4-2i$ [9.2] **37.** 8 **38.** irrational **39.** $\{-2 \pm \sqrt{2}\}$

[9.5] **40.** $\left[-\dfrac{2}{3}, 5\right]$ [10.4] **41.** x-axis **42.** y-axis **43.** origin **44.** $(-\infty, 0)$ [10.5] **45.** 3 [11.1] **46.** $f^{-1}(x) = \dfrac{x^3-5}{3}$

[11.3] **47.** $-\dfrac{3}{4}$ [11.5] **48. (a)** 1.5261 **(b)** -2.4559 [11.6] **49.** $\{2\}$ [12.1] **50.** $x^3+6x^2-11x+13$ [12.2, 12.3] **51. (a)** $-5, 1$

(b) $f(x) = (x+3)(x+5)(x-1)$ **(c)** $(-3, 0), (-5, 0), (1, 0)$; y-intercept: $(0, -15)$ **(d)**

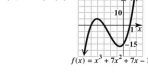

$f(x) = x^3 + 7x^2 + 7x - 15$

[12.4] **52. (a)** $x = -3$ and $x = 3$

(b) $(-2, 0)$ and $(2, 0)$ **(c)** $y = 1$ **(d)**

$f(x) = \dfrac{x^2-4}{x^2-9}$

53.

$C(x) = \dfrac{10x}{49(101-x)}$

CHAPTER **13** CONIC SECTIONS

Section **13.1** (page 805)

Exercises **1. (a)** $(0, 0)$ **(b)** 5 **(c)**

$x^2 + y^2 = 25$

3. B **5.** D **7.** $(x+4)^2 + (y-3)^2 = 4$ **9.** $(x+8)^2 + (y+5)^2 = 5$

11. center: $(-2, -3)$; $r = 2$ **13.** center: $(-5, 7)$; $r = 9$ **15.** center: $(2, 4)$; $r = 4$ **17.** The thumbtack acts as the center and the length of the string acts as the radius. **19.**

$x^2 + y^2 = 9$

21.

$2y^2 = 10 - 2x^2$

23. center: $(-3, 2)$

$(x+3)^2 + (y-2)^2 = 9$

25. center: $(2, 3)$

$x^2 + y^2 - 4x - 6y + 9 = 0$

27.

$\dfrac{x^2}{9} + \dfrac{y^2}{25} = 1$

29.

$\dfrac{x^2}{36} = 1 - \dfrac{y^2}{16}$

31.

$\dfrac{y^2}{25} = 1 - \dfrac{x^2}{49}$

33.

$\dfrac{x^2}{16} + \dfrac{y^2}{4} = 1$

35.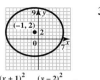

$\dfrac{(x+1)^2}{64} + \dfrac{(y-2)^2}{49} = 1$

37.

$\dfrac{(x-2)^2}{16} + \dfrac{(y-1)^2}{9} = 1$

39. The fixed ends of the string are at the foci, and the constant length of the string represents the sum of the distances from any point on the curve to the foci.

41. By the vertical line test the set is not a function, because a vertical line may intersect the graph of an ellipse in two points.

43. $y_1 = 4 + \sqrt{16 - (x+2)^2}$, $y_2 = 4 - \sqrt{16 - (x+2)^2}$

45.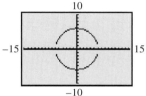

47.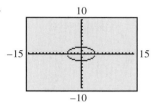

49. (a) 10 m (b) 36 m   **51.** (a) 154.7 million mi

(b) 128.7 million mi (Answers are rounded.)   **53.** $3\sqrt{3}$ units

55. Answers will vary.

Section 13.2   (page 816)

Connections (page 812)   Answers will vary.

Exercises **1.** C   **3.** D   **5.** When written in one of the forms given in the box titled "Equations of Hyperbolas" in this section, it will open up and down if the $-$ sign precedes the x^2-term; it will open left and right if the $-$ sign precedes the y^2-term.   **7.**

$\dfrac{x^2}{16} - \dfrac{y^2}{9} = 1$

9.

$\dfrac{y^2}{9} - \dfrac{x^2}{9} = 1$

11.

$\dfrac{x^2}{25} - \dfrac{y^2}{36} = 1$

13.

$\dfrac{y^2}{16} - \dfrac{x^2}{16} = 1$

15. hyperbola

$x^2 - y^2 = 16$

17. ellipse

$4x^2 + y^2 = 16$

19. circle

$y^2 = 36 - x^2$

21. hyperbola

$9x^2 = 144 + 16y^2$

23. hyperbola

$y^2 = 4 + x^2$

25. domain: $[-4, 4]$; range: $[0, 4]$

$f(x) = \sqrt{16 - x^2}$

27. domain: $[-6, 6]$; range: $[-6, 0]$

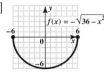

$f(x) = -\sqrt{36 - x^2}$

29. domain: $(-\infty, \infty)$; range: $[3, \infty)$

$\dfrac{y}{3} = \sqrt{1 + \dfrac{x^2}{9}}$

31. $\dfrac{(x-2)^2}{4} - \dfrac{(y+1)^2}{9} = 1$

$(2, -1)$

33. $\frac{y^2}{36} - \frac{(x-2)^2}{49} = 1$

35. (a) 50 m (b) 69.3 m   **37.** for V greater than 4325.68 m per sec   **39.** $y_1 = \sqrt{\frac{x^2}{9} - 1},\ y_2 = -\sqrt{\frac{x^2}{9} - 1}$

41. 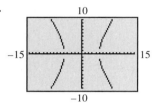   **43.** 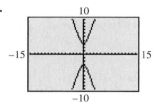   **45.** Answers will vary.

Section 13.3 (page 824)

Exercises **1.** Substitute $x - 1$ for y in the first equation. Then solve for x. Find the corresponding y-values by substituting back into $y = x - 1$. In the first equation, both variables are squared and in the second, both variables are to the first power, so the elimination method is not appropriate.   **3.** one   **5.** none   **7.** 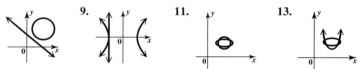   **9.**   **11.**   **13.**   **15.** $\left\{ (0, 0), \left(\frac{1}{2}, \frac{1}{2}\right) \right\}$

17. $\{(-6, 9), (-1, 4)\}$ **19.** $\left\{ \left(-\frac{1}{5}, \frac{7}{5}\right), (1, -1) \right\}$ **21.** $\left\{ (-2, -2), \left(-\frac{4}{3}, -3\right) \right\}$ **23.** $\{(-3, 1), (1, -3)\}$

25. $\left\{ \left(-\frac{3}{2}, -\frac{9}{4}\right), (-2, 0) \right\}$ **27.** $\{(-\sqrt{3}, 0), (\sqrt{3}, 0), (-\sqrt{5}, 2), (\sqrt{5}, 2)\}$ **29.** $\left\{ \left(\frac{i\sqrt{3}}{3}, \frac{-3 + i\sqrt{3}}{6}\right), \left(\frac{-i\sqrt{3}}{3}, \frac{-3 - i\sqrt{3}}{6}\right) \right\}$

31. $\{(-2, 0), (2, 0)\}$ **33.** $\{(\sqrt{3}, 0), (-\sqrt{3}, 0)\}$ **35.** $\{(1, 3), (1, -3), (-1, 3), (-1, -3)\}$ **37.** $\{(-2i\sqrt{2}, -2\sqrt{3}), (-2i\sqrt{2}, 2\sqrt{3}),$
$(2i\sqrt{2}, -2\sqrt{3}), (2i\sqrt{2}, 2\sqrt{3})\}$ **39.** $\{(-\sqrt{5}, -\sqrt{5}), (\sqrt{5}, \sqrt{5})\}$ **41.** $\{(i, 2i), (-i, -2i), (2, -1), (-2, 1)\}$

43. $\{(2, -3), (-3, 2)\}$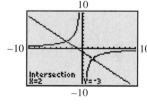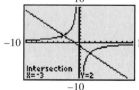

45. length: 12 ft; width: 7 ft
47. $20; \frac{4}{5}$ thousand or 800 calculators
49. 1981; 470 thousand

Section 13.4 (page 833)

Exercises **1.** C   **3.** Answers will vary.   **5.** B   **7.** A   **9.**   **11.**   **13.**

15.

17.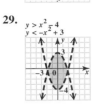

19.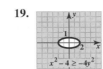

21.

23.

25.

27.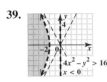

29.

31.

33.

35.

37.

39.

41.

43.

45.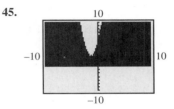

47. maximum: 65; minimum: 8   **49.** maximum: 900; minimum: 0   **51.** maximum of $\frac{42}{5}$ at $\left(\frac{6}{5}, \frac{6}{5}\right)$ **53.** minimum of $\frac{49}{3}$ at $\left(\frac{17}{3}, 5\right)$

55. \$1120 (with 4 pigs, 12 geese)   **57.** 8 of #1 and 3 of #2 (for 100 ft³ of storage)   **59.** 6.4 million gal of gasoline and 3.2 million gal of fuel oil (for \$16,960,000)

Chapter 13 Review Exercises   (page 842)

1. $(x + 2)^2 + (y - 4)^2 = 9$ **3.** $(x - 4)^2 + (y - 2)^2 = 36$ **5.** center: $(4, 1)$; $r = 2$ **7.** center: $(3, -2)$; $r = 5$

9. 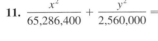   **11.** $\dfrac{x^2}{65{,}286{,}400} + \dfrac{y^2}{2{,}560{,}000} = 1$ **13.**   **15.**   **17.** parabola   **19.** ellipse

21. hyperbola   **23.** $\{(6, -9), (-2, -5)\}$ **25.** $\{(4, 2), (-1, -3)\}$ **27.** $\left\{\left(-\sqrt{2}, 2\right), \left(-\sqrt{2}, -2\right), \left(\sqrt{2}, -2\right), \left(\sqrt{2}, 2\right)\right\}$ **29.** 0, 1, or 2

31.   **33.** 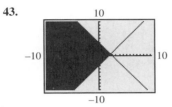   **35.** 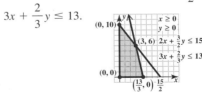   **37.** Let $x = $ number of batches of cakes and $y = $ number of batches of cookies. Then $x \geq 0$, $y \geq 0$, $2x + \dfrac{3}{2}y \leq 15$, and $3x + \dfrac{2}{3}y \leq 13$.

39. 3 batches of cakes and 6 batches of cookies (for maximum profit of \$210)   **41.** $2a + 4b + c = -20$ **42.** $5a + b + c = -26$

43. $-a + b + c = -2$ **44.** $\{(-4, -2, -4)\}$; $x^2 + y^2 - 4x - 2y - 4 = 0$ **45.** center: $(2, 1)$; radius: 3   **47.**

49.

51.

53.

55. (a) 69.8 million km **(b)** 46.0 million km

Chapter 13 Test (page 845)

[13.1] **1.** center: $(2, -3)$; radius: 4

$(x - 2)^2 + (y + 3)^2 = 16$

2. center: $(-4, 1)$; radius: 5   [13.2] **3.**

$f(x) = \sqrt{9 - x^2}$

[13.1] **4.**

$4x^2 + 9y^2 = 36$

[13.2] **5.**

$16y^2 - 4x^2 = 64$

6.

$\dfrac{y}{2} = -\sqrt{1 - \dfrac{x^2}{9}}$

7. ellipse   **8.** hyperbola   **9.** parabola   [13.3] **10.** $\left\{\left(-\dfrac{1}{2}, -10\right), (5, 1)\right\}$

11. $\left\{(-2, -2), \left(\dfrac{14}{5}, -\dfrac{2}{5}\right)\right\}$ **12.** $\{(-\sqrt{22}, -\sqrt{3}), (-\sqrt{22}, \sqrt{3}), (\sqrt{22}, -\sqrt{3}), (\sqrt{22}, \sqrt{3})\}$ [13.4] **13.**

$y < x^2 - 2$

14.

$x^2 + 25y^2 \le 25$
$x^2 + y^2 \le 9$

15. maximum of 48 at $(8, 8)$; minimum of 4 at $(2, 0)$

Cumulative Review Exercises Chapters 1–13 (page 845)

[1.1, 1.2] **1.** -4 [2.1] **2.** $\left\{\dfrac{2}{3}\right\}$ [2.5] **3.** $\left(-\infty, \dfrac{3}{5}\right]$ [2.7] **4.** $\{-4, 4\}$ **5.** $(-\infty, -5) \cup (10, \infty)$ [3.2] **6.** $\dfrac{2}{3}$

[3.3] **7.** $3x + 2y = -13$ [4.1, 4.4, 4.5] **8.** $\{(3, -3)\}$ [4.2, 4.4, 4.5] **9.** $\{(4, 1, -2)\}$ [13.3] **10.** $\left\{(-1, 5), \left(\dfrac{5}{2}, -2\right)\right\}$

[4.3] **11.** 40 mph   [3.5] **12.** $275   [5.4] **13.** $25y^2 - 30y + 9$ **14.** $12r^2 + 40r - 7$ [5.5] **15.** $4x^3 - 4x^2 + 3x + 5 + \dfrac{3}{2x + 1}$

[6.2] **16.** $(3x + 2)(4x - 5)$ **17.** $(2y^2 - 1)(y^2 + 3)$ [6.3] **18.** $(z^2 + 1)(z + 1)(z - 1)$ **19.** $(a - 3b)(a^2 + 3ab + 9b^2)$

[7.1] **20.** $\dfrac{40}{9}$ **21.** $\dfrac{y - 1}{y(y - 3)}$ [7.2] **22.** $\dfrac{3c + 5}{(c + 5)(c + 3)}$ **23.** $\dfrac{1}{p}$ [7.5] **24.** $1\dfrac{1}{5}$ hr   [5.1] **25.** $\dfrac{3}{4}$ **26.** $\dfrac{a^5}{4}$ [8.4] **27.** $2\sqrt[3]{2}$

[8.5] **28.** $\dfrac{3\sqrt{10}}{2}$ [8.7] **29.** $\dfrac{7}{5} + \dfrac{11}{5}i$ [8.6] **30.** $\emptyset$ [6.5] **31.** $\left\{\dfrac{1}{5}, -\dfrac{3}{2}\right\}$ [9.1, 9.2] **32.** $\left\{\dfrac{1 + 2\sqrt{2}}{4}, \dfrac{1 - 2\sqrt{2}}{4}\right\}$

33. $\left\{\dfrac{3 + \sqrt{33}}{6}, \dfrac{3 - \sqrt{33}}{6}\right\}$ [9.3] **34.** $\left\{-\dfrac{\sqrt{6}}{2}, \dfrac{\sqrt{6}}{2}, -\sqrt{7}, \sqrt{7}\right\}$ [9.4] **35.** $v = \dfrac{\pm\sqrt{rFkw}}{kw}$ [10.1] **36. (a)** -1 **(b)** $9x^2 + 18x + 4$

[11.1] **37.** $f^{-1}(x) = \sqrt[3]{x - 4}$ [11.4, 11.5] **38. (a)** 4 **(b)** 7   [11.4] **39.** $\log \dfrac{(3x + 7)^2}{4}$ [11.6] **40.** $\{3\}$ **41. (a)** $12,198.90

(b) $12,214.03   **42. (a)** $16.9 billion **(b)** $75.8 billion   [12.1] **43.** 23   [12.2] **44.** Yes, $x + 2$ is a factor of $f(x)$. The other factor is

$5x^3 + 6x - 4$.   **45.** $-2, -\dfrac{4}{3}, 3$ [3.5] **46.**

$f(x) = -3x + 5$
   [10.2] **47.**

$f(x) = -2(x - 1)^2 + 3$
   [13.4] **48.**

$\dfrac{x^2}{25} + \dfrac{y^2}{16} \le 1$

[13.2] **49.**

$\dfrac{x^2}{4} - \dfrac{y^2}{16} = 1$
   [11.2] **50.**

$f(x) = 3^x$

CHAPTER 14 FURTHER TOPICS IN ALGEBRA

Section 14.1   (page 855)

   1. 2, 3, 4, 5, 6   **3.** $4, \dfrac{5}{2}, 2, \dfrac{7}{4}, \dfrac{8}{5}$ **5.** 3, 9, 27, 81, 243   **7.** $1, \dfrac{1}{4}, \dfrac{1}{9}, \dfrac{1}{16}, \dfrac{1}{25}$ **9.** $-1, 1, -1, 1, -1$ **11.** -70 **13.** $\dfrac{49}{23}$

15. 171   **17.** $4n$ **19.** $\dfrac{1}{3^n}$ **21.** $110, $109, $108, $107, $106, $105; $400   **23.** $6554   **25.** $4 + 5 + 6 + 7 + 8 = 30$

27. $3 + 6 + 11 = 20$ **29.** $-1 + 1 - 1 + 1 - 1 + 1 = 0$ **31.** $0 + 6 + 14 + 24 + 36 = 80$ **33.** $2x + 4x + 6x + 8x + 10x$

35. $x + 2x^2 + 3x^3 + 4x^4 + 5x^5$ **Answers may vary for Exercises 37–41.**   **37.** $\sum\limits_{i=1}^{5} (i + 2)$ **39.** $\sum\limits_{i=1}^{5} \dfrac{1}{i + 1}$ **41.** $\sum\limits_{i=1}^{5} i^2$

43. The similarities are that both are defined by the same linear expression and that points satisfying both lie in a straight line. The difference is that the domain of f consists of all real numbers, but the domain of the sequence is $\{1, 2, 3, \ldots\}$. An example of a similarity is $f(1) = 6$ and $a_1 = 6$. An example of a difference is $f\left(\dfrac{3}{2}\right) = 7$ but $a_{3/2}$ is not allowed.   **45.** A sequence is a list of terms in a specific order, while a series is the indicated sum of the terms of a sequence.   **47.** $\dfrac{59}{7}$ **49.** 5   **51.** 7243   **53.** $\sum\limits_{i=1}^{6} i^2 + \sum\limits_{i=1}^{6} 3i + \sum\limits_{i=1}^{6} 5$ **54.** $3\sum\limits_{i=1}^{6} i$ **55.** $6 \cdot 5 = 30$

56. $\sum\limits_{i=1}^{n} i = \dfrac{n(n + 1)}{2}$ **57.** $\sum\limits_{i=1}^{n} i^2 = \dfrac{n(n + 1)(2n + 1)}{6}$ **58.** $91 + 63 + 30 = 184$ **59.** 572   **60.** -2620

Section 14.2   (page 863)

Exercises **1.** An arithmetic sequence is a sequence (list) of numbers in a specific order such that there is a common difference between any two successive terms. For example, the sequence $1, 5, 9, 13, \ldots$ is arithmetic with difference $d = 5 - 1 = 9 - 5 = 13 - 9 = 4$. As another example, $2, -1, -4, -7, \ldots$ is an arithmetic sequence with $d = -3$.   **3.** $d = 1$ **5.** not arithmetic   **7.** $d = 5$ **9.** 5, 9, 13, 17, 21

11. $-2, -6, -10, -14, -18$ **13.** $a_n = 5n - 3$ **15.** $a_n = \dfrac{3}{4}n + \dfrac{9}{4}$ **17.** $a_n = 3n - 6$ **19.** 76   **21.** 48   **23.** -1 **25.** 16

27. 6   **29.** n represents the number of terms.   **31.** $2S; 101; 100; 10,100$ **32.** $2S = 10,100$ **33.** 5050   **34.** 20,100   **35.** 81

37. -3 **39.** 87   **41.** 390   **43.** 320   **45.** 31,375   **47.** $465   **49.** $2100 per month   **51.** 68; 1100   **53.** no; 3; 9

Section 14.3 (page 872)

Exercises **1.** A geometric sequence is an ordered list of numbers such that each term after the first is obtained by multiplying the previous term by a constant, r, called the common ratio. For example, if the first term is 3 and $r = 4$, then the sequence is 3, 12, 48, 192, ... If the first term is 2 and $r = -1$, then the sequence is 2, -2, 2, -2, ... **3.** $r = 2$ **5.** not geometric **7.** $r = -3$ **9.** $r = -\dfrac{1}{2}$ **11.** $a_n = 5(2)^{n-1}$

13. $a_n = \dfrac{3^{n-1}}{9}$ **15.** $a_n = 10\left(-\dfrac{1}{5}\right)^{n-1}$ **17.** $2(5)^9 = 3{,}906{,}250$ **19.** $\dfrac{1}{2}\left(\dfrac{1}{3}\right)^{11}$ **21.** $2\left(\dfrac{1}{2}\right)^{24} = \dfrac{1}{2^{23}}$ **23.** 2, 6, 18, 54, 162

25. $5, -1, \dfrac{1}{5}, -\dfrac{1}{25}, \dfrac{1}{125}$ **27.** $\dfrac{121}{243}$ **29.** -1.997 **31.** 2.662 **33.** -2.982 **35.** \$66,988.91 **37.** \$130,159.72 **39.** 9

41. $\dfrac{10{,}000}{11}$ **43.** $-\dfrac{9}{20}$ **45.** does not exist **47.** $10\left(\dfrac{3}{5}\right)^4 \approx 1.3$ ft **49.** 3 days; $\dfrac{1}{4}$ g **51. (a)** $1.1(1.06)^5 \approx 1.5$ billion units

(b) approximately 12 yr **53.** $\$50{,}000\left(\dfrac{3}{4}\right)^8 \approx \$5000$ **55.** $.33333\ldots$ **56.** $.66666\ldots$ **57.** $.99999\ldots$

58. $\dfrac{a_1}{1 - r} = \dfrac{.9}{1 - .1} = \dfrac{.9}{.9} = 1$; therefore, $.99999\ldots = 1$ **59.** B

60. $.49999\ldots = .4 + .09999\ldots = \dfrac{4}{10} + \dfrac{1}{10}(.9999\ldots) = \dfrac{4}{10} + \dfrac{1}{10}(1) = \dfrac{5}{10} = \dfrac{1}{2}$

Section 14.4 (page 881)

Connections **(page 877)** **1.** 21 and 34 **2.** 15, 21, 28, 36, 45 **3.** Answers will vary.

Exercises **1.** 720 **3.** 15 **5.** 15 **7.** 1 **9.** 120 **11.** 78 **13.** $m^4 + 4m^3n + 6m^2n^2 + 4mn^3 + n^4$

15. $a^5 - 5a^4b + 10a^3b^2 - 10a^2b^3 + 5ab^4 - b^5$ **17.** $8x^3 + 36x^2 + 54x + 27$ **19.** $\dfrac{x^4}{16} - \dfrac{x^3y}{2} + \dfrac{3x^2y^2}{2} - 2xy^3 + y^4$

21. $m^3x^3 - 3m^2n^2x^2 + 3mn^4x - n^6$ **23.** $r^{12} + 24r^{11}s + 264r^{10}s^2 + 1760r^9s^3$ **25.** $3^{14}x^{14} - 14(3^{13})x^{13}y + 91(3^{12})x^{12}y^2 - 364(3^{11})x^{11}y^3$

27. $t^{20} + 10t^{18}u^2 + 45t^{16}u^4 + 120t^{14}u^6$ **29.** $120(2^7)m^7n^3$ **31.** $\dfrac{7x^2y^6}{16}$ **33.** $36k^7$ **35.** $-160x^6y^3$ **37.** $4320x^9y^4$

Section 14.5 (page 886)

Exercises **1.** positive integers **3.** The statement is not true for $n = 1$—that is, $1 = 1 + 1$ is false. This proof illustrates the situation described in the answer to Exercise 2.

Although we do not usually give proofs, the answers to Exercises 7 and 15 are shown here.

7. *Step 1:* $3(1) = 3$ and $\dfrac{3(1)(1 + 1)}{2} = \dfrac{6}{2} = 3$, so S is true for $n = 1$. *Step 2:* S_k: $3 + 6 + 9 + \cdots + 3k = \dfrac{3(k)(k + 1)}{2}$;

S_{k+1}: $3 + 6 + 9 + \cdots + 3(k + 1) = \dfrac{3(k + 1)[(k + 1) + 1]}{2}$; Add $3(k + 1)$ to each side of S_k and simplify until you obtain S_{k+1}. Since S is

true for $n = 1$ and S is true for $n = k + 1$ when it is true for $n = k$, S is true for every positive integer n.

15. *Step 1:* $\dfrac{1}{1 \cdot 2} = \dfrac{1}{2}$ and $\dfrac{1}{1 + 1} = \dfrac{1}{2}$, so S is true for $n = 1$. *Step 2:* S_k: $\dfrac{1}{1 \cdot 2} + \dfrac{1}{2 \cdot 3} + \dfrac{1}{3 \cdot 4} + \cdots + \dfrac{1}{k(k + 1)} = \dfrac{k}{k + 1}$;

S_{k+1}: $\dfrac{1}{1 \cdot 2} + \dfrac{1}{2 \cdot 3} + \cdots + \dfrac{1}{(k + 1)[(k + 1) + 1]} = \dfrac{k + 1}{(k + 1) + 1}$; Add $\dfrac{1}{(k + 1)[(k + 1) + 1]}$ to each side of S_k and simplify until you obtain

S_{k+1}. Since S is true for $n = 1$ and S is true for $n = k + 1$ when it is true for $n = k$, S is true for every positive integer n.

26. $x^2 + xy + y^2 = \dfrac{x^3 - y^3}{x - y}$ **27.** $x^3 - y^3 = (x - y)(x^2 + xy + y^2)$ **28.** $x^{2n} - x^{2n-1}y + \cdots - xy^{2n-1} + y^{2n} = \dfrac{x^{2n+1} + y^{2n+1}}{x + y}$

29. power; exponents **30.** power; exponents **33.** $\dfrac{4^{n-1}}{3^{n-2}}$ or $3\left(\dfrac{4}{3}\right)^{n-1}$

Section 14.6   (page 895)

Exercises   1. 360   **3.** 72   **5.** 5   **7.** 6   **9.** 1   **11.** 48   **13.** 40,320   **15.** 604,800   **17.** $2.052371412 \times 10^{10}$ **19.** 39,270   **21.** 720; 151,200   **23.** 15,890,700   **25.** 35   **27.** 35   **29.** 84; 324   **31.** A permutation is one of the ways to *arrange r* elements selected from a set of *n* elements. A combination is one of the subsets (*order does not matter*) of *r* elements that can be chosen from a set of *n* elements. Think of a typical choice of *r* elements. According to the problem, if the *r* elements are rearranged, does it represent another different choice?   **33. (a)** permutation **(b)** combination **(c)** combination   **35.** 210   **37. (a)** 56 **(b)** 462 **(c)** 3080 **(d)** 8526   **39.** 210; 5040   **41.** 35   **43. (a)** 220 **(b)** 55 **(c)** 105

Section 14.7   (page 904)

Connections   (page 903)   1. 1,120,529,256   **2.** approximately 806,781,064,300   **3.** $\dfrac{1}{1,120,529,256} \approx .00000000089$

Exercises   1. The probability of event *E* is the number of outcomes in the sample space that belong to *E*, divided by the total number of outcomes in the sample space. For example, suppose a set of cards includes 10 cards, 3 red and 7 black. The probability of randomly drawing a red card is $\dfrac{3}{10}$.   **3.** $S = \{HH, HT, TH, TT\}$ **5.** $S = \{(1, 2), (1, 3), (1, 4), (1, 5), (2, 3), (2, 4), (2, 5), (3, 4), (3, 5), (4, 5)\}$

7. (a) $\{HH, TT\}, \dfrac{1}{2}$ **(b)** $\{HH, HT, TH\}, \dfrac{3}{4}$ **9. (a)** $\{(2, 4)\}, \dfrac{1}{10}$ **(b)** $\{(1, 3), (1, 5), (3, 5)\}, \dfrac{3}{10}$ **(c)** $\emptyset, 0$

(d) $\{(1, 2), (1, 4), (2, 3), (2, 5), (3, 4), (4, 5)\}, \dfrac{3}{5}$ **11.** A probability cannot be greater than 1.   **13. (a)** $\dfrac{1}{5}$ **(b)** $\dfrac{8}{15}$ **(c)** 0 **(d)** 1 to 4 **(e)** 7 to 8

14. 220   **15.** 55   **16.** 55   **17.** 220   **18.** .25   **19.** $\dfrac{3}{13}; \dfrac{10}{13}$ **21.** 499 to 1   **23. (a)** $\dfrac{3}{5}$ **(b)** $\dfrac{7}{10}$ **(c)** $\dfrac{3}{10}$ **25. (a)** $\dfrac{3}{13}$ **(b)** $\dfrac{7}{13}$ **(c)** $\dfrac{3}{13}$

27. (a) $\dfrac{1}{2}$ **(b)** $\dfrac{7}{10}$ **(c)** 0   **29. (a)** .72 **(b)** .70 **(c)** .79   **31.** .90   **33.** .41   **35.** .21   **37.** .79

Chapter 14 Review Exercises   (page 914)

1. $-1, 1, 3, 5$ **3.** 1, 4, 9, 16   **5.** 0, 3, 8, 15   **7.** $2x + 3x^2 + 4x^3 + 5x^4 + 6x^5 + 7x^6$ **9.** 126   **11.** 1888.8 billion dollars

13. arithmetic; $d = 4$ **15.** geometric; $r = -1$ **17.** geometric; $r = \dfrac{1}{2}$ **19.** 73   **21.** $a_n = -5n + 1$ **23.** 15   **25.** 152

27. $a_n = -1(4)^{n-1}$ **29.** $2(-3)^{10} = 118,098$ **31.** $\dfrac{341}{1024}$ **33.** 1   **35.** $32p^5 - 80p^4q + 80p^3q^2 - 40p^2q^3 + 10pq^4 - q^5$

37. $m^2 + 4m\sqrt{mn} + 6mn + 4n\sqrt{mn} + n^2$ **39.** $-18,400(3)^{22}k^3$ **45.** 120   **47.** 35   **49.** The permutations of *n* elements taken *r* at a time are the arrangements of the *r* items. The combinations of *n* elements taken *r* at a time are the (unordered) subsets of *r* items. Each combination may represent several permutations.   **51.** 90   **53.** 456,976,000; 258,336,000   **55. (a)** 4 to 11 **(b)** 3 to 2   **57.** $\dfrac{4}{13}$

59. $\dfrac{3}{4}$ **61.** $a_{40} = 235; S_{10} = 280$ **63.** $a_9 = 6561; S_{10} = -14,762$ **65.** $a_n = 5n - 3$ **67.** $a_n = 27\left(\dfrac{1}{3}\right)^{n-1}$ **69.** 10 sec

71. approximately 42,000   **73. (a)** $\dfrac{5}{10} + \dfrac{5}{10}\left(\dfrac{1}{10}\right) + \dfrac{5}{10}\left(\dfrac{1}{10}\right)^2 + \dfrac{5}{10}\left(\dfrac{1}{10}\right)^3 + \cdots$ **(b)** $\dfrac{1}{10}$ **(c)** $\dfrac{5}{9}$ **75.** No, the terms must be successive, such as the first and second, or the second and third.   **77.** 504

Chapter 14 Test   (page 918)

[14.1] **1.** 0, 2, 0, 2, 0   [14.2] **2.** 4, 6, 8, 10, 12   [14.3] **3.** 48, 24, 12, 6, 3   [14.2] **4.** 0   [14.3] **5.** $\dfrac{64}{3}$ or $-\dfrac{64}{3}$ [14.2] **6.** 75

[14.3] **7.** 124 or 44   [14.1] **8.** 70,439   [14.3] **9.** $137,925.91   **10.** It has a sum if $|r| < 1$.   [14.2] **11.** 70   **12.** 33   **13.** 125,250

[14.3] **14.** 42   **15.** $\dfrac{1}{3}$ **16.** The sum does not exist.   [14.4] **17.** $81k^4 - 540k^3 + 1350k^2 - 1500k + 625$ **18.** $\dfrac{14,080x^8y^4}{9}$

[14.1] **19.** $324 [14.3] **20.** $20(3^{11}) = 3{,}542{,}940$ [14.5] **21.** S_k: $8 + 14 + 20 + \cdots + (6k + 2) = 3k^2 + 5k$

S_{k+1}: $8 + 14 + 20 + \cdots + [6(k + 1) + 2] = 3(k + 1)^2 + 5(k + 1)$

[14.6] **22.** 990 **23.** 45 **24.** 60 **25.** 24,360 **26.** 84 [14.7] **27.** $\dfrac{1}{26}$ **28.** $\dfrac{10}{13}$ **29.** $\dfrac{4}{13}$ **30.** 3 to 10

Cumulative Review Exercises Chapters 1–14 (page 919)

[1.1] **1.** $10, 0, \dfrac{45}{15}$ (or 3), -3 **2.** $-\dfrac{8}{3}, 10, 0, \dfrac{45}{15}$ (or 3), .82, -3 **3.** $\sqrt{13}, -\sqrt{3}$ **4.** all except $\sqrt{-7}$ [1.2, 1.3] **5.** 8 **6.** -35

7. -55 [2.1] **8.** $\left\{\dfrac{1}{6}\right\}$ [2.5] **9.** $[10, \infty)$ [2.7] **10.** $\left\{-\dfrac{9}{2}, 6\right\}$ [2.1] **11.** $\{9\}$ [2.6] **12.** $(-\infty, -3) \cup (4, \infty)$

[2.7] **13.** $(-\infty, -3] \cup [8, \infty)$ [3.2] **14.** $\dfrac{3}{4}$ [3.3] **15.** $3x + y = 4$ [3.1] **16.** [3.4] **17.**

[3.5] **18.** (a) yes (b) $\{-3, -2, 0, 1, 2\}$ (c) $\{2, 6, 4\}$ [4.1] **19.** $\{(3, -5)\}$ **20.** $\{(-1, -2)\}$ [4.4] **21.** $\{(2, 1, 4)\}$ [4.5] **22.** -15

23. $\left\{\left(6, 4, \dfrac{7}{2}\right)\right\}$ [4.3] **24.** 2 lb [5.4] **25.** $20p^2 - 2p - 6$ **26.** $9k^2 - 42k + 49$ [5.2] **27.** $-5m^3 - 3m^2 + 3m + 8$

[5.5] **28.** $2t^3 + 3t^2 - 4t + 2 + \dfrac{3}{3t - 2}$ [6.2] **29.** $z(3z + 4)(2z - 1)$ [6.3] **30.** $(7a^2 + 3b)(7a^2 - 3b)$

31. $(c + 3d)(c^2 - 3cd + 9d^2)$ **32.** $(8r + 3q)^2$ [6.5] **33.** $\left\{-\dfrac{5}{2}, 2\right\}$ [9.5] **34.** $[-2, 3]$ [5.1] **35.** $\dfrac{9}{4}$ **36.** $-\dfrac{27p^2}{10}$

[7.1] **37.** $(-\infty, -9) \cup (-9, 9) \cup (9, \infty)$ **38.** $\dfrac{x + 7}{x - 2}$ [7.2] **39.** $\dfrac{3p - 26}{p(p + 3)(p - 4)}$ [7.4] **40.** $\varnothing$

[9.2] **41.** $\left\{\dfrac{-5 + \sqrt{217}}{12}, \dfrac{-5 - \sqrt{217}}{12}\right\}$ [8.6] **42.** $\{1, 2\}$ [11.2] **43.** $\left\{\dfrac{5}{2}\right\}$ [11.6] **44.** $\{2\}$ [8.4] **45.** $10\sqrt{2}$ [8.7] **46.** 73

[11.1] **47.** $f^{-1}(x) = \dfrac{x - 5}{9}$ [11.2] **48.** 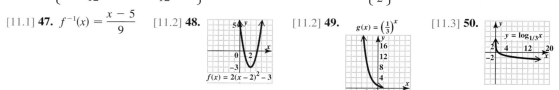 [11.2] **49.** [11.3] **50.** 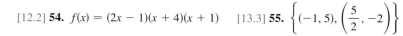 [12.4] **51.**

[13.1] **52.** [13.2] **53.** [12.2] **54.** $f(x) = (2x - 1)(x + 4)(x + 1)$ [13.3] **55.** $\left\{(-1, 5), \left(\dfrac{5}{2}, -2\right)\right\}$

[13.1] **56.** $(x + 5)^2 + (y - 12)^2 = 81$ [14.1] **57.** $-7, -2, 3, 8, 13$ [14.2, 14.3] **58.** (a) 78 (b) $\dfrac{75}{7}$ [14.2] **59.** 30

[14.4] **60.** $32a^5 - 80a^4 + 80a^3 - 40a^2 + 10a - 1$ **61.** $-\dfrac{45x^8y^6}{4}$ [14.6] **63.** (a) 210 (b) 210 [14.7] **64.** $\dfrac{1}{18}$ **65.** $\dfrac{3}{10}$

APPENDIXES

Appendix B   (page 937)

Exercises   1. $w = 3, x = 2, y = -1, z = 4$ **3.** $m = 8, n = -2, z = 2, y = 5, w = 6$ **5.** $a = 2, z = -3, m = 8, k = 1$ **7.** 2×2;
square   **9.** 3×4 **11.** 2×1; column   **13.** To add two matrices of the same size, add corresponding elements. For example,

$[2 \quad 1 \quad -3] + [3 \quad 4 \quad -2] = [5 \quad 5 \quad -5]$. Matrices of different sizes cannot be added.   **15.** $\begin{bmatrix} -2 & -7 & 7 \\ 10 & -2 & 7 \end{bmatrix}$ **17.** $\begin{bmatrix} -6 & 8 \\ 4 & 2 \end{bmatrix}$

19. $\begin{bmatrix} 5x + y & x + y & 7x + y \\ 8x + 2y & x + 3y & 3x + y \end{bmatrix}$ **21.** cannot be added   **23.** $\begin{bmatrix} -4 & 8 \\ 0 & 6 \end{bmatrix}$ **25.** $\begin{bmatrix} 2 & 6 \\ -4 & 6 \end{bmatrix}$ **27.** $\begin{bmatrix} -1 & -3 \\ 2 & -3 \end{bmatrix}$ **29.** no   **31.** $\begin{bmatrix} 13 \\ 25 \end{bmatrix}$

33. $\begin{bmatrix} -17 \\ -1 \end{bmatrix}$ **35.** $\begin{bmatrix} 17 & -10 \\ 1 & 2 \end{bmatrix}$ **37.** $\begin{bmatrix} -2 & 10 \\ 0 & 8 \end{bmatrix}$ **39.** $\begin{bmatrix} -2 & 5 & 0 \\ 6 & 6 & 1 \\ 12 & 2 & -3 \end{bmatrix}$ **41.** $[2 \quad 7 \quad -4]$ **43.** not possible   **45.** 1997–1998;

$\begin{bmatrix} 11{,}375 & 316 & 83{,}000 \\ 6970 & 115 & 73{,}000 \\ 5446 & 159 & 35{,}700 \\ 4534 & 141 & 36{,}700 \\ 4059 & 9 & 27{,}364 \end{bmatrix}$ **47.** $\begin{bmatrix} 100 & 150 \\ 125 & 50 \\ 175 & 200 \end{bmatrix}$; $\begin{bmatrix} 100 & 125 & 175 \\ 150 & 50 & 200 \end{bmatrix}$ **49. (a)** $\begin{bmatrix} 50 & 100 & 30 \\ 10 & 90 & 50 \\ 60 & 120 & 40 \end{bmatrix}$ **(b)** $\begin{bmatrix} 12 \\ 10 \\ 15 \end{bmatrix}$ (If the rows and columns are

interchanged in part (a), this should be a 1×3 matrix.)   **(c)** $\begin{bmatrix} 2050 \\ 1770 \\ 2520 \end{bmatrix}$ (This may be a 1×3 matrix.)   **(d)** $6340

Appendix C   (page 949)

Exercises   1. yes   **3.** no   **5.** no   **7.** yes   **9.** $\begin{bmatrix} -\frac{1}{5} & -\frac{2}{5} \\ \frac{2}{5} & -\frac{1}{5} \end{bmatrix}$ **11.** $\begin{bmatrix} 2 & 1 \\ -\frac{3}{2} & -\frac{1}{2} \end{bmatrix}$ **13.** The inverse does not exist.

15. $\begin{bmatrix} -1 & 1 & 1 \\ 0 & -1 & 0 \\ 2 & -1 & -1 \end{bmatrix}$ **17.** $\begin{bmatrix} 7 & -3 & -3 \\ -1 & 1 & 0 \\ -1 & 0 & 1 \end{bmatrix}$ **19.** $\begin{bmatrix} -\frac{15}{4} & -\frac{1}{4} & -3 \\ \frac{5}{4} & \frac{1}{4} & 1 \\ -\frac{3}{2} & 0 & 1 \end{bmatrix}$ **21.** $\begin{bmatrix} \frac{1}{2} & 0 & \frac{1}{2} & -1 \\ \frac{1}{10} & -\frac{2}{5} & \frac{3}{10} & -\frac{1}{5} \\ -\frac{7}{10} & \frac{4}{5} & -\frac{11}{10} & \frac{12}{5} \\ \frac{1}{5} & \frac{1}{5} & -\frac{2}{5} & \frac{3}{5} \end{bmatrix}$ **23.** $\{(2, 3)\}$ **25.** $\{(-2, 4)\}$

27. $\{(4, -6)\}$ **29.** $\{(10, -1, -2)\}$ **31.** $\{(11, -1, 2)\}$ **33.** $\{(1, 0, 2, 1)\}$ **35.** **(a)** $602.7 = a + 5.543b + 37.14c$
$656.7 = a + 6.933b + 41.30c$
$778.5 = a + 7.638b + 45.62c$

(b) $a \approx -490.547, b = -89, c = 42.71875$ **(c)** $S = -490.547 - 89A + 42.71875B$ **(d)** approximately 843.5   **(e)** $S \approx 1547.5$; Using
only three consecutive years to forecast six years into the future is probably not wise..

Glossary

A

absolute value The absolute value of a number is the distance between 0 and the number on a number line. (Section 1.1)

absolute value equation An absolute value equation is an equation that involves the absolute value of a variable expression. (Section 2.7)

absolute value function The function defined by $f(x) = |x|$ with a graph that includes portions of two lines is called the absolute value function. (Section 10.5)

absolute value inequality An absolute value inequality is an inequality that involves the absolute value of a variable expression. (Section 2.7)

addition property of equality The addition property of equality states that the same number can be added to (or subtracted from) both sides of an equation to obtain an equivalent equation. (Section 2.1)

addition property of inequality The addition property of inequality states that the same number can be added to (or subtracted from) both sides of an inequality without changing the solution set. (Section 2.5)

additive inverse (negative) of a matrix When two matrices are added and a zero matrix results, the matrices are additive inverses (negatives) of each other. (Appendix B)

additive inverse (negative, opposite) Two numbers that are the same distance from 0 on a number line but on opposite sides of 0 are called additive inverses. (Section 1.1)

algebraic expression Any collection of numbers or variables joined by the basic operations of addition, subtraction, multiplication, or division (except by 0), or the operations of raising to powers or taking roots is called an algebraic expression. (Sections 1.3, 5.2)

annuity An annuity is a sequence of equal payments made at equal periods of time. (Section 14.3)

arithmetic mean (average) The arithmetic mean of a group of numbers is the sum of all the numbers divided by the number of numbers. (Section 14.1)

arithmetic sequence (arithmetic progression) An arithmetic sequence is a sequence in which each term after the first differs from the preceding term by a constant amount. (Section 14.2)

array of signs An array of signs is used when evaluating a determinant using expansion by minors. The signs alternate for each row and column, beginning with + in the first row, first column position. (Section 4.5)

associative property of addition The associative property of addition states that the way in which numbers being added are grouped does not change the sum. (Section 1.4)

associative property of multiplication The associative property of multiplication states that the way in which numbers being multiplied are grouped does not change the product. (Section 1.4)

asymptote A line that a graph approaches more and more closely as x approaches a certain value or as the absolute value of x gets larger and larger without bound is called an asymptote of the graph. (Sections 7.4, 11.2, 12.4)

asymptotes of a hyperbola The two intersecting straight lines that the branches of a hyperbola approach are called asymptotes of the hyperbola. (Section 13.2)

augmented matrix An augmented matrix is a matrix that has a vertical bar that separates the columns of the matrix into two groups. (Section 4.4)

axis (axis of symmetry) The axis of a parabola is the vertical or horizontal line through the vertex of the parabola. (Section 10.2)

B

base The base is the number that is a repeated factor when written with an exponent. (Sections 1.3, 5.1)

binomial A binomial is a polynomial with exactly two terms. (Section 5.2)

binomial theorem (general binomial expansion) The binomial theorem is a formula used to expand a binomial raised to a power. (Section 14.4)

boundary In a graph of any inequality in two variables, the boundary is a line or curve that separates the region that satisfies the inequality from the region that does not satisfy the inequality. (Section 13.4)

boundary line In the graph of a linear inequality, the boundary line separates the region that satisfies the inequality from the region that does not satisfy the inequality. (Section 3.4)

C

center of a circle The fixed point that is a fixed distance from all the points that form a circle is the center of the circle. (Section 13.1)

center of an ellipse The center of an ellipse is the fixed point located exactly halfway between the two foci. (Section 13.1)

center-radius form of the equation of a circle The center-radius form of the equation of a circle with center (h, k) and radius r is $(x - h)^2 + (y - k)^2 = r^2$. (Section 13.1)

circle A circle is the set of all points in a plane that lie a fixed distance from a fixed point. (Section 13.1)

coefficient (numerical coefficient) A coefficient is the numerical factor of a term. (Sections 1.4, 5.2)

column matrix A matrix with just one column is called a column matrix. (Appendix B)

column of a matrix A column of a matrix is a group of elements that are read vertically. (Section 4.4)

combination A combination of n elements taken r at a time is one of the ways in which r elements can be chosen from n elements. In combinations, the order of the elements is not important. (Section 14.6)

combined variation If a problem involves a combination of direct variation and inverse

variation, then it is called a combined variation problem. (Section 3.6)

combining like terms Combining like terms is a method of adding or subtracting like terms by using the properties of real numbers. (Section 1.4)

common difference The common difference d is the difference between any two adjacent terms of an arithmetic sequence. (Section 14.2)

common logarithm A common logarithm is a logarithm to base 10. (Section 11.5)

common ratio A common ratio r is the constant multiplier between adjacent terms in a geometric sequence. (Section 14.3)

commutative property of addition The commutative property of addition states that the order of numbers in an addition problem can be changed without changing the sum. (Section 1.4)

commutative property of multiplication The commutative property of multiplication states that the product in a multiplication problem remains the same regardless of the order of the factors. (Section 1.4)

complement In probability, the set of all outcomes in a sample space that do *not* belong to an event is called the complement of the event. (Section 14.7)

complementary angles (complements) Complementary angles are angles whose measures have a sum of 90°. (Section 2.4 Exercises)

completing the square The process of adding to a binomial the number that makes it a perfect square trinomial is called completing the square. (Section 9.1)

complex conjugates The complex conjugate of $a + bi$ is $a - bi$. (Section 8.7)

complex fraction A complex fraction is an expression with one or more fractions in the numerator, denominator, or both. (Section 7.3)

complex number A complex number is any number that can be written in the form $a + bi$, where a and b are real numbers. (Section 8.7)

composite function (composition) A function in which some quantity depends on a variable that, in turn, depends on another variable is called a composite function. (Section 10.1)

compound event In probability, a compound event involves two or more alternative events. (Section 14.7)

compound inequality A compound inequality consists of two inequalities linked by a connective word such as *and* or *or*. (Section 2.6)

comprehensive graph A comprehensive graph of a polynomial function will show the following characteristics: (1) all x-intercepts (zeros); (2) the y-intercept; (3) all turning points; (4) enough of the domain to show the end behavior. (Section 12.3)

conditional equation A conditional equation is true for some replacements of the variable and false for others. (Section 2.1)

conic section When a plane intersects an infinite cone at different angles, the figures formed by the intersections are called conic sections. (Section 13.1)

conjugate The conjugate of $a + b$ is $a - b$. (Section 8.5)

conjugate zeros theorem The conjugate zeros theorem states that if $f(x)$ is a polynomial having only real coefficients and if $a + bi$ is a zero of $f(x)$, where a and b are real numbers, then $a - bi$ is also a zero of $f(x)$. (Section 12.2)

consecutive integers Two integers that differ by one are called consecutive integers. (Section 2.4 Exercises)

consistent system A system of equations with a solution is called a consistent system. (Section 4.1)

constant function A linear function of the form $f(x) = b$, where b is a constant, is called a constant function. (Section 3.5)

constant on an interval A function that is neither increasing nor decreasing on an interval is constant on that interval. (Section 10.4)

constant of variation In the variation equations $y = kx$, or $y = \frac{k}{x}$, or $y = kxz$, the number k is called the constant of variation. (Section 3.6)

constraints In linear programming, the restrictions on a particular situation are called the constraints. (Section 13.4)

contradiction A contradiction is an equation that is never true. It has no solution. (Section 2.1)

coordinate on a number line Each number on a number line is called the coordinate of the point that it labels. (Section 1.1)

coordinates of a point The numbers in an ordered pair are called the coordinates of the corresponding point in the plane. (Section 3.1)

Cramer's rule Cramer's rule uses determinants to solve systems of linear equations. (Section 4.5)

cube root function The function defined by $f(x) = \sqrt[3]{x}$ is called the cube root function. (Section 8.1)

cubing function The polynomial function defined by $f(x) = x^3$ is called the cubing function. (Section 5.3)

D

decreasing function A function f is a decreasing function if its graph goes downward from left to right, that is, $f(x_1) > f(x_2)$ whenever $x_1 < x_2$. (Section 10.4)

degree of a polynomial The degree of a polynomial is the greatest degree of any of the terms in the polynomial. (Section 5.2)

degree of a term The degree of a term is the sum of the exponents on the variables in the term. (Section 5.2)

dependent equations Equations of a system that have the same graph (because they are different forms of the same equation) are called dependent equations. (Section 4.1)

dependent variable In an equation relating x and y, if the value of the variable y depends on the variable x, then y is called the dependent variable. (Section 3.5)

Descartes' rule of signs Descartes' rule of signs is a rule that can help determine the number of positive and the number of negative real zeros of a polynomial function. (Section 12.2 Exercises)

descending powers A polynomial in one variable is written in descending powers of the variable if the degree of the terms of the polynomial decreases from left to right. (Section 5.2)

determinant Associated with every square matrix is a real number called the determinant of the matrix, symbolized by the entries of the matrix placed between two vertical lines. (Section 4.5)

difference The answer to a subtraction problem is called the difference. (Section 1.2)

difference of cubes The difference of cubes, $x^3 - y^3$, can be factored as $x^3 - y^3 = (x - y)(x^2 + xy + y^2)$. (Section 6.3)

difference of squares The difference of squares, $x^2 - y^2$, can be factored as the product of the sum and difference of two terms, or $x^2 - y^2 = (x + y)(x - y)$. (Section 6.3)

difference quotient If the coordinates of point P are $(x, f(x))$ and the coordinates of point Q are $(x + h, f(x + h))$, then the expression $\frac{f(x + h) - f(x)}{h}$ is called the difference quotient. (Section 10.1)

directrix A directrix is a fixed line which, together with the focus (a point not on the line), is used to determine the points that form a parabola. (Section 10.2)

direct variation y varies directly as x if there exists a real number k such that $y = kx$. (Section 3.6)

discriminant The discriminant is the quantity under the radical, $b^2 - 4ac$, in the quadratic formula. (Section 9.2)

distributive property For any real numbers a, b, and c, the distributive property states that $a(b + c) = ab + ac$ and $(b + c)a = ba + ca$. (Section 1.4)

distance The distance between two points on a number line is the absolute value of the difference between the two numbers. (Section 1.2)

division algorithm The division algorithm states that if $f(x)$ and $g(x)$ are polynomials with $g(x)$ of lower degree than $f(x)$ and $g(x)$ of degree one or more, then there exist unique polynomials $q(x)$ and $r(x)$ such that $f(x) = g(x) \cdot q(x) + r(x)$, where either $r(x) = 0$ or the degree of $r(x)$ is less than the degree of $g(x)$. (Section 12.1)

domain The set of all first components (x-values) in the ordered pairs of a relation is the domain. (Section 3.5)

domain of a rational equation The domain of a rational equation is the intersection (overlap) of the domains of the rational expressions in the equation. (Section 7.4)

E

element of a matrix The numbers in a matrix are called the elements of the matrix. (Section 4.4)

elements (members) Elements are the objects that belong to a set. (Section 1.1)

elimination method The elimination method is an algebraic method used to solve a system of equations in which the equations of the system are combined so that one or more variables is eliminated. (Section 4.1)

ellipse An ellipse is the set of all points in a plane the sum of whose distances from two fixed points is constant. (Section 13.1)

empty set (null set) The empty set, denoted by { } or $\emptyset$, is the set containing no elements. (Section 1.1)

equation An equation is a statement that two algebraic expressions are equal. (Section 1.1)

equivalent equations Equivalent equations are equations that have the same solution set. (Section 2.1)

equivalent inequalities Equivalent inequalities are inequalities that have the same solution set. (Section 2.5)

event In probability, an event is any subset of the sample space. (Section 14.7)

expansion by minors A method of evaluating a 3×3 or larger determinant is called expansion by minors. (Section 4.5)

exponent (power) An exponent is a number that indicates how many times a factor is repeated. (Sections 1.3, 5.1)

exponential equation An exponential equation is an equation that has a variable as an exponent. (Section 11.2)

exponential expression A number or letter (variable) written with an exponent is an exponential expression. (Section 1.3)

exponential function An exponential function is a function defined by an expression of the form $f(x) = a^x$, where $a > 0$ and $a \neq 1$ for all real numbers x. (Section 11.2)

extraneous solution A solution to a new equation that does not satisfy the original equation is called an extraneous solution. (Section 8.6)

F

factor A factor of a given number is any number that divides evenly (without remainder) into the given number. (Section 1.3)

factor theorem The factor theorem states that the polynomial $x - k$ is a factor of the

polynomial $f(x)$ if and only if $f(k) = 0$. (Section 12.2)

factoring Writing a polynomial as the product of two or more simpler polynomials is called factoring. (Section 6.1)

factoring by grouping Factoring by grouping is a method of grouping the terms of a polynomial in such a way that the polynomial can be factored even though its greatest common factor is 1. (Section 6.1)

factoring out the greatest common factor Factoring out the greatest common factor is the process of using the distributive property to write a polynomial as a product of the greatest common factor and a simpler polynomial. (Section 6.1)

finite sequence A finite sequence has a domain that includes only the first n positive integers. (Section 14.1)

first-degree equation A first-degree (linear) equation has no term with the variable to a power greater than 1. (Section 2.1)

foci (singular, focus) Foci are fixed points used to determine the points that form a parabola, an ellipse, or a hyperbola. (Sections 10.2, 13.1, 13.2)

FOIL FOIL is a method for multiplying two binomials $(A + B)(C + D)$. Multiply **F**irst terms AC, **O**uter terms AD, **I**nner terms BC, and **L**ast terms BD. Then combine like terms. (Section 5.4)

formula A formula is a mathematical equation in which variables are used to describe a relationship. (Section 2.2)

function A function is a set of ordered pairs (relation) in which each value of the first component x corresponds to exactly one value of the second component y. (Section 3.5)

function notation Function notation $f(x)$ represents the value of the function at x, that is, the y-value that corresponds to x. (Section 3.5)

fundamental principle of counting The fundamental principle of counting states that if one event can occur in m ways and a second event can occur in n ways, then both events can occur in mn ways, provided that the outcome of the first event does not influence the outcome of the second event. (Section 14.6)

fundamental rectangle The asymptotes of a hyperbola are the extended diagonals of its fundamental rectangle, with corners at the points (a, b), $(-a, b)$, $(-a, -b)$, and $(a, -b)$. (Section 13.2)

fundamental theorem of algebra The fundamental theorem of algebra states that every polynomial of degree 1 or more has at least one complex zero. (Section 12.2)

future value of an annuity The future value of an annuity is the sum of the compound amounts of all the payments, compounded to the end of the term. (Section 14.3)

G

general term of a sequence The expression a_n, which defines a sequence, is called the general term of the sequence. (Section 14.1)

geometric sequence (geometric progression) A geometric sequence is a sequence in which each term after the first is a constant multiple of the preceding term. (Section 14.3)

graph of a number The point on a number line that corresponds to a number is its graph. (Section 1.1)

graph of an equation The graph of an equation is the set of all points that correspond to all of the ordered pairs that satisfy the equation. (Section 3.1)

graph of a relation The graph of a relation is the graph of its ordered pairs. (Section 3.5)

greatest common factor (GCF) The greatest common factor of a list of integers is the largest common factor of those integers. The greatest common factor of a polynomial is the largest term that is a factor of all terms in the polynomial. (Section 6.1)

greatest integer function The function defined by $f(x) = [\![x]\!]$, where the symbol $[\![x]\!]$ is used to represent the greatest integer less than or equal to x, is called the greatest integer function. (Section 10.5)

H

horizontal asymptote A horizontal line that a graph approaches as $|x|$ gets larger and larger without bound is called a horizontal asymptote. (Section 12.4)

horizontal line test The horizontal line test states that a function is one-to-one if every horizontal line intersects the graph of the function at most once. (Section 11.1)

hyperbola A hyperbola is the set of all points in a plane such that the absolute value of the difference of the distances from two fixed points is constant. (Section 13.2)

hypotenuse The hypotenuse is the longest side in a right triangle. It is the side opposite the right angle. (Section 8.3)

I

identity An identity is an equation that is true for all replacements of the variable. It has an infinite number of solutions. (Section 2.1)

identity element for addition Since adding 0 to a number does not change the number, 0 is called the identity element for addition. (Section 1.4)

identity element for multiplication Since multiplying a number by 1 does not change the number, 1 is called the identity element for multiplication. (Section 1.4)

identity function The simplest polynomial function is the identity function, defined by $f(x) = x$. (Section 5.3)

identity matrix (multiplicative identity matrix) A square matrix with 1s on the main diagonal and 0s elsewhere is called an identity matrix. If an $n \times n$ matrix is multiplied by the $n \times n$ identity matrix, written I_n, then the original $n \times n$ matrix results. (Appendix C)

identity property The identity properties state that the sum of 0 and any number equals the number, and the product of 1 and any number equals the number. (Section 1.4)

imaginary number A complex number $a + bi$ with $b \neq 0$ is called an imaginary number. (Section 8.7)

imaginary part The imaginary part of the complex number $a + bi$ is b. (Section 8.7)

inconsistent system An inconsistent system of equations is a system with no solution. (Section 4.1)

increasing function A function is an increasing function if its graph goes upward from left to right, that is, $f(x_1) < f(x_2)$ whenever $x_1 < x_2$. (Section 10.4)

independent equations Equations of a system that have different graphs are called independent equations. (Section 4.1)

independent events In probability, if the outcome of one event does not influence the outcome of another, then the events are called independent events. (Section 14.6)

independent variable In an equation relating x and y, if the value of the variable y depends on the variable x, then x is called the independent variable. (Section 3.5)

index (order) In a radical of the form $\sqrt[n]{a}$, n is called the index or order. (Section 8.1)

index of summation When using summation notation, $\sum_{i=1}^{n} f(i)$, the letter i is called the index of summation. (Section 14.1)

inequality An inequality is a statement that two expressions are not equal. (Section 1.1)

infinite sequence An infinite sequence is a function with the set of positive integers as the domain. (Section 14.1)

integers The set of integers is $\{\ldots, -3, -2, -1, 0, 1, 2, 3, \ldots\}$. (Section 1.1)

intersection The intersection of two sets A and B, written $A \cap B$, is the set of elements that belong to both A and B. (Section 2.6)

interval An interval is a portion of a number line. (Section 1.1)

interval notation Interval notation is a simplified notation that uses parentheses ( ) and/or brackets [ ] to describe an interval on a number line. (Section 1.1)

inverse of a function f If f is a one-to-one function, then the inverse of f is the set of all ordered pairs of the form (y, x), where (x, y) belongs to f. (Section 11.1)

inverse property The inverse properties state that a number added to its opposite is 0, and a number multiplied by its reciprocal is 1. (Section 1.4)

inverse variation y varies inversely as x if there exists a real number k such that $y = \frac{k}{x}$. (Section 3.6)

irrational numbers Irrational numbers cannot be written as the quotient of two integers but can be represented by points on the number line. (Section 1.1)

J

joint variation y varies jointly as x and z if there exists a real number k such that $y = kxz$. (Section 3.6)

L

least common denominator (LCD) Given several denominators, the smallest expression that is divisible by all the denominators is called the least common denominator. (Section 7.2)

legs of a right triangle The two shorter sides of a right triangle are called the legs. (Section 8.3)

like terms Terms with exactly the same variables raised to exactly the same powers are called like terms. (Section 1.4)

linear (first-degree) equation in one variable A linear equation in one variable can be written in the form $Ax + B = C$, where A, B, and C are real numbers, with $A \neq 0$. (Section 2.1)

linear equation in two variables A linear equation in two variables is an equation that can be written in the form $Ax + By = C$, where A, B, and C are real numbers and A and B are not both 0. (Section 3.1)

linear function A function defined by an equation of the form $f(x) = mx + b$, for real numbers m and b, is a linear function. (Section 3.5)

linear inequality in one variable A linear inequality in one variable can be written in the form $Ax + B < C$ or $Ax + B > C$ (or with $\leq$ or $\geq$), where A, B, and C are real numbers, with $A \neq 0$. (Section 2.5)

linear inequality in two variables A linear inequality in two variables can be written in the form $Ax + By < C$ or $Ax + By > C$ (or with $\leq$ or $\geq$), where A, B, and C are real numbers and A and B are not both 0. (Section 3.4)

linear programming Linear programming, an application of mathematics to business or social science, is a method for finding an optimum value, for example, minimum cost or maximum profit. (Section 13.4)

linear system (system of linear equations) Two or more linear equations form a linear system. (Section 4.1)

logarithm A logarithm is an exponent; $\log_a x$ is the exponent on the base a that gives the number x. (Section 11.3)

logarithmic equation A logarithmic equation is an equation with a logarithm in at least one term. (Section 11.3)

logarithmic function with base a If a and x are positive numbers with $a \neq 1$, then $f(x) = \log_a x$ defines the logarithmic function with base a. (Section 11.3)

lowest terms A fraction is in lowest terms when there are no common factors in the numerator and denominator (except 1). (Section 7.1)

M

mathematical induction Mathematical induction is a method for proving that a statement S_n is true for every positive integer value of n. In order to prove that S_n is true for every positive integer value of n, you must show that (1) S_1 is true and (2) for any positive integer k, $k \leq n$, if S_k is true, then S_{k+1} is also true. (Section 14.5)

mathematical model In a real-world problem, a mathematical model is one or more equations (or inequalities) that describe the situation. (Section 2.2)

matrix (plural, **matrices**) A matrix is a rectangular array of numbers, consisting of horizontal rows and vertical columns. (Sections 4.4, 4.5)

minor The minor of an element in a 3×3 determinant is the 2×2 determinant remaining when a row and a column of the 3×3 determinant are eliminated. (Section 4.5)

monomial A monomial is a polynomial with only one term. (Section 5.2)

multiplication property of equality The multiplication property of equality states that the same nonzero number can be multiplied by (or divided into) both sides of an equation to obtain an equivalent equation. (Section 2.1)

multiplication property of inequality The multiplication property of inequality states that both sides of an inequality may be multiplied (or divided) by a positive number without changing the direction of the inequality symbol. Multiplying (or dividing) by a negative number reverses the inequality symbol. (Section 2.5)

multiplication property of 0 The multiplication property of 0 states that the product of any real number and 0 is 0. (Section 1.4)

multiplicative inverse of a matrix (inverse matrix) If A is an $n \times n$ matrix, then its multiplicative inverse, written A^{-1}, must satisfy both $AA^{-1} = I_n$ and $A^{-1}A = I_n$. (Appendix C)

multiplicative inverse (reciprocal) The multiplicative inverse of a nonzero real number a is $\frac{1}{a}$. (Section 1.2)

multiplicity of a zero The multiplicity of a zero k of a polynomial $f(x)$ is the number of factors of $x - k$ that appear when the polynomial is written in factored form. (Section 12.2)

mutually exclusive events In probability, two events that cannot occur simultaneously are called mutually exclusive events. (Section 14.7)

N

n-factorial ($n!$) For any positive integer n, $n! = n(n - 1)(n - 2) \cdots (3)(2)(1)$. (Sections 14.4, 14.6)

natural logarithm A natural logarithm is a logarithm to base e. (Section 11.5)

natural numbers (counting numbers) The set of natural numbers includes the numbers used for counting: $\{1, 2, 3, 4, \ldots\}$. (Section 1.1)

negative of a polynomial The negative of a polynomial is that polynomial with every sign changed. (Section 5.2)

nonlinear equation A nonlinear equation is an equation in which some terms have more than one variable or a variable of degree 2 or higher. (Section 13.3)

nonlinear system of equations A nonlinear system of equations is a system that includes at least one nonlinear equation. (Section 13.3)

nonlinear system of inequalities A nonlinear system of inequalities is two or more inequalities to be considered at the same time, at least one of which is nonlinear. (Section 13.4)

number line A number line is a line with a scale that is used to show how numbers relate to each other. (Section 1.1)

number of zeros theorem The number of zeros theorem states that a polynomial of degree n has at most n distinct zeros. (Section 12.2)

numerical coefficient The numerical factor in a term is its numerical coefficient. (Sections 1.4, 5.2)

O

objective function In linear programming, the function to be maximized or minimized is called the objective function. (Section 13.4)

oblique asymptote A nonvertical, nonhorizontal line that a graph approaches as $|x|$ gets larger and larger without bound is called an oblique asymptote. (Section 12.4)

odds The odds in favor of an event is the ratio of the probability of the event to the probability of the complement of the event. (Section 14.7)

one-to-one function A one-to-one function is a function in which each x-value corresponds to only one y-value and each y-value corresponds to only one x-value. (Section 11.1)

ordered pair An ordered pair is a pair of numbers written within parentheses in which the order of the numbers is important. (Section 3.1)

ordered triple A solution of an equation in three variables, written (x, y, z), is called an ordered triple. (Section 4.2)

ordinary annuity An ordinary annuity is an annuity in which the payments are made at the end of each time period and the frequency of payments is the same as the frequency of compounding. (Section 14.3)

origin The point at which the x-axis and y-axis of a rectangular coordinate system intersect is called the origin. (Section 3.1)

outcome In probability, a possible result of each trial in an experiment is called an outcome of the experiment. (Section 14.7)

P

parabola A parabola is the type of curve that is the graph of any quadratic function. Geometrically, a parabola is defined as the set of all points in a plane that are equally distant from a fixed point and a fixed line not containing the point. (Sections 5.3, 10.2)

parallel lines Parallel lines are two lines in the same plane that never intersect. (Section 3.2)

Pascal's triangle Pascal's triangle is a triangular array of numbers that is helpful in expanding binomials. (Section 14.4)

payment period In an annuity, the time between payments is called the payment period. (Section 14.3)

percent Percent, written with the sign %, means "per one hundred." (Section 2.1)

perfect square trinomial A perfect square trinomial is a trinomial that can be factored as the square of a binomial. (Section 6.3)

permutation A permutation of n elements taken r at a time is one of the ways of arranging r elements taken from a set of n elements ($r \leq n$). In permutations, the order of the elements is important. (Section 14.6)

perpendicular lines Perpendicular lines are two lines that intersect to form a right (90°) angle. (Section 3.2)

piecewise linear function A function defined with different linear equations for different parts of its domain is called a piecewise linear function. (Section 10.5)

plot To plot an ordered pair is to locate it on a rectangular coordinate system. (Section 3.1)

point-slope form A linear equation is written in point-slope form if it is in the form $y - y_1 = m(x - x_1)$, where m is the slope of the line and (x_1, y_1) is a point on the line. (Section 3.3)

polynomial A polynomial is a term or a finite sum of terms in which all coefficients are real, all variables have whole number exponents, and no variables appear in denominators. (Section 5.2)

polynomial function A function defined by a polynomial in one variable, consisting of one or more terms, is called a polynomial function. (Section 5.3)

polynomial function of degree n A function defined by $f(x) = a_n x^n + a_{n-1} x^{n-1} + \ldots + a_1 x + a_0$ for complex numbers a_n, $a_{n-1}, \ldots, a_1$, and a_0, where $a_n \neq 0$, is called a polynomial function of degree n. (Section 12.1)

polynomial in x A polynomial containing only the variable x is called a polynomial in x. (Section 5.2)

prime polynomial A prime polynomial is a polynomial that cannot be factored using only integer coefficients. (Section 6.1)

principal root (principal nth root) For even indexes, the symbols $\sqrt{}$, $\sqrt[4]{}$, $\sqrt[6]{}, \ldots, \sqrt[n]{}$, are used for nonnegative roots, which are called principal roots. (Section 8.1)

probability of an event In a sample space with equally likely outcomes, the probability of an event is the ratio of the number of outcomes in the event to the number of outcomes in the sample space. (Section 14.7)

product The answer to a multiplication problem is called the product. (Section 1.2)

product of the sum and difference of two terms The product of the sum and difference of two terms is the difference of the squares of the terms: $(x + y)(x - y) = x^2 - y^2$. (Section 5.4)

proportion A proportion is a statement that two ratios are equal. (Section 7.5)

proportional If y varies directly as x and there exists some number (constant) k such that $y = kx$, then y is said to be proportional to x. (Section 3.6)

Pythagorean formula The Pythagorean formula states that the square of the length of the hypotenuse of a right triangle equals the sum of the squares of the lengths of the two legs. (Section 8.3)

Q

quadrant A quadrant is one of the four regions in the plane determined by a rectangular coordinate system. (Section 3.1)

quadratic equation A quadratic equation is an equation that can be written in the form $ax^2 + bx + c = 0$, where a, b, and c are real numbers, with $a \neq 0$. (Sections 6.5, 9.1)

quadratic formula The quadratic formula is a general formula used to solve any quadratic equation. (Section 9.2)

quadratic function A function defined by an equation of the form $f(x) = ax^2 + bx + c$, for real numbers a, b, and c, with $a \neq 0$, is a quadratic function. (Section 10.2)

quadratic inequality A quadratic inequality can be written in the form $ax^2 + bx + c < 0$ or $ax^2 + bx + c > 0$ (or with $\leq$ or $\geq$), where a, b, and c are real numbers, with $a \neq 0$. (Section 9.5)

quadratic in form An equation that is written in the form, $a[f(x)]^2 + b[f(x)] + c = 0$, for $a \neq 0$, is called quadratic in form. (Section 9.3)

quotient   The answer to a division problem is called the quotient. (Section 1.2)

R

radical   A radical sign with a radicand is called a radical. (Section 8.1)

radical equation   An equation that includes one or more radical expressions with a variable is called a radical equation. (Section 8.6)

radical expression   A radical expression is an algebraic expression that contains radicals. (Section 8.1)

radical sign   The symbol $\sqrt{}$ is called a radical sign. (Section 1.3)

radicand   The number or expression under a radical sign is called the radicand. (Section 8.1)

radius   The radius of a circle is the fixed distance between the center and any point on the circle. (Section 13.1)

range   The set of all second components (y-values) in the ordered pairs of a relation is the range. (Section 3.5)

ratio   A ratio is a comparison of two quantities with the same units. (Section 7.5)

rational expression   The quotient of two polynomials with denominator not 0 is called a rational expression, or algebraic fraction. (Section 7.1)

rational function   A function of the form $f(x) = \frac{P(x)}{Q(x)}$, where $P(x)$ and $Q(x)$ are polynomial functions, $Q(x) \neq 0$, is called a rational function. (Sections 7.1, 12.4)

rational inequality   An inequality that involves fractions is called a rational inequality. (Section 9.5)

rational numbers   Rational numbers can be written as the quotient of two integers, with denominator not 0. (Section 1.1)

rational zeros theorem   The rational zeros theorem states that if $f(x)$ defines a polynomial function with integer coefficients and $\frac{p}{q}$, a rational number written in lowest terms, is a zero of f, then p is a factor of the constant term a_0 and q is a factor of the leading coefficient a_n. (Section 12.2)

rationalizing the denominator   The process of removing radicals from a denominator so that the denominator contains only rational numbers is called rationalizing the denominator. (Section 8.5)

real numbers   Real numbers include all numbers that can be represented by points on the number line, that is, all rational and irrational numbers. (Section 1.1)

real part   The real part of a complex number $a + bi$ is a. (Section 8.7)

reciprocal   Pairs of numbers whose product is 1 are called reciprocals of each other. (Sections 1.2, 7.1)

rectangular (Cartesian) coordinate system   The x-axis and y-axis placed at a right angle at their zero points form a rectangular coordinate system, also called the Cartesian coordinate system. (Section 3.1)

reduced row echelon form   Reduced row echelon form is an extension of row echelon form that has 0s above and below the diagonal of 1s. (Section 4.4)

region of feasible solutions   In linear programming, the region of feasible solutions is the region of the graph that satisfies all of the constraints. (Section 13.4)

relation   A relation is a set of ordered pairs. (Section 3.5)

remainder theorem   The remainder theorem states that if the polynomial $f(x)$ is divided by $x - k$, then the remainder is $f(k)$. (Section 12.1)

rise   Rise is the vertical change between two points on a line, that is, the change in y-values. (Section 3.2)

root (or solution)   A root (or solution) of a polynomial equation $f(x) = 0$ is a number k such that $f(k) = 0$. (Section 12.1)

row echelon form   If a matrix is written with 1s on the diagonal from upper left to lower right and 0s below the 1s, it is said to be in row echelon form. (Section 4.4)

row matrix   A matrix with just one row is called a row matrix. (Appendix B)

row of a matrix   A row of a matrix is a group of elements that are read horizontally. (Section 4.4)

row operations   Row operations are operations on a matrix that produce equivalent matrices leading to the solution of a system of equations. (Section 4.4)

run   Run is the horizontal change between two points on a line, that is, the change in x-values. (Section 3.2)

S

sample space   In probability, the set of all possible outcomes of a given experiment is called the sample space of the experiment. (Section 14.7)

scalar   In work with matrices, a real number is called a scalar to distinguish it from a matrix. (Appendix B)

scientific notation   A number is written in scientific notation when it is expressed in the form $a \times 10^n$, where $1 \leq |a| < 10$ and n is an integer. (Section 5.1)

second-degree inequality   A second-degree inequality is an inequality with at least one variable of degree 2 and no variable with degree greater than 2. (Section 13.4)

sequence   A sequence is a function whose domain is the set of natural numbers. (Section 14.1)

series   The indicated sum of the terms of a sequence is called a series. (Section 14.1)

set   A set is a collection of objects. (Section 1.1)

set-builder notation   Set-builder notation is used to describe a set of numbers without actually having to list all of the elements. (Section 1.1)

signed numbers   Signed numbers are numbers that can be written with a positive or negative sign. (Section 1.1)

simplified radical   A simplified radical meets four conditions:

1. The radicand has no factor raised to a power greater than or equal to the index.

2. The radicand has no fractions.

3. No denominator contains a radical.

4. Exponents in the radicand and the index of the radical have no common factor (except 1).

(Section 8.3)

slope   The ratio of the change in y to the change in x along a line is called the slope of the line. (Section 3.2)

slope-intercept form   A linear equation is written in slope-intercept form if it is in the form $y = mx + b$, where m is the slope and $(0, b)$ is the y-intercept. (Section 3.3)

solution of an equation A solution of an equation is any replacement for the variable that makes the equation true. (Section 2.1)

solution set The solution set of an equation is the set of all solutions of the equation. (Section 2.1)

solution set of a linear system The solution set of a linear system of equations includes all ordered pairs that satisfy all the equations of the system at the same time. (Section 4.1)

solution set of a system of inequalities The solution set of a system of inequalities includes all ordered pairs that make all inequalities of the system true at the same time. (Section 13.4)

square of a binomial The square of a binomial is the sum of the square of the first term, twice the product of the two terms, and the square of the last term: $(x + y)^2 = x^2 + 2xy + y^2$ or $(x - y)^2 = x^2 - 2xy + y^2$. (Section 5.4)

square matrix A square matrix is a matrix that has the same number of rows as columns. (Sections 4.4, 4.5)

square matrix of order n A square matrix of order n is an $n \times n$ matrix. (Appendix B)

square root The opposite of squaring a number is called taking its square root; that is, a number b is a square root of a if $b^2 = a$. (Section 1.3)

square root function The function defined by $f(x) = \sqrt{x}$, with $x \geq 0$, is called the square root function. (Section 8.1)

square root function (extended definition) A function of the form $f(x) = \sqrt{u}$ for an algebraic expression u, with $u \geq 0$, is called a square root function. (Section 13.3)

square root property The square root property states that if $x^2 = k$, then $x = \sqrt{k}$ or $x = -\sqrt{k}$. (Section 9.1)

squaring function The polynomial function defined by $f(x) = x^2$ is called the squaring function. (Section 5.3)

standard form of a complex number The standard form of a complex number is $a + bi$. (Section 8.7)

standard form of a linear equation A linear equation in two variables written in

the form $Ax + By = C$, where A, B, and C are integers with no common factor (except 1) and $A \geq 0$, is in standard form. (Sections 3.1, 3.3)

standard form of a quadratic equation A quadratic equation written in the form $ax^2 + bx + c = 0$, where a, b, and c are real numbers with $a \neq 0$, is in standard form. (Sections 6.5, 9.1)

step function A function with a graph that looks like a series of steps is called a step function. (Section 10.5)

substitution method The substitution method is an algebraic method for solving a system of equations in which one equation is solved for one of the variables and the result is substituted in the other equation. (Section 4.1)

sum The answer to an addition problem is called the sum. (Section 1.2)

sum of cubes The sum of cubes, $x^3 + y^3$, can be factored as $x^3 + y^3 = (x + y) \cdot (x^2 - xy + y^2)$. (Section 6.3)

summation (sigma) notation Summation notation is a compact way of writing a series using the general term of the corresponding sequence. (Section 14.1)

supplementary angles (supplements) Supplementary angles are angles whose measures have a sum of $180°$. (Section 2.4 Exercises)

symmetric with respect to the origin If a graph can be rotated $180°$ about the origin and the result coincides exactly with the original graph, then the graph is symmetric with respect to the origin. (Section 10.4)

symmetric with respect to the x-axis If a graph can be folded in half along the x-axis and the portion of the graph above the x-axis exactly matches the portion below the x-axis, then the graph is symmetric with respect to the x-axis. (Section 10.4)

symmetric with respect to the y-axis If a graph can be folded in half along the y-axis and each half of the graph is the mirror image of the other half, then the graph is symmetric with respect to the y-axis. (Section 10.4)

synthetic division Synthetic division is a shortcut procedure for dividing a polynomial by a binomial of the form $x - k$. (Section 12.1)

system of equations A system of equations consists of two or more equations to be solved at the same time. (Section 4.1)

system of inequalities A system of inequalities consists of two or more inequalities that are considered at the same time. (Section 13.4)

T

term A term is a number, a variable, or the product or quotient of a number and one or more variables raised to powers. (Sections 1.4, 5.2)

term of an annuity The time from the beginning of the first payment period to the end of the last period is called the term of an annuity. (Section 14.3)

terms of a sequence The function values $a_1, a_2, a_3, \ldots$, written in order are called the terms of a sequence. (Section 14.1)

tree diagram A tree diagram is a diagram with branches that is used to systematically list all the outcomes of a counting situation or probability experiment. (Section 14.6)

trial In probability, each repetition of an experiment is called a trial. (Section 14.7)

trinomial A trinomial is a polynomial with exactly three terms. (Section 5.2)

turning points The points on the graph of a function where the function changes from increasing to decreasing or from decreasing to increasing are called turning points. (Section 12.3)

U

union The union of two sets A and B, written $A \cup B$, is the set of elements that belong to either A or B (or both). (Section 2.6)

V

variable A variable is a symbol, usually a letter, used to represent an unknown number. (Section 1.1)

vary directly (is proportional to) y varies directly as x if there exists a real number (constant) k such that $y = kx$. (Section 3.6)

vary inversely y varies inversely as x if there exists a real number (constant) k such that $y = \frac{k}{x}$. (Section 3.6)

vary jointly   If one variable varies as the product of several other variables (sometimes raised to powers), then the first variable is said to vary jointly as the others. (Section 3.6)

Venn diagram   A Venn diagram is a diagram used to illustrate relationships between sets. (Section 14.7)

vertex (corner) point   In linear programming, any optimum value (maximum of minimum) will always occur at a vertex (corner) point of the region of feasible solutions. (Section 13.4)

vertex of a parabola   For a vertical parabola, the vertex is the lowest point if the parabola opens up and the highest point if the parabola opens down. For a horizontal parabola, the vertex is the point farthest to the left if the parabola opens to the right, and the point farthest to the right if the parabola opens to the left. (Section 10.2)

vertical asymptote   A vertical line that a graph approaches, but never touches or intersects, is called a vertical asymptote. (Sections 7.4, 12.4)

vertical line test   The vertical line test states that any vertical line drawn through the graph of a function must intersect the graph in at most one point. (Section 3.5)

W

whole numbers   The set of whole numbers is $\{0, 1, 2, 3, 4, \ldots\}$. (Section 1.1)

X

x-axis   The horizontal number line in a rectangular coordinate system is called the x-axis. (Section 3.1)

x-intercept   A point where a graph intersects the x-axis is called an x-intercept. (Section 3.1)

Y

y-axis   The vertical number line in a rectangular coordinate system is called the y-axis. (Section 3.1)

y-intercept   A point where a graph intersects the y-axis is called the y-intercept. (Section 3.1)

Z

zero matrix   A matrix all of whose elements are 0 is a zero matrix. (Appendix B)

zero of a polynomial function   A zero of a polynomial function f is a value of k such that $f(k) = 0$. (Section 12.2)

zero-factor property   The zero-factor property states that if two numbers have a product of 0, then at least one of the numbers must be 0. (Sections 6.5, 9.1)

Index

Videotape and CD Index

Text/Video/CD Section	Exercise Numbers	Text/Video/CD Section	Exercise Numbers
Section 1.1	85	Section 7.1	65
Section 1.2	none	Section 7.2	57
Section 1.3	75	Section 7.3	25
Section 1.4	27	Section 7.4	3, 31, 35, 41
		Section 7.5	21
Section 2.1	19, 39		
Section 2.2	13, 17, 25, 35, 39	Section 8.1	37
Section 2.3	9, 19, 29, 45, 51, 53	Section 8.2	none
Section 2.4	none	Section 8.3	57, 87
Section 2.5	25	Section 8.4	37, 45
Section 2.6	37	Section 8.5	45, 51
Section 2.7	37, 57	Section 8.6	7, 11, 45, 49, 55, 69
		Section 8.7	63
Section 3.1	47		
Section 3.2	33	Section 9.1	none
Section 3.3	19, 23, 35, 53, 65, 75, 77	Section 9.2	9, 11, 37, 43
		Section 9.3	17, 49, 59
Section 3.4	21	Section 9.4	none
Section 3.5	7, 11, 19, 33, 49, 75(c)	Section 9.5	none
Section 3.6	21, 39		
		Section 10.1	none
Section 4.1	31	Section 10.2	27
Section 4.2	13	Section 10.3	7, 29
Section 4.3	7	Section 10.4	none
Section 4.4	7	Section 10.5	11, 31
Section 4.5	63		
		Section 11.1	19
Section 5.1	15, 17, 19, 81	Section 11.2	17
Section 5.2	41	Section 11.3	47
Section 5.3	19	Section 11.4	41, 43
Section 5.4	55, 63, 99	Section 11.5	37, 49, 53
Section 5.5	31	Section 11.6	37, 49, 53
Section 6.1	15, 59	Section 12.1	9, 33
Section 6.2	73	Section 12.2	none
Section 6.3	35, 41	Section 12.3	none
Section 6.4	21	Section 12.4	none
Section 6.5	41, 45, 65, 79		

Text/Video/CD Section	Exercise Numbers	Text/Video/CD Section	Exercise Numbers
Section 13.1	35, 39	Section 14.3	35
Section 13.2	15, 19, 23	Section 14.4	31
Section 13.3	41	Section 14.5	23
Section 13.4	31	Section 14.6	23, 42
		Section 14.7	19, 21
Section 14.1	21		
Section 14.2	43		